Retroviruses

Molecular Biology, Genomics and Pathogenesis

Edited by

Reinhard Kurth

Robert Koch-Institut
Berlin
Germany

and

Norbert Bannert

Robert Koch-Institut
Zentrum für Biologische Sicherheit
Berlin
Germany

Caister Academic Press

Caister Academic Press
Norfolk, UK

www.caister.com

British Library Cataloguing-in-Publication Data
A catalogue record for this book is available from the British Library

ISBN: 978-1-904455-55-4

Cover image: Cross sectional model of the human immunodeficiency virus (HIV). Created by Dr Stephen Norley of the Robert Koch Institute in Berlin, Germany, using Blender, the free open-source 3D content creation suite (http://www.blender.org/).

Printed and bound in Great Britain
By Cpod, a division of The Cromwell Press Group, Trowbridge, Wiltshire

Contents

Contributors

Norbert Bannert
Centre for Biological Safety 4
Robert Koch Institute
Berlin
Germany

bannertn@rki.de

Jef D. Boeke
The Johns Hopkins University School of Medicine
Baltimore, MD
USA

jboeke@jhmi.edu

Christian J. Buchholz
Paul-Ehrlich-Institut
Department of Medical Biotechnology
Langen
Germany

bucch@pei.de

Klaus Cichutek
Paul-Ehrlich-Institut
Langen
Germany

cickl@pei.de

Xavier Contreras
Box 0703
Depts of Medicine
Microbiology and Immunology
UCSF
San Francisco, CA
USA

xavier.contreras@ucsf.edu

Joachim Denner
Retrovirus Induced Immunosuppression
Robert Koch Institute
Berlin
Germany

dennerj@rki.de

Maribeth V. Eiden
Section on Molecular Virology
Laboratory of Cell and Molecular Regulation
National Institutes of Health
Bethesda, MD
USA

eidenm@mail.nih.gov

Alan Engelman
Department of Cancer Immunology and AIDS
Dana-Farber Cancer Institute
Boston, MA
USA

alan_engelman@dfci.harvard.edu

Volker Erfle
TU München
Institut für Molekulare Virologie
Oberschleißheim
Germany

erfle@gsf.de

Uwe Fiebig
AIDS Immunopathogenesis and Vaccine Development
Robert Koch Institute
Berlin
Germany

fiebigu@rki.de

Robert C. Gallo
Institute of Human Virology and Department of Medicine
School of Medicine University of Maryland
Baltimore, MD
USA

snallo@ihv.umaryland.edu

Nina Gerlach
Niedersächsische Staats- und Universitätsbibliothek
Göttingen
Germany

nina.gerlach@sub.uni-goettingen.de

Heinrich G. Göttlinger
Program in Gene Function and Expression
Program in Molecular Medicine
UMass Medical School
Worcester, MA
USA

heinrich.gottlinger@umassmed.edu

Jonathan Luke Heeney
Department of Veterinary Medicine
Cambridge University
Cambridge
United Kingdom

jlh66@cam.ac.uk

Christine Hohenadl
Veterinärmedizinische Universität Wien
Department für Pathologie
Institut für Virologie
Wien
Austria

christine.hohenadl@vu-wien.ac.at

Oliver Hohn
Centre for Biological Safety 4
Robert Koch Institute
Berlin
Germany

hohno@rki.de

Gerhard Hunsmann
Universitätsmedizin Göttingen
Georg-August-Universität
Institut für Virologie
Göttingen
Germany

virology@medizin.uni-goettingen.de

Reinhard Kurth
Robert Koch-Institut
Berlin
Germany

kurthr@rki.de

Tina Lenasi
Departments of Medicine, Microbiology and Immunology
UCSF
San Francisco, CA
USA
and
University of Helsinki
Department of Virology
Haartman Institute
Helsinki
Finland

tlenasi@yahoo.com

Jeremy Luban
Department of Microbiology and Molecular Medicine
Faculty of Medicine
University of Geneva
Geneva
Switzerland

Jeremy.Luban@unige.ch

Hans Lutz
FVH, FAMH
Resources and Planning
Clinical Laboratory
Vetsuisse Faculty
University of Zurich
Zurich
Switzerland

hlutz@vetclinics.uzh.ch

Walther Mothes
Section of Microbial Pathogenesis,
Yale University School of Medicine
New Haven, CT
USA

walther.mothes@yale.edu

Michael D. Mühlebach
Paul-Ehrlich-Institut
Department of Medical Biotechnology
Langen
Germany

muemi@pei.de

Stephen Norley
Retrovirus Induced Immunosuppression
Robert Koch Institute
Berlin
Germany

norleys@rki.de

Finn Skou Pedersen
Department of Molecular Biology
University of Aarhus
Aarhus
Denmark

fsp@mb.au.dk

B. Matija Peterlin
Departments of Medicine, Microbiology and Immunology
UCSF
San Francisco, CA
USA
and
University of Helsinki
Department of Virology
Haartman Institute
Helsinki
Finland

matija.peterlin@ucsf.edu

Guido Poli
AIDS Immunopathogenesis Unit
Vita-Salute San Raffaele University and Scientific Institute
Milano
Italy

poli.guido@hsr.it

Sandra L. Quackenbush
Department of Microbiology, Immunology, and Pathology
Colorado State University
Fort Collins, CO
USA

Sandra.Quackenbush@colostate.edu

Kathryn Radke
Animal Science Department
University of California
Davis, CA
USA

klradke@ucdavis.edu

Marvin S. Reitz, Jr
Institute of Human Virology and Department of Medicine
School of Medicine University of Maryland
Baltimore, MD
USA

mreitz@ihv.umaryland.edu

Joel Rovnak
Department of Microbiology, Immunology, and Pathology
Colorado State University
Fort Collins, CO
USA

Joel.Rovnak@colostate.edu

Silke Schüle
Paul-Ehrlich-Institut
Department of Medical Biotechnology
Langen
Germany

schsi@pei.de

Jörg Schüpbach
University of Zurich
Institute of Medical Virology
Swiss National Center for Retroviruses
Zurich
Switzerland

Jorg.Schupbach@access.uzh.ch

Matthias Schweizer
Paul-Ehrlich-Institut
Department of Medical Biotechnology
Langen
Germany

schmt@pei.de

Annette Balle Sørensen
The State and University Library
Universitetsparken
Aarhus
Denmark

abs@statsbiblioteket.dk

David E. Symer
Human Cancer Genetics Program
The Ohio State University Comprehensive Cancer Center
Columbus, OH
USA

david.symer@osumc.edu

Pradeep D. Uchil
Section of Microbial Pathogenesis
Yale University School of Medicine
New Haven, CT
USA

pradeep.uchil@yeale.edu

Ernst J. Verschoor
Department of Virology
Biomedical Primate Research Centre
Rijswijk
The Netherlands

verschoor@bprc.nl

Winfried Weissenhorn
Unit for Virus Host–Cell Interactions (UVHCI)
UMR 3265 UJF-EMBL-CNRS
Grenoble
France

weissenhornembl.fr

Preface

For many years the pathogenic and functional properties of retroviruses have been the focus of intense dedicated research. The developments and advances in the field have not only attracted the attention of other virologists but have also significantly enriched our knowledge and the scope of molecular biology, oncology, gene delivery, vaccinology, immunology and evolutionary biology. The discovery of a transmissible agent causing sarcomas in domestic birds by Peyton Rous in 1911 led researchers in the ensuing decades to focus on the oncogenic effects of avian, murine and human retroviruses. However, research efforts shifted dramatically and were strongly intensified with the discovery of a new retrovirus in the early 1980s. The human immunodeficiency virus (HIV), the causative agent of acquired immunodeficiency syndrome (AIDS), slowly erodes the human immune system and without treatment almost inevitably results in death by opportunistic infections or rare forms of tumors. The accumulated knowledge of retroviruses has facilitated the development of effective antiretroviral drugs and therapeutic intervention strategies within a relatively short period of time. Indeed, the first drugs to be licensed in the late 1980s targeted HIV's reverse transcriptase, a fundamental enzyme from which retroviruses get their name, which transcribes the viral genomic RNA into DNA to allow integration into the chromosomal DNA of the host cell.

Retroviruses share a common ancestry with a wide variety of retroelements, a superfamily of more or less mobile genetic sequences that transpose via an RNA intermediate and include retrotransposons and related reverse transcriptase-utilizing entities (see chapter 1 for details). The genetic material of vertebrates is surprisingly rich in retroelements, comprising for example about 42% of the human genome, 8% of which consists of sequences from ancient retroviruses. These so-called endogenous retroviruses are fossils of inherited germ line infections that resulted in vertical transmission from parents to offspring.

Infectious retroviruses are distinct from these retroelements in that their amplification is not confined to those cells already harbouring the integrated proviral element but extends to neighbouring cells and other individuals. The origin of vertebrate retroviruses is yet to be systematically investigated, but due to their similarity in genomic structure and sequences, it is likely that they evolved from Ty3/Gypsy LTR retroelements (see chapter 1 for details). This group of retrotransposons and retrovirus-related elements still exists in plants, fungi and invertebrates.

Retroelements themselves appear to be very ancient and might even have played a role during the switch from the RNA into the DNA world. Retroviruses code for proteins and protein domains that are often homologous to those of eukaryotic cells, some plant and animal DNA viruses and even fungal mitochondria. Although it is well known that retroviruses pick up genetic material from their host cells, the historical origins of retroviruses can nevertheless be traced back to events coincident with the prokaryotic invasion of primitive eukaryotes. In this regard, endogenous retroviruses represent a rich archive of former retroviral infections. The ongoing sequencing of entire species genomes and the use of sophisticated bioinformatic approaches to recognize and compare retroviral sequences will be instrumental in constructing an overall phylogeny that chronicles major aspects of the history and relationships of retroviruses.

Although a comprehensive and rational taxonomy that includes both endogenous and exogenous retroviruses is presently not available, endogenous and exogenous retroviruses of all vertebrates can be grouped into three major classes designated I, II and III based on aspects of homology. This classification is complementary to the latest International Committee on Taxonomy of Viruses (ICTV) grouping of exogenous retroviruses, as class I comprises the genera gamma- and epsilonretroviruses; class II includes lentiviruses, alpha- and betaretroviruses; and class III contains spumaviruses. The taxonomic family *Retroviridae* is presently divided into two subfamilies: orthoretrovirinae and spumavirinae. The spumaretrovirinae contain the genus spumavirus, whereas the orthoretrovirinae comprises the other six genera (Fig. 1).

Although several chapters of the book focus on a single or a limited number of retroviral genera, taxonomy alone is not the basis for the overall structure of this book that consists of 16 chapters. The book begins with chapters reviewing retroelements, endogenous retroviruses and the basic features of retroviruses. Important characteristics of the life cycle of exogenous retroviruses are covered next, followed by aspects of transmission, epidemiology, pathogenesis, host restriction factors and vaccination strategies. Selected retrovirus genera including gammaretroviruses and lentiviruses and their use as gene delivery tools are then discussed. In the final two chapters the focus shifts to simian and human (HTLV and HIV) retroviruses with a discussion of their medical implications.

The book has been written by scientists internationally recognized to be authorities in their fields of research. They provide comprehensive and focused overviews of their respective topics, emphasizing recent advances and future research directions while keeping the content as coherent and plain as possible.

The editors have taken care that this book would not become a compilation of separate chapters but, instead, would offer a broad and consistent overview of the present knowledge of the molecular biology, genomics and pathogenesis of retroviruses. However, each chapter can easily be read and understood independently of the others although this intention inevitably necessitated some degree of redundancy for the sake of clarity. The structure of each chapter is consistent, each having an abstract to explain the concept of the chapter, a conclusion reiterating the most important points, a number of illustrations and tables,

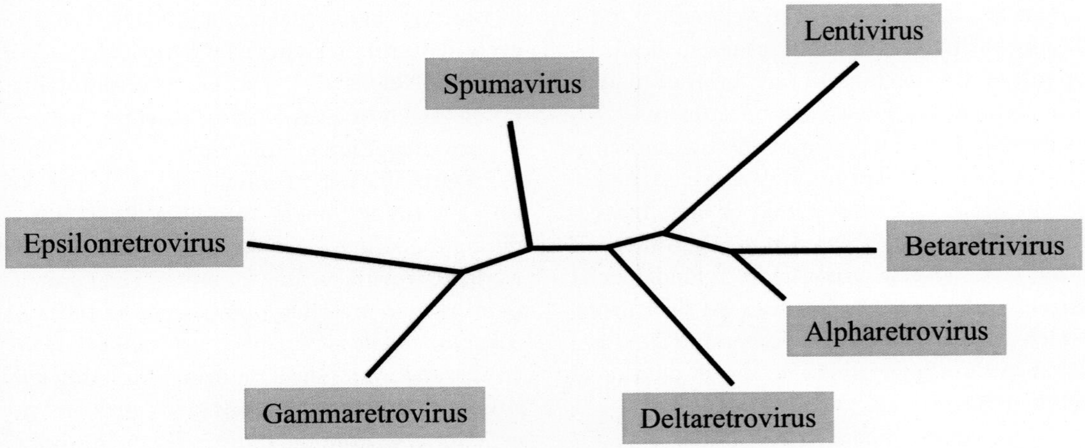

Figure 1 Schematic presentation of the phylogenetic relationships of the *Retroviridae* based on conserved sequences in the polymerase gene.

and finally a selected list of references for further reading. However, despite the extensive bibliography we almost certainly missed a number of publications that other colleagues may consider to be essential and we apologize in advance for such oversights.

We are convinced that this book will help introduce interested students, clinicians, veterinary scientists and virologists to the field and will serve professional retrovirologists as a source of valuable supplementary information and overview.

Finally, we would like to thank Rayk Behrend, Stephen Norley and Laurie von Melchner for their editorial help, Annette Griffin of Horizon Scientific Press for her constant support during the preparation and, of course, all authors for making this book possible through their valuable contributions.

Reinhard Kurth and Norbert Bannert

Dedication

We thank our wives Dr Bärbel-Maria Kurth and Jutta Bannert for their thoughtful support of our enjoyable professional activities. Behind every successful man stands a surprised woman.

Reinhard Kurth
Norbert Bannert

Abbreviations

AAV	adeno-associated virus
Ab	antibodies
Ab-MLV	Abelson murine leukaemia virus
ADA	adenosine deaminase
ADCC	antibody-directed cellular cytotoxicity
AEV	avian erythroblastosis viruses
AGM	African green monkey
AIDS	acquired immunodeficiency syndrome
ALV	avian leukosis virus
A-MLV	amphotropic murine retroviruses
AP	apurinic-apyridiminic
APC	antigen-presenting cells
APOBEC	apolipoprotein B mRNA editing enzyme catalytic polypeptide
ARM	arginine-rich motif
ART	antiretroviral therapy
ASLV	avian sarcoma/leukosis viruses
ATL	adult T-cell leukaemia
AZT	zidovudine (azidothymidine)
B2	rodent-specific SINE retrotransposons
BAAT	bile acid coenzyme A: amino acid N-acetyltransferase
BaEV	baboon endogenous virus
BCG	Bacille Calmette-Guerin
BCR	B-cell receptor
BD	blocking domain
BIV	bovine immunodeficiency virus
BLV	bovine leukaemia virus
BNC	binucleate cells
BP	branch point
Brd4	bromodomain containing protein 4
BRG-1	Brahma-related gene 1
CA	capsid protein
CAEV	caprine arthritis and encephalitis virus
cAMP	cyclic adenosine monophosphate
CBF1	C-promoter binding factor-1
CCD	catalytic core domain
CCR5	chemokine (C-C motif) receptor 5
CD	cluster of differentiation
cDNA	cyclic deoxyribonucleic acid
ChIP	chromatin immunoprecipitation
CHMP	charged multivesicular body protein
CIS	common integration sites
CNPRC	California National Primate Research Center
CNS	central nervous system
c-onc	cellular oncogenes
CPE	cytopathic effect
cPPT	central polypurine tract
CPSF	cleavage polyadenylation specificity factor

cpz	chimpanzee
CR1	chicken repeat 1
CRF	circulating recombinant forms
CRM	chromosome region maintenance
CStF	cleavage stimulation factor
CTD	C-terminal domain
CTE	constitutive transport element
CTL	cytotoxic T-lymphocytes
CTS	central termination sequence
CXCR4	chemokine (C-X-C motif) receptor 4
CypA	cyclophilin A
DC	dendritic cells
ddI	dideoxyinosine
DIS	dimerisation initiation sites
DLS	dimer linkage site
DMRs	differentially methylated regions
DNA	desoxyribonucleic acid
DPF	designated pathogen free
DRB	dichloro-1-b-D-ribofuranosylbenzimidazole
DSIF	DRB sentitivity inducing factor
EDNRB	endothelin B receptor
EGF	epidermal growth factor
EGFR	epidermal growth factor receptor
EIAV	equine infectious anaemia virus
ELISA	enzyme-linked immunosorbent assay
EM	electron microscopy
ENs	endonucleases
ENTV	enzootic nasal tumour virus
Env	envelope glycoproteins
EpoR	erythropoietin receptor
ERM	ezrin, radixin, moesin protein family
ERV	endogenous retrovirus
ES	esterases
ESCRT	endosomal sorting complexes required for transport
ESE	exonic splicing enhancer
ESS	exonic splicing silencer
ET	external domain
EVT	extravillous cytotrophoblast
FACS	flow cytometry
FDA	(US) Food and Drug Administration
FDC	follicular dendritic cells
FeLV	feline leukaemia virus
FeSV	feline sarcoma virus
FGFR	fibroblast growth factor receptor
FIV	feline immunodeficiency virus
FRT	Flp recombinase target
FV	foamy viruses
Fv1	friend virus susceptibility factor 1
FVP	friend MuLV
GA	Gag-encoding region
GALT	gut-associated lymphatic tissues
GaLV	gibbon ape leukaemia virus
GAP	GTPase activating protein
GC	germinal centres
GR boxes	glycine-arginine-rich basic sequences
GTF	general transcription factor
GvHD	graft versus host disease
HA	Haemagglutinin
HAART	highly active antiretroviral therapy
HAR	hyperacute rejection
HAT	histone acetyl transferase
HCG	human chorionic gonadotropin
HDAC	histone deacetylase
HEK	human embryonic kidney cells
hel	HERV-K specific mRNA expressed in lymphocytes
HEPS	highly exposed persistently seronegative individuals
HERV	human endogenous retrovirus
hESC	human embryonic stem cells
HFV	human foamy virus
HHV	human herpes virus
HIV	human immunodeficiency virus
HMBA	hexamethylene bisacetamide

HMG	high mobility group
HMGA1	high mobility group AT-hook protein 1
HNF	hepatocyte nuclear factor
HP1	heterochromatin associated protein 1
HR	heptad repeat
HSC	haematopoietic stem cells
HSV	Herpes simplex virus
HSV-Tk	HSV-1 encoded thymidine kinase
HTDV	human teratocarcinoma derived virus
HTLV	human T-cell lymphotropic virus
ICAM	intra cellular adhesion molecule
ICTVdB	International Committee on Taxonomy of Viruses
IFN	interferon
Ig	Immunoglobulins
IL	interleukin
IL-2	interleukin 2
IN	integrase
indel	insertion/deletion
Inr	initiator
IP	internal promoter
IRES	internal ribosomal entry site
ISS	intronic splicing silencer
IST	inducer of short transcripts
isu-domain	immunosuppressive domain
ITAM	immunoreceptor tyrosine-based activation motif
JSRV	Jaagsiekte sheep retrovirus
kb	kilo bases
kDa	kilo Dalton
KoRV	koala retrovirus
KS	Kaposi's sarcoma
KSHV	Kaposi's sarcoma-associated herpesvirus
LAM	linear amplification-mediated
LCA	Leber's congenital amaurosis
LCMV	lymphocytic choriomeningitis virus
L-Domain	late domain
LEF	lymphocyte enhancer factor
LINE	long interspersed nuclear element
LNX	ligand of numb protein X
LSF	late SV40 factor
LTNP	long-term nonprogression
LTR	long terminal repeat
LTs	long RNA transcripts
m5C	methylated DNA
MA	matrix protein
MAP	mitogen-activated protein
MAPKK	MAP kinase kinase
MAPKKK	MAP kinase kinase kinase
MBP	Maltose binding protein
mDC	Myeloid dendritic cells
MDEV	*Mus dunni* endogenous retrovirus
M-Domain	membrane binding domain
MFSD2	major facilitator superfamily domain containing 2
MHC	major histocompatibility complex
MHR	major homology region
MIR	mammalian interspersed repeat
miRNA	microRNA
MLV	murine leukaemia virus
MMP	Matrix metalloproteinases
MMR	measles-mumps-rubella vaccine
MMS	munich minature swine
MMTV	mouse mammary tumour virus
MMV	mouse minute virus
MoMLV	Moloney murine leukaemia virus
MP	mononuclear phagocytes
MPMV	Mason-Pfizer monkey virus
MSD	major splice donor
MTCT	mother-to-child transmission
mtDNA	mitochondrial DNA
MTOC	microtubule organizing centre
mTRP	melanoma specific tyrosinase related protein
MuLV	murine leukaemia virus

MVA	modified vaccinia virus Ankara
MVB	multivesicular body
MVV	Maedi-Visna virus
NAb	neutralising antibody
NADPH	nicotinamide adenine dinucleotide phosphate
NAIP	neuronal apoptosis inhibitory protein
NC	nucleocapsid protein
Nef	negative factor
NES	nuclear export signal
NF-κB	nuclear factor κB
NHP	non-human primate
NKT	natural killer T-cells
NLS	nuclear localisation sequences
NMR	nuclear magnetic resonance
NPC	nuclear core complex
NSI	non-syncytium inducing
nsSNP	nonsynonymous single nucleotide polymorphisms
NTD	N-terminal domain
N-TEF	negative transcription elongation factor
OPA	ovine pulmonary adenocarcinomas
ORF	open reading frame
PBLs	peripheral blood lymphocytes
PBMC	peripheral blood mononuclear cells
PBS	primer binding site
Pcf11	polyadenylation cleavage factor 11
PCR	polymerase chain reaction
pDC	plasmacytoid dendritic cells
PDGF	platelet-derived growth factor
PEP	post-exposure prevention
PERV	porcine endogenous retrovirus
PHA	phytohaemagglutinin
PHD	plant homeodomain
PHI	primary HIV infection
PI3K	phosphatidylinositide 3-kinase
PIC	preintegration complex
piRNA	piwi protein-associated RNAs
PKC	protein kinase C
PLC	phospholipase C
PLZF	promyelocytic leukaemia zinc finger protein
PMA	phorbol-12-myristate-13-acetate
P-MLV	polytrophic murine retroviruses
POMC	proopiomelanocortin
PPD	PAZ Piwi domain
PPT	polypurine tract
PPT	polypyrimidine tract
PR	protease
PRD	proline-rich domain
PRE	posttranslational regulatory elements
PrEP	pre-exposure prevention
Pro	viral protease
PRR	pattern recognition receptors
P-TEFb	positive transcription elongation factor
PTGS	post-transcriptional gene silencing
PTLV	primate T-lymphotropic virus
PTLV	primate T-cell lymphotrophic viruses
qRT-PCR	quantitative real time PCR
RACK1	receptor for activated C kinase
rasiRNAs	repeat-associated small interfering RNAs
RBD	receptor binding domain
RCR	replication-competent retroviruses
RD	human rhadomyosarcoma
RDR	D-type mammalian retrovirus receptor
RELIK	rabbit endogenous lentivirus type K
REV	reticuloendotheliosis virus
RF	retroperitoneal fibromas
RING	really interesting new gene
RNA	ribonucelic acid
RNAi	RNA interference
RNAPII	RNA polymerase II
RNP	ribonucleoprotein
RP	rapid progression

RRE	Rev responsive element
RSV	Rous sarcoma virus
RT	reverse transcriptase
RTC	reverse transcription complex
Rtl1	retrotransposon like gene
RT-PCR	reverse transcriptase – polymerase chain reaction
rVSV	recombinant VSV
SA	splice acceptor
Sag	superantigen
SAIDS	simian acquired immunodeficiency syndrome
SARS	severe acute respiratory syndrome
SC	small acitve complexes
scAb	single chain antibody
SCID	severe combined immunodeficiency
SD	splice donor
SERV	simian endogenous retroviruses
SFFV	spleen focus-forming virus
sf-Stk	short form of the receptor tyrosine kinase
SFV	simian foamy virus
SFVmar	simian foamy virus from marmosets
SHIV	simian/human immunodeficiency hybrid virus
SI	syncytium inducing
SIN	self-inactivating
siRNA	short interfering RNAs
SIRT1	sirtuin 1
SIV	simian immunodeficiency virus
SIVcpz	SIV of chimpanzees
SIVsmm	SIV of sooty mangabeys
SM	Sooty mangabey
SMRV	squirrel monkey retrovirus
SP	spacer peptide
SP	signal peptide
SRLV	Small ruminant lentiviruses
SRV	Simian retrovirus
SRV	squirrel monkey retrovirus
SSAV	simian sarcoma-associated virus
SSR	start site region
SSSV	salmon swimbladder sarcoma virus
SSV	simian sarcoma virus
STAT	signal transducer and activator of transcription
STAT1	signal transducer of transcription-1
STD	sexually transmitted disease
STEC	spring tumor explant cells
STLV	simian T-cell leukaemia virus
STs	short attenuated transcripts
SU	surface subunit of ENV
TAA	tumour-associated antigen
TAD	transcription activation domain
TAF	TBP-associated factor
TAF9	TATA binding protein-associated factor 9
TAR	transactivation response
Tas	transactivator of spumaviruses
Tat	transactivator (of HIV)
Tax	transactivator of HTLV
TBP	TATA-binding protein
TCR	T-cell receptor
TEM	transmission electron microscopy
TGS	transcriptional gene silencing
Th	T helper cell
TI	transcriptional interference
TLR	Toll-like receptor
TM	transmembrane subunit of Env
TNF	tumour necrosis factor
TPA	12-O-tetradecanoylphorbol 13-acetate
TPRT	target primed reverse transcription
TRAIL	tumor necrosis factor-related apoptosis-inducing ligand
Treg	regulatory T-cell
TRIM	tripartite motif
tRNA	transfer RNA
TSDs	target site duplications
TUNEL	transferase biotin-dUTP nick end labelling

UNAIDS	Joint United Nations Programme on HIV/AIDS
USE	upstream enhancer element
USF	upstream stimulating factor
VEEV	Venezuelan equine encephalitis virus
VEGF	vascular endothelial growth factor
Vif	viral infectivity factor
VLP	virus-like particle
VNTR	variable number of tandem repeats
v-onc	viral oncogenes
vpr	viral protein R
vpr	vacuolar protein sorting
vpu	viral protein U
vpx	viral protein X
VSV	vesicular stomatitis virus
VSV-G	vesicular stomatitis virus glycoprotein
VT	villous cytotrophoblast
VZV	varicella-zoster virus
WDS	walleye dermal sarcoma
WDSV	walleye dermal sarcoma virus
WEH	walleye discreet epidermal hyperplasia
WEHV1/2	walleye epidermal hyperplasia type 1/2
WMSV	woolly monkey sarcoma virus
WMV	wooly monkey virus
WNV	West Nile virus
WPRE	woodchuck hepatitis virus post-transcriptional regulatory element
X-CGD	X-linked chronic granulomatous disease
X-MLV	xenotropic murine leukaemia viruses
XMRV	xenotropic MLV-related virus
YY1	Yin Yang 1
ZDV	Zidovudine

An Everlasting War Dance Between Retrotransposons and their Metazoan Hosts

1

David E. Symer and Jef D. Boeke

Abstract

Many classes of transposons and retrotransposons have invaded and shaped the genomes of their metazoan hosts. Class I transposons (i.e. retrotransposons) include both elements which contain long terminal repeats (LTRs) and many others lacking LTRs. Along with retroviruses, retrotransposons share a fundamental mechanism of mobilization through reverse transcription of an RNA template, via enzymatic activity of a reverse transcriptase. However, unlike retroviruses, retrotransposons remain entirely intracellular during their life cycle. Several distinctive and overlapping genome defence mechanisms have developed over evolutionary time, as host organisms have fought back against these mobile retroelements. The ongoing conflict between these myriad genomic parasites and their widespread host organisms has resulted in positive and deleterious consequences including exaptation, genomic deletions, certain diseases including cancers, and the generation of diversity probably including the formation of new species.

Introduction

In recent years, reference genomic sequences for many metazoan organisms have been assembled and nearly completed. These sequences have highlighted the enormous extent to which many classes of transposons and retrotransposons have invaded and shaped the genomes of their metazoan hosts. Class I transposons (i.e. retrotransposons) include both elements which contain long terminal repeats (LTRs) and many others lacking LTRs. The profound insights of Barbara McClintock, who discovered and characterized several classes of these 'controlling elements' or transposons in maize quite a long time ago (McClintock, 1951, 1984), have been broadened and deepened by decades of research on the myriad retrotransposons and their fossils. These elements are ubiquitous and are generally highly abundant in all eukaryotes.

At the same time, much evidence has accumulated recently that the host organisms have fought back over time against these genomic invaders, through the development of several distinctive and overlapping genome defence mechanisms. A debate has ensued about whether retrotransposons should be considered to be at best neutral genomic parasites, whether they are mostly harmful and/or whether they can provide a dynamic, positive mechanism for adaptation that is useful to their hosts. If the latter is correct, another question is whether such beneficial effects occur only through rare beneficial germ line alterations or also in somatic tissues. This debate has played out even in the terms used to describe transposons, ranging from 'controlling elements' (Davidson and Britten, 1979), 'junk DNA' and 'selfish DNA' (Doolittle and Sapienza, 1980; Orgel and Crick, 1980) to 'itsy-bitsy killing machines' (T. H. Bestor, personal communication) and 'genome's little helper' (J. D. Boeke).

By definition, retrotransposons move from one genomic location to another through an RNA intermediate (Boeke *et al.*, 1985; Garfinkel *et al.*, 1985), and remain entirely intracellular during their entire life cycle. Along with retroviruses, retrotransposons share a fundamental mechanism of mobilization through reverse transcription of an RNA template, via enzymatic activity of a reverse transcriptase. Most of our knowledge about the timing and mechanism of mammalian retrotransposon mobilization has stemmed from careful examination of their distributions in a small number of individuals within a given species, i.e. via germline transmission of the elements. However, substantial recent data have supported the idea that some of these elements can still move within at least some mammalian individuals, i.e. in somatic tissues distinct from and in addition to the germline. For example, careful analysis of several sporadic diseases has revealed that retrotransposition indeed can occur in somatic cells.

Retrotransposons have shaped and changed virtually all eukaryotic genomes, and indeed the biology of their hosts, in many ways. These include insertional mutagenesis of genes and exaptive incorporation of elements into expressed transcripts; contribution of new promoters, terminators, and splice sites across the genome that alter the transcriptome; exon shuffling; genetic and chromosomal instability due to transduction events, the processes of retrotransposition and recombination between repeated elements; and changed epigenetic controls. Extensive evolutionary analysis suggests that retrotransposons generally may have preceded and given rise to retroviruses, the focus of this book, in the distant past. More recently, exogenous retroviruses appear to have infected the germ lines of various metazoans and to have been inherited by vertical transmission, giving rise to endogenous retroviruses.

A recent model addressing the proliferation of retrotransposons in metazoans is based on their coincidence with the appearance of specialized terminal structures at the 5′ and 3′ ends of mRNA transcripts. This model, called the 'cap and tail' hypothesis, suggests that these structures may have played an important mechanistic role in stabilizing and perhaps in mobilizing most retrotransposons' transcripts (Boeke, 2003). Intriguingly, recent evidence suggests that at least some of the host defences directed against retrotransposons may target these terminal cap and tail structures of RNA transcripts for degradation and suppression.

Many families of non-LTR retrotransposons were already identified and characterized extensively in the years before full assembly of eukaryotic genome sequences. Of these, many have been well-described in numerous excellent and comprehensive book chapters, reviews and primary research articles (Beauregard *et al.*, 2008; Belancio *et al.*, 2008; Boeke and Stoye, 1997; Curcio and Derbyshire, 2003; Deininger and Batzer, 2002; Deininger *et al.*, 2003; Eickbush and Jamburuthugoda, 2008; Galun, 2003; Goodier and Kazazian, 2008; Kramerov and Vassetzky, 2005; Ostertag and Kazazian, 2001a; Prak and Kazazian, 2000; Wicker *et al.*, 2007).

In this chapter, we review the major categories of non-LTR and LTR retrotransposons that have been active (and in some cases are still mobilized) in metazoan genomes, with a particular emphasis on mammalian elements (Fig. 1.1). While we cannot begin to describe in detail the incredible diversity and range of retrotransposons that continue to be identified in all metazoans, let alone other eukaryotic organisms, here we provide a detailed summary of recently discovered retrotransposon families and some new information on others. We then discuss the broader impact of some of these elements upon various genomes. A detailed summary of recent studies on host controls of retrotransposons is provided next. Finally, we speculate on the current state of the ongoing conflict between these myriad genomic parasites and their widespread host organisms. Throughout this chapter, we attempt to summarize and synthesize the diverse, sometimes adaptive biological roles that may be played by retrotransposons in the particular contexts of different host species, tissues and cells, genomic locations and epigenetic controls.

Know your enemy: classes of metazoan retrotransposons

Classification by mechanism of retrotransposition

Several classification schemes have been proposed to describe and group the extremely diverse retrotransposons extant in metazoans and other eukaryotic species. Some are based on global structural features, which may include

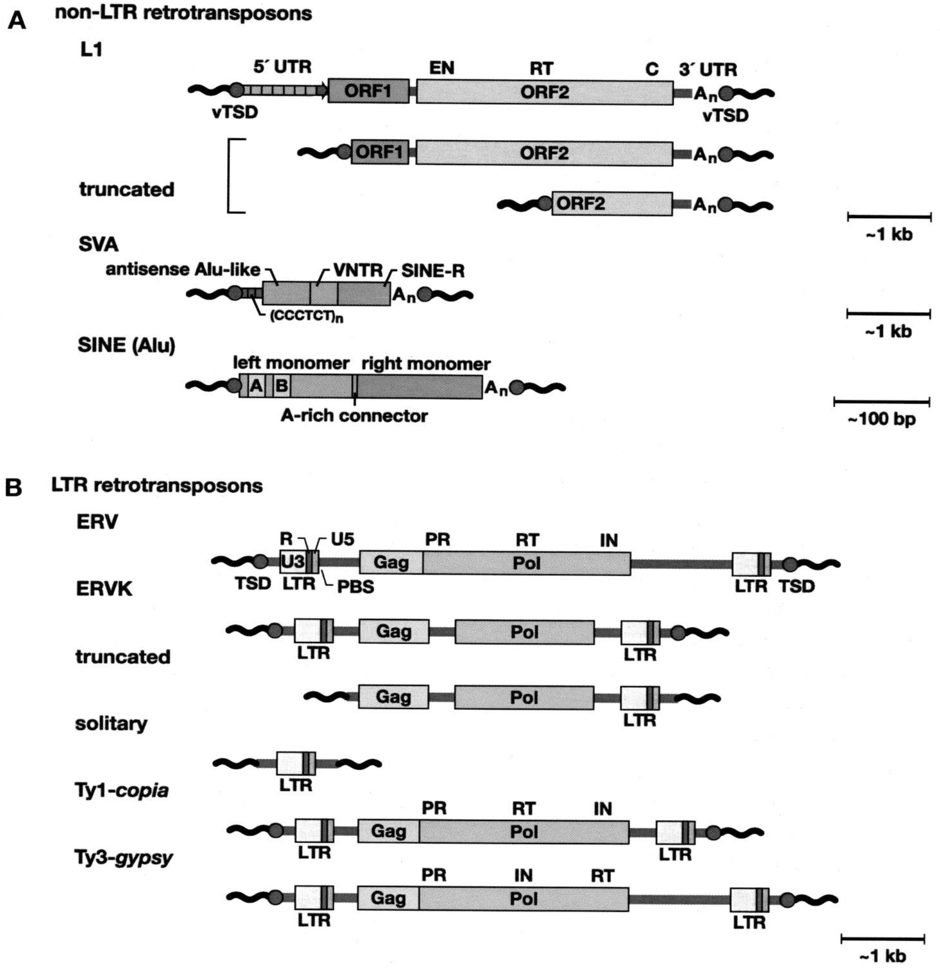

Figure 1.1 Genomic structures of retrotransposons. The overall genomic structures of selected classes of (A) non-LTR retrotransposons and (B) LTR retrotransposons are depicted. (A) Genomic copies of autonomous non-LTR retrotransposons such as (top) Long Interspersed Elements (LINEs such as L1 elements) and (bottom) non-autonomous, non-LTR retrotransposons such as human SVA elements and mammalian SINE elements typically are flanked by target site duplications (TSD, red circles). L1 retrotransposons frequently are truncated from their 5′ ends (bracket). Full-length L1 integrants include 5′ and 3′ untranslated regions (5′ UTR and 3′ UTR, respectively), and open reading frames (ORFs) 1 and 2 which encode an RNA chaperone and endonuclease (EN), reverse transcriptase (RT) and a zinc knuckle, cysteine-rich (C) domains, respectively. Internal RNA polymerase II promoter activity is found in the 5′ UTR. SVA elements typically comprise four regions, including a concatamer of the nucleotide sequence CCCTCT, an antisense Alu-like region, a region with a variable number of tandem repeats (VNTR), and a SINE-R region. Short Interspersed Elements (SINEs) frequently have a two-part structure, and typically do not exceed 500 nt in length. The left monomer of SINEs contains an RNA polymerase III promoter with A and B boxes. It is separated from the right monomer by an A-rich region. A typical SINE element ends with a poly(A) tail at its 3′ end, and is flanked at both ends by a short TSD. (B) LTR retrotransposons are similar to retroviruses in that they typically contain *gag* and *pol* genes. However, *env* genes are defective or absent in retrotransposons, so unlike retroviruses these endogenous elements are not infectious and instead undergo intracellular mobilization. Young genomic integrants are flanked by short TSDs. Each long terminal repeat consists of unique sequences found in encoded transcripts' 3′ (U3) and 5′ (U5) ends, along with a primer binding site (PBS) and R region (R). The *pol* gene typically encodes reverse transcriptase (RT) and integrase (IN) domains, present in either order, and often the *pro* gene is included within *pol*. Frequently, older endogenous retroviruses (ERV retrotransposons) are present as truncated or solitary long terminal repeat elements in genomic DNA sequences. For comparison, Ty1-*copia* and Ty3-*gypsy* retrotransposons also are depicted.

or exclude LTRs and/or other structural features (Fig. 1.1). Other schemes consider the mechanism of transposition by aligning and grouping transposases or reverse transcriptases (Curcio and Derbyshire, 2003; Finnegan, 1989; Kapitonov and Jurka, 2008; Wicker *et al.*, 2007). A recently published classification system considers the site of retrotransposon priming and integration, i.e. extrachromosomal priming versus at the chromosomal target site (Beauregard *et al.*, 2008). Here we will review a classification of retrotransposons based on mechanistic similarities and differences.

A major category of retrotransposons is made up of the 'target priming' retrotransposons, i.e. the non-LTR class, including long interspersed elements (LINEs) (Fig. 1.1) and mobile group II introns (Beauregard *et al.*, 2008). They copy RNA transcripts directly into a genomic target site that has been nicked by one of a variety of endonucleases (ENs) through a process called target primed reverse transcription (TPRT) (Fig. 1.2) (Luan *et al.*, 1993). They contain at least one open reading frame, which encodes at least two distinct enzymatic activities, i.e. reverse transcriptase and endonuclease functions, which can in some cases be separated into two separate functional domains. To date, at least four distinct endonuclease-containing families have been identified, including apurinic–apyridiminic (AP) endonucleases as found in L1 and R1Bm retrotransposons (Feng *et al.*, 1996, 1998); sequence-specific endonucleases loosely related to certain restriction enzymes in R2Bm elements (Yang *et al.*, 1999); GIY-YIB endonucleases encoded by Penelope-like and Athena-like retrotransposons (Arkhipova *et al.*, 2003); and endonucleases encoded by intron-encoded reverse transcriptases/maturases (Zimmerly *et al.*, 1995a,b).

Another major category of retrotransposons is made up of LTR retrotransposons (Fig. 1.1), which are extrachromosomally primed (Beauregard *et al.*, 2008). These elements are related to retroviruses, but lack the *env* gene so they are non-infectious. The integrase encoded by these elements as a part of their *pol* gene product is distinguished by containing the three highly conserved non-contiguous amino acids ($DX_nDX_{35}E$, signifying the sequence Asp followed by a variable number *n* of amino acids, Asp, 35 amino acids and Glu). The three invariant, acidic residues bind metal ions required for catalysis. This same mechanistic and structural feature also is found in DNA transposases and in the integrases encoded by retroviruses such as HIV-1 and avian sarcoma virus (Bujacz *et al.*, 1996a,b; Curcio and Derbyshire, 2003). LTR retrotransposons that utilize a DDE-containing integrase include Ty1/*copia*, Ty3/*gypsy*, BEL elements (Butler *et al.*, 2001a; Frame *et al.*, 2001), and endogenous retroviruses. DDE transposases (integrases) insert the extrachromosomal cDNA copied from the retrotransposon into a new target site, between two staggered nicks on both strands of genomic DNA. Integrase hydrolyses the cDNA phosphodiester backbone at the retrotranposon's ends, resulting in formation of 3′-OH ends, which are used in a transesterification reaction to join the cDNA to the target DNA. Integration of the staggered ends followed by repair presumed to be carried out by host cell machinery generates fixed-length target site duplications (TSDs), a sine qua non of retrotransposition by LTR retrotransposons and other $DX_nDX_{35}E$-encoding elements. Distinct types of LTR elements generate TSDs of different lengths, ranging from 2–9 bp in length. Most LTR retrotransposons and retroviruses generate TSDs of 5±1 bp presumably because cutting

Figure 1.2 Distinct mechanisms of retrotransposition by non-LTR vs. LTR retrotransposons. The life cycles of both autonomous (A) non-LTR and (B) LTR retrotransposons include transcription from genomic templates in the cellular nucleus and translation of open reading frames (ORFs) in the cytoplasm. (A) The full-length transcript from an autonomous, full-length non-LTR retrotransposon is exported to the cytoplasm, where encoded proteins are translated. These proteins are believed to bind to the templating transcript *in cis*, forming ribonucleoprotein complexes that are imported back to the nucleus. An L1-encoded endonuclease (EN) nicks a preferred genomic target site. This nicking generates a free 3′ OH^- directly at the target site that is used by L1-encoded reverse transcriptase (RT) in forming reverse transcribed, single stranded cDNA. This process, called target-primed reverse transcription (TPRT) (Luan *et al.*, 1993), currently is believed to

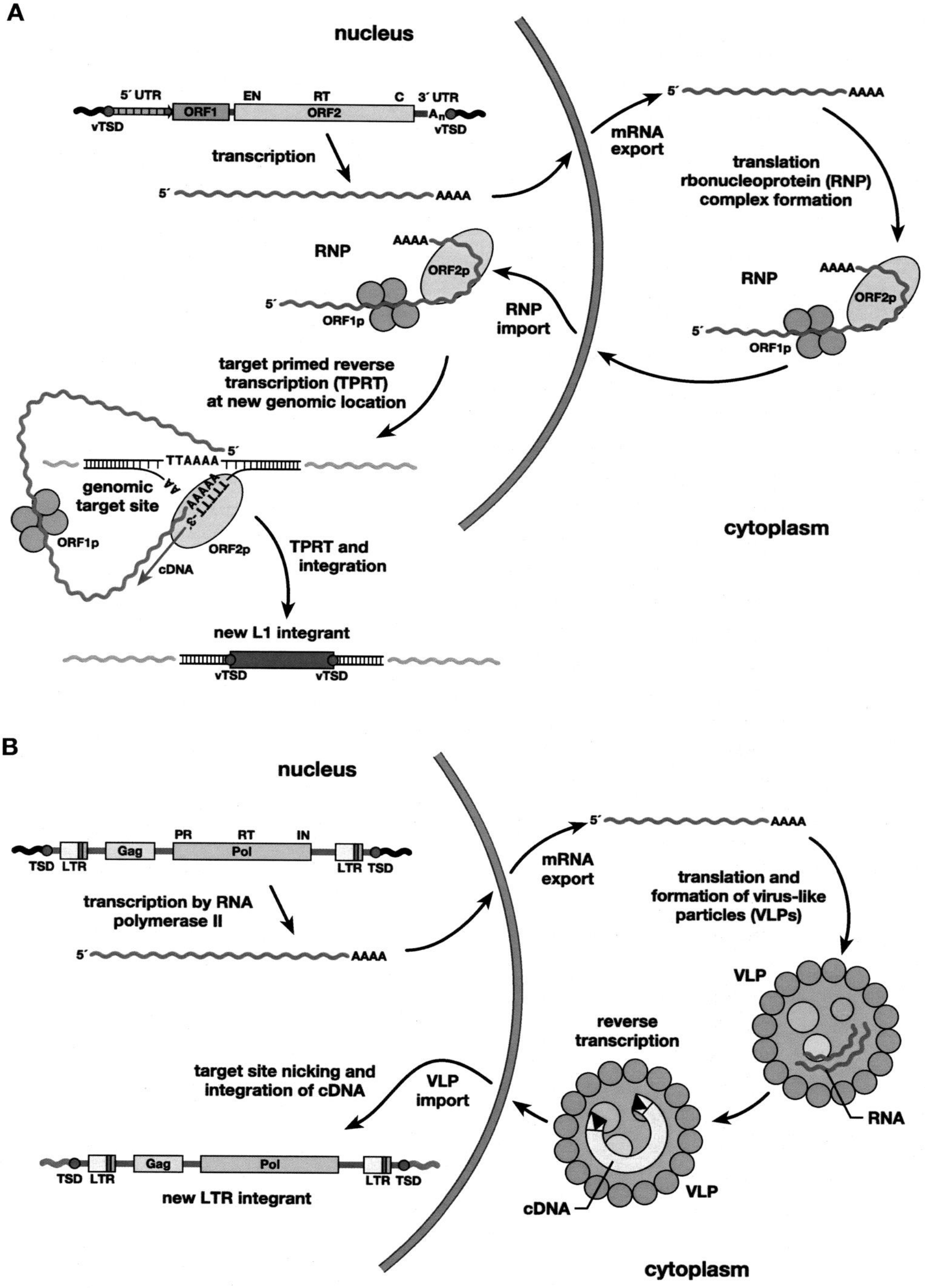

occur for virtually all non-LTR retrotransposition integration events. (B) Similar to non-LTR retrotransposons, the LTR retrotransposons are thought to be transcribed in the nucleus and their transcripts exported to the cytoplasm where encoded proteins are translated. These proteins together with retrotransposon transcripts are assembled into virus-like particles (VLPs). Reverse transcriptase (RT), part of the encoded *pol* gene, first forms a single-stranded cDNA from the RNA template. The PBS hybridizes with particular cellular RNAs such as tRNA, priming the minus-strand cDNA by reverse transcription. Following strand transfer reactions, double-stranded cDNA is formed in the VLPs which then are imported into the nucleus. A genomic target site is nicked by LTR retrotransposon-encoded endonuclease, and integrase catalyses integration of the double-stranded cDNA into a new genomic locus.

orthogonally to the DNA backbone will naturally generate a 5-bp 3′ overhang.

Further subclassifications of the LTR retrotransposons have been proposed (Bowen and McDonald, 1999; Curcio and Derbyshire, 2003; Hull, 1999, 2001; Pringle, 1999; Xiong and Eickbush, 1988, 1990). Probably the most robust sub-classification schemes are based on alignments of the RT protein sequences, as RT is the most highly conserved of the LTR retrotransposon proteins (Xiong and Eickbush, 1990). These schemes agree well with an earlier, simpler subclassification of LTR elements, which was based on the order of the functional domains encoded in the *pol* gene. The *pol* genes are ordered as PR, IN, and then RT/RH in one of the main LTR retrotransposon subgroups, epitomized by the Ty1/*copia* elements in *Saccharomyces cerevisiae* and *Drosophila melanogaster*, respectively. By contrast, the order of *pol* domains found in Ty3/*gypsy* elements, BEL-like retrotransposons (Frame *et al.*, 2001; Kal *et al.*, 1999) and retroviruses is PR, RT/RH, and then IN.

Surprisingly, in certain cases some transposases are used by both DNA transposons and retrotransposons. All DDE DNA transposases and retroviral integrases are part of a larger family of nucleases derived from RNase H enzymes.

Another category of retrotransposons is made up of newly named Y-retrotransposons (Curcio and Derbyshire, 2003), which may form a circular cDNA intermediate by reverse transcription. In turn, this circular cDNA intermediate is presumably integrated into the genomic target by a Y (tyrosine)-containing transposase without forming target site duplications (Curcio and Derbyshire, 2003). This class of unusual retrotransposons includes DIRS1-like (Goodwin and Poulter, 2001) and Kangaroo (Duncan *et al.*, 2002) elements.

Two additional families of transposases, serine transposases and Y2 or rolling circle (RC)-transposases (Curcio and Derbyshire, 2003), appear to be limited to mobilization of DNA transposons. Despite their widespread distribution, DNA transposons appear to be ancient and inactive in most mammalian genomes. However, several families of active DNA transposons, including elements that have been mobilized recently by RC transposase activity, recently have been identified in bats (Pritham and Feschotte, 2007; Ray *et al.*, 2008).

LINEs

The most successful, widely dispersed retrotransposons in most mammals are LINE retroelements. Relatives of LINEs, encompassing the non-LTR retrotransposons, are found in eukaryotes as diverse as plants and some fungi. In primate, rodent, carnivore and ruminant genomes, among others, LINEs are autonomous, i.e. they encode the proteins necessary to mobilize themselves, including reverse transcriptase and endonuclease activities. They appear to be actively mobilized even now in all mammals studied, although a crude comparison of such activities reveals a wide range in such activity in different species and at different times in evolution (International Mouse Sequencing Consortium, 2002; Lander *et al.*, 2001).

There are LINE retrotransposons in vertebrates ranging from fish and reptiles to man. Strikingly, there are even LINE-like retroelements, lacking long terminal repeats like all their family members, present in plants (called Cin4 elements) (Noma *et al.*, 1999) and in various fungi such as *Candida albicans* (Goodwin and Poulter, 2001). The Zorro3 retrotransposon recently has been marked with a retrotransposition reporter construct, allowing a demonstration of its activity *in vivo* (Goodwin *et al.*, 2007). Typical structural features of full-length LINE elements include at least open reading frame (and usually two ORFs), a poly(A) tail and target site duplications (TSDs). Once an element becomes defective through truncation or some other inactivating mutation, these features of new genomic integrants typically are degraded over time, allowing estimation of the age of integrants based upon nucleotide substitution rate compared with consensus sequences, length of TSDs and poly(A) tails, etc. This can be challenging work however, as it is not always possible to determine whether or not full-length elements are active by inspection of the sequence. Functional testing is necessary as was demonstrated elegantly by Kazazian and colleagues (Brouha *et al.*, 2003).

In contrast to the LINE1 (L1) elements typically present in other mammals (Fig. 1.1), there are LINE2 retrotransposons present in high numbers in the recently completed platypus genome (Warren *et al.*, 2008). The latter element is extinct and non-functional in most mammalian lineages. Surprisingly, the overall density of interspersed repeats (about 2 elements per kb of genomic sequence) was found to be higher in platypus than any other metazoan genome analysed to date. The platypus full-length LINE2 element is about 5 kb, while most genomic LINE2 integrants are severely truncated. This element also appears to mobilize a very abundant non-autonomous SINE (short interspersed element) retrotransposon, MIR/Mon-1 in platypus (Warren *et al.*, 2008).

LINE retrotransposons are thought to be mobilized from one chromosomal location to another through an RNA-copying mechanism that occurs at a new (target) genomic location. This mechanism is referred to as target primed reverse transcription (TPRT) (Fig. 1.2A) (Luan *et al.*, 1993), and appears to operate for most non-LTR retrotransposons. The canonical target site for L1 conforms loosely to the consensus 5′ TTTT^AA 3′, consistent with the *in vitro* specificity of the L1 endonuclease (Cost *et al.*, 2002; Feng *et al.*, 1996) and examination of reconstructed target sites of L1 and *Alu* insertions (Gilbert *et al.*, 2002; Jurka, 1997; Symer *et al.*, 2002). This target consensus sequence, which is only loosely conformed to in bona fide new insertions, is thought to reflect nicking specificity by the LINE endonuclease domain encoded at the amino terminus of the second open reading frame (ORF-2) (Cost *et al.*, 2002). A recent structural study revealed that invariant amino acids in other non-LTR retrotransposon endonucleases are conserved in the active site of the L1 endonuclease.

There is at least one endonuclease-independent mechanism for insertion of L1 sequences as well. These occur at high frequencies in cells that are compromised in certain DNA repair pathways, and suggest that retrotransposition mechanisms may in fact exploit elements of native DNA damage repair pathways (Morrish *et al.*, 2002). These results were extended recently by the demonstration that disabled human L1 retrotransposons, with a disrupted endonuclease domain in ORF2, can integrate into dysfunctional telomeres in cells lacking a component of non-homologous end joining DNA repair pathway. The dysfunctional telomeres appear to provide an optimal integration substrate for L1 elements (Morrish *et al.*, 2007).

Genomic copies of LINE retrotransposons typically are truncated from their 5′ ends (Fig. 1.1A) (Akagi *et al.*, 2008; Ostertag and Kazazian, 2001a; Symer *et al.*, 2002; Szak *et al.*, 2002). The mechanism of formation of such truncated integrants initially was attributed to poor processivity by the L1 reverse transcriptase activity encoded in ORF2p. However, recently an alternative proposal was suggested, where ends of integrants appear to be guided to a significant extent by microhomology between the L1 RNA template and the target site. This was initially observed in a tissue culture model wherein marked L1 elements were found to integrate frequently as truncated elements into target sites where the target site sequence shows significant identity to the L1 template sequence itself (Symer *et al.*, 2002; Zingler *et al.*, 2005). Subsequently, additional evidence supporting this model was obtained by inspection of new genomic integrants *in vivo* (Babushok *et al.*, 2006) and those identified from large-scale genome sequencing projects (Akagi *et al.*, 2008; Martin *et al.*, 2005b).

In addition to these common 5′ truncations, about 15% of human L1 integrants contain an inverted segment at the 5′ end, which can occur either in full-length or truncated integrants (Ostertag and Kazazian, 2001b). Thus, despite their huge numbers genome-wide, only a few thousand full-length L1s exist in the human genome, of which approximately 80–100 are thought to be retrotransposition competent. Of these, only a small number are thought to be active donor elements (Brouha *et al.*, 2003). A recent analysis suggested that an element at a given genomic locus active in one individual might be inactive in others, and that allelic differences among specific retrotransposon copies could lead to very different retrotransposition potentials in different individual humans (Seleme *et al.*, 2006).

A parsimonious explanation for L1 integrant inversions and truncations is provided by the elegant twin priming model of Ostertag and Kazazian (Ostertag and Kazazian, 2001b). This model is based on sequential, target-primed reverse transcription reactions (Luan *et al.*, 1993). It suggests that the primer extension reactions at the two free 3′-hydroxyl groups on both DNA strands occur sequentially based on the same RNA template, resulting in an inversion of the newly integrated cDNA (Ostertag and Kazazian, 2001b).

Sense-stranded human L1 transcripts are initiated by an internal promoter within the 5′ UTR of nearly full-length genomic elements (Olovnikov *et al.*, 2007; Ostertag and Kazazian, 2001a). This RNA polymerase II promoter drives expression of bicistronic transcripts encoding two gene products, both required for retrotransposition (Moran *et al.*, 1996). To date, several transcription factors have been identified that regulate L1 expression in a tissue-specific fashion. These include Yin Yang-1 (YY1) (Becker *et al.*, 1993), members of the SRY testis-determining factor family including SOX11 (Tchenio *et al.*, 2000), RUNX3 (Yang *et al.*, 2003) and probably others (Olovnikov *et al.*, 2007).

ORF1p, a ~40-kDa protein, is widely expressed at detectable levels (Goodier *et al.*, 2004; Hohjoh and Singer, 1996; Kulpa and Moran, 2005; Leibold *et al.*, 1990; McMillan and Singer, 1993), while ORF2p, a ~130-kDa protein, has been very difficult to detect until recently (Ergun *et al.*, 2004; Goodier *et al.*, 2004). Both proteins are required for L1 retrotransposition in mammals (Moran *et al.*, 1996), although as few as one molecule of ORF2p per L1 transcript may suffice for retrotransposition (Wei *et al.*, 2001). These proteins appear to bind the L1 transcript *in cis*, forming a ribonucleoprotein complex that is imported from the cytoplasm into the nucleus where new target insertions can occur at genomic target sites (Kulpa and Moran, 2005, 2006).

L1 ORF1p binds nucleic acids and possesses essential nucleic acid chaperone activity (Hohjoh and Singer, 1996, 1997; Kolosha and Martin, 1997, 2003; Martin, 2006; Martin and Bushman, 2001; Martin *et al.*, 2005a). The proximal two-thirds of the protein, starting at its amino terminus, mediates formation of ORF1p homotrimers, through self-assembly of coiled-coil domains. The basic, carboxy-terminal third binds nucleic acids (Martin, 2006; Martin *et al.*, 2003).

ORF2p encodes the endonuclease, reverse transcriptase and zinc finger domains that are required for L1 retrotransposition (Fig. 1.1A). In contrast with relatively low levels of sequence conservation of ORF1 in different mammalian species, ORF2p is much more conserved between human, mouse, rat and other mammalian elements (Li *et al.*, 2006). Its expression is toxic in most cells (Goodier *et al.*, 2004), as its endonuclease activity appears to create double stranded breaks across the genome (Gasior *et al.*, 2006).

To date, the two L1 ORFs have not been shown to be expressed or function as a single fusion protein. Instead, they appear to be translated as separate products from a single, bicistronic L1 transcript. Two recent studies examined the basis for this unconventional expression of two independent proteins, as encoded by both human and mouse genomic elements, from bicistronic L1 transcripts (Alisch *et al.*, 2006; Li *et al.*, 2006). Unlike mechanisms commonly used by LTR retrotransposons and retroviruses, L1 ORF2 (Pol) proteins are not expressed by translational frameshifting or suppression of stop codons. In the full-length mouse L1 transcript, two internal ribosomal entry sites (IRESes) are present upstream of each of the two ORFs, and in particular one is reportedly present in the 3′ end of ORF1 (Li *et al.*, 2006). Surprisingly, though, multiple base changes in the IRES sequences seem to have no deleterious impact on retrotransposition. In both full-length human and mouse L1 transcripts, a short inter-ORF sequence separates the two open reading frames. In both species, this intergenic sequence is dispensable for translation of ORF2. While translation of a non-specific upstream ORF is required for efficient translation of human L1 ORF2, both ORF1-specific sequences and functional ORF1p expression are not required for ORF2 translation. Remarkably, the initiation ATG codon in human ORF2 also is not required for its translation, as it can be replaced by any other codon without compromising retrotransposition. This suggests that the sequence context immediately surrounding the ATG is not as critical as would be expected for internal ribosome reinitiation mediated by an IRES. Thus, human L1 elements lack an essential internal ribosomal entry site (IRES) sequence both in the inter-ORF region and in ORF1, in contrast to mouse L1s (Alisch *et al.*, 2006; McMillan and Singer, 1993). Translation of human L1 ORF2 thus resembles the 'translational coupling' mechanism proposed for SART1 elements, another non-long terminal repeat retrotransposon in silkworm (Kojima *et al.*, 2005).

SINEs

By definition, these short, mobile interspersed elements are less than 500 nt long (Fig. 1.1). A second characteristic feature is that SINEs typically are expressed by RNA polymerase III, unlike other short interspersed repeats such as microsatellites (Deininger *et al.*, 2003). In humans, *Alu* elements are the most common category, with more than 500,000 copies present out of 1 million independent SINEs present in the haploid genome. They are named *Alu* elements because of the presence of an AluI restriction site in many of them. Rodents have B1, B2 and ID elements. Other metazoan genomes have many different, structurally diverse SINE elements (Kramerov and Vassetzky, 2005).

Human *Alu* elements have a two-part structure, where the 5′ part contains an RNA polymerase III promoter with A and B boxes (Fig. 1.1A). The 3′ part is slightly longer than the 5′ part by ~31 nt. These two related monomers are separated by an intervening, central A-rich region that consists of the sequence 5′-A_5TACA_6-3′. A typical element ends with a poly(A) tail at its 3′ end, and is flanked at both ends by a short target site duplication. Characteristics of the 3′ part resemble those of L1 elements, suggesting that *Alu* insertions depend upon L1 retrotransposition machinery.

The two related monomers in human *Alu* and rodent B1 retrotransposons appear to have been derived from 7SL RNA (Quentin, 1992; Ullu and Tschudi, 1984). Most other SINEs originated from tRNA genes, as exemplified by murine B2 elements (Kramerov and Vassetzky, 2005), although a 5S ribosomal transcript appears to have given rise to zebrafish SINE3 elements (Kapitonov and Jurka, 2003a).

SINEs are thought to be mobilized by LINE elements through *trans*-complementation of the SINE transcript by L1-encoded proteins, in particular L1 ORF2p (Boeke and Stoye, 1997; Chen *et al.*, 2007; Dewannieux *et al.*, 2003; Smit *et al.*, 1995). Retrotransposition of human *Alu* elements requires L1 ORF2p but not ORF1p (Dewannieux *et al.*, 2003). Similarly, SVA elements and processed pseudogenes also appear to be mobilized in *trans* by L1 (see below). This is likely despite a preference for L1s to mobilize their own transcripts *in cis* over most other poly(A) transcripts (Wei *et al.*, 2001). SINE transcripts and SVA transcripts might enjoy an advantage over most other polyadenylated transcripts in retrotransposition mediated by L1 proteins, as they share structural features with 7SL RNA (Ullu and Tschudi, 1984), the scaffold of the ribosomal signal recognition particle. These common structural features appear to target them to ribosomes where they purportedly may interact with nascent L1 ORF2p as it is translated (Boeke, 1997; Dewannieux *et al.*, 2003; Sinnett *et al.*, 1991). Although mammalian SINEs typically end in a poly(A) tail, which is required for retrotransposition (Boeke, 1997; Dewannieux and Heidmann, 2005; Roy-Engel *et al.*, 2002), there may be additional factors that explain striking differences in retrotransposition frequencies for old vs. young human *Alu* subfamily members. Based on a plasmid based mobilization assay in cultured cells (Dewannieux *et al.*, 2003), such factors may include the primary sequences, in the 280-nucleotide core region, that distinguish members of different subfamilies (Han *et al.*, 2005) and a differential ability of their expressed transcripts to interact with SRP9/14, resulting in formation of ribonucleoprotein (RNP) complexes (Bennett *et al.*, 2008).

SVA elements

Initially described as a 'composite' retrotransposon in 1994 (Shen *et al.*, 1994), SVA elements were subsequently found to have been mobilized actively in hominoid primates. They were found as new disease-causing insertions resulting in diseases including hereditary elliptocytosis upon insertion into alpha-spectrin (Ostertag *et al.*, 2003) and Fukuyama-type congenital muscular dystrophy upon insertion into Fukutin (Colombo *et al.*, 2000). Most SVA integrants bear hallmarks of L1 retrotransposition, including target site duplications similar in length to those of L1 retrotransposons, a poly(A) tail, and frequent 5′ deletions and/or inversions (Fig. 1.1) (Ostertag *et al.*, 2003). Integrants occasionally include 3′ transduction events (Pickeral *et al.*, 2000) and rarely have 5′ transduced sequences (Lander *et al.*, 2001; Symer *et al.*, 2002). A few instances of

SVA integrants resulting in large genomic deletions have been well described, where HLA-A was deleted directly by an integrated SVA element in individuals of three Japanese families with frequent leukaemias. These genomic deletions along with 3′ and 5′ transduction events are highly reminiscent of events related to L1 retrotransposition (Chen *et al.*, 2005; Gilbert *et al.*, 2002, 2005; Moran *et al.*, 1996; Symer *et al.*, 2002). Each of these genomic features of SVA retrotransposons strongly suggest that L1-encoded proteins (ORF1p and ORF2p) mobilize them in *trans*.

SVA elements are so-named as an acronym, due to the component parts of full-length elements. These include 490 nt of SINE-R sequences derived from the 3′ end of the *env* gene and part of the 3′ long terminal repeat of human endogenous retrovirus-K-10 (HERV-K10) elements, a region containing a variable number of tandem repeats (VNTR) each consisting of 35–50 nucleotides (nt), and *Alu*-like sequences (consisting of antisense sequences together with other sequence of unknown source), together occasionally with a hexameric repeat (CCCTCT) at the 5′ end. There can be more than 10 of these nucleotide hexamers. The name SVA itself stands for 'SINE-R, VNTR, Alu' (Shen *et al.*, 1994) even though the order of these names are opposite to the presumed functional orientation of the element. The VNTR segment can contain 30 or more of the tandem repeats, and ranges from 48 to 2306 nt in overall length (Wang *et al.*, 2005).

While an early analysis found SVA elements (called SINE-R elements at that time) in humans, gorillas and chimpanzees (Zhu *et al.*, 1994), more recent studies showed that SVA elements are present in all hominoid primates (Kim *et al.*, 1999). Subsequent analysis revealed that there are six SVA subfamilies, SVA-A through SVA-F, based upon alignment of the S part (SINE-R) (Wang *et al.*, 2005). Members of subfamilies SVA-E and SVA-F are polymorphic in humans, indicating that they are human-specific retrotransposons. SVA elements were not detected in Old World monkeys such as green monkeys or rhesus macaques (Wang *et al.*, 2005). Since truncated SVA integrants have been identified from all subfamilies, it appears that they originated as a composite element containing the SINE-R, the VNTR and the *Alu*-like sequences from the outset, probably after the divergence of Old World monkeys from hominid primates approximately 25 million years ago. It is possible that full-length elements might have accumulated the hexamer repeat at their 5′ ends more recently, perhaps because of its putative RNA *pol* II promoter activity.

RTE elements

In addition to LINE autonomous non-LTR retrotransposons, RTE retrotransposon family members are present in some mammalian genomes as recently divergent as marsupials (*M. domestica*), along with ruminants and many other non-mammalian metazoan genomes (Malik and Eickbush, 1998). These unusual retrotransposons have a single rather short ORF encoding endonuclease and reverse transcriptase activities. Its sequence seems most similar to the analogous ORF, ORF2, in CR1 autonomous retrotransposons in birds and reptiles (Malik and Eickbush, 1998). The latter are also found in mammalian genomes albeit only in fossil form. Several distinct lineages of SINEs appear to have arisen from RTE elements, and particularly from truncated elements (Malik and Eickbush, 1998).

Retroposed, processed pseudogenes

Processed pseudogenes are a common class of intron-lacking pseudogenes that often contain poly(A) tracts at their 3′ ends, i.e. they look like cDNA copies of mRNAs. At both termini they contain TSDs similar to those of LINES and SINEs, suggesting mobilization by a similar mechanism. Indeed, it has been shown experimentally that both L1 ORF1p and ORF2p are able to mobilize cellular mRNAs in *trans* (Esnault *et al.*, 2000).

CR1 retrotransposons

Chicken repeat 1 (CR1) elements are some of the most abundant and widely dispersed clades of non-LTR retrotransposons. Such elements have been identified in many metazoan genomes including mammals, birds, fish, amphibians, turtles and lizards, and even invertebrates (Haas *et al.*, 1997, 2001; Kapitonov and Jurka, 2003b). Like LINE retrotransposons, these elements typically are truncated extensively from their 5′ ends, but computationally reconstructed, full-length elements contain two ORFs (Haas *et al.*,

1997). Various ORF1 proteins include conserved plant homeodomain (PHD) and esterase (ES) domains, which suggests that they may play a role in chromatin remodelling and/or lipolysis or membrane fusion events (Kapitonov and Jurka, 2003b). The esterase domain is also found in certain RNA virus proteins, where it may play an essential function in infectivity by cleaving sialic acid from its receptor (Herrler *et al.*, 1985). An analogy between various ORF1 proteins and retroviral nucleocapsid proteins, involved in packaging retroviral RNA and particles, has also been made based on the presence of conserved Zn-finger like domains found in some CR1 ORF1 proteins. These conserved domains may bind nucleic acids. CR1 ORF2 is predicted to encode an endonuclease and a reverse transcriptase, suggesting that these retrotransposons could move by target primed reverse transcription (Haas *et al.*, 1997) as proposed for other non-LTR retrotransposons such as R1Bm and L1 elements.

L3 elements are some of the most ancient non-LTR retrotransposons, contain two ORFs, and are members of the CR1 clade (Kapitonov and Jurka, 2003b). Their structure has been reconstructed based on computational analysis. Their ORF1p encodes both ES and PHD domains which are predicted to mediate cell fusion and chromatin remodelling activities (Kapitonov and Jurka, 2003b).

I-factor elements (Drosophila)

These retrotransposons share several structural features with L1 elements. Marked I-factor elements, where reporter genes were inserted into the elements' ORFs, can retrotranspose, albeit inefficiently (Jensen *et al.*, 1994; Pelisson *et al.*, 1991). Like L1s, I-factor elements appear to have a *cis*-preference for their own transcripts. This *cis*-preference appears to be mediated through the 3′ end of the transcripts (Chaboissier *et al.*, 2000).

TART and HeT-A (Drosophila)

The telomeres in *Drosophila melanogaster* and other members of the *Drosophila* genus are made up of a head-to-tail mixed array of two different non-LTR retrotransposons, i.e. TART and HeT-A elements. These two unrelated elements appear to have co-evolved over tens of millions of years, as related elements with certain conserved features are present in other members of the *Drosophila* genus that diverged from *D. melanogaster* approximately 60 million years ago, including *Drosophila virilis* and *Drosophila americana* (Casacuberta and Pardue, 2003). Like LINE elements, TART encodes two open reading frames. ORF1 of both TART and HeT-A encodes a Gag-like protein that includes a CCHC or zinc knuckle motif and the major homology region, both characteristic features of retroviral Gag proteins. TART ORF2 encodes the Pol polypeptide with reverse transcriptase and endonuclease domains. HeT-A elements are about 6 kb in length, while TART elements are ~10 kb. Curiously, HeT-A lacks its own reverse transcriptase, but the HeT-A Gag protein is hypothesized to help direct the TART ORF1/ORF2 proteins to telomeres.

These retrotransposons are not found in euchromatic regions of *Drosophila* species. Instead, they are localized in the telomeric regions of polytene chromosomes. The *Drosophila virilis* chromosomes are acrocentric, and the retrotransposons are enriched in the telomeres of the long chromosomal arms, away from the chromocentre formed by the short arms (Casacuberta and Pardue, 2005).

An unusual feature of these telomeric retrotransposons in *Drosophila* is the frequent presence of CAX repeats in the sense strand of both open reading frames, encoding polyglutamine tracts of unknown significance. Another unusual feature of the TART and HeT-A elements recently identified in *Drosophila virilis* is the X domain in ORF2, again of unknown function. This domain could possibly encode an RNase-H activity that is present in a similar position within ORF2 of another non-LTR retrotransposon in *Drosophila*, I factor (Casacuberta and Pardue, 2003; Malick and Eickbush, 2001).

LTR retrotransposons

LTR retrotransposons comprise two main types, the *copia*/Ty1 and *gypsy*/Ty3 families (Fig. 1.1B), formally known as the Pseudoviridae and Metaviridae, respectively. The *copia* and *gypsy* elements are archetypal elements from *Drosophila* whereas the Ty elements are yeast (*Saccharomyces cerevisiae*) retrotransposons. These four elements represent four specific families (each typically containing dozens of family members), but it is important to recognize that in the larger world of eukaryotic species, this truly represents a tiny tip of a gigantic iceberg. Given that there are dozens of eukaryotic phyla, each probably comprising many genera, and that, on average, perhaps 10 families of these LTR elements are specific to each genus, there are likely to be thousands upon thousands of Pseudoviridae and Metaviridae families. Thus, this is an extremely diverse group of elements with new families discovered every time a eukaryotic genome is sequenced.

The former family differs from retroviruses and the Ty3/*gypsy* family in that the order of functional domains in Pol is PR-IN-RT rather than PR-RT-IN (Fig. 1.1B). Whereas the *copia*/Ty1 family is found in fungi, plants and invertebrates, it is most abundant and diverse in plant lineages. The Ty3/*gypsy* family, which is a likely progenitor to retroviruses, is again represented in all three eukaryotic groups mentioned above. Interestingly, certain members of this group in plants and insects, including *Drosophila gypsy* itself, contain a third open reading frame encoding an Env-like protein, whereas others do not. Furthermore, vertebrate-specific members, both with and without Env-like reading frames, such as the sushi element of puffer fish and the IAP (intracisternal A-particle) elements of rodents (which are more often considered endogenous retroviruses), are now well known. We will briefly review the biology of typical metazoan members of these families.

The process of reverse transcription of LTR retrotransposons and their overall mechanism of retrotransposition is similar to the baroque process employed by retroviruses, in that full-length cDNA is generated from transcripts that lack terminal sequences found at both ends (5′ and 3′) of full-length genomic integrants (Fig. 1.2B). As with retroviruses, LTR retrotransposon transcripts start in the upstream LTR at the repeated region (R) and contain a sequence that is unique to the 5′ end (U5), followed by a primer binding site (PBS). Their transcripts end in the downstream LTR at a terminator that is in the sequence unique to the 3′ end (U3), which is followed by the R region (Figs. 1.1B and 1.2B).

In most but not all *copia* and *gypsy* family elements, a host-encoded tRNA primer hybridizes to the 5′ PBS element in the transcript, priming reverse transcription from it, through the U5 and R sequences to the 5′ end of the transcript (Fig. 1.2B). This results in single-stranded, minus-strand cDNA, forming an RNA/DNA chimera. In copia, the primer tRNA is first cleaved by an unknown mechanism and it is the 5′ half molecule that serves as the primer (Kikuchi *et al.*, 1986). Typically an RNase H activity (encoded by the retrotransposon) progressively degrades the 5′ end of the mRNA template, and the newly formed, single-stranded cDNA hybridizes with the 3′ end of the transcript. This minus-strand exchange thus allows priming and reverse transcription of full-length, minus-strand cDNA, followed by plus-strand synthesis. Ultimately, full-length, double-stranded cDNA is completed, packaged into virus-like particles, imported into the cellular nucleus, and integrated at a new genomic target site by retrotransposon integrase (IN) activity (Fig. 1.2B).

Yet another mechanism is used by the sushi LTR retrotransposons of various fish (as well as related elements in fungi invertebrates and plants), which use 'foldback' priming in which a 5′ fragment of the retrotransposon RNA is used instead of a tRNA (Butler *et al.*, 2001b; Levin, 1995; Lin and Levin, 1997). This may well represent a more primitive mechanism of priming, as it is intramolecular and host function independent. In fact, recent evidence suggests the RNAse H portion of the RT itself makes this cleavage in the related fungal Tf1 element (Hizi, 2008; Levin, 1996). This first strand (minus strand) cDNA is transferred to the 3′ end of the transcript and hybridizes with the R region, while the 5′ end is degraded by RNase H. The short first-strand cDNA is elongated by primer extension, allowing reverse transcription of the rest of the transcript. Meanwhile the transcript is degraded further, and one of the short RNAs primes second strand (plus strand) cDNA synthesis, resulting in an RNA-DNA chimeric molecule that includes the PBS. This plus-strand molecule then anneals to the PBS sequence of the minus-strand cDNA, providing a template for completion of minus-strand synthesis through the primer binding site (PBS), U5, R and U3 to the 5′ end of the retrotransposon. Meanwhile, the plus-strand molecule itself undergoes primer extension, allowing completion of the second strand synthesis by

RT. The entire element is ligated into a staggered target site and presumably undergoes DNA repair of the remaining nicks, generating a *de novo*, full-length LTR retrotransposon integrant with TSDs at its flanks. Most of these steps as outlined above, except for the unique forms of priming, closely resemble the steps in the retroviral life cycle.

Two vertebrate families of LTR elements deserve additional discussion here: IAP of rodents and sushi of vertebrates. IAP retrotransposons (intracisternal A-particle) are of special interest because although they are taxonomically speaking an endogenous retrovirus (Retroviridae), they seem to have adopted a 'retrotransposition lifestyle' in which they move by an infection-free mechanism. This appears to be due to defects in the envelope gene *env* of virtually all IAP elements (Mietz *et al.*, 1987), apparently representing a case of 'devolution' from a classic envelope-containing retrovirus precursor (Ribet *et al.*, 2008). This loss of infectivity, together with a gain of an endoplasmic reticulum targeting signal, led to 'intracellularization' of IAP retrotransposons (Ribet *et al.*, 2008). Another class of LTR retrotransposons in mouse, MusD elements, lack any discernable *env* gene. This latter category is present at approximately 100 copies per mouse genome.

IAP is very important in a review chapter such as this because it represents an extremely well-characterized retroelement, with many papers describing its biology and one in particular documenting its high-frequency retrotransposition in mammalian cells. Several IAP elements from the mouse were tagged with an antisense reporter gene that had been interrupted by an artificial intron, specifically indicating retrotransposition (Boeke *et al.*, 1985; Moran *et al.*, 1996). Using this reporter system, the elements were shown to transpose to efficiencies as high as 10^{-4} to 10^{-3}. This work (Dewannieux *et al.*, 2004) showed that an element lacking an *env* gene was competent for retrotransposition. However, this experiment, even though it is done in tissue culture cells, does not entirely rule out the possibility that the process requires an extracellular *env*-mediated step, with the *env* gene product provided by one of the myriad endogenous retrovirus copies. Interestingly, the highest retrotransposition frequencies were observed in human cells. This result suggests that if there is complementation at a low level by an endogenous retrovirus *env* gene, it cannot be very specific to one type of envelope gene. Thus the case is becoming stronger that the IAP indeed moves via retrotransposons intracellular lifestyle and not by *env* complementation of an infectious process.

The sushi elements are widespread in lower vertebrates (Butler *et al.*, 2001b). Related sequences have been found in mammals. Their unusual mechanism of minus-strand priming of reverse transcription has already been noted above.

Genome biology of retrotransposons

As the genome sequences of many higher eukaryotic species have been completed over the past several years, and more are sequenced virtually every week, the extensive sweep of retrotransposons within all multicellular animals has become very clear. Various classes of retrotransposons have invaded each and every genome repeatedly over evolutionary time, sometimes to quite high numbers. Moreover, sequences from multiple individuals of a given species have been determined recently. This work has allowed detailed comparisons of similarities and differences that distinguish individual genomes, many of which are attributable to direct or indirect effects of retrotransposons (Fig. 1.3). Thus, the era of 'personal genomics' has just started. It seems very likely that retrotransposition has contributed to such individual variation. We speculate that in the near future, even more compelling evidence may be obtained to indicate that retrotransposition occurs actively in the germ line and quite possibly in early normal development (Kano *et al.*, 2009) and perhaps more likely in certain diseases, including human cancers. Indirect effects of these ubiquitous repetitive elements also may play a profound role in many biological functions.

A recent analysis of segmental duplications and copy number variation in various inbred mouse strains indicated that both LINEs and LTR retrotransposons are enriched in the genomic duplications (She *et al.*, 2008). The opposite situation has been observed in the human genome, where segmental duplications are associated with SINEs (Lander *et al.*, 2001). It is possible that these retrotransposons may promote formation

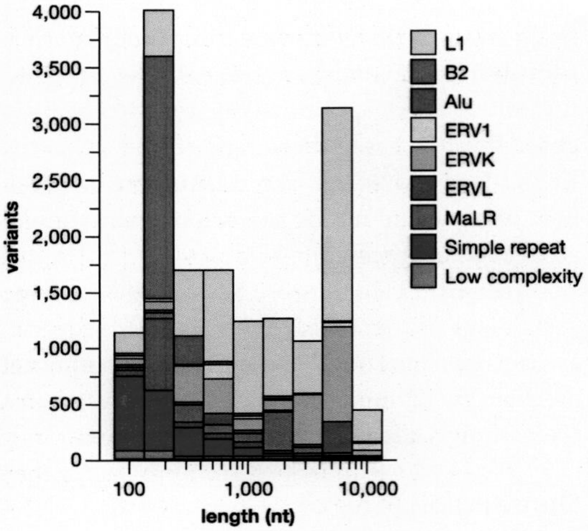

Figure 1.3 Genomic variation between mouse strains due to endogenous retrotransposition. Thousands of mouse strain insertional polymorphisms, ranging from 100 nt to 10 kb, were identified by alignment of whole genome shotgun sequence traces to the C57BL/6J reference genome. Each polymorphic integrant is present in the C57 reference genome and absent from at least one of the strain(s) A/J, DBA/2J, 129S1/SvImJ and 129X1/SvJ. The percentages indicate the relative contribution by each class of retrotransposon, where polymorphic integrants were identified by RepeatMasker to contain >70% RepeatMasker content and range in length from 100 nt to 10 kb. L1 retrotransposition is the most frequent cause of such variation between the mouse strains. Reproduced from Akagi *et al.* (2008) with the permission of Cold Spring Harbor Laboratory Press.

or retention of the segmental duplications, although the mechanistic basis for these phenomena are not known.

LINEs

Approximately 100 retrotransposition-competent L1 retrotransposons reside in the human genome (Brouha *et al.*, 2003; Sassaman *et al.*, 1997). New retrotransposition events have been estimated to occur in about 10% of human sperm cells (Kazazian, 1999; Li *et al.*, 2001). By contrast, approximately 10-fold more retrotransposition-competent elements exist in the mouse genome. The basis for such differences in L1 retrotransposition activities in distinct species has been unclear, given that marked individual human L1 elements appear to retrotranspose more actively than mouse elements in cultured cell or *in vivo* mouse models (Han and Boeke, 2004).

Despite similarities in their primary nucleotide sequences and the presence of intact ORFs, retrotransposition-competent L1s in the human genome appear to undergo retrotransposition at a wide range of frequencies (Brouha *et al.*, 2003; Seleme *et al.*, 2006). In some cases, such differences in retrotransposition may be attributable to single nucleotide polymorphisms within the genomic donor elements, particularly when located within ORF2 (Seleme *et al.*, 2006).

New human and mouse L1 insertions appear to be enriched in genomic regions that have relatively increased densities of A and T nucleotides, probably because of the somewhat promiscuous target specificity of the L1 ORF2p endonuclease (Cost and Boeke, 1998). A typical nicking site sequence for the L1 ORF2p on the bottom DNA strand at an integrant target site is 5′ TTTT^AA 3′, but its nicking activity is quite non-specific, so many other primary sites have been observed (Babushok *et al.*, 2006; Gilbert *et al.*, 2005; Gilbert *et al.*, 2002; Moran *et al.*, 1996; Symer *et al.*, 2002). Even when broader blocks of genomic sequences are considered, recently integrated L1 elements are biased in A/T-rich regions (Akagi *et al.*, 2008; Myers *et al.*, 2002).

Recently, extensive sequencing of additional human and mouse individual- or linear-specific genomes has been accomplished. Analysis of these genomic sequences revealed that active retrotransposition has contributed to structural differences that distinguish the individual genomes (Fig. 1.3). The extent of variation between human individuals due to endogenous L1 mobilization appears to be more modest than that distinguishing mouse strains (Akagi *et al.*, 2008; Kidd *et al.*, 2008; Korbel *et al.*, 2007; Mills *et al.*, 2006a). The movement of L1 elements in mouse has contributed more to the formation of more insertion/ and deletion (indel) sequence variants than have other endogenous retrotransposons, resulting in many thousands of genomic sequence differences that distinguish various mouse strains (Fig. 1.3) (Akagi *et al.*, 2008).

SINEs

Human *Alu* elements appear to be mobilized from a relatively small number of templates corresponding to young subfamilies such as *Alu* Ya5 and *Alu* Yb8 (Bennett *et al.*, 2008; Cordaux *et al.*, 2004), despite the widespread distribution and diversity of *Alu* elements in the genome (Deininger *et al.*, 1992). In particular, the number of active *Alu* copies in the human genome appears to be well over 800, and probably extends into the thousands (Bennett *et al.*, 2008). These so-called 'master donor loci' may be productively transcribed and mobilized due to conserved core nucleotides (Bennett *et al.*, 2008) and/or the presence of surrounding genomic features that favour their expression, such as active promoters or enhancers, efficient terminators, favourable chromatin structure or genomic location, lack of DNA methylation, and other factors (Deininger *et al.*, 1992). They are not randomly distributed throughout the human genome. Instead, they have accumulated in gene-rich regions of chromosomes. When both young, active and ancient, inactive elements are considered, *Alu* retrotransposons account for more than 10% of all human DNA. *Alu* elements continue to retrotranspose actively, as demonstrated by over 40 disease-causing integration events that have been identified to date (Belancio *et al.*, 2008; Chen *et al.*, 2005). A recent estimate of *Alu* mobilization frequencies suggested that 1 in 20 humans is born with a *de novo* Alu integrant (Cordaux *et al.*, 2006).

A relatively similar situation appears to have occurred in the mouse genome (Fig. 1.3), although mobilization of SINE elements has been less active than in human. An elegant model recently has been proposed to explain the gradual but extensive accumulation of these elements in the human genome, whereby *Alu* elements have low levels of retrotransposition activity over long time periods, but certain retrotransposition-competent, 'master' donor elements occasionally can be mobilized at high frequencies, in limited 'bursts' of activity (Han *et al.*, 2005).

During primate evolution, at least 3000 deletion events corresponding to 900 kb of genomic DNA have been attributed to *Alu* retrotransposition (Callinan *et al.*, 2005). It has been proposed that the first strand cDNA generated at the time of *Alu* integration through TPRT (Fig. 1.2A) may be 'hijacked' by microhomology-mediated binding to other chromosomal sequences, leading to genomic deletions (Chen *et al.*, 2007; Symer *et al.*, 2002). Such events have led to diseases such as Apert syndrome, in which a large part of the *FGFR2* gene was deleted by a novel *Alu* integration event (Bochukova *et al.*, 2009).

SVA elements

The genomic distribution of SVA elements is thought to reflect target preferences of L1 endonuclease, i.e. at the same characteristic L1 target sites described above, 5′-TTTT^AA-3′. However, SVA elements, particularly younger integrants, are enriched in G/C-rich and gene-rich regions of the human genome, like SINE retrotransposons are (Wang *et al.*, 2005). They are non-randomly distributed across human chromosomes, as they are enriched significantly on chromosomes 1, 17, 19 and 22, but depleted from 4, 5, 13, 18, 21 and the Y chromosome. Interestingly, SVA elements are neither enriched nor depleted significantly in the X chromosome, unlike LINE retrotransposons themselves. The overall genome-wide distribution of SVA elements resembles that of *Alu* elements, although the youngest subfamilies of SVA elements are found in relatively G/C-rich genomic regions, while the youngest *Alu* elements are localized in A/T-rich regions (Wang *et al.*, 2005). Ono *et al.* estimated that the number of SVA copies per haploid human genome is 4500 (Ono *et al.*, 1987). Batzer and colleagues identified 2762 elements in the draft human genome sequence available in 2004 (Wang *et al.*, 2005), of which 63% are full length.

LTR retrotransposons

Ongoing endogenous retrotransposition by LTR retrotransposons has generated substantial variation between mouse strains (Fig. 1.3) (Akagi *et al.*, 2008; Zhang *et al.*, 2008). It has become clear that endogenous retrovirus-like elements have been much less active in the human genome in comparison with the mouse genome (Bennett *et al.*, 2004; Dewannieux *et al.*, 2006; Maksakova *et al.*, 2006; Mills *et al.*, 2006a,b, 2007).

As is frequently the case for other retroelements in the human and mouse genomes, the orientation of various classes of human LTR retrotransposons located within genes is biased in the antisense direction (van de Lagemaat *et al.*, 2006). This orientation bias presumably reflects selection for or against various effects by the retroelements upon gene expression over evolutionary time (van de Lagemaat *et al.*, 2003).

In *Drosophila*, *gypsy* retrotransposons are organized in specialized insulator bodies, which help to determine the organization of chromatin (Gerasimova *et al.*, 2000).

Processed pseudogenes

Thousands of processed pseudogenes recently have been identified in the human and other eukaryotic genomes, by using computational approaches (Zhang *et al.*, 2004; Zhang and Gerstein, 2004). These sequences appear to have accumulated either from retrotransposition or copied as part of segmental duplications (Zheng *et al.*, 2007). Interestingly, almost one-fifth of all pseudogenes identified in the human genome appear to be transcribed in various cell lines or tissues (Zheng *et al.*, 2007).

A large proportion of human pseudogenes arose from highly transcribed genes such as ribosomal protein-coding and housekeeping genes (Zhang *et al.*, 2003).

The hosts have struck back

Retrotransposons have co-existed with their host metazoan genomes since ancient times. Because of their vast potential to cause widespread mutagenesis, genomic instability and other problems for their hosts, the retrotransposons have been prime targets for various repressive host defences as they both evolved over millions of years.

Cytosine methylation

Methylated cytosines are present in the genomic DNA of many eukaryotes, particularly in vertebrates. These heritable, relatively stable epigenetic marks are less common in eukaryotes with smaller genomes, i.e. less than 10^8 nucleotides, but they have been detected in certain species of all phylogenetic groups of eukaryotes. They involve a single methyl group covalently added at the 5-position of the cytosine base, and thus are called 5-methylcytosine or m^5C.

DNA methylation patterns in eukaryotes are distinct from those in bacteria, which include both m^5C and N-6-methyladenine modifications. However, the eukaryotic DNA methyltransferases that establish and maintain such patterns bear significant homology to the bacterial enzymes, suggesting that a common ancestral form gave rise to all of these methylating enzymes. DNA methylation in bacteria forms a major host defence against bacteriophage infection. Because most m^5C nucleotides in eukaryotes are part of CpG dinucleotides and are localized within repetitive elements such as transposons in the genomic DNA, it has been suggested that a similar host defence mechanism operates against eukaryotic transposons as well (Bestor, 1990; Yoder *et al.*, 1997).

In general, genomic transposons are major targets for DNA methylation in most metazoans' somatic tissues and developmental timepoints. One explanation for this is based on their repetitive distribution in genomes. Transposons' numerous copies could interact by homologous pairing, thereby triggering establishment and/or maintenance of DNA methylation at such hypothetical, 'lined-up' structures (Bender, 1998). Another possibility is that sequence-specific DNA binding proteins could recognize transposons' sequences, thereby directing cytosine methylation (and/or other marks of repressive heterochromatin) specifically to these repetitive elements. Recently, data supporting this second hypothesis have been reported (Cam *et al.*, 2008).

Dnmt3L helps guide and/or activate both *de novo* methyltransferases, Dnmt3a and Dnmt3b, by interacting directly and individually with them, thereby establishing methylation at dense CpG-rich sequences of transposons (Ooi *et al.*, 2007). Dnmt3L oligomerization with Dnmt3a has been found to determine a periodicity of 8–10 bp in methylation of CpG dinucleotides (Jia *et al.*, 2007). There appears to be a similar periodicity in the frequency of CpG sites both in differentially methylated regions of maternally imprinted mouse genes, as well as in sequences genome-wide (Ferguson-Smith and Greally, 2007). However, the basis for the dichotomy observed in methylation at CpG-rich sites, i.e. dense methylation of transposons despite virtually no methylation of the numerous CpG islands upstream of many genes, has not yet been established.

Methylated DNA (in the form of m^5C marks) helps control transcription, recombination and/or integration of genomic retrotransposons in at least three ways. The first of these is that m^5C blocks the binding of certain sequence-specific transcription factors, resulting in transcriptional gene silencing (Hark *et al.*, 2000). A second way by which DNA methylation helps control transposons is that methyl binding domain-containing proteins are recruited to methylated sequences (Hendrich and Bird, 1998). Such proteins in turn may help recruit additional repressive factors and complexes that can bring about further suppressive modifications of chromatin. The third is that m^5C slightly changes the melting temperature of duplex DNA, when compared with similar sequences containing unmethylated cytosines. Together, these factors result in stable, epigenetic silencing of gene expression.

What happens to the epigenetic control of endogenous retrotransposons when DNA methylation is reduced or eliminated in organisms where it is prevalent (Thayer *et al.*, 1993; Woodcock *et al.*, 1997)? The answer appears to depend not only upon the developmental stage, tissue and sex of the organism, but also upon the type of retrotransposon under regulation. Most of the available evidence regarding control of mammalian retroelements has been obtained from knockout mouse models that have been developed over the past several years.

When DNA methylation is disrupted in developing male mouse germ cells by knockout of *Dnmt3L*, methylation at both L1 and IAP elements is decreased, and their transcription is increased very dramatically (Bourc'his and Bestor, 2004; Kato *et al.*, 2007). As noted above, this gene product is not itself an active methyltransferase, but guides the activities of the *de novo* methyltransferases. Interestingly, when either *Dnmt3a* or *Dnmt3b* is knocked out alone, methylation at IAP or L1 retrotransposons in newborn prospermatogonia is only minimally decreased. This result suggests that both *de novo* methyltransferases may act cooperatively to methylate these endogenous mouse retrotransposons, and that their methylation is re-established during the normal development of fetal prospermatogonia. By contrast, knockout of *Dnmt3a* alone results in hypomethylation of SINE B1 elements in newborn prospermatogonia (Kato *et al.*, 2007). Moreover, knockout of *Dnmt3b* results in hypomethylation of both major and minor satellite repeats in newborn prospermatogonia (Kato *et al.*, 2007). A possible association between retrotransposon methylation in the developing male germline and the establishment of imprinting, through methylation of differentially methylated regions (DMRs), has been suggested by recent studies (Bourc'his and Bestor, 2004; Hata *et al.*, 2006; Webster *et al.*, 2005). This link involves establishment of covalent methylcytosine marks at differentially methylated regions (DMRs). By contrast, disruption of *Dnmt3L* in developing female oocytes results in disruption of resulting maternal imprinted marks, without affecting global methylation (Bourc'his *et al.*, 2001).

When the maintenance methyltransferase *Dnmt1* is disrupted in mouse embryos, genomic methylation of IAP retrotransposons is decreased (Walsh *et al.*, 1998). Moreover, transcription of IAP elements is increased markedly, i.e. 50–100 times, in virtually all somatic tissues. Methylation of these LTR retrotransposons is largely erased and then re-established during primordial germ cell development (Kato *et al.*, 2007; Walsh *et al.*, 1998).

By contrast, when DNA methylation is almost completely eliminated in human colorectal cancer cells by double knockout of *DNMT1* and *DNMT3B*, L1 transcripts increase only modestly, i.e. approximately threefold (P. Tiwary, J. Li, K. Akagi and D. E. Symer, manuscript in preparation). Together, these results suggest that differential increases in L1 transcripts with comparable genome-wide hypomethylation may occur in a tissue-specific fashion, perhaps related to presence or absence of tissue-specific factors.

Methylation at individual retrotransposon integrants can vary greatly, even in normal tissues (Reiss and Mager, 2007; Reiss *et al.*, 2007). Some of this variability can be explained by genomic position effects *in cis*, but other unknown factors may also play a role. Such variation has been linked in a few particular cases to changes in gene expression and resulting phenotypic variation (Morgan *et al.*, 1999; Rakyan *et al.*, 2003; Whitelaw and Martin, 2001).

Cytosine methylation also has been identified as a regulator of transcription from *Alu* and other SINE retrotransposons (Liu *et al.*, 1994; Liu and Schmid, 1993). Methylation at SINE elements ranging from plants to mammals may spread into adjacent genes (Arnaud *et al.*, 2000), bringing about aberrant silencing of some tumour-suppressor genes (Graff *et al.*, 1997). In mouse, SINE B1 elements can serve as a target for *de novo* methylation, as methylation centres (Yates *et al.*, 1999).

Recently, other factors in addition to DNA methyltransferases also have been shown to affect methylation across the mammalian genome, and thereby affect transposon silencing. One of these is Lsh, lymphoid specific helicase, a member of the SNF2-helicase family of chromatin remodelling proteins. Homozygous knockout mice lacking *Lsh* have globally reduced CpG methylation (Dennis *et al.*, 2001), particularly at repetitive elements including retrotransposons (Huang *et al.*, 2004). Defective silencing of retrotransposons has been observed in a variety of tissues and developmental contexts including all embryonic tissues and developing female gonads (De La Fuente *et al.*, 2006). The *Lsh* gene product appears to interact directly with *de novo* methyltransferases Dnmt3a and Dnmt3b, bringing about *de novo* methylation (Zhu *et al.*, 2006).

Small RNAs

Extensive recent studies in many animals have indicated that several classes of small RNAs limit the mobilization and other biological effects of retrotransposons *in vivo*, particularly in germline cells, by a variety of elaborate and coordinated defence mechanisms. These small RNA-directed pathways operate both through transcriptional gene silencing and posttranscriptional gene silencing. They are mediated by RNA interference, microRNA-mediated inhibition of protein translation, and Piwi protein-associated silencing of transposons. The recent studies were triggered by pioneering observations made in plants (petunias) and in roundworms (*Caenorhabditis elegans*) (Robert *et al.*, 2004). However, we still lack an understanding of exactly which, when and how various repetitive genomic elements are regulated by particular small RNAs in different species and tissues. We also do not yet know how the Piwi-associated small RNAs suppress transposons, or even how they are generated in detail. The reasons why such small RNA-mediated controls have evolved and persisted in many organisms appear due to strong selective pressure against the threats posed by unchecked retrotransposition in their germlines (Yoder *et al.*, 1997).

Small RNAs interact with, and many of their effects are mediated by, various members of the Argonaute protein family, also called the PAZ Piwi domain (PPD) family. Two broad categories of Argonaute family members are present in most animals, i.e. the Ago clade and the Piwi clade. These proteins typically comprise functionally important parts of protein silencing complexes.

The Ago clade proteins complex with small RNAs of 21–23 nt such as short interfering RNAs (siRNAs) and microRNAs (miRNAs), which are typically processed from double stranded RNA precursors. Formation of these small RNAs depends upon Dicer, an essential RNaseIII endonuclease present in virtually all metazoans (Peters and Meister, 2007). The Ago proteins are expressed ubiquitously in most animal tissues. More recently, Piwi protein-associated RNAs (piRNAs) have been identified. They are longer, i.e. 24–31 nt long, and appear to arise from long, single-stranded precursors (Aravin *et al.*, 2007a; Brennecke *et al.*, 2007), although in mammals such transcripts have not been well characterized to date (O'Donnell and Boeke, 2007). They have been identified in *Drosophila*, mammals, and other multicellular animals, and are not well conserved between species. Expression of the Piwi proteins is restricted to germline tissues.

RNA interference (RNAi), mediated by short interfering RNAs (siRNAs), clearly limits retrotransposition in a variety of organisms. This process brings about posttranscriptional gene silencing through the degradation of expressed transcripts by the Dicer endonuclease as part of the RNA-induced silencing complex endonucleolytic pathway. For example, transposon transcripts, expressed in the *C. elegans* germline, are reduced by this process (Robert *et al.*, 2004, 2005; Sijen and Plasterk, 2003). Double stranded RNAs also inhibit repetitive elements and retrotransposons in *Drosophila* (Aravin *et al.*, 2001; Savitsky *et al.*, 2006; Vagin *et al.*, 2006).

Increased transcription of L1 and IAP retrotransposons has been observed in mouse embryonic stem cells lacking Dicer endonuclease, which have a marked disruption in formation of miRNAs and siRNAs (Kanellopoulou *et al.*, 2005). This strongly suggests that RNA interference and/or translational inhibition due to activity of these small RNAs helps to regulate mammalian retrotransposons at the level of expressed transcripts or proteins. Many human miRNAs are predicted to have specificity for nucleotide sequences that are conserved within retrotransposons, in particular in *Alu* elements (Smalheiser and Torvik, 2005, 2006).

In cultured somatic cells, human L1-specific siRNAs may limit L1 transcripts, but this suppression is modest (Aravin *et al.*, 2007b; Soifer, 2006; Soifer *et al.*, 2005; Yang and Kazazian, 2006). These small RNAs could be processed from transcripts initiated internally within L1 elements. For example, bidirectional promoters existing within the human L1 5′UTR (approximately 600–800 nucleotides apart) may initiate bidirectional transcripts (Matlik *et al.*, 2006; Nigumann *et al.*, 2002; Speek, 2001) that could be processed to small RNAs (Yang and Kazazian, 2006). Similarly, bidirectional promoters have been identified in various LTR transposons (Domansky *et al.*, 2000). Whatever advantage there is to retrotransposons in frequently having these dsRNA-generating promoter arrangement remains mysterious. Perhaps internal bidirectional promoters provide retrotransposons with a built-in, autoregulatory mechanism so that their expression can be regulated or fine-tuned by host defences. It is also possible that heterologous promoters outside of retrotransposons might drive read-through transcripts.

In the germline tissues of *Drosophila* and zebrafish, a large fraction of the small RNAs bound to Piwi family members (i.e. piRNAs) are templated by genomic repetitive elements, particularly defective transposons (Brennecke *et al.*, 2007). *Drosophila* piRNA precursors often are grouped in genomic loci, called piRNA clusters, which appear to serve as master regulators of transposition (Brennecke *et al.*, 2007). These particular small RNAs have been named repeat-associated small interfering RNAs (rasiRNAs), and are a predominant subset of piRNAs in *Drosophila*. They help mediate silencing of transposable elements (Aravin *et al.*, 2003; Houwing *et al.*, 2007; Kim, 2006). By contrast, a much smaller fraction of Piwi-associated small RNAs in mammals align with repetitive element sequences including transposons. There appear to be hundreds of thousands of distinct mammalian piRNAs (Aravin *et al.*, 2007a).

The mechanisms of piRNA expression and regulation are unclear currently. In several diverse species studied to date, distinct pools of piRNAs are bound by distinct Piwi family proteins at various time points of germ cell development. The piRNA sequences typically are either oriented antisense to endogenous transposons and are bound by Piwi and Aubergine, or are biased toward the sense strand of transposons (and are bound by AGO3). A model has been proposed to account for formation and self-amplification of these complementary piRNAs, called the ping-pong amplification loop (Aravin *et al.*, 2007a; Brennecke *et al.*, 2007; Gunawardane *et al.*, 2007). Certain molecular features of piRNAs predicted by this model have been observed in zebrafish (Houwing *et al.*, 2007) and in mouse (Aravin *et al.*, 2007b).

Various Piwi protein family mutations result in increased expression of transposons, observed so far in diverse species including *Drosophila* (Kalmykova *et al.*, 2005; Sarot *et al.*, 2004) and mouse (Aravin *et al.*, 2007b; Carmell *et al.*, 2007). However, detailed mechanisms of piRNA-mediated transposon silencing have been unclear. Both transcriptional gene silencing and posttranscriptional gene silencing have been implicated. Evidence supporting the ping-pong amplification loop model, including strong enrichment of certain nucleotides that mark 5′ ends of putative endonucleolytic cleavage corresponding to both strands of RNAs mapping to transposons, suggests that piRNA-mediated posttranscriptional gene silencing contributes to transposon control in *Drosophila*. There is also a strong association between expression of certain piRNAs and establishment of *de novo* DNA methylation at genomic IAP and L1 retrotransposons in fetal mouse male germ cell development (Kuramochi-Miyagawa *et al.*, 2008).

The phenotypic consequences of certain mouse Piwi protein mutations are remarkably similar to those observed in *Dnmt3L* mutant mice (Aravin and Bourc'his, 2008; Bourc'his and Bestor, 2004), suggesting that suppression of transposons exerted by piRNAs and by *de novo* DNA methylation overlap and may interact extensively in mouse. Human L1-specific small RNAs also could mediate transcriptional gene silencing (TGS) and/or post-transcriptional gene silencing (PTGS) through chromatin modifications,

DNA methylation, and/or other effects directed by Piwi family members (O'Donnell and Boeke, 2007).

Additional cellular factors and organelles involved in silencing transposons through small RNA-mediated pathways recently have been identified. These include Maelstrom (Soper *et al.*, 2008; van der Heijden and Bortvin, 2009), a protein that is a component of the Nuage organelle first identified in *Drosophila* (Lim and Kai, 2007; Lim *et al.*, 2009). Mutant adult male mice lacking Maelstrom have spermatocytes bearing massive DNA damage and chromosomal asynapsis, probably caused by the activity of accumulated L1 ribonucleoproteins (Soper *et al.*, 2008).

Chromatin modifications

Many of the epigenetic processes described above, including small RNAs and DNA methylation, interact with and affect chromatin modifications including histone methylation, acetylation and other modifications; nucleosomal remodelling; and others. It is currently not possible to tease apart or study independently the epigenetic controls of a given transposon integrant in a given metazoan host, since so many of the controls are intertwined at a fundamental level. Perturbations of one type of chromatin modification or epigenetic mark will very likely disrupt or influence other overlapping controls.

Position effects

Expression of human *Alu* elements has been demonstrated to depend upon flanking genomic sequences (Chu *et al.*, 1995; Li and Schmid, 2001; Roy *et al.*, 2000; Ullu and Weiner, 1985), which may reflect differences in chromatin structure, DNA methylation, readthrough transcription, proximity to regulatory elements such as enhancers, promoters and insulators, and perhaps other effects.

APOBECs

Recently, an endogenous human factor, APOBEC3G, was identified as a potent inhibitor of human immunodeficiency virus (HIV) strains lacking a viral infectivity factor (Vif) (see Chapter 11) (Sheehy *et al.*, 2002). Subsequently, over the past few years, a possible role for APOBEC3G and/or other APOBEC family members in controlling endogenous retroviruses and/or retrotransposons has been examined extensively. In general, most endogenous mammalian retrotransposons appear to be inhibited by one or more of the APOBEC family members in various cells and tissues. However, some aspects remain unresolved, and additional work will be needed to clarify mechanistic details of these host defences against retrotransposition.

The APOBEC family of proteins comprises a series of nucleic acid-editing enzymes. The name APOBEC is an acronym for apolipoprotein B mRNA-editing enzyme catalytic polypeptide. APOBEC3 family members are cytosine deaminases (Bishop *et al.*, 2004), which catalyse the conversion of deoxycytidine to deoxyuridine in the elongating minus strand of reverse transcribed cDNA. Uracils are not normally present in DNA, so this enzymatic activity results in editing or degradation of such substrates by uracil DNA deglycosylase. Primates have seven paralogues of APOBEC3, namely APOBEC3A through APOBEC3H, which are abbreviated as A3A, etc. By contrast, there is one *Apobec3* gene in mouse.

The cytosine deaminase motif, present in all members of the APOBEC3 family, is His-X-Glu-X_{23-28}-Pro-Cys-X_{2-4}-Cys. This protein motif occurs once in A3A, A3C and A3H, and twice in A3B, A3F and A3G (Hakata and Landau, 2006; Wedekind *et al.*, 2003).

A3A, A3B, A3C and A3F have been shown to inhibit human L1 retrotransposition in cultured cell assays. A3G inhibits *Alu* but not L1 retrotransposition (Chiu *et al.*, 2006). Surprisingly, the mechanistic basis for inhibition of retrotransposition by APOBEC3 family members appears to be independent of their cytosine deaminase activity. Instead, these gene products may sequester retrotransposon RNA in the cytoplasm or interfere with L1 ORF activities. Both A3A and A3B inhibit human L1 and *Alu* retrotransposition (Bogerd *et al.*, 2006) and mouse LTR (IAP and MusD) retrotransposition (Chen *et al.*, 2006a). A3G also appears to inhibit *Alu* retrotransposition in an L1 ORF1p-dependent fashion, but it does not inhibit L1 retrotransposition *per se* (Hulme *et al.*, 2007; Turelli *et al.*, 2004). It may direct *Alu* transcripts to so-called Stauffen-containing, RNA transport granules in the cytoplasm, thereby sequestering the transcripts away from the L1 proteins needed for their retrotransposition in the nucleus (Chiu *et al.*, 2006).

A double-edged sword: consequences of endogenous retrotransposition

The consequences of endogenous retrotransposition have been both good and bad, and in many cases neutral, for their hosts. The contradictory effects of retrotransposon mobilization, in many organisms, has spawned an ongoing debate about their overall contribution to the hosts' biology and evolution (Belancio *et al.*, 2008).

Many examples of deleterious effects resulting from endogenous retrotransposition have been identified. Various human genetic diseases, mouse phenotypes, etc., that are attributable to endogenous retrotransposition events, have been catalogued recently (Belancio *et al.*, 2008; Callinan and Batzer, 2006; Chen *et al.*, 2006b). These diseases and phenotypes illustrate the various ways by which endogenous retrotransposons can disrupt or modify expression of particular genes, resulting in deleterious phenotypic consequences. In addition, other deleterious products of ongoing retrotransposition probably have been lost over evolutionary time due to purifying negative selection.

Many of the human diseases that are attributable to endogenous retrotransposition have resulted from direct insertional mutagenesis (Callinan and Batzer, 2006). In these cases, the transposon inserted into an exon of a gene, affecting its pattern of expression or structure. Often, an integrant introduces a premature termination codon into the open reading frame of the gene. In this case, a resulting translation product would be foreshortened, so that downstream protein domains would be altered or missing. New integrant events have occurred either in the germline or in somatic cells, resulting in both heritable diseases and in tissue-specific mutations affecting a single individual.

On the other hand, portions of various retrotransposon integrants have been co-opted into useful purposes in their hosts over time. These salutatory events have been called exaptation.

Collateral damage: diseases due to retrotransposition

Extensive compilations of human genetic diseases attributable to endogenous retrotransposition events have been published recently (Belancio *et al.*, 2008; Callinan and Batzer, 2006; Chen *et al.*, 2006b). These extensive summaries illustrate the various ways by which endogenous retrotransposons can disrupt or modify expression of particular genes, resulting in phenotypic consequences. These consequences are manifested as various human diseases. The mechanisms include direct insertional mutagenesis; alteration of promoter, termination or splicing activities; mobilization of exonic sequences through 3′ transduction or 5′ transduction mechanisms; and provision of recombination hotspots.

Adaptive and exaptive responses

Portions of various retrotransposon integrants have been co-opted into useful purposes in their hosts over time; this has been called exaptation. For example, in mammals, recent studies have demonstrated that new genes derived from retrotransposons are expressed and appear to have contributed to genomic innovation, speciation and evolution (Brandt *et al.*, 2005a). Numerous compelling examples now demonstrate that retrotransposition has sparked many forms of evolutionary novelties. These include the formation of new templates for expressed coding and non-coding transcripts, introduction of new regulatory activities including promoters, enhancers and insulators, generation of various forms of genomic variation and instability including structural variations and segmental duplications, and others (Brosius, 2003).

SINE elements have been shown to be 'exapted' by their hosts for various purposes. For example, parts of ancient SINE integrants that are conserved evolutionarily can be expressed in both coding and non-coding transcripts (Bejerano *et al.*, 2006; Kamal *et al.*, 2006; Nishihara *et al.*, 2006; Xie *et al.*, 2006). Several copies of a newly described SINE family have contributed enhancer and ultraconserved exon sequences to expressed genes identified in fish species (Bejerano *et al.*, 2006). A close relative of SINE3 elements (active in zebrafish) has given rise to highly conserved, non-coding sequences containing a central core of approximately 180 nt, found in humans and other organisms as divergent as mammals, birds, zebrafish and coelacanths (Xie *et al.*, 2006). Another example is that up to 20% of human microRNAs may be expressed from *Alu* RNA polymerase III promoters (Borchert *et al.*, 2006).

Recently, bidirectional transcription of a mouse SINE B2 retrotransposon in the growth hormone gene locus on mouse chromosome 11 was found to be associated with expression of that flanking gene in the developing pituitary gland (Lunyak *et al.*, 2007). Both antisense stranded RNA polymerase II (Ferrigno *et al.*, 2001) and sense stranded RNA polymerase III transcripts (Allen *et al.*, 2004) initiated by the SINE element are necessary and sufficient to facilitate remodelling and regulated activation of the genomic locus. The retrotransposon apparently can serve as a boundary element blocking the spread of repressive, heterochromatic marks.

A conserved enhancer sequence that contributes to expression of the proopiomelanocortin (POMC) gene in mammals was found to have originated from an ancient, exapted SINE retrotransposon (Santangelo *et al.*, 2007). The conserved SINE element is a member of the MAR1 family, which in turn is a part of the CORE-SINE superfamily (Ohshima *et al.*, 1996; Smit and Riggs, 1995). Interestingly, several thousand similar copies of the core sequence of the MAR1 family were identified in the human genome, including highly conserved elements. The functional significance of their retention is currently unclear.

Recently a comprehensive analysis of short genomic sequences bound by biologically important transcription factors revealed that several of the factors studied bind to retrotransposon sequences (Fig. 1.4) (Bourque *et al.*, 2008). The particular repeat classes that are enriched in binding these transcription factors include MIR (mammalian interspersed repeat, a SINE retrotransposon), ERVK (mouse endogenous retrovirus K, an LTR retrotransposon), ERV1 (human endogenous retrovirus 1, an LTR retrotransposon) and B2 (rodent-specific SINE retrotransposons) (Fig. 1.4).

There are relatively fewer examples of L1 sequences incorporated into mammalian genes through exaptation, although a few interesting ones have been reported (Burns and Boeke, 2008; Zemojtel *et al.*, 2007). A 3′ transduction event mediated by a new L1 insertion into the human dystrophin gene affected that gene's structure (Holmes *et al.*, 1994). Many other cases of transduction of 3′ flanking sequences were identified subsequently (Goodier *et al.*, 2000; Pickeral *et al.*, 2000). More recently, numerous cases of SVA retrotransposon-mediated transduction of genomic sequences including genes have been described in the human genome (Xing *et al.*, 2006).

Human endogenous retrovirus-E family members with LTRs have introduced promoter activities that are used as alternative promoters that affect expression of human genes such as endothelin B receptor, apolipoprotein C-I, pleiotrophin, amylase and leptin receptor OBR (Medstrand *et al.*, 2001; Schulte *et al.*, 1996; Ting *et al.*, 1992). These effects are typically tissue specific. In some cases, the LTR integrants can also introduce retroelement-specific sequences into transcripts and possibly proteins (Horie *et al.*, 2007; Medstrand *et al.*, 2005).

Recent analysis of mouse and human transcripts revealed that 6–30% of all capped RNAs are initiated from within repetitive elements. In particular, these transcripts very frequently arise from retrotransposon sequences (Faulkner *et al.*, 2009). Certain transposable elements that reside within annotated genes in multiple metazoan species recently were cataloged (Levy *et al.*, 2008).

Perhaps the most interesting aspect of the mammalian Sushi family (also referred to as the MART gene family) is that whereas it seems to be extinct as a retrotransposon, there are about nine copies that appear to represent intriguing cases of retrotransposon domestication. That is, they were exapted and now perform a presumably useful function for the mammalian hosts (Brandt *et al.*, 2005a,b; Youngson *et al.*, 2005). Remarkably, several of these sushi-derived genes have not only acquired introns (suggesting they have been around for a while) but some of the copies occupy syntenic positions in different mammalian genomes; most of them lie on the X chromosome. One of the autosomal members of this gene family, referred to as Rtl1 (retrotransposon like gene), is part of an unusual imprinted locus that is paternally expressed. Remarkably, the

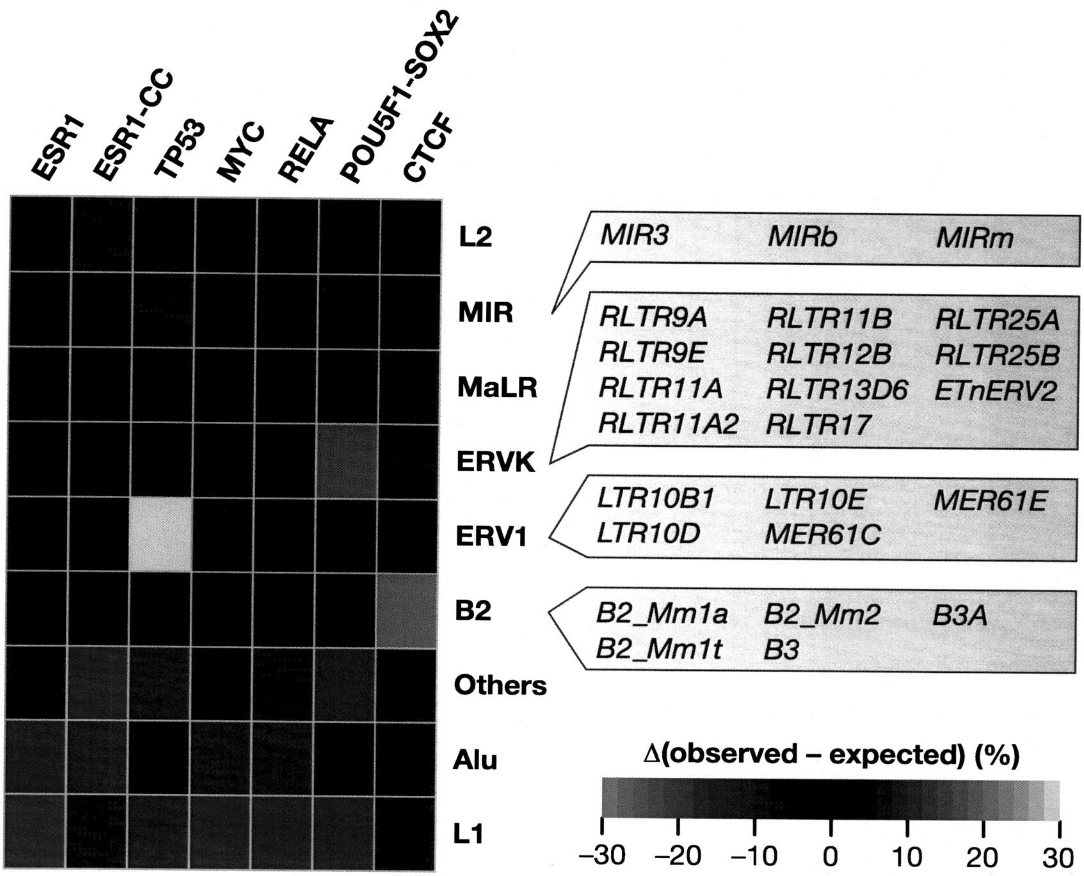

Figure 1.4 Widespread association between transcription factor binding sites and retrotransposons. This heatmap table shows the percentages of instances of retrotransposon family sequences that are bound by the indicated transcription factors, either in excess (yellow) or in deficit (blue) as compared to expected levels from chance alone. Certain DNA sequences corresponding to particular retrotransposon families were bound by transcription factors more than expected due to their prevalence in the genome (yellow). Values computed for seven binding data sets were normalized by subtraction using background data, defined as singleton PETs (for ChIP-PET), randomly selected Affymetrix probes (for ChIP-Chip) or singleton tags (for ChIP-Seq). Background values were subtracted from original data. The percentages are presented as the change in the differences between observed and expected values for experimental vs. background data (Bourque *et al.*, 2008). Reproduced from Bourque *et al.* (2008) with the permission of Cold Spring Harbor Laboratory Press.

antisense strand of the Rtl1 gene region encodes a second transcript, which is the precursor of two imprinted microRNAs, which are themselves maternally expressed (Seitz *et al.*, 2003). A second Sushi-derived gene, Peg10 (paternally expressed gene), is similarly expressed from the paternal allele. While it is conserved in placental mammals (indeed it is highly expressed in the placenta) as well as marsupials, it is absent from the genome of the egg-laying platypus, suggesting an early role in evolution of the more typical mammalian reproductive strategy. Suzuki *et al.* (Suzuki *et al.*, 2007) suggest that heavy cytosine methylation of sushi retrotransposons may have led to the exploitation of methylation as a mechanism for differential expression of maternal and paternal alleles.

Conclusions

The far-reaching insights of Barbara McClintock, who discovered and characterized several classes of maize transposons starting in the 1940s, have been broadened and deepened by decades of research on retrotransposons and their fossils in metazoan genomes. These elements are ubiquitous, are generally highly abundant in all eukaryotes, and have resulted in both advantageous and deleterious consequences for their hosts. It has become clear that many host defence mechanisms, which evolved to suppress endogenous retrotransposons, also play important roles in metazoan development, aging and diseases. In addition, there are many examples of metazoan defences against retrotransposons that are used by their hosts in analogous ways to counteract infections by retroviruses, and vice versa. We conclude that the ongoing war dance between retrotransposons and their metazoan hosts is an everlasting theme of mutual benefit and destruction that will continue to preoccupy our attention – and our genomes.

Acknowledgement

The authors thank Richard Frederickson (Scientific Publications, Graphics and Media, SAIC, NCI-Frederick) for his highly competent help in preparing the figures for this chapter. D.E.S. gratefully acknowledges support by the Intramural Research Program, Center for Cancer Research, National Cancer Institute, National Institutes of Health, during the initial composition of parts of this chapter. Its content does not necessarily reflect the views or policies of the Department of Health and Human Services, nor does mention of trade names, commercial products, or organizations imply endorsement by the US Government.

References

Akagi, K., Li, J., Stephens, R.M., Volfovsky, N., and Symer, D.E. (2008). Extensive variation between inbred mouse strains due to endogenous L1 retrotransposition. Genome Res. *18*, 869–880.

Alisch, R.S., Garcia-Perez, J.L., Muotri, A.R., Gage, F.H., and Moran, J.V. (2006). Unconventional translation of mammalian LINE-1 retrotransposons. Genes Dev. *20*, 210–224.

Allen, T.A., Von Kaenel, S., Goodrich, J.A., and Kugel, J.F. (2004). The SINE-encoded mouse B2 RNA represses mRNA transcription in response to heat shock. Nat. Struct. Mol. Biol. *11*, 816–821.

Aravin, A.A., and Bourc'his, D. (2008). Small RNA guides for *de novo* DNA methylation in mammalian germ cells. Genes Dev. *22*, 970–975.

Aravin, A.A., Hannon, G.J., and Brennecke, J. (2007a). The Piwi-piRNA pathway provides an adaptive defense in the transposon arms race. Science *318*, 761–764.

Aravin, A.A., Lagos-Quintana, M., Yalcin, A., Zavolan, M., Marks, D., Snyder, B., Gaasterland, T., Meyer, J., and Tuschl, T. (2003). The small RNA profile during *Drosophila melanogaster* development. Dev. Cell *5*, 337–350.

Aravin, A.A., Naumova, N.M., Tulin, A.V., Vagin, V.V., Rozovsky, Y.M., and Gvozdev, V.A. (2001). Double-stranded RNA-mediated silencing of genomic tandem repeats and transposable elements in the *D. melanogaster* germline. Curr. Biol. *11*, 1017–1027.

Aravin, A.A., Sachidanandam, R., Girard, A., Fejes-Toth, K., and Hannon, G.J. (2007b). Developmentally regulated piRNA clusters implicate MILI in transposon control. Science *316*, 744–747.

Arkhipova, I.R., Pyatkov, K.I., Meselson, M., and Evgen'ev, M.B. (2003). Retroelements containing introns in diverse invertebrate taxa. Nat. Genet. *33*, 123–124.

Arnaud, P., Goubely, C., Pelissier, T., and Deragon, J.M. (2000). SINE retroposons can be used *in vivo* as nucleation centers for *de novo* methylation. Mol. Cell. Biol. *20*, 3434–3441.

Babushok, D.V., Ostertag, E.M., Courtney, C.E., Choi, J.M., and Kazazian, H.H., Jr. (2006). L1 integration in a transgenic mouse model. Genome Res. *16*, 240–250.

Beauregard, A., Curcio, M.J., and Belfort, M. (2008). The take and give between retrotransposable elements and their hosts. Annu. Rev. Genet. *42*, 587–617.

Becker, K.G., Swergold, G.D., Ozato, K., and Thayer, R.E. (1993). Binding of the ubiquitous nuclear transcription factor YY1 to a *cis* regulatory sequence in the human LINE-1 transposable element. Hum. Mol. Genet. *2*, 1697–1702.

Bejerano, G., Lowe, C.B., Ahituv, N., King, B., Siepel, A., Salama, S.R., Rubin, E.M., Kent, W.J., and Haussler, D. (2006). A distal enhancer and an ultraconserved exon are derived from a novel retroposon. Nature *441*, 87–90.

Belancio, V.P., Hedges, D.J., and Deininger, P. (2008). Mammalian non-LTR retrotransposons: for better or worse, in sickness and in health. Genome Res. *18*, 343–358.

Bender, J. (1998). Cytosine methylation of repeated sequences in eukaryotes: the role of DNA pairing. Trends Biochem. Sci. *23*, 252–256.

Bennett, E.A., Coleman, L.E., Tsui, C., Pittard, W.S., and Devine, S.E. (2004). Natural genetic variation caused by transposable elements in humans. Genetics *168*, 933–951.

Bennett, E.A., Keller, H., Mills, R.E., Schmidt, S., Moran, J.V., Weichenrieder, O., and Devine, S.E. (2008). Active *Alu* retrotransposons in the human genome. Genome Res. *18*, 1875–1883.

Bestor, T.H. (1990). DNA methylation: evolution of a bacterial immune function into a regulator of gene expression and genome structure in higher eukaryotes. Phil. Trans. R. Soc. Lond. B Biol. Sci. *326*, 179–187.

Bishop, K.N., Holmes, R.K., Sheehy, A.M., Davidson, N.O., Cho, S.J., and Malim, M.H. (2004). Cytidine deamination of retroviral DNA by diverse APOBEC proteins. Curr. Biol. *14*, 1392–1396.

Bochukova, E.G., Roscioli, T., Hedges, D.J., Taylor, I.B., Johnson, D., David, D.J., Deininger, P.L., and Wilkie, A.O. (2009). Rare mutations of FGFR2 causing apert syndrome: identification of the first partial gene deletion, and an *Alu* element insertion from a new subfamily. Hum. Mutat. *30*, 204–211.

Boeke, J.D. (1997). LINEs and x – the poly(A) connection. Nat. Genet. *16*, 6–7.

Boeke, J.D. (2003). The unusual phylogenetic distribution of retrotransposons: a hypothesis. Genome Res. *13*, 1975–1983.

Boeke, J.D., Garfinkel, D.J., Styles, C.A., and Fink, G.R. (1985). Ty elements transpose through an RNA intermediate. Cell *40*, 491–500.

Boeke, J.D., and Stoye, J.P. (1997). Retrotransposons, endogenous retroviruses, and the evolution of retroelements. In Retroviruses, Varmus, H., ed. (Cold Spring Harbor, NY, Cold Spring Harbor Laboratory Press), pp. 343–435.

Bogerd, H.P., Wiegand, H.L., Hulme, A.E., Garcia-Perez, J.L., O'Shea, K.S., Moran, J.V., and Cullen, B.R. (2006). Cellular inhibitors of long interspersed element 1 and *Alu* retrotransposition. Proc. Natl. Acad. Sci. U.S.A. *103*, 8780–8785.

Borchert, G.M., Lanier, W., and Davidson, B.L. (2006). RNA polymerase III transcribes human microRNAs. Nat. Struct. Mol. Biol. *13*, 1097–1101.

Bourc'his, D., and Bestor, T.H. (2004). Meiotic catastrophe and retrotransposon reactivation in male germ cells lacking *Dnmt3L*. Nature *431*, 96–99.

Bourc'his, D., Xu, G.L., Lin, C.S., Bollman, B., and Bestor, T.H. (2001). Dnmt3L and the establishment of maternal genomic imprints. Science *294*, 2536–2539.

Bourque, G., Leong, B., Vega, V.B., Chen, X., Lee, Y.L., Srinivasan, K.G., Chew, J.-L., Ruan, Y., Wei, C.-L., Ng, H.H., *et al.* (2008). Evolution of the mammalian transcription factor binding repertoire via transposable elements. Genome Research *18*, 1752–1762.

Bowen, N.J., and McDonald, J.F. (1999). Genomic analysis of Caenorhabditis elegans reveals ancient families of retroviral-like elements. Genome Res. *9*, 924–935.

Brandt, J., Schrauth, S., Veith, A.M., Froschauer, A., Haneke, T., Schultheis, C., Gessler, M., Leimeister, C., and Volff, J.N. (2005a). Transposable elements as a source of genetic innovation: expression and evolution of a family of retrotransposon-derived neogenes in mammals. Gene *345*, 101–111.

Brandt, J., Veith, A.M., and Volff, J.N. (2005b). A family of neofunctionalized Ty3/*gypsy* retrotransposon genes in mammalian genomes. Cytogenet. Genome Res. *110*, 307–317.

Brennecke, J., Aravin, A.A., Stark, A., Dus, M., Kellis, M., Sachidanandam, R., and Hannon, G.J. (2007). Discrete small RNA-generating loci as master regulators of transposon activity in *Drosophila*. Cell *128*, 1089–1103.

Brosius, J. (2003). The contribution of RNAs and retroposition to evolutionary novelties. Genetica *118*, 99–116.

Brouha, B., Schustak, J., Badge, R.M., Lutz-Prigge, S., Farley, A.H., Moran, J.V., and Kazazian, H.H., Jr. (2003). Hot L1s account for the bulk of retrotransposition in the human population. Proc. Natl. Acad. Sci. U.S.A. *100*, 5280–5285.

Bujacz, G., Alexandratos, J., Qing, Z.L., Clement-Mella, C., and Wlodawer, A. (1996a). The catalytic domain of human immunodeficiency virus integrase: ordered active site in the F185H mutant. FEBS Lett. *398*, 175–178.

Bujacz, G., Jaskolski, M., Alexandratos, J., Wlodawer, A., Merkel, G., Katz, R.A., and Skalka, A.M. (1996b). The catalytic domain of avian sarcoma virus integrase: conformation of the active-site residues in the presence of divalent cations. Structure *4*, 89–96.

Burns, K.H., and Boeke, J.D. (2008). Great exaptations. J. Biol. *7*, 5.

Butler, M., Goodwin, T., and Poulter, R. (2001a). An unusual vertebrate LTR retrotransposon from the cod Gadus morhua. Mol. Biol. Evol. *18*, 443–447.

Butler, M., Goodwin, T., Simpson, M., Singh, M., and Poulter, R. (2001b). Vertebrate LTR retrotransposons of the Tf1/sushi group. J. Mol. Evol. *52*, 260–274.

Callinan, P.A., and Batzer, M.A. (2006). Retrotransposable elements and human disease. Genome Dyn. *1*, 104–115.

Callinan, P.A., Wang, J., Herke, S.W., Garber, R.K., Liang, P., and Batzer, M.A. (2005). *Alu* retrotransposition-mediated deletion. J. Mol. Biol. *348*, 791–800.

Cam, H.P., Noma, K., Ebina, H., Levin, H.L., and Grewal, S.I. (2008). Host genome surveillance for retrotransposons by transposon-derived proteins. Nature *451*, 431–436.

Carmell, M.A., Girard, A., van de Kant, H.J., Bourc'his, D., Bestor, T.H., de Rooij, D.G., and Hannon, G.J. (2007). MIWI2 is essential for spermatogenesis and repression of transposons in the mouse male germline. Dev. Cell *12*, 503–514.

Casacuberta, E., and Pardue, M.L. (2003). HeT-A elements in *Drosophila virilis*: retrotransposon telomeres are conserved across the *Drosophila* genus. Proc. Natl. Acad. Sci. U.S.A. *100*, 14091–14096.

Casacuberta, E., and Pardue, M.L. (2005). HeT-A and TART, two *Drosophila* retrotransposons with a bona fide role in chromosome structure for more than 60 million years. Cytogenet. Genome Res. *110*, 152–159.

Chaboissier, M.C., Finnegan, D., and Bucheton, A. (2000). Retrotransposition of the I factor, a non-long terminal repeat retrotransposon of *Drosophila*, generates tandem repeats at the 3′ end. Nucleic Acids Res. *28*, 2467–2472.

Chen, H., Lilley, C.E., Yu, Q., Lee, D.V., Chou, J., Narvaiza, I., Landau, N.R., and Weitzman, M.D. (2006a). APOBEC3A is a potent inhibitor of adeno-associated virus and retrotransposons. Curr. Biol. *16*, 480–485.

Chen, J.M., Ferec, C., and Cooper, D.N. (2006b). LINE-1 endonuclease-dependent retrotranspositional events causing human genetic disease: mutation detection bias and multiple mechanisms of target gene disruption. J. Biomed. Biotechnol. *2006*, 56182.

Chen, J.M., Ferec, C., and Cooper, D.N. (2007). Mechanism of *Alu* integration into the human genome. Genomic Med. *1*, 9–17.

Chen, J.M., Stenson, P.D., Cooper, D.N., and Ferec, C. (2005). A systematic analysis of LINE-1 endonuclease-dependent retrotranspositional events causing human genetic disease. Hum. Genet. *117*, 411–427.

Chiu, Y.L., Witkowska, H.E., Hall, S.C., Santiago, M., Soros, V.B., Esnault, C., Heidmann, T., and Greene,

W.C. (2006). High-molecular-mass APOBEC3G complexes restrict *Alu* retrotransposition. Proc. Natl. Acad. Sci. U.S.A. *103*, 15588–15593.

Chu, W.M., Liu, W.M., and Schmid, C.W. (1995). RNA polymerase III promoter and terminator elements affect *Alu* RNA expression. Nucleic Acids Res. *23*, 1750–1757.

Colombo, R., Bignamini, A.A., Carobene, A., Sasaki, J., Tachikawa, M., Kobayashi, K., and Toda, T. (2000). Age and origin of the FCMD 3′-untranslated-region retrotransposal insertion mutation causing Fukuyama-type congenital muscular dystrophy in the Japanese population. Hum. Genet. *107*, 559–567.

Cordaux, R., Hedges, D.J., and Batzer, M.A. (2004). Retrotransposition of *Alu* elements: how many sources? Trends Genet. *20*, 464–467.

Cordaux, R., Hedges, D.J., Herke, S.W., and Batzer, M.A. (2006). Estimating the retrotransposition rate of human *Alu* elements. Gene *373*, 134–137.

Cost, G.J., and Boeke, J.D. (1998). Targeting of human retrotransposon integration is directed by the specificity of the L1 endonuclease for regions of unusual DNA structure. Biochemistry *37*, 18081–18093.

Cost, G.J., Feng, Q., Jacquier, A., and Boeke, J.D. (2002). Human L1 element target-primed reverse transcription *in vitro*. EMBO J. *21*, 5899–5910.

Curcio, M.J., and Derbyshire, K.M. (2003). The outs and ins of transposition: from mu to kangaroo. Nat. Rev. Mol. Cell. Biol. *4*, 865–877.

Davidson, E.H., and Britten, R.J. (1979). Regulation of gene expression: possible role of repetitive sequences. Science *204*, 1052–1059.

De La Fuente, R., Baumann, C., Fan, T., Schmidtmann, A., Dobrinski, I., and Muegge, K. (2006). Lsh is required for meiotic chromosome synapsis and retrotransposon silencing in female germ cells. Nat. Cell Biol. *8*, 1448–1454.

Deininger, P.L., and Batzer, M.A. (2002). Mammalian retroelements. Genome Res. *12*, 1455–1465.

Deininger, P.L., Batzer, M.A., Hutchison, C.A., 3rd, and Edgell, M.H. (1992). Master genes in mammalian repetitive DNA amplification. Trends Genet. *8*, 307–311.

Deininger, P.L., Moran, J.V., Batzer, M.A., and Kazazian, H.H., Jr. (2003). Mobile elements and mammalian genome evolution. Curr. Opin. Genet. Dev. *13*, 651–658.

Dennis, K., Fan, T., Geiman, T., Yan, Q., and Muegge, K. (2001). Lsh, a member of the SNF2 family, is required for genome-wide methylation. Genes Dev. *15*, 2940–2944.

Dewannieux, M., Dupressoir, A., Harper, F., Pierron, G., and Heidmann, T. (2004). Identification of autonomous IAP LTR retrotransposons mobile in mammalian cells. Nat. Genet. *36*, 534–539.

Dewannieux, M., Esnault, C., and Heidmann, T. (2003). LINE-mediated retrotransposition of marked *Alu* sequences. Nat. Genet. *35*, 41–48.

Dewannieux, M., and Heidmann, T. (2005). Role of poly(A) tail length in *Alu* retrotransposition. Genomics *86*, 378–381.

Dewannieux, M., Harper, F., Richaud, A., Letzelter, C., Ribet, D., Pierron, G., and Heidmann, T. (2006). Identification of an infectious progenitor for the multiple-copy HERV-K human endogenous retroelements. Genome Res *16*, 1548–1556.

Domansky, A.N., Kopantzev, E.P., Snezhkov, E.V., Lebedev, Y.B., Leib-Mosch, C., and Sverdlov, E.D. (2000). Solitary HERV-K LTRs possess bi-directional promoter activity and contain a negative regulatory element in the U5 region. FEBS Lett *472*, 191–195.

Doolittle, W.F., and Sapienza, C. (1980). Selfish genes, the phenotype paradigm and genome evolution. Nature *284*, 601–603.

Duncan, L., Bouckaert, K., Yeh, F., and Kirk, D.L. (2002). kangaroo, a mobile element from *Volvox carteri*, is a member of a newly recognized third class of retrotransposons. Genetics *162*, 1617–1630.

Eickbush, T.H., and Jamburuthugoda, V.K. (2008). The diversity of retrotransposons and the properties of their reverse transcriptases. Virus Res. *134*, 221–234.

Ergun, S., Buschmann, C., Heukeshoven, J., Dammann, K., Schnieders, F., Lauke, H., Chalajour, F., Kilic, N., Stratling, W.H., and Schumann, G.G. (2004). Cell type-specific expression of LINE-1 open reading frames 1 and 2 in fetal and adult human tissues. J. Biol. Chem. *279*, 27753–27763.

Esnault, C., Maestre, J., and Heidmann, T. (2000). Human LINE retrotransposons generate processed pseudogenes. Nat. Genet. *24*, 363–367.

Faulkner, G.J., Kimura, Y., Daub, C.O., Wani, S., Plessy, C., Irvine, K.M., Schroder, K., Cloonan, N., Steptoe, A.L., Lassmann, T., *et al.* (2009). The regulated retrotransposon transcriptome of mammalian cells. Nat. Genet. *41*, 563–571.

Feng, Q., Moran, J.V., Kazazian, H.H., Jr., and Boeke, J.D. (1996). Human L1 retrotransposon encodes a conserved endonuclease required for retrotransposition. Cell *87*, 905–916.

Feng, Q., Schumann, G., and Boeke, J.D. (1998). Retrotransposon R1Bm endonuclease cleaves the target sequence. Proc. Natl. Acad. Sci. U.S.A. *95*, 2083–2088.

Ferguson-Smith, A.C., and Greally, J.M. (2007). Epigenetics: perceptive enzymes. Nature *449*, 148–149.

Ferrigno, O., Virolle, T., Djabari, Z., Ortonne, J.P., White, R.J., and Aberdam, D. (2001). Transposable B2 SINE elements can provide mobile RNA polymerase II promoters. Nat. Genet. *28*, 77–81.

Finnegan, D.J. (1989). Eukaryotic transposable elements and genome evolution. Trends Genet. *5*, 103–107.

Frame, I.G., Cutfield, J.F., and Poulter, R.T. (2001). New BEL-like LTR-retrotransposons in *Fugu rubripes*, *Caenorhabditis elegans*, and *Drosophila* melanogaster. Gene *263*, 219–230.

Galun, E. (2003). Transposable elements: a guide to the perplexed and the novice: with appendices on RNAi, chromatin remodeling and gene tagging (Dordrecht, Kluwer Academic).

Garfinkel, D.J., Boeke, J.D., and Fink, G.R. (1985). Ty element transposition: reverse transcriptase and virus-like particles. Cell *42*, 507–517.

Gasior, S.L., Wakeman, T.P., Xu, B., and Deininger, P.L. (2006). The human LINE-1 retrotransposon creates DNA double-strand breaks. J. Mol. Biol. *357*, 1383–1393.

Gerasimova, T.I., Byrd, K., and Corces, V.G. (2000). A chromatin insulator determines the nuclear localization of DNA. Mol. Cell *6*, 1025–1035.

Gilbert, N., Lutz, S., Morrish, T.A., and Moran, J.V. (2005). Multiple fates of L1 retrotransposition intermediates in cultured human cells. Mol. Cell. Biol. *25*, 7780–7795.

Gilbert, N., Lutz-Prigge, S., and Moran, J.V. (2002). Genomic deletions created upon LINE-1 retrotransposition. Cell *110*, 315–325.

Goodier, J.L., and Kazazian, H.H., Jr. (2008). Retrotransposons revisited: the restraint and rehabilitation of parasites. Cell *135*, 23–35.

Goodier, J.L., Ostertag, E.M., Engleka, K.A., Seleme, M.C., and Kazazian, H.H., Jr. (2004). A potential role for the nucleolus in L1 retrotransposition. Hum. Mol. Genet. *13*, 1041–1048.

Goodier, J.L., Ostertag, E.M., and Kazazian, H.H., Jr. (2000). Transduction of 3′-flanking sequences is common in L1 retrotransposition. Hum. Mol. Genet. *9*, 653–657.

Goodwin, T.J., Busby, J.N., and Poulter, R.T. (2007). A yeast model for target-primed (non-LTR) retrotransposition. BMC Genomics *8*, 263.

Goodwin, T.J., and Poulter, R.T. (2001). The DIRS1 group of retrotransposons. Mol. Biol. Evol. *18*, 2067–2082.

Graff, J.R., Herman, J.G., Myohanen, S., Baylin, S.B., and Vertino, P.M. (1997). Mapping patterns of CpG island methylation in normal and neoplastic cells implicates both upstream and downstream regions in *de novo* methylation. J. Biol. Chem. *272*, 22322–22329.

Gunawardane, L.S., Saito, K., Nishida, K.M., Miyoshi, K., Kawamura, Y., Nagami, T., Siomi, H., and Siomi, M.C. (2007). A slicer-mediated mechanism for repeat-associated siRNA 5′ end formation in *Drosophila*. Science *315*, 1587–1590.

Haas, N.B., Grabowski, J.M., North, J., Moran, J.V., Kazazian, H.H., and Burch, J.B. (2001). Subfamilies of CR1 non-LTR retrotransposons have different 5′UTR sequences but are otherwise conserved. Gene *265*, 175–183.

Haas, N.B., Grabowski, J.M., Sivitz, A.B., and Burch, J.B. (1997). Chicken repeat 1 (CR1) elements, which define an ancient family of vertebrate non-LTR retrotransposons, contain two closely spaced open reading frames. Gene *197*, 305–309.

Hakata, Y., and Landau, N.R. (2006). Reversed functional organization of mouse and human APOBEC3 cytidine deaminase domains. J. Biol. Chem. *281*, 36624–36631.

Han, J.S., and Boeke, J.D. (2004). A highly active synthetic mammalian retrotransposon. Nature *429*, 314–318.

Han, K., Xing, J., Wang, H., Hedges, D.J., Garber, R.K., Cordaux, R., and Batzer, M.A. (2005). Under the genomic radar: the stealth model of *Alu* amplification. Genome Res. *15*, 655–664.

Hark, A.T., Schoenherr, C.J., Katz, D.J., Ingram, R.S., Levorse, J.M., and Tilghman, S.M. (2000). CTCF mediates methylation-sensitive enhancer-blocking activity at the H19/Igf2 locus. Nature *405*, 486–489.

Hata, K., Kusumi, M., Yokomine, T., Li, E., and Sasaki, H. (2006). Meiotic and epigenetic aberrations in *Dnmt3L*-deficient male germ cells. Mol. Reprod. Dev. *73*, 116–122.

Hendrich, B., and Bird, A. (1998). Identification and characterization of a family of mammalian methyl-CpG binding proteins. Mol. Cell. Biol. *18*, 6538–6547.

Herrler, G., Rott, R., Klenk, H.D., Muller, H.P., Shukla, A.K., and Schauer, R. (1985). The receptor-destroying enzyme of influenza C virus is neuraminate-O-acetylesterase. EMBO J. *4*, 1503–1506.

Hizi, A. (2008). The reverse transcriptase of the Tf1 retrotransposon has a specific novel activity for generating the RNA self-primer that is functional in cDNA synthesis. J. Virol. *82*, 10906–10910.

Hohjoh, H., and Singer, M.F. (1996). Cytoplasmic ribonucleoprotein complexes containing human LINE-1 protein and RNA. EMBO J. *15*, 630–639.

Hohjoh, H., and Singer, M.F. (1997). Sequence-specific single-strand RNA binding protein encoded by the human LINE-1 retrotransposon. EMBO J. *16*, 6034–6043.

Holmes, S.E., Dombroski, B.A., Krebs, C.M., Boehm, C.D., and Kazazian, H.H., Jr. (1994). A new retrotransposable human L1 element from the LRE2 locus on chromosome 1q produces a chimaeric insertion. Nat. Genet. *7*, 143–148.

Horie, K., Saito, E.S., Keng, V.W., Ikeda, R., Ishihara, H., and Takeda, J. (2007). Retrotransposons influence the mouse transcriptome: implication for the divergence of genetic traits. Genetics *176*, 815–827.

Houwing, S., Kamminga, L.M., Berezikov, E., Cronembold, D., Girard, A., van den Elst, H., Filippov, D.V., Blaser, H., Raz, E., Moens, C.B., *et al.* (2007). A role for Piwi and piRNAs in germ cell maintenance and transposon silencing in Zebrafish. Cell *129*, 69–82.

Huang, J., Fan, T., Yan, Q., Zhu, H., Fox, S., Issaq, H.J., Best, L., Gangi, L., Munroe, D., and Muegge, K. (2004). Lsh, an epigenetic guardian of repetitive elements. Nucleic Acids Res. *32*, 5019–5028.

Hull, R. (1999). Classification of reverse transcribing elements: a discussion document. Arch. Virol. *144*, 209–213; discussion 213–204.

Hull, R. (2001). Classifying reverse transcribing elements: a proposal and a challenge to the ICTV. International Committee on Taxonomy of Viruses. Arch. Virol. *146*, 2255–2261.

Hulme, A.E., Bogerd, H.P., Cullen, B.R., and Moran, J.V. (2007). Selective inhibition of *Alu* retrotransposition by APOBEC3G. Gene *390*, 199–205.

International Mouse Sequencing Consortium (2002). Initial sequencing and comparative analysis of the mouse genome. Nature *420*, 520–562.

Jensen, S., Cavarec, L., Dhellin, O., and Heidmann, T. (1994). Retrotransposition of a marked *Drosophila* line-like I element in cells in culture. Nucleic Acids Res. *22*, 1484–1488.

Jia, D., Jurkowska, R.Z., Zhang, X., Jeltsch, A., and Cheng, X. (2007). Structure of Dnmt3a bound to Dnmt3L suggests a model for *de novo* DNA methylation. Nature *449*, 248–251.

Jurka, J. (1997). Sequence patterns indicate an enzymatic involvement in integration of mammalian retroposons. Proc. Natl. Acad. Sci. U.S.A. *94*, 1872–1877.

Kal, A.J., van Zonneveld, A.J., Benes, V., van den Berg, M., Koerkamp, M.G., Albermann, K., Strack, N., Ruijter, J.M., Richter, A., Dujon, B., *et al.* (1999). Dynamics of gene expression revealed by comparison of serial analysis of gene expression transcript profiles from yeast grown on two different carbon sources. Mol. Biol. Cell *10*, 1859–1872.

Kalmykova, A.I., Klenov, M.S., and Gvozdev, V.A. (2005). Argonaute protein PIWI controls mobilization of retrotransposons in the *Drosophila* male germline. Nucleic Acids Res. *33*, 2052–2059.

Kamal, M., Xie, X., and Lander, E.S. (2006). A large family of ancient repeat elements in the human genome is under strong selection. Proc. Natl. Acad. Sci. U.S.A. *103*, 2740–2745.

Kanellopoulou, C., Muljo, S.A., Kung, A.L., Ganesan, S., Drapkin, R., Jenuwein, T., Livingston, D.M., and Rajewsky, K. (2005). Dicer-deficient mouse embryonic stem cells are defective in differentiation and centromeric silencing. Genes Dev. *19*, 489–501.

Kano, H., Godoy, I., Courtney, C., Vetter, M.R., Gerton, G.L., Ostertag, E.M., and Kazazian, H.H., Jr. (2009). L1 retrotransposition occurs mainly in embryogenesis and creates somatic mosaicism. Genes Dev. *23*, 1303–1312.

Kapitonov, V.V., and Jurka, J. (2003a). A novel class of SINE elements derived from 5S rRNA. Mol. Biol. Evol. *20*, 694–702.

Kapitonov, V.V., and Jurka, J. (2003b). The esterase and PHD domains in CR1-like non-LTR retrotransposons. Mol. Biol. Evol. *20*, 38–46.

Kapitonov, V.V., and Jurka, J. (2008). A universal classification of eukaryotic transposable elements implemented in Repbase. Nat. Rev. Genet. *9*, 411–412; author reply 414.

Kato, Y., Kaneda, M., Hata, K., Kumaki, K., Hisano, M., Kohara, Y., Okano, M., Li, E., Nozaki, M., and Sasaki, H. (2007). Role of the Dnmt3 family in *de novo* methylation of imprinted and repetitive sequences during male germ cell development in the mouse. Hum. Mol. Genet. *16*, 2272–2280.

Kazazian, H.H., Jr. (1999). An estimated frequency of endogenous insertional mutations in humans. Nat. Genet. *22*, 130.

Kidd, J.M., Cooper, G.M., Donahue, W.F., Hayden, H.S., Sampas, N., Graves, T., Hansen, N., Teague, B., Alkan, C., Antonacci, F., *et al.* (2008). Mapping and sequencing of structural variation from eight human genomes. Nature *453*, 56–64.

Kikuchi, Y., Ando, Y., and Shiba, T. (1986). Unusual priming mechanism of RNA-directed DNA synthesis in *copia* retrovirus-like particles of *Drosophila*. Nature *323*, 824–826.

Kim, H.S., Wadekar, R.V., Takenaka, O., Hyun, B.H., and Crow, T.J. (1999). Phylogenetic analysis of a retroposon family in african great apes. J. Mol. Evol. *49*, 699–702.

Kim, V.N. (2006). Small RNAs just got bigger: Piwi-interacting RNAs (piRNAs) in mammalian testes. Genes Dev. *20*, 1993–1997.

Kojima, K.K., Matsumoto, T., and Fujiwara, H. (2005). Eukaryotic translational coupling in UAAUG stop-start codons for the bicistronic RNA translation of the non-long terminal repeat retrotransposon SART1. Mol. Cell. Biol. *25*, 7675–7686.

Kolosha, V.O., and Martin, S.L. (1997). In vitro properties of the first ORF protein from mouse LINE-1 support its role in ribonucleoprotein particle formation during retrotransposition. Proc. Natl. Acad. Sci. U.S.A. *94*, 10155–10160.

Kolosha, V.O., and Martin, S.L. (2003). High-affinity, non-sequence-specific RNA binding by the open reading frame 1 (ORF1) protein from long interspersed nuclear element 1 (LINE-1). J. Biol. Chem. *278*, 8112–8117.

Korbel, J.O., Urban, A.E., Affourtit, J.P., Godwin, B., Grubert, F., Simons, J.F., Kim, P.M., Palejev, D., Carriero, N.J., Du, L., *et al.* (2007). Paired-end mapping reveals extensive structural variation in the human genome. Science *318*, 420–426.

Kramerov, D.A., and Vassetzky, N.S. (2005). Short retroposons in eukaryotic genomes. Int. Rev. Cytol. *247*, 165–221.

Kulpa, D.A., and Moran, J.V. (2005). Ribonucleoprotein particle formation is necessary but not sufficient for LINE-1 retrotransposition. Hum. Mol. Genet. *14*, 3237–3248.

Kulpa, D.A., and Moran, J.V. (2006). *Cis*-preferential LINE-1 reverse transcriptase activity in ribonucleoprotein particles. Nat. Struct. Mol. Biol. *13*, 655–660.

Kuramochi-Miyagawa, S., Watanabe, T., Gotoh, K., Totoki, Y., Toyoda, A., Ikawa, M., Asada, N., Kojima, K., Yamaguchi, Y., Ijiri, T.W., *et al.* (2008). DNA methylation of retrotransposon genes is regulated by Piwi family members MILI and MIWI2 in murine fetal testes. Genes Dev. *22*, 908–917.

Lander, E.S., Linton, L.M., Birren, B., Nusbaum, C., Zody, M.C., Baldwin, J., Devon, K., Dewar, K., Doyle, M., FitzHugh, W., *et al.* (2001). Initial sequencing and analysis of the human genome. Nature *409*, 860–921.

Leibold, D.M., Swergold, G.D., Singer, M.F., Thayer, R.E., Dombroski, B.A., and Fanning, T.G. (1990). Translation of LINE-1 DNA elements *in vitro* and in human cells. Proc. Natl. Acad. Sci. U.S.A. *87*, 6990–6994.

Levin, H.L. (1995). A novel mechanism of self-primed reverse transcription defines a new family of retroelements. Mol. Cell. Biol. *15*, 3310–3317.

Levin, H.L. (1996). An unusual mechanism of self-primed reverse transcription requires the RNase H domain of reverse transcriptase to cleave an RNA duplex. Mol. Cell. Biol. *16*, 5645–5654.

Levy, A., Sela, N., and Ast, G. (2008). TranspoGene and microTranspoGene: transposed elements influence on the transcriptome of seven vertebrates and invertebrates. Nucleic Acids Res. *36*, D47–D52.

Li, P.W., Li, J., Timmerman, S.L., Krushel, L.A., and Martin, S.L. (2006). The dicistronic RNA from the mouse LINE-1 retrotransposon contains an internal ribosome entry site upstream of each ORF: implications for retrotransposition. Nucleic Acids Res. *34*, 853–864.

Li, T.H., and Schmid, C.W. (2001). Differential stress induction of individual *Alu* loci: implications for transcription and retrotransposition. Gene *276*, 135–141.

Li, X., Scaringe, W.A., Hill, K.A., Roberts, S., Mengos, A., Careri, D., Pinto, M.T., Kasper, C.K., and Sommer, S.S. (2001). Frequency of recent retrotransposition events in the human factor IX gene. Hum. Mutat. *17*, 511–519.

Lim, A.K., and Kai, T. (2007). Unique germ-line organelle, nuage, functions to repress selfish genetic elements in *Drosophila melanogaster*. Proc. Natl. Acad. Sci. U.S.A. *104*, 6714–6719.

Lim, A.K., Tao, L., and Kai, T. (2009). piRNAs mediate posttranscriptional retroelement silencing and localization to pi-bodies in the *Drosophila* germline. J. Cell Biol. *186*, 333–342.

Lin, J.H., and Levin, H.L. (1997). Self-primed reverse transcription is a mechanism shared by several LTR-containing retrotransposons. Rna *3*, 952–953.

Liu, W.M., Maraia, R.J., Rubin, C.M., and Schmid, C.W. (1994). *Alu* transcripts: cytoplasmic localisation and regulation by DNA methylation. Nucleic Acids Res. *22*, 1087–1095.

Liu, W.M., and Schmid, C.W. (1993). Proposed roles for DNA methylation in *Alu* transcriptional repression and mutational inactivation. Nucleic Acids Res. *21*, 1351–1359.

Luan, D.D., Korman, M.H., Jakubczak, J.L., and Eickbush, T.H. (1993). Reverse transcription of R2Bm RNA is primed by a nick at the chromosomal target site: a mechanism for non-LTR retrotransposition. Cell *72*, 595–605.

Lunyak, V.V., Prefontaine, G.G., Nunez, E., Cramer, T., Ju, B.G., Ohgi, K.A., Hutt, K., Roy, R., Garcia-Diaz, A., Zhu, X., *et al.* (2007). Developmentally regulated activation of a SINE B2 repeat as a domain boundary in organogenesis. Science *317*, 248–251.

McClintock, B. (1951). Chromosome organization and genic expression. Cold Spring Harb. Symp. Quant. Biol. *16*, 13–47.

McClintock, B. (1984). The significance of responses of the genome to challenge. Science *226*, 792–801.

McMillan, J.P., and Singer, M.F. (1993). Translation of the human LINE-1 element, L1Hs. Proc. Natl. Acad. Sci. U.S.A. *90*, 11533–11537.

Maksakova, I.A., Romanish, M.T., Gagnier, L., Dunn, C.A., van de Lagemaat, L.N., and Mager, D.L. (2006). Retroviral elements and their hosts: insertional mutagenesis in the mouse germ line. PLoS Genet 2, e2.

Malik, H.S., and Eickbush, T.H. (1998). The RTE class of non-LTR retrotransposons is widely distributed in animals and is the origin of many SINEs. Mol. Biol. Evol. *15*, 1123–1134.

Malik, H.S., and Eickbush, T.H. (2001). Phylogenetic analysis of ribonuclease H domains suggests a late, chimeric origin of LTR retrotransposable elements and retroviruses. Genome Res. *11*, 1187–1197.

Martin, S.L. (2006). The ORF1 Protein Encoded by LINE-1: Structure and Function During L1 Retrotransposition. J. Biomed. Biotechnol. *2006*, 45621.

Martin, S.L., Branciforte, D., Keller, D., and Bain, D.L. (2003). Trimeric structure for an essential protein in L1 retrotransposition. Proc. Natl. Acad. Sci. U.S.A. *100*, 13815–13820.

Martin, S.L., and Bushman, F.D. (2001). Nucleic acid chaperone activity of the ORF1 protein from the mouse LINE-1 retrotransposon. Mol. Cell. Biol. *21*, 467–475.

Martin, S.L., Cruceanu, M., Branciforte, D., Wai-Lun Li, P., Kwok, S.C., Hodges, R.S., and Williams, M.C. (2005a). LINE-1 retrotransposition requires the nucleic acid chaperone activity of the ORF1 protein. J. Mol. Biol. *348*, 549–561.

Martin, S.L., Li, W.L., Furano, A.V., and Boissinot, S. (2005b). The structures of mouse and human L1 elements reflect their insertion mechanism. Cytogenet Genome Res. *110*, 223–228.

Matlik, K., Redik, K., and Speek, M. (2006). L1 antisense promoter drives tissue-specific transcription of human genes. J. Biomed. Biotechnol. *2006*, 71753.

Medstrand, P., Landry, J.R., and Mager, D.L. (2001). Long terminal repeats are used as alternative promoters for the endothelin B receptor and apolipoprotein C-I genes in humans. J. Biol. Chem. *276*, 1896–1903.

Medstrand, P., van de Lagemaat, L.N., Dunn, C.A., Landry, J.R., Svenback, D., and Mager, D.L. (2005). Impact of transposable elements on the evolution of mammalian gene regulation. Cytogenet. Genome Res. *110*, 342–352.

Mietz, J.A., Grossman, Z., Lueders, K.K., and Kuff, E.L. (1987). Nucleotide sequence of a complete mouse intracisternal A-particle genome: relationship to known aspects of particle assembly and function. J. Virol. *61*, 3020–3029.

Mills, R.E., Bennett, E.A., Iskow, R.C., and Devine, S.E. (2007). Which transposable elements are active in the human genome? Trends Genet. *23*, 183–191.

Mills, R.E., Luttig, C.T., Larkins, C.E., Beauchamp, A., Tsui, C., Pittard, W.S., and Devine, S.E. (2006a). An initial map of insertion and deletion (INDEL) variation in the human genome. Genome Res. *16*, 1182–1190.

Mills, R.E., Bennett, E.A., Iskow, R.C., Luttig, C.T., Tsui, C., Pittard, W.S., and Devine, S.E. (2006b). Recently mobilized transposons in the human and chimpanzee genomes. Am. J. Hum. Genet. *78*, 671–679.

Mills, R.E., Bennett, E.A., Iskow, R.C., and Devine, S.E. (2007). Which transposable elements are active in the human genome? Trends Genet. *23*, 183–191.

Moran, J.V., Holmes, S.E., Naas, T.P., DeBerardinis, R.J., Boeke, J.D., and Kazazian, H.H., Jr. (1996). High frequency retrotransposition in cultured mammalian cells. Cell *87*, 917–927.

Morgan, H.D., Sutherland, H.G., Martin, D.I., and Whitelaw, E. (1999). Epigenetic inheritance at the agouti locus in the mouse. Nat. Genet. *23*, 314–318.

Morrish, T.A., Garcia-Perez, J.L., Stamato, T.D., Taccioli, G.E., Sekiguchi, J., and Moran, J.V. (2007). Endonuclease-independent LINE-1 retrotransposition at mammalian telomeres. Nature *446*, 208–212.

Morrish, T.A., Gilbert, N., Myers, J.S., Vincent, B.J., Stamato, T.D., Taccioli, G.E., Batzer, M.A., and Moran, J.V. (2002). DNA repair mediated by endonuclease-independent LINE-1 retrotransposition. Nat. Genet. *31*, 159–165.

Myers, J.S., Vincent, B.J., Udall, H., Watkins, W.S., Morrish, T.A., Kilroy, G.E., Swergold, G.D., Henke, J., Henke, L., Moran, J.V., *et al.* (2002). A comprehensive analysis of recently integrated human Ta L1 elements. Am J. Hum. Genet. *71*, 312–326.

Nigumann, P., Redik, K., Matlik, K., and Speek, M. (2002). Many human genes are transcribed from the antisense promoter of L1 retrotransposon. Genomics *79*, 628–634.

Nishihara, H., Smit, A.F., and Okada, N. (2006). Functional noncoding sequences derived from SINEs in the mammalian genome. Genome Res. *16*, 864–874.

Noma, K., Ohtsubo, E., and Ohtsubo, H. (1999). Non-LTR retrotransposons (LINEs) as ubiquitous components of plant genomes. Mol. Gen. Genet. *261*, 71–79.

O'Donnell, K.A., and Boeke, J.D. (2007). Mighty Piwis defend the germline against genome intruders. Cell *129*, 37–44.

Ohshima, K., Hamada, M., Terai, Y., and Okada, N. (1996). The 3′ ends of tRNA-derived short interspersed repetitive elements are derived from the 3′ ends of long interspersed repetitive elements. Mol. Cell. Biol. *16*, 3756–3764.

Olovnikov, I.A., Adyanova, Z.V., Galimov, E.R., Andreev, D.E., Terenin, I.M., Ivanov, D.S., Prassolov, V.S., and Dmitriev, S.E. (2007). Key role of the internal 5′-UTR segment in the transcription activity of the human L1 retrotransposon. Molec. Biol. *41*, 453–458.

Ono, M., Kawakami, M., and Takezawa, T. (1987). A novel human nonviral retroposon derived from an endogenous retrovirus. Nucleic Acids Res. *15*, 8725–8737.

Ooi, S.K., Qiu, C., Bernstein, E., Li, K., Jia, D., Yang, Z., Erdjument-Bromage, H., Tempst, P., Lin, S.P., Allis, C.D., *et al.* (2007). DNMT3L connects unmethylated lysine 4 of histone H3 to *de novo* methylation of DNA. Nature *448*, 714–717.

Orgel, L.E., and Crick, F.H. (1980). Selfish DNA: the ultimate parasite. Nature *284*, 604–607.

Ostertag, E.M., Goodier, J.L., Zhang, Y., and Kazazian, H.H., Jr. (2003). SVA elements are nonautonomous retrotransposons that cause disease in humans. Am J. Hum. Genet. *73*, 1444–1451.

Ostertag, E.M., and Kazazian, H.H., Jr. (2001a). Biology of mammalian L1 retrotransposons. Annu. Rev. Genet. *35*, 501–538.

Ostertag, E.M., and Kazazian, H.H., Jr. (2001b). Twin priming: a proposed mechanism for the creation of inversions in L1 retrotransposition. Genome Res. *11*, 2059–2065.

Pelisson, A., Finnegan, D.J., and Bucheton, A. (1991). Evidence for retrotransposition of the I factor, a LINE element of *Drosophila* melanogaster. Proc. Natl. Acad. Sci. U.S.A. *88*, 4907–4910.

Peters, L., and Meister, G. (2007). Argonaute proteins: mediators of RNA silencing. Mol. Cell *26*, 611–623.

Pickeral, O.K., Makalowski, W., Boguski, M.S., and Boeke, J.D. (2000). Frequent human genomic DNA transduction driven by LINE-1 retrotransposition. Genome Res. *10*, 411–415.

Prak, E.T., and Kazazian, H.H., Jr. (2000). Mobile elements and the human genome. Nat. Rev. Genet. *1*, 134–144.

Pringle, C.R. (1999). Virus taxonomy at the XIth International Congress of Virology, Sydney, Australia, 1999. Arch. Virol. *144*, 2065–2070.

Pritham, E.J., and Feschotte, C. (2007). Massive amplification of rolling-circle transposons in the lineage of the bat *Myotis lucifugus*. Proc. Natl. Acad. Sci. U.S.A. *104*, 1895–1900.

Quentin, Y. (1992). Origin of the *Alu* family: a family of *Alu*-like monomers gave birth to the left and the right arms of the *Alu* elements. Nucleic Acids Res. *20*, 3397–3401.

Rakyan, V.K., Chong, S., Champ, M.E., Cuthbert, P.C., Morgan, H.D., Luu, K.V., and Whitelaw, E. (2003). Transgenerational inheritance of epigenetic states at the murine Axin(Fu) allele occurs after maternal and paternal transmission. Proc. Natl. Acad. Sci. U.S.A. *100*, 2538–2543.

Ray, D.A., Feschotte, C., Pagan, H.J., Smith, J.D., Pritham, E.J., Arensburger, P., Atkinson, P.W., and Craig, N.L. (2008). Multiple waves of recent DNA transposon activity in the bat, Myotis lucifugus. Genome Res. *18*, 717–728.

Reiss, D., and Mager, D.L. (2007). Stochastic epigenetic silencing of retrotransposons: does stability come with age? Gene *390*, 130–135.

Reiss, D., Zhang, Y., and Mager, D.L. (2007). Widely variable endogenous retroviral methylation levels in human placenta. Nucleic Acids Res. *35*, 4743–4754.

Ribet, D., Harper, F., Dupressoir, A., Dewannieux, M., Pierron, G., and Heidmann, T. (2008). An infectious progenitor for the murine IAP retrotransposon: emergence of an intracellular genetic parasite from an ancient retrovirus. Genome Res. *18*, 597–609.

Robert, V.J., Sijen, T., van Wolfswinkel, J., and Plasterk, R.H. (2005). Chromatin and RNAi factors protect the *C. elegans* germline against repetitive sequences. Genes Dev. *19*, 782–787.

Robert, V.J., Vastenhouw, N.L., and Plasterk, R.H. (2004). RNA interference, transposon silencing, and cosuppression in the Caenorhabditis elegans germ line: similarities and differences. Cold Spring Harb. Symp .Quant. Biol. *69*, 397–402.

Roy, A.M., West, N.C., Rao, A., Adhikari, P., Aleman, C., Barnes, A.P., and Deininger, P.L. (2000). Upstream flanking sequences and transcription of SINEs. J. Mol. Biol. *302*, 17–25.

Roy-Engel, A.M., Salem, A.H., Oyeniran, O.O., Deininger, L., Hedges, D.J., Kilroy, G.E., Batzer, M.A., and Deininger, P.L. (2002). Active *Alu* element 'A-tails': size does matter. Genome Res. *12*, 1333–1344.

Santangelo, A.M., de Souza, F.S., Franchini, L.F., Bumaschny, V.F., Low, M.J., and Rubinstein, M. (2007). Ancient exaptation of a CORE-SINE retroposon into a highly conserved mammalian neuronal enhancer of the proopiomelanocortin gene. PLoS Genet. 3, 1813–1826.

Sarot, E., Payen-Groschene, G., Bucheton, A., and Pelisson, A. (2004). Evidence for a piwi-dependent RNA silencing of the *gypsy* endogenous retrovirus by the *Drosophila* melanogaster flamenco gene. Genetics *166*, 1313–1321.

Sassaman, D.M., Dombroski, B.A., Moran, J.V., Kimberland, M.L., Naas, T.P., DeBerardinis, R.J., Gabriel, A., Swergold, G.D., and Kazazian, H.H., Jr. (1997). Many human L1 elements are capable of retrotransposition. Nat. Genet. *16*, 37–43.

Savitsky, M., Kwon, D., Georgiev, P., Kalmykova, A., and Gvozdev, V. (2006). Telomere elongation is under the control of the RNAi-based mechanism in the *Drosophila* germline. Genes Dev. *20*, 345–354.

Schulte, A.M., Lai, S., Kurtz, A., Czubayko, F., Riegel, A.T., and Wellstein, A. (1996). Human trophoblast and choriocarcinoma expression of the growth factor pleiotrophin attributable to germ-line insertion of an endogenous retrovirus. Proc. Natl. Acad. Sci. U.S.A. *93*, 14759–14764.

Seitz, H., Youngson, N., Lin, S.P., Dalbert, S., Paulsen, M., Bachellerie, J.P., Ferguson-Smith, A.C., and Cavaille, J. (2003). Imprinted microRNA genes transcribed antisense to a reciprocally imprinted retrotransposon-like gene. Nat. Genet. *34*, 261–262.

Seleme, M.C., Vetter, M.R., Cordaux, R., Bastone, L., Batzer, M.A., and Kazazian, H.H., Jr. (2006). Extensive individual variation in L1 retrotransposition capability contributes to human genetic diversity. Proc. Natl. Acad. Sci. U.S.A. *103*, 6611–6616.

She, X., Cheng, Z., Zollner, S., Church, D.M., and Eichler, E.E. (2008). Mouse segmental duplication and copy number variation. Nat. Genet. *40*, 909–914.

Sheehy, A.M., Gaddis, N.C., Choi, J.D., and Malim, M.H. (2002). Isolation of a human gene that inhibits HIV-1 infection and is suppressed by the viral Vif protein. Nature *418*, 646–650.

Shen, L., Wu, L.C., Sanlioglu, S., Chen, R., Mendoza, A.R., Dangel, A.W., Carroll, M.C., Zipf, W.B., and Yu, C.Y. (1994). Structure and genetics of the partially duplicated gene RP located immediately upstream of the complement C4A and the C4B genes in the HLA class III region. Molecular cloning, exon-intron structure, composite retroposon, and breakpoint of gene duplication. J. Biol. Chem. *269*, 8466–8476.

Sijen, T., and Plasterk, R.H. (2003). Transposon silencing in the Caenorhabditis elegans germ line by natural RNAi. Nature *426*, 310–314.

Sinnett, D., Richer, C., Deragon, J.M., and Labuda, D. (1991). *Alu* RNA secondary structure consists of two independent 7 SL RNA-like folding units. J. Biol. Chem. *266*, 8675–8678.

Smalheiser, N.R., and Torvik, V.I. (2005). Mammalian microRNAs derived from genomic repeats. Trends Genet. *21*, 322–326.

Smalheiser, N.R., and Torvik, V.I. (2006). *Alu* elements within human mRNAs are probable microRNA targets. Trends Genet. *22*, 532–536.

Smit, A.F., and Riggs, A.D. (1995). MIRs are classic, tRNA-derived SINEs that amplified before the mammalian radiation. Nucleic Acids Res. *23*, 98–102.

Smit, A.F., Toth, G., Riggs, A.D., and Jurka, J. (1995). Ancestral, mammalian-wide subfamilies of LINE-1 repetitive sequences. J. Mol. Biol. *246*, 401–417.

Soifer, H.S. (2006). Do small RNAs interfere with LINE-1? J. Biomed. Biotechnol. *2006*, 29049.

Soifer, H.S., Zaragoza, A., Peyvan, M., Behlke, M.A., and Rossi, J.J. (2005). A potential role for RNA interference in controlling the activity of the human LINE-1 retrotransposon. Nucleic Acids Res. *33*, 846–856.

Soper, S.F., van der Heijden, G.W., Hardiman, T.C., Goodheart, M., Martin, S.L., de Boer, P., and Bortvin, A. (2008). Mouse maelstrom, a component of nuage, is essential for spermatogenesis and transposon repression in meiosis. Dev. Cell *15*, 285–297.

Speek, M. (2001). Antisense promoter of human L1 retrotransposon drives transcription of adjacent cellular genes. Mol. Cell. Biol. *21*, 1973–1985.

Suzuki, S., Ono, R., Narita, T., Pask, A.J., Shaw, G., Wang, C., Kohda, T., Alsop, A.E., Marshall Graves, J.A., Kohara, Y., *et al.* (2007). Retrotransposon silencing by DNA methylation can drive mammalian genomic imprinting. PLoS Genet. *3*, e55.

Symer, D.E., Connelly, C., Szak, S.T., Caputo, E.M., Cost, G.J., Parmigiani, G., and Boeke, J.D. (2002). Human L1 retrotransposition is associated with genetic instability *in vivo*. Cell *110*, 327–338.

Szak, S.T., Pickeral, O.K., Makalowski, W., Boguski, M.S., Landsman, D., and Boeke, J.D. (2002). Molecular archeology of L1 insertions in the human genome. Genome Biol. *3*, research0052.

Tchenio, T., Casella, J.F., and Heidmann, T. (2000). Members of the SRY family regulate the human LINE retrotransposons. Nucleic Acids Res. *28*, 411–415.

Thayer, R.E., Singer, M.F., and Fanning, T.G. (1993). Undermethylation of specific LINE-1 sequences in human cells producing a LINE-1-encoded protein. Gene *133*, 273–277.

Ting, C.N., Rosenberg, M.P., Snow, C.M., Samuelson, L.C., and Meisler, M.H. (1992). Endogenous retroviral sequences are required for tissue-specific expression of a human salivary amylase gene. Genes Dev. *6*, 1457–1465.

Turelli, P., Vianin, S., and Trono, D. (2004). The innate antiretroviral factor APOBEC3G does not affect human LINE-1 retrotransposition in a cell culture assay. J. Biol. Chem. *279*, 43371–43373.

Ullu, E., and Tschudi, C. (1984). *Alu* sequences are processed 7SL RNA genes. Nature *312*, 171–172.

Ullu, E., and Weiner, A.M. (1985). Upstream sequences modulate the internal promoter of the human 7SL RNA gene. Nature *318*, 371–374.

Vagin, V.V., Sigova, A., Li, C., Seitz, H., Gvozdev, V., and Zamore, P.D. (2006). A distinct small RNA pathway silences selfish genetic elements in the germline. Science *313*, 320–324.

van de Lagemaat, L.N., Landry, J.R., Mager, D.L., and Medstrand, P. (2003). Transposable elements in mammals promote regulatory variation and diversification of genes with specialized functions. Trends Genet. *19*, 530–536.

van de Lagemaat, L.N., Medstrand, P., and Mager, D.L. (2006). Multiple effects govern endogenous retrovirus survival patterns in human gene introns. Genome Biol. *7*, R86.

van der Heijden, G.W., and Bortvin, A. (2009). Transient relaxation of transposon silencing at the onset of mammalian meiosis. Epigenetics *4*, 76–79.

Walsh, C.P., Chaillet, J.R., and Bestor, T.H. (1998). Transcription of IAP endogenous retroviruses is constrained by cytosine methylation. Nat. Genet. *20*, 116–117.

Wang, H., Xing, J., Grover, D., Hedges, D.J., Han, K., Walker, J.A., and Batzer, M.A. (2005). SVA elements: a hominid-specific retroposon family. J. Mol. Biol. *354*, 994–1007.

Warren, W.C., Hillier, L.W., Marshall Graves, J.A., Birney, E., Ponting, C.P., Grutzner, F., Belov, K., Miller, W., Clarke, L., Chinwalla, A.T., *et al.* (2008). Genome analysis of the platypus reveals unique signatures of evolution. Nature *453*, 175–183.

Webster, K.E., O'Bryan, M.K., Fletcher, S., Crewther, P.E., Aapola, U., Craig, J., Harrison, D.K., Aung, H., Phutikanit, N., Lyle, R., *et al.* (2005). Meiotic and epigenetic defects in *Dnmt3L*-knockout mouse spermatogenesis. Proc. Natl. Acad. Sci. U.S.A. *102*, 4068–4073.

Wedekind, J.E., Dance, G.S., Sowden, M.P., and Smith, H.C. (2003). Messenger RNA editing in mammals: new members of the APOBEC family seeking roles in the family business. Trends Genet. *19*, 207–216.

Wei, W., Gilbert, N., Ooi, S.L., Lawler, J.F., Ostertag, E.M., Kazazian, H.H., Boeke, J.D., and Moran, J.V. (2001). Human L1 retrotransposition: *cis* preference versus *trans* complementation. Mol. Cell. Biol. *21*, 1429–1439.

Whitelaw, E., and Martin, D.I. (2001). Retrotransposons as epigenetic mediators of phenotypic variation in mammals. Nat. Genet. *27*, 361–365.

Wicker, T., Sabot, F., Hua-Van, A., Bennetzen, J.L., Capy, P., Chalhoub, B., Flavell, A., Leroy, P., Morgante, M., Panaud, O., *et al.* (2007). A unified classification system for eukaryotic transposable elements. Nat. Rev. Genet. *8*, 973–982.

Woodcock, D.M., Lawler, C.B., Linsenmeyer, M.E., Doherty, J.P., and Warren, W.D. (1997). Asymmetric methylation in the hypermethylated CpG promoter region of the human L1 retrotransposon. J. Biol. Chem. *272*, 7810–7816.

Xie, X., Kamal, M., and Lander, E.S. (2006). A family of conserved noncoding elements derived from an ancient transposable element. Proc. Natl. Acad. Sci. U.S.A. *103*, 11659–11664.

Xing, J., Wang, H., Belancio, V.P., Cordaux, R., Deininger, P.L., and Batzer, M.A. (2006). Emergence of primate genes by retrotransposon-mediated sequence transduction. Proc. Natl. Acad. Sci. U.S.A. *103*, 17608–17613.

Xiong, Y., and Eickbush, T.H. (1988). Similarity of reverse transcriptase-like sequences of viruses, transposable elements, and mitochondrial introns. Mol. Biol. Evol. *5*, 675–690.

Xiong, Y., and Eickbush, T.H. (1990). Origin and evolution of retroelements based upon their reverse transcriptase sequences. EMBO J. *9*, 3353–3362.

Yang, J., Malik, H.S., and Eickbush, T.H. (1999). Identification of the endonuclease domain encoded by R2 and other site-specific, non-long terminal repeat retrotransposable elements. Proc. Natl. Acad. Sci. U.S.A. *96*, 7847–7852.

Yang, N., and Kazazian, H.H., Jr. (2006). L1 retrotransposition is suppressed by endogenously encoded small interfering RNAs in human cultured cells. Nat. Struct. Mol. Biol. *13*, 763–771.

Yang, N., Zhang, L., Zhang, Y., and Kazazian, H.H., Jr. (2003). An important role for RUNX3 in human L1 transcription and retrotransposition. Nucleic Acids Res. *31*, 4929–4940.

Yates, P.A., Burman, R.W., Mummaneni, P., Krussel, S., and Turker, M.S. (1999). Tandem B1 elements located in a mouse methylation center provide a target for *de novo* DNA methylation. J. Biol. Chem. *274*, 36357–36361.

Yoder, J.A., Walsh, C.P., and Bestor, T.H. (1997). Cytosine methylation and the ecology of intragenomic parasites. Trends Genet. *13*, 335–340.

Youngson, N.A., Kocialkowski, S., Peel, N., and Ferguson-Smith, A.C. (2005). A small family of sushi-class retrotransposon-derived genes in mammals and their relation to genomic imprinting. J. Mol. Evol. *61*, 481–490.

Zemojtel, T., Penzkofer, T., Schultz, J., Dandekar, T., Badge, R., and Vingron, M. (2007). Exonization of active mouse L1s: a driver of transcriptome evolution? BMC Genomics *8*, 392.

Zhang, Y., Maksakova, I.A., Gagnier, L., van de Lagemaat, L.N., and Mager, D.L. (2008). Genome-wide assessments reveal extremely high levels of polymorphism of two active families of mouse endogenous retroviral elements. PLoS Genet. *4*, e1000007.

Zhang, Z., Carriero, N., and Gerstein, M. (2004). Comparative analysis of processed pseudogenes in the mouse and human genomes. Trends Genet. *20*, 62–67.

Zhang, Z., and Gerstein, M. (2004). Large-scale analysis of pseudogenes in the human genome. Curr. Opin. Genet. Dev. *14*, 328–335.

Zhang, Z., Harrison, P.M., Liu, Y., and Gerstein, M. (2003). Millions of years of evolution preserved: a comprehensive catalog of the processed pseudogenes in the human genome. Genome Res. *13*, 2541–2558.

Zheng, D., Frankish, A., Baertsch, R., Kapranov, P., Reymond, A., Choo, S.W., Lu, Y., Denoeud, F., Antonarakis, S.E., Snyder, M., *et al.* (2007). Pseudogenes in the ENCODE regions: consensus annotation, analysis of transcription, and evolution. Genome Res. *17*, 839–851.

Zhu, H., Geiman, T.M., Xi, S., Jiang, Q., Schmidtmann, A., Chen, T., Li, E., and Muegge, K. (2006). Lsh is involved in *de novo* methylation of DNA. EMBO J. *25*, 335–345.

Zhu, Z.B., Jian, B., and Volanakis, J.E. (1994). Ancestry of SINE-R.C2 a human-specific retroposon. Hum. Genet. *93*, 545–551.

Zimmerly, S., Guo, H., Eskes, R., Yang, J., Perlman, P.S., and Lambowitz, A.M. (1995a). A group II intron RNA is a catalytic component of a DNA endonuclease involved in intron mobility. Cell *83*, 529–538.

Zimmerly, S., Guo, H., Perlman, P.S., and Lambowitz, A.M. (1995b). Group II intron mobility occurs by target DNA-primed reverse transcription. Cell *82*, 545–554.

Zingler, N., Willhoeft, U., Brose, H.P., Schoder, V., Jahns, T., Hanschmann, K.M., Morrish, T.A., Lower, J., and Schumann, G.G. (2005). Analysis of 5′ junctions of human LINE-1 and *Alu* retrotransposons suggests an alternative model for 5′-end attachment requiring microhomology-mediated end-joining. Genome Res. *15*, 780–789.

Endogenous Retroviruses

Joachim Denner*

2

Abstract

Endogenous retroviruses are genetic elements representing the result of retrovirus infections and integration of the proviruses into the germline of vertebrates including humans. Retroviruses use the enzyme reverse transcriptase (RT) to transcribe their RNA genome into cDNA and incorporate it into the cellular genome. Infections of germ cells result in the presence of these viruses in the genome all cells of the organism and transmission of these sequences to the offspring. Only some endogenous retroviruses are replication competent and produce infectious particles; most are defective. Although the role of endogenous retroviruses during tumour development and autoimmune diseases is still unclear, sufficient evidence has accumulated indicating that retroviruses play an important role in physiological processes. Endogenous retroviruses are involved in placental differentiation and immunosuppression during pregnancy, and retroviral long terminal repeats (LTRs) regulate the expression of cellular genes. During evolution three main processes took place: first, an accumulation of defective proviral DNA ('junk DNA'); second, a development of stronger restriction strategies by the host; and, third, a utilization, 'enslavement' of retroviral genes and LTRs. Since trans-species transmissions of retroviruses are very common, endogenous retrovirus may be important also for the health of other species. For example, pig cells can release porcine endogenous retroviruses that infect human cells and therefore represent a risk for xenotransplantations involving pig cells or organs.

Endogenous retroviruses: evolutionary garbage or well-functioning slaves?

Retroviruses are a large and diverse group of pathogenic and non-pathogenic viruses infecting animals and man. Retroviruses can be transmitted between individuals of the same species or of even to other species (interspecies or trans-species transmission), depending on the presence of one or more specific receptors on the cell surface. They infect their specific target cells, transcribe their RNA genome into DNA using the name giving enzyme reverse transcriptase (RT), and integrate as proviruses into the cellular genome of the target cells (see Chapter 3). For human immunodeficiency virus (HIV)-1 the preferred targets are usually lymphocytes or monocytes that carry both of the cell surface receptors required for infection: CD4 and one of the chemokine receptors CCR5 or CXCR4 (i.e. $CD4^+/CCR5^+$ or $CD4^+/CXCR4^+$ cells). HIV-1 does not usually integrate into the genome of other somatic cells, e.g. liver or kidney cells, and is therefore called an exogenous retrovirus. However, when a retrovirus infects an oocyte, a sperm cell, or precursors of these germ cells, the provirus will fully integrate into the genome of fertilized oocytes. Once this cell starts dividing, the integrated provirus will be present in all cells of the organism.

*This publication is dedicated to my father, Erwin Denner (1928–2009).

Such viruses are called endogenous retroviruses; they behave like normal genes such as albumin or haemoglobin and will be transmitted to the progeny according to Mendelian laws. The transmission of endogenous retroviruses from parent to child is called vertical transmission, whereas transmission from one individual to another is called horizontal transmission. Horizontal transmission is the only way of transmission for exogenous retroviruses, however it is important to note that endogenous retroviruses releasing infectious particles may also be transmitted horizontally. A clear understanding of the nature and transmission of exogenous and endogenous retroviruses is relevant for the understanding of the biological functions of endogenous retroviruses and the potential risk to various populations.

Once a retrovirus becomes integrated into the germ line of a species, i.e. becomes an endogenous retrovirus, there are no mechanisms to eliminate this virus from the genome. However, parts of the virus (with exception of the LTR) may be eliminated via recombination, and deletions, mutations or stop codons may disrupt their open reading frames and prevent expression of functional viral proteins and eventually of infectious viruses (for details see Stoye, 2001). However, most of the proviruses remain part of the genome. As a result, a large portion of the proviruses is defective and unable to replicate. Some 5–8% of human DNA represents retroviral genomes [not including the long interspersed nuclear elements (LINE) and other retrotransposons present in the human genome; see Chapter 1]. Only a few endogenous retroviruses are replication competent and produce infectious particles, and these are mainly found in evolutionarily lower species. These data support the general impression that endogenous retroviruses in the genome are useless – that they represent evolutionary garbage, a burden to the genome that is expensive to maintain. Energy and resources (nucleotides, enzymes) required to replicate up to 8% of the genome are thereby wasted.

In lower species such as mice, koalas and pigs, endogenous retroviruses are often intact and can be activated and expressed as fully infectious particles. In other species, particularly those higher on the evolutionary scale such as non-human primates and man, most of the endogenous retroviruses are defective and unable to produce infectious particles. This suggests that refined strategies of viral restriction evolved in these higher species. There is no doubt that such viruses were infectious when they entered the germ line, but their replication capacity became defective during evolution. Attempts to reconstitute endogenous retroviruses by generating consensus sequences, for example in the case of the human endogenous retrovirus (HERV)-K, resulted in partial (single cycle) restoration of infectious potential (Dewannieux *et al.*, 2006; Lee and Bieniasz, 2007). These studies may help determine whether active mechanisms suppress the replication capacity of endogenous retroviruses and may clarify how host species inactivate unwanted intruders. Numerous cellular restriction factors able to suppress replication of retroviruses – including endogenous retroviruses – have been described (Chiu and Greene, 2008).

On the other hand, some genes of endogenous retroviruses with conserved open reading frames have been described, despite long co-evolution with the host organism. Their functions are only partially understood. Although endogenous retroviruses are often expressed in tumours, it is not known whether genes of these endogenous retroviruses play a role in tumour development. However, a few examples of retroviral genes that benefit the host organism have been reported. For example, Env proteins of endogenous retroviruses drive differentiation of the placental cytotrophoblast into the syncytiotrophoblast and may facilitate immunosuppression; this utilization may be characterized as enslavement (see below).

Although retroviruses can switch between transmission as infectious agents and as host Mendelian elements, and although they can transduce host genes, e.g. viral oncogenes, there are no examples of gene transduction by retroviruses into the germ line of new hosts. Retroviruses could in theory serve to enable horizontal exchange of genetic information. However, other than transporting themselves, endogenous retroviruses do not appear to be purveyors of genes; even the retroviruses that bear oncogenes are not known to be naturally transmitted from host to host.

Since numerous excellent reviews have been published describing endogenous retroviruses

(Baltimore, 1975; Coffin, 1984; Löwer *et al.*, 1996; Boeke and Stoye, 1997; Bannert and Kurth, 2004, 2006; de Parseval and Heidmann, 2005; Weiss, 2006; Voisset *et al.*, 2008) we will concentrate on more recent publications as well as published and unpublished data from our own laboratory. This work focuses mainly on expression of one of the best-characterized human endogenous retroviruses, HERV-K, in normal tissues such as placenta and in tumours. We also describe results on the expression and transmission of porcine endogenous retroviruses (PERVs) in the context of virus safety of xenotransplantation.

Definition and distribution of endogenous retroviruses

One of the most remarkable insights from the human genome project was that more genetic material is devoted to vertically transmitted and defective HERVs than to human protein-coding DNA (Lander *et al.*, 2001). Current estimates place the HERV component of the human genome at 5–8%, comprising at least 31 distinct families, the largest of which, HERV-H, carries approximately 1300 full-length copies (Katzourakis *et al.*, 2005). Similarly high numbers are likely for other mammalian species, indicating that endogenous retroviruses (ERVs), including HERVs, are ubiquitous genomic elements.

Endogenous retroviruses are retroelements characterized by a long terminal repeat (LTR). Retroelements, together with transposons, belong to the group of transposable elements (for details see Bannert and Kurth, 2004). Until recently it was thought that only the simple retroviruses, e.g. alpha-, beta-, gamma- and epsilonretroviruses, become endogenous in their host, whereas those with complex genomes such as the lentiviruses, deltaretroviruses and spumaviruses did not. Recently however, Katzourakis *et al.* (2007) identified 25 full-length but clearly defective copies of a lentivirus called rabbit endogenous lentivirus type K (RELIK) in the genome of European rabbits, although exogenous lentiviruses have not yet been found in rabbits. In contrast, no endogenous lentiviruses have been found in species infected by exogenous lentiviruses. RELIK is estimated to have originated at least 7 million years ago, whereas the exogenous lentiviruses are much younger. Therefore, RELIK may be the origin of all lentiviruses. RELIK encodes for Tat and Rev, but not for Vif, an important protein that counteracts the host-cell mediated deamination process mediated by the apolipoprotein B mRNA editing enzyme, catalytic polypeptide-like (APOBEC) family in mammals. Thus it appears that other lentiviruses such as HIV, simian immunodeficiency virus (SIV), bovine immunodeficiency virus (BIV), and feline immunodeficiency virus (FIV) acquired vif somewhat later.

Endogenous sequences of spumaviruses have also been found (Cordonnier *et al.*, 1995). Investigating expression of retroviral sequences in the placenta, a new class of reverse transcriptase coding sequences and retrovirus-like elements bordered by LTRs and having a potential leucine tRNA primer-binding site were detected, termed HERV-L (human endogenous retrovirus with leucine tRNA primer). In addition, a region with homologies to dUTPase proteins was found downstream from the integrase domain. Amino acid sequence and phylogenetic analysis indicated that the HERV-L *pol* gene is related to that of foamy retroviruses. HERV-L-related sequences were detected in several mammalian species and have expanded to high numbers in primate and mouse genomes (Cordonnier *et al.*, 1995).

Only two families of HERVs have been detected in the genome of vertebrates lower than non-human primates: the HERV-L family in the mouse, rabbit, dog and cow (common ancestor 100 million years) and the HERV-H family in birds, reptiles and fish (common ancestor 400 million years). Interestingly, no element of the ERV-L family possesses an *env* gene, indicating that this family never had an exogenous phase (i.e. replicated horizontally) during its evolutionary history. The genomes of most species exhibit a low number of copies of ERV-L elements (from 10 to 30), while simians (but not prosimians) and mice (but not rats) have undergone bursts resulting in up to 200 copies (Benit *et al.*, 1999). The reason and biological function of this enormous amplification of ERV-L copy numbers in simians and mice is unclear.

Numerous solo-LTRs have been found in the genomes of animals and human. Formation of solo-LTR results from homologous recombination between two LTR flanking the provirus and the subsequent deletion of the internal sequence.

The vast majority of HERV elements in the human gene exist as solo-LTRs (Hughes and Coffin, 2004).

Discovery of endogenous retroviruses

ERVs were first detected in the late 1960s and early 1970s using virological and immunological methods and their existence was later confirmed by nucleic acid detection methods. Since 1904 it had been known that exogenous retroviruses (specifically, the lentivirus equine infectious anaemia virus) are filterable agents. Other retroviruses such as endogenous avian leukaemia viruses (ALV), murine leukaemia virus (MuLV) and mouse mammary tumour virus (MMTV) were described during the following four decades (Table 2.1). When the existence of proviral genes related to exogenous retroviruses was reported in the genome of different species (for review see Weiss, 2006), the interpretation of the earliest-detected ERVs was still impossible. This only became possible after the discovery of reverse transcriptase in 1970 (Baltimore, 1970; Temin and Mizutani, 1970). The name 'retroviruses' was coined in 1974 to emphasize the role of this specific enzyme (Baltimore, 1975). Prior to 1974 such viruses were called oncornaviruses (oncogenic RNA viruses) or were labelled according the electron microscopic classification as type A, B, C and D viruses (Gross, 1980).

Huebner and Todaro (1969) provided the first evidence suggesting the presence of DNA copies of ERVs in the genome. Later studies showed that the genomes of all vertebrate species studied to date are colonized by multiple retroviruses. Phylogenetic studies of viral genomes indicated that the introduction of ERV proceeds in waves, with relatively rapid amplification of copy numbers and dispersal in the host genome.

Research on avian and mouse endogenous retroviruses led to the identification of xenotropic viruses, which could infect only cells from foreign species (Levy, 1973; for review see Levy, 1978). In contrast, ecotropic viruses infect only cells of the same species, amphotropic viruses infect both the same and foreign species, and polytropic viruses infect cells from numerous species. For example, porcine endogenous retrovirus (PERV)-A infects human cells as well as cells from cats, minks, guinea pigs and numerous non-human primates, but does not infect cotton rats, rabbits, rats and mice (for review see Denner, 2008c).

Although retroviral particles were already described in human tissues in the 1970s (e.g. in the placenta: Kalter *et al.*, 1973; Bierwolf *et al.*, 1975) and in human cell lines in the 1980s (Kurth *et al.*, 1980; Löwer *et al.*, 1981), the first sequence evidence was obtained when Ono *et al.* (1986) detected the first HERV-K sequence in the human genome using probes derived from a conserved *pol* region of other retroviruses. These data were later confirmed and extended, and led to the identification of numerous HERV-K on different chromosomes of the humane genome (for review see Bannert and Kurth, 2004).

Activation of endogenous retroviruses

The enhanced expression of viral RNA, proteins and particles – termed *activation* – was already described during the earliest stages of research on ERVs. ERV activation was observed after treatment with chemicals and physical agents (Weiss *et al.*, 1971), including halogenated pyrimidines (Aaronson *et al.*, 1969; Lowy *et al.*, 1971; for review see Denner, 1977) and radiation (Lieberman *et al.*, 1959). Of special interest was the activation of ERVs in immunological reactions *in vivo* as well as after mitogen stimulation *in vitro*. Mitogen-triggered stimulation of peripheral blood mononuclear cells (PBMCs) simulates antigen-specific activation of the immune cells. Release of endogenous retroviruses after stimulation of immune cells was first reported for mice, e.g. after stimulation of lymphocytes with B-cell mitogens *in vitro* (Philipps *et al.*, 1977; De Lamarter *et al.*, 1981; Stoye and Moroni, 1983), during mixed lymphocyte reactions *in vitro*, during the graft-versus-host reaction *in vivo* (Hirsch *et al.*, 1972), as well as after immunization using antigens and different adjuvants *in vivo* (Denner and Dorfman, 1977). Such responses were also observed in other species, including pigs. Differences in the release of PERVs have been observed between different pig strains and even individuals of one strain (Tacke *et al.*, 2000a, 2003) when the mitogen phytohaemagglutinin (PHA) and the tumour promoter 12-*O*-tetradecanoylphorbol

Table 2.1 First description of (A) exogenous and (B) endogenous retroviruses

A

Year	Virus	Authors (for full references, see Vogt, 1997)
Exogenous		
1904	EIAV (equine infectious anaemia virus)	Vallee and Carré
1908	ALV (avian erythroleukaemia virus)	Ellerman and Bang
1911	ASV (avian sarcoma virus)	Rous
1914	ASV (avian sarcoma virus)	Fujinami and Unamoto
1936	MMTV (mouse mammary tumour virus)	Bittner
1951	MuLV (murine leukaemia virus)	Gross
1972	BLV (bovine leukaemia virus)	van der Maaten
1980	HTLV-1 (human T-cell leukaemia virus)	Poiesz
1983	HIV-1 (human immunodeficiency virus)	Barré-Sinoussi, Gallo

B

Year	Particles/sequence	Authors
1966	ALV	Dougherty *et al.* (1966, 1967), Payne *et al.* (1968), Weiss *et al.* (1967)
1969	MuLV	Aaronson *et al.* (1969), Lowy *et al.* (1971)
1969	MMTV	Bentvelzen and Daams (1969), Bentvelzen *et al.* (1970)
1980	HERV-K	Kurth *et al.* (1980), Ono *et al.* (1986)
1995	Foamy virus	Cordonnier *et al.* (1995)
2007	Lentivirus	Katzourakis *et al.* (2007)

13-acetate (TPA) are used for cell stimulation. Although PERVs can be activated by a T-cell mitogen alone, murine endogenous retroviruses are activated only when bromodeoxyuridine is given in addition (De Lamarter *et al.*, 1981; Stoye and Moroni, 1983). The release of PERV particles after stimulation with a biologically active concentrations of the T-cell mitogen was associated with T-cell proliferation and was nearly independent of TPA presence. PERV particles were also released after stimulation with high TPA concentrations in the absence of biologically active PHA concentrations (Tacke *et al.*, 2003). Although the mechanisms of virus activation may be different for PHA and TPA, both agents can activate all three PERV subtypes, PERV-A, PERV-B and the ecotropic PERV-C. Recently, activation of endogenous retroviruses was observed in rat cells after treatment *in vitro* with *Pasteurella pneumotropica* (unpublished data). Treatment of human PBMCs with mitogens did not result in activation of HERV-K. In rare cases, an elevated expression of a 1.5-kb mRNA of HERV-K – designated hel (HERV-K specific mRNA expressed in lymphocytes) – was observed in the mixed lymphocyte reaction (Denner *et al.*, unpublished).

ERVs have also been activated by infection with other viruses, e.g. the LTR directed transcription of the HERV-W was induced by infection with herpes simplex virus (HSV) type 1 (Lee *et al.*, 2003). The effect was partially mediated by the action of HSV-1 immediate early protein 1 and required an Oct-1 binding site in the LTR. The transcription of HERV-K and Alu sequences was also induced by HSV-1. During HIV-1 infection an increased expression of HERV-K *in vitro* was reported (Contreras-Galindo *et al.*, 2006; 2007a; b), although these data are still controversial. Antibodies against HERV-K have been observed in numerous patients with tumours, and particularly those with teratocarcinomas (Vogetseder *et al.*, 1993; Sauter *et al.*, 1995, 1996; Löwer *et al.*, 1996; Boller *et al.*, 1997, Kleinman *et al.*, 2004) or melanomas (Muster *et al.*, 2003; Büscher *et al.*, 2005, 2006). In addition, T-cell responses have been observed in patients with seminomas and melanomas (Schiavetti *et al.*, 2002; Rakoff-Nahoum *et al.*, 2006). Importantly, the sequence of the antigen detected in melanoma patients is an HERV-K-derived open reading frame that is not correlated with the open reading frame used by the virus. Antibodies against HERV-K specific peptides have been found in HIV-1 infected individuals (Löwer *et al.*, 1996); these are not solely due to cross-reacting antibodies against conserved epitopes in the Gag protein, but rather are also directed against the immunodominant epitope in the transmembrane envelope protein of HERV-K (Denner *et al.*, unpublished). For reasons unknown others did not detect HERV-K specific antibodies in HIV-infected individuals (Sauter *et al.*, 1995). An increased HERV-K expression in HIV-1 infected cells has been shown (Johnston *et al.*, 2001; Contreras-Galindo *et al.*, 2006, 2007a,b) and may accompany the increased HERV-K antibodies as has been observed in tumour patients.

Endogenous retroviruses and trans-species transmission of retroviruses

Trans-species transmission is common among retroviruses, although the consequences of the transmission vary with the species involved and the precise retrovirus (for review see Denner, 2007). Some trans-species transmissions are fatal in the new host whereas others remain asymptomatic. HIV-1 and HIV-2 are examples of viruses that are apathogenic in their natural host – non-human primates – but induce fatal acquired immunodeficiency syndrome (AIDS) in humans. Replication-competent endogenous retroviruses release infectious virus particles and all but the ecotropic viruses may infect other species. Again, the outcome may be different for different species: the infection may be aborted or non-pathogenic, may be pathogenic, or the virus may start an endogenization in a new species.

The Koala retrovirus (KoRV) is an excellent example of trans-species transmission and endogenization. In koalas, KoRV induces myeloid leukaemias, lymphomas and immunodeficiencies associated with Chlamydia infections. KoRV is closely related to murine leukaemia viruses and to the gibbon ape leukaemia virus (GaLV) (Hanger *et al.*, 2000). Since GaLV originated from South-East Asian mice such as *Mus caroli* (Lieber *et al.*, 1975) and since gibbons and koalas live on different continents, an intermediate vector may have transmitted the virus to both species. This hypothesis is supported by evidence that the level of sequence divergence between GaLV and KoRV is similar to that of two different strains of GaLV: SEATO and SF (Delassus *et al.*, 1989).

Retroviruses endogenous in DNA viruses

A group of avian retroviruses including the reticuloendotheliosis virus (REV) of turkeys probably had a mammalian origin but had not integrated into avian germ line DNA. Interestingly, these viruses had integrated into the circular DNA of Marek's disease herpesvirus (Isfort *et al.*, 1994) and fowlpox virus (Hertig *et al.*, 1997; Singh *et al.*, 2003). ALV also integrated into Marek's disease herpesvirus (Isfort *et al.*, 1994). On the other hand, integration of a human herpesvirus (HHV)-6 into the human genome and transmission from parent to child has been reported (Tanaka-Taya *et al.*, 2004). It remains unclear whether retroviral enzymes were involved in the integration process.

Endogenization of retroviruses

Identification of the infectious progenitor of an endogenous virus has only rarely been achieved. As mentioned above, the most recent example is the KoRV, which allowed a real-time investigation of retrovirus endogenization. Although KoRV and GaLV obviously have the same progenitor, in contrast to GaLV the infection in koalas is associated with endogenization. As Tarlinton *et al.* (2006) showed recently, koalas in northern Australia carry endogenous sequences, whereas many animals in southern Australia – especially on Kangaroo Island – are not infected at all. One possible scenario to explain this observation is that virus-transmitting rodents from South-East Asia arrived in northern Australian seaports and infected local koalas. This scenario is supported by findings that KoRV infects rat cells *in vitro* and rats *in vivo* (Fiebig *et al.*, 2006). Afterwards, a wave of infection and endogenization moved through Australia, but has not yet reached the southern coast.

PERVs, viruses related to KoRV, have also spread via trans-species transmission. Because similar sequences have been found in mice, but not in rats and hamsters, it was suggested that PERVs originated from murine endogenous retroviruses.

Could HIV-1 become endogenous in humans? Since sperm cells and oocytes do not carry the main receptors for HIV-1, this event seems very unlikely. However, other molecules such as ceramides have been used as cellular receptors for HIV-1, suggesting that cells not expressing CD4 could become infected (Yu *et al.*, 2008). Although the expression of galactosylceramide, binding of HIV-1 to sperm cells, and HIV-1 infection of these cells have been demonstrated (Baccetti *et al.*, 1994; Shevchuk *et al.*, 1998), systematic studies in newborns from HIV-infected parents have not yet been reported.

Coexistence of endogenous and exogenous retroviruses

In many species exogenous and closely related endogenous retroviruses co-exist. Examples include the mouse mammary tumour virus (MMTV) and the murine leukaemia virus (MuLV) in mice, the feline leukaemia virus (FeLV) in cats, the Jaagsiekte sheep retrovirus (JSRV) and the enzootic nasal tumour virus (ENTV) in sheep and goats, PERVs in pigs, ALV in chickens, and KoRV in some, but not all, koalas. In humans, no infectious endogenous retroviruses have been identified so far and for most lentiviruses no related endogenous viruses were found. However, recently in the goat and sheep genome, at least 27 copies of endogenous betaretroviruses related to the exogenous JSRV and ENTV were found. ENTV is the causative agent of respiratory tract carcinomas in sheep. JSRV, causes ovine pulmonary adenocarcinoma. JSRV is a unique betaretrovirus as its envelope glycoprotein is a dominant oncoprotein that induces cell transformation *in vitro*. After interaction with the receptor hyaluronidase 2, a motif in the cytoplasmic tail of the transmembrane envelope protein of JSRV transforms cells by inducing the P13K-AKT and the Raf-MEK-MAPK signalling pathways. The endogenous viruses, enJRSV, play an important role in the evolution of domestic sheep because they block replication of exogenous JRSV and also play a critical role in sheep conceptus development and placental morphogenesis (see below). All enJSRV lack the oncogenic motif in the transmembrane envelope protein and therefore cannot transform cells. EnJSRV blocks JSRV replication by a novel two-step interference mechanism. The early blockade is based on enJSRV expression, which blocks viral entry by receptor interference. The late blockade is based on Gag expression (for review see Palmarini *et al.*, 2004; Arnaud *et al.*, 2008). The Gag of enJSRV carries an arginine to tryptophan mutation in position 21. Co-assembly of the wild and mutated Gag results in multimers unable to traffic to the cell membrane and therefore unable to produce infectious virus (for details see Palmarini *et al.*, 2004).

Immunosuppressive properties of endogenous retroviruses

Most exogenous retroviruses, if not all, induce immunodeficiencies in the infected host (Table 2.2). Regardless of whether it is induced by a gammaretrovirus, betaretrovirus or lentivirus, all immunodeficiencies are characterized by a decrease in the number of $CD4^+$-cells and by opportunistic infections. In all cases, the viral load correlates with progression to disease. The mechanism by which exogenous retroviruses induce immunosuppression is still unclear, however, as summarized below, there is accumulating evidence that the transmembrane envelope protein has a direct immunosuppressive effect (Denner, 1987, 2000).

First, inactivated retrovirus particles can inhibit proliferation of immune competent cells *in vitro*. This is true for HIV-1 (Pahwa *et al.*, 1986; Denner *et al.*, 1996), gammaretroviruses such as FeLV (Olsen *et al.*, 1977), baboon endogenous virus (BaEV) (Denner *et al.*, 1980; Weislow *et al.*, 1981), and PERVs (Denner, 1998; Tacke *et al.*, 2000a,b), as well as for type D retroviruses (Denner *et al.*, 1980; 1985).

Table 2.2 Immunosuppressive exogenous retroviruses (examples)

Genus	Example
Alpharetrovirus	Avian leukosis virus (ALV)
Betaretrovirus	Simian retroviruses 1–3 (SRV 1–3)
Gammaretrovirus	Feline leukaemia virus (FeLV), murine leukaemia virus (MuLV)
Deltaretrovirus	Human T-lymphotropic virus (HTLV)
Lentivirus	Human and simian immunodeficiency viruses (HIV-1, HIV-2, SIVmac)

Second, among all viral proteins tested, the transmembrane envelope proteins are immunosuppressive in *in vitro* assays (Mathes *et al.*, 1979; Denner *et al.*, 1986), and in *in vivo* models. Expression of the transmembrane envelope proteins p15E of murine gamma-retroviruses (Mangeney and Heidmann, 1998), of a type D retrovirus (Blaise *et al.*, 2001) or of the human endogenous retrovirus HERV-H (Mangeney *et al.*, 2001) caused local immunosuppression in immunocompetent animals, thereby preventing rejection of tumour cells (Fig. 2.1). The same effect has been shown for the protein syncytin-1 (but not syncytin-2), which is highly expressed in the human placenta during pregnancy. Syncytin-1 is the envelope protein of HERV-W and syncytin-2 is the envelope protein of HERV-FRD (Blaise *et al.*, 2003). Using the tumour cell assay, it has been shown that syncytin-1 is not immunosuppressive and syncytin-2 is immunosuppressive (Mangeney *et al.*, 2007). Similarly, the murine syncytin-A is not immunosuppressive, but murine syncytin-B is. Using deletion mutants it was shown that a highly conserved domain in the transmembrane envelope proteins – the so-called immunosuppressive domain (isu-domain) – is responsible for the immunosuppressive effect. Mutations of specific amino acids allowed the switch from immunosuppressive to non-immunosuppressive syncytin and vice versa. This experimental approach provided the first demonstration that the immunosuppressive effect is not only local – as previous experiments had indicated (Fig. 2.1E) – but also that the humoral immune response is inhibited. Mice produced more specific antibodies against the non-immunosuppressive syncytin as compared to the immunosuppressive syncytin.

The third body of evidence for the direct immunosuppressive effect of transmembrane envelope protein comes from a C57B1/6 mouse model in which B16 cells are used to induce melanomas (Mangeney *et al.*, 2005) (Fig. 2.2). Use of RNA interference to knock down an endogenous retrovirus induced spontaneously in the B16 melanomas resulted in rejection of the tumour cells in immunocompetent mice, under conditions where control melanoma cells expressing the transmembrane envelope protein p15E grew into lethal tumours. Evidence that the knockdown did not modify the transformed phenotype of the cells was obtained *in vitro* using a soft agar assay and *in vivo* by measurement of tumour cell proliferation in immunoincompetent mice (X-irradiated and severe combined immunodeficiency). Tumour rejection could also be reverted by adoptive transfer of regulatory T-cells (Treg) from control melanoma-engrafted mice, or by re-expression of the envelope gene of the endogenous retrovirus in the knocked down cells (Fig. 2.2). These results show that endogenous retroviruses can be essential for a Treg-mediated subversion of immune surveillance. In addition, retroviral p15E inhibited macrophage accumulation in mice (Cianciolo *et al.*, 1980) and inhibited antibody production against a tumour antigen, thereby increasing tumour incidence *in vivo* (Olsen *et al.*, 1977).

Finally, the fourth body of evidence comes from use of synthetic peptides corresponding to the isu-domain of the transmembrane envelope protein of different retroviruses, which is conserved among all retroviruses (including HIV-1). These peptides are immunosuppressive in several *in vitro* proliferation assays (Cianciolo *et al.*, 1985, 1988; Harrell *et al.*, 1986; Nelson *et al.*, 1989;

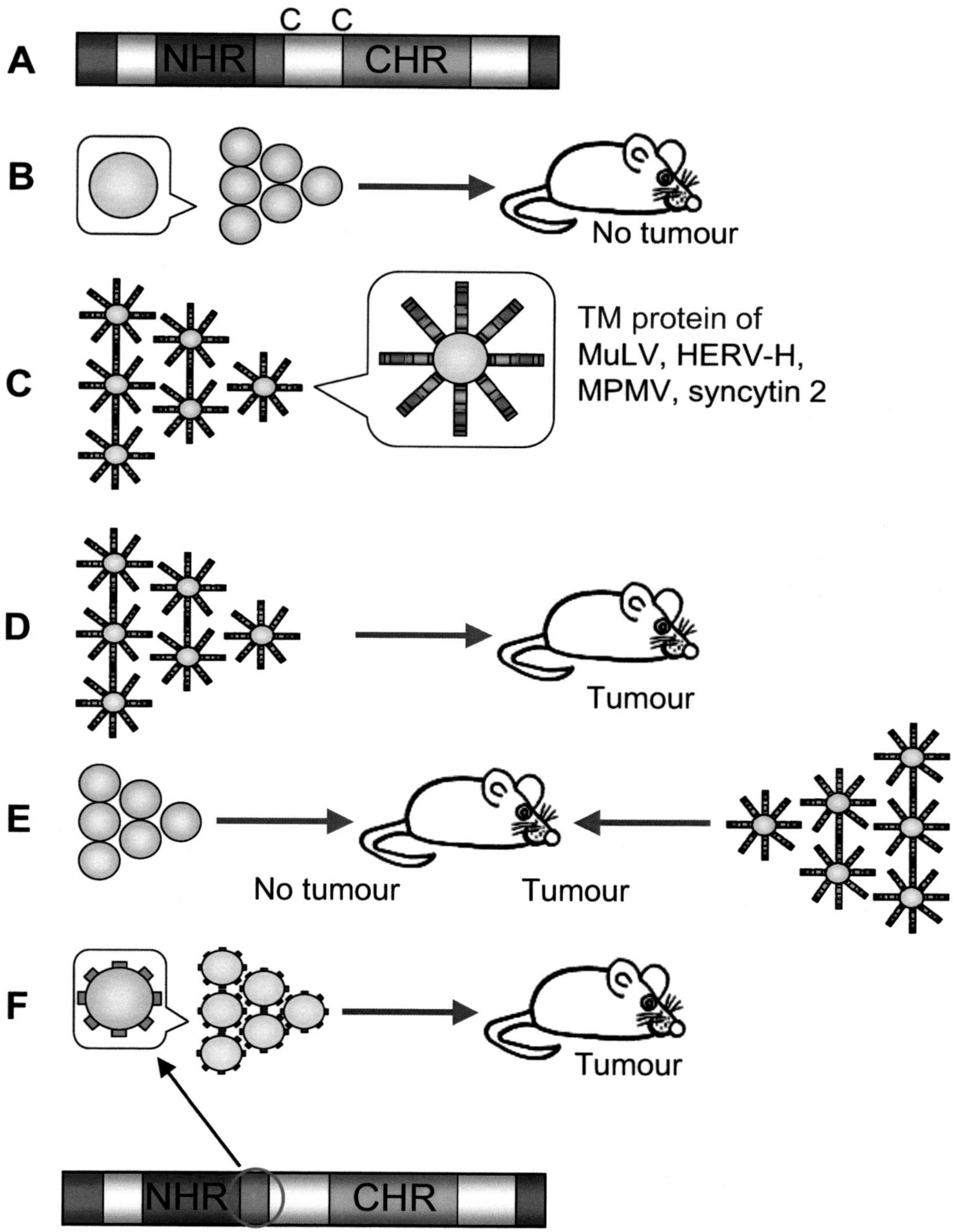

Figure 2.1 Evidence for *in vivo* immunosuppressive properties of TM proteins from different retroviruses (Mangeney and Heidmann, 1998; Blaise *et al.*, 2001; Mangeney *et al.*, 2001, 2007). (A) Schematic presentation of the TM proteins. Red – fusion peptide, green – immunosuppressive (isu) domain. NHR, N-terminal helical region; C-C, cysteine–cysteine loop; CHR, C-terminal region. (B) Injection of tumour cells into immunocompetent mice did not induce tumours. (C), (D) After transfection, these tumour cells expressed the TM protein of the retroviruses MuLV, HERV-H, MPMV, and HERV-FRD (syncytin 2); their inoculation resulted in tumour growth. (E) Inoculation of tumour cells expressing TM proteins resulted in tumour growth, whereas injection of tumour cells not expressing TM *env* into the same animals did not induce tumours. (F) Inoculation of tumour cells expressing only the immunosuppressive domain induced tumours, indicating that this domain is the biological active domain.

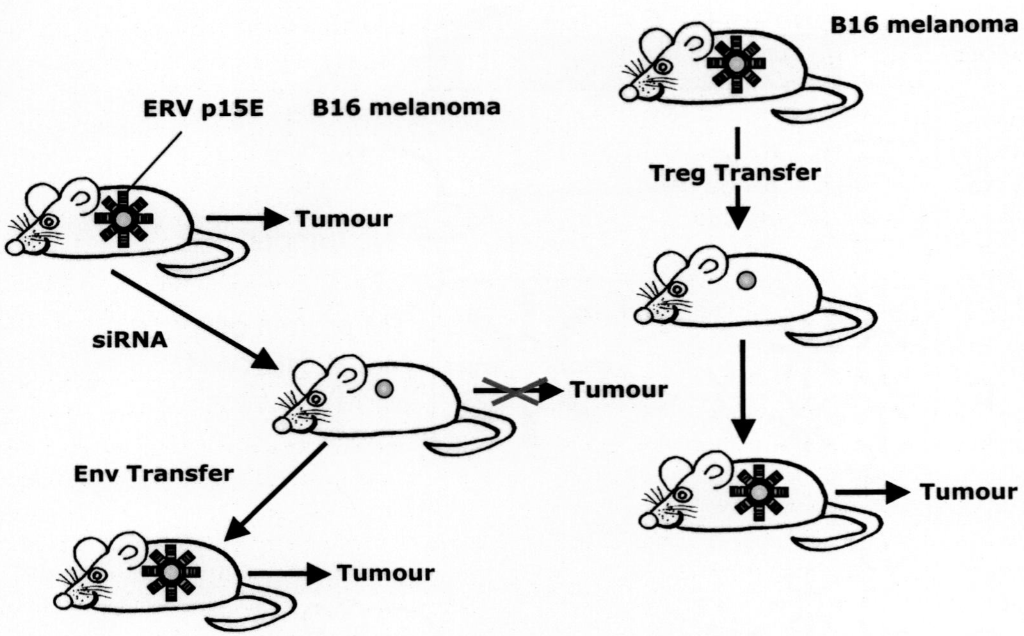

Figure 2.2 Expression of the murine retrovirus TM protein p15E on the cell surface of murine B16 melanoma cells is required for tumour progression *in vivo* (Manganey *et al.*, 2005). Inhibition of TM protein expression by small interfering RNA (siRNA) resulted in an inhibition of tumour growth. Inhibition of tumour growth was reverted by *de novo* transfection and reexpression of the gene encoding the TM protein and upon transfer of regulatory T-cells (Treg) from melanoma-bearing animals resulted again in tumour growth.

Ogasawara *et al.*, 1990; Ruegg *et al.*, 1989a,b; Denner *et al.*, 1994, 1996).

The historical term 'isu-peptide' is slightly confusing as is the term 'fusion peptide' with regard to in the transmembrane envelope proteins of retroviruses. The domains should be called isu-domains and fusion domains, but synthetic peptides derived from the isu-domain may be called isu-peptides. Importantly, the isu-peptides are only biologically active when bound to a carrier protein or after polymerization, presumably because this confers a biologically active conformation or a multiplicity that may be necessary for interactions with target molecules. These isu-peptide–carrier protein conjugates or polymers represent a mono-specific model of the corresponding isu-domain of the retroviral transmembrane envelope protein.

However, it is still unclear whether retroviral transmembrane envelope proteins such as p15E and gp41 with their isu-domains behave similarly *in vivo* as the mono-specific model *in vitro*, which would indicate a role in the immunopathogenesis of retroviral infection. The mechanism by which the corresponding synthetic peptide exerts its immunosuppressive effect *in vitro* is also unclear. The isu-peptides of different retroviruses, including HIV-1, inhibit human T and B-lymphocyte proliferation (Cianciolo *et al.*, 1988; Ruegg *et al.*, 1989a; Denner *et al.*, 1996) and monocyte functions (Harrell *et al.*, 1986; Kleinerman *et al.*, 1987; Tas *et al.*, 1988). Transmembrane envelope proteins and isu-peptides modulate the cytokine production of normal PBMCs: e.g. they increase IL-10 and IFNα production and decrease IL-2 production (Barcova *et al.*, 1998; Denner, 1998; Haraguchi *et al.*, 1992a,b, 1993, 1995a,c; Takeshita *et al.*, 1995; Koutsonikolis *et al.*, 1997; Speth *et al.*, 2000). Cell surface binding proteins (receptors?) have been described – but not fully characterized – for the isu-peptides of gp41 and p15E, which may mediate the effect of the peptides (Qureshi *et al.*, 1990; Kizaki *et al.*, 1991; Chen *et al.*, 1992, 1997, 1998; Denner *et al.*, 1993, 1995; Ebenbichler *et al.*, 1993; Henderson and Qureshi, 1993). In addition, isu-peptides have inhibitory effects on protein kinase C (PKC) and cAMP (Gottlieb *et al.*, 1990; Ruegg *et al.*, 1990a;

Ruegg and Strand, 1990b Kadota *et al.*, 1991; Haraguchi *et al.*, 1995a,b).

Homologies between the immunosuppressive domain of the transmembrane envelope protein of different retroviruses and class I interferons (IFN) have been described (Wegemer *et al.*, 1990; Denner, 1998; Chen *et al.*, 1999). Human class I IFN represent a family of 14 IFNα molecules, one IFNβ molecule and one IFNω molecule. All class I IFNs use the same receptor, which consists of two subunits, IFNαR1 and IFNαR2 (Uze *et al.*, 1994). IFNα itself (Pfeffer, 1987) and peptides corresponding to helical domains of IFN (Ruegg and Strand, 1990c) also exert immunosuppressive effects.

In order to analyse the influence of retroviral transmembrane envelope proteins on cytokine production, cytokine array assays were performed, measuring expression (release in the supernatant) of nearly 100 cytokines by normal human PBMC after exposure to isu-peptides, recombinant transmembrane envelope proteins of different retroviruses (HIV, KoRV, PERV, HERV-K), and purified virus particles. The expression of the cytokines IL-10, IL-6, IL-8, RANTES, MCP-1, MCP-2, TNF-α, MIP-1α, MIP-1β, MIP-3, IL-1β, Gro-(α, β, γ), and Gro-α increased, expression of IL-2 and MIG (CXCL9) decreased, and expression of GCSF, GM-CSF, IL-1α, IL-3, IL-4, IL-5, IL-7, IL-11, IL-12p40, IL-12p70, IL-13, IL-15, IL-17, and many others (altogether over 90 cytokines and other factors) remained unchanged (Denner *et al.*, unpublished). These data indicate that retroviral transmembrane envelope proteins modulate the cytokine production of normal PBMCs and therefore may play a significant role in retrovirus-induced immunopathogenesis. It is important to note that increased IL-6, IL-10, IL-8, Gro-α and TNF values, and decreased IL-2 values, are commonly observed in HIV-1 infected individuals (Shearer and Clerici, 1993).

To study the influence of the isu-domain on gene expression in immune cells, microarray analyses were performed measuring the expression of 29,098 human genes (AB1700 Human Genome Survey Microarray V2.0). Initial data confirmed the influence of retroviral transmembrane envelope proteins on cytokine production, showing the expected elevated expression of different interleukins, e.g. IL-6. However, expression of other genes also changed and the importance of these genes in signal transduction and immunosuppression requires further study. Genes with the highest fold change (altogether 444 transcripts were elevated) include MMP-1, IL-6, IL1A, CXCL13, CXCL and TREM-1 (unpublished data). Similar results were obtained when cells from the same donor were incubated with the recombinant transmembrane envelope protein of HERV-K or the polymer of the isu-peptides of HIV/PERV (unpublished data).

To analyse whether retroviral recombinant transmembrane envelope proteins and isu-peptides induce Tregs, they were incubated with normal human PBMCs and an increase in the number of $CD4^+CD25^+FoxP3^+$ was measured using flow cytometry (FACS) analyses. In addition, real-time RT-PCR analyses showed an increase in the expression of FoxP3 (unpublished data). These results confirmed the involvement of Tregs in retroviral transmembrane envelope protein mediated immunosuppression (Mangeney *et al.*, 2005).

Expression and function of endogenous retroviruses in the placenta

Early findings and functions of the placenta

Expression of retroviral particles in the placenta of different species, including non-human primates and humans, was reported more than 30 years ago (Kalter *et al.*, 1973, 1975a,b; Bierwolf *et al.*, 1975; Seman *et al.*, 1975; Dirksen and Levy, 1977; Feldman, 1979; Stromberg and Benveniste, 1983; Ueno *et al.*, 1983; Feldman *et al.*, 1989). Antigens related to retroviral proteins (Thiry *et al.*, 1981; Derks *et al.*, 1982; Wahlström *et al.*, 1984) and antibodies cross-reacting with primate retroviruses (Thiry *et al.*, 1978) were also described in the human placenta and in pregnant women, respectively. Trophoblast cells isolated from human placentas expressed ERV-3 and secreted an immunosuppressive factor (Boyd *et al.*, 1993). However, the physiological impact of these viruses or their proteins during pregnancy remains unclear.

The placenta is a transient organ that mediates nutrient and gas exchange between mother and foetus during intrauterine life. The placenta is made up of chorionic villi that are covered by a double layer of fetally derived trophoblast cells. The outer fused syncytiotrophoblast is formed from underlying mononuclear cytotrophoblast cells, called villous cytotrophoblast (VT). The fusion of trophoblastic cells into a multinucleated syncytiotrophoblast is a key process of placental morphogenesis. Some of the cytotrophoblast cells grow through the syncytial layer to form columns of extravillous cytotrophoblast (EVT) cells that invade the maternal decidua. These EVT are in close contact with different cell types in the maternal decidua, including immune cells – mainly natural killer cells that are known to be important for the control of trophoblast invasion. Functions of the syncytiotrophoblast include the production of hormones and the supply of the fetal cells with nutrients.

HERV expression in human placenta

A variety of methods have demonstrated that the placenta is a preferential site of HERV expression. Comparison of the expression of fusogenic Env proteins in placenta and 18 other human tissues showed the highest expression in placenta (de Parseval and Heidmann, 2005) (Fig. 2.3). Elevated expression of HERV-W (syncytin-1) (Blond *et al.*, 1999; 2000; Mi *et al.*, 2000), HERV-FRD (syncytin-2) (Malassine *et al.*, 2008), ERV-3 (Kato *et al.*, 1987; Boyd *et al.*, 1993), and HERV-E (Yi and Kim, 2007) have been reported. Some of these endogenous retroviruses are up-regulated hormonally (Prudhomme *et al.*, 2005). Syncytin-1, the envelope protein of HERV-W, mediates the fusion of the villous cytotrophoblast to form the multinucleated syncytiotrophoblast. This observation is supported by evidence that anti-sense RNA specific for syncytin-1 decreases

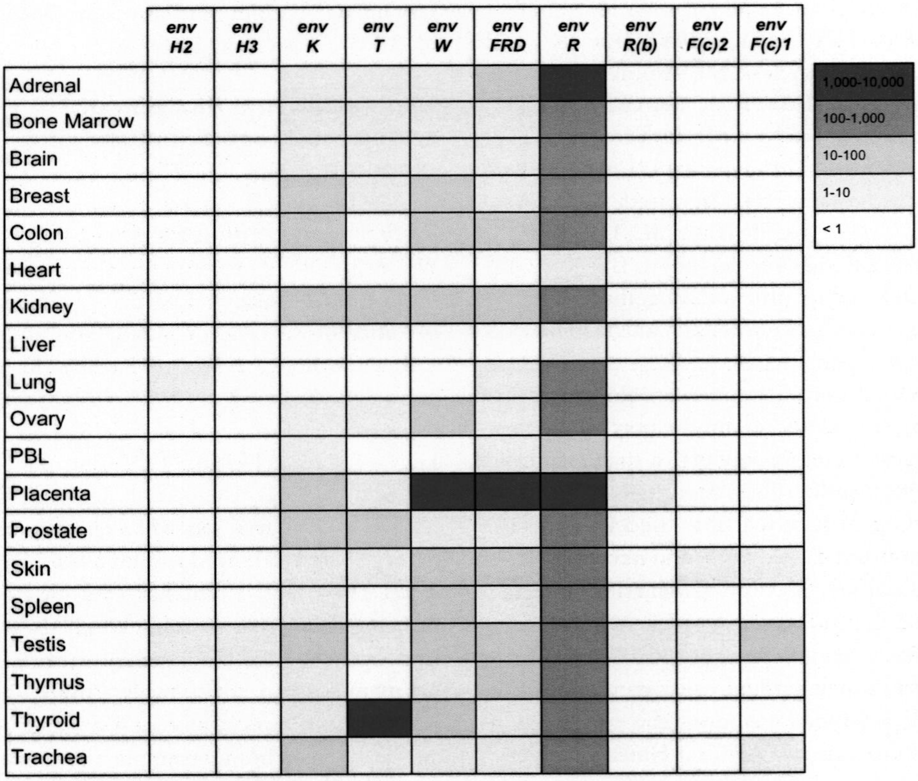

Figure 2.3 Expression of the coding retroviral envelope genes in a panel of 19 healthy human tissues. Values of real-time PCRs are given using 18S RNA as control (for values and details concerning the different viruses and their chromosomal location see de Parseval and Heidmann, 2005). PBL, peripheral blood lymphocytes.

the number and size of syncytiotrophoblast cells, with a concomitant decrease in expression of human chorionic gonadotrophin (HCG) (Frendo *et al.*, 2000, 2003). To induce the formation of syncytia, syncytin-1 interacts with the D type mammalian retrovirus receptor RDR, a neutral amino acid transporter (Blond *et al.*, 2000; Lavilette *et al.*, 2002). Syncytin-2 uses another receptor – major facilitator superfamily domain containing 2 (MFSD2) – which belongs to a large family of presumptive carbohydrate transporters with 10 to 12 membrane-spanning domains (Esnault *et al.*, 2008). A real-time RT-PCR analysis demonstrated specific expression in the placenta. Syncytin-1 and -2 differ in their site of expression within normal placenta, with syncytin-1 localized in both VT and EVT (Frendo *et al.*, 2003; Muir *et al.*, 2006) and syncytin-2 expressed only in VT (Maksakova *et al.*, 2007; Malassine *et al.*, 2007). However, both syncytin-1 and syncytin-2 participate in the fusion of VT to syncytiotrophoblast. Syncytin-1 and -2 are abnormally expressed in several pathological conditions, supporting their relevance in reproduction. For example, in preeclampsia, a condition with pathological trophoblast invasion into the maternal decidua, syncytin-1 is localized in the apical rather than the basal layer of syncytiotrophoblast (Lee *et al.*, 2001; Knerr *et al.*, 2002). Syncytin-2 expression is altered in placenta with trisomy 21, where fusion of VT and maturation of chorionic villi is delayed (Malassine *et al.*, 2008). The strong immunosuppressive effect of syncytin-2, but not of syncytin-1 was discussed above. While the function of HERV-H in placenta is still unknown, ERV-3, a gene originally reported to be responsible for the fusion of spermatozoa with the oocyte, is also thought to participate in the differentiation of the cytotrophoblast into the syncytiotrophoblast, as well as in the immunoregulatory process within the placenta (Venables *et al.*, 1995; Nilsson *et al.*, 1999;).

Using real-time RT-PCR, immunocytochemistry, and Western Blot analyses, expression of HERV-K was studied in normal placental tissue of different gestational ages. The transmembrane envelope protein of HERV-K was found exclusively in VT and EVT-cells, sparing syncytiotrophoblast and other cells (Fig. 2.4) (Kämmerer, U., Germeyer, A., Kapp, M., Stengel, S., Büscher, K., Kurth, R., Denner, J. unpublished). The nature of the invasive EVT-cells was confirmed by counterstaining the tissue with antibodies specific for cytokeratin. The expression increased up to the second trimester and later decreased. These data suggest a potential involvement of HERV-K in placentogenesis and pregnancy. Since retroviral transmembrane envelope proteins, including that of HERV-K, have immunosuppressive properties, expression of HERV-K may contribute to the protection of the embryo.

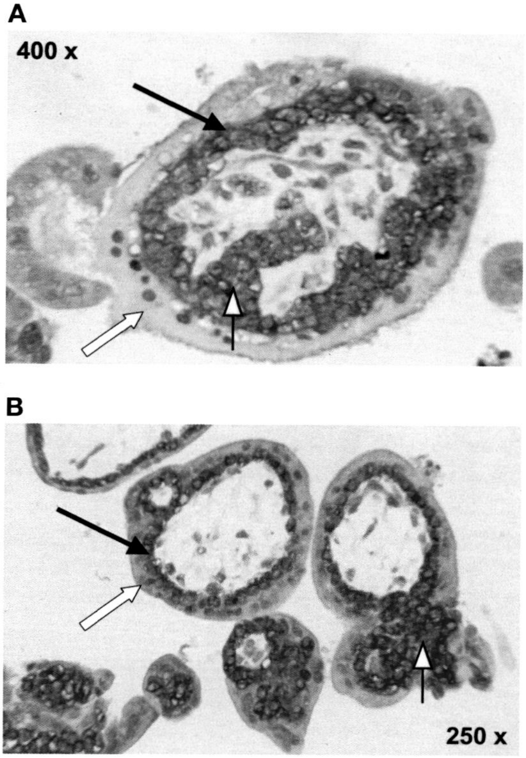

Figure 2.4 Immunohistochemical detection of HERV-K TM protein in formalin-fixed tissue of human placenta. The black arrow indicates villous cytotrophoblasts, the bold white arrow indicates the syncytiotrophoblast, the white head arrow indicates a column of extravillous cytotrophoblasts. (A) Gestational week 6. A strong staining of HERV-K can be seen in all cytotrophoblast cells while the syncytiotrophoblasts are negative. (B) Gestational week 7. A serum specific for the transmembrane envelope protein of HERV-K was used.

enJSRV expression and function in ovine placenta

The endogenous sheep virus enJSRV is expressed in numerous tissues and organs: enJSRV mRNA is abundant in organs of the reproductive tract, in trophectoderm cells and particularly in trophoblast giant binucleate cells (BNC) and multinucleated syncytia which are required for implantation and formation of placentosomes. The latter are required for nutrition of the conceptus. Expression of enJSRV *env* mRNA starts at day 12 after mating. Inhibition of *env* expression by antisense oligonucleotides retards blastocyst growth, inhibits giant BNC differentiation, and results in loss of pregnancy (for review see Arnaud *et al.*, 2008), indicating that sheep reproduction is entirely dependent on enJSRV *env* expression. In the endometrium enJSRV *env* expression increases between day 1 and 13, and as observed in other species (see above), viral particles occur in the epithelia and trophectoderm. It is unclear whether these particles have a biological function. JSRV *env* expression was also detected in sheep foetuses, in the lymphoid cells of the lamina propria of the gut and in the thymus, indicating that it may prevent infection by exogenous JSRV.

Syncytins in mice

Syncytin-A and syncytin-B are two recently identified murine genes which are homologous but are not related to the human syncytin genes (Dupressoir *et al.*, 2005). For both genes the highest expression occurs in the placenta between 9.5 and 14.5 days post coitum. Syncytin-A is not immunosuppressive, whereas syncytin-B is (Mangeney *et al.*, 2007). It is intriguing that in man and mice the two syncytin genes show a similar expression modus and represent different provirus integrations. The entry dates were 20 Myr ago for the murine genes, and 25–49 Myr for the human genes after specification of rodents and primates.

Expression of endogenous retroviruses in tumours

Lessons from mouse models

Leukaemogenesis induced by endogenous murine leukaemia virus in AKR mice is a multistep process leading to recombinant viruses (so called mink cell focus forming viruses) from endogenous viruses (Stocking and Koczak, 1998). *De novo* infection of these recombinant viruses may subsequently activate proto-oncogenes and result in leukaemia. A similar mechanism is involved with the endogenous mouse mammary tumour virus in GR mice (Boeke and Stoye, 1997).

Expression of PERV in melanoma-bearing pigs

An elevated expression of PERV has been observed in melanomas of the Munich miniature swine (MMS) 'Troll' (Dieckhoff *et al.*, 2007b). During cultivation of tumour cells *in vitro* the PERV expression increased and particles were released. While it is unclear whether PERV plays a role in the tumour development, the expression of the immunosuppressive TM protein p15 on the surface of tumour cells suggests that PERV may contribute to tumour progression. Recombinant viruses between PERV-A and PERV-C have also been observed in spleen cells of melanoma-bearing animals, though not in the tumour cells. PERV-A/C recombinant viruses were found *de novo* integrated into the genome of spleen cells, but not of other tissues, and may result from an activation of these viruses due to antitumour immune responses. Such recombinant PERV-A/C are characterized by high titres and may represent a new risk for xenotransplantation (Denner, 2008b).

Expression of HERV-K in human tumours

Although expression of full-length mRNA from deleted and undeleted HERV-K proviruses was detected in all human tissues and cells investigated, expression of spliced *env* and *rec* was detected mainly in tumour cells. Expression of HERV-K was most thoroughly investigated in germ cell tumours and melanomas. Retrovirus-like particles were first detected in human teratocarcinoma cells lines (designated at that time human teratocarcinoma derived virus, HTDV) (Kurth *et al.*, 1980; Löwer *et al.*, 1981). Later it was realized that these particles were encoded by HERV-K(HML-2) sequences (Boller *et al.*, 1993; Löwer *et al.*, 1993b). HML-2 stands for human mouse mammary tumour virus-like-2 according to another classification of these viruses (Bannert and Kurth, 2006). Although the significance of these viruses for the development of germ cell tumours is still unknown, enhanced expression of full-length mRNA, and expression of Env, Rec and Np9 proteins were described in teratocarcinoma cell lines, e.g. in the GH cell line (Löwer *et al.*, 1993a,b, 1995).

Expression of HERV-K was also found in primary melanoma and melanoma-derived cell lines (Muster *et al.*, 2003; Büscher *et al.*, 2005; 2006). Expression of *env* and *rec* mRNA was observed in 45% of the metastatic melanoma biopsies and in 44% of the melanoma cell lines. In normal neonatal melanocytes spliced *rec* was detected but spliced *env* was not detected. Using immunohistochemistry, immunofluorescence, and Western blot analyses with specific antisera, HERV-K proteins were shown to be expressed in primary melanomas, metastases and melanoma cell lines. Sera from 60 melanoma patients, 20 normal blood donors and 20 patients with alopecia were tested by Western blot for antibodies specific to the HERV-K transmembrane envelope protein; 22% (13 from 60) of the melanoma patients' sera, but none of the control sera showed strong responses (Büscher *et al.*, 2005). The frequency of patients producing antibodies was low, suggesting that expression in the other patients was too low for induction of a humoral response. In contrast, up to 85% of patients with seminomas and other germ cell tumours produced antibodies directed against the Env protein, and all Env-positive patients generated high titres of antibodies directed against the transmembrane domain (Sauter *et al.*, 1996). The generation of HERV-K-specific antibodies in tumour patients also indicates a lack of tolerance and suggests that expression during ontogenesis (when discrimination between self and non-self is made) does not occur.

In a second melanoma study, expression of both spliced *env* and *rec* mRNA was detected in 39% of the melanomas and in 40% of the melanoma cell lines and *np9* mRNA was detected in 29% and 21%, respectively. Using antisera specific for Rec and Np9, Rec protein was found in 14% of the melanomas but Np9 was not detected in any samples. Despite the expression of Rec protein, no antibodies specific for Rec were detected. Melanoma cells released virus particles, which were defective and non-infectious (Büscher *et al.*, 2006).

In addition to germ cell tumours and melanomas, data on HERV-K expression in breast cancer were recently reported (Wang-Johanning *et al.*, 2001, 2003; Frank *et al.*, 2008).

Regulation of HERV-K expression by methylation

Evidence that enhanced expression of HERV-K in tumour cells correlates with hypomethylation has been shown for germ cell tumours (Götzinger *et al.*, 1996; Lavie *et al.*, 2005) and melanomas (Stengel *et al.*, unpublished data). Treatment of melanoma cell lines by the demethylating agent 5-aza-2′-deoxycytidine (5-aza-dC) resulted in increased levels of HERV-K expression in cells previously not expressing HERV-K and it was shown that this increase is not the result of transcription factor activation. These results demonstrated that increased HERV-K expression in melanomas is the result of increased promoter activity and demethylation of the 5′LTR (Stengel *et al.*, unpublished data).

HERV-K: accessory genes and their function

Most of the 98,000 HERVs found in the human genome are defective (Belshaw *et al.*, 2005). Among them HERV-K (HML-2) makes up less than 1%. The HERV-K family integrated into the primate genome 30 million years ago and re-integrated either by re-infection or by intracellular transposition processes within the last 200,000 years (Turner *et al.*, 2001). Many HERVs inserted into the human genome after the divergence of humans and chimpanzees six million years ago. Several are insertionally polymorphic: i.e. some humans have the insertion while others do not. Only one, HERV-K113, has open reading frames for all genes (Turner *et al.*, 2001).

HERV-K113 is located on chromosome 19p13.11 and this provirus is not yet fixed in the human population. The prevalence of HERV-K113 was reported to be high in Africa, Asia and Polynesia, but low in Europe and North America (Turner *et al.*, 2001; Moyes *et al.*, 2005). Initial tests of small groups of patients in the United Kingdom showed a higher prevalence of HERV-K113 among patients with multiple

Table 2.3 Prevalence of HERV-K113 and HERV-K115 in different diseases (Moyes *et al.*, 2005; Ruprecht *et al.*, 2008; Burmeister *et al.*, 2004)

	Disease	Sample number	Percentage positive
A	*K113*		
	Multiple sclerosis	109	11.0
	Rheumatoid arthritis	96	5.7
	Sjögren's syndrome	96	15.6
	UK normal	96	4.2
	K115		
	Multiple sclerosis	92	4.4
	Rheumatoid arthritis	96	5.2
	Sjögren's syndrome	96	0
	UK normal	96	1
B	*K113*		
	Seminomas	27	11.1
	Healthy controls	80	18.75
	K115		
	Seminomas	27	7.4
	Healthy controls	80	7.5
C	*K113*		
	Breast cancer		16.7
	Controls		12.7
	K115		
	Breast cancer		4.9
	Controls		9.8

sclerosis or Sjögren's syndrome compared with healthy donors (Moyes *et al.*, 2005). Based on these results HERV-K113 was thought to be a risk factor for some types of autoimmune disorders. However, this result was not confirmed in larger groups of patients (Moyes *et al.*, 2008) (Table 2.3). In addition, there was no correlation between the presence of HERV-K113 and seminomas or breast cancer (Burmeister *et al.*, 2004; Ruprecht *et al.*, 2008). Although particles of HERV-K have been observed in teratocarcinomas and melanomas (Boller *et al.*, 1993, Büscher *et al.*, 2006) and HERV-K113 can produce intact virions (Boller *et al.*, 2008), all HERV-K particles detected up to now were defective – i.e. not infectious. However, using HERV-K consensus sequences based on human specific full-length elements (Dewannieux *et al.*, 2006; Lee and Bieniasz, 2007) or correcting for postinsertional mutations in the youngest HERV-K, HERV-K113 (Beimforde *et al.*, 2008; Bannert *et al.*, personal communication), low levels of infectious particles have been obtained.

In addition to the structural proteins and retrovirus-specific enzymes, genes for two accessory proteins, Rec and Np9, have been described in the genome of HERV-K. Rec encodes a 14-kDa protein localized in the nucleolus of HERV-K particle-producing GH teratocarcinoma cells (Löwer *et al.*, 1995). Rec is translated from a doubly spliced 1.8-kb mRNA and contains an arginine-rich motif specific for RNA- and some DNA-binding proteins. The Rec protein resembles the HIV Rev protein with respect to structural features, intracellular localization and function (Magin *et al.*, 1999). Nuclear localization sequences (NLS) and putative nuclear export signal (NES) sequences have been described and it was clearly demonstrated that Rec supports the export of unspliced HERV-K mRNA depending on Crm1 activity (Magin *et al.*, 1999, 2000). In addition, Rec has tumorigenic potential. It supports tumour growth when expressed in nude mice (Boese *et al.*, 2000), and mice transgenic for *rec* showed a disturbed germ cell development (Galli *et al.*, 2005). Another potential oncoprotein, the 9-kDa protein Np9, is also located in the nucleus. It interacts with the ligand of Numb protein X (LNX) (Armbruester *et al.*, 2004) and has been found in biopsies of mammary carcinomas, in germ cell tumours, and in transformed cell lines, but not in normal cells (Armbruester *et al.*, 2002). LNX is a really interesting new gene (RING)-type E3 ubiquitin ligase that regulates the transcription factor Notch by degrading the Notch antagonist Numb, whereas Rec is encoded by the HERV-K prototype (previously type 2). Np9 is encoded by HERV-K proviruses deleted in the *rec* sequence (type 1). The promyelocytic leukaemia zinc finger protein (PLZF) was identified as an interaction partner of Rec (Boese *et al.*, 2000) and of Np9 (Denne *et al.*, 2007). PLZF is a tumour suppressor and a transcriptional repressor of the c-*myc* proto-oncogene. Abrogation of the transcriptional repression of c-myc resulted in c-myc overexpression and upregulation of p53. Cells stably transfected with PLZF and Rec showed increased cell proliferation (Denne *et al.*, 2007).

The HERV-K(HML-2) *env* gene gives rise to an additional 13-kDa protein by cleavage of the signal peptide (SP) from the Env precursor (Ruggieri *et al.*, 2009). The HML-2 SP represents another functional similarity with the related exogenous mouse mammary tumour virus that likewise encodes a signal peptide which has biological functions in addition to ER-targeting. HML-2 SP and the Rec protein are very similar in sequence, except for the N-terminus. These results strongly indicate that HML-2 SP lacks several functional features previously reported for HML-2 Rec. HML-2 SP is therefore expected to exert a different but still unknown biological function.

Microarrays (gene chips) analyses of human cells that do not normally express *rec* were performed to evaluate changes in the expression of 33,000 cellular genes after transfection and overexpression of *rec*. Results showed an upregulation of numerous proteins including histone deacetylase 3, Sry-box 30, zinc finger protein 256, NFAR 3, and many others (unpublished data). The upregulation was confirmed by real-time RT-PCR. These initial results indicate that Rec expression significantly modulates gene expression in human cells.

Expression of HERV-K in human embryonic stem cells

Using a newly developed quantitative real time PCR the expression of HERV-K was studied in human embryonic stem cells (hESC). The full-length mRNA and all spliced RNA (*env, rec, hel*) were upregulated compared with normal human cells. Expression of the transmembrane envelope protein and Rec was demonstrated using Western blot assays (Büscher, Terstegge, Koch, Itskovitz-Eldor, Brüstle and Denner, unpublished data). After differentiation of hESC to neural precursor cells the HERV-K expression declined to a level characteristic for normal human cells, indicating that HERV-K may be an early differentiation marker. The inverse correlation of HERV-K expression and hESC differentiation provides a basis for further studies on the role of endogenous retroviruses in stem cell proliferation and tumorigenesis.

Porcine endogenous retroviruses and xenotransplantation

Xenotransplantation using porcine cells or organs has been proposed to alleviate the shortage of human donor organs for allotransplantation (Sachs *et al.*, 2001) but it may be associated with a risk of transmission of zoonotic microorganisms. Whereas transmission of most porcine microorganisms can be prevented by using designated pathogen free (DPF) breeding of the animals (Tucker *et al.*, 2002), PERVs cannot be eliminated by this method and therefore pose a potentially high risk for transmission during xenotransplantation (for review see Denner, 1998; 2008c; Blusch *et al.*, 2002; Fishman and Patience, 2004). PERVs belong to the genus gammaretroviruses and are closely related to feline leukaemia viruses (FeLV), murine leukaemia viruses (MuLV), Gibbon ape leukaemia virus (GaLV) and Koala retrovirus (KoRV), all of which induce leukaemia and immunodeficiency in the infected host (Denner, 2008a). More than a hundred proviral copies of PERV are integrated in the pig genome (depending on the pig breed), including three different replication-competent subtypes, PERV-A, PERV-B and PERV-C (Le Tissier *et al.*, 1997; Ericsson *et al.*, 2001; Mang *et al.*, 2001; Patience *et al.*, 2001). Most proviruses are defective and unable to produce replication-competent viruses. Whereas PERV-A and PERV-B are present in the genome of all pigs, PERV-C is not ubiquitous. PERVs are released from normal pig cells (Martin *et al.*, 1998a; Wilson *et al.*, 1998, 2000; Tacke *et al.*, 2000b, 2003; McIntyre *et al.*, 2003) and from pig tumour cell lines (Moennig *et al.*, 1974; Frazier, 1985; Suzuka *et al.*, 1986; Dieckhoff *et al.*, 2007b). Whereas PERV-A and -B infect human cells (human-tropic viruses), PERV-C infects only pig cells (ecotropic virus) (Patience *et al.*, 1997; Takeuchi *et al.*, 1998; Wilson *et al.*, 1998; 2000; Martin *et al.*, 2000; Blusch *et al.*, 2000; Specke *et al.*, 2001a,b, 2002a,b). Shortly after the receptors for PERV-A were identified, it was discovered that mice have a single amino acid mutation which is responsible for the resistance of mice to PERV-A infection (Ericsson *et al.*, 2003). In contrast, the PERV-A receptor is functional in rats, but its expression appears to be too low to support PERV-A infection (Mattiuzzo *et al.*, 2007).

Recombinant PERVs, host range and detection methods

Recently recombinant PERV-A/Cs were described able to replicate in human cells. Extended characterization of these viruses showed an increase in titre after repeated passages on human cells associated with genetic alterations in the LTR of the virus, mainly due to a multimerization of transcription factor NF-Y binding sites (Wilson *et al.*, 2000; Denner *et al.*, 2003). Similar genetic changes in the LTR were also seen in PERV-A passaged on human cells, although the repeats containing the NF-Y binding sites found in the LTR of PERV-A were different from the repeats in the LTR of PERV-C (Scheef *et al.*, 2001; Denner *et al.*, 2003). Some of the recombinant viruses were found *de novo* integrated in spleen cells of miniature pigs or in melanoma-bearing Munich miniature swine (MMS), but not in the germline of these animals (Wood *et al.*, 2004; Martin *et al.*, 2006; Dieckhoff *et al.*, 2007b). However, with regard to xenotransplantation, any increased risk from PERV-A/C recombinant viruses can be easily eliminated by using pigs with germ-lines free of PERV-C, which would prevent recombination with PERV-A. There are two other reasons to avoid the use of PERV-C-containing pigs. First, PERV-C, like other retroviruses, can infect cells not harbouring the specific receptor by receptor-independent infection (Lavillette and Kabat, 2004). Second, recent results indicate that mutations in the C-terminal end of the envelope surface protein (SU) of PERV-C permit binding of PERV-C to human cells and subsequent infection of human cells (Gemeniano *et al.*, 2006).

In conclusion, it is strongly recommended that PERV-C carrying pigs should not be used for the generation of animals for xenotransplantation. PERV-C carrying pigs may increase the risk of generating high-titre PERV-A/C recombinants, or producing receptor independent infections or mutations in the Env protein, either of which could change the tropism of PERV-C towards human cells.

In order to evaluate virus safety in experimental, preclinical and clinical xenotransplantations, specific and sensitive assays have been developed and applied. Retroviral infection may be detected directly by showing the presence of the provirus in the infected cells, showing expression of viral mRNA or viral proteins, or by showing production of infectious viruses. Indirect methods measure either the humoral or the cellular immune response of the infected host. Although no methods for detection of cellular responses against PERVs have yet been developed, numerous methods exist to measure the antibody response (Galbraith *et al.*, 2000; Denner *et al.*, 2008c). Numerous animal sera against PERV proteins have been developed for use as positive controls (Tacke *et al.*, 2001; Fiebig *et al.*, 2003; Irgang *et al.*, 2003). Since antibodies against p27Gag of PERV were detected in some blood assays of xenotransplantation patients and even in sera of normal donors (Paradis *et al.*, 1999; Tacke *et al.*, 2001), well-defined criteria for positive infection should be developed for future clinical testing. Ideally, these criteria will be based on responses against different viral antigens.

Animal models and clinical trials

Despite numerous attempts to develop an animal model to analyse pathogenic consequences of PERV infection and to enable screening of antiviral drugs and neutralizing antibodies *in vivo* (see below), no infection of PERV has been observed in small animals or in non-human primates (Table 2.4). Given that cell lines and primary cells of most of the tested species could be infected *in vitro*, and given that pharmaceutical immunosuppression was successful, these results are surprising and indicate that innate immune responses as well as intracellular restriction factors may prevent infection *in vivo*.

PERV infection was described only for severe combined immunodeficiency (SCID) mice inoculated with foetal pig pancreas cells (Deng *et al.*, 2000; van der Laan *et al.*, 2000). In both studies microchimerism – i.e. the presence of pig cells in different mouse organs – was observed in the infected animals and later evidence showed that pseudotyping with murine endogenous retroviruses was responsible for the presence of PERV proviruses in mouse cells (Martina *et al.*, 2005). Inoculation of cell-free virus did not cause infection of SCID mice (Irgang *et al.*, 2005). Most importantly, inoculation of non-human primates with high doses of high-titre PERV did also not result in PERV infections, although these PERV preparations did infect primary cells from the same animals *in vitro* and although a triple pharmaceutical immunosuppression was applied daily (Specke *et al.*, 2009).

These data correlate with results of preclinical studies involving transplantation of different pig materials to non-human primates (Table 2.5). In the first prospective study, after inoculation of encapsulated pig islet cells into diabetic primates, also no transmission of PERV was observed (Garkavenko *et al.*, 2008). These data indicate that the innate immunity as well as cellular restriction factors such as TRIM5α and APOBEC can prevent PERV transmission to humans and other species. Although the action of intracellular restriction factor activity was observed *in vitro*,

Table 2.4 Animal models of PERV transmission

Species	Immunosuppression	Outcome: Antibodies	Outcome: Provirus	References
Rat	None	Negative	Negative	Specke *et al.* (2001b)
	Cy-A	Negative	Negative	
	Newborn	Negative	Negative	
Mink	None	Negative	Negative	Specke *et al.* (2002a)
SCID mouse		n.t.	Positive? Murine virus mediated	Van der Laan *et al.* (2000), Deng *et al.* (2000)
		Negative	Negative	Irgang *et al.* (2005)
Guinea pig	None	Negative	Negative	Specke *et al.* (2001b)
	None	n.t.	Transient	Argaw *et al.* (2004)
	None	n.t.	Transient	Onion, personal communication
Rhesus monkey, baboon, pig-tailed monkey	Cy-A, RAD, steroids	Negative	Negative	Specke *et al.* (2009)
Lamb	None	Negative	Transient	Popp *et al.* (2007)

Table 2.5 Summary of preclinical data on potential *in vivo* PERV transmission in non-human primates

Animal	Total	Immunosuppression	Reference
Baboon – PAEC, heart (only 1)	15	CyP 15–45mg/kg i.v. (days 0–4)	Martin *et al.* (1998b)
Baboon – pig heart	23	Some baboons	Switzer *et al.* (2001)
Bonnet macaque – pig skin		CsA, MMF, steroids	
STZ-Rhesus – encapsulated pig islets		Some macaque FTBI, CsA, steroids	
STZ-Capuchin – encapsulated pig islets			
Porcine kidney in cynomolgus monkeys with human venous patch	6	CyP, CdA, MPA, steroids, C1inh	Winkler *et al.* (2005)
Cynomolgus monkeys – pig kidney	12	Cyp, CsA, steroids (until day 28)	Loss *et al.* (2001)
Baboons – porcine endothelial cells, mononuclear blood cells, and lungs	15	CyP, MPA	Martin *et al.* (2002)
STZ-cynomolgus monkey – encapsulated islets	12	None	Garkavenko *et al.* (2008)
Baboon – hDAF tg heart (13), hDAF tg kidney (14)	27	GAS914, CyP, CsA, MPA	Moscoso *et al.* (2005)
Baboon – hDAF tg liver perfusion (13)	6	None	Nishitai *et al.* (2005)

C1inh, complement C1 inhibitor; CsA, cyclosporine A; CyP, cyclophosphamide; FTBI, fractionated total body irradiation; hDAF, human decay-accelerating factor; MPA, mycophenolic acid; PAEC, primary aortic endothelial cells; STZ, streptocotozin (induced diabetes).

and productive infection of non-human primate cells is possible, no replication of PERV in these cells was not observed (Ritzhaupt *et al.*, 2002, Denner *et al.*, in preparation). Nevertheless, the primate model is a suitable model to study the infectious risk posed by PERV, because most of the human cells also do not allow replication of PERV *in vitro*. The human embryonic kidney cell line 293 is an exception which well supports PERV replication, This may be explained by the facts, that HEK293 cells are transformed by adenovirus 5 and do not express APOBEC.

Up to now, more than 180 individuals have undergone transplantation of pig cells or tissues, including *ex vivo* perfusion of human blood with pig cells or organs (Table 2.6). PERV transmission was not observed in any of these cases.

Expression of PERV in normal and multitransgenic pigs

Multi-transgenic pigs expressing numerous human genes were generated in an effort to avoid hyperacute rejection (HAR) during xenotransplantation (Petersen *et al.*, 2008). HAR occurs due to preexisting antibodies against 1,3 Gal epitopes present in pigs, but not in humans. Human genes such as TRAIL (Tumour necrosis factor-related apoptosis-inducing ligand), HLA-E/β2m, hDAF (decay-accelerating factor, hCD55), and protectin (hCD59) were introduced in an effort to prevent complement mediated HAR. When multitransgenic pigs were screened for the expression of PERV, a low expression was found in all animals and no differences between transgenic and non-transgenic animals were observed. These results indicate that no transgene was integrated adjacent to or into the locus of a PERV provirus, which could potentially lead to enhanced virus expression (Dieckhoff *et al.*, 2009). The highest expression was found in mini-pigs and crossing other pig lines with mini-pigs resulted in increased PERV expression in the progeny. Evaluation of expression in different organs showed highest expression in the lung and spleen, and very low expression in the pancreas (Dieckhoff *et al.*, 2008).

Table 2.6 Summary of clinical data on potential *in vivo* PERV transmission

Test persons, disease condition	Pig product	No.	Reference
Butchers		44	Tacke *et al.* (2001)
Pig farmers after liver Tx, abattoir workers		63	Hermida-Prieto *et al.* (2007)
Haemophilia	Hyate:C (unheated porcine factor VIII)	88	Heneine *et al.* (2001)
Aortic/mitral valve replacement		18	Moza *et al.* (2001)
Acute liver failure	Bioartificial liver device	6	Kuddus *et al.* (2002)
	Cryopreserved hepatocytes	28	Pitkin and Mullon (1999)
	AMC-BAL	14	Di Nicuolo *et al.* (2005)
	Plasma perfusion though bioreactor, followed by liver Tx + immunosuppression	8	Irgang *et al.* (2003)
	Extracorporeal liver perfusion	2	Levy *et al.* (2000)
Chronic glomerulonephritis	Extracorporeal kidney perfusion	2	Patience *et al.* (1998)
Neurological conditions (Parkinson, Huntington, focal epilepsy)	Fetal pig mesencephalon, lateral ganglionic eminence cells	24	Dinsmore *et al.* (2000)
Diabetes	Encapsulated islets	2	Elliot *et al.* (2000), Garkavenko *et al.* (2004)
	Porcine fetal islets	10	Heneine *et al.* (1998)
Acute liver failure	Cryopreserved hepatocytes (same group as in Pitkin *et al.*, 1999)	28	Paradis *et al.* (1999)*
	Extracorporeal liver perfusion	1	
Burns	Skin	15	
Chronic glomerulonephritis Renal dialysis	Extracorporeal kidney perfusion (same as in Patience *et al.*, 1998)	2	
Various indications	Extracorporeal splenic perfusion	100	
Diabetes	Islets	14	

Abs, antibodies; CsA, cyclosporine; AZA, azathioprine; IS, immunosuppression; pred, prednisolone; PBMC, peripheral blood mononuclear cells; Tx, transplantation.

* Sera from patients with positive reaction against Gag were retested in Tacke *et al.*, 2001, and found negative for PERV transmission despite positive reaction against Gag.

Strategies to prevent PERV transmission

Several approaches have been developed to minimize or prevent PERV transmission to the transplant recipient. First, animals not carrying PERV-C genomes should be selected, thereby preventing generation of high-titre PERV-A/C recombinants. Second, transplant recipients may be vaccinated against PERV. While the development of a vaccine against lentiviruses such as HIV have not yet shown success (for review see Walker and Burton, 2008), effective vaccines against type C retroviruses such as MuLV and FeLV exist (Peters *et al.*, 1975; Lee *et al.*, 1977, Montelaro and Bolognesi, 1995). Immunization of different animal species with the transmembrane envelope protein of PERV successfully induced neutralizing antibodies (Fiebig *et al.*, 2003, Denner *et al.*, in preparation). Interestingly, these antibodies recognize an epitope in the membrane proximal external region (MPER), which is related to an epitope recognized by human antibodies broadly neutralizing HIV-1. Virus-specific neutralizing antibodies recognizing similar epitopes were induced after immunization with the p15E of the closely related gammaretroviruses FeLV (Langhammer *et al.*, 2005a,b) and KoRV (Fiebig *et al.*, 2006). Since no adequate animal model is available to analyse the efficacy of PERV-specific antibodies, cats were immunized with p15E of FeLV. Protection was observed after challenge with infectious FeLV, providing hope that this strategy may also work with PERV (unpublished).

Another strategy is the inhibition of PERV by RNA interference. After selection of suitable small interfering (si)RNA corresponding to sequences in the genome of PERV, inhibition of PERV expression was observed in human cells infected with PERV (Karlas *et al.*, 2004) and in primary pig cells (Miyagawa *et al.*, 2005; Dieckhoff *et al.*, 2006). Based on these studies transgenic pigs have been generated that express a short hairpin (sh) RNA in all cells of the organism and show a reduced expression of PERV (Dieckhoff *et al.*, 2007a).

As for HIV-1, inhibitors of reverse transcriptase such as AZT (zidovudine) and ddI (dideoxyinosine) inhibit PERV replication (Powell *et al.*, 2000; Stephan *et al.*, 2001; Qari *et al.*, 2001). However, RT inhibitors will select for resistant strains soon after the beginning of treatment. Inhibitors of the PERV protease or integrase, which will be necessary for a more effective combination therapy, have yet to be developed.

Shaping the host genome

Proviral inheritance might have numerous consequences for the host (Table 2.7). Some stem from the expression of retroviral proteins such as the reverse transcriptase or the immunosuppressive and syncytia inducing *env* gene. Others stem from the insertion of multiple copies of DNA sequences containing promoters capable of modifying transcription. Transposition of proviruses or replication of endogenous retroviruses and their *de novo* integration may contribute to shaping the genome of the host and influencing gene expression by chromosomal rearrangements. In extreme

Table 2.7 Proviral inheritance: consequences for the host

At the DNA level

Retrotransposition, Insertional mutagenesis

- Insertion into a biologically inert region (e.g., an intron):
  - No expression, but often increase of chromosomal instability (rearrangement)
- Insertion into a biologically active region (e.g., an exon):
  - Regulation of expression
  - Often rapid silencing (DNA methylation, mutations)

At the protein level

- Reverse transcriptase
  - Formation of pseudogenes
- Core protein Gag: Restriction factors
- Surface envelope glycoprotein (SU-Env)
  - Protection from infection
- Transmembrane envelope protein (TM-Env)
  - Fusion (syncytiotrophoblast)
  - Immunosuppression

Table 2.8 HERVs regulate expression of cellular proteins

HERV	Cellular gene	Organ and action	References
HERV	Inserted into promoter of human placenta-specific insulin, ISNL4	LTR-mediated elevated expression during differentiation into syncytiotrophoblast	Bieche *et al.* (2003)
HERV	Azoospermia factor (AZFa) region human Y chromosome	Non-allelic recombinations between HERVs Deletion: Male infertility Duplication: Fertility	Bosch and Jobling (2003)
HERV-E	Amylase	Salivary amylase expression	Samuelson *et al.* (1996)
HERV-E	Pleiotrophin	Trophoblast	Schulte *et al.* (2000)
HERV-E	Endothelin-B receptor	Placenta	Medstrand *et al.* (2001)
HERV-E	Apolipoprotein C1	Liver and other tissues	Medstrand *et al.* (2001)
HERV-E	Opitz syndrome gene Mid1	LTR is strong tissue-specific promoter in placenta and embryo kidney	Landry *et al.* (2002)
HERV-K	Fibroblast growth factor receptor 1 (FGFR1)	Atypical stem cell myeloproliferative disorder after chromosomal translocation	Guasch *et al.* (2003)
HERV-L	β1,3-galactosyltransferase 5	Gastrointestinal tract, mammary gland	Dunn *et al.* (2003)
ERV I	Aromatase P450 (CYP19)	Placental oestrogen synthesis Syncytiotrophoblast	van de Lagemaat *et al.* (2003)
ERV II	BAAT (transferase)	Bile metabolism	van de Lagemaat *et al.* (2003)
ERV III	Carbonic anhydrase 1	Erythroid carbon metabolism	van de Lagemaat *et al.* (2003)
LTR + LINE-2	Chaperonin	McKusick–Kaufmann syndrome	van de Lagemaat *et al.* (2003

cases, integration may activate oncogenes or disrupt tumour suppressor genes and may lead to virally induced tumours; such occurrences have been well documented with certain murine viruses (Boeke and Stoye, 1997).

Retroviral LTR contain the viral promoter, enhancer and polyadenylation signals and can regulate gene expression not only of viral genes but – as has been shown in numerous cases – also of cellular genes. Many LTR-derived promoters are active primarily in the placenta, such as the human endothelin B receptor (EDNRB), Mid1, insulin-like INSL4, pleiotrophin and aromatase (CYP19) genes (Schulte *et al.*, 1996; Medstrand *et al.*, 2001; Landry *et al.*, 2002; Bieche *et al.*, 2003; van de Lagemaat *et al.*, 2003) (Table 2.8).

The LTR of an endogenous retrovirus acts as one of at least two alternative promoters for the human β1,3-galactosyltransferase 5 gene, involved in type 1 Lewis antigen synthesis. This LTR promoter is most active in the gastrointestinal tract and mammary gland. The hepatocyte nuclear factor (HNF)-1 binds to a site within the retroviral promoter and the subsequent expression of HNF-1 and interaction with its binding site correlate with promoter activation. This tissue-specific transcription factor is responsible for the tissue-specific activation of the LTR promoter (Dunn *et al.*, 2003).

Independently acquired LTRs also have regulatory roles for orthologous genes. Multiple, domesticated LTRs of endogenous retroviral elements provide neuronal apoptosis inhibitory protein (NAIP) promoter function in human, mouse, and rat. In humans, an LTR serves as a tissue-specific promoter, which is primarily active in testis. However, in rodents, an ancestral LTR common to all rodent species is the major constitutive promoter for these genes, and a second LTR found in two of the mouse genes is a minor promoter. Such a widespread regulatory role for independently acquired LTRs for orthologous genes in different species represents a remarkable evolutionary scenario (Romanish *et al.*, 2007).

Initially, only sense orientation LTR promoters appeared to regulate human gene expression, although some LTRs were known to possess bidirectional promoter activity *in vitro*. However, it was recently shown that an ERV1 LTR acts as a bidirectional promoter for the human Down syndrome critical region 4 (DSCR4) and DSCR8 genes. While DSCR4 and DSCR8 are essentially co-expressed, their shared LTR promoter is more active in the sense orientation than the antisense orientation; a core region of the promoter required for transcriptional activity in both orientations was identified (Dunn *et al.*, 2006).

Conclusion

Although there is clear evidence from mice and cats that endogenous retroviruses play a role in tumour development, it remains unclear whether this is true for endogenous retroviruses of other species such as non-human primates and man. Whereas endogenous retroviruses are easily released from normal tissues in mice, pigs and koalas, the restriction and inactivation of endogenous retroviruses increased during evolution to primates.

HERV-K(HML-2)s are the best studied human endogenous retroviruses and there are indications that their accessory genes *rec* and *np9* may contribute to tumour development. The expression of the transmembrane envelope protein, which has immunosuppressive properties, may contribute to tumour progression. It will be of great interest to investigate whether therapeutic vaccinations against this immunosuppressive protein have a potential for inhibition of tumour progression.

In contrast to the situation with tumours, there is clear evidence that endogenous retroviruses play a role in physiological processes. First, Env proteins such as syncytin in humans and mice, and Env of enJSRV in sheep, play a role in placental development and therefore in the reproduction of vertebrates. It is important to note that different retroviruses have been recruited in different species, and obviously an utilization, enslavement of retroviruses took place independently several times during evolution of placental animals. In all cases the transmembrane envelope protein plays an important role in generation of the functional placenta, e.g. by fusion of trophoblast cells into the syncytiotrophoblast. Since most of the transmembrane envelope proteins of the endogenous retroviruses involved have immunosuppressive properties, they may contribute to local and distant immunosuppression protecting the semiallotransplant embryo from rejection by the maternal immune system.

The accumulation of junk DNA and the refinement of restriction strategies on the one hand, and the utilization, enslavement of retroviral genes and LTRs on the other, have created an intricate interaction between endogenous retroviruses and the host during the course of evolution.

References

Aaronson, S.A., Hartley, J.W., and Todaro, G.J. (1969). Mouse leukemia virus: 'spontaneous' release by mouse embryo cells after longterm *in vitro* cultivation. Proc. Natl. Acad. Sci. U.S.A. *64*, 87–94.

Argaw, T., Colon-Moran, W., and Wilson, C.A. (2004). Limited infection without evidence of replication by porcine endogenous retrovirus in guinea pigs. J. Gen. Virol. *85*, 15–19.

Armbruester, V., Sauter, M., Krautkraemer, E., Meese, E., Kleiman, A., Best, B., Roemer, K., and Mueller-Lantzsch, N. (2002). A novel gene from the human endogenous retrovirus K expressed in transformed cells. Clin. Cancer Res. *8*, 1800–1807.

Armbruester, V., Sauter, M., Roemer, K., Best, B., Hahn, S., Nty, A., Schmid, A., Philipp, S., Mueller, A., and Mueller-Lantzsch, N. (2004). Np9 protein of human endogenous retrovirus K interacts with ligand of numb protein X. J. Virol. *78*, 10310–10319.

Arnaud, F., Varela, M., Spencer, T.E., and Palmarini, M. (2008). Coevolution of endogenous betaretroviruses of sheep and their host. Cell. Mol. Life. Sci. *65*, 3422–3432.

Baccetti, B., Benedetto, A., Burrini, A.G., Collodel, G., Ceccarini, E.C., Crisà, N., Di Caro, A., Estenoz, M., Garbuglia, A.R., and Massacesi, A.J. (1994). HIV-particles in spermatozoa of patients with AIDS and their transfer into the oocyte. J. Cell Biol. *127*, 903–914.

Baltimore, D. (1970). RNA-dependent DNA polymerase in virions of RNA tumour viruses. Nature *226*, 1209–1211.

Baltimore, D. (1975). Tumor viruses: 1974. Cold Spring Harb. Symp. Quant. Biol. *39*, 1187–1200.

Bannert, N., and Kurth, R. (2004). Retroelements and the human genome: new perspectives on an old relation. Proc. Natl. Acad. Sci. U.S.A. *101*, 14572–14579.

Bannert, N., and Kurth, R. (2006). The evolutionary dynamics of human endogenous retroviral families. Annu. Rev. Genomics Hum. Genet. *7*, 149–173.

Barcova, M., Kacani, L., Speth, C., and Dierich, M. (1998). gp41 envelope protein of human immunodeficiency virus induces interleukin (IL)-10 in monocytes, but not in B, T, or NK cells, leading to reduced IL-2 and interferon-gamma production. J. Infect. Dis. *177*, 905–913.

Belshaw, R., Dawson, A.L., Woolven-Allen, J., Redding, J., Burt, A., and Tristem, M. (2005). Genomewide screening reveals high levels of insertional polymorphism in the human endogenous retrovirus family HERV-K(HML2): implications for present-day activity. J. Virol. *79*, 12507–12514.

Beimforde, N., Hanke, K., Ammar, I., Kurth, R., and Bannert, N. (2008). Molecular cloning and functional characterization of the human endogenous retrovirus K113. Virology *371*, 216–225.

Bénit, L., Lallemand, J.B., Casella, J.F., Philippe, H., and Heidmann, T. (1999). ERV-L elements: a family of endogenous retrovirus-like elements active throughout the evolution of mammals. J. Virol. *73*, 3301–3308.

Bentvelzen, P., and Daams, J.H. (1969). Heredity infections with mammary tumor viruses in mice. J. Natl. Cancer Inst. *43*, 1025–1035.

Bentvelzen, P., Daams, J.H., Hageman, P., and Calafat, J. (1970). Genetic transmission of viruses that incite mammary tumor in mice. Proc. Natl. Acad. Sci. U.S.A. *67*, 377–384.

Bièche, I., Laurent, A., Laurendeau, I., Duret, L., Giovangrandi, Y., Frendo, J.L., Olivi, M., Fausser, J.L., Evain-Brion, D., and Vidaud, M. (2003).Placenta-specific INSL4 expression is mediated by a human endogenous retrovirus element. Biol. Reprod. *68*, 1422–1429.

Bierwolf, D., Rudolph, M., Niezabitowski, A., Jorde, A., Bender, E., Widmaier, R., Graffi, and A. (1975). C-Typ-ähnliche Partikel in normalen menschlichen Plazenten. Arch. Geschwulstforsch. *45*, 628–633.

Blaise, S., Mangeney, M., and Heidmann, T. (2001). The envelope of Mason–Pfizer monkey virus has immunosuppressive properties. J. Gen. Virol. *82*, 1597–1600.

Blaise, S., de Parseval, N., Bénit, L., and Heidmann, T. (2003). Genomewide screening for fusogenic human endogenous retrovirus envelopes identifies syncytin 2, a gene conserved on primate evolution. Proc. Natl. Acad. Sci. U.S.A. *100*, 13013–13018.

Blond, J.L., Beseme, F., Duret, L., Bouton, O., Bedin, F., Perron, H., Mandrand, B., and Mallet, F. (1999). Molecular characterization and placental expression of HERV-W, a new human endogenous retrovirus family. J. Virol. 73, 1175–1185.

Blond, J.L., Lavillette, D., Cheynet, V., Bouton, O., Oriol, G., Chapel-Fernandes, S., Mandrand, B., Mallet, F., and Cosset, F.L. (2000). An envelope glycoprotein of the human endogenous retrovirus HERV-W is expressed in the human placenta and fuses cells expressing the type D mammalian retrovirus receptor. J. Virol. *74*, 3321–3329.

Blusch, J.H., Patience, C., Takeuchi, Y., Templin, C., Roos, C., Von Der Helm, K., Steinhoff, G., and Martin, U. (2000). Infection of nonhuman primate cells by pig endogenous retrovirus. J. Virol. *74*, 7687–7690.

Blusch, J.H., Patience, C., and Martin, U. (2002). Pig endogenous retroviruses and xenotransplantation. Xenotransplantation *9*, 242–251.

Boeke, J.D., and Stoye, J.P. (1997). Retrotransposons, endogenous retroviruses, and the evolution. In Retroviruses, Coffin, J.M. , Hughes, S.H. , Varmus, H.E. , eds. (New York: Cold Spring Harbor Laboratory Press), pp. 343–436.

Boese, A., Sauter, M., Galli, U., Best, B., Herbst, H., Mayer, J., Kremmer, E., Roemer, K., and Mueller-Lantzsch, N. (2000). Human endogenous retrovirus protein cORF supports cell transformation and associates with the promyelocytic leukemia zinc finger protein. Oncogene *19*, 4328–4336.

Boller, K., König, H., Sauter, M., Müller-Lantzsch, N., Löwer, R., Löwer, J., and Kurth, R. (1993). Evidence that HERV-K is the endogenous retrovirus sequence that codes for the human teratocarcinoma-derived retrovirus HTDV. Virology *196*, 349–353.

Boller, K., Janssen, O., Schuldes, H., Tönjes, R.R., and Kurth, R. (1997). Characterization of the antibody response specific for the human endogenous retrovirus HTDV/HERV-K. J. Virol. *71*, 4581–4588.

Boller, K., Schönfeld, K., Lischer, S., Fischer, N., Hoffmann, A., Kurth, R., and Tönjes, R.R. (2008). Human endogenous retrovirus HERV-K113 is capable of producing intact viral particles. J. Gen. Virol. *89*, 567–572.

Bosch, E., and Jobling, M.A. (2003). Duplications of the AZFa region of the human Y chromosome are mediated by homologous recombination between HERVs and are compatible with male fertility. Hum. Mol. Genet. *12*, 341–347.

Boyd, M.T., Bax, C.M.R., Bax, B.E., Bloxam, D.L., and Weiss, R.A. (1993). The human endogenous retrovirus ERV-3 is upregulated in differentiating placental trophoblast cells. Virology *196*, 905–990.

Burmeister, T., Ebert, A.D., Pritze, W., Loddenkemper, C., Schwartz, S., and Thiel, E. (2004). Insertional polymorphisms of endogenous HERV-K113 and HERV-K115 retroviruses in breast cancer patients and age-matched controls. AIDS Res. Hum. Retroviruses *20*, 1223–1229.

Büscher, K., Trefzer, U., Hofmann, M., Sterry, W., Kurth, R., and Denner, J. (2005). Expression of human endogenous retrovirus K in melanomas and melanoma cell lines. Cancer Res. *65*, 4172–4180.

Büscher, K., Hahn, S., Hofmann, M., Trefzer, U., Ozel, M., Sterry, W., Lower, J., Lower, R., Kurth, R., and Denner, J. (2006). Expression of the human endogenous retrovirus-K transmembrane envelope, Rec and

Np9 proteins in melanomas and melanoma cell lines. Melanoma Res. *16*, 223–234.
Chen, Y.H., Ebenbichler, C., Vornhagen, R., Schulz, T., Steindl, F., Bock, G., Katinger, H., and Dierich, M. (1992). HIV-1 gp41 contains two sites for interaction with several proteins on the helper T-lymphoid cell line, H9. AIDS *6*, 533–539.
Chen, Y.H., Stoiber, H., and Dierich, M.P. (1997). Increased levels of antibodies against interferon-alpha in HIV-1 positive individuals may be explained by a common immunological epitope on the human interferon-alpha and HIV-1 gp41. Immunol. Lett. *55*, 15–18.
Chen, Y.H., Speth, C., Wu, W., Stockl, G., Xiao, Y., Yu, T., Ke, Z., Zhao, Y., and Dierich, M. (1998). Antigenic characterization of HIV-1 gp41 binding proteins. Immunol. Lett. *62*, 75–79.
Chen, Y.H., Wu, W., Yang, J., Sui, S., Sun, J., and Dierich, M. (1999). Antibodies against human IFN-alpha and -beta recognized the immunosuppressive domain of HIV-1 gp41 and inhibit gp41-binding to the putative cellular receptor protein p45. Immunol. Lett. *69*, 253–257.
Chiu, Y.L., and Greene, W.C. (2008). The APOBEC3 cytidine deaminases: an innate defensive network opposing exogenous retroviruses and endogenous retroelements. Ann. Rev. Immunol. *26*, 317–353.
Cianciolo, G.J., Matthews, T.J., Bolognesi, D.P., and Snyderman, R. (1980). Macrophage accumulation in mice is inhibited by low molecular weight products from murine leukemia viruses. J. Immunol. *124*, 2900–2905.
Cianciolo, G., Copeland, T., Oroszlan, S., and Snyderman, R. (1985). Inhibition of lymphocyte proliferation by a synthetic peptide homologous to retroviral envelope proteins. Science *230*, 453–455.
Cianciolo, G., Bogerd, H., and Snyderman, R. (1988). Human retrovirus-related synthetic peptides inhibit T-lymphocyte proliferation. Immunol. Lett. *19*, 7–13.
Coffin, J. (1984). Endogenous Viruses. In RNA Tumor Viruses. Molecular Biology of Tumor Viruses. 2nd edn, Weiss, R. , Teich, N. , Varmus, H. , Coffin, J. , eds. (New York, USA: Cold Spring Harbor Laboratory Press), pp. 1109–1204.
Contreras-Galindo, R., González, M., Almodovar-Camacho, S., González-Ramírez, S., Lorenzo, E., and Yamamura, Y. (2006). A new Real-Time-RT-PCR for quantitation of human endogenous retroviruses type K (HERV-K) RNA load in plasma samples: increased HERV-K RNA titers in HIV-1 patients with HAART non-suppressive regimens. J. Virol. Methods *136*, 51–57.
Contreras-Galindo, R., Almodóvar-Camacho, S., González-Ramírez, S., Lorenzo, E., and Yamamura, Y. (2007a). Comparative longitudinal studies of HERV-K and HIV-1 RNA titers in HIV-1-infected patients receiving successful versus unsuccessful highly active antiretroviral therapy. AIDS Res. Hum. Retroviruses. *23*, 1083–1086.
Contreras-Galindo, R., Lopez, P., Velez, R., and Yamamura, Y. (2007b). HIV-1 infection increases the expression of human endogenous retroviruses type K (HERV-K) *in vitro*. AIDS Res. Hum. Retroviruses 23, 116–122.
Cordonnier, A., Casella, J. F., and Heidmann, T. (1995). Isolation of novel human endogenous retrovirus-like elements with foamy virus-related *pol* sequence. J. Virol. *69*, 5890–5897.
De Lamarter, J.F., Monckton, R.P., and Moroni, C. (1981). Transcriptional control of endogenous virus genes in murine lymphocytes. J. Gen. Virol. *52*, 371–375.
Delassus, S., Sonigo, P., and Wain-Hobson, S. (1989). Genetic organization of gibbon ape leukemia virus. Virology *173*, 205–213.
Deng, Y.M., Tuch, B.E., and Rawlinson, W.D. (2000). Transmission of porcine endogenous retroviruses in severe combined immunodeficient mice xenotransplanted with fetal porcine pancreatic cells. Transplantation *70*, 1010–1016.
Denne, M., Sauter, M., Armbruester, V., Licht, J.D., Roemer, K., Mueller-Lantzsch N. (2007). Physical and functional interactions of human endogenous retrovirus proteins Np9 and rec with the promyelocytic leukemia zinc finger protein. J. Virol. *81*, 5607–5616.
Denner, J. (1977). Oncogenic virus induction by halogenated pyrimidines. A review. Arch. Geschwulstforsch. *47*, 151–161.
Denner, J. (1987). Immunosuppression by oncogenic retroviridae. In Immune modulation by infectious agents, Zschiesche, W., ed. (Jena, Fischer Verlag), 114–201.
Denner, J. (1998). Immunosuppression by retroviruses: implications for xenotransplantation. Ann. N.Y. Acad. Sci. *862*, 75–86.
Denner, J. (2000). How does HIV induce AIDS? The virus protein hypothesis. J. Hum. Virol. *3*, 81–82.
Denner, J. (2007). Trans-species transmissions of retroviruses: New cases. Virology *369*, 229–233.
Denner, J. (2008a). Is porcine endogenous retrovirus (PERV) transmission still relevant? Transplant. Proc. *40*, 587–589.
Denner, J. (2008b). Recombinant porcine endogenous retroviruses (PERV-A/C): A new risk for xenotransplantation? Arch. Virol. *153*, 1421–1426.
Denner, J. (2008c). Xenotransplantation: State of the art of biosafety. In Xenotransplantation – ethic, economic, social, cultural and scientific background, E.S. Jansen, and J.W. Simon, eds. (Saarbrücken, Germany, VDM-Verlag), pp. 7–78.
Denner, J., and Dorfman, N. (1977). Small virus-like particles in leukosis-like syndrome induced by certain antigens and immunostimulators. Acta Biol Med German. *36*, 1451–1458.
Denner, J., Wunderlich, V., and Bierwolf, D. (1980). Suppression of human lymphocyte mitogen response by disrupted primate retroviruses of type C (baboon endogenous virus) and type D (PMFV). Acta. Biol. Med. Ger. *39*, 19–26.
Denner, J., Wunderlich, V., and Sydow, G. (1985). Suppression of human lymphocyte mitogen response by retroviruses of type D. I. Action of highly purified intact and disrupted virus. Arch. Virol. *86*, 177–186.
Denner, J., Wunderlich, V., and Bierwolf, D. (1986). Suppression of human lymphocyte mitogen response

by proteins of the type-D retrovirus PMFV. Int. J. Cancer *37*, 311–316.

Denner, J., Vogel, T., Norley, S., Ennen, J., and Kurth, R. (1993). The immunosuppressive (ISU-) peptide of HIV-1: Binding to lymphocyte surface proteins. J. Cancer Res. Clin. Oncol. *119*, 28.

Denner, J., Norley, S., and Kurth, R. (1994). The immunosuppressive peptide of HIV-1: functional domains and immune response in AIDS patients. AIDS *8*, 1063–1072.

Denner, J., Vogel, T., Norley, S., Hoffmann, A., and Kurth, R. (1995). The immunosuppressive (ISU-) peptide of HIV-1: Binding proteins of lymphocytes detected by different methods. J. Cancer Res. Clin. Oncol. *121*, 34.

Denner, J., Persin, C., Vogel, T., Haustein, D., Norley, S., and Kurth, R. (1996). The immunosuppressive peptide of HIV-1 inhibits T and B-lymphocyte stimulation. J. AIDS Hum. Retrovirol. *12*, 442–450.

Denner, J., Specke, V., Thiesen, U., Karlas, A., and Kurth, R. (2003). Genetic alterations of the long terminal repeat of an ecotropic porcine endogenous retrovirus (PERV) during passage in human cells. Virology *314*, 125–133.

de Parseval, N., and Heidmann, T. (2005). Human endogenous retroviruses: from infectious elements to human genes. Cytogenet. Genome Res. *110*, 318–332.

Derks, J.P., Hofmans, L., Bruning, H.W., and von Rood, J.J. (1982). Synthesis of a viral protein with molecular weight of 30,000 (p30) by leukemic cells and antibodies cross-reacting with Simian sarcoma virus p30 in serum of a chronic myeloid leukemia patient. Cancer Res. *42*, 681–686.

Dewannieux, M., Harper, F., Richaud, A., Letzelter, C., Ribet, D., Pierron, G., and Heidmann, T. (2006). Identification of an infectious progenitor for the multiple-copy HERV-K human endogenous retroelements. Genome Res. *16*, 1548–1556.

Dieckhoff, B., Karlas, A., Hofmann, A., Kues, W. A., Petersen, B., Pfeifer, A., Niemann, H., Kurth, R., and Denner, J. (2006). Inhibition of porcine endogenous retroviruses (PERVs) in primary porcine cells by RNA interference using lentiviral vectors. Arch. Virol. *16*, 868–873.

Dieckhoff, B., Karlas, A., Petersen, B., Kues, W.A., Kurth, R., Niemann, H., and Denner, J. (2007a). Production of transgenic pigs expressing PERV-specific shRNA with knock-down of porcine endogenous retroviruses (PERVs). Xenotransplantation *14*, 400–401.

Dieckhoff, B., Puhlmann, J., Büscher, K., Hafner-Marx, A., Herbach, N., Bannert, N., Büttner, M., Wanke, R., Kurth, R., and Denner, J. (2007b). Expression of porcine endogenous retroviruses (PERVs) in melanomas of Munich miniature swine (MMS) Troll. Vet. Microbiol. *123*, 53–68.

Dieckhoff, B., Karlas, A., Petersen, B., Kues, W.A., Kurth, R., Niemann, Denner, J. (2008). Knockdown of porcine endogenous retrovirus (PERV) expression by PERV-specific shRNA in transgenic pigs. Xenotransplantation *15*, 36–45.

Dieckhoff, B., Kessler, B., Jobst, D., Kues, W., Petersen, B., Pfeifer, A., Kurth, R., Niemann, H., Wolf, E., and Denner, J. (2009). Distribution and expression of porcine endogenous retroviruses (PERV) in multi-transgenic pigs generated for xenotransplantation. Xenotransplantation *16*, 64–73.

Di Nicuolo, G., van de Kerkhove, M.P., Hoekstra, R., Beld, M.G., Amoroso, P., Battisti, S., Starace, M., di Florio, E., Scuderi, V., Scala, S., Bracco, A., Mancini, A., Chamuleau, R.A., and Calise, F. (2005). No evidence of *in vitro* and *in vivo* porcine endogenous retrovirus infection after plamapheresis through the AMC-bioartificial liver. Xenotransplantation *12*, 286–292.

Dinsmore, J.H., Manhart, C., Raineri, R., Jacoby, D.B., and Moore, A. (2000). No evidence for infection of human cells with porcine endogenous retrovirus (PERV) after exposure to porcine fetal neuronal cells. Transplantation *70*, 1383–1389.

Dirksen, E.R., and Levy, J.A. (1977). Virus-like particles in placentas from normal individuals and patients with systemic lupus erythematosus. J. Natl. Cancer Inst. *59*, 1187–1192.

Dougherty, R.M., Di Stefano, H.S. (1966).Lack of relationship between infection with avian leukosis virus and the presence of COFAL antigen in chick embryos. Virology *29*, 586–595.

Dougherty, R.M., Di Stefano, H.S., and Roth, F.K. (1967). Virus particles and viral antigens in chicken tissues free of infectious avian leucosis virus. Proc. Natl. Acad. Sci. U.S.A. *58*, 808–817.

Dunn, C.A., Medstrand, P., and Mager, D.L. (2003). An endogenous retroviral long terminal repeat is the dominant promoter for human beta1,3-galactosyltransferase 5 in the colon. Proc. Natl. Acad. Sci. U.S.A. *100*, 12841–12846.

Dunn, C.A., Romanish, M.T., Gutierrez, L.E., van de Lagemaat, L.N., and Mager, D.L. (2006). Transcription of two human genes from a bidirectional endogenous retrovirus promoter. Gene. *366*, 335–342.

Dupressoir, A., Marceau, G., Vernochet, C., Bénit, L., Kanellopoulos, C., Sapin, V., and Heidmann, T. (2005). Syncytin-A and syncytin-B, two fusogenic placenta-specific murine envelope genes of retroviral origin conserved in Muridae. Proc. Natl. Acad. Sci. U.S.A. *102*, 725–730.

Ebenbichler, C., Roder, C., Vornhagen, R., Ratner, L., and Dierich, M. (1993). Cell surface proteins binding to recombinant soluble HIV-1 and HIV-2 transmembrane proteins. AIDS *7*, 489–495.

Elliott, R.B., Escobar, L., Garkavenko, O., Croxson, M.C., Schroeder, B.A., McGregor, M., Ferguson, G., Beckman, N., and Ferguson, S. (2000). No evidence of infection with porcine endogenous retrovirus in recipients of encapsulated porcine islet xenografts. Cell Transplant. *9*, 895–901.

Ericsson, T., Oldmixon, B., Blomberg, J., Rosa, M., Patience, C., and Andersson, G. (2001). Identification of novel porcine endogenous betaretrovirus sequences in miniature swine. J. Virol. *75*, 2765–2770.

Ericsson, T.A., Takeuchi, Y., Templin, C., Quinn, G., Farhadian, S.F., Wood, J.C., Oldmixon, B.A., Suling, K.M., Ishii, J.K., Kitagawa, Y., Miyazawa, T., Salomon, D.R., Weiss, R.A., and Patience, C. (2003). Identification of receptors for pig endogenous retrovirus. Proc. Natl. Acad. Sci. U.S.A. *100*, 6759–6764.

Esnault, C., Priet, S., Ribet, D., Vernochet, C., Bruls, T., Lavialle, C., Weissenbach, J., and Heidmann, T. (2008). A placenta-specific receptor for the fusogenic, endogenous retrovirus-derived, human syncytin-2. Proc. Natl. Acad. Sci. U.S.A. *105*, 17532–17537.

Feldman, D. (1979). Virus particles in the basal plate of rhesus monkey and baboon placenta. Cancer Res. *39*, 1772–1783.

Feldman, D., Valentine, T., Niemann, W.H., Hoar, R.M., Cukierski, M., and Hendrick, A. (1989). C-type virus particles in placentas of rhesus monkeys after maternal treatment with recombinant leukocyte A interferon. J. Exp. Pathol. *4*, 193–198.

Fiebig, U., Stephan, O., Kurth, R., and Denner, J. (2003). Neutralizing antibodies against conserved domains of p15E of porcine endogenous retroviruses (PERVs): basis for a vaccine for xenotransplantation? Virology *307*, 406–413.

Fiebig, U., Hartmann, M. G., Bannert, N., Kurth, R., and Denner, J. (2006). Trans-species transmission of the endogenous koala retrovirus (KoRV). J. Virol. *80*, 5651–5654.

Fishman, J.A., and Patience, C. (2004). Xenotransplantation: infectious risk revisited. Am. J. Transplant. *4*, 1383–1390.

Frank, O., Verbeke, C., Schwarz, N., Mayer, J., Fabarius, A., Hehlmann, R., Leib-Mösch, C., and Seifarth, W. (2008). Variable transcriptional activity of endogenous retroviruses in human breast cancer. J. Virol. *82*, 1808–1818.

Frazier, M.E. (1985). Evidence for retrovirus in miniature swine with radiation-induced leukemia or metaplasia. Arch. Virol. *83*, 83–97.

Frendo, J.L., Vidaud, M., Guibourdenche, J., Luton, D., Muller, F., Bellet, D., Giovagrandi, Y., Tarrade, A., Porquet, D., and Blot, P. (2000). Defect of villous cytotrophoblast differentiation into syncytiotrophoblast in Down's syndrome. J. Clin. Endocrinol. Metab. *85*, 3700–3707.

Frendo, J.L., Olivier, D., Cheynet, V., Blond, J.L., Bouton, O., Vidaud, M., Rabreau, M., Evain-Brion, D., and Mallet, F. (2003). Direct involvement of HERV-W Env glycoprotein in human trophoblast cell fusion and differentiation. Mol. Cell. Biol. *23*, 3566–3574.

Galbraith, D.N., Kelly, H.T., Dyke, A., Reid, G., Haworth, C., Beekman, J., Shepherd, A., and Smith, K.T. (2000). Design and validation of immunological tests for the detection of porcine endogenous retrovirus in biological materials. J. Virol. Methods. *90*, 115–124.

Galli, U.M., Sauter, M., Lecher, B., Maurer, S., Herbst, H., Roemer, K., and Mueller-Lantzsch, N. (2005). Human endogenous retrovirus rec interferes with germ cell development in mice and may cause carcinoma in situ, the predecessor lesion of germ cell tumors. Oncogene *28*, 3223–3228.

Garkavenko, O., Croxson, M.C., Irgang, M., Karlas, A., Denner, J., Elliot, R. B. (2004). Monitoring for presence of potentially xenotic viruses in recipients of pig islet xenotransplantation. J. Clin. Microbiol. *42*, 5353–5356.

Garkavenko, O., Dieckhoff, B., Wynyard, S., Denner, J., Elliott, R.B., Tan, P.L., and Croxson, M.C. (2008). Absence of transmission of potentially xenotic viruses in a prospective pig to primate islet xenotransplantation study. J. Med. Virol. *80*, 2046–2052.

Gemeniano, M., Mpanju, O., Salomon, D.R., Eiden, M.V., and Wilson, C.A. (2006). The infectivity and host range of the ecotropic porcine endogenous retrovirus, PERV-C, is modulated by residues in the C-terminal region of its surface envelope protein. Virology *346*, 108–117.

Gottlieb, R., Kleinerman, E., O'Brian, C., Tsujimoto, S., Cianciolo, G., and Lennarz, W. (1990). Inhibition of protein kinase C by a peptide conjugate homologous to a domain of the retroviral protein p15E. J. Immunol. *145*, 2566–2570.

Götzinger, N., Sauter, M., Roemer, K., and Mueller-Lantzsch, N. (1996). Regulation of human endogenous retrovirus-K Gag expression in teratocarcinoma cell lines and human tumours. J. Gen. Virol. *77*, 2983–2990.

Gross, L. (1980). Oncogenic Viruses , 3rd edn (Oxford: Pergamon).

Guasch, G., Popovici, C., Mugneret, F., Chaffanet, M., Pontarotti, P., Birnbaum, D., and Pébusque, M.J. (2003). Endogenous retroviral sequence is fused to FGFR1 kinase in the 8p12 stem-cell myeloproliferative disorder with t(8;19)(p12;q13.3). Blood *101*, 286–288.

Hanger, J.J., Bromham, L.D., McKee, J.J., O'Brien, T.M., and Robinson, W.F. (2000). The nucleotide sequence of koala (Phascolarctos cinereus) retrovirus: a novel type C endogenous virus related to Gibbon ape leukemia virus. J. Virol. *74*, 4264–4272.

Haraguchi, S., Good, R., Cianciolo, G., and Day, N. (1992a). A synthetic peptide homologous to retroviral envelope protein down-regulates TNF-alpha and IFN-gamma mRNA expression. J. Leukoc. Biol. *52*, 469–472.

Haraguchi, S., Liu, W., Cianciolo, G., Good, R., and Day, N. (1992b). Suppression of human interferon-gamma production by a 17 amino acid peptide homologous to the transmembrane envelope protein of retroviruses: evidence for a primary role played by monocytes. Cell. Immunol. *141*, 388–397.

Haraguchi, S., Good, R., Cianciolo, G., James-Yarish, M., and Day, N. (1993). Transcriptional down-regulation of tumor necrosis factor-alpha gene expression by a synthetic peptide homologous to retroviral envelope protein. J. Immunol. *151*, 2733–2741.

Haraguchi, S., Good, R., and Day, N. (1995a). Immunosuppressive retroviral peptides: cAMP and cytokine patterns. Immunol. Today *16*, 595–603.

Haraguchi, S., Good, R., James-Yarish, M., Cianciolo, G., and Day, N. (1995b). Induction of intracellular cAMP by a synthetic retroviral envelope peptide: a possible mechanism of immunopathogenesis in retroviral infections. Proc. Natl. Acad. Sci. U.S.A. *92*, 5568–5571.

Haraguchi, S., Good, R., James-Yarish, M., Cianciolo, G., and Day, N. (1995c). Differential modulation of Th1- and Th2-related cytokine mRNA expression by a synthetic peptide homologous to a conserved domain within retroviral envelope protein. Proc. Natl. Acad. Sci. U.S.A. *92*, 3611–3615.

Harrell, R.A., Cianciolo, G.J., Copeland, T.D., Oroszlan, S., and Snyderman, R. (1986). Suppression of the respiratory burst of human monocytes by a synthetic peptide homologous to envelope proteins of human and animal retroviruses. J. Immunol. *136*, 3517–3520.

Henderson, L., and Qureshi, M. (1993). A peptide inhibitor of human immunodeficiency virus infection binds to novel human cell surface polypeptides. J. Biol. Chem. *268*, 15291–15297.

Heneine, W., Tibell, A., Switzer, W.M., Sandstrom, P., Rosales, G.V., Matthews, A., Korsgren, O., Chapman, L.E., Folks, T.M., and Groth, C.G. (1998). No evidence of infection with porcine endogenous retrovirus in recipients of porcine islet-cell xenografts. Lancet *352*, 695–699.

Heneine, W., Switzer, W.M., Soucie, J.M., Evatt, B.L., Shanmugam, V., Rosales, G.V., Mattews, A., Sandstrom, P., and Folks, T.M. (2001). Evidence of porcine endogenous retrovirus in porcine factor VIII and evaluation of transmission to recipients with hemophilia. J. Infect. Dis. *183*, 648–652.

Hermida-Prieto, M., Domenech, N., Moscoso, I., Diaz, T., Ishii, J., Salomon, D.R., and Manez, R. (2007). Lack of cross-species transmission of porcine endogenous retrovirus (PERV) to transplant recipients and abattoir workers in contact with pigs. Transplantation *84*, 548–550.

Hertig, C., Coupar, B.E., Gould, A.R., and Boyle, D.B. (1997). Field and vaccine strains of fowlpox virus carry integrated sequences from the avian retrovirus, reticuloendotheliosis virus. Virology *235*, 367–376.

Hirsch, M.S., Phillips, S.M., Solnik, C., Black, P.H., Schwartz, R.S., and Carpenter, C.B. (1972). Activation of leukemia viruses by graft-versus-host and mixed lymphocyte reactions *in vitro*. Proc. Natl. Acad. Sci. U.S.A. *69*, 1069–1072.

Huebner, R.J., and Todaro, G.J. (1969). Oncogenes of RNA tumor viruses as determinants of cancer. Proc. Natl. Acad. Sci. U.S.A. *64*, 1087–1094.

Hughes, J.F., and Coffin, J.M. (2004). Human endogenous retrovirus K solo-LTR formation and insertional polymorphisms: implications for human and viral evolution. Proc. Natl. Acad. Sci. U.S.A. *101*, 1668–1672.

Irgang, M., Sauer, I.M., Karlas, A., Zeilinger, K., Gerlach, J.C., Kurth, R., Neuhaus, P., and Denner, J. (2003). Porcine endogenous retroviruses (PERVs): No infection in patients treated with a bioreactor based on porcine liver cells. J. Clin. Virol. *28*, 141–154.

Irgang, M., Karlas, A., Laue, C., Specke, V., Tacke, S.J., Kurth, R., Schrezenmeir, J., and Denner, J. (2005). Porcine endogenous retroviruses PERV-A and PERV-B infect neither mouse cells *in vitro* nor SCID mice *in vivo*. Intervirology *48*, 167–173.

Isfort, R.J., Qian, Z., Jones, D., Silva, R.F., Witter, R., and Kung, H.J. (1994). Integration of multiple chicken retroviruses into multiple chicken herpesviruses: herpesviral gD as a common target of integration. Virology *203*, 125–133.

Johnston, J.B., Silva, C., Holden, J., Warren, K.G., Clark, A.W., and Power, C. (2001). Monocyte activation and differentiation augment human endogenous retrovirus expression: implications for inflammatory brain diseases. Ann. Neurol. *50*, 434–442.

Kadota, J., Cianciolo, G., and Snyderman, R. (1991). A synthetic peptide homologous to retroviral transmembrane envelope proteins depresses protein kinase C mediated lymphocyte proliferation and directly inactivated protein kinase C: a potential mechanism for immunosuppression. Microbiol. Immunol. *35*, 443–459.

Kalter, S.S., Helmke, R.J., Heberling, R.L., Panigel, M., Fowler, A.K., Strickland, J.E., and Hellmann, A. (1973). C-type particles in normal human placentas. J. Natl. Cancer Inst. *50*, 1081–1084.

Kalter, S.S., Heberling, R.L., Smith, G.C., and Helmke, R.J. (1975a). C-type viruses in chimpanzee (Pan sp.) placentas. J. Natl. Cancer Inst. *55*, 735–736.

Kalter, S.S., Heberling, R.L., Helmke, R.J., Panigel, M., Smith, G.C., Kraemer, D.C., Hellman, A., Fowler, A.K., and Strickland, J.E. (1975b). A comparative study on the presence of C-type viral particles in placentas from primates and other animals. Bibl. Haematol. *40*, 391–401.

Karlas, A., Kurth, R., and Denner, J. (2004). Inhibition of porcine endogenous retroviruses by RNA interference: increasing the safety of xenotransplantation. Virology *325*, 18–23.

Kato, N., Pfeiffer-Ohlson, S., Kato, M., Larsson, E., Rydnert, J., Ohlsson, R., and Cohen, M. (1987). Tissue specific expression of human provirus ERV3 mRNA in human placenta: two of the three ERV3 mRNAs contain human cellular sequences. J. Virol. *61*, 2182–2191.

Katzourakis, A., Rambaut, A., and Pybus, O.G. (2005). The evolutionary dynamics of endogenous retroviruses. Trends Microbiol. *13*, 463–468.

Katzourakis, A., Tristem, M., Pybus, O.G., and Gifford, R.J. (2007). Discovery and analysis of the first endogenous lentivirus. Proc. Natl. Acad. Sci. U.S.A. *104*, 6261–6265.

Kizaki, T., Mitani, M., Cianciolo, G., Ogasawara, M., Good, R., and Day, N. (1991). Specific association of retroviral envelope protein, p15E, with human cell surfaces. Immunol. Lett. *28*, 11–18.

Kleinman, A., Senyuta, N., Tryakin, A., Sauter, M., Karseladze, A., Tjulandin, S., Gurtsevitch, V., and Mueller-Lantzsch, N. (2004). HERV-K(HML-2) GAG/ENV antibodies as indicator for therapy effect in patients with germ cell tumors. Int. J. Cancer *110*, 459–461.

Kleinerman, E., Lachman, L., Knowles, R., Snyderman, R., and Cianciolo, G. (1987). A synthetic peptide homologous to the envelope proteins of retroviruses inhibits monocyte-mediated killing by inactivating interleukin 1. J. Immunol. *139*, 2329–2337.

Knerr, I., Beinder, E., and Rascher, W. (2002). Syncytin, a novel human endogenous retroviral gene in human placenta: evidence for its dysregulation in preeclampsia and HELLP syndrome. Am. J. Obstet. Gynecol. *186*, 210–213.

Koutsonikolis, A., Haraguchi, S., Brigino, E., Owens, U., Good, R., and Day, N. (1997). HIV-1 recombinant gp41 induces IL-10 expression and production in peripheral blood monocytes but not in T-lymphocytes. Immunol. Lett. 55, 109–113.

Kuddus, R., Patzer, J.F., Lopez, R., Mazariegos, G.V., Meighen, B., Kramer, D.J., and Rao, A.S. (2002). Clinical and laboratory evaluation of the safety of a bioartificial liver assist device for potential transmission of porcine endogenous retrovirus. Transplantation 73, 420–429.

Kurth, R., Löwer, R., Löwer, J., Harzmann, R., Pfeiffer, R., Schmidt, C.G., Foghm, J., and Frank, H. (1980). Oncornavirus synthesis in human teratocarcinoma cultures and an increased antiviral immune reactivity in corresponding patients. In: Viruses in naturally occurring cancers. Cold Spring Harbor Conferences on Cell Proliferation, vol. 7, Esssex, M. , Todaro, G. , H. zur Hausen, eds. (Cold Spring Harbor Laboratory), 835–846.

Lander, E.S., Linton, L.M., Birren, B., Nusbaum, C., Zody, M.C., Baldwin, J., Devon, K., Dewar, K., Doyle, M., FitzHugh, W. *et al.* (2001). Initial sequencing and analysis of the human genome. Nature *409*, 860–921.

Landry, J.R., Rouhi, A., Medstrand, P., and Mager, D.L. (2002). The Opitz syndrome gene Mid1 is transcribed from a human endogenous retroviral promoter. Mol. Biol. Evol. *19*, 1934–1942.

Langhammer, S., Fiebig, U., Kurth, R., and Denner, J. (2005a). Neutralising antibodies against the transmembrane protein of feline leukaemia virus (FeLV). Vaccine 23, 3341–3348.

Langhammer, S., Hübner, J., Kurth, R., and Denner, J. (2005b). Antibodies neutralising feline leukemia virus (FeLV) in cats immunised with the transmembrane envelope protein p15E. Immunology *117*, 229–237.

Lavie, L., Kitova, M., Maldener, E., Meese, E., and Mayer, J. (2005). CpG methylation directly regulates transcriptional activity of the human endogenous retrovirus family HERV-K(HML-2). J. Virol. *79*, 876–883.

Lavillette, D., Marin, M., Ruggieri, A., Mallet, F., Cosset, F.L., and Kabat, D. (2002). The envelope glycoprotein of human endogenous retrovirus type W uses a divergent family of amino acid transporters/cell surface receptors. J. Virol. *76*, 6442–6452.

Lavillette, D., and Kabat, D. (2004). Porcine endogenous retroviruses infect cells lacking cognate receptors by an alternative pathway: implications for retrovirus evolution and xenotransplantation. J. Virol. *78*, 8868–8877.

Lee, J.C., Ihle, J.N., and Huebner, R. (1977). The humoral immune response of NIH Swiss and SWR/J mice to vaccination with formalinized AKR or Gross murine leukemia virus. Proc. Natl. Acad. Sci. U.S.A. *74*, 343–347.

Lee, X., Keith, J.C., Stumm, N., Moutsatsos, I., McCoy, J.M., Crum, C.P., Genest, D., Chin, D., Ehrenfels, C., and Pijnenborg, R. (2001). Downregulation of placental syncytin expression and abnormal protein localization in pre-eclampsia. Placenta *22*, 808–812.

Lee, W.J., Kwun, H.J., Kim, H.S., and Jang, K.L. (2003). Activation of the human endogenous retrovirus W long terminal repeat by herpes simplex virus type 1 immediate early protein 1. Mol. Cells *15*, 75–80.

Lee, Y.N., and Bieniasz, P.D. (2007). Reconstitution of an infectious human endogenous retrovirus. PLoS. Pathog. 3, 10.

Levy, J.A. (1973). Xenotropic viruses: murine leukemia viruses associated with NIH Swiss, NZB, and other mouse strains. Science *182*, 1151–1153.

Levy, J.A. (1978). Xenotropic type C viruses. Curr. Top. Microbiol. Immunol. *79*, 111–213.

Levy, M.F., Crippin, J., Sutton, S., Netto, G., McCormack, J., Curiel, T., Goldstein, R.M., Newman, J.T., Gonwa, T.A., Bancherau, J., Diamond, L.E., Byrne, G., Jogan, J., and Klintmalm, G. (2000). Liver allotransplantation after extracorporeal hepatic support with transgenic (hCD55/hCD59) porcine livers: clinical results and lack of pig-to-human transmission of the porcine endogenous retrovirus. Transplantation *69*, 272–280.

Le Tissier, P., Stoye, J.P., Takeuchi, Y., Patience, C., and Weiss, R.A. (1997). Two sets of human-tropic pig retrovirus. Nature *389*, 681–682.

Lieber, M.M., Mahieux, R., Chappey (1975). Isolation from the Asian mouse Mus caroli of an endogenous type C virus related to infectious primate type C viruses. Proc. Natl. Acad. Sci. U.S.A. *72*, 2315–2319.

Lieberman, M., and Kaplan, H.S. (1959). Leukemogenic activity of filtrates from radiation-induced lymphoid tumors of mice. Science *130*, 387–388.

Loss, M., Arends, H., Winkler, M., Przemeck, M., Steinhoff, G., Rensing, S., Kaup, F.J., Hedrich, F.J., Winkler, M.E., and Martin, U. (2001). Analysis of potential porcine endogenous retrovirus (PERV) transmission in a whole-organ xenotransplantation model without interfering microchimerism. Transpl. Int. *14*, 31–37.

Löwer, J., Löwer, R., Stegmann, J., Frank, H., and Kurth, R. (1981). Retrovirus particle production in three of four human teratocarcinoma cell lines. Haematol. Blood Transfus. *26*, 541–544.

Löwer, R., Löwer, J., Tondera-Koch, C., and Kurth, R.A. (1993a). General method for the identification of transcribed retrovirus sequences (R-U5 PCR) reveals the expression of the human endogenous retrovirus loci HERV-H and HERV-K in teratocarcinoma cells. Virology *192*, 501–511.

Löwer, R., Boller, K., Hasenmaier, B., Korbmacher, C., Müller-Lantzsch, N., Löwer, J., and Kurth, R. (1993b). Identification of human endogenous retroviruses with complex mRNA expression and particle formation. Proc. Natl. Acad. Sci. U.S.A. *90*, 4480–4484.

Löwer, R., Tönjes, R.R., Korbmacher, C., Kurth, R., and Löwer, J. (1995). Identification of a rev-related protein by analysis of spliced transcripts of the human endogenous retrovirus HDTV/HERV-K. J. Virol. *69*, 141–149.

Löwer, R., Löwer, J., and Kurth, R. (1996). The viruses in all of us: characteristics and biological significance of human endogenous retrovirus sequences. Proc. Natl. Acad. Sci. U.S.A. *93*, 5177–5184.

Lowy, D.R., Rowe, W.P., Teich, N., and Hartley, J.W. (1971). Murine leukemia virus: high-frequency activation *in vitro* by 5-iododeoxyuridine and 5-bromodeoxyuridine. Science *174*, 155–156.

Magin, C., Löwer, R., and Löwer, J. (1999). cORF and RcRE, the Rev/Rex and RRE/RxRE homologues of the human endogenous retrovirus family HTDV/HERV-K. J. Virol. 73, 9496–9507.

Magin, C., Hesse, J., Löwer, J., and Löwer, R. (2000). Corf, the Rev/Rex homologue of HTDV/HERV-K, encodes an arginine-rich nuclear localization signal that exerts a *trans*-dominant phenotype when mutated. Virology *274*, 11–16.

Maksakova, I.A., Romanish, M.T., Gagnier, L., Dunn, C.A., van de Lagemaat, L.N., Mager, Malassine, A., Blaise, S., Handschuh, K., Lalucque, H., Dupressoir, A., Evain-Brion, D., and Heidmann, T. (2007). Expression of the fusogenic HERV-FRD Env glycoprotein (syncytin 2) in human placenta is restricted to villous cytotrophoblastic cells. Placenta *28*, 185–191.

Malassine, A., Blaise, S., Handschuh, K., Lalucque, H., Dupressoir, A., Evain-Brion, D., and Heidmann, T. (2007). Expression of the fusogenic HERV-FRD Env glycoprotein (syncytin 2) in human placenta is restricted to villous cytotrophoblastic cells. Placenta *28*, 185–191.

Malassiné, A., Frendo, J.L., Blaise, S., Handschuh, K., Gerbaud, P., Tsatsaris, V., Heidmann, T., and Evain-Brion, D. (2008). Human endogenous retrovirus-FRD envelope protein (syncytin 2) expression in normal and trisomy 21-affected placenta. Retrovirology *5*, 6.

Mang, R., Maas, J., Chen, X., Goudsmit, J., and van Der Kuyl, A.C. (2001). Identification of a novel type C porcine endogenous retrovirus: evidence that copy number of endogenous retroviruses increases during host inbreeding. J. Gen. Virol. *82*, 1829–1834.

Mangeney, M., and Heidmann, T. (1998). Tumor cells expressing a retroviral envelope escape immune rejection *in vivo*. Proc. Natl. Acad. Sci. U.S.A. *95*, 14920–14925.

Mangeney, M., de Parseval, N., Thomas, G., and Heidmann, T. (2001). The full-length envelope of an HERV-H human endogenous retrovirus has immunosuppressive properties. J. Gen. Virol. *82*, 2515–2518.

Mangeney, M., Pothlichet, J., Renard, M., Ducos, B., and Heidmann, T. (2005). Endogenous retrovirus expression is required for murine melanoma tumor growth *in vivo*. Cancer Res. *65*, 2588–2591.

Mangeney, M., Renard, M., Schlecht-Louf, G., Bouallaga, I., Heidmann, O., Letzelter, C., Richaud, A., Ducos, B., and Heidmann, T. (2007). Placental syncytins: Genetic disjunction between the fusogenic and immunosuppressive activity of retroviral envelope proteins. Proc. Natl. Acad. Sci. U.S.A. *104*, 20534–20539.

Martin, S.I., Wilkinson, R., and Fishman, J.A. (2006). Genomic presence of recombinant porcine endogenous retrovirus in transmitting miniature swine. Virol. J. 3, 91.

Martin, U., Kiessig, V., Blusch, J.H., Haverich, A., von der Helm, K., Herden, T., and Steinhoff, G. (1998a). Expression of pig endogenous retrovirus by primary porcine endothelial cells and infection of human cells. Lancet 352, 692–694.

Martin, U., Steinhoff, G., Kiessig, V., Chikobava, M., Anssar, M., Morschheuser, T., Lapin, B., and Haverich, A. (1998b). Porcine endogenous retrovirus (PERV) was not transmitted from transplanted porcine endothelial cells to baboons *in vivo*. Transpl. Int. *11*, 247–251.

Martin, U., Winkler, M.E., Id, M., Radeke, H., Arseniev, L., Takeuchi, Y., Simon, A.R., Patience, C., Haverich, A., and Steinhoff, G. (2000). Productive infection of primary human endothelial cells by pig endogenous retrovirus (PERV). Xenotransplantation 7, 138–142.

Martin, U., Tacke, S.J., Simon, A. R., Schröder, C., Wiebe, K., Lapin, B., Haverich, A., Denner, J., and Steinhoff, G. (2002). Absence of PERV specific humoral immune response in baboons after transplantation of porcine cells or organs. Transpl. Int. *15*, 361–368.

Martina, Y., Kurian, S., Cherqui, S., Evanoff, G., Wilson, C., and Salomon, D.R. (2005). Pseudotyping of porcine endogenous retrovirus by xenotropic murine leukemia virus in a pig islet xenotransplantation model. Am. J. Transplant. *5*, 1837–1847.

Mathes, L., Olsen, R., Hebebrand, L., Hoover, E., Schaller, J., Adams, P., and Nichols, W. (1979). Immunosuppressive properties of a virion polypeptide, a 15,000-dalton protein, from feline leukemia virus. Cancer Res. *39*, 950–955.

Mattiuzzo, G., Matouskova, M., and Takeuchi, Y. (2007). Differential resistance to cell entry by porcine endogenous retrovirus subgroup A in rodent species. Retrovirology *14*, 93.

McIntyre, M.C., Kannan, B., Solano-Aguilar, G.I., Wilson, C.A., and Bloom, E.T. (2003). Detection of porcine endogenous retrovirus in cultures of freshly isolated porcine bone marrow cells. Xenotransplantation *10*, 337–342.

Medstrand, P., Landry, J.R., and Mager, D.L. (2001). Long terminal repeats are used as alternative promoters for the endothelin B receptor and apolipoprotein C-I genes in humans. J. Biol Chem. *276*, 1896–1903.

Mi, S., Lee, X., Li, X., Veldman, G.M., Finnerty, H., Racie, L., LaVallie, E., Tang, X.Y., Edouard, P., Howes, S., Keith, J.C., and McCoy, J.M. (2000). Syncytin is a captive retroviral envelope protein involved in human placental morphogenesis. Nature *403*, 785–789.

Miyagawa, S., Nakatsu, S., Nakagawa, T., Kondo, A., Matsunami, K., Hazama, K., Yamada, J., Tomonaga, K., Miyazawa, T., and Shirakura, R. (2005). Prevention of PERV infections in pig to human xenotransplantation by the RNA interference silences gene. J. Biochem. *137*, 503–508.

Moennig, V., Frank, H., Hunsmann, G., Ohms, P., Schwarz, H., and Schäfer, W. (1974). C-type particles produced by a permanent cell line from a leukemic pig. II. Physical, chemical, and serological characterization of the particles. Virology *57*, 179–188.

Montelaro, R., and Bolognesi, D.P. (1995). Vaccines against retroviruses. In The Retroviridae, Vol. 4, Levy, J.A. , ed. (New York, USA: Plenum Press), pp. 605–656.

Moscoso, I., Hermida-Prieto, M., Mañez, R., Lopez-Pelaez, E., Centeno, A., Diaz, T.M., and Domenech, N. (2005). Lack of cross-species transmission of porcine endogenous retrovirus in pig-to-baboon xenotransplantation with sustained depletion of anti-alphagal antibodies. Transplantation *79*, 777–782.

Moyes, D.L., Martin, A., Sawcer, S., Temperton, N., Worthington, J., Griffiths, D.J., and Venables, P.J. (2005). The distribution of the endogenous retroviruses HERV-K113 and HERV-K115 in health and disease. Genomics. *86*, 337–341.

Moyes, D.L., Goris, A., Ban, M., Compston, A., Griffiths, D.J., Sawcer, S., and Venables, P.J. (2008).HERV-K113 is not associated with multiple sclerosis in a large family-based study. AIDS Res. Hum. Retroviruses *24*, 363–365.

Moza, A.K., Mertsching, H., Herden, T., Bader, A., and Haverich, A. (2001). Heart valves from pigs and the porcine endogenous retrovirus: experimentral and clinical data to assess the probability of porcine endogenous retrovirus infection in human subjects. J. Thorac. Cardiovasc. Surg. *121*, 697–701.

Muir, A., Lever, A.M., and Moffett, A. (2006). Human endogenous retrovirus-W envelope (syncytin) is expressed in both villous and extravillous trophoblast populations. J. Gen. Virol. *87*, 2067–2071.

Muster, T., Waltenberger, A., Grassauer, A., Hirschl, S., Caucig, P., Romirer, I., Fodinger, D., Seppele, H., Schanab, O., and Magin-Lachmann, C. (2003). An endogenous retrovirus derived from human melanoma cells. Cancer Res. *63*, 8735–8741.

Nelson, M., Nelson, D., Cianciolo, G., and Snyderman, R. (1989). Effects of CKS-17, a synthetic retroviral envelope peptide, on cell-mediated immunity *in vivo*: immunosuppression, immunogenicity, and relation to immunosuppressive tumor products. Cancer Immunol. Immunother. *30*, 113–118.

Nilsson, B.O., Jin, M., Andersson, A.C., Sundstrom, P., and Larsson, E. (1999). Expression of envelope proteins of endogeneous C-type retrovirus on the surface of mouse and human oocytes at fertilization. Virus Genes *18*, 115–120.

Nishitai, R., Ikai, I., Shiotani, T., Katsura, N., Matsushita, T., Yamanokuchi, S., Matsuo, K., Sugimoto, S., and Yamaoka, Y. (2005). Absence of PERV infection in baboons after transgenic porcine liver perfusion. J. Surg. Res. *124*, 45–51.

Ogasawara, M., Haraguchi, S., Cianciolo, G.J., Mitani, M., Good, R.A., and Day, N.K. (1990). Inhibition of murine cytotoxic T-lymphocyte activity by a synthetic retroviral peptide and abrogation of this activity by IL. J. Immunol. *145*, 456–462.

Olsen, R., Hoover, E., Schaller, J., Mathes, L., and Wolff, L. (1977). Abrogation of resistance to feline oncornavirus disease by immunization with killed feline leukemia virus. Cancer Res. 37, 2082–2085.

Ono, M., Yasunaga, T., Miyata, T., and Ushikubo, H. (1986). Nucleotide sequence of human endogenous retrovirus genome related to the mouse mammary tumor virus genome. J. Virol. *60*, 589–598.

Pahwa, S., Pahwa, R., Good, R.A., Gallo, R.C., and Saxinger, C. (1986). Stimulatory and inhibitory influences of human immunodeficiency virus on normal B-lymphocytes. Proc. Natl. Acad. Sci. U.S.A. *83*, 9124–9128.

Palmarini, M., Mura, M., and Spencer, T.E. (2004). Endogenous betaretroviruses of sheep: teaching new lessons in retroviral interference and adaptation. J. Gen. Virol. *85*, 1–13.

Paradis, K., Langford, G., Long, Z., Heneine, W., Sandstrom, P., Switzer, W.M., Chapman, L.E., Lockey, C., Onions, D., and Otto, E. (1999). Search for cross-species transmission of porcine endogenous retrovirus in patients treated with living pig tissue. The XEN 111 Study Group. Science *285*, 1236–1241.

Patience, C., Takeuchi, Y., and Weiss, R.A. (1997). Infection of human cells by an endogenous retrovirus of pigs. Nat. Med. 3, 282–286.

Patience, C., Patton, G.S., Takeuchi, Y., Weiss, R.A., McClure, M.O., Rydberg, L., and Breimer, M.E. (1998). No evidence of pig DNA or retroviral infection in patients with short-term extracorporeal connection to pig kidneys. Lancet 352, 699–701.

Patience, C., Switzer, W.M., Takeuchi, Y., Griffiths, D.J., Goward, M.E., Heneine, W., Stoye, J.P., and Weiss, R.A. (2001). Multiple groups of novel retroviral genomes in pigs and related species. J. Virol. 75, 2771–2775.

Payne, L.N., and Chubb, R.C. (1968). Studies on the nature and genetic control of an antigen in normal chick embryos which reacts in the COFAL test. J. Gen. Virol. 3, 379–391.

Peters, R.L., Donahoe, R.M., and Kelloff, G.J. (1975). Assay in the mouse for delayed-type hypersensitivity to murine leukemia virus. J. Natl. Cancer Inst. 55, 1089–1095.

Petersen, B., Carnwath, J.W., and Niemann, H. (2008). The perspectives for porcine-to-human xenografts. Comp. Immunol. Microbiol. Infect. Dis. 32, 91–105.

Pfeffer, L. (1987). Cellular effects of interferons. In: Mechanisms of interferon actions, Pfeffer, L.M., ed. (CRC Press, Boca Raton), 1–24.

Phillips, S.M., Stephenson, J.R., and Aaronson, S.A. (1977). Genetic factors infuencing mouse type-C RNA virus induction by naturally occurring B cell mitogens. J. Immunol. *118*, 662–666.

Pitkin, Z., and Mullon, C. (1999). Evidence of absence of porcine endogenous retrovirus (PERV) infection in patients treated with a bioartificial liver support system. Artificial Organs 23, 829–833.

Popp, S.K., Mann, D.A., Milburn, P.J., Gibbs, A.J., McCullagh, P.J., Wilson, J.D., Tönjes, R.R., and Simeonovic, C.J. (2007). Transient transmission of porcine endogenous retrovirus to fetal lambs after pig islet tissue xenotransplantation. Immunol. Cell Biol. *85*, 238–248.

Powell, S.K., Gates, M.E., Langford, G., Gu, M.L., Lockey, C., Long, Z., and Otto, E. (2000). Antiretroviral agents inhibit infection of human cells by porcine endogenous retroviruses. Antimicrob. Agents Chemother. *44*, 3432–3433.

Prudhomme, S., Bonnaud, B., and Mallet, F. (2005). Endogenous retroviruses and animal reproduction. Cytogenet. Genome Res. *110*, 353–364.

Qari, S.H., Magre, S., García-Lerma, J.G., Hussain, A.I., Takeuchi, Y., Patience, C., Weiss, R.A., and Heneine, W. (2001). Susceptibility of the porcine endogenous retrovirus to reverse transcriptase and protease inhibitors. J. Virol. 75, 1048–1053.

Qureshi, N., Coy, D., Garry, R., and Henderson, L. (1990). Characterization of a putative cellular receptor for HIV-1 transmembrane glycoprotein using synthetic peptides. AIDS 4, 553–558.

Rakoff-Nahoum, S., Kuebler, P.J., Heymann, J.J., Sheehy, M.E., Ortiz, G.M., Ogg, G.S., Barbour, J.D., Lenz, J.,

Steinfeld, A.D., and Nixon, D.F. (2006). Detection of T-lymphocytes specific for human endogenous retrovirus K (HERV-K) in patients with seminoma. AIDS Res. Hum. Retroviruses *22*, 52–56.

Ritzhaupt, A., Van Der Laan, L.J., Salomon, D.R., and Wilson, C.A. (2002). Porcine endogenous retrovirus infects but does not replicate in nonhuman primate primary cells and cell lines. J. Virol. *76*, 11312–11320.

Romanish, M.T., Lock, W.M., van de Lagemaat, L.N., Dunn, C.A., and Mager, D.L. (2007). Repeated recruitment of LTR retrotransposons as promoters by the anti-apoptotic locus NAIP during mammalian evolution. PLoS. Genet. *3*, e10.

Ruegg, C., Monell, C., and Strand, M. (1989a). Inhibition of lymphoproliferation by a synthetic peptide with sequence identity to gp41 of human immunodeficiency virus type 1. J. Virol. *63*, 3257–3260.

Ruegg, C., Monell, C., and Strand, M. (1989b). Identification, using synthetic peptides, of the minimum amino acid sequence from the retroviral transmembrane protein p15E required for inhibition of lymphoproliferation and its similarity to gp21 of human T-lymphotropic virus types I and II. J. Virol. *63*, 3250–3256.

Ruegg, C., Clements, J., and Strand, M. (1990a). Inhibition of lymphoproliferation and protein kinase C by synthetic peptides with sequence identity to the transmembrane and Q proteins of visna virus. J. Virol. *64*, 2175–2180.

Ruegg, C., and Strand, M. (1990b). Inhibition of protein kinase C and anti-CD3-induced Ca2+ influx in Jurkat T-cells by a synthetic peptide with sequence identity to HIV-1 gp41. J. Immunol. *144*, 3928–3935.

Ruegg, C., and Strand, M. (1990c). Identification of a decapeptide region of human interferon-alpha with antiproliferative activity and homology to an immunosuppressive sequence of the retroviral transmembrane protein P15E. J. Interferon Res. *10*, 621–626.

Ruggieri, A., Maldener, E., Sauter, M., Mueller-Lantzsch N., Meese, E., Fackler, O.T., and Mayer, J. (2009). Human endogenous retrovirus HERV-K(HML-2) encodes a stable signal peptide with biological properties distinct from Rec. Retrovirology *6*, 17.

Ruprecht, K., Mayer, J., Sauter, M., Roemer, K., and Mueller-Lantzsch, N. (2008). Endogenous retroviruses and cancer. Cell Mol Life Sci. *65*, 3366–3382.

Sachs, D.H., Sykes, M., Robson, S.C., and Cooper, D.K. (2001). Xenotransplantation. Adv. Immunol. *79*, 129–223.

Samuelson, L.C., Phillips, R.S., and Swanberg, L.J. (1996). Amylase gene structures in primates: retroposon insertions and promoter evolution. Mol. Biol. Evol. *13*, 767–779.

Sauter, M., Schommer, S., Kremmer, E., Remberger, K., Dölken, G., Lemm, I., Buck, M., Best, B., Neumann-Haefelin, D., and Mueller-Lantzsch, N. (1995). Human endogenous retrovirus K10: expression of Gag protein and detection of antibodies in patients with seminomas. J. Virol. *69*, 414–421.

Sauter, M., Roemer, K., Best, B., Afting, M., Schommer, S., Seitz, G., Hartmann, M., and Mueller-Lantzsch, N. (1996). Specificity of antibodies directed against Env protein of human endogenous retroviruses in patients with germ cell tumors. Cancer Res. *56*, 4362–4365.

Scheef, G., Fischer, N., Krach, U., and Tönjes, R.R. (2001). The number of a U3 repeat box acting as an enhancer in long terminal repeats of polytropic replication-competent porcine endogenous retroviruses dynamically fluctuates during serial virus passages in human cells. J. Virol. *75*, 6933–6940.

Schiavetti, F., Thonnard, J., Colau, D., Boon, T., and Coulie, P.G. (2002). A human endogenous retroviral sequence encoding an antigen recognized on melanoma by cytolytic T-lymphocytes. Cancer Res. *62*, 5510–5516.

Schulte, A.M., Lai, S., Kurtz, A., Czubayko, F., Riegel, A.T., and Wellstein, A. (1996). Human trophoblast and choriocarcinoma expression of the growth factor pleiotrophin attributable to germ-line insertion of an endogenous retrovirus. Proc. Natl. Acad. Sci. U.S.A. *93*, 14759–14764.

Schulte, A.M., Malerczyk, C., Cabal-Manzano, R., Gajarsa, J.J., List, H.J., Riegel, A.T., and Wellstein, A. (2000). Influence of the human endogenous retrovirus-like element HERV-E.PTN on the expression of growth factor pleiotrophin: a critical role of a retroviral Sp1-binding site. Oncogene *19*, 3988–3998.

Seman, G., Levy, B.M., Panigel, M., and Dmochowski, L. (1975). Type-C virus particles in placenta of the cottontop marmoset (*Saguinus oedipus*). J. Natl. Cancer Inst. *54*, 251–252.

Shearer, G.M., and Clerici, M. (1993). Abnormalities of immune regulation in human immunodeficiency virus infection. Pediatr. Res. 33, 71–74.

Shevchuk, M.M., Nuovo, G.J., and Khalife, G. (1998). HIV in testis: quantitative histology and HIV localization in germ cells. J. Reprod. Immunol. *41*, 69–79.

Singh, P., Schnitzlein, W.M., and Tripathy, D.N. (2003). Reticuloendotheliosis virus sequences within the genomes of field strains of fowlpox virus display variability. J. Virol. *77*, 5855–5562.

Specke, V., Rubant, S., and Denner, J. (2001a). Productive infection of human primary cells and cell lines with porcine endogenous retroviruses (PERVs). Virology *285*, 177–180.

Specke, V., Tacke, S., Boller, K., Schwendemann, J., and Denner, J. (2001b). Porcine endogenous retroviruses (PERVs): In vitro host range and attempts to establish small animal models. J. Gen. Virol. *82*, 837–844.

Specke, V., Plesker, R., Coulibaly, C., Boller, K., and Denner, J. (2002a). Productive infection of a mink cell line with porcine endogenous retroviruses (PERVs) and lack of transmission to minks *in vivo*. Arch. Virol. *147*, 305–319.

Specke, V., Schuurman, H.J., Plesker, R., Coulibaly, C., Özel, M., Langford, G., Kurth, R., and Denner, J. (2002b). Virus safety in xenotransplantation: first exploratory *in vivo* studies in small laboratory animals and non-human primates. Transplant. Immunol. *9*, 281–288.

Specke, V., Plesker, R., Wood, J., Coulibaly, C., Suling, K., Patience, C., Kurth, R., Schuurman, H.J., and Denner, J. (2009). No *in vivo* infection of triple immunosuppressed non-human primates after inoculation with high titers of porcine endogenous retroviruses. Xenotransplantation *16*, 34–44.

Speth, C., Joebstl, B., Barcova, M., and Dierich, M. (2000). HIV-1 envelope protein gp41 modulates expression of interleukin-10 and chemokine receptors on monocytes, astrocytes and neurones. AIDS *14*, 629–636.

Stephan, O., Schwendemann, J., Specke, V., Tacke, J.S., Boller, K., and Denner, J. (2001). Porcine endogenous retroviruses (PERVs): generation of specific antibodies, development of an immunoperoxidase assay (IPA) and inhibition by AZT. Xenotransplantation *8*, 310–316.

Stocking, C., and Kozak, C.A. (2008). Murine endogenous retroviruses. Cell. Mol. Life Sci. *65*, 3383–3398.

Stoye, J.P. (2001). Endogenous retroviruses: still active after all these years? Curr. Biol. *11*, 914–916.

Stoye, J.P., and Moroni, C. (1983). Endogenous retrovirus expression in stimulated murine lymphocytes. Identification of a new locus controlling mitogen induction of a defective virus. J. Exp. Med. *157*, 1660–1674.

Stromberg, K., and Benveniste, R. (1983). Efficient isolation of endogenous rhesus retrovirus from throphoblast. Virology *128*, 518–523.

Suzuka, I., Shimizu, N., Sekiguchi, K., Hoshino, H., Kodama, M., and Shimotohno, K. (1986). Molecular cloning of unintegrated closed circular DNA of porcine retrovirus. FEBS Lett. *198*, 339–343.

Switzer, W.M., Michler, R.E., Shanmugam, V., Matthews, A., Hussain, A.I., Wright, A., Sandstrom, P., Chapman, L.E., Weber, C., Safley, S., Denny, R.R., Navarro, A., Evans, V., Norin, A.J., Kwiatkowski, P., and Heneine, W. (2001). Lack of cross-species transmission of porcine endogenous retrovirus infection to nonhuman primate recipients of porcine cells, tissues, or organs. Transplantation *71*, 959–965.

Tacke, S.J., Kurth, R., and Denner, J. (2000a). Porcine endogenous retrovirus inhibits human immune cells: Risk for xenotransplantation. Virology *268*, 87–93.

Tacke, S.J., Specke, V., Stephan, O., Seibold, E., Bodusch, K., and Denner, J. (2000b). Porcine endogenous retroviruses: diagnostic assays and evidence for immunosuppressive properties. Transplant. Proc. 32, 1166.

Tacke, S.J., Bodusch, K., Berg, A., and Denner, J. (2001). Sensitive and specific immunological detection methods for porcine endogenous retroviruses applicable to experimental and clinical xenotransplantation. Xenotransplantation *8*, 125–135.

Tacke, S.J., Specke, V., and Denner, J. (2003). Differences in release and determination of subtype of porcine endogenous retroviruses (PERV) produced by stimulated normal pig blood cells. Intervirology *46*, 17–24.

Takeshita, S., Breen, E., Ivashchenko, M., Nishanian, P., Kishimoto, T., Vredevoe, D., and Martinez-Maza, O. (1995). Induction of IL-6 and IL-10 production by recombinant HIV-1 envelope glycoprotein 41 (gp41) in the THP-1 human monocytic cell line. Cell. Immunol. *165*, 234–242.

Takeuchi, Y., Patience, C., Magre, S., Weiss, R.A., Banerjee, P.T., Le Tissier, P., and Stoye, J.P. (1998). Host range and interference studies of three classes of pig endogenous retrovirus. J. Virol. *72*, 9986–9991.

Tanaka-Taya, K., Sashihara, J., Kurahashi, H., Amo, K., Miyagawa, H., Kondo, K., Okada, S., and Yamanishi, K. (2004). Human herpesvirus 6 (HHV-6) is transmitted from parent to child in an integrated form and characterization of cases with chromosomally integrated HHV-6 DNA. J. Med. Virol. *73*, 465–473.

Tarlinton, R.E., Meers, J., and Young, P.R. (2006). Retroviral invasion of the koala genome. Nature *442*, 79–81.

Tas, M., Drexhage, H., and Goudsmit, J. (1988). A monocyte chemotaxis inhibiting factor in serum of HIV infected men shares epitopes with the HIV transmembrane protein gp41. Clin. Exp. Immunol. *71*, 13–18.

Temin, H.M., and Mizutani, S. (1970). RNA-dependent DNA polymerase in virions of Rous sarcoma virus. Nature *226*, 1211–1213.

Thiry, L., Sprecher-Goldberger, S., Bossens, M., and Neuray, F. (1978). Immune response to primate oncornaviruses in pre-eclampsia. Lancet *1*, 1268.

Thiry, L., Sprecher-Goldberger, S., Hard, R.C., Bossens, M., and Neuray, F. (1981). Expression of retrovirus-related antigen in pregnancy. I. Antigens cross-reacting with simian retroviruses in human foetal tissues and cord blood lymphocytes. J. Reprod. Immunol. *2*, 309–322.

Tucker, A., Belcher, C., Moloo, B., Bell, J., Mazzulli, T., Humar, A., Hughes, A., McArdle, P., and Talbot, A. (2002). The production of transgenic pigs for potential use in clinical xenotransplantation: microbiological evaluation. Xenotransplantation *9*, 191–202.

Turner, G., Barbulescu, M., Su, M., Jensen-Seaman, M.I., Kidd, K.K., and Lenz, J. (2001). Insertional polymorphisms of full-length endogenous retroviruses in humans. Curr. Biol. *11*, 1531–1535.

Ueno, H., Imamura, M., and Kikuchi, K. (1983). Frequency and antigenicity of type C retrovirus-like particles in human placentas. Virchows Arch. *400*, 31–41.

Uze, G., Di Marco, S., Mouchel-Vielh, E., Monneron, D., Bandu, M.T., Horisberger, M.A., Dorques, A., Lutfalla, G., and Mogensen, K.E. (1994). Domains of interaction between alpha interferon and its receptor components. J. Mol. Biol. *243*, 245–257.

van de Lagemaat, L.N., Landry, J.R., Mager, D.L., and Medstrand, P. (2003). Transposable elements in mammals promote regulatory variation and diversification of genes with specialized functions. Trends Genet. *19*, 530–536.

van der Laan, L.J., Lockey, C., Griffeth, B.C., Frasier, F.S., Wilson, C.A., Onions, D.E., Hering, B.J., Long, Z., Otto, E., Torbett, B.E., and Salomon, D.R. (2000). Infection by porcine endogenous retrovirus after islet xenotransplantation in SCID mice. Nature *407*, 90–94.

Venables, P.J., Brookes, S.M., Griffiths, D., Weiss, R.A., and Boyd, M.T. (1995). Abundance of an endogenous retroviral envelope protein in placental trophoblasts suggests a biological function. Virology *211*, 589–592.

Vogetseder, W., Dumfahrt, A., Mayersbach, P., Schönitzer, D., and Dierich, M.P. (1993). Antibodies in human sera recognizing a recombinant outer membrane protein encoded by the envelope gene of the human endogenous retrovirus K. AIDS Res. Hum. Retroviruses *9*, 687–694.

Vogt, P.K. (1997). Historical introduction to the general properties of retroviruses. In: Retroviruses, Coffin, J.M. , Hughes, S.H. and Varmus, H.E., ed. (New York: Cold Spring Harbor Laboratory), pp. 1–25.

Voisset, C., Weiss, R.A., and Griffiths, D.J. (2008). Human RNA 'rumor' viruses: the search for novel human retroviruses in chronic disease. Microbiol. Mol. Biol. Rev. *72*, 157–196.

Wahlström, T., Nieminen, P., Närvänen, A., Suni, J., Lehtovirta, P., Saksela, E., and Vaheri, A. (1984). Monoclonal antibody defining a human syncytiotrophoblastic polypeptide immunologically related to mammalian retrovirus structural protein p30. Placenta *5*, 465–473.

Walker, B.D., and Burton, D.R. (2008). Toward an AIDS vaccine. Science *320*, 760–764.

Wang-Johanning, F., Frost, A.R., Johanning, G.L., Khazaeli, M.B., LoBuglio, A.F., Shaw, D.R., and Strong, T.V. (2001). Expression of human endogenous retrovirus k envelope transcripts in human breast cancer. Clin. Cancer Res. *7*, 1553–1560.

Wang-Johanning, F., Frost, A.R., Jian, B., Epp, L., Lu, D.W., and Johanning, G.L. (2003). Quantitation of HERV-K *env* gene expression and splicing in human breast cancer. Oncogene *22*, 1528–1535.

Wegemer, D., Kabat, K., and Kloetzer, W. (1990). Biological activities of a synthetic peptide composed of two unlinked domains from a retroviral transmembrane protein sequence. J. Virol. *64*, 1429–1436.

Weislow, O.S., Fisher, O.U., Twardzik, D.R., Hellman, A., and Fowler, A.K. (1981). Depression of mitogen-induced lymphocyte blastogenesis by baboon endogenous retrovirus-associated components. Proc. Soc. Exp. Biol. Med. *166*, 522–527.

Weiss, R.A. (1967). Spontaneous virus production from 'non-virus producing' Rous sarcoma cells. Virology *32*, 719–722.

Weiss, R.A. (2006). The discovery of endogenous retroviruses. Retrovirology *3*, 67.

Weiss, R.A., Friis, R.R., Katz, E., and Vogt, P.K. (1971). Induction of avian tumor viruses in normal cells by physical and chemical carcinogens. Virology *46*, 920–938.

Wilson, C.A., Wong, S., Muller, J., Davidson, C.E., Rose, T.M., and Burd, P. (1998). Type C retrovirus released from porcine primary peripheral blood mononuclear cells infects human cells. J. Virol. *72*, 3082–3087.

Wilson, C.A., Wong, S., VanBrocklin, M., and Federspiel, M.J. (2000). Extended analysis of the *in vitro* tropism of porcine endogenous retrovirus. J. Virol. *74*, 49–56.

Winkler, M.E., Winkler, M., Burian, R., Hecker, J., Loss, M., Przemeck, M., Lorenz, R., Patience, C., Karlas, A., Sommer, S., Denner, J., and Martin, U. (2005). Analysis of pig-to-human porcine endogenous retrovirus transmission in a triple-species kidney xenotransplantation model. Transpl. Int. *17*, 848–858.

Wood, J.C., Quinn, G., Suling, K.M., Oldmixon, B.A., Van Tine, B.A., Cina, R., Arn, S., Huang, C.A., Scobie, L., Onions, D.E., Sachs, D.H., Schuurman, H.J., Fishman, J.A., and Patience, C. (2004). Identification of exogenous forms of human-tropic porcine endogenous retrovirus in miniature Swine. J. Virol. *78*, 2494–2501.

Yi, J.M., and Kim, H.S. (2007). Molecular phylogenetic analysis of the human endogenous retrovirus E (HERV-E) family in human tissues and human cancers. Genes Genet. Syst. *82*, 89–98.

Yu, H., Alfsen, A., Tudor, D., and Bomsel, M. (2008). The binding of HIV-1 gp41 membrane proximal domain to its mucosal receptor, galactosyl ceramide, is structure-dependent. Cell Calcium *43*, 73–82.

Retroviral Particles, Proteins and Genomes

3

Norbert Bannert, Uwe Fiebig and Oliver Hohn

Abstract

Retroviruses, a large group of enveloped viruses named for their typical reverse transcription and integration, comprise seven genera, all of which have the basic proviral genomic structure 5′LTR-*gag-pro-pol-env*-3′LTR. Despite these similarities however, retroviruses have major differences with regard to their genomic organization, protein composition, and architecture. Genera for which differences are limited to the four invariant genes – *gag* (group-specific antigen), *pro* (protease), *pol* (polymerase) and *env* (envelope) – are classified as simple retroviruses, whereas genera that encode accessory proteins are classified as complex retroviruses. Despite matching sequences, the flanking long terminal repeats (LTRs) are dissimilar with regard to most of their functions. This chapter describes the general and the genera-specific facets of retroviral morphology and genomic organization, and their consequences for transcription and protein expression.

Introduction

Retroviruses retrotranscribe their RNA genome into DNA, thereby allowing integration of their 7–12 kb proviral sequences into the chromosomes of infected host cells. All replication competent members of the family have four major genes (5′-*gag-pro-pol-env*-3′) that code for structural proteins and enzymes, which themselves are flanked by LTRs. In addition to these major genes, most retroviruses encode additional proteins that exert influence at various stages of the life cycle and in pathogenesis. Several aspects of retroviral replication are regulated at the DNA or RNA level by specific *cis*- or *trans*-active elements. Transcription is generally initiated by the promoter activity of the 5′LTR and is terminated by a polyadenylation signal in the 3′LTR. Structural and regulatory proteins are translated from full-length and spliced RNAs.

All retrovirus genomes consist of two full-length transcripts. The molecules of the diploid genome are physically linked by hydrogen bonds and have a 5′ cap and 3′ poly-(A), equivalent to cellular mRNAs. The terminal region of the proviral 5′LTR, which drives transcription, and the terminal region of the 3′LTR located downstream of the polyadenylation signal are not present in the genomic RNA. The 'direct repeat' nature of the LTRs allows this region to be duplicated during reverse transcription, re-establishing the complete proviral sequence.

All exogenous retroviruses form spherical particles with a diameter of about 100–150 nm. The viral membrane is studded with glycoprotein spikes needed for attachment and entry into target cells (Fig. 3.1). Electron microscopic studies show two major morphogenetic assembly modes of the viral structures (Bernhard, 1960): cores of B- and D-type viruses – e.g. mouse mammary tumour virus (MMTV), Mason–Pfizer monkey virus (MPMV), and the foamy viruses – assemble in the cytoplasm, whereas C-type viruses – e.g. avian leukaemia virus (ALV), murine leukaemia virus (MLV) and human immunodeficiency virus (HIV) – assemble at the cellular membrane concomitant with the budding process (Fig. 3.2). The retrovirus genome is packaged via an association with the Gag precursor protein. During or soon

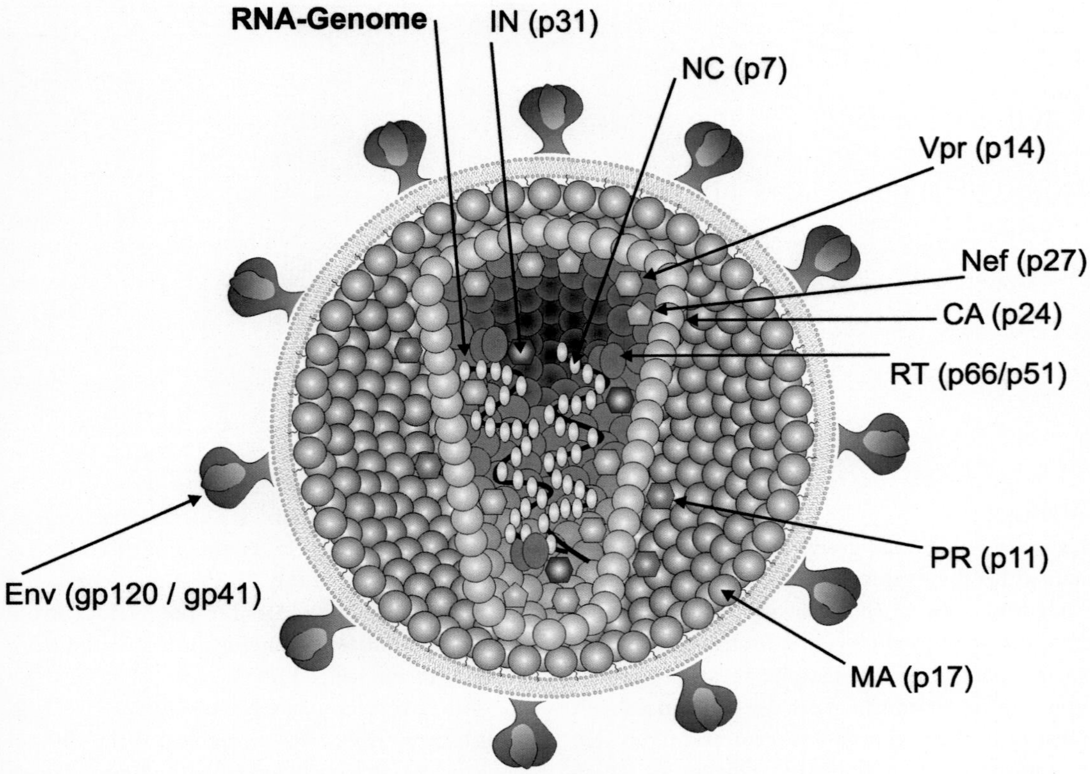

Figure 3.1 Schematic representation of the structure and arrangement of the HIV-1 genome and gene products in the mature HIV-1 particle. The molecular weights of the proteins are indicated.

after the release from the cell, considerable morphological changes occur in most retroviruses, catalysed by the viral protease. With the exception of foamy viruses, the matrix protein remains tethered to the inner leaflet of the membrane, whereas the separated capsid proteins aggregate into a core shell with a characteristic morphology. The capsids enclose the RNA genome with associated nucleocapsid proteins and a cellular tRNA required for priming of retrotranscription. Viral enzymes and a plethora of cellular proteins – many with only partially understood functional relevance – are co-packaged. Foamy viruses, which are grouped in their own subfamily (spumaretrovirinae), incorporate a large amount of reverse transcribed DNA into the virions and have additional distinctive features not shared with the other retroviruses.

In this chapter the genomic organization, the regulatory elements, the role of incorporated cellular proteins and the overall morphology of replication competent retroviruses will be discussed. The functions of the encoded proteins are only described where relevant for the understanding of the morphology and the basic replication and transmission strategies of the diverse members of this virus family. Our aim is to provide a comparative overview of retroviral structure, genomes and proteins and their implications for the life cycle. More detailed descriptions of functional or pathogenic aspects are covered by other chapters in this book.

Retroviral particles

Purification and morphology

In the late 1950s the morphology of retroviruses was visualized by thin section and negative stain electron microscopy (Bernhard, 1958). Fundamental differences in the intracellular assembly site, budding characteristics and core structure of mature retroviral particles were observed (Fig.

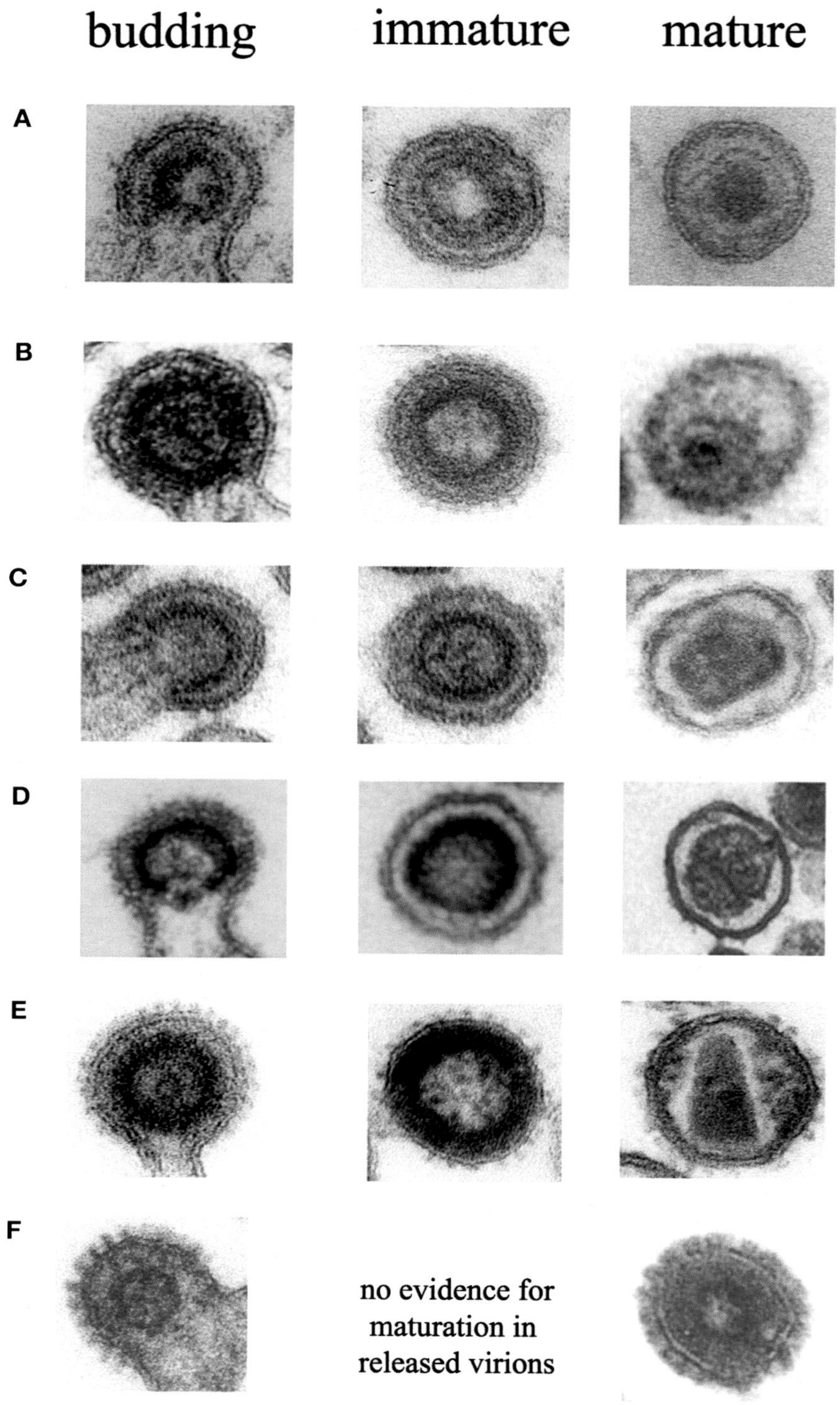

Figure 3.2 Electron micrographs of thin sections representing prototypic exogenous retroviruses at the budding, immature and mature stages. (A) alpharetrovirus (ALV, C-type morphology), (B) betaretrovirus (MMTV, B-type morphology), (C) gammaretrovirus (MLV, C-type morphology), (D) deltaretrovirus (BLV), (E) lentivirus (SIV), (F) spumaretrovirus (SFV). Maturation comparable to orthoretroviruses was not observed in this genus (Wilk *et al.*, 2001). The diameters of retroviruses are in the range of 100–150 nm. (Micrographs are courtesy of Hans R. Gelderblom, Robert Koch Institute).

3.2), and these findings provided the basis for the initial systematic classification of retroviruses.

In order to obtain detailed structural and biochemical information on immature and mature forms, the particles must be isolated and concentrated from cell culture supernatant or from sera into which they were released by the producing cells. Conventional purification and concentration schemes use centrifugation through a 20% (w/w) sucrose cushion or sedimentation to equilibrium in a sucrose gradient. The particles accumulate at about 35% (w/w) sucrose (corresponding to a density of 1.16 g/ml) and sediment at about 600 Sv. A caveat of such isolation methods is the inherent contamination with cellular vesicles that have the same density and therefore cannot be completely removed. This is only a minor problem for electron microscopic (EM) studies of single particles, but is a major problem for biochemical characterizations of particle preparations.

Using biochemical approaches and scanning transmission EM, the dry mass of retroviral particles was determined. Approximately 65% of the particle is made up of protein, most of which is the Gag polyprotein, and the rest consists of 30% lipids, 3% carbohydrates and 2–3% nucleic acids, including the diploid RNA genome and the tRNA that serves as the primer for the reverse transcription (Vogt and Simon, 1999). The lipid composition of retroviral envelopes does not match the usual membrane constitution of the producer cell where the final assembly and egress of the particle takes place; this membrane is enriched in cholesterol, sphingolipids and other constituents of lipid rafts. This membrane microdomain is targeted by the matrix domain of the Gag protein of most retroviruses and sequesters retroviral envelope glycoproteins by a mechanism that is not yet fully understood. The sugars on the envelope protein can contribute more than 50% of the glycoprotein mass.

Early conventional EM studies of retroviral particles clearly showed the spike proteins on the surface of the virions (Fig. 3.2), although their oligomeric composition remained controversial for some time. A trimeric structure of the native spike was suggested by crystallographic and NMR studies of partial glycoprotein subunits and by structure–function analogies to other viruses with fusogenic envelope glycoproteins (Blacklow *et al.*, 1995; Caffrey *et al.*, 1998; Kwong *et al.*, 1998; Lu *et al.*, 1995), and clear electron micrographs of negatively stained human foamy virus particles finally provided definitive evidence for the trimeric nature (Wilk *et al.*, 2000). Direct visualization of complete trilobed propeller-like trimeric HIV and SIV (simian immunodeficiency virus) spikes were later achieved by electron tomography of negatively stained HIV and SIV virions and by cryoelectron tomography with a resolution of about 3 nm (Zanetti *et al.*, 2006; Zhu *et al.*, 2003; Zhu *et al.*, 2006). The latter method was also used to study envelope complexes of MLV (Forster *et al.*, 2005). Although the general trimeric form of retroviral envelopes is now established, there is still substantial controversy concerning several topological aspects of the HIV and SIV trimers (Subramaniam, 2006). Surprisingly, only 14 randomly distributed trimeric complexes were present on the surface of purified HIV and SIV particles (Zhu *et al.*, 2006). Similar numbers of transmembrane and surface subunits were present on the particles, suggesting that surface subunits are not lost by shedding (Zhu *et al.*, 2003). Removal of the cytoplasmic tail of SIV envelopes increased the number of incorporated trimers to an average of 73 complexes per virion. The SIV spikes protrude 12–14 nm and have a diameter of about 10–11 nm (Zanetti *et al.*, 2006; Zhu *et al.*, 2006). Higher numbers of native trimers (up to 400) with comparable dimensions were reported for other retroviruses (Forster *et al.*, 2005; Wilk *et al.*, 2000).

With the exception of the beta-, delta-, and spumaviruses that generally form preassembled capsid structures in the cytoplasm, all exogenous members of the other retroviral genera assemble their immature cores at the cellular membrane prior to the budding process (Fig. 3.2). With the exception of spumaviruses, the structure and appearance of the interior of immature particles changes significantly during or shortly after release from the infected cell in a maturation process involving extensive cleavage of the Gag precursor proteins (Fig. 3.2). The immature non-infectious particles of orthoretroviruses show a spherical morphology with a paracrystalline packing of radially oriented Gag polyproteins, with the N-terminal matrix domain anchored in the lipid bilayer and the RNA genome in contact with the more central nucleocapsid (Briggs *et al.*, 2004; Yeager *et al.*, 1998).

The organization of the genomic RNA in the immature virus is still poorly understood. EM images of immature virions in vitreous ice display characteristic radial densities that can be correlated with distinct Gag domains (Fig. 3.3). The most prominent is the 'railroad track' image that presumably corresponds to the capsid coding domains (Yeager *et al.*, 1998). As exemplified by the average 145 nm immature HIV-1 particle, the spherical lattice is formed by about 5,000 Gag polyproteins. It appears to contain about 700 Vpr molecules, 500 cyclophilin A molecules, and 25 molecules of reverse transcriptase and integrase, in addition to other viral (Vif, Nef) and cellular encoded components (Briggs *et al.*, 2004).

Correct proteolytic processing of the Gag precursors results in the conversion of the particle into a mature state, with an electron dense, roughly spherical, polygonal, conical or cylindrical core. The fine structure and the assembly of retroviral cores were extensively studied, although most of them are quite unstable and tend to disintegrate upon membrane removal (Welker *et al.*, 2000). This is also true for the cone shaped core of HIV, with an average length of 103 nm, a diameter of 52 nm, and an average angle of 21.3°. Ganser and co-workers have proposed that the HIV core adopts the symmetry of a fullerene-type core, forming a curved hexagonal p6 lattice with 12 unevenly distributed pentameric defects – five at the narrow end and seven at the broad end (Ganser *et al.*, 1999; Li *et al.*, 2000). The morphologies of all other observed retroviral cores can be explained by altering the distribution of the 12 pentameric defects (Ganser-Pornillos *et al.*, 2004). High resolution crystallographic and nucleic magnetic resonance (NMR) structural data correspond with the structures observed by cryo-EM reconstructions, and provide clues as to the possible subunit binding interfaces (Gamble *et al.*, 1997; Gitti *et al.*, 1996; Mayo *et al.*, 2003; Rao *et al.*, 1995). In the current model, the HIV core assembly starts after the disassembly of the Gag shell at the narrow end. Genomic RNA or a structure in the membrane might serve as a trigger or template for the process (Briggs *et al.*, 2003). The cone grows towards the distal site with the defined geometry

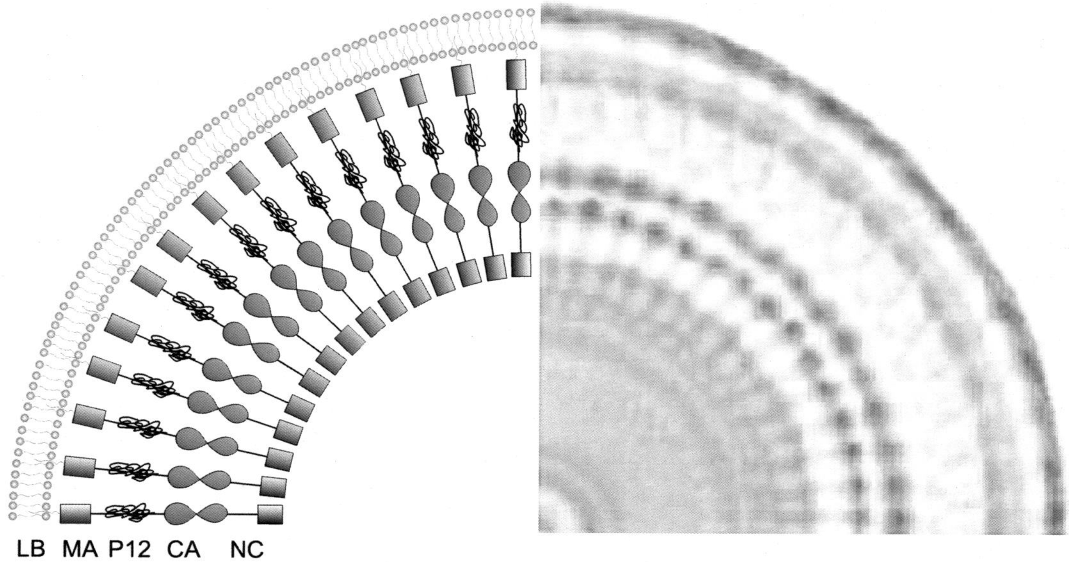

Figure 3.3 Packaging of the Gag polyprotein in immature MLV particles. A schematic depiction (left panel) of a rotationally averaged cryo-EM image of a single particle (right panel) is shown. The track line structure is assigned to the location of the capsid protein but could also include contributions from the nucleocapsid protein together with bound RNA. (LB, lipid bilayer; MA, matrix; P12, phosphoprotein; CA, capsid; NC, nucleocapsid). (Adapted with modifications from Yeager *et al.*, 1998, with permission.)

until limitations in the viral membrane induce a curvature and final closure at the broad end (Zanetti *et al.*, 2006). Less than one-third of the capsid protein released from the Gag precursors in the immature virus is utilized to build the core. The excess seems to be deposited in the space between the core and the membrane, and forms an electron dense structure previously described as lateral bodies (Gelderblom *et al.*, 1987). There is a large degree of variability in all core parameters; only about 60% of HIV particles show cores with regular cone-shape morphology, and irregular tubular, triangular cores are frequently observed. A substantial fraction of HIV particles even contain two cores. These mature particles have an average diameter of 158 nm and are larger than canonical single-core virions. It is not known whether these virions contribute to the large proportion of non-infectious particles.

Incorporated cellular proteins

During the assembly process a number of cytoplasmic proteins become incorporated into retroviral particles (Table 3.1). As already mentioned, the study of these virus-associated cellular proteins is complicated by potential contamination of the virus preparations with microvesicles and proteins that may non-specifically adhere to the virus exterior. Extremely pure virus preparations are therefore required, especially if the cellular

Table 3.1 Host proteins incorporated into retroviral particles

Function	Host molecule	Reference
Adhesions molecules	CD18, CD31; CD62L, ICAM1–3, LFA-1	Bounou *et al.* (2002), Fortin *et al.* (1997)
Antigen presentation	HLA-A, -DR, -DP, -DQ, beta2-microglobulin	Cantin *et al.* (1997a), Cantin *et al.* (2001)
Cell surface receptors	CD2, CD3, CD4, CD8, CD11, CD14, CD19, CD25, CD30, CD44, CD48	Frank *et al.* (1996), Meerloo *et al.* (1992), Orentas and Hildreth (1993)
Complement control	CD46, CD55, CD59, CD21	Montefiori *et al.* (1994), Saifuddin *et al.* (1995, 1997)
Cytosine deaminase	APOBEC 3F/3G	Kao *et al.* (2003), Liddament *et al.* (2004), Mariani *et al.* (2003), Sheehy *et al.* (2002), Wiegand *et al.* (2004)
Enzymes	GAPDH, Tal, Pin1 FKBP12, tRNA synthetase, CD5 (fucosyltransferase), CD39	Barat *et al.* (2007), Briggs *et al.* (1999), Halwani *et al.* (2004), Ott *et al.* (2000, 1996)
Heat shock proteins, chaperone	HSP60; HSP70, HSC70, cyclophilin A	Gurer *et al.* (2002)
Immune response	CD5, CD6; CD43; CD28 CD58, CD90, CD108, CypA	Franke *et al.* (1994), Giguere *et al.* (2002)
Lipids	GM1, cholesterol, phosphatidylserine	Pickl *et al.* (2001)
Membrane proteins	CD63, CD68	Meerloo *et al.* (1992)
RNA-binding protein	Staufen	Mouland *et al.* (2000)
Ser/Thr kinases	MAPK ERK2, NDR1/NDR2	Cartier *et al.* (1997)
Structural proteins	Actin, moesin, ezrin, cofilin	Ott *et al.* (1996)
Uracyl-DNA glycosylase	UNG	Mansky *et al.* (2000), Priet *et al.* (2003)
Transport proteins	Ubiquitin, TSG101, Tal, VPS28, AIP1/ALIX, VPS4B	Amit *et al.* (2004), Bishop and Woodman (2001), Ott *et al.* (2000), VerPlank *et al.* (2001), von Schwedler *et al.* (2003)

proteins in question are incorporated in only low quantities. Establishment of functional roles for these proteins in the viral life cycle requires the specific attraction of proteins into the virion. However, functional roles have not yet been convincingly established for most of the reported co-options. It seems likely that several of these proteins are included simply as a result of their involvement in transport and assembly processes or because they are present in lipid rafts or other sites of particle formation and egress.

In general, cellular proteins interact with internal viral components or are transferred onto the virus as transmembrane or membrane associated proteins. HIV-1 and other lentiviruses are the most extensively studied in this respect. Abundant cellular surface proteins present in retroviral envelopes include the major histocompatibility complex (MHC) proteins class I and II (Cantin *et al.*, 1997b), intracellular adhesion molecule (ICAM)-1 and other adhesion molecules (Fortin *et al.*, 1997), glycosylphosphatidylinositol (GPI)-linked complement regulatory factors (Montefiori *et al.*, 1994; Saifuddin *et al.*, 1995), co-stimulatory molecules, and various enzymes (Barat *et al.*, 2007; Bounou *et al.*, 2001; Giguere *et al.*, 2002). Some of these proteins are immunologically relevant in xeno- and heterogenic settings. An immune response to cellular components on the viral surface may result in neutralization of the virus and complete protection from infection, as was documented for MHC expression in SIV infection (Arthur *et al.*, 1992; Heeney *et al.*, 1994). Some of these host factors, including MHC II, ICAM-1, CD28, CD80 and CD86 molecules, increase viral infectivity if their cognate receptors are expressed on the target cells (Bounou *et al.*, 2002; Cantin *et al.*, 1997a; Fortin *et al.*, 1997; Giguere *et al.*, 2002, 2004).

Among the prominent intracellular components found inside the virus are cytoskeletal proteins such as actin and associated factors [e.g. members of the ezrin, radixin, moesin (ERM) family]. Actin associates directly with the nucleocapsid part of Gag precursor proteins, which may explain its extensive inclusion into nascent particles, with a ratio of up to 1 molecule per 10 molecules of Gag (Damsky *et al.*, 1977; Wilk *et al.*, 1999). Actin was also implicated in several stages of the viral life cycle. The trafficking of retroviral Gag polyproteins to the membrane, as well as some aspects of the subsequent assembly and budding processes, appear to rely on actin polymerization. The role of particle-incorporated actin and associated cytoskeletal proteins is currently unclear (Fackler and Krausslich, 2006). For transport and budding, retroviruses hijack the multivesicular body (MVB) pathway normally used to create vesicles for sorting cellular cargo proteins. Conserved sequences in the Gag polyprotein late (L)-domain become monoubiqutinylated and attract Hrs, Tsg101 and other subunits of the Hrs and ESCRT-complexes (Chapter 7), which organize transport and budding of vesicles. It is therefore not surprising that proteins of the late endosomal sorting machinery are reported to be incorporated into virions during budding (Cantin *et al.*, 2005).

Another Gag binding protein is the cyclic adenosine monophosphate (cAMP)-dependent protein kinase A that, at least in HIV-1 particles, phosphorylates the capsid protein and is required for efficient infectivity (Cartier *et al.*, 2003). A lysyl tRNA synthetase is also engaged in a direct interaction with the HIV-1 capsid protein (Cen *et al.*, 2004; Cen *et al.*, 2001). The current model for the selective packaging of the enzyme involves the formation of a complex with the two viral proteins Gag and Gag-Pro-Pol (Javanbakht *et al.*, 2003). Surprisingly, the packaged synthetase that aminoacylates different $tRNA^{Lys}$ appears to be a mitochondrial version of the protein (Kaminska *et al.*, 2007a; Kaminska *et al.*, 2007b). Again, the function of the enzyme inside the HIV particle is not understood, but limiting cellular expression of the synthetase using small interfering RNA reduces $tRNA^{Lys}$ packaging and viral infectivity (Guo *et al.*, 2003).

Other HIV-1 Gag binding proteins include the translational elongation factor eIF-1 (Cimarelli and Luban, 1999) and cyclophilin A (CypA) (Luban *et al.*, 1993; Thali *et al.*, 1994). The peptidyl-prolyl isomerase cyclophilin A is selectively incorporated at a ratio of one molecule to 10 Gag molecules into HIV-1 and SIVcpz, but not into HIV-2 or other SIV particles. Although CypA is included in HIV-1 virions by virtue of CypA–Gag interactions that occur during virion assembly, the interaction of target cell-derived

CypA with the incoming capsid appears to be the major determinant of CypA's effects on HIV-1 replication (Hatziioannou *et al.*, 2005). A relevant function for this virus associated protein has not yet been firmly established. As discussed in detail in Chapter 11, CypA promotes an early step in the HIV-1 life cycle in human cells (Braaten *et al.*, 1996). In the light of the discovery of the inhibitory effect of rhesus Trim5α on HIV-1 infection (Stremlau *et al.*, 2004), it quickly became clear that target cell CypA is attracted to the incoming HIV-1 capsid to prevent the detrimental inhibition by human Trim5α (Hatziioannou *et al.*, 2005; Sayah *et al.*, 2004). Paradoxically, HIV-1 restriction by Trim5α orthologues from non-human primates, including rhesus monkeys, is dependent on CypA (Berthoux *et al.*, 2005).

Another host strategy to inhibit retroviral replication is the introduction of nonsense mutations into the viral genome by RNA-editing enzymes. APOBEC 3G, a cytosine deaminase, catalyses the change of deoxycytidine (C) to deoxyuridine (U) residues in the minus-strand DNA during reverse transcription. This deamination results in G to A transitions in the newly synthesized HIV-1 plus-strand DNA. The enzyme becomes packaged into HIV-1, equine infectious anaemia virus (EIAV), MLV, and some other retroviruses (Harris *et al.*, 2003; Mangeat *et al.*, 2004). However, the lentiviral Vif protein binds to APOBEG 3G (Chapter 11) in the producer cell and prevents its incorporation into nascent viral particles (Kao *et al.*, 2003). The retrovirus has therefore evolved its own protein as a countermeasure against the incorporation of an inhibitory cellular factor.

As will be discussed later in more detail, retroviruses from several families encode a dUTPase whose activity prevents incorporation of the damaging deoxyuridine into the DNA. Some retroviruses lacking an endogenous dUTPase compensate for this deficiency by attracting a cellular dUTPase via their integrase protein (Chen *et al.*, 2002; Lerner *et al.*, 1995; Priet *et al.*, 2005, 2006).

Finally, chromatin proteins, including the barrier-to-autointegration (BAF) protein (Mansharamani *et al.*, 2003) and RNA-binding factors (Chatel-Chaix *et al.*, 2004), were also detected inside retroviral particles.

Regulatory elements, genes and proteins

Determination of genomic organization

Initial attempts to determine the exact size and gene order of retroviral genomes were carried out in the late 1960s by comparing mutants and recombinants with related retroviruses, and later by applying RNA fingerprinting on isolated viral RNAs (Beemon *et al.*, 1974; Beemon *et al.*, 1976; Joho *et al.*, 1975; Toyoshima and Vogt, 1969; Wang *et al.*, 1976). Data from these experiments and from RNA sedimentation experiments provided strong evidence for the presence of two identical 7–12 kb single-stranded RNA copies in each virus particle.

The major focus at that time was the tumorigenic property of retroviruses. Many experiments were performed with acutely transforming alpha- and gammaretroviruses carrying an oncogene (Chapter 9). By analysing appropriate mutants it quickly became clear that the oncogene is usually dispensable for viral replication (Duesberg and Vogt, 1970). In most cases the oncogene had a cellular origin and rendered provirus replication deficient; this was inferred by the observation that specific oligonucleotides hybridized with sequences in normal cellular DNA (Stehelin *et al.*, 1976). These remarkable findings linked the chromosomal proto-oncogenes with the development of cancers and initiated the modern field of molecular oncology.

The advent of DNA sequencing and cloning and protein sequencing technologies have greatly expanded our knowledge of the precise location of genes and controlling elements. In 1981, the first complete retroviral genome – the Moloney strain of murine leukaemia virus (Mo-MLV) – was published (Shinnick *et al.*, 1981).

The following discussion of the viral sequence is based on the proviral form (i.e. the DNA product of reverse transcription after integration into the DNA of the infected cell), which has the advantage that the promoter is upstream of the coding sequences and there are complete LTRs at both ends of the viral sequence (Fig. 3.4).

Upstream non-coding regions

The proviral sequences preceding the first open reading frame in viruses of this family are referred to as the 5′ long terminal repeat and the 5′ untranslated region. Both regions are packed with regulatory elements that induce and control transcription, splicing, translation, genome dimerization and encapsidation, as well as other fundamental aspects of the viral life cycle. The folding structure of most of these elements, especially at the RNA level, is of chief importance for their functions, and is a prerequisite for efficient interactions with proteins and nucleic acids.

5′ Long terminal repeat

Retrotransposons and retroviruses contain identical direct LTRs at either end of the genomic DNA. The size of these LTRs varies considerably between the retroviral genera; the longest are present in spumaviruses [e.g. simian foamy virus-1 (SFV-1) with 1758 nt] and betaretroviruses (e.g. MMTV with 1332 nt), and the shortest are present in alpharetroviruses (e.g. ALV with 330 nt) and deltaretroviruses (e.g. MPMV with 345 nt). The LTRs of all families can be divided further into three distinct elements: U3 (unique 3′), R (repeated) and U5 (unique 5′).

The U3 region occupies most of the LTR. Common to all retroviruses are *cis*-active elements at the very 5′ terminus of the U3 and at the end of the U5 region, which are involved in the process of integration (Chapter 5). These imperfect inverted repeats, which are characterized by the highly conserved dinucleotides CA or TG, are essential as attachment (*att*) sites for integrase-mediated action or 3′ processing of the proviral substrate for integration (Chow *et al.*, 1992). This CA/TG dinucleotide pair is found exactly two basepairs away from the end of the linear precursor. In HIV-1, 7 to 13 bp adjacent to the highly conserved dinucleotide are required for efficient and specific interaction of the LTR termini with the integrase (LaFemina *et al.*, 1991; Masuda *et al.*, 1998). Only the terminal elements – not the sequences at the borders of the LTRs and the internal proviral sequence – are functional in this respect.

The U3 region in the 5′LTR of the provirus serves as a promoter and drives the transcription by cellular polymerase II. The U3 sequences of even related retroviruses are considerably diverse, consisting of variable arrays of binding motifs of positive and negative regulatory factors (Chapter 6). The core promoter region contains a TATA box bound by the factor TFIIB upstream of the transcriptional start site that defines the U3-R boundary. The *trans*-acting viral regulatory factors encoded by some complex retroviruses are particularly important. These factors include the Tax protein of human T-cell lymphotropic virus-1 (HTLV-1), the Tat protein of HIV-1, and the Tas proteins of SFV-1. Tax requires an element known as TRE-1 (Tax Response Element-1) in the U3 region of HTLV-1, which stimulates transcription (Bantignies *et al.*, 1996). In contrast to Tax and Tas, the Tat proteins of primate lentiviruses are RNA binding proteins. They interact with bulged hairpin structures formed by the first nucleotides of the nascent HIV and SIV transcripts (Feng and Holland, 1988). Binding of Tat to the TAR (trans-activation response) elements induces phosporylation of polymerase II, which increases their processivity and prevents premature termination (Chun and Jeang, 1996). The sequence from the transcriptional initiation site up to the position in the 3′LTR at which the poly(A) sequence is added is defined as the 'R 'region, and is present at both termini of the transcripts. At the beginning of the transcription a regular m7G5′ppp5′G_mp-cap is added to the 5′-end of the nascent transcript by cellular enzymes (Zhou *et al.*, 2003).

The typically 60–200 nt long U5 sequence that follows the R region also contains *cis*- and *trans*-regulatory sequences (Chapter 6). Some of these elements extend into the untranslated region that follows downstream of the 5′LTR.

5′ Untranslated region

One of the most important elements in the 5′ untranslated region (UTR) is the 18 nt long primer binding site (PBS). This sequence (Fig. 3.5) hybridizes with a complementary cellular tRNA that serves as a primer for the initiation of the reverse transcription reaction, beginning with the generation of the short minus-strand strong stop DNA right at the U5 border of the 5′LTR. Retroviruses use tRNAs of different sequence and amino acid specificity for this purpose. For endogenous retroviruses, the amino acid specificity of the tRNA primer was originally used to group the viruses into families.

All retroviruses have a diploid genome, and electron microscopic images show that the two

Alpharetrovirus: RSV

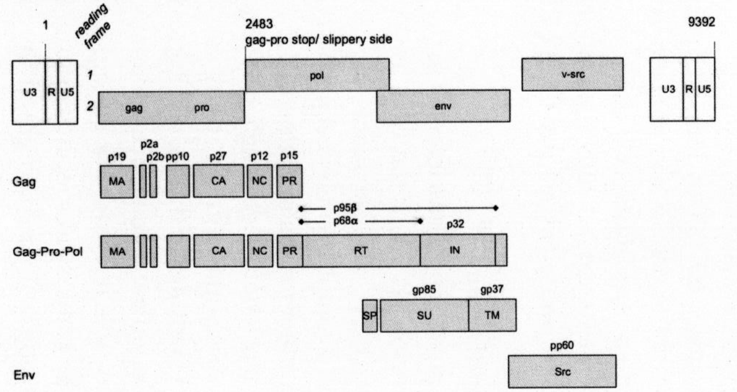

Betaretrovirus: MMTV

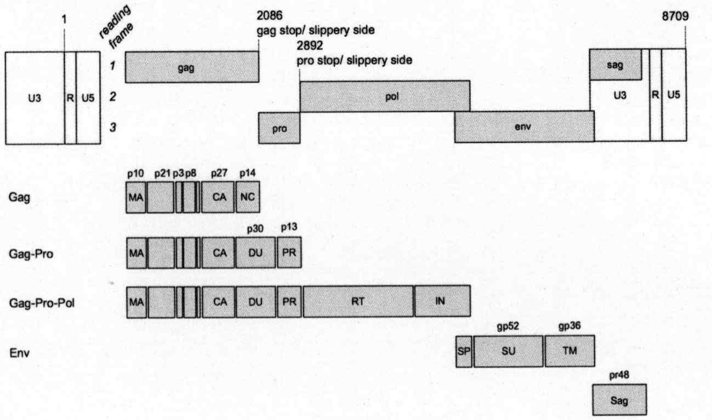

Deltaretrovirus: HTLV-1

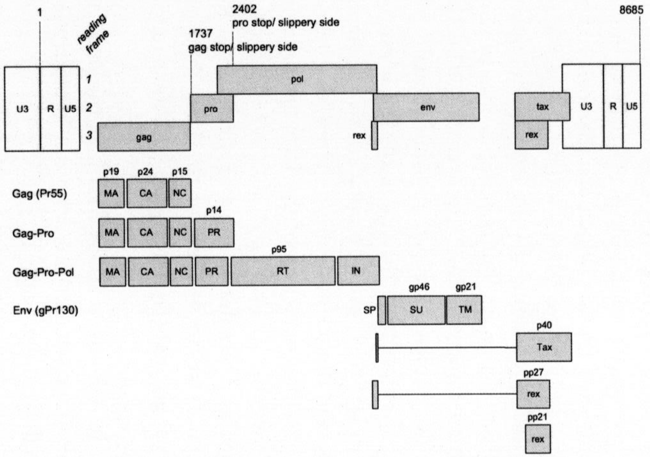

Epsilonretrovirus: WDSV

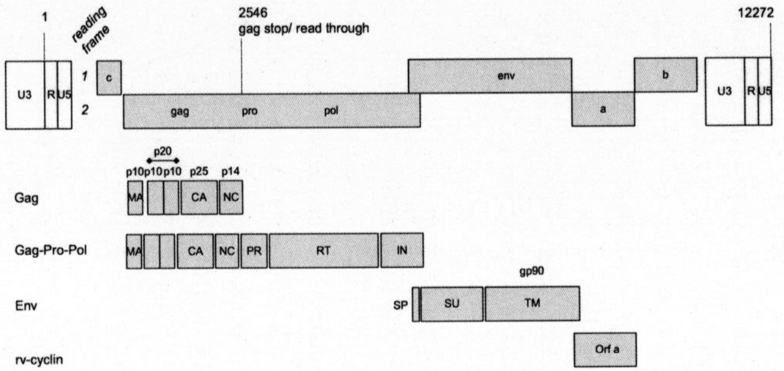

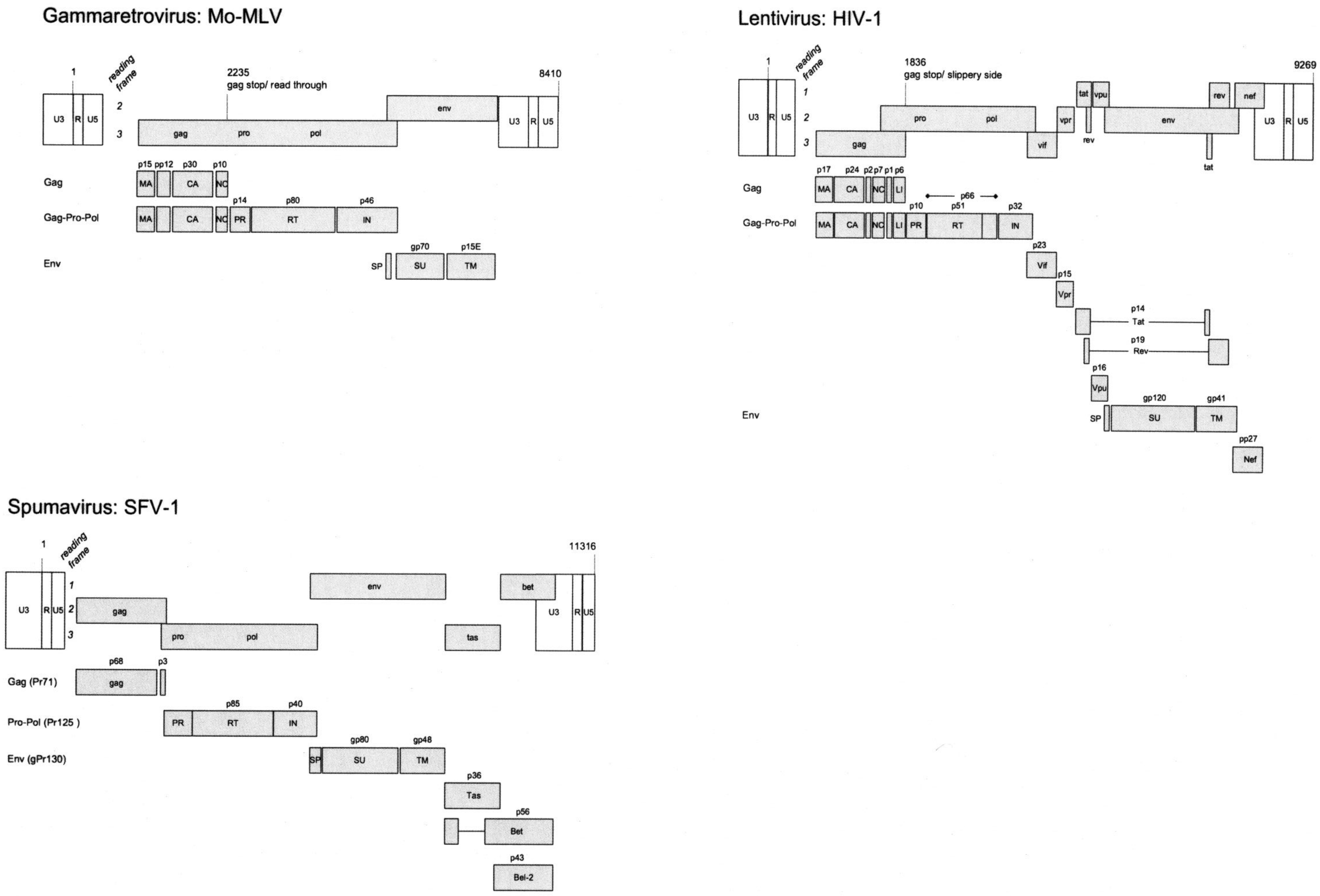

Figure 3.4 Genomic organization of retroviruses. Open reading frames encoded by spliced transcripts are indicated by a line between the boxes.

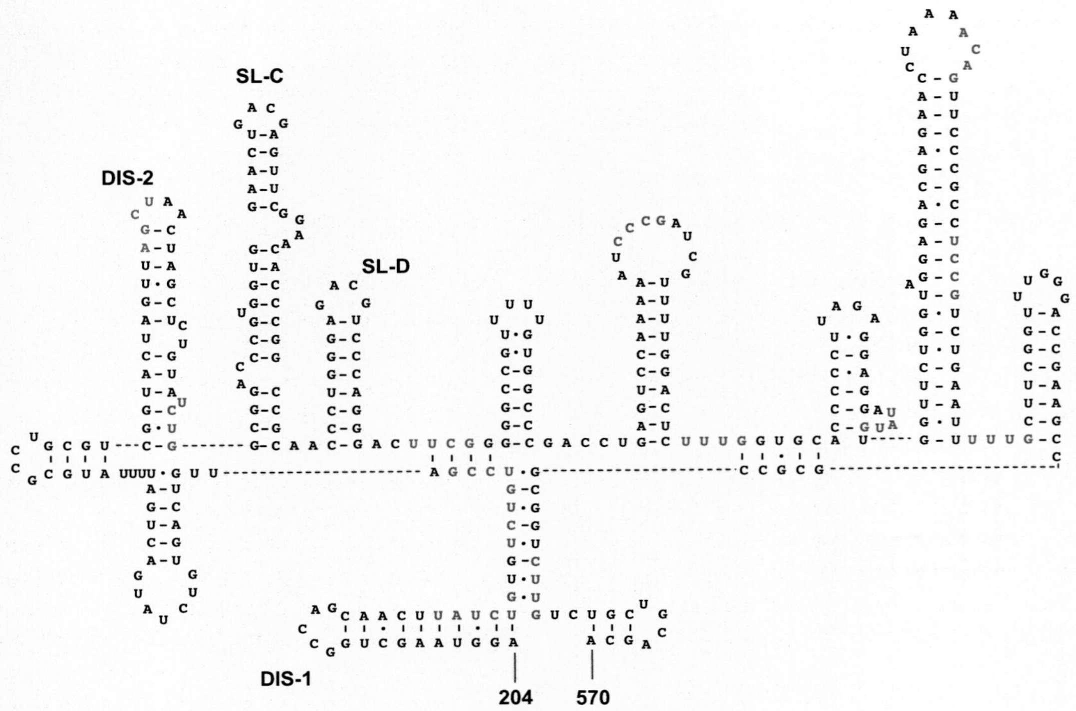

Figure 3.5 Predicted secondary structure of the 5′UTR (residues 204–570) of Mo-MLV. The palindromic AGCU element of DIS-2 which mediates dimerization by interaction with a DIS-2 element of a second RNA is shown in green. Nucleotides involved in binding to the nucleocapsid protein are shown in red. DIS: dimerization initiation site, SL: stem–loop. (Adapted with modifications from D'Souza and Summers, 2005; with permission).

retroviral RNA strands are interconnected at the 5′ site (Bender *et al.*, 1978). The strands appear to be joined at a discrete region, termed the dimer linkage site (DLS). In most retroviruses the DLS resides in the 5′UTR but it can also extend into sequences further downstream. It consists of pseudo-palindromic sequences forming one or two particular hairpins known as dimerization initiation sites (DIS) and other stem–loop structures (Fig. 3.5), including GACG tetraloops and unusual CAG-tri loops (Greatorex, 2004). It was suggested that non-covalent dimerization proceeds through multiple steps, starting with an initial 'kissing complex' and subsequent conversion into the more stable intermolecular duplex found in mature virions (D'Souza and Summers, 2005). Dimerization is physiologically important because deletions and mutations in the DLS sequence lead to impaired maturation and reduced infectivity of the viral particles (Greatorex and Lever, 1998). Why retroviruses have a diploid genome despite severe space limitations is not fully understood and remains a topic of speculation. The special proximity of the linked RNA strands promotes template switching during reverse transcription and facilitates recombination events, which together with the low fidelity of the polymerase increase the genetic diversity of the quasispecies and accelerates evolution of strains (D'Souza and Summers, 2005; Hu and Temin, 1990). Strand transfer may also occur at otherwise deleterious breakpoints, rescuing the reverse transcription by changing to a complementing template. A suitable structure induced by RNA dimerization might also play a role in reverse transcription and other early stages of virus infection (Buxton *et al.*, 2005; Sakuragi *et al.*, 2007).

Dimerization is intimately coupled to an important feature of virus assembly, the encapsidation process of full-length viral RNA. The exact order and mechanistic aspects of dimerization and encapsidation are still a matter of research and debate. The zinc knuckles of the orthoretroviruses' Gag precursor nucleocapsid region play a crucial role in both events. These RNA binding domains catalyse RNA dimerization and are required for efficient packaging (Darlix *et al.*, 1990). Additional nucleocapsid domains and regions of the matrix protein may also be involved in genome packaging of at least some retroviral genera, including deltaretroviruses (Wang *et al.*, 2003). The retroviral RNA dimerization signals generally overlap with elements that are required for efficient RNA packaging. The packaging signals that interact with the zinc knuckles of the nucleocapsid are termed Psi (Ψ)-sites. This interaction probably takes place at the perinuclear region, shortly after the synthesis of the Gag precursor proteins (Poole *et al.*, 2005). For some retroviruses a core encapsidation sequence of about 100 nt – sufficient for incorporation of homo- and heterologous RNAs into virus-like particles – was defined, but sequences upstream and downstream increase the packaging efficiency (Adam and Miller, 1988; Bender *et al.*, 1987). In most retroviruses the Ψ-site extends into the *gag* coding sequence. In HIV-1, up to 300 nt of the 5′*gag* open reading frame (ORF) enhance RNA insertion into particles (McBride *et al.*, 1997). Encapsidation signals in most retroviruses are discontinuous. For example, efficient genome packaging of feline immunodeficiency virus (FIV), a member of the lentivirus genus, requires the first 100 nt of the 5′UTR and the first 100 nt of *gag* (Browning *et al.*, 2003; Kemler *et al.*, 2002). Full-length retroviral genomes are selected for encapsidation with an extraordinary specificity not only from a pool of cellular mRNAs in high excess, but also from subgenomic viral RNAs. The envelopes of all retroviruses and regulatory proteins of complex retroviruses are translated from spliced messenger RNAs. The packaging specificity ensuring Ψ-site has to be excluded from subgenomic viral RNAs or rendered non-functional. In almost all retroviruses the major splice donor site is positioned inside or upstream of crucial packaging motifs, resulting in a loss of functional integrity in all spliced products. In most retroviral genera the splice donor is therefore found in the 5′UTR, and in delta- and spumaretroviruses it is located far upstream, in the R region of the LTR.

The selection principle for full-length transcripts described above appears not to hold true for some alpharetroviruses. For example in avian sarcoma viruses, a highly efficient packaging motif comprising as few as 82 nt is located upstream of the major splice donor (Katz *et al.*, 1986) and is therefore present in spliced transcripts that encode Env. A further peculiarity is the location of the major splice donor site several nucleotides downstream of a putative *gag* start codon. The initial amino acids of Gag and Env proteins are therefore encoded by the same sequence. At present, the precise selection mechanism for full-length RNA in this retroviral genus is still unknown (Banks *et al.*, 1999; Zhou *et al.*, 2007). The general structure of the major splice donor sites of retroviruses follows the regular composition of cellular splice sites (Stoltzfus and Madsen, 2006).

Compared to cellular genes, the 5′UTR of retroviruses is extremely structured and unusually long. Moreover, it contains AUG codons that probably interfere with correct initiation and efficient translation by a regular cap-dependent scanning mechanism, whereby ribosomal subunits bind to the cap and scan downstream until an initiation codon in an appropriate context is encountered. For several members of the *Retroviridae*, including lentiviruses (e.g. HIV, SIV), deltaretroviruses (e.g. HTLV-1), gammaretroviruses (e.g. MLV) and alpharetroviruses [e.g. Rous sarcoma virus (RSV)], initiation by a ribosomal entry site (IRES) located either in the 5′UTR or in the R-U5 region was suggested (Attal *et al.*, 1996; Berlioz and Darlix, 1995; Brasey *et al.*, 2003; Deffaud and Darlix, 2000a,b). However, there are also reports of IRES functionality in a region inside the HIV-1 *gag* sequence (Buck *et al.*, 2001). The presence and relevance of IRES activities in retroviruses presently remains somewhat controversial (Brasey *et al.*, 2003; Miele *et al.*, 1996).

Coding regions: regulatory elements and proteins

Without exception, retroviruses code for four canonical open reading frames: *gag, pro, pol* and *env*. Gag and Pol are usually synthesized as polyproteins that are cleaved by the viral protease into functional units (structural proteins and enzymes). The general strategy for translation and inclusion of the protease and polymerase into particles is the synthesis of Gag fusion proteins, a method that only spumaviruses fail to use. Reading frame changes and redefinition of stop codons are commonly used to achieve an appropriate molecular ratio of the proteins. The envelope protein and most of the auxiliary proteins are usually expressed from spliced transcripts. Coding regions also harbour regulatory elements, including packaging motifs, splice sites and slippery sites for reading frame changes at the ribosome.

Translated sequences upstream of gag

The open reading frame of exogenous retroviruses located furthest upstream usually encodes the Gag polyprotein. Exceptions are the walleye dermal sarcoma virus (WDSV) and related epsilonretroviruses. In WDSV a sequence encoding the 120-amino acid protein Orf C is located just upstream of *gag* in an alternative reading frame (Fig. 3.4). The Orf C protein is targeted to the mitochondria and may be involved in the induction of apoptosis and regression of tumours in infected fish (Nudson *et al.*, 2003). The snakehead fish retrovirus was reported to encode a 14-amino acid long section of the leader peptide belonging to the envelope protein in a region upstream of *gag*. The sequence of the leader peptide is joined with the rest of Env by splicing (Hart *et al.*, 1996).

Expression of the gag *gene*

In all retroviruses Gag is transcribed from unspliced full-length transcripts. The ORFs range from almost 2000 nt in some betaretroviruses (e.g. MPMV) to less than 1200 nt in some deltaretroviruses [e.g. bovine leukaemia virus (BLV)], resulting in proteins of 80–44 kDa. In addition to the regular Gag protein, some murine and feline gammaretroviruses encode an N-terminally extended version. In contrast to the regular form, the transcription of this protein is initiated at an atypical CTG codon. In Mo-MLV this codon is about 264 nucleotides upstream of the ATG start codon for the regular Gag protein. The extra domain encodes a leader peptide directing the protein into the endoplasmic reticulum and Golgi complex, where it becomes glycosylated at several sites. Finally, the glyco-Gag (or $gPr80^{gag}$) reaches the membrane as a type II (extracellular C-terminus) cell surface glycoprotein (Pillemer *et al.*, 1986). This protein is in principle dispensable for viral replication as glygo-Gag negative mutants of Mo-MLV are replication competent and pathogenic, although reversion to wild-type virus is frequently observed *in vivo*. Recently however, a role for the glyco-Gag in a late step of viral budding was suggested (Low *et al.*, 2007). In terms of pathogenicity, the protein seems to be an important determinant of neuro-invasiveness and virulence (Fujisawa *et al.*, 2001).

The regular retroviral Gag polyproteins of most retroviruses are cotranslationally modified by myristoylation. The C_{14} fatty acid is always covalently linked by an amide bond to a glycine following the initiator methionine. Together with hydrophobic and basic 30–90 nt long membrane binding domains (M-domains), it attaches the protein to the inner leaflet of the cell membrane. In orthoretroviruses lacking a myristic acid on their Gag proteins (e.g. avian retroviruses and some lentiviruses), the M-domains are probably potent enough to mediate the membrane anchorage without a fatty acid support. In spumaviruses the leader peptide of the envelope protein is not cleaved off by a signal peptidase, but instead interacts with the N-terminal portion of the Gag protein and may serve the morphogenetic role of targeting preformed capsids to the membrane for budding (Wilk *et al.*, 2001).

The Gag polyprotein of retroviruses of all genera is cleaved by the virus encoded protease during or following release. In spumaviruses the 71-kDa Gag precursor is cleaved once, releasing a 4-kDa C-terminal polypeptide required for efficient particle infectivity (Enssle *et al.*, 1997; Zemba *et al.*, 1998). Accumulating evidence suggests a second cleavage in Gag that occurs shortly after infection and may be essential for the replication, possibly by allowing disassembly of the incoming capsid (Giron *et al.*, 1997; Lehmann-Che *et al.*, 2005). In all other retroviruses except the foamy viruses, three major mature Gag subunits are released: matrix (MA), capsid (CA) and nucleocapsid (NC). These proteins are commonly named after their molecular weight (e.g. HIV-1 p24 for the capsid protein of HIV-1).

The N-terminal matrix protein of about 10–19 kDa contains the M-domains described earlier and remains bound to the viral lipid layer as an internal peripheral membrane protein following release from the precursor. It was suggested that the matrix makes direct contact with the cytoplasmic tail of Env (Belyaev *et al.*, 1994; Freed and Martin, 1995; Rao *et al.*, 1995). However, it remains to be determined whether this proposed interaction is confined to intracellular assembly steps or whether it is permanent and outlasts maturation. The matrix proteins are also known to participate in late entry and uncoating steps as well as in the formation of the preintgration complex and in nuclear import (Hearps and Jans, 2007).

In the Gag precursor protein of many retroviruses, the matrix precedes the major capsid protein domain. The 20–30 kDa capsid proteins released from the inner core of the particle during maturation enclose the incorporated viral enzymes and the viral genome. The capsid contains the most highly conserved sequence among the Gag proteins, known as the major homology region (MHR). The capsid contains the most highly conserved sequence among the Gag proteins, known as the major homology region (MHR). The role of MHR is not fully understood but it seems to be necessary for the assembly process of orthoretroviral particles (Mammano *et al.*, 1994; Provitera *et al.*, 2001).

In all orthoretroviruses the nucleocapsid protein is encoded at the C-terminal end of the capsid domain. The small proteins are highly basic and contain one or two Cys-His motifs (Cys-X_2-Cys-X_4-His-X_4-Cys) which coordinate a single zinc ion (Fig. 3.6). As mentioned earlier, these zinc knuckles are involved in genome packaging. The nucleocapsid protein is also involved in primer tRNA binding during assembly, as well as in the processes of reverse transcription and integration (Bonnet-Mathoniere *et al.*, 1996; Carteau *et al.*, 1999; Remy *et al.*, 1998). Furthermore, nucleocapsid may influence secondary structures of the viral RNA and is needed for efficient movement of the reverse transcriptase along the RNA template. Owing to the binding ability of nucleocapsid to double-stranded nucleic acids, it may remain on the final viral DNA genome for stabilization and protection from degradation (Tanchou *et al.*, 1998). Spumaviruses lack regular Cys-His motifs in their Gag proteins. Instead, the C-terminus of the foamy virus protein contains three glycine-arginine-rich basic sequences called GR boxes. The first box (GR I) binds to viral nucleic acids, allowing fixation of the viral genome to the Gag protein (Lecellier and Saib, 2000).

Members of the alpha-, beta- and gammaretroviruses produce phosphoproteins of approximately 10–20 kDa encoded between the matrix and capsid proteins, but the functional role of these proteins has not been investigated in detail. Several contain one or more of the late assembly domains (described earlier) that mediate virus release (Demirov and Freed, 2004). Sequences with homologous functions are also known for other membrane coated viruses, e.g. vesicular stomatitis Indiana virus (VSV), rabies or Ebola virus (Harty *et al.*, 1999; Irie *et al.*, 2004; Jayakar *et al.*, 2000; Martin-Serrano *et al.*, 2001; Timmins *et al.*, 2003). In equine and primate lentiviruses the L-domains are positioned in the C-terminal region of the Gag polyprotein. In HIV-1 the L-domain consensus sequence PTAP is found in the p6 protein at the very C-terminus of the precursor protein.

The proteolytic cleavage of several Gag precursor proteins produces spacer peptides of different sizes. In HIV-1, the spacer peptides p1 and p2 are released from the junction of CA/NC and NC/p6, respectively. Such spacer peptides released from between the major Gag subunits are

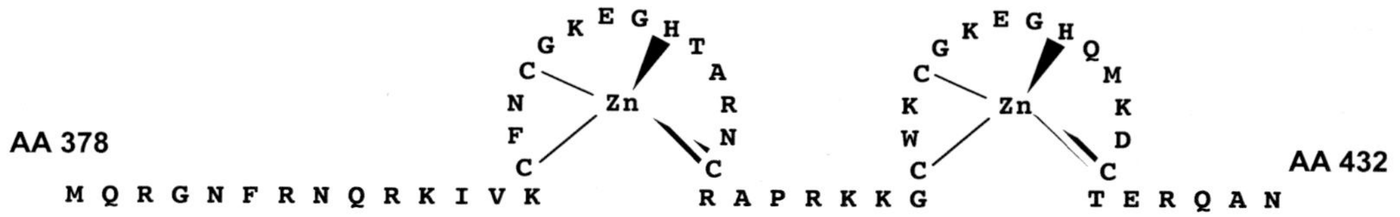

Figure 3.6 Coordination of the zinc ions in the zinc knuckles of the HIV-1 nucleocapsid protein. The sequence corresponds to amino acids 378–432 of the strain HXB2 (for details see (Grigorov *et al.*, 2007).

also known for retroviruses of other genera. They are mostly required in early (disassembly) or late stages (assembly, maturation) of the replication process (Accola *et al.*, 1998; Craven *et al.*, 1993; Krausslich *et al.*, 1995).

Expression of the pro *genes*

In all orthoretrovirinae, Gag, Pro and Pol are translated from the polycistronic full-length mRNA, and Gag, Gag–Pro and/or Gag-Pro-Pol polyproteins are synthesized. This strategy allows Gag to be used to achieve efficient incorporation of the protease and polymerase into nascent viral particles. The polyproteins are subsequently cleaved by the viral protease to generate mature, fully active enzymes. Compared to the structural Gag protein, the Pro and Pol enzymes are required in far lower amounts. To achieve an appropriate translational ratio of these proteins from a single mRNA, retroviruses use two different mechanisms: RNA frameshifting at the ribosome and leaky stops.

A translational –1 (in 5′ direction) frameshift of the mRNA (Jacks and Varmus, 1985) close to the *gag* ORF termination in the *Gag–Pro* overlapping region is necessary in beta-, delta- and lentiviruses to allow the synthesis of the Gag–Pro proteins. In MMTV about 23% of the translations shift –1 close to the *gag–pro* border into the protease ORF (Jacks *et al.*, 1987; Wang *et al.*, 2002). As a result, the amount of protease synthesized is about 4 times less than that of Gag. The general mechanism necessitates specific mRNA signals that induce ribosomes to stall over special homopolymeric 'slippery' sequences (Fig. 3.7), allowing the ribosome-bound amino acids carrying tRNAs to slip one nucleotide upstream (Harger *et al.*, 2002). The pause in translation that stimulates frameshifting at a slippery heptanucleotide is typically caused by a pseudoknot or other hairpin structures a few nucleotides downstream (Giedroc *et al.*, 2000; Jacks *et al.*, 1987; Marcheschi *et al.*, 2007). Without the structured stem–loops or pseudoknots the reading frame change at the slippery site is extremely inefficient.

Of all retroviruses, only members of the alpharetrovirus genus synthesize Gag and protease at an equimolar ratio; in such viruses *gag* has no termination codon and the protease is encoded in the same reading frame. The mature protease is released as a 15-kDa C-terminal processing fragment of the Gag–Pro polyprotein and from the Gag-Pro-Pol precursor (Schatz *et al.*, 1997).

In gamma- and epsilonretroviruses the protease coding sequence immediately follows the *gag*-terminating UAG codon in the same frame. Synthesis of the Gag–Pol proteins results from a translational read-through of the amber stop codon at a frequency of about 5–10%. The UAG codon in this case is translated as a glutamine, resulting in the involvement of a cellular suppressor $tRNA^{Glu}$ in the redefinition process (Yoshinaka *et al.*, 1985). Analogous to the frameshifting course, a pseudoknot structure of the messenger RNA located several nucleotides downstream of the termination codon is mandatory for efficient read-through and continuation of the translation (Fig. 3.7).

The precise mechanism by which retroviruses achieve their high rate of read-through events is not yet fully understood. These structures may have a direct effect on the ribosome movement along the RNA strand, or may influence the binding of the termination factors eRF-1 and eRF-3 to the ribosomes. Furthermore, the availability of reverse transcriptase proteins appears responsible for the high frequency of read-through events; e.g. MLV reverse transcriptase binds to eRF-1 and expression of the enzyme promotes the read-through events of the Gag stop codon (Orlova *et al.*, 2003). Reverse transcriptase molecules therefore enhance the synthesis of Gag-Pro-Pol proteins in a positive feedback loop.

In spumaviruses the protease is translated from a *pro-pol* mRNA. It is generated using the major splice donor and a splice acceptor site in the *gag* coding sequence (Yu *et al.*, 1996). Proteolytic processing of the 125-kDa polyprotein produces an 85-kDa protein with protease and reverse transcriptase activities and a 40-kDa integrase, but a separate protease is not released. The lack of a Gag-Pro-Pol fusion protein raises questions regarding the packaging mechanism for the enzymes. It seems that two encapsidation sequences in the genomic RNA are required to copackage the 125-kDa polyprotein, one located in the 5′ untranslated region of the viral genomic RNA and one in the 3′ region of the *pol* gene (Heinkelein *et al.*, 2000). The C-terminus of Gag is also involved

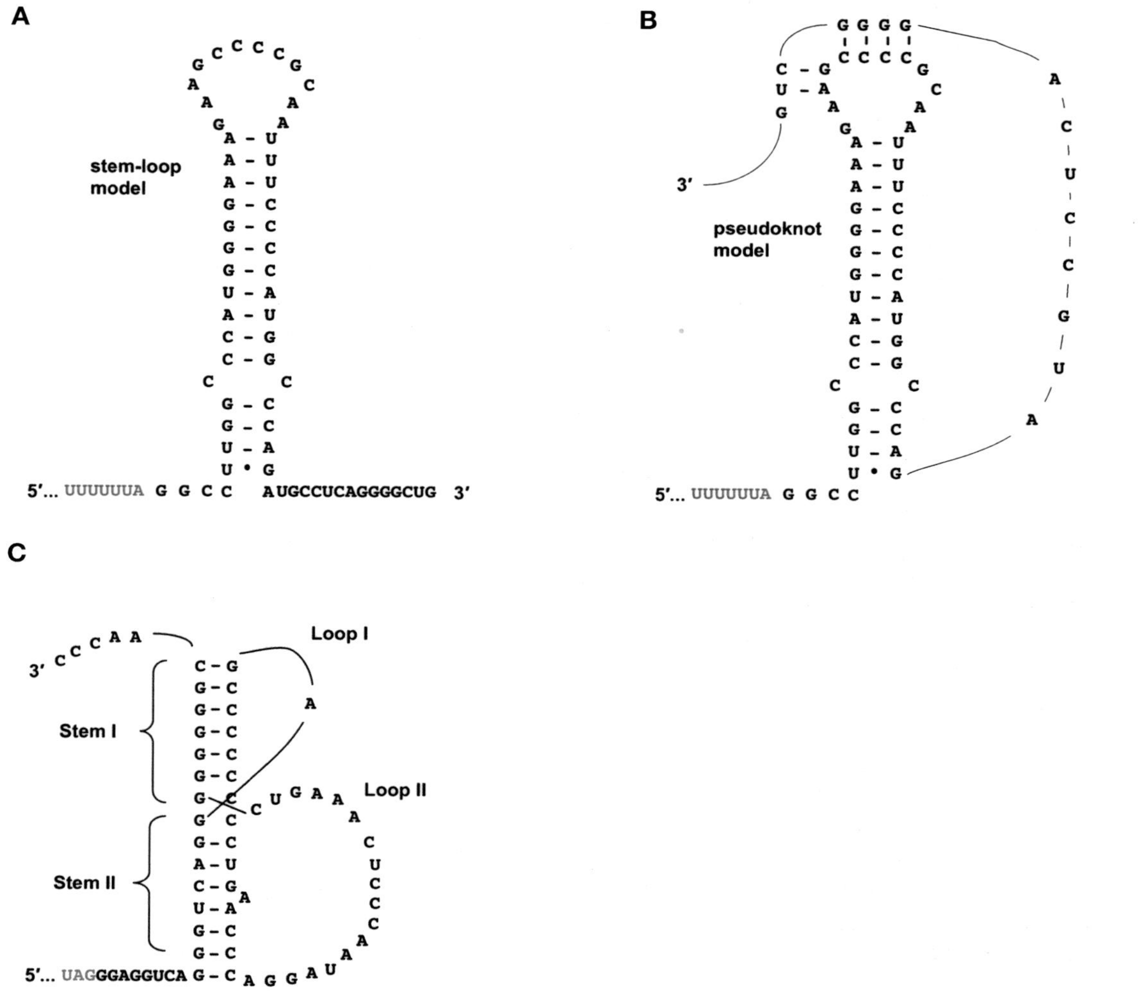

Figure 3.7 Proposed structures of retroviral sequences inducing programmed ribosomal frameshifting. The heptanucleotide slippery site is shown in red. (A) Stem–loop model for the element causing a shift from the *gag* reading frame into the *pro-pol* reading frame in SIV (Marcheschi *et al.*, 2007). (B) An alternative pseudoknot model proposed for the same sequence. (C) Predicted secondary structure responsible for the read-through of the *gag* UAG stop codon (shown in green) of MLV (Alam *et al.*, 1999).

in Pol processing and incorporation (Stenbak and Linial, 2004).

The retroviral protease is a homodimer (Fig. 3.8) of about 14 kDa that exhibits aspartyl protease activity (Li *et al.*, 2005). High resolution three-dimensional structures are known for the enzymes of several retroviruses, including avian sarcoma/leukosis viruses (ASLV), HIV-1, HIV-2, SIV, FIV and EIAV (Lapatto *et al.*, 1989; Miller *et al.*, 1989; Navia *et al.*, 1989; Weber *et al.*, 1989). The substrate is bound into pockets formed by both subunits (Wlodawer and Gustchina, 2000) and the fit into these pockets largely defines the specificity. The process of cleaving the Gag precursor proteins is strictly linked to assembly and early post-release stages of virus particles. Cleavage of the HIV Gag-Pro-Pol protein occurs sequentially at nine different sites. The cleavage rates can differ up to 400-fold, probably reflecting requirements for subsequent steps in virion assembly and maturation (Erickson-Viitanen *et al.*, 1989). The protease activation is not fully understood, but possible mechanisms include the decreased pH after virion release, the high

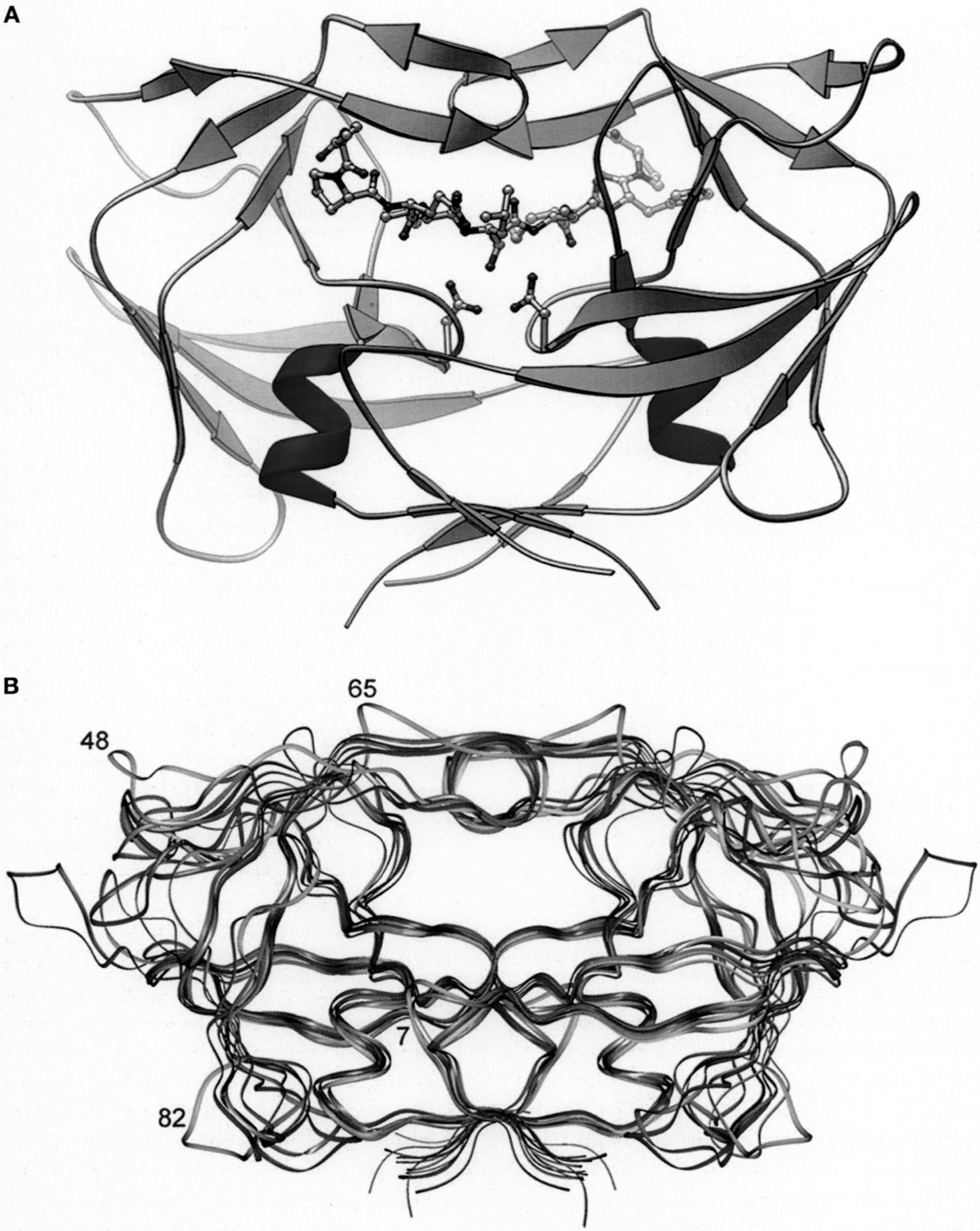

Figure 3.8 The folding structure of HTLV-1 protease (PR) with a bound substrate-based peptide inhibitor and a comparison with other retroviral PRs. (A) Overall view of a dimer. Helices are shown in red and β strands in pale green. The inhibitor and the catalytic aspartates are shown in stick representation. (B) Superposition of seven retroviral enzymes shown in ribbon representation. HTLV-1 PR is coloured blue; HIV-1 PR, green; HIV-2 PR, dark blue; SIV PR, grey; RSV PR, magenta; EIAV PR, yellow; and FIV PR, red. The numbers indicate residues within regions in HTLV-1 PR, with the most pronounced structural differences as compared with other retroviral enzymes. (Reproduced with permission from Li *et al.*, 2005).

concentration of protease molecules needed to achieve dimerization, or even conformational changes in the substrates that abrogate otherwise inhibitory domains.

Expression of the pol *gene*

In addition to the reading frame change at the *gag/pol* border, beta- and deltaretroviruses employ a further −1 frameshift (Fig. 3.4) at the 3′ region of *pro* for the translation of the Gag–Pro–Pol polyprotein. In MMTV, about 8% of the ribosomes shift a second time in the −1 direction to enter the *pol* open reading frame (Jacks *et al.*, 1987; Wang *et al.*, 2002). As a result, about 4 times less protease and 50 times less polymerase are synthesized compared to Gag. In alpharetroviruses the translation of the downstream polymerase also requires a −1 frameshift whereas in all other retroviral genera the *pro* and *pol* coding sequences are in the same open reading frame.

Mature orthoretroviral polymerase enzymes are released from the Gag-Pro-Pol precursor by proteolytic cleavage of the retroviral protease. The C-terminal integrase domain of the Pol polyprotein is also generally separated from the N-terminal reverse transcriptase domain by the action of its own protease.

The reverse transcriptase is an indispensable enzyme in the reproduction cycle of retroviruses. This protein contains an N-terminal polymerase domain and a C-terminal RNase H domain. Although enzymatic activities are present throughout the family, the structure and multimerization of the proteins differ between the genera. In gammaretroviruses for example, reverse transcriptase is a monomer with an N-terminal polymerase and a C-terminal RNase H function. *In vitro* these two domains can be expressed separately and each protein can perform the respective function (Tanese and Goff, 1988). In alpharetroviruses, reverse transcriptase is a heterodimer with a smaller α subunit responsible for polymerase and RNase activity and a larger β subunit with an additional integrase. In HIV-1, reverse transcriptase is also a heterodimer with a larger (p66) and a smaller (p51) subunit resulting from differential cleavage. In both cases, the smaller subunit does not contain the RNase H domain (Ren *et al.*, 1998). Because the reverse transcriptase molecule is an obvious target for antiviral drug development, a large number of studies have evaluated the three-dimensional structure of the HIV-1 enzyme by X-ray crystallography. Owing to dissimilar folding of the p66 and p51, the structure of the reverse transcriptase is asymmetric; illustrations use the form of a human right hand to describe the position of certain domains of the dimer (Chapter 5). The nucleic acid template runs between the fingers and the thumb, whereas the polymerase is located at the base of the palm, and the new double strand continues towards the wrist where the RNase H domain is located (Das and Georgiadis, 2004). The active site of the polymerase often contains a conserved 'YXDD' motif.

Retroviral reverse transcriptase enzymes are characterized by low processivity and low fidelity. The misincorporation rate is in the range of 10^{-3} to 10^{-4} errors per base, depending on the primer, template and type of assay. Unlike most host DNA polymerases, reverse transcriptase shows no proofreading nuclease activity (Battula and Loeb, 1976). However, studies with HIV-1 mutants resistant to the antiviral nucleoside analogue zidovudine (AZT) show a potential to remove the incorporated AZT and rescue a terminated chain for continued elongation (Meyer *et al.*, 1998). However, it should be noted that a high frequency of mutated HIV virions *in vivo* also results from the host RNA polymerase II which translates the new viral genomes.

The retroviral enzyme can perform strand displacement reactions but not nick-translation reactions. During reverse transcription the protein can shift between both RNA strands of a viral genome, thereby creating new mosaic genomes (copy-choice recombination). In HIV, double infections with particles belonging to different subgroups can result in new viruses, which are known as circulating recombinant forms (CRF).

Similar to the reverse transcriptase, the RNase H domain of the polymerase requires divalent cations for activity. The template must be a nucleic acid duplex – normally the RNA:DNA hybrid – but there is also evidence that HIV-1 and MLV enzymes can degrade RNA:RNA duplexes (Ben-Artzi *et al.*, 1992). The RNase H domain generally acts as a nuclease together with the reverse transcriptase, degrading the RNA template about 17–18 nt behind the growing 3′ end (Gopalakrishnan *et al.*, 1992).

The dUTPase

About 20 years after the demonstration of protease and polymerase another enzymatically active protein, a deoxyuridine triphosphate pyrophosphatase (dUTPase) that cleaves the alpha-beta phosphodiester of dUTP to form pyrophosphate and dUMP, was identified in betaretroviruses and non-primate lentiviruses, and later in various endogenous retroviruses (Koppe *et al.*, 1994; Mayer and Meese, 2002; McGeoch, 1990). The enzyme prevents detrimental uracil incorporation during reverse transcription (Elder *et al.*, 1992; Koppe *et al.*, 1994). Several studies showed that this activity enhances productive viral replication, particularly in non-dividing cells in which the lower dNTP levels delay DNA synthesis and support dUTP incorporation (Lichtenstein *et al.*, 1995; Payne and Elder, 2001; Steagall *et al.*, 1995).

In betaretroviruses and non-primate lentiviruses the dUTPase genes are located at different genomic locations (Baldo and McClure, 1999). In betaretroviruses the dUTPase coding region is situated between *gag* and *pro*, so translation requires the −1 frameshift mentioned earlier. In the case of MMTV, cleavage of the transframe Gag–Pro or Gag–Pro–Pol fusion proteins releases the mature p30 (30 kDa) dUTPase enzyme. This consists of the almost complete nucleocapsid protein plus a 154 amino acid extension from the *pro* cistron enclosing the enzymatically active site (Bergman *et al.*, 1994; Koppe *et al.*, 1994). The nucleocapsid RNA binding domains (zinc knuckle motifs) might serve to tether the homotrimeric dUTPase to the genomic RNA in an optimal position for preventing (by hydrolysing dUTP and maintaining a low level of dUTP/dTTP) uracil incorporation into the DNA strands during reverse transcription (Barabas *et al.*, 2006; Payne and Elder, 2001).

In non-primate lentiviruses the dUTPase coded by the *dut* open reading frame is part of the polymerase and is located between reverse transcriptase and integrase. The mature active enzyme is released by processing of the Gag-Pro-Pol polyprotein. Interestingly, the lentiviral dUTPase also forms homotrimers, whereas protease and reverse transcriptase are dimers (Prasad *et al.*, 1996).

Genes and proteins of the central region of retroviral genomes

Downstream of the *pol* reading frame, lenti- and deltaretroviruses encode several auxiliary proteins in the central part of their genomes (or at least their first exons). In lentiviruses the entire ORFs of a number of accessory proteins are located in this region (Fig. 3.4). These are translated from singly spliced subgenomic mRNAs created using the major splice donor site and appropriate splice acceptor sites upstream of the respective translational start codon.

Of these proteins, the viral infectivity factor (Vif) is one of the most prominent. With the exception of EIAV, all lentiviruses carry a gene coding for a 20- to 30-kDa Vif protein. It is the open reading frame located furthest upstream and usually partially overlaps with *pol*. Vif is packaged into retroviral particles by association with the viral RNA and a component of the nucleocapsid complex. Its antiviral activity occurs through binding of cytosine deaminases and induction of protein degradation (Bishop *et al.*, 2004); which will be discussed in more detail in Chapter 11.

In addition to *vif*, the central region of primate lentiviruses contains the *vpr, vpu and vpx* genes, although *vpx* is only present in HIV-2/SIVmac/SIVmn and *vpu* is unique to HIV-1 and HIVcpz (Fig. 3.9).

Vpr (12–14 kDa) is translated from a spliced RNA that overlaps with the *vif* and *tat* genes in HIV-1. The protein is predominantly expressed at late stages of the viral life cycle and is found in association with the p6 Gag protein in the viral core. The Vpr protein has multiple functions, including induction of transport of the viral pre-integration complex to the nucleus, stimulation of the expression of viral genes by transactivation of the long terminal repeat, induction of apoptosis, and arrest of infected cells in the G2-M phase of the cell cycle (Moon and Yang, 2006).

Vpu is a membrane phosphoprotein of about 16 kDa that is translated from an alternative reading frame of the bicistronic mRNA encoding the envelope protein. A weak ATG initiation codon upstream of the *env* gene is used to start translation of *vpu* and the protein is expressed late in the

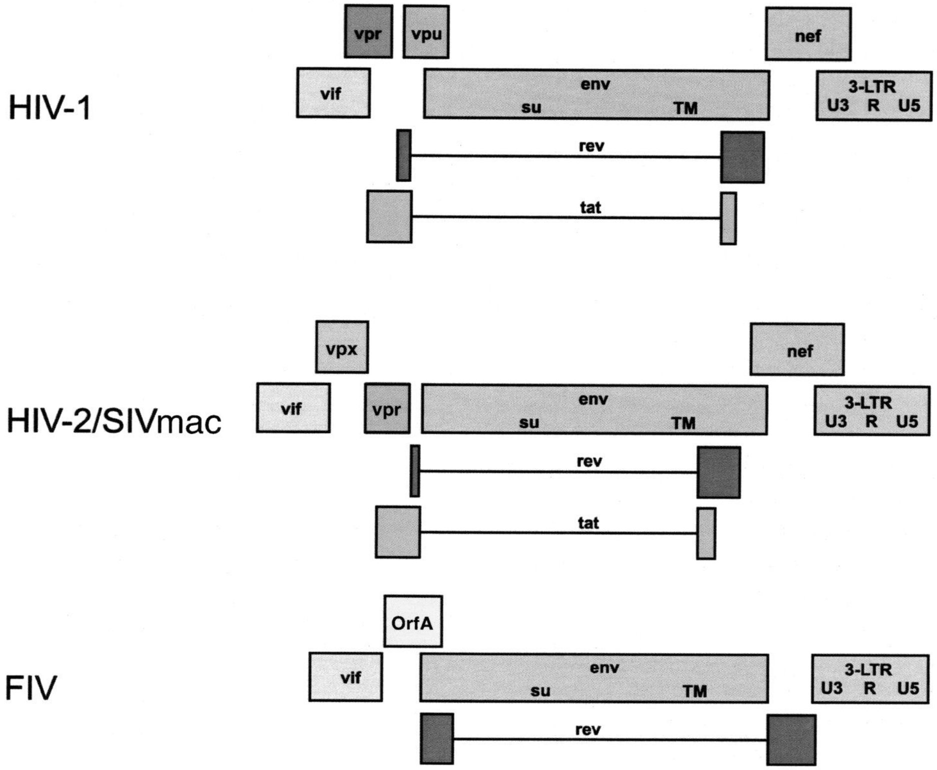

Figure 3.9 Organization of coding sequences in the downstream half of lentiviral genomes.

virus production cycle. Vpu increases virus replication through formation of membrane pores in the lipid bilayers. A Vpu-mediated downregulation of CD4 receptor expression by degradation of newly synthesized CD4 in the rough endoplasmic reticulum (resulting in an elevated release of enveloped particles) was also reported (Wildum *et al.*, 2006; Willey *et al.*, 1992). Recently, a role of Vpu in the detachment of viral particles from the producer cell line by antagonism of the adhesion to tetherin (CD317), was reported (Neil *et al.*, 2008).

Additional factors encoded in the central region are the inadequately investigated *vpw* and *vpy* genes of the bovine lentivirus group, *orf A* of FIV (which seems to have a weak transactivator function), and the *S2* protein of EIAV. Transcription of *S2* presumably occurs by ribosomal leaky scanning of a tricistronic mRNA message encoding Tat, S2, and Env or from a bicistronic transcript encoding S2 and Env (Fagerness *et al.*, 2006; Molina *et al.*, 2002; Schiltz *et al.*, 1992).

The transactivator (*tat*) gene of the equine and ovine/caprine lentiviral groups and the *rev* gene of EIAV are translated from a singly spliced mRNA and are also completely encoded in the central region. The Tat and Rev proteins in all other lentiviruses, as well as the deltaretrovirus homologues Tax and Rex, are translated from doubly spliced mRNAs in which a second splicing event removes most of the envelope coding sequence as an intron and the second exon is positioned at the 3′ end of (or behind) the *env* open reading frame. The Tat proteins of primate lentiviruses are small nucleoproteins that contain zinc finger-like motifs. The 37- to 40-kDa Tax proteins of HTLV are also found in the nucleus of infected cells where they drive transcription from the LTR promoter and promote expression of a number of cellular genes.

The single primary transcript of most retroviruses directs synthesis of many viral proteins. The coordination and regulation of splicing is a key requirement for balanced protein synthesis, genome encapsidation and efficient replication. Retroviruses make use of cellular transport pathways, allowing nuclear export of intron-containing RNAs that are normally restricted to the nucleus (Malim *et al.*, 1990; Pasquinelli *et al.*, 1997). This enhanced transport is mediated by highly structured *cis*-acting sequences of about 200 nt that are present in the *env* open reading frame, the adjacent 3′UTR or in the 3′LTR (Bray *et al.*, 1994; Itoh *et al.*, 1989; Magin-Lachmann *et al.*, 2001; Yang *et al.*, 1999).

In complex retroviruses these *cis*-active sites interact with a virus encoded accessory protein. Of these, the Rev protein of HIV-1 that binds to the Rev responsive element (RRE) in the *env* gene of HIV, and the Rex-1 protein of HTLV-1 that interacts with the RxRE element in the 3′LTR of the virus, were studied most extensively. Both interactions promote export of unspliced or incompletely spliced transcripts (Hidaka *et al.*, 1988; Zapp and Green, 1989). The 13- to 27-kDa Rev and Rex proteins have nuclear localization and export signals that enable them to shuttle between the cytoplasm and the nucleus.

The Rev/RRE system of HIV-1 also underlies a temporal switch from the early phase of regulatory protein expression of completely spliced RRE-less transcripts (e.g. Tat, Rev, Nef) to the late phase of virion production. The later phase requires a higher ratio of structural proteins and unspliced genomic RNA for encapsidation. In addition to lenti- and deltaretroviruses, the human endogenous retrovirus-K (HERV-K), a typical betaretrovirus, also regulates its nuclear RNA export through the related Rec/RcRE system (Magin-Lachmann *et al.*, 2001). Alongside the viral proteins mentioned, key cellular factors of nuclear export are also recruited to the target sequence. The protein complex guides the viral RNA into the CRM1-dependent pathway normally used for the export of cellular proteins and spleisosomal or ribosomal RNAs. With this mechanism it bypasses restrictions imposed by splice site determinants and *cis*-active CRS/INS inhibitory elements that were postulated in the structural genes (Gruter *et al.*, 1998; Schneider *et al.*, 1997).

In simple retroviruses such as the betaretrovirus Mason Pfizer monkey virus (MPMV), the functionally and spatially related constitutive transport element (CTE) avoids retention of intron-containing transcripts. The CTE does not make use of the Crm1-RanGTP pathway, but directly recruits the cellular mRNA export receptor transactivation response (TAP) and its cofactor Nxt to the viral RNA (Cullen, 2003; Gruter *et al.*, 1998).

The two post-transcriptional regulatory systems (Rev/RRE and CTE) reveal retroviruses' efficient use of cellular transport pathways to overcome inherent genomic obstacles, thereby maximizing production of infectious progeny virus.

Expression of the env *gene*

The envelope protein of all retroviruses is translated from a spliced transcript. Typically, almost the entire *gag-pro-pol* sequences of these subgenomic RNAs are spliced out. The envelope glycoprotein mediates viral attachment and subsequent entry into target cells by binding to receptors on the cell surface. Env is the only viral protein localized on the surface of infected cells and particles, and is the main determinant of viral tropism and the main target for neutralizing antibodies. It is a class I protein (cytoplasmic C-terminus), and is structurally and functionally homologous to similar proteins of other viruses (e.g. filoviruses, paramyxoviruses or coronaviruses). The native precursor undergoes co- and posttranslational modifications in the endoplasmic reticulum and in the Golgi complex, including extensive glycosylation, formation of disulphide bonds, oligomerization, and cleavage into two functionally distinct domains: an N-terminal surface subunit (SU) and a C-terminal transmembrane subunit (TM). This cleavage, usually mediated by a cellular furin protease, is essential for fusiogenic activity and thus for virus infectivity. The SU region of HIV-1 consists of five variable regions (V1 to V5), and five conserved regions (C1 to C5) that form an inner domain, an outer domain and a bridging sheet (Huang *et al.*, 2005; Kwong *et al.*, 1998; Wyatt *et al.*, 1998). Transmembrane and surface units remain non-covalently associated and form trimers. The receptor binding domain is located in the surface subunit and minor changes in this region can result in the use of new or alternative receptors; glycosylation and oligomerization frequently mask the receptor binding site, making access difficult for neutralizing antibodies.

Interaction with the specific receptor induces conformational changes that initiate fusion between the viral and cellular membrane (Chapter 4).

The TM subunit contains an N-terminal hydrophobic fusion peptide of approximately 20 amino acids, which is inserted into the target cell membrane as an early event in the membrane fusion process. All retroviruses contain two leucine-zipper-like regions (heptad repeats) C-terminal to the fusion peptide, which are capable of forming a coiled-coil structure. This structure is a triple-stranded coiled coil, with each strand formed by an N-terminal heptad repeat (HR1) from one of the three TM subunits. The second heptad repeat (HR2) is located behind a central disulphide-bonded loop. During the fusion process, both heptad repeats interact by forming a stable six helix bundle (Chapter 4).

Regulatory genes at the 3′ end of the genomes

Several of the complex retroviruses encode additional accessory proteins in the region downstream of the *env* gene. The Nef (negative factor) protein of primate lentiviruses is one of the most fascinating retroviral proteins. The HIV-1 *nef* gene overlaps partially with the 3′ end of the *env* gene and the 3′LTR. Like Tat and Rev it is synthesized from a doubly spliced transcript without the RRE, but in contrast to these two proteins, its complete coding sequence is in the second exon. Nef is produced at high levels shortly after infection and is post-translationally modified by phosphorylation and by the attachment of myristic acid to its N-terminus, which ensures a tight association with the cellular membrane. The 10- to 34-kDa lentiviral Nef proteins are also incorporated at low levels in viral particles. The protein fulfils a plethora of functions important for efficient replication *in vivo*, which has implications for the development of AIDS in humans and other non-natural hosts (Guy *et al.*, 1987; Schindler *et al.*, 2006). For example, the HIV-1 Nef protein promotes viral infection and the development of AIDS by activating T-cells (Schindler *et al.*, 2006). Nef proteins can also downregulate the CD4-receptor on the cell surface, thereby preventing interference with the budding process and virus spread (Garcia and Miller, 1991). To accomplish this, Nef links CD4 to components of the clathrin-dependent trafficking pathways at the plasma membrane, which induces internalization and delivery of CD4 to lysosomes for degradation (Lindwasser *et al.*, 2007; Piguet *et al.*, 1998). Moreover, Nef protects infected cells from cytotoxic T-lymphocytes by reducing the expression of MHC-I on the cell surface (Schwartz *et al.*, 1996), and Nef binding to p53 blocks apoptosis (Greenway *et al.*, 2002).

Epsilonretroviruses also encode accessory proteins downstream of their *env* open reading frame. WDSV has two genes in this region, *orf a* and *orf b* (Holzschu *et al.*, 1995); *orf b* appears to be the product of an *orf a* gene duplication (LaPierre *et al.*, 1998). Transcripts encoding these proteins result from a pattern of alternative splice site selection, and six alternative transcripts are known to encode variant forms of the Orf A protein (Quackenbush *et al.*, 1997). These proteins show some homology to cyclins A and D, which play important roles in cell cycle progression and are therefore termed retroviral cyclins (Holzschu *et al.*, 1995; LaPierre *et al.*, 1998; Zhang *et al.*, 1999). They are suspected to be involved in the development and regression of seasonal hyperplastic dermal lesions in walleye fish and in the regulation of viral gene expression (Holzschu *et al.*, 2003; Rovnak *et al.*, 2001). The tumorigenic potential of retroviral cyclins also extends to mammalian cells, since Orf a can induce skin lesions in transgenic mice (Lairmore *et al.*, 2000).

The Sag superantigen is a protein encoded in a region overlapping the *env* and the 3′LTR of exogenous and endogenous MMTV (Choi *et al.*, 1991; Marrack *et al.*, 1991; Woodl, *et al.*, 1991) but is absent from other members of the betaretrovirus genus. The protein is expressed on the surface of B-cells and antigen-presenting cells, and activates T-cells by interaction with specific T-cell receptor β-chains (Acha-Orbea and MacDonald, 1995; Pullen *et al.*, 1990). The resulting T-cell activation in turn stimulates the infected B-cells to proliferate, thereby amplifying the number of virus-infected cells (Held *et al.*, 1993). Expression of superantigen early in life leads to clonal deletion of responsive T-cells in the thymus (Simpson *et al.*, 1993). In activated lymphocytes Sag expression is regulated by internal enhancer elements (Elliott *et al.*, 1988; Miller *et al.*, 1992; Reuss and Coffin, 1998).

Spumaviruses also have an additional promoter/enhancer activity (Lochelt *et al.*, 1993): a sequence within the *env* region directs expression of two accessory proteins, Tas (transactivator of spumaviruses, previously called Bel-1 for between *env* and LTR) and Bet. Soon after infection, transcription from the internal promoter leads to translation of the Tas protein. The DNA-binding transactivator protein enhances its own production from the internal promoter and also acts on the prototypical 5′LTR promoter, for which the affinity is lower and whose activation requires higher Tas concentrations. The LTR-driven expression of structural genes reflects the switching from the early to the late phase of infection. Activation of the internal promoter also induces transcription of the Bet protein, which shares its first 88 amino acids with Tas. Although the Bet protein is efficiently expressed, its functional significance is only partially understood and seems to be extremely complex and cell type dependent. In contrast to Tas, Bet appears to support the establishment and maintenance of viral persistence *in vivo*. The protein is efficiently released from infected cells and taken up by surrounding naïve cells where it enters the nucleus (Lecellier *et al.*, 2002). It was postulated that Bet mediates superinfection resistance by interfering with an early post entry step (Bock *et al.*, 1998) and inactivates cytidine deaminases of the APOBEC3 family (Lochelt *et al.*, 2005).

Downstream non-coding regions

Regulatory elements in this region of the retroviral genome are predominantly involved in processes of reverse transcription and termination of the transcripts at the polyadenylation signal. All retroviral transcripts are polyadenylated, regardless of whether they are destined for protein synthesis or encapsidation.

The polypurine track and regulatory elements in the 3′UTR

Successful reverse transcription of the retroviral RNA requires primers to initiate minus and plus-strand DNA synthesis. As mentioned earlier in this chapter, a cellular tRNA bound to the primer binding site (PBS) downstream of the 5′LTR initiates minus-strand DNA synthesis in all retroviruses. At a later stage, 10–30 min after initiation of reverse transcription, an internal RNase H resistant region consisting of about 10–20 purines, annealed to the nascent minus-strand DNA, primes the plus-strand DNA synthesis (Boone and Skalka, 1981; Myers *et al.*, 1980). The polypurine tract (PPT) is located just upstream of the 3′LTR. Considerable evidence indicates that immediately after minus-strand synthesis has passed beyond the PPT, a second reverse transcriptase enzyme binds the nascent DNA-RNA hybrid and its RNase H activity cleaves at the 3′ terminus of the PPT (Huber and Richardson, 1990; Luo *et al.*, 1990; Luo and Taylor, 1990; Rausch and Le Grice, 2004; Wohrl and Moelling, 1990). The precise position of this cleavage is critical since it determines the start of the plus-strand synthesis which defines the 5′ end of the viral DNA. The typical PPT consists of two motifs with different aspects of PPT function: an upstream A-tract and a downstream G-tract. The A-tract probably plays a role in correct positioning of the reverse transcriptase for cleavage at the PPT-U3 junction (McWilliams *et al.*, 2003), whereas the G-tract enhances extension of the nascent primer leading to the synthesis of the plus-strand strong stop DNA (Powell and Levin, 1996). This PPT is immediately preceded by a U-rich sequence common to most retroviruses that may be involved in the selection and removal of the plus-strand primer (Bacharach *et al.*, 2000).

Certain retroviruses, including HIV-1, FIV, EIAV and SFV-1 (Charneau *et al.*, 1992; Charneau and Clavel, 1991; Stetor *et al.*, 1999; Whitwam *et al.*, 2001), contain a perfect copy of the PPT upstream in a central region of the genome. In HIV-1 the central PPT (cPPT), which resides in the integrase gene, is used as an additional primer site for plus-strand DNA synthesis. Its functionality is required for efficient replication (Hungnes *et al.*, 1992). The resulting plus-strand discontinuity stops the polymerase at the central termination sequence (CTS), a specific position approximately 90 nt behind the cPPT, within a duplex characterized by A-tracts. (Charneau *et al.*, 1994). Therefore the outcome of the plus-strand synthesis is a final product with a 90 nt overlap or ‘flap’ in the middle of the genome. The reason for this flap is not fully understood but it was suggested that it enhances flexibility in the centre of the DNA, thereby simplifying passage of the preintegration complex through the pore complex (Zennou *et al.*, 2000). The flap is likely to be resolved and repaired by a host encoded flap endonuclease 1 and ligase activity (Rumbaugh *et al.*, 1998).

3′ Long terminal repeat

Although identical in sequence, the two LTRs have many different functions in the life cycle of a retrovirus. The most prominent disparity is the initiation of transcription in the upstream repeat and its termination in the downstream repeat. The latter process relies on *cis*-active elements, structural determinants and host cell factors. As for almost all eukaryotic mRNAs, mature retroviral transcripts end with a poly(A) tail of 200–300 nucleotides that are not encoded by the chromosomal DNA (Guntaka, 1993). The adenine tail serves to stabilize the mRNA and influence transport, splicing and translation (Colgan and Manley, 1997; Scott and Imperiale, 1996). It usually starts about 15–25 nt downstream of a conserved hexameric AAUAAA or AGUAAA poly(A) signal. One exception is HTLV-1 and other deltaretroviruses, in which the poly(A) signal and poly(A) tail are separated by approximately 260 nt. The poly(A) sequence is synthesized by the poly(A) tail polymerase after cleavage of the primary transcript (pre-mRNA) that extends well beyond the consensus hexanucleotide located either in the U3 or in the R region of the LTR (Guntaka, 1993). The poly(A) signal site is bound by the cleavage polyadenylation specificity factor (CPSF) (Jenny and Keller, 1995; Keller, 1995). An additional regulatory region known as the upstream enhancer element (USE), located in the U3 region, is essential for this interaction and efficient polyadenylation (Valsamakis *et al.*, 1991). It was proposed that this element initially interacts with the CPSF and functions as an entry site for CPSF (Klasens *et al.*, 1999). Another well-characterized *cis*-active RNA element that controls termination is a GU-rich or U-rich sequence recognized by the cleavage stimulation factor (CStF). Together with CPSF and other cellular factors, it guides cleavage and polyadenylation (Bohnlein *et al.*, 1989; Colgan and Manley, 1997). The GU- or U-rich element is located in the U5 region, usually 1–40 nt downstream of the poly(A) site. In HTLV and BLV, the AAUAAA motif and the GU-rich element are more than 250 nt apart, explaining the long distance between the poly(A) signal and the poly(A) site mentioned earlier. However, formation of a secondary structure seems to juxtapose both motifs, allowing formation of the termination machinery and polyadenylation at this remote site (Bar-Shira *et al.*, 1991; Bardwell *et al.*, 1991).

In most retroviruses the polyadenylation signal lies within the R region of the LTR. Consequently, it is present twice in the transcript and a premature termination at the upstream poly(A) signal in the 5′LTR must be prevented. It is not completely clear how this is achieved. The missing upstream enhancer element and the proximity of the polyadenylation signal to the close terminal cap structure and to the downstream major splice donor site were reported to be critical parameters in this respect (Ashe *et al.*, 1997; Cherrington and Ganem, 1992; Klasens *et al.*, 1999). However, structural studies revealed a dominant occlusion of the hexamer motif in a hairpin structure and tertiary interactions that form a long-distance pseudoknot or branched multiple hairpin structure (Gee *et al.*, 2006; Paillart *et al.*, 2004a,b; Paillart *et al.*, 2002). *In vitro* studies showed that destabilization of the hairpin structures enables premature polyadenylation in this region (Das *et al.*, 1997). On the other hand, the U5 region of the 3′LTR is presumably a dominant linear and more flexible structure with a large asymmetric loop amenable to the binding of CPSF and other factors (Gee *et al.*, 2006). Local structural characteristics of the nascent transcript are obviously therefore major determinants that prevent recognition of the polyadenylation signal in the 5′LTR, but trigger it in the 3′LTR.

Conclusion

Retroviruses have been the most intensively studied family of viruses over the past two decades. The lability of the particles and certain features of their proteins make structural analysis a challenging task. Recent advances in crystallographic techniques, NMR and cryo-tomography have helped to unravel many of the structural details of entire particles, their proteins and their maturation. Although crucial information about some aspects of the envelope protein architecture (e.g. the membrane proximal region and its dynamics during the fusion process) is urgently needed, impressive insights have been achieved in this field. High resolution structural data for retroviral proteins have enabled and complemented the development of new viral inhibitors and antiviral drugs.

Moreover, the analysis of endogenous proviruses provides a glimpse into ancient retroviral precursors and the strategies that have led to the successful evolution of retroviruses over millions of years. The resulting insights provide invaluable information with regard to modern viral diseases. Understanding the general concepts of retroviral behaviour and the many alternative strategies used by subgroups of retroviruses aids in the identification of potential targets for interference with retroviral replication, and, at a more general level, teaches us about cell biology. In addition, insights into genome structure and the genetic mechanisms of retroviruses have enabled development of approaches for the transfer and integration of therapeutic genes into human chromosomes (Chapter 13). Continued efforts to further our understanding of the structural and regulatory elements of retroviruses provide the basis for a deeper understanding of retrovirus biology and for the development of new and finely tuned strategies for interfering with their often devastating effects.

References

Accola, M.A., Hoglund, S., and Gottlinger, H.G. (1998). A putative alpha-helical structure which overlaps the capsid-p2 boundary in the human immunodeficiency virus type 1 Gag precursor is crucial for viral particle assembly. J. Virol. *72*, 2072–2078.

Acha-Orbea, H., and MacDonald, H.R. (1995). Superantigens of mouse mammary tumor virus. Annu. Rev. Immunol. *13*, 459–486.

Adam, M.A., and Miller, A.D. (1988). Identification of a signal in a murine retrovirus that is sufficient for packaging of nonretroviral RNA into virions. J. Virol. *62*, 3802–3806.

Alam, S.L., Wills, N.M., Ingram, J.A., Atkins, J.F., and Gesteland, R.F. (1999). Structural studies of the RNA pseudoknot required for readthrough of the *gag*-termination codon of murine leukemia virus. J. Mol. Biol. *288*, 837–852.

Amit, I., Yakir, L., Katz, M., Zwang, Y., Marmor, M.D., Citri, A., Shtiegman, K., Alroy, I., Tuvia, S., Reiss, Y., Roubini, E., Cohen, M., Wides, R., Bacharach, E., Schubert, U., and Yarden, Y. (2004). Tal, a Tsg101-specific E3 ubiquitin ligase, regulates receptor endocytosis and retrovirus budding. Genes Dev. *18*, 1737–1752.

Arthur, L.O., Bess, J.W., Jr., Sowder, R.C., 2nd, Benveniste, R.E., Mann, D.L., Chermann, J.C., and Henderson, L.E. (1992). Cellular proteins bound to immunodeficiency viruses: implications for pathogenesis and vaccines. Science *258*, 1935–1938.

Ashe, M.P., Pearson, L.H., and Proudfoot, N.J. (1997). The HIV-1, 5′-LTR poly(A) site is inactivated by U1 snRNP interaction with the downstream major splice donor site. EMBO J. *16*, 5752–5763.

Attal, J., Theron, M.C., Taboit, F., Cajero-Juarez, M., Kann, G., Bolifraud, P., and Houdebine, L.M. (1996). The RU5 ('R') region from human leukaemia viruses (HTLV-1) contains an internal ribosome entry site (IRES)-like sequence. FEBS Lett. *392*, 220–224.

Bacharach, E., Gonsky, J., Lim, D., and Goff, S.P. (2000). Deletion of a short, untranslated region adjacent to the polypurine tract in Moloney murine leukemia virus leads to formation of aberrant 5′ plus-strand DNA ends *in vivo*. J. Virol. *74*, 4755–4764.

Baldo, A.M., and McClure, M.A. (1999). Evolution and horizontal transfer of dUTPase-encoding genes in viruses and their hosts. J. Virol. *73*, 7710–7721.

Banks, J.D., Kealoha, B.O., and Linial, M.L. (1999). An Mpsi-containing heterologous RNA, but not *env* mRNA, is efficiently packaged into avian retroviral particles. J. Virol. *73*, 8926–8933.

Bantignies, F., Rousset, R., Desbois, C., and Jalinot, P. (1996). Genetic characterization of transactivation of the human T-cell leukemia virus type 1 promoter: Binding of Tax to Tax-responsive element 1 is mediated by the cyclic AMP-responsive members of the CREB/ATF family of transcription factors. Mol. Cell. Biol. *16*, 2174–2182.

Bar-Shira, A., Panet, A., and Honigman, A. (1991). An RNA secondary structure juxtaposes two remote genetic signals for human T-cell leukemia virus type I RNA 3′-end processing. J. Virol. *65*, 5165–5173.

Barabas, O., Nemeth, V., and Vertessy, B.G. (2006). Crystallization and preliminary X-ray studies of dUTPase from Mason–Pfizer monkey retrovirus. Acta Crystallogr. *62*, 399–401.

Barat, C., Martin, G., Beaudoin, A.R., Sevigny, J., and Tremblay, M.J. (2007). The nucleoside triphosphate diphosphohydrolase-1/CD39 is incorporated into human immunodeficiency type 1 particles, where it remains biologically active. J. Mol. Biol. *371*, 269–282.

Bardwell, V.J., Wickens, M., Bienroth, S., Keller, W., Sproat, B.S., and Lamond, A.I. (1991). Site-directed ribose methylation identifies 2′-OH groups in polyadenylation substrates critical for AAUAAA recognition and poly(A) addition. Cell *65*, 125–133.

Battula, N., and Loeb, L.A. (1976). On the fidelity of DNA replication. Lack of exodeoxyribonuclease activity and error-correcting function in avian myeloblastosis virus DNA polymerase. J. Biol. Chem. *251*, 982–986.

Beemon, K., Duesberg, P., and Vogt, P. (1974). Evidence for crossing-over between avian tumor viruses based on analysis of viral RNAs. Proc. Natl. Acad. Sci. U.S.A. *71*, 4254–4258.

Beemon, K.L., Faras, A.J., Hasse, A.T., Duesberg, P.H., and Maisel, J.E. (1976). Genomic complexities of murine leukemia and sarcoma, reticuloendotheliosis, and visna viruses. J. Virol. *17*, 525–537.

Belyaev, A.S., Stuart, D., Sutton, G., and Roy, P. (1994). Crystallization and preliminary X-ray investigation of recombinant simian immunodeficiency virus matrix protein. J. Mol. Biol. *241*, 744–746.

Ben-Artzi, H., Zeelon, E., Gorecki, M., and Panet, A. (1992). Double-stranded RNA-dependent RNase activity associated with human immunodeficiency virus type 1 reverse transcriptase. Proc. Natl. Acad. Sci. U.S.A. *89*, 927–931.

Bender, M.A., Palmer, T.D., Gelinas, R.E., and Miller, A.D. (1987). Evidence that the packaging signal of Moloney murine leukemia virus extends into the *gag* region. J. Virol. *61*, 1639–1646.

Bender, W., Chien, Y.H., Chattopadhyay, S., Vogt, P.K., Gardner, M.B., and Davidson, N. (1978). High-molecular-weight RNAs of AKR, NZB, and wild mouse viruses and avian reticuloendotheliosis virus all have similar dimer structures. J. Virol. *25*, 888–896.

Bergman, A.C., Bjornberg, O., Nord, J., Nyman, P.O., and Rosengren, A.M. (1994). The protein p30, encoded at the Gag–Pro junction of mouse mammary tumor virus, is a dUTPase fused with a nucleocapsid protein. Virology *204*, 420–424.

Berlioz, C., and Darlix, J.L. (1995). An internal ribosomal entry mechanism promotes translation of murine leukemia virus *gag* polyprotein precursors. J. Virol. *69*, 2214–2222.

Bernhard, W. (1958). Electron microscopy of tumor cells and tumor viruses; a review. Cancer Res. *18*, 491–509.

Bernhard, W. (1960). The detection and study of tumor viruses with the electron microscope. Cancer Res. *20*, 712–727.

Berthoux, L., Sebastian, S., Sokolskaja, E., and Luban, J. (2005). Cyclophilin A is required for TRIM5{alpha}-mediated resistance to HIV-1 in Old World monkey cells. Proc. Natl. Acad. Sci. U.S.A. *102*, 14849–14853.

Bishop, K.N., Holmes, R.K., Sheehy, A.M., and Malim, M.H. (2004). APOBEC-mediated editing of viral RNA. Science *305*, 645.

Bishop, N., Woodman, P. (2001). TSG101/mammalian VPS23 and mammalian VPS28 interact directly and are recruited to VPS4-induced endosomes. J. Biol. Chem. *276*, 11735–11742.

Blacklow, S.C., Lu, M., and Kim, P.S. (1995). A trimeric subdomain of the simian immunodeficiency virus envelope glycoprotein. Biochemistry *34*, 14955–14962.

Bock, M., Heinkelein, M., Lindemann, D., and Rethwilm, A. (1998). Cells expressing the human foamy virus (HFV) accessory Bet protein are resistant to productive HFV superinfection. Virology *250*, 194–204.

Bohnlein, S., Hauber, J., and Cullen, B.R. (1989). Identification of a U5-specific sequence required for efficient polyadenylation within the human immunodeficiency virus long terminal repeat. J. Virol. *63*, 421–424.

Bonnet-Mathoniere, B., Girard, P.M., Muriaux, D., and Paoletti, J. (1996). Nucleocapsid protein 10 activates dimerization of the RNA of Moloney murine leukaemia virus *in vitro*. Eur. J. Biochem./FEBS *238*, 129–135.

Boone, L.R., and Skalka, A.M. (1981). Viral DNA synthesized *in vitro* by avian retrovirus particles permeabilized with melittin. II. Evidence for a strand displacement mechanism in plus-strand synthesis. J. Virol. *37*, 117–126.

Bounou, S., Dumais, N., and Tremblay, M.J. (2001). Attachment of human immunodeficiency virus-1 (HIV-1) particles bearing host-encoded B7-2 proteins leads to nuclear factor-kappa B- and nuclear factor of activated T-cells-dependent activation of HIV-1 long terminal repeat transcription. J. Biol. Chem. *276*, 6359–6369.

Bounou, S., Leclerc, J.E., and Tremblay, M.J. (2002). Presence of host ICAM-1 in laboratory and clinical strains of human immunodeficiency virus type 1 increases virus infectivity and CD4(+)-T-cell depletion in human lymphoid tissue, a major site of replication *in vivo*. J. Virol. *76*, 1004–1014.

Braaten, D., Franke, E.K., and Luban, J. (1996). Cyclophilin A is required for an early step in the life cycle of human immunodeficiency virus type 1 before the initiation of reverse transcription. J. Virol. *70*, 3551–3560.

Brasey, A., Lopez-Lastra, M., Ohlmann, T., Beerens, N., Berkhout, B., Darlix, J.L., and Sonenberg, N. (2003). The leader of human immunodeficiency virus type 1 genomic RNA harbors an internal ribosome entry segment that is active during the G2/M phase of the cell cycle. J. Virol. *77*, 3939–3949.

Bray, M., Prasad, S., Dubay, J.W., Hunter, E., Jeang, K.T., Rekosh, D., and Hammarskjold, M.L. (1994). A small element from the Mason–Pfizer monkey virus genome makes human immunodeficiency virus type 1 expression and replication Rev-independent. Proc. Natl. Acad. Sci. U.S.A. *91*, 1256–1260.

Briggs, C.J., Ott, D.E., Coren, L.V., Oroszlan, S., and Tözsér, J. (1999). Comparison of the effect of FK506 and cyclosporin A on virus production in H9 cells chronically and newly infected by HIV-1. Arch. Virol. *144*, 2151–2160.

Briggs, J.A., Simon, M.N., Gross, I., Krausslich, H.G., Fuller, S.D., Vogt, V.M., and Johnson, M.C. (2004). The stoichiometry of Gag protein in HIV-1. Nature Struct. Mol. Biol. *11*, 672–675.

Briggs, J.A., Wilk, T., Welker, R., Krausslich, H.G., and Fuller, S.D. (2003). Structural organization of authentic, mature HIV-1 virions and cores. EMBO J. *22*, 1707–1715.

Browning, M.T., Mustafa, F., Schmidt, R.D., Lew, K.A., and Rizvi, T.A. (2003). Delineation of sequences important for efficient packaging of feline immunodeficiency virus RNA. J. Gen. Virol. *84*, 621–627.

Buck, C.B., Shen, X., Egan, M.A., Pierson, T.C., Walker, C.M., and Siliciano, R.F. (2001). The human immunodeficiency virus type 1 *gag* gene encodes an internal ribosome entry site. J. Virol. *75*, 181–191.

Buxton, P., Tachedjian, G., and Mak, J. (2005). Analysis of the contribution of reverse transcriptase and integrase proteins to retroviral RNA dimer conformation. J. Virol. *79*, 6338–6348.

Caffrey, M., Cai, M., Kaufman, J., Stahl, S.J., Wingfield, P.T., Covell, D.G., Gronenborn, A.M., and Clore, G.M. (1998). Three-dimensional solution structure of the 44 kDa ectodomain of SIV gp41. EMBO J. *17*, 4572–4584.

Cantin, R., Fortin, J.F., Lamontagne, G., and Tremblay, M. (1997a). The acquisition of host-derived major histocompatibility complex class II glycoproteins by human immunodeficiency virus type 1 accelerates the process of virus entry and infection in human T-lymphoid cells. Blood *90*, 1091–1100.

Cantin, R., Fortin, J.F., Lamontagne, G., and Tremblay, M. (1997b). The presence of host-derived HLA-DR1 on human immunodeficiency virus type 1 increases viral infectivity. J. Virol. *71*, 1922–1930.

Cantin, R., Methot, S., and Tremblay, M.J. (2005). Plunder and stowaways: incorporation of cellular proteins by enveloped viruses. J. Virol. *79*, 6577–6587.

Cantin, R., Martin, G., and Tremblay, M.J. (2001). A novel virus capture assay reveals a differential acquisition of host HLA-DR by clinical isolates of human immunodeficiency virus type 1 expanded in primary human cells depending on the nature of producing cells and the donor source. J. Gen. Virol. *82*, 2979–2987.

Carteau, S., Gorelick, R.J., and Bushman, F.D. (1999). Coupled integration of human immunodeficiency virus type 1 cDNA ends by purified integrase *in vitro*: stimulation by the viral nucleocapsid protein. J. Virol. *73*, 6670–6679.

Cartier, C., Deckert, M., Grangeasse, C., Trauger, R., Jensen, F., Bernard, A., Cozzone, A., Desgranges, C., and Boyer, V. (1997). Association of ERK2 mitogen-activated protein kinase with human immunodeficiency virus particles. J. Virol. *71*, 4832–4837.

Cartier, C., Hemonnot, B., Gay, B., Bardy, M., Sanchiz, C., Devaux, C., and Briant, L. (2003). Active cAMP-dependent protein kinase incorporated within highly purified HIV-1 particles is required for viral infectivity and interacts with viral capsid protein. J. Biol. Chem. *278*, 35211–35219.

Cen, S., Javanbakht, H., Niu, M., and Kleiman, L. (2004). Ability of wild-type and mutant lysyl-tRNA synthetase to facilitate tRNA(Lys) incorporation into human immunodeficiency virus type 1. J. Virol. *78*, 1595–1601.

Cen, S., Khorchid, A., Javanbakht, H., Gabor, J., Stello, T., Shiba, K., Musier-Forsyth, K., and Kleiman, L. (2001). Incorporation of lysyl-tRNA synthetase into human immunodeficiency virus type 1. J. Virol. *75*, 5043–5048.

Charneau, P., Alizon, M., and Clavel, F. (1992). A second origin of DNA plus-strand synthesis is required for optimal human immunodeficiency virus replication. J. Virol. *66*, 2814–2820.

Charneau, P., and Clavel, F. (1991). A single-stranded gap in human immunodeficiency virus unintegrated linear DNA defined by a central copy of the polypurine tract. J. Virol. *65*, 2415–2421.

Charneau, P., Mirambeau, G., Roux, P., Paulous, S., Buc, H., and Clavel, F. (1994). HIV-1 reverse transcription. A termination step at the center of the genome. J. Mol. Biol. *241*, 651–662.

Chatel-Chaix, L., Clement, J.F., Martel, C., Beriault, V., Gatignol, A., DesGroseillers, L., and Mouland, A.J. (2004). Identification of Staufen in the human immunodeficiency virus type 1 Gag ribonucleoprotein complex and a role in generating infectious viral particles. Mol. Cell. Biol. *24*, 2637–2648.

Chen, R., Wang, H., and Mansky, L.M. (2002). Roles of uracil-DNA glycosylase and dUTPase in virus replication. J. Gen. Virol. *83*, 2339–2345.

Cherrington, J., and Ganem, D. (1992). Regulation of polyadenylation in human immunodeficiency virus (HIV): contributions of promoter proximity and upstream sequences. EMBO J. *11*, 1513–1524.

Choi, Y., Kappler, J.W., and Marrack, P. (1991). A superantigen encoded in the open reading frame of the 3′ long terminal repeat of mouse mammary tumour virus. Nature *350*, 203–207.

Chow, S.A., Vincent, K.A., Ellison, V., and Brown, P.O. (1992). Reversal of integration and DNA splicing mediated by integrase of human immunodeficiency virus. Science *255*, 723–726.

Chun, R.F., and Jeang, K.T. (1996). Requirements for RNA polymerase II carboxyl-terminal domain for activated transcription of human retroviruses human T-cell lymphotropic virus I and HIV-1. J. Biol. Chem. *271*, 27888–27894.

Cimarelli, A., and Luban, J. (1999). Translation elongation factor 1-alpha interacts specifically with the human immunodeficiency virus type 1 Gag polyprotein. J. Virol. *73*, 5388–5401.

Colgan, D.F., and Manley, J.L. (1997). Mechanism and regulation of mRNA polyadenylation. Genes Dev. *11*, 2755–2766.

Craven, R.C., Leure-duPree, A.E., Erdie, C.R., Wilson, C.B., and Wills, J.W. (1993). Necessity of the spacer peptide between CA and NC in the Rous sarcoma virus *gag* protein. J. Virol. *67*, 6246–6252.

Cullen, B.R. (2003). Nuclear mRNA export: insights from virology. Trends Biochem. Sci. *28*, 419–424.

Damsky, C.H., Sheffield, J.B., Tuszynski, G.P., and Warren, L. (1977). Is there a role for actin in virus budding? J. Cell. Biol. *75*, 593–605.

Darlix, J.L., Gabus, C., Nugeyre, M.T., Clavel, F., and Barre-Sinoussi, F. (1990). *Cis* elements and *trans*-acting factors involved in the RNA dimerization of the human immunodeficiency virus HIV-1. J. Mol. Biol. *216*, 689–699.

Das, A.T., Klaver, B., Klasens, B.I., van Wamel, J.L., and Berkhout, B. (1997). A conserved hairpin motif in the R-U5 region of the human immunodeficiency virus type 1 RNA genome is essential for replication. J. Virol. *71*, 2346–2356.

Das, D., and Georgiadis, M.M. (2004). The crystal structure of the monomeric reverse transcriptase from Moloney murine leukemia virus. Structure *12*, 819–829.

Deffaud, C., and Darlix, J.L. (2000a). Characterization of an internal ribosomal entry segment in the 5′ leader of murine leukemia virus *env* RNA. J. Virol. *74*, 846–850.

Deffaud, C., and Darlix, J.L. (2000b). Rous sarcoma virus translation revisited: characterization of an internal ribosome entry segment in the 5′ leader of the genomic RNA. J. Virol. *74*, 11581–11588.

Demirov, D.G., and Freed, E.O. (2004). Retrovirus budding. Virus Res. *106*, 87–102.

D'Souza, V., and Summers, M.F. (2005). How retroviruses select their genomes. Nature Rev. 3, 643–655.

Duesberg, P.H., and Vogt, P.K. (1970). Differences between the ribonucleic acids of transforming and nontransforming avian tumor viruses. Proc. Natl. Acad. Sci. U.S.A. *67*, 1673–1680.

Elder, J.H., Lerner, D.L., Hasselkus-Light, C.S., Fontenot, D.J., Hunter, E., Luciw, P.A., Montelaro, R.C., and Phillips, T.R. (1992). Distinct subsets of retroviruses encode dUTPase. J. Virol. *66*, 1791–1794.

Elliott, J.F., Pohajdak, B., Talbot, D.J., Shaw, J., and Paetkau, V. (1988). Phorbol diester-inducible, cyclosporine-suppressible transcription from a novel promoter within the mouse mammary tumor virus *env* gene. J. Virol. *62*, 1373–1380.

Enssle, J., Fischer, N., Moebes, A., Mauer, B., Smola, U., and Rethwilm, A. (1997). Carboxy-terminal cleavage of the human foamy virus Gag precursor molecule is an essential step in the viral life cycle. J. Virol. *71*, 7312–7317.

Erickson-Viitanen, S., Manfredi, J., Viitanen, P., Tribe, D.E., Tritch, R., Hutchison, C.A., 3rd, Loeb, D.D., and Swanstrom, R. (1989). Cleavage of HIV-1 *gag* polyprotein synthesized *in vitro*: sequential cleavage by the viral protease. AIDS Res. Hum. Retroviruses 5, 577–591.

Fackler, O.T., and Krausslich, H.G. (2006). Interactions of human retroviruses with the host cell cytoskeleton. Curr. Opin. Microbiol. *9*, 409–415.

Fagerness, A.J., Flaherty, M.T., Perry, S.T., Jia, B., Payne, S.L., and Fuller, F.J. (2006). The S2 accessory gene of equine infectious anemia virus is essential for expression of disease in ponies. Virology *349*, 22–30.

Feng, S., and Holland, E.C. (1988). HIV-1 *tat trans*-activation requires the loop sequence within tar. Nature *334*, 165–167.

Forster, F., Medalia, O., Zauberman, N., Baumeister, W., and Fass, D. (2005). Retrovirus envelope protein complex structure in situ studied by cryo-electron tomography. Proc. Natl. Acad. Sci. U.S.A. *102*, 4729–4734.

Fortin, J.F., Cantin, R., Lamontagne, G., and Tremblay, M. (1997). Host-derived ICAM-1 glycoproteins incorporated on human immunodeficiency virus type 1 are biologically active and enhance viral infectivity. J. Virol. *71*, 3588–3596.

Frank, I., Stoiber, H., Godar, S., Stockinger, H., Steindl, F., Katinger, H.W., and Dierich, M.P. (1996). Acquisition of host cell-surface-derived molecules by HIV-1. AIDS *10*, 1611–1620.

Franke, E.K., Yuan, H.E., Luban, J. (1994). Specific incorporation of cyclophilin A into HIV-1 virions. Nature *372*, 359–362.

Freed, E.O., and Martin, M.A. (1995). Virion incorporation of envelope glycoproteins with long but not short cytoplasmic tails is blocked by specific, single amino acid substitutions in the human immunodeficiency virus type 1 matrix. J. Virol. *69*, 1984–1989.

Fujisawa, R., McAtee, F.J., Favara, C., Hayes, S.F., and Portis, J.L. (2001). N-terminal cleavage fragment of glycosylated Gag is incorporated into murine oncornavirus particles. J. Virol. *75*, 11239–11243.

Gamble, T.R., Yoo, S., Vajdos, F.F., von Schwedler, U.K., Worthylake, D.K., Wang, H., McCutcheon, J.P., Sundquist, W.I., and Hill, C.P. (1997). Structure of the carboxyl-terminal dimerization domain of the HIV-1 capsid protein. Science *278*, 849–853.

Ganser-Pornillos, B.K., von Schwedler, U.K., Stray, K.M., Aiken, C., and Sundquist, W.I. (2004). Assembly properties of the human immunodeficiency virus type 1 CA protein. J. Virol. *78*, 2545–2552.

Ganser, B.K., Li, S., Klishko, V.Y., Finch, J.T., and Sundquist, W.I. (1999). Assembly and analysis of conical models for the HIV-1 core. Science *283*, 80–83.

Garcia, J.V., and Miller, A.D. (1991). Serine phosphorylation-independent downregulation of cell-surface CD4 by nef. Nature *350*, 508–511.

Gee, A.H., Kasprzak, W., and Shapiro, B.A. (2006). Structural differentiation of the HIV-1 poly(A) signals. J. Biolmolec. Struct. Dynamics *23*, 417–428.

Gelderblom, H.R., Hausmann, E.H., Ozel, M., Pauli, G., and Koch, M.A. (1987). Fine structure of human immunodeficiency virus (HIV) and immunolocalization of structural proteins. Virology *156*, 171–176.

Giedroc, D.P., Theimer, C.A., and Nixon, P.L. (2000). Structure, stability and function of RNA pseudoknots involved in stimulating ribosomal frameshifting. J. Mol. Biol. *298*, 167–185.

Giguere, J.F., Bounou, S., Paquette, J.S., Madrenas, J., and Tremblay, M.J. (2004). Insertion of host-derived costimulatory molecules CD80 (B7.1) and CD86 (B7.2) into human immunodeficiency virus type 1 affects the virus life cycle. J. Virol. *78*, 6222–6232.

Giguere, J.F., Paquette, J.S., Bounou, S., Cantin, R., and Tremblay, M.J. (2002). New insights into the functionality of a virion-anchored host cell membrane protein: CD28 versus HIV type 1. J. Immunol. *169*, 2762–2771.

Giron, M.L., Colas, S., Wybier, J., Rozain, F., and Emanoil-Ravier, R. (1997). Expression and maturation of human foamy virus Gag precursor polypeptides. J. Virol. *71*, 1635–1639.

Gitti, R.K., Lee, B.M., Walker, J., Summers, M.F., Yoo, S., and Sundquist, W.I. (1996). Structure of the amino-terminal core domain of the HIV-1 capsid protein. Science *273*, 231–235.

Gopalakrishnan, V., Peliska, J.A., and Benkovic, S.J. (1992). Human immunodeficiency virus type 1 reverse transcriptase: spatial and temporal relationship between the polymerase and RNase H activities. Proc. Natl. Acad. Sci. U.S.A. *89*, 10763–10767.

Greatorex, J. (2004). The retroviral RNA dimer linkage: different structures may reflect different roles. Retrovirology *1*, 22.

Greatorex, J., and Lever, A. (1998). Retroviral RNA dimer linkage. J. Gen. Virol. *79*, 2877–2882.

Greenway, A.L., McPhee, D.A., Allen, K., Johnstone, R., Holloway, G., Mills, J., Azad, A., Sankovich, S., and Lambert, P. (2002). Human immunodeficiency virus type 1 Nef binds to tumor suppressor p53 and protects cells against p53-mediated apoptosis. J. Virol. *76*, 2692–2702.

Grigorov, B., Decimo, D., Smagulova, F., Pechoux, C., Mougel, M., Muriaux, D., and Darlix, J.L. (2007). Intracellular HIV-1 Gag localization is impaired by mutations in the nucleocapsid zinc fingers. Retrovirology *4*, 54.

Gruter, P., Tabernero, C., von Kobbe, C., Schmitt, C., Saavedra, C., Bachi, A., Wilm, M., Felber, B.K., and Izaurralde, E. (1998). TAP, the human homolog of Mex67p, mediates CTE-dependent RNA export from the nucleus. Mol. Cell *1*, 649–659.

Guntaka, R.V. (1993). Transcription termination and polyadenylation in retroviruses. Microbiol. Rev. *57*, 511–521.

Guo, F., Cen, S., Niu, M., Javanbakht, H., and Kleiman, L. (2003). Specific inhibition of the synthesis of human lysyl-tRNA synthetase results in decreases in tRNA(Lys) incorporation, tRNA(3)(Lys) annealing to viral RNA, and viral infectivity in human immunodeficiency virus type 1. J. Virol. *77*, 9817–9822.

Gurer, C., Cimarelli, A., and Luban, J. (2002). Specific incorporation of heat shock protein 70 family members into primate lentiviral virions. J. Virol. *76*, 4666–4670.

Guy, B., Kieny, M.P., Riviere, Y., Le Peuch, C., Dott, K., Girard, M., Montagnier, L., and Lecocq, J.P. (1987). HIV F/3′ orf encodes a phosphorylated GTP-binding protein resembling an oncogene product. Nature *330*, 266–269.

Halwani, R., Cen, S., Javanbakht, H., Saadatmand, J., Kim, S., Shiba, K., and Kleiman, L. (2004). Cellular distribution of Lysyl-tRNA synthetase and its interaction with Gag during human immunodeficiency virus type 1 assembly. J. Virol. *78*, 7553–7564.

Harger, J.W., Meskauskas, A., and Dinman, J.D. (2002). An 'integrated model' of programmed ribosomal frameshifting. Trends Biochem. Sci. *27*, 448–454.

Harris, R.S., Bishop, K.N., Sheehy, A.M., Craig, H.M., Petersen-Mahrt, S.K., Watt, I.N., Neuberger, M.S., and Malim, M.H. (2003). DNA deamination mediates innate immunity to retroviral infection. Cell *113*, 803–809.

Hart, D., Frerichs, G.N., Rambaut, A., and Onions, D.E. (1996). Complete nucleotide sequence and transcriptional analysis of snakehead fish retrovirus. J. Virol. *70*, 3606–3616.

Harty, R.N., Paragas, J., Sudol, M., and Palese, P. (1999). A proline-rich motif within the matrix protein of vesicular stomatitis virus and rabies virus interacts with WW domains of cellular proteins: implications for viral budding. J. Virol. 73, 2921–2929.

Hatziioannou, T., Perez-Caballero, D., Cowan, S., and Bieniasz, P.D. (2005). Cyclophilin interactions with incoming human immunodeficiency virus type 1 capsids with opposing effects on infectivity in human cells. J. Virol. *79*, 176–183.

Hearps, A.C., and Jans, D.A. (2007). Regulating the functions of the HIV-1 matrix protein. AIDS Res. Hum. Retroviruses 23, 341–346.

Heeney, J.L., van Els, C., de Vries, P., ten Haaft, P., Otting, N., Koornstra, W., Boes, J., Dubbes, R., Niphuis, H., Dings, M., *et al.* (1994). Major histocompatibility complex class I-associated vaccine protection from simian immunodeficiency virus-infected peripheral blood cells. The Journal of experimental medicine *180*, 769–774.

Heinkelein, M., Thurow, J., Dressler, M., Imrich, H., Neumann-Haefelin, D., McClure, M.O., and Rethwilm, A. (2000). Complex effects of deletions in the 5′ untranslated region of primate foamy virus on viral gene expression and RNA packaging. J. Virol. *74*, 3141–3148.

Held, W., Waanders, G.A., Shakhov, A.N., Scarpellino, L., Acha-Orbea, H., and MacDonald, H.R. (1993). Superantigen-induced immune stimulation amplifies mouse mammary tumor virus infection and allows virus transmission. Cell *74*, 529–540.

Hidaka, M., Inoue, J., Yoshida, M., and Seiki, M. (1988). Post-transcriptional regulator (rex) of HTLV-1 initiates expression of viral structural proteins but suppresses expression of regulatory proteins. EMBO J. *7*, 519–523.

Holzschu, D., Lapierre, L.A., and Lairmore, M.D. (2003). Comparative pathogenesis of epsilonretroviruses. J. Virol. *77*, 12385–12391.

Holzschu, D.L., Martineau, D., Fodor, S.K., Vogt, V.M., Bowser, P.R., and Casey, J.W. (1995). Nucleotide sequence and protein analysis of a complex piscine retrovirus, walleye dermal sarcoma virus. J. Virol. *69*, 5320–5331.

Hu, W.S., and Temin, H.M. (1990). Genetic consequences of packaging two RNA genomes in one retroviral particle: pseudodiploidy and high rate of genetic recombination. Proc. Natl. Acad. Sci. U.S.A. *87*, 1556–1560.

Huang, C.C., Tang, M., Zhang, M.Y., Majeed, S., Montabana, E., Stanfield, R.L., Dimitrov, D.S., Korber, B., Sodroski, J., Wilson, I.A., *et al.* (2005). Structure of a V3-containing HIV-1 gp120 core. Science *310*, 1025–1028.

Huber, H.E., and Richardson, C.C. (1990). Processing of the primer for plus strand DNA synthesis by human immunodeficiency virus 1 reverse transcriptase. J. Biol. Chem. *265*, 10565–10573.

Hungnes, O., Tjotta, E., and Grinde, B. (1992). Mutations in the central polypurine tract of HIV-1 result in delayed replication. Virology *190*, 440–442.

Irie, T., Licata, J.M., McGettigan, J.P., Schnell, M.J., and Harty, R.N. (2004). Budding of PPxY-containing rhabdoviruses is not dependent on host proteins TGS101 and VPS4A. J. Virol. *78*, 2657–2665.

Itoh, M., Inoue, J., Toyoshima, H., Akizawa, T., Higashi, M., and Yoshida, M. (1989). HTLV-1 rex and HIV-1 *rev* act through similar mechanisms to relieve suppression of unspliced RNA expression. Oncogene *4*, 1275–1279.

Jacks, T., Townsley, K., Varmus, H.E., and Majors, J. (1987). Two efficient ribosomal frameshifting events are required for synthesis of mouse mammary tumor virus *gag*-related polyproteins. Proc. Natl. Acad. Sci. U.S.A. *84*, 4298–4302.

Jacks, T., and Varmus, H.E. (1985). Expression of the Rous sarcoma virus *pol* gene by ribosomal frameshifting. Science *230*, 1237–1242.

Javanbakht, H., Halwani, R., Cen, S., Saadatmand, J., Musier-Forsyth, K., Gottlinger, H., and Kleiman, L. (2003). The interaction between HIV-1 Gag and human lysyl-tRNA synthetase during viral assembly. J. Biol. Chem. *278*, 27644–27651.

Jayakar, H.R., Murti, K.G., and Whitt, M.A. (2000). Mutations in the PPPY motif of vesicular stomatitis virus matrix protein reduce virus budding by inhibiting a late step in virion release. J. Virol. *74*, 9818–9827.

Jenny, A., and Keller, W. (1995). Cloning of cDNAs encoding the 160kDa subunit of the bovine cleavage and polyadenylation specificity factor. Nucleic acids research *23*, 2629–2635.

Joho, R.H., Billeter, M.A., and Weissmann, C. (1975). Mapping of biological functions on RNA of avian tumor viruses: location of regions required for transformation and determination of host range. Proc. Natl. Acad. Sci. U.S.A. *72*, 4772–4776.

Kaminska, M., Francin, M., Shalak, V., and Mirande, M. (2007a). Role of HIV-1 Vpr-induced apoptosis on the release of mitochondrial lysyl-tRNA synthetase. FEBS Lett. *581*, 3105–3110.

Kaminska, M., Shalak, V., Francin, M., and Mirande, M. (2007b). Viral hijacking of mitochondrial lysyl-tRNA synthetase. J. Virol. *81*, 68–73.

Kao, S., Khan, M.A., Miyagi, E., Plishka, R., Buckler-White, A., and Strebel, K. (2003). The human immunodeficiency virus type 1 Vif protein reduces intracellular expression and inhibits packaging of APOBEC3G (CEM15), a cellular inhibitor of virus infectivity. J. Virol. 77, 11398–11407.

Katz, R.A., Terry, R.W., and Skalka, A.M. (1986). A conserved *cis*-acting sequence in the 5′ leader of avian sarcoma virus RNA is required for packaging. J. Virol. *59*, 163–167.

Keller, W. (1995). No end yet to messenger RNA 3′ processing! Cell *81*, 829–832.

Kemler, I., Barraza, R., and Poeschla, E.M. (2002). Mapping the encapsidation determinants of feline immunodeficiency virus. J. Virol. *76*, 11889–11903.

Klasens, B.I., Huthoff, H.T., Das, A.T., Jeeninga, R.E., and Berkhout, B. (1999). The effect of template RNA structure on elongation by HIV-1 reverse transcriptase. Biochimica et biophysica acta *1444*, 355–370.

Koppe, B., Menendez-Arias, L., and Oroszlan, S. (1994). Expression and purification of the mouse mammary tumor virus Gag–Pro transframe protein p30 and characterization of its dUTPase activity. J. Virol. *68*, 2313–2319.

Krausslich, H.G., Facke, M., Heuser, A.M., Konvalinka, J., and Zentgraf, H. (1995). The spacer peptide between human immunodeficiency virus capsid and nucleocapsid proteins is essential for ordered assembly and viral infectivity. J. Virol. *69*, 3407–3419.

Kwong, P.D., Wyatt, R., Robinson, J., Sweet, R.W., Sodroski, J., and Hendrickson, W.A. (1998). Structure of an HIV gp120 envelope glycoprotein in complex with the CD4 receptor and a neutralizing human antibody. Nature *393*, 648–659.

LaFemina, R.L., Callahan, P.L., and Cordingley, M.G. (1991). Substrate specificity of recombinant human immunodeficiency virus integrase protein. J. Virol. *65*, 5624–5630.

Lairmore, M.D., Stanley, J.R., Weber, S.A., and Holzschu, D.L. (2000). Squamous epithelial proliferation induced by walleye dermal sarcoma retrovirus cyclin in transgenic mice. Proc. Natl. Acad. Sci. U.S.A. *97*, 6114–6119.

Lapatto, R., Blundell, T., Hemmings, A., Overington, J., Wilderspin, A., Wood, S., Merson, J.R., Whittle, P.J., Danley, D.E., Geoghegan, K.F., *et al.* (1989). X-ray analysis of HIV-1 proteinase at 2.7 A resolution confirms structural homology among retroviral enzymes. Nature *342*, 299–302.

LaPierre, L.A., Casey, J.W., and Holzschu, D.L. (1998). Walleye retroviruses associated with skin tumors and hyperplasias encode cyclin D homologs. J. Virol. *72*, 8765–8771.

Lecellier, C.H., and Saib, A. (2000). Foamy viruses: between retroviruses and pararetroviruses. Virology *271*, 1–8.

Lecellier, C.H., Vermeulen, W., Bachelerie, F., Giron, M.L., and Saib, A. (2002). Intra- and intercellular trafficking of the foamy virus auxiliary bet protein. J. Virol. *76*, 3388–3394.

Lehmann-Che, J., Giron, M.L., Delelis, O., Lochelt, M., Bittoun, P., Tobaly-Tapiero, J., de The, H., and Saib, A. (2005). Protease-dependent uncoating of a complex retrovirus. J. Virol. *79*, 9244–9253.

Lerner, D.L., Wagaman, P.C., Phillips, T.R., Prospero-Garcia, O., Henriksen, S.J., Fox, H.S., Bloom, F.E., and Elder, J.H. (1995). Increased mutation frequency of feline immunodeficiency virus lacking functional deoxyuridine-triphosphatase. Proc. Natl. Acad. Sci. U.S.A. *92*, 7480–7484.

Li, M., Laco, G.S., Jaskolski, M., Rozycki, J., Alexandratos, J., Wlodawer, A., and Gustchina, A. (2005). Crystal structure of human T-cell leukemia virus protease, a novel target for anticancer drug design. Proc. Natl. Acad. Sci. U.S.A. *102*, 18332–18337.

Li, S., Hill, C.P., Sundquist, W.I., and Finch, J.T. (2000). Image reconstructions of helical assemblies of the HIV-1 CA protein. Nature *407*, 409–413.

Lichtenstein, D.L., Rushlow, K.E., Cook, R.F., Raabe, M.L., Swardson, C.J., Kociba, G.J., Issel, C.J., and Montelaro, R.C. (1995). Replication *in vitro* and *in vivo* of an equine infectious anemia virus mutant deficient in dUTPase activity. J. Virol. *69*, 2881–2888.

Liddament, M.T., Brown, W.L., Schumacher, A.J., and Harris, R.S. (2004). APOBEC3F properties and hypermutation preferences indicate activity against HIV-1 *in vivo*. Curr. Biol. *14*, 1385–1391.

Lindwasser, O.W., Smith, W.J., Chaudhuri, R., Yang, P., Hurley, J.H., and Bonifacino, J.S. (2008). A diacidic motif in HIV-1 Nef is a novel determinant of binding to AP2. J. Virol. *82*, 1166–1174.

Lochelt, M., Muranyi, W., and Flugel, R.M. (1993). Human foamy virus genome possesses an internal, Bel-1-dependent and functional promoter. Proc. Natl. Acad. Sci. U.S.A. *90*, 7317–7321.

Lochelt, M., Romen, F., Bastone, P., Muckenfuss, H., Kirchner, N., Kim, Y.B., Truyen, U., Rosler, U., Battenberg, M., Saib, A., *et al.* (2005). The antiretroviral activity of APOBEC3 is inhibited by the foamy virus accessory Bet protein. Proc. Natl. Acad. Sci. U.S.A. *102*, 7982–7987.

Low, A., Datta, S., Kuznetsov, Y., Jahid, S., Kothari, N., McPherson, A., and Fan, H. (2007). Mutation in the glycosylated *gag* protein of murine leukemia virus results in reduced *in vivo* infectivity and a novel defect in viral budding or release. J. Virol. *81*, 3685–3692.

Lu, M., Blacklow, S.C., and Kim, P.S. (1995). A trimeric structural domain of the HIV-1 transmembrane glycoprotein. Nature structural biology *2*, 1075–1082.

Luban, J., Bossolt, K.L., Franke, E.K., Kalpana, G.V., and Goff, S.P. (1993). Human immunodeficiency virus type 1 Gag protein binds to cyclophilins A and B. Cell *73*, 1067–1078.

Luo, G.X., Sharmeen, L., and Taylor, J. (1990). Specificities involved in the initiation of retroviral plus-strand DNA. J. Virol. *64*, 592–597.

Luo, G.X., and Taylor, J. (1990). Template switching by reverse transcriptase during DNA synthesis. J. Virol. *64*, 4321–4328.

McBride, M.S., Schwartz, M.D., and Panganiban, A.T. (1997). Efficient encapsidation of human immunodeficiency virus type 1 vectors and further characterization of *cis* elements required for encapsidation. J. Virol. *71*, 4544–4554.

McGeoch, D.J. (1990). Protein sequence comparisons show that the 'pseudoproteases' encoded by poxviruses and certain retroviruses belong to the deoxyuridine triphosphatase family. Nucleic acids research *18*, 4105–4110.

McWilliams, M.J., Julias, J.G., Sarafianos, S.G., Alvord, W.G., Arnold, E., and Hughes, S.H. (2003). Mutations in the 5′ end of the human immunodeficiency virus type 1 polypurine tract affect RNase H cleavage specificity and virus titer. J. Virol. 77, 11150–11157.

Magin-Lachmann, C., Hahn, S., Strobel, H., Held, U., Lower, J., and Lower, R. (2001). Rec (formerly Corf) function requires interaction with a complex, folded RNA structure within its responsive element rather than binding to a discrete specific binding site. J. Virol. *75*, 10359–10371.

Malim, M.H., Tiley, L.S., McCarn, D.F., Rusche, J.R., Hauber, J., and Cullen, B.R. (1990). HIV-1 structural gene expression requires binding of the Rev *trans*-activator to its RNA target sequence. Cell *60*, 675–683.

Mammano, F., Ohagen, A., Hoglund, S., and Gottlinger, H.G. (1994). Role of the major homology region of human immunodeficiency virus type 1 in virion morphogenesis. J. Virol. *68*, 4927–4936.

Mangeat, B., Turelli, P., Liao, S., and Trono, D. (2004). A single amino acid determinant governs the species-specific sensitivity of APOBEC3G to Vif action. J. Biol. Chem. *279*, 14481–14483.

Mansharamani, M., Graham, D.R., Monie, D., Lee, K.K., Hildreth, J.E., Siliciano, R.F., and Wilson, K.L. (2003). Barrier-to-autointegration factor BAF binds p55 Gag and matrix and is a host component of human immunodeficiency virus type 1 virions. J. Virol. 77, 13084–13092.

Mansky, L.M., Preveral, S., Selig, L., Benarous, R., and Benichou, S. (2000). The interaction of vpr with uracil DNA glycosylase modulates the human immunodeficiency virus type 1 In vivo mutation rate. J. Virol. *74*, 7039–7047.

Marcheschi, R.J., Staple, D.W., and Butcher, S.E. (2007). Programmed ribosomal frameshifting in SIV is induced by a highly structured RNA stem–loop. J. Mol. Biol. *373*, 652–663.

Mariani, R., Chen, D., Schröfelbauer, B., Navarro, F., König, R., Bollman, B., Münk, C., Nymark-McMahon H., and Landau, N.R. (2003). Species-specific exclusion of APOBEC3G from HIV-1 virions by Vif. Cell *114*, 21–31.

Marrack, P., Kushnir, E., and Kappler, J. (1991). A maternally inherited superantigen encoded by a mammary tumour virus. Nature *349*, 524–526.

Martin-Serrano, J., Zang, T., and Bieniasz, P.D. (2001). HIV-1 and Ebola virus encode small peptide motifs that recruit Tsg101 to sites of particle assembly to facilitate egress. Nature medicine *7*, 1313–1319.

Masuda, T., Kuroda, M.J., and Harada, S. (1998). Specific and independent recognition of U3 and U5 att sites by human immunodeficiency virus type 1 integrase *in vivo*. J. Virol. 72, 8396–8402.

Mayer, J., and Meese, E.U. (2002). The human endogenous retrovirus family HERV-K(HML-3). Genomics *80*, 331–343.

Mayo, K., Huseby, D., McDermott, J., Arvidson, B., Finlay, L., and Barklis, E. (2003). Retrovirus capsid protein assembly arrangements. J. Mol. Biol. *325*, 225–237.

Meerloo, T., Parmentier, H.K., Osterhaus, A.D., Goudsmit, J., and Schuurman, H.J. (1992). Modulation of cell surface molecules during HIV-1 infection of H9 cells. An immunoelectron microscopic study. AIDS *6*, 1105–1116.

Meyer, P.R., Matsuura, S.E., So, A.G., and Scott, W.A. (1998). Unblocking of chain-terminated primer by HIV-1 reverse transcriptase through a nucleotide-dependent mechanism. Proc. Natl. Acad. Sci. U.S.A. *95*, 13471–13476.

Miele, G., Mouland, A., Harrison, G.P., Cohen, E., and Lever, A.M. (1996). The human immunodeficiency virus type 1, 5′ packaging signal structure affects translation but does not function as an internal ribosome entry site structure. J. Virol. *70*, 944–951.

Miller, C.L., Garner, R., and Paetkau, V. (1992). An activation-dependent, T-lymphocyte-specific transcriptional activator in the mouse mammary tumor virus *env* gene. Mol. Cell. Biol. *12*, 3262–3272.

Miller, M., Jaskolski, M., Rao, J.K., Leis, J., and Wlodawer, A. (1989). Crystal structure of a retroviral protease proves relationship to aspartic protease family. Nature *337*, 576–579.

Molina, R.P., Matukonis, M., Paszkiet, B., Zhang, J., Kaleko, M., and Luo, T. (2002). Mapping of the bovine immunodeficiency virus packaging signal and RRE and incorporation into a minimal gene transfer vector. Virology *304*, 10–23.

Montefiori, D.C., Cornell, R.J., Zhou, J.Y., Zhou, J.T., Hirsch, V.M., and Johnson, P.R. (1994). Complement control proteins, CD46, CD55, and CD59, as common surface constituents of human and simian immunodeficiency viruses and possible targets for vaccine protection. Virology *205*, 82–92.

Moon, H.S., and Yang, J.S. (2006). Role of HIV Vpr as a regulator of apoptosis and an effector on bystander cells. Molecules and cells *21*, 7–20.

Mouland, A.J., Mercier, J., Luo, M., Bernier, L., DesGroseillers L., and Cohen, E.A. (2000). The double-stranded RNA-binding protein Staufen is incorporated in human immunodeficiency virus type 1: evidence for a role in genomic RNA encapsidation. J. Virol. *74*, 5441–5451.

Myers, J.C., Dobkin, C., and Spiegelman, S. (1980). RNA primer used in synthesis of anticomplementary DNA by reverse transcriptase of avian myeloblastosis virus. Proc. Natl. Acad. Sci. U.S.A. *77*, 1316–1320.

Navia, M.A., Fitzgerald, P.M., McKeever, B.M., Leu, C.T., Heimbach, J.C., Herber, W.K., Sigal, I.S., Darke, P.L., and Springer, J.P. (1989). Three-dimensional structure of aspartyl protease from human immunodeficiency virus HIV-1. Nature *337*, 615–620.

Neil, S.J., Zang, T., and Bieniasz, P.D. (2008). Tetherin inhibits retrovirus release and is antagonized by HIV-1 Vpu. Nature *451*, 425–430.

Nudson, W.A., Rovnak, J., Buechner, M., and Quackenbush, S.L. (2003). Walleye dermal sarcoma virus Orf C is targeted to the mitochondria. J. Gen. Virol. *84*, 375–381.

Orentas, R.J., and Hildreth, J.E. (1993). Association of host cell surface adhesion receptors and other membrane proteins with HIV and SIV. AIDS Res. Hum. Retroviruses *9*, 1157–1165.

Orlova, M., Yueh, A., Leung, J., and Goff, S.P. (2003). Reverse transcriptase of Moloney murine leukemia virus binds to eukaryotic release factor 1 to modulate suppression of translational termination. Cell *115*, 319–331.

Ott, D.E., Coren, L.V., Chertova, E.N., Gagliardi, T.D., and Schubert, U. (2000). Ubiquitination of HIV-1 and MuLV Gag. Virology *278*, 111–121.

Ott, D.E., Coren, L.V., Kane, B.P., Busch, L.K., Johnson, D.G., Sowder, R.C 2nd, Chertova, E.N., Arthur, L.O., and Henderson, L.E. (1996). Cytoskeletal proteins inside human immunodeficiency virus type 1 virions. J. Virol. *70*, 7734–7743.

Paillart, J.C., Dettenhofer, M., Yu, X.F., Ehresmann, C., Ehresmann, B., and Marquet, R. (2004a). First snapshots of the HIV-1 RNA structure in infected cells and in virions. J. Biol. Chem. *279*, 48397–48403.

Paillart, J.C., Shehu-Xhilaga, M., Marquet, R., and Mak, J. (2004b). Dimerization of retroviral RNA genomes: an inseparable pair. Nature Rev. *2*, 461–472.

Paillart, J.C., Skripkin, E., Ehresmann, B., Ehresmann, C., and Marquet, R. (2002). In vitro evidence for a long range pseudoknot in the 5′-untranslated and matrix coding regions of HIV-1 genomic RNA. J. Biol. Chem. *277*, 5995–6004.

Pasquinelli, A.E., Ernst, R.K., Lund, E., Grimm, C., Zapp, M.L., Rekosh, D., Hammarskjold, M.L., and Dahlberg, J.E. (1997). The constitutive transport element (CTE) of Mason–Pfizer monkey virus (MPMV) accesses a cellular mRNA export pathway. EMBO J. *16*, 7500–7510.

Payne, S.L., and Elder, J.H. (2001). The role of retroviral dUTPases in replication and virulence. Curr. Protein Peptide Sci. *2*, 381–388.

Pickl, W.F., Pimentel-Muiños, F.X., and Seed, B. (2001). Lipid rafts and pseudotyping. J. Virol. *75*, 7175–7183.

Piguet, V., Chen, Y.L., Mangasarian, A., Foti, M., Carpentier, J.L., and Trono, D. (1998). Mechanism of Nef-induced CD4 endocytosis: Nef connects CD4 with the mu chain of adaptor complexes. EMBO J. *17*, 2472–2481.

Pillemer, E.A., Kooistra, D.A., Witte, O.N., and Weissman, I.L. (1986). Monoclonal antibody to the amino-terminal L sequence of murine leukemia virus glycosylated *gag* polyproteins demonstrates their unusual orientation in the cell membrane. J. Virol. *57*, 413–421.

Poole, E., Strappe, P., Mok, H.P., Hicks, R., and Lever, A.M. (2005). HIV-1 Gag–RNA interaction occurs at a perinuclear/centrosomal site; analysis by confocal microscopy and FRET. Traffic (Copenhagen, Denmark) *6*, 741–755.

Powell, M.D., and Levin, J.G. (1996). Sequence and structural determinants required for priming of plus-strand DNA synthesis by the human immunodeficiency virus type 1 polypurine tract. J. Virol. *70*, 5288–5296.

Prasad, G.S., Stura, E.A., McRee, D.E., Laco, G.S., Hasselkus-Light, C., Elder, J.H., and Stout, C.D. (1996). Crystal structure of dUTP pyrophosphatase from feline immunodeficiency virus. Protein Sci *5*, 2429–2437.

Priet, S., Gros, N., Navarro, J.M., Boretto, J., Canard, B., Querat, G., and Sire, J. (2005). HIV-1-associated uracil DNA glycosylase activity controls dUTP misincorporation in viral DNA and is essential to the HIV-1 life cycle. Mol. Cell *17*, 479–490.

Priet, S., Navarro, J.M., Gros, N., Quérat, G., and Sire, J. (2003). Differential incorporation of uracil DNA glycosylase UNG2 into HIV-1, HIV-2, and SIV(MAC) viral particles. Virology *307*, 283–289.

Priet, S., Sire, J., and Querat, G. (2006). Uracils as a cellular weapon against viruses and mechanisms of viral escape. Curr. HIV Res. *4*, 31–42.

Provitera, P., Goff, A., Harenberg, A., Bouamr, F., Carter, C., and Scarlata, S. (2001). Role of the major homology region in assembly of HIV-1 Gag. Biochemistry *40*, 5565–5572.

Pullen, A.M., Wade, T., Marrack, P., and Kappler, J.W. (1990). Identification of the region of T-cell receptor beta chain that interacts with the self-superantigen MIs-1a. Cell *61*, 1365–1374.

Quackenbush, S.L., Holzschu, D.L., Bowser, P.R., and Casey, J.W. (1997). Transcriptional analysis of walleye dermal sarcoma virus (WDSV). Virology *237*, 107–112.

Rao, Z., Belyaev, A.S., Fry, E., Roy, P., Jones, I.M., and Stuart, D.I. (1995). Crystal structure of SIV matrix antigen and implications for virus assembly. Nature *378*, 743–747.

Rausch, J.W., and Le Grice, S.F. (2004). 'Binding, bending and bonding': polypurine tract-primed initiation of plus-strand DNA synthesis in human immunodeficiency virus. Int. J. Biochem. Cell Biol. *36*, 1752–1766.

Remy, E., de Rocquigny, H., Petitjean, P., Muriaux, D., Theilleux, V., Paoletti, J., and Roques, B.P. (1998). The annealing of tRNA3Lys to human immunodeficiency virus type 1 primer binding site is critically dependent on the NCp7 zinc fingers structure. J. Biol. Chem. *273*, 4819–4822.

Ren, J., Esnouf, R.M., Hopkins, A.L., Jones, E.Y., Kirby, I., Keeling, J., Ross, C.K., Larder, B.A., Stuart, D.I., and Stammers, D.K. (1998). 3′-Azido-3′-deoxythymidine drug resistance mutations in HIV-1 reverse transcriptase can induce long range conformational changes. Proc. Natl. Acad. Sci. U.S.A. *95*, 9518–9523.

Reuss, F.U., and Coffin, J.M. (1998). Mouse mammary tumor virus superantigen expression in B cells is regulated by a central enhancer within the *pol* gene. J. Virol. *72*, 6073–6082.

Rovnak, J., Casey, J.W., and Quackenbush, S.L. (2001). Intracellular targeting of walleye dermal sarcoma virus Orf A (rv-cyclin). Virology *280*, 31–40.

Rumbaugh, J.A., Fuentes, G.M., and Bambara, R.A. (1998). Processing of an HIV replication intermediate by the human DNA replication enzyme FEN1. J. Biol. Chem. *273*, 28740–28745.

Saifuddin, M., Hedayati, T., Atkinson, J.P., Holguin, M.H., Parker, C.J., and Spear, G.T. (1997). Human immunodeficiency virus type 1 incorporates both glycosyl phosphatidylinositol-anchored CD55 and CD59 and integral membrane CD46 at levels that protect from complement-mediated destruction. J. Gen. Virol. *78*, 1907–1911.

Saifuddin, M., Parker, C.J., Peeples, M.E., Gorny, M.K., Zolla-Pazner, S., Ghassemi, M., Rooney, I.A., Atkinson, J.P., and Spear, G.T. (1995). Role of virion-associated glycosylphosphatidylinositol-linked proteins CD55 and CD59 in complement resistance of cell line-derived and primary isolates of HIV-1. J. Exp. Med. *182*, 501–509.

Sakuragi, J., Sakuragi, S., and Shioda, T. (2007). Minimal region sufficient for genome dimerization in the human immunodeficiency virus type 1 virion and its potential roles in the early stages of viral replication. J. Virol. *81*, 7985–7992.

Sayah, D.M., Sokolskaja, E., Berthoux, L., and Luban, J. (2004). Cyclophilin A retrotransposition into TRIM5 explains owl monkey resistance to HIV-1. Nature *430*, 569–573.

Schatz, G., Pichova, I., and Vogt, V.M. (1997). Analysis of cleavage site mutations between the NC and PR Gag domains of Rous sarcoma virus. J. Virol. *71*, 444–450.

Schiltz, R.L., Shih, D.S., Rasty, S., Montelaro, R.C., and Rushlow, K.E. (1992). Equine infectious anemia virus gene expression: characterization of the RNA splicing pattern and the protein products encoded by open reading frames S1 and S2. J. Virol. *66*, 3455–3465.

Schindler, M., Munch, J., Kutsch, O., Li, H., Santiago, M.L., Bibollet-Ruche, F., Muller-Trutwin, M.C., Novembre, F.J., Peeters, M., Courgnaud, V., *et al.* (2006). Nef-mediated suppression of T-cell activation was lost in a lentiviral lineage that gave rise to HIV-1. Cell *125*, 1055–1067.

Schneider, R., Campbell, M., Nasioulas, G., Felber, B.K., and Pavlakis, G.N. (1997). Inactivation of the human immunodeficiency virus type 1 inhibitory elements allows Rev-independent expression of Gag and Gag/protease and particle formation. J. Virol. *71*, 4892–4903.

Schwartz, O., Marechal, V., Le Gall, S., Lemonnier, F., and Heard, J.M. (1996). Endocytosis of major histocompatibility complex class I molecules is induced by the HIV-1 Nef protein. Nat. Med. *2*, 338–342.

Scott, J.M., and Imperiale, M.J. (1996). Reciprocal effects of splicing and polyadenylation on human immunodeficiency virus type 1 pre-mRNA processing. Virology *224*, 498–509.

Sheehy, A.M., Gaddis, N.C., Choi, J.D., and Malim, M.H. (2002). Isolation of a human gene that inhibits HIV-1 infection and is suppressed by the viral Vif protein. Nature *418*, 646–650.

Shinnick, T.M., Lerner, R.A., and Sutcliffe, J.G. (1981). Nucleotide sequence of Moloney murine leukaemia virus. Nature *293*, 543–548.

Simpson, E., Dyson, P.J., Knight, A.M., Robinson, P.J., Elliott, J.I., and Altmann, D.M. (1993). T-cell receptor repertoire selection by mouse mammary tumor viruses and MHC molecules. Immunol. Rev. *131*, 93–115.

Steagall, W.K., Robek, M.D., Perry, S.T., Fuller, F.J., and Payne, S.L. (1995). Incorporation of uracil into viral DNA correlates with reduced replication of EIAV in macrophages. Virology *210*, 302–313.

Stehelin, D., Varmus, H.E., Bishop, J.M., and Vogt, P.K. (1976). DNA related to the transforming gene(s) of avian sarcoma viruses is present in normal avian DNA. Nature *260*, 170–173.

Stenbak, C.R., and Linial, M.L. (2004). Role of the C terminus of foamy virus Gag in RNA packaging and Pol expression. J. Virol. *78*, 9423–9430.

Stetor, S.R., Rausch, J.W., Guo, M.J., Burnham, J.P., Boone, L.R., Waring, M.J., and Le Grice, S.F. (1999). Characterization of (+) strand initiation and termination sequences located at the center of the equine infectious anemia virus genome. Biochemistry *38*, 3656–3667.

Stoltzfus, C.M., and Madsen, J.M. (2006). Role of viral splicing elements and cellular RNA binding proteins in regulation of HIV-1 alternative RNA splicing. Curr. HIV Res. *4*, 43–55.

Stremlau, M., Owens, C.M., Perron, M.J., Kiessling, M., Autissier, P., and Sodroski, J. (2004). The cytoplasmic body component TRIM5alpha restricts HIV-1 infection in Old World monkeys. Nature *427*, 848–853.

Subramaniam, S. (2006). The SIV surface spike imaged by electron tomography: one leg or three? PLoS Pathogens 2, e91.

Tanchou, V., Decimo, D., Pechoux, C., Lener, D., Rogemond, V., Berthoux, L., Ottmann, M., and Darlix, J.L. (1998). Role of the N-terminal zinc finger of human immunodeficiency virus type 1 nucleocapsid protein in virus structure and replication. J. Virol. *72*, 4442–4447.

Tanese, N., and Goff, S.P. (1988). Domain structure of the Moloney murine leukemia virus reverse transcriptase: mutational analysis and separate expression of the DNA polymerase and RNase H activities. Proc. Natl. Acad. Sci. U.S.A. *85*, 1777–1781.

Thali, M., Bukovsky, A., Kondo, E., Rosenwirth, B., Walsh, C.T., Sodroski, J., and Gottlinger, H.G. (1994).

Functional association of cyclophilin A with HIV-1 virions. Nature *372*, 363–365.

Timmins, J., Schoehn, G., Ricard-Blum, S., Scianimanico, S., Vernet, T., Ruigrok, R.W., and Weissenhorn, W. (2003). Ebola virus matrix protein VP40 interaction with human cellular factors Tsg101 and Nedd4. J. Mol. Biol. *326*, 493–502.

Toyoshima, K., and Vogt, P.K. (1969). Temperature sensitive mutants of an avian sarcoma virus. Virology *39*, 930–931.

Valsamakis, A., Zeichner, S., Carswell, S., and Alwine, J.C. (1991). The human immunodeficiency virus type 1 polyadenylylation signal: a 3′ long terminal repeat element upstream of the AAUAAA necessary for efficient polyadenylylation. Proc. Natl. Acad. Sci. U.S.A. *88*, 2108–2112.

VerPlank L., Bouamr, F., LaGrassa T.J., Agresta, B., Kikonyogo, A., Leis, J., and Carter, C.A. (2001). Tsg101, a homologue of ubiquitin-conjugating (E2) enzymes, binds the L domain in HIV type 1 Pr55(Gag). Proc. Natl. Acad. Sci. U.S.A. *98*, 7724–7729.

Vogt, V.M., and Simon, M.N. (1999). Mass determination of rous sarcoma virus virions by scanning transmission electron microscopy. J. Virol. *73*, 7050–7055.

von Schwedler U.K., Stuchell, M., Müller, B., Ward, D.M., Chung, H.Y., Morita, E., Wang, H.E., Davis, T., He, G.P., Cimbora, D.M., Scott, A., Kräusslich, H.G., Kaplan, J., Morham, S.G., and Sundquist, W.I. (2003). The protein network of HIV budding. Cell *114*, 701–713.

Wang, H., Norris, K.M., and Mansky, L.M. (2003). Involvement of the matrix and nucleocapsid domains of the bovine leukemia virus Gag polyprotein precursor in viral RNA packaging. J. Virol. *77*, 9431–9438.

Wang, L., Galehouse, D., Mellon, P., Duesberg, P., Mason, W.S., and Vogt, P.K. (1976). Mapping oligonucleotides of Rous sarcoma virus RNA that segregate with polymerase and group-specific antigen markers in recombinants. Proc. Natl. Acad. Sci. U.S.A. *73*, 3952–3956.

Wang, Y., Wills, N.M., Du, Z., Rangan, A., Atkins, J.F., Gesteland, R.F., and Hoffman, D.W. (2002). Comparative studies of frameshifting and non-frameshifting RNA pseudoknots: a mutational and NMR investigation of pseudoknots derived from the bacteriophage T2 gene 32 mRNA and the retroviral Gag–Pro frameshift site. RNA (New York, NY *8*, 981–996.

Weber, I.T., Miller, M., Jaskolski, M., Leis, J., Skalka, A.M., and Wlodawer, A. (1989). Molecular modeling of the HIV-1 protease and its substrate binding site. Science (New York, NY *243*, 928–931.

Welker, R., Hohenberg, H., Tessmer, U., Huckhagel, C., and Krausslich, H.G. (2000). Biochemical and structural analysis of isolated mature cores of human immunodeficiency virus type 1. J. Virol. *74*, 1168–1177.

Whitwam, T., Peretz, M., and Poeschla, E. (2001). Identification of a central DNA flap in feline immunodeficiency virus. J. Virol. 75, 9407–9414.

Wiegand, H.L., Doehle, B.P., Bogerd, H.P., and Cullen, B.R. (2004). A second human antiretroviral factor, APOBEC3F, is suppressed by the HIV-1 and HIV-2 Vif proteins. EMBO J. *23*, 2451–2458.

Wildum, S., Schindler, M., Munch, J., and Kirchhoff, F. (2006). Contribution of Vpu, Env, and Nef to CD4 down-modulation and resistance of human immunodeficiency virus type 1-infected T-cells to superinfection. J. Virol. *80*, 8047–8059.

Wilk, T., de Haas, F., Wagner, A., Rutten, T., Fuller, S., Flugel, R.M., and Lochelt, M. (2000). The intact retroviral Env glycoprotein of human foamy virus is a trimer. J. Virol. *74*, 2885–2887.

Wilk, T., Geiselhart, V., Frech, M., Fuller, S.D., Flugel, R.M., and Lochelt, M. (2001). Specific interaction of a novel foamy virus Env leader protein with the N-terminal Gag domain. J. Virol. *75*, 7995–8007.

Wilk, T., Gowen, B., and Fuller, S.D. (1999). Actin associates with the nucleocapsid domain of the human immunodeficiency virus Gag polyprotein. J. Virol. *73*, 1931–1940.

Willey, R.L., Maldarelli, F., Martin, M.A., and Strebel, K. (1992). Human immunodeficiency virus type 1 Vpu protein induces rapid degradation of CD4. J. Virol. *66*, 7193–7200.

Wlodawer, A., and Gustchina, A. (2000). Structural and biochemical studies of retroviral proteases. Biochimica et biophysica acta *1477*, 16–34.

Wohrl, B.M., and Moelling, K. (1990). Interaction of HIV-1 ribonuclease H with polypurine tract containing RNA-DNA hybrids. Biochemistry *29*, 10141–10147.

Woodland, D.L., Happ, M.P., Gollob, K.J., and Palmer, E. (1991). An endogenous retrovirus mediating deletion of alpha beta T-cells? Nature *349*, 529–530.

Wyatt, R., Kwong, P.D., Desjardins, E., Sweet, R.W., Robinson, J., Hendrickson, W.A., and Sodroski, J.G. (1998). The antigenic structure of the HIV gp120 envelope glycoprotein. Nature 393, 705–711.

Yang, J., Bogerd, H.P., Peng, S., Wiegand, H., Truant, R., and Cullen, B.R. (1999). An ancient family of human endogenous retroviruses encodes a functional homolog of the HIV-1 Rev protein. Proc. Natl. Acad. Sci. U.S.A. *96*, 13404–13408.

Yeager, M., Wilson-Kubalek, E.M., Weiner, S.G., Brown, P.O., and Rein, A. (1998). Supramolecular organization of immature and mature murine leukemia virus revealed by electron cryo-microscopy: implications for retroviral assembly mechanisms. Proc. Natl. Acad. Sci. U.S.A. *95*, 7299–7304.

Yoshinaka, Y., Katoh, I., Copeland, T.D., and Oroszlan, S. (1985). Murine leukemia virus protease is encoded by the Gag–Pol gene and is synthesized through suppression of an amber termination codon. Proc. Natl. Acad. Sci. U.S.A. *82*, 1618–1622.

Yu, S.F., Baldwin, D.N., Gwynn, S.R., Yendapalli, S., and Linial, M.L. (1996). Human foamy virus replication: a pathway distinct from that of retroviruses and hepadnaviruses. Science (New York, NY *271*, 1579–1582.

Zanetti, G., Briggs, J.A., Grunewald, K., Sattentau, Q.J., and Fuller, S.D. (2006). Cryo-electron tomographic structure of an immunodeficiency virus envelope complex in situ. PLoS pathogens *2*, e83.

Zapp, M.L., and Green, M.R. (1989). Sequence-specific RNA binding by the HIV-1 Rev protein. Nature *342*, 714–716.

Zemba, M., Wilk, T., Rutten, T., Wagner, A., Flugel, R.M., and Lochelt, M. (1998). The carboxy-terminal p3Gag domain of the human foamy virus Gag precursor is required for efficient virus infectivity. Virology *247*, 7–13.

Zennou, V., Petit, C., Guetard, D., Nerhbass, U., Montagnier, L., and Charneau, P. (2000). HIV-1 genome nuclear import is mediated by a central DNA flap. Cell *101*, 173–185.

Zhang, Z., Kim, E., and Martineau, D. (1999). Functional characterization of a piscine retroviral promoter. J. Gen. Virol. *80*, 3065–3072.

Zhou, J., Bean, R.L., Vogt, V.M., and Summers, M. (2007). Solution structure of the Rous sarcoma virus nucleocapsid protein: muPsi RNA packaging signal complex. J. Mol. Biol. *365*, 453–467.

Zhou, M., Deng, L., Kashanchi, F., Brady, J.N., Shatkin, A.J., and Kumar, A. (2003). The Tat/TAR-dependent phosphorylation of RNA polymerase II C-terminal domain stimulates cotranscriptional capping of HIV-1 mRNA. Proc. Natl. Acad. Sci. U.S.A. *100*, 12666–12671.

Zhu, P., Chertova, E., Bess, J., Jr., Lifson, J.D., Arthur, L.O., Liu, J., Taylor, K.A., and Roux, K.H. (2003). Electron tomography analysis of envelope glycoprotein trimers on HIV and simian immunodeficiency virus virions. Proc. Natl. Acad. Sci. U.S.A. *100*, 15812–15817.

Zhu, P., Liu, J., Bess, J., Jr., Chertova, E., Lifson, J.D., Grise, H., Ofek, G.A., Taylor, K.A., and Roux, K.H. (2006). Distribution and three-dimensional structure of AIDS virus envelope spikes. Nature *441*, 847–852.

Retroviral Entry and Uncoating

4

Walther Mothes and Pradeep D. Uchil

Abstract

Retroviruses form small 100-nm particles of simple composition, yet are able to replicate, spread and cause severe diseases. This is possible, because throughout their replication cycle, retroviruses utilize host factors and hijack cellular pathways. In addition, retroviruses have to overcome a strong innate and adaptive immune response. As such, retroviral replication is the result of a complex co-evolution of viral biology, the cell biology of the host and immune evasion. This complex nature of retroviral infections also applies to the subject of this chapter, how retroviruses enter cells, uncoat to reverse transcribe and to deliver their genomes into the nucleus of the cell. We will discuss the viral aspects of entry, cover cell biological aspects of viral trafficking and deal with innate cellular factors targeting incoming viruses. Finally, we will review virus entry in the context of retroviral pathogenesis and discuss how virus entry and budding are coordinated at sites of cell–cell contact during a spreading infection.

Retroviral fusion machines

Retroviruses, like other enveloped viruses, enter cells by fusion of viral and cellular membranes (Goff, 2007; Hernandez *et al.*, 1996; Young, 2001). The fusion machine that mediates this step lies within the viral envelope glycoprotein (Env). Env folds into a 'loaded' energy-rich metastable conformation on the viral surface that responds to a cellular trigger to adopt its thermodynamically stable conformation that brings membranes together for fusion (Carr *et al.*, 1997; Ruigrok *et al.*, 1986; Smith *et al.*, 2004; Wallin *et al.*, 2005). The biosynthesis of a metastable protein poses an interesting thermodynamic problem. Any *de novo* synthesized protein would by default fold into its energy minimum. To solve this problem, Env is first synthesized as a precursor protein that folds into a trimeric structure in the endoplasmic reticulum (Einfeld and Hunter, 1988). The metastable conformation is then generated by proteolytic cleavage. Cellular proteases such *trans*-Golgi-localized furins cleave Env into surface (SU) and transmembrane (TM) subunits yielding a metastable trimer of heterodimers (Hallenberger *et al.*, 1997; Hunter and Swanstrom, 1990).

The metastable Env protein can be activated for fusion by interacting with cellular factors such as receptor or by exposure to low pH. An interaction with a receptor at the cell surface may suffice to induce fusion at the plasma membrane. In contrast, because low pH is found in endosomes, low pH activated fusion often correlates with internalization.

It is critical for the directionality of the viral replication cycle that Env does not come into contact with its activating factors while being synthesized in the producer cell. Downregulation of receptor in infected cells is of central importance and the Human immunodeficiency virus (HIV) encodes for three proteins devoted to the downregulation of CD4 (Vpu, Nef, Env) (Aiken *et al.*, 1994; Lama *et al.*, 1999; Levesque *et al.*, 2003; Wildum *et al.*, 2006). In the case of low-pH activated fusion proteins, the situation is further complicated because of the low-pH environment of the trans-Golgi (Demaurex *et al.*, 1998). Viral fusion proteins could be prematurely activated while passing through this compartment. Therefore, fusion machines often require additional 'cues' to be activated for fusion at the right location and the right time. Influenza A virus haemagglutinin can be activated by proteases in the lung following inhalation (Klenk and Garten, 1994). The fusiogenicity of Env proteins of the murine leukaemia virus (MLV) and HIV are further enhanced when the cytoplasmic tail of Env is processed by the viral protease following virus release (Melikyan *et al.*, 2000a; Rein *et al.*, 1994; Wyma *et al.*, 2004). The Env protein of most avian leukosis viruses (ALV) only becomes responsive to low pH after an initial priming step induced by receptor (Mothes *et al.*, 2000; Narayan *et al.*, 2003). HIV Env is activated for fusion by interaction with receptor as well as co-receptor (Doms and Trono, 2000). Thus, in general, multifactorial activation allows viruses to better orchestrate entry events and induce membrane fusion at the desired location.

Conformational changes leading to membrane fusion

Interaction of Env with receptor/co-receptor or exposure to low pH induces the conformational changes that lead to membrane fusion. Current insights into the fusion reaction are based on the determination of various structural intermediates. For an increasing number of viral glycoproteins, X-ray structures of the precursor protein, the proteolytically activated glycoprotein and the final post-fusion conformation have been determined. Based on these structures, we distinguish three types of viral fusion machines at the time of writing (Harrison, 2005; Kielian and Rey, 2006). Type I fusion proteins are exemplified by influenza A virus haemagglutinin (HA) and use a series of alpha helix and loop rearrangements to promote fusion (Bullough *et al.*, 1994; Carr and Kim, 1993; Wilson *et al.*, 1981). The post-fusion conformation of most retroviruses resembles that of HA, suggesting that retroviral Env belong to type I fusion machines (Skehel and Wiley, 1998). In contrast to type I fusion machines, type II fusion machines, represented by alphavirus E1 and flavivirus E proteins, rearrange entire protein domains rather than alpha helixes and loops (Bressanelli *et al.*, 2004; Gibbons *et al.*, 2004; Modis *et al.*, 2004). The G protein of vesicular stomatitis virus (VSV) is the founding member of a third subgroup that exhibits features of both types of fusion machines (Roche *et al.*, 2006; Roche *et al.*, 2007).

Using influenza A virus haemagglutinin as a model protein for retroviral fusion the following individual fusion steps can be imagined (Skehel and Wiley, 2000): Proteolytic processing of the precursor HA0 would cleave close to a hydrophobic region called the fusion peptide generating the metastable HA1/2 dimer (Fig. 4.1A). By binding to sialic acid on surface proteins or glycolipids, HA would then be internalized to a low pH endosome. Binding to receptor sialic acid is not known to induce any conformational changes. However, exposure to endosomal low pH would subsequently destabilize the globular head of the HA1 trimer and trigger the 'spring-loaded' transmembrane HA2 subunit to expose the N-terminal fusion peptide (Fig. 4.1B) (Carr *et al.*, 1997; Durrer *et al.*, 1996; Stegmann *et al.*, 1990). This is accomplished by a conversion of a loop into a long continuous alpha helix thereby exposing the fusion peptide at the tip for insertion into the target membrane (Fig. 4.1B). Fusion is probably then driven by a second wave of conformational changes, whereby a helix turns into a loop, causing an anti-parallel backfolding of the protein (Fig. 4.1B and C) (Bullough *et al.*, 1994). These conformational changes lead to the formation of a highly stable coiled-coil structure in which the fusion peptide and the transmembrane anchor are located on the same side of the molecule (Fig. 4.1B and C). The resulting trimer of TM hairpins forms a highly stable hexamer of alpha helixes, which is the defining feature of type I fusion machines.

The parallel lipid reaction proceeds through a hemifusion intermediate with merged outer lipid leaflets, followed by the fusion of inner leaflets to generate a fusion pore (Fig. 4.1C) (Kemble *et al.*, 1994; Melikyan *et al.*, 2005; Melikyan *et al.*, 1995). The widening of an initially flickering fusion pore probably represents the critical step in the fusion reaction and requires cooperativity between several HA trimers (Chernomordik *et al.*, 2006; Cohen and Melikyan, 2004; Markovic *et al.*, 2001). Viral and cellular lipids such as ceramide, cholesterol and sphingomyelin, probably play an important role

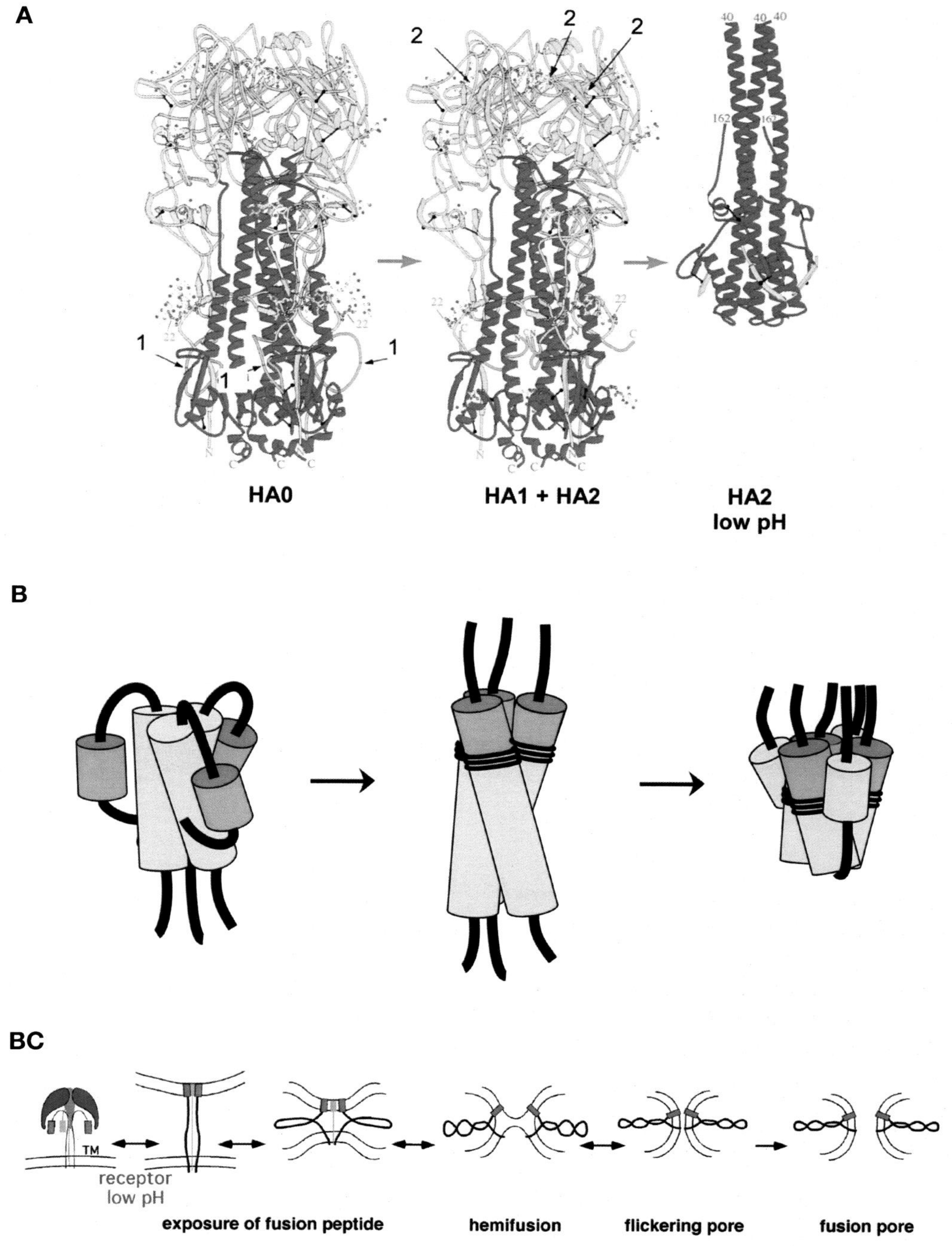

Figure 4.1 Membrane fusion mediated by type I fusion proteins. (A) Conformation changes that take place in influenza A haemagglutinin (HA), a prototype type I fusion protein. Reprinted from Skehel and Wiley (2000) with permission from Annual Reviews. Depicted snapshots represent the precursor HA protein (HA0), the activated metastable HA1/2 dimer after proteolytic cleavage in the surface loop (1) and the post-fusion conformation of HA2 at low pH. The globular head of HA1 binds receptor sialic acid (2). (B) Scheme to illustrate the basic underlying loop to helix and helix to loop rearrangements of HA2. (C) The spring-loaded exposure of the fusion peptide followed by the back folding of the molecule to a structure where fusion peptide and transmembrane anchor to the same side of the molecule represents a plastic working model of how viral type I fusion proteins force membranes together for fusion. Membrane fusion proceeds through a series of events involving a hemifusion intermediate, a flickering pore, fusion pore and pore enlargement.

in the fusion reaction (Chernomordik *et al.*, 2006; Rawat *et al.*, 2003).

Structural as well as sequence comparison of TM proteins of retroviruses suggest that most Env proteins are type I fusion machines (Caffrey *et al.*, 1998; Chan *et al.*, 1997; Fass *et al.*, 1996; Kobe *et al.*, 1999; Malashkevich *et al.*, 1998; Tan *et al.*, 1997; Weissenhorn *et al.*, 1997). However, in contrast to HA, Env of HIV and most gamma retroviruses are activated by interaction with receptor and not exposure to low pH (Hernandez *et al.*, 1996; Kolokoltsov and Davey, 2004; Maddon *et al.*, 1988; McClure *et al.*, 1988; Mothes *et al.*, 2000; Stein *et al.*, 1987; Wallin *et al.*, 2004). In response to receptor binding, conformational changes in SU lead to the activation of the underlying TM protein for fusion. In the case of HIV, Env is activated by an ordered multi-step process that at the same time protects Env from neutralizing antibodies (Fig. 4.2A) (Kwong, 2005; Moore and Doms, 2003). The primary receptor CD4 recognizes a constitutively exposed protein region on the surface of the highly glycosylated Env (Chen *et al.*, 2005; Kwong *et al.*, 1998). Multiple CD4 molecules are required to establish an affinity high enough for this region of low immunogenicity (Zhou *et al.*, 2007). However, once binding is established half of the Env protein dramatically refolds. As a result, a bridging sheet is formed that now exposes the binding site for the HIV co-receptor (Kwong *et al.*, 1998; Rizzuto *et al.*, 1998; Zhou *et al.*, 2007). In the case of CCR5, sulphated tyrosine residues at the N-terminus of the receptor are critical for the protein–protein interaction (Farzan *et al.*, 1999). Co-receptor binding probably then leads to shedding of SU, thereby activating the underlying TM for membrane fusion.

In the case of the murine leukaemia virus (MLV), HTLV and other gamma retroviruses, SU consists of two domains separated by a proline-rich linker; the N-terminal receptor-binding domain (RBD) and the C-terminal domain (Davey *et al.*, 1997; Fass *et al.*, 1997; Kim *et al.*, 2000; Manel *et al.*, 2003). Interestingly, all these viruses preferentially use channel proteins as cellular receptors (Young, 2001). In response to receptor binding, RBD changes its conformation and can then communicate with the C-terminal half of SU and the underlying TM (Barnett and Cunningham, 2001; Lavillette *et al.*, 2001). Interestingly, critical to the structural rearrangements induced by receptor is a CXXC motif in SU (Pinter *et al.*, 1997). CXXC motifs are typically found in disulphide isomerases and the presence of such motifs suggests that reshuffling of disulphide bridges contributes to the activation of the fusion protein (Pinter *et al.*, 1997). Initially, after proteolytic activation of Env, MLV SU and TM remain tethered by a disulphide bridge between a cysteine in the CXXC motif and a conserved cysteine in the CX6CC motif of TM. However, in response to receptor binding, Env reacts like a disulphide isomerase and reshuffles the cysteine bridge so that the two cysteines within the CXXC motif are now linked (Fig. 4.2B) (Wallin *et al.*, 2004). As a consequence, oxidized SU is released from TM, thereby inducing the conformational changes that lead to fusion. Consistent with the concept of metastability, these disulphide bridge rearrangements can also be induced by heat, urea or detergent (Wallin *et al.*, 2004, 2005). Thus, in the case of gamma retroviruses, the metastable TM is suppressed by SU and activated after its release. In the case of HA, low pH probably plays a similar role by destabilizing the globular head of HA1 and releasing the 'TM' equivalent HA2 (Huang *et al.*, 2002).

Finally, an interesting twist to the way viral Env proteins are activated for fusion comes from the ALV that uses a combination of receptor and low pH (Barnard *et al.*, 2004; Matsuyama *et al.*, 2004; Mothes *et al.*, 2000; Smith *et al.*, 2004). Neither low pH nor soluble receptor alone can activate ALV Env. In contrast, in the combined presence of soluble receptor and low pH, Env is fully activated and the formation of the highly stable Env trimer is observed (Fig. 4.3A and B). Because the reaction is performed *in vitro* in the absence of target membranes the irreversible triggering of Env by receptor and low pH results in complete inactivation of the virus (Fig. 4.3B). Last, but not least, cell–cell fusion (syncytia formation) is observed in tissue culture in response to low pH only when Env expressing cells can interact with cell expressing cognate receptor (Fig. 4.3C). In closing, the discussed examples of influenza HA, HIV, MLV and ALV demonstrate how related type I fusion machines can be activated by different cellular triggers.

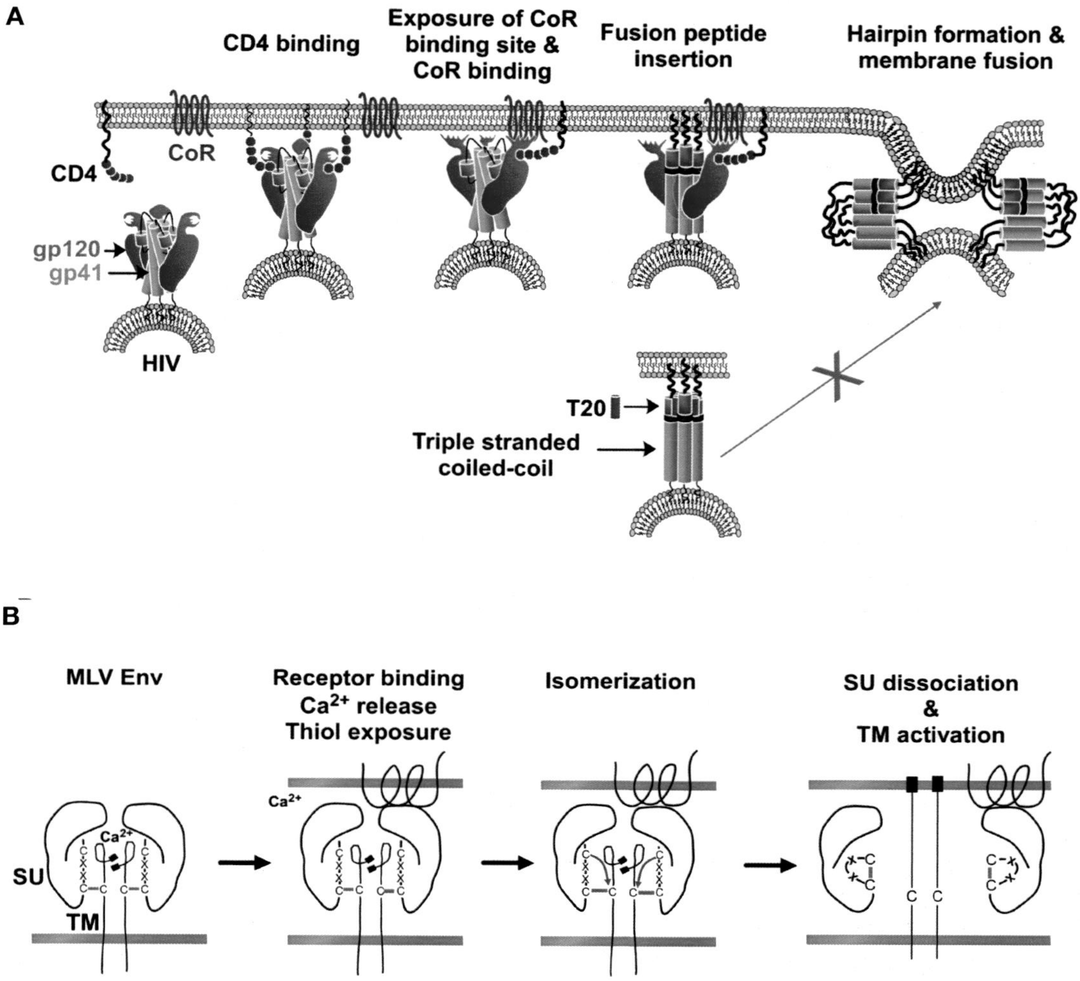

Figure 4.2 Conformational changes in HIV and MLV Env leading to membrane fusion. (A) Binding of CD4 to SU (gp120) of HIV Env results in formation of a bridging sheet, which exposes a conserved co-receptor (CoR) binding site. Co-receptor binding then probably triggers conformational changes leading to exposure of fusion peptide in the TM (gp41). The fusion peptide inserts into the target cell membrane and is followed by formation of a helical hairpin structure in which gp41 folds back on itself thereby bringing both membranes together for membrane fusion. The bottom portion of the figure displays how addition of enfuvirtide (T20 peptide) blocks membrane fusion by preventing the formation of the hairpin structure. Figure, courtesy of Bob Doms. (B) Schematic model for MLV fusion. Isomerization of the SU–TM disulphide-bond controls the fusion function in MLV Env. Two SU–TM complexes of the trimeric and Ca^{++} stabilized Env are shown. Each consists of the membrane-anchored TM subunit, with the fusion peptide (arrow), and the peripheral SU subunit. The SU–TM disulphide-bond and the associated CXXC motif in SU are indicated. Binding to receptor leads to the isomerization of the CXXC, the release of oxidized SU, thereby inducing the conformational changes within TM that lead to the insertion of the fusion peptide into the target membrane. Adapted from Wallin *et al.* (2004) by permission from Macmillan Publishers Ltd.

Entry under the pressure of adaptive immune response

A critical challenge for viral fusion machines is that conformational changes have to take place in the presence of neutralizing antibodies (Burton *et al.*, 2004; Kwong, 2005). While retroviruses can mutate their envelope proteins to escape antibodies, the host likewise mutates the variable region of immunoglobulin proteins to generate new neutralizing antibodies. The cat and mouse

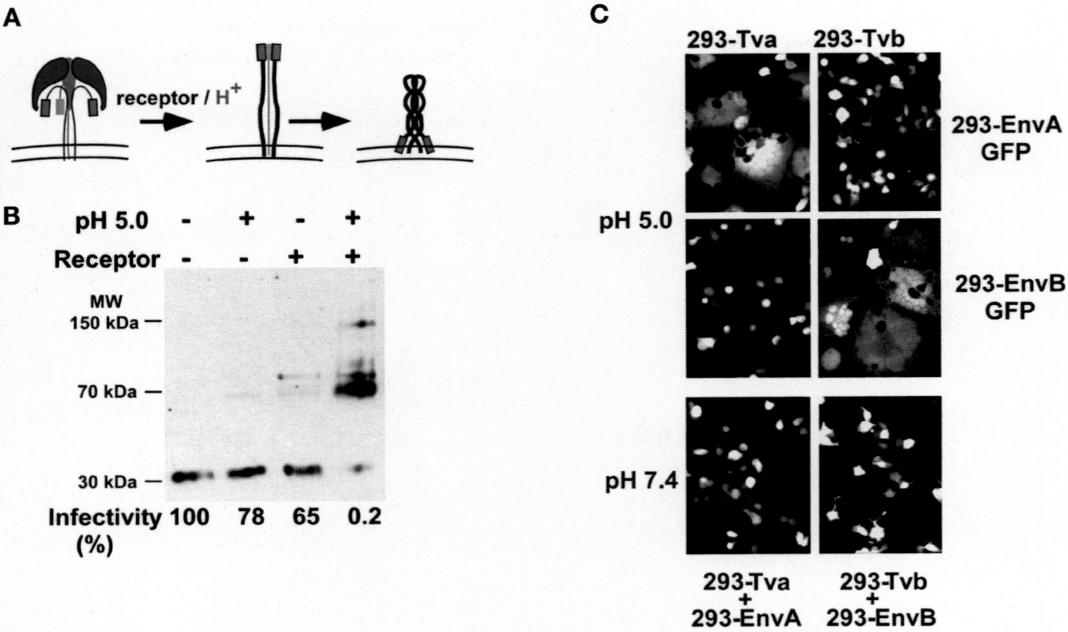

Figure 4.3 ALV enters cells by receptor priming and low pH triggering. (A) Schematic representation of the conformational changes in ALV TM induced by receptor and low pH. When performed with purified virions in the absence of target membranes, TM activation leads to the insertion into the viral membrane thereby inactivating the protein. (B) Conversion of monomeric TM to high molecular weight species, which probably represents the post-fusion conformation of ALV Env as examined by western blot using TM-specific antibodies. The blot shows that both receptor interaction and low-pH induction individually do not activate ALV Env (lanes 1–3). In contrast, majority of the Env proteins are activated when both receptor and low-pH are applied (lane 4). The formation of the activated high molecular weight species of TM correlates with the complete inactivation of ALV (per cent infectivity below the blot). (C) Syncytium formation with human 293 cells. Transfected 293 cells that express Tva or Tvb were mixed with transfected 293 cells expressing GFP and EnvA or EnvB. Cell mixtures were treated at a pH of 5.0 or 7.4 and images were recorded with a FITC filter. The panels show that activation of ALV Env as monitored by synctia formation depends on both specific receptor and low pH. (B) and (C) were reprinted from Mothes *et al.* (2000) with permission from Elsevier.

game is further complicated by the fact that the binding affinity between Env and its activating receptors must follow identical biochemical principles as the interaction between Env and neutralizing antibodies. Hydrophobic interfaces or salt-bridges can generate enough affinity to promote a protein–protein interaction. Viral Env proteins have adopted several strategies to evade neutralizing antibodies, but still generate enough affinity to induce the conformational changes. First, Env proteins are heavily glycosylated effectively shielding the protein surface (Wei *et al.*, 2003; Wyatt *et al.*, 1998). Second, HIV sterically limits physical access of antibodies to receptor/co-receptor binding sites (Kwong *et al.*, 1998). Only artificial Fab fragments have access to the important co-receptor binding region of HIV. Third, conformational changes are perfectly timed in sequential steps to only transiently expose the critical regions to receptors (Melikyan *et al.*, 2000b). Fourth, viruses use cooperativity between multiple Env trimers, whereby multiple low-affinity interactions generate enough critical mass to induce conformational changes (Markovic *et al.*, 2001). Consequently, the few isolated neutralizing antibodies against HIV recognize their target often in unusual ways, for instance by binding to glycans, recognizing short-lived intermediates or by carrying sulphated tyrosine residues in molecular mimicry of the N-terminus of CCR5 (Burton *et al.*, 2004; Choe *et al.*, 2003; Huang *et al.*, 2007). Whether these antibodies will allow an effective antiviral therapy is still an open question (Montefiori *et al.*, 2007).

In contrast to HIV, a master of antibody evasion, MMTV and MLV are antibody-controlled in adult mice. The virus is passed on from mother to pup prior to the development of a full immune system in the offspring (Finke and Acha-Orbea, 2001).

Entry inhibitors

Progress in the field of retroviral entry has led to new antiviral therapies that interfere with this step. In general, a prerequisite for a successful inhibitor is that it should be specific for the virus with little or no side effects on the cell. Thus, commonly used anti-HIV therapies are based on inhibitors targeting the three viral enzymes reverse transcriptase, protease and more recently integrase. Env must also be viewed as a specific viral enzyme mediating membrane fusion. Indeed, enfuvirtide (T-20) has joined the FDA-approved list of HIV inhibitors (Kilby *et al.*, 1998; Wild *et al.*, 1993). This peptide inhibitor is homologous to one of the helix of the six-helix bundle. By intercalating into the TM structure with high affinity, it prevents the formation of the antiparallel six-helix bundle (Fig. 4.2A). Because the six-helix bundle is the defining feature of type I fusion machines and the most conserved property of the highly variable HIV Env, it represents a good target (Moore and Doms, 2003; Skehel and Wiley, 1998). In addition, several small molecule inhibitors target every single step of HIV entry, e.g. the binding of CD4 to Env and the engagement of co-receptor (Moore and Doms, 2003). Most of these inhibitors target less conserved regions and are more strain-specific as compared to enfuvirtide. Inhibitors that block the interaction of Env with receptor and co-receptor should – if possible – not interfere with the recognition of Env by neutralizing antibodies. However, because the thermodynamics of protein–protein or protein–drug interactions follow similar basic rules, these motifs are often targeted by drugs as well as antibodies (Zhou *et al.*, 2007). Despite these difficulties, it is exciting that entry inhibitors are entering the clinics.

Cell biology of retroviral entry

In this section we will explore the entry process from a cellular perspective. In principle, any existing cellular entry pathway can be utilized by viruses and as such, viruses have been instrumental tools in characterizing various forms of endocytosis (Brandenburg and Zhuang, 2007; Marsh and Helenius, 2006; Pelkmans and Helenius, 2003). Viruses entering cells via endocytosis will first engage a receptor at the plasma membrane that in response to virus binding will mediate their transport to endosomal and lysosomal compartments (Fig. 4.4). Within endosomes, many viruses utilize low pH to trigger their fusion machine. Other endocytosis-dependent viruses can use different endosomal features to trigger fusion. Besides low pH, endosomes are rich in proteases, specific reductases and oxidative species such as nitric oxide. Any of these 'cues' could in principle be used with the identical outcome that fusion is induced in endosomes. Ebola and SARS viruses, for instance, use endosomal cathepsins to be activated for fusion in endosomes (Chandran *et al.*, 2005; Simmons *et al.*, 2005). Both viruses only appear to be 'pH dependent', because cathepsins require low pH for full activity. Thus, viruses can evolve to utilize any factor found in a specific cellular compartment or microenvironment as the trigger for its fusion machine. Having illustrated the cell biological point of view, we will now focus on individual steps beginning with binding to cell receptors.

Virus surfing towards entry sites

Initial attachment to cells is often mediated by non-specific low affinity receptors (Mondor *et al.*, 1998; Pizzato *et al.*, 1999). The inefficiency of this binding reaction is major cause of the observed low infectivity to particle ratio of many viruses (Damico and Bates, 2000; O'Doherty *et al.*, 2000; Roan and Greene, 2007; van der Schaar *et al.*, 2007). After binding to cellular protrusions such as filopodia and ruffles, viral particles then engage specific receptors and move towards cellular entry sites in an actin-dependent process (Fig. 4.4) (Ewers *et al.*, 2005; Lehmann *et al.*, 2005; Rust *et al.*, 2004). Receptor oligomerization in response to virus binding engages the underlying retrograde flow of actin, causing viruses to

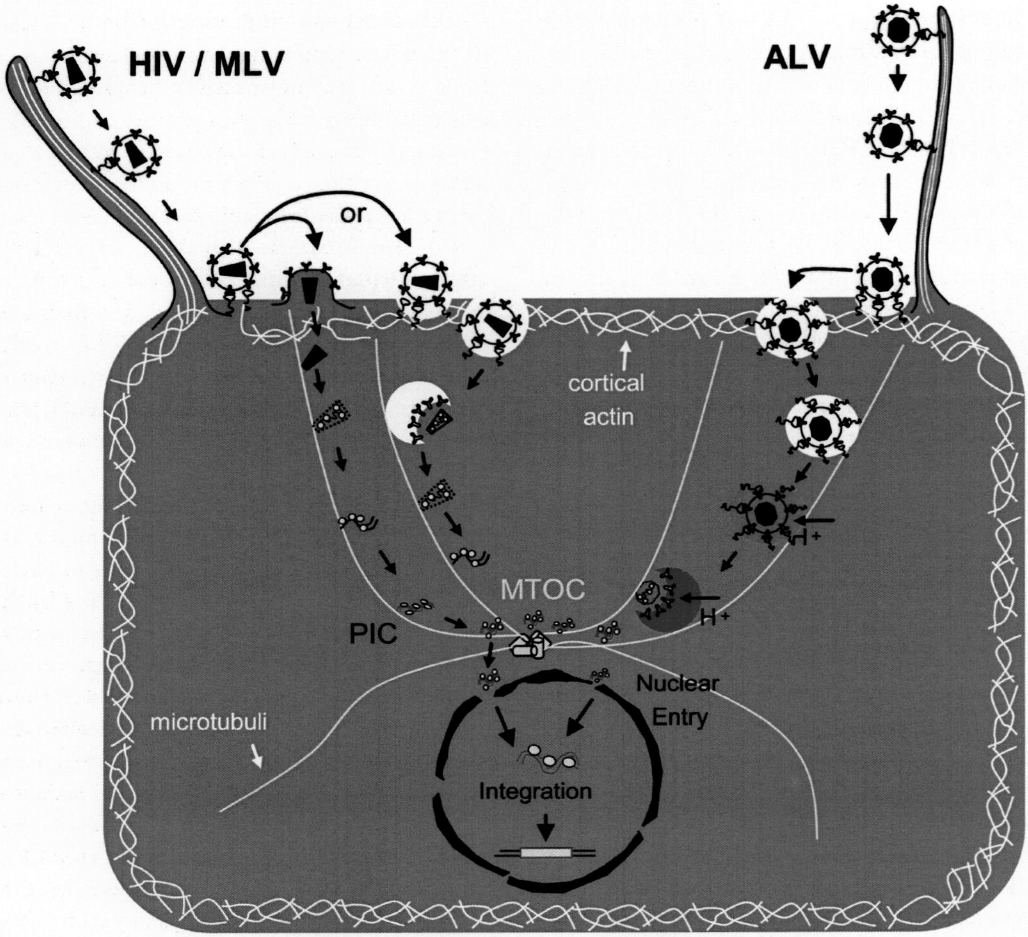

Figure 4.4 Cell biology of retroviral entry. Virus entry probably begins by binding to cellular protrusions such a filopodia. This is followed by receptor engagement and signalling to establish a link to the underlying retrograde actin flow. Viruses then surf along filopodia to reach entry sites where they can cross the cortical actin cytoskeleton. In the case of HIV and MLV, fusion can occur either at the plasma membrane or via endocytosis. In contrast, ALV fusion is dependent on both receptor priming and acidification of endocytic compartment. Fusion leads to release of viral cores into the cytoplasm. This is followed by uncoating/reverse transcription and formation of pre-integration complex (PIC). These complexes use microtubules to reach the microtubule organizing centre (MTOC) before entering the nucleus to finally integrate into the host chromosome.

surf towards endocytic hot spots at the cell body (Lehmann *et al.*, 2005). A similar movement of the epidermal growth factor (EGF) along the surface of filopodia depends on active signalling and the kinase activity within the cytoplasmic tail of EGF receptor (Lidke *et al.*, 2004, 2005). Most retroviruses, including pH-independent HIV and MLV move towards these hot spots despite the fact that interaction with receptor is believed to be sufficient to induce fusion. Membrane fusion into filopodia is likely delayed in the periphery of the cell because a tight tethering of the membrane to the underlying cortical actin creates a high membrane tension (Sheetz, 2001). In contrast, membrane mixing of viral and cellular membranes is probably facilitated at the cell body where membrane tension is low.

Cell entry and crossing the cortical actin cytoskeleton

Following actin-dependent motility towards entry sites, viruses are then either endocytosed or interaction with receptor induces membrane fusion at the plasma membrane (Fig. 4.4). HIV must recruit CD4 and co-receptors from distinct microdomains in an actin-dependent process (Del Real *et al.*, 2002; Gomez and Hope, 2005; Iyengar *et al.*, 1998; Jimenez-Baranda *et al.*, 2007; Steffens and Hope, 2004). Other viruses such as ALV are endocytosed (Mothes *et al.*, 2000). At present, six different forms of endocytosis are being distinguished, a number that is likely to grow (Marsh and Helenius, 2006). Depending of the viral receptor isoform, subgroup A of ALV uses either dynamin or caveolin-dependent endocytosis (Narayan *et al.*, 2003). In contrast, HIV and MLV may enter cells in a pH-independent manner (Kolokoltsov and Davey, 2004; Maddon *et al.*, 1988; McClure *et al.*, 1988; Stein *et al.*, 1987; Wallin *et al.*, 2004). For both viruses, expression of Env on the surface of cells induces strong syncytia in tissue culture at neutral pH, entry cannot be inhibited by lysomotropic agents (de Duve *et al.*, 1974) and low pH cannot irreversibly inactivate purified virions. However, despite the pH-independence, both viruses enter certain cell types by endocytosis (Daecke *et al.*, 2005; Katen *et al.*, 2001; Kizhatil and Albritton, 1997; Marechal *et al.*, 2001; McClure *et al.*, 1990; Miyauchi *et al.*, 2009; Vidricaire *et al.*, 2004). The fact, that the literature on the entry pathway of HIV and MLV is controversial may reflect a real flexibility on the viral side, differences between various cell types as well as a certain promiscuity on the cellular side. This is also corroborated by the observation, that retroviruses can be 'pseudotyped' with the Env proteins of other viruses. For instance, all retroviral capsids can incorporate VSVG into their envelope and infect cells with high efficiency using a low pH-dependent endocytic pathway (Aiken, 1997; Chazal *et al.*, 2001).

The opposite may not necessarily be possible. Low pH and/or endocytosis-dependent viruses such as ALV cannot efficiently enter cells when fusion is induced at the cell surface, e.g. by treatment of cells with bound virus with low pH (Brandenburg *et al.*, 2007; Marsh and Bron, 1997; Miyauchi *et al.*, 2009; Mothes *et al.*, 2000; Tscherne *et al.*, 2006). Viruses that evolved to enter cells via endocytosis may need to travel through the cortical actin cytoskeleton within an endosome and later use the microtubule-dependent motility of endosomes to reach a perinuclear location. Thus, these viruses exploit the trafficking routes of endosomal cargo containers for their journey into the cell (Fig. 4.4). Endosomes are pushed through the peripheral actin cytoskeleton by the formation of short actin tails (Merrifield *et al.*, 2002; Merrifield *et al.*, 1999; Pelkmans *et al.*, 2002; Taunton *et al.*, 2000). Dynamin plays a role in actin nucleation during this process and myosin 6, a minus-end actin motor, mediates minus-end motility of vesicles from the cell periphery towards the nucleus (Aschenbrenner *et al.*, 2003; Lee and De Camilli, 2002; Orth *et al.*, 2002; Sun and Whittaker, 2007).

In contrast, capsids of viruses that can enter cells at the plasma membrane must have acquired the ability to cross the cortical actin cytoskeleton and travel within the cytoplasm (Fig. 4.4). The initial recruitment of receptor to viruses at the cell surface probably already induces subsequent changes in the cortical actin cytoskeleton (Yoder *et al.*, 2008). One factor modulating the ability of HIV to cross the cortical actin could be nef. The infectivity of HIV lacking nef is significantly reduced (Chowers *et al.*, 1994; Miller *et al.*, 1994). This defect can be rescued in two ways. First, the infectivity is restored when HIV is sent into cells via an endocytic route – e.g. by using VSVG protein as an Env protein (Aiken, 1997; Chazal *et al.*, 2001). Second, the defect can be overcome by dissolving the cortical actin cytoskeleton with lactrunculin B (Campbell *et al.*, 2004). Thus, nef probably confers HIV capsids the ability to travel through the dense cortical actin cytoskeleton. Interestingly, nef must be expressed in the producer cells to manifest its effect on virus entry (Aiken and Trono, 1995; Chowers *et al.*, 1995; Miller *et al.*, 1995). While the exact mechanism by which nef enhances HIV infectivity remains to be determined, recent data demonstrate that nef must associate with dynamin-2 in the producer cell (Pizzato *et al.*, 2007).

Cytoplasmic trafficking and nuclear entry

Having crossed the cortical actin, retroviral capsids then probably switch from actin to microtubule mediated motility to move towards the nucleus (Fig. 4.4) (Goff, 2007; Greber and

Way, 2006). HIV and Foamy viruses, like herpes and adenoviruses use a dynein-dependent movement along microtubules to accumulate in a perinuclear region around the microtubule organizing centre (MTOC) (Dohner *et al.*, 2002; McDonald *et al.*, 2002; Petit *et al.*, 2003; Saib *et al.*, 1997; Sodeik *et al.*, 1997; Suomalainen *et al.*, 1999). How capsids then move from the MTOC to the nucleus is not known, though the process may involve an actin-mediated step (Arhel *et al.*, 2006). The preintegration complex of HIV then travels through the nuclear pore (see Chapter 5). Retroviruses fundamentally differ in their ability to enter the nucleus of non-dividing cells (Katz *et al.*, 2005; Yamashita and Emerman, 2006). HIV and to some extent ALV can infect non-dividing cells, but MLV and Foamy virus entry into the nucleus depends on the progression of the cell cycle (Bieniasz *et al.*, 1995; Bukrinsky *et al.*, 1992; Goff, 2007; Hatziioannou and Goff, 2001; Nisole and Saib, 2004; Roe *et al.*, 1993). While entry into the nucleus is one of the most controversial fields in HIV research (see Chapter 5), recent evidence demonstrates that the determinants lie in the capsid protein. HIV carrying a MLV capsid protein becomes cell cycle dependent (Yamashita and Emerman, 2004). In fact, just two amino acids substitutions in the HIV capsid protein turn HIV into an MLV-like retrovirus that has lost its ability to enter non-dividing cells and now depends on the cell cycle (Yamashita *et al.*, 2007).

Reverse transcription and uncoating

At some point after fusion and before integration into chromosomal DNA of the host, viral capsids uncoat and the RNA genome is reverse transcribed into DNA. Reverse transcription may proceed as soon as intact capsids have access to nucleotides following fusion. *In vitro*, purified virions undergo efficient reverse transcription in the presence of nucleotides (Baltimore, 1970; Retzel *et al.*, 1980; Temin and Mizutani, 1970). However, completion of reverse transcription may depend on nuclear factors (Arhel *et al.*, 2007; Narayan and Young, 2004).

Very little is known about the uncoating of retroviral capsids, but it is predicted to depend on specific cellular factors and ATP (Goff, 2007; Narayan and Young, 2004). Before we review the few data reported on uncoating, it is helpful to revisit how the capsid is initially generated. During assembly and budding of HIV, the Gag polyprotein precursor protein assembles into a highly stable and spherical immature capsid (see Chapter 7). During or shortly after completion of budding, activation of the viral protease initiates the maturation of the Gag precursor protein into its individual subunits. Matrix protein remains bound to the lipid bilayer, nucleocapsid binds to the genome and a subset of monomeric capsid molecules reassemble *de novo* into the conical HIV core that is so characteristic of the mature and infectious HIV. Importantly, the resulting mature capsids are more fragile (Freed, 1998; Gottlinger, 2001; Morita and Sundquist, 2004; Wiegers *et al.*, 1998). The instability of the mature capsid can be viewed as the first critical step towards the uncoating reaction. Similar to the generation of a metastable Env protein that can be triggered for fusion, capsid maturation may generate a metastable shell that can be triggered for uncoating once it reaches the cytoplasm.

Where and when HIV capsids uncoat is controversial. At present, two main working models exist. In one model, uncoating would precede reverse transcription (Forshey *et al.*, 2002; Zhang *et al.*, 2000). Support for this model comes from the observation that mutant HIV capsids with altered stability are impaired in reverse transcription (Forshey *et al.*, 2002). Following uncoating and reverse transcription, the preintegration complex (PIC) would form. Biochemical evidence suggests that most of the capsid proteins are removed from the HIV PIC (Fassati and Goff, 2001; Miller *et al.*, 1997), but stay associated with the MLV PIC (Bowerman *et al.*, 1989; Fassati and Goff, 1999). In contrast, the second model proposes that uncoating occurs directly at the nuclear pore (Arhel *et al.*, 2007). The formation of a 'DNA Flap' during the completion of reverse transcription would trigger uncoating and subsequent nuclear entry directly at the nuclear pore (Arhel *et al.*, 2007). HIV cores carrying capsid protein accumulated at the nuclear pore when the DNA Flap was mutated. Support for a role in nuclear factors in late events of reverse transcription of ALV were also observed in a cell-free uncoating assay (Narayan and Young, 2004). While current experimental evidence is apparently contradictory, an emerging role of capsid in the nuclear entry of HIV may indicate that common ground will be found (Dismuke and Aiken, 2006; McDonald *et al.*, 2002; Yamashita and Emerman, 2004; Yamashita *et al.*, 2007). Finally, nuclear import is probably regulated by phosphorylation and SUMOylation (Yueh and Goff, 2003; Yueh *et al.*, 2006).

In contrast to HIV, for foamy virus, reverse transcription already occurs during assembly and budding in the producer cell (Yu *et al.*, 1996). Foamy virus capsids are highly stable and uncoating is linked to its release from the MTOC (Lehmann-Che *et al.*, 2007; Saib *et al.*, 1997). In quiescent T-cells, Foamy viruses can stay stably at the MTOC for weeks. Release of particles from the MTOC and subsequent uncoating only occurs upon activation of cells (Lehmann-Che *et al.*, 2007).

Innate cellular factors targeting incoming viruses

Classical innate immunity describes the ability of cells to recognize foreign patterns such as genomes, capsids, envelopes and lipids and to mount an antiviral response (Honda *et al.*, 2006; Iwasaki and Medzhitov, 2004; Kawai and Akira, 2006; Kunzi and Pitha, 2003). In the case of retroviruses, genome and capsid shell probably represent structures that are perceived as 'foreign' by cells. Indeed, TRIM5, an E3 ubiquitin ligase, can recognize incoming retroviral capsids, directly bind to them and induce their premature disassembly prior to reverse transcription (Sayah *et al.*, 2004; Sebastian and Luban, 2005; Stremlau *et al.*, 2004; Stremlau *et al.*, 2006). APOBEC3G, is a cytidine deaminase that is packaged into retroviruses and exerts its antiviral effect during reverse transcription (Harris and Liddament, 2004; Okeoma *et al.*, 2007). As such, early retroviral entry steps have evolved under the pressure to escape an innate or intrinsic cellular response (Bieniasz, 2004; Goff, 2004; Nisole *et al.*, 2005; Sokolskaja and Luban, 2006). Consequently, the virus has an interest to shield and protect its genome and fortify its capsid while still interfacing with essential cellular factors. Innate immunity to retroviruses is extensively covered in Chapter 11. Here it is important to realize that the ability of cells to recognize and fight incoming genomes and capsids affects our thinking about retroviral entry and uncoating. Drastic uncoating steps that expose the vulnerable genome are unlikely. Direct uncoating and injection of genomes at the nuclear pore as observed for Herpes and adenovirus and proposed for HIV would represent an appealing solution to the problem (Arhel *et al.*, 2007). In this context, it is also intriguing that the highly stable foamy capsid lacks most lysines. Most structurally important positive charges are replaced by arginines. The vulnerability of retroviral capsids to ubiquitination and proteasome-dependent degradation could be a consequence of an innate cellular defence (Schubert *et al.*, 2000; Wu *et al.*, 2006). On the other hand, viruses mutate quickly and can adapt to cellular responses. HIV is not recognized by the human TRIM5 protein, but is vulnerable to TRIM5 proteins from different species (Sawyer *et al.*, 2005; Stremlau *et al.*, 2004). The viral protein vif counteracts the antiviral APOBEC3G (Bieniasz, 2004; Goff, 2004; Harris and Liddament, 2004). Thus, the high mutation rate allows the virus to evade the innate immune system. Retroviruses even have the potential to use the antiviral response towards its advantage and to integrate it into their life cycle. The incorporation of NFkB sites into its LTR (Alcami *et al.*, 1995; Hiscott *et al.*, 2001) represents a particularly clever move by HIV because most innate immunity signalling pathways induce NFkB (Honda *et al.*, 2006; Iwasaki and Medzhitov, 2004; Kawai and Akira, 2006; Kunzi and Pitha, 2003). It is also intriguing to speculate that TRIM5-like activities could be utilized by retroviruses to promote uncoating of retroviral capsids (Stremlau *et al.*, 2004).

The recent progress in the field of innate immunity to retroviruses may lead to novel antiviral therapies that aim to strengthen the innate recognition of retroviruses. Small molecules may be isolated that either restore the antiviral activity of human TRIM5 to recognize HIV or interfere with the inhibitory effect of vif on APOBEC3G. In the see-saw game between immune evasion/innate immunity, antiviral therapy would augment the host giving protection an advantage.

Virus entry in the context of retroviral pathogenesis

Towards the end of this chapter we would like to place the individual aspects into the broader context of retroviral pathogenesis. First, HIV is predominantly a sexually transmitted disease. Therefore, the first entry event into cells occurs at vaginal and rectal mucosal surfaces (Bomsel and Alfsen, 2003). Interestingly, semen contains a component that strongly enhances the binding of HIV to cells (Munch *et al.*, 2007). In the case of mother to child transmission via breast milk HIV is transmitted via the tonsillar and gastrointestinal mucosa (Kourtis *et al.*, 2003; Miotti *et al.*, 1999). Given the importance of mucosal infection in retroviral pathogenesis, astonishingly little is known about the mechanism by which HIV crosses the epithelial barrier. HIV and MLV probably first bind to microvilli at the apical surface of polarized epithelial cells to move to the base of microvilli, are then endocytosed and transcytosed to the basolateral side of the cell layer (Alfsen *et al.*, 2005; Bomsel and Alfsen, 2003; Lehmann *et al.*, 2005).

Once retroviruses reach the basolateral side, they can infect underlying lymphocytes or even infect the mucosal cell layer through the basolateral face (Bomsel and Alfsen, 2003). The tropism of each retroviral infection is then determined by multiple factors such as receptor expression, presence of restricting or essential host factors and cell cycle progression. In addition, as retroviruses must adapt to innate immunity factors at the cellular level they also take advantage of the natural defence of the whole organism against new invading pathogens. Plasmacytoid and Langerhans dendritic cells play a critical role in the first line of defence to invading pathogens (Colonna *et al.*, 2004; Iwasaki and Medzhitov, 2004; Liu, 2001). As such, retroviral spreading is critically dependent on their ability to replicate in these cells – although often at very low levels (Granelli-Piperno *et al.*, 1999; Lin *et al.*, 2000; Pope *et al.*, 1995; Turville *et al.*, 2004). The recognition of retroviruses by the innate immune system leads to the activation and maturation of dendritic cells (DC) that in response migrate towards lymph nodes (Iwasaki and Medzhitov, 2004). Retroviruses exploit the migratory behaviour of DCs to infect T and B-cells in lymph nodes (Cameron *et al.*, 1992; Courreges *et al.*, 2007; Geijtenbeek *et al.*, 2000; McDonald *et al.*, 2003). The switch from replication in antigen-presenting cells (M-tropic virus using CCR5) to replication in T-cells (T-tropic virus using CXCR4) is slow and represents a critical point in the pathogenesis of AIDS (Doranz *et al.*, 1996; Moore and Doms, 2003).

Efficient coordination of virus entry and release at sites of cell–cell contact

When a retroviral infection spreads through an organism, viruses probably spread predominantly by direct cell–cell contact. Indeed, virus spreading is 100- to 1000-fold more efficient when budding can be directly coupled to entry at sites of cell–cell contact (Carr *et al.*, 1999; Dimitrov *et al.*, 1993; Gupta *et al.*, 1989; Johnson and Huber, 2002; Li and Burrell, 1992; Phillips, 1994). The contribution of cell-free virus to viral spreading is minimal in static cultures (Sourisseau *et al.*, 2007). This has been most dramatically demonstrated for HTLV-1, which is poorly infectious in cell-free form but spreads efficiently among T-cells (Bangham, 2003). Morphological analyses have revealed that, during transmission, HTLV-1 budding and entry are polarized to cell–cell contact zones. These zones are specifically enriched in microtubules, actin and adhesions factors (Barnard *et al.*, 2005; Igakura *et al.*, 2003; Nejmeddine *et al.*, 2005). Such a characteristic contact zone has been named the 'virological' or 'infectious' synapse (Fig. 4.5A)(Igakura *et al.*, 2003; McDonald *et al.*, 2003) due to its resemblance to the immunological synapse formed between T-cells and antigen-presenting cells (Huppa and Davis, 2003). In the case of HIV, infection of T-cells is dramatically stimulated by the addition of dendritic cells (Cameron *et al.*, 1992; Pope *et al.*, 1994; Pope *et al.*, 1995). Dendritic cells capture HIV via surface lectins prior to contact-mediated delivery of HIV to T-cells (Cavrois *et al.*, 2007; Geijtenbeek *et al.*, 2000; Turville *et al.*, 2002). A visual analysis of uninfected T-cells conjugated to HIV-pulsed DCs has revealed the polarized delivery of virus to zones of tight cell–cell contact (McDonald *et al.*, 2003). Further work has noted that DC-captured HIV is trafficked together with the tetraspanins CD9 and CD81, but not CD63 or MHC class II molecules to the infectious synapse (Garcia *et al.*, 2005). Thus, HIV may utilize the formation of immunological cell–cell interfaces

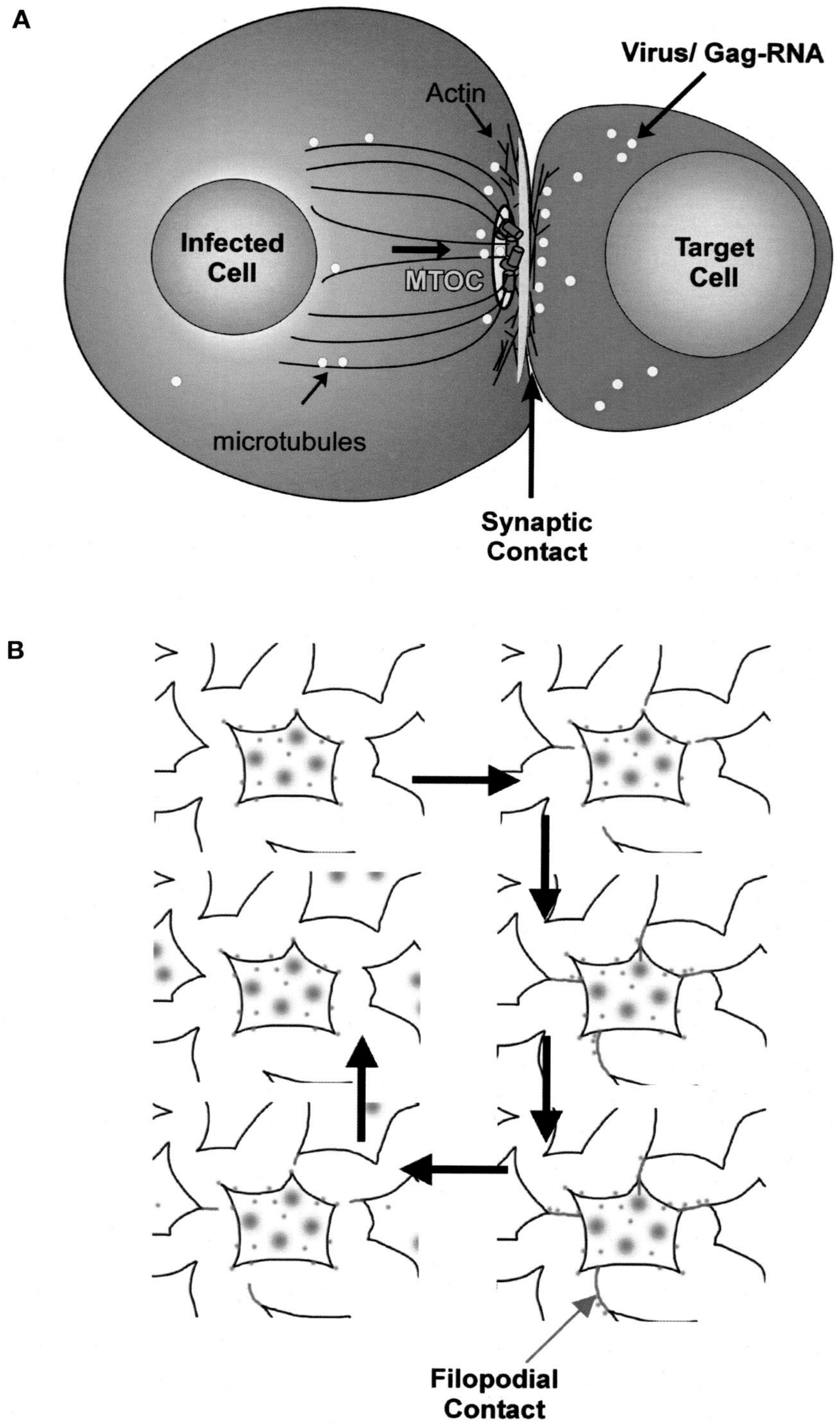

Figure 4.5 Viral spreading at sites of cell–cell contact. (A) Retrovirus spreading from an infected cell to a target cell via synaptic cell–cell contacts, called virological or infectious synapses. Synaptic contacts exhibit a polarization of the MTOC and actin cytoskeleton towards the contact zone. Adapted from Jolly and Sattentau (2004). (B) Retroviral spreading in fibroblasts as observed by time-lapse microscopy. Infected cells would recruit filopodia from neighbouring non-infected cells to establish filopodial bridges. Viruses would then move along the surface of these cell–cell bridges to infect neighbouring cells.

between antigen presenting cells and T-cells for the purpose of viral spreading (McDonald *et al.*, 2003; Sol-Foulon *et al.*, 2007). Spreading via these tight interfaces also carries the advantage that it is resistant to neutralizing antibodies (Chen *et al.*, 2007; Ganesh *et al.*, 2004). Interestingly, HIV expresses nef to simultaneously interfere with the immunological functions of synapses such as antigen presentation (Fackler *et al.*, 2007; Schindler *et al.*, 2006; Sol-Foulon *et al.*, 2007; Thoulouze *et al.*, 2006).

HIV can also induce tight cell–cell contacts between infected and uninfected T-cells (Fig. 4.5A). Similar to immunological synapses, talin concentrates at the cell–cell interface. Formation of these contacts depends on Env–receptor interaction and requires functional actin and myosin (Igakura *et al.*, 2003; Jolly *et al.*, 2004; Nejmeddine *et al.*, 2005; Pearce-Pratt *et al.*, 1994). Consistent with a general role for cell–cell contact in retroviral spread, numerous studies have also noted that HIV budding is polarized to regions of cell–cell contact (Bourinbaiar and Phillips, 1991; Deschambeault *et al.*, 1999; Johnson and Huber, 2002; Phillips, 1994; Phillips and Bourinbaiar, 1992; Phillips *et al.*, 1998).

In contrast to HIV and HTLV that replicate in lymphocytes, other retroviruses such as MLV have a broader tissue range (Finke and Acha-Orbea, 2001). To gain insights into the mechanism of MLV spreading in fibroblasts, our laboratory co-cultured infected cells generating fluorescent MLV virions with receptor expressing target cells. Time-lapse imaging resulted in the first documentation of cell-to-cell transmission of retroviruses in living cells (Sherer *et al.*, 2007). These results revealed a novel mode of viral spreading via thin filopodial bridges (Fig. 4.5B). These filopodia originate from non-infected cells and interact through their tips with infected cells. A strong association of the viral envelope glycoprotein (Env) in an infected cell with the receptor molecules in a target cell generates a stable bridge. Viruses then specifically bud at these sites and move along the outer surface of the filopodial bridge toward the target cell. The entire process of assembly, transfer and fusion can proceed in as little as 60min (Jin *et al.*, 2009).

Stable filopodial bridges, designated as cytonemes or nanotubules, have previously been described in fruit flies and in dendritic cells, which are thought to only transmit HIV via synaptic contacts (Onfelt *et al.*, 2006; Ramirez-Weber and Kornberg, 1999; Rustom *et al.*, 2004). The transmission of retroviruses via cytonemes and synapses probably follows a similar underlying molecular mechanism (Hope, 2007). Interestingly, filopodia play an important role during the formation of neurological synapses (Dent *et al.*, 2007; Niell *et al.*, 2004; Ziv and Smith, 1996). Thus, transmission of viruses via cytonemes/nanotubules and synapses both present the ability of viruses to utilize the ability of cells to communicate with each other to spread efficiently.

Conclusions

Much progress has been made in the field of retroviral entry and as a result the first HIV entry inhibitors are now part of antiviral therapies. However, despite the recent progress, areas such as capsid uncoating and nuclear entry remain poorly understood. New approaches such as genome-wide RNAi screens will reveal host factors involved in these processes (Goff, 2008). It is the hope that future insights into the cell biology of HIV replication can lead to novel antiviral inhibitors targeting cellular factors essential for HIV replication. With the failure of the first HIV vaccine trials our hope is redirected towards understanding the basic cell biology and innate immunity of HIV. As we study each individual process using a reductionist approach, we will also need to revisit findings in the broader context of retroviral pathogenesis. The roles of mucosa in HIV entry and of cell–cell contact in spreading are predicted to yield important new insights. In closing, retroviruses will continue to amaze us in their ability to utilize cellular pathways and evade the immune system.

References

Aiken, C. (1997). Pseudotyping human immunodeficiency virus type 1 (HIV-1) by the glycoprotein of vesicular stomatitis virus targets HIV-1 entry to an endocytic pathway and suppresses both the requirement for Nef and the sensitivity to cyclosporin A. J. Virol. *71*, 5871–5877.

Aiken, C., Konner, J., Landau, N.R., Lenburg, M.E., and Trono, D. (1994). Nef induces CD4 endocytosis: requirement for a critical dileucine motif in the membrane-proximal CD4 cytoplasmic domain. Cell *76*, 853–864.

Aiken, C., and Trono, D. (1995). Nef stimulates human immunodeficiency virus type 1 proviral DNA synthesis. J. Virol. *69*, 5048–5056.

Alcami, J., Lain de Lera, T., Folgueira, L., Pedraza, M.A., Jacque, J.M., Bachelerie, F., Noriega, A.R., Hay, R.T., Harrich, D., Gaynor, R.B., *et al.* (1995). Absolute dependence on kappa B responsive elements for initiation and Tat-mediated amplification of HIV transcription in blood CD4 T-lymphocytes. EMBO J. *14*, 1552–1560.

Alfsen, A., Yu, H., Magerus-Chatinet, A., Schmitt, A., and Bomsel, M. (2005). HIV-1-infected blood mononuclear cells form an integrin- and agrin-dependent viral synapse to induce efficient HIV-1 transcytosis across epithelial cell monolayer. Mol. Biol. Cell *16*, 4267–4279.

Arhel, N., Genovesio, A., Kim, K. A., Miko, S., Perret, E., Olivo-Marin, J. C., Shorte, S., and Charneau, P. (2006). Quantitative four-dimensional tracking of cytoplasmic and nuclear HIV-1 complexes. Nat. Methods 3, 817–824.

Arhel, N. J., Souquere-Besse, S., Munier, S., Souque, P., Guadagnini, S., Rutherford, S., Prevost, M. C., Allen, T. D., and Charneau, P. (2007). HIV-1 DNA Flap formation promotes uncoating of the pre-integration complex at the nuclear pore. EMBO J. *26*, 3025–3037.

Aschenbrenner, L., Lee, T., and Hasson, T. (2003). Myo6 facilitates the translocation of endocytic vesicles from cell peripheries. Mol. Biol. Cell *14*, 2728–2743.

Baltimore, D. (1970). RNA-dependent DNA polymerase in virions of RNA tumour viruses. Nature *226*, 1209–1211.

Bangham, C.R. (2003). Human T-lymphotropic virus type 1 (HTLV-1): persistence and immune control. Int. J. Hematol. *78*, 297–303.

Barnard, A.L., Igakura, T., Tanaka, Y., Taylor, G.P., and Bangham, C.R. (2005). Engagement of specific T-cell surface molecules regulates cytoskeletal polarization in HTLV-1-infected lymphocytes. Blood *106*, 988–995.

Barnard, R.J., Narayan, S., Dornadula, G., Miller, M.D., and Young, J. A. (2004). Low pH is required for avian sarcoma and leukosis virus Env-dependent viral penetration into the cytosol and not for viral uncoating. J. Virol. *78*, 10433–10441.

Barnett, A.L., and Cunningham, J.M. (2001). Receptor binding transforms the surface subunit of the mammalian C-type retrovirus envelope protein from an inhibitor to an activator of fusion. J. Virol. *75*, 9096–9105.

Bieniasz, P.D. (2004). Intrinsic immunity: a front-line defense against viral attack. Nat. Immunol 5, 1109–1115.

Bieniasz, P.D., Weiss, R.A., and McClure, M.O. (1995). Cell cycle dependence of foamy retrovirus infection. J. Virol. *69*, 7295–7299.

Bomsel, M., and Alfsen, A. (2003). Entry of viruses through the epithelial barrier: pathogenic trickery. Nat. Rev. Mol. Cell. Biol. *4*, 57–68.

Bourinbaiar, A.S., and Phillips, D.M. (1991). Transmission of human immunodeficiency virus from monocytes to epithelia. J. Acquir. Immune Defic. Syndr. *4*, 56–63.

Bowerman, B., Brown, P.O., Bishop, J.M., and Varmus, H.E. (1989). A nucleoprotein complex mediates the integration of retroviral DNA. Genes Dev. *3*, 469–478.

Brandenburg, B., Lee, L.Y., Lakadamyali, M., Rust, M.J., Zhuang, X., and Hogle, J.M. (2007). Imaging poliovirus entry in live cells. PLoS Biol. *5*, e183.

Brandenburg, B., and Zhuang, X. (2007). Virus trafficking – learning from single-virus tracking. Nat. Rev. Microbiol. *5*, 197–208.

Bressanelli, S., Stiasny, K., Allison, S. L., Stura, E. A., Duquerroy, S., Lescar, J., Heinz, F. X., and Rey, F. A. (2004). Structure of a flavivirus envelope glycoprotein in its low-pH-induced membrane fusion conformation. EMBO J. *23*, 728–738.

Bukrinsky, M.I., Sharova, N., Dempsey, M.P., Stanwick, T.L., Bukrinskaya, A.G., Haggerty, S., and Stevenson, M. (1992). Active nuclear import of human immunodeficiency virus type 1 preintegration complexes. Proc. Natl. Acad. Sci. U.S.A. *89*, 6580–6584.

Bullough, P.A., Hughson, F.M., Treharne, A.C., Ruigrok, R. W., Skehel, J.J., and Wiley, D.C. (1994). Crystals of a fragment of influenza haemagglutinin in the low pH induced conformation. J. Mol. Biol. *236*, 1262–1265.

Burton, D.R., Desrosiers, R.C., Doms, R.W., Koff, W.C., Kwong, P.D., Moore, J.P., Nabel, G.J., Sodroski, J., Wilson, I. A., and Wyatt, R.T. (2004). HIV vaccine design and the neutralizing antibody problem. Nat. Immunol. *5*, 233–236.

Caffrey, M., Cai, M., Kaufman, J., Stahl, S. J., Wingfield, P.T., Covell, D.G., Gronenborn, A.M., and Clore, G.M. (1998). Three-dimensional solution structure of the 44 kDa ectodomain of SIV gp41. EMBO J. *17*, 4572–4584.

Cameron, P.U., Freudenthal, P. S., Barker, J. M., Gezelter, S., Inaba, K., and Steinman, R. M. (1992). Dendritic cells exposed to human immunodeficiency virus type-1 transmit a vigorous cytopathic infection to $CD4^+$ T-cells. Science *257*, 383–387.

Campbell, E.M., Nunez, R., and Hope, T.J. (2004). Disruption of the actin cytoskeleton can complement the ability of Nef to enhance human immunodeficiency virus type 1 infectivity. J. Virol. *78*, 5745–5755.

Carr, C.M., Chaudhry, C., and Kim, P.S. (1997). Influenza hemagglutinin is spring-loaded by a metastable native conformation. Proc. Natl. Acad. Sci. U.S.A. *94*, 14306–14313.

Carr, C.M., and Kim, P.S. (1993). A spring-loaded mechanism for the conformational change of influenza hemagglutinin. Cell 73, 823–832.

Carr, J.M., Hocking, H., Li, P., and Burrell, C.J. (1999). Rapid and efficient cell-to-cell transmission of human immunodeficiency virus infection from monocyte-derived macrophages to peripheral blood lymphocytes. Virology *265*, 319–329.

Cavrois, M., Neidleman, J., Kreisberg, J. F., and Greene, W.C. (2007). In vitro derived dendritic cells *trans*-Infect CD4 T-cells primarily with surface-bound HIV-1 Virions. PLoS Pathog. *3*, e4.

Chan, D. C., Fass, D., Berger, J. M., and Kim, P.S. (1997). Core structure of gp41 from the HIV envelope glycoprotein. Cell *89*, 263–273.

Chandran, K., Sullivan, N.J., Felbor, U., Whelan, S.P., and Cunningham, J.M. (2005). Endosomal proteolysis of the ebola virus glycoprotein is necessary for infection. Science *308*, 1643–1645.

Chazal, N., Singer, G., Aiken, C., Hammarskjold, M.L., and Rekosh, D. (2001). Human immunodeficiency virus type 1 particles pseudotyped with envelope proteins that fuse at low pH no longer require Nef for optimal infectivity. J. Virol. *75*, 4014–4018.

Chen, B., Vogan, E. M., Gong, H., Skehel, J.J., Wiley, D.C., and Harrison, S.C. (2005). Structure of an unliganded simian immunodeficiency virus gp120 core. Nature *433*, 834–841.

Chen, P., Hubner, W., Spinelli, M. A., and Chen, B.K. (2007). Predominant mode of human immunodeficiency virus transfer between T-cells is mediated by sustained *env*-dependent neutralization-resistant virological synapses. J. Virol. *81*, 12582–12595.

Chernomordik, L.V., Zimmerberg, J., and Kozlov, M.M. (2006). Membranes of the world unite! J. Cell Biol. *175*, 201–207.

Choe, H., Li, W., Wright, P. L., Vasilieva, N., Venturi, M., Huang, C. C., Grundner, C., Dorfman, T., Zwick, M. B., Wang, L., *et al.* (2003). Tyrosine sulfation of human antibodies contributes to recognition of the CCR5 binding region of HIV-1 gp120. Cell *114*, 161–170.

Chowers, M.Y., Pandori, M. W., Spina, C. A., Richman, D.D., and Guatelli, J.C. (1995). The growth advantage conferred by HIV-1 nef is determined at the level of viral DNA formation and is independent of CD4 downregulation. Virology *212*, 451–457.

Chowers, M.Y., Spina, C.A., Kwoh, T.J., Fitch, N.J., Richman, D.D., and Guatelli, J.C. (1994). Optimal infectivity *in vitro* of human immunodeficiency virus type 1 requires an intact nef gene. J. Virol. *68*, 2906–2914.

Cohen, F.S., and Melikyan, G.B. (2004). The energetics of membrane fusion from binding, through hemifusion, pore formation, and pore enlargement. J. Membr. Biol. *199*, 1–14.

Colonna, M., Trinchieri, G., and Liu, Y.J. (2004). Plasmacytoid dendritic cells in immunity. Nat. Immunol. *5*, 1219–1226.

Courreges, M. C., Burzyn, D., Nepomnaschy, I., Piazzon, I., and Ross, S.R. (2007). Critical role for dendritic cells in mouse mammary tumor virus in in vivo infection. J. Virol. *81*, 3769–3777.

Daecke, J., Fackler, O.T., Dittmar, M.T., and Krausslich, H.G. (2005). Involvement of clathrin-mediated endocytosis in human immunodeficiency virus type 1 entry. J. Virol. *79*, 1581–1594.

Damico, R., and Bates, P. (2000). Soluble receptor-induced retroviral infection of receptor-deficient cells. J. Virol. *74*, 6469–6475.

Davey, R.A., Hamson, C. A., Healey, J.J., and Cunningham, J.M. (1997). In vitro binding of purified murine ecotropic retrovirus envelope surface protein to its receptor, MCAT-1. J. Virol. *71*, 8096–8102.

de Duve, C., de Barsy, T., Poole, B., Trouet, A., Tulkens, P., and Van Hoof, F. (1974). Commentary. Lysosomotropic agents. Biochem. Pharmacol. *23*, 2495–2531.

Del Real, G., Jimenez-Baranda, S., Lacalle, R. A., Mira, E., Lucas, P., Gomez-Mouton, C., Carrera, A. C., Martinez, A. C., and Manes, S. (2002). Blocking of HIV-1 infection by targeting CD4 to nonraft membrane domains. J. Exp. Med. *196*, 293–301.

Demaurex, N., Furuya, W., D'Souza, S., Bonifacino, J. S., and Grinstein, S. (1998). Mechanism of acidification of the *trans*-Golgi network (TGN). In situ measurements of pH using retrieval of TGN38 and furin from the cell surface. J. Biol. Chem. *273*, 2044–2051.

Dent, E.W., Kwiatkowski, A.V., Mebane, L.M., Philippar, U., Barzik, M., Rubinson, D.A., Gupton, S., Van Veen, J.E., Furman, C., Zhang, J., *et al.* (2007). Filopodia are required for cortical neurite initiation. Nat. Cell Biol. *9*, 1347–1359.

Deschambeault, J., Lalonde, J. P., Cervantes-Acosta, G., Lodge, R., Cohen, E. A., and Lemay, G. (1999). Polarized human immunodeficiency virus budding in lymphocytes involves a tyrosine-based signal and favors cell-to-cell viral transmission. J. Virol. *73*, 5010–5017.

Dimitrov, D.S., Willey, R.L., Sato, H., Chang, L.J., Blumenthal, R., and Martin, M.A. (1993). Quantitation of human immunodeficiency virus type 1 infection kinetics. J. Virol. *67*, 2182–2190.

Dismuke, D.J., and Aiken, C. (2006). Evidence for a functional link between uncoating of the human immunodeficiency virus type 1 core and nuclear import of the viral preintegration complex. J. Virol. *80*, 3712–3720.

Dohner, K., Wolfstein, A., Prank, U., Echeverri, C., Dujardin, D., Vallee, R., and Sodeik, B. (2002). Function of dynein and dynactin in herpes simplex virus capsid transport. Mol. Biol. Cell *13*, 2795–2809.

Doms, R.W., and Trono, D. (2000). The plasma membrane as a combat zone in the HIV battlefield. Genes Dev. *14*, 2677–2688.

Doranz, B.J., Rucker, J., Yi, Y., Smyth, R. J., Samson, M., Peiper, S.C., Parmentier, M., Collman, R.G., and Doms, R.W. (1996). A dual-tropic primary HIV-1 isolate that uses fusin and the beta-chemokine receptors CKR-5, CKR-3, and CKR-2b as fusion cofactors. Cell *85*, 1149–1158.

Durrer, P., Galli, C., Hoenke, S., Corti, C., Gluck, R., Vorherr, T., and Brunner, J. (1996). H^+-induced membrane insertion of influenza virus hemagglutinin involves the HA2 amino-terminal fusion peptide but not the coiled coil region. J. Biol. Chem. *271*, 13417–13421.

Einfeld, D., and Hunter, E. (1988). Oligomeric structure of a prototype retrovirus glycoprotein. Proc. Natl. Acad. Sci. U.S.A. *85*, 8688–8692.

Ewers, H., Smith, A.E., Sbalzarini, I. F., Lilie, H., Koumoutsakos, P., and Helenius, A. (2005). Single-particle tracking of murine polyoma virus-like particles on live cells and artificial membranes. Proc. Natl. Acad. Sci. U.S.A. *102*, 15110–15115.

Fackler, O.T., Alcover, A., and Schwartz, O. (2007). Modulation of the immunological synapse: a key to HIV-1 pathogenesis? Nat Rev Immunol 7, 310–317.

Farzan, M., Mirzabekov, T., Kolchinsky, P., Wyatt, R., Cayabyab, M., Gerard, N. P., Gerard, C., Sodroski, J., and Choe, H. (1999). Tyrosine sulfation of the amino terminus of CCR5 facilitates HIV-1 entry. Cell *96*, 667–676.

Fass, D., Davey, R. A., Hamson, C. A., Kim, P. S., Cunningham, J. M., and Berger, J. M. (1997). Structure of a murine leukemia virus receptor-binding glycoprotein at 2.0 angstrom resolution. Science *277*, 1662–1666.

Fass, D., Harrison, S. C., and Kim, P.S. (1996). Retrovirus envelope domain at 1.7 angstrom resolution. Nat Struct Biol *3*, 465–469.

Fassati, A., and Goff, S.P. (1999). Characterization of intracellular reverse transcription complexes of Moloney murine leukemia virus. J. Virol. *73*, 8919–8925.

Fassati, A., and Goff, S.P. (2001). Characterization of intracellular reverse transcription complexes of human immunodeficiency virus type 1. J. Virol. *75*, 3626–3635.

Finke, D., and Acha-Orbea, H. (2001). Immune response to murine and feline retroviruses. In Retroviral Immunology, Pantaleo, G. , and Walker, B. D. , eds. (Totowa, NJ: Humana Press), pp. 125–157.

Forshey, B.M., von Schwedler, U., Sundquist, W. I., and Aiken, C. (2002). Formation of a human immunodeficiency virus type 1 core of optimal stability is crucial for viral replication. J. Virol. *76*, 5667–5677.

Freed, E.O. (1998). HIV-1 *gag* proteins: diverse functions in the virus life cycle. Virology *251*, 1–15.

Ganesh, L., Leung, K., Lore, K., Levin, R., Panet, A., Schwartz, O., Koup, R. A., and Nabel, G. J. (2004). Infection of specific dendritic cells by CCR5-tropic human immunodeficiency virus type 1 promotes cell-mediated transmission of virus resistant to broadly neutralizing antibodies. J. Virol. *78*, 11980–11987.

Garcia, E., Pion, M., Pelchen-Matthews, A., Collinson, L., Arrighi, J. F., Blot, G., Leuba, F., Escola, J. M., Demaurex, N., Marsh, M., and Piguet, V. (2005). HIV-1 trafficking to the dendritic cell-T-cell infectious synapse uses a pathway of tetraspanin sorting to the immunological synapse. Traffic *6*, 488–501.

Geijtenbeek, T.B., Kwon, D.S., Torensma, R., van Vliet, S. J., van Duijnhoven, G.C., Middel, J., Cornelissen, I.L., Nottet, H.S., KewalRamani, V.N., Littman, D.R., *et al.* (2000). DC-SIGN, a dendritic cell-specific HIV-1-binding protein that enhances *trans*-infection of T-cells. Cell *100*, 587–597.

Gibbons, D.L., Vaney, M.C., Roussel, A., Vigouroux, A., Reilly, B., Lepault, J., Kielian, M., and Rey, F. A. (2004). Conformational change and protein–protein interactions of the fusion protein of Semliki Forest virus. Nature *427*, 320–325.

Goff, S.P. (2004). Retrovirus restriction factors. Mol. Cell *16*, 849–859.

Goff, S.P. (2007). Host factors exploited by retroviruses. Nat. Rev. Microbiol. *5*, 253–263.

Goff, S.P. (2008). Knockdown screens to knockout HIV-1. Cell *135*, 417–420.

Gomez, C., and Hope, T. J. (2005). The ins and outs of HIV replication. Cell. Microbiol. *7*, 621–626.

Gottlinger, H.G. (2001). The HIV-1 assembly machine. Aids *15* Suppl. *5*, S13–20.

Granelli-Piperno, A., Finkel, V., Delgado, E., and Steinman, R.M. (1999). Virus replication begins in dendritic cells during the transmission of HIV-1 from mature dendritic cells to T-cells. Curr. Biol. *9*, 21–29.

Greber, U.F., and Way, M. (2006). A superhighway to virus infection. Cell *124*, 741–754.

Gupta, P., Balachandran, R., Ho, M., Enrico, A., and Rinaldo, C. (1989). Cell-to-cell transmission of human immunodeficiency virus type 1 in the presence of azidothymidine and neutralizing antibody. J. Virol. *63*, 2361–2365.

Hallenberger, S., Moulard, M., Sordel, M., Klenk, H.D., and Garten, W. (1997). The role of eukaryotic subtilisin-like endoproteases for the activation of human immunodeficiency virus glycoproteins in natural host cells. J. Virol. *71*, 1036–1045.

Harris, R. S., and Liddament, M.T. (2004). Retroviral restriction by APOBEC proteins. Nat. Rev. Immunol. *4*, 868–877.

Harrison, S. C. (2005). Mechanism of membrane fusion by viral envelope proteins. Adv Virus Res. *64*, 231–261.

Hatziioannou, T., and Goff, S.P. (2001). Infection of nondividing cells by Rous sarcoma virus. J. Virol. *75*, 9526–9531.

Hernandez, L.D., Hoffman, L.R., Wolfsberg, T. G., and White, J. M. (1996). Virus-cell and cell–cell fusion. Annu. Rev. Cell Dev. Biol. *12*, 627–661.

Hiscott, J., Kwon, H., and Genin, P. (2001). Hostile takeovers: viral appropriation of the NF-kappaB pathway. J. Clin. Invest. *107*, 143–151.

Honda, K., Takaoka, A., and Taniguchi, T. (2006). Type I interferon [corrected] gene induction by the interferon regulatory factor family of transcription factors. Immunity *25*, 349–360.

Hope, T.J. (2007). Bridging efficient viral infection. Nat. Cell Biol. *9*, 243–244.

Huang, C.C., Lam, S. N., Acharya, P., Tang, M., Xiang, S. H., Hussan, S. S., Stanfield, R. L., Robinson, J., Sodroski, J., Wilson, I. A., *et al.* (2007). Structures of the CCR5 N terminus and of a tyrosine-sulfated antibody with HIV-1 gp120 and CD4. Science *317*, 1930–1934.

Huang, Q., Opitz, R., Knapp, E.W., and Herrmann, A. (2002). Protonation and stability of the globular domain of influenza virus hemagglutinin. Biophys J. *82*, 1050–1058.

Hunter, E., and Swanstrom, R. (1990). Retrovirus envelope glycoproteins. Curr. Top. Microbiol. Immunol. *157*, 187–253.

Huppa, J.B., and Davis, M.M. (2003). T-cell-antigen recognition and the immunological synapse. Nat. Rev. Immunol. *3*, 973–983.

Igakura, T., Stinchcombe, J.C., Goon, P.K., Taylor, G.P., Weber, J.N., Griffiths, G.M., Tanaka, Y., Osame, M., and Bangham, C.R. (2003). Spread of HTLV-I between lymphocytes by virus-induced polarization of the cytoskeleton. Science *299*, 1713–1716.

Iwasaki, A., and Medzhitov, R. (2004). Toll-like receptor control of the adaptive immune responses. Nat. Immunol. *5*, 987–995.

Iyengar, S., Hildreth, J.E., and Schwartz, D.H. (1998). Actin-dependent receptor colocalization required for human immunodeficiency virus entry into host cells. J. Virol. 72, 5251–5255.

Jimenez-Baranda, S., Gomez-Mouton, C., Rojas, A., Martinez-Prats, L., Mira, E., Ana Lacalle, R., Valencia, A., Dimitrov, D. S., Viola, A., Delgado, R., *et al.* (2007). Filamin-A regulates actin-dependent clustering of HIV receptors. Nat. Cell Biol. *9*, 838–846.

Jin, J., Sherer, N. M., Heidecker, G., Derse, D., and Mothes, W. (2009). Assembly of the murine leukemia virus is directed towards sites of cell-cell contact. PLoS Biol. *7*, e1000163.

Johnson, D.C., and Huber, M.T. (2002). Directed egress of animal viruses promotes cell-to-cell spread. J. Virol. *76*, 1–8.

Jolly, C., Kashefi, K., Hollinshead, M., and Sattentau, Q.J. (2004). HIV-1 cell to cell transfer across an Env-induced, actin-dependent synapse. J. Exp. Med. *199*, 283–293.

Jolly, C., and Sattentau, Q.J. (2004). Retroviral spread by induction of virological synapses. Traffic *5*, 643–650.

Katen, L.J., Januszeski, M.M., Anderson, W.F., Hasenkrug, K.J., and Evans, L.H. (2001). Infectious entry by amphotropic as well as ecotropic murine leukemia viruses occurs through an endocytic pathway. J. Virol. *75*, 5018–5026.

Katz, R.A., Greger, J.G., and Skalka, A.M. (2005). Effects of cell cycle status on early events in retroviral replication. J. Cell Biochem *94*, 880–889.

Kawai, T., and Akira, S. (2006). Innate immune recognition of viral infection. Nat. Immunol. *7*, 131–137.

Kemble, G.W., Danieli, T., and White, J.M. (1994). Lipid-anchored influenza hemagglutinin promotes hemifusion, not complete fusion. Cell *76*, 383–391.

Kielian, M., and Rey, F.A. (2006). Virus membrane-fusion proteins: more than one way to make a hairpin. Nat. Rev. Microbiol. *4*, 67–76.

Kilby, J.M., Hopkins, S., Venetta, T. M., DiMassimo, B., Cloud, G. A., Lee, J. Y., Alldredge, L., Hunter, E., Lambert, D., Bolognesi, D., *et al.* (1998). Potent suppression of HIV-1 replication in humans by T-20, a peptide inhibitor of gp41-mediated virus entry. Nat. Med. *4*, 1302–1307.

Kim, F.J., Seiliez, I., Denesvre, C., Lavillette, D., Cosset, F.L., and Sitbon, M. (2000). Definition of an amino-terminal domain of the human T-cell leukemia virus type 1 envelope surface unit that extends the fusogenic range of an ecotropic murine leukemia virus. J. Biol. Chem. *275*, 23417–23420.

Kizhatil, K., and Albritton, L.M. (1997). Requirements for different components of the host cell cytoskeleton distinguish ecotropic murine leukemia virus entry via endocytosis from entry via surface fusion. J. Virol. *71*, 7145–7156.

Klenk, H.D., and Garten, W. (1994). Host cell proteases controlling virus pathogenicity. Trends Microbiol. *2*, 39–43.

Kobe, B., Center, R.J., Kemp, B.E., and Poumbourios, P. (1999). Crystal structure of human T-cell leukemia virus type 1 gp21 ectodomain crystallized as a maltose-binding protein chimera reveals structural evolution of retroviral transmembrane proteins. Proc. Natl. Acad. Sci. U.S.A. *96*, 4319–4324.

Kolokoltsov, A. A., and Davey, R. A. (2004). Rapid and sensitive detection of retrovirus entry by using a novel luciferase-based content-mixing assay. J. Virol. *78*, 5124–5132.

Kourtis, A.P., Butera, S., Ibegbu, C., Beled, L., and Duerr, A. (2003). Breast milk and HIV-1: vector of transmission or vehicle of protection? Lancet Infect Dis 3, 786–793.

Kunzi, M.S., and Pitha, P. M. (2003). Interferon targeted genes in host defense. Autoimmunity *36*, 457–461.

Kwong, P.D. (2005). Human immunodeficiency virus: refolding the envelope. Nature *433*, 815–816.

Kwong, P.D., Wyatt, R., Robinson, J., Sweet, R.W., Sodroski, J., and Hendrickson, W.A. (1998). Structure of an HIV gp120 envelope glycoprotein in complex with the CD4 receptor and a neutralizing human antibody. Nature *393*, 648–659.

Lama, J., Mangasarian, A., and Trono, D. (1999). Cell-surface expression of CD4 reduces HIV-1 infectivity by blocking Env incorporation in a Nef- and Vpu-inhibitable manner. Curr. Biol. *9*, 622–631.

Lavillette, D., Boson, B., Russell, S.J., and Cosset, F.L. (2001). Activation of membrane fusion by murine leukemia viruses is controlled *in cis* or in *trans* by interactions between the receptor-binding domain and a conserved disulfide loop of the carboxy terminus of the surface glycoprotein. J. Virol. *75*, 3685–3695.

Lee, E., and De Camilli, P. (2002). Dynamin at actin tails. Proc. Natl. Acad. Sci. U.S.A. *99*, 161–166.

Lehmann, M.J., Sherer, N. M., Marks, C.B., Pypaert, M., and Mothes, W. (2005). Actin- and myosin-driven movement of viruses along filopodia precedes their entry into cells. J. Cell Biol. *170*, 317–325.

Lehmann-Che, J., Renault, N., Giron, M.L., Roingeard, P., Clave, E., Tobaly-Tapiero, J., Bittoun, P., Toubert, A., de The, H., and Saib, A. (2007). Centrosomal latency of incoming foamy viruses in resting cells. PLoS Pathog 3, e74.

Levesque, K., Zhao, Y.S., and Cohen, E.A. (2003). Vpu exerts a positive effect on HIV-1 infectivity by down-modulating CD4 receptor molecules at the surface of HIV-1-producing cells. J. Biol. Chem. *278*, 28346–28353.

Li, P., and Burrell, C.J. (1992). Synthesis of human immunodeficiency virus DNA in a cell-to-cell transmission model. AIDS Res. Hum. Retroviruses *8*, 253–259.

Lidke, D.S., Lidke, K.A., Rieger, B., Jovin, T.M., and Arndt-Jovin, D.J. (2005). Reaching out for signals: filopodia sense EGF and respond by directed retrograde transport of activated receptors. J. Cell. Biol. *170*, 619–626.

Lidke, D.S., Nagy, P., Heintzmann, R., Arndt-Jovin, D.J., Post, J. N., Grecco, H.E., Jares-Erijman, E.A., and Jovin, T M. (2004). Quantum dot ligands provide new insights into erbB/HER receptor-mediated signal transduction. Nat. Biotechnol. *22*, 198–203.

Lin, C.L., Sewell, A.K., Gao, G.F., Whelan, K.T., Phillips, R.E., and Austyn, J.M. (2000). Macrophage-tropic HIV induces and exploits dendritic cell chemotaxis. J. Exp. Med. *192*, 587–594.

Liu, Y. J. (2001). Dendritic cell subsets and lineages, and their functions in innate and adaptive immunity. Cell *106*, 259–262.

McClure, M.O., Marsh, M., and Weiss, R.A. (1988). Human immunodeficiency virus infection of CD4-bearing cells occurs by a pH-independent mechanism. EMBO J. *7*, 513–518.

McClure, M.O., Sommerfelt, M.A., Marsh, M., and Weiss, R.A. (1990). The pH independence of mammalian retrovirus infection. J. Gen. Virol. *71*, 767–773.

McDonald, D., Vodicka, M. A., Lucero, G., Svitkina, T.M., Borisy, G.G., Emerman, M., and Hope, T.J. (2002). Visualization of the intracellular behavior of HIV in living cells. J. Cell Biol. *159*, 441–452.

McDonald, D., Wu, L., Bohks, S.M., KewalRamani, V.N., Unutmaz, D., and Hope, T.J. (2003). Recruitment of HIV and its receptors to dendritic cell-T-cell junctions. Science *300*, 1295–1297.

Maddon, P.J., McDougal, J. S., Clapham, P.R., Dalgleish, A. G., Jamal, S., Weiss, R.A., and Axel, R. (1988). HIV infection does not require endocytosis of its receptor, CD4. Cell *54*, 865–874.

Malashkevich, V.N., Chan, D.C., Chutkowski, C.T., and Kim, P.S. (1998). Crystal structure of the simian immunodeficiency virus (SIV) gp41 core: conserved helical interactions underlie the broad inhibitory activity of gp41 peptides. Proc. Natl. Acad. Sci. U.S.A. *95*, 9134–9139.

Manel, N., Kim, F.J., Kinet, S., Taylor, N., Sitbon, M., and Battini, J. L. (2003). The ubiquitous glucose transporter GLUT-1 is a receptor for HTLV. Cell *115*, 449–459.

Marechal, V., Prevost, M. C., Petit, C., Perret, E., Heard, J. M., and Schwartz, O. (2001). Human immunodeficiency virus type 1 entry into macrophages mediated by macropinocytosis. J. Virol. *75*, 11166–11177.

Markovic, I., Leikina, E., Zhukovsky, M., Zimmerberg, J., and Chernomordik, L. V. (2001). Synchronized activation and refolding of influenza hemagglutinin in multimeric fusion machines. J. Cell. Biol. *155*, 833–844.

Marsh, M., and Bron, R. (1997). SFV infection in CHO cells: cell-type specific restrictions to productive virus entry at the cell surface. J. Cell. Sci. *110*, 95–103.

Marsh, M., and Helenius, A. (2006). Virus entry: open sesame. Cell *124*, 729–740.

Matsuyama, S., Delos, S.E., and White, J.M. (2004). Sequential roles of receptor binding and low pH in forming prehairpin and hairpin conformations of a retroviral envelope glycoprotein. J. Virol. *78*, 8201–8209.

Melikyan, G.B., Barnard, R.J., Abrahamyan, L.G., Mothes, W., and Young, J.A. (2005). Imaging individual retroviral fusion events: from hemifusion to pore formation and growth. Proc. Natl. Acad. Sci. U.S.A. *102*, 8728–8733.

Melikyan, G.B., Markosyan, R.M., Brener, S.A., Rozenberg, Y., and Cohen, F.S. (2000a). Role of the cytoplasmic tail of ecotropic moloney murine leukemia virus Env protein in fusion pore formation. J. Virol. *74*, 447–455.

Melikyan, G.B., Markosyan, R.M., Hemmati, H., Delmedico, M. K., Lambert, D.M., and Cohen, F.S. (2000b). Evidence that the transition of HIV-1 gp41 into a six-helix bundle, not the bundle configuration, induces membrane fusion. J. Cell Biol. *151*, 413–423.

Melikyan, G.B., White, J. M., and Cohen, F.S. (1995). GPI-anchored influenza hemagglutinin induces hemifusion to both red blood cell and planar bilayer membranes. J. Cell Biol. *131*, 679–691.

Merrifield, C.J., Feldman, M. E., Wan, L., and Almers, W. (2002). Imaging actin and dynamin recruitment during invagination of single clathrin-coated pits. Nat. Cell Biol. *4*, 691–698.

Merrifield, C.J., Moss, S.E., Ballestrem, C., Imhof, B.A., Giese, G., Wunderlich, I., and Almers, W. (1999). Endocytic vesicles move at the tips of actin tails in cultured mast cells. Nat. Cell Biol. *1*, 72–74.

Miller, M.D., Farnet, C. M., and Bushman, F.D. (1997). Human immunodeficiency virus type 1 preintegration complexes: studies of organization and composition. J. Virol. *71*, 5382–5390.

Miller, M.D., Warmerdam, M. T., Gaston, I., Greene, W. C., and Feinberg, M.B. (1994). The human immunodeficiency virus-1 nef gene product: a positive factor for viral infection and replication in primary lymphocytes and macrophages. J. Exp. Med. *179*, 101–113.

Miller, M.D., Warmerdam, M. T., Page, K.A., Feinberg, M. B., and Greene, W.C. (1995). Expression of the human immunodeficiency virus type 1 (HIV-1) nef gene during HIV-1 production increases progeny particle infectivity independently of gp160 or viral entry. J. Virol. *69*, 579–584.

Miotti, P.G., Taha, T.E., Kumwenda, N.I., Broadhead, R., Mtimavalye, L. A., Van der Hoeven, L., Chiphangwi, J. D., Liomba, G., and Biggar, R. J. (1999). HIV transmission through breastfeeding: a study in Malawi. Jama *282*, 744–749.

Miyauchi, K., Kim, Y., Latinovic, O., Morozov, V., and Melikyan, G.B. (2009). HIV enters cells via endocytosis and dynamin-dependent fusion with endosomes. Cell *137*, 433–444.

Modis, Y., Ogata, S., Clements, D., and Harrison, S. C. (2004). Structure of the dengue virus envelope protein after membrane fusion. Nature *427*, 313–319.

Mondor, I., Ugolini, S., and Sattentau, Q. J. (1998). Human immunodeficiency virus type 1 attachment to HeLa CD4 cells is CD4 independent and gp120 dependent and requires cell surface heparans. J. Virol. *72*, 3623–3634.

Montefiori, D., Sattentau, Q., Flores, J., Esparza, J., and Mascola, J. (2007). Antibody-based HIV-1 vaccines: recent developments and future directions. PLoS Med *4*, e348.

Moore, J.P., and Doms, R.W. (2003). The entry of entry inhibitors: a fusion of science and medicine. Proc. Natl. Acad. Sci. U.S.A. *100*, 10598–10602.

Morita, E., and Sundquist, W.I. (2004). Retrovirus budding. Annu Rev Cell Dev Biol *20*, 395–425.

Mothes, W., Boerger, A.L., Narayan, S., Cunningham, J.M., and Young, J.A. (2000). Retroviral entry mediated by receptor priming and low pH triggering of an envelope glycoprotein. Cell *103*, 679–689.

Munch, J., Rucker, E., Standker, L., Adermann, K., Goffinet, C., Schindler, M., Wildum, S., Chinnadurai,

R., Rajan, D., Specht, A., *et al.* (2007). Semen-derived amyloid fibrils drastically enhance HIV infection. Cell *131*, 1059–1071.

Narayan, S., Barnard, R.J., and Young, J.A. (2003). Two retroviral entry pathways distinguished by lipid raft association of the viral receptor and differences in viral infectivity. J. Virol. *77*, 1977–1983.

Narayan, S., and Young, J.A. (2004). Reconstitution of retroviral fusion and uncoating in a cell-free system. Proc. Natl. Acad. Sci. U.S.A. *101*, 7721–7726.

Nejmeddine, M., Barnard, A.L., Tanaka, Y., Taylor, G.P., and Bangham, C.R. (2005). Human T-lymphotropic virus, type 1, tax protein triggers microtubule reorientation in the virological synapse. J. Biol. Chem. *280*, 29653–29660.

Niell, C.M., Meyer, M. P., and Smith, S. J. (2004). In vivo imaging of synapse formation on a growing dendritic arbor. Nat. Neurosci. *7*, 254–260.

Nisole, S., and Saib, A. (2004). Early steps of retrovirus replicative cycle. Retrovirology *1*, 9.

Nisole, S., Stoye, J. P., and Saib, A. (2005). TRIM family proteins: retroviral restriction and antiviral defence. Nat. Rev. Microbiol. *3*, 799–808.

O'Doherty, U., Swiggard, W.J., and Malim, M.H. (2000). Human immunodeficiency virus type 1 spinoculation enhances infection through virus binding. J. Virol. *74*, 10074–10080.

Okeoma, C.M., Lovsin, N., Peterlin, B.M., and Ross, S.R. (2007). APOBEC3 inhibits mouse mammary tumour virus replication *in vivo*. Nature *445*, 927–930.

Onfelt, B., Nedvetzki, S., Benninger, R.K., Purbhoo, M.A., Sowinski, S., Hume, A. N., Seabra, M.C., Neil, M.A., French, P.M., and Davis, D.M. (2006). Structurally distinct membrane nanotubes between human macrophages support long-distance vesicular traffic or surfing of bacteria. J. Immunol. *177*, 8476–8483.

Orth, J.D., Krueger, E.W., Cao, H., and McNiven, M.A. (2002). The large GTPase dynamin regulates actin comet formation and movement in living cells. Proc. Natl. Acad. Sci. U.S.A. *99*, 167–172.

Pearce-Pratt, R., Malamud, D., and Phillips, D.M. (1994). Role of the cytoskeleton in cell-to-cell transmission of human immunodeficiency virus. J. Virol. *68*, 2898–2905.

Pelkmans, L., and Helenius, A. (2003). Insider information: what viruses tell us about endocytosis. Curr Opin Cell Biol *15*, 414–422.

Pelkmans, L., Puntener, D., and Helenius, A. (2002). Local actin polymerization and dynamin recruitment in SV40-induced internalization of caveolae. Science *296*, 535–539.

Petit, C., Giron, M.L., Tobaly-Tapiero, J., Bittoun, P., Real, E., Jacob, Y., Tordo, N., De The, H., and Saib, A. (2003). Targeting of incoming retroviral Gag to the centrosome involves a direct interaction with the dynein light chain 8. J. Cell Sci. *116*, 3433–3442.

Phillips, D.M. (1994). The role of cell-to-cell transmission in HIV infection. Aids *8*, 719–731.

Phillips, D.M., and Bourinbaiar, A.S. (1992). Mechanism of HIV spread from lymphocytes to epithelia. Virology *186*, 261–273.

Phillips, D.M., Tan, X., Perotti, M.E., and Zacharopoulos, V.R. (1998). Mechanism of monocyte-macrophage-mediated transmission of HIV. AIDS Res. Hum. Retroviruses *14* Suppl. *1*, S67–70.

Pinter, A., Kopelman, R., Li, Z., Kayman, S.C., and Sanders, D.A. (1997). Localization of the labile disulfide bond between SU and TM of the murine leukemia virus envelope protein complex to a highly conserved CWLC motif in SU that resembles the active-site sequence of thiol-disulfide exchange enzymes. J. Virol. *71*, 8073–8077.

Pizzato, M., Helander, A., Popova, E., Calistri, A., Zamborlini, A., Palu, G., and Gottlinger, H.G. (2007). Dynamin 2 is required for the enhancement of HIV-1 infectivity by Nef. Proc. Natl. Acad. Sci. U.S.A. *104*, 6812–6817.

Pizzato, M., Marlow, S. A., Blair, E.D., and Takeuchi, Y. (1999). Initial binding of murine leukemia virus particles to cells does not require specific Env–receptor interaction. J. Virol. *73*, 8599–8611.

Pope, M., Betjes, M.G., Romani, N., Hirmand, H., Cameron, P.U., Hoffman, L., Gezelter, S., Schuler, G., and Steinman, R.M. (1994). Conjugates of dendritic cells and memory T-lymphocytes from skin facilitate productive infection with HIV-1. Cell *78*, 389–398.

Pope, M., Gezelter, S., Gallo, N., Hoffman, L., and Steinman, R. M. (1995). Low levels of HIV-1 infection in cutaneous dendritic cells promote extensive viral replication upon binding to memory $CD4^+$ T-cells. J. Exp. Med. *182*, 2045–2056.

Ramirez-Weber, F. A., and Kornberg, T.B. (1999). Cytonemes: cellular processes that project to the principal signaling center in *Drosophila* imaginal discs. Cell *97*, 599–607.

Rawat, S.S., Viard, M., Gallo, S. A., Rein, A., Blumenthal, R., and Puri, A. (2003). Modulation of entry of enveloped viruses by cholesterol and sphingolipids (Review). Mol. Membr. Biol. *20*, 243–254.

Rein, A., Mirro, J., Haynes, J. G., Ernst, S. M., and Nagashima, K. (1994). Function of the cytoplasmic domain of a retroviral transmembrane protein: p15E-p2E cleavage activates the membrane fusion capability of the murine leukemia virus Env protein. J. Virol. *68*, 1773–1781.

Retzel, E.F., Collett, M.S., and Faras, A.J. (1980). Enzymatic synthesis of deoxyribonucleic acid by the avian retrovirus reverse transcriptase *in vitro*: optimum conditions required for transcription of large ribonucleic acid templates. Biochemistry *19*, 513–518.

Rizzuto, C.D., Wyatt, R., Hernandez-Ramos, N., Sun, Y., Kwong, P.D., Hendrickson, W.A., and Sodroski, J. (1998). A conserved HIV gp120 glycoprotein structure involved in chemokine receptor binding. Science *280*, 1949–1953.

Roan, N.R., and Greene, W.C. (2007). A Seminal Finding for Understanding HIV Transmission. Cell *131*, 1044–1046.

Roche, S., Bressanelli, S., Rey, F.A., and Gaudin, Y. (2006). Crystal structure of the low-pH form of the vesicular stomatitis virus glycoprotein G. Science *313*, 187–191.

Roche, S., Rey, F.A., Gaudin, Y., and Bressanelli, S. (2007). Structure of the prefusion form of the vesicular stomatitis virus glycoprotein G. Science *315*, 843–848.

Roe, T., Reynolds, T.C., Yu, G., and Brown, P.O. (1993). Integration of murine leukemia virus DNA depends on mitosis. EMBO J. *12*, 2099–2108.

Ruigrok, R.W., Martin, S.R., Wharton, S.A., Skehel, J. J., Bayley, P.M., and Wiley, D.C. (1986). Conformational changes in the hemagglutinin of influenza virus which accompany heat-induced fusion of virus with liposomes. Virology *155*, 484–497.

Rust, M.J., Lakadamyali, M., Zhang, F., and Zhuang, X. (2004). Assembly of endocytic machinery around individual influenza viruses during viral entry. Nat. Struct. Mol. Biol. *11*, 567–573.

Rustom, A., Saffrich, R., Markovic, I., Walther, P., and Gerdes, H.H. (2004). Nanotubular highways for intercellular organelle transport. Science *303*, 1007–1010.

Saib, A., Puvion-Dutilleul, F., Schmid, M., Peries, J., and de The, H. (1997). Nuclear targeting of incoming human foamy virus Gag proteins involves a centriolar step. J. Virol. *71*, 1155–1161.

Sawyer, S.L., Wu, L. I., Emerman, M., and Malik, H.S. (2005). Positive selection of primate TRIM5alpha identifies a critical species-specific retroviral restriction domain. Proc. Natl. Acad. Sci. U.S.A. *102*, 2832–2837.

Sayah, D.M., Sokolskaja, E., Berthoux, L., and Luban, J. (2004). Cyclophilin A retrotransposition into TRIM5 explains owl monkey resistance to HIV-1. Nature *430*, 569–573.

Schindler, M., Munch, J., Kutsch, O., Li, H., Santiago, M. L., Bibollet-Ruche, F., Muller-Trutwin, M. C., Novembre, F. J., Peeters, M., Courgnaud, V., *et al.* (2006). Nef-mediated suppression of T-cell activation was lost in a lentiviral lineage that gave rise to HIV-1. Cell *125*, 1055–1067.

Schubert, U., Ott, D. E., Chertova, E.N., Welker, R., Tessmer, U., Princiotta, M. F., Bennink, J. R., Krausslich, H.G., and Yewdell, J. W. (2000). Proteasome inhibition interferes with *gag* polyprotein processing, release, and maturation of HIV-1 and HIV-2. Proc. Natl. Acad. Sci. U.S.A. *97*, 13057–13062.

Sebastian, S., and Luban, J. (2005). TRIM5alpha selectively binds a restriction-sensitive retroviral capsid. Retrovirology 2, 40.

Sheetz, M.P. (2001). Cell control by membrane-cytoskeleton adhesion. Nat. Rev. Mol. Cell. Biol. *2*, 392–396.

Sherer, N.M., Lehmann, M. J., Jimenez-Soto, L. F., Horensavitz, C., Pypaert, M., and Mothes, W. (2007). Retroviruses can establish filopodial bridges for efficient cell-to-cell transmission. Nat. Cell Biol. *9*, 310–315.

Simmons, G., Gosalia, D.N., Rennekamp, A. J., Reeves, J. D., Diamond, S. L., and Bates, P. (2005). Inhibitors of cathepsin L prevent severe acute respiratory syndrome coronavirus entry. Proc. Natl. Acad. Sci. U.S.A. *102*, 11876–11881.

Skehel, J.J., and Wiley, D.C. (1998). Coiled coils in both intracellular vesicle and viral membrane fusion. Cell *95*, 871–874.

Skehel, J.J., and Wiley, D.C. (2000). Receptor binding and membrane fusion in virus entry: the influenza hemagglutinin. Annu. Rev. Biochem. *69*, 531–569.

Smith, J.G., Mothes, W., Blacklow, S.C., and Cunningham, J.M. (2004). The mature avian leukosis virus subgroup A envelope glycoprotein is metastable, and refolding induced by the synergistic effects of receptor binding and low pH is coupled to infection. J. Virol. *78*, 1403–1410.

Sodeik, B., Ebersold, M. W., and Helenius, A. (1997). Microtubule-mediated transport of incoming herpes simplex virus 1 capsids to the nucleus. J. Cell Biol. *136*, 1007–1021.

Sokolskaja, E., and Luban, J. (2006). Cyclophilin, TRIM5, and innate immunity to HIV-1. Curr. Opin. Microbiol. *9*, 404–408.

Sol-Foulon, N., Sourisseau, M., Porrot, F., Thoulouze, M.I., Trouillet, C., Nobile, C., Blanchet, F., di Bartolo, V., Noraz, N., Taylor, N., *et al.* (2007). ZAP-70 kinase regulates HIV cell-to-cell spread and virological synapse formation. EMBO J. *26*, 516–526.

Sourisseau, M., Sol-Foulon, N., Porrot, F., Blanchet, F., and Schwartz, O. (2007). Inefficient human immunodeficiency virus replication in mobile lymphocytes. J. Virol. *81*, 1000–1012.

Steffens, C.M., and Hope, T.J. (2004). Mobility of the human immunodeficiency virus (HIV) receptor CD4 and coreceptor CCR5 in living cells: implications for HIV fusion and entry events. J. Virol. *78*, 9573–9578.

Stegmann, T., White, J. M., and Helenius, A. (1990). Intermediates in influenza induced membrane fusion. EMBO J. *9*, 4231–4241.

Stein, B.S., Gowda, S. D., Lifson, J.D., Penhallow, R.C., Bensch, K.G., and Engleman, E.G. (1987). pH-independent HIV entry into CD4-positive T-cells via virus envelope fusion to the plasma membrane. Cell *49*, 659–668.

Stremlau, M., Owens, C.M., Perron, M.J., Kiessling, M., Autissier, P., and Sodroski, J. (2004). The cytoplasmic body component TRIM5alpha restricts HIV-1 infection in Old World monkeys. Nature *427*, 848–853.

Stremlau, M., Perron, M., Lee, M., Li, Y., Song, B., Javanbakht, H., Diaz-Griffero, F., Anderson, D., Sundquist, W.I., and Sodroski, J. (2006). Specific recognition and accelerated uncoating of retroviral capsids by the TRIM5alpha restriction factor. Proc. Natl. Acad. Sci. U.S.A. *103*, 5514–5519.

Sun, X., and Whittaker, G.R. (2007). Role of the actin cytoskeleton during influenza virus internalization into polarized epithelial cells. Cell Microbiol *9*, 1672–1682.

Suomalainen, M., Nakano, M.Y., Keller, S., Boucke, K., Stidwill, R.P., and Greber, U.F. (1999). Microtubule-dependent plus- and minus end-directed motilities are competing processes for nuclear targeting of adenovirus. J. Cell Biol. *144*, 657–672.

Tan, K., Liu, J., Wang, J., Shen, S., and Lu, M. (1997). Atomic structure of a thermostable subdomain of HIV-1 gp41. Proc. Natl. Acad. Sci. U.S.A. *94*, 12303–12308.

Taunton, J., Rowning, B.A., Coughlin, M.L., Wu, M., Moon, R.T., Mitchison, T.J., and Larabell, C.A. (2000). Actin-dependent propulsion of endosomes and lysosomes by recruitment of N-WASP. J. Cell Biol. *148*, 519–530.

Temin, H.M., and Mizutani, S. (1970). RNA-dependent DNA polymerase in virions of Rous sarcoma virus. Nature *226*, 1211–1213.

Thoulouze, M.I., Sol-Foulon, N., Blanchet, F., Dautry-Varsat, A., Schwartz, O., and Alcover, A. (2006). Human immunodeficiency virus type-1 infection impairs the formation of the immunological synapse. Immunity *24*, 547–561.

Tscherne, D.M., Jones, C.T., Evans, M.J., Lindenbach, B.D., McKeating, J.A., and Rice, C.M. (2006). Time- and temperature-dependent activation of hepatitis C virus for low-pH-triggered entry. J. Virol. *80*, 1734–1741.

Turville, S.G., Cameron, P.U., Handley, A., Lin, G., Pohlmann, S., Doms, R.W., and Cunningham, A.L. (2002). Diversity of receptors binding HIV on dendritic cell subsets. Nat. Immunol. *3*, 975–983.

Turville, S.G., Santos, J.J., Frank, I., Cameron, P.U., Wilkinson, J., Miranda-Saksena, M., Dable, J., Stossel, H., Romani, N., Piatak, M., Jr., *et al.* (2004). Immunodeficiency virus uptake, turnover, and 2-phase transfer in human dendritic cells. Blood *103*, 2170–2179.

van der Schaar, H.M., Rust, M. J., Waarts, B.L., van der Ende-Metselaar, H., Kuhn, R. J., Wilschut, J., Zhuang, X., and Smit, J. M. (2007). Characterization of the early events in dengue virus cell entry by biochemical assays and single-virus tracking. J. Virol. *81*, 12019–12028.

Vidricaire, G., Imbeault, M., and Tremblay, M.J. (2004). Endocytic host cell machinery plays a dominant role in intracellular trafficking of incoming human immunodeficiency virus type 1 in human placental trophoblasts. J. Virol. *78*, 11904–11915.

Wallin, M., Ekstrom, M., and Garoff, H. (2004). Isomerization of the intersubunit disulphide-bond in Env controls retrovirus fusion. EMBO J. *23*, 54–65.

Wallin, M., Ekstrom, M., and Garoff, H. (2005). The fusion-controlling disulfide bond isomerase in retrovirus Env is triggered by protein destabilization. J. Virol. *79*, 1678–1685.

Wei, X., Decker, J. M., Wang, S., Hui, H., Kappes, J.C., Wu, X., Salazar-Gonzalez, J.F., Salazar, M.G., Kilby, J. M., Saag, M. S., *et al.* (2003). Antibody neutralization and escape by HIV-1. Nature *422*, 307–312.

Weissenhorn, W., Dessen, A., Harrison, S.C., Skehel, J. J., and Wiley, D.C. (1997). Atomic structure of the ectodomain from HIV-1 gp41. Nature *387*, 426–430.

Wiegers, K., Rutter, G., Kottler, H., Tessmer, U., Hohenberg, H., and Krausslich, H.G. (1998). Sequential steps in human immunodeficiency virus particle maturation revealed by alterations of individual Gag polyprotein cleavage sites. J. Virol. *72*, 2846–2854.

Wild, C., Greenwell, T., and Matthews, T. (1993). A synthetic peptide from HIV-1 gp41 is a potent inhibitor of virus-mediated cell–cell fusion. AIDS Res. Hum. Retroviruses *9*, 1051–1053.

Wildum, S., Schindler, M., Munch, J., and Kirchhoff, F. (2006). Contribution of Vpu, Env, and Nef to CD4 down-modulation and resistance of human immunodeficiency virus type 1-infected T-cells to superinfection. J. Virol. *80*, 8047–8059.

Wilson, I. A., Skehel, J.J., and Wiley, D.C. (1981). Structure of the haemagglutinin membrane glycoprotein of influenza virus at 3 A resolution. Nature *289*, 366–373.

Wu, X., Anderson, J.L., Campbell, E.M., Joseph, A.M., and Hope, T.J. (2006). Proteasome inhibitors uncouple rhesus TRIM5alpha restriction of HIV-1 reverse transcription and infection. Proc. Natl. Acad. Sci. U.S.A. *103*, 7465–7470.

Wyatt, R., Kwong, P.D., Desjardins, E., Sweet, R. W., Robinson, J., Hendrickson, W. A., and Sodroski, J. G. (1998). The antigenic structure of the HIV gp120 envelope glycoprotein. Nature *393*, 705–711.

Wyma, D.J., Jiang, J., Shi, J., Zhou, J., Lineberger, J.E., Miller, M.D., and Aiken, C. (2004). Coupling of human immunodeficiency virus type 1 fusion to virion maturation: a novel role of the gp41 cytoplasmic tail. J. Virol. *78*, 3429–3435.

Yamashita, M., and Emerman, M. (2004). Capsid is a dominant determinant of retrovirus infectivity in nondividing cells. J. Virol. *78*, 5670–5678.

Yamashita, M., and Emerman, M. (2006). Retroviral infection of non-dividing cells: old and new perspectives. Virology *344*, 88–93.

Yamashita, M., Perez, O., Hope, T. J., and Emerman, M. (2007). Evidence for direct involvement of the capsid protein in HIV infection of nondividing cells. PLoS Pathog. *3*, 1502–1510.

Yoder, A., Yu, D., Dong, L., Iyer, S.R., Xu, X., Kelly, J., Liu, J., Wang, W., Vorster, P.J., Agulto, L., *et al.* (2008). HIV envelope-CXCR4 signaling activates cofilin to overcome cortical actin restriction in resting CD4 T cells. Cell *134*, 782–792.

Young, J. A. (2001). Virus entry and uncoating, In Field's Virology, Knipe, D.M. , ed. (Philadelphia: Lippincott Williams & Wilkins), pp. 87–103.

Yu, S. F., Baldwin, D.N., Gwynn, S.R., Yendapalli, S., and Linial, M.L. (1996). Human foamy virus replication: a pathway distinct from that of retroviruses and hepadnaviruses. Science *271*, 1579–1582.

Yueh, A., and Goff, S.P. (2003). Phosphorylated serine residues and an arginine-rich domain of the moloney murine leukemia virus p12 protein are required for early events of viral infection. J. Virol. *77*, 1820–1829.

Yueh, A., Leung, J., Bhattacharyya, S., Perrone, L. A., de los Santos, K., Pu, S. Y., and Goff, S.P. (2006). Interaction of moloney murine leukemia virus capsid with Ubc9 and PIASy mediates SUMO-1 addition required early in infection. J. Virol. *80*, 342–352.

Zhang, H., Dornadula, G., Orenstein, J., and Pomerantz, R.J. (2000). Morphologic changes in human immunodeficiency virus type 1 virions secondary to intravirion reverse transcription: evidence indicating that reverse transcription may not take place within the intact viral core. J. Hum Virol *3*, 165–172.

Zhou, T., Xu, L., Dey, B., Hessell, A.J., Van Ryk, D., Xiang, S.H., Yang, X., Zhang, M.Y., Zwick, M.B., Arthos, J., *et al.* (2007). Structural definition of a conserved neutralization epitope on HIV-1 gp120. Nature *445*, 732–737.

Ziv, N.E., and Smith, S.J. (1996). Evidence for a role of dendritic filopodia in synaptogenesis and spine formation. Neuron *17*, 91–102.

Reverse Transcription and Integration

5

Alan Engelman

Abstract

Retroviruses are unique among animal viruses in that their replication requires the recombination of their own genetic material with that of the infected host cell. Two virus-encapsulated enzymes, reverse transcriptase and integrase, are dedicated to provirus formation. Reverse transcriptase, using a packaged cellular tRNA primer to initiate DNA synthesis from the viral RNA template, generates linear double-stranded DNA within the context of the reverse transcription nucleoprotein complex. The integrase enzyme processes the neo-synthesized DNA ends as the preintegration complex moves toward the cell nucleus. After finding a suitable chromatin acceptor site, the integrase recombines the processed DNA ends with a cell chromosome. This chapter focuses on the mechanisms of viral DNA synthesis, its transport to the nucleus, and the resulting chromosomal DNA integration.

Introduction

Reverse transcription and integration are key components of the early phase of the retroviral lifecycle. As these processes are carried out by viral enzymes with known active sites, they are highly sought after targets for pharmacological intervention of viral-induced diseases like HIV/AIDS (Pomerantz and Horn, 2003; Reeves and Piefer, 2005). Understanding the molecular mechanisms of reverse transcription and integration therefore not only deepens our understanding of these basic biological processes but helps to formulate optimized pharmaceuticals to control the spread of the devastating AIDS pandemic.

Overview of the early events in retroviral replication

The infection begins when viral envelope glycoproteins engage specific cellular receptors (Fig. 5.1, step 1). The ribonucleoprotein core components of the viral particle that include the RNA genome in association with the viral nucleocapsid protein and reverse transcriptase and integrase enzymes are subsequently exposed to the cell cytoplasm as the viral and cellular membranes fuse (Fig. 5.1, step 2). It is important to appreciate that the intracellular events up to and including integration occur in the context of a nucleoprotein complex that is derived from the viral core. Though numerous conformational changes including structural rearrangements and interactions with important cellular co-factors occur between virus entry and integration, a large nucleoprotein structure is maintained (Brown *et al.*, 1987; Bowerman *et al.*, 1989; Ellison *et al.*, 1990; Farnet and Haseltine, 1990; Karageorgos *et al.*, 1993; Miller *et al.*, 1997; Wei *et al.*, 1997; Chen *et al.*, 1999; Fassati and Goff, 1999, 2001).

After entry the virus core undergoes a conformational change(s) that is operationally defined as uncoating because it is required for the productive initiation of reverse transcription (Fig. 5.1, step 3; see Chapter 4 for details) (Crawford and Goff, 1984; Alin and Goff, 1996; Forshey *et al.*, 2002; Narayan and Young, 2004; Auewarakul *et al.*, 2005). DNA synthesis ensues within the context of the reverse transcription complex (RTC) (Karageorgos *et al.*, 1993; Fassati and Goff, 1999, 2001) in association with actin microfilaments (Bukrinskaya *et al.*, 1998). The

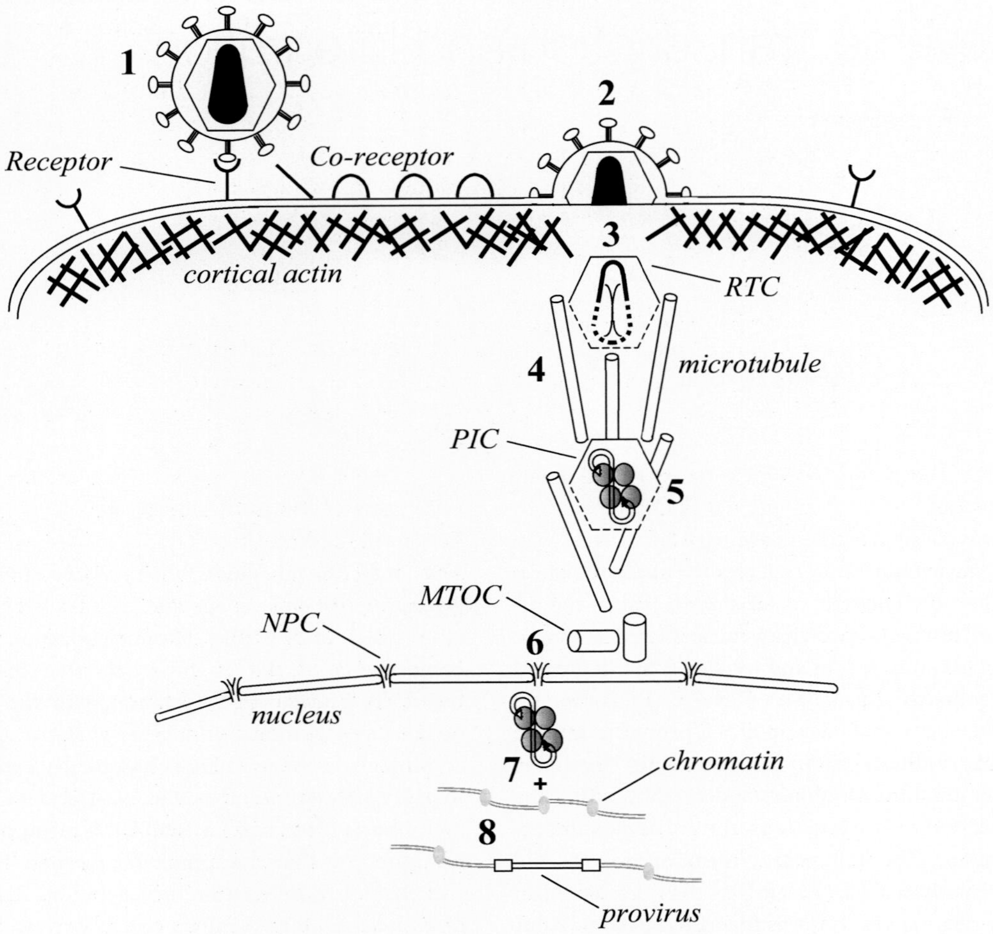

Figure 5.1 The early events of the retroviral life cycle. Infection begins with virus binding, when virion envelope surface glycoproteins interact with specific receptors on the surface of the cell (step 1). A series of protein conformational changes results in the fusion of the virion and cellular membranes and subsequent entry of the viral core into the cell cytoplasm (step 2; see Chapter 4 for details). The core partially uncoats, resulting in the initiation of reverse transcription in the context of the reverse transcription complex (RTC; step 3). The RTC travels along microtubules (step 4) and matures into the preintegration complex (PIC) as DNA synthesis is completed and the 3′ ends of the reverse transcript become processed by the integrase enzyme (step 5). Particles display affinity for the microtubule organizing centre (MTOC) at the periphery of the nuclear membrane as they approach the nucleus. Lentiviruses are actively transported via the nuclear pore complex (NPC; step 6) whereas other retroviruses await nuclear membrane breakdown during mitosis to gain access to chromosomes for integration. After locating a suitable chromatin acceptor site, the integrase covalently joins the 3′ ends of viral cDNA to host DNA (step 7). The resulting DNA recombination intermediate is repaired by host cell machinery to yield the integrated provirus (step 8).

RTC moves toward the cell nucleus in association with microtubules (Fig. 5.1, step 4) (McDonald *et al.*, 2002; Arhel *et al.*, 2006a) and matures into the preintegration complex (PIC) as DNA synthesis nears completion and the 3′ ends of the reverse transcript are processed by integrase (Fig. 5.1, step 5) (Brown *et al.*, 1989; Roth *et al.*, 1989; Pauza, 1990; Miller *et al.*, 1997). The PIC must gain access to cell chromosomes before integrase can catalyse integration, and different viruses use different mechanisms to ensure this outcome. Lentiviruses are actively transported into nuclei (Bukrinsky *et al.*, 1992), presumably through nuclear core complexes (NPCs) that reside within the nuclear membrane (Fig. 5.1, step 6). Productive infection by γ-retroviruses like Moloney murine leukaemia virus (MoMLV) in contrast requires cells to pass through the M phase of the cell cycle when the nuclear membrane breaks down (Roe *et al.*, 1993; Lewis and Emerman, 1994), reflecting the relative inability of these viruses to infect non-dividing cells (Lewis *et al.*, 1992). After locating a suitable target site within chromatin, the viral integrase joins the 3′ ends of viral cDNA to the 5′ phosphates of a double-stranded cut in cell DNA (Fig. 5.1, step 7), which yields a recombination intermediate containing single-stranded DNA gaps (Fujiwara and Mizuuchi, 1988; Brown *et al.*, 1989; Lee and Coffin, 1991). Repair of the gaps, which is presumably mediated by host cell enzymes (Brin *et al.*, 2000; Yoder and Bushman, 2000), yields the integrated provirus (Fig. 5.1, step 8).

Though the overview of post-entry uncoating leading to the initiation of cDNA synthesis suffices as a general introduction, it is important to note that it does not equally apply to all retroviruses. The Retroviridae family consists of two subfamilies, Orthoretrovirinae, comprising α-, β-, δ-, ε-, γ-, and *Lentivirus* genera, and Spumaretrovirinae, made up of the sole *Spumaretrovirus* genus. The taxonomic distinction reflects the atypical behaviour of the spumaviruses (Rethwilm, 2003; Delelis *et al.*, 2004). In particular, reverse transcription is accomplished in the virus producer cell before particle release, as compared to in the target cell after virus infection (Yu *et al.*, 1996; Moebes *et al.*, 1997). The reverse transcriptase and integrase enzymes of most retroviruses are furthermore synthesized as part of a Gag–Pol polyprotein that is cleaved into separate components during virus budding and maturation, whereas spumaviruses express their enzymes separately from Gag, as a Pol polyprotein (Netzer *et al.*, 1993; Enssle *et al.*, 1996; Lochelt and Flugel, 1996).

Reverse transcription

Retroviral reverse transcriptase enzymes are RNA- and DNA-dependent DNA polymerases with an associated RNase H activity that specifically degrades the RNA component of RNA/DNA hybrids. Virus particles house two copies of a single-stranded, plus-sense RNA genome, the end regions of which harbour sequence elements that play essential roles in reverse transcription. The termini comprise an identical repeat or R sequence, whereas unique sequences abutting R at the 5′ and 3′ ends of the RNA are called U5 and U3, respectively. Immediately internal to U5 is the primer binding site (PBS) essential for priming the synthesis of the DNA minus-strand, whereas the polypurine tract (PPT) that primes plus-strand synthesis is located downstream, abutting U3 (Fig. 5.2).

The mechanism of cDNA synthesis

The mechanistic details of reverse transcription were deciphered by analysing the structures of replication intermediates isolated from acutely infected cells or synthesized *in vitro* using virions under non-denaturing lysis conditions in so-called endogenous reverse transcription reactions (Haseltine *et al.*, 1976, 1979; Varmus *et al.*, 1978; Gilboa *et al.*, 1979; Boone and Skalka, 1980, 1981; Zack *et al.*, 1990; Borroto-Esoda and Boone, 1991). DNA synthesis initiates via a partially unfolded cellular tRNA primer that anneals to the PBS and primes minus-strand synthesis (Fig. 5.2). Numerous factors contribute to the selective incorporation of specific tRNA species into virions (Table 5.1) (Waters *et al.*, 1980; Jiang *et al.*, 1993; Mak and Kleiman, 1997) as well their placement onto the viral RNA genome. Reverse transcriptase is the primary viral determinant governing the specificity of incorporation (Peters and Hu, 1980; Levin and Seidman, 1981; Mak *et al.*, 1994) and some enzymes, like those from α-retroviruses (Araya *et al.*, 1979) and human immunodeficiency virus type 1 (HIV-1) (Barat *et al.*, 1993), contribute to primer annealing as well. The viral nucleocapsid protein also contributes to tRNA positioning (Barat *et al.*, 1989, 1993). Certain viruses, such as HIV-1 and Rous sarcoma virus (RSV), in addition package the cognate aminoacyl-tRNA synthetase enzyme (Cen *et al.*, 2001, 2002). The lack of specific prolyl-tRNA synthetase incorporation into MoMLV particles may represent a less stringent requirement for a specific tRNA in priming γ-retroviral DNA synthesis (Cen *et al.*, 2002). Interactions between the tRNA primer and the viral genome outside of the PBS also contribute to proper initiation of reverse transcription (Aiyar *et al.*, 1992; Isel *et al.*, 1995; Kang and Morrow, 1999; Beerens *et al.*, 2001).

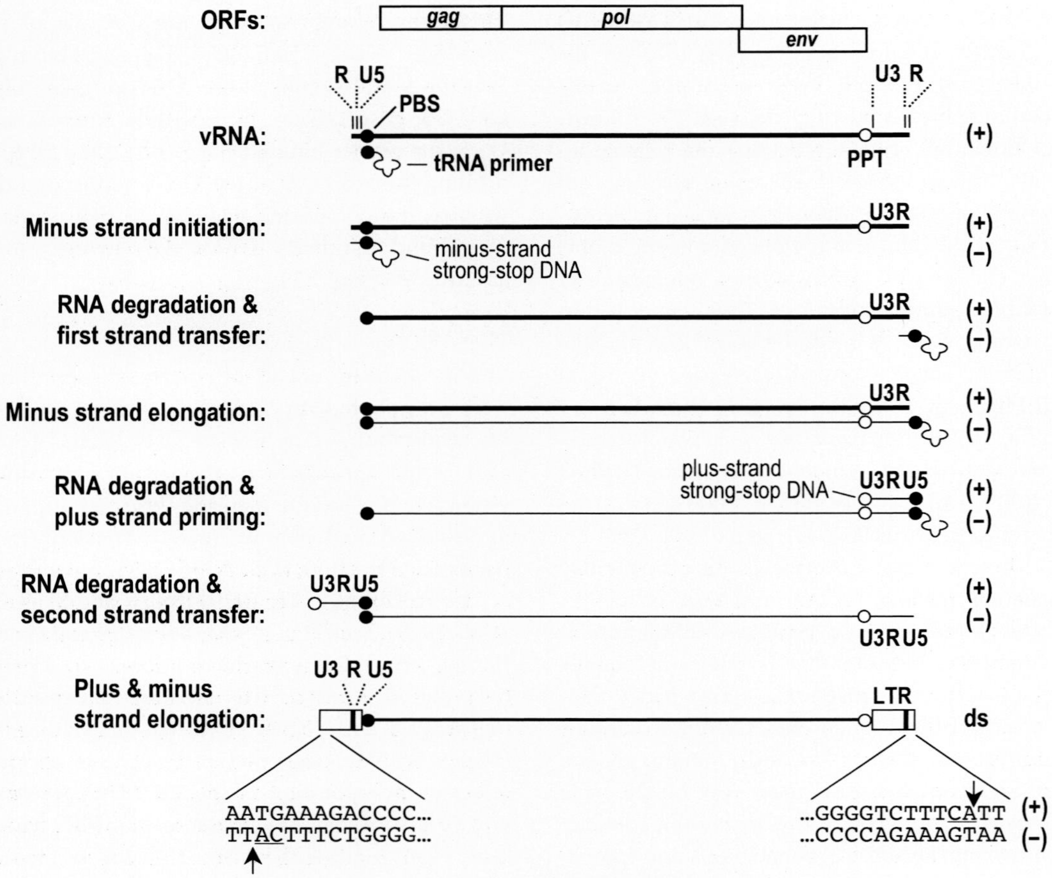

Figure 5.2 Mechanism of reverse transcription. Shown above the viral RNA (vRNA) genome are the *gag*, *pol*, and *env* open reading frames (ORFs) present in all replication-competent retroviruses (see Chapter 3 for details). DNA synthesis is primed by a cellular tRNA that anneals to complementary sequences in the viral RNA primer binding site (PBS), resulting in the synthesis of minus-strand strong-stop DNA containing R and U5 sequences. This DNA becomes annealed to the 3′ end of the viral genome following RNA degradation and the first strand transfer. Following minus-strand extension and RNA degradation, the polypurine tract (PPT) RNA primer is used to prime plus-strand synthesis. The resultant plus-strand strong-stop DNA becomes annealed to the 3′ region of the extended minus-strand following RNA degradation and the second strand transfer. Elongation of the plus and minus strands results in double-stranded (ds) cDNA with a copy of the long terminal repeat (LTR) at each end. The sequences of the Moloney murine leukaemia virus U3 and U5 DNA attachments sites important for integration are shown underneath the LTRs, with the phylogenetically conserved CA/GT dinucleotides and phosphodiester bonds cleaved by integrase during 3′ processing indicated by underlines and vertical arrows, respectively. The polarity of nucleic acid strands (+ or –) is indicated on the right.

Extension of the nascent minus-strand to the 5′ end of the genome yields so-called minus-strand strong-stop DNA comprising R and U5 sequences (Fig. 5.2). RNase H activity plays a key role in the transfer of minus-strand strong-stop DNA to the 3′ end of the viral genome (Tanese *et al.*, 1991). After strand transfer, the minus strand is extended to copy the remaining RNA sequence, followed by RNase H degradation of the majority of the RNA genome. The PPT in contrast resists degradation, resulting in the formation of the primer for plus-strand synthesis (Finston and Champoux, 1984; Omer *et al.*, 1984; Smith *et al.*, 1984a,b). The plus-strand is extended (rightward in Fig. 5.2) until reverse transcriptase encounters 1-methy-adenosine, eighteen nucleotides into the tRNA primer, which fails to support DNA synthesis (Ben-Artzi *et al.*, 1996; Auxilien *et*

Table 5.1 tRNA primer for initiation of reverse transcription of select retroviruses

Genus	Representative virus	tRNA	Reference
α	Rous sarcoma virus	Trp	Harada *et al.* (1975)
β	Mouse mammary tumour virus Mason–Pfizer monkey virus	Lys-3 Lys-1,2	Peters and Glover (1980), Sonigo *et al.* (1986)
γ	Moloney murine leukaemia virus	Pro	Peters *et al.* (1977), Harada *et al.* (1979)
δ	Human T-lymphotropic virus type 1	Pro	Seiki *et al.* (1983)
ε	Walleye dermal sarcoma virus	His	Holzschu *et al.* (1995)
Spuma	Simian foamy virus	Lys-1,2	Kupiec *et al.* (1991)
Lenti	Human immunodeficiency virus type 1	Lys-3	Ratner *et al.* (1985)
	Maedi-visna virus	Lys-1,2	Sonigo *et al.* (1985)

al., 1999) and results in the formation of plus-strand strong-stop DNA. Following RNase H removal of the tRNA primer from the 5′ end of the minus-strand, plus-strand strong-stop DNA undergoes the second strand transfer to anneal with the 3′ end of the minus strand. The plus and minus strands are then extended to yield a linear double-stranded (ds) cDNA containing a copy of U3-R-U5 sequences, also known as the long terminal repeat (LTR), at each end (Fig. 5.2). In addition to terminal region sequence homology, the viral nucleocapsid protein participates in the proper transfer of DNA strands during reverse transcription (Allain *et al.*, 1994; Peliska *et al.*, 1994; Rodriguez-Rodriguez *et al.*, 1995).

The resulting minus-strand is an intact polynucleotide, whereas the plus strands of many viruses contain gaps due to discontinuous DNA synthesis (Boone and Skalka, 1981; Harris *et al.*, 1981; Kung *et al.*, 1981; Hsu and Taylor, 1982; Kupiec *et al.*, 1988; Schweizer *et al.*, 1989; Charneau and Clavel, 1991; Hungnes *et al.*, 1991; Miller *et al.*, 1995). In particular, lenti- and spumaretroviruses contain a second origin of plus-strand synthesis known as the central PPT (cPPT) (Kupiec *et al.*, 1988; Charneau and Clavel, 1991; Whitwam *et al.*, 2001). Approximately 100 nucleotides downstream from the HIV-1 cPPT lies a termination signal for reverse transcription, such that the transferred and extended plus-strand effectively terminates ~100 nucleotides downstream from the cPPT (Charneau *et al.*, 1994). The resulting centrally located, triple-stranded region in the neo-synthesized lentiviral genome is referred to as the central DNA flap (Zennou *et al.*, 2000). Numerous groups have investigated the importance of this highly conserved structure on HIV-1 replication, with fairly different results reported. Certain DNA flap mutant strains were significantly perturbed in their ability to replicate (Zennou *et al.*, 2000; Ao *et al.*, 2004; Arhel *et al.*, 2006b; De Rijck and Debyser, 2006) whereas others showed little to no effect (Charneau *et al.*, 1992; Hungnes *et al.*, 1992; Dvorin *et al.*, 2002; Limón *et al.*, 2002b; Arhel *et al.*, 2006b; Marsden and Zack, 2007). Yeast retrotransposons are close relatives that share many of the intracellular replication steps utilized by their retroviral cousins (refer to Chapter 1 for details), and reverse transcription of the Ty1 genome also yields a central DNA flap structure. Consistent with a relatively modest role in lentiviral replication, flap-deficient Ty1 retrotransposition was reduced approximately twofold (Heyman *et al.*, 2003).

Enzyme structural organization

The subunit compositions of reverse transcriptase enzymes differ markedly among retroviral genera. The MoMLV enzyme for example functions as a single monomeric polypeptide that contains both polymerase and RNase H activities (Fig. 5.3A) (Moelling, 1974; Roth *et al.*, 1985). The δ-retroviral bovine leukaemia virus (Perach and Hizi, 1999) and β-retroviral mouse mammary tumour virus (Taube *et al.*, 1998) proteins have likewise been characterized as functional monomers.

The lentiviral and α-retroviral enzymes by contrast function as obligate dimers. The larger β subunit of the avian viral enzyme retains the integrase coding region at its C-terminus; processing by the viral protease yields the smaller α subunit (Fig. 5.3A) (Gibson and Verma, 1974). Function is delegated asymmetrically, as the α subunit of the αβ heterodimer catalyses the majority of DNA polymerase and RNase H activities (Werner and Wohrl, 2000). The HIV-1 heterodimer also arises via asymmetric processing, whereby the vast majority of the RNase H domain is removed from the larger p66 subunit to create p51 (Fig. 5.3A) (Mizrahi *et al.*, 1989; Bathurst *et al.*, 1990; Graves *et al.*, 1990). By contrast to α-retroviruses, DNA polymerase and RNase H activities of HIV-1 reverse transcriptase are catalysed by the larger subunit of the heterodimer (Le Grice *et al.*, 1991; Hostomsky *et al.*, 1992; Lederer *et al.*, 1992).

Crystal structures of reverse transcriptase shed significant light on the asymmetry of catalytic function among the two HIV-1 polypeptides: although the N-terminal 440 residues of p66 are identical to p51, their secondary structural elements fold into grossly different tertiary structures (Fig. 5.3B) (Kohlstaedt *et al.*, 1992; Jacobo-Molina *et al.*, 1993; Huang *et al.*, 1998). The p66 subunit contains five subdomains whose overall structure resembles that of a right hand, leading to the fingers, palm, and thumb subdomain terminology (in addition to connection and RNase H domains; Fig. 5.3B). The palm subdomain harbours active site residues Asp-110, Asp-185, and Asp-186 (Fig. 5.3A), and its constituent α helices and β strands adopt a fold that is analogous to the active site domains of other DNA polymerases (Kohlstaedt *et al.*, 1992). The crystal structure of the MoMLV polymerase domain revealed a positively charged groove that probably accommodates substrate nucleic acid (Das and Georgiadis, 2004). The non-catalytic p51 subunit contributed to forming the analogous HIV-1 surface, helping to explain the requirement for an obligate lentiviral dimer in contrast to the active monomeric γ-retroviral enzyme (Huang *et al.*, 1998; Das and Georgiadis, 2004).

Structures of HIV-1 reverse transcriptase in complex with modelled primer:templates have shed significant light on the molecular details of DNA synthesis (Jacobo-Molina *et al.*, 1993; Huang *et al.*, 1998). Markedly kinked DNA (green in Fig. 5.3B) binds in the positively charged groove that extends from the polymerase active site to the RNase H active site. In the vicinity of the polymerase active site the DNA adopts A conformation, whereas the structure becomes more B-like as the duplex extends toward the RNase H domain.

The RTC and cytoplasmic transport

Retroviral particles possess buoyant densities of approximately 1.16 to 1.18 g/ml (Toplin, 1967; Sarkar and Moore, 1974; Matheka *et al.*, 1976; Mergener *et al.*, 1992). Within 1 h of infection, MoMLV and HIV-1 subcellular particles displaying densities of 1.34–1.35 g/mL are detected in the cytoplasm (Fassati and Goff, 1999, 2001). Because these particles both co-sediment with newly synthesized viral DNA and continue to synthesize endogenous viral DNA *in vitro* when supplied with dNTPs, they are referred to as RTCs (Lee and Coffin, 1991; Fassati and Goff, 1999, 2001; Iordanskiy *et al.*, 2006). The sizes and/or shapes of HIV-1 and MoMLV RTCs change as a function of time post-infection as evidenced by variable sedimentation velocities through sucrose (Karageorgos *et al.*, 1993; Fassati and Goff, 1999, 2001). These changes may reflect maturation steps that take place as RTCs morph into PICs and/or move from the cytoplasm into the nucleus (Karageorgos *et al.*, 1993; Fassati and Goff, 1999, 2001; Iordanskiy *et al.*, 2006; Arhel *et al.*, 2007).

The cytoplasm presents a relatively viscous environment (Luby-Phelps, 2000), precluding retroviruses from reaching the cell nucleus in the absence of active transport (Campbell and Hope, 2005). Insight into the mechanism of retrovirus trafficking has been gleaned from microscopic analyses of fluorescently labelled particles in live cells (McDonald *et al.*, 2002; Arhel *et al.*, 2006a). HIV-1 and human foamy virus engage the dynein-mediated microtubule motor complex to traffic toward the microtubule organizing centre (MTOC) at the nuclear periphery (McDonald *et al.*, 2002; Petit *et al.*, 2003; Arhel *et al.*, 2006a) though of note, polymerized microtubules seem dispensable under certain conditions of HIV-1 (Bukrinskaya *et al.*, 1998; Kootstra *et al.*, 2000; Groschel and Bushman, 2005) and α-retroviral

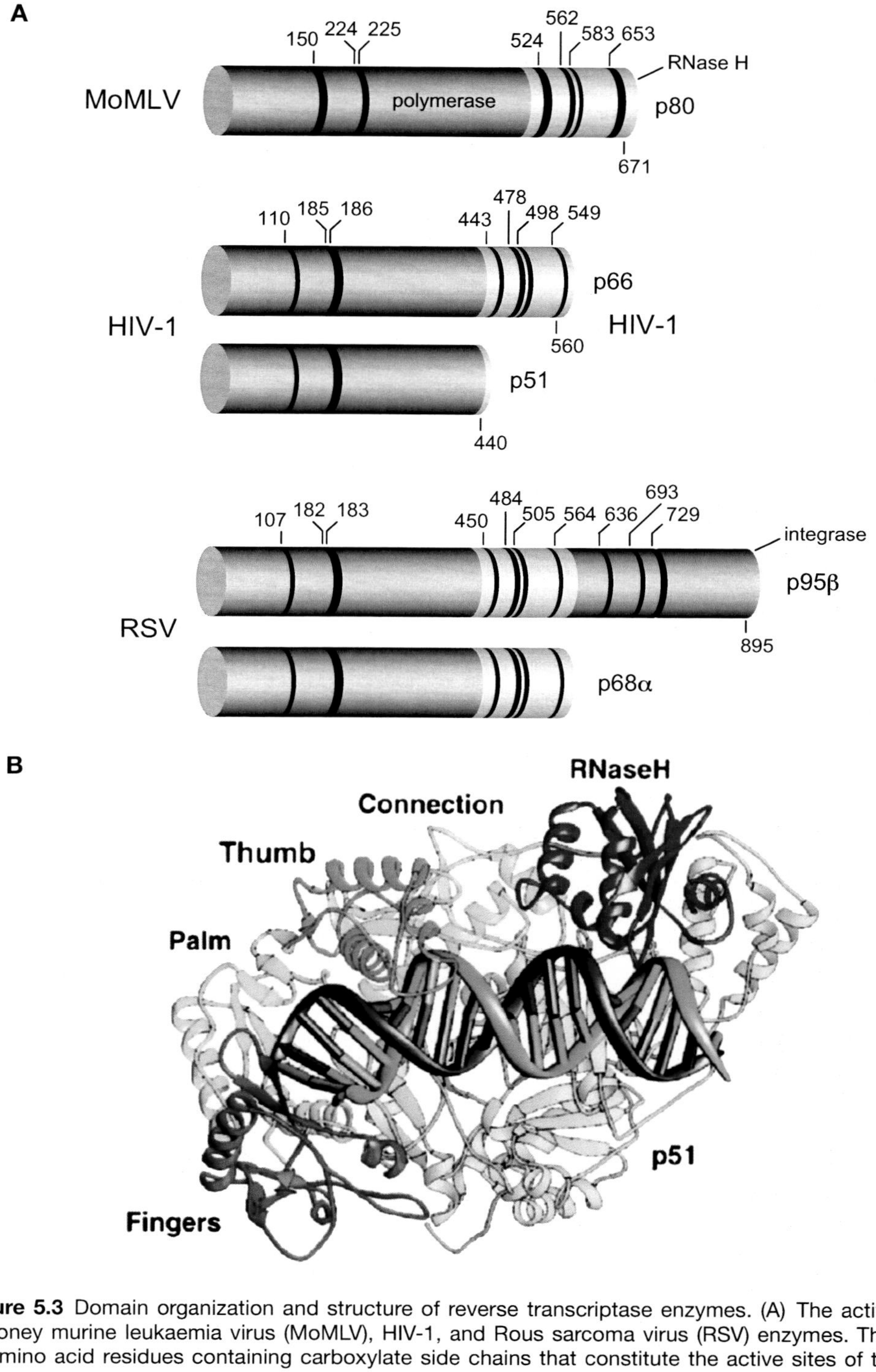

Figure 5.3 Domain organization and structure of reverse transcriptase enzymes. (A) The active forms of Moloney murine leukaemia virus (MoMLV), HIV-1, and Rous sarcoma virus (RSV) enzymes. The positions of amino acid residues containing carboxylate side chains that constitute the active sites of the different enzymes are highlighted in black. Four such residues are conserved in the RNase H enzymes, though only the first three appear to be essential for catalysis (Kanaya *et al.*, 1990). (B) Crystal structure of a HIV-1 p66/p51 primer–template complex (Protein Data Base accession code 1RTD). The nucleoprotein structure was covalently trapped via a disulphide bond created between an engineered Cys at position 258 in the thumb domain (orange) and a thiol-substituted guanine nucleotide in the template strand. Gold, dTTP substrate nucleotide bound at the polymerase active site. Panel B, from Huang *et al.*, 1998, is reprinted with permission from AAAS.

(Katz *et al.*, 2002) infection. The RTC matures into the PIC as it moves toward the nucleus (Fig. 5.1, step 5). PICs are perhaps best defined operationally as nucleoprotein complexes isolated from acutely infected cells that integrate neo-synthesized cDNA into an exogenously added target DNA *in vitro* (Brown *et al.*, 1987, 1989; Fujiwara and Mizuuchi, 1988; Bowerman *et al.*, 1989; Ellison *et al.*, 1990; Farnet and Haseltine, 1990; Lee and Coffin, 1991). Integrase, the viral enzyme that accomplishes integration, possesses two catalytic activities, 3′ processing and DNA strand transfer (Fig. 5.4). The LTR ends of HIV-1 cDNA can be processed by integrase soon after their synthesis in the cell cytoplasm (Miller *et al.*, 1997), suggesting that integrase 3′ processing activity may be linked to the completion of DNA end synthesis. Consistent with this notion, HIV-1 reverse transcriptase and integrase enzymes physically interact with each other (Wu *et al.*, 1999; Tasara *et al.*, 2001; Hehl *et al.*, 2004; Zhu *et al.*, 2004) and genetic complementation experiments moreover suggest that the two enzymes remain associated during the formation of the Ty1 PIC (Wilhelm and Wilhelm, 2006).

Accomplishing end processing in the cell cytoplasm well before encountering the chromosomal targets of integration within the nucleus presents a particular challenge for retroviruses. A variety of footprinting approaches has revealed that the terminal regions of preintegrative DNA are specifically protected from DNA cleavage reagents *in vitro* (Bowerman *et al.*, 1989; Miller *et al.*, 1997; Wei *et al.*, 1997; Chen *et al.*, 1999; Khiytani and Dimmock, 2002). Exposed internal regions in relatively high concentration to 3′ ends that are poised for DNA strand transfer provide an ideal scenario for integrase-mediated autointegration of the DNA ends into interior regions of the reverse transcript. Autointegration, which occurs during infection (Shoemaker *et al.*, 1981; Li *et al.*, 1991) and Ty1 retrotransposition (Garfinkel *et al.*, 2006), is suicidal, as the neo-synthesized cDNA either becomes inverted or cut in two (Shoemaker *et al.*, 1981; Lee and Craigie, 1994). One therefore envisions that retroviruses have evolved a mechanism to suppress this unwanted outcome, and the barrier-to-autointegration factor (BAF), a small, non-histone DNA binding protein, was discovered via its ability to suppress MoMLV PIC autointegration activity *in vitro* (Lee and Craigie, 1994, 1998). Follow-up studies utilizing RNA interference to reduce cellular protein levels have yielded conflicting results as to the importance of BAF during HIV-1 and MoMLV infection: one study reported as much as a tenfold reduction in the ability of knockdown cells to become infected (Jacque and Stevenson, 2006) whereas a separate study saw little if any effects (Shun *et al.*, 2007a). Additional work is therefore required to ascertain if BAF plays a role in protecting preintegrative cDNA from autointegration and/or helping PICs to accomplish chromosomal integration (Chen and Engelman, 1998; Suzuki and Craigie, 2002; Suzuki *et al.*, 2004; Jacque and Stevenson, 2006).

Nuclear localization

PICs must access chromosomes for integration to occur. Different retroviruses appear to accomplish this task by using different means, though the precise mechanisms of retroviral nuclear localization are some of the most contentious and least understood aspects of the viral lifecycles (see Fassati, 2006, Yamashita and Emerman, 2006, and Suzuki and Craigie, 2007, for recent reviews).

Cellular and viral phenotypes

All retroviruses are seemingly prohibited from productively infecting quiescent cells due to limited nucleotide precursor concentrations affecting the extent of reverse transcription and RTC formation (Fritsch and Temin, 1977; Varmus *et al.*, 1977; Stevenson *et al.*, 1990b; Zack *et al.*, 1990; Gao *et al.*, 1993; Goulaouic *et al.*, 1994; Meyerhans *et al.*, 1994; O'Brien *et al.*, 1994; Spina *et al.*, 1995; Korin and Zack, 1998; Kootstra *et al.*, 2000; Pierson *et al.*, 2002; Triques and Stevenson, 2004; Plesa *et al.*, 2007). Different viruses however display marked preferences for infecting dividing cells versus growth-arrested cells that are nonetheless metabolically active. HIV-1 and other lentiviruses efficiently infect non-dividing primary cells and growth-arrested cell lines (Weinberg *et al.*, 1991; Lewis *et al.*, 1992) whereas MoMLV requires cells to be actively cycling (Harel *et al.*, 1981; Miller *et al.*, 1990; Lewis *et al.*, 1992). Results with α- and spumaretroviruses have yielded more intermediate phenotypes. Avian viruses have been reported to transduce non-dividing cells at approximately

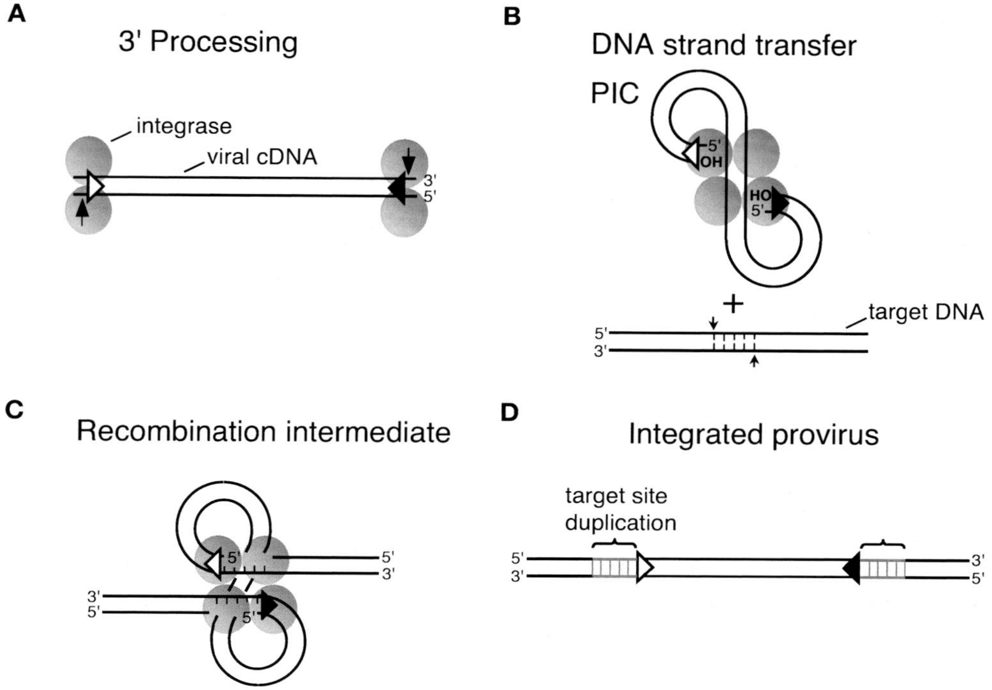

Figure 5.4 Mechanism of retroviral DNA integration. (A) The active integrase multimer (a dimer is drawn for simplicity) hydrolyses each cDNA end adjacent to the phylogenetically conserved sequence CA (marked by short vertical arrows; see Fig. 5.2 for attachment site sequences) during the 3′ processing reaction. (B) After gaining nuclear entry and locating a suitable chromatin acceptor site for integration, integrase uses the hydroxyl groups at the recessed 3′ ends to cut the opposing strands of target DNA in a staggered fashion, which at the same time joins the viral ends to the resulting 5′ phosphates. (C) The product of DNA strand transfer is a recombination intermediate harbouring single-strand gaps adjacent to the viral 5′ ends. (D) Repair of the gaps, which is presumably mediated by host enzymes, yields the integrated provirus flanked by a short duplication (grey) whose sequence is defined by the identity of the double-stranded DNA cut in B. The open and filled triangles demarcate the U3 and U5 DNA attachment sites in the upstream and downstream LTR, respectively, important for integrase function.

3–30% of the levels of cycling cell controls (Hatziioannou and Goff, 2001; Katz *et al.*, 2002). Foamy viruses, like γ-retroviruses, are blocked from productively infecting growth-arrested cell lines (Bieniasz *et al.*, 1995; Trobridge and Russell, 2004) though, akin to lentiviruses, a foamy viral vector efficiently transduced serum-starved cells that were subsequently stimulated to divide (Trobridge and Russell, 2004). An affinity of foamy viral (Saib *et al.*, 1997; Lehmann-Che *et al.*, 2007) or HIV-1 (McDonald *et al.*, 2002; Zamborlini *et al.*, 2007) RTCs for the cellular MTOC may very well contribute to the preintegration latency of these viruses.

HIV-1 PICs are actively transported into cell nuclei (Bukrinsky *et al.*, 1992) whereas cells are required to pass through the M phase of their growth cycle to support MoMLV infection (Roe *et al.*, 1993; Lewis and Emerman, 1994). Such studies helped to solidify the hypothesis that lentiviral PICs contain nuclear localization signals (NLSs) to impart active transport in non-dividing cells, whereas γ-retroviral complexes lack such signals and therefore wait for nuclear membrane dissolution to access chromosomes. Such dogma however is under constant re-evaluation for the following reasons. Studies to identity lentiviral NLSs that specifically function under conditions of growth arrest led to proposed roles for HIV-1 matrix (Bukrinsky *et al.*, 1993; von Schwedler *et al.*, 1994; Gallay *et al.*, 1995a, 1995b; Bukrinskaya *et al.*, 1996; Haffar *et al.*, 2000), Vpr (Heinzinger *et al.*, 1994; Popov *et al.*, 1998a,b; Iijima *et al.*, 2004; Nitahara-Kasahara *et al.*, 2007), and integrase

(Gallay *et al.*, 1997; Bouyac-Bertoia *et al.*, 2001; Ikeda *et al.*, 2004; Ao *et al.*, 2005), yet a number of reports at the same time seemingly discounted important roles for each of these factors (Freed *et al.*, 1995, 1997; Fouchier *et al.*, 1997; Reil *et al.*, 1998; Kootstra and Schuitemaker, 1999; Mannioui *et al.*, 2005 in terms of matrix; Connor *et al.*, 1995; Reil *et al.*, 1998; Kootstra and Schuitemaker, 1999; Bouyac-Bertoia *et al.*, 2001; Trobridge and Russell, 2004 for Vpr; Petit *et al.*, 2000; Dvorin *et al.*, 2002; Limón *et al.*, 2002a; Lu *et al.*, 2004 for integrase). A multiply mutated HIV-1 derivative lacking Vpr, the central DNA flap, and carrying MoMLV matrix and integrase moreover infected growth arrested cells as efficiently as dividing cells, discounting essential roles for each of these HIV-1 factors in nuclear import under these conditions (Yamashita and Emerman, 2005). Experiments designed to uncover factors that function specifically in growth arrested cells may be misguided: HIV-1 appears to utilize an active transport mechanism in cycling as well as in non-dividing cells (Katz *et al.*, 2003) and the addition of functional NLSs to γ-retroviral vectors in some cases failed to significantly increase the efficiency at which they infected growth-arrested cells (Seamon *et al.*, 2002; Caron and Caruso, 2005). Yet another study has brought into question the extent of the block to γ-retroviral infection in non-dividing primary human cells (Jarrosson-Wuilleme *et al.*, 2006).

HIV-1/MoMLV chimeric viruses carrying the MoMLV capsid protein were specifically blocked from infecting growth-arrested cells, leading to the hypothesis that the γ-retroviral capsid protein may act as an effective inhibitor of nuclear transport (Yamashita and Emerman, 2004). Accordingly, MoMLV RTCs and PICs seem to retain more of the virion complement of capsid protein than their HIV-1 counterparts (Bowerman *et al.*, 1989; Karageorgos *et al.*, 1993; Miller *et al.*, 1997; Fassati and Goff, 1999, 2001). Moreover, cycling cells supported normal levels of HIV-1 Q63A/Q67A capsid mutant reverse transcription but the resulting PICs, which retained an unusually high level of capsid protein, were defective for nuclear import (Dismuke and Aiken, 2006). Consistent with a role for capsid protein shedding in regulating nuclear import, central DNA flap mutant RTCs, which also retained an elevated level of capsid protein, accumulated at the nuclear periphery (Arhel *et al.*, 2007).

Additional analyses of HIV-1 capsid mutants indicated that the block to infection might occur after nuclear localization, hinting that protein shedding could mediate a relatively late structural conformation required for proper intranuclear trafficking as compared to entry into the nucleus (Yamashita *et al.*, 2007). Certain mutations in the MoMLV Gag p12 (Yuan *et al.*, 1999, 2002; Yueh and Goff, 2003) and capsid (Yueh *et al.*, 2006) proteins yielded an analogous phenotype whereby virus infection was apparently blocked at entry into the nucleus or soon thereafter to obscure proper trafficking. The post nuclear entry events that occur to ensure chromosomal engagement for integration are not well understood. BAF and its cellular binding partner lamina-associated polypeptide 2α (LAP2α) were reported to play critical roles for MoMLV whereas emerin was additionally implicated for HIV-1 (Jacque and Stevenson, 2006). However, a follow-up study failed to confirm important roles for these inner nuclear membrane-associated factors during infection (Shun *et al.*, 2007a). Additional work is therefore required to glean insight into intranuclear trafficking mechanisms – continued analysis of viral mutants that are apparently blocked at this step(s) is expected to help decipher salient details.

Host factors and nuclear localization

Considering the lack of consensus surrounding the viral determinants important for PIC nuclear localization, it is not surprising that a clear picture is also lacking for which cell factors mediate the process. Briefly, classical nuclear import occurs via an NLS-bearing protein first interacting with importin-α or a similar karyopherin in the cytosol, followed by their combined interaction with importin-β. This complex then docks with nucleoporins located on the cytoplasmic side of the NPC to affect energy-dependent nuclear translocation (see Stewart, 2007, for a recent review). Ectopic expression of HIV-1 Vpr can lead to its accumulation at the nuclear periphery (Fouchier *et al.*, 1998; Vodicka *et al.*, 1998; Depienne *et al.*, 2000; Le Rouzic *et al.*, 2002) and the protein was moreover shown to interact with various import factors including importin-α (Popov *et al.*, 1998b;

Vodicka *et al.*, 1998; Nitahara-Kasahara *et al.*, 2007), importin-β (Fouchier *et al.*, 1998), and nucleoporins (Fouchier *et al.*, 1998; Le Rouzic *et al.*, 2002). Vpx, an accessory protein common to simian immunodeficiency virus (SIV) and HIV-2 that shares Vpr's ability to affect nuclear import (Fletcher 3rd *et al.*, 1996), interacts with heat shock protein (HSP) 40, and this interaction can influence PIC nuclear import (Cheng *et al.*, 2008).

Integrase is an attractive PIC karyophile because it is the one viral protein that must remain associated with the reverse transcript until integration. Moreover, the Ty1 (Kenna *et al.*, 1998; Moore *et al.*, 1998) and Ty3 (Lin *et al.*, 2001) proteins regulate retrotransposon PIC nuclear translocation. Lentiviral integrases are karyophilic (Petit *et al.*, 1999; Pluymers *et al.*, 1999; Cherepanov *et al.*, 2000; Depienne *et al.*, 2000; Devroe *et al.*, 2003; Woodward *et al.*, 2003) and similar to Vpr, HIV-1 integrase has been shown to interact with importin-α (Gallay *et al.*, 1997; Fassati *et al.*, 2003; Hearps and Jans, 2006), importin-β (Fassati *et al.*, 2003; Hearps and Jans, 2006), as well as importin-7 (Fassati *et al.*, 2003; Ao *et al.*, 2007) and transportin-1 (Fassati *et al.*, 2003). Cells knocked down for importin-7 expression supported 40–100% and 10–40% of the levels of MoMLV and HIV-1 infectivity, respectively, as compared to unsilenced control cells (Fassati *et al.*, 2003), indicating a potential role for importin-7 in lentiviral nuclear import. Follow up studies however questioned the overall relevance of importin-7 in this process (Zielske and Stevenson, 2005; Ao *et al.*, 2007).

The extent to which importin–integrase interactions contribute to the karyophilic properties of the lentiviral proteins is an ongoing area of investigation. Lens epithelium-derived growth factor (LEDGF)/p75 is a dominant lentiviral integrase binding protein (Cherepanov *et al.*, 2003; Cherepanov, 2007) that plays a critical role in targeting PICs to active genes during integration (see below). As lentiviral integrases re-localized to the cytoplasm upon LEDGF/p75 silencing (Maertens *et al.*, 2003; Llano *et al.*, 2004b; Emiliani *et al.*, 2005), their karyophilic properties are strongly influenced by this normally chromatin-associated protein (Maertens *et al.*, 2004; Vanegas *et al.*, 2005). HIV-1 integrase is targeted for degradation by the cellular proteasome (Mulder and Muesing, 2000; Devroe at al., 2003) and LEDGF/p75 significantly protects it from this fate (Llano *et al.*, 2004a). Knockdown cells treated with proteasome inhibitors unveiled a nuclear population of HIV-1 integrase, indicating that the lentiviral proteins may access nuclei via a LEDGF/p75-independent mechanism (Emiliani *et al.*, 2005). MoMLV integrase is predominantly cytoplasmic when expressed in cells (Llano *et al.*, 2004b) whereas α-retroviral integrase is karyophilic (Kukolj *et al.*, 1997). Certain mutations that disrupted integrase localization also perturbed α-retroviral replication, though the part(s) of the lifecycle affected by the mutations was not investigated (Kukolj *et al.*, 1998). Additional work may address whether the karyophilic property of avian viral integrase contributes to the modest abilities for these viruses to infect non-dividing cells (Hatziioannou and Goff, 2001; Katz *et al.*, 2002).

Defective tRNAs lacking 3′ CCA ends affected the efficiency of HIV-1 RTC nuclear localization in *in vitro* transport assays under conditions wherein MoMLV transport was unaffected (Zaitseva *et al.*, 2006), indicating that host nucleic acid might play a physiologically relevant role in lentiviral PIC nuclear import. More recently a genome-wide screen to identify host factors important for HIV-1 replication highlighted transportin-3 as a potential player in PIC nuclear localization: knockdown cells were infected normally by MoMLV, whereas HIV-1 was blocked downstream from reverse transcription, perhaps at PIC nuclear import or integration (Brass *et al.*, 2008).

Integration

Integration proceeds via the integrase protein acting on the DNA attachment sites comprising the U3 and U5 ends of the nascent reverse transcript (Figs. 5.2 and 5.4) (Panganiban and Temin, 1983, 1984; Donehower and Varmus, 1984; Schwartzberg *et al.*, 1984; Quinn and Grandgenett, 1988; Stevenson *et al.*, 1990a; LaFemina *et al.*, 1992; Enssle *et al.*, 1999). The salient DNA cutting and joining steps were deciphered by analysing sequences of integrated virus-host junctions (Dhar *et al.*, 1980; Ju and Skalka, 1980; Shimotohno *et al.*, 1980; Hishinuma *et al.*, 1981; Hughes *et al.*,

1981; Majors and Varmus, 1981; Vincent *et al.*, 1990; Vink *et al.*, 1990; Neves *et al.*, 1998) as well as the end structures of replication intermediates isolated from acutely infected cells (Brown *et al.*, 1989; Roth *et al.*, 1989; Pauza, 1990; Lee and Coffin, 1991) and *in vitro* recombination intermediates formed during PIC integration assays (Fujiwara and Mizuuchi, 1988; Brown *et al.*, 1989; Lee and Coffin, 1991). Several additional criteria solidified the concept that retroviral integration is mechanistically analogous to the reactions carried out by a number of prokaryotic DNA transposase proteins as well as the RAG1/2 recombinase that mediates rearrangements of immunoglobulin and T-cell receptor genes in higher eukaryotes (see Craig *et al.*, 2002, for several pertinent chapters).

Molecular mechanism

The hallmark of retroviral attachment sites is an invariant CA/GT dinucleotide located near each 3′ end of unintegrated DNA (Fig. 5.2) (reviewed in Skalka, 1993 and Katzman and Katz, 1999), and *in vitro* assays utilizing purified proteins and model attachment site substrates were invaluable for deciphering the biochemical mechanisms of integrase function (Katzman *et al.*, 1989; Craigie *et al.*, 1990; Katz *et al.*, 1990; Sherman and Fyfe, 1990; Vora *et al.*, 1990; Bushman and Craigie, 1991; Pahl and Flugel, 1993). Integrase uses water to hydrolyse the phosphodiester bond 3′ of the conserved adenine during 3′ processing, which in many cases liberates a terminal dinucleotide from each LTR (Fig. 5.2 and 5.4) (Engelman *et al.*, 1991; Vink *et al.*, 1991, 1994; Dotan *et al.*, 1995; Engelman, 1996; Skinner *et al.*, 2001). Other oxygen-containing nucleophiles, like the hydroxyl groups in glycerol or at the 3′ end of the dinucleotide that is to be cleaved, can surprisingly substitute for water during 3′ processing (Engelman *et al.*, 1991; Vink *et al.*, 1991, 1994; Dotan *et al.*, 1995; Skinner *et al.*, 2001). Utilization of the 3′-OH yields a cyclic dinucleotide product, which afforded the monitoring of the number of chiral inversions of a participating phosphorothioate placed at the scissile bond. As chirality was inverted in the resulting cyclic dinucleotide, 3′ processing probably occurred via a single transesterification (hydrolysis in the usual case) in the absence of an integrase-viral DNA covalent intermediate (Engelman *et al.*, 1991). Monitoring the analogous reaction catalysed by the bacteriophage MuA transposase protein likewise revealed inversion of the participating phosphorothioate upon hydrolysis (Mizuuchi *et al.*, 1999). The placement of chiral phosphorothioate centres in the backbones of the target DNA strands used during integration and transposition also revealed that DNA strand transfer yields inversion of chirality, again indicating that this reaction occurs via a single transesterification in the absence of a protein–DNA covalent intermediate (Engelman *et al.*, 1991; Mizuuchi and Adzuma, 1991). Therefore, the 3′-OH generated during hydrolysis in the cytoplasm is used by integrase in the nucleus to cut the target DNA backbone, which at the same time joins the viral DNA end to the cell chromosome (Fig. 5.4B and C). As the foamy viral U3 DNA attachment site is not processed by integrase (Juretzek *et al.*, 2004), the 3′-OH nucleophile for DNA strand transfer in this case is generated by reverse transcriptase.

The final step in the integration process, gap repair (Fig. 5.4C and D), is probably completed by host cell enzymes. Experiments using substrates that modelled the single-strand gap structures created by DNA strand transfer revealed the requirement for three enzymatic activities *in vitro*: polymerization, which could be fulfilled by DNA polymerase β, δ, or reverse transcriptase; ligation, accomplished by ligase I, III, or IV; and flap endonucleolytic cleavage, catalysed by flap endonuclease 1 (Brin *et al.*, 2000; Yoder and Bushman, 2000). Genetic approaches implicated potential roles for poly(ADP-ribose) polymerase 1 (PARP-1) (Gaken *et al.*, 1996; Ha *et al.*, 2001) as well as several kinases involved in DNA repair including DNA-dependent protein kinase (DNA-PK) (Daniel *et al.*, 1999, 2004), ataxia-telangiectasia mutated (ATM) (Daniel *et al.*, 2001), and ATM- and Rad3-related (ATR) (Daniel *et al.*, 2003) during virus infection. However, follow-up studies failed to support important roles for these factors in integration gap repair (Baekelandt *et al.*, 2000; Siva and Bushman, 2002; Ariumi *et al.*, 2005; DeHart *et al.*, 2005). The redundant functionality of certain repair factors *in vitro* as well as the pleiotropic effects of disrupting such factors in cells perhaps contributed to these contradictory findings. Other DNA repair factors, including Rad52 (Lau *et al.*, 2004) and XPB and

XPD DNA helicases (Yoder *et al.*, 2006), can antagonize PIC function prior to integration and thus appear to contribute to the intrinsic cellular antiviral response (refer to Chapter 11 for subject overview).

Enzyme structure and functional organization

Results of limited proteolysis (Engelman and Craigie, 1992), deletion analyses (Bushman *et al.*, 1993; Vink *et al.*, 1993; Bushman and Wang, 1994; Kulkosky *et al.*, 1995; Shibagaki *et al.*, 1997), complementation assays that resurrected enzyme function by mixing otherwise defective mutant proteins together (Engelman *et al.*, 1993; van Gent *et al.*, 1993; Pahl and Flugel, 1995; Ellison *et al.*, 1995; Jonsson *et al.*, 1996; Diamond and Bushman, 2005), and structural biology (Dyda *et al.*, 1994; Bujacz *et al.*, 1995, 1996; Eijkelenboom *et al.*, 1995, 1997; Lodi *et al.*, 1995; Cai *et al.*, 1997; Chen *et al.*, 2000a,b; Yang *et al.*, 2000; Wang *et al.*, 2001) revealed that the integrase proteins comprise three functional domains: the N-terminal domain (NTD), catalytic core domain (CCD), and C-terminal domain (CTD) (Fig. 5.5). The CCD, which adopts an RNase H-like fold, houses invariant Asp and Glu residues that comprise the DDE active site motif (Drelich *et al.*, 1992; Engelman and Craigie, 1992; Kulkosky *et al.*, 1992; van Gent *et al.*, 1992; Bushman *et al.*, 1993; Leavitt *et al.*, 1993). Catalysis is believed to require two magnesium ions bound by the active site residues. Unlike reverse transcription, a major limitation in understanding the mechanistic details of retroviral integration is the lack of a holoenzyme structure bound to substrate DNA(s). Insight has nevertheless been gleaned from analysing related Tn5 transposase (Davies *et al.*, 2000) and RNase H (Nowotny *et al.*, 2005) nucleoprotein structures. The active site metal ions probably participate as Lewis acids. One activates the attacking nucleophile, whereas the second activates the chemical leaving group. The 3′ processing nucleophile – water – is quite different from the 3′-OH used during DNA strand transfer, yet the same active site catalyses both reactions (Drelich *et al.*, 1992; Engelman and Craigie, 1992; Kulkosky *et al.*, 1992; van Gent *et al.*, 1992; Leavitt *et al.*, 1993). To accommodate such active site gymnastics, the metal ion that activates the incoming hydroxyl group during 3′ processing is proposed to activate the 3′-OH leaving group during DNA strand transfer, in what has been termed a ping-pong reaction mechanism (Kennedy *et al.*, 2000; Nowotny *et al.*, 2005). Disruption of metal ion binding is the proposed antiviral mechanism of clinically relevant HIV-1 integrase inhibitors (Grobler *et al.*, 2002).

Kinetic (Jones *et al.*, 1992), photo cross-linking (Engelman *et al.*, 1993; Heuer and Brown, 1997; Faure *et al.*, 2005; Zhao *et al.*, 2007), yeast two-hybrid (Kalpana and Goff, 1993; Kalpana *et al.*, 1999; Berthoux *et al.*, 2007), and complementation analyses revealed that 3′ processing and DNA strand transfer are catalysed by an integrase multimer *in vitro* (Engelman *et al.*, 1993; van Gent *et al.*, 1993; Ellison *et al.*, 1995; Pahl and Flugel, 1995; Jonsson *et al.*, 1996) and during virus infection (Fletcher 3rd *et al.*, 1997; Wu *et al.*, 1997; Kalpana *et al.*, 1999; Lu *et al.*, 2004; Berthoux *et al.*, 2007). Owing to the lack of relevant structures, the precise make-up of the active multimer(s) is unknown. A dimer suffices *in vitro* to process 3′ ends (Faure *et al.*, 2005; Guiot *et al.*, 2006) though a tetramer may be required for efficient activity (Bosserman *et al.*, 2007). A tetramer appears to be the basic catalytic unit for DNA strand transfer activity (Bao *et al.*, 2003; Faure *et al.*, 2005; Li *et al.*, 2006). In addition to harbouring the enzyme active site, the CCD interacts specifically with the viral DNA attachment site (Du *et al.*, 1997; Gerton and Brown, 1997; Heuer and Brown, 1997; Jenkins *et al.*, 1997; Esposito and Craigie, 1998; Gerton *et al.*, 1998; Chen *et al.*, 2006; Johnson *et al.*, 2006; Zhao *et al.*, 2007) and non-specifically with target DNA (van Gent *et al.*, 1992; Heuer and Brown, 1997; Appa *et al.*, 2001; Dirac and Kjems, 2001; Harper *et al.*, 2001; Diamond and Bushman, 2005; Lu *et al.*, 2005). The NTD contains two conserved His and Cys residues that coordinate a zinc atom (Fig. 5.5A and B) (Burke *et al.*, 1992; McEuen *et al.*, 1992; Bushman *et al.*, 1993; Cai *et al.*, 1997; Eijkelenboom *et al.*, 1997) and zinc binding promotes enzyme multimerization and catalysis (Lee and Han, 1996; Zheng *et al.*, 1996; Lee *et al.*, 1997). The NTD also interacts with viral DNA (van den Ent *et al.*, 1999; Zhao *et al.*, 2007). The CTD, which is the least conserved of the protein domains (Fig. 5.5A), likewise

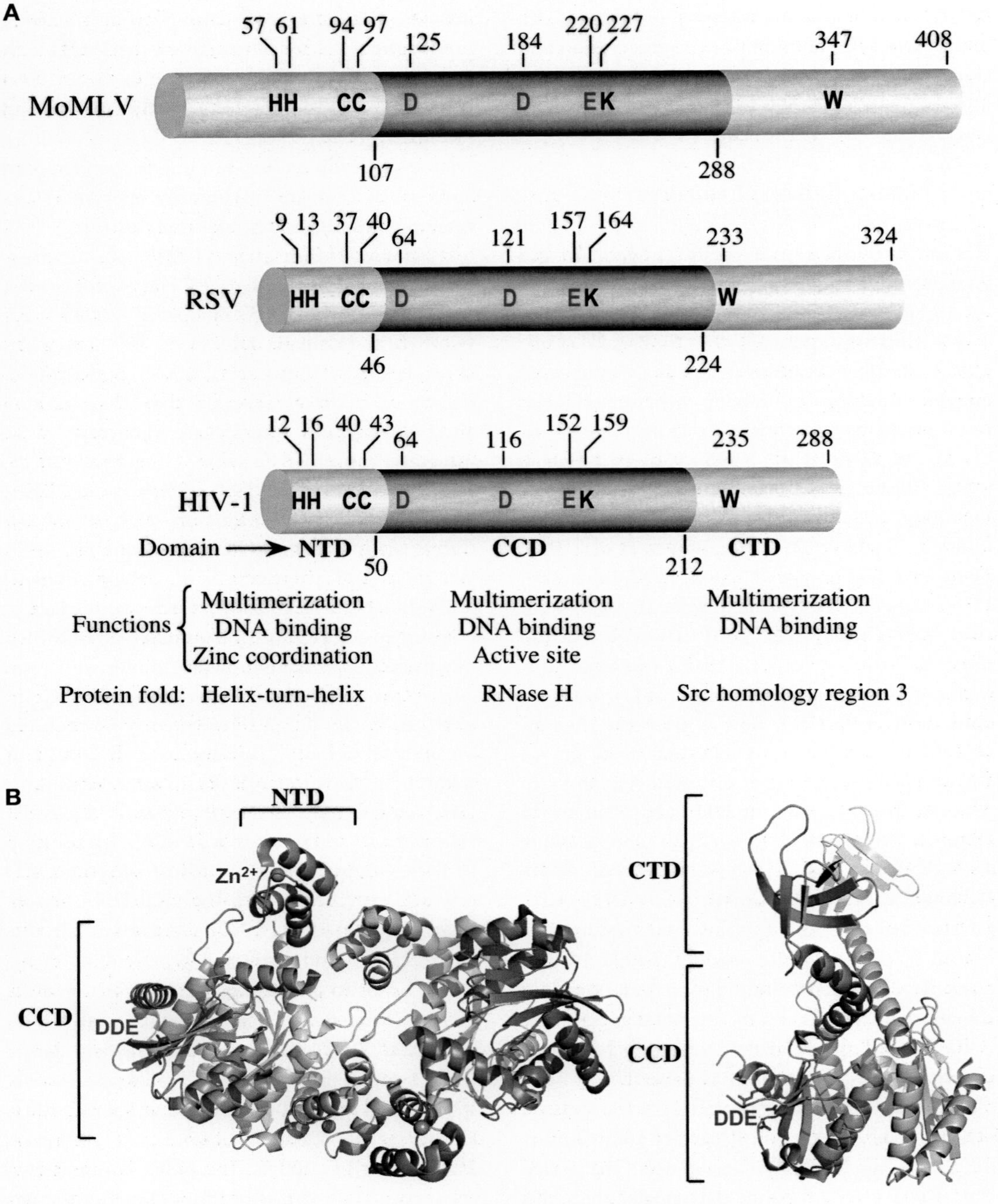

Figure 5.5 Integrase domain organization and HIV-1 crystal structures. (A) The N-terminal domain (NTD), catalytic core domain (CCD), and C-terminal domain (CTD) of monomeric Moloney murine leukaemia virus (MoMLV), Rous sarcoma virus (RSV), and HIV-1 integrase are shown highlighting amino acid residues conserved among all retroviruses. Known domain functions as well as three dimensional protein folds are also indicated. The Asp and Glu residues in the CCDs that constitute the enzyme active sites are highlighted in red. (B) The HIV-1 NTD-CCD (Protein Data Base accession code 1K6Y) and CCD-CTD (accession code 1EX4) two domain structures are shown, highlighting the separate domains. The Asp and Glu residues that constitute the enzyme active site are shown as red sticks; the zinc atoms bound to the NTDs are shown as grey spheres. (B) reproduced from Vandegraaff and Engelman (2007) with permission from Cambridge University Press.

contributes to multimerization (Andrake and Skalka, 1995; Jenkins *et al.*, 1996; Puras Lutzke and Plasterk, 1998) as well as specific (Gao *et al.*, 2001; Zhao *et al.*, 2007) and non-specific (Mumm and Grandgenett, 1991; Woerner *et al.*, 1992; Woerner and Marcus-Sekura, 1993; Vink *et al.*, 1993; Engelman *et al.*, 1994; Puras Lutzke *et al.*, 1994; Puras Lutzke and Plasterk, 1998) DNA binding. Based on available data, a number of models have been proposed for the active nucleoprotein complex (Heuer and Brown, 1998; Gao *et al.*, 2001; Wang *et al.*, 2001; Karki *et al.*, 2004; De Luca *et al.*, 2005; Wielens *et al.*, 2005; Chen *et al.*, 2006; Zhao *et al.*, 2007).

The integrase tetramer bound to processed U3 and U5 DNA attachment sites is thought to cut opposing strands of target DNA in a 'coupled' or 'concerted' reaction (Fig. 5.4B and C) (Fitzgerald *et al.*, 1992; Goodarzi *et al.*, 1995; Vora and Grandgenett, 1995; Carteau *et al.*, 1999; Hindmarsh *et al.*, 1999; Yang and Roth, 2001; Sinha *et al.*, 2002; Vora *et al.*, 2004; Bera *et al.*, 2005; Li and Craigie, 2005; Sinha and Grandgenett, 2005) though the integration of the two ends may actually occur sequentially as compared to simultaneously. Mutating the invariant CA/GTs within both DNA attachment sites rendered HIV-1 non-infectious (Masuda *et al.*, 1998; Brown *et al.*, 1999) whereas the titres of single-end HIV-1, foamy, and α-retroviral mutants were reduced as little as 24% but at most by ~10-fold (Masuda *et al.*, 1998; Juretzek *et al.*, 2004; Oh *et al.*, 2006, 2008). Sequencing resultant α-retroviral-host junctions revealed that defective U5 ends were joined by a non-viral mechanism(s) (Oh *et al.*, 2006) whereas approximately 60% of U3 end mutant proviruses formed normally (Oh *et al.*, 2008). The joining of one DNA attachment site might therefore suffice to commit the second DNA end for integration. Consistent with this interpretation, the HIV-1 integrase tetramer catalysed the sequential integration of two modelled DNA attachment sites *in vitro* (Li *et al.*, 2006).

Host factor roles

A variety of host cell factors have been proposed to participate in integration, though confirmatory evidence for important roles during virus infection has been advanced for a scant few, principally LEDGF/p75. Owing to a number of recent reviews (Van Maele *et al.*, 2006; Engelman, 2007; Vandegraaff and Engelman, 2007; Engelman and Cherepanov, 2008; Poeschla, 2008), this section will briefly overview the role of LEDGF/p75 in lentiviral DNA integration.

HIV-1 integrase expressed in human 293T-cells interacted tightly with endogenous LEDGF/p75, and recombinant LEDGF/p75 purified following its expression in bacteria moreover significantly stimulated integrase activities *in vitro* (Cherepanov *et al.*, 2003). The host–virus interaction was independently discovered by surveying human proteins for binding to HIV-1 integrase in HeLa cells (Turlure *et al.*, 2004) and in yeast (Emiliani *et al.*, 2005). Subsequent work revealed the LEDGF/p75 interaction specific to lentiviral integrases (Llano *et al.*, 2004b; Busschots *et al.*, 2005; Cherepanov, 2007). Considering this specificity, two different roles for LEDGF/p75 in retroviral replication could be envisaged. One function was nuclear import, and the other was targeting PICs to active genes for integration.

Retroviral integration occurs basically throughout the cell genome, though subtle biases at the levels of local DNA sequence (Stevens and Griffith, 1996; Carteau *et al.*, 1998; Holman and Coffin, 2005; Wu *et al.*, 2005; Derse *et al.*, 2007) and genetic structure (reviewed in Bushman *et al.*, 2005, Engelman, 2007, and Vandegraaff and Engelman, 2007) exist. Each bias moreover appears genus-specific. Lentiviruses disfavour non-gene regions and significantly favour active genes, integrating into transcription units nearly equally along their lengths (Schroder *et al.*, 2002; Hematti *et al.*, 2004; Mitchell *et al.*, 2004; Crise *et al.*, 2005; Hacker *et al.*, 2006; Kang *et al.*, 2006; MacNeil *et al.*, 2006). MoMLV by contrast modestly prefers genes and transcriptional activity, but fully one-fourth of all integrations occur within a 10 kb window of gene start sites (Wu *et al.*, 2003; Mitchell *et al.*, 2004; Shun *et al.*, 2007b). Spumaretroviruses display yet a third pattern wherein genes are slightly disfavoured though transcriptional start sites and associated CpG islands are nonetheless targeted significantly above random (Nowrouzi *et al.*, 2006; Trobridge *et al.*, 2006). Alpha- and δ-retroviruses display the least overall preference for genomic features, targeting genes at frequencies similar to MoMLV with more modest preferences for promoters and CpG islands (Mitchell *et al.*, 2004; Narezkina *et al.*, 2004; Derse *et al.*, 2007).

Initial attempts to gauge the importance of LEDGF/p75 in lentiviral replication using RNA interference yielded mixed results (Llano *et al.*, 2004b; Vandegraaff *et al.*, 2006; Vandekerckhove *et al.*, 2006; Zielske and Stevenson, 2006). Subsequent work revealed that endogenous LEDGF/p75 levels far exceed those required for normal levels of integration (Llano *et al.*, 2006; Shun *et al.*, 2007b). Thus clean genetic systems emerged via over-expressing in target cells the part of LEDGF/p75 that binds integrase (De Rijck *et al.*, 2006; Llano *et al.*, 2006; Hombrouck *et al.*, 2007) and/or generating deep knockdown (Llano *et al.*, 2006) or genetic knockout (Marshall *et al.*, 2007; Shun *et al.*, 2007b) cell lines. HIV-1 PIC nuclear localization was unabated (Llano *et al.*, 2006; Shun *et al.*, 2007b) yet lentiviral integration was reduced approximately 5- to 50-fold in cells expressing little or no LEDGF/p75 (Llano *et al.*, 2006; Marshall *et al.*, 2007; Shun *et al.*, 2007b). The proviruses that formed under these conditions revealed novel distributions: genes and transcriptional activity were now targeted at frequencies mimicking those normally observed for γ- and α-retroviruses, with concomitant integration surges in the vicinities of promoters and CpG islands (Marshall *et al.*, 2007; Shun *et al.*, 2007b). Integrase catalytic function appeared unaffected under these conditions: 3′ processing proceeded normally, and HIV-1 PICs derived from knockout cell nuclei moreover supported normal levels of DNA strand transfer activity *in vitro* (Shun *et al.*, 2007b). Accordingly, the weakly conserved target DNA consensus sequences at sites of integration were maintained in the absence of LEDGF/p75 (Marshall *et al.*, 2007; Shun *et al.*, 2007b). Thus, LEDGF/p75 plays a critical role in connecting lentiviral PICs to active genes for integration. The biochemical mechanism awaits clarification, though an affinity of LEDGF/p75 for active genes, perhaps in association with elongating RNA polymerase II or as a splicing factor, might be expected. Future work will undoubtedly attempt to target the LEDGF/p75–integrase interaction (Cherepanov *et al.*, 2005) for the development of novel antiviral drugs (Al-Mawsawi and Neamati, 2007; Engelman, 2007) as well as alter the specificity of lentiviral integration via modulating the distribution of LEDGF/p75 along human chromosomes.

Conclusions

Reverse transcription and integration define the signature replication steps of retroviruses, and much is known about their mechanisms of action. Drugs that inhibit HIV-1 reverse transcriptase activity have been in the clinic for approximately 20 years (Pomerantz and Horn, 2003) whereas the first integrase inhibitor was licensed in late 2007 (Evering and Markowitz, 2007). Owing to the relatively high rate at which HIV-1 acquires drug resistance mutations (Lucas, 2005), there is a continual need to develop new classes of antiviral drugs (Piacenti, 2006) and detailed knowledge of enzyme reaction mechanism is an important component of these efforts. Future work to decipher currently unknown aspects of reverse transcriptase and integrase function is therefore central to the long-term goal of overcoming the immense societal burdens associated with the current AIDS pandemic. Virus–host interactions critical for retroviral replication bring to the table a relatively new set of targets for antiviral drug development (Reeves and Piefer, 2005; Rice and Sutton, 2007). Research aimed at discovering novel virus–host interactions that play important roles in reverse transcription, RTC trafficking, PIC nuclear import, and integration are therefore expected in the coming years to significantly increase the diversity of molecules that might one day make their way to the clinic to help counteract the spread of HIV/AIDS.

Acknowledgements

Work in my laboratory is supported by grants from the US National Institutes of Health. I am grateful to Abraham Brass, Robert Craigie, Eric Freed, and Alice Telesnitsky for their critical comments on the manuscript.

References

Aiyar, A., Cobrinik, D., Ge, Z., Kung, H.J., and Leis, J. (1992). Interaction between retroviral U5 RNA and the T psi C loop of the tRNA(Trp) primer is required for efficient initiation of reverse transcription. J. Virol. *66*, 2464–2472.

Al-Mawsawi, L.Q., and Neamati, N. (2007). Blocking interactions between HIV-1 integrase and cellular cofactors: an emerging anti-retroviral strategy. Trends Pharmacol. Sci. *28*, 526–535.

Alin, K., and Goff, S.P. (1996). Amino acid substitutions in the CA protein of Moloney murine leukemia virus that block early events in infection. Virology *222*, 339–351.

Allain, B., Lapadat-Tapolsky, M., Berlioz, C., and Darlix, J.L. (1994). Transactivation of the minus-strand DNA transfer by nucleocapsid protein during reverse transcription of the retroviral genome. EMBO J. *13*, 973–981.

Andrake, M.D., and Skalka, A.M. (1995). Multimerization determinants reside in both the catalytic core and C terminus of avian sarcoma virus integrase. J. Biol. Chem. *270*, 29299–29306.

Ao, Z., Fowke, K., Cohen, E., and Yao, X. (2005). Contribution of the C-terminal tri-lysine regions of human immunodeficiency virus type 1 integrase for efficient reverse transcription and viral DNA nuclear import. Retrovirology *2*, 62.

Ao, Z., Huang, G., Yao, H., Xu, Z., Labine, M., Cochrane, A.W., and Yao, X. (2007). Interaction of human immunodeficiency virus type 1 integrase with cellular nuclear import receptor importin 7 and its impact on viral replication. J. Biol. Chem. *282*, 13456–13467.

Ao, Z., Yao, X., and Cohen, E.A. (2004). Assessment of the role of the central DNA flap in human immunodeficiency virus type 1 replication by using a single-cycle replication system. J. Virol. *78*, 3170–3177.

Appa, R.S., Shin, C.-G., Lee, P., and Chow, S.A. (2001). Role of the nonspecific DNA-binding region and alpha helices within the core domain of retroviral integrase in selecting target DNA sites for integration. J. Biol. Chem. *276*, 45848–45855.

Araya, A., Sarih, L., and Litvak, S. (1979). Reverse transcriptase mediated binding of primer tRNA to the viral genome. Nucleic Acids Res. *6*, 3831–3844.

Arhel, N., Genovesio, A., Kim, K.A., Miko, S., Perret, E., Olivo-Marin, J.C., Shorte, S., and Charneau, P. (2006a). Quantitative four-dimensional tracking of cytoplasmic and nuclear HIV-1 complexes. Nat. Methods *3*, 817–824.

Arhel, N., Munier, S., Souque, P., Mollier, K., and Charneau, P. (2006b). Nuclear import defect of human immunodeficiency virus type 1 DNA flap mutants is not dependent on the viral strain or target cell type. J. Virol. *80*, 10262–10269.

Arhel, N.J., Souquere-Besse, S., Munier, S., Souque, P., Guadagnini, S., Rutherford, S., Prévost, M.C., Allen, T.D., and Charneau, P. (2007). HIV-1 DNA Flap formation promotes uncoating of the pre-integration complex at the nuclear pore. EMBO J. *26*, 3025–3037.

Ariumi, Y., Turelli, P., Masutani, M., and Trono, D. (2005). DNA damage sensors ATM, ATR, DNA-PKcs, and PARP-1 are dispensable for human immunodeficiency virus type 1 integration. J. Virol. *79*, 2973–2978.

Auewarakul, P., Wacharapornin, P., Srichatrapimuk, S., Chutipongtanate, S., and Puthavathana, P. (2005). Uncoating of HIV-1 requires cellular activation. Virology 337, 93–101.

Auxilien, S., Keith, G., Le Grice, S.F., and Darlix, J.-L. (1999). Role of post-transcriptional modifications of primer tRNALys,3 in the fidelity and efficacy of plus strand DNA transfer during HIV-1 reverse transcription. J. Biol. Chem. *274*, 4412–4420.

Baekelandt, V., Claeys, A., Cherepanov, P., De Clercq, E., De Strooper, B., Nuttin, B., and Debyser, Z. (2000). DNA-dependent protein kinase is not required for efficient lentivirus integration. J. Virol. *74*, 11278–11285.

Bao, K.K., Wang, H., Miller, J.K., Erie, D.A., Skalka, A.M., and Wong, I. (2003). Functional oligomeric state of avian sarcoma virus integrase. J. Biol. Chem. *278*, 1323–1327.

Barat, C., Lullien, V., Schatz, O., Keith, G., Nugeyre, M.T., Grüninger-Leitch, F., Barré-Sinoussi, F., LeGrice, S.F., and Darlix, J.L. (1989). HIV-1 reverse transcriptase specifically interacts with the anticodon domain of its cognate primer tRNA. EMBO J. *8*, 3279–3285.

Barat, C., Schatz, O., Le Grice, S., and Darlix, J.L. (1993). Analysis of the interactions of HIV1 replication primer tRNA(Lys,3) with nucleocapsid protein and reverse transcriptase. J. Mol. Biol. *231*, 185–190.

Bathurst, I.C., Moen, L.K., Lujan, M.A., Gibson, H.L., Feucht, P.H., Pichuantes, S., Craik, C.S., Santi, D.V., and Barr, P.J. (1990). Characterization of the human immunodeficiency virus type-1 reverse transcriptase enzyme produced in yeast. Biochem. Biophys. Res. Commun. *171*, 589–595.

Beerens, N., Groot, F., and Berkhout, B. (2001). Initiation of HIV-1 reverse transcription is regulated by a primer activation signal. J. Biol. Chem. *276*, 31247–31256.

Ben-Artzi, H., Shemesh, J., Zeelon, E., Amit, B., Kleiman, L., Gorecki, M., and Panet, A. (1996). Molecular analysis of the second template switch during reverse transcription of the HIV RNA template. Biochemistry *35*, 10549–10557.

Bera, S., Vora, A.C., Chiu, R., Heyduk, T., and Grandgenett, D.P. (2005). Synaptic complex formation of two retrovirus DNA attachment sites by integrase: a fluorescence energy transfer study. Biochemistry *44*, 15106–15114.

Berthoux, L., Sebastian, S., Muesing, M.A., and Luban, J. (2007). The role of lysine 186 in HIV-1 integrase multimerization. Virology *364*, 227–236.

Bieniasz, P.D., Weiss, R.A., and McClure, M.O. (1995). Cell cycle dependence of foamy retrovirus infection. J. Virol. *69*, 7295–7299.

Boone, L.R., and Skalka, A. (1980). Two species of full-length cDNA are synthesized in high yield by melittin-treated avian retrovirus particles. Proc. Natl. Acad. Sci. U.S.A. *77*, 847–851.

Boone, L.R., and Skalka, A. (1981). Viral DNA synthesized *in vitro* by avian retrovirus particles permeabilized with melittin. J. Virol. *37*, 109–116.

Borroto-Esoda, K., and Boone, L.R. (1991). Equine infectious anemia virus and human immunodeficiency virus DNA synthesis *in vitro*: characterization of the endogenous reverse transcriptase reaction. J. Virol. *65*, 1952–1959.

Bosserman, M.A., O'Quinn, D.F., and Wong, I. (2007). Loop202–208 in avian sarcoma virus integrase mediates tetramer assembly and processing activity. Biochemistry *46*, 11231–11239.

Bouyac-Bertoia, M., Dvorin, J.D., Fouchier, R.A.M., Jenkins, Y., Meyer, B.E., Wu, L.I., Emerman, M., and Malim, M.H. (2001). HIV-1 infection requires a functional integrase NLS. Mol. Cell *7*, 1025–1035.

Bowerman, B., Brown, P.O., Bishop, J.M., and Varmus, H.E. (1989). A nucleoprotein complex mediates the integration of retroviral DNA. Genes Dev. *3*, 469–478.

Brass, A.L., Dykxhoorn, D.M., Benita, Y., Yan, N., Engelman, A., Xavier, R.J., Lieberman, J., and Elledge, S.J. (2008). Identification of host proteins required for HIV infection through a functional genomic screen. Science *319*, 921–926.

Brin, E., Yi, J., Skalka, A.M., and Leis, J. (2000). Modeling the late steps in HIV-1 retroviral integrase-catalyzed DNA integration. J. Biol. Chem. *275*, 39287–39295.

Brown, H.E.V., Chen, H., and Engelman, A. (1999). Structure-based mutagenesis of the human immunodeficiency virus type 1 DNA attachment site: effects on integration and cDNA synthesis. J. Virol. *73*, 9011–9020.

Brown, P.O., Bowerman, B., Varmus, H.E., and Bishop, J.M. (1987). Correct integration of retroviral DNA *in vitro*. Cell *49*, 347–356.

Brown, P.O., Bowerman, B., Varmus, H.E., and Bishop, J.M. (1989). Retroviral integration: structure of the initial covalent product and its precursor, and a role for the viral IN protein. Proc. Natl. Acad. Sci. U.S.A. *86*, 2525–2529.

Bujacz, G., Alexandratos, J., Qing, Z.L., Clement-Mella, C., and Wlodawer, A. (1996). The catalytic domain of human immunodeficiency virus integrase: ordered active site in the F185H mutant. FEBS Lett. *398*, 175–178.

Bujacz, G., Jaskolski, M., Alexandratos, J., Wlodawer, A., Merkel, G., Katz, R.A., and Skalka, A.M. (1995). High-resolution structure of the catalytic domain of avian sarcoma virus integrase. J. Mol. Biol. *253*, 333–346.

Bukrinskaya, A., Brichacek, B., Mann, A., and Stevenson, M. (1998). Establishment of a functional human immunodeficiency virus type 1 (HIV-1) reverse transcription complex involves the cytoskeleton. J. Exp. Med. *188*, 2113–2125.

Bukrinskaya, A.G., Ghorpade, A., Heinzinger, N.K., Smithgall, T.E., Lewis, R.E., and Stevenson, M. (1996). Phosphorylation-dependent human immunodeficiency virus type 1 infection and nuclear targeting of viral DNA. Proc. Natl. Acad. Sci. U.S.A. *93*, 367–371.

Bukrinsky, M., Haggerty, S., Dempsey, M., Sharova, N., Adzhubel, A., Spitz, L., Lewis, P., Goldfarb, D., Emerman, M., and Stevenson, M. (1993). A nuclear localization signal within HIV-1 matrix protein that governs infection of non-dividing cells. Nature *365*, 666–669.

Bukrinsky, M., Sharova, N., Dempsey, M., Stanwick, T., Bukrinskaya, A., Haggerty, S., and Stevenson, M. (1992). Active nuclear import of human immunodeficiency virus type 1 preintegration complexes. Proc. Natl. Acad. Sci. U.S.A. *89*, 6580–6584.

Burke, C.J., Sanyal, G., Bruner, M.W., Ryan, J.A., LaFemina, R.L., Robbins, H.L., Zeft, A.S., Middaugh, C.R., and Cordingley, M.G. (1992). Structural implications of spectroscopic characterization of a putative zinc finger peptide from HIV-1 integrase. J. Biol. Chem. *267*, 9639–9644.

Bushman, F., Lewinski, M., Ciuffi, A., Barr, S., Leipzig, J., Hannenhalli, S., and Hoffmann, C. (2005). Genome-wide analysis of retroviral DNA integration. Nat. Rev. Microbiol. *3*, 848–858.

Bushman, F.D., and Craigie, R. (1991). Activities of human immunodeficiency virus (HIV) integration protein *in vitro*: Specific cleavage and integration of HIV DNA. Proc. Natl. Acad. Sci. U.S.A. *88*, 1339–1343.

Bushman, F.D., Engelman, A., Palmer, I., Wingfield, P., and Craigie, R. (1993). Domains of the integrase protein of human immunodeficiency virus type 1 responsible for polynucleotidyl transfer and zinc binding. Proc. Natl. Acad. Sci. U.S.A. *90*, 3428–3432.

Bushman, F.D., and Wang, B. (1994). Rous sarcoma virus integrase protein: mapping functions for catalysis and substrate binding. J. Virol. *68*, 2215–2223.

Busschots, K., Vercammen, J., Emiliani, S., Benarous, R., Engelborghs, Y., Christ, F., and Debyser, Z. (2005). The interaction of LEDGF/p75 with integrase is lentivirus-specific and promotes DNA binding. J. Biol. Chem. *280*, 17841–17847.

Cai, M., Zheng, R., Caffrey, M., Craigie, R., Clore, G.M., and Gronenborn, A.M. (1997). Solution structure of the N-terminal zinc binding domain of HIV-1 integrase. Nat. Struct. Biol. *4*, 567–577.

Campbell, E.M., and Hope, T.J. (2005). Gene therapy progress and prospects: Viral trafficking during infection. Gene Ther. *12*, 1353–1359.

Caron, M.-C., and Caruso, M. (2005). A nuclear localization signal in the matrix of spleen necrosis virus (SNV) does not allow efficient gene transfer into quiescent cells with SNV-derived vectors. Virology *338*, 292–296.

Carteau, S., Gorelick, R.J., and Bushman, F.D. (1999). Coupled integration of human immunodeficiency virus type 1 cDNA ends by purified integrase *in vitro*: Stimulation by the viral nucleocapsid protein. J. Virol. *73*, 6670–6679.

Carteau, S., Hoffmann, C., and Bushman, F. (1998). Chromosome structure and human immunodeficiency virus type 1 cDNA integration: Centromeric alphoid repeats are a disfavored target. J. Virol. *72*, 4005–4014.

Cen, S., Javanbakht, H., Kim, S., Shiba, K., Craven, R., Rein, A., Ewalt, K., Schimmel, P., Musier-Forsyth, K., and Kleiman, L. (2002). Retrovirus-specific packaging of aminoacyl-tRNA synthetases with cognate primer tRNAs. J. Virol. *76*, 13111–13115.

Cen, S., Khorchid, A., Javanbakht, H., Gabor, J., Stello, T., Shiba, K., Musier-Forsyth, K., and Kleiman, L. (2001). Incorporation of lysyl-tRNA synthetase into human immunodeficiency virus type 1. J. Virol. *75*, 5043–5048.

Charneau, P., Alizon, M., and Clavel, F. (1992). A second origin of DNA plus-strand synthesis is required for optimal human immunodeficiency virus replication. J. Virol. *66*, 2814–2820.

Charneau, P., and Clavel, F. (1991). A single-stranded gap in human immunodeficiency virus unintegrated linear DNA defined by a central copy of the polypurine tract. J. Virol. *65*, 2415–2421.

Charneau, P., Mirambeau, G., Roux, P., Paulous, S., Buc, H., and Clavel, F. (1994). HIV-1 reverse transcription. A termination step at the center of the genome. J. Mol. Biol. *241*, 651–662.

Chen, A., Weber, I.T., Harrison, R.W., and Leis, J. (2006). Identification of amino acids in HIV-1 and avian sarcoma virus integrase subsites required for specific recognition of the long terminal repeat ends. J. Biol. Chem. *281*, 4173–4182.

Chen, H., and Engelman, A. (1998). The barrier-to-autointegration protein is a host factor for HIV type 1 integration. Proc. Natl. Acad. Sci. U.S.A. *95*, 15270–15274.

Chen, H., Wei, S.-Q., and Engelman, A. (1999). Multiple integrase functions are required to form the native structure of the human immunodeficiency virus type I intasome. J. Biol. Chem. *274*, 17358–17364.

Chen, J.C.-H., Krucinski, J., Miercke, L.J.W., Finer-Moore, J.S., Tang, A.H., Leavitt, A.D., and Stroud, R.M. (2000a). Crystal structure of the HIV-1 integrase catalytic core and C-terminal domains: A model for viral DNA binding. Proc. Natl. Acad. Sci. U.S.A. *97*, 8233–8238.

Chen, Z., Yan, Y., Munshi, S., Li, Y., Zugay-Murphy, J., Xu, B., Witmer, M., Felock, P., Wolfe, A., and Sardana, V. (2000b). X-ray structure of simian immunodeficiency virus integrase containing the core and C-terminal domain (residues 50–293) – an initial glance of the viral DNA binding platform. J. Mol. Biol. *296*, 521–533.

Cheng, X., Belshan, M., and Ratner, L. (2008). Hsp40 facilitates HIV-2 vpx-mediated preintegration complex nuclear import. J. Virol. *82*, 1229–1237.

Cherepanov, P. (2007). LEDGF/p75 interacts with divergent lentiviral integrases and modulates their enzymatic activity *in vitro*. Nucleic Acids Res. *35*, 113–124.

Cherepanov, P., Ambrosio, A.L.B., Rahman, S., Ellenberger, T., and Engelman, A. (2005). From the Cover: Structural basis for the recognition between HIV-1 integrase and transcriptional coactivator p75. Proc. Natl. Acad. Sci. U.S.A. *102*, 17308–17313.

Cherepanov, P., Maertens, G., Proost, P., Devreese, B., Van Beeumen, J., Engelborghs, Y., De Clercq, E., and Debyser, Z. (2003). HIV-1 integrase forms stable tetramers and associates with LEDGF/p75 protein in human cells. J. Biol. Chem. *278*, 372–381.

Cherepanov, P., Pluymers, W., Claeys, A., Proost, P., De Clercq, E., and Debyser, Z. (2000). High-level expression of active HIV-1 integrase from a synthetic gene in human cells. FASEB J. *14*, 1389–1399.

Connor, R.I., Chen, B.K., Choe, S., and Landau, N.R. (1995). Vpr is required for efficient replication of human immunodeficiency virus type-1 in mononuclear phagocytes. Virology *206*, 935–944.

Craig, N.L., Craigie, R., Gellert, M., and Lambowitz, A.M. (2002). Mobile DNA II (Washington, D.C., ASM Press).

Craigie, R., Fujiwara, T., and Bushman, F. (1990). The IN protein of Moloney murine leukemia virus processes the viral DNA ends and accomplishes their integration *in vitro*. Cell *62*, 829–837.

Crawford, S., and Goff, S.P. (1984). Mutations in *gag* proteins P12 and P15 of Moloney murine leukemia virus block early stages of infection. J. Virol. *49*, 909–917.

Crise, B., Li, Y., Yuan, C., Morcock, D.R., Whitby, D., Munroe, D.J., Arthur, L.O., and Wu, X. (2005). Simian immunodeficiency virus integration preference is similar to that of human immunodeficiency virus type 1. J. Virol. *79*, 12199–12204.

Daniel, R., Greger, J.G., Katz, R.A., Taganov, K.D., Wu, X., Kappes, J.C., and Skalka, A.M. (2004). Evidence that stable retroviral transduction and cell survival following DNA integration depend on components of the nonhomologous end joining repair pathway. J. Virol. *78*, 8573–8581.

Daniel, R., Kao, G., Taganov, K., Greger, J.G., Favorova, O., Merkel, G., Yen, T.J., Katz, R.A., and Skalka, A.M. (2003). Evidence that the retroviral DNA integration process triggers an ATR-dependent DNA damage response. Proc. Natl. Acad. Sci. U.S.A. *100*, 4778–4783.

Daniel, R., Katz, R.A., Merkel, G., Hittle, J.C., Yen, T.J., and Skalka, A.M. (2001). Wortmannin potentiates integrase-mediated killing of lymphocytes and reduces the efficiency of stable transduction by retroviruses. Mol. Cell Biol. *21*, 1164–1172.

Daniel, R., Katz, R.A., and Skalka, A.M. (1999). A role for DNA-PK in retroviral DNA integration. Science *284*, 644–647.

Das, D., and Georgiadis, M.M. (2004). The crystal structure of the monomeric reverse transcriptase from Moloney murine leukemia virus. Structure *12*, 819–829.

Davies, D.R., Goryshin, I.Y., Reznikoff, W.S., and Rayment, I. (2000). Three-dimensional structure of the Tn5 synaptic complex transposition intermediate. Science *289*, 77–85.

De Luca, L., Vistoli, G., Pedretti, A., Barreca, M.L., and Chimirri, A. (2005). Molecular dynamics studies of the full-length integrase–DNA complex. Biochem. Biophys. Res. Commun. *336*, 1010–1016.

De Rijck, J., and Debyser, Z. (2006). The central DNA flap of the human immunodeficiency virus type 1 is important for viral replication. Biochem. Biophys. Res. Commun. *349*, 1100–1110.

De Rijck, J., Vandekerckhove, L., Gijsbers, R., Hombrouck, A., Hendrix, J., Vercammen, J., Engelborghs, Y., Christ, F., and Debyser, Z. (2006). Overexpression of the lens epithelium-derived growth factor/p75 integrase binding domain inhibits human immunodeficiency virus replication. J. Virol. *80*, 11498–11509.

DeHart, J.L., Andersen, J.L., Zimmerman, E.S., Ardon, O., An, D.S., Blackett, J., Kim, B., and Planelles, V. (2005). The ataxia telangiectasia-mutated and Rad3-related protein is dispensable for retroviral integration. J. Virol. *79*, 1389–1396.

Delelis, O., Lehmann-Che, J., and Saib, A. (2004). Foamy viruses – a world apart. Curr. Opin. Microbiol. *7*, 400–406.

Depienne, C., Roques, P., Creminon, C., Fritsch, L., Casseron, R., Dormont, D., Dargemont, C., and Benichou, S. (2000). Cellular distribution and karyophilic properties of matrix, integrase, and Vpr proteins from the human and simian immunodeficiency viruses. Exp. Cell Res. *260*, 387–395.

Derse, D., Crise, B., Li, Y., Princler, G., Lum, N., Stewart, C., McGrath, C.F., Hughes, S.H., Munroe, D.J., and Wu, X. (2007). HTLV-1 integration target sites in the human genome: comparison with other retroviruses. J. Virol. *81*, 6731–6741.

Devroe, E., Engelman, A., and Silver, P.A. (2003). Intracellular transport of human immunodeficiency virus type 1 integrase. J. Cell Sci. *116*, 4401–4408.

Dhar, R., McClements, W.L., Enquist, L.W., and Vande Woude, G.F. (1980). Nucleotide sequences of integrated Moloney sarcoma provirus long terminal repeats and their host and viral junctions. Proc. Natl. Acad. Sci. U.S.A. *77*, 3937–3941.

Diamond, T.L., and Bushman, F.D. (2005). Division of labor within human immunodeficiency virus integrase complexes: Determinants of catalysis and target DNA capture. J. Virol. *79*, 15376–15387.

Dirac, A.M.G., and Kjems, J. (2001). Mapping DNA-binding sites of HIV-1 integrase by protein footprinting. Eur. J. Biochem. *268*, 743–751.

Dismuke, D.J., and Aiken, C. (2006). Evidence for a functional link between uncoating of the human immunodeficiency virus type 1 core and nuclear import of the viral preintegration complex. J. Virol. *80*, 3712–3720.

Donehower, L.A., and Varmus, H.E. (1984). A mutant murine leukemia virus with a single missense codon in *pol* is defective in a function affecting integration. Proc. Natl. Acad. Sci. U.S.A. *81*, 6461–6465.

Dotan, I., Scottoline, B.P., Heuer, T.S., and Brown, P.O. (1995). Characterization of recombinant murine leukemia virus integrase. J. Virol. *69*, 456–468.

Drelich, M., Wilhelm, R., and Mous, J. (1992). Identification of amino acid residues critical for endonuclease and integration activities of HIV-1 IN protein *in vitro*. Virology *188*, 459–468.

Du, Z., Ilyinskii, P.O., Lally, K., Desrosiers, R.C., and Engelman, A. (1997). A mutation in integrase can compensate for mutations in the simian immunodeficiency virus att site. J. Virol. *71*, 8124–8132.

Dvorin, J.D., Bell, P., Maul, G.G., Yamashita, M., Emerman, M., and Malim, M.H. (2002). Reassessment of the roles of integrase and the central DNA flap in human immunodeficiency virus type 1 nuclear import. J. Virol. *76*, 12087–12096.

Dyda, F., Hickman, A.B., Jenkins, T.M., Engelman, A., Craigie, R., and Davies, D.R. (1994). Crystal structure of the catalytic domain of HIV-1 integrase: similarity to other polynucleotidyl transferases. Science *266*, 1981–1986.

Eijkelenboom, A.P., van den Ent, F.M., Vos, A., Doreleijers, J.F., Hard, K., Tullius, T.D., Plasterk, R.H., Kaptein, R., and Boelens, R. (1997). The solution structure of the amino-terminal HHCC domain of HIV-2 integrase: a three-helix bundle stabilized by zinc. Curr. Biol. *7*, 739–746.

Eijkelenboom, A.P.A.M., Puras Lutzke, R.A., Boelens, R., Plasterk, R.H.A., Kaptein, R., and Hård, K. (1995). The DNA-binding domain of HIV-1 integrase has an SH3-like fold. Nat. Struct. Biol. *2*, 807–810.

Ellison, V., Abrams, H., Roe, T., Lifson, J., and Brown, P. (1990). Human immunodeficiency virus integration in a cell-free system. J. Virol. *64*, 2711–2715.

Ellison, V., Gerton, J., Vincent, K.A., and Brown, P.O. (1995). An essential interaction between distinct domains of HIV-1 integrase mediates assembly of the active multimer. J. Biol. Chem. *270*, 3320–3326.

Emiliani, S., Mousnier, A., Busschots, K., Maroun, M., Van Maele, B., Tempe, D., Vandekerckhove, L., Moisant, F., Ben-Slama, L., Witvrouw, M., *et al.* (2005). Integrase mutants defective for interaction with LEDGF/p75 are impaired in chromosome tethering and HIV-1 replication. J. Biol. Chem. *280*, 25517–25523.

Engelman, A. (1996). Biochemical characterization of recombinant equine infectious anemia virus integrase. Protein Expression Purif. *8*, 299–304.

Engelman, A. (2007). Host cell factors and HIV-1 integration. Future HIV Ther. *1*, 415–426.

Engelman, A., Bushman, F.D., and Craigie, R. (1993). Identification of discrete functional domains of HIV-1 integrase and their organization within an active multimeric complex. EMBO J. *12*, 3269–3275.

Engelman, A., and Cherepanov, P. (2008). The lentiviral integrase binding protein LEDGF/p75 and HIV-1 replication. PLoS Pathog. *4*, e1000046.

Engelman, A., and Craigie, R. (1992). Identification of conserved amino acid residues critical for human immunodeficiency virus type 1 integrase function *in vitro*. J. Virol. *66*, 6361–6369.

Engelman, A., Hickman, A.B., and Craigie, R. (1994). The core and carboxyl-terminal domains of the integrase protein of human immunodeficiency virus type 1 each contribute to nonspecific DNA binding. J. Virol. *68*, 5911–5917.

Engelman, A., Mizuuchi, K., and Craigie, R. (1991). HIV-1 DNA integration: mechanism of viral DNA cleavage and DNA strand transfer. Cell *67*, 1211–1221.

Enssle, J., Jordan, I., Mauer, B., and Rethwilm, A. (1996). Foamy virus reverse transcriptase is expressed independently from the Gag protein. Proc. Natl. Acad. Sci. U.S.A. *93*, 4137–4141.

Enssle, J., Moebes, A., Heinkelein, M., Panhuysen, M., Mauer, B., Schweizer, M., Neumann-Haefelin, D., and Rethwilm, A. (1999). An active foamy virus integrase is required for virus replication. J. Gen. Virol. *80*, 1445–1452.

Esposito, D., and Craigie, R. (1998). Sequence specificity of viral end DNA binding by HIV-1 integrase reveals critical regions for protein–DNA interaction. EMBO J. *17*, 5832–5843.

Evering, T.H., and Markowitz, M. (2007). Raltegravir (MK-0518): An integrase inhibitor for the treatment of HIV-1. Drugs Today (Barc.) *43*, 865–877.

Farnet, C.M., and Haseltine, W.A. (1990). Integration of human immunodeficiency virus type 1 DNA *in vitro*. Proc. Natl. Acad. Sci. U.S.A. *87*, 4164–4168.

Fassati, A. (2006). HIV infection of non-dividing cells: a divisive problem. Retrovirology 3, 74.

Fassati, A., and Goff, S.P. (1999). Characterization of intracellular reverse transcription complexes of Moloney murine leukemia virus. J. Virol. 73, 8919–8925.

Fassati, A., and Goff, S.P. (2001). Characterization of intracellular reverse transcription complexes of human immunodeficiency virus type 1. J. Virol. *75*, 3626–3635.

Fassati, A., Gorlich, D., Harrison, I., Zaytseva, L., and Mingot, J.M. (2003). Nuclear import of HIV-1 intracellular reverse transcription complexes is mediated by importin 7. EMBO J. *22*, 3675–3685.

Faure, A., Calmels, C., Desjobert, C., Castroviejo, M., Caumont-Sarcos, A., Tarrago-Litvak, L., Litvak, S., and Parissi, V. (2005). HIV-1 integrase crosslinked oligomers are active *in vitro*. Nucleic Acids Res. *33*, 977–986.

Finston, W.I., and Champoux, J.J. (1984). RNA-primed initiation of Moloney murine leukemia virus plus strands by reverse transcriptase *in vitro*. J. Virol. *51*, 26–33.

Fitzgerald, M.L., Vora, A.C., Zeh, W.G., and Grandgenett, D.P. (1992). Concerted integration of viral DNA termini by purified avian myeloblastosis virus integrase. J. Virol. *66*, 6257–6263.

Fletcher 3rd, T.M., Brichacek, B., Sharova, N., Newman, M.A., Stivahtis, G., Sharp, P.M., Emerman, M., Hahn, B.H., and Stevenson, M. (1996). Nuclear import and cell cycle arrest functions of the HIV-1 Vpr protein are encoded by two separate genes in HIV-2/SIV(SM). EMBO J. *15*, 6155–6165.

Fletcher 3rd, T.M., Soares, M.A., McPhearson, S., Hui, H., Wiskerchen, M., Muesing, M.A., Shaw, G.M., Leavitt, A.D., Boeke, J.D., and Hahn, B.H. (1997). Complementation of integrase function in HIV-1 virions. EMBO J. *16*, 5123–5138.

Forshey, B.M., von Schwedler, U., Sundquist, W.I., and Aiken, C. (2002). Formation of a human immunodeficiency virus type 1 core of optimal stability is crucial for viral replication. J. Virol. *76*, 5667–5677.

Fouchier, R.A.M., Meyer, B.E., Simon, J.H.M., Fischer, U., Albright, A.V., Gonzalez-Scarano, F., and Malim, M.H. (1998). Interaction of the human immunodeficiency virus type 1 Vpr protein with the nuclear pore complex. J. Virol. *72*, 6004–6013.

Fouchier, R.A.M., Meyer, B.E., Simon, J.H.M., Fischer, U., and Malim, M.H. (1997). HIV-1 infection of non-dividing cells: evidence that the amino-terminal basic region of the viral matrix protein is important for Gag processing but not for post-entry nuclear import. EMBO J. *16*, 4531–4539.

Freed, E.O., Englund, G., Maldarelli, F., and Martin, M.A. (1997). Phosphorylation of residue 131 of HIV-1 matrix is not required for macrophage infection. Cell *88*, 171–173.

Freed, E.O., Englund, G., and Martin, M.A. (1995). Role of the basic domain of human immunodeficiency virus type 1 matrix in macrophage infection. J. Virol. *69*, 3949–3954.

Fritsch, E.F., and Temin, H.M. (1977). Inhibition of viral DNA synthesis in stationary chicken embryo fibroblasts infected with avian retroviruses. J. Virol. *24*, 461–469.

Fujiwara, T., and Mizuuchi, K. (1988). Retroviral DNA integration: structure of an integration intermediate. Cell *54*, 497–504.

Gaken, J.A., Tavassoli, M., Gan, S.U., Vallian, S., Giddings, I., Darling, D.C., Galea-Lauri, J., Thomas, M.G., Abedi, H., Schreiber, V., *et al.* (1996). Efficient retroviral infection of mammalian cells is blocked by inhibition of poly(ADP-ribose) polymerase activity. J. Virol. *70*, 3992–4000.

Gallay, P., Hope, T., Chin, D., and Trono, D. (1997). HIV-1 infection of nondividing cells through the recognition of integrase by the importin/karyopherin pathway. Proc. Natl. Acad. Sci. U.S.A. *94*, 9825–9830.

Gallay, P., Swingler, S., Aiken, C., and Trono, D. (1995a). HIV-1 infection of nondividing cells: C-terminal tyrosine phosphorylation of the viral matrix protein is a key regulator. Cell *80*, 379–388.

Gallay, P., Swingler, S., Song, J., Bushman, F., and Trono, D. (1995b). HIV nuclear import is governed by the phosphotyrosine-mediated binding of matrix to the core domain of integrase. Cell *83*, 569–576.

Gao, K., Butler, S.L., and Bushman, F. (2001). Human immunodeficiency virus type 1 integrase: arrangement of protein domains in active cDNA complexes. EMBO J. *20*, 3565–3576.

Gao, W., Cara, A., Gallo, R.C., and Lori, F. (1993). Low levels of deoxynucleotides in peripheral blood lymphocytes: A strategy to inhibit human immunodeficiency virus type 1 replication. Proc. Natl. Acad. Sci. U.S.A. *90*, 8925–8928.

Garfinkel, D.J., Stefanisko, K.M., Nyswaner, K.M., Moore, S.P., Oh, J., and Hughes, S.H. (2006). Retrotransposon suicide: Formation of Ty1 circles and autointegration via a central DNA flap. J. Virol. *80*, 11920–11934.

Gerton, J.L., and Brown, P.O. (1997). The core domain of HIV-1 integrase recognizes key features of its DNA substrates. J. Biol. Chem. *272*, 25809–25815.

Gerton, J.L., Ohgi, S., Olsen, M., DeRisi, J., and Brown, P.O. (1998). Effects of mutations in residues near the active site of human immunodeficiency virus type 1 integrase on specific enzyme–substrate interactions. J. Virol. *72*, 5046–5055.

Gibson, W., and Verma, I.M. (1974). Studies on the reverse transcriptase of RNA tumor viruses. Structural relatedness of two subunits of avian turmor viruses. Proc. Natl. Acad. Sci. U.S.A. *71*, 4991–4994.

Gilboa, E., Mitra, S.W., Goff, S., and Baltimore, D. (1979). A detailed model of reverse transcription and tests of crucial aspects. Cell *18*, 93–100.

Goodarzi, G., Im, G.J., Brackmann, K., and Grandgenett, D. (1995). Concerted integration of retrovirus-like DNA by human immunodeficiency virus type 1 integrase. J. Virol. *69*, 6090–6097.

Goulaouic, H., Subra, F., Mouscadet, J.F., Carteau, S., and Auclair, C. (1994). Exogenous nucleosides promote the completion of MoMLV DNA synthesis in G0-arrested Balb c/3T3 fibroblasts. Virology *200*, 87–97.

Graves, M.C., Meidel, M.C., Pan, Y.-C.E., Manneberg, M., Lahm, H.-W., and Gruninger-Leitch, F. (1990). Identification of a human immunodeficiency virus-1 protease cleavage site within the 66,000 dalton subunit of reverse transcriptase. Biochem. Biophys. Res. Commun. *168*, 30–36.

Grobler, J.A., Stillmock, K., Hu, B., Witmer, M., Felock, P., Espeseth, A.S., Wolfe, A., Egbertson, M., Bourgeois,

M., Melamed, J., *et al.* (2002). Diketo acid inhibitor mechanism and HIV-1 integrase: Implications for metal binding in the active site of phosphotransferase enzymes. Proc. Natl. Acad. Sci. U.S.A. *99*, 6661–6666.

Groschel, B., and Bushman, F. (2005). Cell cycle arrest in G2/M promotes early steps of infection by human immunodeficiency virus. J. Virol. *79*, 5695–5704.

Guiot, E., Carayon, K., Delelis, O., Simon, F., Tauc, P., Zubin, E., Gottikh, M., Mouscadet, J.-F., Brochon, J.-C., and Deprez, E. (2006). Relationship between the oligomeric status of HIV-1 integrase on DNA and enzymatic activity. J. Biol. Chem. *281*, 22707–22719.

Ha, H.C., Juluri, K., Zhou, Y., Leung, S., Hermankova, M., and Snyder, S.H. (2001). Poly(ADP-ribose) polymerase-1 is required for efficient HIV-1 integration. Proc. Natl. Acad. Sci. U.S.A. *98*, 3364–3368.

Hacker, C.V., Vink, C.A., Wardell, T.W., Lee, S., Treasure, P., Kingsman, S.M., Mitrophanous, K.A., and Miskin, J.E. (2006). The integration profile of EIAV-based vectors. Mol. Ther. *14*, 536–545.

Haffar, O.K., Popov, S., Dubrovsky, L., Agostini, I., Tang, H., Pushkarsky, T., Nadler, S.G., and Bukrinsky, M. (2000). Two nuclear localization signals in the HIV-1 matrix protein regulate nuclear import of the HIV-1 pre-integration complex. J. Mol. Biol. *299*, 359–368.

Harada, F., Peters, G.G., and Dahlberg, J.E. (1979). The primer tRNA for Moloney murine leukemia virus DNA synthesis. Nucleotide sequence and aminoacylation of tRNAPro. J. Biol. Chem. *254*, 10979–10985.

Harada, F., Sawyer, R.C., and Dahlberg, J.E. (1975). A primer ribonucleic acid for initiation of *in vitro* Rous sarcoma virus deoxyribonucleic acid synthesis. J. Biol. Chem. *250*, 3487–3497.

Harel, J., Rassart, E., and Jolicoeur, P. (1981). Cell cycle dependence of synthesis of unintegrated viral DNA in mouse cells newly infected with murine leukemia virus. Virology *110*, 202–207.

Harper, A.L., Skinner, L.M., Sudol, M., and Katzman, M. (2001). Use of patient-derived human immunodeficiency virus type 1 integrases to identify a protein residue that affects target site selection. J. Virol. *75*, 7756–7762.

Harris, J.D., Scott, J.V., Traynor, B., Brahic, M., Stowring, L., Ventura, P., Haase, A.T., and Peluso, R. (1981). Visna virus DNA: discovery of a novel gapped structure. Virology *113*, 573–583.

Haseltine, W.A., Coffin, J.M., and Hageman, T.C. (1979). Structure of products of the Moloney murine leukemia virus endogenous DNA polymerase reaction. J. Virol. *30*, 375–383.

Haseltine, W.A., Kleid, D.G., Panet, A., Rothenberg, E., and Baltimore, D. (1976). Ordered transcription of RNA tumor virus genomes. J. Mol. Biol. *106*, 109–131.

Hatziioannou, T., and Goff, S.P. (2001). Infection of nondividing cells by Rous sarcoma virus. J. Virol. *75*, 9526–9531.

Hearps, A.C., and Jans, D.A. (2006). HIV-1 integrase is capable of targeting DNA to the nucleus via an Importin alpha/beta dependent mechanism. Biochem. J. *398*, 475–484.

Hehl, E.A., Joshi, P., Kalpana, G.V., and Prasad, V.R. (2004). Interaction between human immunodeficiency virus type 1 reverse transcriptase and integrase proteins. J. Virol. *78*, 5056–5067.

Heinzinger, N., Bukrinsky, M., Haggerty, S., Ragland, A., Kewalramani, V., Lee, M., Gendelman, H., Ratner, L., Stevenson, M., and Emerman, M. (1994). The Vpr protein of human immunodeficiency virus type 1 influences nuclear localization of viral nucleic acids in nondividing host cells. Proc. Natl. Acad. Sci. U.S.A. *91*, 7311–7315.

Hematti, P., Hong, B.-K., Ferguson, C., Adler, R., Hanawa, H., Sellers, S., Holt, I.E., Eckfeldt, C.E., Sharma, Y., Schmidt, M., *et al.* (2004). Distinct genomic integration of MLV and SIV vectors in primate hematopoietic stem and progenitor cells. PLoS Biol. *2*, e423.

Heuer, T.S., and Brown, P.O. (1997). Mapping features of HIV-1 integrase near selected sites on viral and target DNA molecules in an active enzyme–DNA complex by photo-cross-linking. Biochemistry *36*, 10655–10665.

Heuer, T.S., and Brown, P.O. (1998). Photo-cross-linking studies suggest a model for the architecture of an active human immunodeficiency virus type 1 integrase–DNA complex. Biochemistry *37*, 6667–6678.

Heyman, T., Wilhelm, M., and Wilhelm, F.X. (2003). The central PPT of the yeast retrotransposon Ty1 is not essential for transposition. J. Mol. Biol. *331*, 315–320.

Hindmarsh, P., Ridky, T., Reeves, R., Andrake, M., Skalka, A.M., and Leis, J. (1999). HMG protein family members stimulate human immunodeficiency virus type 1 and avian sarcoma virus concerted DNA integration *in vitro*. J. Virol. 73, 2994–3003.

Hishinuma, F., DeBona, P.J., Astrin, S., and Skalka, A.M. (1981). Nucleotide sequence of acceptor site and termini of integrated avian endogenous provirus ev1: integration creates a 6 bp repeat of host DNA. Cell *23*, 155–164.

Holman, A.G., and Coffin, J.M. (2005). Symmetrical base preferences surrounding HIV-1, avian sarcoma/leukosis virus, and murine leukemia virus integration sites. Proc. Natl. Acad. Sci. U.S.A. *102*, 6103–6107.

Holzschu, D.L., Martineau, D., Fodor, S.K., Vogt, V.M., Bowser, P.R., and Casey, J.W. (1995). Nucleotide sequence and protein analysis of a complex piscine retrovirus, walleye dermal sarcoma virus. J. Virol. *69*, 5320–5331.

Hombrouck, A., De Rijck, J., Hendrix, J., Vandekerckhove, L., Voet, A., Maeyer, M.D., Witvrouw, M., Engelborghs, Y., Christ, F., Gijsbers, R., *et al.* (2007). Virus evolution reveals an exclusive role for LEDGF/p75 in chromosomal tethering of HIV. PLoS Pathog. 3, e47.

Hostomsky, Z., Hostomska, Z., Fu, T.B., and Taylor, J. (1992). Reverse transcriptase of human immunodeficiency virus type 1: functionality of subunits of the heterodimer in DNA synthesis. J. Virol. *66*, 3179–3182.

Hsu, T.W., and Taylor, J.M. (1982). Single-stranded regions on unintegrated avian retrovirus DNA. J. Virol. *44*, 47–53.

Huang, H., Chopra, R., Verdine, G.L., and Harrison, S.C. (1998). Structure of a covalently trapped catalytic complex of HIV-1 reverse transcriptase: implications for drug resistance. Science *282*, 1669–1675.

Hughes, S.H., Mutschler, A., Bishop, J.M., and Varmus, H.E. (1981). A Rous sarcoma virus provirus is flanked by short direct repeats of a cellular DNA sequence present in only one copy prior to integration. Proc. Natl. Acad. Sci. U.S.A. *78*, 4299–4303.

Hungnes, O., Tjotta, E., and Grinde, B. (1991). The plus strand is discontinuous in a subpopulation of unintegrated HIV-1 DNA. Arch. Virol. *116*, 133–141.

Hungnes, O., Tjøtta, E., and Grinde, B. (1992). Mutations in the central polypurine tract of HIV-1 result in delayed replication. Virology *190*, 440–442.

Iijima, S., Nitahara-Kasahara, Y., Kimata, K., Zhong Zhuang, W., Kamata, M., Isogai, M., Miwa, M., Tsunetsugu-Yokota, Y., and Aida, Y. (2004). Nuclear localization of Vpr is crucial for the efficient replication of HIV-1 in primary $CD4^+$ T-cells. Virology *327*, 249–261.

Ikeda, T., Nishitsuji, H., Zhou, X., Nara, N., Ohashi, T., Kannagi, M., and Masuda, T. (2004). Evaluation of the functional involvement of human immunodeficiency virus type 1 integrase in nuclear import of viral cDNA during acute infection. J. Virol. *78*, 11563–11573.

Iordanskiy, S., Berro, R., Altieri, M., Kashanchi, F., and Bukrinsky, M. (2006). Intracytoplasmic maturation of the human immunodeficiency virus type 1 reverse transcription complexes determines their capacity to integrate into chromatin. Retrovirology 3, 4.

Isel, C., Ehresmann, C., Keith, G., Ehresmann, B., and Marquet, R. (1995). Initiation of reverse transcripion of HIV-1: Secondary structure of the HIV-1 RNA/tRNALys3 (template/primer) complex. J. Mol. Biol. *247*, 236–250.

Jacobo-Molina, A., Ding, J., Nanni, R.G., Clark, A.D., Jr., Lu, X., Tantillo, C., Williams, R.L., Kamer, G., Ferris, A.L., and Clark, P. (1993). Crystal structure of human immunodeficiency virus type 1 reverse transcriptase complexed with double-stranded DNA at 3.0 A resolution shows bent DNA. Proc. Natl. Acad. Sci. U.S.A. *90*, 6320–6324.

Jacque, J.M., and Stevenson, M. (2006). The inner-nuclear-envelope protein emerin regulates HIV-1 infectivity. Nature *441*, 641–645.

Jarrosson-Wuilleme, L., Goujon, C., Bernaud, J., Rigal, D., Darlix, J.-L., and Cimarelli, A. (2006). Transduction of nondividing human macrophages with gammaretrovirus-derived vectors. J. Virol. *80*, 1152–1159.

Jenkins, T.M., Engelman, A., Ghirlando, R., and Craigie, R. (1996). A soluble active mutant of HIV-1 integrase: involvement of both the core and the C-terminal domains in multimerization. J. Biol. Chem. *271*, 7712–7718.

Jenkins, T.M., Esposito, D., Engelman, A., and Craigie, R. (1997). Critical contacts between HIV-1 integrase and viral DNA identified by structure-based analysis and photo-crosslinking. EMBO J. *16*, 6849–6859.

Jiang, M., Mak, J., Ladha, A., Cohen, E., Klein, M., Rovinski, B., and Kleiman, L. (1993). Identification of tRNAs incorporated into wild-type and mutant human immunodeficiency virus type 1. J. Virol. *67*, 3246–3253.

Johnson, A.A., Santos, W., Pais, G.C.G., Marchand, C., Amin, R., Burke, T.R., Jr., Verdine, G., and Pommier, Y. (2006). Integration requires a specific interaction of the donor DNA terminal 5′-cytosine with glutamine 148 of the HIV-1 integrase flexible loop. J. Biol. Chem. *281*, 461–467.

Jones, K.S., Coleman, J., Merkel, G.W., Laue, T.M., and Skalka, A.M. (1992). Retroviral integrase functions as a multimer and can turn over catalytically. J. Biol. Chem. *267*, 16037–16040.

Jonsson, C., Donzella, G., Gaucan, E., Smith, C., and Roth, M. (1996). Functional domains of Moloney murine leukemia virus integrase defined by mutation and complementation analysis. J. Virol. *70*, 4585–4597.

Ju, G., and Skalka, A.M. (1980). Nucleotide sequence analysis of the long terminal repeat (LTR) of avian retroviruses: structural similarities with transposable elements. Cell *22*, 379–386.

Juretzek, T., Holm, T., Gartner, K., Kanzler, S., Lindemann, D., Herchenroder, O., Picard-Maureau, M., Rammling, M., Heinkelein, M., and Rethwilm, A. (2004). Foamy virus integration. J. Virol. *78*, 2472–2477.

Kalpana, G.V., and Goff, S.P. (1993). Genetic analysis of homomeric interactions of human immunodeficiency virus type 1 integrase using the yeast ywo-hybrid system. Proc. Natl. Acad. Sci. U.S.A. *90*, 10593–10597.

Kalpana, G.V., Reicin, A., Cheng, G.S., Sorin, M., Paik, S., and Goff, S.P. (1999). Isolation and characterization of an oligomerization-defective mutant of HIV-1 integrase. Virology *259*, 274–285.

Kanaya, S., Kohara, A., Miura, Y., Sekiguchi, A., Iwai, S., Inoue, H., Ohtsuka, E., and Ikehara, M. (1990). Identification of the amino acid residues involved in an active site of Escherichia coli ribonuclease H by site-directed mutagenesis. J. Biol. Chem. *265*, 4615–4621.

Kang, S.-M., and Morrow, C.D. (1999). Genetic analysis of a unique human immunodeficiency virus type 1 (HIV-1) with a primer binding site complementary to tRNAMet supports a role for U5-PBS stem–loop RNA structures in initiation of HIV-1 reverse transcription. J. Virol. 73, 1818–1827.

Kang, Y., Moressi, C.J., Scheetz, T.E., Xie, L., Tran, D.T., Casavant, T.L., Ak, P., Benham, C.J., Davidson, B.L., and McCray, P.B., Jr. (2006). Integration site choice of a feline immunodeficiency virus vector. J. Virol. *80*, 8820–8823.

Karageorgos, L., Li, P., and Burrel, C. (1993). Characterization of HIV replication complexes early after cell-to-cell infection. AIDS Res. Hum. Retroviruses *9*, 817–823.

Karki, R.G., Tang, Y., Burke, T.R.J., and Nicklaus, M.C. (2004). Model of full-length HIV-1 integrase complexed with viral DNA as template for anti-HIV drug design. J. Comput. Aided Mol. Des. *18*, 739–760.

Katz, R.A., Greger, J.G., Boimel, P., and Skalka, A.M. (2003). Human immunodeficiency virus type 1 DNA nuclear import and integration are mitosis independent in cycling cells. J. Virol. *77*, 13412–13417.

Katz, R.A., Greger, J.G., Darby, K., Boimel, P., Rall, G.F., and Skalka, A.M. (2002). Transduction of interphase cells by avian sarcoma virus. J. Virol. *76*, 5422–5434.

Katz, R.A., Merkel, G., Kulkosky, J., Leis, J., and Skalka, A.M. (1990). The avian retroviral IN protein is both necessary and sufficient for integrative recombination *in vitro*. Cell *63*, 87–95.

Katzman, M., and Katz, R.A. (1999). Substrate recognition by retroviral integrases. Adv. Virus Res. *52*, 371–395.

Katzman, M., Katz, R.A., Skalka, A.M., and Leis, J. (1989). The avian retroviral integration protein cleaves the terminal sequences of linear viral DNA at the *in vivo* sites of integration. J. Virol. *63*, 5319–5327.

Kenna, M.A., Brachmann, C.B., Devine, S.E., and Boeke, J.D. (1998). Invading the yeast nucleus: a nuclear localization signal at the C terminus of Ty1 integrase is required for transposition *in vivo*. Mol. Cell Biol. *18*, 1115–1124.

Kennedy, A.K., Haniford, D.B., and Mizuuchi, K. (2000). Single active site catalysis of the successive phosphoryl transfer steps by DNA transposases: Insights from phosphorothioate stereoselectivity. Cell *101*, 295–305.

Khiytani, D.K., and Dimmock, N.J. (2002). Characterization of a human immunodeficiency virus type 1 pre-integration complex in which the majority of the cDNA is resistant to DNase I digestion. J. Gen. Virol. *83*, 2523–2532.

Kohlstaedt, L.A., Wang, J., Friedman, J.M., Rice, P.A., and Steitz, T.A. (1992). Crystal structure at 3.5 A resolution of HIV-1 reverse transcriptase complexed with an inhibitor. Science *256*, 1783–1790.

Kootstra, N.A., and Schuitemaker, H. (1999). Phenotype of HIV-1 lacking a functional nuclear localization signal in matrix protein of Gag and Vpr is comparable to wild-type HIV-1 in primary macrophages. Virology *253*, 170–180.

Kootstra, N.A., Zwart, B.M., and Schuitemaker, H. (2000). Diminished human immunodeficiency virus type 1 reverse transcription and nuclear transport in primary macrophages arrested in early G1 phase of the cell cycle. J. Virol. *74*, 1712–1717.

Korin, Y.D., and Zack, J.A. (1998). Progression to the G1b phase of the cell cycle is required for completion of human immunodeficiency virus type 1 reverse transcription in T-cells. J. Virol. *72*, 3161–3168.

Kukolj, G., Jones, K., and Skalka, A. (1997). Subcellular localization of avian sarcoma virus and human immunodeficiency virus type 1 integrases. J. Virol. *71*, 843–847.

Kukolj, G., Katz, R.A., and Skalka, A.M. (1998). Characterization of the nuclear localization signal in the avian sarcoma virus integrase. Gene *223*, 157–163.

Kulkosky, J., Jones, K.S., Katz, R.A., Mack, J.P., and Skalka, A.M. (1992). Residues critical for retroviral integrative recombination in a region that is highly conserved among retroviral/retrotransposon integrases and bacterial insertion sequence transposases. Mol. Cell Biol. *12*, 2331–2338.

Kulkosky, J., Katz, R.A., Merkel, G., and Skalka, A.M. (1995). Activities and substrate specificity of the evolutionarily conserved central domain of retroviral integrase. Virology *206*, 448–456.

Kung, H.J., Fung, Y.K., Majors, J.E., Bishop, J.M., and Varmus, H.E. (1981). Synthesis of plus strands of retroviral DNA in cells infected with avian sarcoma virus and mouse mammary tumor virus. J. Virol. *37*, 127–138.

Kupiec, J.-J., Tobaly-Tapiero, J., Canivet, M., Santillana-Hayat, M., Flugel, R.M., Peries, J., and Emanoil-Ravier, R. (1988). Evidence for a gapped linear duplex DNA intermediate in the replicative cycle of human and simian spumavirases. Nucleic Acids Res. *16*, 9557–9565.

Kupiec, J.J., Kay, A., Hayat, M., Ravier, R., Périès, J., and Galibert, F. (1991). Sequence analysis of the simian foamy virus type 1 genome. Gene *101*, 2903–2909.

LaFemina, R., Schneider, C., Robbins, H., Callahan, P., LeGrow, K., Roth, E., Schleif, W., and Emini, E. (1992). Requirement of active human immunodeficiency virus type 1 integrase enzyme for productive infection of human T-lymphoid cells. J. Virol. *66*, 7414–7419.

Lau, A., Kanaar, R., Jackson, S.P., and O'Conner, M.J. (2004). Suppression of retroviral infection by the RAD52 DNA repair protein. EMBO J. *23*, 3421–3429.

Le Grice, S.F., Naas, T., Wohlgensinger, B., and Schatz, O. (1991). Subunit-selective mutagenesis indicates minimal polymerase activity in heterodimer-associated p51 HIV-1 reverse transcriptase. EMBO J. *10*, 3905–3911.

Le Rouzic, E., Mousnier, A., Rustum, C., Stutz, F., Hallberg, E., Dargemont, C., and Benichou, S. (2002). Docking of HIV-1 Vpr to the nuclear envelope is mediated by the interaction with the nucleoporin hCG1. J. Biol. Chem. *277*, 45091–45098.

Leavitt, A.D., Shiue, L., and Varmus, H.E. (1993). Site-directed mutagenesis of HIV-1 integrase demonstrates differential effects on integrase functions *in vitro*. J. Biol. Chem. *268*, 2113–2119.

Lederer, H., Schatz, O., May, R., Crespi, H., Darlix, J.L., Le Grice, S.F., and Heumann, H. (1992). Domain structure of the human immunodeficiency virus reverse transcriptase. EMBO J. *11*, 1131–1139.

Lee, M., and Craigie, R. (1994). Protection of retroviral DNA from autointegration: Involvement of a cellular factor. Proc. Natl. Acad. Sci. U.S.A. *91*, 9823–9827.

Lee, M.S., and Craigie, R. (1998). A previously unidentified host protein protects retroviral DNA from autointegration. Proc. Natl. Acad. Sci. U.S.A. *95*, 1528–1533.

Lee, S.P., and Han, M.K. (1996). Zinc stimulates Mg2+-dependent 3′-processing activity of human immunodeficiency virus type 1 integrase *in vitro*. Biochemistry *35*, 3837–3844.

Lee, S.P., Xiao, J., Knutson, J.R., Lewis, M.S., and Han, M.K. (1997). Zn2+ promotes the self-association of human immunodeficiency virus type-1 integrase *in vitro*. Biochemistry *36*, 173–180.

Lee, Y.M., and Coffin, J.M. (1991). Relationship of avian retrovirus DNA synthesis to integration *in vitro*. Mol. Cell Biol. *11*, 1419–1430.

Lehmann-Che, J., Renault, N., Giron, M.L., Roingeard, P., Clave, E., Tobaly-Tapiero, J., Bittoun, P., Toubert, A., The, H., and Saib, A. (2007). Centrosomal latency of incoming foamy viruses in resting cells. PLoS Pathog. 3, e74.

Levin, J.G., and Seidman, J.G. (1981). Effect of polymerase mutations on packaging of primer tRNAPro during murine leukemia virus assembly. J. Virol. *38*, 403–408.

Lewis, P., Hensel, M., and Emerman, M. (1992). Human immunodeficiency virus infection of cells arrested in the cell cycle. EMBO J. *11*, 3053–3058.

Lewis, P.F., and Emerman, M. (1994). Passage through mitosis is required for oncoretroviruses but not for the human immunodeficiency virus. J. Virol. *68*, 510–516.

Li, M., and Craigie, R. (2005). Processing of viral DNA ends channels the HIV-1 integration reaction to concerted integration. J. Biol. Chem. *280*, 29334–29339.

Li, M., Mizuuchi, M., Burke, T.R.J., and Craigie, R. (2006). Retroviral DNA integration: reaction pathway and critical intermediates. EMBO J. *25*, 1295–1304.

Li, Y., Kappes, J.C., Conway, J.A., Price, R.W., Shaw, G.M., and Hahn, B.H. (1991). Molecular characterization of human immunodeficiency virus type 1 cloned directly from uncultured human brain tissue: identification of replication-competent and -defective viral genomes. J. Virol. *65*, 3973–3985.

Limón, A., Devroe, E., Lu, R., Ghory, H.Z., Silver, P.A., and Engelman, A. (2002a). Nuclear localization of human immunodeficiency virus type 1 preintegration complexes (PICs): V165A and R166A are pleiotropic integrase mutants primarily defective for integration, not PIC nuclear import. J. Virol. *76*, 10598–10607.

Limón, A., Nakajima, N., Lu, R., Ghory, H.Z., and Engelman, A. (2002b). Wild-type levels of nuclear localization and human immunodeficiency virus type 1 replication in the absence of the central DNA flap. J. Virol. *76*, 12078–12086.

Lin, S.S., Nymark-McMahon, M.H., Yieh, L., and Sandmeyer, S.B. (2001). Integrase mediates nuclear localization of Ty3. Mol. Cell Biol. *21*, 7826–7838.

Llano, M., Delgado, S., Vanegas, M., and Poeschla, E.M. (2004a). Lens epithelium-derived growth factor/p75 prevents proteasomal degradation of HIV-1 integrase. J. Biol. Chem. *279*, 55570–55577.

Llano, M., Saenz, D.T., Meehan, A., Wongthida, P., Peretz, M., Walker, W.H., Teo, W., and Poeschla, E.M. (2006). An essential role for LEDGF/p75 in HIV integration. Science *314*, 461–464.

Llano, M., Vanegas, M., Fregoso, O., Saenz, D., Chung, S., Peretz, M., and Poeschla, E.M. (2004b). LEDGF/p75 determines cellular trafficking of diverse lentiviral but not murine oncoretroviral integrase proteins and is a component of functional lentiviral preintegration complexes. J. Virol. *78*, 9524–9537.

Lochelt, M., and Flugel, R.M. (1996). The human foamy virus *pol* gene is expressed as a Pro-Pol polyprotein and not as a Gag–Pol fusion protein. J. Virol. *70*, 1033–1040.

Lodi, P.J., Ernst, J.A., Kuszewski, J., Hickman, A.B., Engelman, A., Craigie, R., Clore, G.M., and Gronenborn, A.M. (1995). Solution structure of the DNA binding domain of HIV-1 integrase. Biochemistry *34*, 9826–9833.

Lu, R., Limón, A., Devroe, E., Silver, P.A., Cherepanov, P., and Engelman, A. (2004). Class II integrase mutants with changes in putative nuclear localization signals are primarily blocked at a post-nuclear entry step of human immunodeficiency virus type 1 replication. J. Virol. *78*, 12735–12746.

Lu, R., Limon, A., Ghory, H.Z., and Engelman, A. (2005). Genetic analyses of DNA-binding mutants in the catalytic core domain of human immunodeficiency virus type 1 integrase. J. Virol. *79*, 2493–2505.

Luby-Phelps, K. (2000). Cytoarchitecture and physical properties of cytoplasm: volume, viscosity, diffusion, intracellular surface area. Int. Rev. Cytol. *192*, 189–221.

Lucas, G.M. (2005). Antiretroviral adherence, drug resistance, viral fitness and HIV disease progression: a tangled web is woven. J. Antimicrob. Chemother. *55*, 413–416.

McDonald, D., Vodicka, M.A., Lucero, G., Svitkina, T.M., Borisy, G.G., Emerman, M., and Hope, T.J. (2002). Visualization of the intracellular behavior of HIV in living cells. J. Cell Biol. *159*, 441–452.

McEuen, A.R., Edwards, B., Koepke, K.A., Ball, A.E., Jennings, B.A., Wolstenholme, A.J., Danson, M.J., and Hough, D.W. (1992). Zinc binding by retroviral integrase. Biochem. Biophys. Res. Commun. *189*, 813–818.

MacNeil, A., Sankale, J.L., Meloni, S.T., Sarr, A.D., Mboup, S., and Kanki, P. (2006). Genomic sites of human immunodeficiency virus type 2 (HIV-2) integration: similarities to HIV-1 *in vitro* and possible differences *in vivo*. J. Virol. *80*, 7316–7321.

Maertens, G., Cherepanov, P., Debyser, Z., Engelborghs, Y., and Engelman, A. (2004). Identification and characterization of a functional nuclear localization signal in the HIV-1 integrase interactor LEDGF/p75. J. Biol. Chem. *279*, 33421–33429.

Maertens, G., Cherepanov, P., Pluymers, W., Busschots, K., De Clercq, E., Debyser, Z., and Engelborghs, Y. (2003). LEDGF/p75 is essential for nuclear and chromosomal targeting of HIV-1 integrase in human cells. J. Biol. Chem. *278*, 33528–33539.

Majors, J.E., and Varmus, H.E. (1981). Nucleotide sequences at host-proviral junctions for mouse mammary tumour virus. Nature *289*, 253–258.

Mak, J., Jiang, M., Wainberg, M.A., Hammarskjold, M.L., Rekosh, D., and Kleiman, L. (1994). Role of Pr160Gag–Pol in mediating the selective incorporation of tRNA(Lys) into human immunodeficiency virus type 1 particles. J. Virol. *68*, 2065–2072.

Mak, J., and Kleiman, L. (1997). Primer tRNAs for reverse transcription. J. Virol. *71*, 8087–8095.

Mannioui, A., Nelson, E., Schiffer, C., Felix, N., Le Rouzic, E., Benichou, S., Gluckman, J.C., and Canque, B. (2005). Human immunodeficiency virus type 1 KK26–27 matrix mutants display impaired infectivity, circularization and integration but not nuclear import. Virology 339, 21–30.

Marsden, M.D., and Zack, J.A. (2007). Human immunodeficiency virus bearing a disrupted central DNA flap is pathogenic *in vivo*. J. Virol. *81*, 6146–6150.

Marshall, H.M., Ronen, K., Berry, C., Llano, M., Sutherland, H., Saenz, D., Bickmore, W., Poeschla, E., and Bushman, F.D. (2007). Role of PSIP1/LEDGF/

p75 in lentiviral infectivity and integration targeting. PLoS ONE *2*, e1340.

Masuda, T., Kuroda, M.J., and Harada, S. (1998). Specific and independent recognition of U3 and U5 att sites by human immunodeficiency virus type 1 integrase *in vivo*. J. Virol. *72*, 8396–8402.

Matheka, H.D., Coggins, L., Shively, J.N., and Norcross, N.L. (1976). Purification and characterization of equine infectious anemia virus. Arch. Virol. *51*, 107–114.

Mergener, K., Fäcke, M., Welker, R., Brinkmann, V., Gelderblom, H.R., and Kräusslich, H.G. (1992). Analysis of HIV particle formation using transient expression of subviral constructs in mammalian cells. Virology *186*, 25–39.

Meyerhans, A., Vartanian, J.P., Hultgren, C., Plikat, U., Karlsson, A., Wang, L., Eriksson, S., and Wain-Hobson, S. (1994). Restriction and enhancement of human immunodeficiency virus type 1 replication by modulation of intracellular deoxynucleoside triphosphate pools. J. Virol. *68*, 535–540.

Miller, D.G., Adam, M.A., and Miller, A.D. (1990). Gene transfer by retrovirus vectors occurs only in cells that are actively replicating at the time of infection. Mol. Cell Biol. *10*, 4239–4242.

Miller, M., Farnet, C., and Bushman, F. (1997). Human immunodeficiency virus type 1 preintegration complexes: studies of organization and composition. J. Virol. *71*, 5382–5390.

Miller, M.D., Wang, B., and Bushman, F.D. (1995). Human immunodeficiency virus type 1 preintegration complexes containing discontinuous plus strands are competent to integrate *in vitro*. J. Virol. *69*, 3938–3944.

Mitchell, R.S., Beitzel, B.F., Schroder, A.R.W., Shinn, P., Chen, H., Berry, C.C., Ecker, J.R., and Bushman, F.D. (2004). Retroviral DNA integration: ASLV, HIV, and MLV show distinct target site preferences. PLoS Biol. *2*, e234.

Mizrahi, V., Lazarus, G.M., Miles, L.M., Meyers, C.A., and Debouck, C. (1989). Recombinant HIV-1 reverse transcriptase: purification, primary structure, and polymerase/ribonuclease H activities. Arch. Biochem. Biophys. *273*, 347–358.

Mizuuchi, K., and Adzuma, K. (1991). Inversion of the phosphate chirality at the target site of Mu DNA strand transfer: evidence for a one-step transesterification mechanism. Cell *66*, 129–140.

Mizuuchi, K., Nobbs, T.J., Halford, S.E., Adzuma, K., and Qin, J. (1999). A new method for determining the stereochemistry of DNA cleavage reactions: application to the SfiI and HpaII restriction endonucleases and to the MuA transposase. Biochemistry *38*, 4640–4648.

Moebes, A., Enssle, J., Bieniasz, P.D., Heinkelein, M., Lindemann, D., Bock, M., McClure, M.O., and Rethwilm, A. (1997). Human foamy virus reverse transcription that occurs late in the viral replication cycle. J. Virol. *71*, 7305–7311.

Moelling, K. (1974). Characterization of reverse transcriptase and RNase H from friend-murine leukemia virus. Virology *62*, 46–59.

Moore, S.P., Rinckel, L.A., and Garfinkel, D.J. (1998). A Ty1 integrase nuclear localization signal required for retrotransposition. Mol. Cell Biol. *18*, 1105–1114.

Mulder, L.C.F., and Muesing, M.A. (2000). Degradation of HIV-1 integrase by the N-end rule pathway. J. Biol. Chem. *275*, 29749–29753.

Mumm, S.R., and Grandgenett, D.P. (1991). Defining nucleic acid-binding properties of avian retrovirus integrase by deletion analysis. J. Virol. *65*, 1160–1167.

Narayan, S., and Young, J.A.T. (2004). Reconstitution of retroviral fusion and uncoating in a cell-free system. Proc. Natl. Acad. Sci. U.S.A. *101*, 7721–7726.

Narezkina, A., Taganov, K.D., Litwin, S., Stoyanova, R., Hayashi, J., Seeger, C., Skalka, A.M., and Katz, R.A. (2004). Genome-wide analyses of avian sarcoma virus integration sites. J. Virol. *78*, 11656–11663.

Netzer, K.-O., Schliephake, A., Maurer, B., Watanabe, R., Aguzzi, A., and Rethwilm, A. (1993). Identification of *pol*-related gene products of human foamy virus. Virology *192*, 336–338.

Neves, M., Périès, J., and Saïb, A. (1998). Study of human foamy virus proviral integration in chronically infected murine cells. Res. Virol. *149*, 393–401.

Nitahara-Kasahara, Y., Kamata, M., Yamamoto, T., Zhang, X., Miyamoto, Y., Muneta, K., Iijima, S., Yoneda, Y., Tsunetsugu-Yokota, Y., and Aida, Y. (2007). Novel nuclear import of Vpr promoted by importin alpha is crucial for human immunodeficiency virus type 1 replication in macrophages. J. Virol. *81*, 5284–5293.

Nowotny, M., Gaidamakov, S.A., Crouch, R.J., and Yang, W. (2005). Crystal structures of RNase H bound to an RNA/DNA hybrid: substrate specificity and metal-dependent catalysis. Cell *121*, 1005–1016.

Nowrouzi, A., Dittrich, M., Klanke, C., Heinkelein, M., Rammling, M., Dandekar, T., von Kalle, C., and Rethwilm, A. (2006). Genome-wide mapping of foamy virus vector integrations into a human cell line. J. Gen. Virol. *87*, 1339–1347.

O'Brien, W.A., Namazi, A., Kalhor, H., Mao, S.H., Zack, J.A., and Chen, I.S. (1994). Kinetics of human immunodeficiency virus type 1 reverse transcription in blood mononuclear phagocytes are slowed by limitations of nucleotide precursors. J. Virol. *68*, 1258–1263.

Oh, J., Chang, K.W., and Hughes, S.H. (2006). Mutations in the U5 sequences adjacent to the primer binding site do not affect tRNA cleavage by Rous sarcoma virus RNase H but do cause aberrant integrations *in vivo*. J. Virol. *80*, 451–459.

Oh, J., Chang, K.W., Wierzchoslawski, R., Alvord, W.G., and Hughes, S.H. (2008). Rous sarcoma virus (RSV) integration *in vivo*: a CA dinucleotide is not required in U3, and RSV linear DNA does not autointegrate. J. Virol. *82*, 503–512.

Omer, C., Resnick, R., and Faras, A.J. (1984). Evidence for involvement of an RNA primer in initiation of strong-stop plus DNA synthesis during reverse transcription *in vitro*. J. Virol. *50*, 465–470.

Pahl, A., and Flugel, R.M. (1993). Endonucleolytic cleavages and DNA-joining activities of the integration protein of human foamy virus. J. Virol. *67*, 5426–5434.

Pahl, A., and Flugel, R.M. (1995). Characterization of the human spuma retrovirus integrase by site-directed mutagenesis, by complementation analysis, and by swapping the zinc finger domain of HIV-1. J. Biol. Chem. *270*, 2957–2966.

Panganiban, A.T., and Temin, H.M. (1983). The terminal nucleotides of retrovirus DNA are required for integration but not virus production. Nature *306*, 155–160.

Panganiban, A.T., and Temin, H.M. (1984). The retrovirus *pol* gene encodes a product required for DNA integration: Identification of a retrovirus int locus. Proc. Natl. Acad. Sci. U.S.A. *81*, 7885–7889.

Pauza, C. (1990). Two bases are deleted from the termini of HIV-1 linear DNA during integrative recombination. Virology *179*, 886–889.

Peliska, J.A., Balasubramanian, S., Giedroc, D.P., and Benkovic, S.J. (1994). Recombinant HIV-1 nucleocapsid protein accelerates HIV-1 reverse transcriptase catalyzed DNA strand transfer reactions and modulates RNase H activity. Biochemistry *33*, 13817–13823.

Perach, M., and Hizi, A. (1999). Catalytic features of the recombinant reverse transcriptase of bovine leukemia virus expressed in bacteria. Virology *259*, 176–189.

Peters, G., and Glover, C. (1980). tRNA's and priming of RNA-directed DNA synthesis in mouse mammary tumor virus. J. Virol. *35*, 31–40.

Peters, G., Harada, F., Dahlberg, J.E., Panet, A., Haseltine, W.A., and Baltimore, D. (1977). Low-molecular-weight RNAs of Moloney murine leukemia virus: identification of the primer for RNA-directed DNA synthesis. J. Virol. *21*, 1031–1041.

Peters, G.G., and Hu, J. (1980). Reverse transcriptase as the major determinant for selective packaging of tRNA's into Avian sarcoma virus particles. J. Virol. *36*, 692–700.

Petit, C., Giron, M.-L., Tobaly-Tapiero, J., Bittoun, P., Real, E., Jacob, Y., Tordo, N., de The, H., and Saib, A. (2003). Targeting of incoming retroviral Gag to the centrosome involves a direct interaction with the dynein light chain 8. J. Cell Sci. *116*, 3433–3442.

Petit, C., Schwartz, O., and Mammano, F. (1999). Oligomerization within virions and subcellular localization of human immunodeficiency virus type 1 integrase. J. Virol. *73*, 5079–5088.

Petit, C., Schwartz, O., and Mammano, F. (2000). The karyophilic properties of human immunodeficiency virus type 1 integrase are not required for nuclear import of proviral DNA. J. Virol. *74*, 7119–7126.

Piacenti, F.J. (2006). An update and review of antiretroviral therapy. Pharmacotherapy *26*, 1111–1133.

Pierson, T.C., Zhou, Y., Kieffer, T.L., Ruff, C.T., Buck, C., and Siliciano, R.F. (2002). Molecular characterization of preintegration latency in human immunodeficiency virus type 1 infection. J. Virol. *76*, 8518–8531.

Plesa, G., Dai, J., Baytop, C., Riley, J.L., June, C.H., and O'Doherty, U. (2007). Addition of deoxynucleosides enhances human immunodeficiency virus type 1 integration and 2LTR formation in resting $CD4^+$ T-cells. J. Virol. *81*, 13938–13942.

Pluymers, W., Cherepanov, P., Schols, D., De Clercq, E., and Debyser, Z. (1999). Nuclear localization of human immunodeficiency virus type 1 integrase expressed as a fusion protein with green fluorescent protein. Virology *258*, 327–332.

Poeschla, E.M. (2008). Integrase, LEDGF/p75 and HIV replication. Cell. Mol. Life Sci. *65*, 1403–1424.

Pomerantz, R.J., and Horn, D.L. (2003). Twenty years of therapy for HIV-1 infection. Nat. Med. *9*, 867–873.

Popov, S., Rexach, M., Ratner, L., Blobel, G., and Bukrinsky, M. (1998a). Viral protein R regulates docking of the HIV-1 preintegration complex to the nuclear pore complex. J. Biol. Chem. *273*, 13347–13352.

Popov, S., Rexach, M., Zybarth, G., Reiling, N., Lee, M.-A., Ratner, L., Lane, C.M., Moore, M.S., Blobel, G., and Bukrinsky, M. (1998b). Viral protein R regulates nuclear import of the HIV-1 pre-integration complex. EMBO J. *17*, 909–917.

Puras Lutzke, R.A., and Plasterk, R.H.A. (1998). Structure-based mutational analysis of the C-terminal DNA-binding domain of human immunodeficiency virus type 1 integrase: critical residues for protein oligomerization and DNA binding. J. Virol. *72*, 4841–4848.

Puras Lutzke, R.A., Vink, C., and Plasterk, R.H.A. (1994). Characterization of the minimal DNA-binding domain of the HIV integrase protein. Nucleic Acids Res. *22*, 4125–4131.

Quinn, T.P., and Grandgenett, D.P. (1988). Genetic evidence that the avian retrovirus DNA endonuclease domain of *pol* is necessary for viral integration. J. Virol. *62*, 2307–2312.

Ratner, L., Haseltine, W., Patarca, R., Livak, K.J., Starcich, B., Josephs, S.F., Doran, E.R., Rafalski, J.A., Whitehorn, E.A., Baumeister, K., *et al.* (1985). Complete nucleotide sequence of the AIDS virus, HTLV-III. Nature *313*, 277–284.

Reeves, J.D., and Piefer, A.J. (2005). Emerging drug targets for antiretroviral therapy. Drugs *65*, 1747–1766.

Reil, H., Bukovsky, A.A., Gelderblom, H.R., and Gottlinger, H.G. (1998). Efficient HIV-1 replication can occur in the absence of the viral matrix protein. EMBO J. *17*, 2699–2708.

Rethwilm, A. (2003). The replication strategy of foamy viruses. Curr. Top. Microbiol. Immunol. *277*, 1–26.

Rice, A.P., and Sutton, R.E. (2007). Targeting protein–protein interactions for HIV therapeutics. Future HIV Ther. *1*, 369–385.

Rodriguez-Rodriguez, L., Tsuchihashi, Z., Fuentes, G.M., Bambara, R.A., and Fay, P.J. (1995). Influence of human immunodeficiency virus nucleocapsid protein on synthesis and strand transfer by the reverse transcriptase *in vitro*. J. Biol. Chem. *270*, 15005–15011.

Roe, T., Reynolds, T.C., Yu, G., and Brown, P.O. (1993). Integration of murine leukemia virus DNA depends on mitosis. EMBO J. *12*, 2099–2108.

Roth, M.J., Schwartzberg, P.L., and Goff, S.P. (1989). Structure of the termini of DNA intermediates in the integration of retroviral DNA: dependence on IN function and terminal DNA sequence. Cell *58*, 47–54.

Roth, M.J., Tanese, N., and Goff, S.P. (1985). Purification and characterization of murine retroviral reverse transcriptase expressed in Escherichia coli. J. Biol. Chem. *260*, 9326–9335.

Saib, A., Puvion-Dutilleul, F., Schmid, M., Peries, J., and de The, H. (1997). Nuclear targeting of incoming human foamy virus Gag proteins involves a centriolar step. J. Virol. *71*, 1155–1161.

Sarkar, N.H., and Moore, D.H. (1974). Separation of B and C type virions by centrifugation in gentle density gradients. J. Virol. *13*, 1143–1147.

Schroder, A.R.W., Shinn, P., Chen, H., Berry, C., Ecker, J.R., and Bushman, F. (2002). HIV-1 integration in the human genome favors active genes and local hotspots. Cell *110*, 521–529.

Schwartzberg, P., Colicelli, J., and Goff, S.P. (1984). Construction and analysis of deletion mutations in the *pol* gene of moloney murine leukemia virus: A new viral function required for productive infection. Cell *37*, 1043–1052.

Schweizer, M., Renne, R., and Neumann-Haefelin, D. (1989). Structural analysis of proviral DNA in simian foamy virus (LK-3)-infected cells. Arch. Virol. *109*, 103–114.

Seamon, J.A., Jones, K.S., Miller, C., and Roth, M.J. (2002). Inserting a nuclear targeting signal into a replication-competent Moloney murine leukemia virus affects viral export and is not sufficient for cell cycle-independent infection. J. Virol. *76*, 8475–8484.

Seiki, M., Hattori, S., Hirayama, Y., and Yoshida, M. (1983). Human adult T-cell leukemia virus: Complete nucleotide sequence of the provirus genome integrated in leukemia cell DNA. Proc. Natl. Acad. Sci. U.S.A. *80*, 3618–3622.

Sherman, P.A., and Fyfe, J.A. (1990). Human immunodeficiency virus integration protein expressed in Escherichia coli possesses selective DNA cleaving activity. Proc. Natl. Acad. Sci. U.S.A. *87*, 5119–5123.

Shibagaki, Y., Holmes, M.L., Appa, R.S., and Chow, S.A. (1997). Characterization of feline immunodeficiency virus integrase and analysis of functional domains. Virology *230*, 1–10.

Shimotohno, K., Mizutani, S., and Temin, H.M. (1980). Sequence of retrovirus provirus resembles that of bacterial transposable elements. Nature *285*, 550–554.

Shoemaker, C., Hoffman, J., Goff, S.P., and Baltimore, D. (1981). Intramolecular integration within Moloney murine leukemia virus DNA. J. Virol. *40*, 164–172.

Shun, M.-C., Daigle, J.E., Vandegraaff, N., and Engelman, A. (2007a). Wild-type levels of human immunodeficiency virus type 1 infectivity in the absence of cellular emerin protein. J. Virol. *81*, 166–172.

Shun, M.-C., Raghavendra, N.K., Vandegraaff, N., Daigle, J.E., Hughes, S., Kellam, P., Cherepanov, P., and Engelman, A. (2007b). LEDGF/p75 functions downstream from preintegration complex formation to effect gene-specific HIV-1 integration. Genes Dev. *21*, 1767–1778.

Sinha, S., and Grandgenett, D.P. (2005). Recombinant human immunodeficiency virus type 1 integrase exhibits a capacity for full-site integration *in vitro* that is comparable to that of purified preintegration complexes from virus-infected cells. J. Virol. *79*, 8208–8216.

Sinha, S., Pursley, M.H., and Grandgenett, D.P. (2002). Efficient concerted integration by recombinant human immunodeficiency virus type 1 integrase without cellular or viral cofactors. J. Virol. *76*, 3105–3113.

Siva, A.C., and Bushman, F. (2002). Poly(ADP-ribose) polymerase 1 is not strictly required for infection of murine cells by retroviruses. J. Virol. *76*, 11904–11910.

Skalka, A.M. (1993). Retroviral DNA integration: lessons for transposon shuffling. Gene *135*, 175–182.

Skinner, L.M., Sudol, M., Harper, A.L., and Katzman, M. (2001). Nucleophile selection for the endonuclease activities of human, ovine, and avian retroviral integrases. J. Biol. Chem. *276*, 114–124.

Smith, J.K., Cywinski, A., and Taylor, J.M. (1984a). Initiation of plus-strand DNA synthesis during reverse transcription of an avian retrovirus genome. J. Virol. *49*, 200–204.

Smith, J.K., Cywinski, A., and Taylor, J.M. (1984b). Specificity of initiation of plus-strand DNA by Rous sarcoma virus. J. Virol. *52*, 314–319.

Sonigo, P., Alizon, M., Staskus, K., Klatzmann, D., Cole, S., Danos, O., Retzel, E., Tiollais, P., Haase, A., and Wain-Hobson, S. (1985). Nucleotide sequence of the visna lentivirus: relationship to the AIDS virus. Cell *42*, 369–382.

Sonigo, P., Barker, C., Hunter, E., and Wain-Hobson, S. (1986). Nucleotide sequence of Mason–Pfizer monkey virus: an immunosuppressive D-type retrovirus. Cell *45*, 375–385.

Spina, C.A., Guatelli, J.C., and Richman, D.D. (1995). Establishment of a stable, inducible form of human immunodeficiency virus type 1 DNA in quiescent CD4 lymphocytes *in vitro*. J. Virol. *69*, 2977–2988.

Stevens, S.W., and Griffith, J.D. (1996). Sequence analysis of the human DNA flanking sites of human immunodeficiency virus type 1 integration. J. Virol. *70*, 6459–6462.

Stevenson, M., Haggerty, S., Lamonica, C.A., Meier, C.M., Welch, S.K., and Wasiak, A.J. (1990a). Integration is not necessary for expression of human immunodeficiency virus type 1 protein products. J. Virol. *64*, 2421–2425.

Stevenson, M., Stanwick, T.L., Dempsey, M.P., and Lamonica, C.A. (1990b). HIV-1 replication is controlled at the level of T-cell activation and proviral integration. EMBO J. *9*, 1551–1560.

Stewart, M. (2007). Molecular mechanism of the nuclear protein import cycle. Nat. Rev. Mol. Cell Biol. *8*, 195–208.

Suzuki, Y., and Craigie, R. (2002). Regulatory mechanisms by which barrier-to-autointegration factor blocks autointegration and stimulates intermolecular integration of Moloney murine leukemia virus preintegration complexes. J. Virol. *76*, 12376–12380.

Suzuki, Y., and Craigie, R. (2007). The road to chromatin – nuclear entry of retroviruses. Nat. Rev. Microbiol. *5*, 187–196.

Suzuki, Y., Yang, H., and Craigie, R. (2004). LAP2alpha and BAF collaborate to organize the Moloney murine leukemia virus preintegration complex. EMBO J. *23*, 4670–4680.

Tanese, N., Telesnitsky, A., and Goff, S.P. (1991). Abortive reverse transcription by mutants of Moloney murine leukemia virus deficient in the reverse transcriptase-associated RNase H function. J. Virol. *65*, 4387–4397.

Tasara, T., Maga, G., Hottiger, M.O., and Hubscher, U. (2001). HIV-1 reverse transcriptase and integrase enzymes physically interact and inhibit each other. FEBS Lett. *507*, 39–44.

Taube, R., Loya, S., Avidan, O., Perach, M., and Hizi, A. (1998). Reverse transcriptase of mouse mammary tumour virus: expression in bacteria, purification and biochemical characterization. Biochem. J. *332*, 807–808.

Toplin, I. (1967). Purification of the Moloney and Rauscher murine leukemia viruses by use of zonal ultracentrifuge systems. Appl. Microbiol. *15*, 582–589.

Triques, K., and Stevenson, M. (2004). Characterization of restrictions to human immunodeficiency virus type 1 infection of monocytes. J. Virol. *78*, 5523–5527.

Trobridge, G., and Russell, D.W. (2004). Cell cycle requirements for transduction by foamy virus vectors compared to those of oncovirus and lentivirus vectors. J. Virol. *78*, 2327–2335.

Trobridge, G.D., Miller, D.G., Jacobs, M.A., Allen, J.M., Kiem, H.-P., Kaul, R., and Russell, D.W. (2006). Foamy virus vector integration sites in normal human cells. Proc. Natl. Acad. Sci. U.S.A. *103*, 1498–1503.

Turlure, F., Devroe, E., Silver, P.A., and Engelman, A. (2004). Human cell proteins and human immunodeficiency virus DNA integration. Front. Biosci. *9*, 3187–3208.

van den Ent, F.M.I., Vos, A., and Plasterk, R.H.A. (1999). Dissecting the role of the N-terminal domain of human immunodeficiency virus integrase by *trans*-complementation analysis. J. Virol. *73*, 3176–3183.

van Gent, D.C., Groeneger, A.A.M.O., and Plasterk, R.H.A. (1992). Mutational analysis of the integrase protein of human immunodeficiency virus type 2. Proc. Natl. Acad. Sci. U.S.A. *89*, 9598–9602.

van Gent, D.C., Vink, C., Groeneger, A.A.M.O., and Plasterk, R.H.A. (1993). Complementation between HIV integrase proteins mutated in different domains. EMBO J. *12*, 3261–3267.

Van Maele, B., Busschots, K., Vandekerckhove, L., Christ, F., and Debyser, Z. (2006). Cellular co-factors of HIV-1 integration. Trends Biochem. Sci. *31*, 98–105.

Vandegraaff, N., Devroe, E., Turlure, F., Silver, P.A., and Engelman, A. (2006). Biochemical and genetic analyses of integrase-interacting proteins lens epithelium-derived growth factor (LEDGF)/p75 and hepatoma-derived growth factor related protein 2 (HRP2) in preintegration complex function and HIV-1 replication. Virology *346*, 415–426.

Vandegraaff, N., and Engelman, A. (2007). Molecular mechanism of HIV integration and therapeutic intervention. Expert Rev. Mol. Med. *9*, 1–19.

Vandekerckhove, L., Christ, F., Van Maele, B., De Rijck, J., Gijsbers, R., Van den Haute, C., Witvrouw, M., and Debyser, Z. (2006). Transient and stable knockdown of the integrase cofactor LEDGF/p75 reveals its role in the replication cycle of human immunodeficiency virus. J. Virol. *80*, 1886–1896.

Vanegas, M., Llano, M., Delgado, S., Thompson, D., Peretz, M., and Poeschla, E. (2005). Identification of the LEDGF/p75 HIV-1 integrase-interaction domain and NLS reveals NLS-independent chromatin tethering. J. Cell Sci. *118*, 1733–1743.

Varmus, H.E., Heasley, S., Kung, H.J., Oppermann, H., Smith, V.C., Bishop, J.M., and Shank, P.R. (1978). Kinetics of synthesis, structure and purification of avian sarcoma virus-specific DNA made in the cytoplasm of acutely infected cells. J. Mol. Biol. *120*, 55–82.

Varmus, H.E., Padgett, T., Heasley, S., Simon, G., and Bishop, J.M. (1977). Cellular functions are required for the synthesis and integration of avian sarcoma virus-specific DNA. Cell *11*, 307–319.

Vincent, K.A., York-Higgins, D., Quiroga, M., and Brown, P.O. (1990). Host sequences flanking the HIV provirus. Nucleic Acids Res. *18*, 6045–6047.

Vink, C., Groenink, M., Elgersma, Y., Fouchier, R.A., Tersmette, M., and Plasterk, R.H. (1990). Analysis of the junctions between human immunodeficiency virus type 1 proviral DNA and human DNA. J. Virol. *64*, 5626–5627.

Vink, C., Oude Groeneger, A.M., and Plasterk, R.H.A. (1993). Identification of the catalytic and DNA-binding region of the human immunodeficiency virus type I integrase protein. Nucleic Acids Res. *21*, 1419–1425.

Vink, C., van der Linden, K.H., and Plasterk, R.H. (1994). Activities of the feline immunodeficiency virus integrase protein produced in Escherichia coli. J. Virol. *68*, 1468–1474.

Vink, C., Yeheskiely, E., van der Marel, G.A., Van Boom, J.H., and Plasterk, R.H.A. (1991). Site-specific hydrolysis and alcoholysis of human immunodeficiency virus DNA termini mediated by the viral integrase protein. Nucleic Acids Res. *19*, 6691–6698.

Vodicka, M.A., Koepp, D.M., Silver, P.A., and Emerman, M. (1998). HIV-1 Vpr interacts with the nuclear transport pathway to promote macrophage infection. Genes Dev. *12*, 175–185.

von Schwedler, U., Kornbluth, R.S., and Trono, D. (1994). The nuclear localization signal of the matrix protein of human immunodeficiency virus type 1 allows the establishment of infection in macrophages and quiescent T-lymphocytes. Proc. Natl. Acad. Sci. U.S.A. *91*, 6992–6996.

Vora, A., Bera, S., and Grandgenett, D. (2004). Structural organization of avian retrovirus integrase in assembled intasomes mediating full-site integration. J. Biol. Chem. *279*, 18670–18678.

Vora, A.C., Fitzgerald, M.L., and Grandgenett, D.P. (1990). Removal of 3′-OH-terminal nucleotides from blunt-ended long terminal repeat termini by the avian retrovirus integration protein. J. Virol. *64*, 5656–5659.

Vora, A.C., and Grandgenett, D.P. (1995). Assembly and catalytic properties of retrovirus integrase–DNA complexes capable of efficiently performing concerted integration. J. Virol. *69*, 7483–7488.

Wang, J.-Y., Ling, H., Yang, W., and Craigie, R. (2001). Structure of a two-domain fragment of HIV-1 integrase: implications for domain organization in the intact protein. EMBO J. *20*, 7333–7343.

Waters, L.C., Mullin, B.C., Bailiff, E.G., and Popp, R.A. (1980). Differential association of transfer RNAs with the genomes of murine, feline and primate retroviruses. Biochim. Biophys. Acta *608*, 112–126.

Wei, S.-Q., Mizuuchi, K., and Craigie, R. (1997). A large nucleoprotein assembly at the ends of the viral DNA mediates retroviral DNA integration. EMBO J. *16*, 7511–7520.

Weinberg, J.B., Matthews, T.J., Cullen, B.R., and Malim, M.H. (1991). Productive human immunodeficiency virus type 1 (HIV-1) infection of nonproliferating human monocytes. J. Exp. Med. *174*, 1477–1482.

Werner, S., and Wohrl, B.M. (2000). Asymmetric subunit organization of heterodimeric Rous sarcome virus reverse transcriptase alphabeta: localization of the polymerase and RNase H active sites in the alpha subunit. J. Virol. *74*, 3245–3252.

Whitwam, T., Peretz, M., and Poeschla, E. (2001). Identification of a central DNA flap in feline immunodeficiency virus. J. Virol. *75*, 9407–9414.

Wielens, J., Crosby, I.T., and Chalmers, D.K. (2005). A three-dimensional model of the human immunodeficiency virus type 1 integration complex. J. Comput. Aided Mol. Des. *19*, 301–317.

Wilhelm, M., and Wilhelm, F.X. (2006). Cooperation between reverse transcriptase and integrase during reverse transcription and formation of the preintegrative complex of Ty1. Eukaryot. Cell *5*, 1760–1769.

Woerner, A.M., Klutch, M., Levin, J.G., and Marcus-Sekura, C.J. (1992). Localization of DNA binding activity of HIV-1 integrase to the C-terminal half of the protein. AIDS Res. Hum. Retroviruses *8*, 297–304.

Woerner, A.M., and Marcus-Sekura, C.J. (1993). Characterization of a DNA binding domain in the C-terminus of HIV-1 integrase by deletion mutagenesis. Nucleic Acids Res. *21*, 3507–3511.

Woodward, C.L., Wang, Y., Dixon, W.J., Htun, H., and Chow, S.A. (2003). Subcellular localization of feline immunodeficiency virus integrase and mapping of its karyophilic determinant. J. Virol. *77*, 4516–4527.

Wu, X., Li, Y., Crise, B., and Burgess, S.M. (2003). Transcription start regions in the human genome are favored targets for MLV integration. Science *300*, 1749–1751.

Wu, X., Li, Y., Crise, B., Burgess, S.M., and Munroe, D.J. (2005). Weak palindromic consensus sequences are a common feature found at the integration target sites of many retroviruses. J. Virol. *79*, 5211–5214.

Wu, X., Liu, H., Xiao, H., Conway, J.A., Hehl, E., Kalpana, G.V., Prasad, V., and Kappes, J.C. (1999). Human immunodeficiency virus type 1 integrase protein promotes reverse transcription through specific interactions with the nucleoprotein reverse transcription complex. J. Virol. *73*, 2126–2135.

Wu, X., Liu, H., Xiao, H., Conway, J.A., Hunter, E., and Kappes, J.C. (1997). Functional RT and IN incorporated into HIV-1 particles independently of the Gag/Pol precursor protein. EMBO J. *16*, 5113–5122.

Yamashita, M., and Emerman, M. (2004). Capsid is a dominant determinant of retrovirus infectivity in nondividing cells. J. Virol. *78*, 5670–5678.

Yamashita, M., and Emerman, M. (2005). The cell cycle independence of HIV infections is not determined by known karyophilic viral elements. PLoS Pathog. *1*, e18.

Yamashita, M., and Emerman, M. (2006). Retroviral infection of non-dividing cells: Old and new perspectives. Virology *344*, 88–93.

Yamashita, M., Perez, O., Hope, T.J., and Emerman, M. (2007). Evidence for direct involvement of the capsid protein in HIV infection of nondividing cells. PLoS Pathog. 3, e156.

Yang, F., and Roth, M.J. (2001). Assembly and catalysis of concerted two-end integration events by Moloney murine leukemia virus integrase. J. Virol. 75, 9561–9670.

Yang, Z.-N., Mueser, T.C., Bushman, F.D., and Hyde, C.C. (2000). Crystal structure of an active two-domain derivative of rous sarcoma virus integrase. J. Mol. Biol. *296*, 535–548.

Yoder, K., Sarasin, A., Kraemer, K., McIlhatton, M., Bushman, F., and Fishel, R. (2006). The DNA repair genes XPB and XPD defend cells from retroviral infection. Proc. Natl. Acad. Sci. U.S.A. *103*, 4622–4627.

Yoder, K.E., and Bushman, F.D. (2000). Repair of gaps in retroviral DNA integration intermediates. J. Virol. *74*, 11191–11200.

Yu, S.F., Baldwin, D.N., Gwynn, S.R., Yendapalli, S., and Linial, M.L. (1996). Human foamy virus replication: a pathway distinct from that of retroviruses and hepadnaviruses. Science *271*, 1579–1582.

Yuan, B., Fassati, A., Yueh, A., and Goff, S.P. (2002). Characterization of Moloney murine leukemia virus p12 mutants blocked during early events of infection. J. Virol. *76*, 10801–10810.

Yuan, B., Li, X., and Goff, S.P. (1999). Mutations altering the moloney murine leukemia virus p12 Gag protein affect virion production and early events of the virus life cycle. EMBO J. *18*, 4700–4710.

Yueh, A., and Goff, S.P. (2003). Phosphorylated serine residues and an arginine-rich domain of the Moloney murine leukemia virus p12 protein are required for early events of viral infection. J. Virol. *77*, 1820–1829.

Yueh, A., Leung, J., Bhattacharyya, S., Perrone, L.A., de los Santos, K., Pu, S.-y., and Goff, S.P. (2006). Interaction of Moloney murine leukemia virus capsid with Ubc9 and PIASy mediates SUMO-1 addition required early in infection. J. Virol. *80*, 342–352.

Zack, J.A., Arrigo, S.J., Weitsman, S.R., Go, A.S., Haislip, A., and Chen, I.S. (1990). HIV-1 entry into quiescent primary lymphocytes: molecular analysis reveals a labile, latent viral structure. Cell *61*, 213–222.

Zaitseva, L., Myers, R., and Fassati, A. (2006). tRNAs promote nuclear import of HIV-1 intracellular reverse transcription complexes. PLoS Biol. *4*, e332.

Zamborlini, A., Lehmann-Che, J., Clave, E., Giron, M.-L., Tobaly-Tapiero, J., Roingeard, P., Emiliani, S., Toubert, A., de The, H., and Saib, A. (2007). Centrosomal pre-integration latency of HIV-1 in quiescent cells. Retrovirology *4*, 63.

Zennou, V., Petit, C., Guetard, D., Nerhbass, U., Montagnier, L., and Charneau, P. (2000). HIV-1 genome nuclear import is mediated by a central DNA flap. Cell *101*, 173–185.

Zhao, Z., McKee, C.J., Kessl, J.J., Santos, W.L., Daigle, J.E., Engelman, A., Verdine, G., and Kvaratskhelia, M. (2007). Subunit specific protein footprinting reveals significant structural rearrangements and a role for N-terminal LYS-14 of HIV-1 integrase during viral DNA binding. J. Biol. Chem. *283*, 5632–5641.

Zheng, R., Jenkins, T.M., and Craigie, R. (1996). Zinc folds the N-terminal domain of HIV-1 integrase, promotes multimerization, and enhances catalytic activity. Proc. Natl. Acad. Sci. U.S.A. *93*, 13659–13664.

Zhu, K., Dobard, C., and Chow, S.A. (2004). Requirement for integrase during reverse transcription of human immunodeficiency virus type 1 and the effect of cysteine mutations of integrase on its interactions with reverse transcriptase. J. Virol. *78*, 5045–5055.

Zielske, S.P., and Stevenson, M. (2005). Importin 7 may be dispensable for human immunodeficiency virus type 1 and simian immunodeficiency virus infection of primary macrophages. J. Virol. *79*, 11541–11546.

Zielske, S.P., and Stevenson, M. (2006). Modest but reproducible inhibition of human immunodeficiency virus type 1 infection in macrophages following LEDGFp75 silencing. J. Virol. *80*, 7275–7280.

Transcription, Splicing and Transport of Retroviral RNA

6

Tina Lenasi, Xavier Contreras and B. Matija Peterlin

Abstract

Studies of retroviruses have contributed greatly to our understanding of mechanisms that regulate eukaryotic gene expression. They include transcription, processing of nascent transcripts and transport of mRNA species from the nucleus to the cytoplasm. For example, analyses of viral promoters and enhancers revealed important aspects of initiation and elongation of transcription by RNA polymerase II. Sites of integration further emphasized contributions of chromatin and distal interactions between *cis*-acting sequences to the expression of viral genes and those of nearby oncogenes that lead to the transformation of target cells. At the level of DNA, they also introduced the concept of transcriptional interference for the silencing of viral 3′ long terminal repeats, where their transcription terminates and nascent transcripts become polyadenylated. Next, studies of their complex splicing patterns revealed suboptimal splice donor and acceptor sites, splicing enhancers and silencers, as well as the competition between splicing and export of incompletely spliced retroviral mRNA species from the nucleus to the cytoplasm. There, they defined RNA and protein export mechanisms for cellular and viral macromolecules. Finally, current studies of the silencing of retroviral genomes promise to elucidate mechanism for turning on and off the expression of eukaryotic genes. Of all mammalian retroviruses, HIV has been studied the most and forms the basis of this chapter. However, lessons learned from this primate lentivirus inform all other retroviruses.

Structure of the LTR

Three regions have been defined in the LTR. They contain untranslated 3′ (U3), repeated (R), and untranslated 5′ (U5) sequences (Fig. 6.1). The start site of transcription is between U3 and R regions. Whereas promoter and enhancer reside in U3, TAR and polyadenylation sequences are in R and the packaging signal as well as the major splice donor site are in U5 regions, respectively.

Promoter and enhancer sequences

The LTR is highly responsive to mitogenic stimuli and to Tat. From the 3′ to 5′ direction, the promoter contains the Initiator (Inr), TATA box and Sp1-binding sites. Next, the compact enhancer binds NF-κB, nuclear factor of activated T-cells (NF-AT) (Fig. 6.1). Additional extended enhancer sequences between positions −130 and −201 are important for HIV transcription in peripheral blood lymphocytes (PBLs) and in some T-cell lines (Kim *et al.*, 1993). They bind two activators that are highly enriched in T-cells: the lymphocyte enhancer factor (LEF), which is a lymphocyte-specific high mobility group (HMG) protein in immature B and T-cells and in mature T-cells (Waterman *et al.*, 1991; Waterman and Jones, 1990) and the thymocyte-enriched Ets-1 protein. In addition, the upstream stimulating factor (USF) binds to an E-box motif (Fig. 6.1) (Gaynor, 1992). As with many HMG proteins, LEF generates a strong (130°) bend in the DNA that could influence the structure of the promoter (Giese *et al.*, 1992). In addition, LEF contains a potent transcription activation domain (TAD)

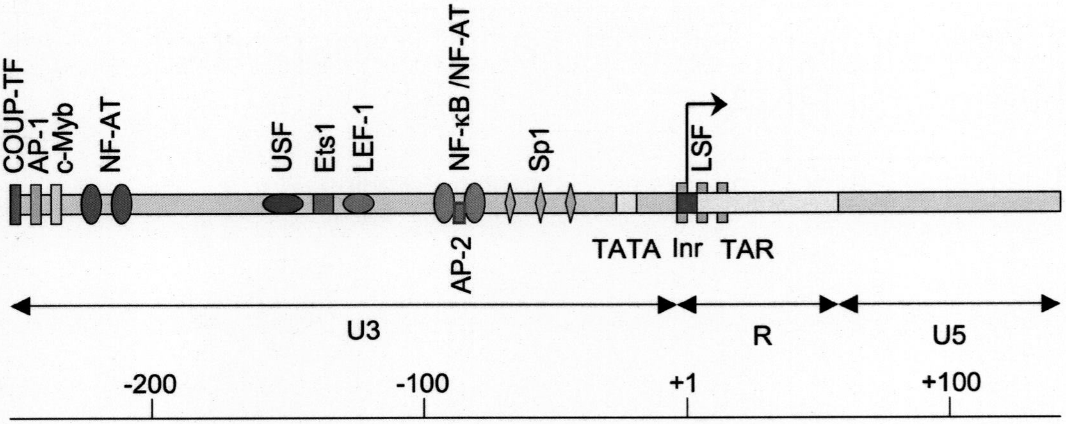

Figure 6.1 Structure of the core LTR of HIV-1. LTR is composed of three regions: 3′ untranslated region (U3) (green), repeated region (grey) and 5′ untranslated region (U5) (blue). The following host transcription factors bind to the LTR: COUP-TF (dark red), AP-1 (light blue), c-Myb (light pink), NF-AT (brown), USF (dark brown), Ets1 (purple), LEF-1 (blue), NF-κB (orange), AP-2 (blue), Sp1 (yellow) and LSF (yellow-green). TATA box and initiator element (Inr) are marked in grey and red, respectively. Arrow (shown in light blue)marks the transcription start site.

(Carlsson *et al.*, 1993). Interestingly, this TAD is preferentially active in T-cells and is strongly influenced by the context of its binding site, indicating that it may act in concert with other lymphoid-specific proteins such as Ets-1. This motif in the LTR is similar to the enhancer for the α-chain of the T-cell antigen receptor, which also binds LEF and Ets-1 (Giese *et al.*, 1992). Since they influence the relaxation of chromatin structure on the provirus, they could also ensure that the promoter is in an open configuration, or to counter repressive effects of sequences flanking the integration site. In this extended enhancer, other sequences bind additionally NF-AT, activated protein-1 (AP-1), chicken ovalbumin upstream promoter transcription factor (COUP-TF), and c-Myb (Fig. 6.1).

Downstream sequences

Sequences immediately downstream of the transcription start site contain at least three different elements: the start site region (SSR), which includes the initiator (Inr), the inducer of short transcripts (IST) (Ratnasabapathy *et al.*, 1990; Sheldon *et al.*, 1993), and TAR (Fig. 6.1). TAR forms an RNA stem–loop that recruits Tat to the LTR (see below). The IST element (Ratnasabapathy *et al.*, 1990; Sheldon *et al.*, 1993) is a DNA sequence closely overlapping TAR (–5 to +80), and mediates the synthesis of short attenuated transcripts (STs) that accumulate in the absence of Tat (Adams *et al.*, 1994; Feinberg *et al.*, 1991; Kessler and Mathews, 1992; Laspia *et al.*, 1989, 1990). They measure 55–60 nucleotides, and include the entire TAR secondary structure. Mutations in IST interfere specifically with the synthesis of STs and do not affect full-length long RNA transcripts (LTs) (Sheldon *et al.*, 1993). Fusing the IST to heterologous promoters confers upon them the ability to synthesize STs in cells (Ratnasabapathy *et al.*, 1990). Late SV40 factor (LSF) binds to this R region. Its binding to a weak site that overlaps the TATA box is inhibitory to HIV transcription (Kato *et al.*, 1991). There, LSF competes for the recruitment of TFIID to the LTR. Moreover, it cooperates with Yin Yang 1 (YY1) to inhibit viral transcription (see below).

Importance of chromatin

Like other genes, the HIV genome is packed into chromatin via nucleosomes, which are formed by four distinct histones. Two nucleosomes, Nuc-0 and Nuc-1 are positioned on the LTR from positions –413 to –253 and +1 to +155 (Van Lint *et al.*, 1994; Verdin *et al.*, 1993). Since chromatin can restrict access to activators, PIC, and/or the movement of RNAPII, the regulation of these nucleosomes is critical for HIV transcription. This regulation occurs via post-translational modifications of histones. Importantly, their reversible acetylation is associated with relaxation and tightening of chromatin. Histone acetyl transferases (HATs) and histone deacetylases (HDACs) regulate levels of acetylation of histones. Whereas HATs acetylate histones, which lead to relaxed chromatin, HDACs have the opposite effect.

Activation of HIV transcription

On the LTR, Sp1 could recruit a HAT, the p300/CREB binding protein (p300/CBP). Indeed, Sp1-mediated activation of p21 requires p300/CBP (Xiao *et al.*, 2000). In addition, mutations within Sp1 sites of the Gα (i2) promoter prevent transcriptional activation mediated by HDAC inhibitors (Yang *et al.*, 2001a). NF-κB also recruits p300/CBP for some of its transcriptional effects (Chen *et al.*, 2005).

Further, HATs, such as p300/CBP and p300/CBP-associated factor (PCAF) are recruited to the LTR by Tat (Benkirane *et al.*, 1998). However, in this context, p300/CBP acetylates Tat and dissociates it from TAR. In addition, the recruitment of HATs by the acetylated Tat protein increases histone acetylation during HIV transcription (Dorr *et al.*, 2002).

After the acetylation of histones, the ATP-dependent chromatin remodelling complex SWI/SNF is recruited to the LTR (Angelov *et al.*, 2000). In the presence of the high mobility group AT-hook protein 1 (HMGA1), the activating transcription factor 3 (ATF-3) also recruits the Brahma-related gene 1 (BRG-1) to the 3 AP1 sites located at the 3′ end of Nuc-1. Next, SWI/SNF disrupts Nuc-1, which renders HIV chromatin more accessible to other transcription factors and also facilitates transcription elongation (Henderson *et al.*, 2000; Henderson *et al.*, 2004).

Repression of HIV transcription

YY1, in cooperation with LSF represses HIV transcription (Romerio *et al.*, 1997). Indeed, LSF and YY1 act cooperatively (Coull *et al.*, 2000). YY1 also recruits HDAC1 via its glycine/alanine domain. Tat overcomes these repressive effect of YY1. Thus, whereas the expression of YY1 decreases histone acetylation, Tat increases it (He and Margolis, 2002).

CBF1 (C-promoter binding factor-1) binds to positions +148 to +153 on the LTR of latently infected cells (Tyagi and Karn, 2007). It also recruits HDACs and corepressor complexes to the LTR in newly infected cells. Additionally, in activated T-cells, where HIV transcription is most active, CBF1 levels are low.

Histone methylation also regulates chromatin remodelling. Indeed, the histone methyltransferase Suv39H1 trimethylates K9 and K27 in histone H3 and recruits the heterochromatin associated protein 1 (HP1), which marks transcriptionally inactive proviruses (du Chene *et al.*, 2007; Marban *et al.*, 2007). Thus, when levels of HP1 are decreased by RNA interference, HIV transcription is increased in latently infected cells (du Chene *et al.*, 2007). In accordance with these results, 5′-azacytidine, which demethylates DNA, also promotes HIV transcription (Tanaka *et al.*, 2003).

Initiation and early events

TFIIH and assembly of the PIC

HIV mRNA synthesis is a complex multistage process that requires cooperative actions of viral and cellular proteins. Multiple checkpoints have to be overcome for the efficient expression of viral genes. Transcription begins with initiation. At this stage, the TATA-Binding Protein (TBP) recognizes the TATA box. This binding initiates the recruitment of the TBP-associated factors (TAFs), which assemble into the multisubunit TFIID complex. Following this recruitment, additional general transcription factors (GTFs), which include TFIIA, TFIIB, TFIIE, TFIIF, and TFIIH, associate with RNAPII to form the PIC (Fig. 6.2A).

The function of RNAPII is regulated via its C-terminal domain (CTD), which in humans contains 52 conserved heptapeptide repeats (YSPTSPS) (Garriga and Grana, 2004; Price, 2000). For initiation, the CTD is not phosphorylated and binds the Mediator complex. For elongation, the CTD must be heavily phosphorylated. This posttranslational modification is the consequence of a complicated process. During promoter clearance and early RNA synthesis, Cdk7 from THIIH phosphorylates serines at position 5 (S5) in the CTD (Fig. 6.2B) (Serizawa *et al.*, 1995; Shiekhattar *et al.*, 1995). TFIIH also mediates ATP-dependent strand separation via its 3′-5′ helicase activity at the transcription start site

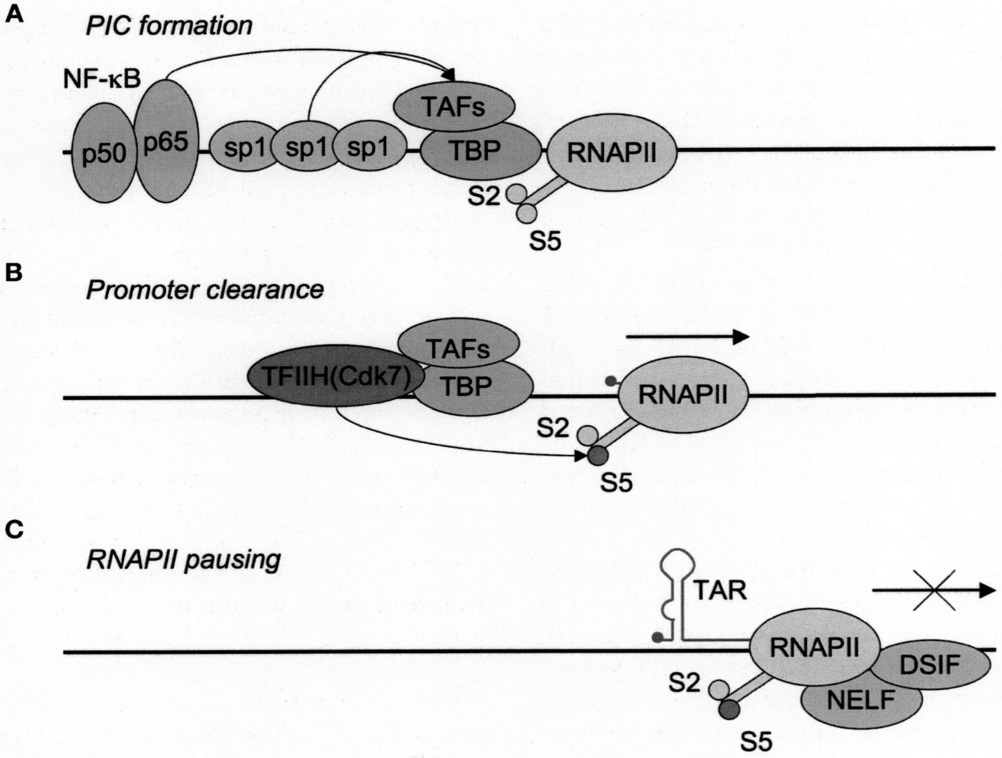

Figure 6.2 Initiation of transcription. Initiation of transcription on the HIV genome is characterized by 3 major steps. (A) First, NF-κB composed of p50 and p65 subunits and Sp1 are required for the formation of the pre-initiation complex (PIC) and the recruitment of TBP and TAFs. (B) Then, TFIIH binds to the promoter and phosphorylates S5 in the CTD of RNAPII. This event stimulates promoter clearance and synthesis of the first nucleotides of RNA. (C) Finally, the negative transcription elongation factors NELF and DSIF are recruited and arrest RNAPII in close proximity to the promoter

to open DNA from positions –9 to +2 (Holstege *et al.*, 1997; Holstege *et al.*, 1996). In addition, it inhibits the arrest of early transcription intermediates (Moreland *et al.*, 1999; Spangler *et al.*, 2001). S5 phosphorylation of the CTD also facilitates the recruitment of capping enzymes (see below for details).

NF-κB

Activation of target cells is critical for HIV transcription. Indeed, the LTR contains binding sites that recruit several transcription factors, which are induced in activated T-lymphocytes or tissue macrophages: NF-AT, AP-1, NF-κB, among others. NF-κB, which is composed of p50 and p65, is sequestered in the cytoplasm by IκB. The activation of target cells induces the phosphorylation, ubiquitylation and proteasomal degradation of IκB and thus translocation of NF-κB to the nucleus, where it binds to the enhancer and stimulates initiation and elongation of HIV transcription (Nabel and Baltimore, 1987).

Sp1

In addition to HATs and HDACs, Sp1 binds several TAFs and GTFs (Dynlacht *et al.*, 1991; Gill *et al.*, 1994; Hoey *et al.*, 1993). Moreover, Sp1 interacts with other DNA-bound proteins, including distal Sp1 monomers, YY-1, and NF-κB. These interactions promote the formation of the PIC.

RNAPII stalling and N-TEF

After promoter clearance, N-TEF induces RNAPII stalling and inhibits elongation of transcription (Fig. 6.2C). N-TEF contains two subcomplexes. First, dichloro-1-b-D-ribofuranosylbenzimidazole (DRB)-sensitivity inducing factor (DSIF) is required for effects of DRB, which inhibits the kinase activity of P-TEFb to block RNAPII elongation (Wada *et al.*, 1998; Yamaguchi *et al.*, 1999). It contains two subunits similar to yeast transcription factors Spt4 and Spt5. The negative elongation factor (NELF) is also necessary for effects of DSIF. Although neither alone displays major effects, together they arrest RNAPII in *in vitro* transcription assays (Renner *et al.*, 2001). Thus, N-TEF blocks RNAPII elongation in close proximity to the promoter.

Premature termination and Pcf11

Polyadenylation cleavage factor 11 (Pcf11) is another factor that contributes to early blocks in transcription elongation. It promotes the dissociation of engaged RNAPII from the DNA template (Buratowski, 2005; Rosonina *et al.*, 2006). Indeed, its depletion by siRNA enhances HIV transcription (Zhang *et al.*, 2007). In addition, chromatin immunoprecipitation (ChIP) assays demonstrated that Pcf11 is recruited to the LTR. These studies suggested that the pause induced by DSIF and NELF allows Pcf11 to dismantle the elongation complex.

Elongation of transcription

Studies with Tat revealed that P-TEFb is required to overcome early blocks after promoter clearance thus allowing for efficient elongation of HIV transcription. In this regard, two major steps exist during viral transcription and they are defined by differential recruitment of P-TEFb to the LTR. In the early phase, DNA-bound activators, especially NF-κB, recruit P-TEFb in activated cells and lead to an inefficient transcription (Fig. 6.3B). Next, Tat recruits P-TEFb to TAR in the late, more productive phase of HIV transcription (Fig. 6.3C).

P-TEFb

P-TEFb is composed of a regulatory C-type cyclin subunit (CycT1, CycT2 or CycK) and a catalytic subunit, the cyclin-dependent kinase 9 (Cdk9). In addition, different isoforms of CycT2 and Cdk9 exist in cells. Once recruited to the transcription complex, Cdk9 phosphorylates serines at position 2 (S2) in the CTD of RNAPII (Fig. 6.3). This post-translational modification, along with S5 phosphorylation, increases the diameter and rigidity of the CTD. During this process, most subunits of PIC, which include the Mediator and GTFs, are released from RNAPII. This phosphorylation also acts as a scaffold for the recruitment of splicing and polyadenylation factors. Indeed, the phosphorylation of S2 by P-TEFb marks the transition from initiation to elongation of transcription. Of note, P-TEFb also phosphorylates the C terminal repeats (CTRs) in Spt5 (Fig. 6.3) (Kim and Sharp, 2001). Thus modified, DSIF functions as a positive elongation factor (Yamada *et al.*, 2006). Finally, P-TEFb phosphorylates the RD subunit of NELF (NELF-E) (Fig. 6.3). This phosphorylation releases NELF from double stranded RNA and results in increased transcription elongation (Fujinaga *et al.*, 2004).

Regulation of P-TEFb

P-TEFb is found in two complexes in cells (Fig. 6.3A) (Nguyen *et al.*, 2001; Yang *et al.*, 2001b). The small, active complex (SC) is composed of CycT1 and Cdk9 alone or together with the bromodomain-containing protein 4 (Brd4). In contrast, the large inactive complex (LC) of about 500 kDa contains Hexamethylene bisacetamide (HMBA)-inducible protein 1 or 2 (HEXIM1 or HEXIM2) and 7SK snRNA in addition to P-TEFb (Michels *et al.*, 2003; Yik *et al.*, 2003).

Brd4 is a member of the BET family of proteins that contain two tandem bromodomains (BDI and BDII) and an external (ET) domain (Wu and Chiang, 2007). Since Brd4 interacts with the active form of P-TEFb (Yang *et al.*, 2005), this finding suggested that Brd4 is involved in activation of RNAPII elongation. Indeed, Brd4 increases S2 phosphorylation by Cdk9 and enhances HIV transcription. Brd4 can recruit P-TEFb to the LTR by two distinct mechanisms. First, it enhances the recruitment of active P-TEFb to acetylated chromatin. Indeed, pre-treatment of cells with HDAC inhibitors facilitated the recruitment of Brd4 and Cdk9 to an integrated LTR (Jang *et al.*, 2005). Second, Brd4 is found in the Mediator complex. Indeed, one of its subunits, TRAP220 could be found only in pull-down complexes containing the

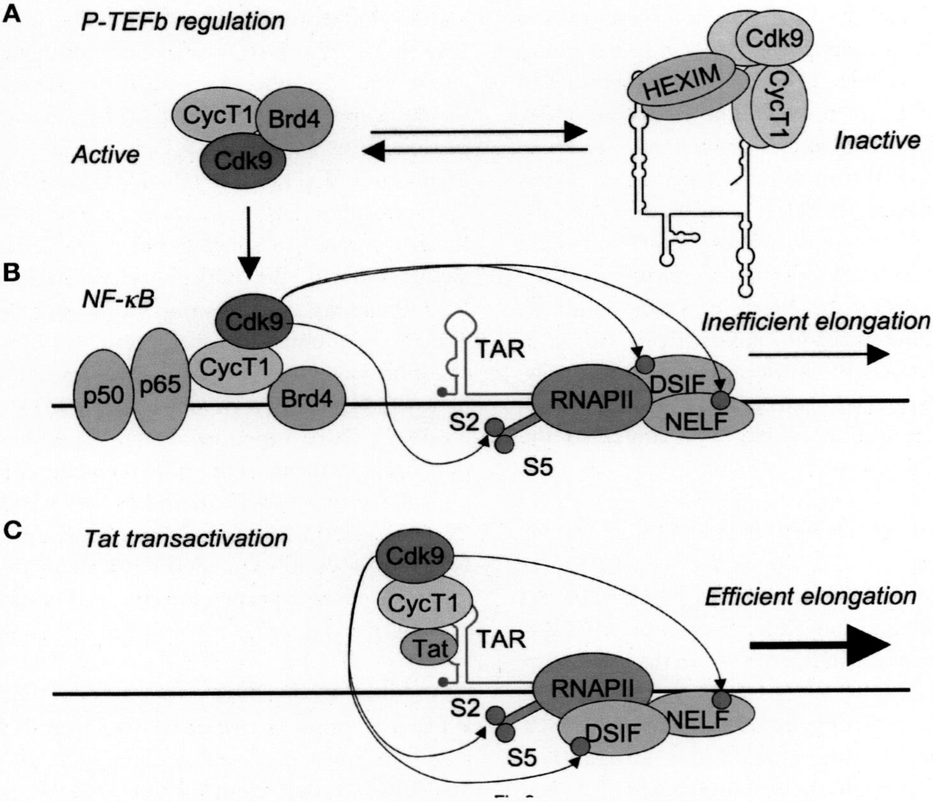

Figure 6.3 Elongation of transcription. P-TEFb, composed of CycT1 and Cdk9 is required for elongation of transcription on the viral genome. (A) P-TEFb exists in two complexes. The inactive form, bound to HEXIM1/2 and 7SK snRNA and the active form, which is free or bound to Brd4. (B) In early steps of the viral replicative cycle, the active form of P-TEFb is recruited to the LTR via Brd4 and NF-κB. This recruitment leads to the inefficient elongation of transcription. (C) Later the recruitment of Tat and P-TEFb to TAR leads to more efficient transcription elongation. P-TEFb phosphorylates S2 in the CTD of RNAPII and also phosphorylates subunits of DSIF and NELF. Thus phosphorylated, NELF disengages from RNAPII and phosphorylated DSIF acts as a positive elongation factor.

wild-type Cdk9 protein but not its mutated S175D derivative, which is catalytically active but does not bind Brd4, suggesting that these interactions depend on Brd4 (Yang *et al.*, 2005). It is possible that Brd4 also helps to recruit the Mediator to the LTR. After the initiation of transcription and S5 phosphorylation by TFIIH and RNAPII arrest, the Mediator could then recruit P-TEFb to enhance S2 phosphorylation and transcription elongation. Indeed, since transcription from the LTR using Cdk9-depleted nuclear extracts was restored using the wild-type Cdk9 but not mutant Cdk9-S175D proteins, it requires Brd4 in the absence of NF-κB and Tat (Yang *et al.*, 2005).

In conclusion, Brd4 can mediate the recruitment of P-TEFb to DNA and could account for transcription elongation observed in the absence of NF-κB and Tat. However, large amounts of Brd4 can also compete with Tat for binding to P-TEFb. Indeed, expression of a peptide corresponding to the P-TEFb binding domain of Brd4 interfered with effects of Tat on the LTR (Bisgrove *et al.*, 2007).

In the LC, Cdk9 is enzymatically inactive. HEXIM1 was originally identified as the most induced gene following the administration of the powerful differentiation agent HMBA to vascular smooth muscle cells (VSMC) (Kusuhara *et al.*, 1999). Another isoform, HEXIM2 is also found in the LC (Byers *et al.*, 2005). 7SK snRNA is one of the most abundant RNA species in cells. However, its function remained unknown for 10 years. Interestingly, the inactivation of P-TEFb by HEXIM1 and 7SK snRNA contributes significantly to the control of cell growth and differentiation. For example, cell growth signals release P-TEFb from the LC during cardiac hypertrophy, a disease characterized by the enlargement of myocytes due to a global increase of RNA synthesis (Sano *et al.*, 2002). In addition, following stress signals such as UV, actinomycin D or DRB, the LC falls apart and releases P-TEFb, which stimulates transcription (Nguyen *et al.*, 2001; Yang *et al.*, 2001b). It was also demonstrated that HMBA induces a transient release of active P-TEFb (He *et al.*, 2006). This transient release is due to the phosphorylation of HEXIM1 via the Akt pathway and is accompanied by the recruitment of active P-TEFb to the LTR (Contreras *et al.*, 2007). Importantly, the overexpression of HEXIM1 inhibits HIV transcription and replication, which underlines the importance of P-TEFb in viral replication (Fraldi *et al.*, 2005; Shimizu *et al.*, 2007). Of interest, Tat can interact with P-TEFb directly in the LC and recruits active P-TEFb to TAR (Barboric *et al.*, 2007).

Recently, two new components of the LC have been identified. One study revealed that P-TEFb is associated with a network of proteins including a previously uncharacterized protein BCDIN3 (Bicoid-interacting protein 3) (Jeronimo *et al.*, 2007). BCDIN3 is a methyltransferase that caps 7SK snRNA. Thus, it is also called the methyl phosphate capping enzyme, MEPCE. In addition, LARP7 (or PIP7S) was shown to be associated with the LC (He *et al.*, 2008; Krueger *et al.*, 2008). LARP7 provides RNA 3′ end protection and thus stabilizes the binding of HEXIM1 to 7SK snRNA. Knockdown of this protein leads to the disruption of the LC and a rapid degradation of 7SK snRNA. The release of P-TEFb induced by LARP7 knock-down then increases Tat-independent and -dependent HIV transcription. In addition, the dissociation of the LC leads to the formation of a different complex, where 7SK snRNA interacts with several hnRNPs. Thus, in addition to LARP7, levels of BCDIN3 and RNPs should influence HIV transcription.

Role of NF-κB

The ability of NF-κB to stimulate elongation of transcription depends on P-TEFb (Fig. 6.3B), whose activity can be blocked by a dominant-negative Cdk9 protein and chemical inhibitors of Cdk9 (Barboric *et al.*, 2001). In addition, ChIPs revealed that NF-κB recruits P-TEFb and increases transcriptional elongation (Williams *et al.*, 2007). NF-κB is particularly critical for early HIV transcription and for the synthesis of Tat.

Role of Tat

Tat is expressed from multiply spliced viral transcripts and recruits P-TEFb to TAR. This results in amplification of effects of NF-κB and sustains high levels of viral replication. For this reason, Tat is an indispensable viral protein (Jeang and Gatignol, 1994). When Tat is mutated, no detectable progeny virions are produced. Tat is a nuclear protein of 14 kDa containing of up to 101 residues encoded by two exons. A shorter 72 residues 'one-exon' Tat possesses all the transcriptional activating properties of the full-length protein. It is composed of several functional domains: cysteine-rich (aa 20 to 31) and core domains (aa 31 to 47) involved in binding to CycT1 and transcription activation, a basic arginine-rich motif (ARM) (aa 48 to 57), which binds TAR and contains a nuclear localization signal (Mann and Frankel, 1991; Vives *et al.*, 1997) and a tripeptide arginine-glycine-aspartic acid motif (RGD, aa 78 to 80), which interacts with integrins (Zocchi *et al.*, 1997).

3D structure of Tat was studied by nuclear magnetic resonance (NMR) (Bayer *et al.*, 1995). This work demonstrated that the N-terminal region (V4, P14) is found near the hydrophobic core and the glutamic acid-rich domain. Cysteine-rich and basic domains are very flexible and form loops (Bayer *et al.*, 1995; Peloponese *et al.*, 2000). Thus, Tat is rather unstructured. However, it might become more structured with its interacting partners (Long and Crothers, 1999; Seewald *et al.*, 1998). For example, interaction with TAR is accompanied by the formation of a short α-helix in the ARM of Tat (Loret *et al.*, 1992).

Tat binds to the 5′ bulge in TAR via its ARM from positions 49 to 57. But this interaction is not sufficient for transcriptional activation *in vivo* (Fig. 6.3C). N-terminal cysteine-rich and core regions bind CycT1 from P-TEFb. In CycT1, there is the Tat-TAR recognition motif (TRM), which also binds TAR and where a cysteine at position 261 plays a critical role for binding Tat (Garber *et al.*, 1998). Thus Tat and P-TEFb bind to TAR cooperatively. In murine cells, which are not permissive for HIV transcription, CycT1 lacks this C261 (Wei *et al.*, 1998).

The assembly of the RNA–protein complex between Tat, TAR and P-TEFb is regulated *in vivo*. Indeed, Tat is expressed in very small amounts in infected cells. The stoichiometry is of about 100 to 1000 viral transcripts containing TAR to one Tat molecule (Kiernan *et al.*, 1999). In its unmodified form, the half-life of the Tat–TAR complex is of about 41 seconds *in vitro*, which is not compatible with the rapid kinetics required to allow HIV transcription. Actually, p300/CBP and PCAF can acetylate Tat on lysines at positions 28 (by PCAF) and 50 (by P300/CBP) (Chan and La Thangue, 2001; Kiernan *et al.*, 1999). These two modifications have different functional consequences. Whereas the acetylation of K28 can dissociate Tat from P-TEFb, that of K50 weakens interactions between Tat and TAR (Kiernan *et al.*, 1999). Acetylated Tat, which is released from P-TEFb, recruits PCAF to the elongating RNAPII, thus facilitating chromatin remodelling (Bres *et al.*, 2002; Dorr *et al.*, 2002). At termination, Tat can be deacetylated by sirtuin 1 (SIRT1) (Pagans *et al.*, 2005) which allows for its recycling and reassembly with TAR. In addition, the phosphorylation of Cdk9 reinforces the RNA–protein complex between P-TEFb, Tat and TAR (Garber *et al.*, 2000).

Role of other viral factors

Vpr

Vpr is an accessory protein of HIV, whose function is important for efficient viral replication in non-dividing cells, notably macrophages (Ayyavoo *et al.*, 1997; Connor *et al.*, 1995; Eckstein *et al.*, 2001). Vpr associates with budding virions and blocks newly infected cells in the G2 phase of the cell cycle, which provides the best conditions for LTR-directed transcription and which cooperates with effects of Tat (Hrimech *et al.*, 1999; Jowett *et al.*, 1995). Vpr can bind to p300/CBP (Felzien *et al.*, 1998; Kino *et al.*, 2002), the glucocorticoid receptor (GR) (Sherman *et al.*, 2000) and even Tat and/or CycT1 to activate transcription (Sawaya *et al.*, 2000). In addition, Vpr can activate NF-κB (Varin *et al.*, 2005).

Nef

Nef is another accessory protein of HIV. It contains a myristoylated residue, which allows its localization to the membrane. Thus localized, Nef potentiates several signalling pathways, notably those within the virological synapse, which also leads to the activation of NF-κB, AP-1 (Varin *et al.*, 2003), NF-AT (Manninen *et al.*, 2000) and signal transducer of transcription-1 (STAT1) (Federico *et al.*, 2001).

Transcription from unintegrated DNA

Unintegrated DNA accumulates in T-lymphocytes, lymphoid and brain tissues, as well as in cultured cells. During the asymptomatic phase of HIV infection, levels of non-integrated proviruses can reach 99% of total viral DNA. The non-integrated HIV DNA exists in three different forms: 1-LTR and 2-LTR circles and linear DNA. Thus, transcription from unintegrated DNA could account for a non-negligible part of viral gene expression. However, the synthesis of viral proteins is inefficient in absence of integration. Indeed, only small amounts of viral proteins were detected from this episomal viral DNA (Spina *et al.*, 1995; Wu and Marsh, 2001). Other studies demonstrated that circular forms of DNA can persist in non-dividing cells and sustain reporter gene expression to a level equivalent to wild-type vectors (Kelly *et al.*, 2008; Saenz *et al.*, 2004).

Transcription in different cell types

HIV transcription differs considerably in different cell types (Rohr *et al.*, 2003). In each cell type, different combinations of *cis*-acting sequences and *trans*-acting factors on the LTR influence the expression of viral genes. Cellular activators and inhibitors as well as components of the microenvironment such as cytokines and the differentiation state of infected cells are key components for the modulation of the early phase of HIV transcription.

In T-lymphocytes, cellular activation, followed by nuclear translocation of NF-κB and

NF-AT are essential for viral replication. The absence of transcription in quiescent memory T-cells during proviral latency is in great part due to the absence of such factors (see below). In thymocytes, various cytokines stimulate NF-κB which is required for LTR-driven transcription (Chene *et al.*, 1999).

The state of differentiation of monocytes, macrophages and dendritic cells (DCs) is also critical for HIV transcription. Whereas mature DCs and monocytes (Sonza *et al.*, 1996) are not permissive, immature DCs and macrophages are permissive for viral replication (Granelli-Piperno *et al.*, 1998). This block in mature DC is at the level of transcription and is independent of NF-κB and Sp1 (Bakri *et al.*, 2001). Indeed, during macrophage differentiation, levels of CycT1 increase and thus facilitate HIV transcription (Liou *et al.*, 2002). Nuclear factor interleukin-6 (NF-IL6) and USF also play important roles in these cells.

Interestingly, in microglial cells, which support viral replication in the brain, Sp1 plays an essential role, anchoring NF-IL6, CREB/ATF, and COUP-TF. Differentiation by a combination of cytokines such as IFN-γ, IL-1α, and TNF-α, is also required for viral replication, which confirms the important role of NF-κB in these cells (Janabi *et al.*, 1998).

Processing of HIV pre-mRNA

Capping of nascent transcripts

Similar to the host genes transcribed by RNAPII, HIV nascent transcripts undergo processing including capping, splicing and polyadenylation. These processes occur mostly cotranscriptionally by actions of corresponding host factors. First, pre-mRNA is capped on its 5′ end as soon as its length has attained 19–22 nucleotides. This step of cotranscriptional processing is facilitated by RNAPII pausing due to N-TEF. The pausing allows capping enzymes, RNA guanylyltransferase and RNA (guanine-N7) methyltransferase, to synthesize m7G(5′)ppp(5′)N cap in three subsequent enzymatic reactions (Furuichi and Shatkin, 1977). The capping reaction is also stimulated by Tat via TAR-dependent phosphorylation of the CTD by P-TEFb (Chiu *et al.*, 2002; Zhou *et al.*, 2003).

Splicing of nascent transcripts

Alternative use of splice acceptor and splice donor sites

Further processing of viral pre-mRNA gives rise to a variety of transcripts and consequently, many proteins due to the alternative use of splice sites in the 9-kb-long primary transcript, which contains nine open reading frames (ORFs) (Fig. 6.4). HIV uses the splicing machinery of the host cell for the splicing of its pre-mRNA. Thus, the appropriate *cis*-elements are present in nascent HIV transcripts. These include 5′ splice donor (SD), 3′ splice acceptor (SA) and branch point (BP) sites. The first two are positioned at the 5′ and 3′ ends of introns, respectively, and the last one is some 10 nucleotides upstream of the SA site. Preceding the SA site is another important signal, the polypyrimidine tract (PPT), which is composed of a stretch of pyrimidines. The splicing of introns from the HIV pre-mRNA is sequential, with the 5′ intron being removed before its 3′ neighbour (Beyer and Osheim, 1988). It has been demonstrated that the removal of introns in the order of their appearance has the same directionality even after the insertion of heterologous exons in different positions of the viral genome (Bohne *et al.*, 2005).

9kb long nascent HIV mRNA contains five SD and nine SA sites (Fig. 6.4). Spliced viral transcripts are divided into three major classes according to the frequency of splicing (Purcell and Martin, 1993). The first is a class of unspliced mRNAs, the second is a class of singly spliced mRNAs (4 kb size class) lacking *gag–pol* coding region and the third is a class of multiply spliced mRNAs lacking *gag–pol* and *env* coding regions (1.8 kb size class). The unspliced mRNA serves as the viral genomic RNA for encapsidation and as the template for the expression of Gag–Pol polyproteins. The unspliced mRNAs together with 4 kb size class belong to Rev-dependent and the 1.8 kb size class to Rev-independent mRNA species (Fig. 6.4). In contrast to Rev-dependent, Rev-independent transcripts do not require Rev for their export to the cytoplasm.

The major SD site is located upstream of the *gag* ORF and the first SA site at the end of the *pol* ORF, which results in the removal of the *gag–pol* region as an intron from singly and multiply spliced mRNAs (Pollard and Malim, 1998). The

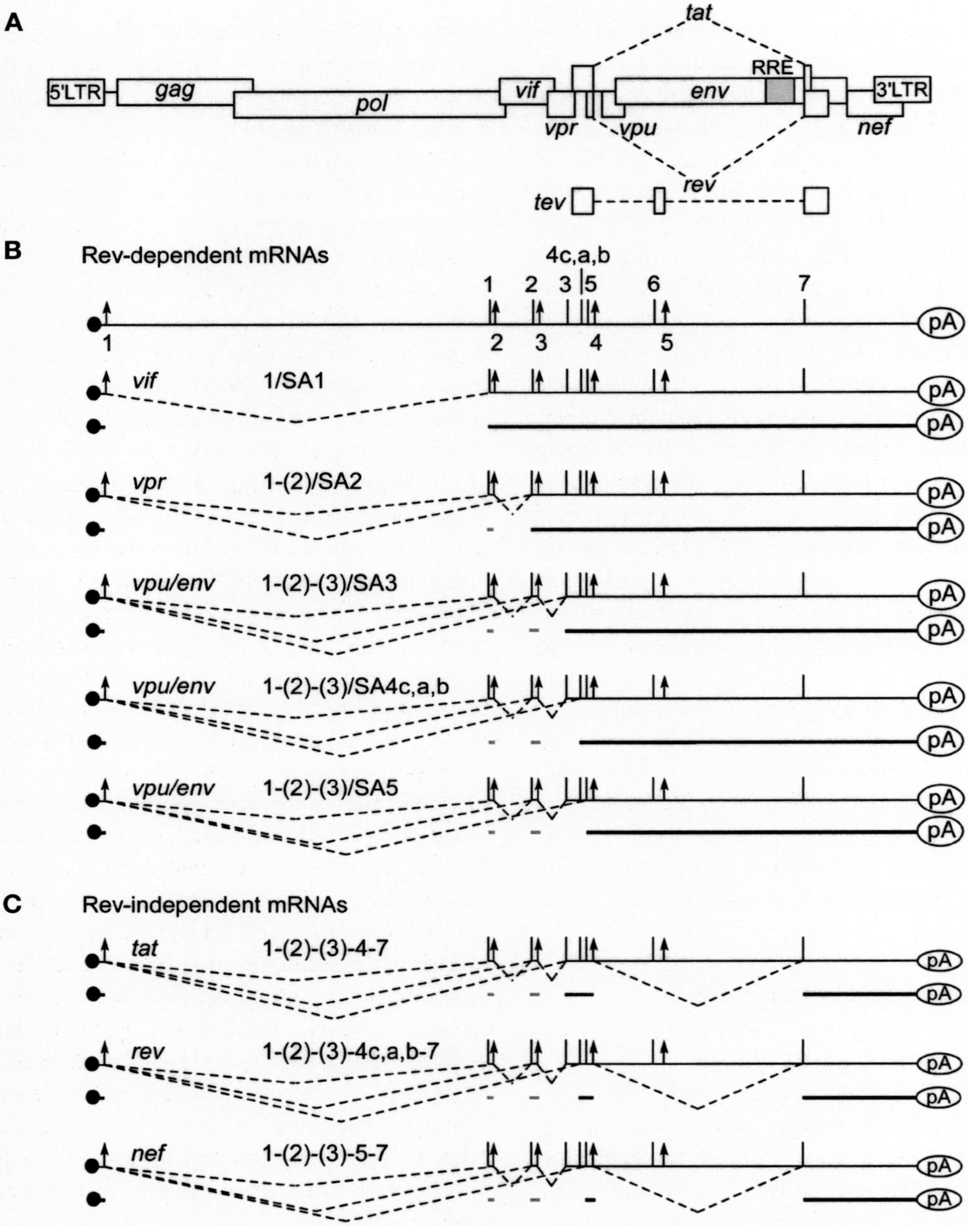

Figure 6.4 Splicing of the HIV pre-mRNA. (A) Schematic representation of the genomic organization of HIV. 5′ LTR, 3′LTR and ORFs are depicted as white boxes. Grey box represents the RRE. (B) and (C) Black circle represents the 5′ cap and circled pA represents the poly(A) tail. Splice donor (SD) and splice acceptor (SA) sites are illustrated as vertical arrows and vertical lines, respectively. Dotted lines encompass regions, which are spliced out. Identities of mRNAs resulting from individual splicing patterns are indicated above each scheme. The exon inclusion in processed mRNAs is described as follows: the numbers represent included exons before the final splicing event, which is designated in the case of incompletely spliced mRNAs with a dash followed by the last SA site used. After this splice site, the rest of nascent mRNA is included in the processed mRNA molecule. Numbers in brackets represent alternatively included exons 2 and 3. Below the splicing schemes are the corresponding processed mRNAs illustrated with thick black and grey lines representing included regions and alternatively included exons 2 and 3, respectively. (B) Rev-independent mRNAs resulting from incomplete splicing of nascent mRNA. The upper panel represents unspliced pre-mRNA with numbers above the scheme corresponding to SA sites and numbers below the scheme to SD sites. (C) Rev-independent mRNAs resulting from complete splicing of nascent mRNA.

use of other SD and SA sites depends on their strength. Specifically, SD sites contain mostly a conserved consensus sequence. However, SA sites in HIV are weak due to their short PPTs with interspersed purines and non-consensus BPs (Amendt *et al.*, 1995; Dyhr-Mikkelsen and Kjems, 1995; O'Reilly *et al.*, 1995; Si *et al.*, 1997; Staffa and Cochrane, 1994). The least efficiently used is the first SA site (SA1) while the other SA sites are of similar strength (O'Reilly *et al.*, 1995). Two of the splice sites (SD5 and SA6) are present only in HXB2 strain of HIV. The use of these splice sites results in the inclusion of exon 6, which produces transcripts containing the first exon of *tat*, a small portion of *env* and the second exon of *rev* (Benko *et al.*, 1990; Salfeld *et al.*, 1990) (Fig. 6.4). This hybrid mRNA encodes a protein Tev, expression of which is regulated during the viral life cycle similarly to the expression of other regulatory viral proteins. In the study of Salfeld *et al.*, Tev exhibited only Tat activity (Salfeld *et al.*, 1990), in contrast, Benko *et al.* demonstrated its Tat and Rev activities (Benko *et al.*, 1990).

Three SA sites (SA4a, SA4b and SA4c) located upstream of the *rev* ORF and one SA site (SA5) located at the beginning of the *rev* ORF (Fig. 6.4) are preceded by eight BPs positioned close to one another (Swanson and Stoltzfus, 1998). The close proximity of BPs and other *cis*-acting elements involved in the splicing process results in the competition for binding of splicing factors and consequently, in alternative use of the four splice sites.

To increase the complexity of HIV transcripts, exon 2 (between SA1 and SD2) and exon 3 (between SA2 and SD3) are alternatively included in the processed mRNAs from 4 kb as well as from 1.8kb size class (Fig. 6.4). This occurrence is due to non-consensus SD and SA sites (Furtado *et al.*, 1991; Muesing *et al.*, 1987; Robert-Guroff *et al.*, 1990). Exons 2 and 3 are non-coding leader exons and by testing expression vectors with different leader exons in HeLa-CD4$^+$ cells it has been demonstrated that the presence of exon 2 stimulated mRNA and protein expression. On the other hand, the presence of exon 3 reduced it (Krummheuer *et al.*, 2001). Furthermore, the authors attributed these effects to the exon-specific nuclear degradation of non-polyadenylated mRNA species. The outcome of alternative inclusion of these two exons in mRNAs together with alternative use of other splice sites result in over 40 differentially spliced HIV mRNAs (Fig. 6.4).

Splicing enhancer and silencer elements

The efficiency of intron removal does not depend solely on the strength of their splice sites, but also on the presence of various *cis*-acting elements in weak exons and neighbouring introns. These elements either enhance splicing by promoting the use of weak splice sites or inhibit splicing by preventing the use of weak splice sites, which are in most cases upstream SA sites. In the first case, they are called exonic splicing enhancers (ESEs) and in the second case exonic splicing silencers (ESSs) or intronic splicing silencers (ISSs). While ESEs are divided in the two major types, which are the more common purine-rich and non-purine-rich, ESSs and ISSs are highly diverse (Zheng, 2004). All three classes of splicing elements reside at different locations in the HIV genome and they are functional when transcribed into pre-mRNA.

ESEs bind SR proteins, which are required for the removal of constitutively spliced exons, but they also regulate alternative splicing. They contain N-terminal RNA binding and C-terminal RS (serine-arginine rich) domains for interactions with other proteins. By binding to ESEs in weak exons, SR proteins facilitate non-consensus SD or SA site to be recognized by the spliceosome (Graveley, 2000). On the other hand, ESSs rapidly associate with proteins from hnRNP A/B family, among which hnRNP A1 is most often described to repress splicing by preventing the proper spliceosomal complex assembly (Burd and Dreyfuss, 1994).

ESEs and ESSs in the HIV genome were identified according to their specific sequence and by mutating parts of the viral genome. The first ESE was discovered in the terminal *tat–rev* exon and its structure revealed a typical splicing enhancer sequence (GAA)3 (Staffa and Cochrane, 1995). Another ESE termed ESE3 was found upstream of the (GAA)3 element in the same exon (Tange *et al.*, 2001). Bidirectional ESE (GAR element) was identified downstream of SA5 site (Caputi *et al.*, 2004). The authors demonstrated that the GAR element was required for the recruitment of U1 snRNP to the SD4 site, a step necessary for splicing of Tat, Rev and Nef mRNAs. Similar to splicing enhancer elements, splicing silencers were discovered in different exons in the HIV genome. The first ESS (ESS2) was found in the first Tat exon and it inhibited the use of the upstream SA3 site (Amendt *et al.*, 1994). Later another ESS (ESS2p) was found in the same exon and it also repressed splicing at SA3 site (Jacquenet *et al.*, 2001). However, unlike ESS2 and other ESS in HIV genome, which were shown to bind proteins from hnRNP A/B family (Bilodeau *et al.*, 2001; Caputi *et al.*, 1999; Yang *et al.*, 2001b), ESS2p inhibited splicing by binding hnRNP H. Next, an ESS in exon 3 termed ESSV was shown to repress splicing at SA2 site, resulting in reduced levels of Vpr mRNA (Bilodeau *et al.*, 2001). Inactivation of ESSV by mutation resulted in 95% decreased viral replication due to increased inclusion of exon 3 and decreased accumulation of unspliced viral transcripts (Madsen and Stoltzfus, 2005).

Elements, which promote splicing, often reside close to the ones that inhibit it. This situation was demonstrated for the first Tat exon downstream of SA3 site. Here, an ESS was found to counteract a juxtaposed ESE (ESE2). Binding of hnRNP A1 to the ESS prevents the binding of SR proteins to ESE2 (Zahler *et al.*, 2004). A similar case was described for the terminal exon, where they identified a bipartite ESS (ESS3) just downstream of above described ESEs (Staffa and Cochrane, 1995; Tange *et al.*, 2001). Furthermore, it was demonstrated that hnRNP A1 inhibits exon inclusion by binding to the high-affinity site in ESS3 followed by its association with the upstream region containing ESE3. Although ESE3 is recognized by SR proteins SC35 and SF2/ASF, only the latter prevents the binding of hnRNP A1 to ESS3 (Zhu *et al.*, 2001). In addition, an ISS was identified upstream of SA7 site in the 3′ terminal part of the last intron (Tange *et al.*, 2001). Later, a more global picture of interactions between enhancer and silencer elements around SA7 site was illustrated (Marchand *et al.*, 2002). Here, the authors discovered a novel hnRNP A1 binding site that overlaps ESE3/(GAA)3 region. In addition, they established that ISS, ESE3/(GAA)3 and ESS3 are located in three different pre-mRNA stem–loops, where the positions of the *cis*-acting elements allow for the cooperative binding of hnRNP A1 to both repressor elements and competition of positive and negative binding factors for their binding sites. The region containing ESS overlapping ESE3/(GAA)3 enhances or represses the use of SA7 site depending on the cellular ratio of the SF2/ASF and hnRNP A1. The authors therefore named it a Janus splicing regulator. Another such example was described for exon 2, where an ESS (GGGG motif) and an ESE proximal to SD2 were identified. The former decreases exon 2 inclusion and consequently Vif expression and the latter facilitates exon 2 inclusion resulting in increased Vif expression (Exline *et al.*, 2008). Competition between both elements allows for the proper ratio of Vif mRNA to unspliced mRNAs in infected cells.

Changed balance of cellular splicing factors during viral replication

Studies identifying cellular genes differentially expressed during HIV infection revealed changes in levels of certain splicing factors. For example, 2 days after infection of H9 cells, 2- to 3-fold increased SC35 mRNA and concomitant increased SC35 protein levels were detected (Maldarelli *et al.*, 1998). Another group investigated differentially expressed genes 60 hours after the infection of MT-4 cells with NL4–3 strain of HIV (Ryo *et al.*, 2000). Among other down-regulated mRNAs, they identified another member of the SR protein family, splicing factor 9G8. Next, Fukuhara *et al.* demonstrated that HIV gene expression in Flp-In293 cells decreased overall SR protein activity (Fukuhara *et al.*, 2006), which correlated with low levels of HIV replication in these cells. The study by Dowling *et al.* focused on changes in the expression of different cellular splicing factors in infected compared to uninfected macrophage cultures (Dowling *et al.*, 2008). They detected great up-regulation of the SR protein SC35 and down-regulation of hnRNPs of the A/B and H families in the first or second week after infection. In addition, they observed increased cytoplasmic SC35 expression after long-term infection. In conclusion, viral infection triggers changes in levels of splicing factors that probably contribute to the establishment of a proper environment for viral persistence in infected cells.

Polyadenylation of nascent transcripts

The final step in the processing of nascent transcripts before they are transported to the cytoplasm is polyadenylation. Similar to capping and splicing, host cellular proteins are required for endonucleolytic cleavage and polyadenylation of viral pre-mRNAs. *Cis*-acting elements involved in these processes are highly conserved polyadenylation (poly(A)) signal (AAUAAA) located 10–30 nucleotides upstream of the cleavage site and GU-rich region without a consensus sequence immediately downstream of the cleavage site (Levitt *et al.*, 1989). The first step in the polyadenylation process is binding of CPSF (cleavage polyadenylation specificity factor) to the AAUAAA sequence, followed by binding of CstF (cleavage stimulation factor) to the downstream element (Bienroth *et al.*, 1991; Takagaki *et al.*, 1990).

HIV genome contains two identical LTRs and each of them has a poly(A) site. For a proper expression of the virus, the poly(A) site in the 5′LTR has to be ignored or transcription would not proceed further into the viral genome. Several studies investigated the mechanism resulting in the poor use of the upstream poly(A) site. Some of them favour a hypothesis that the proximity of transcription start site or Cap complex blocks the use of this site (Cherrington and Ganem, 1992; Weichs an der Glon *et al.*, 1991). In addition, Cherrington *et al.* propose that the sequence upstream of the initiation site in the U3 region contributes to the efficiency of polyadenylation in the 3′LTR (Cherrington and Ganem, 1992). Owing to the absence of this sequence in the pre-mRNA originating from the 5′ LTR, the corresponding poly(A) site is less efficiently recognized. Some other *in vitro* and *in vivo* studies support this notion (Ashe *et al.*, 1995; Gilmartin *et al.*, 1992). Later on, another hypothesis proposed a different mechanism for the inhibition of the upstream poly(A) site (Ashe *et al.*, 1995). In this case, the major SD site (MSD) located 200 bp downstream of the poly(A) site blocked polyadenylation. Upon mutating this site, the poly(A) site in the 5′LTR was activated. They further demonstrated the existence of a stem–loop structure near the MSD that binds U1 snRNP, which is associated with the blocking of this poly(A) site (Ashe *et al.*, 2000; Ashe *et al.*, 1997). Most likely, combined effects of promoter proximity and MSD contribute to the inefficient use of the poly(A) site in the 5′LTR.

Export of unspliced and partially spliced transcripts into the cytoplasm

Additional diversity of mature transcripts results from the export of unspliced and partially spliced transcripts to the cytoplasm. These mRNA molecules represent about half of all viral transcripts and include mRNAs that encode Gag, Pol, Env, Vpu, Vif and Vpr (reviewed in Cullen, 1992). The transport of unspliced and singly spliced transcripts from the nucleus to the cytoplasm is mediated by Rev. Levels of Rev in the nucleus determine the proportions of unprocessed mRNA species (Rev-dependent mRNAs) relative to fully processed mRNA species (Rev-independent mRNAs) in the cytoplasm (Feinberg *et al.*, 1986; Knight *et al.*, 1987; Sodroski *et al.*, 1986). To these ends, viral regulatory proteins (Tat, Rev and Nef) encoded by Rev-independent mRNAs are expressed early in the viral replicative cycle. Thus, increased synthesis of structural and accessory proteins encoded by Rev-dependent mRNAs is observed later.

Rev functions by binding a highly conserved *cis*-acting sequence, the RRE RNA stem–loop in the *env* gene, which is present in RNA coding for Env (Emerman *et al.*, 1989; Malim *et al.*, 1989; Rosen *et al.*, 1988; Zapp and Green, 1989). Shuttling of Rev between the nucleus and the cytoplasm is dependent on two signals in Rev, the arginine-rich motif (ARM) that is also its nuclear localization signal (NLS) and the leucine-rich nuclear export signal (NES) (Bohnlein *et al.*, 1991; Malim *et al.*, 1991). After its synthesis in the cytoplasm, Rev is transported to the nucleus through binding the nuclear import factor importin β via its NLS (Truant and Cullen, 1999) (Fig. 6.5, step 1). In the nucleus, Rev interacts with RRE via a region overlapping the NLS (Fig. 6.5, step 2), thus masking this signal and exposing the NES (Henderson and Percipalle, 1997). Binding of a single Rev molecule is followed by a cooperative binding of up to nine molecules, which results in the formation of a multimeric complex

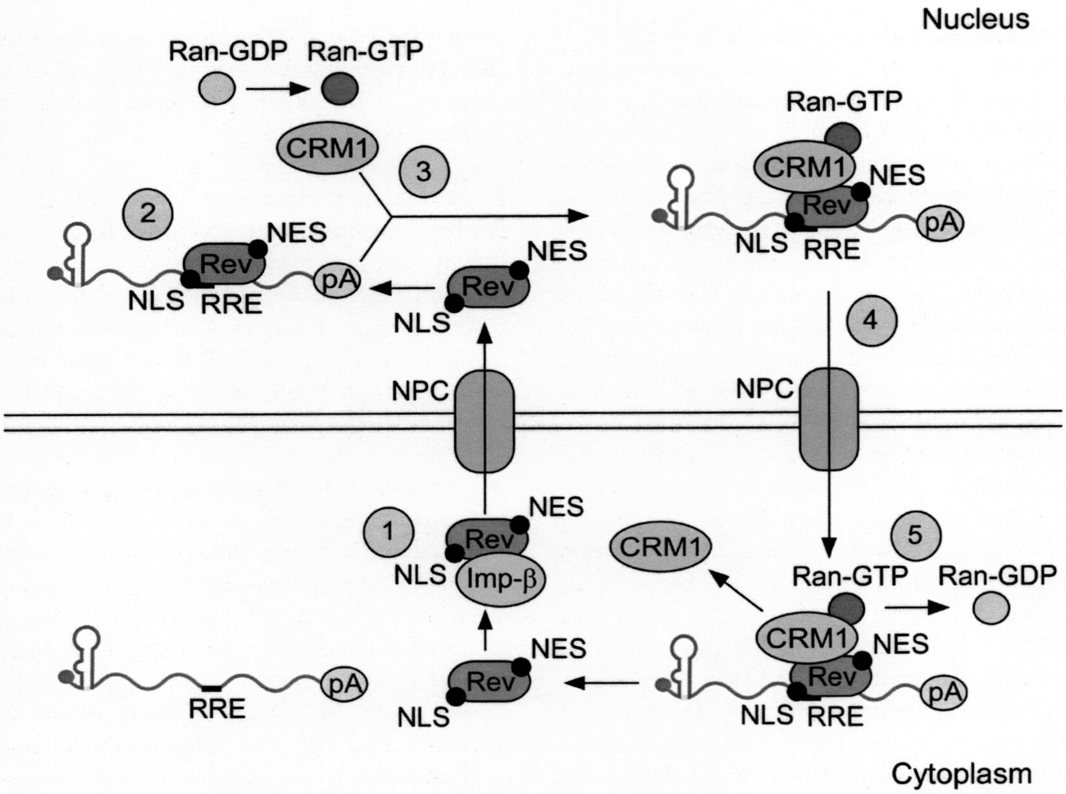

Figure 6.5 Rev regulates export of partially spliced HIV mRNAs from nucleus to cytoplasm. Step 1: in the cytoplasm, Rev binds importin β (imp-β) and is translocated to the nucleus through the nuclear pore complex (NLC). Step 2: Rev binds RRE in unspliced and singly spliced HIV mRNAs via a region overlapping the NLS. This binding is followed by a conformational change that masks the NLS and exposes the NES. Step 3: Crm1 and Ran-GTP associate with Rev bound to mRNA. Step 4: the complex is exported to the cytoplasm through the NPC. Step 5: after the hydrolysis of Ran-GTP to Ran-GDP, Crm1 is released from the complex. Consequently, Rev and mRNA dissociate. Rev with unmasked NLS binds imp-β and is again imported to the nucleus.

(Charpentier *et al.*, 1997; Heaphy *et al.*, 1991; Kjems *et al.*, 1991; Malim and Cullen, 1991; Mann *et al.*, 1994). Exposed NES interacts with a cellular protein Crm1, a member of importin/exportin family of nucleoplasmic transport factors (Fornerod *et al.*, 1997; Neville *et al.*, 1997) (Fig. 6.5, step 3). With the recruitment of Crm1, the export pathway ensues (Bogerd *et al.*, 1998; Fischer *et al.*, 1995) (Fig. 6.5, step 4). Interactions between Rev and Crm1 are regulated by Ran, a cellular GTPase (Izaurralde *et al.*, 1997; Stade *et al.*, 1997). Rev-bound Crm1 interacts with Ran-GTP. In the cytoplasm, the latter is hydrolysed to Ran-GDP (Fig. 6.5, step 5), a step that catalyses the release of Crm1 from the Rev–RRE complex. Dissociation of Rev from Crm1 and consequently from the RRE allows again for the binding of importin β to Rev, which is followed by its import to the nucleus (Fig. 6.5, step 1).

It has been proposed that in addition to its role in the export, Rev also inhibits the splicing of HIV transcripts to enhance the export of unprocessed mRNA species to the cytoplasm (Kjems *et al.*, 1991; Powell *et al.*, 1997).

Viral latency

One of the characteristics of the HIV provirus is that it can reside in a latent form, in which it is not expressed and thus not recognized by the immune system. Although the pool of latently infected cells in the patients with AIDS is very small (~10^6 per individual), it is the major obstacle in curing AIDS. Latent provirus remains replication-competent and cannot be purged from patients even when treated for long periods of time with highly active antiretroviral therapy (HAART). The latent reservoir consists primarily of latently infected resting memory CD4$^+$ T-lymphocytes (Finzi *et al.*, 1999). The other cell types contributing to the reservoir are microglia in the central nervous system (CNS) (Takahashi *et al.*, 1996), dendritic cells (Geijtenbeek *et al.*, 2000), resting monocytes and macrophages (Igarashi *et al.*, 2003). Owing to the long lifespan of the resting CD4$^+$ T-lymphocytes, the decay rate of the pool of latently infected cells is extremely slow with a suggested half-life of 44 months (Finzi *et al.*, 1999).

Integration state of viral DNA determines the type of latency that can be designated as pre-integration or post-integration latency. In resting CD4$^+$ T-lymphocytes, the viral genome is predominantly in the non-integrated form resulting in a pre-integration latency. Nevertheless, this type of latency is of minor importance for the maintenance of the viral reservoir because of the short half-life of the non-integrated viral DNA. Most likely, this latency results from the deficiency of ATP needed for the energy-dependent import of the viral cDNA and pre-integration complex to the nucleus of resting cells (Bukrinsky *et al.*, 1992).

Post-integration latency occurs when the integration of HIV DNA into the genome of CD4$^+$ T-lymphocytes is followed by the regression of infected cells to a resting state (Chun *et al.*, 2003). The establishment and maintenance of post-integration latency is dictated by a combination of different mechanisms.

Inefficient transcriptional initiation and/or elongation from the 5′-LTR

At the level of transcription, the major contributors to the post-integration latency are transcriptional interference (TI), the chromatin environment, lack of key host transcription factors and lack of Tat.

TI is observed when two promoters with the same or the opposite orientation are in a relative close proximity (Adhya and Gottesman, 1982). Initiation of transcription from a downstream promoter is suppressed by ongoing transcription from an upstream promoter. When HIV integrates into an actively transcribed host gene, 5′LTR is inhibited or suppressed (Fig. 6.6A). Based on the studies demonstrating that the HIV genome integrates preferentially into actively transcribed genes (Han *et al.*, 2004; Lewinski *et al.*, 2005; Schroder *et al.*, 2002), it is very likely that TI is an important cause of latency. Indeed, recent observations demonstrated that viral integration into the actively transcribed host genes led to TI caused by the elongating RNAPII transcribing through the viral promoter (Lenasi *et al.*, 2008). Two approaches to overcome TI include the decreased transcription from the upstream promoter and/or the activation of the downstream promoter. Accordingly, to activate the 5′ LTR, one could use TNF-α, prostratin, a phorbol ester that activates cells (Korin *et al.*, 2002; Kulkosky *et al.*, 2001) and broad activators of T-lymphocytes (IL-2) (Chun *et al.*, 1999; Stellbrink *et al.*, 2002) (Fig. 6.6B).

Although proviruses integrate preferentially into active transcription units, repressive chromatin structure has been observed on the transcriptionally silent LTR (Steger and Workman, 1997). In the repressed state, the LTR is occupied by HDAC1 that deacetylates surrounding histones and promotes the formation of a silent chromatin (Zhong *et al.*, 2002) (Fig. 6.7A). This chromatin structure is de-repressed with HDAC inhibitors (valproic acid, TSA) (Fig. 6.7B) or certain host proteins (NF-κB) that recruit chromatin remodelling complexes to allow the formation of PIC (Williams *et al.*, 2006) (Fig. 6.7C).

Another feature contributing to the post-integration latency is availability of key host transcriptional activators like NF-κB and NF-AT. In resting cells, levels of active NF-κB are often too low for productive transcription. In this case, transcription is initiated but blocked at the level of elongation. Activation of resting cells with IL-2 or TNF-α elevates levels of active NF-κB, which binds to the LTR (Nabel and Baltimore, 1987). Recruitment of HATs (p300/CBP) and P-TEFb by NF-κB consequently surmounts the elongation block (Williams *et al.*, 2007; Zhong *et al.*, 1998) (Fig. 6.7C). Since Tat is responsible for productive transcription elongation, mutations in the Tat gene can also maintain proviral latency in infected cells.

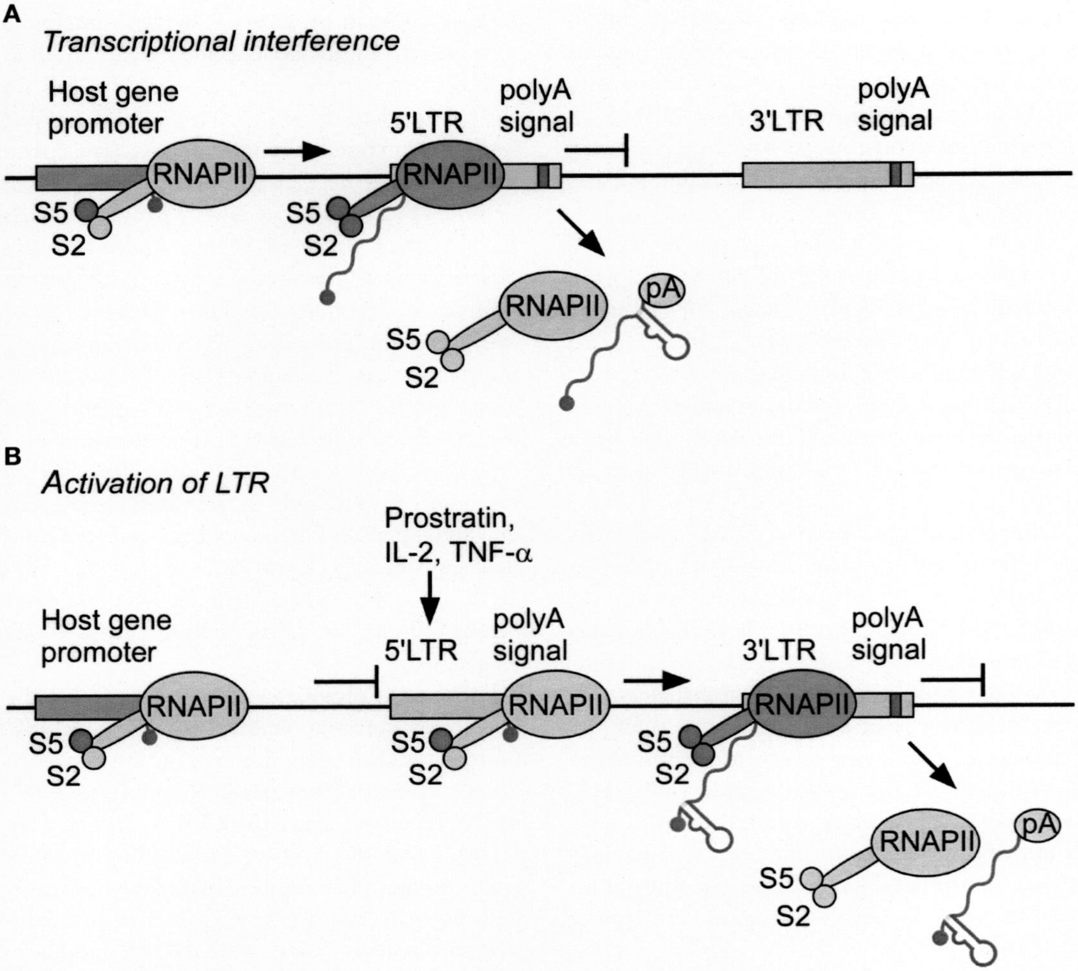

Figure 6.6 Transcriptional interference from the host promoter suppresses transcription from the 5′LTR. (A) Elongating RNAPII from an upstream promoter reads through the 5′LTR and prevents formation of the PIC on this promoter. The majority of transcription terminates at the poly(A) site in this LTR, which gives rise to truncated transcripts. (B) Sufficient activation of the 5′LTR prevents upstream RNAPII from reading through the HIV genome. These induced activators facilitate the formation of PIC on the 5′ LTR, which is followed by transcription elongation. Now, transcription terminates at the poly(A) site in the 3′LTR, which results in the synthesis of viral mRNAs.

Inhibition of viral expression at the post-transcriptional level

Inhibition of post-transcriptional processes contributes less significantly to HIV latency and reservoir. Mechanisms operating at this stage are lack of Rev, lack of host polypyrimidine tract binding protein (PTB), and RNA interference. The first two result in inefficient export of HIV mRNA species from the nucleus to the cytoplasm. Whereas the deficiency of Rev prevents the export of incompletely spliced mRNAs (see above), low levels of PTB result in the accumulation of incompletely spliced mRNAs in the nucleus (Lassen *et al.*, 2006). Overexpression of viral Rev or host PTB, respectively, reverts this post-transcriptional block. The third mechanism that inhibits viral gene expression post-transcriptionally involves host- and virus-derived microRNAs (miRNAs), which probably target HIV transcripts. Consequently, they are degraded by the cellular RNAi machinery (reviewed in Weinberg and Morris, 2006).

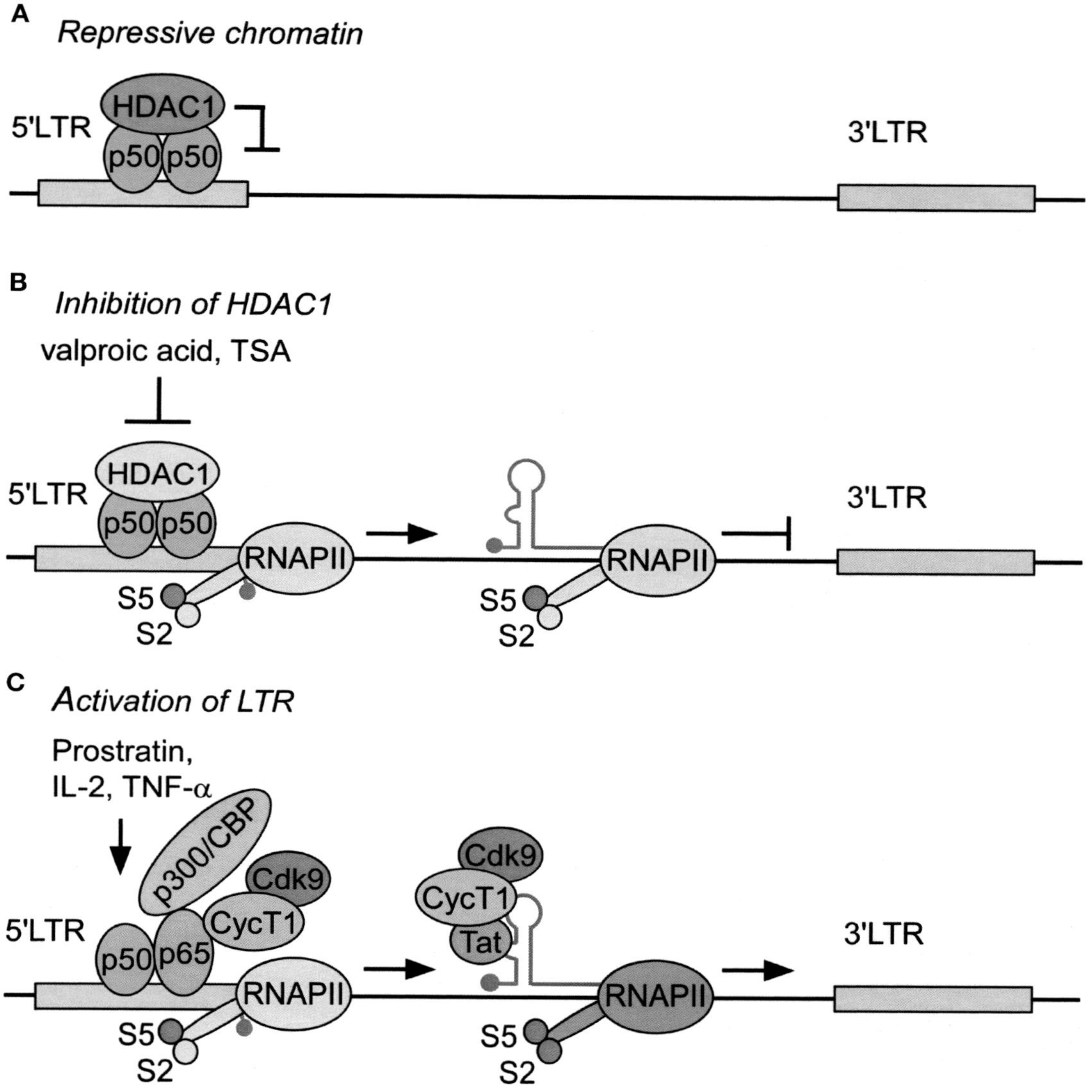

Figure 6.7 Repressive chromatin structure and lack of transcriptional activators prevent transcription of the provirus. (A) Recruitment of HDAC1 to p50 homodimer, which is bound to the LTR, promotes the deacetylation of histones and the formation of transcriptionally non-permissive chromatin. (B) HDAC inhibitors relieve the repressed chromatin structure, thus allowing the formation of PIC and transcription initiation. Insufficient levels of transcription activators result in abortive transcription after the synthesis of TAR. (C) Activation of the 5′LTR results in the recruitment of active NF-κB (p50-p65), which in turn recruits CBP/p300 and P-TEFb for efficient transcription elongation.

Conclusions

In this chapter, we covered aspects of HIV molecular biology that have extensive resonance with the expression of other retroviruses and host eukaryotic genes. Other retroviruses contain LTRs as well as compact, complex genomes and require that incompletely and completely spliced viral transcripts be transported from the nucleus to the cytoplasm. Some of the simpler retroviruses, such as the murine leukaemia virus do not encode Tat or Rev but rely more fully on host cell proteins to accomplish these tasks. In this light, it is important to note that it was the study of Tat and Rev that brought P-TEFb and the control of transcription elongation as well as Crm1 and active export of macromolecules from the nucleus to the cytoplasm to eukaryotic biology. In addition, most other retroviruses persist in the organism and other lentiviruses establish similar reservoirs to that of HIV in the infected host.

In all these processes, much is known but many questions remain. For example, we know little of structures of complexes between viral and cellular proteins. How are they changed in different environments of resting versus activated cells? How is the precise balance between cotranscriptional and posttranscriptional processing of viral transcripts achieved? Does the export by Rev also dictate further steps in translation and RNA encapsidation? In all these studies, can we find differences between viral and cellular processes that will lead to improved therapies of retroviral infections in the future? Clearly, studies of retroviruses in general and HIV in particular have enriched greatly the knowledge of eukaryotic biology. The future should reveal additional insights into this uneasy symbiosis between retroviruses and their mammalian hosts.

References

Adams, M., Sharmeen, L., Kimpton, J., Romeo, J.M., Garcia, J.V., Peterlin, B.M., Groudine, M., and Emerman, M. (1994). Cellular latency in human immunodeficiency virus-infected individuals with high CD4 levels can be detected by the presence of promoter-proximal transcripts. Proc. Natl. Acad. Sci. U.S.A. *91*, 3862–3866.

Adhya, S., and Gottesman, M. (1982). Promoter occlusion: transcription through a promoter may inhibit its activity. Cell *29*, 939–944.

Amendt, B.A., Hesslein, D., Chang, L.J., and Stoltzfus, C.M. (1994). Presence of negative and positive *cis*-acting RNA splicing elements within and flanking the first *tat* coding exon of human immunodeficiency virus type 1. Mol. Cell. Biol. *14*, 3960–3970.

Amendt, B.A., Si, Z.H., and Stoltzfus, C.M. (1995). Presence of exon splicing silencers within human immunodeficiency virus type 1 *tat* exon 2 and *tat*-rev exon 3: evidence for inhibition mediated by cellular factors. Mol. Cell. Biol. *15*, 4606–4615.

Angelov, D., Charra, M., Seve, M., Cote, J., Khochbin, S., and Dimitrov, S. (2000). Differential remodeling of the HIV-1 nucleosome upon transcription activators and SWI/SNF complex binding. J. Mol. Biol. *302*, 315–326.

Ashe, M.P., Furger, A., and Proudfoot, N.J. (2000). Stem–loop 1 of the U1 snRNP plays a critical role in the suppression of HIV-1 polyadenylation. Rna *6*, 170–177.

Ashe, M.P., Griffin, P., James, W., and Proudfoot, N.J. (1995). Poly(A) site selection in the HIV-1 provirus: inhibition of promoter-proximal polyadenylation by the downstream major splice donor site. Genes Dev. *9*, 3008–3025.

Ashe, M.P., Pearson, L.H., and Proudfoot, N.J. (1997). The HIV-1, 5′-LTR poly(A) site is inactivated by U1 snRNP interaction with the downstream major splice donor site. EMBO J. *16*, 5752–5763.

Ayyavoo, V., Mahalingam, S., Rafaeli, Y., Kudchodkar, S., Chang, D., Nagashunmugam, T., Williams, W.V., and Weiner, D.B. (1997). HIV-1 viral protein R (Vpr) regulates viral replication and cellular proliferation in T-cells and monocytoid cells in vitro. J. Leukoc. Biol. *62*, 93–99.

Bakri, Y., Schiffer, C., Zennou, V., Charneau, P., Kahn, E., Benjouad, A., Gluckman, J.C., and Canque, B. (2001). The maturation of dendritic cells results in postintegration inhibition of HIV-1 replication. J. Immunol. *166*, 3780–3788.

Barboric, M., Nissen, R.M., Kanazawa, S., Jabrane-Ferrat, N., and Peterlin, B.M. (2001). NF-kappaB binds P-TEFb to stimulate transcriptional elongation by RNA polymerase II. Mol. Cell *8*, 327–337.

Barboric, M., Yik, J.H., Czudnochowski, N., Yang, Z., Chen, R., Contreras, X., Geyer, M., Matija Peterlin, B., and Zhou, Q. (2007). Tat competes with HEXIM1 to increase the active pool of P-TEFb for HIV-1 transcription. Nucleic Acids Res. 35, 2003–2012.

Bayer, P., Kraft, M., Ejchart, A., Westendorp, M., Frank, R., and Rosch, P. (1995). Structural studies of HIV-1 Tat protein. J. Mol. Biol. *247*, 529–535.

Benkirane, M., Chun, R.F., Xiao, H., Ogryzko, V.V., Howard, B.H., Nakatani, Y., and Jeang, K.T. (1998). Activation of integrated provirus requires histone acetyltransferase. p300 and P/CAF are coactivators for HIV-1 Tat. J. Biol. Chem. *273*, 24898–24905.

Benko, D.M., Schwartz, S., Pavlakis, G.N., and Felber, B.K. (1990). A novel human immunodeficiency virus type 1 protein, tev, shares sequences with tat, *env*, and *rev* proteins. J. Virol. *64*, 2505–2518.

Beyer, A.L., and Osheim, Y.N. (1988). Splice site selection, rate of splicing, and alternative splicing on nascent transcripts. Genes Dev. 2, 754–765.

Bienroth, S., Wahle, E., Suter-Crazzolara, C., and Keller, W. (1991). Purification of the cleavage and polyadenylation factor involved in the 3′-processing of messenger RNA precursors. J. Biol. Chem. *266*, 19768–19776.

Bilodeau, P.S., Domsic, J.K., Mayeda, A., Krainer, A.R., and Stoltzfus, C.M. (2001). RNA splicing at human immunodeficiency virus type 1, 3′ splice site A2 is regulated by binding of hnRNP A/B proteins to an exonic splicing silencer element. J. Virol. 75, 8487–8497.

Bisgrove, D.A., Mahmoudi, T., Henklein, P., and Verdin, E. (2007). Conserved P-TEFb-interacting domain of BRD4 inhibits HIV transcription. Proc. Natl. Acad. Sci. U.S.A. *104*, 13690–13695.

Bogerd, H.P., Echarri, A., Ross, T.M., and Cullen, B.R. (1998). Inhibition of human immunodeficiency virus Rev and human T-cell leukemia virus Rex function, but not Mason–Pfizer monkey virus constitutive transport element activity, by a mutant human nucleoporin targeted to Crm1. J. Virol. 72, 8627–8635.

Bohne, J., Wodrich, H., and Krausslich, H.G. (2005). Splicing of human immunodeficiency virus RNA is position-dependent suggesting sequential removal of introns from the 5′ end. Nucleic Acids Res. 33, 825–837.

Bohnlein, E., Berger, J., and Hauber, J. (1991). Functional mapping of the human immunodeficiency virus type 1 Rev RNA binding domain: new insights into the domain structure of Rev and Rex. J. Virol. *65*, 7051–7055.

Bres, V., Tagami, H., Peloponese, J.M., Loret, E., Jeang, K.T., Nakatani, Y., Emiliani, S., Benkirane, M., and Kiernan, R.E. (2002). Differential acetylation of Tat coordinates its interaction with the co-activators cyclin T1 and PCAF. EMBO J. *21*, 6811–6819.

Bukrinsky, M.I., Sharova, N., Dempsey, M.P., Stanwick, T.L., Bukrinskaya, A.G., Haggerty, S., and Stevenson, M. (1992). Active nuclear import of human immunodeficiency virus type 1 preintegration complexes. Proc. Natl. Acad. Sci. U.S.A. *89*, 6580–6584.

Buratowski, S. (2005). Connections between mRNA 3′ end processing and transcription termination. Curr. Opin. Cell Biol. *17*, 257–261.

Burd, C.G., and Dreyfuss, G. (1994). RNA binding specificity of hnRNP A1: significance of hnRNP A1 high-affinity binding sites in pre-mRNA splicing. EMBO J. *13*, 1197–1204.

Byers, S.A., Price, J.P., Cooper, J.J., Li, Q., and Price, D.H. (2005). HEXIM2, a HEXIM1-related protein, regulates positive transcription elongation factor b through association with 7SK. J. Biol. Chem. *280*, 16360–16367.

Caputi, M., Freund, M., Kammler, S., Asang, C., and Schaal, H. (2004). A bidirectional SF2/ASF- and SRp40-dependent splicing enhancer regulates human immunodeficiency virus type 1 rev, *env*, vpu, and nef gene expression. J. Virol. *78*, 6517–6526.

Caputi, M., Mayeda, A., Krainer, A.R., and Zahler, A.M. (1999). hnRNP A/B proteins are required for inhibition of HIV-1 pre-mRNA splicing. EMBO J. *18*, 4060–4067.

Carlsson, P., Waterman, M.L., and Jones, K.A. (1993). The hLEF/TCF-1 alpha HMG protein contains a context-dependent transcriptional activation domain that induces the TCR alpha enhancer in T-cells. Genes Dev. *7*, 2418–2430.

Chan, H.M., and La Thangue, N.B. (2001). p300/CBP proteins: HATs for transcriptional bridges and scaffolds. J. Cell Sci. *114*, 2363–2373.

Charpentier, B., Stutz, F., and Rosbash, M. (1997). A dynamic *in vivo* view of the HIV-I Rev–RRE interaction. J. Mol. Biol. *266*, 950–962.

Chen, L.F., Williams, S.A., Mu, Y., Nakano, H., Duerr, J.M., Buckbinder, L., and Greene, W.C. (2005). NF-kappaB RelA phosphorylation regulates RelA acetylation. Mol. Cell. Biol. *25*, 7966–7975.

Chene, L., Nugeyre, M.T., Barre-Sinoussi, F., and Israel, N. (1999). High-level replication of human immunodeficiency virus in thymocytes requires NF-kappaB activation through interaction with thymic epithelial cells. J. Virol. *73*, 2064–2073.

Cherrington, J., and Ganem, D. (1992). Regulation of polyadenylation in human immunodeficiency virus (HIV): contributions of promoter proximity and upstream sequences. EMBO J. *11*, 1513–1524.

Chiu, Y.L., Ho, C.K., Saha, N., Schwer, B., Shuman, S., and Rana, T.M. (2002). Tat stimulates cotranscriptional capping of HIV mRNA. Mol. Cell *10*, 585–597.

Chun, T.W., Engel, D., Mizell, S.B., Hallahan, C.W., Fischette, M., Park, S., Davey, R.T., Jr., Dybul, M., Kovacs, J.A., Metcalf, J.A., *et al.* (1999). Effect of interleukin-2 on the pool of latently infected, resting $CD4^+$ T-cells in HIV-1-infected patients receiving highly active anti-retroviral therapy. Nat. Med. *5*, 651–655.

Chun, T.W., Justement, J.S., Lempicki, R.A., Yang, J., Dennis, G., Jr., Hallahan, C.W., Sanford, C., Pandya, P., Liu, S., McLaughlin, M., *et al.* (2003). Gene expression and viral prodution in latently infected, resting $CD4^+$ T-cells in viremic versus aviremic HIV-infected individuals. Proc. Natl. Acad. Sci. U.S.A. *100*, 1908–1913.

Connor, R.I., Chen, B.K., Choe, S., and Landau, N.R. (1995). Vpr is required for efficient replication of human immunodeficiency virus type-1 in mononuclear phagocytes. Virology *206*, 935–944.

Contreras, X., Barboric, M., Lenasi, T., and Peterlin, B.M. (2007). HMBA releases P-TEFb from HEXIM1 and 7SK snRNA via PI3K/Akt and activates HIV transcription. PLoS Pathog. 3, 1459–1469.

Coull, J.J., Romerio, F., Sun, J.M., Volker, J.L., Galvin, K.M., Davie, J.R., Shi, Y., Hansen, U., and Margolis, D.M. (2000). The human factors YY1 and LSF repress the human immunodeficiency virus type 1 long terminal repeat via recruitment of histone deacetylase 1. J. Virol. *74*, 6790–6799.

Cullen, B.R. (1992). Mechanism of action of regulatory proteins encoded by complex retroviruses. Microbiol. Rev. *56*, 375–394.

Dorr, A., Kiermer, V., Pedal, A., Rackwitz, H.R., Henklein, P., Schubert, U., Zhou, M.M., Verdin, E., and Ott, M. (2002). Transcriptional synergy between Tat and PCAF is dependent on the binding of acetylated Tat to the PCAF bromodomain. EMBO J. *21*, 2715–2723.

Dowling, D., Nasr-Esfahani, S., Tan, C.H., O'Brien, K., Howard, J.L., Jans, D.A., Purcell, D.F., Stoltzfus, C.M., and Sonza, S. (2008). HIV-1 infection induces changes in expression of cellular splicing factors that regulate alternative viral splicing and virus production in macrophages. Retrovirology *5*, 18.

du Chene, I., Basyuk, E., Lin, Y.L., Triboulet, R., Knezevich, A., Chable-Bessia, C., Mettling, C., Baillat, V., Reynes, J., Corbeau, P., *et al.* (2007). Suv39H1 and HP1gamma are responsible for chromatin-mediated HIV-1 transcriptional silencing and post-integration latency. EMBO J. *26*, 424–435.

Dyhr-Mikkelsen, H., and Kjems, J. (1995). Inefficient spliceosome assembly and abnormal branch site selection in splicing of an HIV-1 transcript *in vitro*. J. Biol. Chem. *270*, 24060–24066.

Dynlacht, B.D., Hoey, T., and Tjian, R. (1991). Isolation of coactivators associated with the TATA-binding protein that mediate transcriptional activation. Cell *66*, 563–576.

Eckstein, D.A., Sherman, M.P., Penn, M.L., Chin, P.S., De Noronha, C.M., Greene, W.C., and Goldsmith, M.A. (2001). HIV-1 Vpr enhances viral burden by facilitating infection of tissue macrophages but not nondividing $CD4^+$ T-cells. J. Exp. Med. *194*, 1407–1419.

Emerman, M., Vazeux, R., and Peden, K. (1989). The *rev* gene product of the human immunodeficiency virus affects envelope-specific RNA localization. Cell *57*, 1155–1165.

Exline, C.M., Feng, Z., and Stoltzfus, C.M. (2008). Negative and positive mRNA splicing elements act competitively to regulate human immunodeficiency virus type 1 vif gene expression. J. Virol. *82*, 3921–3931.

Federico, M., Percario, Z., Olivetta, E., Fiorucci, G., Muratori, C., Micheli, A., Romeo, G., and Affabris, E. (2001). HIV-1 Nef activates STAT1 in human monocytes/macrophages through the release of soluble factors. Blood *98*, 2752–2761.

Feinberg, M.B., Baltimore, D., and Frankel, A.D. (1991). The role of Tat in the human immunodeficiency virus life cycle indicates a primary effect on transcriptional elongation. Proc. Natl. Acad. Sci. U.S.A. *88*, 4045–4049.

Feinberg, M.B., Jarrett, R.F., Aldovini, A., Gallo, R.C., and Wong-Staal, F. (1986). HTLV-III expression and production involve complex regulation at the levels of splicing and translation of viral RNA. Cell *46*, 807–817.

Felzien, L.K., Woffendin, C., Hottiger, M.O., Subbramanian, R.A., Cohen, E.A., and Nabel, G.J. (1998). HIV transcriptional activation by the accessory protein, VPR, is mediated by the p300 co-activator. Proc. Natl. Acad. Sci. U.S.A. *95*, 5281–5286.

Finzi, D., Blankson, J., Siliciano, J.D., Margolick, J.B., Chadwick, K., Pierson, T., Smith, K., Lisziewicz, J., Lori, F., Flexner, C., *et al.* (1999). Latent infection of $CD4^+$ T-cells provides a mechanism for lifelong persistence of HIV-1, even in patients on effective combination therapy. Nat. Med. *5*, 512–517.

Fischer, U., Huber, J., Boelens, W.C., Mattaj, I.W., and Luhrmann, R. (1995). The HIV-1 Rev activation domain is a nuclear export signal that accesses an export pathway used by specific cellular RNAs. Cell *82*, 475–483.

Fornerod, M., Ohno, M., Yoshida, M., and Mattaj, I.W. (1997). CRM1 is an export receptor for leucine-rich nuclear export signals. Cell *90*, 1051–1060.

Fraldi, A., Varrone, F., Napolitano, G., Michels, A.A., Majello, B., Bensaude, O., and Lania, L. (2005). Inhibition of Tat activity by the HEXIM1 protein. Retrovirology *2*, 42.

Fujinaga, K., Irwin, D., Huang, Y., Taube, R., Kurosu, T., and Peterlin, B.M. (2004). Dynamics of human immunodeficiency virus transcription: P-TEFb phosphorylates RD and dissociates negative effectors from the transactivation response element. Mol. Cell. Biol. *24*, 787–795.

Fukuhara, T., Hosoya, T., Shimizu, S., Sumi, K., Oshiro, T., Yoshinaka, Y., Suzuki, M., Yamamoto, N., Herzenberg, L.A., Herzenberg, L.A., and Hagiwara, M. (2006). Utilization of host SR protein kinases and RNA-splicing machinery during viral replication. Proc. Natl. Acad. Sci. U.S.A. *103*, 11329–11333.

Furtado, M.R., Balachandran, R., Gupta, P., and Wolinsky, S.M. (1991). Analysis of alternatively spliced human immunodeficiency virus type-1 mRNA species, one of which encodes a novel *tat-env* fusion protein. Virology *185*, 258–270.

Furuichi, Y., and Shatkin, A.J. (1977). 5′-termini of reovirus mRNA: ability of viral cores to form caps post-transcriptionally. Virology 77, 566–578.

Garber, M.E., Mayall, T.P., Suess, E.M., Meisenhelder, J., Thompson, N.E., and Jones, K.A. (2000). CDK9 autophosphorylation regulates high-affinity binding of the human immunodeficiency virus type 1 *tat*-P–TEFb complex to TAR RNA. Mol. Cell. Biol. *20*, 6958–6969.

Garber, M.E., Wei, P., KewalRamani, V.N., Mayall, T.P., Herrmann, C.H., Rice, A.P., Littman, D.R., and Jones, K.A. (1998). The interaction between HIV-1 Tat and human cyclin T1 requires zinc and a critical cysteine residue that is not conserved in the murine CycT1 protein. Genes Dev. *12*, 3512–3527.

Garriga, J., and Grana, X. (2004). Cellular control of gene expression by T-type cyclin/CDK9 complexes. Gene *337*, 15–23.

Gaynor, R. (1992). Cellular transcription factors involved in the regulation of HIV-1 gene expression. Aids *6*, 347–363.

Geijtenbeek, T.B., Kwon, D.S., Torensma, R., van Vliet, S.J., van Duijnhoven, G.C., Middel, J., Cornelissen, I.L., Nottet, H.S., KewalRamani, V.N., Littman, D.R., *et al.* (2000). DC-SIGN, a dendritic cell-specific HIV-1-binding protein that enhances trans-infection of T-cells. Cell *100*, 587–597.

Giese, K., Cox, J., and Grosschedl, R. (1992). The HMG domain of lymphoid enhancer factor 1 bends DNA and facilitates assembly of functional nucleoprotein structures. Cell *69*, 185–195.

Gill, G., Pascal, E., Tseng, Z.H., and Tjian, R. (1994). A glutamine-rich hydrophobic patch in transcription factor Sp1 contacts the dTAFII110 component of the *Drosophila* TFIID complex and mediates transcriptional activation. Proc. Natl. Acad. Sci. U.S.A. *91*, 192–196.

Gilmartin, G.M., Fleming, E.S., and Oetjen, J. (1992). Activation of HIV-1 pre-mRNA 3′ processing *in vitro* requires both an upstream element and TAR. EMBO J. *11*, 4419–4428.

Granelli-Piperno, A., Delgado, E., Finkel, V., Paxton, W., and Steinman, R.M. (1998). Immature dendritic cells selectively replicate macrophagetropic (M-tropic) human immunodeficiency virus type 1, while mature cells efficiently transmit both M- and T-tropic virus to T-cells. J. Virol. 72, 2733–2737.

Graveley, B.R. (2000). Sorting out the complexity of SR protein functions. Rna *6*, 1197–1211.

Han, Y., Lassen, K., Monie, D., Sedaghat, A.R., Shimoji, S., Liu, X., Pierson, T.C., Margolick, J.B., Siliciano, R.F., and Siliciano, J.D. (2004). Resting $CD4^+$ T-cells from human immunodeficiency virus type 1 (HIV-1)-infected individuals carry integrated HIV-1 genomes within actively transcribed host genes. J. Virol. *78*, 6122–6133.

He, G., and Margolis, D.M. (2002). Counterregulation of chromatin deacetylation and histone deacetylase occupancy at the integrated promoter of human immunodeficiency virus type 1 (HIV-1) by the HIV-1 repressor YY1 and HIV-1 activator Tat. Mol. Cell. Biol. *22*, 2965–2973.

He, N., Jahchan, N.S., Hong, E., Li, Q., Bayfield, M.A., Maraia, R.J., Luo, K., and Zhou, Q. (*2008*). A La-Related Protein Modulates 7SK snRNP Integrity to Suppress P-TEFb-Dependent Transcriptional Elongation and Tumorigenesis. Mol. Cell *29*, 588–599.

He, N., Pezda, A.C., and Zhou, Q. (2006). Modulation of a P-TEFb functional equilibrium for the global control of cell growth and differentiation. Mol. Cell. Biol. *26*, 7068–7076.

Heaphy, S., Finch, J.T., Gait, M.J., Karn, J., and Singh, M. (1991). Human immunodeficiency virus type 1 regulator of virion expression, rev, forms nucleoprotein filaments after binding to a purine-rich 'bubble' located within the rev-responsive region of viral mRNAs. Proc. Natl. Acad. Sci. U.S.A. *88*, 7366–7370.

Henderson, A., Bunce, M., Siddon, N., Reeves, R., and Tremethick, D.J. (2000). High-mobility-group protein I can modulate binding of transcription factors to the U5 region of the human immunodeficiency virus type 1 proviral promoter. J. Virol. *74*, 10523–10534.

Henderson, A., Holloway, A., Reeves, R., and Tremethick, D.J. (2004). Recruitment of SWI/SNF to the human immunodeficiency virus type 1 promoter. Mol. Cell. Biol. *24*, 389–397.

Henderson, B.R., and Percipalle, P. (1997). Interactions between HIV Rev and nuclear import and export factors: the Rev nuclear localisation signal mediates specific binding to human importin-beta. J. Mol. Biol. *274*, 693–707.

Hoey, T., Weinzierl, R.O., Gill, G., Chen, J.L., Dynlacht, B.D., and Tjian, R. (1993). Molecular cloning and functional analysis of *Drosophila* TAF110 reveal properties expected of coactivators. Cell *72*, 247–260.

Holstege, F.C., Fiedler, U., and Timmers, H.T. (1997). Three transitions in the RNA polymerase II transcription complex during initiation. EMBO J. *16*, 7468–7480.

Holstege, F.C., van der Vliet, P.C., and Timmers, H.T. (1996). Opening of an RNA polymerase II promoter occurs in two distinct steps and requires the basal transcription factors IIE and IIH. EMBO J. i, 1666–1677.

Hrimech, M., Yao, X.J., Bachand, F., Rougeau, N., and Cohen, E.A. (1999). Human immunodeficiency virus type 1 (HIV-1) Vpr functions as an immediate-early protein during HIV-1 infection. J. Virol. *73*, 4101–4109.

Igarashi, T., Imamichi, H., Brown, C.R., Hirsch, V.M., and Martin, M.A. (2003). The emergence and characterization of macrophage-tropic SIV/HIV chimeric viruses (SHIVs) present in $CD4^+$ T-cell-depleted rhesus monkeys. J. Leukoc. Biol. *74*, 772–780.

Izaurralde, E., Kutay, U., von Kobbe, C., Mattaj, I.W., and Gorlich, D. (1997). The asymmetric distribution of the constituents of the Ran system is essential for transport into and out of the nucleus. EMBO J. *16*, 6535–6547.

Jacquenet, S., Mereau, A., Bilodeau, P.S., Damier, L., Stoltzfus, C.M., and Branlant, C. (2001). A second exon splicing silencer within human immunodeficiency virus type 1 *tat* exon 2 represses splicing of Tat mRNA and binds protein hnRNP H. J. Biol. Chem. *276*, 40464–40475.

Janabi, N., Di Stefano, M., Wallon, C., Hery, C., Chiodi, F., and Tardieu, M. (1998). Induction of human immunodeficiency virus type 1 replication in human glial cells after proinflammatory cytokines stimulation: effect of IFNgamma, IL1beta, and TNFalpha on differentiation and chemokine production in glial cells. Glia 23, 304–315.

Jang, M.K., Mochizuki, K., Zhou, M., Jeong, H.S., Brady, J.N., and Ozato, K. (2005). The bromodomain protein Brd4 is a positive regulatory component of P-TEFb and stimulates RNA polymerase II-dependent transcription. Mol. Cell *19*, 523–534.

Jeang, K.T., and Gatignol, A. (1994). Comparison of regulatory features among primate lentiviruses. Curr. Top. Microbiol. Immunol. *188*, 123–144.

Jeronimo, C., Forget, D., Bouchard, A., Li, Q., Chua, G., Poitras, C., Therien, C., Bergeron, D., Bourassa, S., Greenblatt, J., *et al.* (2007). Systematic analysis of the protein interaction network for the human transcription machinery reveals the identity of the 7SK capping enzyme. Mol. Cell *27*, 262–274.

Jowett, J.B., Planelles, V., Poon, B., Shah, N.P., Chen, M.L., and Chen, I.S. (1995). The human immunodeficiency virus type 1 vpr gene arrests infected T-cells in the G2 + M phase of the cell cycle. J. Virol. *69*, 6304–6313.

Kato, H., Horikoshi, M., and Roeder, R.G. (1991). Repression of HIV-1 transcription by a cellular protein. Science *251*, 1476–1479.

Kelly, J., Beddall, M.H., Yu, D., Iyer, S.R., Marsh, J.W., and Wu, Y. (2008). Human macrophages support persistent transcription from unintegrated HIV-1 DNA. Virology *372*, 300–312.

Kessler, M., and Mathews, M.B. (1992). Premature termination and processing of human immunodeficiency virus type 1-promoted transcripts. J. Virol. *66*, 4488–4496.

Kiernan, R.E., Vanhulle, C., Schiltz, L., Adam, E., Xiao, H., Maudoux, F., Calomme, C., Burny, A., Nakatani, Y., Jeang, K.T., *et al.* (1999). HIV-1 *tat* transcriptional activity is regulated by acetylation. EMBO J. *18*, 6106–6118.

Kim, J.B., and Sharp, P.A. (2001). Positive transcription elongation factor B phosphorylates hSPT5 and RNA polymerase II carboxyl-terminal domain independently of cyclin-dependent kinase-activating kinase. J. Biol. Chem. *276*, 12317–12323.

Kim, J.Y., Gonzalez-Scarano, F., Zeichner, S.L., and Alwine, J.C. (1993). Replication of type 1 human immunodeficiency viruses containing linker substitution mutations in the –201 to –130 region of the long terminal repeat. J. Virol. *67*, 1658–1662.

Kino, T., Gragerov, A., Slobodskaya, O., Tsopanomichalou, M., Chrousos, G.P., and Pavlakis, G.N. (2002). Human immunodeficiency virus type 1 (HIV-1) accessory protein Vpr induces transcription of the HIV-1 and glucocorticoid-responsive promoters by binding directly to p300/CBP coactivators. J. Virol. *76*, 9724–9734.

Kjems, J., Brown, M., Chang, D.D., and Sharp, P.A. (1991). Structural analysis of the interaction between the human immunodeficiency virus Rev protein and the Rev response element. Proc. Natl. Acad. Sci. U.S.A. *88*, 683–687.

Klaver, B., and Berkhout, B. (1994). Comparison of 5′ and 3′ long terminal repeat promoter function in human immunodeficiency virus. J. Virol. *68*, 3830–3840.

Knight, D.M., Flomerfelt, F.A., and Ghrayeb, J. (1987). Expression of the art/trs protein of HIV and study of its role in viral envelope synthesis. Science *236*, 837–840.

Korin, Y.D., Brooks, D.G., Brown, S., Korotzer, A., and Zack, J.A. (2002). Effects of prostratin on T-cell activation and human immunodeficiency virus latency. J. Virol. *76*, 8118–8123.

Krueger, B.J., Jeronimo, C., Roy, B.B., Bouchard, A., Barrandon, C., Byers, S.A., Searcey, C.E., Cooper, J.J., Bensaude, O., Cohen, E.A., *et al.* (2008). LARP7 is a stable component of the 7SK snRNP while P-TEFb, HEXIM1 and hnRNP A1 are reversibly associated. Nucleic Acids Res. *36*, 2219–2229.

Krummheuer, J., Lenz, C., Kammler, S., Scheid, A., and Schaal, H. (2001). Influence of the small leader exons 2 and 3 on human immunodeficiency virus type 1 gene expression. Virology *286*, 276–289.

Kulkosky, J., Culnan, D.M., Roman, J., Dornadula, G., Schnell, M., Boyd, M.R., and Pomerantz, R.J. (2001). Prostratin: activation of latent HIV-1 expression suggests a potential inductive adjuvant therapy for HAART. Blood *98*, 3006–3015.

Kusuhara, M., Nagasaki, K., Kimura, K., Maass, N., Manabe, T., Ishikawa, S., Aikawa, M., Miyazaki, K., and Yamaguchi, K. (1999). Cloning of hexamethylene-bis-acetamide-inducible transcript, HEXIM1, in human vascular smooth muscle cells. Biomed. Res. *20*, 273–279.

Laspia, M.F., Rice, A.P., and Mathews, M.B. (1989). HIV-1 Tat protein increases transcriptional initiation and stabilizes elongation. Cell *59*, 283–292.

Laspia, M.F., Rice, A.P., and Mathews, M.B. (1990). Synergy between HIV-1 Tat and adenovirus E1A is principally due to stabilization of transcriptional elongation. Genes Dev. *4*, 2397–2408.

Lassen, K.G., Ramyar, K.X., Bailey, J.R., Zhou, Y., and Siliciano, R.F. (2006). Nuclear retention of multiply spliced HIV-1 RNA in resting CD4$^+$ T-cells. PLoS Pathog. *2*, e68.

Lenasi, T., Contreras, X., and Peterlin, B.M. (2008). Transcriptional interference antagonizes proviral gene expression to promote HIV latency. Cell Host Microbe *4*, 89–91.

Levitt, N., Briggs, D., Gil, A., and Proudfoot, N.J. (1989). Definition of an efficient synthetic poly(A) site. Genes Dev. *3*, 1019–1025.

Lewinski, M.K., Bisgrove, D., Shinn, P., Chen, H., Hoffmann, C., Hannenhalli, S., Verdin, E., Berry, C.C., Ecker, J.R., and Bushman, F.D. (2005). Genome-wide analysis of chromosomal features repressing human immunodeficiency virus transcription. J. Virol. *79*, 6610–6619.

Liou, L.Y., Herrmann, C.H., and Rice, A.P. (2002). Transient induction of cyclin T1 during human macrophage differentiation regulates human immunodeficiency virus type 1 Tat transactivation function. J. Virol. *76*, 10579–10587.

Long, K.S., and Crothers, D.M. (1999). Characterization of the solution conformations of unbound and Tat peptide-bound forms of HIV-1 TAR RNA. Biochemistry *38*, 10059–10069.

Loret, E.P., Georgel, P., Johnson, W.C., Jr., and Ho, P.S. (1992). Circular dichroism and molecular modeling yield a structure for the complex of human immunodeficiency virus type 1 *trans*-activation response RNA and the binding region of Tat, the *trans*-acting transcriptional activator. Proc. Natl. Acad. Sci. U.S.A. *89*, 9734–9738.

Madsen, J.M., and Stoltzfus, C.M. (2005). An exonic splicing silencer downstream of the 3′ splice site A2 is required for efficient human immunodeficiency virus type 1 replication. J. Virol. *79*, 10478–10486.

Maldarelli, F., Xiang, C., Chamoun, G., and Zeichner, S.L. (1998). The expression of the essential nuclear splicing factor SC35 is altered by human immunodeficiency virus infection. Virus Res. *53*, 39–51.

Malim, M.H., and Cullen, B.R. (1991). HIV-1 structural gene expression requires the binding of multiple Rev monomers to the viral RRE: implications for HIV-1 latency. Cell *65*, 241–248.

Malim, M.H., Hauber, J., Le, S.Y., Maizel, J.V., and Cullen, B.R. (1989). The HIV-1 *rev trans*-activator acts through a structured target sequence to activate nuclear export of unspliced viral mRNA. Nature *338*, 254–257.

Malim, M.H., McCarn, D.F., Tiley, L.S., and Cullen, B.R. (1991). Mutational definition of the human immunodeficiency virus type 1 Rev activation domain. J. Virol. *65*, 4248–4254.

Mann, D.A., and Frankel, A.D. (1991). Endocytosis and targeting of exogenous HIV-1 Tat protein. EMBO J. *10*, 1733–1739.

Mann, D.A., Mikaelian, I., Zemmel, R.W., Green, S.M., Lowe, A.D., Kimura, T., Singh, M., Butler, P.J., Gait, M.J., and Karn, J. (1994). A molecular rheostat. Co-operative *rev* binding to stem I of the rev-response element modulates human immunodeficiency virus type-1 late gene expression. J. Mol. Biol. *241*, 193–207.

Manninen, A., Renkema, G.H., and Saksela, K. (2000). Synergistic activation of NFAT by HIV-1 nef and the Ras/MAPK pathway. J. Biol. Chem. *275*, 16513–16517.

Marban, C., Suzanne, S., Dequiedt, F., de Walque, S., Redel, L., Van Lint, C., Aunis, D., and Rohr, O. (2007). Recruitment of chromatin-modifying enzymes by CTIP2 promotes HIV-1 transcriptional silencing. EMBO J. *26*, 412–423.

Marchand, V., Mereau, A., Jacquenet, S., Thomas, D., Mougin, A., Gattoni, R., Stevenin, J., and Branlant, C. (2002). A Janus splicing regulatory element modulates HIV-1 *tat* and *rev* mRNA production by coordination of hnRNP A1 cooperative binding. J. Mol. Biol. 323, 629–652.

Michels, A.A., Nguyen, V.T., Fraldi, A., Labas, V., Edwards, M., Bonnet, F., Lania, L., and Bensaude, O. (2003). MAQ1 and 7SK RNA interact with CDK9/cyclin T complexes in a transcription-dependent manner. Mol. Cell. Biol. *23*, 4859–4869.

Moreland, R.J., Tirode, F., Yan, Q., Conaway, J.W., Egly, J.M., and Conaway, R.C. (1999). A role for the TFIIH XPB DNA helicase in promoter escape by RNA polymerase II. J. Biol. Chem. *274*, 22127–22130.

Muesing, M.A., Smith, D.H., and Capon, D.J. (1987). Regulation of mRNA accumulation by a human immunodeficiency virus *trans*-activator protein. Cell *48*, 691–701.

Nabel, G., and Baltimore, D. (1987). An inducible transcription factor activates expression of human immunodeficiency virus in T-cells. Nature *326*, 711–713.

Neville, M., Stutz, F., Lee, L., Davis, L.I., and Rosbash, M. (1997). The importin-beta family member Crm1p bridges the interaction between Rev and the nuclear pore complex during nuclear export. Curr. Biol. *7*, 767–775.

Nguyen, V.T., Kiss, T., Michels, A.A., and Bensaude, O. (2001). 7SK small nuclear RNA binds to and inhibits the activity of CDK9/cyclin T complexes. Nature *414*, 322–325.

O'Reilly, M.M., McNally, M.T., and Beemon, K.L. (1995). Two strong 5′ splice sites and competing, suboptimal 3′ splice sites involved in alternative splicing of human immunodeficiency virus type 1 RNA. Virology *213*, 373–385.

Pagans, S., Pedal, A., North, B.J., Kaehlcke, K., Marshall, B.L., Dorr, A., Hetzer-Egger, C., Henklein, P., Frye, R., McBurney, M.W., *et al.* (2005). SIRT1 regulates HIV transcription via Tat deacetylation. PLoS Biol. 3, e41.

Peloponese, J.M., Jr., Gregoire, C., Opi, S., Esquieu, D., Sturgis, J., Lebrun, E., Meurs, E., Collette, Y., Olive, D., Aubertin, A.M., *et al.* (2000). 1H-13C nuclear magnetic resonance assignment and structural characterization of HIV-1 Tat protein. C. R. Acad. Sci. III *323*, 883–894.

Pollard, V.W., and Malim, M.H. (1998). The HIV-1 Rev protein. Annu. Rev. Microbiol. *52*, 491–532.

Powell, D.M., Amaral, M.C., Wu, J.Y., Maniatis, T., and Greene, W.C. (1997). HIV Rev-dependent binding of SF2/ASF to the Rev response element: possible role in Rev-mediated inhibition of HIV RNA splicing. Proc. Natl. Acad. Sci. U.S.A. *94*, 973–978.

Price, D.H. (2000). P-TEFb, a cyclin-dependent kinase controlling elongation by RNA polymerase II. Mol. Cell. Biol. *20*, 2629–2634.

Purcell, D.F., and Martin, M.A. (1993). Alternative splicing of human immunodeficiency virus type 1 mRNA modulates viral protein expression, replication, and infectivity. J. Virol. *67*, 6365–6378.

Ratnasabapathy, R., Sheldon, M., Johal, L., and Hernandez, N. (1990). The HIV-1 long terminal repeat contains an unusual element that induces the synthesis of short RNAs from various mRNA and snRNA promoters. Genes Dev. *4*, 2061–2074.

Renner, D.B., Yamaguchi, Y., Wada, T., Handa, H., and Price, D.H. (2001). A highly purified RNA polymerase II elongation control system. J. Biol. Chem. *276*, 42601–42609.

Robert-Guroff, M., Popovic, M., Gartner, S., Markham, P., Gallo, R.C., and Reitz, M.S. (1990). Structure and expression of *tat*-, rev-, and nef-specific transcripts of human immunodeficiency virus type 1 in infected lymphocytes and macrophages. J. Virol. *64*, 3391–3398.

Rohr, O., Marban, C., Aunis, D., and Schaeffer, E. (2003). Regulation of HIV-1 gene transcription: from lymphocytes to microglial cells. J. Leukoc. Biol. *74*, 736–749.

Romerio, F., Gabriel, M.N., and Margolis, D.M. (1997). Repression of human immunodeficiency virus type 1 through the novel cooperation of human factors YY1 and LSF. J. Virol. *71*, 9375–9382.

Rosen, C.A., Terwilliger, E., Dayton, A., Sodroski, J.G., and Haseltine, W.A. (1988). Intragenic *cis*-acting art gene-responsive sequences of the human immunodeficiency virus. Proc. Natl. Acad. Sci. U.S.A. *85*, 2071–2075.

Rosonina, E., Kaneko, S., and Manley, J.L. (2006). Terminating the transcript: breaking up is hard to do. Genes Dev. *20*, 1050–1056.

Ryo, A., Suzuki, Y., Arai, M., Kondoh, N., Wakatsuki, T., Hada, A., Shuda, M., Tanaka, K., Sato, C., Yamamoto, M., and Yamamoto, N. (2000). Identification and characterization of differentially expressed mRNAs in HIV type 1-infected human T-cells. AIDS Res. Hum. Retroviruses *16*, 995–1005.

Saenz, D.T., Loewen, N., Peretz, M., Whitwam, T., Barraza, R., Howell, K.G., Holmes, J.M., Good, M., and Poeschla, E.M. (2004). Unintegrated lentivirus DNA persistence and accessibility to expression in nondividing cells: analysis with class I integrase mutants. J. Virol. *78*, 2906–2920.

Salfeld, J., Gottlinger, H.G., Sia, R.A., Park, R.E., Sodroski, J.G., and Haseltine, W.A. (1990). A tripartite HIV-1 *tat-env*-rev fusion protein. EMBO J. *9*, 965–970.

Sano, M., Abdellatif, M., Oh, H., Xie, M., Bagella, L., Giordano, A., Michael, L.H., DeMayo, F.J., and Schneider, M.D. (2002). Activation and function of cyclin T-Cdk9 (positive transcription elongation factor-b) in cardiac muscle-cell hypertrophy. Nat. Med. *8*, 1310–1317.

Sawaya, B.E., Khalili, K., Gordon, J., Taube, R., and Amini, S. (2000). Cooperative interaction between HIV-1 regulatory proteins Tat and Vpr modulates transcription of the viral genome. J. Biol. Chem. *275*, 35209–35214.

Schroder, A.R., Shinn, P., Chen, H., Berry, C., Ecker, J.R., and Bushman, F. (2002). HIV-1 integration in the human genome favors active genes and local hotspots. Cell *110*, 521–529.

Seewald, M.J., Metzger, A.U., Willbold, D., Rosch, P., and Sticht, H. (1998). Structural model of the HIV-1 Tat(46–58)–TAR complex. J. Biomol. Struct. Dyn. *16*, 683–692.

Serizawa, H., Makela, T.P., Conaway, J.W., Conaway, R.C., Weinberg, R.A., and Young, R.A. (1995). Association of Cdk-activating kinase subunits with transcription factor TFIIH. Nature *374*, 280–282.

Sheldon, M., Ratnasabapathy, R., and Hernandez, N. (1993). Characterization of the inducer of short transcripts, a human immunodeficiency virus type 1 transcriptional element that activates the synthesis of short RNAs. Mol. Cell. Biol. *13*, 1251–1263.

Sherman, M.P., de Noronha, C.M., Pearce, D., and Greene, W.C. (2000). Human immunodeficiency virus type 1 Vpr contains two leucine-rich helices that mediate glucocorticoid receptor coactivation independently of its effects on G(2) cell cycle arrest. J. Virol. *74*, 8159–8165.

Shiekhattar, R., Mermelstein, F., Fisher, R.P., Drapkin, R., Dynlacht, B., Wessling, H.C., Morgan, D.O., and Reinberg, D. (1995). Cdk-activating kinase complex is a component of human transcription factor TFIIH. Nature *374*, 283–287.

Shimizu, S., Urano, E., Futahashi, Y., Miyauchi, K., Isogai, M., Matsuda, Z., Nohtomi, K., Onogi, T., Takebe, Y., Yamamoto, N., and Komano, J. (2007). Inhibiting lentiviral replication by HEXIM1, a cellular negative regulator of the CDK9/cyclin T complex. Aids *21*, 575–582.

Si, Z., Amendt, B.A., and Stoltzfus, C.M. (1997). Splicing efficiency of human immunodeficiency virus type 1 *tat* RNA is determined by both a suboptimal 3′ splice site and a 10 nucleotide exon splicing silencer element located within *tat* exon 2. Nucleic Acids Res. *25*, 861–867.

Sodroski, J., Goh, W.C., Rosen, C., Dayton, A., Terwilliger, E., and Haseltine, W. (1986). A second post-transcriptional *trans*-activator gene required for HTLV-III replication. Nature *321*, 412–417.

Sonza, S., Maerz, A., Deacon, N., Meanger, J., Mills, J., and Crowe, S. (1996). Human immunodeficiency virus type 1 replication is blocked prior to reverse transcription and integration in freshly isolated peripheral blood monocytes. J. Virol. *70*, 3863–3869.

Spangler, L., Wang, X., Conaway, J.W., Conaway, R.C., and Dvir, A. (2001). TFIIH action in transcription initiation and promoter escape requires distinct regions of downstream promoter DNA. Proc. Natl. Acad. Sci. U.S.A. *98*, 5544–5549.

Spina, C.A., Guatelli, J.C., and Richman, D.D. (1995). Establishment of a stable, inducible form of human immunodeficiency virus type 1 DNA in quiescent CD4 lymphocytes *in vitro*. J. Virol. *69*, 2977–2988.

Stade, K., Ford, C.S., Guthrie, C., and Weis, K. (1997). Exportin 1 (Crm1p) is an essential nuclear export factor. Cell *90*, 1041–1050.

Staffa, A., and Cochrane, A. (1994). The *tat/rev* intron of human immunodeficiency virus type 1 is inefficiently spliced because of suboptimal signals in the 3′ splice site. J. Virol. *68*, 3071–3079.

Staffa, A., and Cochrane, A. (1995). Identification of positive and negative splicing regulatory elements within the terminal *tat*-rev exon of human immunodeficiency virus type 1. Mol. Cell. Biol. *15*, 4597–4605.

Steger, D.J., and Workman, J.L. (1997). Stable co-occupancy of transcription factors and histones at the HIV-1 enhancer. EMBO J. *16*, 2463–2472.

Stellbrink, H.J., van Lunzen, J., Westby, M., O'Sullivan, E., Schneider, C., Adam, A., Weitner, L., Kuhlmann, B., Hoffmann, C., Fenske, S., *et al.* (2002). Effects of interleukin-2 plus highly active antiretroviral therapy on HIV-1 replication and proviral DNA (COSMIC trial). Aids *16*, 1479–1487.

Swanson, A.K., and Stoltzfus, C.M. (1998). Overlapping *cis* sites used for splicing of HIV-1 *env*/nef and *rev* mRNAs. J. Biol. Chem. *273*, 34551–34557.

Takagaki, Y., Manley, J.L., MacDonald, C.C., Wilusz, J., and Shenk, T. (1990). A multisubunit factor, CstF, is required for polyadenylation of mammalian pre-mRNAs. Genes Dev. *4*, 2112–2120.

Takahashi, K., Wesselingh, S.L., Griffin, D.E., McArthur, J.C., Johnson, R.T., and Glass, J.D. (1996). Localization of HIV-1 in human brain using polymerase chain reaction/in situ hybridization and immunocytochemistry. Ann. Neurol. *39*, 705–711.

Tanaka, J., Ishida, T., Choi, B.I., Yasuda, J., Watanabe, T., and Iwakura, Y. (2003). Latent HIV-1 reactivation in transgenic mice requires cell cycle -dependent demethylation of CREB/ATF sites in the LTR. Aids *17*, 167–175.

Tange, T.O., Damgaard, C.K., Guth, S., Valcarcel, J., and Kjems, J. (2001). The hnRNP A1 protein regulates HIV-1 *tat* splicing via a novel intron silencer element. EMBO J. *20*, 5748–5758.

Truant, R., and Cullen, B.R. (1999). The arginine-rich domains present in human immunodeficiency virus type 1 Tat and Rev function as direct importin beta-dependent nuclear localization signals. Mol. Cell. Biol. *19*, 1210–1217.

Tyagi, M., and Karn, J. (2007). CBF-1 promotes transcriptional silencing during the establishment of HIV-1 latency. EMBO J. *26*, 4985–4995.

Van Lint, C., Ghysdael, J., Paras, P., Jr., Burny, A., and Verdin, E. (1994). A transcriptional regulatory element is associated with a nuclease-hypersensitive site in the *pol* gene of human immunodeficiency virus type 1. J. Virol. *68*, 2632–2648.

Varin, A., Decrion, A.Z., Sabbah, E., Quivy, V., Sire, J., Van Lint, C., Roques, B.P., Aggarwal, B.B., and Herbein, G. (2005). Synthetic Vpr protein activates activator protein-1, c-Jun N-terminal kinase, and NF-kappaB and stimulates HIV-1 transcription in promonocytic cells and primary macrophages. J. Biol. Chem. *280*, 42557–42567.

Varin, A., Manna, S.K., Quivy, V., Decrion, A.Z., Van Lint, C., Herbein, G., and Aggarwal, B.B. (2003). Exogenous Nef protein activates NF-kappa B, AP-1, and c-Jun N-terminal kinase and stimulates HIV transcription in promonocytic cells. Role in AIDS pathogenesis. J. Biol. Chem. *278*, 2219–2227.

Verdin, E., Paras, P., Jr., and Van Lint, C. (1993). Chromatin disruption in the promoter of human immunodeficiency virus type 1 during transcriptional activation. EMBO J. *12*, 3249–3259.

Vives, E., Brodin, P., and Lebleu, B. (1997). A truncated HIV-1 Tat protein basic domain rapidly translocates through the plasma membrane and accumulates in the cell nucleus. J. Biol. Chem. *272*, 16010–16017.

Wada, T., Takagi, T., Yamaguchi, Y., Watanabe, D., and Handa, H. (1998). Evidence that P-TEFb alleviates the negative effect of DSIF on RNA polymerase

II-dependent transcription *in vitro*. EMBO J. *17*, 7395–7403.

Waterman, M.L., Fischer, W.H., and Jones, K.A. (1991). A thymus-specific member of the HMG protein family regulates the human T-cell receptor C alpha enhancer. Genes Dev. *5*, 656–669.

Waterman, M.L., and Jones, K.A. (1990). Purification of TCF-1 alpha, a T-cell-specific transcription factor that activates the T-cell receptor C alpha gene enhancer in a context-dependent manner. New Biol. *2*, 621–636.

Wei, P., Garber, M.E., Fang, S.M., Fischer, W.H., and Jones, K.A. (1998). A novel CDK9-associated C-type cyclin interacts directly with HIV-1 Tat and mediates its high-affinity, loop-specific binding to TAR RNA. Cell *92*, 451–462.

Weichs an der Glon, C., Monks, J., and Proudfoot, N.J. (1991). Occlusion of the HIV poly(A) site. Genes Dev. *5*, 244–253.

Weinberg, M.S., and Morris, K.V. (2006). Are viral-encoded microRNAs mediating latent HIV-1 infection? DNA Cell Biol. *25*, 223–231.

Williams, S.A., Chen, L.F., Kwon, H., Ruiz-Jarabo, C.M., Verdin, E., and Greene, W.C. (2006). NF-kappaB p50 promotes HIV latency through HDAC recruitment and repression of transcriptional initiation. EMBO J. *25*, 139–149.

Williams, S.A., Kwon, H., Chen, L.F., and Greene, W.C. (2007). Sustained induction of NF-kappa B is required for efficient expression of latent human immunodeficiency virus type 1. J. Virol. *81*, 6043–6056.

Wu, S.Y., and Chiang, C.M. (2007). The double bromodomain-containing chromatin adaptor Brd4 and transcriptional regulation. J. Biol. Chem. *282*, 13141–13145.

Wu, Y., and Marsh, J.W. (2001). Selective transcription and modulation of resting T-cell activity by preintegrated HIV DNA. Science *293*, 1503–1506.

Xiao, H., Hasegawa, T., and Isobe, K. (2000). p300 collaborates with Sp1 and Sp3 in p21(waf1/cip1) promoter activation induced by histone deacetylase inhibitor. J. Biol. Chem. *275*, 1371–1376.

Yamada, T., Yamaguchi, Y., Inukai, N., Okamoto, S., Mura, T., and Handa, H. (2006). P-TEFb-mediated phosphorylation of hSpt5 C-terminal repeats is critical for processive transcription elongation. Mol. Cell *21*, 227–237.

Yamaguchi, Y., Wada, T., Watanabe, D., Takagi, T., Hasegawa, J., and Handa, H. (1999). Structure and function of the human transcription elongation factor DSIF. J. Biol. Chem. *274*, 8085–8092.

Yang, J., Kawai, Y., Hanson, R.W., and Arinze, I.J. (2001a). Sodium butyrate induces transcription from the G alpha(i2) gene promoter through multiple Sp1 sites in the promoter and by activating the MEK-ERK signal transduction pathway. J. Biol. Chem. *276*, 25742–25752.

Yang, Z., Yik, J.H., Chen, R., He, N., Jang, M.K., Ozato, K., and Zhou, Q. (2005). Recruitment of P-TEFb for stimulation of transcriptional elongation by the bromodomain protein Brd4. Mol. Cell *19*, 535–545.

Yang, Z., Zhu, Q., Luo, K., and Zhou, Q. (2001b). The 7SK small nuclear RNA inhibits the CDK9/cyclin T1 kinase to control transcription. Nature *414*, 317–322.

Yik, J.H., Chen, R., Nishimura, R., Jennings, J.L., Link, A.J., and Zhou, Q. (2003). Inhibition of P-TEFb (CDK9/Cyclin T) kinase and RNA polymerase II transcription by the coordinated actions of HEXIM1 and 7SK snRNA. Mol. Cell *12*, 971–982.

Zahler, A.M., Damgaard, C.K., Kjems, J., and Caputi, M. (2004). SC35 and heterogeneous nuclear ribonucleoprotein A/B proteins bind to a juxtaposed exonic splicing enhancer/exonic splicing silencer element to regulate HIV-1 *tat* exon 2 splicing. J. Biol. Chem. *279*, 10077–10084.

Zapp, M.L., and Green, M.R. (1989). Sequence-specific RNA binding by the HIV-1 Rev protein. Nature *342*, 714–716.

Zhang, Z., Klatt, A., Henderson, A.J., and Gilmour, D.S. (2007). Transcription termination factor Pcf11 limits the processivity of Pol II on an HIV provirus to repress gene expression. Genes Dev. *21*, 1609–1614.

Zheng, Z.M. (2004). Regulation of alternative RNA splicing by exon definition and exon sequences in viral and mammalian gene expression. J. Biomed. Sci. *11*, 278–294.

Zhong, H., May, M.J., Jimi, E., and Ghosh, S. (2002). The phosphorylation status of nuclear NF-kappa B determines its association with CBP/p300 or HDAC-1. Mol. Cell *9*, 625–636.

Zhong, H., Voll, R.E., and Ghosh, S. (1998). Phosphorylation of NF-kappa B p65 by PKA stimulates transcriptional activity by promoting a novel bivalent interaction with the coactivator CBP/p300. Mol. Cell *1*, 661–671.

Zhou, M., Deng, L., Kashanchi, F., Brady, J.N., Shatkin, A.J., and Kumar, A. (2003). The Tat/TAR-dependent phosphorylation of RNA polymerase II C-terminal domain stimulates cotranscriptional capping of HIV-1 mRNA. Proc. Natl. Acad. Sci. U.S.A. *100*, 12666–12671.

Zhu, J., Mayeda, A., and Krainer, A.R. (2001). Exon identity established through differential antagonism between exonic splicing silencer-bound hnRNP A1 and enhancer-bound SR proteins. Mol. Cell *8*, 1351–1361.

Zocchi, M.R., Poggi, A., and Rubartelli, A. (1997). The RGD-containing domain of exogenous HIV-1 Tat inhibits the engulfment of apoptotic bodies by dendritic cells. Aids *11*, 1227–1235.

Assembly and Release

7

Heinrich G. Göttlinger and Winfried Weissenhorn

Abstract

Retroviral assembly and release are both mediated by the viral Gag polyprotein. Proteolytic processing of Gag within the immature virus particle yields the internal structural proteins of the mature virion, of which matrix (MA), capsid (CA), and nucleocapsid (NC) are common to all retroviruses. Within the context of the unprocessed polyprotein, the MA domain is primarily required for Gag membrane targeting and for the incorporation of the viral surface glycoproteins into progeny virions. The CA domain of Gag provides the major driving force for the assembly of immature particles. After rearranging into a different type of lattice subsequent to the proteolytic processing of Gag, CA forms the core of the mature virion. NC nucleates immature particle assembly through the concentration of Gag on RNA molecules, and also plays an essential role in the encapsidation of the viral RNA genome. Retroviral Gag proteins also harbour conserved motifs that co-opt a cellular budding machinery to promote the separation of the lipid envelope of the nascent virion from the cell surface, and thus the release of an extracellular virion. The distinct roles of the various Gag domains in virus morphogenesis are the subject of this review.

Introduction

HIV-1 and other orthoretroviruses egress from infected cells by budding from the plasma membrane. This non-lytic mode of release leads to the acquisition of a plasma membrane-derived lipid envelope that surrounds the viral capsid and anchors the viral transmembrane envelope (Env) glycoprotein spikes. The major structural proteins within the virion's lipid envelope are derived from the *gag* gene product, which is encoded by all retroviruses. Gag is synthesized as a polyprotein precursor that is sufficient for the assembly of immature virus-like particles (VLP). Because other virion components have little effect on the efficiency of viral particle production, Gag can be considered the engine that drives assembly. However, the assembly of infectious particles requires the co-assembly of Gag with Env, genomic viral RNA and Gag–Pol, another virally encoded precursor protein that contains the essential viral enzymes protease (PR), reverse transcriptase (RT), and integrase (IN). The Gag–Pol product is generated by occasional ribosomal frameshifting or by suppression of the Gag termination codon, which causes translation to proceed from the *gag* frame into the adjacent *pol* gene. Gag–Pol is incorporated into nascent viral particles through its N-terminal Gag portion, which engages in Gag–Gag interactions.

The assembly of Gag and Gag–Pol molecules at a ratio reflecting their different levels of synthesis initially leads to the formation of a spherical particle of immature morphology. Immature particles are characterized by the presence of a relatively thick protein shell underneath the viral lipid envelope that is predominantly composed of radially arranged Gag precursor molecules. For most retroviruses, including HIV-1, the assembly of immature particles occurs at the plasma membrane, which serves as a scaffold to concentrate Gag precursor molecules. But in the case of HIV-1, there is additional biochemical evidence that Gag assembly intermediates begin to form in the cytosol (Lee *et al.*, 1999; Lingappa *et al.*, 1997; Nermut *et al.*, 2003; Tritel and Resh, 2000). However, the first assembly product readily visualized by electron microscopy is an electron-dense patch of Gag underneath the plasma membrane. Through the lateral addition of Gag precursor molecules, this patch grows and deforms the plasma membrane into a spherical bud that increasingly protrudes from the cell surface. The bud ultimately pinches off, releasing an extracellular virion. However, the Gag proteins of some retroviruses – e.g. those of Mason–Pfizer monkey virus (MPMV) and mouse mammary tumour virus (MMTV) – do not require the help of a membrane to complete assembly. Consequently, these viruses assemble immature particles in the cytosol. These preassembled structures are then transported to the cell surface, where they become enveloped by budding through the plasma membrane.

Regardless of the mode of assembly, immature particles are non-infectious, presumably because their unprocessed Gag protein shell is too rigid to allow disassembly following entry into a new target cell. Their conversion into infectious particles requires the processing of the Gag and Gag–Pol polyproteins by PR, which results in the liberation of structural components of the mature virion and of essential viral enzymes. The cleavage of Gag by PR also leads to the morphological maturation of the virion due to the rearrangement of individual Gag processing products. As reflected by an increased sensitivity to detergents (Wang and Barklis, 1993), this morphological transformation results in a less stable viral protein core, which primes the viral particle for uncoating.

All retroviruses produce three major conserved Gag processing products, called matrix (MA), capsid (CA), and nucleocapsid (NC), which are arranged in the Gag precursor in that order. In contrast, other Gag cleavage products are only found in certain retroviral genera or subgroups. For example, primate lentiviruses share a p6 domain that follows NC in the Gag precursor and occurs only in these viruses. MA is the only Gag cleavage product that remains associated with the internal surface of the viral lipid envelope, where it probably forms the outer protein shell of the mature virion. CA condenses into another protein shell called the capsid (or core), which gives the mature virion its characteristic morphology, and NC locates within the mature capsid in a complex with the genomic viral RNA and Gag-Pol-derived viral enzymes.

As detailed in the following sections, the various Gag domains play distinct roles during virus morphogenesis, with MA predominantly responsible for Gag membrane targeting and Env incorporation, CA (together with NC) providing the principal force for assembly, and NC mediating the recognition and encapsidation of the viral genome. In addition, retroviral Gag polyproteins harbour specialized domains that promote virus release after completion of assembly by recruiting a host machinery that normally functions in a topologically related endosomal budding pathway.

MA and Gag membrane binding

MA constitutes the N-terminal domain of the Gag precursor in all orthoretroviruses. During particle assembly, MA is the primary determinant responsible for Gag membrane targeting. MA remains bound to the viral lipid membrane following virus maturation.

Role of myristylation

The MA domains of most retroviruses undergo cotranslational modification via the attachment of myristic acid, a 14-carbon saturated fatty acid, to an N-terminal glycine residue after removal of the initiation methionine. The myristylation of Gag is essential for the assembly of viral particles by the gammaretrovirus Moloney murine leukaemia virus (Mo-MLV) and the lentivirus human immunodeficiency virus (HIV)-1 (Bryant and Ratner, 1990; Gottlinger *et al.*, 1989; Rein *et al.*, 1986), presumably because it is required for the stable association of Gag with membranes (Spearman *et al.*, 1994). Consistent with this notion, the membrane-independent intracellular assembly of the betaretrovirus MPMV does not require Gag myristylation, but the association of the preassembled capsids with the plasma membrane is prevented if Gag myristylation cannot occur (Rhee and Hunter, 1987). Remarkably, HIV-1 particle production can remain efficient even if the entire MA domain is deleted and replaced by a myristylation signal (Lee and Linial, 1994; Reil *et al.*, 1998). Indeed, if a heterologous myristylation signal is appended to the N-terminus of Gag, under certain conditions HIV-1 can replicate efficiently in the absence of the entire MA domain (Reil *et al.*, 1998). These observations indicate that the myristyl group provides the dominant membrane targeting signal in HIV-1 MA. They also indicate that Gag binds to membranes in a highly cooperative manner, because the membrane binding energy provided by a single myristate appears too low to allow the stable attachment of a myristylated protein to a membrane (Blenis and Resh, 1993).

Role of basic residues

While the myristylation signal is the only portion of HIV-1 MA that is strictly required for membrane-driven Gag assembly, conserved basic regions in the MA domains of HIV-1 and various other retroviruses also play a role. A bipartite membrane-targeting motif was first reported for HIV-1 MA, where the N-terminal myristylation signal functioned together with a N-proximal basic region; mutations of all positively charged residues in this region severely impaired budding (Zhou *et al.*, 1994). However, it appears that the role of these basic residues is redundant, since more limited substitutions had only modest effects on viral particle production (Freed *et al.*, 1995). Basic residues in several viral and cellular membrane proteins are thought to contribute to membrane binding through electrostatic interactions with acidic phospholipids, which are concentrated at the inner leaflet of the plasma membrane (Resh, 1994).

A role of N-proximal basic MA residues in membrane targeting is supported by the crystal structures of simian immunodeficiency virus (SIV) and HIV-1 MA, both of which showed a trimeric assembly with a flat composite surface enriched in conserved basic residues (Hill *et al.*, 1996; Rao *et al.*, 1995). The folded globular domain of HIV-1 MA is composed of five major helices (Hill *et al.*, 1996; Massiah *et al.*, 1994), and despite limited sequence homology the topologies of several other retroviral MA proteins are rather similar (Conte and Matthews, 1998).

The clustering of basic residues into a putative membrane binding patch is another widely conserved feature. A particularly interesting case is an N-terminal portion of Rous sarcoma virus (RSV) MA called the membrane-binding (M) domain, because RSV Gag is not myristylated and also does not depend on acetylation of the N-terminal methionine for budding (Erdie and Wills, 1990). Like the MA domains of other retroviruses, the RSV-M domain possesses a strongly cationic surface patch, and in the absence of a requirement for myristylation this patch is thought to largely account for membrane binding via electrostatic interactions (McDonnell *et al.*, 1998).

Role of MA in conferring cooperativity on Gag membrane binding

Curiously, large deletions in HIV-1 MA that removed the entire globular domain very significantly increased the efficiency of viral particle production (Reil *et al.*, 1998). Since this is contrary to what the model of Gag membrane binding outlined above would have predicted, it indicates that the model is incomplete. The positive effect on HIV-1 particle production of removing the globular head of MA was particularly pronounced at low Gag expression levels, and provided the first indication that the intact MA domain has an inhibitory effect that needs to be overcome during assembly. A membrane flotation and microscopic analysis subsequently revealed that the globular head of HIV-1 MA inhibits a plasma membrane-specific targeting signal contained within the six amino-terminal MA residues (Perez-Caballero *et al.*, 2004). Gag membrane binding in the presence – but not in the absence – of the MA globular head was dramatically dependent on Gag expression levels, implying a high degree of cooperativity. Consistent with this notion, Gag multimerization was required to overcome the inhibitory effect of MA. Together, these observations indicated that Gag multimerization profoundly increases the affinity of Gag for membranes by unmasking a membrane binding site that coincides with the MA myristylation signal (Perez-Caballero *et al.*, 2004).

The myristyl switch model of Gag membrane binding

An inhibitory effect of the intact MA domain on particle assembly is entirely compatible with a myristyl switch model of Gag membrane binding. Such a model was originally proposed to explain the observation that the membrane-targeting signal appears hidden if the HIV-1 MA domain is expressed by itself rather than in the context of the Gag precursor (Zhou and Resh, 1996). While this finding could also reflect the well-documented relationship between Gag multimerization and Gag membrane binding (Li *et al.*, 2007; Ono *et al.*, 2000a; Perez-Caballero *et al.*, 2004), there is now considerable evidence to support the myristyl switch model. For instance, deletions within the α-helical globular head of HIV-1 MA cause a dramatic increase in its overall membrane binding *in vivo* (Spearman *et al.*, 1997). Because the membrane binding by MA *in vivo* depends on myristic acid, this result suggests that deletions in the globular head of the molecule can make the N-terminal myristyl group more accessible (Spearman *et al.*, 1997). Other studies have shown that point mutations near the N-terminus of HIV-1 MA, such as V7R, L8A and L8I, block Gag membrane binding without affecting myristylation, consistent with the interpretation that such mutations prevent the exposure of the myristyl group (Ono and Freed, 1999; Paillart and Gottlinger, 1999). Furthermore, the resulting severe defects in particle production were completely rescued by second-site mutations in the α-helical core of MA, which by themselves appeared to increase Gag membrane binding, again arguing for a role of the MA globular head in sequestering the myristyl group prior to membrane binding (Ono and Freed, 1999; Paillart and Gottlinger, 1999). Subsequent nuclear magnetic resonance (NMR) studies confirmed that myristylated HIV-1 MA adopts myristate-exposed and myristate-sequestered states, as anticipated (Tang *et al.*, 2004). This work also indicated that myristate exposure is triggered by MA trimerization (Tang *et al.*, 2004), possibly providing an explanation for the observation that the association of Gag with membranes *in vivo* is highly dependent on Gag expression levels (Perez-Caballero *et al.*, 2004). NMR studies also confirmed that the V7R, L8A and L8I mutations in HIV-1 MA stabilize the myristate-sequestered MA species, as expected (Saad *et al.*, 2007).

A particularly attractive aspect of the myristyl switch model is that it provides a mechanism for Gag's discrimination between different cellular membranes. Consistent with this notion, a large deletion in HIV-1 MA that presumably led to constitutive exposure of the myristyl group, redirected virus assembly from the plasma membrane to the endoplasmic reticulum (Facke *et al.*, 1993). A possible interpretation of this result was that Gag lacking the globular head of MA simply associated with the most abundant cellular membrane (Facke *et al.*, 1993). Interestingly, point mutations in the basic domain of HIV-1 MA also retargeted virus assembly to intracellular sites (Ono *et al.*, 2000b), implying that the basic domain confers selectivity to Gag membrane binding, possibly by regulating the exposure of the myristyl group.

It subsequently emerged that selective membrane binding by HIV-1 MA involves the acidic phospholipid phosphatidylinositol 4,5-bisphosphate [$PI(4,5)P_2$], which is concentrated at the inner leaflet of the plasma membrane (Ono *et al.*, 2004). Manipulating the levels or localization of $PI(4,5)P_2$ altered the subcellular localization of Gag, indicating that $PI(4,5)P_2$ regulates the targeting of Gag to the plasma membrane (Ono *et al.*, 2004). Structural studies then revealed that $PI(4,5)P_2$ binds directly to HIV-1 MA and is likely to function both as an allosteric trigger for myristate exposure and as a direct membrane anchor (Fig. 7.1) (Saad *et al.*, 2006). Thus, myristate exposure by HIV-1 MA appears to be regulated both by Gag multimerization and by $PI(4,5)P_2$, and the relative contribution of these factors to Gag membrane targeting remains to be determined.

Surprisingly, a very recent study indicates that HIV-2 MA does not exhibit concentration-dependent exposure of the myristyl group, and $PI(4,5)P_2$ analogues that bind HIV-2 MA also do not trigger myristate exposure (Saad *et al.*, 2008). On the other hand, as in the case of HIV-1, enzymes that altered $PI(4,5)P_2$ localization affected the site of HIV-2 assembly (Saad *et al.*, 2008). Thus, although the models of myristate exposure and Gag membrane targeting proposed for HIV-1 are compelling, they may not be applicable for all retroviruses.

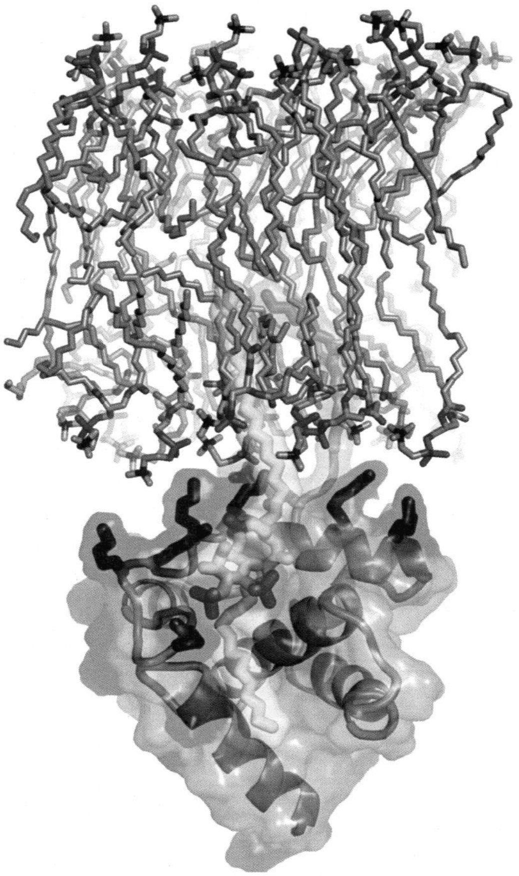

Figure 7.1 Model of membrane-associated HIV-1 MA with bound phosphatidylinositol 4,5-bisphosphate (yellow, with red phosphates). The myristyl group at the N-terminus of MA is coloured in green, and basic MA residues are coloured in blue (courtesy of Michael F. Summers).

Role of the endosomal accumulation of Gag

A particularly controversial issue is the site where Gag initially associates with membranes and nucleates assembly. Although it has long been considered an established fact that retroviral Gag proteins are generally targeted to the plasma membrane, several recent studies have suggested that infectious HIV-1 assembly in macrophages occurs in late endosomes or multivesicular bodies (MVB) (Nguyen *et al.*, 2003; Ono and Freed, 2004; Pelchen-Matthews *et al.*, 2003; Raposo *et al.*, 2002). Since MVBs are sites of exosome formation and can fuse with the plasma membrane, a so-called Trojan exosome hypothesis was proposed, which states that retroviruses use the pre-existing exosome biogenesis pathway for the formation and release of infectious particles (Gould *et al.*, 2003). This hypothesis was attractive because it is well established that retroviruses hijack the cellular protein network required for the formation of exosomes to promote virus release (Adamson and Freed, 2007; Bieniasz, 2006; Morita and Sundquist, 2004). Numerous studies then suggested that retrovirus morphogenesis on endosomal membranes is not limited to macrophages, and that the Gag proteins of HIV-1, Mo-MLV, and human T-cell lymphotrophic virus (HTLV)-I use MVBs as trafficking intermediates and as sites of virus assembly in all cells (Blot *et al.*, 2004; Dong *et al.*, 2005; Houzet *et al.*, 2006; Nydegger *et al.*, 2003; Perlman and Resh, 2006; Sherer *et al.*, 2003). However, it has recently become clear that much if not all of the Gag protein associated with endosomes gets there through the re-endocytosis of trapped virions and budding structures from the plasma membrane (Harila *et al.*, 2006; Neil *et al.*, 2006).

Interestingly, the endocytosis of nascent retroviral particles from the cell surface, and consequently the endosomal accumulation of Gag, is prevented by Vpu (Harila *et al.*, 2006; Neil *et al.*, 2006), an accessory protein of HIV-1 that has long been known to promote retroviral release (Gottlinger *et al.*, 1993; Klimkait *et al.*, 1990). The reuptake of nascent virions from the plasma membrane that is counteracted by Vpu depends on a cellular tethering activity that is constitutively active in some cell types and can be induced by interferon-α in others (Neil *et al.*, 2007). In groundbreaking work, the tethering factor antagonized by Vpu was identified as the interferon-inducible cell surface protein CD317/BST-2/HM1.24, which was renamed tetherin (Neil *et al.*, 2008; Van Damme *et al.*, 2008).

Consistent with the theory that the internalization of Gag accounts for its presence in endosomes, several recent studies revealed that newly synthesized HIV-1 Gag initially travels to the plasma membrane and reaches endocytic compartments only later (Finzi *et al.*, 2007; Harila *et al.*, 2006; Jouvenet *et al.*, 2006). Also, the endosomal accumulation of HIV-1 Gag is prevented if endocytosis is inhibited, whereas viral particle production is unaffected (Jouvenet *et al.*, 2006).

Finally, a novel ultrastructural approach recently revealed that the plasma membrane is the primary site of HIV-1 budding even in macrophages (Welsch *et al.*, 2007). These recent findings suggest that the endosomal localization of Gag is more a manifestation of an innate antiviral mechanism than indicative of a role in retrovirus assembly.

MA and Env incorporation

For HIV-1 and presumably all other lentiviruses, MA has an essential role in the incorporation of the Env glycoprotein spikes into assembling virions. The Env glycoprotein spikes are trimeric complexes formed by the surface (SU) and transmembrane (TM) glycoproteins, which are derived from a common precursor that is cleaved by a cellular protease within the secretory pathway. Following cleavage, SU and TM remain associated with each other, and the complex becomes anchored in the viral lipid bilayer via TM. Although retroviral Env proteins are not required for viral particle assembly per se, their incorporation is essential for the assembly of infectious particles.

Role of MA in different retroviral genera

An early study showed that the MA protein of the avian RSV can be chemically cross-linked to TM, indicating that these virion components interact with each other (Gebhardt *et al.*, 1984). However, further results showed that RSV virions lacking the cytoplasmic tail of TM are infectious, which argues against an essential role of the MA–TM interaction in the incorporation of Env during virus assembly (Perez *et al.*, 1987). Similarly, the cytoplasmic tail of Mo-MLV Env is not strictly required for its incorporation into virions (Januszeski *et al.*, 1997). Nevertheless, the situation appeared to be different for lentiviruses, because mutagenic analyses of HIV-1 showed that the globular head of MA is absolutely essential for the incorporation of Env (Dorfman *et al.*, 1994b; Yu *et al.*, 1992). Even some single amino acid substitutions in MA blocked the incorporation of HIV-1 Env (Freed and Martin, 1995), confirming the essential role of MA. Consistent with these findings, other studies demonstrated that truncations of the long cytoplasmic tail of HIV-1 TM can impair Env incorporation (Dubay *et al.*, 1992; Yu *et al.*, 1993). Furthermore, HIV-1 Env was not incorporated into VLP formed by the Gag protein of the lentivirus visna virus (Dorfman *et al.*, 1994b). In contrast, HIV-1 Env was efficiently incorporated if the MA domain of visna virus was replaced by that of HIV-1 (Dorfman *et al.*, 1994b). Collectively, these observations appeared to support the notion that HIV-1 Env incorporation depends on a specific interaction between MA and TM.

Role of MA in accommodating the cytoplasmic tail of Env

The inability of visna capsids to incorporate HIV-1 Env was unexpected, because pseudotype formation can readily occur between much more distantly related retroviruses (Dorfman *et al.*, 1994b). For instance, HIV-1 can be efficiently pseudotyped with the Env proteins of murine retroviruses (Kimpton and Emerman, 1992; Lusso *et al.*, 1990; Spector *et al.*, 1990), which are essentially unrelated. It is even possible to pseudotype HIV-1 and other retroviral particles with cellular receptors such as Tva or CD4 and CXCR4, which confer infectivity for target cells expressing the cognate Env proteins (Balliet and Bates, 1998; Endres *et al.*, 1997). Thus, there was an unexplained discrepancy between the remarkable ease with which HIV-1 particles could be pseudotyped and the nearly complete inability of HIV-1 Env to pseudotype retroviruses such as Mo-MLV (Mammano *et al.*, 1997; Wilson *et al.*, 1989). Further analysis revealed that the length of the cytoplasmic domain of Env accounts for this difference. The cytoplasmic tails of lentiviral TM proteins are unusually long; for instance, the HIV-1 tail comprises about 150 amino acids. In contrast, viral glycoproteins that efficiently pseudotype HIV-1 and other retroviruses have cytoplasmic tails of less than 50 amino acids. Remarkably, the removal of the long cytoplasmic tail of HIV-1 Env was sufficient to allow its efficient incorporation into Mo-MLV particles, leading to the formation of infectious pseudotypes (Mammano *et al.*, 1997; Schnierle *et al.*, 1997). The removal of the cytoplasmic domain of TM also allowed the efficient incorporation of HIV-1 Env into HIV-1 MA mutants that were unable to incorporate full-length HIV-1 Env, and even rescued virus replication in highly permissive

MT-4 cells (Freed and Martin, 1995; Mammano *et al.*, 1995). Indeed, it was demonstrated that the removal of the cytoplasmic tail of TM to restore Env incorporation allows the efficient replication of HIV-1 in MT-4 cells, even if the entire MA domain is replaced by a myristylation signal (Reil *et al.*, 1998). Taken together, these observations imply that the primary role of HIV-1 MA in Env incorporation is to accommodate the bulky cytoplasmic domain of HIV-1 TM. In contrast, MA is not required to accommodate Env proteins with short cytoplasmic domains; these can be incorporated in a MA-independent manner if expressed at sufficiently high levels.

Cell type-dependent role of MA

While the cytoplasmic tails of retroviral Envs are clearly not essential for their incorporation into virions, there is nevertheless evidence for a cell type-dependent requirement for an interaction between HIV-1 MA and TM (Murakami and Freed, 2000). Specifically, the cytoplasmic domain of TM, although largely dispensable for Env incorporation in MT-4 or HeLa cells, is required in the majority of T-cell lines and in natural target cells for HIV-1 (Murakami and Freed, 2000). These findings strengthen the case for a specific interaction between HIV-1 MA and TM, and this notion is further supported by direct biochemical evidence (Cosson, 1996; Wyma *et al.*, 2000). Indirect evidence for an interaction between HIV-1 TM and Gag were obtained much earlier through the observation that HIV-1 Env determines the site of virus release in polarized epithelial cells (Owens *et al.*, 1991). In addition, polarized HIV-1 budding depended on the cytoplasmic domain of TM and was abolished by mutations in MA that prevented the incorporation of Env (Lodge *et al.*, 1994). It subsequently became clear that the effect of TM on the polarized budding of HIV-1, which may favour cell-to-cell transmission, is dependent on a membrane-proximal tyrosine-based motif (Deschambeault *et al.*, 1999; Lodge *et al.*, 1997). This tyrosine-based motif is highly conserved among retroviruses, and also directs the endocytosis of Env, thereby limiting its exposure on the cell surface (Rowell *et al.*, 1995; Sauter *et al.*, 1996). Interestingly, HIV-1 Env internalization is suppressed by Gag in a MA-dependent manner, which provides an elegant mechanism for the removal of excess cell surface Env that does not participate in viral particle assembly (Egan *et al.*, 1996).

The cell type-dependent requirement for the cytoplasmic tail of TM in the incorporation of HIV-1 Env led to the suggestion that a host factor that interacts with TM may be involved (Murakami and Freed, 2000). One candidate is TIP47, which was reported to interact simultaneously with HIV-1 Gag and Env and to play an essential role in Env incorporation (Lopez-Verges *et al.*, 2006). It was also reported that cholesterol-rich lipid rafts support not only HIV-1 budding (Nguyen and Hildreth, 2000; Ono and Freed, 2001), but also the incorporation of Env, which apparently requires the Gag-mediated association of Env with detergent-resistant membranes (Bhattacharya *et al.*, 2006). Another very recent study established that HIV-1 can escape from the entry-inhibiting effects of a cholesterol-binding compound via cleavage of TM by PR (Waheed *et al.*, 2007). Since the escape mutants were fully replication competent even though much of the cytoplasmic tail of TM was cleaved off after assembly (Waheed *et al.*, 2007), this study provides some of the most compelling evidence to date that the primary role of the cytoplasmic tail of HIV-1 TM is in Env incorporation.

CA and assembly

Although multiple Gag regions are probably involved in the assembly of an immature retroviral particle, the CA domain is of special importance. As shown first for HIV-1, CA is composed of two largely α-helical domains (Gamble *et al.*, 1997; Gitti *et al.*, 1996): the N-terminal domain (NTD) and the C-terminal domain (CTD). The smaller CTD begins with the major homology region (MHR), a stretch of about 20 amino acids that is conserved in all replication-competent orthoretroviruses (Gamble *et al.*, 1997). Overall, the CTD is the most conserved domain of Gag.

In early genetic studies, mutations in the CA domains of different retroviruses had remarkably different effects on immature particle assembly. Whereas Mo-MLV assembly was blocked by linker insertions throughout CA (Hansen *et al.*, 1990; Schwartzberg *et al.*, 1984), most of the CA domain of RSV, including the MHR, was dispensable for assembly (Wills and Craven, 1991). In the case of HIV-1, mutations in the NTD often disrupt the assembly of conical mature capsids and abolish virion infectivity, but generally have no effect on the assembly of immature particles by Gag (Borsetti *et al.*, 1998; Dorfman *et al.*, 1994a; Reicin *et al.*, 1995, 1996; Wang and Barklis, 1993). In contrast, many mutations in the smaller CTD dramatically impair immature particle assembly (Dorfman *et al.*, 1994a; McDermott *et al.*, 1996; Reicin *et al.*, 1995). Notably, HIV-1 particle assembly is absolutely dependent on the integrity of the MHR at the N-terminus of the CTD, where even conservative single amino acid substitutions often have dramatic effects (Chang *et al.*, 2007; Mammano *et al.*, 1994; von Schwedler *et al.*, 2003a). Overall, these genetic studies implied that HIV-1 CA harbours two functionally distinct domains, an N-terminal domain that is dispensable for immature particle assembly and a required C-terminal domain beginning with the MHR. It later emerged that these functionally defined regions of CA coincide with independently folded domains that are connected via a hinge region just proximal of the MHR (Gamble *et al.*, 1997; Gitti *et al.*, 1996).

Role of the CA NTD

Image reconstructions of HIV-1 CA tubes assembled *in vitro* indicated that the mature retroviral capsid is composed of hexameric rings formed by the CA NTDs, and that these rings are linked via the CTDs (Li *et al.*, 2000). The NTD of HIV-1 CA is composed of seven α-helices and shaped like an arrowhead (Gitti *et al.*, 1996), and despite very limited sequence homology the overall fold of the NTD is conserved across retroviral genera (Cornilescu *et al.*, 2001; Jin *et al.*, 1999; Khorasanizadeh *et al.*, 1999; Kingston *et al.*, 2000; Mortuza *et al.*, 2004). In the case of HIV-1, helices 4 and 5 of the NTD are connected by a long loop that binds to the cellular prolyl isomerase cyclophilin A (Gamble *et al.*, 1996). This interaction is of considerable interest because drugs that bind cyclophilin A with high affinity and prevent its incorporation into assembling virions significantly inhibit HIV-1 replication (Franke *et al.*, 1994b; Thali *et al.*, 1994). However, the interaction is not conserved even among closely related lentiviruses, and is dispensable for HIV-1 assembly. A feature of the NTD that is conserved across orthoretroviruses is an N-terminal β-hairpin structure that is stabilized by a salt bridge between the charged proline at the processed N-terminus of CA and an invariant buried aspartate (Gitti *et al.*, 1996; Kingston *et al.*, 2000; Mortuza *et al.*, 2004). Consequently, it is thought that the β-hairpin forms only after the proteolytic processing of immature Gag, and that this rearrangement of the CA N-terminus triggers the assembly of the mature core (Gitti *et al.*, 1996; Mortuza *et al.*, 2004; von Schwedler *et al.*, 1998). Consistent with this conformational switch model, the addition of short N-terminal extensions to HIV-1 CA causes a switch from tube to sphere formation during *in vitro* assembly assays (Gross *et al.*, 1998; von Schwedler *et al.*, 1998), thus changing assembly from a mature- to an immature-like mode. Furthermore, as predicted by the model, point mutations in the putative new CA–CA interface generated by the refolding of the CA N-terminus allow immature particle assembly *in vivo*, but prevent the assembly of mature cores and thus abolish virion infectivity (von Schwedler *et al.*, 1998). On the other hand, it has been shown that an HIV-1 Gag-derived protein with an unprocessed CA domain can assemble into either spheres or tubes *in vitro*, depending upon the pH of the reaction (Gross *et al.*, 2000). Since CA has an immature N-terminus in this context, the formation of a salt bridge by the processed N-terminus of CA is not absolutely essential for the assembly of structures resembling mature cores.

Role of the CA CTD

The HIV-1 CTD is a globular domain composed of a 3_{10} helix, an extended strand, and four α-helices (Gamble *et al.*, 1997; Worthylake *et al.*, 1999). Intact HIV-1 CA has long been known to self-associate, although with low affinity (Rose *et al.*, 1992), and the isolated CTD dimerizes with a similar affinity as intact CA and is therefore referred to as the dimerization domain (Gamble *et al.*, 1997; Worthylake *et al.*, 1999). Dimerization of the CTD is considered essential for its role in connecting the hexameric rings formed by the NTD (Li *et al.*, 2000). It is therefore curious that in contrast to HIV-1 CA, HTLV-I and RSV CA remain monomeric even at high protein concentrations (Campos-Olivas *et al.*,

2000; Khorasanizadeh *et al.*, 1999). The first crystal structures of the HIV-1 CTD indicated that the MHR is distinct from the dimer interface (Gamble *et al.*, 1997; Worthylake *et al.*, 1999), leaving the critical role of the MHR in virus assembly largely unexplained. In these models, the principal dimer interface is formed by helix 2 of the HIV-1 CTD, which packs in a parallel manner against helix 2 of another monomer (Gamble *et al.*, 1997; Worthylake *et al.*, 1999). However, while mutations at the putative interface prevented HIV-1 CA dimerization in solution, they did not completely abolish HIV-1 assembly (von Schwedler *et al.*, 2003a).

Recently, a different model of HIV-1 CTD dimerization was described, in which the MHR forms a major part of the dimerization interface (Ivanov *et al.*, 2007). These authors previously found that the SCAN (SRE-ZBP, CT-fin-51, AW-1, and number 18 cDNA) domain, a dimerization module in mammalian zinc finger proteins, is a close structural homolog of the HIV-1 CTD (Ivanov *et al.*, 2005). However, the SCAN domain uses an entirely different dimerization interface that is formed by swapping the region corresponding to the MHR between the monomers (Ivanov *et al.*, 2005). This led to the suggestion that an analogous MHR-swapped dimer forms during retroviral assembly, thereby providing a straightforward explanation for the unusual conservation and critical role of the MHR (Ivanov *et al.*, 2005). Direct evidence for this concept comes from a recent X-ray structure of a HIV-1 CTD mutant, which demonstrated that the CTD is indeed capable of adopting a domain-swapped architecture in which the MHR forms a major part of the dimer interface (Ivanov *et al.*, 2007). In order to obtain sufficient amounts of material for analysis, domain-swapped dimerization was triggered by deleting Ala-177 in the linker region that connects the MHR with the rest of the CTD. However, the authors suggest that in the intact Gag precursor, Gag modules adjacent to the CTD may help to trigger formation of the domain-swapped dimer by inducing proximity (Ivanov *et al.*, 2007). Notably, although ΔAla-177 deletion mutants can assemble into both spherical VLPs and tubes *in vitro* (Ivanov *et al.*, 2007), virus particle assembly in human cells is completely abolished (Sergei Popov and Heinrich Göttlinger, unpublished observation). Thus, the biological relevance of the ΔAla-177-induced domain-swapped dimer must still be fully verified.

Role of the CA-NC boundary

In all structures of the HIV-1 CTD, C-terminal residues that are critical for immature particle assembly were disordered (Gamble *et al.*, 1997; Worthylake *et al.*, 1999). Functional studies indicated that these C-terminal CTD residues form part of a distinct assembly domain that extends into the adjacent SP1 peptide (also called p2), which separates CA from NC (Accola *et al.*, 1998; Gottlinger *et al.*, 1989; Krausslich *et al.*, 1995). Although SP1 is sometimes considered a spacer peptide, its first four amino-acids in fact constitute a critical assembly determinant (Accola *et al.*, 1998; Krausslich *et al.*, 1995; Liang *et al.*, 2002). This N-terminal region of SP1 is highly conserved and is predicted to form part of an α-helix that begins in the CTD (Accola *et al.*, 1998). Interestingly, a mutagenic analysis supports the idea that the propensity of the CA-SP1 boundary to adopt an α-helical conformation is essential for its role in viral particle assembly *in vivo* (Accola *et al.*, 1998). This putative α-helix appears to be mainly required for the induction of curvature, and thus for an immature as opposed to a mature assembly mode, because its disruption invariably led to the assembly of large sheets of Gag underneath the plasma membrane (Accola *et al.*, 1998; Gottlinger *et al.*, 1989; Krausslich *et al.*, 1995). Furthermore, the disruption of the helix by PR is probably required for the assembly of mature capsids, because cleavage site mutations at the CA-SP1 junction prevent the formation of mature capsids (Accola *et al.*, 1998; Krausslich *et al.*, 1995). These concepts are supported by *in vitro* assembly studies, which confirmed that SP1 functions as a switch that allows the formation of immature-like spherical capsids, whereas only mature-like tubes and cones are assembled in its absence (Gross *et al.*, 2000).

Like HIV-1, lentiviruses in general have a CA-NC spacer region; this spacer is also highly critical for the assembly of bovine immunodeficiency virus (BIV) (Guo *et al.*, 2004). The avian alpharetrovirus RSV has a 12-residue spacer peptide (SP), which together with the last eight residues of CA and the first four residues of

NC forms a distinct morphology-determining assembly element (Keller *et al.*, 2008). If this element is disrupted, mature-like tubes rather than immature spheres are formed (Keller *et al.*, 2008), which is analogous to the requirement for SP1 in the assembly of spherical HIV-1 particles. Retroviruses that lack a CA-NC spacer region may nevertheless harbour a functionally equivalent assembly element, as illustrated by the role of the recently identified 'charged assembly helix' at the C-terminus of the CTD of Murine leukaemia virus (MLV) (Cheslock *et al.*, 2003). Intriguingly, the function of this motif in MLV assembly appears mainly dependent on its α-helical conformation and not on its primary sequence (Cheslock *et al.*, 2003), which is again reminiscent of the role of HIV-1 SP1. Overall, these parallels between widely divergent retroviruses indicate that the region between the folded portions of the CTD and of NC is specifically required for the immature assembly mode.

Arrangement of Gag in immature capsids

Immature HIV-1 particles are heterogeneous in size and ultrastructure, precluding their analysis by crystallography. However, cryo-electron microscopy combined with radial density analysis confirmed the radial arrangement of HIV-1 Gag molecules in immature particles, and revealed the presence of concentric layers of density that are thought to correspond to the separately folded MA, NTD, CTD, and NC domains (Fuller *et al.*, 1997; Wilk *et al.*, 2001). Furthermore, image processing of cryo-electron micrographs of authentic immature HIV-1 particles indicated the presence of a hexagonal lattice at the same radius as the density inferred to be the NTD (Briggs *et al.*, 2004). This hexagonal order was reminiscent of that seen within helical tubes assembled from purified HIV-1 CA, where the NTDs of CA formed hexameric rings that were connected to each other via the CTDs (Li *et al.*, 2000). Surprisingly, electron cryotomography recently revealed that the ordered Gag lattice in immature HIV-1 virions is not continuous, but rather is interrupted by substantial regions of disorder (Wright *et al.*, 2007). In areas with an ordered CA lattice, radial projections showed patches of hexagonal order in both the NTD and CTD layers (Wright *et al.*, 2007). In contrast, no such order was apparent in the MA and NC layers, even though it was recently shown that HIV-1 MA assembles on membranes as a hexamer (Alfadhli *et al.*, 2007). Intriguingly, a distinct hexagonal lattice with a slightly smaller spacing, inferred to be composed of the final CA residues and of the SP1 spacer peptide, was observed between the CTD and NC layers (Wright *et al.*, 2007). The peaks of the SP1 lattice are located directly below the holes in the CA rings, and based on the observed density below the CTD rings the authors proposed that SP1 forms a six-helix bundle in immature virions (Wright *et al.*, 2007). In the immature lattice, the SP1 bundle is proposed to hold CA hexamers together from below (Wright *et al.*, 2007). In contrast, in the mature lattice it is the NTDs that drive hexamer formation by CA from above. If confirmed, these results could nicely explain why the HIV-1 NTD is dispensable for the assembly of immature particles but absolutely essential for the formation of mature conical capsids (Borsetti *et al.*, 1998).

Assembly of mature capsids

After proteolytic processing of the immature Gag shell, the assembly of mature capsids is believed to occur *de novo* rather than through condensation of the pre-existing CA lattice, because virions with two or more mature capsids are sometimes observed (Benjamin *et al.*, 2005). Like immature virions, mature capsids differ significantly in size and shape (Briggs *et al.*, 2003). To explain this variability, it was suggested that the assembly of the mature HIV-1 capsid begins at the narrow end, and that capsid elongation proceeds through the interior of the virion until it hits the viral membrane on the other side (Briggs *et al.*, 2006). This model is based on the finding that neither the size of the narrow cap nor the cone angle vary with virion size, whereas the diameter of the broad end of the mature capsid correlates with the virion diameter (Briggs *et al.*, 2006). Only about a third of the available CA ultimately contributes to the formation of the mature capsid (Benjamin *et al.*, 2005; Briggs *et al.*, 2004), which may facilitate uncoating once the free CA is released upon entry into a new target cell (Mortuza *et al.*, 2004).

Pure recombinant HIV-1 CA-NC can by itself assemble into cones that resemble mature capsids, demonstrating that no other viral or cellular factors are required (Ganser *et al.*, 1999). The synthetic capsids exhibited quantized cone angles, indicating that they were composed of a closed hexagonal lattice (Ganser *et al.*, 1999). This led to the proposal that HIV-1 capsids are 'fullerene cones', which are formed through the closure of a hexagonal lattice via the inclusion of 12 pentons (Fig. 7.2). Specifically, the inclusion of 5 pentons at the narrow end yields a cone of the shape most often seen in mature HIV-1 virions, and the cylindrical and spherical capsids seen in other retroviruses can be created through a different spacing of the pentons. Image reconstructions of HIV-1 CA tubes and authentic mature capsids assembled *in vitro* revealed that these are indeed helical assemblies, providing support for the fullerene cone model (Briggs *et al.*, 2003; Li *et al.*, 2000). Molecular modelling of *in vitro* assemblies of HIV-1 or RSV CA indicated that the NTDs form hexameric rings and the CTDs form dimeric linkers that interconnect the hexamers (Li *et al.*, 2000; Mayo *et al.*, 2002). In support of this concept, a hexameric form of the NTD of MLV CA was directly observed by X-ray crystallography (Muriaux *et al.*, 2004).

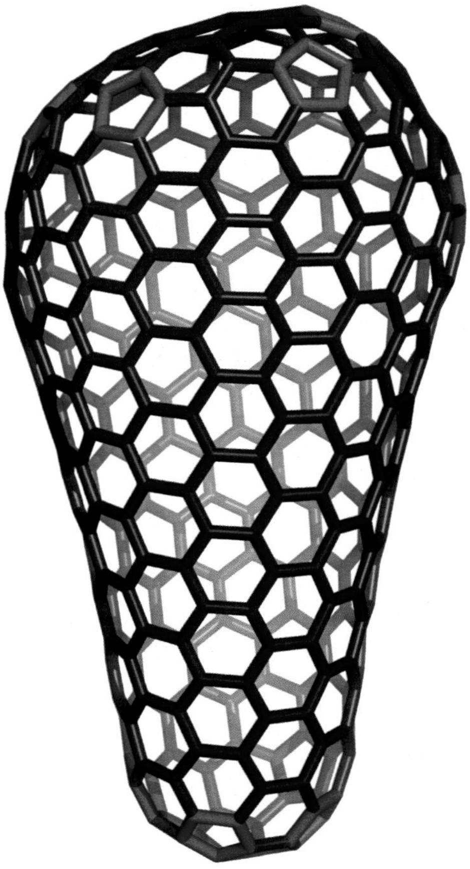

Figure 7.2 Hexagonal lattice model of the mature HIV-1 capsid. Pentameric defects are in red (courtesy of Wesley Sundquist).

A hexameric lattice assembled from full-length HIV-1 CA was recently visualized by electron cryocrystallography (Ganser-Pornillos *et al.*, 2007). This study confirmed that the NTDs form the inner rings of the CA hexamers (Fig. 7.3A), and indicated that the CTDs link neighbouring hexamers by forming conventional rather than domain-swapped dimers (Fig. 7.3B) (Ganser-Pornillos *et al.*, 2007). Importantly, the study also revealed that interactions between the NTDs and CTDs of neighbouring CA molecules in the same hexamer create a third interface that stabilizes the hexagonal CA lattice (Fig. 7.3A and C) (Ganser-Pornillos *et al.*, 2007). This novel interface is in part formed by NTD helix 4 and CTD helices 8 and 9 (Ganser-Pornillos *et al.*, 2007), which is in remarkable agreement with earlier chemical cross-linking experiments showing that lysine residues in helices 4 and 9 are in close proximity in tubes of HIV-1 CA assembled *in vitro* (Lanman *et al.*, 2003). The first evidence for a NTD–CTD interaction came from genetic studies of RSV, which unexpectedly revealed that the assembly defect of a mutant with a change in helix 9 within the CTD can be suppressed by a second-site mutation in helix 4 of the NTD (Bowzard *et al.*, 2001). Furthermore, the existence of the NTD–CTD interaction was suggested by *in vitro* assembly studies, in which the CTD-mediated inhibition of HIV-1 CA assembly was alleviated by addition of the NTD (Lanman *et al.*, 2002). A particularly exciting aspect of the new model for full-length HIV-1 CA is that it clarifies the mechanism by which the 12-residue peptide CAI (Sticht *et al.*, 2005) and the small molecule *N*-(3-chloro-4-methylphenyl)-*N'*-{2-[({5-[(dimethylamino)-methyl]-2-furyl}-methyl)-sulphanyl]ethyl}urea (CAP-1) (Tang *et al.*, 2003) inhibit HIV-1 CA assembly. The model indicates that both inhibitors act by disrupting the NTD–CTD interface (Fig. 7.3C), and thus highlights this interaction

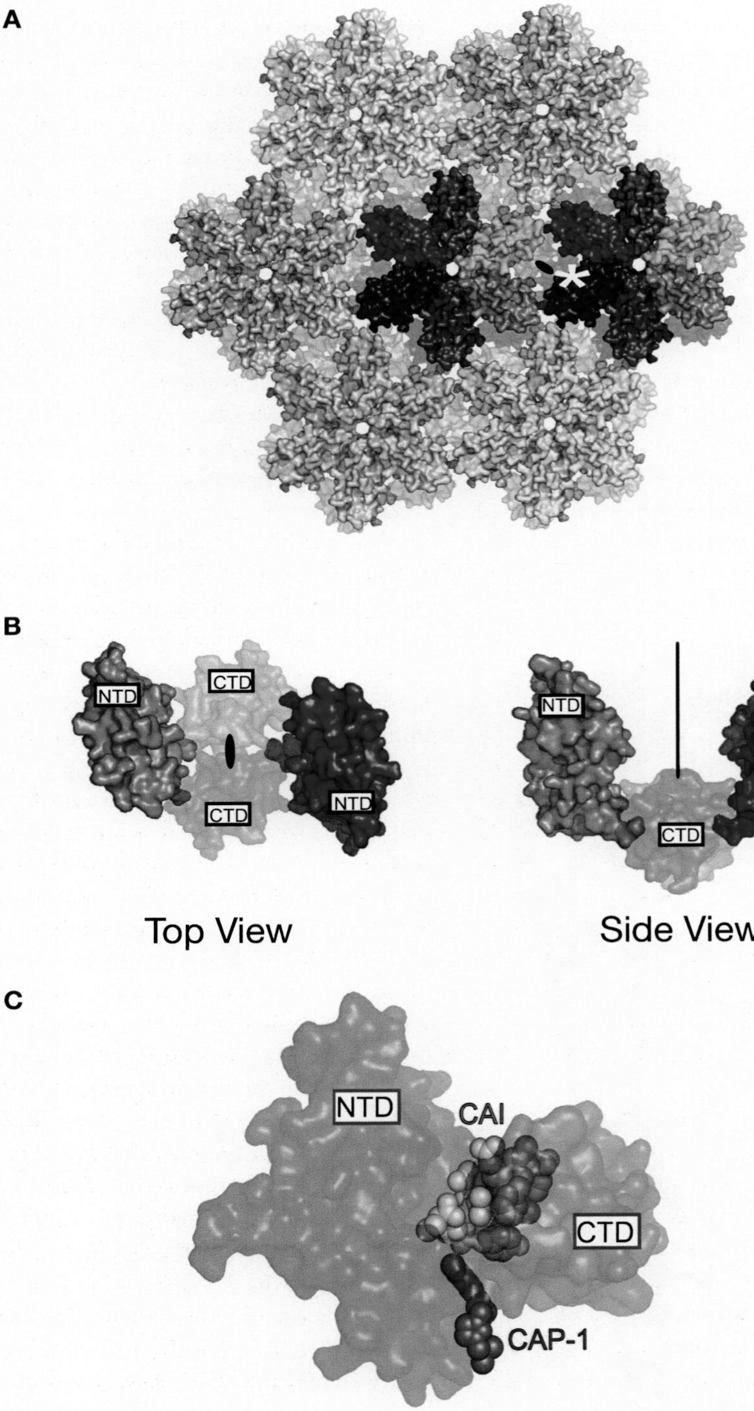

Figure 7.3 (A) Molecular model of the hexameric lattice formed by HIV-1 CA, as seen from the outside of the mature capsid. The CA monomers within two of the hexamers are coloured individually, with individual NTDs and CTDs depicted in bright and light shades, respectively. An CTD–CTD and a NTD–CTD interface are highlighted by a black dyad and a white asterisk, respectively. (B) Expanded view of a CA dimer. (C) Model illustrating how the capsid assembly inhibitors CAI and CAP-1 might interfere with the formation of the NTD–CTD interface between adjacent CA molecules within a CA hexamer. (Reprinted from WI Sundquist and CP Hill (2007). How to Assemble a Capsid. Cell *131*, 17–19, with permission from Elsevier.)

as a promising target for a new class of antiviral compounds (Ganser-Pornillos *et al.*, 2007).

The interconnected roles of NC in RNA encapsidation and assembly

In the Gag proteins of all orthoretroviruses, the NC domain follows CA, either directly or separated by a spacer peptide. A major essential function of NC during assembly is the specific encapsidation of the dimeric viral genomic RNA into nascent virions (Berkowitz *et al.*, 1996). An invariant feature of NC in orthoretroviruses is the presence of one or two copies of a CCHC (Cys-Cys-His-Cys) array that coordinates zinc (Summers *et al.*, 1992). The two CCHC arrays in HIV-1 NC mediate sequence specific interactions with the genomic viral RNA by binding to exposed guanosines in the packaging signal (Amarasinghe *et al.*, 2000; De Guzman *et al.*, 1998), the region of the viral genome that is required *in cis* for its specific encapsidation. In the case of MLV, which harbours only a single CCHC array, the binding of NC to the packaging signal is regulated by a RNA conformational switch mechanism (D'Souza and Summers, 2004). Specifically, NC-binding elements in the packaging signal are sequestered by base-pairing in the monomeric RNA and become exposed for high affinity NC–binding upon dimerization (D'Souza and Summers, 2004). The highly basic NC protein also binds non-specifically to nucleic acids and promotes various essential annealing reactions in the retroviral life cycle.

A role for NC in assembly became apparent when it was noted that a RSV mutant with a deletion in NC showed a strong defect in virus particle release (Dupraz and Spahr, 1992). Also, point mutations in HIV-1 NC that disrupted both CCHC arrays simultaneously significantly reduced HIV-1 particle formation in human cells (Dorfman *et al.*, 1993). Furthermore, a yeast two hybrid-based Gag multimerization assay supported the proposal that NC is involved in Gag–Gag interactions (Franke *et al.*, 1994a). A linkage between RNA binding and assembly was then suggested by reports showing that an N-terminal basic region of HIV-1 NC is particularly critical for assembly (Dawson and Yu, 1998; Sandefur *et al.*, 2000; Zhang and Barklis, 1997). Mutagenic studies also revealed an inverse relationship between the number of basic NC residues that were substituted and the level of viral particle assembly (Bowzard *et al.*, 1998; Cimarelli and Luban, 2000; Cimarelli *et al.*, 2000), arguing for a critical role of the non-specific RNA-binding activity of NC in assembly. Consistent with this view, an *in vitro* assay for Gag–Gag interactions revealed that NC and RNA are both required, indicating that the NC-mediated interaction between Gag monomers occurs through a RNA bridge (Burniston *et al.*, 1999). This bridging function can be exerted both by viral and cellular RNAs, and probably serves to concentrate Gag molecules to trigger the formation of protein–protein contacts involving CA.

Support for this model comes from the observation that retroviral particles always contain roughly similar amounts of nucleic acid, and that this is achieved through the packaging of cellular RNA if genomic RNA is not available (Muriaux *et al.*, 2001). Furthermore, RNA appears to play a role in the maintenance of MLV capsids, since detergent-treated immature capsids were disrupted by treatment with RNase (Muriaux *et al.*, 2001). Thus, while it has long been known that the presence of genomic RNA is not required for retroviral assembly, RNA is nevertheless a structural element of retroviral particles (Muriaux *et al.*, 2001).

In a ground-breaking study, Campbell and Vogt showed that RSV and HIV-1 CA-NC fragments can assemble into particle-like structures *in vitro*, and that the efficiency of assembly is dramatically increased in the presence of heterologous RNA (Campbell and Vogt, 1995). These authors also reported that the length of the mature-like tubes that were formed was proportional to the length of the input RNA, suggesting that RNA may function as a scaffold in the assembly process (Campbell and Vogt, 1995). The *in vitro* assembly of HIV-1 Gag into small spherical particles also required nucleic acid, and removal of the nucleic acid disrupted the assembly products (Campbell and Rein, 1999). Together, these studies confirmed the role of RNA in the assembly and maintenance of both mature and immature retroviral structures.

Although there is an absolute requirement for nucleic acid, the *in vitro* assembly of HIV-1

Gag occurs both on RNA and DNA, and even on short oligonucleotides (Campbell and Rein, 1999). The minimum length requirement for these oligonucleotides is about twice the binding site of NC, indicating that the primary role of nucleic acid in assembly is to induce Gag dimerization (Campbell and Rein, 1999; Ma and Vogt, 2002). Indeed, oligonucleotide-induced RSV Gag dimerization could be directly observed if *in vitro* assembly was arrested at a basic pH, and Gag dimerization and assembly shared the same requirement for oligonucleotide length (Ma and Vogt, 2004). Interestingly, in this model system, nucleic acid of a minimal length was required for Gag dimerization, but not for the maintenance of these dimers or for their ability to polymerize into higher order structures. Thus, while RSV CA remains a monomer even at very high protein concentrations, it appears that the proximity induced by the NC-nucleic acid interaction triggers the exposure of interfaces in CA that mediate Gag dimerization and subsequent assembly (Ma and Vogt, 2004).

If the role of NC and of nucleic acid in assembly is limited to the induction of Gag dimerization, then unrelated dimerization domains might be able to substitute for the role of NC in assembly, and this was indeed observed. Barklis and co-workers were the first to show that the replacement of HIV-1 NC with unrelated polypeptides that form interprotein contacts allows efficient assembly even in the absence of detectable RNA incorporation (Zhang *et al.*, 1998). Efficient particle production was even observed by minimal HIV-1 Gag constructs in which NC was replaced by a leucine zipper dimerization domain, and that also lacked most of MA and the NTD (Accola *et al.*, 2000). Thus, the remaining CTD-SP1 fragment of HIV-1 Gag in combination with a heterologous dimerization domain was sufficient to drive assembly. In a recent *in vitro* assembly study, the replacement of HIV-1 NC by a reactive thiol allowed the initiation of assembly via a cross-linking mechanism, further supporting the view that the role of NC in assembly is to promote the initial pairing of Gag molecules (Alfadhli *et al.*, 2005). A foreign dimerization domain could also mimic the function of RSV NC in the formation and budding of VLP, allowing nucleic acid-independent assembly (Johnson *et al.*, 2002). Thus, the role of the NC–RNA interaction in promoting the formation of assembly-competent Gag dimers is probably universal among retroviruses. As was pointed out earlier (Johnson *et al.*, 2002), the fact that NC can be functionally replaced by completely unrelated dimerization domains argues against a scaffolding role for RNA in retroviral assembly, and instead suggests a model in which the NC and nucleic acid-mediated juxtaposition triggers complex conformational changes that convert Gag into an assembly-competent form.

Gag domains involved in virus release

Since retroviruses become enveloped by budding through the plasma membrane, a membrane fission event ultimately becomes necessary to separate the virion envelope from the cell membrane. It is now well established that Gag actively promotes this late assembly step by engaging components of a cellular budding pathway.

Virus release involves L domains

The first evidence that the detachment of retroviruses from the cell surface and from each other does not occur spontaneously but rather involves specific Gag sequences was obtained for HIV-1, where virus release depends on the C-terminal p6 domain of Gag, which has no structural role (Gottlinger *et al.*, 1991). Mutations within a short sequence motif in HIV-1 p6 arrested virus assembly at a late budding stage (Gottlinger *et al.*, 1991; Huang *et al.*, 1995); a similar late budding defect was subsequently observed for RSV upon mutation of the p2b spacer peptide of Gag (Wills *et al.*, 1994). It then became apparent that most if not all orthoretroviruses contain functionally equivalent regions in Gag, which are now commonly referred to as 'late assembly' or L domains.

To date, three different types of L domains have been characterized in detail, and these all map to short, highly conserved proline-containing motifs. Remarkably, different types of L domains are functionally interchangeable between unrelated viruses, and their position within Gag can vary (Parent *et al.*, 1995), although some context dependency is conserved (Martin-Serrano *et al.*, 2004; Strack *et al.*, 2002). The primary L domain of HIV-1 maps to a P(T/S)AP motif near the

N-terminus of p6 (Gottlinger *et al.*, 1991; Huang *et al.*, 1995). The requirement for the P(T/S)AP L domain is cell type-dependent; in adherent cells such as macrophages the P(T/S)AP motif is required for virus–cell detachment, whereas in T-cell lines and primary lymphocytes it primarily promotes virion–virion detachment (Demirov *et al.*, 2002). However, the P(T/S)AP L domain is crucial for HIV-1 replication regardless of cell type (Demirov *et al.*, 2002).

The P(T/S)AP motif is highly conserved among most primate lentiviruses, even though p6 is otherwise the most variable Gag domain among this group of viruses. Although other lentiviruses lack a p6 domain, the P(T/S)AP motif is conserved and occupies an equivalent location in Gag. A notable exception is equine infectious anaemia virus (EIAV), which harbours a p9 domain at the C-terminus of Gag that is unrelated to the p6 domains of primate lentiviruses. Nevertheless, the EIAV p9 domain functions in virus release (Parent *et al.*, 1995), and the L domain in p9 was mapped to a conserved LYPxL motif (Puffer *et al.*, 1997). Interestingly, HIV-1 harbours a related motif with the consensus sequence LYPx_nL, which acts as an auxiliary L domain (Strack *et al.*, 2003).

In contrast to the L domains of lentiviruses, those of other retroviruses are located between the MA and CA domains of Gag. The first L domain of this type that was mapped precisely was that of RSV, which consists of the sequence PPPPY (Wills *et al.*, 1994; Xiang *et al.*, 1996). Several other retroviruses were subsequently shown to harbour L domains with the consensus sequence PPxY, including MLV, MPMV, HTLV-I, and bovine leukaemia virus (Le Blanc *et al.*, 2002; Wang *et al.*, 2002; Yasuda and Hunter, 1998; Yuan *et al.*, 1999). L domains occur either alone or in combination, and were also documented in the filovirus matrix protein VP40 (Harty *et al.*, 2000; Martin-Serrano *et al.*, 2001; Strack *et al.*, 2000), the rhabdovirus matrix protein M (Craven *et al.*, 1999), and the Lassa virus Z protein (Perez *et al.*, 2003; Strecker *et al.*, 2003).

L domains engage the MVB pathway

Numerous studies have shown that L domains serve as entry points into a network of so-called class E vacuolar protein sorting (Vps) proteins, which normally function in multi-vesicular body (MVB) formation. MVB formation is morphologically similar to virus budding in that the endosomal membrane must curve and bud away from the cytoplasm, which explains why enveloped viruses make use of this conserved cellular machinery to escape from cells. The connection between the MVB pathway and L domain function became apparent when it was shown that the P(T/S)AP-type L domain binds to the class E Vps protein Tsg101 (Garrus *et al.*, 2001; Martin-Serrano *et al.*, 2001; VerPlank *et al.*, 2001). Similarly, the LYPx_nL-type L domain mediates binding to the Tsg101 interaction partner ALIX/AIP1 (ALG-2-interacting protein X), whose yeast homolog Bro1 also functions in the class E Vps pathway (Martin-Serrano *et al.*, 2003a; Strack *et al.*, 2003; Vincent *et al.*, 2003; von Schwedler *et al.*, 2003b). The third type of L domain (PPxY) mediates interactions with proteins that contain WW (Trp-Trp) domains, in particular with HECT (homologous to E6-associated protein C-terminus) domain-containing E3 ubiquitin ligases of the Nedd4 family, which are part of an enzymatic cascade in which ubiquitin moieties are transferred to lysine residues on the protein substrate (Bouamr *et al.*, 2003; Harty *et al.*, 2000; Kikonyogo *et al.*, 2001; Martin-Serrano *et al.*, 2005; Strack *et al.*, 2000; Yasuda *et al.*, 2002). At least some of the Nedd4 family ubiquitin ligases are recruited to the aberrant endosomal structures that develop if the MVB pathway is blocked (Martin-Serrano *et al.*, 2005), so it appears that all three L domain types provide a link to the pathway catalysing MVB biogenesis.

MVB biogenesis

The protein network involved in MVB formation was first identified in *Saccharomyces cerevisiae*, and implicated in membrane protein trafficking from the Golgi or plasma membrane to the vacuole/lysosome for degradation. The ubiquitination of transmembrane proteins serves as a signal for their transport to endosomal membrane microdomains, ultimately leading to membrane invagination and vesicle budding, and thus to the delivery of cargo from the limiting membrane of the endosome into the lumen of the organelle (Gruenberg and Stenmark, 2004; Katzmann *et al.*, 2002). In *Saccharomyces cerevisiae* this process is largely mediated by ~17 non-essential proteins known as class E Vps proteins. Mutations that disrupt the function of any of the yeast class E Vps proteins lead to the formation of so-called class E compartments, which are enlarged endosomal structures that cannot mature into MVBs (Katzmann *et al.*, 2002). Although the mammalian system is more complex, the framework is similar, and at least one human orthologue is known for each yeast class E Vps protein (Babst, 2005). Most class E Vps proteins participate in the formation of three distinct heteromeric endosomal sorting complexes required for transport (ESCRT). These complexes, known as ESCRT-I, -II, and –III, are transiently recruited to endosomal membranes, and are thought to function in a sequential manner (Babst, 2005). Endosomal sorting also depends on the AAA-type ATPase Vps4, which disassembles ESCRT-III and dissociates all ESCRT complexes from endosomes (Babst *et al.*, 1997; Babst *et al.*, 1998; Katzmann *et al.*, 2002).

ESCRT-I

The initial selection of ubiquitinated cargo on endosomal membranes is mediated by a complex consisting of the class E Vps proteins Hrs and STAM, also known as ESCRT-0 (Williams and Urbe, 2007). This leads to the recruitment of ESCRT-I to the endosomal membrane, mediated in part by an interaction between ESCRT-I component Tsg101 and a PSAP motif in Hrs (Bache *et al.*, 2003; Katzmann *et al.*, 2003; Lu *et al.*, 2003; Pornillos *et al.*, 2003). Thus, the P(T/S)AP-type L domain of HIV-1 and other retroviruses may mimic the Tsg101-recruiting activity of Hrs (Pornillos *et al.*, 2003). Tsg101 contains an ubiquitin-binding UEV (ubiquitin E2 variant) domain, and NMR studies revealed that the HIV-1 PTAP (Pro-Thr-Ala-Pro) motif binds at a conserved hydrophobic groove in the Tsg101 UEV domain (Pornillos *et al.*, 2002). The Tsg101 UEV domain is related to E2 ubiquitin-conjugating enzymes but lacks a catalytic site, and crystallographic studies showed that Tsg101 UEV binds ubiquitin differently from other E2 enzymes (Sundquist *et al.*, 2004), allowing the simultaneous binding to the PTAP motif. This may be relevant for the function of Tsg101 in virus budding, because a Tsg101 mutant with a deletion in the ubiquitin binding pocket exerted a strong dominant-negative effect on the release of HIV-1 Gag (Goff *et al.*, 2003).

The UEV domain is dispensable for Tsg101's role in HIV-1 budding if the rest of the protein is recruited to HIV-1 Gag by alternative means (Martin-Serrano *et al.*, 2001). The critical C-terminal portion of Tsg101 binds Vps28, another ESCRT-I component that is required for HIV-1 release (Martin-Serrano *et al.*, 2003b; Stuchell *et al.*, 2004). Tsg101 also binds to another ESCRT-I subunit called Vps37, of which human cells express four different versions (Stuchell *et al.*, 2004). One of these versions rescues the budding of an HIV-1 PTAP mutant if recruited to the site of assembly, indicating that Vps37 can also function in virus release (Stuchell *et al.*, 2004). More recently, a fourth member of ESCRT-I, MVB12, was identified and shown to play an auxiliary role in HIV-1 budding (Morita *et al.*, 2007a). Thus, all components of ESCRT-I appear to contribute in some way to its function in virus release. However, the exact mechanism by which ESCRT-I catalyses retrovirus budding remains elusive.

ESCRT-II

In yeast, ESCRT-I functions upstream of ESCRT-II, a complex composed of Vps22p, Vps25p and Vps36p (Babst *et al.*, 2002b). Each of the three yeast components of ESCRT-II has a human orthologue, and these bind to ubiquitin, Tsg101, and ESCRT-III in a similar manner as the yeast proteins (Langelier *et al.*, 2006). Also, human ESCRT-II, like the yeast complex, transiently associates with endosomal membranes (Langelier *et al.*, 2006). As expected, the human ESCRT-II complex has a role in MVB vesicle formation, because the depletion of one of its subunits inhibited the lysosomal targeting of the epidermal growth factor receptor (Langelier *et al.*, 2006). However, the depletion of ESCRT-II did not inhibit HIV-1 release or infectivity, indicating that viral L domains somehow bypass ESCRT-II (Langelier *et al.*, 2006).

ESCRT-III

Whereas ESCRT-I and ESCRT-II are preformed complexes, ESCRT-III assembles on endosomal membranes from monomeric cytosolic subunits

(Babst *et al.*, 2002a; Saksena *et al.*, 2007). ESCRT-III is composed of structurally related class E Vps proteins that have highly basic N-terminal and highly acidic C-terminal halves. Yeast expresses six such proteins: Chmp1p through Chmp6p (*Ch*arged *m*ultivesicular body *p*rotein) (Howard *et al.*, 2001). Human cells contain eleven CHMP proteins, ten of which correspond to the six orthologues found in yeast (Horii *et al.*, 2006; Howard *et al.*, 2001; Kranz *et al.*, 2001). The myristylated CHMP6/Vps20 component of ESCRT-III interacts with ESCRT-II EAP20/Vps25, thus linking ESCRT-II to ESCRT-III (Teo *et al.*, 2004). The binding of ESCRT-II to two copies of CHMP6 is thought to initiate ESCRT-III assembly (Saksena *et al.*, 2007).

The structure of C-terminally truncated CHMP3 shows a four helical bundle core that contains two regions responsible for homodimerization and polymerization *in vitro*, both of which also appear to play a role in the heterodimerization and polymerization of CHMP proteins *in vivo* (Muziol *et al.*, 2006). CHMP proteins exist in a closed cytosolic conformation (Lata *et al.*, 2008a; Zamborlini *et al.*, 2006), and the removal of C-terminal autoinhibitory regions leads to spontaneous membrane targeting and polymerization (Lin *et al.*, 2005; Muziol *et al.*, 2006; Shim *et al.*, 2007).

ESCRT-III has been implicated in retrovirus budding, since the overexpression of catalytically inactive Vps4, the ATPase that disassembles ESCRT-III (Saksena *et al.*, 2007), potently inhibited the release of both HIV-1 and MLV (Garrus *et al.*, 2001). Since dominant-negative Vps4 caused the accumulation of fully assembled viral particles at the cell surface (Garrus *et al.*, 2001), ESCRT-III and Vps4 are thought to act at a final step in budding. The fact that ESCRT-III and VPS4 are also required for cytokinesis (Carlton and Martin-Serrano, 2007; Morita *et al.*, 2007b) supports the hypothesis that ESCRT-III and VPS4 constitute the core of the membrane fission machinery mediating the release of intraluminal vesicles at the endosomal membrane, the abscission of daughter cells during cytokinesis, and the release of enveloped viruses such as HIV-1.

A role of ESCRT-III in viral budding is also indicated by the finding that the fusion of CHMP proteins to a bulky tag, or the removal of autoinhibitory regions, converts them into highly potent inhibitors of retrovirus release (Martin-Serrano *et al.*, 2003a; Muziol *et al.*, 2006; Strack *et al.*, 2003; von Schwedler *et al.*, 2003b; Zamborlini *et al.*, 2006). This produces fully assembled virions that stay attached to the plasma membrane and fail to pinch off (Strack *et al.*, 2003; von Schwedler *et al.*, 2003b), similar to what is seen with dominant-negative Vps4 (Garrus *et al.*, 2001). Although it appears that all CHMP proteins can be converted into inhibitors of HIV-1 budding, the three human CHMP4 isoforms differ with regard to the level of inhibition that can be achieved, which could indicate functional diversification (Carlton *et al.*, 2008). Furthermore, despite their potency as dominant-negative inhibitors, clearly not all CHMP proteins are required for retroviral budding. In particular, HIV-1 release from cells depleted of CHMP5 or CHMP6 was not reduced (Langelier *et al.*, 2006; Ward *et al.*, 2005). In contrast, the CHMP5 binding partner LIP5, which regulates the catalytic activity of Vps4 (Azmi *et al.*, 2008), is required for HIV-1 budding (Ward *et al.*, 2005).

Conserved C-terminal amphipathic 'MIM1' helical peptide motifs, which are present in human CHMP1A and 2B and in the yeast Vps2 and Did2 proteins, interact with the MIT (microtubule interacting and trafficking) domains of human and yeast Vps4 (Obita *et al.*, 2007; Stuchell-Brereton *et al.*, 2007). A similar binding mode was proposed for the Vps4–CHMP3 interaction (Lata *et al.*, 2008b), whereas CHMP4 and CHMP6 use a different 'MIM2' motif to recruit Vps4 (Kieffer *et al.*, 2008). Mutations within the CHMP–VPS4 interaction motifs arrest HIV-1 budding, highlighting the importance of these interactions for retroviral release (Kieffer *et al.*, 2008; Stuchell-Brereton *et al.*, 2007).

When overexpressed in mammalian cells, CHMP4A and CHMP4B assemble into membrane-associated circular filaments, and in the presence of dominant-negative Vps4 these filaments deform the plasma membrane into buds and tubules protruding from the cell surface (Hanson *et al.*, 2008). CHMP4 isoforms may thus play a special role in retroviral budding. Furthermore, recent evidence showed that CHMP2A and CHMP3 can copolymerize into helical tubular structures (Fig. 7.4) that contain

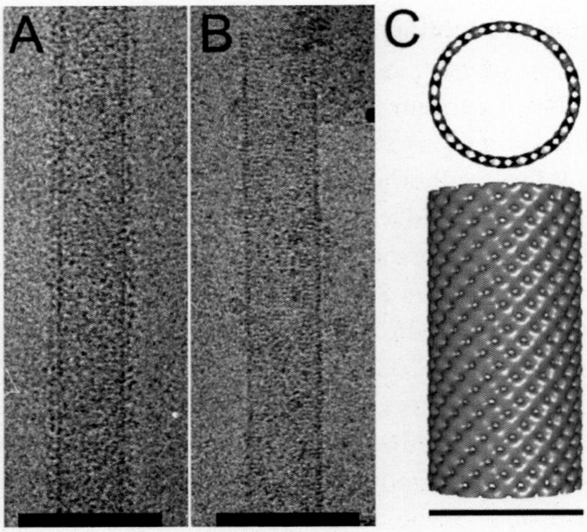

Figure 7.4 (A) and (B) Cryo-electron micrographs of tubular structures formed by C-terminally truncated versions of CHMP2 and CHMP3. The tubes had a fuzzy surface if maltose binding protein (MBP) was left attached to CHMP2 (A), but exhibited a smooth surface after MBP was removed (B). (C) EM reconstruction model of the lattice formed by CHMP2 and CHMP3. (Reprinted with permission from Lata *et al.*, 2008b.)

the membrane binding surface on the outside and the VPS4 interaction motif on the inside. The polymer formation is cooperatively enhanced on lipid bilayers, and the polymers are disassembled by Vps4 *in vitro* (Lata *et al.*, 2008b). Thus, the CHMP2A-CHMP3 helical structure may assemble on the inside of a budding vesicle or virion. The removal of individual subunits from helical CHMP polymers by Vps4 might then induce membrane constriction and abscission of two membranes, leading to virus release (Kieffer *et al.*, 2008; Lata *et al.*, 2008b).

ALIX

All three human CHMP4 isoforms directly interact with ALIX/AIP1, the cellular binding partner for the LYPx_nL-type L domains of HIV-1 and EIAV (Katoh *et al.*, 2004; Martin-Serrano *et al.*, 2003a; Strack *et al.*, 2003; von Schwedler *et al.*, 2003b). ALIX also interacts with Tsg101, and thus could provide a direct link between ESCRT-I and ESCRT-III (Strack *et al.*, 2003; von Schwedler *et al.*, 2003b). Although ALIX was first considered to be the orthologue of yeast Bro1, which functions in the MVB pathway (Odorizzi *et al.*, 2003), the role of Bro1 in receptor down regulation in yeast was not confirmed for ALIX in mammalian cells (Cabezas *et al.*, 2005; Doyotte *et al.*, 2008; Schmidt *et al.*, 2004). Instead, HD-PTP (His-domain-containing protein tyrosine phosphatase), a homologue of ALIX that also interacts with CHMP4B (Ichioka *et al.*, 2007), is required for receptor sorting (Doyotte *et al.*, 2008). Thus, HD-PTP seems to constitute the functional homologue of yeast Bro1 (Doyotte *et al.*, 2008). Nevertheless, HD-PTP does not support LYPx_nL-mediated virus budding (Yoshiko Usami and Heinrich Göttlinger, unpublished observation).

Structural analyses revealed that ALIX is composed of three distinct regions: two of these regions are folded domains and the third is a C-terminal proline-rich domain (PRD) that harbours docking sites for cellular binding partners such as Tsg101. The banana-shaped N-terminal Bro1 domains of human ALIX and of yeast Bro1 both contain a core of tetratricopeptide repeats (Fisher *et al.*, 2007; Kim *et al.*, 2005). The concave surface of the ALIX Bro1 domain interacts with C-terminal amphipathic helices in CHMP4A, B and C (McCullough *et al.*, 2008). The Bro1 domain of ALIX is linked to a V-shaped middle domain. The two arms of the V domain are composed of three-helix bundles, and a hydrophobic groove on arm 2 constitutes the binding site for LYPx_nL-type L domains (Fisher *et al.*, 2007; Lee *et al.*, 2007; Zhai *et al.*, 2008). Notably, EIAV p9, which does not interact with Tsg101, binds ALIX about 60-fold more tightly than HIV-1 p6 (Zhai *et al.*, 2008). Thus, retroviruses such as EIAV that depend on ALIX (Fisher *et al.*, 2007; Strack *et al.*, 2003) possess a high-affinity binding site, whereas viruses such as HIV-1 that rely primarily on Tsg101 bind ALIX with much lower affinity.

Although ALIX is much less important for HIV-1 than for EIAV budding, dominant-negative ALIX fragments potently inhibit HIV-1 release (Munshi *et al.*, 2007; Strack *et al.*, 2003). Also, ALIX is required for the budding of a minimal HIV-1 Gag construct (Strack *et al.*, 2003). Furthermore, the overexpression of wild type ALIX potently rescues the severe budding defect of HIV-1 mutants that lack a functional PTAP L domain (Fisher *et al.*, 2007; Usami *et al.*, 2007). As expected, this rescue depends on the LYPx_nL-type L domain in HIV-1 p6 (Fisher *et al.*,

2007; Usami *et al.*, 2007; Zhai *et al.*, 2008) and on the integrity of the ligand binding site in the V domain of ALIX (Fisher *et al.*, 2007). The ability of ALIX to rescue Tsg101 binding site mutants of HIV-1 is also absolutely dependent on its interaction with CHMP4 proteins (Fisher *et al.*, 2007; McCullough *et al.*, 2008; Usami *et al.*, 2007); this observation provides the most convincing evidence to date that CHMP4 plays a pivotal role in retroviral budding. In contrast, the PRD-mediated interaction of ALIX with endophilin, CD2AP, CIN85, or CEP55 is dispensable for its function in retrovirus budding, even though the PRD is essential (Carlton *et al.*, 2008; Fisher *et al.*, 2007; Usami *et al.*, 2007). Furthermore, a PSAP motif in the PRD mediating binding to Tsg101 is also dispensable (Fisher *et al.*, 2007; Usami *et al.*, 2007). Analysis of a series of truncation mutants showed that the very C-terminus of the ~ 150 amino acid PRD is essential for ALIX's ability to promote virus budding, even though this region is poorly conserved (Usami *et al.*, 2007). Interestingly, the same C-terminal region is also required for ALIX multimerization, and the forced oligomerization of ALIX restores its ability to promote virus release (Carlton *et al.*, 2008). One explanation for these observations is that ALIX multimerization is required for its interaction with Tsg101 (Carlton *et al.*, 2008).

A recent study revealed that HIV-1 Gag contains a second binding site for ALIX in NC (Popov *et al.*, 2008). In contrast to HIV-1 p6, which binds to the V domain of ALIX, NC interacts with the N-terminal Bro1 domain through its two CCHC arrays (Popov *et al.*, 2008). ALIX's ability to rescue PTAP L domain mutants depends on intact CCHC motifs in NC, and the disruption of the CCHC motifs led to a phenotype resembling that of L domain mutants (Popov *et al.*, 2008). These results support the notion that HIV-1 NC and its interaction with ALIX play a role in virus release. Notably, a L domain-like budding defect was also observed after deletion of either of the two CCHC arrays in the NC domain of RSV (Lee and Linial, 2006), which suggests a more general role of NC in retroviral budding.

Ubiquitin and retrovirus budding

Retroviral particles contain considerable amounts of free ubiquitin, as well as a small amount of ubiquitin covalently bound to Gag (Ott *et al.*, 2000; Ott *et al.*, 1998; Putterman *et al.*, 1990). For HIV-1, the mono-ubiquitination of Gag can occur in all of its domains and is highly dependent on the membrane association of Gag (Gottwein and Krausslich, 2005; Jager *et al.*, 2007). Evidence that proteasome inhibitors interfere with virus release – probably by depleting the levels of free ubiquitin – suggested a role for ubiquitin in retroviral budding (Patnaik *et al.*, 2000; Schubert *et al.*, 2000; Strack *et al.*, 2000). Although these observations may be due to an indirect effect of ubiquitin depletion on the class E Vps pathway, PPxY-type L domains also induced a robust mono-ubiquitination of Gag constructs to which they were appended (Strack *et al.*, 2000). Furthermore, the induction of Gag ubiquitination by PPxY-type L domains correlated perfectly with L domain function (Strack *et al.*, 2000). This finding suggested that PPxY-type L domains function by recruiting Nedd4 family ubiquitin ligases, which bind PPxY motifs through their multiple WW domains. Indeed, numerous subsequent studies implicated Nedd4 and related HECT domain ubiquitin ligases in the function of PPxY-type L domains (Blot *et al.*, 2004; Bouamr *et al.*, 2003; Harty *et al.*, 2000; Kikonyogo *et al.*, 2001; Martin-Serrano *et al.*, 2005; Segura-Morales *et al.*, 2005; Timmins *et al.*, 2003; Vana *et al.*, 2004; Yasuda *et al.*, 2002,2003). Perhaps most compelling is the observation that the overexpression of several catalytically active Nedd4 family ubiquitin ligases rescued release defects caused by 'leaky' PPxY-motif mutations or by inhibitory fragments derived from these enzymes (Martin-Serrano *et al.*, 2005). The same study also showed that Nedd4 family ubiquitin ligases are recruited to aberrant endosomal compartments induced by catalytically inactive Vps4, thus establishing a potential link between PPxY-type L domains and the MVB pathway (Martin-Serrano *et al.*, 2005).

Interestingly, a recent study showed that overexpression of a Nedd4 family ubiquitin ligase markedly enhanced PPxY-dependent budding in the absence of ubiquitin acceptors in Gag (Zhadina *et al.*, 2007). This effect was abrogated by mutating the catalytic site of the ubiquitin ligase, implying that the conjugation of ubiquitin to a *trans*-acting factor is critical for the effect of

Nedd4 ubiquitin ligases on virus release (Zhadina *et al.*, 2007).

Because HIV-1 Gag lacks PPxY motifs, Nedd4 family ubiquitin ligases were not expected to affect HIV-1 budding. However, two recent studies showed that HIV-1 budding defects can be rescued in a remarkably potent manner through the overexpression of a specific splice variant of Nedd4–2 (also called Nedd4L) that lacks most of the N-terminal C2 domain (Chung *et al.*, 2008; Usami *et al.*, 2008). This rescue depended on a catalytically active HECT domain (Chung *et al.*, 2008; Usami *et al.*, 2008) and on the naturally truncated C2 domain, which accounts for the ability of the ubiquitin ligase to interact with HIV-1 Gag (Usami *et al.*, 2008). It appears that the Nedd4–2 splice variant interacts with the CTD of CA, since all other HIV-1 Gag domains were dispensable for its effect (Chung *et al.*, 2008; Usami *et al.*, 2008). This interaction may contribute to viral budding at endogenous expression levels in cell types that express limiting amounts of Tsg101, because depletion of the ubiquitin ligase further reduced the release of HIV-1 L domain mutants (Chung *et al.*, 2008). Interestingly, the stimulation of HIV-1 budding by Nedd4-2 required Tsg101 (Chung *et al.*, 2008). This finding lends further support to the notion that all types of L domains ultimately connect to the same cellular budding machinery.

References

Accola, M.A., Hoglund, S., and Gottlinger, H.G. (1998). A putative alpha-helical structure which overlaps the capsid-p2 boundary in the human immunodeficiency virus type 1 Gag precursor is crucial for viral particle assembly. J. Virol. 72, 2072–2078.

Accola, M.A., Strack, B., and Gottlinger, H.G. (2000). Efficient particle production by minimal *gag* constructs which retain the carboxy-terminal domain of human immunodeficiency virus type 1 capsid-p2 and a late assembly domain. J. Virol. *74*, 5395–5402.

Adamson, C.S., and Freed, E.O. (2007). Human immunodeficiency virus type 1 assembly, release, and maturation. Adv Pharmacol 55, 347–387.

Alfadhli, A., Dhenub, T.C., Still, A., and Barklis, E. (2005). Analysis of human immunodeficiency virus type 1 Gag dimerization-induced assembly. J. Virol. *79*, 14498–14506.

Alfadhli, A., Huseby, D., Kapit, E., Colman, D., and Barklis, E. (2007). Human immunodeficiency virus type 1 matrix protein assembles on membranes as a hexamer. J. Virol. *81*, 1472–1478.

Amarasinghe, G.K., De Guzman, R.N., Turner, R.B., Chancellor, K.J., Wu, Z.R., and Summers, M.F. (2000). NMR structure of the HIV-1 nucleocapsid protein bound to stem–loop SL2 of the psi-RNA packaging signal. Implications for genome recognition. J. Mol. Biol. *301*, 491–511.

Azmi, I.F., Davies, B.A., Xiao, J., Babst, M., Xu, Z., and Katzmann, D.J. (2008). ESCRT-III family members stimulate Vps4 ATPase activity directly or via Vta1. Dev. Cell *14*, 50–61.

Babst, M. (2005). A protein's final ESCRT. Traffic *6*, 2–9.

Babst, M., Katzmann, D. J., Estepa-Sabal, E. J., Meerloo, T., and Emr, S. D. (2002a). Escrt-III: an endosome-associated heterooligomeric protein complex required for mvb sorting. Dev. Cell 3,271–282.

Babst, M., Katzmann, D.J., Snyder, W.B., Wendland, B., and Emr, S.D. (2002b). Endosome-associated complex, ESCRT-II, recruits transport machinery for protein sorting at the multivesicular body. Dev. Cell 3, 283–289.

Babst, M., Sato, T.K., Banta, L.M., and Emr, S.D. (1997). Endosomal transport function in yeast requires a novel AAA-type ATPase, Vps4p. EMBO J. *16*, 1820–1831.

Babst, M., Wendland, B., Estepa, E.J., and Emr, S.D. (1998). The Vps4p AAA ATPase regulates membrane association of a Vps protein complex required for normal endosome function. EMBO J. *17*, 2982–2993.

Bache, K.G., Brech, A., Mehlum, A., and Stenmark, H. (2003). Hrs regulates multivesicular body formation via ESCRT recruitment to endosomes. J. Cell Biol. *162*, 435–442.

Balliet, J. W., and Bates, P. (1998). Efficient infection mediated by viral receptors incorporated into retroviral particles. J. Virol. 72, 671–676.

Benjamin, J., Ganser-Pornillos, B.K., Tivol, W.F., Sundquist, W.I., and Jensen, G.J. (2005). Three-dimensional structure of HIV-1 virus-like particles by electron cryotomography. J. Mol. Biol. *346*, 577–588.

Berkowitz, R., Fisher, J., and Goff, S.P. (1996). RNA packaging. Curr. Top. Microbiol. Immunol. *214*, 177–218.

Bhattacharya, J., Repik, A., and Clapham, P.R. (2006). Gag regulates association of human immunodeficiency virus type 1 envelope with detergent-resistant membranes. J. Virol. *80*, 5292–5300.

Bieniasz, P.D. (2006). Late budding domains and host proteins in enveloped virus release. Virology *344*, 55–63.

Blenis, J., and Resh, M.D. (1993). Subcellular localization specified by protein acylation and phosphorylation. Curr. Opin. Cell Biol. *5*, 984–989.

Blot, V., Perugi, F., Gay, B., Prevost, M.C., Briant, L., Tangy, F., Abriel, H., Staub, O., Dokhelar, M.C., and Pique, C. (2004). Nedd4.1-mediated ubiquitination and subsequent recruitment of Tsg101 ensure HTLV-1 Gag trafficking towards the multivesicular body pathway prior to virus budding. J. Cell Sci. *117*, 2357–2367.

Borsetti, A., Ohagen, A., and Gottlinger, H. G. (1998). The C-terminal half of the human immunodeficiency virus type 1 Gag precursor is sufficient for efficient particle assembly. J. Virol. 72, 9313–9317.

Bouamr, F., Melillo, J.A., Wang, M.Q., Nagashima, K., de Los Santos, M., Rein, A., and Goff, S.P. (2003).

PPPYVEPTAP motif is the late domain of human T-cell leukemia virus type 1 Gag and mediates its functional interaction with cellular proteins Nedd4 and Tsg101. J. Virol. *77*, 11882–11895.

Bowzard, J.B., Bennett, R.P., Krishna, N.K., Ernst, S.M., Rein, A., and Wills, J. W. (1998). Importance of basic residues in the nucleocapsid sequence for retrovirus Gag assembly and complementation rescue. J. Virol. *72*, 9034–9044.

Bowzard, J.B., Wills, J.W., and Craven, R.C. (2001). Second-site suppressors of Rous sarcoma virus Ca mutations: evidence for interdomain interactions. J. Virol. *75*, 6850–6856.

Briggs, J.A., Grunewald, K., Glass, B., Forster, F., Krausslich, H.G., and Fuller, S.D. (2006). The mechanism of HIV-1 core assembly: insights from three-dimensional reconstructions of authentic virions. Structure *14*, 15–20.

Briggs, J.A., Simon, M.N., Gross, I., Krausslich, H.G., Fuller, S.D., Vogt, V.M., and Johnson, M.C. (2004). The stoichiometry of Gag protein in HIV-1. Nat. Struct. Mol. Biol. *11*, 672–675.

Briggs, J.A., Wilk, T., Welker, R., Krausslich, H.G., and Fuller, S.D. (2003). Structural organization of authentic, mature HIV-1 virions and cores. EMBO J. *22*, 1707–1715.

Bryant, M., and Ratner, L. (1990). Myristoylation-dependent replication and assembly of human immunodeficiency virus 1. Proc. Natl. Acad. Sci. U.S.A. *87*, 523–527.

Burniston, M.T., Cimarelli, A., Colgan, J., Curtis, S.P., and Luban, J. (1999). Human immunodeficiency virus type 1 Gag polyprotein multimerization requires the nucleocapsid domain and RNA and is promoted by the capsid- dimer interface and the basic region of matrix protein. J. Virol. 73, 8527–8540.

Cabezas, A., Bache, K. G., Brech, A., and Stenmark, H. (2005). Alix regulates cortical actin and the spatial distribution of endosomes. J. Cell Sci. *118*, 2625–2635.

Campbell, S., and Rein, A. (1999). In vitro assembly properties of human immunodeficiency virus type 1 Gag protein lacking the p6 domain. J. Virol. 73, 2270–2279.

Campbell, S., and Vogt, V.M. (1995). Self-assembly *in vitro* of purified CA-NC proteins from Rous sarcoma virus and human immunodeficiency virus type 1. J. Virol. *69*, 6487–6497.

Campos-Olivas, R., Newman, J.L., and Summers, M.F. (2000). Solution structure and dynamics of the Rous sarcoma virus capsid protein and comparison with capsid proteins of other retroviruses. J. Mol. Biol. *296*, 633–649.

Carlton, J.G., Agromayor, M., and Martin-Serrano, J. (2008). Differential requirements for Alix and ESCRT-III in cytokinesis and HIV-1 release. Proc. Natl. Acad. Sci. U.S.A. *105*, 10541–10546.

Carlton, J.G., and Martin-Serrano, J. (2007). Parallels between cytokinesis and retroviral budding: a role for the ESCRT machinery. Science *316*, 1908–1912.

Chang, Y.F., Wang, S. M., Huang, K. J., and Wang, C. T. (2007). Mutations in capsid major homology region affect assembly and membrane affinity of HIV-1 Gag. J. Mol. Biol. *370*, 585–597.

Cheslock, S.R., Poon, D.T., Fu, W., Rhodes, T. D., Henderson, L.E., Nagashima, K., McGrath, C.F., and Hu, W.S. (2003). Charged assembly helix motif in murine leukemia virus capsid: an important region for virus assembly and particle size determination. J. Virol. *77*, 7058–7066.

Chung, H.Y., Morita, E., von Schwedler, U., Muller, B., Krausslich, H.G., and Sundquist, W. I. (2008). NEDD4L overexpression rescues the release and infectivity of human immunodeficiency virus type 1 constructs lacking PTAP and YPXL late domains. J. Virol. *82*, 4884–4897.

Cimarelli, A., and Luban, J. (2000). Human immunodeficiency virus type 1 virion density is not determined by nucleocapsid basic residues. J. Virol. *74*, 6734–6740.

Cimarelli, A., Sandin, S., Hoglund, S., and Luban, J. (2000). Basic residues in human immunodeficiency virus type 1 nucleocapsid promote virion assembly via interaction with RNA. J. Virol. *74*, 3046–3057.

Conte, M.R., and Matthews, S. (1998). Retroviral matrix proteins: a structural perspective. Virology *246*, 191–198.

Cornilescu, C.C., Bouamr, F., Yao, X., Carter, C., and Tjandra, N. (2001). Structural analysis of the N-terminal domain of the human T-cell leukemia virus capsid protein. J. Mol. Biol. *306*, 783–797.

Cosson, P. (1996). Direct interaction between the envelope and matrix proteins of HIV-1. EMBO J. *15*, 5783–5788.

Craven, R.C., Harty, R.N., Paragas, J., Palese, P., and Wills, J.W. (1999). Late domain function identified in the vesicular stomatitis virus M protein by use of rhabdovirus-retrovirus chimeras. J. Virol. 73, 3359–3365.

D'Souza, V., and Summers, M. F. (2004). Structural basis for packaging the dimeric genome of Moloney murine leukaemia virus. Nature *431*, 586–590.

Dawson, L., and Yu, X.F. (1998). The role of nucleocapsid of HIV-1 in virus assembly. Virology *251*, 141–157.

De Guzman, R.N., Wu, Z.R., Stalling, C.C., Pappalardo, L., Borer, P.N., and Summers, M.F. (1998). Structure of the HIV-1 nucleocapsid protein bound to the SL3 psi-RNA recognition element. Science *279*, 384–388.

Demirov, D.G., Orenstein, J.M., and Freed, E.O. (2002). The late domain of human immunodeficiency virus type 1, p6 promotes virus release in a cell type-dependent manner. J. Virol. *76*, 105–117.

Deschambeault, J., Lalonde, J. P., Cervantes-Acosta, G., Lodge, R., Cohen, E. A., and Lemay, G. (1999). Polarized human immunodeficiency virus budding in lymphocytes involves a tyrosine-based signal and favors cell-to-cell viral transmission. J. Virol. *73*, 5010–5017.

Dong, X., Li, H., Derdowski, A., Ding, L., Burnett, A., Chen, X., Peters, T.R., Dermody, T. S., Woodruff, E., Wang, J.J., and Spearman, P. (2005). AP-3 directs the intracellular trafficking of HIV-1 Gag and plays a key role in particle assembly. Cell *120*, 663–674.

Dorfman, T., Bukovsky, A., Ohagen, A., Hoglund, S., and Gottlinger, H.G. (1994a). Functional domains of the capsid protein of human immunodeficiency virus type 1. J. Virol. *68*, 8180–8187.

Dorfman, T., Luban, J., Goff, S. P., Haseltine, W. A., and Gottlinger, H.G. (1993). Mapping of functionally important residues of a cysteine-histidine box in the human immunodeficiency virus type 1 nucleocapsid protein. J. Virol. *67*, 6159–6169.

Dorfman, T., Mammano, F., Haseltine, W. A., and Gottlinger, H.G. (1994b). Role of the matrix protein in the virion association of the human immunodeficiency virus type 1 envelope glycoprotein. J. Virol. *68*, 1689–1696.

Doyotte, A., Mironov, A., McKenzie, E., and Woodman, P. (2008). The Bro1-related protein HD-PTP/PTPN23 is required for endosomal cargo sorting and multivesicular body morphogenesis. Proc. Natl. Acad. Sci. U.S.A. *105*, 6308–6313.

Dubay, J. W., Roberts, S.J., Hahn, B.H., and Hunter, E. (1992). Truncation of the human immunodeficiency virus type 1 transmembrane glycoprotein cytoplasmic domain blocks virus infectivity. J. Virol. *66*, 6616–6625.

Dupraz, P., and Spahr, P.F. (1992). Specificity of Rous sarcoma virus nucleocapsid protein in genomic RNA packaging. J. Virol. *66*, 4662–4670.

Egan, M.A., Carruth, L.M., Rowell, J. F., Yu, X., and Siliciano, R. F. (1996). Human immunodeficiency virus type 1 envelope protein endocytosis mediated by a highly conserved intrinsic internalization signal in the cytoplasmic domain of gp41 is suppressed in the presence of the Pr55gag precursor protein. J. Virol. *70*, 6547–6556.

Endres, M.J., Jaffer, S., Haggarty, B., Turner, J.D., Doranz, B.J., O'Brien, P.J., Kolson, D.L., and Hoxie, J.A. (1997). Targeting of HIV- and SIV-infected cells by CD4-chemokine receptor pseudotypes. Science *278*, 1462–1464.

Erdie, C. R., and Wills, J.W. (1990). Myristylation of Rous sarcoma virus Gag protein does not prevent replication in avian cells. J. Virol. *64*, 5204–5208.

Facke, M., Janetzko, A., Shoeman, R. L., and Krausslich, H. G. (1993). A large deletion in the matrix domain of the human immunodeficiency virus *gag* gene redirects virus particle assembly from the plasma membrane to the endoplasmic reticulum. J. Virol. *67*, 4972–4980.

Finzi, A., Orthwein, A., Mercier, J., and Cohen, E.A. (2007). Productive human immunodeficiency virus type 1 assembly takes place at the plasma membrane. J. Virol. *81*, 7476–7490.

Fisher, R.D., Chung, H.Y., Zhai, Q., Robinson, H., Sundquist, W.I., and Hill, C.P. (2007). Structural and biochemical studies of ALIX/AIP1 and its role in retrovirus budding. Cell *128*, 841–852.

Franke, E.K., Yuan, H. E., Bossolt, K.L., Goff, S.P., and Luban, J. (1994a). Specificity and sequence requirements for interactions between various retroviral Gag proteins. J. Virol. *68*, 5300–5305.

Franke, E.K., Yuan, H.E., and Luban, J. (1994b). Specific incorporation of cyclophilin A into HIV-1 virions. Nature *372*, 359–362.

Freed, E.O., Englund, G., and Martin, M.A. (1995). Role of the basic domain of human immunodeficiency virus type 1 matrix in macrophage infection. J. Virol. *69*, 3949–3954.

Freed, E. O., and Martin, M. A. (1995). Virion incorporation of envelope glycoproteins with long but not short cytoplasmic tails is blocked by specific, single amino acid substitutions in the human immunodeficiency virus type 1 matrix. J. Virol. *69*, 1984–1989.

Fuller, S.D., Wilk, T., Gowen, B.E., Krausslich, H. G., and Vogt, V.M. (1997). Cryo-electron microscopy reveals ordered domains in the immature HIV-1 particle. Curr. Biol. *7*, 729–738.

Gamble, T.R., Vajdos, F.F., Yoo, S., Worthylake, D.K., Houseweart, M., Sundquist, W I., and Hill, C.P. (1996). Crystal structure of human cyclophilin A bound to the amino-terminal domain of HIV-1 capsid. Cell *87*, 1285–1294.

Gamble, T.R., Yoo, S., Vajdos, F.F., von Schwedler, U.K., Worthylake, D.K., Wang, H., McCutcheon, J.P., Sundquist, W.I., and Hill, C.P. (1997). Structure of the carboxyl-terminal dimerization domain of the HIV-1 capsid protein. Science *278*, 849–853.

Ganser-Pornillos, B.K., Cheng, A., and Yeager, M. (2007). Structure of full-length HIV-1 CA: a model for the mature capsid lattice. Cell *131*, 70–79.

Ganser, B.K., Li, S., Klishko, V.Y., Finch, J. T., and Sundquist, W.I. (1999). Assembly and analysis of conical models for the HIV-1 core. Science *283*, 80–83.

Garrus, J.E., von Schwedler, U.K., Pornillos, O.W., Morham, S.G., Zavitz, K.H., Wang, H.E., Wettstein, D.A., Stray, K.M., Cote, M., Rich, R L., *et al.* (2001). Tsg101 and the vacuolar protein sorting pathway are essential for HIV-1 budding. Cell *107*, 55–65.

Gebhardt, A., Bosch, J.V., Ziemiecki, A., and Friis, R.R. (1984). Rous sarcoma virus p19 and gp35 can be chemically crosslinked to high molecular weight complexes. An insight into virus assembly. J. Mol. Biol. *174*, 297–317.

Gitti, R.K., Lee, B.M., Walker, J., Summers, M.F., Yoo, S., and Sundquist, W.I. (1996). Structure of the amino-terminal core domain of the HIV-1 capsid protein. Science 273, 231–235.

Goff, A., Ehrlich, L.S., Cohen, S.N., and Carter, C.A. (2003). Tsg101 control of human immunodeficiency virus type 1 Gag trafficking and release. J. Virol. *77*, 9173–9182.

Gottlinger, H.G., Dorfman, T., Cohen, E.A., and Haseltine, W.A. (1993). Vpu protein of human immunodeficiency virus type 1 enhances the release of capsids produced by *gag* gene constructs of widely divergent retroviruses. Proc. Natl. Acad. Sci. U.S.A. *90*, 7381–7385.

Gottlinger, H.G., Dorfman, T., Sodroski, J.G., and Haseltine, W.A. (1991). Effect of mutations affecting the p6 *gag* protein on human immunodeficiency virus particle release. Proc. Natl. Acad. Sci. U.S.A. *88*, 3195–3199.

Gottlinger, H.G., Sodroski, J.G., and Haseltine, W.A. (1989). Role of capsid precursor processing and myristoylation in morphogenesis and infectivity of human immunodeficiency virus type 1. Proc. Natl. Acad. Sci. U.S.A. *86*, 5781–5785.

Gottwein, E., and Krausslich, H.G. (2005). Analysis of human immunodeficiency virus type 1 Gag ubiquitination. J. Virol. *79*, 9134–9144.

Gould, S.J., Booth, A. M., and Hildreth, J.E. (2003). The Trojan exosome hypothesis. Proc. Natl. Acad. Sci. U.S.A. *100*, 10592–10597.

Gross, I., Hohenberg, H., Huckhagel, C., and Krausslich, H. G. (1998). N-Terminal extension of human immunodeficiency virus capsid protein converts the *in vitro* assembly phenotype from tubular to spherical particles. J. Virol. *72*, 4798–4810.

Gross, I., Hohenberg, H., Wilk, T., Wiegers, K., Grattinger, M., Muller, B., Fuller, S., and Krausslich, H.G. (2000). A conformational switch controlling HIV-1 morphogenesis. EMBO J. *19*, 103–113.

Gruenberg, J., and Stenmark, H. (2004). The biogenesis of multivesicular endosomes. Nat. Rev. Mol. Cell. Biol. *5*, 317–323.

Guo, X., Hu, J., Whitney, J.B., Russell, R.S., and Liang, C. (2004). Important role for the CA-NC spacer region in the assembly of bovine immunodeficiency virus Gag protein. J. Virol. *78*, 551–560.

Hansen, M., Jelinek, L., Whiting, S., and Barklis, E. (1990). Transport and assembly of *gag* proteins into Moloney murine leukemia virus. J. Virol. *64*, 5306–5316.

Hanson, P.I., Roth, R., Lin, Y., and Heuser, J.E. (2008). Plasma membrane deformation by circular arrays of ESCRT-III protein filaments. J. Cell Biol. *180*, 389–402.

Harila, K., Prior, I., Sjoberg, M., Salminen, A., Hinkula, J., and Suomalainen, M. (2006). Vpu and Tsg101 regulate intracellular targeting of the human immunodeficiency virus type 1 core protein precursor Pr55gag. J. Virol. *80*, 3765–3772.

Harty, R.N., Brown, M.E., Wang, G., Huibregtse, J., and Hayes, F.P. (2000). A PPxY motif within the VP40 protein of Ebola virus interacts physically and functionally with a ubiquitin ligase: implications for filovirus budding. Proc. Natl. Acad. Sci. U.S.A. *97*, 13871–13876.

Hill, C.P., Worthylake, D., Bancroft, D. P., Christensen, A.M., and Sundquist, W.I. (1996). Crystal structures of the trimeric human immunodeficiency virus type 1 matrix protein: implications for membrane association and assembly. Proc. Natl. Acad. Sci. U.S.A. *93*, 3099–3104.

Horii, M., Shibata, H., Kobayashi, R., Katoh, K., Yorikawa, C., Yasuda, J., and Maki, M. (2006). CHMP7, a novel ESCRT-III-related protein, associates with CHMP4b and functions in the endosomal sorting pathway. Biochem J. *400*, 23–32.

Houzet, L., Gay, B., Morichaud, Z., Briant, L., and Mougel, M. (2006). Intracellular assembly and budding of the Murine Leukemia Virus in infected cells. Retrovirology 3, 12.

Howard, T.L., Stauffer, D.R., Degnin, C.R., and Hollenberg, S.M. (2001). CHMP1 functions as a member of a newly defined family of vesicle trafficking proteins. J. Cell Sci. *114*, 2395–2404.

Huang, M., Orenstein, J.M., Martin, M.A., and Freed, E.O. (1995). p6Gag is required for particle production from full-length human immunodeficiency virus type 1 molecular clones expressing protease. J. Virol. *69*, 6810–6818.

Ichioka, F., Takaya, E., Suzuki, H., Kajigaya, S., Buchman, V.L., Shibata, H., and Maki, M. (2007). HD-PTP and Alix share some membrane-traffic related proteins that interact with their Bro1 domains or proline-rich regions. Arch. Biochem. Biophys. *457*, 142–149.

Ivanov, D., Stone, J. R., Maki, J.L., Collins, T., and Wagner, G. (2005). Mammalian SCAN domain dimer is a domain-swapped homolog of the HIV capsid C-terminal domain. Mol. Cell *17*, 137–143.

Ivanov, D., Tsodikov, O.V., Kasanov, J., Ellenberger, T., Wagner, G., and Collins, T. (2007). Domain-swapped dimerization of the HIV-1 capsid C-terminal domain. Proc. Natl. Acad. Sci. U.S.A. *104*, 4353–4358.

Jager, S., Gottwein, E., and Krausslich, H.G. (2007). Ubiquitination of human immunodeficiency virus type 1 Gag is highly dependent on Gag membrane association. J. Virol. *81*, 9193–9201.

Januszeski, M.M., Cannon, P.M., Chen, D., Rozenberg, Y., and Anderson, W.F. (1997). Functional analysis of the cytoplasmic tail of Moloney murine leukemia virus envelope protein. J. Virol. *71*, 3613–3619.

Jin, Z., Jin, L., Peterson, D.L., and Lawson, C.L. (1999). Model for lentivirus capsid core assembly based on crystal dimers of EIAV p26. J. Mol. Biol. *286*, 83–93.

Johnson, M.C., Scobie, H.M., Ma, Y.M., and Vogt, V.M. (2002). Nucleic acid-independent retrovirus assembly can be driven by dimerization. J. Virol. *76*, 11177–11185.

Jouvenet, N., Neil, S.J., Bess, C., Johnson, M.C., Virgen, C.A., Simon, S.M., and Bieniasz, P.D. (2006). Plasma membrane is the site of productive HIV-1 particle assembly. PLoS Biol. *4*, e435.

Katoh, K., Shibata, H., Hatta, K., and Maki, M. (2004). CHMP4b is a major binding partner of the ALG-2-interacting protein Alix among the three CHMP4 isoforms. Arch. Biochem. Biophys. *421*, 159–165.

Katzmann, D.J., Odorizzi, G., and Emr, S.D. (2002). Receptor downregulation and multivesicular-body sorting. Nat Rev Mol. Cell. Biol. 3, 893–905.

Katzmann, D.J., Stefan, C.J., Babst, M., and Emr, S.D. (2003). Vps27 recruits ESCRT machinery to endosomes during MVB sorting. J. Cell Biol. *162*, 413–423.

Keller, P.W., Johnson, M.C., and Vogt, V.M. (2008). Mutations in the spacer peptide and adjoining sequences in Rous sarcoma virus Gag lead to tubular budding. J. Virol. *82*, 6788–6797.

Khorasanizadeh, S., Campos-Olivas, R., and Summers, M. F. (1999). Solution structure of the capsid protein from the human T-cell leukemia virus type-I. J. Mol. Biol. *291*, 491–505.

Kieffer, C., Skalicky, J. J., Morita, E., De Domenico, I., Ward, D. M., Kaplan, J., and Sundquist, W. I. (2008). Two distinct modes of ESCRT-III recognition are required for VPS4 functions in lysosomal protein targeting and HIV-1 budding. Dev. Cell *15*, 62–73.

Kikonyogo, A., Bouamr, F., Vana, M.L., Xiang, Y., Aiyar, A., Carter, C., and Leis, J. (2001). Proteins related to the Nedd4 family of ubiquitin protein ligases interact with the L domain of Rous sarcoma virus and are required for *gag* budding from cells. Proc. Natl. Acad. Sci. U.S.A. *98*, 11199–11204.

Kim, J., Sitaraman, S., Hierro, A., Beach, B.M., Odorizzi, G., and Hurley, J.H. (2005). Structural basis for endosomal targeting by the Bro1 domain. Dev. Cell *8*, 937–947.

Kimpton, J., and Emerman, M. (1992). Detection of replication-competent and pseudotyped human immunodeficiency virus with a sensitive cell line on the basis of activation of an integrated beta-galactosidase gene. J. Virol. *66*, 2232–2239.

Kingston, R.L., Fitzon-Ostendorp, T., Eisenmesser, E. Z., Schatz, G. W., Vogt, V.M., Post, C.B., and Rossmann, M.G. (2000). Structure and self-association of the Rous sarcoma virus capsid protein. Structure *8*, 617–628.

Klimkait, T., Strebel, K., Hoggan, M.D., Martin, M.A., and Orenstein, J.M. (1990). The human immunodeficiency virus type 1-specific protein vpu is required for efficient virus maturation and release. J. Virol. *64*, 621–629.

Kranz, A., Kinner, A., and Kolling, R. (2001). A family of small coiled-coil-forming proteins functioning at the late endosome in yeast. Mol. Biol. Cell *12*, 711–723.

Krausslich, H.G., Facke, M., Heuser, A.M., Konvalinka, J., and Zentgraf, H. (1995). The spacer peptide between human immunodeficiency virus capsid and nucleocapsid proteins is essential for ordered assembly and viral infectivity. J. Virol. *69*, 3407–3419.

Langelier, C., von Schwedler, U.K., Fisher, R.D., De Domenico, I., White, P.L., Hill, C. P., Kaplan, J., Ward, D., and Sundquist, W.I. (2006). Human ESCRT-II complex and its role in human immunodeficiency virus type 1 release. J. Virol. *80*, 9465–9480.

Lanman, J., Lam, T.T., Barnes, S., Sakalian, M., Emmett, M.R., Marshall, A.G., and Prevelige, P.E., Jr. (2003). Identification of novel interactions in HIV-1 capsid protein assembly by high-resolution mass spectrometry. J. Mol. Biol. *325*, 759–772.

Lanman, J., Sexton, J., Sakalian, M., and Prevelige, P.E., Jr. (2002). Kinetic analysis of the role of intersubunit interactions in human immunodeficiency virus type 1 capsid protein assembly *in vitro*. J. Virol. *76*, 6900–6908.

Lata, S., Roessle, M., Solomons, J., Jamin, M., Gottlinger, H.G., Svergun, D.I., and Weissenhorn, W. (2008a). Structural basis for autoinhibition of ESCRT-III CHMP3. J. Mol. Biol. *378*, 818–827.

Lata, S., Schoehn, G., Jain, A., Pires, R., Piehler, J., Gottlinger, H.G., and Weissenhorn, W. (2008b). Helical Structures of ESCRT-III Are Disassembled by VPS4. Science *321*, 1354–1357.

Le Blanc, I., Prevost, M.C., Dokhelar, M. C., and Rosenberg, A.R. (2002). The PPPY motif of human T-cell leukemia virus type 1 Gag protein is required early in the budding process. J. Virol. *76*, 10024–10029.

Lee, E.G., and Linial, M.L. (2006). Deletion of a Cys-His motif from the Alpharetrovirus nucleocapsid domain reveals late domain mutant-like budding defects. Virology *347*, 226–233.

Lee, P.P., and Linial, M.L. (1994). Efficient particle formation can occur if the matrix domain of human immunodeficiency virus type 1 Gag is substituted by a myristylation signal. J. Virol. *68*, 6644–6654.

Lee, S., Joshi, A., Nagashima, K., Freed, E.O., and Hurley, J.H. (2007). Structural basis for viral late-domain binding to Alix. Nat. Struct. Mol. Biol. *14*, 194–199.

Lee, Y.M., Liu, B., and Yu, X.F. (1999). Formation of virus assembly intermediate complexes in the cytoplasm by wild-type and assembly-defective mutant human immunodeficiency virus type 1 and their association with membranes. J. Virol. *73*, 5654–5662.

Li, H., Dou, J., Ding, L., and Spearman, P. (2007). Myristoylation is required for human immunodeficiency virus type 1 Gag–Gag multimerization in mammalian cells. J. Virol. *81*, 12899–12910.

Li, S., Hill, C.P., Sundquist, W.I., and Finch, J.T. (2000). Image reconstructions of helical assemblies of the HIV-1 CA protein. Nature *407*, 409–413.

Liang, C., Hu, J., Russell, R.S., Roldan, A., Kleiman, L., and Wainberg, M.A. (2002). Characterization of a putative alpha-helix across the capsid-SP1 boundary that is critical for the multimerization of human immunodeficiency virus type 1 gag. J. Virol. *76*, 11729–11737.

Lin, Y., Kimpler, L. A., Naismith, T. V., Lauer, J.M., and Hanson, P.I. (2005). Interaction of the mammalian endosomal sorting complex required for transport (ESCRT) III protein hSnf7–1 with itself, membranes, and the AAA+ ATPase SKD1. J. Biol. Chem. *280*, 12799–12809.

Lingappa, J.R., Hill, R.L., Wong, M.L., and Hegde, R.S. (1997). A multistep, ATP-dependent pathway for assembly of human immunodeficiency virus capsids in a cell-free system. J. Cell Biol. *136*, 567–581.

Lodge, R., Gottlinger, H., Gabuzda, D., Cohen, E.A., and Lemay, G. (1994). The intracytoplasmic domain of gp41 mediates polarized budding of human immunodeficiency virus type 1 in MDCK cells. J. Virol. *68*, 4857–4861.

Lodge, R., Lalonde, J.P., Lemay, G., and Cohen, E.A. (1997). The membrane-proximal intracytoplasmic tyrosine residue of HIV-1 envelope glycoprotein is critical for basolateral targeting of viral budding in MDCK cells. EMBO J. *16*, 695–705.

Lopez-Verges, S., Camus, G., Blot, G., Beauvoir, R., Benarous, R., and Berlioz-Torrent, C. (2006). Tail-interacting protein TIP47 is a connector between Gag and Env and is required for Env incorporation into HIV-1 virions. Proc. Natl. Acad. Sci. U.S.A. *103*, 14947–14952.

Lu, Q., Hope, L.W., Brasch, M., Reinhard, C., and Cohen, S.N. (2003). TSG101 interaction with HRS mediates endosomal trafficking and receptor down-regulation. Proc. Natl. Acad. Sci. U.S.A. *100*, 7626–7631.

Lusso, P., di Marzo Veronese, F., Ensoli, B., Franchini, G., Jemma, C., DeRocco, S. E., Kalyanaraman, V. S., and Gallo, R. C. (1990). Expanded HIV-1 cellular tropism by phenotypic mixing with murine endogenous retroviruses. Science *247*, 848–852.

Ma, Y.M., and Vogt, V.M. (2002). Rous sarcoma virus Gag protein–oligonucleotide interaction suggests a critical role for protein dimer formation in assembly. J. Virol. *76*, 5452–5462.

Ma, Y.M., and Vogt, V.M. (2004). Nucleic acid binding-induced Gag dimerization in the assembly of Rous sarcoma virus particles *in vitro*. J. Virol. *78*, 52–60.

McCullough, J., Fisher, R.D., Whitby, F.G., Sundquist, W.I., and Hill, C.P. (2008). ALIX–CHMP4 interactions in the human ESCRT pathway. Proc. Natl. Acad. Sci. U.S.A. *105*, 7687–7691.

McDermott, J., Farrell, L., Ross, R., and Barklis, E. (1996). Structural analysis of human immunodeficiency virus type 1 Gag protein interactions, using cysteine-specific reagents. J. Virol. *70*, 5106–5114.

McDonnell, J.M., Fushman, D., Cahill, S.M., Zhou, W., Wolven, A., Wilson, C.B., Nelle, T.D., Resh, M.D., Wills, J., and Cowburn, D. (1998). Solution structure and dynamics of the bioactive retroviral M domain from Rous sarcoma virus. J. Mol. Biol. *279*, 921–928.

Mammano, F., Kondo, E., Sodroski, J., Bukovsky, A., and Gottlinger, H. G. (1995). Rescue of human immunodeficiency virus type 1 matrix protein mutants by envelope glycoproteins with short cytoplasmic domains. J. Virol. *69*, 3824–3830.

Mammano, F., Ohagen, A., Hoglund, S., and Gottlinger, H.G. (1994). Role of the major homology region of human immunodeficiency virus type 1 in virion morphogenesis. J. Virol. *68*, 4927–4936.

Mammano, F., Salvatori, F., Indraccolo, S., De Rossi, A., Chieco-Bianchi, L., and Gottlinger, H.G. (1997). Truncation of the human immunodeficiency virus type 1 envelope glycoprotein allows efficient pseudotyping of Moloney murine leukemia virus particles and gene transfer into $CD4^+$ cells. J. Virol. *71*, 3341–3345.

Martin-Serrano, J., Eastman, S.W., Chung, W., and Bieniasz, P.D. (2005). HECT ubiquitin ligases link viral and cellular PPXY motifs to the vacuolar protein-sorting pathway. J. Cell Biol. *168*, 89–101.

Martin-Serrano, J., Perez-Caballero, D., and Bieniasz, P.D. (2004). Context-dependent effects of L domains and ubiquitination on viral budding. J. Virol. *78*, 5554–5563.

Martin-Serrano, J., Yarovoy, A., Perez-Caballero, D., and Bieniasz, P.D. (2003a). Divergent retroviral late-budding domains recruit vacuolar protein sorting factors by using alternative adaptor proteins. Proc. Natl. Acad. Sci. U.S.A. *100*, 12414–12419.

Martin-Serrano, J., Zang, T., and Bieniasz, P.D. (2001). HIV-1 and Ebola virus encode small peptide motifs that recruit Tsg101 to sites of particle assembly to facilitate egress. Nat. Med. *7*, 1313–1319.

Martin-Serrano, J., Zang, T., and Bieniasz, P.D. (2003b). Role of ESCRT-I in retroviral budding. J. Virol. *77*, 4794–4804.

Massiah, M.A., Starich, M.R., Paschall, C., Summers, M.F., Christensen, A.M., and Sundquist, W.I. (1994). Three-dimensional structure of the human immunodeficiency virus type 1 matrix protein. J. Mol. Biol. *244*, 198–223.

Mayo, K., Vana, M.L., McDermott, J., Huseby, D., Leis, J., and Barklis, E. (2002). Analysis of Rous sarcoma virus capsid protein variants assembled on lipid monolayers. J. Mol. Biol. *316*, 667–678.

Morita, E., Sandrin, V., Alam, S. L., Eckert, D. M., Gygi, S. P., and Sundquist, W. I. (2007a). Identification of human MVB12 proteins as ESCRT-I subunits that function in HIV budding. Cell Host Microbe *2*, 41–53.

Morita, E., Sandrin, V., Chung, H.Y., Morham, S. G., Gygi, S.P., Rodesch, C.K., and Sundquist, W.I. (2007b). Human ESCRT and ALIX proteins interact with proteins of the midbody and function in cytokinesis. EMBO J. *26*, 4215–4227.

Morita, E., and Sundquist, W. I. (2004). Retrovirus budding. Annu. Rev. Cell Dev. Biol. *20*, 395–425.

Mortuza, G.B., Haire, L.F., Stevens, A., Smerdon, S.J., Stoye, J.P., and Taylor, I.A. (2004). High-resolution structure of a retroviral capsid hexameric amino-terminal domain. Nature *431*, 481–485.

Munshi, U.M., Kim, J., Nagashima, K., Hurley, J.H., and Freed, E.O. (2007). An Alix fragment potently inhibits HIV-1 budding: Characterization of binding to retroviral YPXL late domains. J. Biol. Chem. *282*, 3847–3855.

Murakami, T., and Freed, E.O. (2000). The long cytoplasmic tail of gp41 is required in a cell type-dependent manner for HIV-1 envelope glycoprotein incorporation into virions. Proc. Natl. Acad. Sci. U.S.A. *97*, 343–348.

Muriaux, D., Costes, S., Nagashima, K., Mirro, J., Cho, E., Lockett, S., and Rein, A. (2004). Role of murine leukemia virus nucleocapsid protein in virus assembly. J. Virol. *78*, 12378–12385.

Muriaux, D., Mirro, J., Harvin, D., and Rein, A. (2001). RNA is a structural element in retrovirus particles. Proc. Natl. Acad. Sci. U.S.A. *98*, 5246–5251.

Muziol, T., Pineda-Molina, E., Ravelli, R. B., Zamborlini, A., Usami, Y., Gottlinger, H., and Weissenhorn, W. (2006). Structural basis for budding by the ESCRT-III factor CHMP3. Dev. Cell *10*, 821–830.

Neil, S.J., Eastman, S.W., Jouvenet, N., and Bieniasz, P.D. (2006). HIV-1 Vpu promotes release and prevents endocytosis of nascent retrovirus particles from the plasma membrane. PLoS Pathog. *2*, e39.

Neil, S.J., Sandrin, V., Sundquist, W.I., and Bieniasz, P.D. (2007). An interferon-alpha-induced tethering mechanism inhibits HIV-1 and Ebola virus particle release but is counteracted by the HIV-1 Vpu protein. Cell Host Microbe 2, 193–203.

Neil, S.J., Zang, T., and Bieniasz, P.D. (2008). Tetherin inhibits retrovirus release and is antagonized by HIV-1 Vpu. Nature *451*, 425–430.

Nermut, M. V., Zhang, W. H., Francis, G., Ciampor, F., Morikawa, Y., and Jones, I. M. (2003). Time course of Gag protein assembly in HIV-1-infected cells: a study by immunoelectron microscopy. Virology *305*, 219–227.

Nguyen, D.G., Booth, A., Gould, S. J., and Hildreth, J.E. (2003). Evidence that HIV budding in primary macrophages occurs through the exosome release pathway. J. Biol. Chem. *278*, 52347–52354.

Nguyen, D.H., and Hildreth, J.E. (2000). Evidence for budding of human immunodeficiency virus type 1 selectively from glycolipid-enriched membrane lipid rafts. J. Virol. *74*, 3264–3272.

Nydegger, S., Foti, M., Derdowski, A., Spearman, P., and Thali, M. (2003). HIV-1 egress is gated through late endosomal membranes. Traffic *4*, 902–910.

Obita, T., Saksena, S., Ghazi-Tabatabai, S., Gill, D. J., Perisic, O., Emr, S.D., and Williams, R.L. (2007). Structural basis for selective recognition of ESCRT-III by the AAA ATPase Vps4. Nature *449*, 735–739.

Odorizzi, G., Katzmann, D. J., Babst, M., Audhya, A., and Emr, S.D. (2003). Bro1 is an endosome-associated protein that functions in the MVB pathway in Saccharomyces cerevisiae. J. Cell Sci. *116*, 1893–1903.

Ono, A., Ablan, S.D., Lockett, S.J., Nagashima, K., and Freed, E.O. (2004). Phosphatidylinositol (4,5) bisphosphate regulates HIV-1 Gag targeting to the plasma membrane. Proc. Natl. Acad. Sci. U.S.A. *101*, 14889–14894.

Ono, A., Demirov, D., and Freed, E.O. (2000a). Relationship between human immunodeficiency virus type 1 Gag multimerization and membrane binding. J. Virol. *74*, 5142–5150.

Ono, A., and Freed, E.O. (1999). Binding of human immunodeficiency virus type 1 Gag to membrane: role of the matrix amino terminus. J. Virol. *73*, 4136–4144.

Ono, A., and Freed, E.O. (2001). Plasma membrane rafts play a critical role in HIV-1 assembly and release. Proc. Natl. Acad. Sci. U.S.A. *98*, 13925–13930.

Ono, A., and Freed, E.O. (2004). Cell-type-dependent targeting of human immunodeficiency virus type 1 assembly to the plasma membrane and the multivesicular body. J. Virol. *78*, 1552–1563.

Ono, A., Orenstein, J. M., and Freed, E. O. (2000b). Role of the Gag matrix domain in targeting human immunodeficiency virus type 1 assembly. J. Virol. *74*, 2855–2866.

Ott, D.E., Coren, L.V., Chertova, E.N., Gagliardi, T. D., and Schubert, U. (2000). Ubiquitination of HIV-1 and MuLV Gag. Virology *278*, 111–121.

Ott, D.E., Coren, L.V., Copeland, T.D., Kane, B.P., Johnson, D.G., Sowder, R. C., 2nd, Yoshinaka, Y., Oroszlan, S., Arthur, L.O., and Henderson, L. E. (1998). Ubiquitin is covalently attached to the p6Gag proteins of human immunodeficiency virus type 1 and simian immunodeficiency virus and to the p12Gag protein of Moloney murine leukemia virus. J. Virol. *72*, 2962–2968.

Owens, R.J., Dubay, J.W., Hunter, E., and Compans, R.W. (1991). Human immunodeficiency virus envelope protein determines the site of virus release in polarized epithelial cells. Proc. Natl. Acad. Sci. U.S.A. *88*, 3987–3991.

Paillart, J.C., and Gottlinger, H.G. (1999). Opposing effects of human immunodeficiency virus type 1 matrix mutations support a myristyl switch model of *gag* membrane targeting. J. Virol. *73*, 2604–2612.

Parent, L. J., Bennett, R. P., Craven, R. C., Nelle, T. D., Krishna, N. K., Bowzard, J. B., Wilson, C. B., Puffer, B. A., Montelaro, R. C., and Wills, J. W. (1995). Positionally independent and exchangeable late budding functions of the Rous sarcoma virus and human immunodeficiency virus Gag proteins. J. Virol. *69*, 5455–5460.

Patnaik, A., Chau, V., and Wills, J.W. (2000). Ubiquitin is part of the retrovirus budding machinery. Proc. Natl. Acad. Sci. U.S.A. *97*, 13069–13074.

Pelchen-Matthews, A., Kramer, B., and Marsh, M. (2003). Infectious HIV-1 assembles in late endosomes in primary macrophages. J. Cell Biol. *162*, 443–455.

Perez-Caballero, D., Hatziioannou, T., Martin-Serrano, J., and Bieniasz, P. D. (2004). Human immunodeficiency virus type 1 matrix inhibits and confers cooperativity on *gag* precursor–membrane interactions. J. Virol. *78*, 9560–9563.

Perez, L.G., Davis, G.L., and Hunter, E. (1987). Mutants of the Rous sarcoma virus envelope glycoprotein that lack the transmembrane anchor and cytoplasmic domains: analysis of intracellular transport and assembly into virions. J. Virol. *61*, 2981–2988.

Perez, M., Craven, R. C., and de la Torre, J. C. (2003). The small RING finger protein Z drives arenavirus budding: implications for antiviral strategies. Proc. Natl. Acad. Sci. U.S.A. *100*, 12978–12983.

Perlman, M., and Resh, M.D. (2006). Identification of an intracellular trafficking and assembly pathway for HIV-1 gag. Traffic *7*, 731–745.

Popov, S., Popova, E., Inoue, M., and Gottlinger, H.G. (2008). Human immunodeficiency virus type 1 Gag engages the Bro1 domain of ALIX/AIP1 through the nucleocapsid. J. Virol. *82*, 1389–1398.

Pornillos, O., Alam, S.L., Davis, D.R., and Sundquist, W.I. (2002). Structure of the Tsg101 UEV domain in complex with the PTAP motif of the HIV-1, p6 protein. Nat Struct Biol *9*, 812–817.

Pornillos, O., Higginson, D. S., Stray, K. M., Fisher, R.D., Garrus, J.E., Payne, M., He, G.P., Wang, H.E., Morham, S.G., and Sundquist, W. (2003). HIV Gag mimics the Tsg101-recruiting activity of the human Hrs protein. J. Cell Biol. *162*, 425–434.

Puffer, B.A., Parent, L.J., Wills, J. W., and Montelaro, R.C. (1997). Equine infectious anemia virus utilizes a YXXL motif within the late assembly domain of the Gag p9 protein. J. Virol. *71*, 6541–6546.

Putterman, D., Pepinsky, R.B., and Vogt, V.M. (1990). Ubiquitin in avian leukosis virus particles. Virology *176*, 633–637.

Rao, Z., Belyaev, A.S., Fry, E., Roy, P., Jones, I.M., and Stuart, D.I. (1995). Crystal structure of SIV matrix antigen and implications for virus assembly. Nature *378*, 743–747.

Raposo, G., Moore, M., Innes, D., Leijendekker, R., Leigh-Brown, A., Benaroch, P., and Geuze, H. (2002). Human macrophages accumulate HIV-1 particles in MHC II compartments. Traffic *3*, 718–729.

Reicin, A.S., Ohagen, A., Yin, L., Hoglund, S., and Goff, S. P. (1996). The role of Gag in human immunodeficiency virus type 1 virion morphogenesis and early steps of the viral life cycle. J. Virol. *70*, 8645–8652.

Reicin, A.S., Paik, S., Berkowitz, R.D., Luban, J., Lowy, I., and Goff, S.P. (1995). Linker insertion mutations in the human immunodeficiency virus type 1 *gag* gene: effects on virion particle assembly, release, and infectivity. J. Virol. *69*, 642–650.

Reil, H., Bukovsky, A.A., Gelderblom, H.R., and Gottlinger, H.G. (1998). Efficient HIV-1 replication can occur in the absence of the viral matrix protein. EMBO J. *17*, 2699–2708.

Rein, A., McClure, M.R., Rice, N.R., Luftig, R.B., and Schultz, A.M. (1986). Myristylation site in Pr65gag is essential for virus particle formation by Moloney murine leukemia virus. Proc. Natl. Acad. Sci. U.S.A. *83*, 7246–7250.

Resh, M. D. (1994). Myristylation and palmitylation of Src family members: the fats of the matter. Cell *76*, 411–413.

Rhee, S. S., and Hunter, E. (1987). Myristylation is required for intracellular transport but not for assembly of D-type retrovirus capsids. J. Virol. *61*, 1045–1053.

Rose, S., Hensley, P., O'Shannessy, D. J., Culp, J., Debouck, C., and Chaiken, I. (1992). Characterization of HIV-1, p24 self-association using analytical affinity chromatography. Proteins *13*, 112–119.

Rowell, J. F., Stanhope, P. E., and Siliciano, R. F. (1995). Endocytosis of endogenously synthesized HIV-1 envelope protein. Mechanism and role in processing for association with class II MHC. J. Immunol. *155*, 473–488.

Saad, J.S., Ablan, S.D., Ghanam, R.H., Kim, A., Andrews, K., Nagashima, K., Soheilian, F., Freed, E.O., and Summers, M.F. (2008). Structure of the myristylated human immunodeficiency virus type 2 matrix protein and the role of phosphatidylinositol-(4,5)-bisphosphate in membrane targeting. J. Mol Biol. *382*, 434–447.

Saad, J.S., Loeliger, E., Luncsford, P., Liriano, M., Tai, J., Kim, A., Miller, J., Joshi, A., Freed, E.O., and Summers, M.F. (2007). Point mutations in the HIV-1 matrix protein turn off the myristyl switch. J. Mol. Biol. *366*, 574–585.

Saad, J.S., Miller, J., Tai, J., Kim, A., Ghanam, R. H., and Summers, M. F. (2006). Structural basis for targeting HIV-1 Gag proteins to the plasma membrane for virus assembly. Proc. Natl. Acad. Sci. U.S.A. *103*, 11364–11369.

Saksena, S., Sun, J., Chu, T., and Emr, S. D. (2007). ESCRTing proteins in the endocytic pathway. Trends Biochem. Sci. 32, 561–573.

Sandefur, S., Smith, R. M., Varthakavi, V., and Spearman, P. (2000). Mapping and characterization of the N-terminal I domain of human immunodeficiency virus type 1 Pr55(Gag). J. Virol. *74*, 7238–7249.

Sauter, M.M., Pelchen-Matthews, A., Bron, R., Marsh, M., LaBranche, C.C., Vance, P.J., Romano, J., Haggarty, B.S., Hart, T.K., Lee, W.M., and Hoxie, J.A. (1996). An internalization signal in the simian immunodeficiency virus transmembrane protein cytoplasmic domain modulates expression of envelope glycoproteins on the cell surface. J. Cell Biol. *132*, 795–811.

Schmidt, M.H., Hoeller, D., Yu, J., Furnari, F.B., Cavenee, W. K., Dikic, I., and Bogler, O. (2004). Alix/AIP1 antagonizes epidermal growth factor receptor downregulation by the Cbl-SETA/CIN85 complex. Mol. Cell. Biol. *24*, 8981–8993.

Schnierle, B.S., Stitz, J., Bosch, V., Nocken, F., Merget-Millitzer, H., Engelstadter, M., Kurth, R., Groner, B., and Cichutek, K. (1997). Pseudotyping of murine leukemia virus with the envelope glycoproteins of HIV generates a retroviral vector with specificity of infection for CD4-expressing cells. Proc. Natl. Acad. Sci. U.S.A. *94*, 8640–8645.

Schubert, U., Ott, D.E., Chertova, E.N., Welker, R., Tessmer, U., Princiotta, M. F., Bennink, J. R., Krausslich, H.G., and Yewdell, J.W. (2000). Proteasome inhibition interferes with *gag* polyprotein processing, release, and maturation of HIV-1 and HIV-2. Proc. Natl. Acad. Sci. U.S.A. *97*, 13057–13062.

Schwartzberg, P., Colicelli, J., Gordon, M.L., and Goff, S.P. (1984). Mutations in the *gag* gene of Moloney murine leukemia virus: effects on production of virions and reverse transcriptase. J. Virol. *49*, 918–924.

Segura-Morales, C., Pescia, C., Chatellard-Causse, C., Sadoul, R., Bertrand, E., and Basyuk, E. (2005). Tsg101 and Alix interact with murine leukemia virus Gag and cooperate with Nedd4 ubiquitin ligases during budding. J. Biol. Chem. *280*, 27004–27012.

Sherer, N.M., Lehmann, M.J., Jimenez-Soto, L. F., Ingmundson, A., Horner, S. M., Cicchetti, G., Allen, P.G., Pypaert, M., Cunningham, J.M., and Mothes, W. (2003). Visualization of retroviral replication in living cells reveals budding into multivesicular bodies. Traffic *4*, 785–801.

Shim, S., Kimpler, L.A., and Hanson, P.I. (2007). Structure/function analysis of four core ESCRT-III Proteins reveals common regulatory role for extreme C-terminal domain. Traffic *8*, 1068–1079.

Spearman, P., Horton, R., Ratner, L., and Kuli-Zade, I. (1997). Membrane binding of human immunodeficiency virus type 1 matrix protein *in vivo* supports a conformational myristyl switch mechanism. J. Virol. *71*, 6582–6592.

Spearman, P., Wang, J. J., Vander Heyden, N., and Ratner, L. (1994). Identification of human immunodeficiency virus type 1 Gag protein domains essential to membrane binding and particle assembly. J. Virol. *68*, 3232–3242.

Spector, D.H., Wade, E., Wright, D.A., Koval, V., Clark, C., Jaquish, D., and Spector, S.A. (1990). Human immunodeficiency virus pseudotypes with expanded cellular and species tropism. J. Virol. *64*, 2298–2308.

Sticht, J., Humbert, M., Findlow, S., Bodem, J., Muller, B., Dietrich, U., Werner, J., and Krausslich, H.G. (2005). A peptide inhibitor of HIV-1 assembly *in vitro*. Nat. Struct. Mol. Biol. *12*, 671–677.

Strack, B., Calistri, A., Accola, M.A., Palu, G., and Gottlinger, H.G. (2000). A role for ubiquitin ligase recruitment in retrovirus release. Proc. Natl. Acad. Sci. U.S.A. *97*, 13063–13068.

Strack, B., Calistri, A., Craig, S., Popova, E., and Gottlinger, H.G. (2003). AIP1/ALIX is a binding partner for HIV-1, p6 and EIAV p9 functioning in virus budding. Cell *114*, 689–699.

Strack, B., Calistri, A., and Gottlinger, H.G. (2002). Late assembly domain function can exhibit context dependence and involves ubiquitin residues implicated in endocytosis. J. Virol. *76*, 5472–5479.

Strecker, T., Eichler, R., Meulen, J., Weissenhorn, W., Dieter Klenk, H., Garten, W., and Lenz, O. (2003).

Lassa virus Z protein is a matrix protein and sufficient for the release of virus-like particles. J. Virol. *77*, 10700–10705.

Stuchell-Brereton, M.D., Skalicky, J. J., Kieffer, C., Karren, M.A., Ghaffarian, S., and Sundquist, W.I. (2007). ESCRT-III recognition by VPS4 ATPases. Nature *449*, 740–744.

Stuchell, M.D., Garrus, J. E., Muller, B., Stray, K.M., Ghaffarian, S., McKinnon, R., Krausslich, H. G., Morham, S.G., and Sundquist, W. I. (2004). The human endosomal sorting complex required for transport (ESCRT-I) and its role in HIV-1 budding. J. Biol. Chem. *279*, 36059–36071.

Summers, M.F., Henderson, L. E., Chance, M.R., Bess, J. W., Jr., South, T. L., Blake, P.R., Sagi, I., Perez-Alvarado, G., Sowder, R.C., 3rd, Hare, D.R., *et al.* (1992). Nucleocapsid zinc fingers detected in retroviruses: EXAFS studies of intact viruses and the solution-state structure of the nucleocapsid protein from HIV-1. Protein Sci. *1*, 563–574.

Sundquist, W.I., Schubert, H.L., Kelly, B.N., Hill, G.C., Holton, J.M., and Hill, C.P. (2004). Ubiquitin recognition by the human TSG101 protein. Mol. Cell *13*, 783–789.

Tang, C., Loeliger, E., Kinde, I., Kyere, S., Mayo, K., Barklis, E., Sun, Y., Huang, M., and Summers, M.F. (2003). Antiviral inhibition of the HIV-1 capsid protein. J. Mol. Biol. *327*, 1013–1020.

Tang, C., Loeliger, E., Luncsford, P., Kinde, I., Beckett, D., and Summers, M.F. (2004). Entropic switch regulates myristate exposure in the HIV-1 matrix protein. Proc. Natl. Acad. Sci. U.S.A. *101*, 517–522.

Teo, H., Perisic, O., Gonzalez, B., and Williams, R.L. (2004). ESCRT-II, an endosome-associated complex required for protein sorting: crystal structure and interactions with ESCRT-III and membranes. Dev. Cell *7*, 559–569.

Thali, M., Bukovsky, A., Kondo, E., Rosenwirth, B., Walsh, C. T., Sodroski, J., and Gottlinger, H. G. (1994). Functional association of cyclophilin A with HIV-1 virions. Nature *372*, 363–365.

Timmins, J., Schoehn, G., Ricard-Blum, S., Scianimanico, S., Vernet, T., Ruigrok, R.W., and Weissenhorn, W. (2003). Ebola virus matrix protein VP40 interaction with human cellular factors Tsg101 and Nedd4. J. Mol. Biol. *326*, 493–502.

Tritel, M., and Resh, M.D. (2000). Kinetic analysis of human immunodeficiency virus type 1 assembly reveals the presence of sequential intermediates. J. Virol. *74*, 5845–5855.

Usami, Y., Popov, S., and Gottlinger, H.G. (2007). Potent rescue of human immunodeficiency virus type 1 late domain mutants by ALIX/AIP1 depends on its CHMP4 binding site. J. Virol. *81*, 6614–6622.

Usami, Y., Popov, S., Popova, E., and Gottlinger, H. G. (2008). Efficient and specific rescue of human immunodeficiency virus type 1 budding defects by a Nedd4-like ubiquitin ligase. J. Virol. *82*, 4898–4907.

Van Damme, N., Goff, D., Katsura, C., Jorgenson, R. L., Mitchell, R., Johnson, M.C., Stephens, E.B., and Guatelli, J. (2008). The interferon-induced protein BST-2 restricts HIV-1 release and is downregulated from the cell surface by the viral Vpu protein. Cell Host Microbe 3, 245–252.

Vana, M. L., Tang, Y., Chen, A., Medina, G., Carter, C., and Leis, J. (2004). Role of Nedd4 and ubiquitination of Rous sarcoma virus Gag in budding of virus-like particles from cells. J. Virol. *78*, 13943–13953.

VerPlank, L., Bouamr, F., LaGrassa, T. J., Agresta, B., Kikonyogo, A., Leis, J., and Carter, C. A. (2001). Tsg101, a homologue of ubiquitin-conjugating (E2) enzymes, binds the L domain in HIV type 1 Pr55(Gag). Proc. Natl. Acad. Sci. U.S.A. *98*, 7724–7729.

Vincent, O., Rainbow, L., Tilburn, J., Arst, H.N., Jr., and Penalva, M.A. (2003). YPXL/I is a protein interaction motif recognized by aspergillus PalA and its human homologue, AIP1/Alix. Mol. Cell. Biol. *23*, 1647–1655.

von Schwedler, U.K., Stemmler, T.L., Klishko, V.Y., Li, S., Albertine, K. H., Davis, D.R., and Sundquist, W.I. (1998). Proteolytic refolding of the HIV-1 capsid protein amino-terminus facilitates viral core assembly. EMBO J. *17*, 1555–1568.

von Schwedler, U.K., Stray, K.M., Garrus, J. E., and Sundquist, W.I. (2003a). Functional surfaces of the human immunodeficiency virus type 1 capsid protein. J. Virol. *77*, 5439–5450.

von Schwedler, U. K., Stuchell, M., Muller, B., Ward, D.M., Chung, H.Y., Morita, E., Wang, H.E., Davis, T., He, G.P., Cimbora, D.M., *et al.* (2003b). The protein network of HIV budding. Cell *114*, 701–713.

Waheed, A. A., Ablan, S.D., Roser, J.D., Sowder, R. C., Schaffner, C.P., Chertova, E., and Freed, E.O. (2007). HIV-1 escape from the entry-inhibiting effects of a cholesterol-binding compound via cleavage of gp41 by the viral protease. Proc. Natl. Acad. Sci. U.S.A. *104*, 8467–8471.

Wang, C.T., and Barklis, E. (1993). Assembly, processing, and infectivity of human immunodeficiency virus type 1 *gag* mutants. J. Virol. *67*, 4264–4273.

Wang, H., Norris, K.M., and Mansky, L. M. (2002). Analysis of bovine leukemia virus *gag* membrane targeting and late domain function. J. Virol. *76*, 8485–8493.

Ward, D.M., Vaughn, M.B., Shiflett, S.L., White, P.L., Pollock, A.L., Hill, J., Schnegelberger, R., Sundquist, W.I., and Kaplan, J. (2005). The role of LIP5 and CHMP5 in multivesicular body formation and HIV-1 budding in mammalian cells. J. Biol. Chem. *280*, 10548–10555.

Welsch, S., Keppler, O.T., Habermann, A., Allespach, I., Krijnse-Locker, J., and Krausslich, H.G. (2007). HIV-1 buds predominantly at the plasma membrane of primary human macrophages. PLoS Pathog. 3, e36.

Wilk, T., Gross, I., Gowen, B. E., Rutten, T., de Haas, F., Welker, R., Krausslich, H. G., Boulanger, P., and Fuller, S. D. (2001). Organization of immature human immunodeficiency virus type 1. J. Virol. *75*, 759–771.

Williams, R.L., and Urbe, S. (2007). The emerging shape of the ESCRT machinery. Nat Rev Mol. Cell. Biol. *8*, 355–368.

Wills, J.W., Cameron, C.E., Wilson, C.B., Xiang, Y., Bennett, R.P., and Leis, J. (1994). An assembly domain of the Rous sarcoma virus Gag protein required late in budding. J. Virol. *68*, 6605–6618.

Wills, J.W., and Craven, R.C. (1991). Form, function, and use of retroviral *gag* proteins. Aids 5, 639–654.

Wilson, C., Reitz, M.S., Okayama, H., and Eiden, M.V. (1989). Formation of infectious hybrid virions with gibbon ape leukemia virus and human T-cell leukemia virus retroviral envelope glycoproteins and the *gag* and *pol* proteins of Moloney murine leukemia virus. J. Virol. *63*, 2374–2378.

Worthylake, D. K., Wang, H., Yoo, S., Sundquist, W.I., and Hill, C.P. (1999). Structures of the HIV-1 capsid protein dimerization domain at 2.6 A resolution. Acta Crystallogr.D Biol. Crystallogr. *55*, 85–92.

Wright, E.R., Schooler, J. B., Ding, H.J., Kieffer, C., Fillmore, C., Sundquist, W. I., and Jensen, G. J. (2007). Electron cryotomography of immature HIV-1 virions reveals the structure of the CA and SP1 Gag shells. EMBO J. *26*, 2218–2226.

Wyma, D.J., Kotov, A., and Aiken, C. (2000). Evidence for a stable interaction of gp41 with Pr55(Gag) in immature human immunodeficiency virus type 1 particles. J. Virol. *74*, 9381–9387.

Xiang, Y., Cameron, C.E., Wills, J.W., and Leis, J. (1996). Fine mapping and characterization of the Rous sarcoma virus Pr76gag late assembly domain. J. Virol. *70*, 5695–5700.

Yasuda, J., and Hunter, E. (1998). A proline-rich motif (PPPY) in the Gag polyprotein of Mason–Pfizer monkey virus plays a maturation-independent role in virion release. J. Virol. *72*, 4095–4103.

Yasuda, J., Hunter, E., Nakao, M., and Shida, H. (2002). Functional involvement of a novel Nedd4-like ubiquitin ligase on retrovirus budding. EMBO Rep 3, 636–640.

Yasuda, J., Nakao, M., Kawaoka, Y., and Shida, H. (2003). Nedd4 regulates egress of Ebola virus-like particles from host cells. J. Virol. *77*, 9987–9992.

Yu, X., Yuan, X., Matsuda, Z., Lee, T. H., and Essex, M. (1992). The matrix protein of human immunodeficiency virus type 1 is required for incorporation of viral envelope protein into mature virions. J. Virol. *66*, 4966–4971.

Yu, X., Yuan, X., McLane, M. F., Lee, T.H., and Essex, M. (1993). Mutations in the cytoplasmic domain of human immunodeficiency virus type 1 transmembrane protein impair the incorporation of Env proteins into mature virions. J. Virol. *67*, 213–221.

Yuan, B., Li, X., and Goff, S. P. (1999). Mutations altering the moloney murine leukemia virus p12 Gag protein affect virion production and early events of the virus life cycle. EMBO J. *18*, 4700–4710.

Zamborlini, A., Usami, Y., Radoshitzky, S. R., Popova, E., Palu, G., and Gottlinger, H. (2006). Release of autoinhibition converts ESCRT-III components into potent inhibitors of HIV-1 budding. Proc. Natl. Acad. Sci. U.S.A. *103*, 19140–19145.

Zhadina, M., McClure, M.O., Johnson, M.C., and Bieniasz, P.D. (2007). Ubiquitin-dependent virus particle budding without viral protein ubiquitination. Proc. Natl. Acad. Sci. U.S.A. *104*, 20031–20036.

Zhai, Q., Fisher, R.D., Chung, H. Y., Myszka, D.G., Sundquist, W.I., and Hill, C.P. (2008). Structural and functional studies of ALIX interactions with YPX(n) L late domains of HIV-1 and EIAV. Nat. Struct. Mol. Biol. *15*, 43–49.

Zhang, Y., and Barklis, E. (1997). Effects of nucleocapsid mutations on human immunodeficiency virus assembly and RNA encapsidation. J. Virol. *71*, 6765–6776.

Zhang, Y., Qian, H., Love, Z., and Barklis, E. (1998). Analysis of the assembly function of the human immunodeficiency virus type 1 *gag* protein nucleocapsid domain. J. Virol. *72*, 1782–1789.

Zhou, W., Parent, L.J., Wills, J.W., and Resh, M.D. (1994). Identification of a membrane-binding domain within the amino-terminal region of human immunodeficiency virus type 1 Gag protein which interacts with acidic phospholipids. J. Virol. *68*, 2556–2569.

Zhou, W., and Resh, M.D. (1996). Differential membrane binding of the human immunodeficiency virus type 1 matrix protein. J. Virol. *70*, 8540–8548.

Transmission and Epidemiology

8

Hans Lutz, Gerhard Hunsmann and Jörg Schüpbach

Abstract

Although the prevalence of feline retroviruses has decreased significantly during the last 20 years, they still occur worldwide and in some areas they are still of veterinary importance. In Europe and the USA, EIAV infection has almost been eradicated. As BIV does not cause disease, it is not studied widely and little information is available on its prevalence. BLV occurs in many countries and is of considerable economic importance. The small ruminant retroviruses CAEV and VMV occur worldwide and in some areas at high frequency. Lentiviruses collectively called SIV affect both non-primates and primates. They are naturally present in Africa but not in Asia, North and South America. In 2006, HIV-1 and HIV-2 infection was estimated by the United Nations Program on HIV/AIDS to have affected 39.5 million people. The worldwide prevalence among adults was estimated to be 1%. Most affected by the AIDS epidemic is the population living in sub-Saharan Africa.

FeLV, BIV, BLV and VMV are usually transmitted by direct or even indirect contact. FIV is predominantly transmitted by bites and via milk. EIAV and CAEV are transmitted via milk, by fomites and iatrogenically. SIV and HIV are transmitted through sexual contact and by contaminated needles and blood.

Cross-species transmissions of FeLV and VMV occur occasionally. The HIV epidemic is the result of the zoonotic transmissions of SIV from chimpanzees.

Retroviruses of non-primate species

Natural hosts and prevalence

Feline leukaemia virus (FeLV)

FeLV occurs worldwide and affects domestic cats and some small wild felids in their natural habitat. Wild cats affected by FeLV include European wild cats (*Felis silvestris silvestris*), yaguarondis (*Puma yaguarondi*), and Geoffrey's cats (*Oncifelas geoffroyi*) (Boid *et al.*, 1991; Filoni *et al.*, 2003; Leutenegger *et al.*, 1999; McOrist *et al.*, 1991). With the exception of lions (Hofmann-Lehmann *et al.*, 1996), most larger felids can also become infected by FeLV when they come in contact with infected domestic cats. For example, FeLV infection was shown in a captive leopard (Rasheed and Gardner, 1981), a captive cheetah (Marker *et al.*, 2003), and in Florida panthers (Nolen, 2004).

Whereas the clinical signs of FeLV infection in wild felids and cheetahs are similar to those of the domestic cat, the European wild cat shows no clear clinical signs of pathogenicity (Leutenegger *et al.*, 1999). In the domestic cat, FeLV infection is usually diagnosed via detection of the FeLV p27 protein in plasma by enzyme-linked immunosorbent assay (ELISA) or immunochromatography assays. Serology is not reliable for indirect detection of FeLV infection because many cats have antibodies to endogenous FeLV sequences. The prevalence of FeLV infection in domestic cats depends on their population density and on other characteristics such as age, gender, health status, and lifestyle (Hoover *et al.*, 1976; Levy *et al.*, 2006). FeLV infection rates usually range between 1% and 10%, but can reach 30% under crowded conditions (Archambault *et al.*, 1993; Fromont *et al.*, 1997). Detection of exogenous proviral FeLV DNA using polymerase chain reaction (PCR) usually gives a prevalence rate about 10% higher than detection via ELISA; these results suggest that about 10% of infected domestic cats have immune responses that enable them to overcome overt infection (Arjona *et al.*, 2007; Hofmann-Lehmann *et al.*, 2001).

Feline immunodeficiency virus (FIV)

FIV affects a number of domestic and wild felid species and occurs worldwide (Archambault *et al.*, 1993; Brown *et al.*, 1994; Hofmann-Lehmann *et al.*, 1996; Lutz *et al.*, 1992; Spencer *et al.*, 1992; Troyer *et al.*, 2005). Domestic cats, pumas (*Puma concolor*), lions (*Panthera leo*), leopards (*Panthera pardus*), and Pallas' cats (*Otocolobus manul*) are infected by species-specific strains. Hyaenidae and cheetahs are also endemically infected (Troyer *et al.*, 2005), whereas the large Asian felids do not show endemic infections (Archambault *et al.*, 1993; Brown *et al.*, 1994; Lutz *et al.*, 1992).

In the domestic cat the prevalence of FIV varies greatly; in some geographical areas up to 20% or more of the cats may be infected while in other areas the virus is rarely present (Hitt *et al.*, 1992; Ishida *et al.*, 1989; Lee *et al.*, 2002; Levy *et al.*, 2006; Lutz *et al.*, 1988; Malik *et al.*, 1997; Norris *et al.*, 2007). Interestingly, the prevalence is high in the north and south of Europe, but is low in Germany, Austria and Switzerland (Bandecchi *et al.*, 1992; Courchamp and Pontier, 1994; Hosie *et al.*, 1989; Lutz *et al.*, 1988; Ueland and Lutz, 1992). FIV is most prevalent in non-castrated male cats because their territorial behaviour increases individual exposure (Ishida *et al.*, 1989). FIV strains in domestic cats show considerable genetic variation, thus enabling the distinction of several clades (i.e. subtypes) that occur with varying frequencies (Carpenter *et al.*, 1998; Matteucci *et al.*, 2000; Sodora *et al.*, 1994).

Feline sarcoma virus (FeSV)

FeSVs are replication-defective FeLVs found extremely rarely in cats infected with FeLV. They arise from FeLVs that acquire cellular oncogenes and subsequently induce multiple fibrosarcomas (Majumder *et al.*, 1990; Sarma *et al.*, 1971). Owing to the general decrease in the prevalence of FeLV infection, FeSVs have rarely been observed in recent years.

Equine infectious anaemia virus (EIAV)

The natural host of EIAV are horses, donkeys and mules (Coggins, 1984). Thirty years ago EIAV occurred frequently and with a worldwide distribution, with prevalence greater in warm climates (Akiyama *et al.*, 1967; Coggins and Auchnie, 1977; Etcheverrigaray *et al.*, 1978; Issel and Adams, 1979; McGuire *et al.*, 1974; Yaoi *et al.*, 1959). It is still present in many countries – including several states of the USA – but control measures have successfully decreased its prevalence (Bicout *et al.*, 2006; Loftin *et al.*, 1990; Nagarajan and Simard, 2007). EIAV nearly disappeared from Europe during the late 1990s, but several cases were observed in Italy and Croatia between 2002 and 2007 (http://www.oie.int/wahid-prod/public.php?page=home).

Bovine immunodeficiency virus (BIV)

BIV infects cattle (*Bos taurus taurus*) and water buffalo (*Bubalus bubalis*) (Meas *et al.*, 2000), and it occurs worldwide at varying prevalence (Amills *et al.*, 2002; Belloc *et al.*, 1996; Cavirani *et al.*, 1998; Cho *et al.*, 1999; Cockerell *et al.*, 1992; Hidalgo *et al.*, 1995; Hirai *et al.*, 1996; Horzinek *et al.*, 1991; Muluneh, 1994; Whetstone *et al.*, 1990). Because BIV does not cause pathognomonic clinical signs and because it does not appear to disturb the immune system, very little information on BIV has been published.

Small ruminant lentiviruses (caprine arthritis and encephalitis virus (CAEV) of goat and maedi-visna virus (MVV) of sheep)

Small ruminant lentiviruses (SRLV) occur worldwide and are highly prevalent in Europe (Cutlip *et al.*, 1992; Madewell *et al.*, 1987; Peterhans *et al.*, 2004). Countries with especially high prevalence of CAEV infection include the United States, France, Norway and Canada, and – until recently – Switzerland (Smith and Sherman, 1994). The natural hosts of CAEV are goats and the natural hosts of MVV are sheep, although some interspecies transmissions were reported for both (Pisoni *et al.*, 2005; Shah *et al.*, 2004b); for example, CAEV can be experimentally transmitted to calves (Morin *et al.*, 2003). So far there is no evidence for CAEV or MVV infection in deer (Chomel *et al.*, 1994).

Bovine leukaemia virus (BLV)
Bovine leukaemia virus (BLV) occurs worldwide and is of great economic importance (Monti and Frankena, 2005; Pelzer, 1997) (http://www.oie.int/wahid-prod/public.php?page=disease_status_lists&disease_id=35). Many species are susceptible to BLV, including sheep, goats and rabbits (Burny *et al.*, 1985).

Modes of transmission

Feline leukaemia virus (FeLV)
FeLV is mainly transmitted horizontally by close contact between susceptible cats and infected cats which shed FeLV in their saliva in large amounts (Caldwell *et al.*, 1975; Gomes-Keller *et al.*, 2006). Therefore, FeLV occurs most commonly among socially well adapted cats. In addition, due to the shedding of FeLV in saliva, the infection is also transmitted by bites incurred during mating or territorial aggression (Heath, 1971). Transmission by milk to uninfected kittens is also well documented (Pacitti *et al.*, 1986). FeLV is occasionally observed in non-viraemic, latently infected queens because latent FeLV mammary gland infection may occur during development of the mammary gland (Pacitti *et al.*, 1986). FeLV is also shed in urine and faeces, but this is of lesser relevance for transmission. In addition, FeLV can be transmitted by parasites and fomites (Vobis *et al.*, 2003). Kittens and younger cats are especially susceptible to infection (Hoover *et al.*, 1976). Vertical transmission (*in utero*) is possible but rare because the frequency of FeLV infection has decreased markedly and because infection in utero usually results in abortion or early death of the offspring (Hardy *et al.*, 1976).

Feline immunodeficiency virus (FIV)
In the domestic cat, FIV is usually transmitted by bites (Ishida *et al.*, 1989). Among socially well adapted cats that do not show aggressive behaviour, FIV is rarely transmitted. Transmission to offspring is rare when the queen is asymptomatic (Callanan *et al.*, 1991). The efficacy of vertical transmission depends on the FIV strain and the phase of the infection (i.e. the viral load of the queen). Under experimental conditions and using a strain of the B clade, vertical transmission in utero and via milk affected up to 20% of offspring (Allison and Hoover, 2003; O'Neil *et al.*, 1996).

Transmission of lion lentivirus to offspring is less well understood; discordant mother-cup antibody status suggests that maternal transmission is infrequent (Brown *et al.*, 1994).

Feline sarcoma virus (FeSV)
FeSV is replication deficient, so transmission between cats is doubtful and has never been demonstrated.

Equine infectious anaemia virus (EIAV)
EIAV is horizontally transmitted mechanically by fomites, iatrogenically, and via insects such as tabanids (Hawkins *et al.*, 1976; Issel and Foil, 1984). Transmission by needles has been clearly demonstrated (Williams *et al.*, 1981). In an experimental study, EIAV did not readily replicate in cell cultures of 4 insect species, thus suggesting that transmission is purely mechanical (Shen *et al.*, 1978). Depending on the virulence of the strain and the viral load, EIAV is transmitted from the mare to the foal with varying degrees of efficacy (Kemen and Coggins, 1972).

Bovine immunodeficiency virus (BIV)
BIV is most likely transmitted by direct contact (Belloc *et al.*, 1996; Bouillant and Archambault, 1990). In addition, transmission *in utero* has been documented (Meas *et al.*, 2002; Moody *et al.*, 2002; Scholl *et al.*, 2000). Transmission may occur via colostrum (Moody *et al.*, 2002) and via semen, although the latter source has been disputed (Burger *et al.*, 2000; Nash *et al.*, 1995). Although there are no reports of iatrogenic transmission of BIV, it seems likely that BIV could be transmitted by contaminated needles, which has been shown for many other lentivirus infections.

Small ruminant lentiviruses (SRLV)
In most cases, SRLV infection occurs via ingestion of infected colostrum or milk, although transmission can also occur via inhalation or ingestion of respiratory secretions. VMV is readily transmitted

during close contact between infected and non-infected animals or through contact with aerosols, whereas CAEV is rarely transmitted under such conditions (Berriatua *et al.*, 2003; Blacklaws *et al.*, 2004; Gudnadottir and Palsson, 1965). The degree to which SRLV infection is transmitted horizontally by semen or in utero is unclear (Blacklaws *et al.*, 2004).

Bovine leukaemia virus (BLV)

For many years it has been known that BLV is transmitted by blood sucking insects such as tabanids (Bech-Nielsen *et al.*, 1978; Manet *et al.*, 1989). In addition, iatrogenic transmission, e.g. during routine vaccinations or during dehorning, is highly efficacious (Hopkins and DiGiacomo, 1997; Lassauzet *et al.*, 1990). Direct contact between susceptible and infected cattle is also a major source of spread (Hopkins and DiGiacomo, 1997). Vertical transmission either *in utero* or by ingestion of colostrum was observed, although this appears to be infrequent under natural conditions (Ferrer and Piper, 1981).

Cross-species infections (zoonoses)

Feline leukaemia virus (FeLV)

Under natural conditions, FeLV is not known to affect species other than felids. However, experimental transmission of cell-free extracts of FeLV (Rickard strain) induced tumour formation in dogs (Rickard *et al.*, 1973). Owing to the close contact between domestic cats and humans, interspecies transmission of FeLV from cats to humans is of special interest. Although it was speculated that FeLV is of zoonotic importance, no evidence of transmission to humans has been documented (Jarrett, 1971; Loar, 1987). A more recent study investigated FeLV specific DNA in blood samples from human leukaemia patients, and concluded that none of the cases involved FeLV-specific DNA sequences (Nowotny *et al.*, 1995). From these and several other (non-cited) studies, it can be concluded that FeLV is of no zoonotic importance.

Feline immunodeficiency virus (FIV)

FIV was successfully transmitted from large felids to domestic cats under experimental conditions (Lutz *et al.*, 1992; VandeWoude *et al.*, 1997), although this success came only after several failures. Under special conditions such as in zoos, FIV can be transmitted from domestic cats to wild felids (Archambault *et al.*, 1993; Carpenter *et al.*, 1996; Lutz *et al.*, 1992). The transmission of FIV from domestic cat to a free-ranging felid was shown for the Tsushima cat (*Felis bengalensis euptilura*) (Nishimura *et al.*, 1999). However, determination of proviral sequences showed that most wild felid populations are affected by FIV strains distinct from those of other species, which suggests that although FIV transfer between various feline species has occurred in the past, it is infrequent today (Troyer *et al.*, 2005). The zoonotic potential of FIV appears to be extremely low, as antibodies to FIV have not been found in veterinarians and animal care workers at high risk of exposure to FIV (Butera *et al.*, 2000). However, FIV can infect primary human cells *in vitro*, and experimental transfer of infected cells to non-human primates was shown to induce clinical signs (Johnston *et al.*, 2001).

Feline sarcoma virus (FeSV)

Under experimental conditions FeSV can induce sarcomas in dogs and can transform baboon cells (Gardner *et al.*, 1971; Melnick *et al.*, 1973). However, there are no reports on the transmission of FeSV to other species.

Equine infectious anaemia virus

No information is available on the potential of EIAV to infect humans or other species.

Bovine immunodeficiency virus (BIV)

BIV can be experimentally transmitted to rabbits, where it causes clinical and pathological signs similar to those seen in human immunodeficiency virus (HIV)-infected human patients (Archambault *et al.*, 1993; Gonda *et al.*, 1994; Walder *et al.*, 2001). Infection of sheep and goats has been reported (Archambault *et al.*, 1993). Infection of human cells by BIV and BIV vectors is possible *in vitro*, but appears to be of low efficacy (Berkowitz *et al.*, 2001; DiGiacomo and Hopkins, 1997).

Small ruminant lentiviruses (SRLV)

Interspecies transmission of SRLV between sheep and goats and vice versa has been demonstrated (Shah *et al.*, 2004a). Recent observations suggest that transmission of MVV between sheep and goats is rare but possible in both directions, not only by direct contact but also indirectly in stables used for housing MVV infected sheep before CAEV-free goats were housed in the same stable (Shah *et al.*, 2004b).

Bovine leukaemia virus (BLV)

BLV is readily transmitted to sheep and goats, where it causes similar clinical signs as in cattle (Burny *et al.*, 1985). Using a highly sensitive PCR approach, BLV was determined to be not responsible for a variety of human T-cell derived neoplasms (Burmeister *et al.*, 2007).

Lentiviruses of primates

In addition to their potency in non-primate species, lentiviruses also infect primates. Interestingly, only simian Old World anthropoid primates (suborder *Anthropoidea*) from sub-Saharan Africa seem to host this group of viruses, which are collectively called the simian immunodeficiency viruses (SIV) (VandeWoude and Apetrei, 2006). Up to now such viruses have not been reported in prosimians, New World monkeys (*Platyrhinae*), or Asian anthropoids, and appear to be limited to Hominidae and African *Cercipithecidae* (Table 8.1).

Serological techniques are the first choice for study of viral infections. In the studies described below, most laboratories screened non human primate sera with ELISA and Western blot tests developed for the detection of HIV-1 or HIV-2 antibodies in human sera. Antigens employed in commercial HIV-1 assays are derived from the widely used type B subtypes designated 'LAI' and 'NM', whereas HIV-2 assays are based on the group A strain designated 'ROD' isolate. When species-specific secondary antibodies are not available, these tests rely on secondary antibodies that react with human immunoglobulins and have limited interspecies cross reactivity. Thus, the current census depends on the detection of interspecies cross-reactive antibodies with assay formats optimized for human samples, and may therefore underestimate infection rates.

Another constraint of these studies is the limited sample size for animals living in zoological gardens or special sanctuaries, although some limitations were overcome by the development of techniques that enabled detection of antibodies in urine and faecal samples, as well as assays for viral RNA from feral apes and monkeys (Santiago *et al.*, 2002). This latter approach delivered partial sequence information from individual isolates that could be used to determine evolutionary relationships, and also enabled reconstruction of full-length molecular viral clones competent for infection and replication *in vitro* and *in vivo*. The results of these studies shed new light on the prevalence and transmission of SIV under natural conditions, and provided information about SIV infectivity, potential for interspecies transmission, pathogenicity, and zoonotic potential.

Evidence for SIV infection was reported for 9 out of 10 species representing the 10 families of the African *Cercopithecinae* genus; most of this evidence was based on serological methods. Interestingly, macaques living in wide geographic ranges from Morocco (Barbary macaques; *M. sylvana*) to the north of Japan (Japanese monkey; *M. fuscata*) are not naturally infected with SIV-type lentiviruses.

The seroprevalence in the African monkey samples was quite variable, ranging from a few per cent to over 60% in mandrills (*Mandrillus sphinx*), grivets (*Clorocebus ethiops*) and blue monkeys (*Cercopithecus mitis*) living in Central Africa. High prevalence in a species is indicative of high intra-species infectivity and may also be a prerequisite for natural inter-species transmission.

Baboons (*Papio anubis* and *Papio hamadryas*) and Patas monkeys (*Erythrocebus patas*) can be infected with SIVagm from sympatric vervet monkeys (*Chlorocebus pygerythrus*). SIVs of chimpanzees originated from a recombination of viruses from the red-capped mangabey (*Cercocebus torquatus*) and the Greater spot-nosed monkey (*Cercopithecus nictitans*): the former provided the 5′-end and the latter provided the 3′-end containing a *vpu* gene (Courgnaud *et al.*, 2002). Cross-species transmission can also occur among captive animals; one chimpanzee (*p. t. vellerosus*) acquired SIV from a cage mate of the *p. t. troglodytes* subspecies (Corbet *et al.*, 2000). Finally, the origin of AIDS in breeding colonies of several US primate centres was traced to transmission of an SIV from sooty mangabeys (*Cercocebus atys*) and adaptation by serial passage in Rhesus monkeys (*Macaca mulatta*) kept in the same enclosure (VandeWoude and Apetrei, 2006).

Table 8.1 SIV infection in old world primates (catarrhini)

Family and genus			
Scientific[a]	Common	Distribution	SIV status
Hominoidea	ape, hominoid		
Hominidae	gorillas, chimpanzees, humans		
Homo	humans	world-wide	
Pan	bonobo, chimpanzee	Africa	+
Gorilla	gorilla	Africa	+
Pongo	orang-utan	Asia	-
Hylobatidae	lesser ape, gibbon	Asia	-
Cercopithecoidea	macaque, baboon, guenon, colobine		
Cercopithecidae	cheek pouch monkey		
Macaca	macaque	Asia[b]	-
Papio	baboon	Africa	+
Mandrillus	drill, mandrill	Africa	+
Cercocebus	mangabey	Africa	+
Lophocebus	mangabey	Africa	+
Cercopithecus	guenon	Africa	+
Allenopithecus	Allen's swamp monkey	Africa	+
Miopithecus	dwarf guenon	Africa	+
Erythrocebus	patas monkey	Africa	+
Chlorocebus	vervet	Africa	+
Colobidae[c]	leaf-eating monkey		
Procolobus	olive and red colobus monkey	Africa	
Colobus	colobus monkey	Africa	+
Presbytis	leaf monkey	Asia	

a bold, superfamily; unbold, family; italic, genus

b 23 species of macaques are found in Asia and one in North Africa.

c no evidence has been found of SIV infection in Asian colobine genera of Semnopithecus, Trachypithecus, Pygathrix, Rhinopithecus, and Nasalis.

Among the three ape families (superfamily Hominoidae) SIV-type lentiviruses were verified in African great apes and humans (Hominidae), but not in orang-utans (Pongidae) or gibbons (Hylobatidae) that live in restricted ranges of South-East Asia. The chimpanzee species (*Pan troglodytes*) comprises four subspecies (as defined by mtDNA) that live in non-overlapping geographic ranges in West and Central Africa: *P. t. verus* (Guinea to Ghana), *P. t. vellerosus* (north of the Sanaga River in northern Cameroon and bordering Nigeria), *P. t. troglodytes* (south of the Sanaga and north of the Congo River), and *P. t. schweinfurtii* (Eastern-Central Africa, primarily Uganda and Tanzania). For central and eastern chimpanzees, all SIV except the western *P. t. verus* are harboured with prevalence below 10% (VandeWoude and Apetrei, 2006). Interestingly,

SIV-type lentiviruses have not been found in lesser chimpanzees (bonobo, *pan paniscus*) living south of the Congo River. Three distinct SIV strains closely related to group O HIV-1 were recently discovered in 6/213 faecal samples collected in Cameroon from wild-living gorillas (*G. g. gorilla*), which is the most western of the four gorilla subspecies of western and Central Africa (Van Heuverswyn *et al.*, 2006).

Cluster analysis of SIV sequences obtained from different primate species showed that some viruses were transmitted through species barriers. The evidence for cross-species transmission to humans is compelling for HIV-1 and HIV-2, and suggests that these viruses originated from ancestral SIV closely related to viruses found in chimpanzee (*P. t. troglodytes*) of West–Central (for HIV-1), or sooty mangabey (*Cercocebus atys*) of West Africa (for HIV-2). HIV-1 was transmitted at least seven times, thereby giving rise to the three lineages of M, N, and O viruses (Keele *et al.*, 2006). The 'original' M-group virus, which is still most prevalent in Africa, was most likely transmitted only once, from a chimpanzee to a human, in about 1930 (Mokili *et al.*, 2005). Thereafter the M-group viruses diversified into 9 clades and numerous circulating recombinant forms (CRF), and gave rise to the AIDS pandemic.

N and O clade viruses also have chimpanzee cousins. HIV-1 O infects about 5% of HIV carriers in Cameroon, whereas N viruses have only been found in seven individuals from West-Central Africa. This region is the most likely source of the AIDS pandemic since (i) the diversity of circulating HIV-1 is highest in this part of Africa, (ii) chimpanzees and gorillas harbour viruses closely related to all three HIV-1 lineages, (iii) archival serum samples from human residents of this geographic area or from frequent visitors contained HIV-1 sequences most closely related to current chimpanzee and gorilla viruses, and (iv) retrospective examination of medical records found AIDS cases in Central Africa in the early 1960s.

The closest relative of HIV-2 is found in sooty mangabeys of West Africa (Kanki *et al.*, 1986). Molecular epidemiological studies identified 10 independent transmissions to humans. The epidemic strains are of the A and B subgroup, whereas the non-epidemic C to H subgroups occur only in a very limited number of individuals (VandeWoude and Apetrei, 2006). The C to H viruses appear to be less infectious and pathogenic for humans.

SIV type lentiviruses circulate in primates living predominantly in West-central Africa. Phylogenetic evidence indicates that an ancestral SIV was introduced into the primate lineage about 35 million years ago. Most SIVs spread within their host species but few of them successfully cross species barriers. The primary infectivity after cross-species transmission seems to be very low and many infections are abortive – i.e. they do not produce chronic virus carriers. The higher levels of replication needed for transmission may require adaptation and thus may be dependent on the duration of circulation in the species. Evidence for pathogenicity of these chronic retroviral infections in their natural well-adapted hosts is scarce.

The origin and spread of SIVmac, HIV-1 and HIV-2 resulted from human ignorance, with deleterious consequences for the new hosts. HIV evolution in humans is spurred by replication errors of reverse transcriptase and by multiple infections that allow for recombination, which enable the virus to evade immunological and chemical attacks. Owing to this genomic plasticity of HIV, no protective vaccine can be expected in the near future. Therefore, a combination of prevention and treatment will be required to contain the HIV pandemic.

Human immunodeficiency virus (HIV)

Epidemiology and transmission of human immunodeficiency viruses

The human immunodeficiency viruses HIV-1 and HIV-2 are the causative agents of acquired immunodeficiency syndrome (AIDS). Both viruses replicate primarily in $CD4^+$ T-lymphocytes. In the absence of effective antiretroviral therapy, the progressive destruction of these important immune regulatory cells leads to a gradually increasing immunodeficiency which finally results in opportunistic infections, tumour formation, and death.

HIV-1, first isolated in 1983 (Barre-Sinoussi *et al.*, 1983), is the more aggressive virus and is responsible for the AIDS pandemic. HIV-2, discovered in 1986 (Clavel *et al.*, 1986), is less pathogenic. Rates of heterosexual and mother-to-child transmission of HIV-2 are low. HIV-2 rarely causes AIDS and latency dominates the clinical picture (Bock and Markovitz, 2001; Jaffar *et al.*, 2004; Schim van der Loeff and Aaby, 1999).

According to United Nations Program on HIV/AIDS (UNAIDS) estimates, a total of 39.5 million people were living with AIDS in 2006, and the global prevalence among adults aged 15–49 years was estimated at 1% (Fig. 8.1). New infections in 2006 amounted to 4.3 million, and children younger than 15 years accounted for 530,000 (12.3%) of these new infections. The number of deaths attributable to AIDS in 2006 was 2.9 million, including about 380,000 children (13.1%).

The region most affected by the epidemic is sub-Saharan Africa, where the overall prevalence among adults is 5.9% (as compared to about 1% worldwide). Other regions with a high overall prevalence among adults include the Caribbean (1.2%), Eastern Europe and Central Asia (0.9%), and North America (0.8%). A lower prevalence is found in South and South-East Asia (0.6%), Latin America (0.5%), Oceania (0.4%), Western and Central Europe (0.3%), Middle East and North Africa (0.2%), and East Asia (0.1%).

Almost two-thirds (24.7 million people, 63%) of the global population of persons infected with HIV live in sub-Saharan Africa. An estimated 2.8 million adults and children living in this region became infected with HIV in 2006, which is more than in all other regions of the world combined. The 2.1 million AIDS deaths in sub-Saharan Africa in 2006 accounted for 72% of global AIDS deaths. Across this region, women bear a disproportionately large part of the AIDS burden: not only because they are more likely than men to be infected with HIV, but in most countries they are also more likely to be caring for people infected with HIV. Southern Africa is the epicentre of the epidemic, accounting for 32% of the world's total HIV population and 34% of global AIDS deaths. HIV prevalence in adults is around 20–30% in Zimbabwe, South Africa, Botswana, Lesotho and Namibia. The highest prevalence of any country is in Swaziland, where 33.4% of adults are infected. In some of these countries the HIV epidemic has already led to a dramatic reduction of average life expectancy. Although surveillance

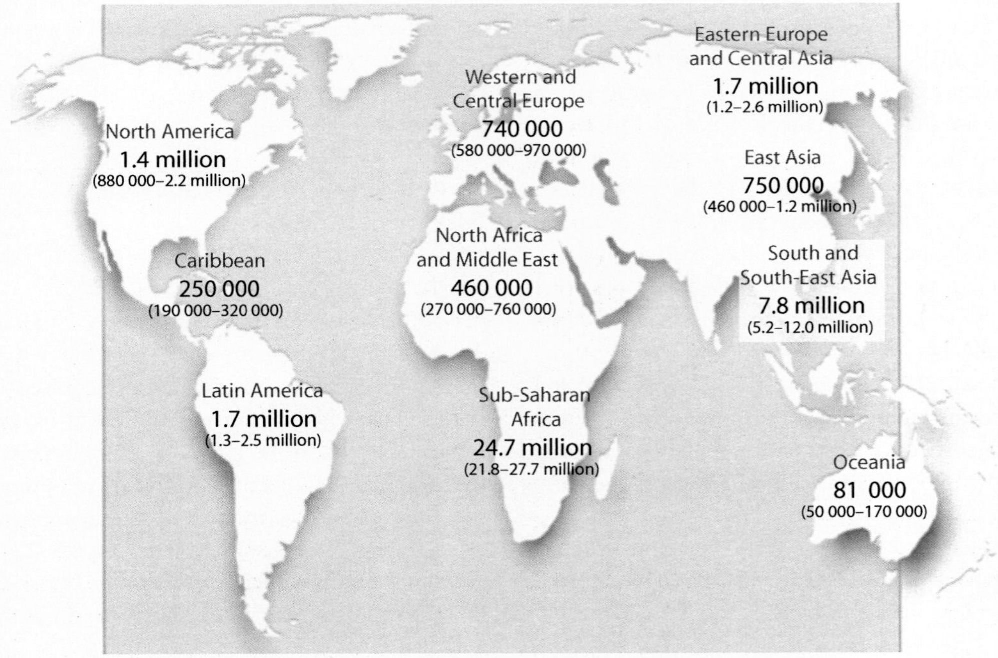

Figure 8.1 Adults and children estimated to be living with HIV infection in 2006. Reprinted with permission from UNAIDS (2006).

data suggest that the epidemic may have peaked in some countries, it is still on the rise in others (UNAIDS 2006). Thus, more than 25 years after the recognition of AIDS as a new disease entity and more than 20 years after the identification of its cause, effective and sustainable measures to halt the epidemic and reverse its course have not yet been implemented on a global scale.

Origin of HIV

As described above, the SIVs naturally infect non-primate species as well as various species of Old World monkeys and the chimpanzee (Fig. 8.2A). The primate lentiviruses are categorized into five major lineages. Lineage 1 contains the various isolates of HIV-1, which are subclassified into three groups: M (main), O (outlier), and N (Simon *et al.*, 1998). From the phylogenetic tree it is evident that group M isolates (e.g. HIV-1/LAI) are more closely related to two isolates from chimpanzee (SIVcpzGAB1 and SIVcpzUS), than to isolates of HIV-1 group O (HIV-1/ANT70) or to another chimpanzee isolate (SIVcpzANT).

These data indicate that the HIV-1 epidemic is the result of zoonotic virus transmissions from chimpanzee, subspecies *Pan troglodytes troglodytes*, to human (Gao *et al.*, 1999). The origin of group M diversification, i.e. the beginning of the HIV-1 pandemic, is placed around 1930 (Korber *et al.*, 2000; Salemi *et al.*, 2001). Recent investigations involving HIV serology and reverse transcription polymerase chain reaction (RT-PCR) performed on faecal samples collected in big ape habitats in Cameroon demonstrated a wide variety of SIVcpz isolates, which are organized in phylogenetic clades restricted to the respective habitats. SIVcpz prevalence in some habitats is as high as 23–35%, while in others it is absent or affects only a small per cent of the population. Phylogenetic analysis of SIVcpz together with HIV-1 isolates clearly showed that HIV-1 group M originated from SIVcpz isolates that are prevalent

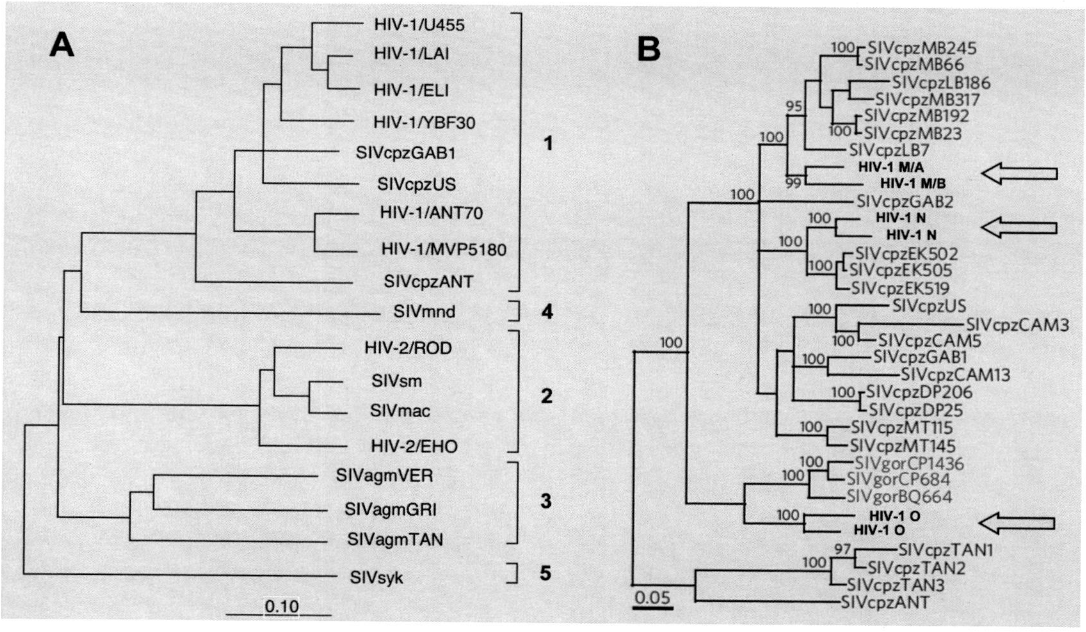

Figure 8.2 Origin of HIV-1 and HIV-2. (A) Phylogenetic tree of primate lentiviruses, derived from Pol protein sequences. Numbers 1 to 5 indicate the five major lineages. HIV-1/U455 is a group M, subtype A isolate. ELI is of group M subtype D, and LAI is of group M subtype B. ANT70 and MVP5180 represent group O, and YBF30 is group N. ROD and EHO represent different subtypes of HIV-2. SIVcpzGAB1, SIVcpzUS, and SIVcpzANT are chimpanzee isolates, mnd, mandrill; agm, African green monkey; syk, Sykes' monkey; sm, sooty mangabey. The bar at the bottom denotes genomic diversity. (B) Phylogenetic tree showing the relationship of HIV-1 groups M, N and O to chimpanzee and, respectively, gorilla lentiviruses (gor). Combined and modified from references (Gao *et al.*, 1999; Sharp *et al.*, 1994) and (Van Heuverswyn *et al.*, 2006)

in two *P. t. troglodytes* populations living in the extreme south-east of Cameroon. HIV-1 group N originated from SIVcpz isolates from *P. t. troglodytes* living in a different area located about 250 km to the west–north-west (Keele *et al.*, 2006). These results indicate that wild chimpanzees act as a reservoir for HIV-1 groups M and N. Viruses closely related to HIV-1 group O were isolated from gorillas living 400 km apart from each other in forest habitats of Cameroon (Van Heuverswyn *et al.*, 2006). Phylogenetic analysis demonstrated that both HIV-1 group O and SIVgor originated from chimpanzee viruses (Fig. 8.2B). Whether chimpanzees transmitted HIV-1 group O viruses to gorillas and humans independently, or whether they were transmitted from chimpanzees to gorillas, which then transmitted it to humans secondarily, is unknown.

Lineage 2 of primate lentiviruses contains the various isolates of HIV-2, which are related to viruses infecting sooty mangabeys (SIV_{sm}). SIV_{sm} was also transmitted naturally to macaques. HIV-2 strain ROD differs less from SIV_{sm} or SIV_{mac} than it does from another human isolate, HIV-2/EHO (Fig. 8.2A). This, together with other similar examples, led to the conclusion that the HIV-2 epidemic was effected via multiple simian-to-human cross-species transmissions. Transmission of the epidemic subtypes HIV-2 A and B may have occurred around 1940 (Lemey *et al.*, 2003).

HIV groups and subtypes and their global distribution

The extraordinary variability of HIV that results from rapid mutation and recombination has led to the development and distinct geographical distribution of various HIV clades (McCutchan, 2000; Peeters and Sharp, 2000). As noted previously, HIV-1 is composed of three phylogenetic groups, M, N, and O. Group M is divided into subtypes A, B, C, D, F, G, H, J and K, which show genetic variation of about 15–20%. The overall variation between subtypes is approximately 25–35%, depending on the subtypes and genome regions examined (Korber *et al.*, 2001). Viral recombination, a consequence of multiple viral infections within a single individual (co-infection or superinfection), has produced a great variety of CRFs that increasingly dominate the epidemic. To date more than 20 CRFs have been defined based on their identification in at least three epidemiologically unlinked individuals and characterization of the full-length sequence. According to a WHO study involving 23,874 HIV-1 samples from 70 countries, subtype C accounted for 50% of all infections worldwide in 2004, and subtypes A, B, D and G accounted for 12%, 10%, 3% and 6%, respectively. Subtypes F, H, J. and K together accounted for 1%. The circulating recombinant forms CRF01_AE and CRF02_AG each accounted for 5%, and CRF03_AB for 0.1%. Other recombinants accounted for the remaining 8% of infections. Altogether, recombinant forms were responsible for 18% of all infections (Hemelaar *et al.*, 2006). Isolates of HIV-1 group O, which are almost exclusively restricted to persons originating from Cameroon, Gabon and Equatorial Guinea, differ as much from one another as do viruses from different subtypes of group M, but their limited number has so far precluded a definition of distinct subtypes.

HIV-1 group N viruses were isolated from only a few individuals from Cameroon (Simon *et al.*, 1998). A total of seven subtypes of HIV-2, two of which are epidemic (A and B) and five non-epidemic (C to G), have been defined, each of which resulted from separate simian-to-human transmissions (Lemey *et al.*, 2003).

Of all HIV-1 infections worldwide, 63% are present in sub-Saharan Africa. In 2004, 56% of infections in that region were caused by subtype C, with smaller proportions caused by subtypes A (14%) or G (10%), CRF02_AG (7%), or other recombinants (9%). Subtype C accounts for more than 97% of the infections in Southern Africa, Ethiopia and India, and accounts for significant proportions in East, North and Central Africa. Subtype A is responsible for about 30% of infections in East and Central Africa, 20% in West Africa, and 80% in Eastern Europe and Central Asia. Subtype B, which until two decades ago was solely responsible for the epidemic in North America, the Caribbean, Latin America, Europe and Australia, now accounts for 75–95% of HIV infections in these regions. Subtype D accounts for 10–15% of infections in Central and East Africa and about 50% in North Africa. Subtype G accounts for about 30% of infections in West Africa and more than 10% in Central Africa.

Subtypes F, H, J and K remain a minority among populations in all world regions (Hemelaar *et al.*, 2006).

Recombinant forms of HIV are of increasing relevance. CRF01_AE and CRF02_AG have produced heterosexual epidemics in Asia and West Africa, respectively. CRF01_AE is responsible for 85% of new infections in South and South-East Asia and 16% of new infections in East Asia. CRF02_AG is responsible for about 30% of new infections in West Africa and about 6.7% in Central Africa (Njai *et al.*, 2006). Non-B subtypes account for an increasing proportion of newly diagnosed HIV-1 infections in Europe (Böni *et al.*, 1999; Lot *et al.*, 2004).

HIV transmission and establishment of infection

HIV is transmitted predominantly by sexual intercourse, connatally from mother to child, post-natally by breast feeding, or by parenteral inoculation – i.e. intravenous drug injection. Globally, the most frequent route of transmission is by sexual intercourse. The probability of HIV-1 transmission per unprotected coital act is estimated at 1/10–1/1600 for male-to-male transmission, 1/200–1/2000 for male-to-female transmission, and 1/200–1/10,000 for female-to-male transmission. The average risk is 0.5–1% per injected drug use, 12–50% for connatal mother-to-child transmission, 12% for breast-feeding, 90% for contaminated blood transfusion, and 0.1–1.0% for nosocomial transmission (reviewed in Levy, 1997). In general, the risk is proportional to the viral load determined by RT-PCR for HIV-1 RNA (Quinn *et al.*, 2000; Wawer *et al.*, 2005). The virus is not transmitted through casual contact in household settings, and there is no evidence for transmission by non-human vectors.

Sexual transmission is mediated by infectious HIV-1 particles and/or virus-infected cells in the semen or mucosal secretions. The relative transmissibility of cell-free versus cell-associated virus is unknown (Gupta and Klasse, 2006). The risk of transmitting or acquiring infection varies greatly, and epidemiologic studies indicate that transmission is linked to viral shedding, i.e. the amount of infectious virus in genital fluids. This in turn is linked to the disease stage and is highest during acute infection and late-stage AIDS (Wawer *et al.*, 2005). Highly active antiretroviral therapy (HAART) can reduce HIV-1 shedding in semen and the female genital tract to undetectable levels, but virions can sometimes be found in semen even when they are undetectable in the blood plasma. Although some untreated infected individuals pose a low transmission risk – notably, no virus transmission was observed from individuals with less than 1500 copies of HIV-1 RNA per millilitre plasma or serum (Quinn *et al.*, 2000), others may be 'super-shedders' and thus highly infectious. Acutely infected individuals pose a particular risk (Wawer *et al.*, 2005). Moreover, other sexually transmitted diseases (STDs) markedly increase both viral shedding and the risk of acquiring HIV-1 infection (reviewed by Galvin and Cohen, 2004; Kaul *et al.*, 2007).

In order for HIV to be sexually transmitted, virions or infected cells must cross the epithelial barriers of the female or male genital tract (reviewed by Gupta and Klasse, 2006; Kaul *et al.*, 2007; Shattock and Moore, 2003). The multiple layers of stratified squamous epithelium that line the most exposed regions of the female and male genital mucosa (vagina and ectocervix in women; inner foreskin, penile glans and fossa navicularis in men) constitute a significant physical barrier. However, these barriers may be penetrated via physical breaches or by infection of intra-epithelial Langerhans cells. The single-layered columnar epithelium that lines the endocervix is more fragile than the stratified epithelium, especially when present as cervical ectopy located on the exocervix and exposed directly to physical stress. The single-layered rectal epithelium likewise provides little protection against potential trauma during intercourse, which may facilitate HIV-1 access to the underlying target cells and even the systemic circulation. Moreover, the rectum, unlike the genital tract, is populated with organized lymphoid tissues (lymphoid follicles) and the rectal epithelium contains specialized M-cells that are capable of binding and presenting HIV-1 to the underlying lymphoid tissue. Such physiological and anatomical differences could account for the markedly increased risk of acquiring HIV-1 infection during anal intercourse.

Both the genital and rectal subepithelial stromal tissues are densely populated with dendritic cells, macrophage, and T-cells that express the virus receptor, the CD4 molecule, the co-receptor CCR5 and, to a lesser extent, the alternative co-receptor CXCR4, and are thus susceptible to HIV-1 infection. Any break in epithelial integrity permits virions a direct access to these target cells, allowing the establishment of infection in mucosal sites. Infection of these cells can be detected within 1 hour of the addition of SIV to the macaque vagina and is most commonly observed where the epithelium is abraded (reviewed in Miller and Shattock, 2003).

Dendritic cells (DC) appear to play an important role in ferrying the virus to regional lymph nodes. Tissue culture studies showed that DC can capture and transmit HIV to $CD4^+$ T-cells, mainly through the membrane-located DC-SIGN molecule, which interacts with gp120 (Geijtenbeek and van Kooyk, 2003). *In vivo*, the immature DC with the captured HIV migrate to the regional lymphoid tissues where they transmit the virus to activated $CD4^+$ T-lymphocytes (reviewed by Teleshova *et al.*, 2003; Wu and KewalRamani, 2006).

The per oral route of infection is involved in many mother-to-child transmissions via breastfeeding, but whether the site of actual virus transmission is within the oral cavity or the small intestine is unclear (Herzberg *et al.*, 2006). The per oral route of infection may also be responsible for intra-partum transmission, as HIV-1 RNA is frequently detected in gastric aspirates of neonates born to HIV-infected mothers (Mandelbrot *et al.*, 1999). Oral transmission was implicated in adult infections in which the only risk factor was receptive oral intercourse (Campo *et al.*, 2006; Syrjanen, 2006). In parenteral infections, the likely primary target cells of intravenously inoculated virus consist of dendritic cells, which further transmit the virus to circulating $CD4^+$ T-lymphocytes (Cameron *et al.*, 2007).

The availability of densely packed $CD4^+$ T-cells in the absence of an efficient immune response in early infection results in large-scale virus production within the regional lymphoid tissues. As a consequence, free virus and virus-infected cells leave the lymph node by the efferent lymphatics to infect lymph node stations further downstream and enter the blood. This leads to generalized infection of all organs including the gut-associated lymphatic tissues (GALT) and the central nervous system. The SIV model has shown that this initial propagation is very rapid: infection of DC in the lamina propria of the vagina and the regional lymph nodes can be detected within 2 days, and plasma viraemia was demonstrated 5 days after inoculation (Spira *et al.*, 1996). In the absence of a specific immune response, replication of HIV within the lymphatics (which harbour 98% of the total number of lymphocytes in the body) causes a rapid increase in the production and release of viral particles and in the number of virus-infected cells distributed throughout the body, including the genital organs and secretions. Individuals in the acute stage of HIV infection are thus highly infectious (Pilcher *et al.*, 2004). Phylogenetic studies suggest that nearly half of all sexual transmissions occur during this early phase of infection (Brenner *et al.*, 2007; Pillay and Fisher, 2007).

Prevention of HIV transmission

Any measure that effectively reduces exposure to the viral inoculum present in genital secretions will reduce the extent of sexual transmission of HIV. Condoms, one of the most effective measures, reduce the incidence of new HIV infections by 80–95% when used consistently (Pinkerton and Abramson, 1997; Weller and Davis, 2002). Early identification, counselling and treatment of infected individuals is important, particularly for individuals in the highly infectious acute stage of infection or those identified by contact tracing (Holtgrave, 2007; Marks *et al.*, 2005; Wawer *et al.*, 2005). Prompt treatment of STDs reduces inflammation and thereby reduces the number of infected, activated and virus producing $CD4^+$ T-lymphocytes and macrophages in genital secretions (Galvin and Cohen, 2004). Use of HAART is effective in reducing HIV-1 shedding in genital secretions (Barroso *et al.*, 2000; Vernazza *et al.*, 2000), and a reduction in HIV-1 transmission from infected patients receiving HAART was observed in several studies (Bunnell *et al.*, 2006; Castilla *et al.*, 2005; Fang *et al.*, 2004; Porco *et al.*, 2004). Expanded access to HAART as a cost-efficient means to curb the growth of the HIV pandemic has been proposed (Montaner *et al.*, 2006).

Preventative measures that reduce the risk of HIV sexual transmission include the condom or its female-controlled equivalent, the femidom. Behavioural changes such as abstinence and limitation of partners are of similar importance (Green *et al.*, 2006; Shelton *et al.*, 2004). Male circumcision reduces the infectable epithelial area of the penis, and is associated with a 60% risk reduction (Auvert *et al.*, 2005; Bailey *et al.*, 2007; Gray *et al.*, 2007).

Microbicides that inactivate the virus within the female genital tract are urgently needed, but the clinical efficacy studies conducted so far were disappointing. Some microbicides, notably nonoxynol-9 and cellulose sulphate, even increase the risk of HIV infection (Check, 2007; Van Damme *et al.*, 2002). However, other microbicide approaches, in particular those involving topical application of antiretroviral drugs, may be more promising. Post-exposure prevention (PEP) of HIV transmission by systemic antiretroviral drug treatment has proven effective and is a standard procedure in developed countries (Cardo *et al.*, 1997). Pre-exposure prevention (PrEP) in persons at an increased risk of HIV infection is now being evaluated in clinical trials with antiretroviral drugs like Tenofovir, a nucleotide reverse transcriptase inhibitor of high efficacy (CDC, 2007). Vaccine development has not made significant progress and is years away from routine application.

Given the high proportion of HIV transmissions that occur from individuals in the stage of acute infection, the rate of change of sexual partners –especially concurrent partners –is a crucial determinant in the spread of HIV. As infectiousness is dramatically higher during the acute stage, transmission is particularly heightened by partner change among newly infected individuals. The risk for transmission of other STD, which further increases viral shedding or susceptibility to HIV infection, further magnifies the spread of HIV when multiple partnerships are involved. Avoidance of multi-partner sexual relationships should thus be a particular focus for controlling the HIV spread in sexually dominated epidemics (Shelton *et al.*, 2004).

Mother-to-child transmission of HIV

In the absence of any intervention, the risk of mother-to-child-transmission (MTCT) of HIV is around 15–30% if the mother does not breastfeed the child. It is estimated that about 20% of connatal infections occur in utero and the rest occur through exposure to blood in the birth channel. The rate of connatal transmission correlates with the viral load in maternal plasma. While no transmission is seen at a plasma viral load below 1000 copies/ml, women with a HIV-1 RNA concentration above 100 000 and no antiretroviral treatment transmit the virus at rates above 50% (Garcia *et al.*, 1999; Mofenson *et al.*, 1999). Late postnatal transmission by breastfeeding presents a significant further risk, accounting for about 42% of all MTCT cases (Coutsoudis *et al.*, 2004; Taha *et al.*, 2007). The overall risk of late postnatal transmission was estimated at about 10% (8.9 transmissions/100 child-years of breast-feeding). The risk is generally constant throughout breastfeeding, but transmission is increased with a lower maternal $CD4^+$ T-cell count and male sex of the child (Coutsoudis *et al.*, 2004). In developed countries, the combination of elective Caesarean section, HAART of the mother combined with a short-time intervention in the newborn, and avoidance of breast feeding have reduced the vertical transmission rates to around 1–2%. Both Caesarean section and HAART are, however, not widely available in low -and middle-income countries. In these countries, various easier and less costly antiretroviral regimens have been offered to pregnant women or their newborn babies, or to both. While it is not entirely clear which regimen is best, a combination of ZDV and 3TC given to mothers in the antenatal, intrapartum and postpartum periods, and to babies for a week after delivery, or a regimen involving a single dose of nevirapine given to mothers in labour and babies immediately after birth, seem to be effective and feasible (Volmink *et al.*, 2007).

Conclusion

In conclusion, HIV is still increasing in prevalence and impact. In contrast, due to better understanding of the mode of transmission, the development of diagnostic tools in combination with separation- and removal-programmes in infected populations, retrovirus infections of animals have lost much of their importance. In addition, vaccines against FeLV infection have contributed to overcome this infection.

References

Akiyama, Y., Yamamoto, H., Yoshino, T., Ishitani, R., and Watanabe, S. (1967). Equine infectious anemia occurring in Hokkaido, Japan– its histopathology and a critical view of the occurrence and diagnosis of this disease. Natl. Inst. Anim. Health Q. (Tokyo) 7, 95–106.

Allison, R.W., and Hoover, E.A. (2003). Feline immunodeficiency virus is concentrated in milk early in lactation. AIDS Res. Hum. Retroviruses *19*, 245–253.

Amills, M., Ramiya, V., Norimine, J., Olmstead, C.A., and Lewin, H.A. (2002). Reduced IL-2 and IL-4 mRNA expression in $CD4^+$ T-cells from bovine leukemia virus-infected cows with persistent lymphocytosis. Virology *304*, 1–9.

Archambault, D., Nadin-Davis, S., Lutze-Wallace, C., and Bouillant, A.M. (1993). The bovine immunodeficiency virus: 1990–1992 update. Vet. Res. *24*, 179–187.

Arjona, A., Barquero, N., Domenech, A., Tejerizo, G., Collado, V.M., Toural, C., Martin, D., and Gomez-Lucia, E. (2007). Evaluation of a novel nested PCR for the routine diagnosis of feline leukemia virus (FeLV) and feline immunodeficiency virus (FIV). J. Feline Med. Surg. *9*, 14–22.

Auvert, B., Taljaard, D., Lagarde, E., Sobngwi-Tambekou, J., Sitta, R., and Puren, A. (2005). Randomized, controlled intervention trial of male circumcision for reduction of HIV infection risk: the ANRS (1265) Trial. PLoS Med. *2*, e298.

Bailey, R.C., Moses, S., Parker, C.B., Agot, K., Maclean, I., Krieger, J.N., Williams, C.F., Campbell, R.T., and Ndinya-Achola, J.O. (2007). Male circumcision for HIV prevention in young men in Kisumu, Kenya: a randomised controlled trial. Lancet *369*, 643–656.

Bandecchi, P., Matteucci, D., Baldinotti, F., Guidi, G., Abramo, F., Tozzini, F., and Bendinelli, M. (1992). Prevalence of feline immunodeficiency virus and other retroviral infections in sick cats in Italy. Vet. Immunol. Immunopathol. *31*, 337–345.

Barre-Sinoussi, F., Chermann, J.C., Rey, F., Nugeyre, M.T., Chamaret, S., Gruest, J., Dauguet, C., Axler-Blin, C., Vezinet-Brun, F., Rouzioux, C. *et al.* (1983). Isolation of a T-lymphotropic retrovirus from a patient at risk for acquired immune deficiency syndrome (AIDS). Science *220*, 868–871.

Barroso, P.F., Schechter, M., Gupta, P., Melo, M.F., Vieira, M., Murta, F.C., Souza, Y., and Harrison, L.H. (2000). Effect of antiretroviral therapy on HIV shedding in semen. Ann. Intern. Med. *133*, 280–284.

Bech-Nielsen, S., Piper, C.E., and Ferrer, J.F. (1978). Natural mode of transmission of the bovine leukemia virus: role of bloodsucking insects. Am. J. Vet. Res. *39*, 1089–1092.

Belloc, C., Polack, B., Schwartz-Cornil, I., Brownlie, J., and Levy, D. (1996). Bovine immunodeficiency virus: facts and questions. Vet. Res. *27*, 395–402.

Berkowitz, R., Ilves, H., Lin, W.Y., Eckert, K., Coward, A., Tamaki, S., Veres, G., and Plavec, I. (2001). Construction and molecular analysis of gene transfer systems derived from bovine immunodeficiency virus. J. Virol. *75*, 3371–3382.

Berriatua, E., Alvarez, V., Extramiana, B., Gonzalez, L., Daltabuit, M., and Juste, R. (2003). Transmission and control implications of seroconversion to Maedi-Visna virus in Basque dairy-sheep flocks. Prev. Vet. Med. *60*, 265–279.

Bicout, D.J., Carvalho, R., Chalvet-Monfray, K., and Sabatier, P. (2006). Distribution of equine infectious anemia in horses in the north of Minas Gerais State, Brazil. J. Vet. Diagn. Invest. *18*, 479–482.

Blacklaws, B.A., Berriatua, E., Torsteinsdottir, S., Watt, N.J., de Andres, D., Klein, D., and Harkiss, G.D. (2004). Transmission of small ruminant lentiviruses. Vet. Microbiol. *101*, 199–208.

Bock, P.J., and Markovitz, D.M. (2001). Infection with HIV-2. Aids *15* Suppl. *5*, S35–S45.

Boid, R., McOrist, S., Jones , T.W., Easterbee, N., Hubbard, A.L., and Jarrett, O. (1991). Isolation of FeLV from a wild felid (*Felis silvestris*). Vet. Rec. *128*, 256.

Böni, J., Pyra, H., Gebhardt, M., Perrin, L., Bürgisser, P., Matter, L., Fierz,W., Erb, P., Piffaretti, J.C., Minder, E., *et al.* (1999). High frequency of non-B subtypes in newly diagnosed HIV-1 infections in Switzerland. J. Acquir. Immune Defic. Syndr. Hum. Retrovirol. *22*, 174–179.

Bouillant, A.M., and Archambault, D. (1990). [Bovine immunodeficiency virus: short review. Ann. Rech. Vet. *21*, 239–250.

Brenner, B.G., Roger, M., Routy, J.P., Moisi, D., Ntemgwa, M., Matte, C., Baril, J.G., Thomas, R., Rouleau,D., and Bruneau, J. *et al.* (2007). High rates of forward transmission events after acute/early HIV-1 infection. J. Infect. Dis. *195*, 951–959.

Brown, E.W., Yuhki, N., Packer, C., and O'Brien, S.J. (1994). A lion lentivirus related to feline immunodeficiency virus: epidemiologic and phylogenetic aspects. J. Virol. *68*, 5953–5968.

Bunnell, R., Ekwaru, J.P., Solberg, P., Wamai, N., Bikaako-Kajura, W., Were, W., Coutinho, A., Liechty, C., Madraa, E., Rutherford, G. *et al.* (2006). Changes in sexual behavior and risk of HIV transmission after antiretroviral therapy and prevention interventions in rural Uganda. Aids *20*, 85–92.

Burger, R.A., Nelson, P.D., Kelly-Quagliana, K., and Coats, K.S. (2000). Failure to detect bovine immunodeficiency virus contamination of stud bull spermatozoa, blood leukocytes, or semen leukocytes in samples supplied by artificial insemination centers. Am. J. Vet. Res. *61*, 816–819.

Burmeister, T., Schwartz, S., Hummel, M., Hoelzer, D., and Thiel, E. (2007). No genetic evidence for involvement of Deltaretroviruses in adult patients with precursor and mature T-cell neoplasms. Retrovirology *4*, 11.

Burny, A., Bruck, C., Cleuter, Y., Couez, D., Deschamps,J., Ghysdael, J., Gregoire, D., Kettmann, R., Mammerickx, M., Marbaix, G. *et al.* (1985). Bovine leukemia virus, a versatile agent with various pathogenic effects in various animal species. Cancer Res. *45*(9 Suppl.), 4578s–4582s.

Butera, S.T., Brown, J., Callahan, M.E., Owen, S.M., Matthews, A.L., Weigner, D.D., Chapman, L.E., and Sandstrom, P.A. (2000). Survey of veterinary conference attendees for evidence of zoonotic infection by feline retroviruses. J. Am. Vet. Med. Assoc. *217*, 1475–1479.

Caldwell, G.G., Baumgartener, L., Carter, C., Cotter, S., Currier, R., Essex, M., Hardy, W., Olson, C., and Olsen, R. (1975). Seroepidemiologic testing in man for evidence of antibodies to feline leukemia virus and bovine leukemia virus. Bibl. Haematol. *43*, 238–241.

Callanan, J.J., Hosie, M.J., and Jarrett, O. (1991). Transmission of feline immunodeficiency virus from mother to kitten. Vet. Rec. *128*, 332–333.

Cameron, P.U., Handley, A.J., Baylis, D.C., Solomon, A.E., Bernard, N., Purcell, D.F., and Lewin, S.R. (2007). Preferential infection of dendritic cells during human immunodeficiency virus type 1 infection of blood leukocytes. J. Virol. *81*, 2297–2306.

Campo, J., Perea, M.A., del Romero, J., Cano, J., Hernando, V., and Bascones, A. (2006). Oral transmission of HIV, reality or fiction? An update. Oral Dis. *12*, 219–228.

Cardo, D.M., Culver, D.H., Ciesielski, C.A., Srivastava, P.U., Marcus, R., Abiteboul, D., Heptonstall, J., Ippolito, G., Lot, F., McKibben, P.S. *et al.* (1997). A case-control study of HIV seroconversion in health care workers after percutaneous exposure. Centers for Disease Control and Prevention Needlestick Surveillance Group. N. Engl. J. Med. 337, 1485–1490.

Carpenter, M.A., Brown, E.W., Culver, M., Johnson, W.E., Pecon-Slattery, J., Brousset, D., and O'Brien S.J. (1996). Genetic and phylogenetic divergence of feline immunodeficiency virus in the puma (*Puma concolor*). J. Virol. *70*, 6682–6693.

Carpenter, M.A., Brown, E.W., MacDonald, D.W., and O'Brien, S.J. (1998). Phylogeographic patterns of feline immunodeficiency virus genetic diversity in the domestic cat. Virology *251*, 234–243.

Castilla, J., Del Romero, J., Hernando, V., Marincovich, B., Garcia, S., and Rodriguez, C. (2005). Effectiveness of highly active antiretroviral therapy in reducing heterosexual transmission of HIV. J. Acquir. Immune Defic. Syndr. *40*, 96–101.

Cavirani, S., Donofrio, G., Chiocco, D., Foni, E., Martelli, P., Allegri, G., Cabassi, C.S., De Iaco, B., and Flammini, C.F. (1998). Seroprevalence to bovine immunodeficiency virus and lack of association with leukocyte counts in Italian dairy cattle. Prev. Vet. Med. *37*, 147–157.

C.D.C. (2007). CDC Trials of Pre-Exposure Prophylaxis for HIV Prevention. (Atlanta, GA: CDC.)

Check, E. (2007). Scientists rethink approach to HIV gels. Nature *446*, 12.

Cho, K.O., Meas, S., Park, N.Y., Kim, Y.H., Lim, Y.K., Endoh, D., Lee, S.I., Ohashi, K., Sugimoto, C., and Onuma, M. (1999). Seroprevalence of bovine immunodeficiency virus in dairy and beef cattle herds in Korea. J. Vet. Med. Sci. *61*, 549–551.

Chomel, B.B., Carniciu, M.L., Kasten, R.W., Castelli, P.M., Work, T.M., and Jessup, D.A. (1994). Antibody prevalence of eight ruminant infectious diseases in California mule and black-tailed deer (*Odocoileus hemionus*). J. Wildl. Dis. *30*, 51–59.

Clavel, F., Guyader, M., Guetard, D., Salle, M., Montagnier, L., and Alizon, M. (1986). Molecular cloning and polymorphism of the human immune deficiency virus type 2. Nature *324*, 691–695.

Cockerell, G.L., Jensen, W.A., Rovnak, J., Ennis ,W.H., and Gonda, M.A. (1992). Seroprevalence of bovine immunodeficiency-like virus and bovine leukemia virus in a dairy cattle herd. Vet. Microbiol. *31*, 109–116.

Coggins, L. (1984). Carriers of equine infectious anemia virus. J. Am. Vet. Med. Assoc. *184*, 279–281.

Coggins, L., and Auchnie, J.A. (1977). Control of equine infectious anemia in horses in Hong Kong. J. Am. Vet. Med. Assoc. *170*, 1299–1301.

Corbet, S., Muller-Trutwin, M.C., Versmisse, P., Delarue, S., Ayouba, A., Lewis, J., Brunak, S., Martin, P., Brun-Vezinet, F., Simon, F., *et al.* (2000). *env* sequences of simian immunodeficiency viruses from chimpanzees in Cameroon are strongly related to those of human immunodeficiency virus group N from the same geographic area. J. Virol. *74*, 529–534.

Courchamp, F., and Pontier, D. (1994). Feline immunodeficiency virus: an epidemiological review. C. R. Acad. Sci. III *317*, 1123–1134.

Courgnaud, V., Salemi, M., Pourrut, X., Mpoudi-Ngole, E., Abela, B., Auzel, P., Bibollet-Ruche, F., Hahn, B., Vandamme, A.M., Delaporte, E. *et al.* (2002). Characterization of a novel simian immunodeficiency virus with a vpu gene from greater spot-nosed monkeys (*Cercopithecus nictitans*) provides new insights into simian/human immunodeficiency virus phylogeny. J. Virol. *76*, 8298–8309.

Coutsoudis, A., Dabis, F., Fawzi, W., Gaillard, P., Haverkamp, G., Harris, D.R., Jackson, J.B., Leroy, V., Meda, N., Msellati ,P., *et al.* (2004). Late postnatal transmission of HIV-1 in breast-fed children: an individual patient data meta-analysis. J. Infect. Dis. *189*, 2154–2166.

Cutlip, R.C., Lehmkuhl, H.D., Sacks, J.M., and Weaver, A.L. (1992). Prevalence of antibody to caprine arthritis-encephalitis virus in goats in the United States. J. Am. Vet. Med. Assoc. *200*, 802–805.

DiGiacomo, R.F., and Hopkins, S.G. (1997). Food animal and poultry retroviruses and human health. Vet. Clin. North Am. Food Anim. Pract. *13*, 177–190.

Etcheverrigaray, M.E., Oliva, G.A., and Zabala Suarez, J.E. (1978). Immunodiffusion serologic study of equine infectious anemia in the Province of Buenos Aires, Argentina. Rev. Assoc. Argent Microbiol. *10*, 20–23.

Fang, C.T., Hsu, H.M., Twu, S.J., Chen, M.Y., Chang, Y.Y., Hwang, J.S., Wang, J.D., and Chuang, C.Y. (2004). Decreased HIV transmission after a policy of providing free access to highly active antiretroviral therapy in Taiwan. J. Infect. Dis. *190*, 879–885.

Ferrer, J.F., and Piper, C.E. (1981). Role of colostrum and milk in the natural transmission of the bovine leukemia virus. Cancer Res. *41*, 4906–4909.

Filoni, C., Adania, C.H., Durigon, E.L., and Catao-Dias, J.L. (2003). Serosurvey for feline leukemia virus and lentiviruses in captive small neotropic felids in Sao Paulo state, Brazil. J. Zoo. Wildl. Med. *34*, 65–68.

Fromont, E., Artois, M., Langlais, M., Courchamp, F., and Pontier, D. (1997). Modelling the feline leukemia virus (FeLV) in natural populations of cats (*Felis catus*). Theor. Popul. Biol. *52*, 60–70.

Galvin, S.R., and Cohen, M.S. (2004). The role of sexually transmitted diseases in HIV transmission. Nat. Rev. Microbiol. 2, 33–42.

Gao, F., Bailes, E., Robertson, D.L., Chen, Y.L., Rodenburg, C.M., Michael, S.F., Cummins, L.B., Arthur, L.O., Peeters, M., Shaw, G.M *et al.* (1999). Origin of HIV-1 in the chimpanzee Pan troglodytes troglodytes. Nature *397*, 436–441.

Garcia, P.M., Kalish, L.A., Pitt, J., Minkoff, H., Quinn, T.C., Burchett, S.K., Kornegay, J., Jackson, B., Moye, J., Hanson C *et al.* (1999). Maternal levels of plasma human immunodeficiency virus type 1 RNA and the risk of perinatal transmission. Women and Infants Transmission Study Group. [see comments]. N. Engl. J. Med. *341*, 394–402.

Gardner, M.B., Arnstein, P., Johnson, E., Rongey, R.W., Charman, H.P., and Huebner, R.J. (1971). Feline sarcoma virus tumor induction in cats and dogs. J. Am. Vet. Med. Assoc. *158* (Suppl. 2), 1046–1053.

Geijtenbeek, T.B., van Kooyk Y. (2003). DC-SIGN: a novel HIV receptor on DCs that mediates HIV-1 transmission. Curr. Top. Microbiol. Immunol. *276*, 31–54.

Gomes-Keller M.A., Tandon, R., Gonczi, E., Meli, M.L., Hofmann-Lehmann R., and Lutz, H. (2006). Shedding of feline leukemia virus RNA in saliva is a consistent feature in viremic cats. Vet. Microbiol. *112*, 11–21.

Gonda, M.A., Luther, D.G., Fong, S.E., and Tobin, G.J. (1994). Bovine immunodeficiency virus: molecular biology and virus–host interactions. Virus Res. *32*, 155–181.

Gray, R.H., Kigozi, G., Serwadda, D., Makumbi, F., Watya, S., Nalugoda, F., Kiwanuka, N., Moulton, L.H., Chaudhary, M.A., Chen, M.Z *et al.* (2007). Male circumcision for HIV prevention in men in Rakai, Uganda: a randomised trial. Lancet *369*, 657–666.

Green, E.C., Halperin, D.T., Nantulya, V., and Hogle, J.A. (2006). Uganda's HIV prevention success: the role of sexual behavior change and the national response. AIDS Behav. *10*, 335–46; discussion 347–350.

Gudnadottir, M., and Palsson, P.A. (1965). Successful transmission of visna by intrapulmonary inoculation. J. Infect. Dis. *115*, 217–225.

Gupta, K., and Klasse, P.J. (2006). How do viral and host factors modulate the sexual transmission of HIV? Can transmission be blocked? PLoS Med. 3, e79.

Hardy, W.D., Jr., Hess, P.W., MacEwen E.G., McClelland A.J., Zuckerman, E.E., Essex, M., Cotter, S.M., and Jarrett, O. (1976). Biology of feline leukemia virus in the natural environment. Cancer Res. *36*: 582–588.

Hawkins, J.A., Adams, W.V., Jr., Wilson, B.H., Issel, C.J., and Roth, E.E. (1976). Transmission of equine infectious anemia virus by *Tabanus fuscicostatus*. J. Am. Vet. Med. Assoc. *168*, 63–64.

Heath, C.W., Jr. (1971). Epidemiologic implications of feline leukemia virus. J. Am. Vet. Med. Assoc. *158*, Suppl. 2: (1119)+.

Hemelaar, J., Gouws, E., Ghys, P.D., and Osmanov, S. (2006). Global and regional distribution of HIV-1 genetic subtypes and recombinants in 2004. AIDS *20*, W13–W23.

Herzberg, M.C., Weinberg, A., and Wahl, S.M. (2006). (C3) The oral epithelial cell and first encounters with HIV-1. Adv. Dent. Res. *19*, 158–166.

Hidalgo, G., Flores, M., and Bonilla, J.A. (1995). Detection and isolation of bovine immunodeficiency-like virus (BIV) in dairy herds of Costa Rica. Zentralbl. Veterinarmed. B *42*, 155–161.

Hirai, N., Kabeya, H., Ohashi, K., Sugimoto, C., and Onuma, M. (1996). Detection of antibodies against bovine immunodeficiency-like virus in daily cattle in Hokkaido. J. Vet. Med. Sci. *58*, 455–457.

Hitt, M.E., Spangler, L., McCarville C. (1992). Prevalence of feline immunodeficiency virus in submissions of feline serum to a diagnostic laboratory in Atlantic Canada. Can. Vet. J. *33*, 723–726.

Hofmann-Lehmann R., Fehr, D., Grob, M., Elgizoli, M., Packer, C., Martenson, J.S., O'Brien S.J., and Lutz, H. (1996). Prevalence of antibodies to feline parvovirus, calicivirus, herpesvirus, coronavirus, and immunodeficiency virus and of feline leukemia virus antigen and the interrelationship of these viral infections in free-ranging lions in east Africa. Clin. Diagn. Lab. Immunol. *3*, 554–562.

Hofmann-Lehmann R., Huder, J.B., Gruber, S., Boretti, F., Sigrist, B., and Lutz, H. (2001). Feline leukaemia provirus load during the course of experimental infection and in naturally infected cats. J. Gen. Virol. *82*, 1589–1596.

Holtgrave, D.R. (2007). Costs and Consequences of the US Centers for Disease Control and Prevention's Recommendations for Opt-Out HIV Testing. PLoS Med. *4*, e194.

Hoover, E.A., Olsen, R.G., Hardy, W.D., Jr., Schaller, J.P., and Mathes, L.E. (1976). Feline leukemia virus infection: age-related variation in response of cats to experimental infection. J. Natl. Cancer. Inst. *57*, 365–369.

Hopkins, S.G., DiGiacomo R.F. (1997). Natural transmission of bovine leukemia virus in dairy and beef cattle. Vet. Clin. North Am. Food Anim. Pract. *13*, 107–128.

Horzinek, M., Keldermans, L., Stuurman, T., Black, J., Herrewegh, A., Sillekens, P., and Koolen, M. (1991). Bovine immunodeficiency virus: immunochemical characterization and serological survey. J. Gen. Virol. *72*, 2923–2928.

Hosie, M.J., Robertson, C., and Jarrett, O. (1989). Prevalence of feline leukaemia virus and antibodies to feline immunodeficiency virus in cats in the United Kingdom. Vet. Rec. *125*, 293–297.

Ishida, T., Washizu, T., Toriyabe, K., Motoyoshi, S., Tomoda, I., and Pedersen, N.C. (1989). Feline immunodeficiency virus infection in cats of Japan. J. Am. Vet. Med. Assoc. *194*, 221–225.

Issel, C.J., Adams, W.V., Jr. (1979). Serologic survey for equine infectious anemia virus in Louisiana. J. Am. Vet. Med. Assoc. *174*, 286–288.

Issel, C.J., and Foil, L.D. (1984). Studies on equine infectious anemia virus transmission by insects. J. Am. Vet. Med. Assoc. *184*, 293–297.

Jaffar, S., Grant, A.D., Whitworth, J., Smith, P.G., and Whittle, H. (2004). The natural history of HIV-1 and HIV-2 infections in adults in Africa: a literature review. Bull. World Hlth Org. *82*, 462–469.

Jarrett, O. (1971). Virology and host range of feline leukemia virus. J. Am. Vet. Med. Assoc. *158* (Suppl. 2), 1032.

Johnston, J. B., Olson, M. E., Rud, E. W., and Power, C. (2001). Xenoinfection of nonhuman primates by feline immunodeficiency virus. Curr. Biol. *11*, 1109–1113.

Kanki, P.J., Barin, F., M'Boup S., Allan, J.S., Romet-Lemonne J.L., Marlink, R., McLane M.F., Lee, T.H., Arbeille, B., Denis F *et al.* (1986). New human T-lymphotropic retrovirus related to simian T-lymphotropic virus type III (STLV-IIIAGM). Science *232*, 238–243.

Kaul, R., Pettengell, C., Sheth, P.M., Sunderji, S., Biringer, A., Macdonald, K., Walmsley, S., and Rebbapragada, A. (2007). The genital tract immune milieu: An important determinant of HIV susceptibility and secondary transmission. J. Reprod. Immunol. *77*, 32–40.

Keele, B.F., Van Heuverswyn F., Li, Y., Bailes, E., Takehisa, J., Santiago, M.L., Bibollet-Ruche F., Chen, Y., Wain, L.V., Liegeois F *et al.* (2006). Chimpanzee reservoirs of pandemic and nonpandemic HIV-1. Science *313*, 523–526.

Kemen, M.J., Jr., Coggins, L. (1972). Equine infectious anemia: transmission from infected mares to foals. J. Am. Vet. Med. Assoc. *161*, 496–499.

Korber, B., Gaschen, B., Yusim, K., Thakallapally, R., Kesmir, C., and Detours, V. (2001). Evolutionary and immunological implications of contemporary HIV-1 variation. Br. Med. Bull. *58*, 19–42.

Korber, B., Muldoon, M., Theiler, J., Gao, F., Gupta, R., Lapedes, A., Hahn, B.H., Wolinsky, S., and Bhattacharya, T. (2000). Timing the ancestor of the HIV-1 pandemic strains. Science *288*, 1789–1796.

Lassauzet, M.L., Thurmond, M.C., Johnson, W.O., Stevens, F., and Picanso, J.P. (1990). Effect of brucellosis vaccination and dehorning on transmission of bovine leukemia virus in heifers on a California dairy. Can J. Vet. Res. *54*, 184–189.

Lee, I.T., Levy, J.K., Gorman, S.P., Crawford, P.C., and Slater, M.R. (2002). Prevalence of feline leukemia virus infection and serum antibodies against feline immunodeficiency virus in unowned free-roaming cats. J. Am. Vet. Med. Assoc. *220*, 620–622.

Lemey, P., Pybus, O.G., Wang, B., Saksena, N.K., Salemi, M., and Vandamme, A.M. (2003). Tracing the origin and history of the HIV-2 epidemic. Proc. Natl. Acad. Sci. U.S.A. *100*, 6588–6592.

Leutenegger, C., Hofmann-Lehmann R., Riols, C., Liberek, M., Worel, G., Lups, P., Fehr, D., Hartmann, M., Weilenmann, P., and Lutz, H. (1999). Viral infections in free-living populations of the European wildcat. J. Wildlife Dis. *35*, 678–686.

Levy, J.E. (1997). HIV and the Pathogenesis of AIDS (2nd edn): ASM Press, Washington, DC.

Levy, J.K., Scott, H.M., Lachtara, J.L., and Crawford, P.C. (2006). Seroprevalence of feline leukemia virus and feline immunodeficiency virus infection among cats in North America and risk factors for seropositivity. J. Am. Vet. Med. Assoc. *228*, 371–376.

Loar, A.S. (1987). The zoonotic potential of feline leukemia virus. Vet. Clin. North Am. Small Anim. Pract. *17*, 105–115.

Loftin, M.K., Levine, J.F., McGinn T., and Coggins, L. (1990). Distribution of equine infectious anemia in equids in southeastern United States. J. Am. Vet. Med. Assoc. *197*, 1018–20.

Lot, F., Semaille, C., Cazein, F., Barin, F., Pinget, R., Pillonel, J., and Desenclos, J.C. (2004). Preliminary results from the new HIV surveillance system in France. Euro Surveill. *9*, 34–37.

Lutz, H., Egberink, H., Arnold, P., Winkler, G., Wolfensberger, C., Jarrett, O., Parodi, A.L., Pedersen, N.C., and Horzinek, M.C. (1988). Felines T-lymphotropes Lentivirus (FTLV): Experimentelle Infektion und Vorkommen in einigen Ländern Europas. Kleintierpraxis 33, 455–459, 455–459.

Lutz, H., Isenbugel, E., Lehmann, R., Sabapara, R.H., and Wolfensberger, C. (1992). Retrovirus infections in non-domestic felids: serological studies and attempts to isolate a lentivirus. Vet. Immunol. Immunopathol. *35*, 215–224.

McCutchan F.E. (2000). Understanding the genetic diversity of HIV-1. AIDS *14* (suppl. 3): S31–S44.

McGuire T.C., Crawford, T.B., and Henson, J.B. (1974). Prevalence of antibodies to herpesvirus types 1 and 2, arteritis and infectious anemia viral antigens in equine serum. Am. J. Vet. Res. *35*, 181–185.

McOrist S., Boid, R., Jones, T.W., Easterbee, N., Hubbard, A.L., and Jarrett, O. (1991). Some viral and protozool diseases in the European wildcat (Felis silvestris). J. Wildl. Dis. *27*, 693–696.

Madewell, B.R., Ameghino, E., Rivera, H., Inope, L., De Martini J. (1987). Seroreactivity of Peruvian sheep and goats to small ruminant lentivirus-ovine progressive pneumonia virus. Am. J. Vet. Res. *48*, 372–374.

Majumder, S., Ray, P., and Besmer, P. (1990). Tyrosine protein kinase activity of the HZ4-feline sarcoma virus P80gag-kit-transforming protein. Oncogene Res. *5*, 329–335.

Malik, R., Kendall, K., Cridland, J., Coulston, S., Stuart, A.J., Snow, D., and Love, D.N. (1997). Prevalences of feline leukaemia virus and feline immunodeficiency virus infections in cats in Sydney. Aust. Vet. J. *75*, 323–327.

Mandelbrot, L., Burgard, M., Teglas, J.P., Benifla, J.L., Khan, C., Blot, P., Vilmer, E., Matheron, S., Firtion, G., Blanche S *et al.* (1999). Frequent detection of HIV-1 in the gastric aspirates of neonates born to HIV-infected mothers. AIDS *13*, 2143–9.

Manet, G., Guilbert, X., Roux, A., Vuillaume, A., and Parodi, A.L. (1989). Natural mode of horizontal transmission of bovine leukemia virus (BLV): the potential role of tabanids (*Tabanus* spp.). Vet. Immunol. Immunopatho.l *22*, 255–263.

Marker, L., Munson, L., Basson, P.A., and Quackenbush, S. (2003). Multicentric T-cell lymphoma associated with feline leukemia virus infection in a captive namibian cheetah (Acinonyx jubatus). J. Wildlife Dis. *39*, 690–695.

Marks, G., Crepaz, N., Senterfitt, J.W., and Janssen, R.S. (2005). Meta-analysis of high-risk sexual behavior in persons aware and unaware they are infected with HIV in the United States: implications for HIV prevention programs. J. Acquir. Immune Defic. Syndr. *39*, 446–453.

Matteucci, D., Poli, A., Mazzetti, P., Sozzi, S., Bonci, F., Isola, P., Zaccaro, L., Giannecchini, S., Calandrella, M., Pistello M *et al.* (2000). Immunogenicity of an anti-clade B feline immunodeficiency fixed-cell virus vaccine in field cats. J. Virol. *74*, 10911–10919.

Meas, S., Ohashi, K., Tum, S., Chhin, M., Te, K., Miura, K., Sugimoto, C., and Onuma, M. (2000). Seroprevalence of bovine immunodeficiency virus and bovine leukemia virus in draught animals in Cambodia. J. Vet. Med. Sci. *62*, 779–781.

Meas, S., Usui, T., Ohashi, K., Sugimoto, C., and Onuma, M. (2002). Vertical transmission of bovine leukemia virus and bovine immunodeficiency virus in dairy cattle herds. Vet. Microbiol. *84*, 275–282.

Melnick, J.L., Altenburg, B., Arnstein, P., Mirkovic, R., and Tevethia, S.S. (1973). Transformation of baboon cells with feline sarcoma virus. Intervirology *1*, 386–398.

Miller, C.J., and Shattock, R.J. (2003). Target cells in vaginal HIV transmission. Microbes Infect. *5*, 59–67.

Mofenson, L.M., Lambert, J.S., Stiehm, E.R., Bethel, J., Meyer, W.A., 3rd, Whitehouse, J., Moye, J., Jr., Reichelderfer, P., Harris, D.R., Fowler, M.G, *et al.* (1999). Risk factors for perinatal transmission of human immunodeficiency virus type 1 in women treated with zidovudine. Pediatric AIDS Clinical Trials Group Study 185 Team. N. Engl. J. Med. *341*, 385–393.

Mokili, J., and Korber, B. (2005). The speed of HIV in Africa. J. Neurovirol. *11* Suppl. *1*, 66–75.

Montaner, J.S., Hogg, R., Wood, E., Kerr, T., Tyndall, M., Levy, A.R., and Harrigan, P.R. (2006). The case for expanding access to highly active antiretroviral therapy to curb the growth of the HIV epidemic. Lancet *368*, 531–536.

Monti, G.E., and Frankena, K. (2005). Survival analysis on aggregate data to assess time to sero-conversion after experimental infection with Bovine Leukemia virus. Prev. Vet. Med. *68*, 241–262.

Moody, C.A., Pharr, G.T., Murphey, J., Hughlett, M.B., Weaver, C.C., Nelson, P.D., and Coats, K.S. (2002). Confirmation of vertical transmission of bovine immunodeficiency virus in naturally infected dairy cattle using the polymerase chain reaction. J. Vet. Diagn. Invest. *14*, 113–119.

Morin, T., Guiguen, F., Bouzar, B.A., Villet, S., Greenland, T., Grezel, D., Gounel, F., Gallay, K., Garnier, C., Durand J. *et al.* (2003). Clearance of a productive lentivirus infection in calves experimentally inoculated with caprine arthritis-encephalitis virus. J. Virol. *77*, 6430–7.

Muluneh, A. (1994). Seroprevalence of bovine immunodeficiency-virus (BIV) antibodies in the cattle population in Germany. Zentralbl. Veterinarmed. B *41*, 679–684.

Nagarajan, M.M., and Simard, C. (2007). Gag genetic heterogeneity of equine infectious anemia virus (EIAV) in naturally infected horses in Canada. Virus Res. *129*, 228–235.

Nash, J.W., Hanson, L.A., St Cyr Coats K. (1995). Bovine immunodeficiency virus in stud bull semen. Am. J. Vet. Res. *56*, 760–763.

Nishimura, Y., Goto, Y., Yoneda, K., Endo, Y., Mizuno, T., Hamachi, M., Maruyama, H., Kinoshita, H., Koga, S., Komori M *et al.* (1999). Interspecies transmission of feline immunodeficiency virus from the domestic cat to the Tsushima cat (Felis bengalensis euptilura) in the wild. J. Virol. *73*, 7916–7921.

Njai, H.F., Gali, Y., Vanham, G., Clybergh, C., Jennes, W., Vidal, N., Butel, C., Mpoudi-Ngolle E., Peeters, M., and Arien, K.K. (2006). The predominance of Human Immunodeficiency Virus type 1 (HIV-1) circulating recombinant form 02 (CRF02_AG) in West Central Africa may be related to its replicative fitness. Retrovirology 3, 40.

Nolen, R.S. (2004). Feline leukemia virus threatens endangered panthers. J. Am. Vet. Med. Assoc. *224*, 1721–2.

Norris, J.M., Bell, E.T., Hales, L., Toribio, J.A., White, J.D., Wigney, D.I., Baral, R.M., and Malik, R. (2007). Prevalence of feline immunodeficiency virus infection in domesticated and feral cats in eastern Australia. J. Feline Med. Surg. *9*, 300–308.

Nowotny, N., Uthman, A., Haas, O.A., Borkhardt, A., Lechner, K., Egberink, H.F., Mostl, K., and Horzinek, M.C. (1995). Is it possible to catch leukemia from a cat? Lancet *346*, 252–253.

O'Neil L.L., Burkhard, M.J., and Hoover, E.A. (1996). Frequent perinatal transmission of feline immunodeficiency virus by chronically infected cats. J. Virol. *7*, (2894)–901.

Pacitti, A.M., Jarrett, O., and Hay, D. (1986). Transmission of feline leukaemia virus in the milk of a non-viraemic cat. Vet. Rec. *118*, 381–384.

Peeters, M., and Sharp, P.M. (2000). Genetic diversity of HIV-1: the moving target. AIDS *14* (Suppl. 3), S129–S140.

Pelzer, K.D. (1997). Economics of bovine leukemia virus infection. Vet. Clin. North Am. Food Anim. Pract. *13*, 129–141.

Peterhans, E., Greenland, T., Badiola, J., Harkiss, G., Bertoni, G., Amorena, B., Eliaszewicz, M., Juste, R.A., Krassnig, R., Lafont, J.P *et al.* (2004). Routes of transmission and consequences of small ruminant lentiviruses (SRLVs) infection and eradication schemes. Vet. Res. *35*, 257–274.

Pilcher, C.D., Tien, H.C., Eron, J.J., Jr., Vernazza, P.L., Leu, S.Y., Stewart, P.W., Goh, L.E., and Cohen, M.S. (2004). Brief but efficient: acute HIV infection and the sexual transmission of HIV. J. Infect. Dis. *189*, 1785–1792.

Pillay, D., and Fisher, M. (2007). Primary HIV Infection, phylogenetics, and antiretroviral prevention. J. Infect. Dis. *195*, 924–926.

Pinkerton, S.D., and Abramson, P.R. (1997). Effectiveness of condoms in preventing HIV transmission. Soc. Sci. Med. *44*, 1303–12.

Pisoni, G., Quasso, A., and Moroni, P. (2005). Phylogenetic analysis of small-ruminant lentivirus subtype B1 in mixed flocks: evidence for natural transmission from goats to sheep. Virology *339*, 147–152.

Porco, T.C., Martin, J.N., Page-Shafer K.A., Cheng, A., Charlebois, E., Grant, R.M., and Osmond, D.H. (2004). Decline in HIV infectivity following the introduction of highly active antiretroviral therapy. AIDS *18*, 81–88.

Quinn, T.C., Wawer, M.J., Sewankambo, N., Serwadda, D., Li, C., Wabwire-Mangen F., Meehan, M.O., Lutalo, T., and Gray, R.H. (2000). Viral load and heterosexual transmission of human immunodeficiency virus type 1. Rakai Project Study Group. N. Engl. J. Med. *342*, 921–929.

Rasheed, S., and Gardner, M.B. (1981). Isolation of feline leukemia virus from a leopard cat cell line and search for retrovirus in wild felidae. J. Natl. Cancer. Inst. *67*, 929–933.

Rickard, C.G., Post, J.E., Noronha, F., and Barr, L.M. (1973). Interspecies infection by feline leukemia virus: serial cell-free transmission in dogs of malignant lymphomas induced by feline leukemia virus. Bibl. Haematol. *39*, 102–112.

Salemi, M., Strimmer, K., Hall, W.W., Duffy, M., Delaporte, E., Mboup, S., Peeters, M., and Vandamme, A.M. (2001). Dating the common ancestor of SIVcpz and HIV-1 group M and the origin of HIV-1 subtypes using a new method to uncover clock-like molecular evolution. FASEB J. *15*, 276–278.

Santiago, M.L., Rodenburg, C.M., Kamenya, S., Bibollet-Ruche F., Gao, F., Bailes, E., Meleth, S., Soong, S.J., Kilby, J.M., Moldoveanu Z *et al.* (2002). SIVcpz in wild chimpanzees. Science *295*, 465.

Sarma, P.S., Log, T., and Theilen, G.H. (1971). ST feline sarcoma virus. Biological characteristics and *in vitro* propagation. Proc. Soc. Exp. Biol. Med. *137*, 1444–1448.

Schim van der Loeff M.F., and Aaby, P. (1999). Towards a better understanding of the epidemiology of HIV-2. AIDS *13*, S69–84.

Scholl, D.T., Truax, R.E., Baptista, J.M., Ingawa, K., Orr, K.A., O'Reilly K.L., and Jenny, B.F. (2000). Natural transplacental infection of dairy calves with bovine immunodeficiency virus and estimation of effect on neonatal health. Prev. Vet. Med. *43*, 239–252.

Shah, C., Boni, J., Huder, J.B., Vogt, H.R., Muhlherr, J., Zanoni, R., Miserez, R., Lutz, H., and Schupbach, J. (2004)a. Phylogenetic analysis and reclassification of caprine and ovine lentiviruses based on 104 new isolates: evidence for regular sheep-to-goat transmission and worldwide propagation through livestock trade. Virology *319*, 12–26.

Shah, C., Huder, J.B., Boni, J., Schonmann, M., Muhlherr, J., Lutz, H., and Schupbach, J. (2004)b. Direct evidence for natural transmission of small-ruminant lentiviruses of subtype A4 from goats to sheep and vice versa. J. Virol. *78*, 7518–7522.

Shattock, R.J., and Moore, J.P. (2003). Inhibiting sexual transmission of HIV-1 infection. Nat. Rev. Microbiol. *1*, 25–34.

Shelton, J.D., Halperin, D.T., Nantulya, V., Potts, M., Gayle, H.D., and Holmes, K.K. (2004). Partner reduction is crucial for balanced 'ABC' approach to HIV prevention. BMJ *328*, 891–893.

Shen, D.T., Gorham, J.R., Jones, R.H., and Crawford, T.B. (1978). Failure to propagate equine infectious anemia virus in mosquitoes and Culicoides variipennis. Am. J. Vet. Res. *39*, 875–876.

Simon, F., Mauclere, P., Roques, P., Loussertajaka, I., Mullertrutwin, M.C., Saragosti, S., Georgescourbot, M.C., Barresinoussi, F., and Brunvezinet, F. (1998). Identification of a new human immunodeficiency virus type 1 distinct from group M and Ggroup O. Nat. Med. *4*, 1032–1037.

Smith, M.C., and Sherman, D.M. (1994). Goat Medicine (Philadelphia: Lea & Febiger), pp. 73–79.

Sodora, D.L., Shpaer, E.G., Kitchell, B.E., Dow, S.W., Hoover, E.A., and Mullins, J.I. (1994). Identification of three feline immunodeficiency virus (FIV) *env* gene subtypes and comparison of the FIV and human immunodeficiency virus type 1 evolutionary patterns. J. Virol. *68*, 2230–8.

Spencer, J.A., Van Dijk A.A., Horzinek, M.C., Egberink, H.F., Bengis, R.G., Keet, D.F., Morikawa, S., and Bishop, D.H. (1992). Incidence of feline immunodeficiency virus reactive antibodies in free-ranging lions of the Kruger National Park and the Etosha National Park in southern Africa detected by recombinant FIV p24 antigen. Onderstepoort J. Vet. Res. *59*, 315–322.

Spira, A.I., Marx, P.A., Patterson, B.K., Mahoney, J., Koup, R.A., Wolinsky, S.M., and Ho, D.D. (1996). Cellular targets of infection and route of viral dissemination after an intravaginal inoculation of simian immunodeficiency virus into rhesus macaques. J. Exp. Med. *183*, 215–225.

Syrjanen, S. (2006). PL7 Oral viral infections that could be transmitted oro-genitally. Oral Dis. *12* Suppl. *1*, 2.

Taha, T.E., Hoover, D.R., Kumwenda, N.I., Fiscus, S.A., Kafulafula, G., Nkhoma, C., Chen, S., Piwowar, E., Broadhead, R.L., Jackson, J.B *et al.* (2007). Late postnatal transmission of HIV-1 and associated factors. J. Infect. Dis. *196*, 10–14.

Teleshova, N., Frank, I., and Pope, M. (2003). Immunodeficiency virus exploitation of dendritic cells in the early steps of infection. J. Leukoc. Biol. *74*, 683–690.

Troyer, J.L., Pecon-Slattery J., Roelke, M.E., Johnson, W., VandeWoude S., Vazquez-Salat N., Brown, M., Frank, L., Woodroffe, R., Winterbach C, *et al.* (2005). Seroprevalence and genomic divergence of circulating strains of feline immunodeficiency virus among Felidae and Hyaenidae species. J. Virol. *79*, 8282–8294.

Ueland, K., and Lutz, H. (1992). Prevalence of feline leukemia virus and antibodies to feline immunodeficiency virus in cats in Norway. Zentralbl. Veterinarmed. B *39*, 53–58.

UNAIDS (2006). UNAIDS Report on the global AIDS epidemic. (Geneva: UNAIDS).

Van Damme L., Ramjee, G., Alary, M., Vuylsteke, B., Chandeying, V., Rees, H., Sirivongrangson, P., Mukenge-Tshibaka L., Ettiegne-Traore V., Uaheowitchai C *et al.* (2002). Effectiveness of COL-(1492), a nonoxynol-9 vaginal gel, on HIV-1 transmission in female sex workers: a randomised controlled trial. Lancet *360*, 971–977.

Van Heuverswyn F., Li, Y., Neel, C., Bailes, E., Keele, B.F., Liu, W., Loul, S., Butel, C., Liegeois, F., Bienvenue Y *et al.* (2006). Human immunodeficiency viruses: SIV infection in wild gorillas. Nature *444*, 164.

VandeWoude S., and Apetrei, C. (2006). Going wild: lessons from naturally occurring T-lymphotropic lentiviruses. Clin .Microbiol. Rev. *19*, 728–762.

VandeWoude S., O'Brien S.J., and Hoover, E.A. (1997). Infectivity of lion and puma lentiviruses for domestic cats. J. Gen. Virol. *78*, 795–800.

Vernazza, P.L., Troiani, L., Flepp, M.J., Cone, R.W., Schock, J., Roth, F., Boggian, K., Cohen, M.S., Fiscus, S.A., and Eron, J.J. (2000). Potent antiretroviral treatment of HIV-infection results in suppression of the seminal shedding of HIV. The Swiss HIV Cohort Study. AIDS *14*, 117–121.

Vobis, M., D'Haese J., Mehlhorn, H., and Mencke, N. (2003). Evidence of horizontal transmission of feline leukemia virus by the cat flea (*Ctenocephalides felis*). Parasitol. Res. *91*, 467–470.

Volmink, J., Siegfried, N.L., van der Merwe L., and Brocklehurst, P. (2007). Antiretrovirals for reducing the risk of mother-to-child transmission of HIV infection. Cochrane Database Syst. Rev. (1), CD003510.

Walder, R., Kalvatchev, L., Perez, F., Garzaro, D., and Barrios, M. (2001). Bovine immunodeficiency virus in experimentally infected rabbit: tropism for lymphoid and nonlymphoid tissues. Comp. Immunol. Microbiol. Infect. Dis. *24*, 1–20.

Wawer, M.J., Gray, R.H., Sewankambo, N.K., Serwadda, D., Li, X., Laeyendecker, O., Kiwanuka, N., Kigozi, G., Kiddugavu, M., Lutalo T *et al.* (2005). Rates of HIV-1 transmission per coital act, by stage of HIV-1 infection, in Rakai, Uganda. J. Infect. Dis. *191*, 1403–9.

Weller, S., and Davis, K. (2002). Condom effectiveness in reducing heterosexual HIV transmission. Cochrane Database Syst. Rev. 1, CD003255.

Whetstone, C.A., VanDerMaaten M.J., and Black, J.W. (1990). Humoral immune response to the bovine immunodeficiency-like virus in experimentally and naturally infected cattle. J. Virol. *64*, 3557–3561.

Williams, D.L., Issel, C.J., Steelman, C.D., Adams, W.V., Jr., Benton, C.V. (1981). Studies with equine infectious anemia virus: transmission attempts by mosquitoes and survival of virus on vector mouthparts and hypodermic needles, and in mosquito tissue culture. Am. J. Vet. Res. *42*, 1469–1473.

Wu, L., KewalRamani V.N. (2006). Dendritic-cell interactions with HIV: infection and viral dissemination. Nat. Rev. Immunol. *6*, 859–868.

Yaoi, H., Nagata, A., Goto, N., and Saito, K. (1959). Studies on the virus of equine infectious anemia. I. Retransmission of Arakawa's virus to horse. Yokohama Med. Bull. *10*, 1–10.

Pathogenesis of Oncoviral Infections

9

Finn Skou Pedersen and Annette Balle Sørensen

Abstract

Retroviruses cause cancer in natural or laboratory settings by a variety of mechanisms. The acutely transforming or transducing retroviruses induce tumours in animals within days to weeks. They harbour a host-cell derived gene, an oncogene, which infiltrates signalling cascades that regulate cell growth and survival. The oncoprotein encoded by the viral oncogene is activated to dominant signalling, either by deregulated expression or as a result of a modified protein structure that uncouples downstream signalling from upstream physiological signals. The *cis*-acting or non-acutely transforming viruses cause disease with latency periods of months. These viruses work as insertional mutagens to promote multi-step oncogenesis, and large-scale mapping of proviral insertions in tumour DNAs provides a rich source of candidate genes with a potential role in cancer of non-retroviral aetiology. Viral proteins may also stimulate target cells to proliferate. One example is the mouse mammary tumour virus which stimulates lymphocytes via a virus-encoded superantigen. Other examples are the mitogenic stimulation of erythrocyte precursor cells by the defective envelope protein of the mouse spleen focus-forming virus, and the direct oncogenic affect of the envelope protein of the Jaagsiekte sheep retrovirus.

Overview of oncogenic retroviruses

The family *Retroviridae* provides a multitude of examples of virus isolates that are able to induce disease in laboratory or natural settings. Disease types include immunodeficiencies and neurological disorders, as well as non-malignant and malignant proliferative diseases. The term oncoviruses or oncoretroviruses typically refers to oncogenic retroviruses of the alpharetroviruses, betaretroviruses, or gammaretroviruses genera. The focus of this chapter will be on how oncoviral infections can cause malignant cancers.

Oncogenic retroviruses can be divided into three main groups by their mechanism of cell transformation (Rosenberg and Jolicoeur, 1997; Nevins, 2007) (Fig. 9.1).

Transducing retroviruses

These are acutely transforming retroviruses with a host-derived oncogene. They are typically able to cause oncogenic transformation of cells in culture and cause the induction of specific groups of tumours in animals with latency periods of days to weeks. With the exception of Rous sarcoma virus (RSV), the oncogenic agent of viruses of this group is a defective virus that has incorporated a host-derived oncogene at the expense of one or more viral genes needed for replication. Such defective viruses have been isolated as mixtures with helper viruses that are able to complement the defective virus with the missing viral proteins. These viruses are represented in the genera alpharetroviruses and gammaretroviruses.

Cis-acting retroviruses

These are non-acutely transforming retroviruses; they are typically not able to cause oncogenic transformation of cells in culture and are all replication-competent. They cause disease in animals

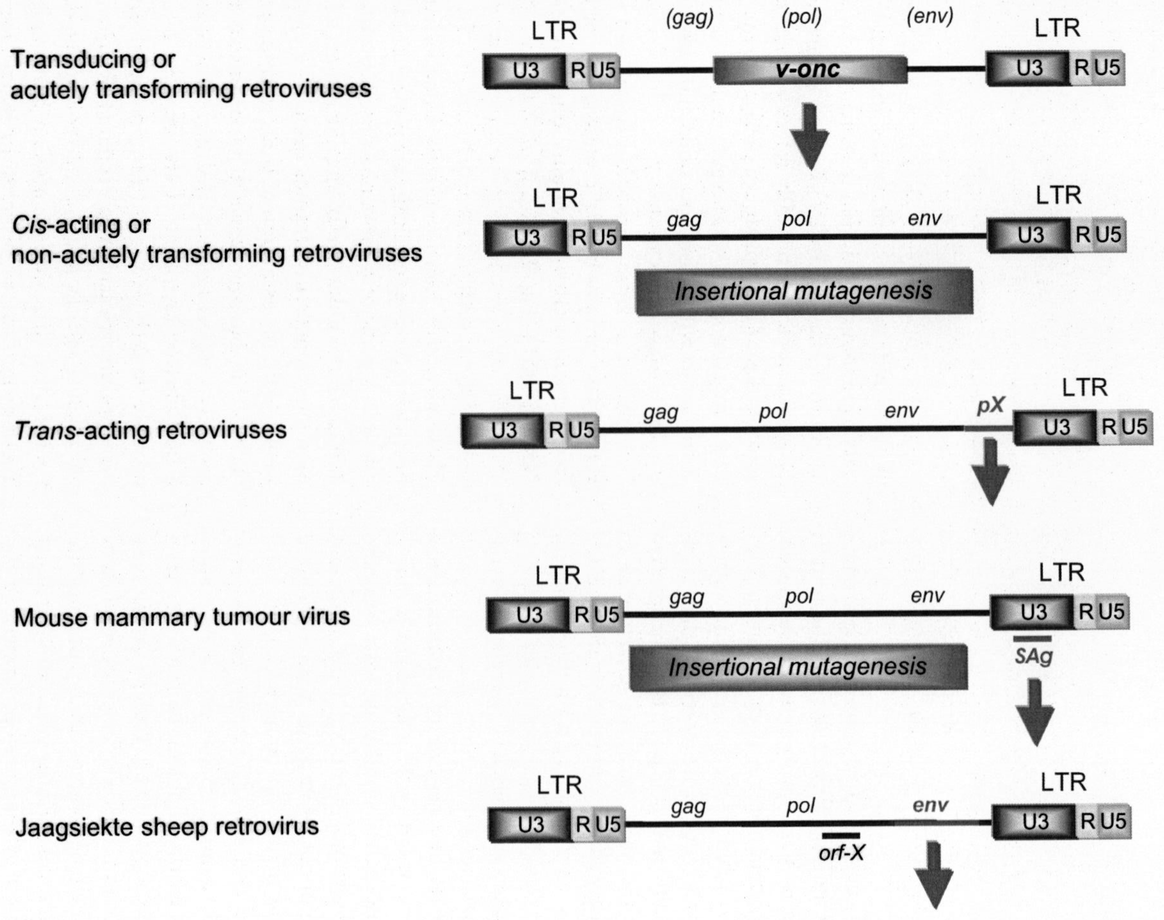

Figure 9.1 Oncogenic retroviruses. Schematic representation of the genomic organization within the three major groups; the transducing, the *cis*-acting, and the *trans*-acting retroviruses. In addition, two special cases are shown, the mouse mammary tumour virus (MMTV) and the Jaagsiekte sheep retrovirus (JSRV). The red boxes, lines, and arrows indicate the major players in oncogenesis.

after latency periods of a few months to more than a year. Mutagenesis of the host genome caused by provirus insertions during multiple rounds of infection in the animal is a key component in oncogenesis. Members of this group are found in the genera alpharetroviruses, betaretroviruses, and gammaretroviruses.

Trans-acting retroviruses

These are oncogenic retroviruses with a viral gene product directly implicated in tumorigenesis. In this mixed group we are dealing with non-defective viruses in which one or more viral gene products also required for the infection process of the virus have a quite diverse range of effects on tumorigenesis. Some viruses of this group are able to transform cells in culture. For this group of viruses the critical viral gene has no counterpart in the host cell, which distinguishes them from the acutely transforming retroviruses, which all carry a host-derived oncogene. Well-established examples are found in the genera betaretroviruses, deltaretroviruses, and epsilonretroviruses. For viral determinants of oncogenesis in deltaretroviruses and epsilonretroviruses, the reader is referred to Chapters 14, 15 and 16.

Isolation of oncogenic retroviruses

Today, modern researchers working with oncogenesis by retroviruses are in much debt to classic isolates with diverse biological properties. The history goes back to 1908 when the Danish researchers Ellermann and Bang reported that leukosis in chicken could be caused by a virus (Ellerman and Bang, 1908). The disease described in this work is now recognized as a form of leukaemia or lymphoma, and the type of viruses responsible for the disease as avian leukosis virus (ALV). In 1911, the New York-based researcher Peyton Rous, also working with chicken, reported that solid tumours, sarcomas, could be serially transmitted by a cell-free filtrate (Rous, 1911). The virus discovered by Rous, Rous sarcoma virus (RSV), served as a prototype of a transducing, acutely transforming retrovirus for research on transformation in cell culture and tumorigenesis in animals during most of the century. ALV and RSV both belong to the genus of alpharetroviruses and are now commonly referred to as avian sarcoma/leukosis viruses (ASLV). In the following years, many other acutely transforming retroviruses were isolated from chicken (Table 9.1), some with novel tumorigenic specificities, such as the avian erythroblastosis virus and the avian leukaemia virus MC29 that target different lineages of the haematopoietic system.

In mammals, the findings of retroviral tumour induction came from studies of inbred mouse strains with characteristic patterns of increased incidence of certain types of cancers. In one mouse strain characterized by high incidence of mammary cancer it was suggested that the oncogenic agent could be transmitted through the milk from mother to offspring, because the foster nursing of pups by mothers of a low-incidence strain led to a reduction in disease incidence. In 1942 John Bittner published direct evidence of such mammary carcinoma transmission by a cell-free agent from milk (Bittner, 1942), an agent that was later identified as mouse mammary tumour virus (MMTV). The two modes of transmission of MMTV in mice – vertically through the germ line and horizontally through the milk – made the picture quite complex until the concept of endogenous retroviruses was established. A clarification of this dichotomy of transmission of MMTV-induced mammary carcinoma awaited the demonstration of germ-line inherited MMTV (Bentvelzen and Daams, 1969). A similar complex picture with contributions from endogenous viruses came from studies of virus-induced leukaemia in mice. Ludwik Gross found that cell-free extracts of leukemic thymuses from the high leukaemia incidence strain AKR could induce thymic leukaemia in a low-leukaemia incidence strain. Moreover, cell-free extracts of increased virulence (Gross leukaemia virus) were derived by serial passage in the low-leukaemogenic C3H strain (Gross, 1957). Gross discovered that virus inoculation should take place within the first few days after birth for disease development to take place. This helps the virus to avoid a strong immune response and establish an infection. A classic retroviral disease in mice described by Charlotte Friend in 1957 (Friend, 1957) is characterized by spleen enlargement and erythroleukaemia (*i.e.* leukaemia characterized by the proliferation of erythrocyte precursor cells). The mechanism of disease induction by this virus, the Friend virus

Table 9.1 Examples of oncogenic retroviruses

Genus	Virus	Species	Disease	Oncogene
Alpha-retrovirus	Avian leukosis virus (ALV)	Chicken	Bursal lymphoma	None
	Rous sarcoma virus (RSV)	Chicken	Sarcoma	*src*
	Fujinami sarcoma virus (FuSV)	Chicken	Sarcoma	*fps*
	Avian myelocytoma virus MH2	Chicken	Myelocytoma	*myc* plus *mil*
	Avian erythroblastosis virus (AEV)	Chicken	Erythroblastosis	*erbB* (plus *erbA*)
	Avian myeloblastosis virus (AMV)	Chicken	Myeloblastosis	*myb*
	Avian myeloblastosis erythroblastosis virus E26	Chicken	Erythroblastosis/ myeloblastosis	*myb* plus *ets*
	Avian myelocytoma MC29	Chicken	Myelocytoma	*myc*
	Avian sarcoma virus 16	Chicken	Haemangiosarcomas	*p3k*
	Avian sarcoma virus 17	Chicken	Sarcoma	*jun*
Beta-retrovirus	Mouse mammary tumour virus (MMTV)	Mouse	Mammary carcinoma	None
	Jaagsiekte sheep retrovirus (JSRV)	Sheep	Pulmonary adenocarcinoma	Viral *env*
	Enzootic nasal tumour virus (ENTV)	Sheep/goat	Nasal adenocarcinoma	Viral *env*
Gamma-retrovirus	Gross murine leukaemia virus (Gross MLV)	Mouse	Thymic lymphoma	None
	Friend murine leukaemia virus (Fr-MLV)	Mouse	Acute erythroleukaemia	None
	Radiation leukaemia virus (RadLV)	Mouse	Thymic lymphoma	None
	Moloney murine leukaemia virus (Mo-MLV)	Mouse	Thymic/non-thymic lymphoma	None
	Feline leukaemia virus (FeLV)	Cat	Thymic lymphoma	None
	Reticuloendotheliosis virus strain T (Rev-T)	Turkey	Lymphomas	*rel*
	Moloney murine sarcoma virus (Mo-MSV)	Mouse	Sarcoma	*mos*
	Harvey murine sarcoma virus (Ha-MSV)	Rat	Sarcoma/ erythroleukaemia	H-*ras*
	Kirsten murine sarcoma virus (Ki-MSV)	Rat	Sarcoma/ erythroleukaemia	K-*ras*
	Abelson murine leukaemia virus (Ab-MLV)	Mouse	Pre-B-lymphoma	*abl*
	Gibbon ape leukaemia virus (GALV)	Primate	Lymphomas, others	None
	McDonough feline sarcoma virus (SM-FeSV)	Cat	Sarcoma	*fms*
	Simian sarcoma virus (SSV)	Woolly monkey	Sarcoma	*sis*
	Murine sarcoma virus 3611	Mouse	Sarcoma	*raf*
	AKT8	Mouse	T-cell lymphoma	*akt*

complex, has remained an important research topic ever since. The Moloney strain of MLV is another classic leukaemogenic mouse retrovirus isolate, which is a potent inducer of lymphomas in several inbred mouse strains (Moloney, 1960). In particular during the 1960s and 1970s, several other retroviruses that cause malignant disease in mammalian species such as mice and cats as well as various primates were also isolated. Some of these (Table 9.1) are non-acutely transforming retroviruses that induce leukaemias/lymphomas with latency periods of months, e.g. the feline leukaemia virus (FeLV) (Jarrett *et al.*, 1964) and the gibbon ape leukaemia virus (GALV) (Theilen *et al.*, 1971). Others are acutely transforming retroviruses with short latency periods, representing several interesting tumorigenic specificities, such as the pre-B-lymphocyte-transforming Abelson murine leukaemia virus (Ab-MLV) (Abelson and Rabstein, 1970).

For more than half a century after their initial discovery, oncogenic retroviruses were detected, quantified and characterized in animal hosts, also including eggs in case of the bird viruses. By the 1950s cell culture became an important tool for the propagation and studies of animal viruses. A hallmark in retrovirology was the demonstration by Temin and Rubin (1958) of a quantitative focus-formation assay for the oncogenic transformation of chicken embryo fibroblasts by Rous sarcoma virus. The transformed foci were characterized by an increase in saturation density and anchorage independence. Transformed foci produced viruses with focus-forming ability. A second major advancement came with the introduction of recombinant DNA and sequencing technologies into retroviral research about 1980, which allowed the complete genetic characterization and directed mutational analysis of the rich harvest of retroviral isolates of diverse biological properties from the preceding decades.

Transducing retroviruses

In the 1950s and 1960s, genetic studies of viral replication and cell transformation in chicken embryo fibroblasts led to the identification of four Rous sarcoma virus (RSV) genes. The *gag*, *pol*, and *env* genes were required for viral growth, whereas the fourth gene, *src*, was needed for cell transformation but dispensable for viral replication (Martin, 2004). An RSV mutant without a functional *src* gene would replicate but not transform. RSV mutants with an impaired ability of viral growth could be scored by their ability to make transformed foci from which no transforming virus was produced. From such cells, however, a virus with transforming properties could be rescued by superinfection by a helper virus (Hanafusa *et al.*, 1963; Temin, 1963) (Fig. 9.2). In a further development, temperature-sensitive mutants of Rous sarcoma virus were isolated, which were unable to establish or maintain oncogenic transformation at the non-permissive temperature (Martin, 1970). A change to the permissive temperature would re-establish transformation. These results showed that a functional Src protein is needed to induce transformation and to maintain the transformed phenotype. Stehelin *et al.* (1976) used defined mutants of RSV to generate and select cDNAs specific for the *src* gene. Hybridization reactions using these cDNAs as probes demonstrated that nucleotide sequences closely related to *src* were present in the genomic DNA of normal chicken DNA and that similar sequences were present in the genome of a variety of avian species. In its origin, *src* was therefore not a viral gene, but a cellular gene captured and transferred by a retrovirus.

The pioneering work on Rous sarcoma virus was followed by intense studies of a large number of transducing retroviruses from birds and mammals. The major discoveries that the *src* gene of Rous sarcoma virus plays no role in retroviral replication and that it is derived from a cellular gene, turned out to be true for retroviral oncogenes in general. Table 9.2 lists examples of oncogenes captured in transducing retrovirus isolates with transforming properties. Some genes have been captured independently in more than one isolate of the same species, such as the *myc* gene in several viruses from chicken, and orthologues of the same gene have been captured in different species, e.g. *raf* in the mouse and *mil* in the chicken. Some transducing retroviruses, such as the avian erythroblastosis viruses AEV-ES4 and AEV-R as well as MH2 avian myelocytoma virus, have captured not only one, but two distinct cellular genes. The cellular genes which gave rise to the viral oncogenes (v-*onc*) are termed

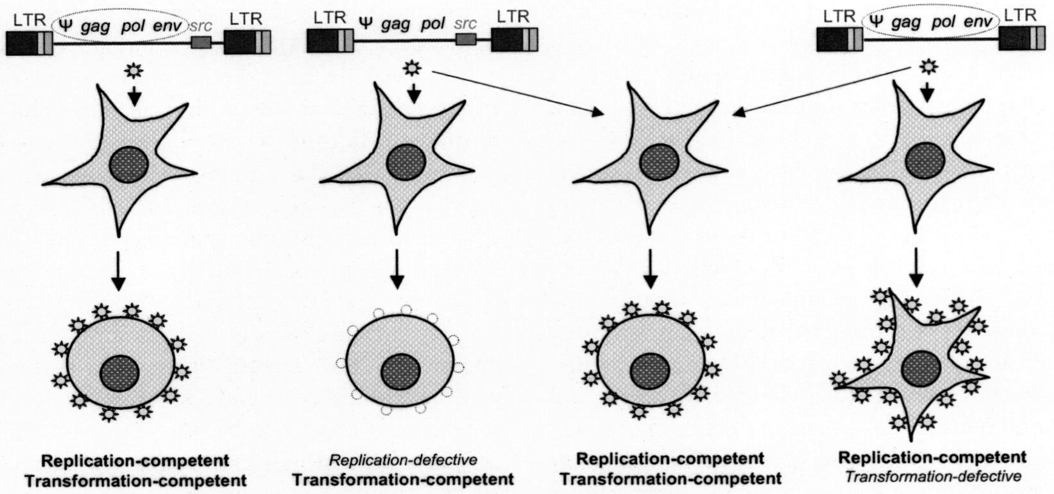

Figure 9.2 Replication and transformation competence of Rous sarcoma virus (RSV) mutants. The left column shows the replication- and transformation-competent RSV, which contains all *cis*-acting signals (such as the packaging signal Ψ) and the genes necessary for replication (*gag*, *pol*, and *env*) as well as the transforming v-*src* gene. The second column shows the replication-defective virus in which the transduced v-*src* has replaced part of the coding sequences. Although the virus can be integrated and transcribed, and consequently the transforming oncoprotein (v-Src) expressed, no viral particles can be produced in this case. However, such a virus can be rescued – as shown in columns 3 – by superinfection by a replication-competent helper virus (column 4). The defective genome, which still contains the packaging signal Ψ, is transcribed and packaged into viral particles assembled by proteins provided from the helper virus (in fact, the virus particles produced in this case are a mixture of particles containing the defective, v-*src*-including, genome, the complete helper virus genome, or both).

proto-oncogenes or cellular oncogenes (c-*onc*) and their protein products, oncoproteins.

Retroviral capture of oncogenes

Since oncogene capture is a rare event, a model cannot be subjected to rigorous experimental testing in a quantitative setting. Prevailing models hold that a provirus becomes integrated upstream of a cellular oncogene in a co-transcriptional orientation (Felder *et al.*, 1991; Coffin, 1992). A chimeric RNA is generated by transcription from the 5′-end of the provirus into the flanking oncogene, possibly facilitated by deletion of the 3′-part of the provirus. This chimeric RNA may undergo splicing events that will remove some or all of the introns of the proto-oncogene, but it must retain the retroviral packaging signal. The model further proposes that the same cell expresses a complete retrovirus, and that the chimeric retroviral-oncogene RNA will be encapsidated as a heterodimer into a retroviral particle linked to a complete retroviral genome. After infection of another cell by such a virus, template switching by the retroviral reverse transcriptase may lead to the formation of a chimeric proviral genome with an oncogene flanked by 5′ and 3′ parts of the virus. In formal terms this model proposes two recombination events, the first one at the double-stranded DNA level by integration of the provirus near the oncogene, the second one during cDNA synthesis by template switching of reverse transcriptase between viral and oncogene sequences. In many cases the transducing virus has captured additional sequences or undergone other changes. The resulting transducing virus contains all *cis*-acting sequences needed for the transcription, packaging, reverse transcription, and integration. Some have also retained *cis*-acting sequences for splicing. With the rare exception of Rous sarcoma virus, transducing retroviruses have lost part of the coding sequences for viral protein. Such viruses that lack the coding capacity for one or more viral proteins have been isolated as mixtures with replication-competent viruses

Table 9.2 Examples of oncogenes captured in transducing retroviruses

Functional class	Oncogene	Virus	Origin	Oncogene product	Mode of oncogenic activation
Growth factor	sis	SSV	Monkey, sarcoma	Env-Sis	Deregulated expression
Receptor phosphotyrosine kinase	erbB	AEV-ES4	Chicken, erythroblastosis	v-ErbB	Modified protein with constitutive signalling
	fms	SM-FeSV	Cat, sarcoma	Gag-Fms	Modified protein with constitutive signalling
Non-receptor phosphotyrosine kinase	src	RSV	Chicken, sarcoma	v-Src	Modified protein with constitutive signalling
	abl	Ab-MLV	Mouse, pre-B-lymphoma	Gag-Abl	Modified protein with constitutive signalling
G-protein	H-ras	Ha-MSV	Rat, sarcoma/ erythroleukaemia	v-Hras	Modified protein with constitutive signalling
	K-ras	Ki-MSV	Rat, sarcoma/ erythroleukaemia	v-Kras	Modified protein with constitutive signalling
Phosphatidylinositol kinase	p3k	ASV16	Chicken, sarcoma	Gag-PI3K	Protein localization to membrane
Ser/Thr kinase	raf	3611-MSV	Mouse, sarcoma	Gag-Raf	Modified protein with deregulated properties
	mos	Mo-MSV	Mouse, sarcoma	Env-Mos	Deregulated expression
	akt	AKT8	Mouse, thymoma	Gag-Akt	Protein localization to membrane
Adapter	crk	CT10	Chicken, sarcoma	Gag-Crk	Modified protein with increased stimulation of signalling
Transcription factor	myc	MC29	Chicken, sarcoma	Gag-Myc	Mutated protein/deregulated expression
	jun	ASV-17	Chicken, sarcoma	Gag-Jun	Protein deregulation by mutation

that provide the missing proteins (Fig. 9.2). Many oncogenes of the replication-defective viruses are expressed as fusion proteins with viral Gag or Env sequences, often with N-terminally located viral polypeptides (Table 9.2). In Rous sarcoma virus, which is unique among transducing retroviruses in having retained *gag*, *pol*, and *env*, the oncoprotein Src is translated from a specific subgenomic mRNA with its own splice-acceptor signal.

Function of oncogenes

The discovery of transducing retroviruses with cell-derived genes that are able to transform cells in a dominant manner has had an immense impact on molecular cancer research. The oncogenic potential of a cellular counterpart of a viral oncogene can also be activated by diverse genetic changes not involving retroviruses such as point mutations, gene amplifications, and chromosomal rearrangements that affect the structure and expression of oncoproteins. Moreover, conserved over long evolutionary distances and thus appearing to possess fundamental regulatory functions, oncogenes have also had a major impact on research of basic cell biology from yeast to man.

Retroviral transduction of an oncogene causes the transformation of appropriate target cells within a very short time. Tumours in animals inoculated with a transducing virus appear within days and are usually polyclonal in origin. In contrast to most cancers that develop after a multi-step transformation process, the tumour induction by retroviral transduction of an oncogene is probably a single-step process. To understand how the expression of a single v-*onc* gene can infiltrate and substantially change the normal regulatory program in a cell, it is important to uncover the normal function of the corresponding c-*onc* to see how this function is deregulated or corrupted in the v-*onc*. While the function of many cellular oncoproteins has been determined, such understanding is still missing or only tentative for others. The general picture is that the cellular counterparts of viral oncogenes are part of signalling networks that are involved in the regulation of cell growth and survival (Fig. 9.3). These networks receive input from the outside of the cell and propagate signals to the nucleus where they are converted into patterns of gene regulation. The extracellular growth signal is typically a polypeptide growth factor that binds to the extracellular domain of a specific receptor at the cell surface. The growth factor receptor may be an integral membrane protein with tyrosine-specific kinase activity located in its cytoplasmic part. Ligand binding may cause receptor-dimerization that in turn activates the kinase and leads to autophosphorylation of specific tyrosine residues of the receptor. This signal, thus transferred across the cell-surface membrane by the receptor, can then be recognized through specific protein binding of phosphotyrosine-containing peptide motifs, and further propagated by complex steps of sequential protein–protein interactions and/or enzymatic reactions. These signal cascades finally modify transcription factors that control the expression of specific sets of genes involved in growth control.

Activation of oncogenes in transducing retroviruses

What distinguishes the effect of an oncogene transduced by a retrovirus from that of its normal counterpart, already present in the genome of the infected cell? One important point is the deregulation of expression imposed upon the oncogene by its location in the context of the provirus. Transcription is driven by viral promoter-enhancer sequences of the LTR, and any steps of normal regulation of RNA processing or export are expected to be lost by the placement of an intronless gene in the context of retroviral RNA sequences. Control of oncoprotein translation is in many cases expected to exploit exclusively viral signals. Such deregulation of expression does in fact appear to be sufficient for the activation of transforming properties of some oncoproteins, since their normal cellular counterpart is also transforming when expressed in a retroviral context or overexpressed by other means (Table 9.2). In many cases, however, the deregulated expression of a structurally normal oncoprotein is in itself not sufficient for cell transformation. Structural changes in the protein itself are also needed. Transduced oncogenes often carry deletions at one or both ends as well as internal deletions and point mutations. Below are provided well-studied examples of how specific structural changes of an oncoprotein may uncouple its downstream signalling from stimulation by physiological upstream

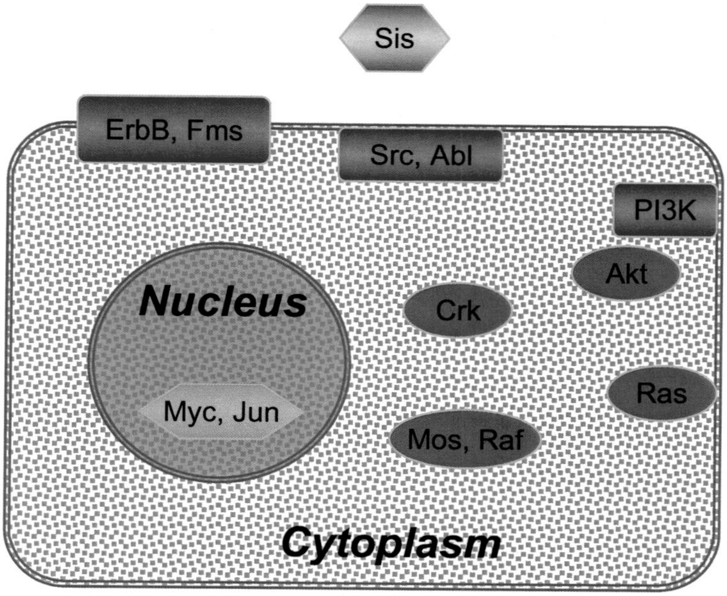

Figure 9.3 Examples of oncoproteins in signalling pathways. Oncoproteins deregulate signalling cascades which leads to growth stimulation and survival irrespective of the presence of physiologic growth-activating signals. Sis is an example of a growth-factor-like oncoprotein, which stimulates signalling through binding to a growth factor receptor in a predominantly autocrine manner. The v-ErbB and v-Fms oncoproteins are examples of modified growth factor receptor tyrosine kinases with constitutive signalling activity. The v-Src and v-Abl oncoproteins are examples of membrane-bound non-receptor tyrosine kinases with deregulated signalling. The viral Ras-oncoproteins are mutationally activated forms of GTPases with central roles in the propagation of growth stimulatory signals towards the nucleus. Crk is an adaptor protein with no catalytic activity, but with specific protein–protein interaction domains that serve to relocate proteins into active signal transduction complexes. Examples of oncoprotein members of two pathways involved in the further propagation of growth and survival signals towards the nucleus are Raf and Mos of the MAP kinase pathway, and PI3K and Akt of the PI3K/Akt pathway. Myc and Jun are examples of two nuclear oncoproteins with transcription factor activity.

signals. When oncoproteins are fused with viral protein sequences such as deleted forms of Gag protein, the viral protein part of fusion proteins is not directly implicated in the transformation process. However, the viral sequences may have important modulatory effects on protein stability or subcellular localization.

Growth factor stimulation

The *sis* oncogene was identified in the transducing retrovirus simian sarcoma virus (SSV), isolated from a woolly monkey kept as a pet animal (Theilen *et al.*, 1971). The Sis oncoprotein shows about 88% identity with the B-chain of the human platelet-derived growth factor, PDGF-B (Doolittle *et al.*, 1983). PDGF-B forms homodimers or heterodimers with other PDGF family members and transmits mitogenic stimuli through PDGF receptors, which are expressed on a variety of cell types. The Sis oncoprotein is expressed as a fusion protein with an N-terminal signal peptide derived from the viral envelope protein. This signal peptide sustains the translocation of the oncoprotein into vesicles of the endoplasmic reticulum, where it is dimerized and proteolytically cleaved into a homodimer of similar structure to the normal PDGF-B homodimer. Only a small fraction of the Sis oncoprotein is secreted. The product of the *sis* gene is functionally similar to PDGF-B (Westermark and Heldin, 1991). It binds to PDGF receptors, induces internalization of the receptor–ligand complex, induces receptor phosphorylation on tyrosine activities, and transmits mitogenic stimuli to the cell. Most of the PDGF–receptor interaction is believed to be initiated in intracellular vesicles where also the PDGF-receptor is processed. This PDGF-like activity of Sis appears to be fully responsible for its transforming ability, which is dependent upon the presence of PDGF receptors in the target cells. The overexpression of PDGF-B on PDGF receptor-bearing cells is also able to cause cell transformation. Hence, the Sis oncoprotein exerts its transforming ability through PDGF-like mitogenic stimuli in a predominantly autocrine manner (Westermark and Heldin, 1991). PDGF receptors function in the cell through activation of a variety of downstream effectors, such as phospholipase C (PLC), the adapter protein Grb2 in complex with the nucleotide releasing factor SOS, the phosphatidylinositide 3-kinase (PI3K), the GTPase activating protein (GAP), and the signal transducers and activators of transcription (STATs).

Receptor tyrosine kinase signalling

The oncogene *erbB* was first identified in avian erythroblastosis virus (AEV), named ES-4, which is a virus that induces erythroblastosis (proliferation of erythrocyte precursor cells) and fibrosarcomas (solid tumours of fibroblast-like cells) in chicken (Engelbreth-Holm and Meyer, 1932). The classic AEV isolates carry two oncogenes *erbA* and *erbB*, while more recent isolates carry only *erbB*. ErbB is both necessary and sufficient to induce the pathogenic effects characteristic of AEV (Frykberg *et al.*, 1983). The viral *erbB* oncogene shows close homology to the epidermal growth factor receptor (EGFR) gene. Four major structural and functional domains have been identified in EGFR: an extracellular N-terminal domain with affinity for the growth factor ligand, a transmembrane part, a cytoplasmic kinase domain, and a cytoplasmic carboxyl-terminal regulatory part with five tyrosine phosphorylation sites. In normal growth stimulation through EGFR, ligand binding is believed to induce an allosteric change that favours receptor oligomerization (Schlessinger, 1986). As a result of this, the kinase activity of EGFR, which is essential for the mitogenic effect of EGF, becomes elevated and causes the autophosphorylation of tyrosines in its C-terminal part. This in turn promotes the docking of EGFR substrates and adaptors that sustain the further propagation of mitogenic signals in an only partially understood manner. EGFR partners of potential importance in downstream signalling overlap with those mentioned above for PDGFR. ErbB differs from EGFR by having lost most of the extracellular growth-factor-binding domain and by having various strain-different deletions in the cytoplasmic part. Studies of various isolates of AEV and of conditional and non-conditional mutants have led to the conclusion that membrane localization and the presence of the kinase domain are essential for the transforming activity of ErbB. Moreover, the C-terminal deletions that differ among isolates have an influence on the transforming potential and disease specificity (Raines *et al.*, 1988). These results are compatible with the notion that the ErbB oncoprotein functions as a constitutively activated EGFR that transmits mitogenic signals into the cell irrespective of EGF stimulation. *Fms*, the oncogene of the McDonough strain of feline sarcoma virus (McDonough *et al.*, 1971) encodes a constitutively activated version of another growth-factor receptor tyrosine kinase, the receptor for the macrophage colony stimulating factor 1 (Sherr, 1988).

Non-receptor tyrosine kinase signalling

The *src* gene was the first oncogene discovered and genetically defined by its transforming properties. It was also the first oncogene of a transduced retrovirus for which its cellular origin was established and for which a protein product was identified (Brugge and Erikson, 1977). The Src oncoprotein was shown to be a protein kinase and the founding member of a large family of tyrosine-specific protein kinases with roles in signal transduction (Hunter and Sefton, 1980). In spite of these early front-running manifestations of *src* research, its role in cell transformation is still only partially understood. Src is a membrane-bound non-receptor tyrosine-specific kinase. The cellular Src protein has several well-defined domains. An N-terminal myristylation domain mediates membrane binding. The SH2 and SH3 (Src homology regions 2 and 3, respectively) domains are protein–protein interaction domains common to tyrosine-specific protein kinases as well as other proteins. SH2 domains bind to phosphotyrosine motifs in a context-specific manner. The catalytic kinase domain is followed by a regulatory C-terminal domain. This domain organization is maintained in the viral Src protein, but the C-terminal regulatory regions harbour deletions (Cooper and Howell, 1993). These structural changes of the viral oncoprotein are important for its transforming properties, since overexpression of the cellular form is insufficient for cell transformation. Increased kinase activity is needed for the transforming activity. The kinase activity of cellular Src is kept in an inactive state by multiple intramolecular interactions, and regulated positively and negatively by the phosphorylation of specific tyrosine residues in the C-terminal region. Specifically, phosphorylation of tyrosine 527 regulates the kinase activity of cellular Src negatively, most likely by an intramolecular binding of the SH2 domain with an effect on protein folding and catalytic activity (Cooper and Howell, 1993). Significantly, viral Src lacks tyrosine 527 and is constitutively active as a kinase. Numerous proteins have been suggested as downstream targets of Src signalling. One example of an important partner is the focal adhesion kinase, a tyrosine kinase located at the points of focal adhesion of an adherent tissue-culture cell to the surface of the culture flask. The focal adhesion kinase is activated by binding to the SH2 domain of Src. This in turn results in the activation of several downstream pathways which contributes to the reduced adhesion and morphological changes characteristic of transformed fibroblasts (Parsons, 2003). Another well-studied viral oncoprotein (Abl) with membrane-bound tyrosine kinase activity was derived from the Abelson murine leukaemia virus, which is an effective transforming agent for early stages of the B-lymphocyte lineage (Chen *et al.*, 1994).

Activated Ras proteins

Oncogenes of the *ras* family have been captured independently in a number of transducing retroviruses. The best studied examples are the Harvey sarcoma virus (Harvey, 1964) with a viral oncogene derived from the cellular H-*ras* gene, and the Kirsten sarcoma virus (Kirsten and Mayer, 1967) with a viral oncogene derived from the cellular K-*ras* gene. The third mammalian *ras* gene, N-*ras*, has not been naturally captured in a transducing retrovirus. The cellular Ras proteins are GTPases that act at the cell membrane as nodal points in the transmission of mitogenic signals (Takuwa and Takuwa, 2001; Mitin *et al.*, 2005). They receive input from several transmission paths including signals from tyrosine kinases of growth factor receptors such as EGFR and membrane-bound signalling proteins such as Src. The inputs are converted to signals transmitted through protein kinase cascades towards the nucleus. Ras proteins belong to a superfamily of GTPase proteins with properties as a molecular switch, where the GTP-bound form is in the on-state and the GDP-bound form is in the off-state. These molecular switch properties of Ras proteins are supported by detailed structural information. In a normal cell the transition between the signalling-active GTP form and the signalling-inactive GDP is regulated by several factors. Among these, the guanine nucleotide exchange factors activate Ras through an exchange of Ras-bound GDP for GTP. Sos is an important guanine nucleotide exchange factor that is recruited to activate Ras by activated growth factor receptors. In contrast, the GTPase activating proteins (GAPs) have a dampening effect on Ras signalling by stimulating the Ras-catalysed hydrolysis of GTP to GDP. Activated Ras oncoproteins encoded in transducing retroviruses harbour characteristic point mutations in codons 12, 13, or 61 that inhibit the binding of GAP proteins to Ras in the GTP-bound form (Tabin and Weinberg, 1985). Thereby these mutations inhibit the stimulation of the GTPase activity, and the oncoprotein remains in the activated GTP-bound form, which results in the constitutive propagation of growth-stimulatory signals towards the nucleus through various pathways. One of these is the mitogen-activated protein (MAP) kinase cascade discussed in the next paragraph. The activating mutations detected in viral *ras* oncogenes are also commonly found in human cancer.

Activation of the MAP kinase pathway

The *raf/mil* oncogene was captured by a murine (*raf*) (Rapp *et al.*, 1983) as well as an avian (*mil*) transducing retrovirus, the latter also containing *myc* (Coll *et al.*, 1983). Raf/Mil acts through a cascade of protein kinases of the MAP kinase pathway, which relays signals from extracellular stimuli into specific intracellular responses in the form of phosphorylation events. Raf is a cytoplasmic serine/threonine kinase that is positively and negatively regulated by phosphorylation and protein–protein interactions in specific domains. It is emerging as a nodal propagator of growth signals and a key downstream effector of Ras. In mammalian cells, there are several distinct MAP kinase signalling pathways. These signalling cascades are typically based upon a core element of three sequentially acting kinases termed MAP kinase kinase kinase (MAPKKK), which phosphorylates MAP kinase kinase (MAPKK), which in turn phosphorylates the MAP kinase (Raman *et al.*, 2007). Signal transduction through this core is regulated by positive as well as negative input. Docking interactions of MAP kinase pathway members as well as scaffolding interactions contribute to the efficiency and specificity of signal transduction. Once activated, the MAP kinase translocates to the nucleus and modifies downstream effectors by phosphorylation on serine or threonine residues. Raf/Mil is a MAPKKK of the RAF-MEK-ERK pathway. In the transducing viruses, the cellular *raf* sequences are truncated at the 5′ end, which contributes to its deregulated signalling properties and oncogenic potential. The mammalian genome contains three *raf* genes (A-*raf*, B-*raf*, and C-*raf*). In particular B-*raf* is frequently mutated in human cancers (Rapp *et al.*, 2006). The *mos* gene is another example of a MAPKKK gene captured by a transducing retrovirus. *Mos* (Wood *et al.*, 1984) is an intronless gene expressed at very low level in most tissues with the exception of germ cells where it is highly expressed. Overexpression of the product of the *c-mos* gene is sufficient for cell transformation, hence in this case deregulation of expression does appear to be sufficient for oncogenic activation.

Activation of the PI3K/Akt kinase cascade

Another signalling cascade that receives stimuli from growth factors is the phosphatidylinositide 3-kinase (PI3K)/c-Akt kinase cascade. PI3K lies at a nodal point of multiple signal cascades, and signalling through this pathway has been particularly implicated in the promotion of cell growth and survival. Receptor activation results in the recruitment of PI3K isoforms to the inner surface of the plasma membrane; here the PI3Ks catalyse the transfer of phosphate from ATP to the D-3 position of the inositol ring of membrane bound phosphoinositides. These lipids then function as transducers of signals to downstream pathways through their binding to specific protein motifs (PH domains). PI3K-generated phospholipids at the membrane stimulate the serine/threonine kinase Akt (protein kinase B), which has a critical role in the propagation of survival signals from PI3K. Protein–protein interaction at the inner face of the plasma membrane appears to play key roles in the further signalling through Akt. Interestingly, the genes encoding PI3K and Akt were captured in transducing retroviruses avian sarcoma virus 16 (Chang *et al.*, 1997) and murine T-cell lymphoma-derived virus AKT8 (Bellacosa *et al.*, 1993), respectively. The viral form of PI3K (v-p3k) differs from its cellular counterpart by harbouring an N-terminal Gag-fusion, by a 14-amino acids deletion in the N-terminal part and by four point mutations. The fusion of Gag sequences to the catalytic subunit of PI3K stimulates its membrane binding through the myristylated N-terminus of Gag. Increased membrane binding *per se* is critical for activation of the transforming properties of PI3K, and can also be mediated by engineering of other membrane-targeting motifs onto the cellular PI3K protein. The transforming activity of such modified PI3K proteins was found to correlate with their ability to induce activating phosphorylation of Akt (Aoki *et al.*, 2000). The viral Akt oncoprotein, which is also a Gag fusion protein, differs from its cellular counterpart with respect to subcellular distribution and posttranslational modifications. The transforming properties of the Akt oncoprotein depend upon increased localization at the plasma membrane and require its kinase activity (Aoki *et al.*, 1998).

Activation of signal transduction by Crk family adaptors

The v-*crk* gene was identified in the chicken sarcoma virus isolate CT10 (Mayer *et al.*, 1988). It encoded a Gag-Crk fusion protein with no apparent catalytic domain, but with two domains homologous to other signalling proteins now recognized as the SH2 and SH3 domains involved in protein–protein interactions. The transformation of chicken embryo fibroblasts by this viral oncoprotein caused an increase in protein tyrosine phosphorylation in spite of the absence of a kinase domain in the oncoprotein. When the cellular Crk protein was overexpressed, neither cell transformation nor increased phosphorylation was observed (Fajardo *et al.*, 1993). Crk is the founding member of a widely expressed family of adaptor proteins that act through protein–protein interaction sites to relocate proteins and activate signal transduction complexes (Feller, 2001).

Nuclear oncoproteins

Many oncogenes of transducing retroviruses encode nuclear proteins with transcription regulatory activity. A prominent example is the *myc* gene that has been captured in several independent isolates and also plays an important role in many human cancers, such as haematopoietic malignancies. The Myc protein regulates transcription by combining with a small dimerization partner Max via a helix-loop-helix and a leucine zipper motif to form a DNA-binding complex with binding specificity for a DNA sequence CACGTG (which represents an E-box motif CANNTG) (Luscher, 2001). Myc has an N-terminal transactivating domain. Dimerization with Max is required not only for transcriptional activation, but also for oncogenic transformation as well as for Myc-induced apoptosis (Amati *et al.*, 1993). Phosphorylation-site mutation may contribute to the oncogenic activation of Myc. The downstream target genes critical for oncogenic transformation are now beginning to be elucidated. Microarray hybridization experiments reveal that Myc affects up to 15% of all genes (Dang *et al.*, 2006) with the consistent representation of gene classes associated with metabolism, protein biosynthesis, cell cycle regulation, cell adhesion, and the cytoskeleton. It also remains a distinct possibility that transcription-independent effects are involved in the downstream effects of Myc.

Jun is another example of an oncoprotein with transcription regulatory activity. The *jun* gene was identified as the insert in avian sarcoma virus 17. Like Myc, Jun has a leucine zipper motif and exerts its functions in various dimeric protein complexes. The transcription factor AP-1 is a dimeric complex between Jun and Fos (Bohmann *et al.*, 1987; Angel *et al.*, 1988), a product of an oncogene also captured in transducing retroviruses. Besides Fos proteins, Jun proteins can participate in heterodimeric complexes with proteins of the ATF family of transcription factors and with other leucine zipper proteins. Jun has an N-terminal transcriptional transactivating domain, a basic domain involved in DNA binding, and a leucine zipper dimerization domain that are all needed for the oncogenic transformation of chicken cells. The viral Jun oncoprotein differs from its cellular counterpart by a 27-amino acid deletion in the transactivation domain as well as two regulatory point mutations; these changes contribute to the transforming potential of viral Jun (Vogt and Bader, 2005).

The *cis*-acting retroviruses

The majority of the retroviruses that trigger the development of neoplasms in their hosts do not themselves harbour oncogenes. These non-acutely transforming or *cis*-acting retroviruses, comprising members from the three genera alpharetroviruses, betaretroviruses, and gammaretroviruses, as a minimum contain the retroviral structural *gag, pol,* and *env* genes in addition to necessary *cis*-acting signals, thus ensuring their replication-competence. Owing to the lack of a directly transforming mediator such as a transduced oncogene, this group of viruses is generally not capable of transforming cells in culture, which is mirrored in the prolonged latency period of tumour induction compared with that of the transducing retroviruses (months/years compared to weeks). However, though quite different for the separate virus/host systems, the latency period within a specific system is remarkably constant, probably reflecting that the underlying mechanisms such as the number of hits in genes or pathways for each system generally are rather similar from time to time. This is supported by studies with transgenic or knock-out animals. If one hit is fixed like e.g. overexpression of a target oncogene in a transgenic animal, the latency period is shortened, thus corresponding to a reduction in number of hits.

The *cis*-acting retroviruses, being endogenous or exogenous, cause many different diseases, most of which are haematopoietic in origin, but appearance of cancer in brain (Iwata *et al.*, 2002) and mammary tissues is observed as well (Table 9.1). Although the complex process of disease induction and progression may vary considerably, depending on the characteristics of the inducing virus, some common features can be identified.

Insertional mutagenesis

Important clues to mechanisms underlying tumorigenesis induced by the non-oncogene bearing retroviruses came from studies conducted more than 25 years ago. These studies revealed that the retrovirally induced tumours were clonal or oligoclonal descendants of a single infected cell, as judged by their proviral integration pattern. Such observations implied that great importance should be attached to insertional mutagenesis with regard to understanding the development of tumours induced by the *cis*-acting retroviruses. The reality of this supposition was definitely demonstrated in 1981 by the analyses performed by William S. Hayward, Benjamin G. Neel, and Susan M. Astrin (Hayward *et al.*, 1981). They screened avian leukosis virus (ALV)-induced bursal lymphomas in chicken and found that in 31 out of 37 lymphomas a provirus had integrated upstream of the c-*myc* gene in a co-transcriptional orientation. Subsequently, similar observations were obtained in Suzanne Cory's laboratory with murine leukaemia virus (MLV)-induced T-cell lymphomas. Here it was found that 8 of 32 lymphomas displayed a provirus insertion into the c-*myc* gene, however in a distinctively different pattern, as six of the integrations were found upstream of c-*myc* in opposite transcriptional orientation, while two were found downstream of the c-*myc* gene (Corcoran *et al.*, 1984). Still in the same period, parallel studies in Anton Berns' laboratory identified in MLV-induced T-cell lymphomas another common target site, denoted *pim*-1 for proviral integration site MLV (Cuypers *et al.*, 1984; Selten *et al.*, 1985), and Roel Nusse and Harold E. Varmus identified a common mouse mammary tumour virus (MMTV) provirus integration region in mouse mammary tumours. They named this region MMTV *Int1* (Nusse and Varmus, 1982).

Since these opening and pivotal studies, huge numbers of equivalent studies have followed, greatly facilitated by the invention and refinement of suitable procedures, first and foremost founded on the PCR technique. As a result, thousands of retroviral integration sites (RIS) have been identified, the majority of which have been assembled in the Retrovirus Tagged Cancer Gene Database (RTCGD), (http://RTCGD.ncifcrf.gov) (Akagi *et al.*, 2004), which at present (October 2007) contains almost 3400 RISs. Around 500 of these have been defined as common integration sites (CIS), meaning that these sites have been targeted more than on one occasion, and hence may point to potential cancer genes.

It is important to keep in mind that provirus integration, being an obligatory step in the normal retroviral replication cycle mediated by the viral IN protein, is essentially random with respect to target sites (Chapter 5). Many studies have demonstrated that certain chromatin structures, e.g. actively transcribed regions may be favoured by the integration machinery (Rohdewohld *et al.*, 1987; Scherdin *et al.*, 1990). Likewise, different types of retroviruses appear to have an inclination for specific integration site targets relative to an arbitrary transcription unit; thus MLV vectors seem to prefer promoter regions and first intron sequences, while HIV vectors prefer integration anywhere in the transcriptional unit, but not upstream of the transcriptional start site, and as opposed to this, ASLV vectors show only a weak preference for active genes and no preference for transcription start regions (Wu *et al.*, 2003; Mitchell *et al.*, 2004; Narezkina *et al.*, 2004; De Palma *et al.*, 2005). Such observations contribute to an understanding of the molecular mechanisms behind the integration process, but do not change the basic concept that in view of the total genome, retroviral integration is fundamentally a random process, although certain local DNA structures may be favoured. When a provirus has been inserted by chance near a gene that is involved in cellular growth and differentiation, this cell may gain a selective growth advantage. Consequently, expansion of this cell may eventually lead to the formation of a clonal tumour in which each cell contains a provirus integration at this particular site. The probability of hitting a certain cellular gene is of course extremely low on a single cell basis, but with a sufficient number of infected cells in a tissue, the probability of a specific locus being targeted will approach the value one. As mentioned above, more than one hit/mutation is needed for a malignant tumour to develop. These supplementary hits can be introduced by additional proviral integrations or by other kinds of mutations of appropriate target genes/pathways. Whether or not a cell will gain a selective growth advantage by an integrated provirus will at the outset depend on the capability of that provirus to exert a significant effect on the target gene(s). This will again among other things depend on the composition of transcription-factor binding sites in the proviral promoter and enhancer sequences of the LTR, – if these sites match the transcription-factor profile in the particular cell type, thus in the cellular context providing a strong promoter and enhancer.

Mechanisms of altered gene expression by insertional mutagenesis

The studies mentioned in the previous paragraph have altogether identified two principal ways of affecting expression of the target cellular gene by insertional mutagenesis, namely by activation/deregulation via chimeric transcripts or by activation/deregulation of the normal cellular transcripts (Fig. 9.4).

Activation/deregulation via chimeric transcripts

By this mode of activation, the resulting oncotranscript will contain viral sequences in all cases in addition to cellular target-gene sequences. The oncotranscript will either have been initiated in the proviral promoter or have been terminated at the viral poly(A) site.

In the cases of viral promoter utilization, the provirus has frequently integrated upstream of the cellular target gene, but also integrations within an intron, often the first or second ones, are common. In any event, the integrations must be positioned in a co-transcriptional orientation for the promoter to function. Transcription may start within the viral 3′LTR (often referred to as promoter activation) or within the 5′LTR (often referred to as read-through activation). When transcription starts from the 3′LTR promoter, a transcript containing viral R and U5 regions will be generated (Fig. 9.4A). The integrated proviruses often suffer internal deletions, and occasionally the 5′LTR has been removed. The majority of the provirus integrations within the *c-myc* gene in ALV induced lymphomas are representatives of this kind of promoter activation (Hayward *et al.*, 1981; Robinson and Gagnon, 1986; Goodenow and Hayward, 1987), which has otherwise been reported only in a few cases, one example being the provirus integrations of the Akv MLV variant Akv1–99 into the N-ras gene (Martin-Hernandez *et al.*, 2001). Alternatively, transcription starts from the promoter located within the 5′LTR and proceeds into the virus and downstream into the neighbouring cellular gene, using a viral splice donor site together with a splice acceptor site in the flanking exons (Fig. 9.4B). An example of this mode of activation includes Abelson virus-induced lymphosarcomas, in which the helper virus Moloney MLV has inserted upstream of exon1 in the *c-myb* gene (Shen-Ong *et al.*, 1986). An unusual read-through transcript including the viral *env* region has been described by Patriotis and Tsichlis (1994). Analyses of Moloney MLV-induced rat T-cell lymphomas revealed a transcript, which after initiation at the 5′LTR, was spliced to the viral *env* gene using the normal retroviral splice donor and acceptor sites and then, using a cryptic splice donor site of the virus, the 5′part of *env* was spliced to the downstream *Mlvi-4* sequence. A recent example of promoter utilization is represented by the proviral insertions found within *Evi12* in Cas-Br-MuLV and AKXD induced haematopoietic tumours (van den Akker *et al.*, 2007). In this study, transcripts originating from the viral 5′LTR as well as the 3′LTR were identified. Most of the transcripts encoded truncated forms of Grp94 protein, however some transcripts contained viral *gag* sequences as well, and these transcripts encoded Gag/Grp94 fusion products.

Another way of insertional activation of cellular genes via chimeric transcripts is the usage of the viral poly(A) site. In such cases, the proviruses are found inserted within a gene, often in the 3′-non-coding region, and in the same transcriptional orientation, but before the normal cellular poly(A) site (Fig. 9.4C). This insertion pattern can lead to expression of an altered protein product or removal of regulating sequences. A well-known example of the latter are the integrations found downstream of *pim-1* in MLV-induced T-cell lymphomas. Here, the insertions result in removal of 3′-non-coding AU-rich motifs, which confer instability to the normal *pim-1* mRNA. This elimination of destabilizing sequences results in elevated and abnormal levels

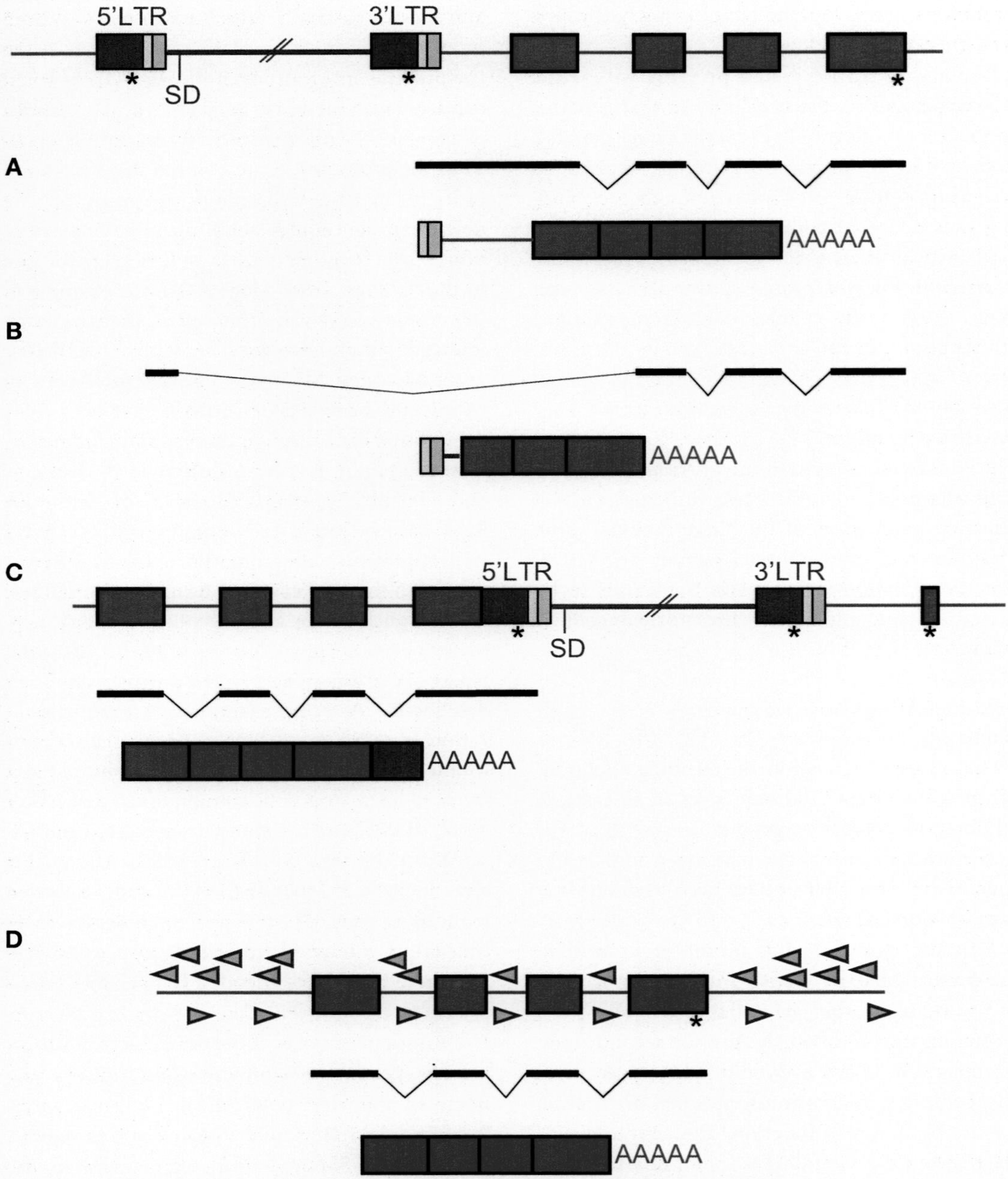

Figure 9.4 Mechanisms of altered gene expression by insertional mutagenesis. (A), (B) and (C) show examples of the activation or deregulation through production of chimeric transcripts, while panel (D) illustrates the activation or deregulation of normal cellular transcripts. (A), (B) 3′LTR and 5′LTR promoter (or read-through) activation, respectively. In both cases the resulting transcript contains as a minimum viral R and U5 in addition to c-*onc* exon sequences. (C) An example of exploiting the viral poly(A) site. In this case the transcript is put together by c-onc exons joined to viral U3 sequences. (D) The blue arrowheads provide a picture of possible sites and orientations of integrated proviruses relative to the cellular target gene. The resulting transcript contains no sequences of viral origin. Red boxes depict exons of the cellular target gene. SD, the canonical *env* splice donor site. *Poly(A) signal. AAAAA represents the poly(A) tail.

of *pim-1* mRNA (Selten *et al.*, 1986; Shaw and Kamen, 1986).

Activation/deregulation of the normal cellular transcripts

This is the insertion pattern observed in by far the majority of tumours induced by the *cis*-acting retroviruses, which probably relates to the fact that no stringent position or orientation requirement is attached. Provirus integrations have been observed both upstream, within introns, and downstream of a cellular target gene, in both orientations and at distances of more than 50 kb away from the target gene (Fig. 9.4D). The structure of the target gene transcript is not altered, and even the level of transcription appears in several cases unaltered at the time of measurement (end-stage tumours). This pattern of insertions is often referred to as 'enhancer activation' because it is conceivable that enhancers in the proviral LTRs contribute to the effect on target gene expression. However, the precise mechanisms of activation, which may be rather complex, are not clarified. There are innumerable examples of this kind of insertion pattern, and in a wide range of tumour systems (Li *et al.*, 1999; Lund *et al.*, 2002; Mikkers *et al.*, 2002; Suzuki *et al.*, 2002; Erkel, *et al.*, 2004; Johansson *et al.*, 2004; Theodorou *et al.*, 2007); however, in many cases there has been no positive demonstration of effect on gene expression, so the accurate target gene(s) have in fact not clearly been pointed out. Accordingly, in such cases it also remains untested if alternative or cryptic promoters are activated. Nevertheless, several independent integrations into the same chromosomal region strongly indicate that that region plays a role in tumorigenesis.

Insertional mutagenesis as a tool to understanding cancer

For retroviruses implicated in human cancers, such as human T-cell leukaemia virus (HTLV), insertional mutagenesis does not in general appear to contribute to oncogenesis. Recently, however, the very first example of a human replication-competent gammaretrovirus, termed xenotropic MLV-related virus (XMRV), was identified in prostate cancer tissue from patients homozygous for a reduced-activity variant of the antiviral enzyme RNase L, and primary integration site analyses were performed. These analyses revealed provirus integrations into the *CREB5*, *NFATc3* and *APPBP2* genes (Urisman *et al.*, 2006; Dong *et al.*, 2007). Yet, it still remains to be examined whether these integrations do play a role at all in the progression of human prostate cancer.

Insertional mutagenesis is, on the other hand, a common mechanism in rodent, feline and avian retroviruses, where the provirus – as described above – inserts itself into the host genome and affects the expression of the neighbouring genes. It has turned out that genes affected by insertional mutagenesis in animal tumours are often the same that are deregulated in human cancers, and hence, the application of *cis*-acting retroviral mutagenesis as a tool for discovering genes involved in human neoplasms has proven superior to many other approaches. In particular, use of different mouse models has found favour and resulted in the before-mentioned assembly of the approximately 3400 RIS in the RTCGD database. In addition, another public web database called VISION (viral insertion sites identifying oncogenes) (Weiser *et al.*, 2007) containing about 500 retroviral insertion sites specifically related to murine B-cell tumours has very recently been published (http://www.mouse-genome.bcm.tmc.edu/vision). Not only mammals, but also more distant species have proven valuable as *cis*-acting retroviral mutagenesis models for the identification of genes potentially involved in the formation of human cancers. A recent example is the mapping of provirus integration sites in MAV-2 induced nephroblastoma (tumour of the kidneys) in chicken, which revealed several novel common integration sites, and demonstrated that human orthologues of the tagged chicken genes are in many cases deregulated in human renal tumours (Pajer *et al.*, 2006).

It is also noteworthy that although many of the oncogenes identified by retroviral mutagenesis models are the same as those captured by the transducing retroviruses, there are genes that appear 'model-specific'. An example is the N-*ras* gene which has been identified as a target in Akv1–99 MLV-induced tumours in NMRI mice (Martin-Hernandez *et al.*, 2001), but never observed as captured by a transducing retrovirus.

Recently, insertional mutagenesis was found to contribute to the development of malignant disease in a human gene therapy trial for X-linked severe combined immunodeficiency (Hacein-Bey-Abina *et al.*, 2003; McCormack and Rabbitts, 2004). An MLV-derived retroviral vector was used to transfer a functional copy of the γc chain of cytokine receptors into haematopoietic precursor cells of affected children. While the treatment was successfully able to restore a functional immune system, two of the treated children developed a leukaemia-like lymphoproliferative disease, in which the retroviral vector has acted as an insertional mutation at a specific locus harbouring the LMO2 gene, which has also been found as a target for insertional mutagenesis in MLV-induced lymphomas in mice.

Characteristics of retroviral mutagenesis target genes

The oncogenic transformation of a cell requires several mutations conferring abnormal growth or differentiation properties. These mutations should somehow together give rise to acquisition of a capacity for unlimited proliferation and independence from signals regulating cell growth. The genes responsible for such properties are typically divided into two major classes, the proto-oncogenes and the tumour-suppressor genes. The number of proto-oncogenes identified by *cis*-acting retroviral mutagenesis exceeds by far the number of tumour-suppressor genes. This is not surprising, since a tumour-suppressor gene will provide a selective advantage to a cell only if both alleles are inactivated. This means that both alleles in a single cell should be targeted by a provirus, which is quite unlikely to happen. Alternatively, the insertion into one of the alleles should give rise to haploinsufficiency or a dominant negative protein. The *p53* is beyond compare the best-known example of a tumour suppressor gene that has been inactivated by an integrated provirus. The p53 protein is a transcription factor essential for the prevention of cancer formation. It has a major role in the cellular response to a diverse range of signals, like DNA damage, hypoxia, or oncogenic activation (Gomez-Lazaro *et al.*, 2004; Stiewe, 2007). The *p53* gene has been shown to be inactivated in the majority of erythroleukaemias induced by Friend virus – either by deletions, point mutations, or provirus insertions. Occasionally, even both *p53* alleles have been targeted by provirus integrations (Ben-David and Bernstein, 1991; Ney and D'Andrea, 2000). In order to enhance the possibility of identifying tumour-suppressor genes, Suzuki *et al.* (2006) carried out a study where they performed an insertional mutagenesis screen in *Blm*-deficient mice, which have increased frequencies of mitotic recombination due to a mutation in the RecQ protein-like-3 helicase gene. The rationale behind this approach was that if one allele of a tumour-suppressor gene was targeted by a provirus, such an integration might be duplicated to the other allele at an increased frequency in *Blm*-deficient mice because of an increased frequency of non-sister chromatid exchange, leading to provirus integration homozygosity at this locus. If an essential tumour-suppressor gene was hereby impaired, this would contribute to tumour development. By using homozygosity as a marker, they succeeded in identifying already characterized as well as novel tumour-suppressor genes.

A recently identified class of non-protein-coding small RNAs, the microRNAs or miRNAs, appear to be correlated with development of cancer (Zhang *et al.*, 2007). Interestingly, avian BIC, a non-coding RNA gene found to be activated by proviral insertions in avian leukosis virus-induced lymphomas (Tam *et al.*, 1997), is now known to encode miR-155. The miR-17–92 miRNA cluster located on mouse chromosome 14 has been targeted both in SL3-3 MLV-induced T-cell lymphomas (Wang *et al.*, 2006) and in Friend MLV-induced erythroleukaemia cells (Cui *et al.*, 2007). The corresponding miRNA cluster in humans has been correlated with lymphomas and lung cancer (Ota *et al.*, 2004; Hayashita *et al.*, 2005). In radiation leukaemia virus (RadLV) induced T-cell lymphomas, proviral integration sites mapped to another miRNA cluster, miR-106–363, located on mouse chromosome X (Landais *et al.*, 2007), a cluster not previously correlated with cancer development.

Viral elements influencing pathogenesis by the *cis*-acting retroviruses

Retroviruses can infect many tissues *in vivo*, but generally, the tumours induced by each virus type predominantly originate from a particular cell type. For instance, Moloney MLV induces mainly T-cell lymphomas, whereas Friend MLV induces erythroleukaemia in susceptible mouse strains. Such preference is determined by both retroviral and cellular features. Here, the most essential viral elements for pathogenesis will be described.

Role of the LTR

Containing both promoter and enhancer sequences as well as a polyadenylation site, the LTR constitutes the central regulator of retroviral expression (described in detail in Chapter 6), and therefore has drawn much attention as an attractive candidate for playing an important role in oncogenesis. The earliest proofs that LTR is indeed a central determinant in both disease induction and specificity came with experiments performed in the 1980s using recombinant viruses in which the LTR from one virus was exchanged with the LTR from another. These experiments showed that the induced disease matched the origin of the LTR and not the body of the virus containing the structural genes (Celander and Haseltine, 1984; Lenz *et al.*, 1984; Weber and Schaffner, 1985). Later studies revealed that the specificity could be located mainly to the enhancer region in U3, and further narrowed down to sequences defining different transcription-factor binding sites. In line with this, it has been demonstrated by numerous examples that a few changes in the U3 can have a dramatic effect on the oncogenic potential of a virus, for instance mutation of the Runx sites in the T-lymphomagenic MLVs, Moloney and SL3-3, significantly alter the pathogenic picture. In the case of Moloney MLV the disease specificity was changed from T-cell lymphoma to erythroleukaemia, while in the case of SL3-3 MLV the outcome depended on the number and characteristics of the introduced mutations; some mutations increased the latency period of disease onset, while others almost completely abolished the oncogenic potential (Speck *et al.*, 1990a,b; Hallberg *et al.*, 1991). Other examples include modification of the distribution of multiple classes of B-cell lymphomas in mice when deleting one copy of the 99-bp enhancer repeat, or mutating various transcription-factor sites in the B-lymphomagenic Akv MLV (Sørensen *et al.*, 2007b), and within the avian system a study by Khelifi-Younes *et al.* (2003) showed that introduction of deletions of U3 in the LTR of myeloblastosis-associated virus (MAV) type 1(N), which normally induces nephroblastoma in chicken, changes the pathogenic spectrum and significantly reduces the tumorigenic potential of this virus.

In spite of a predominant role of the LTR in pathogenesis by the *cis*-acting retroviruses, also other viral sequences have been shown to be important for the ability and potency of a particular virus to induce cancer, as described below.

Role of the env *region*

In addition to their classical role in mediating viral entry and being involved in the phenomenon of interference (see Chapter 4), Env proteins may play both direct and indirect roles in pathogenesis by the *cis*-acting retroviruses. Thus, a significant role of *env* in retrovirally induced lymphomagenesis in mice and cats is the contribution to the production of recombinant polytropic viruses possessing the ability to superinfect cells. The broadened host range of these recombinant viruses – in the murine context often named mink cell focus-forming (MCF) viruses because they form cytopathic foci in mink cells – allows them to infect cells that already express ecotropic viruses (Sourvinos *et al.*, 2000). This re-infection step greatly enhances the probability that provirus insertions into genes or loci contributing to increased growth of the cell will occur.

Besides the ability to overcome superinfection resistance some Env proteins are able to bind to growth-factor receptors on the cell surface and trigger a growth-stimulating signal by mimicking normal ligand–receptor interaction. Jing-Po Li and David Baltimore showed in 1991 that the Env SU protein of Moloney or Friend MCF was able to bind to the erythropoietin receptor (EpoR) like the gp55 of the spleen focus-forming virus, SFFV (see 'Friend virus' below), and that this interaction induced growth-factor-independent proliferation of an IL-3-dependent cell line (Li and Baltimore, 1991). A few years later Philip N. Tsichlis and co-workers demonstrated that infection with MCF viruses of an IL-2-dependent rat cell line led to rapid induction of IL-9 gene expression. This, combined with the withdrawal of IL-2, rapidly selected for cells in which the IL-9 receptor was up-regulated through insertional mutagenesis in the locus *Gfi-2* (Flubacher *et al.*, 1994). Such studies suggest that part of the contribution of the polytropic recombinant viruses to leukaemogenesis in some settings may involve an interaction between the Env protein and a cellular receptor, which will in turn initiate mitogenic signals, thus expanding the available pool of target cells.

Role of the gag *region*

In addition to the *env* gene, the viral *gag* gene sequences have also proven to play a role in retroviral pathogenesis, at least in some cases. Thus it has been demonstrated that introduction of only three synonymous nucleotide mutations in the capsid-coding gene of Moloney MLV changed the oncogenic properties of this virus. While wild-type Moloney MLV induces T-cell lymphomas in 100% of the inoculated mice, the *gag* mutant exhibited a much broader specificity, inducing both erythroid and myelomonocytic leukaemias in addition to the expected T-cell tumours. In contrast, the equivalent mutations in a Friend MLV background did not seem to influence the pathogenic potential of this virus at all. Both wild-type and mutant Friend MLVs induced exclusively the characteristic erythroleukaemia (Audit *et al.*, 1999; Dejardin *et al.*, 2000; Houzet *et al.*, 2003). The introduced mutations were located at a well-conserved alternative splice donor site (SD'), and mutating this same site and/or an alternative splice acceptor site, also located in the *gag* region, in the B-lymphomagenic Akv MLV resulted in a modification of the oncogenic properties of this virus, changing the distribution of the different B-cell types as well as generating tumours of additional specificities such as *de novo* diffuse large B-cell lymphoma and histiocytic sarcoma (Sørensen *et al.*, 2007a).

The *gag* SD' site has also been found to be used for generating the oncogenic *gag–myb* fusion RNAs in promonocytic leukaemias induced by Moloney MLV in pristane-treated BALB/c mice (Nason-Burchenal and Wolff, 1993). When the SD' site was mutated in this model, the proportion of myeloid leukaemia decreased significantly, while the proportion of lymphoid leukaemia increased. Moreover, the typical 5'LTR promotion of c-*myb* was impaired, thereby signifying a specific requirement of the SD' site for this mechanism (Ramirez *et al.*, 2004). These studies indicate that the role of *gag* SD'-site in oncogenesis, although the site is conserved among many species, seems to be strongly dependent on the virus type and experimental model.

Specific examples

Mouse mammary tumour virus – a role for a viral superantigen

Mouse mammary tumour virus (MMTV) belongs to the genus betaretroviruses, which induces mammary tumours in mice. MMTV in many ways resembles the other *cis*-acting viruses portrayed here. It carries the typical *gag*, *pol*, and *env* genes, and it can be transmitted as a stably integrated, endogenous provirus (Mtv), or as an exogenous virus through the milk of infected female mice to newborn pups (Fig. 9.5). Regarding the underlying mechanisms of tumour induction, insertional mutagenesis is an essential element, and as mentioned previously, already as early as in 1982 the first common integration site (Int1) in mouse mammary tumour tissue was isolated (Nusse and Varmus, 1982). However, until very recently only quite few genes involved in mouse mammary tumorigenesis have been isolated. Then in 2007 a high-throughput screen of MMTV-induced mouse mammary tumours was published (Theodorou *et al.*, 2007). This screen identified 664 retroviral integration sites (RISs) from a panel of 160 independent MMTV-induced tumours. Of this pool of RISs, a total of 33 CISs could be recognized, many of which represent novel candidate genes for breast cancer development.

Like for other *cis*-acting retroviruses, the LTR has been shown to play an important role in MMTV-induced tumorigenesis. The various hormone-responsive and tissue-specific elements contained in this region are strong contributors to the high expression of the virus in the mammary gland cells in the adult female mouse (Beato, 1996; Truss *et al.*, 1996). In addition to the LTR, sequences within the *gag* region seem to support mammary gland cell transformation (Hook *et al.*, 2000; Swanson *et al.*, 2006).

However, a striking difference between MMTV and all other *cis*-acting retroviruses is

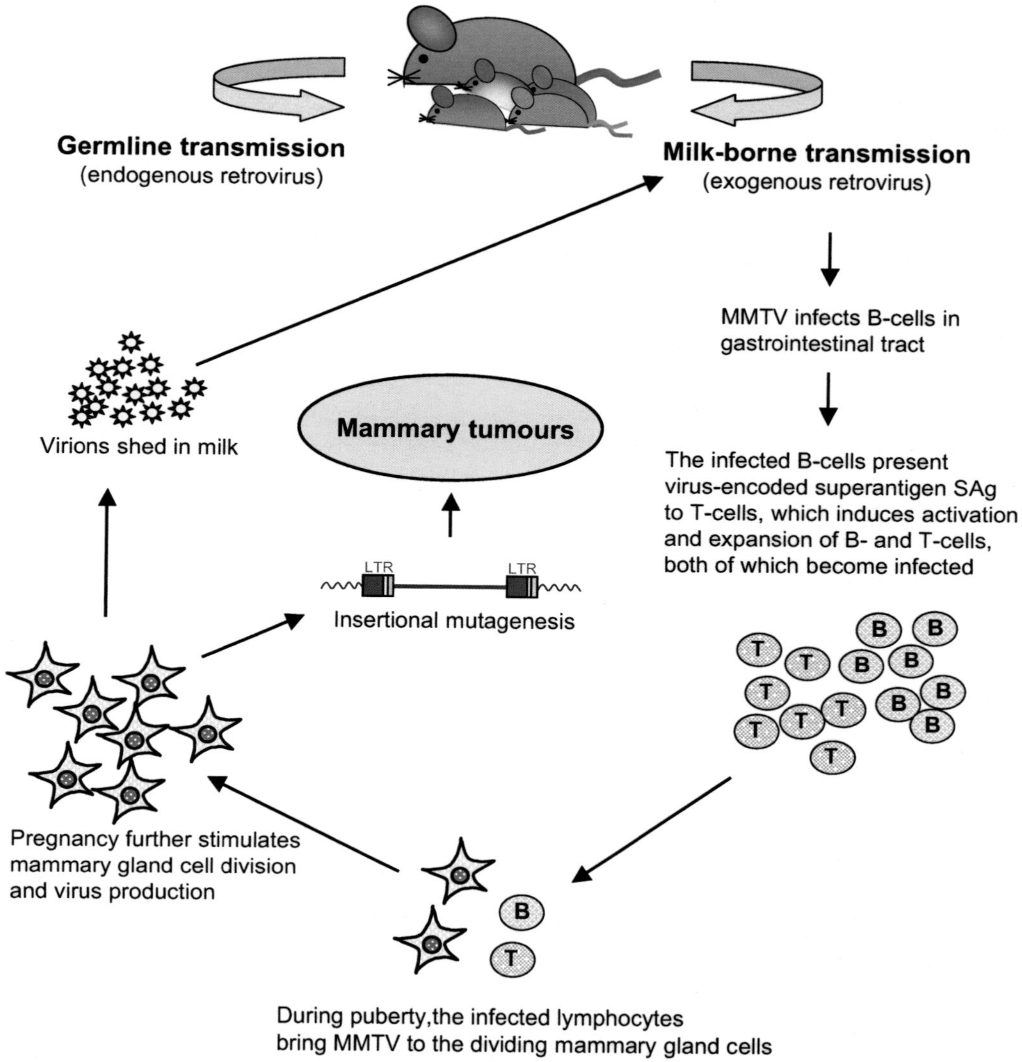

Figure 9.5 Pathway of milk-borne MMTV transmission (see text for explanation).

the presence of an additional gene, *sag*, located within the 3′LTR (Fig. 9.1), and encoding a viral superantigen called SAg (Choi *et al.*, 1991). The primary targets for exogenous MMTV are B-cells as well as dendritic cells located in Peyer's patches of the gastrointestinal tract of neonatally infected pups (Fig. 9.5). The infected B-cells subsequently present the MMTV-encoded SAg molecule in association with major histocompatibility complex (MHC) class II to cognate T-cells holding a particular Vβ chain of the T-cell receptor. This results in activation of these T-cells and concomitant stimulation of neighbouring lymphocytes. The overall effect is a production of a reservoir of lymphocytes set for infection. During sexual maturation of the mouse, the infected lymphocytes bring the virus to the dividing mammary gland epithelial cells. These cells are regarded as the major targets for MMTV in adult mice, as secretion of MMTV into the milk to be transmitted to the offspring is a critical step in the replication cycle. The multiple rounds of reinfections and provirus integrations in the mammary-gland cell population eventually lead to development of carcinomas (Czarneski *et al.*, 2003; Acha-Orbea *et al.*, 2007).

It has been demonstrated that a functional SAg protein is necessary to establish MMTV infection (Golovkina *et al.*, 1992; Held *et al.*, 1993; Beutner *et al.*, 1996; Golovkina *et al.*, 1998), although both a SAg-dependent and a SAg-independent infection step seem to exist. Thus, it has been shown that the initial infection in naturally proliferating cells in neonates appears to be independent of SAg function, while the following SAg-dependent activation of T-cells and resulting stimulation of B-cell proliferation is an absolute requirement for infection to proceed (Pobezinskaya *et al.*, 2004). Moreover, as a means to dealing with the immune system, MMTV subverts the innate immune response via Toll-like receptor-4-mediated production of IL-10 by B-cells, which suppresses the antivirus adaptive immune response (Jude *et al.*, 2003).

Friend virus complex

Friend virus is made up of a complex consisting of a replication-competent Friend MLV (a genuine *cis*-acting retrovirus) and a replication-defective spleen focus-forming virus, SFFV. The latter virus is the reason that Friend virus is capable of inducing leukaemia in adult mice, in contrast to the *cis*-acting retroviruses which induce diseases only when introduced into newborn mice. There are at least two naturally occurring variants of SFFV, which are called $SFFV_P$ (polycythaemia strain) and $SFFV_A$ (anaemia strain). Mice infected with the viral complex of $SFFV_P$ and Friend MuLV (denoted FVP) become polycythaemic (increase in red cell mass caused by increased erythropoiesis), while mice infected with $SFFV_A$ and Friend MuLV (denoted FVA) develop anaemia (Ney and D'Andrea, 2000). Overall, Friend disease development is characterized by two distinct stages, where a massive splenic enlargement is seen in the first stage, followed in the second stage by the emergence of fully transformed cells, at the outset in the spleen, but eventually in blood, bone marrow, and liver as well (Fig. 9.6). Mice infected with FVP often die during the first stage of disease because of splenic rupture, but if they survive this stage, they will develop leukaemia as well.

The polyclonal expansion observed in stage one is primarily a result of constitutive activation of the erythropoietin receptor, EpoR; an activation that is caused by binding of the SFFV Env-product (gp55) to EpoR. SFFV behaves as a transducing virus, but it does not contain any transduced cellular genes. Rather, the equivalent pathogenic component in this virus is a defective *env* gene, encoding gp55, which in the N-terminus is related to MCF Env and in the C-terminus is related to ecotropic MLV Env. The gp55 protein has been shown to bind EpoR at the cell surface and thereby trigger the proliferation of erythroid progenitor cells, resulting in the splenic enlargement typical of the first disease stage (Ben-David and Bernstein, 1991; Ferro *et al.*, 1993; Wang *et al.*, 1993). The Env product of $SFFV_P$ induces Epo-independent proliferation and differentiation of erythroid cells, whereas Env of $SFFV_A$ induces Epo-independent proliferation, but requires Epo for differentiation. In both cases, SFFV Env-activated EpoR signalling in erythroid cells depends upon the interaction with a short form of the receptor tyrosine kinase (sf-Stk) (Fig. 9.6) encoded by the cellular *Fv-2* gene (Lilly, 1970; Persons *et al.*, 1999). Interestingly, SFFV has also been shown able to transform fibroblast engineered to express sf-Stk (Nishigaki *et al.*, 2005; Jelacic *et al.*, 2007).

The following emergence of malignant clones in stage two requires additional events, including insertional mutagenesis (Fig. 9.6). The insertional mutagenesis may involve both inactivation of the tumour-suppressor genes *p53* or *p45 NFE2* and activation of oncogenes belonging to the *ets* family (Lee *et al.*, 2003). The p53 gene has been shown to be inactivated by retroviral integrations, deletions, and/or point mutations in the majority of the erythroleukaemia cell lines induced by Friend virus or Friend MLV strains. Besides the inactivation of the *p53* gene, almost all erythroleukaemias induced by the Friend virus display proviral integrations into the *Spi-1* locus, encoding the DNA-binding transcriptional activator PU.1.

The replication-competent part of the Friend virus, Friend MLV, can – when injected into newborn mice – induce a variety of haematopoietic neoplasms, including erythroleukaemia, lymphomas, and myeloblastic leukaemias, where the type of neoplasm induced depends in large part on the mouse strain. The obvious question to put now

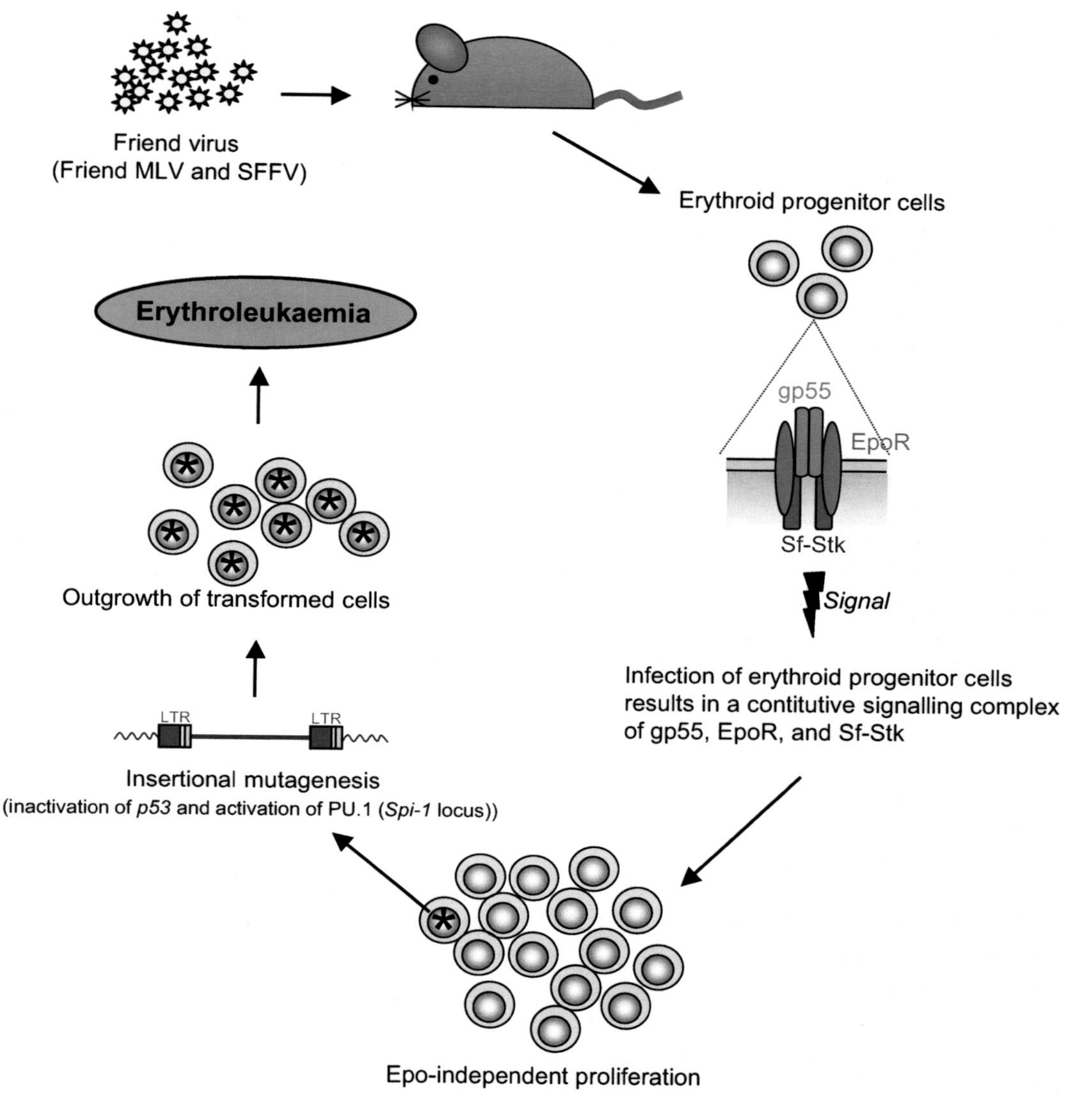

Figure 9.6 Friend virus induced erythroleukaemia (see text for explanation).

will be: do the events in the progress of these diseases resemble the ones induced by the Friend virus complex? And, yes, to some extent they do. After infection with Friend MLV recombinant MCF viruses emerge, and as mentioned previously, the MCF Env protein is related to gp55 and has been shown to bind EpoR and at least in some cases induce growth factor independence. Thus, the early stages of Friend MLV and Friend virus-complex induced diseases may seem quite similar. In the late event, the same distinct pattern of inactivation of *p53* (by point mutations, deletions, provirus insertions) and activation of *Spi-1* by provirus insertions does not appear in the Friend MLV induced erythroleukaemia. Interestingly however, another member of the *ets* gene family, *Fli-1*, is frequently targeted, indicating that the *ets* gene family does play a central role in retrovirally induced erythroleukaemia (Truong and Ben-David, 2000; Lee *et al.*, 2003). Other diseases induced by Friend MLV, such as myeloid and lymphoid tumours do not show this preference for *Fli-1* proviral insertions. In these cases a number of other target genes have been identified

(Sola *et al.*, 1988; Askew *et al.*, 1991; Kone *et al.*, 2002; Yatsula *et al.*, 2006).

Jaagsiekte sheep retrovirus – a paradigm for an oncogenic Env protein

The Jaagsiekte sheep retrovirus (JSRV) is a betaretrovirus responsible for the development of ovine pulmonary adenocarcinomas (OPA), a form of lung cancer, which causes significant veterinary problems in many countries (Leroux *et al.*, 2007). The target cells are differentiated lung epithelial cells, *i.e.* type II pneumocytes in the alveoli, and Clara cells in the bronchioli. The virus is transmitted between animals by close contact, and naturally infected sheep exhibit incubation periods of 2–4 years. In experimental settings multifocal lesions typical of OPA may be observed in a few weeks following inoculation of JSRV, which points in the direction of an oncogenic element in the JSRV genome. Moreover, transfection of JSRV DNA induces transformed foci in NIH 3T3 fibroblasts (Maeda *et al.*, 2001). However, no transduced gene of cellular origin is present in the JSRV genome, which contains the normal battery of retroviral genes, *gag*, *pol*, and *env* (Fig. 9.1). An alternative open reading frame in *pol*, *orf*-x, has been suspected as a contributor to transformation. However, disruption of *orf*-x has no effect on transformation (Maeda *et al.*, 2001) or *in vivo* tumorigenesis (Cousens *et al.*, 2007), and *orf*-x bears no resemblance to known oncogenes. It is now clear that the Env-protein of JSRV has cell-transforming properties in addition to its function in retroviral entry (Liu and Miller, 2007). It is also clear that the JSRV envelope is sufficient to induce lung tumours *in vivo* similar to those induced by JSRV in sheep. This was shown in immunodeficient mice using a replication-incompetent adeno-associated virus vector for the delivery and expression of JSRV Env (Wootton *et al.*, 2005). A similar result was obtained in sheep using a replication-defective retrovirus carrying JSRV *env* under JSRV LTR control (Caporale *et al.*, 2006). The envelope protein of a related agent, enzootic nasal tumour virus (ENTV), which is responsible for nasal adenocarcinomas that affect sheep and goats has been shown to display similar properties of cell transformation and tumour induction as JSRV Env. The rapid experimental induction of tumours by JSRV Env clearly documents an oncogenic role for this viral protein, but is also in sharp contrast to the latency periods of disease induction of months to years in natural settings. This emphasizes the potential importance of additional steps such as insertional mutagenesis under conditions of natural infections. The presence of common insertion sites for JSRV in the sheep genome in OPA tissue has been investigated, and one common insertion site was found on sheep chromosome 16 (Cousens *et al.*, 2004). Altogether, the possible role of specific insertion sites remains unclear.

Retroviral Env proteins mediate cell entry by the binding to specific receptors on the cell surface. The specific receptor for JSRV and ENTV is hyalurinidase 2 (Hyal2), a protein attached to the cell surface through a glycosylphosphatidylinositol (GPI)-anchor (Rai *et al.*, 2001). Evidence that the SU subunit of JSRV may affect cell transformation by interaction with the Hyal2 receptor comes from studies in BEAS-2 human bronchial epithelial cells (Danilkovitch-Miagkova *et al.*, 2003). Interestingly, Hyal2 in complex with a receptor tyrosine kinase, RON, inhibits RON signalling. JSRV expression leads to Hyal 2 proteasomal degradation, which leads to reactivation of RON and tyrosine kinase signalling. However, these findings have not been reproduced in other cell lines. Specific modifications of the surface subunit (SU) of Env also abolish transformation in some settings (Hofacre and Fan, 2004).

The cytoplasmic tail of TM of JSRV Env is clearly necessary for transformation in multiple cell lines since exchange with that of other retroviruses completely abolished transformation (Liu and Miller, 2007). However, reverse exchange experiments have not provided conclusive evidence as to whether this cytoplasmic tail is in itself sufficient for transformation. Much attention has been given to an YXXM motif, conserved in the cytoplasmic TM tail of all transforming JSRV and ENTV strains. If phosphorylated at tyrosine, this is a putative binding site for the regulatory subunit of PI3K. Mutation of the YXXM motif abolished transformation in rodent fibroblasts (Hull and Fan, 2006). It has been found that the activity of the PI3K/Akt signalling pathway is elevated in JSRV-transformed rodent fibroblasts. However, how JSRV Env triggers signalling is uncertain, since PI3K-dependent Akt-signalling was independent of the YXXM motif in some settings, and since it has not been possible to obtain direct evidence of a YXXM docking site function (Liu and Miller, 2007). Altogether, no unifying picture of how JSRV Env transforms different cells has been obtained. Recent evidence also points to activation of the Raf-MEK-ERK pathway in JSRV transformation, but it is unclear how this pathway becomes activated (Liu *et al.*, 2003; Maeda *et al.*, 2005).

Research on the mechanism of oncogenesis by JSRV has created a new paradigm for a retroviral envelope protein with transforming properties. This is reminiscent of other *trans* effects of complete or modified retroviral envelope proteins as discussed for the spleen focus-forming virus and MCF viruses earlier in this chapter. MMTV was recently found to cause transformation-like morphological changes in mammary epithelial cell lines (Katz *et al.*, 2005). This effect was found to depend upon an immunoreceptor tyrosine-based activation motif (ITAM) encoded in the cytoplasmic tail of the TM subunit. ITAM motifs are potential docking sites for SH2 domains and associated with various cellular effects. Interestingly, this motif was also shown to play a role in virus-induced mammary tumours (Ross *et al.*, 2006).

Another *trans*-acting retroviral envelope is that of the avian haemangioma retrovirus (Alian *et al.*, 2000), which was isolated during an epidemic of haemangioma (vascular tumours) in hens. The envelope gene of the avian haemangioma virus has been found to cause either apoptosis or cell proliferation in cell culture depending on cell type.

Conclusions

A multitude of viral isolates from birds and mammals provide a solid foundation for research on oncogenic retroviruses in diverse biological models of tumour induction as well as in cell-culture models of transformation. The finding that oncogenes in transducing viruses are captured from their host cells has had an enormous impact on modern tumour biology. Some of the genes first identified because of their association with a retrovirus are in fact among the most intensively studied mammalian genes. They work in cascades of the transmission of signals for cell growth and survival. The oncogenes of the transducing viruses alter the normal growth and survival programs of the cell as a result of deregulated expression or structural changes of the encoded oncoprotein. The discovery that many oncogenic retroviruses cause tumours by insertional mutagenesis has given access to an additional pool of equally interesting genes that are targeted by proviral insertion during tumorigenesis. Post-genomic screenings for such insertion targets with a function in multistep oncogenesis in a range of disease models has now led to huge numbers of genes or loci for further studies. Novel ways by which proteins encoded by oncogenic retroviruses play a role in pathogenesis by targeting specific host molecules continue to be discovered. The superantigen encoded by mouse mammary tumour virus provides an example of how a retrovirus may exploit the host immune system during the establishment of infection. Another example is the defective envelope protein encoded by the spleen focus-forming virus, which has the ability to stimulate the erythropoietin receptor. Unexpectedly, the ability of the Jaagsiekte sheep retrovirus to cause lung cancer is mediated by specific interaction between the envelope protein of this virus and the target cell. Altogether, there is no reason to believe that the major role played by oncogenic retroviruses in research in the molecular and cellular biology of tumorigenesis has come to an end.

References

Abelson, H.T., and Rabstein, L.S. (1970). Lymphosarcoma: virus-induced thymic-independent disease in mice. Cancer Res. *30*, 2213–2222.

Acha-Orbea, H., Shakhov, A.N., and Finke, D. (2007). Immune response to MMTV infection. Front. Biosci. *12*, 1594–1609.

Akagi, K., Suzuki, T., Stephens, R.M., Jenkins, N.A., and Copeland, N.G. (2004). RTCGD: retroviral tagged cancer gene database. Nucleic Acids Res. *32*, D523–7.

Alian, A., Sela-Donenfeld, D., Panet, A., and Eldor, A. (2000). Avian hemangioma retrovirus induces cell proliferation via the envelope (*env*) gene. Virology *276*, 161–168.

Amati, B., Brooks, M.W., Levy, N., Littlewood, T.D., Evan, G.I., and Land, H. (1993). Oncogenic activity of the c-Myc protein requires dimerization with Max. Cell *72*, 233–245.

Angel, P., Allegretto, E.A., Okino, S.T., Hattori, K., Boyle, W.J., Hunter, T., and Karin, M. (1988). Oncogene jun encodes a sequence-specific *trans*-activator similar to AP-1. Nature 332, 166–171.

Aoki, M., Batista, O., Bellacosa, A., Tsichlis, P., and Vogt, P.K. (1998). The akt kinase: molecular determinants of oncogenicity. Proc. Natl. Acad. Sci. U.S.A. *95*, 14950–14955.

Aoki, M., Schetter, C., Himly, M., Batista, O., Chang, H.W., and Vogt, P.K. (2000). The catalytic subunit of phosphoinositide 3-kinase: requirements for oncogenicity. J. Biol. Chem. *275*, 6267–6275.

Askew, D.S., Bartholomew, C., Buchberg, A.M., Valentine, M.B., Jenkins, N.A., Copeland, N.G., and Ihle, J.N. (1991). His-1 and His-2: identification and chromosomal mapping of two commonly rearranged sites of viral integration in a myeloid leukemia. Oncogene *6*, 2041–2047.

Audit, M., Dejardin, J., Hohl, B., Sidobre, C., Hope, T.J., Mougel, M., and Sitbon, M. (1999). Introduction of a *cis*-acting mutation in the capsid-coding gene of moloney murine leukemia virus extends its leukemogenic properties. J. Virol. *73*, 10472–10479.

Beato, M. (1996). Chromatin structure and the regulation of gene expression: remodeling at the MMTV promoter. J. Mol. Med. *74*, 711–724.

Bellacosa, A., Franke, T.F., Gonzalez-Portal, M.E., Datta, K., Taguchi, T., Gardner, J., Cheng, J.Q., Testa, J.R., and Tsichlis, P.N. (1993). Structure, expression and chromosomal mapping of c-akt: relationship to v-akt and its implications. Oncogene *8*, 745–754.

Ben-David, Y., and Bernstein, A. (1991). Friend virus-induced erythroleukemia and the multistage nature of cancer. Cell *66*, 831–834.

Bentvelzen, P., and Daams, J.H. (1969). Hereditary infections with mammary tumor viruses in mice. J. Natl. Cancer Inst. *43*, 1025–1035.

Beutner, U., McLellan, B., Kraus, E., and Huber, B.T. (1996). Lack of MMTV superantigen presentation in MHC class II-deficient mice. Cell. Immunol. *168*, 141–147.

Bittner, J.J. (1942). The Milk-Influence of Breast Tumors in Mice. Science *95*, 462–463.

Bohmann, D., Bos, T.J., Admon, A., Nishimura, T., Vogt, P.K., and Tjian, R. (1987). Human proto-oncogene c-jun encodes a DNA binding protein with structural and functional properties of transcription factor AP-1. Science *238*, 1386–1392.

Brugge, J.S., and Erikson, R.L. (1977). Identification of a transformation-specific antigen induced by an avian sarcoma virus. Nature *269*, 346–348.

Caporale, M., Cousens, C., Centorame, P., Pinoni, C., De las Heras, M., and Palmarini, M. (2006). Expression of the jaagsiekte sheep retrovirus envelope glycoprotein is sufficient to induce lung tumors in sheep. J. Virol. *80*, 8030–8037.

Celander, D., and Haseltine, W.A. (1984). Tissue-specific transcription preference as a determinant of cell tropism and leukaemogenic potential of murine retroviruses. Nature *312*, 159–162.

Chang, H.W., Aoki, M., Fruman, D., Auger, K.R., Bellacosa, A., Tsichlis, P.N., Cantley, L.C., Roberts, T.M., and Vogt, P.K. (1997). Transformation of chicken cells by the gene encoding the catalytic subunit of PI 3-kinase. Science *276*, 1848–1850.

Chen, Y.Y., Wang, L.C., Huang, M.S., and Rosenberg, N. (1994). An active v-abl protein tyrosine kinase blocks immunoglobulin light-chain gene rearrangement. Genes Dev. *8*, 688–697.

Choi, Y., Kappler, J.W., and Marrack, P. (1991). A superantigen encoded in the open reading frame of the 3′ long terminal repeat of mouse mammary tumour virus. Nature *350*, 203–207.

Coffin, J.M. (1992). Genetic diversity and evolution of retroviruses. Curr. Top. Microbiol. Immunol. *176*, 143–164.

Coll, J., Righi, M., Taisne, C.D., Dissous, C., Gegonne, A., and Stehelin, D. (1983). Molecular cloning of the avian acute transforming retrovirus MH2 reveals a novel cell-derived sequence (v-mil) in addition to the myc oncogene. EMBO J. *2*, 2189–2194.

Cooper, J.A., and Howell, B. (1993). The when and how of Src regulation. Cell *73*, 1051–1054.

Corcoran, L.M., Adams, J.M., Dunn, A.R., and Cory, S. (1984). Murine T lymphomas in which the cellular myc oncogene has been activated by retroviral insertion. Cell *37*, 113–122.

Cousens, C., Bishop, J.V., Philbey, A.W., Gill, C.A., Palmarini, M., Carlson, J.O., DeMartini, J.C., and Sharp, J.M. (2004). Analysis of integration sites of Jaagsiekte sheep retrovirus in ovine pulmonary adenocarcinoma. J. Virol. *78*, 8506–8512.

Cousens, C., Maeda, N., Murgia, C., Dagleish, M.P., Palmarini, M., and Fan, H. (2007). In vivo tumorigenesis by Jaagsiekte sheep retrovirus (JSRV) requires Y590 in Env TM, but not full-length orfX open reading frame. Virology *367*, 413–421.

Cui, J.W., Li, Y.J., Sarkar, A., Brown, J., Tan, Y.H., Premyslova, M., Michaud, C., Iscove, N., Wang, G.J., and Ben-David, Y. (2007). Retroviral insertional activation of the Fli-3 locus in erythroleukemias encoding a cluster of microRNAs that convert Epo-induced differentiation to proliferation. Blood *110*, 2631–2640.

Cuypers, H.T., Selten, G., Quint, W., Zijlstra, M., Maandag, E.R., Boelens, W., van Wezenbeek, P., Melief, C., and Berns, A. (1984). Murine leukemia virus-induced T-cell lymphomagenesis: integration of proviruses in a distinct chromosomal region. Cell *37*, 141–150.

Czarneski, J., Rassa, J.C., and Ross, S.R. (2003). Mouse mammary tumor virus and the immune system. Immunol. Res. *27*, 469–480.

Dang, C.V., O'Donnell, K.A., Zeller, K.I., Nguyen, T., Osthus, R.C., and Li, F. (2006). The c-Myc target gene network. Semin. Cancer Biol. *16*, 253–264.

Danilkovitch-Miagkova, A., Duh, F.M., Kuzmin, I., Angeloni, D., Liu, S.L., Miller, A.D., and Lerman, M.I. (2003). Hyaluronidase 2 negatively regulates RON receptor tyrosine kinase and mediates transformation of epithelial cells by jaagsiekte sheep retrovirus. Proc. Natl. Acad. Sci. U.S.A. *100*, 4580–4585.

De Palma, M., Montini, E., Santoni de Sio, F.R., Benedicenti, F., Gentile, A., Medico, E., and Naldini, L. (2005). Promoter trapping reveals significant differences in integration site selection between MLV and HIV vectors in primary hematopoietic cells. Blood *105*, 2307–2315.

Dejardin, J., Bompard-Marechal, G., Audit, M., Hope, T.J., Sitbon, M., and Mougel, M. (2000). A novel subgenomic murine leukemia virus RNA transcript results from alternative splicing. J. Virol. *74*, 3709–3714.

Dong, B., Kim, S., Hong, S., Das Gupta, J., Malathi, K., Klein, E.A., Ganem, D., Derisi, J.L., Chow, S.A., and Silverman, R.H. (2007). An infectious retrovirus susceptible to an IFN antiviral pathway from human prostate tumors. Proc. Natl. Acad. Sci. U.S.A. *104*, 1655–1660.

Doolittle, R.F., Hunkapiller, M.W., Hood, L.E., Devare, S.G., Robbins, K.C., Aaronson, S.A., and Antoniades, H.N. (1983). Simian sarcoma virus onc gene, v-sis, is derived from the gene (or genes) encoding a platelet-derived growth factor. Science *221*, 275–277.

Ellerman, V., and Bang, O. (1908). Experimentelle Leukämie bei Hühnern. Zentralbl. Bakteriol. Parasitenkd. Infectionskr. Hyg. Abt. Orig. *46*, 595–609.

Engelbreth-Holm, J.A., and Meyer, R. (1932). Bericht über neue Erfahrungen mit einem Stamm Hühner-Erythroleukose Acta Pathol. Microbiol. Scand. *9*, 293–332.

Erkeland, S.J., Valkhof, M., Heijmans-Antonissen, C., van Hoven-Beijen, A., Delwel, R., Hermans, M.H., and Touw, I.P. (2004). Large-scale identification of disease genes involved in acute myeloid leukemia. J. Virol. *78*, 1971–1980.

Fajardo, J.E., Birge, R.B., and Hanafusa, H. (1993). A 31-amino-acid N-terminal extension regulates c-Crk binding to tyrosine-phosphorylated proteins. Mol. Cell. Biol. *13*, 7295–7302.

Felder, M.P., Eychene, A., Barnier, J.V., Calogeraki, I., Calothy, G., and Marx, M. (1991). Common mechanism of retrovirus activation and transduction of c-mil and c-Rmil in chicken neuroretina cells infected with Rous-associated virus type 1. J. Virol. *65*, 3633–3640.

Feller, S.M. (2001). Crk family adaptors-signalling complex formation and biological roles. Oncogene *20*, 6348–6371.

Ferro, F.E.,Jr, Kozak, S.L., Hoatlin, M.E., and Kabat, D. (1993). Cell surface site for mitogenic interaction of erythropoietin receptors with the membrane glycoprotein encoded by Friend erythroleukemia virus. J. Biol. Chem. *268*, 5741–5747.

Flubacher, M.M., Bear, S.E., and Tsichlis, P.N. (1994). Replacement of interleukin-2 (IL-2)-generated mitogenic signals by a mink cell focus-forming (MCF) or xenotropic virus-induced IL-9-dependent autocrine loop: implications for MCF virus-induced leukemogenesis. J. Virol. *68*, 7709–7716.

Friend, C. (1957). Cell-free transmission in adult Swiss mice of a disease having the character of a leukemia. J. Exp. Med. *105*, 307–318.

Frykberg, L., Palmieri, S., Beug, H., Graf, T., Hayman, M.J., and Vennstrom, B. (1983). Transforming capacities of avian erythroblastosis virus mutants deleted in the erbA or erbB oncogenes. Cell 32, 227–238.

Golovkina, T.V., Chervonsky, A., Dudley, J.P., and Ross, S.R. (1992). Transgenic mouse mammary tumor virus superantigen expression prevents viral infection. Cell *69*, 637–645.

Golovkina, T.V., Dudley, J.P., and Ross, S.R. (1998). B and T-cells are required for mouse mammary tumor virus spread within the mammary gland. J. Immunol. *161*, 2375–2382.

Gomez-Lazaro, M., Fernandez-Gomez, F.J., and Jordan, J. (2004). P53: Twenty Five Years Understanding the Mechanism of Genome Protection. J. Physiol. Biochem. *60*, 287–307.

Goodenow, M.M., and Hayward, W.S. (1987). 5′ long terminal repeats of Myc-associated proviruses appear structurally intact but are functionally impaired in tumors induced by avian leukosis viruses. J. Virol. *61*, 2489–2498.

Gross, L. (1957). Development and serial cellfree passage of a highly potent strain of mouse leukemia virus. Proc. Soc. Exp. Biol. Med. *94*, 767–771.

Hacein-Bey-Abina, S., Von Kalle, C., Schmidt, M., McCormack, M.P., Wulffraat, N., Leboulch, P., Lim, A., Osborne, C.S., Pawliuk, R., Morillon, E. *et al.* (2003). LMO2-associated clonal T-cell proliferation in two patients after gene therapy for SCID-X1. Science *302*, 415–419.

Hallberg, B., Schmidt, J., Luz, A., Pedersen, F.S., and Grundstrom, T. (1991). SL3-3 enhancer factor 1 transcriptional activators are required for tumor formation by SL3-3 murine leukemia virus. J. Virol. *65*, 4177–4181.

Hanafusa, H., Hanafusa, T., and Rubin, H. (1963). The defectiveness of Rous sarcoma virus. Proc. Natl. Acad. Sci. U.S.A. *49*, 572–580.

Harvey, J.J. (1964). An unidentified virus which causes the rapid production of tumours in mice. Nature *204*, 1104–1105.

Hayashita, Y., Osada, H., Tatematsu, Y., Yamada, H., Yanagisawa, K., Tomida, S., Yatabe, Y., Kawahara, K., Sekido, Y., and Takahashi, T. (2005). A polycistronic microRNA cluster, miR-17-92, is overexpressed in human lung cancers and enhances cell proliferation. Cancer Res. *65*, 9628–9632.

Hayward, W.S., Neel, B.G., and Astrin, S.M. (1981). Activation of a cellular onc gene by promoter insertion in ALV-induced lymphoid leukosis. Nature *290*, 475–480.

Held, W., Shakhov, A.N., Izui, S., Waanders, G.A., Scarpellino, L., MacDonald, H.R., and Acha-Orbea, H. (1993). Superantigen-reactive $CD4^+$ T-cells are required to stimulate B cells after infection with mouse mammary tumor virus. J. Exp. Med. *177*, 359–366.

Hofacre, A., and Fan, H. (2004). Multiple domains of the Jaagsiekte sheep retrovirus envelope protein are required for transformation of rodent fibroblasts. J. Virol. 78, 10479–10489.

Hook, L.M., Agafonova, Y., Ross, S.R., Turner, S.J., and Golovkina, T.V. (2000). Genetics of mouse mammary tumor virus-induced mammary tumors: linkage of tumor induction to the *gag* gene. J. Virol. *74*, 8876–8883.

Houzet, L., Battini, J.L., Bernard, E., Thibert, V., and Mougel, M. (2003). A new retroelement constituted by a natural alternatively spliced RNA of murine replication-competent retroviruses. EMBO J. *22*, 4866–4875.

Hull, S., and Fan, H. (2006). Mutational analysis of the cytoplasmic tail of jaagsiekte sheep retrovirus envelope protein. J. Virol. *80*, 8069–8080.

Hunter, T., and Sefton, B.M. (1980). Transforming gene product of Rous sarcoma virus phosphorylates tyrosine. Proc. Natl. Acad. Sci. U.S.A. *77*, 1311–1315.

Iwata, N., Ochiai, K., Hayashi, K., Ohashi, K., and Umemura, T. (2002). Avian retrovirus infection causes naturally occurring glioma: isolation and transmission of a virus from so-called fowl glioma. Avian Pathol. *31*, 193–199.

Jarrett, W.F., Crawford, E.M., Martin, W.B., and Davie, F. (1964). A virus-like particle associated with leukemia (lymphosarcoma). Nature *202*, 567–569.

Jelacic, T.M., Thompson, D., Hanson, C., Cmarik, J., Nishigaki, K., and Ruscetti, S. (2007). The tyrosine kinase Sf-Stk and its downstream signals are required for maintenance of Friend spleen focus-forming virus-induced fibroblast transformation. J. Virol. *82*, 419–427.

Johansson, F.K., Brodd, J., Eklof, C., Ferletta, M., Hesselager, G., Tiger, C.F., Uhrbom, L., and Westermark, B. (2004). Identification of candidate cancer-causing genes in mouse brain tumors by retroviral tagging. Proc. Natl. Acad. Sci. U.S.A. *101*, 11334–11337.

Jude, B.A., Pobezinskaya, Y., Bishop, J., Parke, S., Medzhitov, R.M., Chervonsky, A.V., and Golovkina, T.V. (2003). Subversion of the innate immune system by a retrovirus. Nat. Immunol. *4*, 573–578.

Katz, E., Lareef, M.H., Rassa, J.C., Grande, S.M., King, L.B., Russo, J., Ross, S.R., and Monroe, J.G. (2005). MMTV Env encodes an ITAM responsible for transformation of mammary epithelial cells in three-dimensional culture. J. Exp. Med. *201*, 431–439.

Khelifi-Younes, C., Dambrine, G., Cherel, Y., Soubieux, D., Li, C.L., and Perbal, B. (2003). Deletions within the U3 long terminal repeat alter the tumorigenic potential of myeloblastosis associated virus type 1(N). Virology *316*, 84–89.

Kirsten, W.H., and Mayer, L.A. (1967). Morphologic responses to a murine erythroblastosis virus. J. Natl. Cancer Inst. *39*, 311–355.

Kone, J., Arroyo, J., Savinelli, T., Lin, S., Boyd, K., Wu, Y., Nimmakayalu, M., Copeland, N.G., Jenkins, N.A., Qumsiyeh, M. *et al.* (2002). F-MuLV acceleration of myelomonocytic tumorigenesis in SV40 large T antigen transgenic mice is accompanied by retroviral insertion at Fli1 and a novel locus, Fim4. Leukemia *16*, 1827–1834.

Landais, S., Landry, S., Legault, P., and Rassart, E. (2007). Oncogenic potential of the miR-106–363 cluster and its implication in human T-cell leukemia. Cancer Res. *67*, 5699–5707.

Lee, C.R., Cervi, D., Truong, A.H., Li, Y.J., Sarkar, A., and Ben-David, Y. (2003). Friend virus-induced erythroleukemias: a unique and well-defined mouse model for the development of leukemia. Anticancer Res. *23*, 2159–2166.

Lenz, J., Celander, D., Crowther, R.L., Patarca, R., Perkins, D.W., and Haseltine, W.A. (1984). Determination of the leukaemogenicity of a murine retrovirus by sequences within the long terminal repeat. Nature *308*, 467–470.

Leroux, C., Girard, N., Cottin, V., Greenland, T., Mornex, J.F., and Archer, F. (2007). Jaagsiekte Sheep Retrovirus (JSRV): from virus to lung cancer in sheep. Vet. Res. *38*, 211–228.

Li, J., Shen, H., Himmel, K.L., Dupuy, A.J., Largaespada, D.A., Nakamura, T., Shaughnessy, J.D.,Jr, Jenkins, N.A., and Copeland, N.G. (1999). Leukaemia disease genes: large-scale cloning and pathway predictions. Nat. Genet. *23*, 348–353.

Li, J.P., and Baltimore, D. (1991). Mechanism of leukemogenesis induced by mink cell focus-forming murine leukemia viruses. J. Virol. *65*, 2408–2414.

Lilly, F. (1970). Fv-2: identification and location of a second gene governing the spleen focus response to Friend leukemia virus in mice. J. Natl. Cancer Inst. *45*, 163–169.

Liu, S.L., Lerman, M.I., and Miller, A.D. (2003). Putative phosphatidylinositol 3-kinase (PI3K) binding motifs in ovine betaretrovirus Env proteins are not essential for rodent fibroblast transformation and PI3K/Akt activation. J. Virol. *77*, 7924–7935.

Liu, S.L., and Miller, A.D. (2007). Oncogenic transformation by the jaagsiekte sheep retrovirus envelope protein. Oncogene *26*, 789–801.

Lund, A.H., Turner, G., Trubetskoy, A., Verhoeven, E., Wientjens, E., Hulsman, D., Russell, R., DePinho, R.A., Lenz, J., and van Lohuizen, M. (2002). Genome-wide retroviral insertional tagging of genes involved in cancer in Cdkn2a-deficient mice. Nat. Genet. *32*, 160–165.

Luscher, B. (2001). Function and regulation of the transcription factors of the Myc/Max/Mad network. Gene *277*, 1–14.

Maeda, N., Fu, W., Ortin, A., de las Heras, M., and Fan, H. (2005). Roles of the Ras-MEK-mitogen-activated protein kinase and phosphatidylinositol 3-kinase-Akt-mTOR pathways in Jaagsiekte sheep retrovirus-induced transformation of rodent fibroblast and epithelial cell lines. J. Virol. *79*, 4440–4450.

Maeda, N., Palmarini, M., Murgia, C., and Fan, H. (2001). Direct transformation of rodent fibroblasts by jaagsiekte sheep retrovirus DNA. Proc. Natl. Acad. Sci. U.S.A. *98*, 4449–4454.

Martin, G.S. (2004). The road to Src. Oncogene *23*, 7910–7917.

Martin, G.S. (1970). Rous sarcoma virus: a function required for the maintenance of the transformed state. Nature *227*, 1021–1023.

Martin-Hernandez, J., Sørensen, A.B., and Pedersen, F.S. (2001). Murine leukemia virus proviral insertions between the N-ras and unr genes in B-cell lymphoma DNA affect the expression of N-ras only. J. Virol. *75*, 11907–11912.

Mayer, B.J., Hamaguchi, M., and Hanafusa, H. (1988). A novel viral oncogene with structural similarity to phospholipase C. Nature *332*, 272–275.

McCormack, M.P., and Rabbitts, T.H. (2004). Activation of the T-cell oncogene LMO2 after gene therapy for X-linked severe combined immunodeficiency. N. Engl. J. Med. *350*, 913–922.

McDonough, S.K., Larsen, S., Brodey, R.S., Stock, N.D., and Hardy, W.D.,Jr. (1971). A transmissible feline fibrosarcoma of viral origin. Cancer Res. *31*, 953–956.

Mikkers, H., Allen, J., Knipscheer, P., Romeijn, L., Hart, A., Vink, E., and Berns, A. (2002). High-throughput retroviral tagging to identify components of specific signaling pathways in cancer. Nat. Genet. *32*, 153–159.

Mitchell, R.S., Beitzel, B.F., Schroder, A.R., Shinn, P., Chen, H., Berry, C.C., Ecker, J.R., and Bushman, F.D. (2004). Retroviral DNA integration: ASLV, HIV, and MLV show distinct target site preferences. PLoS Biol. *2*, E234.

Mitin, N., Rossman, K.L., and Der, C.J. (2005). Signaling interplay in Ras superfamily function. Curr. Biol. *15*, R563–74.

Moloney, J.B. (1960). Biological studies on a lymphoid-leukemia virus extracted from sarcoma 37. I. Origin and introductory investigations. J. Natl. Cancer Inst. *24*, 933–951.

Narezkina, A., Taganov, K.D., Litwin, S., Stoyanova, R., Hayashi, J., Seeger, C., Skalka, A.M., and Katz, R.A. (2004). Genome-wide analyses of avian sarcoma virus integration sites. J. Virol. *78*, 11656–11663.

Nason-Burchenal, K., and Wolff, L. (1993). Activation of c-myb is an early bone-marrow event in a murine model for acute promonocytic leukemia. Proc. Natl. Acad. Sci. U.S.A. *90*, 1619–1623.

Nevins, J.R. (2007). Cell Transformation by Viruses. In D.N. Knipe, and P.M. Howley, eds, Fields Virology (Philadelphia: Lippincott–Raven Press), pp. 209–248.

Ney, P.A., and D'Andrea, A.D. (2000). Friend erythroleukemia revisited. Blood *96*, 3675–3680.

Nishigaki, K., Hanson, C., Jelacic, T., Thompson, D., and Ruscetti, S. (2005). Friend spleen focus-forming virus transforms rodent fibroblasts in cooperation with a short form of the receptor tyrosine kinase Stk. Proc. Natl. Acad. Sci. U.S.A. *102*, 15488–15493.

Nusse, R., and Varmus, H.E. (1982). Many tumors induced by the mouse mammary tumor virus contain a provirus integrated in the same region of the host genome. Cell *31*, 99–109.

Ota, A., Tagawa, H., Karnan, S., Tsuzuki, S., Karpas, A., Kira, S., Yoshida, Y., and Seto, M. (2004). Identification and characterization of a novel gene, C13orf25, as a target for 13q31–q32 amplification in malignant lymphoma. Cancer Res. *64*, 3087–3095.

Pajer, P., Pecenka, V., Kralova, J., Karafiat, V., Prukova, D., Zemanova, Z., Kodet, R., and Dvorak, M. (2006). Identification of potential human oncogenes by mapping the common viral integration sites in avian nephroblastoma. Cancer Res. *66*, 78–86.

Parsons, J.T. (2003). Focal adhesion kinase: the first ten years. J. Cell. Sci. *116*, 1409–1416.

Patriotis, C., and Tsichlis, P.N. (1994). The activated Mlvi-4 locus in Moloney murine leukemia virus-induced rat T-cell lymphomas encodes an *env*/Mlvi-4 fusion protein. J. Virol. *68*, 7927–7932.

Persons, D.A., Paulson, R.F., Loyd, M.R., Herley, M.T., Bodner, S.M., Bernstein, A., Correll, P.H., and Ney, P.A. (1999). Fv2 encodes a truncated form of the Stk receptor tyrosine kinase. Nat. Genet. *23*, 159–165.

Pobezinskaya, Y., Chervonsky, A.V., and Golovkina, T.V. (2004). Initial stages of mammary tumor virus infection are superantigen independent. J. Immunol. *172*, 5582–5587.

Rai, S.K., Duh, F.M., Vigdorovich, V., Danilkovitch-Miagkova, A., Lerman, M.I., and Miller, A.D. (2001). Candidate tumor suppressor HYAL2 is a glycosylphosphatidylinositol (GPI)-anchored cell-surface receptor for jaagsiekte sheep retrovirus, the envelope protein of which mediates oncogenic transformation. Proc. Natl. Acad. Sci. U.S.A. *98*, 4443–4448.

Raines, M.A., Maihle, N.J., Moscovici, C., Moscovici, M.G., and Kung, H.J. (1988). Molecular characterization of three erbB transducing viruses generated during avian leukosis virus-induced erythroleukemia: extensive internal deletion near the kinase domain activates the fibrosarcoma- and hemangioma-inducing potentials of erbB. J. Virol. *62*, 2444–2452.

Raman, M., Chen, W., and Cobb, M.H. (2007). Differential regulation and properties of MAPKs. Oncogene *26*, 3100–3112.

Ramirez, J.M., Houzet, L., Koller, R., Bies, J., Wolff, L., and Mougel, M. (2004). Activation of c-myb by 5′ retrovirus promoter insertion in myeloid neoplasms is dependent upon an intact alternative splice donor site (SD′) in gag. Virology *330*, 398–407.

Rapp, U.R., Gotz, R., and Albert, S. (2006). BuCy RAFs drive cells into MEK addiction. Cancer. Cell. *9*, 9–12.

Rapp, U.R., Reynolds, F.H.,Jr, and Stephenson, J.R. (1983). New mammalian transforming retrovirus: demonstration of a polyprotein gene product. J. Virol. *45*, 914–924.

Robinson, H.L., and Gagnon, G.C. (1986). Patterns of proviral insertion and deletion in avian leukosis virus-induced lymphomas. J. Virol. *57*, 28–36.

Rohdewohld, H., Weiher, H., Reik, W., Jaenisch, R., and Breindl, M. (1987). Retrovirus integration and chromatin structure: Moloney murine leukemia proviral integration sites map near DNase I-hypersensitive sites. J. Virol. *61*, 336–343.

Rosenberg, N., and Jolicoeur, P. (1997). Retroviral pathogenesis. In Retroviruses, Coffin, J.M., Hughes, S.H., and Varmus, H.E. eds., Cold Spring Harbor Laboratory Press) pp. 475–585.

Ross, S.R., Schmidt, J.W., Katz, E., Cappelli, L., Hultine, S., Gimotty, P., and Monroe, J.G. (2006). An immunoreceptor tyrosine activation motif in the mouse mammary tumor virus envelope protein plays a role in virus-induced mammary tumors. J. Virol. *80*, 9000–9008.

Rous, P. (1911). A sarcoma of the fowl transmissible by an agent separable from the tumor cells J. Exp. Med. *13*, 397–411.

Scherdin, U., Rhodes, K., and Breindl, M. (1990). Transcriptionally active genome regions are preferred targets for retrovirus integration. J. Virol. *64*, 907–912.

Schlessinger, J. (1986). Allosteric regulation of the epidermal growth factor receptor kinase. J. Cell Biol. *103*, 2067–2072.

Selten, G., Cuypers, H.T., and Berns, A. (1985). Proviral activation of the putative oncogene Pim-1 in MuLV induced T-cell lymphomas. EMBO J. *4*, 1793–1798.

Selten, G., Cuypers, H.T., Boelens, W., Robanus-Maandag, E., Verbeek, J., Domen, J., van Beveren, C., and Berns, A. (1986). The primary structure of the putative oncogene pim-1 shows extensive homology with protein kinases. Cell *46*, 603–611.

Shaw, G., and Kamen, R. (1986). A conserved AU sequence from the 3′ untranslated region of GM-CSF mRNA mediates selective mRNA degradation. Cell *46*, 659–667.

Shen-Ong, G.L., Morse, H.C.,3rd, Potter, M., and Mushinski, J.F. (1986). Two modes of c-myb activation in virus-induced mouse myeloid tumors. Mol. Cell. Biol. *6*, 380–392.

Sherr, C.J. (1988). The fms oncogene. Biochim. Biophys. Acta *948*, 225–243.

Sola, B., Simon, D., Mattei, M.G., Fichelson, S., Bordereaux, D., Tambourin, P.E., Guenet, J.L., and Gisselbrecht, S. (1988). Fim-1, Fim-2/c-fms, and Fim-3, three common integration sites of Friend murine leukemia virus in myeloblastic leukemias, map to mouse chromosomes 13, 18, and 3, respectively. J. Virol. *62*, 3973–3978.

Sørensen, A.B., Lund, A.H., Kunder, S., Quintanilla-Martinez, L., Schmidt, J., Wang, B., Wabl, M., and Pedersen, F.S. (2007). Impairment of alternative splice sites defining a novel gammaretroviral exon within *gag* modifies the oncogenic properties of Akv murine leukemia virus. Retrovirology *4*, 46.

Sørensen, K.D., Kunder, S., Quintanilla-Martinez, L., Sørensen, J., Schmidt, J., and Pedersen, F.S. (2007). Enhancer mutations of Akv murine leukemia virus inhibit the induction of mature B-cell lymphomas and shift disease specificity towards the more differentiated plasma cell stage. Virology *362*, 179–191.

Sourvinos, G., Tsatsanis, C., and Spandidos, D.A. (2000). Mechanisms of retrovirus-induced oncogenesis. Folia Biol. (Praha) *46*, 226–232.

Speck, N.A., Renjifo, B., Golemis, E., Fredrickson, T.N., Hartley, J.W., and Hopkins, N. (1990). Mutation of the core or adjacent LVb elements of the Moloney murine leukemia virus enhancer alters disease specificity. Genes Dev. *4*, 233–242.

Speck, N.A., Renjifo, B., and Hopkins, N. (1990). Point mutations in the Moloney murine leukemia virus enhancer identify a lymphoid-specific viral core motif and 1,3-phorbol myristate acetate-inducible element. J. Virol. *64*, 543–550.

Stehelin, D., Varmus, H.E., Bishop, J.M., and Vogt, P.K. (1976). DNA related to the transforming gene(s) of avian sarcoma viruses is present in normal avian DNA. Nature *260*, 170–173.

Stiewe, T. (2007). The p53 family in differentiation and tumorigenesis. Nat. Rev. Cancer. *7*, 165–168.

Suzuki, T., Minehata, K., Akagi, K., Jenkins, N.A., and Copeland, N.G. (2006). Tumor suppressor gene identification using retroviral insertional mutagenesis in Blm-deficient mice. EMBO J. *25*, 3422–3431.

Suzuki, T., Shen, H., Akagi, K., Morse, H.C., Malley, J.D., Naiman, D.Q., Jenkins, N.A., and Copeland, N.G. (2002). New genes involved in cancer identified by retroviral tagging. Nat. Genet. *32*, 166–174.

Swanson, I., Jude, B.A., Zhang, A.R., Pucker, A., Smith, Z.E., and Golovkina, T.V. (2006). Sequences within the *gag* gene of mouse mammary tumor virus needed for mammary gland cell transformation. J. Virol. *80*, 3215–3224.

Tabin, C.J., and Weinberg, R.A. (1985). Analysis of viral and somatic activations of the cHa-ras gene. J. Virol. *53*, 260–265.

Takuwa, N., and Takuwa, Y. (2001). Regulation of cell cycle molecules by the Ras effector system. Mol. Cell. Endocrinol. *177*, 25–33.

Tam, W., Ben-Yehuda, D., and Hayward, W.S. (1997). bic, a novel gene activated by proviral insertions in avian leukosis virus-induced lymphomas, is likely to function through its noncoding RNA. Mol. Cell Biol. *17*, 1490–1502.

Temin, H.M. (1963). Further evidence for a converted, non-virus-producing state of Rous sarcoma virus-infected cells. Virology *20*, 235–245.

Temin, H.M., and Rubin, H. (1958). Characteristics of an assay for Rous sarcoma virus and Rous sarcoma cells in tissue culture. Virology *6*, 669–688.

Theilen, G.H., Gould, D., Fowler, M., and Dungworth, D.L. (1971). C-type virus in tumor tissue of a woolly monkey (Lagothrix spp.) with fibrosarcoma. J. Natl. Cancer Inst. *47*, 881–889.

Theodorou, V., Kimm, M.A., Boer, M., Wessels, L., Theelen, W., Jonkers, J., and Hilkens, J. (2007). MMTV insertional mutagenesis identifies genes, gene families and pathways involved in mammary cancer. Nat. Genet. *39*, 759–769.

Truong, A.H., and Ben-David, Y. (2000). The role of Fli-1 in normal cell function and malignant transformation. Oncogene *19*, 6482–6489.

Truss, M., Bartsch, J., Mows, C., Chavez, S., and Beato, M. (1996). Chromatin structure of the MMTV promoter and its changes during hormonal induction. Cell. Mol. Neurobiol. *16*, 85–101.

Urisman, A., Molinaro, R.J., Fischer, N., Plummer, S.J., Casey, G., Klein, E.A., Malathi, K., Magi-Galluzzi, C., Tubbs, R.R., Ganem, D., Silverman, R.H., and DeRisi, J.L. (2006). Identification of a novel Gammaretrovirus in prostate tumors of patients homozygous for R462Q RNASEL variant. PLoS Pathog. 2, e25.

van den Akker, E., Aarts, L.H., and Delwel, R. (2007). Viral insertion in Evi12 causes expression of aberrant Grp94 mRNAs containing the viral *gag* myristylation motif. Virology *366*, 227–233.

Vogt, P.K., and Bader, A.G. (2005). Jun: stealth, stability, and transformation. Mol. Cell *19*, 432–433.

Wang, C.L., Wang, B.B., Bartha, G., Li, L., Channa, N., Klinger, M., Killeen, N., and Wabl, M. (2006). Activation of an oncogenic microRNA cistron by provirus integration. Proc. Natl. Acad. Sci. U.S.A. *103*, 18680–18684.

Wang, Y., Kayman, S.C., Li, J.P., and Pinter, A. (1993). Erythropoietin receptor (EpoR)-dependent mitogenicity of spleen focus-forming virus correlates with viral pathogenicity and processing of *env* protein but not with formation of gp52–EpoR complexes in the endoplasmic reticulum. J. Virol. *67*, 1322–1327.

Weber, F., and Schaffner, W. (1985). Enhancer activity correlates with the oncogenic potential of avian retroviruses. EMBO J. *4*, 949–956.

Weiser, K.C., Liu, B., Hansen, G.M., Skapura, D., Hentges, K.E., Yarlagadda, S., Morse Iii, H.C., and Justice, M.J. (2007). Retroviral insertions in the VISION database identify molecular pathways in mouse lymphoid leukemia and lymphoma. Mamm. Genome *18*, 701–722.

Westermark, B., and Heldin, C.H. (1991). Platelet-derived growth factor in autocrine transformation. Cancer Res. *51*, 5087–5092.

Wood, T.G., McGeady, M.L., Baroudy, B.M., Blair, D.G., and Vande Woude, G.F. (1984). Mouse c-mos oncogene activation is prevented by upstream sequences. Proc. Natl. Acad. Sci. U.S.A. *81*, 7817–7821.

Wootton, S.K., Halbert, C.L., and Miller, A.D. (2005). Sheep retrovirus structural protein induces lung tumours. Nature *434*, 904–907.

Wu, X., Li, Y., Crise, B., and Burgess, S.M. (2003). Transcription start regions in the human genome are favored targets for MLV integration. Science *300*, 1749–1751.

Yatsula, B., Galvao, C., McCrann, M., and Perkins, A.S. (2006). Assessment of F-MuLV-induced tumorigenesis reveals new candidate tumor genes including Pecam1, St7, and Prim2. Leukemia *20*, 162–165.

Zhang, B., Pan, X., Cobb, G.P., and Anderson, T.A. (2007). microRNAs as oncogenes and tumor suppressors. Dev. Biol. *302*, 1–12.

Pathogenesis of Immunodeficiency Virus Infections

10

Guido Poli and Volker Erfle

Abstract

Extensive analysis of naturally occurring simian immunodeficiency viruses (SIVs) and comparative phylogenetic studies with human immunodeficiency viruses (HIVs) suggests that the latter are close relatives of the SIVcpz viruses of chimpanzees (HIV-1) or the SIVsmm viruses of sooty mangabeys (HIV-2). Crossing of species barriers resulted in adaptation to the human host and subsequent acquisition of a pathogenic phenotype. Naturally occurring T-lymphocyte-tropic lentiviral infections are highly prevalent and productive but are not usually pathogenic for native hosts. Crossing species barriers may produce an abortive infection or, as in the case of the HIVs, may enhance virulence after several cycles of transmission. The large number of species carrying these viruses may suggest that infection confers an evolutionary advantage to the host. The virulent T-lymphocyte-tropic lentiviruses have a similar genomic structure and exhibit comparable replication strategies. Their major targets are lymphocytes populating lymphoid organs and tissues, and antigen-presenting cells (dendritic cells, mononuclear phagocytes). Within these targets the virus can replicate to very high titres and thereby exhaust $CD4^+$ T-cells, producing profound immunodeficiency. Although the infection of lymphoid organs and tissue is the pathologic hallmark of HIV infection, this virus also infects cells of the central nervous system.

This chapter discusses various pathogenic mechanisms involved in immune activation and dysregulation, and summarizes characteristics of HIV/SIV gene–host factor interaction in the immune and central nervous systems.

Introduction

The human immunodeficiency viruses (HIV-1 and -2) belong to the Lentiviridae subfamily of retroviruses. Even before the human viruses were identified, several animal species were shown to harbour species-specific viruses that cause various degrees of immunodeficiency. (For this reason, all such viruses will be hereafter referred to as 'immunodeficiency viruses'.) The most prominent examples of immunodeficiency viruses are the simian immunodeficiency viruses (SIV) that infect various monkey species, including chimpanzees (which are very close to humans in terms of genetic make up), sooty mangabeys (SM), and African green monkeys (AGM). Furthermore, other mammalian species such as cats and cattle harbour the feline and bovine immunodeficiency viruses (FIV and BIV, respectively) (Bendinelli *et al.*, 1995; Suarez *et al.*, 1993; Stump and VandeWoude, 2007).

In the absence of antiretroviral therapy (ART), infection with HIV-1 or HIV-2 usually leads to the development of symptoms collectively called acquired immunodeficiency syndrome (AIDS). Although immunodeficiency is the most prominent pathogenic outcome of this infection, the AIDS definition also includes profound neurological abnormalities, and secondary infections or malignancies such as B-cell lymphomas and Kaposi's sarcoma (KS). KS was later found to be caused by another virus of the herpesviridae family [KSHV or human herpes virus (HHV)-8] (Huang *et al.*, 2001).

A fatal disease similar to AIDS in humans infected with HIV-1 has been observed in rhesus macaques experimentally infected with SIV. Therefore, SIV infection of rhesus macaques is considered a good model for the pathogenesis of HIV infection in humans. However, this model does not reflect natural SIV infection in other monkey species such as AGM or SM (VandeWoude and Apetrei, 2006; Beer *et al.*, 1998, 1999). SIV infection has been detected in more than 40 different non-human primate (NHP) species, but these naturally infected NHP generally show no signs of disease. Therefore, the SIV strain(s) pathogenic for rhesus and cynomolgus monkeys can be considered to have only laboratory – not clinical – relevance. Pathogenic SIVs have a common ancestor that was isolated from a SM; infections with the modern day-descendant of this ancestor do not induce disease despite high levels of virus replication in the peripheral blood compartment, tissues and organs. Because the SIVsm virus model in macaques has an ancestor in infected chimpanzees (Pandrea *et al.*, 2008), it may be a suitable model for HIV-1 in humans. The SIV strains isolated from chimpanzees in the wild (i.e. SIVcpz) do not appear to be pathogenic in their natural host, although this may not be an absolute rule (Keele, 2009). However, virus transmission from NHP to humans produces adaptation to the human host and subsequent acquisition of a pathogenic phenotype (Sharp *et al.*, 1999; Hahn *et al.*, 2000; Wain *et al.*, 2007).

Like HIV in humans (and like the experimental infection of macaques by SIVsmm), naturally transmitted FIV infections in cats are life threatening infections (Bendinelli *et al.*, 1995). In contrast, the infection of cattle with BIV usually has no health consequences for the infected animal (St-Louis *et al.*, 2004, 2005).

Amongst all immunodeficiency viruses in simian, feline and bovine species, two main patterns of host–virus interaction are observed during infection, and these are described below.

Within this category of infections the predominant pattern is a life-long infection and replication without obvious pathological consequences. This pattern has been described in most NHP species in Africa (Holzammer *et al.*, 2001), in feline species such as pumas and lions, and in cattle with 'natural' infections (VandeWoude and Apetrei, 2006). Only single animals show signs of illness, generally after very long incubation periods (Ling *et al.*, 2004). These 'semi-pathogenic' immunodeficiency viruses almost always arise from trans-species infection. But trans-species transmission only leads to the characteristic disease phenotype in rare cases; clearly, many rounds of replication are required to select the pathogenic phenotype.

First round transmissions, such as the infection of domestic cats with Puma FIV, results in an asymptomatic infection and even protects the infected cats against a superinfection with a pathogenic FIV (VandeWoude *et al.*, 2003). The 'semi-pathogenic' type of infection is typically associated with chronic progressive diseases involving long incubation periods. The chronic infection is characterized by high replication rates of the virus, and typically produces a mild acute illness, often involving some lymphoproliferative changes at the early stages of infection. After long latency periods the onset of AIDS is preceded by a lymphadenopathy syndrome. Immunodeficiency characterized by a progressive loss of thymic and peripheral CD4 T-cells and neurological disorders (primarily encephalopathy) are the pathological endpoints. The final clinical symptomatology of AIDS is due to opportunistic infections and tumours (Apetrei *et al.*, 2004). However, there are exemptions from this typical clinical pattern; namely, two cases of acute immunodeficiency viruses have been identified: a lethal variant from SIVsmm, designated SIVsmm-pbj14 (Fultz, 1994) and a BIV variant, the Jembrana disease virus in Bali cattle (Chadwick *et al.*, 1995a,b). These two viruses induce clinical signs of acute lymphadenopathy and lymphopenia, often with a fatal outcome, even after a short incubation period of several days.

Thus, the pathogenic potential of these different immunodeficiency viruses is dependent on many different parameters such as the virus and its genes, the host species, the target organ and/or the target cell, and the quality and breadth of the immune response to infection (Miller *et al.*, 2000; Norley *et al.*, 1999; Kirchhoff and Silvestri, 2008; Kirchhoff *et al.*, 2008). The fact that many immunodeficiency viruses do not usually cause problems in their natural host raises the question of the role of these viruses in the animal kingdom, and may provide hints for the development of protective strategies in species where immunodeficiency viruses become pathogenic.

Immunodeficiency viruses exhibit a typical retroviral genomic makeup. This genome contains the codes for the structural proteins necessary to form the typical architecture of the virus particle, for proteins that perform essential functions in the life cycle of the virus (including viral enzymes), and for proteins that regulate the virus–cell interaction during viral replication (regulatory and accessory proteins). Besides their enabling functions on behalf of the virus, these proteins often influence the host's cellular functions, thereby inducing aberrant responses in infected cells or organs that can lead to immunodeficiency or other pathogenic phenotypes (Table 10.1).

All the genes encoding for the structural proteins and the viral enzymes are shared among HIVs, SIVs, FIVs and BIVs; although their sequences are different, these genes perform equivalent functions in the viral life cycle. This functional similarity was confirmed by experiments in which either portions of genes or entire individual genes (e.g. *tat*) were inserted into a different viral genomic environment. The major differences between the gene sequences are related to regulatory proteins, for which NHP immunodeficiency viruses (i.e. SIV) are characterized by the presence of a *nef* gene that is not present in FIV or BIV. Vpu is part of the genome of HIV-1, HIV-2 and the SIVs, whereas Vpr is present in HIV-1 and HIV-2, which has in addition a similar protein, VpX. The overall sequence of HIV-2 is very similar to most SIVs.

Pathogenesis by immunodeficiency viruses cannot be attributed to just one or two single genes or proteins, and a significant body of literature describes both direct and indirect mechanisms with pathogenic potential; non-structural proteins in particular may be involved in pathogenesis. For example, multiple mechanisms have been proposed to explain the death and dysfunction of $CD4^+$ T-lymphocytes, the hallmark of AIDS (Fauci, 1996). Multiple mechanisms are also likely involved in the apoptosis of various cell types that commonly occur in immunodeficiency virus infections (e.g. apoptotic neuronal death in HIV encephalopathy). Investigation of the HIV-1 genes *tat*, *nef*, *vpr* and also *env* have yielded significant information on signalling pathways leading to apoptosis in the various affected cell types (Acheampong *et al.*, 2005b), and this is described in more detail below.

- Vpr is an HIV-1 accessory protein that transactivates the long terminal repeat (LTR), induces G_2 arrest, and also induces apoptosis in infected T-cells (Ayyavoo *et al.*, 1997; Andersen *et al.*, 2008). However, the magnitude of T-lymphocyte apoptosis cannot be accounted for by the death of infected cells alone, suggesting that bystander T-cells must be profoundly affected by viral proteins such as gp120, Tat, Nef or VpR (Finkel *et al.*, 1995).
- The Tat protein is released from infected cells and this extracellular Tat can be taken up by non-infected T-cells (Frankel and Pabo, 1988; Ensoli *et al.*, 1993). Working alone or in concert with gp120, Tat protein accelerates or induces apoptosis in the exposed T-cells (Westendorp *et al.*, 1995; Noonan and Albini, 2000). Interestingly, chimpanzee cells are resistant to HIV-1 *tat* protein induced apoptosis (Ehret *et al.*, 1996).
- The first insights into the possible role of the HIV-1 *nef* gene came from studies with *nef* deleted SIV strains (Kestler *et al.*, 1991; Daniel *et al.*, 1992) and from patients accidentally infected with a HIV-1 harbouring a truncated *nef* gene (Brambilla *et al.*, 1999; Greenough *et al.*, 1999; Rhodes *et al.*, 2000; Hofmann-Lehmann *et al.*, 2003 Verity *et al.*, 2007). Both non-human animals and the infected humans maintained extremely low virus burdens and normal $CD4^+$ T-lymphocyte concentrations, and have remained healthy for a significantly longer time than human or non-human animals infected with SIV or HIV with a functional *nef* gene. In a murine model for *nef* mediated pathogenesis – transgenic CD4 promoter/Nef mice – the animals develop a severe AIDS-like disease characterized by pre-mature death, most likely via apoptosis of thymic and peripheral $CD4^+$ and $CD8^+$ T-cells (Priceputu *et al.*, 2005, 2007).

Table 10.1 Functions and cellular effects of HIV proteins

Viral protein	Main function	Effects on target cells	Key references
Structural proteins			
p24/30 Gag	Virion core formation	?	Ganser-Pornillos *et al.* (2008)
p17/19 Gag	Core structure stabilization	?	Ganser-Pornillos *et al.* (2008), Hearps and Jans (2007)
p6 Gag	Virion release	Interaction with tetherin and other membrane associated proteins	Freed (2003), Fujii *et al.* (2007)
gp120 Env	Receptor/co-receptor binding	Down-modulation of CD4 from the plasma membrane and induction of cell signalling	Phogat and Wyatt (2007), Roux and Taylor (2007), Sirois *et al.* (2007)
gp41 Env	Virion-cell fusion	?	Phogat and Wyatt (2007)
Viral enzymes			
RT	RNA-dependent DNA polymerase	?	Camarasa *et al.* (2006)
RNase H	Removal of RNA template from RNA/DNA hybrid during RT	?	Camarasa *et al.* (2006)
Protease	Cleavage of Gag p55 precursor protein	?	Camarasa *et al.* (2006)
Integrase	Integration of proviral DNA into cellular DNA	?	Camarasa *et al.* (2006)
Regulatory and accessory proteins			
Tat	Enhancement of viral transcription	Several effects due to interaction with cell surface receptors upon active or passive release and cellular uptake (T-cell activation and apoptosis, chemotactic activity)	Ammosova *et al.* (2006), Gatignol (2007)
Rev	Nuclear export of viral RNA	?	Ammosova *et al.* (2006), Rossi *et al.* (2007)
Nef	Enhancement of virus replication	Downmodulation of MHC class I antigens and CD4 from the plasma membrane; T-cell activation and apoptosis, chemotactic activity	Foster and Garcia (2007)
VpR/VpX	Enhancement of virus replication in macrophage	Functional interaction with intracellular glucocorticoid receptors	Dehart and Planelles, (2008), Andersen *et al.* (2008)
Vif	Virion infectivity factor	Counteraction of APOBEC-3G and -3F	Dehart *et al.* (2008), Goila-Gaur and Strebel (2008)
VpU	Late events in the virus life cycle	Interaction with tetherin, related to morphogenesis, maturation and release of new progeny virions	Strebel (2007) McNatt *et al.* (2009)

The apoptotic potential of Nef has also been demonstrated in human T-lymphocytes with recombinant Nef protein (Zauli *et al.*, 1999).

- The apoptotic potential of extracellular Tat and Nef in T-cells, together with the known chemotactic activity of these two HIV proteins for macrophages and T-cells, opens a new view on the viral dissemination strategy (Albini *et al.*, 1998; Koedel *et al.*, 1999; Lehmann *et al.*, 2006). The secretion of one or both of these proteins from infected cells provides protection against T-cell attacks only so long as virus production is ongoing. Protein secretion also actively attracts new target cells. This combined effect may help to explain the delayed or restricted *in vivo* disease that occurs in SIVs and HIVs in which *nef* has been deleted or truncated. Interference with these activities of Tat and Nef may be an attractive therapeutic target, and has potential to prolong the disease free period in infected individuals.
- Vpu has recently been linked to inactivation of tetherin, a host protein controlling the release of virions from infected cells (McNatt *et al.*, 2009).

General features of the immune response to immunodeficiency virus infection

Most of the information on the immunological response to exogenous retroviruses comes from studies of HIV-1 and, to a lesser extent, HIV-2. The analysis of the adaptive and innate immune responses to these pathogenic human retroviruses sets the stage for the analysis and comparison of responses to other retroviruses.

To provide a basis for understanding the peculiarities of the immune response to HIVs, some general features of the immunological activation following viral infections are summarized below. The two major arms of the immune response are labelled 'innate' (i.e. 'natural') and 'adaptive' (i.e. 'specific') (Schoenborn and Wilson, 2007; van Kooyk and Rabinovich, 2008). These two systems should not be considered independent responses to the microbial invader, but rather as two multifactorial components strategically evolved with the goal of preserving the integrity of the host from invaders. That these invaders can modify the genomic fidelity of the host itself is exemplified by the existence and significant accumulation of sequences of endogenous retroviruses.

The adaptive immune response targets both the type of virus and the specific virus (or microbe) that has infected the host; in contrast the innate immune response triggers a set of general events typical for many pathogens, with no strict specificity for the particular invader. An adaptive response depends on the specificity of recognition and affinity of the T-cell receptor (TCR) and B-cell receptor (BCR), which are expressed on the surface of T-lymphocytes (either $CD4^+$ or $CD8^+$) and B-lymphocytes, respectively. Innate immune responses include an array of both cell-associated and soluble factors that are rapidly mobilized by the invasion (or attempted invasion) of a microorganism (Schoenborn and Wilson, 2007; van Kooyk and Rabinovich, 2008). Cell surface receptors recognizing motifs of the intruder are also a fundamental component of the innate immune response, as exemplified by the so-called Toll-like receptors (TLR) and, more generally, by pattern recognition receptors (PRR) that recognize common structural motifs of the pathogen rather than strict sequences (as in the case of the adaptive response) (An *et al.*, 2008; Gilliet *et al.*, 2008).

The temporal difference between the innate (rapid) and adaptive (slow) immune response is a fundamental distinction between these two types of reactions, which work together to protect the host. While the innate immune response contains further spread of the invading virus, the adaptive immune response specifically attacks the invader. Through the production of specific immunoglobulins (Ig) or antibodies (Ab), B-cells attack native antigens on the surface of the invading microorganism; the antigens are recognized either by their amino acid sequence or their conformation (the BCR is a membrane Ig). In contrast, T-lymphocytes can only recognize peptides of the microorganism after their intracellular processing and fragmentation. Viral peptides are coupled to molecules of the major histocompatibility complex (MHC) in order to enable presentation to the T-cells, which is an absolute prerequisite for triggering a specific cell-mediated immune response (Vyas *et al.*, 2008). Every host cell except neurons expresses MHC Class I molecules; if these molecules are loaded with viral peptides the infected cell can be recognized by CD8$^+$ T-lymphocytes and eliminated by cytolysis. For this reason, activated CD8$^+$ T-cells are frequently referred to as cytotoxic T-lymphocytes (CTL) (Schoenborn and Wilson, 2007). A restricted number of cells, namely myeloid dendritic cells (mDC), mononuclear phagocytes (MP) including circulating monocytes and tissue macrophages, and B-lymphocytes express MHC Class II molecules and, for this reason, they are commonly referred to as antigen presenting cells (APC). If these molecules are loaded with microbial peptides, they can be recognized by CD4$^+$ T-lymphocytes that usually do not function as CTL (although a CD4$^+$ CTL response clearly exists), but rather as 'helper' cells for other effector cells such as CD8$^+$ T-lymphocytes and B-cells. The nature of CD4$^+$ T-cell activity is not completely understood, but it certainly involves the production of several cytokines and chemokines that orchestrate cell activation and recruitment at sites of infection. Depending on the type of microbial invader, T-cells can be oriented towards a 'type 1' (Th1) or 'type 2' (Th2) immune response (Bettelli *et al.*, 2008). Th1 cells are typically induced by intracellular pathogens, including bacteria and viruses that promote the synthesis and release of interleukin(IL)-12 from macrophages. IL-12, together with other factors, induces the activation and differentiation of naïve CD4$^+$ T-lymphocytes (Th0) to become Th1 cells, which secrete *pro*-inflammatory cytokines and chemokines. The cytokine interferon (IFN)-γ initiates a feed-back that promotes further macrophage activation (Bettelli *et al.*, 2008). This *pro*-inflammatory immune response is optimal for the development and maintenance of a robust CD8$^+$ CTL response. Studies in non-human animals clearly showed that in the absence of CD4$^+$ T-cell reactions, CTLs develop in response to a viral infection (every cell expressed MHC Class I molecules), but the response fades with time.

In contrast, extra-cellular pathogens (typically parasites) trigger the release of IL-4 from cells of the innate immune system, such as natural killer T (NKT) cells and DC, and this cytokine (as well as macrophage-derived IL-13) promotes the differentiation of Th2 cells (Stock and Akbari, 2008). Th2 cells secrete IL-4 (an autocrine factor that drives Th2 cell polarization), which induces an isotypic switch from IgG to IgA and IgE (i.e. the substitution of the stem of the Ab, named Fc, but leaving the same variable and hypervariable regions responsible for specific target recognition). These Ab play a fundamental protective role in mucosal immunity (IgA) and are fundamental in the anti-parasite immune response (IgE) both directly and after binding to eosinophils and mast cells (Galli *et al.*, 2008a,b). The binding of Ab occurs via the Fcε portion, which allows degranulation of lytic factors against the invading pathogen. Other Th2 cytokines such as IL-5 also activate these cells. Finally, the Th2 response can be associated with an anti-inflammatory effect (D'Ambrosio *et al.*, 2000). Particularly in the mouse, Th2 cells selectively secrete IL-10, a potent anti-inflammatory cytokine that is also secreted by both Th1 and Th2 cells in humans (D'Ambrosio *et al.*, 2000).

Recently, a novel subset of polarized T-cell responses, named Th17 cells, has been defined (Korn *et al.*, 2009), while polarization of macrophages into M1 and M2 cells has been linked to control of HIV infection (Cassol *et al.*, 2009).

Immunopathogenesis of HIV and SIV infections

Given the ultimate immunosuppressive and CD4$^+$ T-cell depleting effects of HIV infection, the fact that a strong and robust immune response follows the primary HIV infection (PHI) (Harari *et al.*, 2002) seems counter-intuitive. However, both CD4$^+$ and CD8$^+$ T-cell responses develop rapidly, in a matter of a few weeks, and evolve with conventionally high levels of specificity. Viral spreading is initially carried out by a homogeneous virus that almost invariably uses CCR5 (R5) as entry co-receptor (Liu *et al.*, 2006). Viral diversification and eventual emergence (i.e. for subtype B) of CXCR4-using strains (usually duotropic R5X4, more rarely monotropic X4 or utilizing additional chemokine co-receptors such as CCR2 and CCR3) emerges only later on in the disease, during the immunodeficient stage of infection (Berger *et al.*, 1998; Koot *et al.*, 1993, 1999). The Ab response, which includes the development of neutralizing Ab (NAb), appears after the PHI (Polonis *et al.*, 2008).

The insurgence of a robust adaptive immune response to HIV is believed to play a major role in the partial containment of virus replication and reduction of the number of infected cells. Macaques experimentally infected with SIV (closely related to HIV-2) and transiently depleted of $CD8^+$ cells (as well as depletion of a substantial fraction of NK cells and T-lymphocytes) show a burst of viraemia and $CD4^+$ T-cell depletion; however, the $CD4^+$ T-cells partially recover when $CD8^+$ T-cells return to their usual physiological levels (Barouch *et al.*, 2000). Observations in humans characterized either by rapid progression (RP) or long-term nonprogression (LTNP) of HIV disease demonstrate the important role of the $CD8^+$ T-cell response in combating the pathogenic consequences of unchecked viral replication. In humans with low levels of $CD8^+$ T-cells, an immune response to RP is almost absent and is overwhelmed by untamed virus replication. Conversely, humans with good $CD8^+$ CTL responses show a stronger control of HIV infection.

However, the 'chicken vs. egg' question has not yet been solved in these extreme examples of natural infection. Are LTNP actually protected by the $CD8^+$ T-cell response (along with other features of the immune response to HIV) or this is rather the result of a relatively benign infection, as determined by other essential features of host/virus interactions (e.g. the role of the CCR5Δ32 heterozygotic allelic configuration, which is frequently observed in LTNP) (Navis *et al.*, 2008; Saksena *et al.*, 2007; Schellens *et al.*, 2008)? With regard to RP, it should be noted that natural apathogenic infection of SM and AGM is also characterized by very high levels of virus replication coupled with the host's minimal immunological response to the virus (Pandrea *et al.*, 2008). However, the emergence of viral variants despite cell-mediated responses and Nab indicates that the CTL response certainly influences HIV replication and *in vivo* evolution. When studied in later stages of disease such as during the AIDS phase, CTL are often still detectable but they bear features of imperfect activation or functional exhaustion as compared, e.g. to the CTL response against cytomegalovirus (Harari *et al.*, 2002).

A distinctive feature of $CD8^+$ T-cells of HIV+ individuals is the infiltration of the germinal centres (GC) of lymph nodes (a B-cell area) (Emilie *et al.*, 1990). This morphological feature is peculiar to HIV infection (i.e. not observed in individuals with autoimmune diseases), and indicates an immunopathogenic role of CTL in HIV disease analogous to the role of CTL in hepatitis B virus infected liver, where the tissue damage is not consequent to viral replication per se, but rather to the $CD8^+$-dependent T-cell response against infected cells (Guidotti and Chisari, 2001). The GCs of the lymph nodes are architectural structures that organize the B-cell response to infectious agents and are supported by a dense network of follicular DC (FDC) (Burton *et al.*, 1997; Pantaleo *et al.*, 1998). These cells, which are of different origin from circulating mDC or plasmacytoid DC (pDC), are crucial for the B-cell response because they fix antigens in their native configuration and thereby preserve entire virions that maintain their infectivity for weeks and months after infection. *In situ* hybridization studies reveal that HIV virions decorate the network of FDC in the GC, creating an extracellular reservoir of the virus; the fibrotic involution of GC in late stage disease (following a florid phase earlier in infection) suggests a loss of their filter function, which may increase the peripheral viral load (i.e. viraemia) (Burton *et al.*, 1997; Pantaleo *et al.*, 1998).

CTL of late stage patients can secrete IFN-γ in response to HIV peptides, but do not secrete IL-2, other cytokines, or perforin, all of which are fundamental for inducing a lytic effect on infected cells (Harari *et al.*, 2002).

The $CD4^+$ T-cell response to HIV infection plays a crucial role in the organization of the immune response, and in the pathogenesis of disease progression (Rosenberg *et al.*, 2000). These cells are the primary (though not exclusive) targets of HIV, and their physical depletion and functional impairment critically contributes to the progression of HIV infection towards AIDS.

It was recently shown that a substantial loss of $CD4^+$ T-cells coupled with high levels of virus replication occurs in the gut-associated lymphoid tissue (GALT) within days to weeks after PHI, regardless of the infection's port of entry (Brenchley *et al.*, 2004; Li *et al.*, 2005; Mattapallil *et al.*, 2005). According to these studies, HIV infection and disease may last for several years before the opportunistic infections and tumours typical of the AIDS phase emerge. While the AIDS phase is the most dramatic stage of the infection with regard to patient health, it only occurs after a rapid pathogenic depletion of $CD4^+$ T-cells shortly after PHI. The selective infection and depletion of HIV-specific $CD4^+$ T-cells has been shown only *ex vivo*. Although clinical trials with highly active antiretroviral therapy (HAART) showed that substantial immunological functional reconstitution of $CD4^+$ T-cells occurs in the GALT (Wilcox and Saag, 2008), these results indicated that mucosal-associated immune responses are a fundamental feature of the natural history of HIV infection and should be considered by anyone who undertakes the difficult task of developing a vaccine against this infection. Recently, an interaction between gp120 Env of certain R5 HIV-1 strains was shown to interact with and promote cell signalling through the $\alpha4\beta7$ integrin, the homing receptor for $CD4^+$ T-lymphocytes in the gut (Arthos *et al.*, 2008). Since such an interaction can promote more efficient virus replication it is conceivable that it plays a role in both HIV spreading in the gut and in the skewed replication compared with CXCR4-using viruses (Arthos *et al.*, 2008).

Another important aspect of the immunopathogenesis of HIV infection is the role of 'resting' $CD4^+$ T-cells that are either uninfected or latently infected (Finzi *et al.*, 1999). *In vitro* studies demonstrated that resting $CD4^+$ T-cells can be infected but they support HIV replication only to a limited extent due to a lack of dideoxynucleotides and other fundamental co-factors that are required for completion of the reverse transcription process that precedes the integration of proviral DNA into host chromosomes (Lori *et al.*, 1994). Nonetheless, *in vivo* studies in macaques experimentally infected with SIV by non-traumatic exposure of the virus in the genital mucosa, as well as studies of infected humans, have suggested that $CD4^+$ T-cells in a resting state are the first cell type to be productively infected (Zhang *et al.*, 1999). Activated $CD4^+$ T-cells and macrophages are infected several days later and amplify the infectious process, since activated T-cells produce significantly more virus on a per cell basis than resting T-cells. However, activated T-cells survive for a much shorter time than resting T-cells (Zhang *et al.*, 1999). If these studies are confirmed, they will bear profound implications for identification of host factors that enable replication and for the development of novel antiviral agents.

Infected $CD4^+$ T-lymphocytes with a 'resting memory' phenotype after an immunological encounter with the virus also play an important role in the pathogenesis of HIV infection (Chun *et al.*, 2000; Chun *et al.*, 2001; Finzi *et al.*, 1997). The initial euphoria following the success of the first HAART trials was significantly dampened by the observation that suspension of very long duration antiviral regimens almost always leads to resurgence of virus replication and $CD4^+$ T-cell depletion to pre-therapy levels. Even therapy interruption protocols that are specifically designed to boost the host's immune response have been essentially abandoned (Lori *et al.*, 2002) (although salvage therapy for patients infected with multi-resistant viruses remains a therapeutic option). It appears that a small number of latently infected $CD4^+$ T-cells can quickly reignite the infection *in vivo* after therapy suspension. In addition, other cell types such as monocytes, macrophages and astrocytes in the central nervous system (CNS) are chronically infected *in vivo* and may contribute to reigniting the infection (Brack-Werner, 1999; Fulcher *et al.*, 2004).

Both the host immune response and cytokines help to control the balance between latent and productive HIV infection. In general, *pro*-inflammatory cytokines play an inductive role in virus replication by activating host transcription factors such as NF-kB and AP-1 that can directly bind to the HIV LTR, thereby triggering or potentiating virus replication (Alfano *et al.*, 2008); AP-1 has also been shown to bind to a functional intragenic enhancer (Verdin *et al.*, 1990). Other cytokine-regulated host transcription factors affecting HIV transcription and expression in T-cells and/or macrophages include NFAT, IRF3 and STAT5 (Alfano *et al.*, 2008; Williams and Greene, 2007).

Several infected individuals show evidence of a cleavage of STAT5 at the C-terminus, which generates a dominant STAT5Δ isoform that can bind to one or two sites in the HIV LTR (in proximity of the NF-kB binding sites) but acts as a repressor rather than an activator of virus expression (Crotti *et al.*, 2007). Furthermore, cytokines and IFNs control HIV replication at post-transcriptional levels and may therefore participate in the regulation of the final phases of viral assembly and release from the plasma membrane, as shown in the case of IFN-α/β (Poli *et al.*, 1989), recently linked to tetherin (McNatt *et al.*, 2009). In macrophages this event is also regulated by the accumulation of virions in intracellular vacuoles, although the nature of this process is still controversial. Cytokines such as IFN-γ and chemokines like CCL2 have the capacity to increase the intracellular accumulation of virions in macrophages *in vitro* (Biswas *et al.*, 1992; Fantuzzi *et al.*, 2003), a feature that has also been observed *in vivo* in the brains of individuals with AIDS dementia complex (Koenig *et al.*, 1986; Orenstein and Jannotta, 1988).

Neuropathogenesis of HIV and SIV infections

A histological hallmark of HIV-1 infection of the CNS is the presence of accumulations of microglial cells and, occasionally, giant cells located close to blood vessels. These pathologies occur mostly in late stages of the HIV-1 infection. That these infected microglial nodules replicate the virus has been shown using immunohistological staining with antibodies against structural proteins such as Gag (Budka, 1991; Lawrence and Major, 2002). The susceptibility of human microglial cells to HIV-1 infection was also confirmed *in vitro*. Microglial cells are known to express CD4 and CCR5.

Infected macrophages may also be present in the microglial nodules, and it is possible that the infected monocytes transport the virus through the blood–brain barrier, from the peripheral blood stream to the CNS. This view of CNS infection dynamics is supported by experiments in SIV infected rhesus monkeys, in which SIV Gag-expressing macrophages were detected in the extracellular space between blood vessels and the surrounding astrocytic layer (Patrick *et al.*, 2002). The typical microglial nodules were also present in brains of SIV infected monkeys.

Astrocytes, the most numerous of all CNS cell types, normally function as neuronal supporters and guardians. Their possible role as major targets of HIV-1 infection was first suspected from *in vitro* experiments, which showed a limited susceptibility of this cell type to HIV-1, and also a very low replication of the virus. Further detailed *in vitro* analysis of the infection and replication dynamics revealed that HIV-1 infected astrocytes show a large presence of ribonucleic acid (RNA) for multiple spliced messages and only few RNA molecules for single spliced and unspliced messages, which indicates a high expression of regulatory proteins (predominantly Nef) and low expression of structural proteins. Virus progeny were rarely detected in the supernatants of these cultures, but infectious virus was rescued from the HIV harbouring astrocytes by co-cultivation with macrophages. The reason for this phenotype is an astrocyte-specific interference with the HIV Rev function (Brack-Werner, 1999).

Furthermore, evaluation of brain sections revealed that astrocytes are a major target cell for HIV-1 infection *in vivo* (Bagasra *et al.*, 1996). This was demonstrated using immunostaining with Nef specific antibodies, in situ hybridization with probes for multiple spliced messages, and polymerase chain reaction (PCR) for integrated proviruses. It is noteworthy that the number of HIV-1 infected cells in the brain of a patient in an advanced state of infection is of the same magnitude as the number of infected cells in the haematopoietic system. Studies in macaques showed that Nef and Rev expressing astrocytes were present within the first three weeks after primary viral infection (Guillemin *et al.*, 2000). The infection of neurons *in vitro* is anecdotal and requires additional *in vivo* proof.

The facts described above regarding the HIV and SIV infection of the CNS may indicate that the CNS infection follows a biphasic pattern. First, shortly after the individual is infected, the virus is transferred via infected monocytes through the blood–brain barrier, thereby enabling infection of astrocytes, which may lead to astrogliosis and early signs of encephalitis. Second, the infected blood–brain barrier becomes leaky and allows CNS invasion by immune cells that activate and infect microglial cells, resulting in the clinical pathology that is typical of AIDS associated neuropathy.

Thus infected astrocytes and microglial cells are likely sources for virus-mediated effectors in AIDS related neuropathy. The Nef protein is the major HIV component expressed in infected astrocytes. Nef overexpression in human astrocytes can be easily achieved, and leads to characteristic signs of both astrogliosis (Kohleisen *et al.*, 1999) and apoptosis of astrocytes, neurons and brain microvascular endothelial cells (Van Marle *et al.*, 2004; Acheampong *et al.*, 2005a). Nef protein overexpression modulates the sphingomyelinase signalling pathway triggered by tumour necrosis factor α (TNF-α), thus leading to important modifications in the activation and proliferation of glial cells (Richard *et al.*, 1997), which may explain the widespread reactive astrogliosis observed in AIDS associated neuropathy. Furthermore, Nef protein induces blood–brain barrier disruption in the rat via activation of matrix metalloproteinase-9, one of several factors that has been implicated in lentivirus induced brain disease (Sporer *et al.*, 2000; Johnston *et al.*, 2002). In addition, Tat is expressed in astrocytes with low HIV levels and is reported to have numerous effects on the function and viability of neuronal cells. These effects include induction of apoptosis, discrete changes in intracellular signalling, and alteration of the blood–brain barrier (Brack-Werner, 1999; Patrick *et al.*, 2002).

Microglial cells actively replicate immunodeficiency viruses, giving rise to extracellular Env glycoprotein that is regarded as a major factor in neuropathogenesis. Transgenic mice expressing gp120 Env manifest neuropathological features resembling the neuropathology in brains of AIDS patients (Toggas *et al.*, 1994).

Thus, both *in vitro* and *in vivo*, HIV-1 gp120 Env produces injury or apoptosis in both rodent and human neurons. This effect depends predominantly on activation of chemokine receptors on microglial cells or macrophages, but may also be induced by a direct binding of gp120 Env to CXCR4 and CCR5 receptors present on human neurons upon interaction *in trans* with CD4 (Catani *et al.*, 2000; Kaul and Lipton, 1999; Kaul and Lipton, 2004). The role of active virus production in cells of the microglial/macrophage lineage as cause for late AIDS dementia is also supported by the significant reduction of HIV neuropathy under HAART.

Conclusion

Prior to the AIDS epidemic it was assumed that severely pathogenic human retroviruses would be easily eliminated by the hosts' immune systems via mechanisms such as complement-mediated lysis of virions or of infected cells. Therefore, the emergence of HIV at pandemic proportions was not anticipated by most scientists. This assumption has of course been proven wrong. The global challenge of the HIV/AIDS epidemic, as well as minor outbreaks due to other pathogenic retroviruses such as the first exogenous human T-lymphotropic pathogenic retrovirus (HTLV-1) (Suzuki and Gojobori, 1998), severe acute respiratory syndrome (SARS) (Vicenzi *et al.*, 2004), or deadly avian flu viruses (e.g. H5N1), emphasizes the importance of continuing and expanding research on all potential pathogenic microbes, including even those that might at first appear to be irrelevant.

Acknowledgements

The authors wish to thank their faithful friends and collaborators for intellectual input during many years of professional research on virus–cell interaction. This work was supported in part by a grant of the VI National Program for research against AIDS of the Istituto Superiore di Sanità, Rome, Italy (to G.P.), and the Deutsche Forschungsgemeinschaft (SFB Programme), as well as the Bundesforschungsministerium für Bildung und Forschung (AIDS Programme) (to V.E.).

References

Acheampong, E.A., Parveen, Z., Muthoga, L.W., Kalayeh, M., Mukhtar, M., and Pomerantz, R.J. (2005a). Human immunodeficiency virus type 1 Nef potentially induces apoptosis in primary human brain microvascular endothelial cells via the activation of caspases. J. Virol. 79, 4257–4269.

Acheampong, E.A., Parveen, Z., Muthoga, L.W., Wasmuth-Peroud, V., Kalayeh, M., Bashir, A., Diecidue, R., Mukhtar, M., and Pomerantz, R J. (2005b). Molecular interactions of human immunodeficiency virus type 1 with primary human oral keratinocytes. J. Virol. 79, 8440–8453.

Albini, A., Ferrini, S., Benelli, R., Sforzini, S., Giunciuglio, D., Aluigi, M. G., Proudfoot, A.E., Alouani, S., Wells, T.N., Mariani, G., Rabin, R.L., Farber, J.M., and Noonan, D.M. (1998). HIV-1 Tat mimicry of chemokines. Proc. Natl. Acad. Sci. U.S.A. 95, 13153 –13158.

Alfano, M., Crotti, A., Vicenzi, E., and Poli, G. (2008). New players in cytokine control of HIV infection. Curr. HIV/AIDS Rep. 5, 27–32.

Ammosova, T., Berro, R., Jerebtsova, M., Jackson, A., Charles, S., Klase, Z., Southerland, W., Gordeuk, V. R., Kashanchi, F., and Nekhai, S. (2006). Phosphorylation of HIV-1 Tat by CDK2 in HIV-1 transcription. Retrovirology 3, 78.

An, H., Hou, J., Zhou, J., Zhao, W., Xu, H., Zheng, Y., Yu, Y., Liu, S., and Cao, X. (2008). Phosphatase SHP-1 promotes TLR- and RIG-I-activated production of type I interferon by inhibiting the kinase IRAK1. Nat. Immunol. *9*, 542–550.

Andersen, J.L., Le Rouzic, E., and Planelles, V. (2008). HIV-1 Vpr: mechanisms of G2 arrest and apoptosis. Exp. Mol. Pathol. *85*, 2–10.

Apetrei, C., Robertson, D.L., and Marx, P.A. (2004). The history of SIVS and AIDS: epidemiology, phylogeny and biology of isolates from naturally SIV infected non-human primates (NHP) in Africa. Front. Biosci. *9*, 225–254.

Arthos, J., Cicala, C., Martinelli, E., Macleod, K., Van Ryk, D., Wei, D., Xiao, Z., Veenstra, T. D., Conrad, T. P., Lempicki, R.A., *et al.* (2008). HIV-1 envelope protein binds to and signals through integrin alpha4beta7, the gut mucosal homing receptor for peripheral T-cells. Nat. Immunol. *9*, 301–309.

Ayyavoo, V., Mahboubi, A., Mahalingam, S., Ramalingam, R., Kudchodkar, S., Williams, W.V., Green, D.R., and Weiner, D.B. (1997). HIV-1 Vpr suppresses immune activation and apoptosis through regulation of nuclear factor kappa B [see comments]. Nat. Med. *3*, 1117–1123.

Bagasra, O., Lavi, E., Bobroski, L., Khalili, K., Pestaner, J.P., Tawadros, R., and Pomerantz, R.J. (1996). Cellular reservoirs of HIV-1 in the central nervous system of infected individuals: identification by the combination of in situ polymerase chain reaction and immunohistochemistry. Aids *10*, 573–585.

Barouch, D.H., Santra, S., Schmitz, J.E., Kuroda, M.J., Fu, T.M., Wagner, W., Bilska, M., Craiu, A., Zheng, X.X., Krivulka, G.R., *et al.* (2000). Control of viremia and prevention of clinical AIDS in rhesus monkeys by cytokine-augmented DNA vaccination. Science *290*, 486–492.

Beer, B., Denner, J., Brown, C. R., Norley, S., zur Megede, J., Coulibaly, C., Plesker, R., Holzammer, S., Baier, M., Hirsch, V. M., and Kurth, R. (1998). Simian immunodeficiency virus of African green monkeys is apathogenic in the newborn natural host. J. Acquir. Immune Defic. Syndr. Hum. Retroviruses *18*, 210–220.

Beer, B.E., Bailes, E., Goeken, R., Dapolito, G., Coulibaly, C., Norley, S.G., Kurth, R., Gautier, J.P., Gautier-Hion, A., Vallet, D., *et al.* (1999). Simian immunodeficiency virus (SIV) from sun-tailed monkeys (*Cercopithecus solatus*): evidence for host-dependent evolution of SIV within the *C. lhoesti* superspecies. J. Virol. *73*, 7734–7744.

Bendinelli, M., Pistello, M., Lombardi, S., Poli, A., Garzelli, C., Matteucci, D., Ceccherini-Nelli, L., Malvaldi, G., and Tozzini, F. (1995). Feline immunodeficiency virus: an interesting model for AIDS studies and an important cat pathogen. Clin. Microbiol. Rev. *8*, 87–112.

Berger, E.A., Doms, R.W., Fenyo, E.M., Korber, B.T., Littman, D.R., Moore, J.P., Sattentau, Q.J., Schuitemaker, H., Sodroski, J., and Weiss, R.A. (1998). A new classification for HIV-1 [letter]. Nature *391*, 240.

Bettelli, E., Korn, T., Oukka, M., and Kuchroo, V.K. (2008). Induction and effector functions of T(H)17 cells. Nature *453*, 1051–1057.

Biswas, P., Poli, G., Kinter, A.L., Justement, J.S., Stanley, S.K., Maury, W.J., Bressler, P., Orenstein, J.M., and Fauci, A.S. (1992). Interferon gamma induces the expression of human immunodeficiency virus in persistently infected promonocytic cells (U1) and redirects the production of virions to intracytoplasmic vacuoles in phorbol myristate acetate-differentiated U1 cells. J. Exp. Med. *176*, 739–750.

Brack-Werner, R. (1999). Astrocytes: HIV cellular reservoirs and important participants in neuropathogenesis [editorial]. Aids *13*, 1–22.

Brambilla, A., Turchetto, L., Gatti, A., Bovolenta, C., Veglia, F., Santagostino, E., Gringeri, A., Clementi, M., Poli, G., Bagnarelli, P., and Vicenzi, E. (1999). Defective *nef* alleles in a cohort of hemophiliacs with progressing and nonprogressing HIV-1 infection. Virology *259*, 349–368.

Brenchley, J.M., Schacker, T.W., Ruff, L.E., Price, D.A., Taylor, J.H., Beilman, G.J., Nguyen, P.L., Khoruts, A., Larson, M., Haase, A.T., and Douek, D.C. (2004). $CD4^+$ T-cell depletion during all stages of HIV disease occurs predominantly in the gastrointestinal tract. J. Exp. Med. *200*, 749–759.

Budka, H. (1991). Neuropathology of human immunodeficiency virus infection. Brain Pathol. *1*, 163–175.

Burton, G.F., Masuda, A., Heath, S.L., Smith, B.A., Tew, J.G., and Szakal, A.K. (1997). Follicular dendritic cells (FDC) in retroviral infection: host/pathogen perspectives. Immunol. Rev. *156*, 185–197.

Camarasa, M.J., Velazquez, S., San-Felix, A., Perez-Perez, M.J., and Gago, F. (2006). Dimerization inhibitors of HIV-1 reverse transcriptase, protease and integrase: a single mode of inhibition for the three HIV enzymes? Antiviral Res. *71*, 260–267.

Cassol E., Cassetta, L., Rizzi, C., Alfano, M., and Poli, G. (2009). M1 and M2a polarization of human monocyte-derived macrophages inhibits HIV-1 replication by distinct mechanisms. J. Immunol. *182*, 6237–6246.

Catani, M.V., Corasaniti, M.T., Navarra, M., Nistico, G., Finazzi-Agro, A., and Melino, G. (2000). gp120 induces cell death in human neuroblastoma cells through the CXCR4 and CCR5 chemokine receptors. J. Neurochem. *74*, 2373–2379.

Chadwick, B.J., Coelen, R. J., Sammels, L.M., Kertayadnya, G., and Wilcox, G.E. (1995a). Genomic sequence analysis identifies Jembrana disease virus as a new bovine lentivirus. J. Gen. Virol. *76*, 189–192.

Chadwick, B.J., Coelen, R.J., Wilcox, G. E., Sammels, L.M., and Kertayadnya, G. (1995b). Nucleotide sequence analysis of Jembrana disease virus: a bovine lentivirus associated with an acute disease syndrome. J. Gen. Virol. *76*, 1637–1650.

Chun, T.W., Davey, R.T., Jr., Ostrowski, M., Shawn Justement, J., Engel, D., Mullins, J.I., and Fauci, A.S. (2000). Relationship between pre-existing viral reservoirs and the re-emergence of plasma viremia after discontinuation of highly active anti-retroviral therapy. Nat. Med. *6*, 757–761.

Chun, T.W., Justement, J.S., Moir, S., Hallahan, C.W., Ehler, L.A., Liu, S., McLaughlin, M., Dybul, M., Mican, J. M., and Fauci, A.S. (2001). Suppression of HIV replication in the resting CD4$^+$ T-cell reservoir by autologous CD8$^+$ T-cells: implications for the development of therapeutic strategies. Proc. Natl. Acad. Sci. U.S.A. *98*, 253–258.

Crotti, A., Lusic, M., Lupo, R., Lievens, P.M., Liboi, E., Chiara, G.D., Tinelli, M., Lazzarin, A., Patterson, B.K., Giacca, M., *et al.* (2007). Naturally occurring C-terminally truncated STAT5 is a negative regulator of HIV-1 expression. Blood *109*, 5380–5389.

D'Ambrosio, D., Iellem, A., Colantonio, L., Clissi, B., Pardi, R., and Sinigaglia, F. (2000). Localization of Th-cell subsets in inflammation: differential thresholds for extravasation of Th1 and Th2 cells. Immunol Today *21*, 183–186.

Daniel, M. D., Kirchhoff, F., Czajak, S.C., Sehgal, P. K., and Desrosiers, R.C. (1992). Protective effects of a live attenuated SIV vaccine with a deletion in the *nef* gene. Science *258*, 1938–1941.

DeHart, J.L., Bosque, A., Harris, R., and Planelles, V. (2008). Human immunodeficiency virus Type 1 Vif induces cell cycle delay via recruitment of the same E3 ubiquitin ligase complex that targets APOBEC3 proteins for degradation. J. Virol. *82*, 9265–9272.

DeHart, J.L., and Planelles, V. (2008). Human immunodeficiency virus type 1 Vpr links proteasomal degradation and checkpoint activation. J. Virol. *82*, 1066–1072.

Ehret, A., Westendorp, M.O., Herr, I., Debatin, K.M., Heeney, J.L., Frank, R., and Krammer, P.H. (1996). Resistance of chimpanzee T-cells to human immunodeficiency virus type 1 Tat-enhanced oxidative stress and apoptosis. J. Virol. *70*, 6502–6507.

Emilie, D., Peuchmaur, M., Maillot, M.C., Crevon, M. C., Brousse, N., Delfraissy, J.F., Dormont, J., and Galanaud, P. (1990). Production of interleukins in human immunodeficiency virus-1- replicating lymph nodes. J. Clin. Invest. *86*, 148–159.

Ensoli, B., Buonaguro, L., Barillari, G., Fiorelli, V., Gendelman, R., Morgan, R.A., Wingfield, P., and Gallo, R.C. (1993). Release, uptake, and effects of extracellular human immunodeficiency virus type 1 Tat protein on cell growth and viral transactivation. J. Virol. *67*, 277–287.

Fantuzzi, L., Spadaro, F., Vallanti, G., Canini, I., Ramoni, C., Vicenzi, E., Belardelli, F., Poli, G., and Gessani, S. (2003). Endogenous CCL2 (monocyte chemotactic protein-1) modulates human immunodeficiency virus type-1 replication and affects cytoskeleton organization in human monocyte-derived macrophages. Blood. *102*, 2334-2337.

Fauci, A.S. (1996). Host factors and the pathogenesis of HIV-induced disease. Nature *384*, 529–534.

Finkel, T.H., Tudor-Williams, G., Banda, N.K., Cotton, M.F., Curiel, T., Monks, C., Baba, T. W., Ruprecht, R.M., and Kupfer, A. (1995). Apoptosis occurs predominantly in bystander cells and not in productively infected cells of HIV- and SIV-infected lymph nodes. Nat. Med. *1*, 129–134.

Finzi, D., Blankson, J., Siliciano, J.D., Margolick, J.B., Chadwick, K., Pierson, T., Smith, K., Lisziewicz, J., Lori, F., Flexner, C., *et al.* (1999). Latent infection of CD4$^+$ T-cells provides a mechanism for lifelong persistence of HIV-1, even in patients on effective combination therapy. Nat. Med. *5*, 512–517.

Finzi, D., Hermankova, M., Pierson, T., Carruth, L.M., Buck, C., Chaisson, R.E., Quinn, T.C., Chadwick, K., Margolick, J., Brookmeyer, R., *et al.* (1997). Identification of a reservoir for HIV-1 in patients on highly active antiretroviral therapy. Science *278*, 1295–1300.

Foster, J.L., and Garcia, J.V. (2007). Role of Nef in HIV-1 replication and pathogenesis. Adv. Pharmacol. *55*, 389–409.

Frankel, A.D., and Pabo, C.O. (1988). Cellular uptake of the *tat* protein from human immunodeficiency virus. Cell *55*, 1189–1193.

Freed, E.O. (2003). The HIV–TSG101 interface: recent advances in a budding field. Trends Microbiol. *11*, 56–59.

Fujii, K., Hurley, J.H., and Freed, E.O. (2007). Beyond Tsg101: the role of Alix in 'ESCRTing' HIV-1. Nat. Rev. Microbiol. *5*, 912–916.

Fulcher, J.A., Hwangbo, Y., Zioni, R., Nickle, D., Lin, X., Heath, L., Mullins, J. I., Corey, L., and Zhu, T. (2004). Compartmentalization of human immunodeficiency virus type 1 between blood monocytes and CD4$^+$ T-cells during infection. J. Virol. *78*, 7883–7893.

Fultz, P.N. (1994). SIVsmmPBj14: an atypical lentivirus. Curr. Top. Microbiol. Immunol. *188*, 65–76.

Galli, S.J., Grimbaldeston, M., and Tsai, M. (2008a). Immunomodulatory mast cells: negative, as well as positive, regulators of immunity. Nat. Rev. Immunol. *8*, 478–486.

Galli, S. J., Tsai, M., and Piliponsky, A. M. (2008b). The development of allergic inflammation. Nature *454*, 445–454.

Ganser-Pornillos, B. K., Yeager, M., and Sundquist, W. I. (2008). The structural biology of HIV assembly. Curr. Opin. Struct. Biol. *18*, 203–217.

Gatignol, A. (2007). Transcription of HIV: Tat and cellular chromatin. Adv. Pharmacol. *55*, 137–159.

Gilliet, M., Cao, W., and Liu, Y.J. (2008). Plasmacytoid dendritic cells: sensing nucleic acids in viral infection and autoimmune diseases. Nat. Rev. Immunol. *8*, 594–606.

Goila-Gaur, R., and Strebel, K. (2008). HIV-1 Vif, APOBEC, and intrinsic immunity. Retrovirology *5*, 51.

Greenough, T.C., Sullivan, J.L., and Desrosiers, R.C. (1999). Declining CD4 T-cell counts in a person infected with nef-deleted HIV-1. N. Engl. J. Med. *340*, 236–237.

Guidotti, L.G., and Chisari, F.V. (2001). Noncytolytic control of viral infections by the innate and adaptive immune response. Annu. Rev. Immunol. *19*, 65–91.

Guillemin, G., Croitoru, J., Le Grand, R.L., Franck-Duchenne, M., Dormont, D., and Boussin, F.D. (2000). Simian immunodeficiency virus mac251 infection of astrocytes. J. Neurovirol. *6*, 173–186.

Hahn, B.H., Shaw, G.M., De Cock, K.M., and Sharp, P.M. (2000). AIDS as a zoonosis: scientific and public health implications. Science *287*, 607–614.

Harari, A., Rizzardi, G.P., Ellefsen, K., Ciuffreda, D., Champagne, P., Bart, P.A., Kaufmann, D., Telenti, A., Sahli, R., Tambussi, G., *et al.* (2002). Analysis of HIV-1- and CMV-specific memory CD4 T-cell responses during primary and chronic infection. Blood *100*, 1381–1387.

Hearps, A.C., and Jans, D.A. (2007). Regulating the functions of the HIV-1 matrix protein. AIDS Res. Hum. Retroviruses 23, 341–346.

Hofmann-Lehmann, R., Vlasak, J., Williams, A.L., Chenine, A. L., McClure, H.M., Anderson, D.C., O'Neil, S., and Ruprecht, R.M. (2003). Live attenuated, nef-deleted SIV is pathogenic in most adult macaques after prolonged observation. Aids *17*, 157–166.

Holzammer, S., Holznagel, E., Kaul, A., Kurth, R., and Norley, S. (2001). High virus loads in naturally and experimentally SIVagm-infected African green monkeys. Virology *283*, 324–331.

Huang, L.M., Chao, M.F., Chen, M.Y., Shih, H., Chiang, Y.P., Chuang, C.Y., and Lee, C.Y. (2001). Reciprocal regulatory interaction between human herpesvirus 8 and human immunodeficiency virus type 1. J. Biol. Chem. *276*, 13427–13432.

Johnston, J.B., Silva, C., and Power, C. (2002). Envelope gene-mediated neurovirulence in feline immunodeficiency virus infection: induction of matrix metalloproteinases and neuronal injury. J. Virol. *76*, 2622–2633.

Kaul, M., and Lipton, S.A. (1999). Chemokines and activated macrophages in HIV gp120-induced neuronal apoptosis. Proc. Natl. Acad. Sci. U.S.A. *96*, 8212–8216.

Kaul, M., and Lipton, S.A. (2004). Signaling pathways to neuronal damage and apoptosis in human immunodeficiency virus type 1-associated dementia: Chemokine receptors, excitotoxicity, and beyond. J. Neurovirol .*10* Suppl. *1*, 97–101.

Keele B.F., Jones, J.H., Terio, K.A., Estes, J.D., Rudicell, R.S., Wilson, M.L., Li, Y., Learn, G.H., Beasley, T.M., Schumacher-Stankey J., *et al.* (2009). Increased mortality and AIDS-like immunopathology in wild chimpanzees infected with SIVcpz. Nature *460*, 515–519.

Kestler, H.W., Ringler, D.J., Mori, K., Panicali, D.L., Sehgal, P. K., Daniel, M.D., and Desrosiers, R.C. (1991). Importance of the *nef* gene for maintenance of high virus loads and for development of AIDS. Cell *65*, 651–662.

Kirchhoff, F., Schindler, M., Specht, A., Arhel, N., and Munch, J. (2008). Role of Nef in primate lentiviral immunopathogenesis. Cell. Mol. Life Sci. *65*, 2621–2636.

Kirchhoff, F., and Silvestri, G. (2008). Is Nef the elusive cause of HIV-associated hematopoietic dysfunction? J. Clin. Invest. *118*, 1622–1625.

Koedel, U., Kohleisen, B., Sporer, B., Lahrtz, F., Ovod, V., Fontana, A., Erfle, V., and Pfister, H.W. (1999). HIV type 1 Nef protein is a viral factor for leukocyte recruitment into the central nervous system. J. Immunol. *163*, 1237–1245.

Koenig, S., Gendelman, H.E., Orenstein, J.M., Dal Canto, M.C., Pezeshkpour, G.H., Yungbluth, M., Janotta, F., Aksamit, A., Martin, M.A., and Fauci, A.S. (1986). Detection of AIDS virus in macrophages in brain tissue from AIDS patients with encephalopathy. Science *223*, 1089–1093.

Kohleisen, B., Shumay, E., Sutter, G., Foerster, R., Brack-Werner, R., Nuesse, M., and Erfle, V. (1999). Stable expression of HIV-1 Nef induces changes in growth properties and activation state of human astrocytes. Aids *13*, 2331–2341.

Koot, M., Keet, I.P., Vos, A.H., de Goede, R.E., Roos, M.T., Coutinho, R.A., Miedema, F., Schellekens, P.T., and Tersmette, M. (1993). Prognostic value of HIV-1 syncytium-inducing phenotype for rate of $CD4^+$ cell depletion and progression to AIDS. Ann. Intern. Med. *118*, 681–688.

Koot, M., van Leeuwen, R., de Goede, R.E., Keet, I.P., Danner, S., Eeftinck Schattenkerk, J.K., Reiss, P., Tersmette, M., Lange, J.M., and Schuitemaker, H. (1999). Conversion rate towards a syncytium-inducing (SI) phenotype during different stages of human immunodeficiency virus type 1 infection and prognostic value of SI phenotype for survival after AIDS diagnosis. J. Infect. Dis. *179*, 254–258.

Korn T., Bettelli, E., Oukka, M., and Kuchroo, V.K. (2009). IL-17 and Th17 Cells. Annu. Rev. Immunol. *27*, 485–517.

Lawrence, D.M., and Major, E.O. (2002). HIV-1 and the brain: connections between HIV-1-associated dementia, neuropathology and neuroimmunology. Microbes Infect. *4*, 301–308.

Lehmann, M.H., Walter, S., Ylisastigui, L., Striebel, F., Ovod, V., Geyer, M., Gluckman, J.C., and Erfle, V. (2006). Extracellular HIV-1 Nef increases migration of monocytes. Exp Cell Res *312*, 3659–3668.

Li, Q., Duan, L., Estes, J.D., Ma, Z.M., Rourke, T., Wang, Y., Reilly, C., Carlis, J., Miller, C.J., and Haase, A.T. (2005). Peak SIV replication in resting memory $CD4^+$ T-cells depletes gut lamina propria $CD4^+$ T-cells. Nature *434*, 1148–1152.

Ling, B., Apetrei, C., Pandrea, I., Veazey, R.S., Lackner, A.A., Gormus, B., and Marx, P.A. (2004). Classic AIDS in a sooty mangabey after an 18-year natural infection. J. Virol. *78*, 8902–8908.

Liu, Y., McNevin, J., Cao, J., Zhao, H., Genowati, I., Wong, K., McLaughlin, S., McSweyn, M.D., Diem, K., Stevens, C.E., *et al.* (2006). Selection on the human immunodeficiency virus type 1 proteome following primary infection. J. Virol. *80*, 9519–9529.

Lori, F., Foli, A., and Lisziewicz, J. (2002). Structured treatment interruptions as a potential alternative therapeutic regimen for HIV-infected patients: a review of recent clinical data and future prospects. J. Antimicrob. Chemother. *50*, 155–160.

Lori, F., Malykh, A., Cara, A., Sun, D., Weinstein, J.N., Lisziewicz, J., and Gallo, R.C. (1994). Hydroxyurea as an inhibitor of human immunodeficiency virus-type 1 replication. Science *266*, 801–805.

McNatt M.W., Zang, T., Hatziioannou, T., Bartlett, M., Fofana, I.B., Johnson, W.E., Neil, S.J., and Bieniasz, P.D. (2009). Species-specific activity of HIV-1 Vpu and positive selection of tetherin transmembrane domain variants. PLoS Pathog. *5*, e1000300.

Mattapallil, J.J., Douek, D. C., Hill, B., Nishimura, Y., Martin, M., and Roederer, M. (2005). Massive infection and loss of memory $CD4^+$ T-cells in multiple tissues during acute SIV infection. Nature *434*, 1093–1097.

Miller, R.J., Cairns, J.S., Bridges, S., and Sarver, N. (2000). Human immunodeficiency virus and AIDS: insights from animal lentiviruses. J. Virol. *74*, 7187–7195.

Navis, M., Schellens, I.M., van Swieten, P., Borghans, J.A., Miedema, F., Kootstra, N.A., van Baarle, D., and Schuitemaker, H. (2008). A nonprogressive clinical course in HIV-infected individuals expressing human leukocyte antigen B57/5801 is associated with preserved $CD8^+$ T-lymphocyte responsiveness to the HW9 epitope in Nef. J. Infect. Dis. *197*, 871–879.

Noonan, D., and Albini, A. (2000). From the outside in: extracellular activities of HIV Tat. Adv. Pharmacol. *48*, 229–250.

Norley, S., Beer, B., Holzammer, S., zur Megede, J., and Kurth, R. (1999). Why are the natural hosts of SIV resistant to AIDS? Immunol Lett. *66*, 47–52.

Orenstein, J.M., and Jannotta, F. (1988). Human immunodeficiency virus and papovavirus infections in acquired immunodeficiency syndrome: an ultrastructural study of three cases. Hum. Pathol. *19*, 350–361.

Pandrea, I., Sodora, D.L., Silvestri, G., and Apetrei, C. (2008). Into the wild: simian immunodeficiency virus (SIV) infection in natural hosts. Trends Immunol. *29*, 419–428.

Pantaleo, G., Cohen, O.J., Schacker, T., Vaccarezza, M., Graziosi, C., Rizzardi, G.P., Kahn, J., Fox, C.H., Schnittman, S.M., Schwartz, D.H., *et al.* (1998). Evolutionary pattern of human immunodeficiency virus (HIV) replication and distribution in lymph nodes following primary infection: implications for antiviral therapy. Nat. Med. *4*, 341–345.

Patrick, M.K., Johnston, J.B., and Power, C. (2002). Lentiviral neuropathogenesis: comparative neuroinvasion, neurotropism, neurovirulence, and host neurosusceptibility. J. Virology *76*, 7923–7931.

Phogat, S., and Wyatt, R. (2007). Rational modifications of HIV-1 envelope glycoproteins for immunogen design. Curr. Pharm. Des. *13*, 213–227.

Poli, G., Orenstein, J. M., Kinter, A., Folks, T. M., and Fauci, A. S. (1989). Interferon-alpha but not AZT suppresses HIV expression in chronically infected cell lines. Science *244*, 575–577.

Polonis, V.R., Brown, B.K., Rosa Borges, A., Zolla-Pazner, S., Dimitrov, D. S., Zhang, M.Y., Barnett, S.W., Ruprecht, R.M., Scarlatti, G., Fenyo, E.M., *et al.* (2008). Recent advances in the characterization of HIV-1 neutralization assays for standardized evaluation of the antibody response to infection and vaccination. Virology *375*, 315–320.

Priceputu, E., Hanna, Z., Hu, C., Simard, M. C., Vincent, P., Wildum, S., Schindler, M., Kirchhoff, F., and Jolicoeur, P. (2007). Primary human immunodeficiency virus type 1 *nef* alleles show major differences in pathogenicity in transgenic mice. J. Virol. *81*, 4677–4693.

Priceputu, E., Rodrigue, I., Chrobak, P., Poudrier, J., Mak, T.W., Hanna, Z., Hu, C., Kay, D.G., and Jolicoeur, P. (2005). The Nef-mediated AIDS-like disease of CD4C/human immunodeficiency virus transgenic mice is associated with increased Fas/FasL expression on T-cells and T-cell death but is not prevented in Fas-, FasL-, tumor necrosis factor receptor 1-, or interleukin-1beta-converting enzyme-deficient or Bcl2-expressing transgenic mice. J. Virol. *79*, 6377–6391.

Rhodes, D.I., Ashton, L., Solomon, A., Carr, A., Cooper, D., Kaldor, J., and Deacon, N. (2000). Characterization of three nef-defective human immunodeficiency virus type 1 strains associated with long-term nonprogression. Australian Long-Term Nonprogressor Study Group. J. Virol. *74*, 10581–10588.

Richard, A., Robichaud, G., Lapointe, R., Bourgoin, S., Darveau, A., and Poulin, L. (1997). Interference of HIV-1 Nef in the sphingomyelin transduction pathway activated by tumour necrosis factor-alpha in human glial cells. Aids *11*, F1–7.

Rosenberg, E. S., Altfeld, M., Poon, S. H., Phillips, M. N., Wilkes, B. M., Eldridge, R.L., Robbins, G.K., D'Aquila, R.T., Goulder, P.J., and Walker, B.D. (2000). Immune control of HIV-1 after early treatment of acute infection. Nature *407*, 523–526.

Rossi, J. J., June, C.H., and Kohn, D.B. (2007). Genetic therapies against HIV. Nat. Biotechnol. *25*, 1444–1454.

Roux, K.H., and Taylor, K. A. (2007). AIDS virus envelope spike structure. Curr. Opin. Struct. Biol. *17*, 244–252.

Saksena, N.K., Rodes, B., Wang, B., and Soriano, V. (2007). Elite HIV controllers: myth or reality? AIDS Rev. *9*, 195–207.

Schellens, I.M., Borghans, J.A., Jansen, C.A., De Cuyper, I.M., Geskus, R.B., van Baarle, D., and Miedema, F. (2008). Abundance of early functional HIV-specific $CD8^+$ T-cells does not predict AIDS-free survival time. PLoS ONE 3, e2745.

Schoenborn, J. R., and Wilson, C.B. (2007). Regulation of interferon-gamma during innate and adaptive immune responses. Adv. Immunol. *96*, 41–101.

Sharp, P.M., Bailes, E., Robertson, D.L., Gao, F., and Hahn, B.H. (1999). Origins and evolution of AIDS viruses. Biol. Bull. *196*, 338–342.

Sirois, S., Touaibia, M., Chou, K.C., and Roy, R. (2007). Glycosylation of HIV-1 gp120 V3 loop: towards the rational design of a synthetic carbohydrate vaccine. Curr. Med. Chem. *14*, 3232–3242.

Sporer, B., Koedel, U., Paul, R., Kohleisen, B., Erfle, V., Fontana, A., and Pfister, H. W. (2000). Human immunodeficiency virus type-1 Nef protein induces blood–brain barrier disruption in the rat: role of matrix metalloproteinase-9. J. Neuroimmunol. *102*, 125–130.

St-Louis, M.C., Abed, Y., and Archambault, D. (2005). The bovine immunodeficiency virus: cloning of a *tat*/*rev* cDNA encoding a novel Tat protein with enhanced transactivation activity. Arch. Virol. *150*, 1529–1547.

St-Louis, M.C., Cojocariu, M., and Archambault, D. (2004). The molecular biology of bovine immunodeficiency virus: a comparison with other lentiviruses. Anim. Health Res. Rev. *5*, 125–143.

Stock, P., and Akbari, O. (2008). Recent advances in the role of NKT-cells in allergic diseases and asthma. Curr. Allergy Asthma Rep. *8*, 165–170.

Strebel, K. (2007). HIV accessory genes Vif and Vpu. Adv. Pharmacol. *55*, 199–232.

Stump, D.S., and VandeWoude, S. (2007). Animal models for HIV AIDS: a comparative review. Comp. Med. *57*, 33–43.

Suarez, D.L., VanDerMaaten, M.J., Wood, C., and Whetstone, C.A. (1993). Isolation and characterization of new wild-type isolates of bovine lentivirus. J. Virol. *67*, 5051–5055.

Suzuki, Y., and Gojobori, T. (1998). The origin and evolution of human T-cell lymphotropic virus types I and II. Virus Genes *16*, 69–84.

Toggas, S.M., Masliah, E., Rockenstein, E. M., Rall, G.F., Abraham, C.R., and Mucke, L. (1994). Central nervous system damage produced by expression of the HIV-1 coat protein gp120 in transgenic mice. Nature *367*, 188–193.

van Kooyk, Y., and Rabinovich, G. A. (2008). Protein–glycan interactions in the control of innate and adaptive immune responses. Nat. Immunol. *9*, 593–601.

van Marle, G., Henry, S., Todoruk, T., Sullivan, A., Silva, C., Rourke, S.B., Holden, J., McArthur, J.C., Gill, M.J., and Power, C. (2004). Human immunodeficiency virus type 1 Nef protein mediates neural cell death: a neurotoxic role for IP-10. Virology *329*, 302–318.

VandeWoude, S., Hageman, C.L., and Hoover, E.A. (2003). Domestic cats infected with lion or puma lentivirus develop anti-feline immunodeficiency virus immune responses. J. Acquir.Immune Defic.Syndr. *34*, 20–31.

VandeWoude, S., and Apetrei, C. (2006). Going wild: lessons from naturally occurring T-lymphotropic lentiviruses. Clin. Microbiol. Rev. *19*, 728–762.

Verdin, E., Becker, N., Bex, F., Droogmans, L., and Burny, A. (1990). Identification and characterization of an enhancer in the coding region of the genome of human immunodeficiency virus type 1. Proc. Natl. Acad. Sci. U.S.A. *87*, 4874–4878.

Verity, E.E., Zotos, D., Wilson, K., Chatfield, C., Lawson, V. A., Dwyer, D.E., Cunningham, A., Learmont, J., Dyer, W., Sullivan, J., *et al.* (2007). Viral phenotypes and antibody responses in long-term survivors infected with attenuated human immunodeficiency virus type 1 containing deletions in the *nef* and long terminal repeat regions. J. Virol. *81*, 9268–9278.

Vicenzi, E., Canducci, F., Pinna, D., Mancini, N., Carletti, S., Lazzarin, A., Bordignon, C., Poli, G., and Clementi, M. (2004). Coronaviridae and SARS-associated coronavirus strain HSR1. Emerg. Infect. Dis. *10*, 413–418.

Vyas, J.M., Van der Veen, A.G., and Ploegh, H.L. (2008). The known unknowns of antigen processing and presentation. Nat. Rev. Immunol. *8*, 607–618.

Wain, L.V., Bailes, E., Bibollet-Ruche, F., Decker, J.M., Keele, B. F., Van Heuverswyn, F., Li, Y., Takehisa, J., Ngole, E.M., Shaw, G.M., *et al.* (2007). Adaptation of HIV-1 to its human host. Mol. Biol. Evol. *24*, 1853–1860.

Westendorp, M. O., Frank, R., Ochsenbauer, C., Stricker, K., Dhein, J., Walczak, H., Debatin, K. M., and Krammer, P. H. (1995). Sensitization of T-cells to CD95-mediated apoptosis by HIV-1 Tat and gp120. Nature *375*, 497–500.

Wilcox, C.M., and Saag, M.S. (2008). Gastrointestinal complications of HIV infection: changing priorities in the HAART era. Gut *57*, 861–870.

Williams, S.A., and Greene, W.C. (2007). Regulation of HIV-1 latency by T-cell activation. Cytokine *39*, 63–74.

Zauli, G., Gibellini, D., Secchiero, P., Dutartre, H., Olive, D., Capitani, S., and Collette, Y. (1999). Human immunodeficiency virus type 1 Nef protein sensitizes CD4(+) T lymphoid cells to apoptosis via functional upregulation of the CD95/CD95 ligand pathway. Blood *93*, 1000–1010.

Zhang, Z., Schuler, T., Zupancic, M., Wietgrefe, S., Staskus, K. A., Reimann, K.A., Reinhart, T.A., Rogan, M., Cavert, W., Miller, C.J., *et al.* (1999). Sexual transmission and propagation of SIV and HIV in resting and activated $CD4^+$ T-cells. Science *286*, 1353–1357.

Retroviral Restriction Factors

11

Jeremy Luban

Abstract

Over the course of the retrovirus replication cycle, viral cDNA is inserted into host chromosomal DNA to establish the provirus. This process results in a permanent insertion mutation in the host cell genome. Host cells have evolved intracellular factors that block the spread of retroviral infection. Some of these antiviral factors act prior to integration and therefore also block the mutagenic potential of infection. Several such factors have been identified, including Fv1, the APOBEC3 complex, and TRIM5. These host factors potently block HIV-1 and other retroviruses from establishment of the provirus. Here we will review current understanding of Fv1, APOBEC3, and TRIM5 proteins, with particular emphasis on TRIM5.

Introduction

Retroviral integration

Integration, an essential step in the retroviral life cycle, results in ligation of a cDNA copy of the viral genome into host chromosomal DNA (Fig. 11.1). The retroviral integration machinery, or preintegration complex, includes viral cDNA and the viral integrase protein (IN) (Bowerman *et al.*, 1989; Farnet and Bushman, 1997; Farnet and Haseltine, 1991; Li *et al.*, 1998; Miller *et al.*, 1997). The resulting insertion of viral nucleic acid is inherited by all daughter cells. Retroviral insertion sites are located throughout the host genome, though retroviral preintegration complexes exhibit particular preferences with respect to target nucleotide sequence, chromatin packing, or the presence of cellular DNA binding proteins and other epigenetic markers (Pryciak and Varmus, 1992b; Wang *et al.*, 2007).

Integration is mutagenic

From the perspective of the virus, the site of integration is important since local features influence subsequent expression of the viral promoter (Jordan *et al.*, 2003; Lewinski *et al.*, 2005). From the perspective of the host, integration has the potential to disrupt open reading frames or to alter gene expression via effects on transcription or splicing. In this context it is important to consider the impact on host evolution of retroviruses and other mobile elements (retrotransposons, see Chapter 1) that replicate via reverse transcription. Retrotransposons have been found in all eukaryotic genomes. 64% to 73% of the maize genome consists of dispersed repetitive sequences, probably retrotransposon in origin (Meyers *et al.*, 2001). 35% to 45% of the human genome consists of transposable elements and 17% of the human genome consists of 500,000 copies of the non-LTR retrotransposon LINE-1 (L1) (Lander *et al.*, 2001; Venter *et al.*, 2001). Though most L1s are incapable of retrotransposition, as many as 100 are capable of retrotransposition (Brouha *et al.*, 2003; Sassaman *et al.*, 1997) and such mobilization can cause human disease (Kazazian *et al.*, 1988). It is therefore not surprising that host cells have evolved intracellular proteins that limit the retroviral replication cycle and block its mutagenic potential.

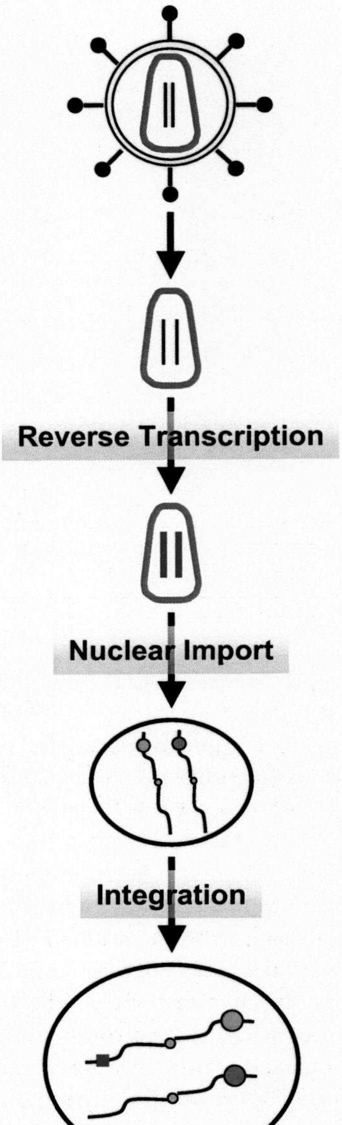

Figure 11.1 Early events in HIV-1 infection. Idealized presentation of the essential steps by which an extracellular HIV-1 virion establishes a provirus. The virion membrane fuses with the host cell membrane, releasing the virion core (blue cone) into the host cell cytoplasm. The virion core is a hexameric lattice of the capsid (CA) protein. Reverse transcription then probably occurs within the virion core, converting the viral RNA (2 black lines) to double-stranded viral cDNA (two thicker red lines). The viral cDNA then gains access to the nucleus via the nuclear pore where it is ligated to host chromosomal DNA, establishing the provirus (red box shown on the chromosome with the yellow marker).

Fv1

Capsid-specific inhibition of murine retroviruses

Resistance of particular mouse strains to a transmissible leukaemia-like illness was observed by Charlotte Friend in the 1950s (Friend, 1957). Subsequently, mouse breeding experiments performed by Frank Lilly identified *Fv1* (Friend virus susceptibility factor 1)(Lilly, 1967; Steeves and Lilly, 1977), a locus encoding an inhibitory activity that acts against particular strains of murine leukaemia virus (MLV). Murine leukaemia viruses are the only viruses known to be inhibited by it (Baumann *et al.*, 2004; Hatziioannou *et al.*, 2004a). Mouse strains bear Fv1 alleles that exhibit specificity for different strains of virus (Fig. 11.2). The Fv1 alleles exhibit co-dominance and each inhibits a different subset of viruses. Mice carrying the *Fv1n* allele, for example, are resistant to infection by B-tropic strains of MLV, but are susceptible to N-tropic strains. The viral determinant for susceptibility to this inhibition is the capsid (CA) protein (Bock *et al.*, 2000; Boone *et al.*, 1983; DesGroseillers and Jolicoeur, 1983; Kozak and Chakraborti, 1996; Lassaux *et al.*, 2005; Qi *et al.*, 1998; Rommelaere *et al.*, 1979; Stevens *et al.*, 2004), which constitutes a hexameric lattice that surrounds the viral RNA genome in the virion, and probably in the cytoplasm of newly infected cells (Briggs *et al.*, 2003; Ganser *et al.*, 2003; Ganser-Pornillos *et al.*, 2007; Li *et al.*, 2000; Mayo *et al.*, 2003; Mortuza *et al.*, 2004). The barrier to retroviral infection by Fv1 occurs after reverse transcription but sometime before integration of the viral genome; the precise step that is inhibited has been difficult to pinpoint. Fv1 is generally said to act after the preintegration complex enters the nucleus (Pryciak and Varmus, 1992a), though under different conditions, effects of Fv1 on reverse transcription have also been detected (Yang *et al.*, 1980).

Cloning of Fv1

In 1996 *Fv1* was cloned and found to be the remnant of an endogenous virus with greatest sequence homology to the *gag* of HERV-L and MERV-L (Best *et al.*, 1996). It has only been detected in the mouse genome. The codominant antiviral activity of the different *Fv1* alleles, along

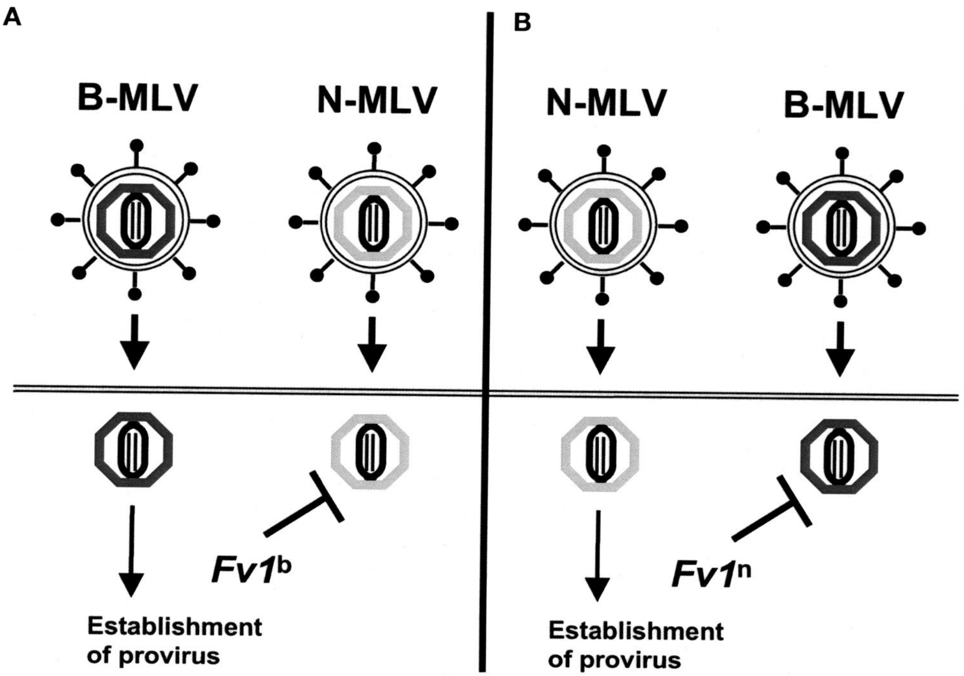

Figure 11.2 Fv1 restriction. B-tropic MLV (blue core) and N-tropic MLV (yellow cores) are shown infecting cells bearing the Fv1^b allele (A) or cells bearing the Fv1^n allele (B). Virions are shown entering target cells in which the plasma membrane is represented by the double horizontal line. When infection is successful the virions gain access to chromosomal DNA in the nucleus. The Fv1^b allele present in some strains of mice (A) restricts infection by N-tropic MLV, which bear an arginine at CA residue 110 (yellow core). B-tropic MLV has a glutamine at CA residue 110 (blue core) and is not restricted by Fv1^b. Mouse strains bearing the Fv1^n allele (B) restrict infection by B-MLV but not by N-MLV.

with the CA-determinants for sensitivity to restriction, suggests that *Fv1* encodes a protein that binds directly to the CA protein of sensitive viruses. An interaction between restrictive Fv1 protein and the CA of a susceptible MLV has never been detected. Though the mechanism by which retroviral infection is blocked by *Fv1* is still unknown, studies of Fv1 established CA as a retroviral tropism determinant and as a target of cellular factors that regulate retroviral infectivity.

APOBEC3

The lentiviral protein Vif

Most lentiviruses, including HIV-1, encode a protein called Vif, that promotes virion infectivity (Dettenhofer *et al.*, 2000; Fisher *et al.*, 1987; Gabuzda *et al.*, 1992; Gibbs *et al.*, 1994; Harmache *et al.*, 1996; Kan *et al.*, 1986; Lee *et al.*, 1986; Michaels *et al.*, 1993; Park *et al.*, 1994; Simon and Malim, 1996; Sodroski *et al.*, 1986; Sova and Volsky, 1993; Strebel *et al.*, 1987; Tomonaga *et al.*, 1992; von Schwedler *et al.*, 1993). Most experiments indicate that Vif acts in the virion producer cell during virus assembly and that virions produced in its absence are not infectious. Analysis of virions produced in the absence of Vif indicated that reverse transcription during entry into target cells is impaired (Nascimbeni *et al.*, 1998; Sova and Volsky, 1993; von Schwedler *et al.*, 1993). Vif is essential for HIV-1 replication in H9 cells, CEM cells, and in primary blood cells, but not in Jurkat, CEM-SS, or SupT1 cells. The fact that Vif is essential in some cell types but not others suggested that Vif either inhibited or substituted for a cellular factor. The generation of heterokaryons by fusion of permissive with non-permissive cells indicated that the non-permissive phenotype is dominant (Madani and Kabat, 1998; Simon *et al.*, 1998), and, therefore, that Vif blocks a cellular, lentiviral inhibitor that acts during virion assembly. The effect of the inhibitor is only manifest during infection of a new target cell. In 2002, a functional expression screen revealed APOBEC3G to be the vif-sensitive, antiviral factor (Sheehy *et al.*, 2002).

Inhibition of HIV-1 and other retroelements by APOBEC3 proteins

Apolipoprotein B mRNA-editing catalytic polypeptides (APOBEC) are a family of cytidine deaminases that include APOBEC1, AID, APOBEC2, and a group of APOBEC3 proteins. APOBEC3 genes are clustered on human chromosome 22 and include APOBEC3A, 3B, 3C, 3DE, 3F, 3G, and 3H (Jarmuz *et al.*, 2002). Human APOBEC3G, and subsequently 3F and 3B, were found to block *vif*-defective human immunodeficiency virus type-1 (HIV-1) and related viruses (Conticello *et al.*, 2003; Harris *et al.*, 2003; Kao *et al.*, 2003; Mangeat *et al.*, 2003; Mariani *et al.*, 2003; Marin *et al.*, 2003; Sheehy *et al.*, 2002; Shindo *et al.*, 2003; Stopak *et al.*, 2003; Yu *et al.*, 2003; Zhang *et al.*, 2003). Hepatitis B virus and HTLV1 are also inhibited by these factors (Derse *et al.*, 2004; Rosler *et al.*, 2005; Sasada *et al.*, 2005; Turelli *et al.*, 2004). APOBEC3A does not inhibit HIV-1 but blocks replication of adeno-associated virus and retrotransposition of endogenous elements such as LINE-1 (Bogerd *et al.*, 2006a; Bogerd *et al.*, 2006b; Chen *et al.*, 2006; Muckenfuss *et al.*, 2006).

Cytidine deaminase activity and retroviral cDNA

The initial body of work from many laboratories indicated that deamination activity is critical for APOBEC3-mediated inhibition of HIV-1 replication. Sequencing of viral cDNA revealed that A3B, A3G and A3F caused extensive mutagenesis of *vif*-defective HIV-1 proviruses (Bishop *et al.*, 2004; Doehle *et al.*, 2005; Harris *et al.*, 2003; Lecossier *et al.*, 2003; Liddament *et al.*, 2004; Mangeat *et al.*, 2003; Mariani *et al.*, 2003; Wiegand *et al.*, 2004; Yu *et al.*, 2004; Zhang *et al.*, 2003; Zheng *et al.*, 2004). Viral cDNA becomes hypermutated when Vif mutant HIV-1 is produced in the presence of APOBEC3G presumably due to C-to-U modification during minus-strand viral DNA synthesis. APOBEC3 from virion-producer cells is packaged into nascent virions, but exerts its negative effect on viral replication after the viral RNA is reverse transcribed in the target cell. APOBEC3 then hypermutates the viral genome, leading to aborted infection. HIV-1 blocks virion incorporation of APOBEC3G (Kao *et al.*, 2003; Mariani *et al.*, 2003; Marin *et al.*, 2003; Sheehy *et al.*, 2003; Stopak *et al.*, 2003; Yu *et al.*, 2003), perhaps by targeting the cellular factor to a ubiquitin ligase complex, and thereby prevents hypermutation of the viral genome.

Non-catalytic mechanisms of viral inhibition

Recent research has challenged, or extended these models concerning APOBEC3 antiviral activity. Several investigators have reported that APOBEC3 mutants lacking deoxycytidine deaminase activity retain the ability to inhibit virus infectivity (Bishop *et al.*, 2006; Bogerd *et al.*, 2006b; Chen *et al.*, 2006; Holmes *et al.*, 2007; Jonsson *et al.*, 2006; Muckenfuss *et al.*, 2006; Newman *et al.*, 2005; Turelli *et al.*, 2004). Additionally, target cell APOBEC3G, as opposed to producer cell APOBEC3G, has been reported to inhibit HIV-1 infection of certain cell types, including resting T-cells and monocytes (Chiu *et al.*, 2005). In these particular cell types, APOBEC3G is found in a low molecular weight complex that is active against HIV-1 (Chiu *et al.*, 2005). It was recently reported that APOBEC3G can interact with cellular mRNAs (Chiu *et al.*, 2006; Gallois-Montbrun *et al.*, 2007; Kozak *et al.*, 2006; Wichroski *et al.*, 2006) to form high-molecular-mass (HMM) complexes (Chiu *et al.*, 2005; Kozak *et al.*, 2006; Luo *et al.*, 2004) and to associate with stress granules, staufen granules, or P bodies.

In more commonly studied cell types, APOBEC3G is bound up in Staufen-containing RNA transport granules and Ro ribonucleoprotein complexes, along with Alu and small Y RNAs (Chiu *et al.*, 2006). These high molecular weight complexes are inactive against HIV-1 but capable of inhibiting Alu retrotransposition (Chiu *et al.*, 2006). These observations are perhaps reconciled with the previous ones in that newly synthesized APOBEC3G is incorporated into HIV-1 virions where it is maintained enzymatically inactive until RNAseH activation upon infection of a new target cell (Soros *et al.*, 2007). APOBEC3A associates with LINE-1 RNA to form high molecular weight complexes and to inhibit retrotransposition of these retroelements (Niewiadomska *et al.*, 2007).

TRIM5

Lv1 and Ref1 activity

Primate lentiviruses, of which HIV-1 is a member, constitute a large family of viruses that are found in over 30 primate species in sub-Saharan Africa (Sharp *et al.*, 2005). HIV-1 replicates efficiently in human cells but its replication is potently blocked in cells from many species of Old World monkeys (Balzarini *et al.*, 1997; Besnier *et al.*, 2003; Cowan *et al.*, 2002; Himathongkham and Luciw, 1996; Hofmann *et al.*, 1999; Li *et al.*, 1992; Munk *et al.*, 2002; Shibata *et al.*, 1991). In contrast, other primate lentiviruses such as SIVmac239 replicate well in cells from Old World monkeys but poorly in cells from New World monkeys (Hofmann *et al.*, 1999). Owl monkeys are unusual New World monkey species that defy this convention: cells from these animals restrict HIV-1 but not SIV (Hofmann *et al.*, 1999; Towers *et al.*, 2003).

In each of these cases, the species-specific block to infection occurs early in infection, generally prior to reverse transcription (Besnier *et al.*, 2002; Cowan *et al.*, 2002; Hatziioannou *et al.*, 2003; Himathongkham and Luciw, 1996; Hofmann *et al.*, 1999; Munk *et al.*, 2002; Shibata *et al.*, 1995; Towers *et al.*, 2003). The existence of a specific antiviral factor responsible for these blocks to infection was suggested by the fact that the block was saturable, and by cell-fusion experiments indicating that the block to infection was dominant (Besnier *et al.*, 2002; Cowan *et al.*, 2002; Munk *et al.*, 2002).

Another important clue was provided by the fact that N-tropic MLV, and not B-MLV, is inhibited by an Fv1-like antiviral activity in cells from humans and other mammals (Towers *et al.*, 2000). By analogy with Fv1 this activity was called Ref1, and the anti-HIV-1 activity in non-human primate cells was called Lv1. As is the case with Fv1 and Ref1, the viral determinant for sensitivity to the Lv1 antiviral factor was found to be CA. This was supported by the generation of chimeric viruses in which components from HIV-1 were swapped with those from SIV (Berthoux *et al.*, 2004; Bukovsky *et al.*, 1997; Cowan *et al.*, 2002; Dorfman and Gottlinger, 1996; Hatziioannou *et al.*, 2004b; Ikeda *et al.*, 2004; Kootstra *et al.*, 2003; Owens *et al.*, 2003, 2004; Sayah *et al.*, 2004; Shibata *et al.*, 1991; Shibata *et al.*, 1995; Stremlau *et al.*, 2004; Towers *et al.*, 2003).

The CA-binding protein cyclophilin A and Lv1 activity

Other clues about the importance of CA, and about the identity of Lv1, came from the finding that HIV-1 CA binds the ubiquitous, cytoplasmic protein, cyclophilin A (CypA) (Luban *et al.*, 1993). CypA binds a surface-exposed loop on HIV-1 CA (Franke *et al.*, 1994; Gamble *et al.*, 1996; Luban *et al.*, 1993). This ability to bind to CypA was not observed with other retroviruses such as MLV or $SIV_{MAC}239$, but it was true of essentially all HIV-1 isolates. More recently it has been shown that CypA also interacts with the CA of SIV strains isolated from particular African green monkey species (*Cercopithecus tantalus*), as well as FIV (Diaz-Griffero *et al.*, 2006b; Lin and Emerman, 2006; Zhang *et al.*, 2006). CypA interaction with HIV-1 CA promotes an early step in the infection of human cells (Braaten *et al.*, 1996a; Braaten *et al.*, 1996b; Braaten and Luban, 2001; Franke and Luban, 1996; Franke *et al.*, 1994; Hatziioannou *et al.*, 2005; Sokolskaja *et al.*, 2004; Thali *et al.*, 1994), the same point in the retroviral life cycle where Lv1 was acting. Analysis of the retroviral replication cycle using RNAi to disrupt CypA indicated that target cell CypA, and not producer cell CypA, increases HIV-1 infectivity (Hatziioannou *et al.*, 2005; Sokolskaja *et al.*, 2004).

Competitive inhibitors and genetic modifications that block the HIV-1 CA–CypA interaction were shown to decrease HIV-1 sensitivity to Lv1 restriction activity in simian cells (Berthoux *et al.*, 2004; Berthoux *et al.*, 2005b; Kootstra *et al.*, 2003; Sayah *et al.*, 2004; Towers *et al.*, 2003). Evaluation of chimeric viruses generated from HIV-1 and simian immunodeficiency virus (SIV) sequences identified the CypA-binding region of CA as the viral determinant for species-specific tropism (Berthoux *et al.*, 2004; Bukovsky *et al.*, 1997; Cowan *et al.*, 2002; Dorfman and Gottlinger, 1996; Hatziioannou *et al.*, 2004b; Ikeda *et al.*, 2004; Kootstra *et al.*, 2003; Owens *et al.*, 2004; Owens *et al.*, 2003; Sayah *et al.*, 2004; Shibata *et al.*, 1991, 1995; Stremlau *et al.*, 2004; Towers *et al.*, 2003). This raised the possibility that CypA was somehow relevant to CA-specific restriction of HIV-1, but analysis of CypA coding sequences failed to identify species-specific polymorphisms that account for the block to HIV-1 infection in non-human primates, and simian CypA cDNA was insufficient to transfer the HIV-1 restriction activity to otherwise permissive human cells (Sayah *et al.*, 2004; Yin *et al.*, 1998). It was even found that the anti-HIV-1 activity in owl monkey cells was eliminated by any of three different shRNAs specific for CypA but restriction activity in the CypA knockdown cells was not restored when CypA protein was reintroduced (Sayah *et al.*, 2004).

Identification of TRIM5α as Lv1 in macaque cells

Screens using cDNA from two different primate species revealed TRIM5 to be responsible for Lv1 activity (Fig. 11.3). The first screened a rhesus macaque library for cDNAs that conferred HIV-1 resistance to otherwise permissive human cells (Stremlau *et al.*, 2004). Macaque TRIM5α cDNA was found among cells that remained uninfected after challenge with HIV-1. The rhesus macaque TRIM5 gene produces several isoforms via differential splicing, but TRIM5α was the only isoform that conferred HIV-1 restriction activity. It blocks HIV-1 after entry into the cell but before reverse transcription is completed. Like Fv1 it was found to be CA-specific.

TRIMCyp identified as the HIV-1 inhibitor in owl monkey cells

The second screen that found TRIM5 had the goal of identifying the source of Lv1 activity in owl monkey cells (Sayah *et al.*, 2004). As mentioned above, owl monkeys are the only New World primates that potently restrict HIV-1 but not $SIV_{MAC}239$ (Hofmann *et al.*, 1999). The Lv1 activity in owl monkey cells was suppressed nearly completely by cyclosporine or mutations

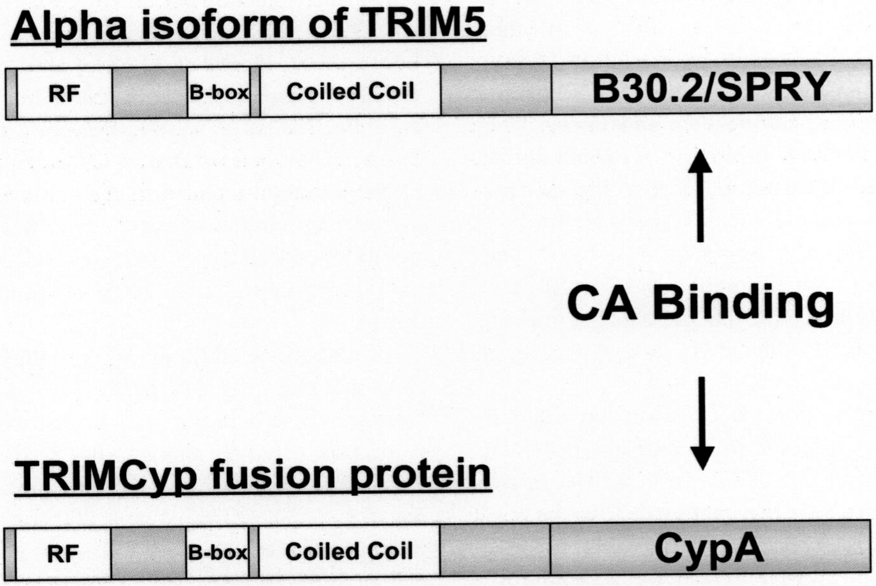

Figure 11.3 TRIM5α and TRIMCyp. Schematic diagram showing the structural motifs present in the two forms of TRIM5 that exhibit retroviral restriction activity. The TRIM5α isoform is found in most primates, in rabbits and in cows. TRIMCyp fusion genes are found in all species of *Aotus* (New World owl monkeys) and in three species of *Macaca* (Old World macaques). The ring finger (RF), B-box motif, and coiled-coil domains are common to these two proteins, and are the elements that define a protein as a tripartite motif family member. The two proteins are distinguished from each other by their very different CA-specific recognition domains: PRYSPRY or B30.2 domain; CypA, cyclophilin A domain.

that disrupt the CA–CypA interaction (Towers *et al.*, 2003). Similarly, three different shRNAs specific for CypA suppressed Lv1 activity in these cells (Sayah *et al.*, 2004). Surprisingly, restoration of CypA protein to the shRNA-expressing owl monkey cells failed to restore restriction activity and the owl monkey CypA cDNA was not sufficient to transfer Lv1 activity to human cells. Screen of an owl monkey cDNA library for clones with homology to CypA, but size-selected to be at least twice the size of the CypA mRNA, revealed a remarkable mRNA fusion consisting of TRIM5 and CypA (Sayah *et al.*, 2004) (Fig. 11.3). This unusual TRIM5 variant was sufficient to transfer Lv1 activity to human cells (Nisole *et al.*, 2004; Sayah *et al.*, 2004).

Generation of the TRIMCyp fusion gene

TRIMCyp is the only TRIM5 allele found in all ten species of owl monkey (Nisole *et al.*, 2004; Ribeiro *et al.*, 2005; Sayah *et al.*, 2004; Song *et al.*, 2005b). It arose via retrotransposition of a complete CypA cDNA into TRIM5 intron 7 (Fig. 11.4) and has all the hallmarks of a cDNA integration event mediated by the enzymatic machinery of a LINE-1 element. TRIMCyp presumably conferred a potent survival advantage to the animal in which it arose, though the selective pressure on these South American primates is unknown. Recently, additional TRIMCyp-sensitive retroviruses were identified, HIV-2_{ROD}, an SIV from *Cercopithecus tantalus*, and FIV (Diaz-Griffero *et al.*, 2006b; Lin and Emerman, 2006; Virgen *et al.*, 2008; Zhang *et al.*, 2006). The CA of these retroviruses binds CypA and TRIMCyp and the restriction of these viruses is blocked by the competitive inhibitor cyclosporine (Lin and Emerman, 2006; Takeuchi *et al.*, 2007).

Convergent evolution: a TRIMCyp fusion in *Macaca* spp.

Though not as ubiquitous as in *Aotus* species, a TRIMCyp fusion gene has also been found in three species of the genus *Macaca* (Brennan *et al.*, 2008; Newman *et al.*, 2008; Virgen *et al.*, 2008; Wilson *et al.*, 2008). As with *Aotus*, insertion of the *Macaca* CypA cDNA into the TRIM5 locus was most likely catalysed by L1 machinery, but the *Macaca* fusion gene was clearly generated independently of the *Aotus* fusion gene. Rather than being inserted into TRIM5 intron 7, the *Macaca* cDNA was inserted 3′ to exon 8 (Fig. 11.4). In the transcript from the *Macaca* TRIM5 fusion gene, the CypA cDNA is spliced to the end of exon 6.

The Macaca TRIMCyp fusion gene has several interesting features. The usual 3′ splice acceptor site before exon 7 has been disrupted (Newman *et al.*, 2008); this partly explains how exons 7 and 8 are bypassed to generate the fusion with the downstream CypA. Several non-synonymous amino acid changes are found when

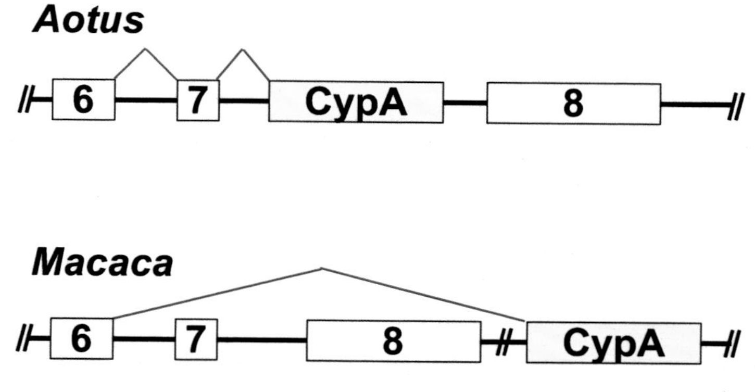

Figure 11.4 Structure of TRIMCyp genes from *Aotus* and *Macaca*. Exons 6, 7, and 8 are shown (numbered white boxes), as they are arranged within Aotus and Macaca. New exons were created independently in *Aotus* and in *Macaca* by LINE-1-catalysed retroinsertion of cyclophilin A cDNA; the positions of these exons are shown (yellow boxes). The red lines show how the exons are linked by splicing to generate mRNAs encoding retroviral restriction factors.

the CypA domain of the fusion gene is compared with the parental *Macaca* CypA. One of these changes, R69H, renders the TRIMCyp fusion unable to bind to HIV-1 CA (Virgen *et al.*, 2008), and thus offers an explanation for the observation that it does not restrict HIV-1 (Newman *et al.*, 2008; Virgen *et al.*, 2008; Wilson *et al.*, 2008). Even more interesting is the fact that this fusion protein retains restriction activity against HIV-2 and FIV. As with the *Aotus* fusion gene, the selective pressure that led to the retention of this peculiar allele is unknown.

TRIM5α orthologues restrict a wide range of retroviruses

Macaque TRIM5α inhibits FIV and EIAV, as well as HIV-1 (Hatziioannou *et al.*, 2004c; Keckesova *et al.*, 2004; Saenz *et al.*, 2005) (Table 11.1). Human TRIM5α possesses modest activity against HIV-1 and potently inhibits N-tropic MLV, thus accounting for Ref1 activity (Hatziioannou *et al.*, 2004c; Keckesova *et al.*, 2004; Perron *et al.*, 2004; Sokolskaja *et al.*, 2006; Yap *et al.*, 2004). Many other primate species have been examined and found to have TRIM5 orthologues that inhibit a wide-range of retroviruses (Ohkura *et al.*, 2006; Song *et al.*, 2005b). The African green monkey *Cercopithecus tantalus*, for example, blocks HIV-1, SIVmac, N-tropic MLV, and equine infectious anaemia virus (EIAV). In most cases, it has been shown that the viral determinant for sensitivity to restriction by a given TRIM5α orthologue is the retroviral CA (Cowan *et al.*, 2002; Dodding *et al.*, 2005; Stremlau *et al.*, 2004).

Rabbits encode a TRIM5 with antiretroviral activity (Schaller *et al.*, 2007), as do cows (Si *et al.*, 2006; Ylinen *et al.*, 2006). Cows in fact have multiple TRIM5 genes (Sawyer *et al.*, 2007). TRIM5α orthologues from these different species now account for restriction activity against a broad range of retroviruses, including HIV-1, HIV-2, SIVs from several non-human primate species, N-MLV, and EIAV (Hatziioannou *et al.*, 2004c; Keckesova *et al.*, 2004; Perron *et al.*, 2004; Si *et al.*, 2006; Song *et al.*, 2005b; Yap *et al.*, 2004; Ylinen *et al.*, 2005). Mice and cats lack a close TRIM5 homologue (Song *et al.*, 2005b) and dogs have lost the TRIM5 gene (Sawyer *et al.*, 2007).

Restriction of HIV-1 by human TRIM5α

When it comes to inhibiting HIV-1, human TRIM5α is much less potent than the rhesus

Table 11.1 TRIM5α orthologues and the retroviruses that they restrict

Host species	Restricted viruses
Homo sapiens	N-MLV, EIAV, HIV-1 (weak)
Pan troglodytes (chimpanzee)	N-MLV
Pongo pygmateus (orang-utan)	N-MLV, SIVmac, N-MLV
Macaca mulatta (rhesus macaque)	HIV-1, N-MLV, SIVagm, EIAV, SIVmac (weak)
Cercopithecus tantalus (AGM)	HIV-1, SIVmac, N-MLV, EIAV
Cercopithecus aethiops (AGM vero)	HIV-1, N-MLV, EIAV, SIVagm (weak)
Aotus trivirgatus (owl monkey)	HIV-1, HIV2rod, FIV, SIVagmTAN
Saimiri boliviensis (squirrel monkey)	SIVmac
Ateles geoffroyi (spider monkey)	HIV-1, SIVmac, SIVagm
Bos taurus (cow)	HIV-1, HIV-2, SIVmac, N-MLV
Oryctolagus cuniculus (rabbit)	HIV-1, HIV-2, FIV, EIAV, M-MLV

macaque orthologue. Nonetheless, endogenous human TRIM5α does indeed restrict HIV-1 to a modest extent, as demonstrated by knockdown of TRIM5 expression in HeLa cells using RNAi (Keckesova *et al.*, 2006; Sokolskaja *et al.*, 2006; Stremlau *et al.*, 2006b). This modest anti-HIV-1 activity potentially decreases the rates of virus transmission or contributes to the delay in immunosuppression – median of 10 years – that is observed in most HIV-1 infected people.

Several studies have attempted to determine if polymorphisms within the human TRIM5α gene show a correlation with susceptibility to HIV-1 infection (Goldschmidt *et al.*, 2006; Javanbakht *et al.*, 2006a; Sawyer *et al.*, 2006; Speelmon *et al.*, 2006; van Manen *et al.*, 2008). Non-synonymous single nucleotide polymorphisms (nsSNP) have been found in human populations but efforts to correlate susceptibility to infection or towards AIDS progression have not been straight forward. One non-synonymous SNP, H43Y, shows decreased *in vitro* restriction activity against both N-MLV and HIV-1 (Sawyer *et al.*, 2006) and is associated with accelerated disease progression among individuals in the Amsterdam Cohort who were homozygous for this variant (van Manen *et al.*, 2008). One study found an association of R136Q with increased susceptibility to HIV-1 infection in European Americans (Speelmon *et al.*, 2006) but another study found a slight protective effect of R136Q in African Americans (Javanbakht *et al.*, 2006a). Another group offers the possible explanation that the effects of R136Q are context dependent (van Manen *et al.*, 2008). These investigators observed a protective effect of the 136Q variant, but only after the emergence in infected subjects of CXCR4-using HIV-1 variants. Interestingly, naive CD4 T-cells, which are selectively targeted by X4 HIV-1, express more Trim5α than do memory $CD4^+$ T-cells. In contrast, when the 136Q allele was present in combination with a particular variant in the 5′UTR, accelerated disease progression was observed. Though conclusions are not yet clear and simple, it seems that polymorphisms in Trim5 probably influence the clinical course of HIV-1 infection *in vivo*, perhaps in a manner that is only detectable in the context of particular cell types, viral strains, or other host genetic factors.

Innate immunity and the tripartite motif family of proteins

TRIM5 is a member of the very large tripartite motif family of genes of which there are roughly 70 family members (Reymond *et al.*, 2001). TRIM family proteins are defined by the presence of ring finger, B-box, and coiled-coil domains. Coding diversity is enhanced further by the fact that many of the family members encode multiple splice variants. TRIM19 (PML), for example, has nine isoforms, and TRIM5 has three isoforms (http://www.ncbi.nlm.nih.gov). The alpha isoform is the only TRIM5 isoform that possesses a PRYSPRY domain at the C-terminus, and, aside from the unusual CypA-fusions found in some species, is the only isoform found to have antiviral activity.

Among the TRIM family members that have been tested, modest retroviral restriction activity has also been detected with TRIM1 and TRIM34 (Si *et al.*, 2006; Song *et al.*, 2005a; Yap *et al.*, 2004, 2006; Zhang *et al.*, 2006). Several other TRIM proteins are involved in immunity, in various ways. TRIM25 (Gack *et al.*, 2007), for example, is required for RIG-I-mediated antiviral activity. The SPRY domain of TRIM25 interacts with the caspase recruitment domain of RIG-I. This results in ubiquitination of RIG-I and signalling via IRF-3. TRIM19, otherwise known as PML, has antiviral activity targeting several herpesviruses (Everett and Chelbi-Alix, 2007). It has been suggested that TRIM19 exhibits activity against HIV-1 and other retroviruses (Turelli *et al.*, 2001) but others have failed to confirm this (Berthoux *et al.*, 2003; Zhang *et al.*, 2006). Recently, TRIM28 was found to be an essential component of the protein complex that recognizes the MLV primer site and silences transcription from this virus in embryonic cells (Wolf and Goff, 2007).

TRIM5α structure

Studies by many laboratories have sought to elucidate the importance of each of the TRIM5α domains for retroviral restriction activity. No domain is dispensable for full restriction activity. The following sections will review what is known about the structure and roles of each of the TRIM5α domains in retroviral restriction.

The ring finger domain

The amino terminus of TRIM5α possesses a ring finger domain of about 40 amino acids. Ring finger domains function as E3 ubiquitin ligases that mediate the transfer of ubiquitin from E2 enzymes to specific substrate molecules. Indeed, *in vitro*, TRIM5 accepts ubiquitin from the E2 enzyme UbcH5B (Xu *et al.*, 2003). An obvious target of TRIM5α-mediated ubiquitination is retroviral CA, though no lab has yet reported an effect of TRIM5 on CA ubiquitination. TRIM5α protein is rapidly turned over via a proteasome-dependent mechanism and a mutant lacking the ring domain has a much prolonged half-life (Diaz-Griffero *et al.*, 2006a). However, the importance of the ring domain for TRIM5α-mediated restriction activity has been a matter of great controversy. Ring finger mutants retain partial restriction activity (Javanbakht *et al.*, 2005; Perez-Caballero *et al.*, 2005a; Perron *et al.*, 2004; Stremlau *et al.*, 2004), demonstrating that the ring domain is not absolutely required for antiviral potential. Another study showed that E1 ubiquitin-activating activity is not required for TRIM5-mediated restriction activity (Perez-Caballero *et al.*, 2005b). Proteasome inhibitors only minimally counteract restriction activity though they increase viral cDNA levels in the cytoplasm (Anderson *et al.*, 2006; Wu *et al.*, 2006). Some in the field now believe that restriction activity occurs via a multistep, redundant mechanism (see below). This explains how E3 ubiquitin ligase activity might be an important, though non-essential component of restriction activity.

The B-box domain

The B-box domain, a defining element of the TRIM family of proteins, is located between the ring finger and coiled-coil motifs (Reymond *et al.*, 2001). It binds zinc and has an overall structure similar to a ring finger motif (Borden *et al.*, 1995; Massiah *et al.*, 2007; Massiah *et al.*, 2006). In most experiments reported to date, the TRIM5α B-box was found to be essential for restriction activity (Javanbakht *et al.*, 2005; Perez-Caballero *et al.*, 2005a). The function is not known but there is some evidence that it influences CA-binding by the PRYSPRY domain, perhaps by changing the orientation of the latter (Li *et al.*, 2007).

The coiled-coil domain

Two blocks of amino acid residues C-terminal to the TRIM5 B-box motif have a high probability of forming a coiled-coil according to standard algorithms (Berger *et al.*, 1995). The first of the two blocks is 38 residues in length. The second block is 31 residues in length. The coiled-coil domain was shown to mediate multimerization of TRIM proteins generally (Reymond *et al.*, 2001), and specifically of TRIM5 (Berthoux *et al.*, 2005a). Consistent with this multimerization activity, TRIM5 fragments lacking restriction activity exhibit dominant negative activity against the wild-type protein, so long as the coiled-coil domain is intact (Berthoux *et al.*, 2005a; Javanbakht *et al.*, 2005; Perez-Caballero *et al.*, 2005a; Stremlau *et al.*, 2004). Cross-linking studies demonstrated that TRIM5α exists as a trimer (Mische *et al.*, 2005) and that the coiled-coil domain and sequences immediately C-terminal to it are necessary and sufficient for trimerization (Javanbakht *et al.*, 2006b). Interestingly, replacement of the coiled-coil domain with a heterologous trimerization domain from GCN4 recapitulated the trimerization of TRIM5α but did not restore restriction activity (Javanbakht *et al.*, 2006b). This suggests that either the coiled-coil domain is responsible for more than simple oligomerization or that the placement of the TRIM5α trimer has precise spatial requirements for restriction activity.

The PRYSPRY or B30.2 domain

The carboxy terminus of the alpha isoform of TRIM5 is a B30.2 or PRYSPRY domain. This nomenclature requires some explanation. A search for coding sequences within the human major histocompatibility complex (MHC) class I region (chromosome 6p21.3) identified an exon called B30-2 encoding a 166 amino acid peptide (Vernet *et al.*, 1993). This peptide was found to have homology to several proteins, including some proteins that were later identified as TRIM family members, and this class of proteins have come to be called B30.2. Independently, computer algorithms identified a closely related domain named SPRY in the splA kinase from *Dictyostelium discoideum* and in mammalian ryanodine receptors (Ponting *et al.*, 1997). Sequence alignments revealed the SPRY domain to be contained within the slightly larger B30.2 domain. The literature sometimes refers to a SPRY domain which is really a PRY segment of 61 amino acids followed by a 139 amino acid-long SPRY segment (Apweiler *et al.*, 2004). Some have chosen to call the approximately 200-amino-acid-long homologous sequences found in all the above-mentioned proteins the PRYSPRY domain, and this is the term we will use here (Apweiler *et al.*, 2004; Grutter *et al.*, 2006).

The structures of several PRYSPRY domains have been solved (Grutter *et al.*, 2006; James *et al.*, 2007; Masters *et al.*, 2006; Woo *et al.*, 2006) and based on this information the three dimensional structure of the TRIM5α PRYSPRY domain has been modelled. PRYSPRY forms one compact domain of a 13-stranded β-sandwich, containing a hydrophobic core and a putative ligand-binding pocket. Deletion of the PRYSPRY domain of several proteins abolishes interaction with their respective binding partners, suggesting that, in each case, PRYSPRY acts as a protein-interacting module. In the case of TRIM5α, the PRYSPRY domain is essential for restriction activity and constitutes the main determinant for virus specificity (Hatziioannou *et al.*, 2004c; Keckesova *et al.*, 2004; Nakayama *et al.*, 2005; Perez-Caballero *et al.*, 2005a; Perron *et al.*, 2004; Sawyer *et al.*, 2005; Song *et al.*, 2005a,b; Stremlau *et al.*, 2004, 2005; Yap *et al.*, 2004, 2005). TRIM5α hybrid proteins, for example, that contain the human TRIM5 ring finger, B-box, and coiled-coil domains fused to the rhesus macaque PRYSPRY domain are able to potently restrict HIV-1, thereby taking on the restriction phenotype of rhesus TRIM5α (Perez-Caballero *et al.*, 2005a).

PRYSPRY domain and retroviral restriction specificity

The carboxy-terminal CypA domain of owl monkey TRIMCyp and the similarly placed SPRY domain of TRIM5α are required for species-specific differences in restriction activity (Berthoux *et al.*, 2005a; Hatziioannou *et al.*, 2004c; Keckesova *et al.*, 2004; Nakayama *et al.*, 2005; Perez-Caballero *et al.*, 2005a; Perron *et al.*, 2004; Sawyer *et al.*, 2005; Sayah *et al.*, 2004; Song *et al.*, 2005a,b; Stremlau *et al.*, 2004, 2005; Yap *et al.*, 2004, 2005). Sequence analysis revealed several variable regions within the TRIM5α PRYSPRY domain (Sawyer *et al.*, 2005; Song *et al.*, 2005a). In particular, comparison of TRIM5α orthologues identified a 13-amino-acid patch in the PRYSPRY domain where non-synonymous amino acid changes clustered (Sawyer *et al.*, 2005; Song *et al.*, 2005a). Based on these sequence comparisons chimeric TRIM5α proteins were generated and tested for restriction activity. One fruitful series of experiments was done with chimeras generated with human and rhesus TRIM5α orthologues; the rhesus protein potently restricts HIV-1, while the human protein does not. Assessment of the restriction activity of chimeric rhesus-human TRIM5α proteins demonstrated that HIV-1 restriction specificity was largely determined by the amino acid residues in the PRYSPRY variable hotspot (Perez-Caballero *et al.*, 2005a; Sawyer *et al.*, 2005; Stremlau *et al.*, 2005). Ultimately, conversion of a single amino acid in human TRIM5α (residue R332) greatly increased anti-HIV-1 activity (Li *et al.*, 2006; Yap *et al.*, 2005). The mapping of specificity determinants for N-MLV restriction by human TRIM5α also emphasized the importance of residues in the first variable loop, but pointed out the importance of other variable regions in PRYSPRY (Maillard *et al.*, 2007; Ohkura *et al.*, 2006; Perron *et al.*, 2006). Tandem sequence duplications and triplications in the variable region of African green monkey and spider monkey TRIM5α perhaps explain the wide variety of retroviruses restricted by these orthologues (Song *et al.*, 2005a). All these studies suggest that this variable region of PRYSPRY provides a flexible surface that can readily evolve to interact with diverse CA ligands.

PRYSPRY domain and retroviral CA binding

As discussed above, the TRIM5α PRYSPRY domain has been replaced by the CA-binding protein cyclophilin A, in owl monkeys and certain species of macaques (Brennan *et al.*, 2008; Newman *et al.*, 2008; Nisole *et al.*, 2004; Sayah *et al.*, 2004; Virgen *et al.*, 2008; Wilson *et al.*, 2008). Along with the demonstrated CA-specificity of TRIM5α restriction, these unusual TRIM5 isoforms indicated that the TRIM5α PRYSPRY domain must bind directly to CA of restricted retroviruses. Nonetheless, though conventional methods for detecting protein–protein interactions such as the yeast two-hybrid system and coimmunoprecipitation readily detect TRIMCyp interaction with HIV-1 CA, attempts to detect TRIM5α interaction with CA using these methods have failed (Berthoux *et al.*, 2005a; Sebastian and Luban, 2005).

The activity of CA-specific restriction factors such as Fv1 can be saturated by loading target cells with CA from restriction-sensitive retroviruses (Bassin *et al.*, 1980; Boone *et al.*, 1990; Duran-Troise *et al.*, 1981; Hartley *et al.*, 1970). The same is true of TRIM5α-mediated restriction (Besnier *et al.*, 2002; Cowan *et al.*, 2002; Munk *et al.*, 2002). In either case, though, saturation of restriction activity requires delivery of a mature virion CA: neither CA protein associated with protease-defective virions, nor free CA protein expressed directly within the target cell, will overcome restriction (Cowan *et al.*, 2002; Dodding *et al.*, 2005; Towers *et al.*, 2003). These observations led to the hypothesis that the TRIM5α trimer recognizes a higher-order structure of CA multimers (Javanbakht *et al.*, 2006b; Mische *et al.*, 2005; Sebastian and Luban, 2005; Stremlau *et al.*, 2006a), perhaps the CA lattice which is modelled as a hexameric unit in retrovirion cores (Briggs *et al.*, 2003; Ganser *et al.*, 2003; Ganser-Pornillos *et al.*, 2007; Li *et al.*, 2000; Mayo *et al.*, 2003; Mortuza *et al.*, 2004).

Based on the above arguments, CA-binding assays have been developed that recapitulate the restriction specificity seen in cells. Detergent-stripped MLV virions were mixed with GST-tagged human TRIM5α in the first assay. Pull-down of TRIM5α on glutathione-conjugated Sepharose beads resulted in the co-sedimentation of restricted N-tropic but not of unrestricted B-tropic MLV and the interaction was dependent on the PRYSPRY domain (Sebastian and Luban, 2005). This method has not worked using rhesus TRIM5α and HIV-1 CA, perhaps because HIV-1 virion cores are less stable than those of MLV (Bowerman *et al.*, 1989; Fassati and Goff, 1999; Fassati and Goff, 2001; Welker *et al.*, 2000).

The second approach that showed TRIM5α interaction with CA used the reverse strategy (Stremlau *et al.*, 2006a). Retroviral core-like complexes were assembled *in vitro* from recombinant HIV-1 CA-NC fusion protein. These complexes were incubated with cell lysates containing TRIM5α and the complexes were accelerated through a sucrose cushion. The CA complex pellet was then probed for TRIM5α. Using this assay, it was shown that the ability of TRIM5α orthologues to associate with HIV-1 CA correlated with the ability to restrict HIV-1.

Mechanism of TRIM5-mediated restriction

The block to retroviral infection caused by TRIM5 occurs very early after virion fusion with the target cell membrane (Fig. 11.5), prior to detectable accumulation of viral cDNA (Berthoux *et al.*, 2004; Berthoux *et al.*, 2003; Cowan *et al.*, 2002; Munk *et al.*, 2002; Perez-Caballero *et al.*, 2005b; Stremlau *et al.*, 2004; Towers *et al.*, 2003). Under particular conditions effects can be detected later in the replication cycle, either at the time of nuclear transport or after the virus has gained access to the nucleus (Anderson *et al.*, 2006; Berthoux *et al.*, 2004; Wu *et al.*, 2006; Yap *et al.*, 2006); these experiments involved assessment of TRIM5 chimeras and mutants or of infection in the presence of pharmacologic agents, including proteasome inhibitors, arsenic, or cyclosporine. Though each of these conditions is unphysiological, these results nonetheless point out that, following recognition of a susceptible retroviral capsid, TRIM5 may block all subsequent steps in the viral replication pathway.

Requirement for CA binding

There is general agreement that retroviral restriction requires direct binding of the susceptible retroviral CA by TRIM5 (Fig. 11.5). As discussed above, the major CA-recognition element is found at the TRIM5 C-terminus and is either a PRYSPRY domain or the CA-binding element cyclophilin A. To date, any manipulation that disrupts interaction of a particular retroviral CA with these recognition elements will disrupt restriction activity against that retrovirus (Berthoux *et al.*, 2005a; Hatziioannou *et al.*, 2004c; Keckesova *et al.*, 2004; Nakayama *et al.*, 2005; Owens *et al.*, 2003, 2004; Perez-Caballero *et al.*, 2005a; Perron *et al.*, 2004; Sawyer *et al.*, 2005; Sayah *et al.*, 2004; Song *et al.*, 2005a,b; Stremlau *et al.*, 2004, 2005; Towers *et al.*, 2003; Yap *et al.*, 2004, 2005). Other

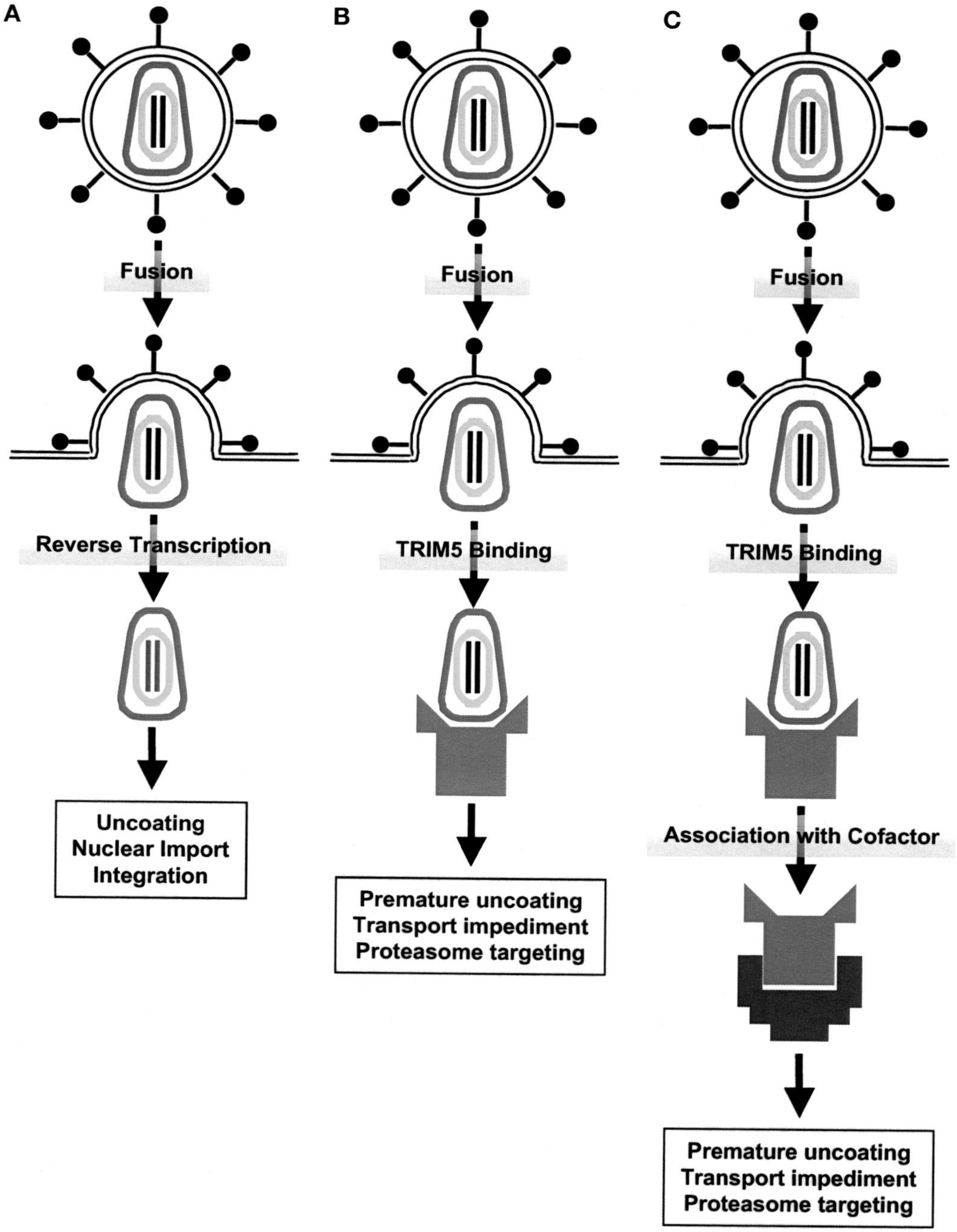

Figure 11.5 Alternative models for the mechanism by which TRIM5 restricts retroviruses. Retrovirus entry into a target cells is shown schematically as in Fig. 11.1. In the unrestricted condition (A), the virion core containing viral cDNA is able to establish the provirus. This requires reverse transcription, nuclear import and integration. Transformation of the viral RNA genome (pair of vertical black lines) to viral cDNA (pair of vertical red lines) is shown. The virion core may undergo an uncoating step prior to nuclear entry. In the simplest models for the mechanism of retroviral restriction (B), TRIM5 binding to the virion core is sufficient to block infection. Binding of TRIM5 (green polygon) to the virion core might promote premature virion uncoating, targeting of virion proteins for degradation by the proteasome, or mistargeting of the virion core within the cytoplasm. Some experimental data invoke more complex models (C) in which retroviral restriction by TRIM5 requires recruitment of additional host factors (blue polygon) that then promote premature virion uncoating, target virion proteins for degradation, or mistarget the virion core.

than this binding event, the requirement for any additional factors/cofactors in TRIM5-mediated restriction is currently a matter of intense debate.

In some cases, TRIM5 binding to retroviral CA may be sufficient to block infection. The CA-binding protein CypA exhibits anti-HIV-1 restriction activity when it is fused to a heterologous multimerization domain (Javanbakht *et al.*, 2007; Yap *et al.*, 2007). Furthermore, FIV restriction by owl monkey *Aotus* TRIMCyp can be detected in the absence of the ring finger and B-box domains (Diaz-Griffero *et al.*, 2007). Hence, there is evidence that a multivalent CA-binding domain is sufficient for TRIM5-mediated restriction. Direct interaction with the virion core might block association with an essential viral replication cofactor present in the target cell cytoplasm or cause the viral nucleoprotein complex to mislocalize within the cell. In contrast to the above findings with TRIMCyp, TRIM5α deletion mutants that retain the coiled-coil multimerization domain and PRYSPRY domain are sufficient for virion core binding (Stremlau *et al.*, 2006a) but lack restriction activity (Javanbakht *et al.*, 2005; Perez-Caballero *et al.*, 2005a; Stremlau *et al.*, 2004).

The hypothesis of retroviral uncoating

By analogy with other viruses, it has been proposed that retroviral infection requires that the CA protein lattice disassemble, a hypothetical process that has been called uncoating. Whether uncoating is required for reverse transcription or for nuclear transport, or whether it is required at all, has been difficult to ascertain. Technical obstacles associated with analysis of the early events in the retroviral life cycle include the fact that only a small minority of the retroviral particles that enter a cell succeed in establishing a provirus (McDonald *et al.*, 2002; Piatak *et al.*, 1993); a significant proportion of the viral mass that is taken up by cells enters a non-productive, endocytic pathway (Marechal *et al.*, 1998); and viral replication complexes appear to be relatively unstable and perhaps inaccessible to biochemical analysis (Farnet and Haseltine, 1991; Fassati and Goff, 1999, 2001; Welker *et al.*, 2000). For example, CA generally does not remain associated with reverse transcription complexes isolated from acutely infected cells (Farnet and Haseltine, 1991; Fassati and Goff, 1999, 2001) yet, microscopic examination of infected cells shows that reverse transcription probably occurs within an intact retrovirion core (Arhel *et al.*, 2007; McDonald *et al.*, 2002). Reverse transcription can be completed *in vitro* within isolated CA cores (Warrilow *et al.*, 2007) and CA mutants have been identified with replication defects that are evident after reverse transcription has been completed (Dismuke and Aiken, 2006; Yamashita *et al.*, 2007).

Nonetheless, interesting results were obtained recently, in which physical methods were used to separate intact cytoplasmic virion cores from soluble viral proteins and viral protein present in the endocytic compartment (Stremlau *et al.*, 2006a). Good correlation was made between retroviral restriction activity by TRIM5 and premature conversion of particulate capsids to soluble CA protein. This was demonstrated with HIV-1 restriction by rhesus macaque TRIM5α (Stremlau *et al.*, 2006a), N-MLV restriction by human TRIM5α (Perron *et al.*, 2007), and by owl monkey (*Aotus*) TRIMCyp restriction of HIV-1 and FIV (Diaz-Griffero *et al.*, 2007). Based on these experiments it has been suggested that simple interaction of TRIM5α with virion cores is sufficient to block retroviral infection; by doing so, TRIM5α causes the premature disassembly of virion cores, before reverse transcription is completed. Some observations with this experimental system suggest that the mechanism must be more complicated than this, though. Proteasome inhibitors increase the yield of particulate HIV-1 CA in the cytoplasm of cells expressing either rhesus macaque TRIM5alpha or owl monkey TRIMCyp (Diaz-Griffero *et al.*, 2007). Additionally, particulate HIV-1 CA is detected in cells expressing TRIMCyp constructs that bear B-box domain mutations, even though these mutants are able to restrict HIV-1 (Diaz-Griffero *et al.*, 2007). In other words, there is not a perfect correlation between retroviral restriction activity and the ability to disrupt particulate CA.

TRIM5 ubiquitin ligase activity

The presence of the ring finger in TRIM5 suggests that E3 ubiquitin ligase activity is required for retroviral restriction activity. Consistent with the importance of this activity, TRIM5 receives ubiquitin from an E2 transferase enzyme *in vitro* (Xu *et al.*, 2003), mutation of ring finger sequences decreases restriction activity (Javanbakht *et al.*, 2005; Perez-Caballero *et al.*, 2005a; Perez-Caballero *et al.*, 2005b; Stremlau *et al.*, 2004), and proteasome inhibitors increase viral titre, though to only a modest extent at best (Butler *et al.*, 2002; Perez-Caballero *et al.*, 2005b; Schwartz *et al.*, 1998). Nonetheless ubiquitin is dispensable for restriction activity (Javanbakht *et al.*, 2005; Perez-Caballero *et al.*, 2005b; Stremlau *et al.*, 2006a) and interpretation of ubiquitin experiments is complicated further by the fact that TRIM5 itself is polyubiquitinated and its half-life is increased by proteasome inhibitors (Diaz-Griffero *et al.*, 2006a). What is clear is that proteasome inhibitors completely rescue the inhibitory effect of TRIM5 on reverse transcription, but not the effect of TRIM5 on retroviral transduction (Wu *et al.*, 2006). This apparent discrepancy, along with other observations described below, indicates that in addition to blocking reverse transcription, TRIM5 also blocks subsequent steps in the retroviral life cycle.

Two-step mechanism of TRIM5-mediated restriction

In addition to blocking viral cDNA accumulation in restrictive cells, TRIM5 presents a block to the nuclear import of viral cDNA. This was first observed during the analysis of the effects of As_2O_3 (Berthoux *et al.*, 2004), a drug which increases the titre of HIV-1 and other retroviruses (Turelli *et al.*, 2001). The stimulatory effect of As_2O_3 on retroviral transduction is attributable to effects of the drug on TRIM5-mediated restriction (Berthoux *et al.*, 2004; Berthoux *et al.*, 2003; Sebastian *et al.*, 2006). The drug has no effect on the restriction activity of Fv1 (Berthoux *et al.*, 2003). As_2O_3 seems not to act directly on TRIM5, since the drug suppresses restriction activity in some cell lines but not others, even though both cells express the same TRIM5 cDNA (Sebastian *et al.*, 2006). The effect of As_2O_3 is to rescue the level of viral cDNA synthesis, without significant change in the levels of viral cDNA in the nucleus (Berthoux *et al.*, 2004). In contrast, in the context of rhesus macaque TRIM5α, cyclosporine increases the nuclear forms of HIV-1 cDNA with minimal effect on total cDNA synthesis (Berthoux *et al.*, 2004). Thus, TRIM5 seems to impose a block at the level of reverse transcription and at the level of nuclear transport. As discussed above, subsequent experiments with proteasome inhibitors confirmed that TRIM5 has additional blocks downstream of reverse transcription (Wu *et al.*, 2006). Additionally, fusions of CypA with the ring finger, B-box, and coiled-coil domains taken from different TRIM family members indicate that TRIM-mediated restriction can occur with blocks apparent either before or after reverse transcription, depending upon the TRIM family member (Yap *et al.*, 2006).

Conclusions

Integration, an essential step in the life cycle of exogenous retroviruses, endogenous retroviruses, or other retro-elements, is inherently mutagenic to the chromosomal DNA target. To protect themselves, host cells elaborate proteins that are commonly called restriction factors. Among these restriction factors are Fv1 and members of the APOBEC and TRIM family of proteins. Restriction factors block the critical replicative steps of retroelements – such as reverse transcription – that ultimately culminate in integration. Detailed understanding of the mechanisms by which they act is lacking. Fv1 and TRIM5 target retroviruses in a CA-specific manner. Fv1 is homologous to endogenous retroviral Gag polyproteins and its mechanism of action remains extremely mysterious. This protein is only encoded in mice and particular murine retroviruses are the only viruses known to be restricted. In contrast, a wide- range of retroviruses are restricted by different orthologues of TRIM5α. In primates, the gene encoding TRIM5α has undergone extremely rapid diversifying selection, presumably in response to lethal challenges by exogenous retroviruses. TRIM5α restriction activity requires multimerization and direct interaction with the CA-hexameric lattice of the retroviral virion core. Some believe that it simply acts by prematurely disrupting the structure of the virion core. The functional relevance for restriction of its ring-finger E3 ubiquitin ligase activity has been questioned. Others believe that there are unidentified cellular cofactors required for activity. The APOBEC3 gene family has expanded 10-fold from mice to humans. Most of the encoded proteins have been found to restrict retroviruses or retroelements in some context. Here too there is much to learn about mechanism. Almost certainly, the adenosine deaminase activity is important for restriction activity but non-catalytic, antiviral activities have been reported. These factors are transcriptionally responsive to interferon. This fact, along with other observations, suggests that in the near future these restriction factors may come to be considered components of more classical innate immunity.

References

Anderson, J.L., Campbell, E.M., Wu, X., Vandegraaff, N., Engelman, A., and Hope, T.J. (2006). proteasome inhibition reveals that a functional preintegration complex intermediate can be generated during restriction by diverse TRIM5 proteins. J. Virol. *80*, 9754–9760.

Apweiler, R., Bairoch, A., Wu, C.H., Barker, W.C., Boeckmann, B., Ferro, S., Gasteiger, E., Huang, H., Lopez, R., Magrane, M., *et al.* (2004). UniProt: the Universal Protein knowledgebase. Nucleic Acids Res. *32*, D115–119.

Arhel, N.J., Souquere-Besse, S., Munier, S., Souque, P., Guadagnini, S., Rutherford, S., Prevost, M.C., Allen, T.D., and Charneau, P. (2007). HIV-1 DNA Flap formation promotes uncoating of the pre-integration complex at the nuclear pore. EMBO J. *26*, 3025–3037.

Balzarini, J., De Clercq, E., and Uberla, K. (1997). SIV/HIV-1 hybrid virus expressing the reverse transcriptase gene of HIV-1 remains sensitive to HIV-1-specific reverse transcriptase inhibitors after passage in rhesus macaques. J. Acquir. Immune Defic. Syndr. Hum. Retroviruses *15*, 1–4.

Bassin, R.H., Gerwin, B.I., Levin, J.G., Duran-Troise, G., Benjers, B.M., and Rein, A. (1980). Macromolecular requirements for abrogation of Fv-1 restriction by murine leukemia viruses. J. Virol. *35*, 287–297.

Baumann, J.G., Unutmaz, D., Miller, M.D., Breun, S.K., Grill, S M., Mirro, J., Littman, D.R., Rein, A., and KewalRamani, V.N. (2004). Murine T-cells potently restrict human immunodeficiency virus infection. J. Virol. *78*, 12537–12547.

Berger, B., Wilson, D.B., Wolf, E., Tonchev, T., Milla, M., and Kim, P.S. (1995). Predicting coiled coils by use of pairwise residue correlations. Proc. Natl. Acad. Sci. U.S.A. *92*, 8259–8263.

Berthoux, L., Towers, G.J., Gurer, C., Salomoni, P., Pandolfi, P.P., and Luban, J. (2003). As_2O_3 enhances retroviral reverse transcription and counteracts Ref1 antiviral activity. J. Virol. *77*, 3167–3180.

Berthoux, L., Sebastian, S., Sokolskaja, E., and Luban, J. (2004). Lv1 inhibition of human immunodeficiency virus type 1 is counteracted by factors that stimulate synthesis or nuclear translocation of viral cDNA. J. Virol. *78*, 11739–11750.

Berthoux, L., Sebastian, S., Sayah, D.M., and Luban, J. (2005a). Disruption of human TRIM5alpha antiviral activity by nonhuman primate orthologues. J. Virol. *79*, 7883–7888.

Berthoux, L., Sebastian, S., Sokolskaja, E., and Luban, J. (2005b). Cyclophilin A is required for TRIM5α-mediated resistance to HIV-1 in Old World monkey cells. Proc. Natl. Acad. Sci. U.S.A. *102*, 14849–14853.

Besnier, C., Takeuchi, Y., and Towers, G. (2002). Restriction of lentivirus in monkeys. Proc. Natl. Acad. Sci. U.S.A. *99*, 11920–11925.

Besnier, C., Ylinen, L., Strange, B., Lister, A., Takeuchi, Y., Goff, S.P., and Towers, G.J. (2003). Characterization of murine leukemia virus restriction in mammals. J. Virol. *77*, 13403–13406.

Best, S., Le Tissier, P., Towers, G., and Stoye, J. P. (1996). Positional cloning of the mouse retrovirus restriction gene Fv1. Nature *382*, 826–829.

Bishop, K.N., Holmes, R.K., and Malim, M. H. (2006). Antiviral potency of APOBEC proteins does not correlate with cytidine deamination. J. Virol. *80*, 8450–8458.

Bishop, K.N., Holmes, R.K., Sheehy, A.M., Davidson, N.O., Cho, S.J., and Malim, M.H. (2004). Cytidine deamination of retroviral DNA by diverse APOBEC proteins. Curr. Biol. *14*, 1392–1396.

Bock, M., Bishop, K.N., Towers, G., and Stoye, J.P. (2000). Use of a transient assay for studying the genetic determinants of Fv1 restriction. J. Virol. *74*, 7422–7430.

Bogerd, H.P., Wiegand, H.L., Doehle, B.P., Lueders, K.K., and Cullen, B.R. (2006a). APOBEC3A and APOBEC3B are potent inhibitors of LTR-retrotransposon function in human cells. Nucleic Acids Res. *34*, 89–95.

Bogerd, H.P., Wiegand, H.L., Hulme, A. E., Garcia-Perez, J.L., O'Shea, K. S., Moran, J.V., and Cullen, B.R. (2006b). Cellular inhibitors of long interspersed element 1 and Alu retrotransposition. Proc. Natl. Acad. Sci. U.S.A. *103*, 8780–8785.

Boone, L.R., Innes, C.L., and Heitman, C. K. (1990). Abrogation of Fv-1 restriction by genome-deficient virions produced by a retrovirus packaging cell line. J. Virol. *64*, 3376–3381.

Boone, L.R., Myer, F.E., Yang, D.M., Ou, C.Y., Koh, C.K., Roberson, L.E., Tennant, R.W., and Yang, W.K. (1983). Reversal of Fv-1 host range by *in vitro* restriction endonuclease fragment exchange between molecular clones of N-tropic and B-tropic murine leukemia virus genomes. J. Virol. *48*, 110–119.

Borden, K.L., Lally, J.M., Martin, S.R., O'Reilly, N. J., Etkin, L.D., and Freemont, P.S. (1995). Novel topology of a zinc-binding domain from a protein involved in regulating early Xenopus development. EMBO J. *14*, 5947–5956.

Bowerman, B., Brown, P.O., Bishop, J.M., and Varmus, H. (1989). A nucleoprotein complex mediates the integration of retroviral DNA. Genes Dev. 3, 469–478.

Braaten, D., Aberham, C., Franke, E.K., Yin, L., Phares, W., and Luban, J. (1996a). Cyclosporine A-resistant human immunodeficiency virus type 1 mutants demonstrate that Gag encodes the functional target of cyclophilin A. J. Virol. *70*, 5170–5176.

Braaten, D., Franke, E.K., and Luban, J. (1996b). Cyclophilin A is required for an early step in the life cycle of human immunodeficiency virus type 1 before the initiation of reverse transcription. J. Virol. *70,* 3551–3560.

Braaten, D., and Luban, J. (2001). Cyclophilin A regulates HIV-1 infectivity, as demonstrated by gene targeting in human T-cells. EMBO J. *20,* 1300–1309.

Brennan, G., Kozyrev, Y., and Hu, S.-L. (2008). TRIMCyp expression in Old World primates *Macaca nemestrina* and *Macaca fascicularis.* Proc. Natl. Acad. Sci. U.S.A. *105,* 3569–3574.

Briggs, J.A., Wilk, T., Welker, R., Krausslich, H.G., and Fuller, S.D. (2003). Structural organization of authentic, mature HIV-1 virions and cores. EMBO J. *22,* 1707–1715.

Brouha, B., Schustak, J., Badge, R.M., Lutz-Prigge, S., Farley, A.H., Moran, J.V., and Kazazian, H.H., Jr. (2003). Hot L1s account for the bulk of retrotransposition in the human population. Proc. Natl. Acad. Sci. U.S.A. *100,* 5280–5285.

Bukovsky, A.A., Weimann, A., Accola, M.A., and Gottlinger, H.G. (1997). Transfer of the HIV-1 cyclophilin-binding site to simian immunodeficiency virus from Macaca mulatta can confer both cyclosporin sensitivity and cyclosporin dependence. Proc. Natl. Acad. Sci. U.S.A. *94,* 10943–10948.

Butler, S.L., Johnson, E.P., and Bushman, F.D. (2002). Human immunodeficiency virus cDNA metabolism: notable stability of two-long terminal repeat circles. J. Virol. *76,* 3739–3747.

Chen, H., Lilley, C.E., Yu, Q., Lee, D.V., Chou, J., Narvaiza, I., Landau, N.R., and Weitzman, M.D. (2006). APOBEC3A is a potent inhibitor of adeno-associated virus and retrotransposons. Curr. Biol. *16,* 480–485.

Chiu, Y. L., Soros, V. B., Kreisberg, J. F., Stopak, K., Yonemoto, W., and Greene, W. C. (2005). Cellular APOBEC3G restricts HIV-1 infection in resting $CD4^+$ T-cells. Nature *435,* 108–114.

Chiu, Y. L., Witkowska, H. E., Hall, S. C., Santiago, M., Soros, V. B., Esnault, C., Heidmann, T., and Greene, W. C. (2006). High-molecular-mass APOBEC3G complexes restrict Alu retrotransposition. Proc. Natl. Acad. Sci. U.S.A. *103,* 15588–15593.

Conticello, S. G., Harris, R. S., and Neuberger, M. S. (2003). The Vif protein of HIV triggers degradation of the human antiretroviral DNA deaminase APOBEC3G. Curr. Biol. *13,* 2009–2013.

Cowan, S., Hatziioannou, T., Cunningham, T., Muesing, M. A., Gottlinger, H. G., and Bieniasz, P. D. (2002). Cellular inhibitors with Fv1-like activity restrict human and simian immunodeficiency virus tropism. Proc. Natl. Acad. Sci. U.S.A. *99,* 11914–11919.

Derse, D., Heidecker, G., Mitchell, M., Hill, S., Lloyd, P., and Princler, G. (2004). Infectious transmission and replication of human T-cell leukemia virus type 1. Front Biosci 9, 2495–2499.

DesGroseillers, L., and Jolicoeur, P. (1983). Physical mapping of the Fv-1 tropism host range determinant of BALB/c murine leukemia viruses. J. Virol. *48,* 685–696.

Dettenhofer, M., Cen, S., Carlson, B.A., Kleiman, L., and Yu, X.F. (2000). Association of human immunodeficiency virus type 1 Vif with RNA and its role in reverse transcription. J. Virol. *74,* 8938–8945.

Diaz-Griffero, F., Kar, A., Lee, M., Stremlau, M., Poeschla, E., and Sodroski, J. (2007). Comparative requirements for the restriction of retrovirus infection by TRIM5alpha and TRIMCyp. Virology *369,* 400–410.

Diaz-Griffero, F., Li, X., Javanbakht, H., Song, B., Welikala, S., Stremlau, M., and Sodroski, J. (2006a). Rapid turnover and polyubiquitylation of the retroviral restriction factor TRIM5. Virology *349,* 300–315.

Diaz-Griffero, F., Vandegraaff, N., Li, Y., McGee-Estrada, K., Stremlau, M., Welikala, S., Si, Z., Engelman, A., and Sodroski, J. (2006b). Requirements for capsid-binding and an effector function in TRIMCyp-mediated restriction of HIV-1. Virology *351,* 404–419.

Dismuke, D.J., and Aiken, C. (2006). Evidence for a functional link between uncoating of the human immunodeficiency virus type 1 core and nuclear import of the viral preintegration complex. J. Virol. *80,* 3712–3720.

Dodding, M.P., Bock, M., Yap, M.W., and Stoye, J.P. (2005). Capsid processing requirements for abrogation of Fv1 and Ref1 restriction. J. Virol. *79,* 10571–10577.

Doehle, B.P., Schafer, A., and Cullen, B.R. (2005). Human APOBEC3B is a potent inhibitor of HIV-1 infectivity and is resistant to HIV-1 Vif. Virology *339,* 281–288.

Dorfman, T., and Gottlinger, H.G. (1996). The human immunodeficiency virus type 1 capsid p2 domain confers sensitivity to the cyclophilin-binding drug SDZ NIM 811. J. Virol. *70,* 5751–5757.

Duran-Troise, G., Bassin, R.H., Wallace, B.F., and Rein, A. (1981). Balb/3T3 cells chronically infected with N-tropic murine leukemia virus continue to express Fv-1b restriction. Virology *112,* 795–799.

Everett, R.D., and Chelbi-Alix, M.K. (2007). PML and PML nuclear bodies: implications in antiviral defence. Biochimie *89,* 819–830.

Farnet, C.M., and Bushman, F.D. (1997). HIV-1 cDNA integration: requirement of HMG I(Y) protein for function of preintegration complexes *in vitro.* Cell *88,* 483–492.

Farnet, C.M., and Haseltine, W.A. (1991). Determination of viral proteins present in the human immunodeficiency virus type 1 preintegration complex. J. Virol. *65,* 1910–1915.

Fassati, A., and Goff, S.P. (1999). Characterization of intracellular reverse transcription complexes of Moloney murine leukemia virus. J. Virol. 73, 8919–8925.

Fassati, A., and Goff, S.P. (2001). Characterization of intracellular reverse transcription complexes of human immunodeficiency virus type 1. J. Virol. *75,* 3626–3635.

Fisher, A.G., Ensoli, B., Ivanoff, L., Chamberlain, M., Petteway, S., Ratner, L., Gallo, R.C., and Wong-Staal, F. (1987). The sor gene of HIV-1 is required for efficient virus transmission *in vitro.* Science *237,* 888–893.

Franke, E.K., and Luban, J. (1996). Inhibition of HIV-1 replication by cyclosporine A or related compounds correlates with the ability to disrupt the Gag-cyclophilin A interaction. Virology *222,* 279–282.

Franke, E.K., Yuan, H.E., and Luban, J. (1994). Specific incorporation of cyclophilin A into HIV-1 virions. Nature *372,* 359–362.

Friend, C. (1957). Cell-free transmission in adult Swiss mice of a disease having the character of a leukemia. J. Exp. Med. *105*, 307–318.

Gabuzda, D.H., Lawrence, K., Langhoff, E., Terwilliger, E., Dorfman, T., Haseltine, W.A., and Sodroski, J. (1992). Role of vif in replication of human immunodeficiency virus type 1 in CD4+ T-lymphocytes. J. Virol. *66*, 6489–6495.

Gack, M.U., Shin, Y.C., Joo, C.H., Urano, T., Liang, C., Sun, L., Takeuchi, O., Akira, S., Chen, Z., Inoue, S., and Jung, J.U. (2007). TRIM25 RING-finger E3 ubiquitin ligase is essential for RIG-I-mediated antiviral activity. Nature *446*, 916–920.

Gallois-Montbrun, S., Kramer, B., Swanson, C.M., Byers, H., Lynham, S., Ward, M., and Malim, M.H. (2007). Antiviral protein APOBEC3G localizes to ribonucleoprotein complexes found in P bodies and stress granules. J. Virol. *81*, 2165–2178.

Gamble, T.R., Vajdos, F.F., Yoo, S., Worthylake, D.K., Houseweart, M., Sundquist, W.I., and Hill, C.P. (1996). Crystal structure of human cyclophilin A bound to the amino-terminal domain of HIV-1 capsid. Cell *87*, 1285–1294.

Ganser, B.K., Cheng, A., Sundquist, W.I., and Yeager, M. (2003). Three-dimensional structure of the M-MuLV CA protein on a lipid monolayer: a general model for retroviral capsid assembly. EMBO J. *22*, 2886–2892.

Ganser-Pornillos, B. K., Cheng, A., and Yeager, M. (2007). Structure of full-length HIV-1 CA: a model for the mature capsid lattice. Cell *131*, 70–79.

Gibbs, J.S., Regier, D.A., and Desrosiers, R.C. (1994). Construction and *in vitro* properties of SIVmac mutants with deletions in 'nonessential' genes. AIDS Res. Hum. Retroviruses *10*, 607–616.

Goldschmidt, V., Bleiber, G., May, M.T., Martinez, R., Ortiz, M., and Telenti, A. (2006). Role of common human TRIM5alpha variants in HIV-1 disease progression. Retrovirology *3*, 54.

Grutter, C., Briand, C., Capitani, G., Mittl, P.R., Papin, S., Tschopp, J., and Grutter, M.G. (2006). Structure of the PRYSPRY-domain: implications for autoinflammatory diseases. FEBS Lett. *580*, 99–106.

Harmache, A., Russo, P., Guiguen, F., Vitu, C., Vignoni, M., Bouyac, M., Hieblot, C., Pepin, M., Vigne, R., and Suzan, M. (1996). Requirement of caprine arthritis encephalitis virus vif gene for *in vivo* replication. Virology *224*, 246–255.

Harris, R.S., Bishop, K.N., Sheehy, A.M., Craig, H. M., Petersen-Mahrt, S.K., Watt, I.N., Neuberger, M.S., and Malim, M.H. (2003). DNA deamination mediates innate immunity to retroviral infection. Cell *113*, 803–809.

Hartley, J.W., Rowe, W.P., and Huebner, R.J. (1970). Host-range restrictions of murine leukemia viruses in mouse embryo cell cultures. J. Virol. *5*, 221–225.

Hatziioannou, T., Cowan, S., and Bieniasz, P.D. (2004a). Capsid-dependent and -independent postentry restriction of primate lentivirus tropism in rodent cells. J. Virol. *78*, 1006–1011.

Hatziioannou, T., Cowan, S., Goff, S.P., Bieniasz, P.D., and Towers, G.J. (2003). Restriction of multiple divergent retroviruses by Lv1 and Ref1. EMBO J. *22*, 385–394.

Hatziioannou, T., Cowan, S., Von Schwedler, U.K., Sundquist, W.I., and Bieniasz, P.D. (2004b). Species-specific tropism determinants in the human immunodeficiency virus type 1 capsid. J. Virol. *78*, 6005–6012.

Hatziioannou, T., Perez-Caballero, D., Cowan, S., and Bieniasz, P.D. (2005). Cyclophilin interactions with incoming human immunodeficiency virus type 1 capsids with opposing effects on infectivity in human cells. J. Virol. *79*, 176–183.

Hatziioannou, T., Perez-Caballero, D., Yang, A., Cowan, S., and Bieniasz, P.D. (2004c). Retrovirus resistance factors Ref1 and Lv1 are species-specific variants of TRIM5alpha. Proc. Natl. Acad. Sci. U.S.A. *101*, 10774–10779.

Himathongkham, S., and Luciw, P.A. (1996). Restriction of HIV-1 (subtype B) replication at the entry step in rhesus macaque cells. Virology *219*, 485–488.

Hofmann, W., Schubert, D., LaBonte, J., Munson, L., Gibson, S., Scammell, J., Ferrigno, P., and Sodroski, J. (1999). Species-specific, postentry barriers to primate immunodeficiency virus infection. J. Virol. *73*, 10020–10028.

Holmes, R.K., Koning, F.A., Bishop, K.N., and Malim, M.H. (2007). APOBEC3F can inhibit the accumulation of HIV-1 reverse transcription products in the absence of hypermutation. Comparisons with APOBEC3G. J. Biol. Chem. *282*, 2587–2595.

Ikeda, Y., Ylinen, L. M., Kahar-Bador, M., and Towers, G.J. (2004). Influence of *gag* on human immunodeficiency virus type 1 species-specific tropism. J. Virol. *78*, 11816–11822.

James, L.C., Keeble, A.H., Khan, Z., Rhodes, D.A., and Trowsdale, J. (2007). Structural basis for PRYSPRY-mediated tripartite motif (TRIM) protein function. Proc. Natl. Acad. Sci. U.S.A. *104*, 6200–6205.

Jarmuz, A., Chester, A., Bayliss, J., Gisbourne, J., Dunham, I., Scott, J., and Navaratnam, N. (2002). An anthropoid-specific locus of orphan C to U RNA-editing enzymes on chromosome 22. Genomics *79*, 285–296.

Javanbakht, H., An, P., Gold, B., Petersen, D.C., O'Huigin, C., Nelson, G.W., O'Brien, S.J., Kirk, G.D., Detels, R., Buchbinder, S., *et al.* (2006a). Effects of human TRIM5alpha polymorphisms on antiretroviral function and susceptibility to human immunodeficiency virus infection. Virology *354*, 15–27.

Javanbakht, H., Diaz-Griffero, F., Stremlau, M., Si, Z., and Sodroski, J. (2005). The contribution of RING and B-box 2 domains to retroviral restriction mediated by monkey TRIM5alpha. J. Biol. Chem. *280*, 26933–26940.

Javanbakht, H., Diaz-Griffero, F., Yuan, W., Yeung, D.F., Li, X., Song, B., and Sodroski, J. (2007). The ability of multimerized cyclophilin A to restrict retrovirus infection. Virology *367*, 19–29.

Javanbakht, H., Yuan, W., Yeung, D.F., Song, B., Diaz-Griffero, F., Li, Y., Li, X., Stremlau, M., and Sodroski, J. (2006b). Characterization of TRIM5alpha trimerization and its contribution to human immunodeficiency virus capsid binding. Virology *353*, 234–246.

Jonsson, S.R., Hache, G., Stenglein, M.D., Fahrenkrug, S.C., Andresdottir, V., and Harris, R.S. (2006). Evolutionarily conserved and non-conserved retrovirus restriction activities of artiodactyl APOBEC3F proteins. Nucleic Acids Res. *34*, 5683–5694.

Jordan, A., Bisgrove, D., and Verdin, E. (2003). HIV reproducibly establishes a latent infection after acute infection of T-cells *in vitro*. EMBO J. *22*, 1868–1877.

Kan, N.C., Franchini, G., Wong-Staal, F., DuBois, G.C., Robey, W.G., Lautenberger, J.A., and Papas, T.S. (1986). Identification of HTLV-III/LAV sor gene product and detection of antibodies in human sera. Science *231*, 1553–1555.

Kao, S., Khan, M. A., Miyagi, E., Plishka, R., Buckler-White, A., and Strebel, K. (2003). The human immunodeficiency virus type 1 Vif protein reduces intracellular expression and inhibits packaging of APOBEC3G (CEM15), a cellular inhibitor of virus infectivity. J. Virol. *77*, 11398–11407.

Kazazian, H.H., Jr., Wong, C., Youssoufian, H., Scott, A.F., Phillips, D. G., and Antonarakis, S.E. (1988). Haemophilia A resulting from *de novo* insertion of L1 sequences represents a novel mechanism for mutation in man. Nature *332*, 164–166.

Keckesova, Z., Ylinen, L.M., and Towers, G.J. (2004). The human and African green monkey TRIM5alpha genes encode Ref1 and Lv1 retroviral restriction factor activities. Proc. Natl. Acad. Sci. U.S.A. *101*, 10780–10785.

Keckesova, Z., Ylinen, L. M., and Towers, G.J. (2006). Cyclophilin A renders human immunodeficiency virus type 1 sensitive to old world monkey but not human TRIM5{alpha} antiviral activity. J. Virol. *80*, 4683–4690.

Kootstra, N.A., Munk, C., Tonnu, N., Landau, N.R., and Verma, I.M. (2003). Abrogation of postentry restriction of HIV-1-based lentiviral vector transduction in simian cells. Proc. Natl. Acad. Sci. U.S.A. *100*, 1298–1303.

Kozak, C.A., and Chakraborti, A. (1996). Single amino acid changes in the murine leukemia virus capsid protein gene define the target of Fv1 resistance. Virology *225*, 300–305.

Kozak, S.L., Marin, M., Rose, K.M., Bystrom, C., and Kabat, D. (2006). The anti-HIV-1 editing enzyme APOBEC3G binds HIV-1 RNA and messenger RNAs that shuttle between polysomes and stress granules. J. Biol. Chem. *281*, 29105–29119.

Lander, E.S., Linton, L.M., Birren, B., Nusbaum, C., Zody, M.C., Baldwin, J., Devon, K., Dewar, K., Doyle, M., FitzHugh, W., *et al.* (2001). Initial sequencing and analysis of the human genome. Nature *409*, 860–921.

Lassaux, A., Sitbon, M., and Battini, J.L. (2005). Residues in the murine leukemia virus capsid that differentially govern resistance to mouse Fv1 and human Ref1 restrictions. J. Virol. *79*, 6560–6564.

Lecossier, D., Bouchonnet, F., Clavel, F., and Hance, A.J. (2003). Hypermutation of HIV-1 DNA in the absence of the Vif protein. Science *300*, 1112.

Lee, T.H., Coligan, J.E., Allan, J.S., McLane, M. F., Groopman, J.E., and Essex, M. (1986). A new HTLV-III/LAV protein encoded by a gene found in cytopathic retroviruses. Science *231*, 1546–1549.

Lewinski, M.K., Bisgrove, D., Shinn, P., Chen, H., Hoffmann, C., Hannenhalli, S., Verdin, E., Berry, C.C., Ecker, J.R., and Bushman, F.D. (2005). Genome-wide analysis of chromosomal features repressing human immunodeficiency virus transcription. J. Virol. *79*, 6610–6619.

Li, J., Lord, C.I., Haseltine, W., Letvin, N.L., and Sodroski, J. (1992). Infection of cynomolgus monkeys with a chimeric HIV-1/SIVmac virus that expresses the HIV-1 envelope glycoproteins. J. Acquir. Immune Defic. Syndr. *5*, 639–646.

Li, L., Farnet, C.M., Anderson, W.F., and Bushman, F.D. (1998). Modulation of activity of Moloney murine leukemia virus preintegration complexes by host factors *in vitro*. J. Virol. *72*, 2125–2131.

Li, S., Hill, C.P., Sundquist, W.I., and Finch, J. T. (2000). Image reconstructions of helical assemblies of the HIV-1 CA protein. Nature *407*, 409–413.

Li, X., Song, B., Xiang, S. H., and Sodroski, J. (2007). Functional interplay between the B-box 2 and the B30.2(SPRY) domains of TRIM5alpha. Virology *366*, 234–244.

Li, Y., Li, X., Stremlau, M., Lee, M., and Sodroski, J. (2006). Removal of arginine 332 allows human TRIM5alpha to bind human immunodeficiency virus capsids and to restrict infection. J. Virol. *80*, 6738–6744.

Liddament, M.T., Brown, W.L., Schumacher, A.J., and Harris, R.S. (2004). APOBEC3F properties and hypermutation preferences indicate activity against HIV-1 *in vivo*. Curr. Biol. *14*, 1385–1391.

Lilly, F. (1967). Susceptibility to two strains of Friend leukemia virus in mice. Science *155*, 461–462.

Lin, T.Y., and Emerman, M. (2006). Cyclophilin A interacts with diverse lentiviral capsids. Retrovirology *3*, 70.

Luban, J., Bossolt, K.L., Franke, E.K., Kalpana, G.V., and Goff, S.P. (1993). Human immunodeficiency virus type 1 Gag protein binds to cyclophilins A and B. Cell *73*, 1067–1078.

Luo, K., Liu, B., Xiao, Z., Yu, Y., Yu, X., Gorelick, R., and Yu, X. F. (2004). Amino-terminal region of the human immunodeficiency virus type 1 nucleocapsid is required for human APOBEC3G packaging. J. Virol. *78*, 11841–11852.

McDonald, D., Vodicka, M. A., Lucero, G., Svitkina, T. M., Borisy, G. G., Emerman, M., and Hope, T. J. (2002). Visualization of the intracellular behavior of HIV in living cells. J. Cell Biol. *159*, 441–452.

Madani, N., and Kabat, D. (1998). An endogenous inhibitor of human immunodeficiency virus in human lymphocytes is overcome by the viral Vif protein. J. Virol. 72, 10251–10255.

Maillard, P.V., Reynard, S., Serhan, F., Turelli, P., and Trono, D. (2007). Interfering residues narrow the spectrum of MLV restriction by human TRIM5alpha. PLoS Pathog 3, e200.

Mangeat, B., Turelli, P., Caron, G., Friedli, M., Perrin, L., and Trono, D. (2003). Broad antiretroviral defence by human APOBEC3G through lethal editing of nascent reverse transcripts. Nature *424*, 99–103.

Marechal, V., Clavel, F., Heard, J.M., and Schwartz, O. (1998). Cytosolic Gag p24 as an index of productive entry of human immunodeficiency virus type 1. J. Virol. 72, 2208–2212.

Mariani, R., Chen, D., Schrofelbauer, B., Navarro, F., Konig, R., Bollman, B., Munk, C., Nymark-McMahon, H., and Landau, N.R. (2003). Species-specific exclusion of APOBEC3G from HIV-1 virions by Vif. Cell *114*, 21–31.

Marin, M., Rose, K.M., Kozak, S.L., and Kabat, D. (2003). HIV-1 Vif protein binds the editing enzyme APOBEC3G and induces its degradation. Nat. Med. *9*, 1398–1403.

Massiah, M.A., Matts, J.A., Short, K.M., Simmons, B.N., Singireddy, S., Yi, Z., and Cox, T C. (2007). Solution structure of the MID1 B-box2 CHC(D/C)C(2)H(2) zinc-binding domain: insights into an evolutionarily conserved RING fold. J. Mol. Biol. *369*, 1–10.

Massiah, M.A., Simmons, B.N., Short, K.M., and Cox, T.C. (2006). Solution structure of the RBCC/TRIM B-box1 domain of human MID1: B-box with a RING. J. Mol. Biol. *358*, 532–545.

Masters, S.L., Yao, S., Willson, T.A., Zhang, J.G., Palmer, K.R., Smith, B.J., Babon, J.J., Nicola, N.A., Norton, R.S., and Nicholson, S.E. (2006). The SPRY domain of SSB-2 adopts a novel fold that presents conserved Par-4-binding residues. Nat. Struct. Mol. Biol. *13*, 77–84.

Mayo, K., Huseby, D., McDermott, J., Arvidson, B., Finlay, L., and Barklis, E. (2003). Retrovirus capsid protein assembly arrangements. J. Mol. Biol. *325*, 225–237.

Meyers, B.C., Tingey, S.V., and Morgante, M. (2001). Abundance, distribution, and transcriptional activity of repetitive elements in the maize genome. Genome Res. *11*, 1660–1676.

Michaels, F.H., Hattori, N., Gallo, R.C., and Franchini, G. (1993). The human immunodeficiency virus type 1 (HIV-1) vif protein is located in the cytoplasm of infected cells and its effect on viral replication is equivalent in HIV-2. AIDS Res. Hum. Retroviruses *9*, 1025–1030.

Miller, M.D., Farnet, C.M., and Bushman, F.D. (1997). Human immunodeficiency virus type 1 preintegration complexes: studies of organization and composition. J. Virol. *71*, 5382–5390.

Mische, C.C., Javanbakht, H., Song, B., Diaz-Griffero, F., Stremlau, M., Strack, B., Si, Z., and Sodroski, J. (2005). Retroviral restriction factor TRIM5alpha is a trimer. J. Virol. *79*, 14446–14450.

Mortuza, G.B., Haire, L. F., Stevens, A., Smerdon, S.J., Stoye, J.P., and Taylor, I.A. (2004). High-resolution structure of a retroviral capsid hexameric amino-terminal domain. Nature *431*, 481–485.

Muckenfuss, H., Hamdorf, M., Held, U., Perkovic, M., Lower, J., Cichutek, K., Flory, E., Schumann, G. G., and Munk, C. (2006). APOBEC3 proteins inhibit human LINE-1 retrotransposition. J. Biol. Chem. *281*, 22161–22172.

Munk, C., Brandt, S. M., Lucero, G., and Landau, N.R. (2002). A dominant block to HIV-1 replication at reverse transcription in simian cells. Proc. Natl. Acad. Sci. U.S.A. *99*, 13843–13848.

Nakayama, E.E., Miyoshi, H., Nagai, Y., and Shioda, T. (2005). A specific region of 37 amino acid residues in the SPRY (B30.2) domain of African green monkey TRIM5alpha determines species-specific restriction of simian immunodeficiency virus SIVmac infection. J. Virol. *79*, 8870–8877.

Nascimbeni, M., Bouyac, M., Rey, F., Spire, B., and Clavel, F. (1998). The replicative impairment of Vif- mutants of human immunodeficiency virus type 1 correlates with an overall defect in viral DNA synthesis. J. Gen. Virol. *79 (Pt 8)*, 1945–1950.

Newman, E.N., Holmes, R.K., Craig, H. M., Klein, K.C., Lingappa, J.R., Malim, M.H., and Sheehy, A.M. (2005). Antiviral function of APOBEC3G can be dissociated from cytidine deaminase activity. Curr. Biol. *15*, 166–170.

Newman, R., Hall, L., Kirmaier, A., Pozzi, L., Pery, E., Farzan, M., O'Neil, S., and Johnson, W. (2008). Evolution of a Trim5-CypA Splice Isoform in Old World Monkeys. PLoS Pathog *4*, e1000003.

Niewiadomska, A.M., Tian, C., Tan, L., Wang, T., Sarkis, P.T., and Yu, X.F. (2007). Differential inhibition of long interspersed element 1 by APOBEC3 does not correlate with high-molecular-mass-complex formation or P-body association. J. Virol. *81*, 9577–9583.

Nisole, S., Lynch, C., Stoye, J.P., and Yap, M.W. (2004). A Trim5-cyclophilin A fusion protein found in owl monkey kidney cells can restrict HIV-1. Proc. Natl. Acad. Sci. U.S.A. *101*, 13324–13328.

Ohkura, S., Yap, M.W., Sheldon, T., and Stoye, J.P. (2006). All Three Variable Regions of the TRIM5{alpha} B30.2 domain can contribute to the specificity of retrovirus restriction. J. Virol. *80*, 8554–8565.

Owens, C.M., Song, B., Perron, M. J., Yang, P.C., Stremlau, M., and Sodroski, J. (2004). Binding and susceptibility to postentry restriction factors in monkey cells are specified by distinct regions of the human immunodeficiency virus type 1 capsid. J. Virol. *78*, 5423–5437.

Owens, C.M., Yang, P.C., Gottlinger, H., and Sodroski, J. (2003). Human and simian immunodeficiency virus capsid proteins are major viral determinants of early, postentry replication blocks in simian cells. J. Virol. *77*, 726–731.

Park, I.W., Myrick, K., and Sodroski, J. (1994). Effects of vif mutations on cell-free infectivity and replication of simian immunodeficiency virus. J. Acquir Immune Defic Syndr 7, 1228–1236.

Perez-Caballero, D., Hatziioannou, T., Yang, A., Cowan, S., and Bieniasz, P.D. (2005a). Human tripartite motif 5alpha domains responsible for retrovirus restriction activity and specificity. J. Virol. *79*, 8969–8978.

Perez-Caballero, D., Hatziioannou, T., Zhang, F., Cowan, S., and Bieniasz, P.D. (2005b). Restriction of human immunodeficiency virus type 1 by TRIM-CypA occurs with rapid kinetics and independently of cytoplasmic bodies, ubiquitin, and proteasome activity. J. Virol. *79*, 15567–15572.

Perron, M.J., Stremlau, M., Lee, M., Javanbakht, H., Song, B., and Sodroski, J. (2007). The human TRIM5alpha restriction factor mediates accelerated uncoating of the N-tropic murine leukemia virus capsid. J. Virol. *81*, 2138–2148.

Perron, M.J., Stremlau, M., and Sodroski, J. (2006). Two surface-exposed elements of the B30.2/SPRY domain as potency determinants of N-tropic murine leukemia virus restriction by human TRIM5alpha. J. Virol. *80*, 5631–5636.

Perron, M.J., Stremlau, M., Song, B., Ulm, W., Mulligan, R.C., and Sodroski, J. (2004). TRIM5alpha mediates the postentry block to N-tropic murine leukemia viruses in human cells. Proc. Natl. Acad. Sci. U.S.A. *101*, 11827–11832.

Piatak, M., Jr., Saag, M.S., Yang, L.C., Clark, S.J., Kappes, J. C., Luk, K.C., Hahn, B. H., Shaw, G.M., and Lifson, J.D. (1993). High levels of HIV-1 in plasma during all stages of infection determined by competitive PCR. Science *259*, 1749–1754.

Ponting, C., Schultz, J., and Bork, P. (1997). SPRY domains in ryanodine receptors (Ca^{2+}-release channels). Trends Biochem. Sci. *22*, 193–194.

Pryciak, P.M., and Varmus, H.E. (1992a). Fv-1 restriction and its effects on murine leukemia virus integration *in vivo* and *in vitro*. J. Virol. *66*, 5959–5966.

Pryciak, P.M., and Varmus, H.E. (1992b). Nucleosomes, DNA-binding proteins, and DNA sequence modulate retroviral integration target site selection. Cell *69*, 769–780.

Qi, C.F., Bonhomme, F., Buckler-White, A., Buckler, C., Orth, A., Lander, M.R., Chattopadhyay, S.K., and Morse, H.C., 3rd (1998). Molecular phylogeny of Fv1. Mamm. Genome *9*, 1049–1055.

Reymond, A., Meroni, G., Fantozzi, A., Merla, G., Cairo, S., Luzi, L., Riganelli, D., Zanaria, E., Messali, S., Cainarca, S., *et al.* (2001). The tripartite motif family identifies cell compartments. EMBO J. *20*, 2140–2151.

Ribeiro, I.P., Menezes, A.N., Moreira, M. A., Bonvicino, C.R., Seuanez, H.N., and Soares, M.A. (2005). Evolution of cyclophilin A and TRIMCyp retrotransposition in New World primates. J. Virol. *79*, 14998–15003.

Rommelaere, J., Donis-Keller, H., and Hopkins, N. (1979). RNA sequencing provides evidence for allelism of determinants of the N-, B- or NB-tropism of murine leukemia viruses. Cell *16*, 43–50.

Rosler, C., Kock, J., Kann, M., Malim, M. H., Blum, H.E., Baumert, T.F., and von Weizsacker, F. (2005). APOBEC-mediated interference with hepadnavirus production. Hepatology *42*, 301–309.

Saenz, D.T., Teo, W., Olsen, J.C., and Poeschla, E.M. (2005). Restriction of feline immunodeficiency virus by Ref1, Lv1, and primate TRIM5alpha proteins. J. Virol. *79*, 15175–15188.

Sasada, A., Takaori-Kondo, A., Shirakawa, K., Kobayashi, M., Abudu, A., Hishizawa, M., Imada, K., Tanaka, Y., and Uchiyama, T. (2005). APOBEC3G targets human T-cell leukemia virus type 1. Retrovirology *2*, 32.

Sassaman, D.M., Dombroski, B. A., Moran, J.V., Kimberland, M. L., Naas, T.P., DeBerardinis, R.J., Gabriel, A., Swergold, G.D., and Kazazian, H.H., Jr. (1997). Many human L1 elements are capable of retrotransposition. Nat. Genet. *16*, 37–43.

Sawyer, S.L., Emerman, M., and Malik, H.S. (2007). Discordant Evolution of the Adjacent Antiretroviral Genes TRIM22 and TRIM5 in Mammals. PLoS Pathog *3*, e197.

Sawyer, S.L., Wu, L.I., Akey, J. M., Emerman, M., and Malik, H.S. (2006). High-frequency persistence of an impaired allele of the retroviral defense gene TRIM5alpha in humans. Curr. Biol. *16*, 95–100.

Sawyer, S.L., Wu, L.I., Emerman, M., and Malik, H.S. (2005). Positive selection of primate TRIM5alpha identifies a critical species-specific retroviral restriction domain. Proc. Natl. Acad. Sci. U.S.A. *102*, 2832–2837.

Sayah, D.M., Sokolskaja, E., Berthoux, L., and Luban, J. (2004). Cyclophilin A retrotransposition into TRIM5 explains owl monkey resistance to HIV-1. Nature *430*, 569–573.

Schaller, T., Hue, S., and Towers, G.J. (2007). An active TRIM5 protein in rabbits indicates a common antiviral ancestor for mammalian TRIM5 proteins. J. Virol. *81*, 11713–11721.

Schwartz, O., Marechal, V., Friguet, B., Arenzana-Seisdedos, F., and Heard, J.M. (1998). Antiviral activity of the proteasome on incoming human immunodeficiency virus type 1. J. Virol. *72*, 3845–3850.

Sebastian, S., and Luban, J. (2005). TRIM5alpha selectively binds a restriction-sensitive retroviral capsid. Retrovirology *2*, 40.

Sebastian, S., Sokolskaja, E., and Luban, J. (2006). Arsenic counteracts human immunodeficiency virus type 1 restriction by various TRIM5 orthologues in a cell type-dependent manner. J. Virol. *80*, 2051–2054.

Sharp, P.M., Shaw, G. M., and Hahn, B.H. (2005). Simian immunodeficiency virus infection of chimpanzees. J. Virol. *79*, 3891–3902.

Sheehy, A.M., Gaddis, N. C., Choi, J.D., and Malim, M.H. (2002). Isolation of a human gene that inhibits HIV-1 infection and is suppressed by the viral Vif protein. Nature *418*, 646–650.

Sheehy, A.M., Gaddis, N.C., and Malim, M.H. (2003). The antiretroviral enzyme APOBEC3G is degraded by the proteasome in response to HIV-1 Vif. Nat. Med. *9*, 1404–1407.

Shibata, R., Kawamura, M., Sakai, H., Hayami, M., Ishimoto, A., and Adachi, A. (1991). Generation of a chimeric human and simian immunodeficiency virus infectious to monkey peripheral blood mononuclear cells. J. Virol. *65*, 3514–3520.

Shibata, R., Sakai, H., Kawamura, M., Tokunaga, K., and Adachi, A. (1995). Early replication block of human immunodeficiency virus type 1 in monkey cells. J. Gen. Virol. *76 (Pt 11)*, 2723–2730.

Shindo, K., Takaori-Kondo, A., Kobayashi, M., Abudu, A., Fukunaga, K., and Uchiyama, T. (2003). The enzymatic activity of CEM15/Apobec-3G is essential for the regulation of the infectivity of HIV-1 virion but not a sole determinant of its antiviral activity. J. Biol. Chem. *278*, 44412–44416.

Si, Z., Vandegraaff, N., O'Huigin, C., Song, B., Yuan, W., Xu, C., Perron, M., Li, X., Marasco, W. A., Engelman, A., *et al.* (2006). Evolution of a cytoplasmic tripartite motif (TRIM) protein in cows that restricts retroviral infection. Proc. Natl. Acad. Sci. U.S.A. *103*, 7454–7459.

Simon, J.H., Gaddis, N.C., Fouchier, R. A., and Malim, M H. (1998). Evidence for a newly discovered cellular anti-HIV-1 phenotype. Nat. Med. *4*, 1397–1400.

Simon, J.H., and Malim, M.H. (1996). The human immunodeficiency virus type 1 Vif protein modulates the postpenetration stability of viral nucleoprotein complexes. J. Virol. *70*, 5297–5305.

Sodroski, J., Goh, W. C., Rosen, C., Tartar, A., Portetelle, D., Burny, A., and Haseltine, W. (1986). Replicative and cytopathic potential of HTLV-III/LAV with sor gene deletions. Science *231*, 1549–1553.

Sokolskaja, E., Berthoux, L., and Luban, J. (2006). Cyclophilin A and TRIM5alpha independently regulate human immunodeficiency virus type 1 infectivity in human cells. J. Virol. *80*, 2855–2862.

Sokolskaja, E., Sayah, D.M., and Luban, J. (2004). Target cell cyclophilin A modulates human immunodeficiency virus type 1 infectivity. J. Virol. *78*, 12800–12808.

Song, B., Gold, B., O'Huigin, C., Javanbakht, H., Li, X., Stremlau, M., Winkler, C., Dean, M., and Sodroski, J. (2005a). The B30.2(SPRY) domain of the retroviral restriction factor TRIM5alpha exhibits lineage-specific length and sequence variation in primates. J. Virol. *79*, 6111–6121.

Song, B., Javanbakht, H., Perron, M., Park, D.H., Stremlau, M., and Sodroski, J. (2005b). Retrovirus restriction by TRIM5alpha variants from Old World and New World primates. J. Virol. *79*, 3930–3937.

Soros, V.B., Yonemoto, W., and Greene, W.C. (2007). Newly synthesized APOBEC3G is incorporated into HIV virions, inhibited by HIV RNA, and subsequently activated by RNase H. PLoS Pathog 3, e15.

Sova, P., and Volsky, D.J. (1993). Efficiency of viral DNA synthesis during infection of permissive and nonpermissive cells with vif-negative human immunodeficiency virus type 1. J. Virol. *67*, 6322–6326.

Speelmon, E.C., Livingston-Rosanoff, D., Li, S. S., Vu, Q., Bui, J., Geraghty, D.E., Zhao, L.P., and McElrath, M.J. (2006). Genetic association of the antiviral restriction factor TRIM5alpha with human immunodeficiency virus type 1 infection. J. Virol. *80*, 2463–2471.

Steeves, R., and Lilly, F. (1977). Interactions between host and viral genomes in mouse leukemia. Annu. Rev. Genet. *11*, 277–296.

Stevens, A., Bock, M., Ellis, S., LeTissier, P., Bishop, K.N., Yap, M.W., Taylor, W., and Stoye, J.P. (2004). Retroviral capsid determinants of Fv1 NB and NR tropism. J. Virol. *78*, 9592–9598.

Stopak, K., de Noronha, C., Yonemoto, W., and Greene, W.C. (2003). HIV-1 Vif blocks the antiviral activity of APOBEC3G by impairing both its translation and intracellular stability. Mol. Cell *12*, 591–601.

Strebel, K., Daugherty, D., Clouse, K., Cohen, D., Folks, T., and Martin, M.A. (1987). The HIV 'A' (sor) gene product is essential for virus infectivity. Nature *328*, 728–730.

Stremlau, M., Owens, C. M., Perron, M. J., Kiessling, M., Autissier, P., and Sodroski, J. (2004). The cytoplasmic body component TRIM5alpha restricts HIV-1 infection in Old World monkeys. Nature *427*, 848–853.

Stremlau, M., Perron, M., Lee, M., Li, Y., Song, B., Javanbakht, H., Diaz-Griffero, F., Anderson, D. J., Sundquist, W. I., and Sodroski, J. (2006a). Specific recognition and accelerated uncoating of retroviral capsids by the TRIM5α restriction factor. Proc. Natl. Acad. Sci. U.S.A. *103*, 5514–5519.

Stremlau, M., Perron, M., Welikala, S., and Sodroski, J. (2005). Species-specific variation in the B30.2(SPRY) domain of TRIM5α determines the potency of human immunodeficiency virus restriction. J. Virol. *79*, 3139–3145.

Stremlau, M., Song, B., Javanbakht, H., Perron, M., and Sodroski, J. (2006b). Cyclophilin A: An auxiliary but not necessary cofactor for TRIM5alpha restriction of HIV-1. Virology *351*, 112–120.

Takeuchi, H., Buckler-White, A., Goila-Gaur, R., Miyagi, E., Khan, M.A., Opi, S., Kao, S., Sokolskaja, E., Pertel, T., Luban, J., and Strebel, K. (2007). Vif counteracts a cyclophilin A-imposed inhibition of simian immunodeficiency viruses in human cells. J. Virol. *81*, 8080–8090.

Thali, M., Bukovsky, A., Kondo, E., Rosenwirth, B., Walsh, C.T., Sodroski, J., and Gottlinger, H.G. (1994). Functional association of cyclophilin A with HIV-1 virions. Nature *372*, 363–365.

Tomonaga, K., Norimine, J., Shin, Y.S., Fukasawa, M., Miyazawa, T., Adachi, A., Toyosaki, T., Kawaguchi, Y., Kai, C., and Mikami, T. (1992). Identification of a feline immunodeficiency virus gene which is essential for cell-free virus infectivity. J. Virol. *66*, 6181–6185.

Towers, G., Bock, M., Martin, S., Takeuchi, Y., Stoye, J.P., and Danos, O. (2000). A conserved mechanism of retrovirus restriction in mammals. Proc. Natl. Acad. Sci. U.S.A. *97*, 12295–12299.

Towers, G.J., Hatziioannou, T., Cowan, S., Goff, S.P., Luban, J., and Bieniasz, P.D. (2003). Cyclophilin A modulates the sensitivity of HIV-1 to host restriction factors. Nat. Med. *9*, 1138–1143.

Turelli, P., Doucas, V., Craig, E., Mangeat, B., Klages, N., Evans, R., Kalpana, G., and Trono, D. (2001). Cytoplasmic recruitment of INI1 and PML on incoming HIV preintegration complexes: interference with early steps of viral replication. Mol. Cell 7, 1245–1254.

Turelli, P., Mangeat, B., Jost, S., Vianin, S., and Trono, D. (2004). Inhibition of hepatitis B virus replication by APOBEC3G. Science *303*, 1829.

van Manen, D., Rits, M.A.N., Beugeling, C., van Dort, K., Schuitemaker, H., and Kootstra, N.A. (2008). The effect of Trim5 polymorphisms on the clinical course of HIV-1 infection. PLoS Pathog. *4*, e18.

Venter, J.C., Adams, M.D., Myers, E.W., Li, P.W., Mural, R.J., Sutton, G.G., Smith, H.O., Yandell, M., Evans, C.A., Holt, R.A., *et al.* (2001). The sequence of the human genome. Science *291*, 1304–1351.

Vernet, C., Boretto, J., Mattei, M.G., Takahashi, M., Jack, L.J., Mather, I.H., Rouquier, S., and Pontarotti, P. (1993). Evolutionary study of multigenic families mapping close to the human MHC class I region. J. Mol. Evol. 37, 600–612.

Virgen, C., Kratovac, Z., Bieniasz, P. D., and Hatziioannou, T. (2008). Independent genesis of chimeric TRIM5-Cyclophilin proteins in two primate species. Proc. Natl. Acad. Sci. U.S.A. *105*, 3563–3568.

von Schwedler, U., Song, J., Aiken, C., and Trono, D. (1993). Vif is crucial for human immunodeficiency virus type 1 proviral DNA synthesis in infected cells. J. Virol. *67*, 4945–4955.

Wang, G.P., Ciuffi, A., Leipzig, J., Berry, C.C., and Bushman, F.D. (2007). HIV integration site selection: analysis by massively parallel pyrosequencing reveals association with epigenetic modifications. Genome Res. *17*, 1186–1194.

Warrilow, D., Stenzel, D., and Harrich, D. (2007). Isolated HIV-1 core is active for reverse transcription. Retrovirology *4*, 77.

Welker, R., Hohenberg, H., Tessmer, U., Huckhagel, C., and Krausslich, H. G. (2000). Biochemical and structural analysis of isolated mature cores of human immunodeficiency virus type 1. J. Virol. *74*, 1168–1177.

Wichroski, M.J., Robb, G.B., and Rana, T.M. (2006). Human retroviral host restriction factors APOBEC3G and APOBEC3F localize to mRNA processing bodies. PLoS Pathog *2*, e41.

Wiegand, H.L., Doehle, B.P., Bogerd, H.P., and Cullen, B.R. (2004). A second human antiretroviral factor, APOBEC3F, is suppressed by the HIV-1 and HIV-2 Vif proteins. EMBO J. *23*, 2451–2458.

Wilson, S., Webb, B., Ylinen, L., Verschoor, E., Heeney, J., and Towers, G. (2008). Independent evolution of an antiviral TRIMCyp in Rhesus Macaques. Proc. Natl. Acad. Sci. U.S.A. *105*, 3557–3562.

Wolf, D., and Goff, S.P. (2007). TRIM28 mediates primer-binding site targeted silencing of murine leukemia virus in embryonic cells. Cell *131*, 46–57.

Woo, J.S., Imm, J.H., Min, C.K., Kim, K.J., Cha, S.S., and Oh, B.H. (2006). Structural and functional insights into the B30.2/SPRY domain. EMBO J. *25*, 1353–1363.

Wu, X., Anderson, J.L., Campbell, E.M., Joseph, A.M., and Hope, T.J. (2006). Proteasome inhibitors uncouple rhesus TRIM5alpha restriction of HIV-1 reverse transcription and infection. Proc. Natl. Acad. Sci. U.S.A. *103*, 7465–7470.

Xu, L., Yang, L., Moitra, P. K., Hashimoto, K., Rallabhandi, P., Kaul, S., Meroni, G., Jensen, J. P., Weissman, A.M., and D'Arpa, P. (2003). BTBD1 and BTBD2 colocalize to cytoplasmic bodies with the RBCC/tripartite motif protein, TRIM5delta. Exp Cell Res *288*, 84–93.

Yamashita, M., Perez, O., Hope, T.J., and Emerman, M. (2007). Evidence for direct involvement of the capsid protein in HIV infection of nondividing cells. PLoS Pathog *3*, 1502–1510.

Yang, W.K., Kiggans, J.O., Yang, D.M., Ou, C.Y., Tennant, R.W., Brown, A., and Bassin, R.H. (1980). Synthesis and circularization of N- and B-tropic retroviral DNA Fv-1 permissive and restrictive mouse cells. Proc. Natl. Acad. Sci. U.S.A. *77*, 2994–2998.

Yap, M.W., Dodding, M. P., and Stoye, J. P. (2006). Trim-cyclophilin A fusion proteins can restrict human immunodeficiency virus type 1 infection at two distinct phases in the viral life cycle. J. Virol. *80*, 4061–4067.

Yap, M.W., Mortuza, G. B., Taylor, I.A., and Stoye, J.P. (2007). The design of artificial retroviral restriction factors. Virology *365*, 302–314.

Yap, M.W., Nisole, S., Lynch, C., and Stoye, J. P. (2004). Trim5alpha protein restricts both HIV-1 and murine leukemia virus. Proc. Natl. Acad. Sci. U.S.A. *101*, 10786–10791.

Yap, M.W., Nisole, S., and Stoye, J.P. (2005). A single amino acid change in the SPRY domain of human Trim5alpha leads to HIV-1 restriction. Curr. Biol. *15*, 73–78.

Yin, L., Boussard, S., Allan, J., and Luban, J. (1998). The HIV type 1 replication block in nonhuman primates is not explained by differences in cyclophilin A primary structure. AIDS Res. Hum. Retroviruses *14*, 95–97.

Ylinen, L.M., Keckesova, Z., Webb, B.L., Gifford, R.J., Smith, T.P., and Towers, G.J. (2006). Isolation of an active Lv1 gene from cattle indicates that tripartite motif protein-mediated innate immunity to retroviral infection is widespread among mammals. J. Virol. *80*, 7332–7338.

Ylinen, L. M., Keckesova, Z., Wilson, S. J., Ranasinghe, S., and Towers, G J. (2005). Differential restriction of human immunodeficiency virus type 2 and simian immunodeficiency virus SIVmac by TRIM5alpha alleles. J. Virol. *79*, 11580–11587.

Yu, Q., Chen, D., Konig, R., Mariani, R., Unutmaz, D., and Landau, N.R. (2004). APOBEC3B and APOBEC3C are potent inhibitors of simian immunodeficiency virus replication. J. Biol. Chem. *279*, 53379–53386.

Yu, X., Yu, Y., Liu, B., Luo, K., Kong, W., Mao, P., and Yu, X.F. (2003). Induction of APOBEC3G ubiquitination and degradation by an HIV-1 Vif–Cul5–SCF complex. Science *302*, 1056–1060.

Zhang, F., Hatziioannou, T., Perez-Caballero, D., Derse, D., and Bieniasz, P.D. (2006). Antiretroviral potential of human tripartite motif-5 and related proteins. Virology *353*, 396–409.

Zhang, H., Yang, B., Pomerantz, R. J., Zhang, C., Arunachalam, S.C., and Gao, L. (2003). The cytidine deaminase CEM15 induces hypermutation in newly synthesized HIV-1 DNA. Nature *424*, 94–98.

Zheng, Y.H., Irwin, D., Kurosu, T., Tokunaga, K., Sata, T., and Peterlin, B.M. (2004). Human APOBEC3F is another host factor that blocks human immunodeficiency virus type 1 replication. J. Virol. *78*, 6073–6076.

Molecular Vaccines and Correlates of Protection

12

Stephen Norley and Reinhard Kurth

Abstract

The failure of 'classical' vaccines to induce protection to the most important of all retroviruses, HIV, has led to the development of a huge variety of 'molecular vaccines', i.e. vaccines produced using modern molecular biological techniques. Such vaccines range from simple plasmid DNA coding for the genes of choice, through recombinant viruses carrying such genes to engineered bacteria designed to deliver HIV genes to the mucosal immune system. Evaluation of such vaccines in animal models has resulted in sporadic successes and many failures and the few human clinical trials have been, at best, negative. However, the relative success of molecular vaccines in combating other retroviral infections and the continuing refinement of HIV/SIV vaccines showing some efficacy suggests that a molecular AIDS vaccine may be achievable. In the end, the HIV/AIDS pandemic will only be defeated by the development of an effective, stable, and inexpensive vaccine.

Introduction

Although desirable at the level of basic science, the driving force behind the study of human pathogens is ultimately to facilitate the development of strategies for effective prevention, therapy and vaccination. Vaccines in particular are an essential part of our arsenal to combat infectious diseases and it is thanks to vaccines that many of the crippling and lethal diseases of history are now no longer of concern. Starting with Jenner and the smallpox vaccine, the development of prophylactic vaccines was, until relatively recently, a rather hit-or-miss affair, being based on the fortuitous existence or generation of a related apathogenic or attenuated microbe or on the chemical inactivation of the pathogen itself. With the dawn of modern recombinant DNA technology, it became possible to produce vaccines in an informed and targeted manner – by selecting the genes or proteins of interest or by deliberately deleting genes necessary for replication or virulence to produce 'molecular vaccines'.

Of course, the term 'molecular vaccines' could in principle be used to describe any form of vaccine (all vaccines are, after all, composed of molecules). For the purposes of this chapter, however, we will limit ourselves to vaccines developed using molecular biological techniques as opposed to those more traditional forms of vaccines that have been used for decades or even centuries (Fig. 12.1), primarily because these traditional approaches have failed. We will also concentrate predominantly on the efforts to develop and evaluate molecular vaccines to combat HIV infection, for it is in this area that by far the greatest efforts have been, and are being, made.

Background

If one studies the list of currently ongoing clinical trials for AIDS vaccines, it becomes obvious that all present-day vaccine candidates fall into the category of 'molecular vaccines'. On the face of it, this is curious, because there are relatively few licensed 'molecular vaccines' for other diseases and the vast majority of successful vaccines in the past (and indeed, at present) have been produced

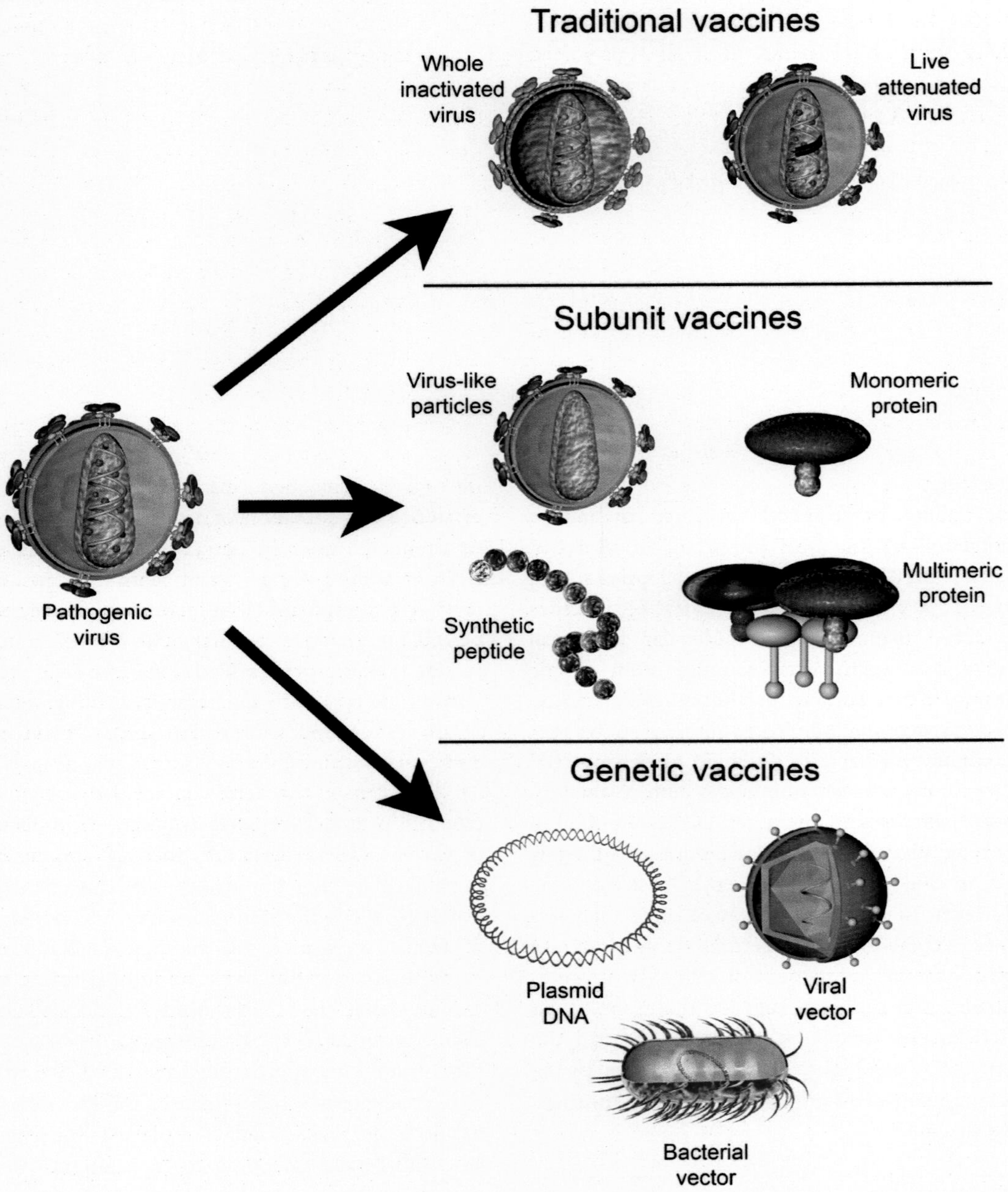

Figure 12.1 Forms of antiretroviral vaccines. Traditional forms of vaccines such as whole inactivated virus or live attenuated virus have been shown to be ineffective or inappropriate for an AIDS vaccines. Subunit vaccines are usually used to induce neutralizing antibodies but have so far failed to be effective in human trials. Most AIDS vaccine development currently concentrates on the use of genetic vaccines designed to deliver viral genes either directly or via a vector organism (replication competent or incompetent) to allow *de novo* synthesis of viral proteins and the stimulation of both humoral and cellular immune responses.

using 'traditional' methodologies (whole-inactivated virus or live attenuated virus). In order to understand why present-day AIDS vaccines concentrate almost exclusively on modern molecular technologies, it is necessary to first review the reasons for abandoning the traditional forms of vaccine production that have been so successful in the past.

Whole-inactivated virus vaccines

Many successful vaccines have been produced by simply growing and purifying the pathogen in question and subjecting the microbe to inactivation using chemicals that abolish infectivity while preserving antigenic structures. Vaccines such as the Salk polio vaccine and present-day influenza vaccines are produced in this way and have been, to a certain degree, successful. Despite initial concerns about the safety of such vaccines, it was therefore logical that attempts should be made to develop and test whole inactivated virus vaccines. As vaccination/challenge experiments for an AIDS vaccine can obviously not be performed in human volunteers, most of these studies used the simian immunodeficiency virus (SIV)/rhesus macaque animal model (Table 12.1). SIVmac and its relatives are immunodeficiency viruses that, like HIV-2, are derived from a virus (SIVsm) endemic in African sooty mangabeys (Hirsch *et al.*, 1989). SIVmac infection of Asian macaques results in a pattern of infection and disease very similar to that seen in HIV-1 infected humans (Naidu *et al.*, 1988; Kestler *et al.* ,1989; Marthas *et al.*, 1989).

In the late 1980s and early 1990s, many groups carried out experiments testing the efficacy of whole inactivated SIV as vaccine against SIVmac challenge and the results were highly promising. In most cases a small number of immunizations with chemically inactivated SIV in the presence of a good adjuvant was sufficient to induce a state of sterilizing immunity in recipient macaques (Carlson *et al.*, 1990; Gardner *et al.*, 1991; Hartung *et al.*, 1992) because upon challenge with the live virus, vaccinees appeared to be absolutely protected from infection. Owing to understandable safety concerns, however, it was considered important to determine which of the viral proteins (or combinations thereof) were responsible for this protection in the hope that such knowledge would allow the production of vaccines not involving the use of live, infectious virus. It was during the course of such experiments that it became clear that the protection induced by whole inactivated vaccines in the SIV/macaque system was entirely due to an unexpected artefact in the system. Both the virus used to produce the vaccine and that used to challenge the vaccinated animals was produced in transformed human T-cell lines, and both viruses carried human cell membrane proteins such as MHC-molecules incorporated during budding from the host cells. It was the antibody response to these xenogeneic proteins that was found to be eliminating the live virus upon challenge (Stott *et al.*, 1990; Chan *et*

Table 12.1 Animal models for AIDS vaccine development

Species	Virus	Pros	Cons
Chimpanzee	HIV-1	Closely related to humans Infectible with HIV-1	Very expensive. Major ethical problems. Lack of pathogenesis
Rhesus macaque Cynomolgus macaque Pig-tailed macaque	SIVmac (and SIVmne, SIVsm) HIV-2	Induction of disease closely resembling human AIDS. Results of preclinical trials tend to predict outcome of clinical trials.	All viruses are derivatives of SIVsm and therefore resemble HIV-2, not HIV-1
Rhesus macaque Cynomolgus macaque Pig-tailed macaque	Simian/Human Immunodeficiency Virus (SHIV)	Virus expresses (usually) the HIV-1 envelope glycoprotein, allowing Env-based vaccines to be tested.	Protection readily achieved but does not predict outcome of human trials
Transgenic or humanized mice	HIV-1	Inexpensive. Easy handling. High throughput.	Highly artificial systems of unknown relevance to the situation in humans
Cats	FIV	Relatively inexpensive. Natural infection allowing large scale trials in the wild.	Unknown relevance to the situation in humans

al., 1992), either by direct binding and neutralization or by induction of effector mechanisms such as complement-mediated virolysis (Spear *et al.*, 1993). Indeed, whereas animals vaccinated with such human-cell derived inactivated virus vaccines were fully protected from infection with human-cell derived challenge virus, they were totally susceptible to infection with the same virus passaged once through rhesus cells (and therefore bearing rhesus cellular proteins instead of human proteins; Norley *et al.*, 1998). Unfortunately, there is so far no practical way to take advantage of this artefact as a basis for an effective AIDS vaccine in humans and the results strongly indicate that an antibody response to an inactivated HIV vaccines would be ineffective in terms of protection.

Live attenuated virus vaccines

Live attenuated viruses have historically been the most effective form of vaccines, the most well-known examples of which are the smallpox vaccine, the Sabin polio vaccine and the MMR (measles–mumps–rubella) vaccine. Use of the smallpox vaccine has led to eradication of what until relatively recently was a major cause of death and suffering in the world and use of the polio vaccine has led us to the brink of eradicating poliomyelitis (Cochi and Kew, 2008). Such vaccines were usually produced either by repeatedly passaging the pathogenic virus through cell culture until loss of its pathogenic properties (e.g. for MMR) or by using a related but non-pathogenic virus. Such procedures were not available for HIV/SIV, but by the targeted deletion of non-essential genes from the SIV genome it was possible to produce a virus able to replicate in the rhesus host and induce antiviral immune responses but which did not appear capable of inducing the simian AIDS inevitably resulting from infection with the full-length virus (Kestler *et al.*, 1991). Furthermore, animals 'vaccinated' with such deleted viruses were soon shown to be fully protected against infection with the pathogenic wild-type virus (Daniel *et al.*, 1992). The degree of infection was so impressive that by the mid-1990s, plans were under way to test a multiply deleted form of HIV as a vaccine in human volunteers. These plans, and indeed the whole concept of using live attenuated vaccines in humans, came crashing down when it was shown that newborn macaques infected with the deleted SIV 'vaccine' rapidly succumbed to AIDS caused by the vaccine virus within a short period of time, presumably due to their immature immune system (Baba *et al.*, 1999). Furthermore, adult macaques infected for many years with the attenuated vaccine viruses started to develop AIDS symptoms (Baba *et al.*, 1999), as did humans accidentally infected through blood transfusion for over 12 years with a naturally occurring deletion mutant of HIV (Learmont *et al.*, 1999). These results indicated that the deleted genes were wholly unnecessary for disease induction in newborns and that even in adults disease progression was simply delayed, not abrogated. The hopes of developing a protective live attenuated HIV vaccine were therefore quickly dashed.

Many groups continue to work with live attenuated viruses in the macaque model because if an immunological mechanism for the protection induced by these viruses could be identified, it might be possible to induce similar protective immune responses using a safe form of vaccine. However, there is reason to believe that such investigations are doomed to failure, that the protection afforded by live attenuated SIV vaccines is not based on the immune response but rather on competition between the vaccine virus and the challenge virus for suitable target cells.

The failure of the whole inactivated virus vaccines to induce protection in the macaque animal model convinced many researchers that vaccines based on recombinant subunits of the virus were also likely to be ineffective. During the late 1980s and early 1990s, much effort had been put into cloning the *env* gene of HIV and to use this to produce purified envelope protein by recombinant DNA technology. It was known that the outer envelope protein, gp120, contains epitopes recognized by neutralizing antibodies and it was quickly shown that laboratory animals immunized with gp120 produced in bacteria or in eukaryotic cells responded by the production of antibodies able to neutralize laboratory isolates of HIV *in vitro* (Matthews *et al.*, 1986; Putney *et al.*, 1986). Indeed, using the only available animal model for HIV infection, it was possible to induce protective immunity in chimpanzees against infection with such laboratory strains of HIV (Berman *et al.*, 1990; Girard *et al.*, 1991). Based on such promising results, clinical trials using purified recombinant envelope protein were started and, as expected, resulted in the induction of neutralizing antibodies (Francis *et al.*, 1998). However, during the phase II clinical trial it became clear that although the neutralizing antibodies being induced could indeed prevent infection of T-cell lines in the laboratory with the corresponding laboratory isolates of HIV, primary isolates of HIV grown in human PBMC were largely unaffected (Hanson, 1994; Matthews, 1994; Mascola *et al.*, 1996). Fears that the vaccines might be leading researchers up a blind alley were further realized when a number of volunteers participating in the phase II clinical trial became infected with HIV at a time at which they had high titres of neutralizing antibodies to laboratory strains of HIV in circulation. Despite scepticism amongst the scientific community, the planned phase III trial involving many thousands of volunteers went ahead and the results were, as expected, disappointingly negative (Pitisuttithum *et al.*, 2006).

Molecular vaccines against HIV

The failure of vaccines designed to induce protective neutralizing antibodies more or less forced most AIDS vaccine development in the direction of immunogens able to induce both cellular (T-cell) and humoral (antibody) responses (Table 12.2). Although antibodies can be induced readily with protein-based immunogens, the induction of virus-specific cytotoxic T-cells usually requires the *de novo* synthesis of viral proteins within the host cells (as occurs during normal infection) with the corresponding endogenous processing and expression in the context of the MHC-I molecule at the cell surface to be recognized by the T-cell receptor, (although recently methods have been developed that sometimes allow this requirement to be circumvented). The loss of the live attenuated virus vaccine as an avenue of vaccine development meant that ways had to be found to induce viral gene expression in the vaccinee, a problem for which there are, thankfully, a wide range of solutions. These range from inoculation with DNA corresponding to the gene(s) of choice to the use of recombinant viruses and bacteria that carry and express the HIV genes (Table 12.3). Each of these approaches

Table 12.2 Potentially protective immune responses induced by candidate AIDS vaccines

Effector mechanism	Target antigen	Induced by	Efficacy *in vivo*
Neutralizing antibodies	Envelope glycoprotein (gp120/gp41)	Whole inactivated virus Live attenuated virus Subunit protein VLPs Peptides Genetic vaccines	Protection mediated by passive transfer of particular antibodies
Antibody-dependent cellular cytotoxicity	Envelope glycoprotein (gp120/gp41)	Whole inactivated virus Live attenuated virus Subunit protein VLPs Peptides Genetic vaccines	Not demonstrated
Complement-mediated lysis of virions or infected cells	Envelope glycoprotein (gp120/gp41)	Live attenuated virus Subunit protein	Not demonstrated Possible role in enhancement
Cytotoxic T-cells	Any viral protein	Live attenuated virus Modified peptides VLPs Genetic vaccines	Demonstrated by $CD8^+$ T-cell depletion and by vaccination

Table 12.3 Examples of molecular vaccines

Naked DNA – polyepitope strings
Naked DNA – whole viral genes
Poxvirus vectors (vaccinia, MVA, canarypox)
Adenovirus vectors
Adeno-associated virus
Alphavirus vectors (SFV, Sindbis)
Herpesvirus vectors (HSV, VZV)
Measles virus vectors
Rhabdovirus vectors (rabies virus, VSV)
Poliovirus vectors
Single-round infectious viruses
Bacterial vectors (*Salmonella*, *Shigella*, *Listeria*, BCG)

has its own advantages and limitations as will be described in more detail.

DNA vaccines

The concept that inoculation of 'naked DNA' corresponding to viral genes into a human or other animal could result in the *de novo* synthesis of proteins and the stimulation of both cellular and humoral immune responses was first developed in the late 1980s and early 1990s. Margaret Liu and her team succeeded in protecting mice from lethal infection with influenza virus by intramuscular injection of DNA corresponding to the influenza nucleoprotein and haemagglutinin genes (Montgomery *et al.*, 1993; Ulmer *et al.*, 1994). Because protection could be mediated by the induction of CTLs recognizing epitopes in the relatively conserved viral nucleoprotein instead of the induction of highly type-specific neutralizing antibodies recognizing the variable haemagglutin protein, the DNA vaccine could have been broadly effective in the scope of protection induced. Incidentally, although the whole field of DNA immunization took off in the 1990s, it had been known (but largely forgotten) since the early 1960s that administration of purified viral DNA (in this case papillomavirus DNA) could result in productive infection of an animal (Chambers and Ito, 1964).

The first demonstration of the induction of anti-HIV neutralizing antibody responses (in both mice and in primates) using DNA was achieved by Wang *et al.* (1993) using vaccine constructs coding for HIV gp120. This was followed by the induction of HIV-specific CTL by DNA immunization and the ability of such responses to protect mice from 'challenge' with tumour cells expressing the HIV gene (Wang *et al.*, 1994). Subsequently, cellular immune responses to HIV proteins were successfully induced in primates by a number of groups (Shiver *et al.*, 1995; Wang *et al.*, 1995; Yasutomi *et al.*, 1996; Lekutis *et al.*, 1997; Shiver *et al.*, 1997). Promising experiments in chimpanzees in which immune responses and even protection from heterologous HIV challenge were induced by DNA immunization (Boyer *et al.*, 1997; Ugen *et al.*, 1997; Bagarazzi *et al.*, 1998), plus similar data in macaques challenged with a hybrid simian/human immunodeficiency virus (SHIV) bearing the HIV envelope protein (Boyer *et al.*, 1996) led to the first phase I clinical trials in humans of such vaccines (Ugen *et al.*, 1998). As expected, the vaccines were found to be very safe. However, despite the promising results using DNA vaccines, it soon became clear that in humans and other primates such constructs were very weakly immunogenic, inducing only very low levels of antiviral CTLs and antibodies. Efforts therefore concentrated on improving the immunogenicity of naked DNA.

The initial studies of DNA vaccines used simple inoculation, usually intramuscular or intradermal and although it was possible to demonstrate long-term gene expression *in vivo* of the foreign gene, the efficiency was very low and it was unclear whether the immune system was being effectively involved. In order to recruit the dendritic cells patrolling the skin, methods were developed to administer the genes directly into the dermal cells at a high efficiency. For example, systems have been developed that blast particle-associated DNA under high pressure into the skin and achieves an efficiency far higher than can be achieved by injection of soluble DNA. Microscopic gold spheres coated in plasmid DNA shot directly into the cell's cytoplasm resulted in a higher efficiency of expression (Haynes *et al.*, 1994; Fuller *et al.*, 1996) and a reduction in the amount of DNA required to achieve measurable immune responses. More recently, methods have been developed to allow *in vivo* electroporation of DNA across the cell membrane (Titomirov *et al.*, 1991; Vanbever and Preat, 1999; Hirao *et al.*, 2008) and at a rather lower level of technical sophistication, the application of DNA using a standard tattooing machine was shown to be highly effective (Pokorna *et al.*, 2008).

In addition to improving the method of delivery, the efficiency of expression *in vivo* can be dramatically increased by modifying the nucleotide sequence of the chosen gene. By designing and synthesizing a gene using codons corresponding to those most commonly used in mammalian genes and by eliminating inhibitory sequences from the gene sequence without affecting the translated protein sequence, expression efficiencies and immune responses upon inoculation can be increased by orders of magnitude (Gao *et al.*, 2003).

The immune response induced by a DNA vaccine can also be dramatically enhanced by co-administration of DNA coding for immunomodulatory cytokine genes. For example, DNA coding for IL-2, IL-12, GMCSF and others have been used to direct and amplify the immune response in vaccinees. (Kim *et al.*, 1997, 2000; Svanholm *et al.*, 1997; Barouch *et al.*, 2000, 2002; Boyer *et al.*, 2005).

One of the advantages of a molecular vaccine designed to induce CTLs is that it is not necessary to maintain the structure of the protein that is to be expressed. Problems of conformation, correct glycosylation, multivalent structures etc. do not arise when all that is required is the production of a protein containing the correct epitope(s) because the cell that takes up foreign DNA will automatically select and correctly express the relevant protein epitopes in the context of its MHC. Indeed, it is not necessary to provide the DNA for the whole protein at all. Some groups have taken the approach of synthesizing genes that code for an artificial protein consisting of a string of CTL, T-helper and antibody epitopes in tandem, epitopes taken from different genes or even from different viruses (Hanke *et al.*, 2002).

Unfortunately, despite promising results in laboratory animals, clinical trials using such multiepitope DNA vaccines were abandoned when the strength and frequency of responses did not meet the required level (Guimaraes-Walker *et al.*, 2008). It is, however, possible that an approach such as this using codon-optimized DNA delivered bioballistically in conjunction with immunomodulatory cytokine DNA might be more successful.

Despite advances such as codon-optimization and the use of cytokine genes as 'adjuvants', the relatively poor immunogenicity of DNA vaccines in humans has meant that many vaccine strategies use DNA to prime the immune system in preparation for subsequent boosting by other forms of vaccine. As the only genes expressed as a result of DNA immunization are those intended to induce a protective immune response against the pathogen, priming with DNA does not result in vector-specific immune responses that might interfere with subsequent immunizations. Indeed, as will be discussed later, the use of different vectors whose only commonality is the gene of choice is an effective way to avoid the problems of anti-vector immunity (Fig. 12.2).

Recombinant viral vectors

Poxvirus vectors

One of the first candidate HIV vaccine to be used in humans was a replication-competent recombinant version of vaccinia virus (the smallpox vaccine) carrying the HIV envelope gene either alone or in combination with a recombinant protein boost (Zagury *et al.*, 1987, 1988a,b). This was shown to induce HIV-specific T- and B-cell

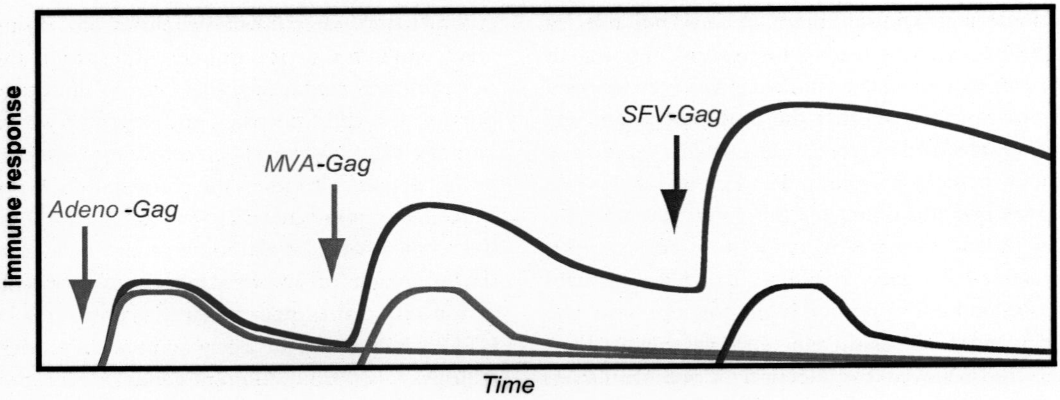

Figure 12.2 The prime-boost vaccination strategy. Immunization with recombinant vectors will result in the generation of immune responses to the vector (as well as to the protein of interest) that can limit the use of the same vector as a booster immunization. The use of multiple, different vectors expressing the same 'vaccine gene' will result in a boosting of the desired immune responses but not those directed at the vectors. In this hypothetical example, sequential immunization of adenovirus, MVA and SFV-based vectors all expressing the same HIV *gag* gene results in repeated boosting of the anti-Gag immune response but only transient induction of the vector-specific immune responses.

responses (Berzofsky *et al.*, 1988; Zagury *et al.*, 1988) and initially seemed a promising approach to vaccine development. However, it was quickly and tragically realized that the use of a live, replicating vaccinia vector in a population that may already encompass HIV-infected immunocompromised individuals could result in death from disseminated, systemic vaccinia infection under conditions of immunosuppression (Redfield *et al.*, 1987; Zagury, 1991). The use of *any* live, replicating pathogenic vector was essentially ruled out by this result, despite successful vaccination/challenge experiments in macaques (Hu *et al.*, 1992, 1993) and focus shifted to attenuated vectors that could be used to deliver the HIV gene(s) of choice to the host cell but which could not themselves replicate.

Attenuated vaccinia vectors (e.g. NYVAC) or vectors based on avian poxviruses (e.g. ALVAC) incapable of productive replication in mammals were developed (Cox *et al.*, 1993; Egan *et al.*, 1995; Pialoux *et al.*, 1995) and despite sporadic results in macaques (Franchini *et al.*, 1995; Myagkikh *et al.*, 1996) and chimpanzees (Girard *et al.*,1997), small-scale human trials were soon carried out (Egan *et al.*, 1995; Pialoux *et al.*, 1995; Ferrari *et al.*, 1997; Belshe *et al.*, 1998).

Not all poxvirus vectors are equal, however: each has its advantages and disadvantages. One popular poxvirus-based non-replicating vector is based on modified vaccinia virus Ankara (MVA). This virus, produced by multiple passage in cell culture and used as a smallpox vaccine in 150,000 Europeans without side-effects has a relatively late stop in virus production that allows for high-level gene expression. The virus was characterized and developed as a general vaccine vector (Meyer *et al.*, 1991; Sutter and Moss, 1992) before being adopted for HIV (Belyakov *et al.*, 1998; Hanke *et al.*, 1998; Seth *et al.*, 1998).

Owing to its potential as a component of an AIDS vaccine, MVA-based vectors have been used extensively, either alone or in combination with other vectors (usually DNA) to assess their protective efficacy in the macaque animal model for AIDS and reductions in viral load were achieved (Hirsch *et al.*, 1996; Ourmanov *et al.*, 2000). Promising results were also achieved using the more 'natural' mucosal challenge route (Amara *et al.*, 2001; Verrier *et al.*, 2002) and there is evidence that protection correlates with day-of-challenge antiviral $CD4^+$ T-cell responses and preservation of $CD4^+$ T-cell memory (Manrique *et al.*, 2008). Indeed, mucosal immunization using MVA was shown to be effective in inducing mucosal immunity and in reducing post-challenge virus loads (Bertley *et al.*, 2004; Wang *et al.*, 2004; Gherardi and Esteban, 2005). Furthermore, by using a challenge system in which very low levels of virus were repeatedly administered intrarectally (a scenario more relevant to real-world situations than single high-dose challenges) it was possible to show protection of some immunized macaques from systemic infection (Ellenberger *et al.*, 2006; Aidoo *et al.*, 2007).

Head-to-head comparisons of vaccine vectors based on NYVAC and MVA have shown that, despite differences in the levels of antigen expression and the nature of the immune responses induced (Gomez *et al.*, 2007) the protective efficacies against challenge in the SIV/macaque animal models were comparable (Mooij *et al.*, 2008).

The poxvirus vectors have been used to make a wide range of vaccines carrying various HIV genes and many have been or are being tested in clinical trials. Of the 23 current clinical trials, five involve MVA, three ALVAC and one NYVAC. Indeed, at the time of writing, the only candidate AIDS vaccine in phase III efficacy trials is based on a recombinant canarypox (ALVAC) expressing the *env* and *gag/pol* genes of HIV followed by an Env protein boost (Vanichseni *et al.*, 2004; Watanaveeradej *et al.*, 2006).

One potential disadvantage of using poxvirus vectors as part of an AIDS vaccine is the preexisting immunity to vaccinia in those old enough (approximately 40 years of age or older at the time of writing) to have received the smallpox vaccine. Vaccinia-specific antibodies could theoretically block the entry of the vector virus into the host cells and therefore prevent expression of the HIV genes. Certainly, the induction of anti-SIV T-cell responses in macaques previously immunized with another MVA construct was greatly diminished, suggesting at the very least that multiple boosts with MVA might be ineffective (Sharpe *et al.*, 2001). Fortunately, the majority of those most in need of an AIDS vaccine are too young to have been exposed to vaccinia (or smallpox) and are therefore immunologically naïve. As will be discussed later, the problem of pre-existing immunity poses a greater obstacle to those vectors based on viruses that are still in circulation.

Adenovirus vectors

Until relatively recently, vaccine vectors based on replication deficient adenoviruses seemed to offer one of the most promising avenues of development. However, as is the case with replication-defective poxvirus vectors, much of the earlier work was carried out with recombinant viruses able to replicate in the vaccinee. Adenovirus constructs expressing the HIV envelope gp120 were found to be immunogenic in laboratory animals (Natuk *et al.*, 1992) and in chimpanzees (Natuk *et al.*, 1993). Indeed, combining live adenoviral vectors expressing Env with a boost using purified gp120 protein gave solid protection against high-dose HIV challenge in chimpanzees (Lubeck *et al.*, 1997). In the more rigorous rhesus macaque/SIVmac challenge system, a similar strategy using adenovirus expressing SIV gp120 alone or in combination with gp120 protein, while not protecting against vaginal challenge, reduced the ensuing viral loads in vaccine recipients (Buge *et al.*, 1997).

However, in the studies using replication-competent adenovirus vectors, it was usually possible to reisolate the vaccine virus from the stools of infected animals (Buge *et al.*, 1997). Recognizing that an infectious genetically altered organism carrying HIV genes and able to spread to others was unlikely to receive public acceptance or licensing, attention turned to replication-defective variants of adenovirus as a basis of vaccine development. Like the deficient poxvirus vectors, replication-defective adenoviruses were able to deliver the gene of interest (e.g. HIV-1 *env*) to the host cells, resulting in the *de novo* synthesis of the protein and the stimulation of both cellular and humoral immune responses in mice (Bruce *et al.*, 1999). However, it was a study carried out by Shiver *et al.* (2002) comparing the immunogenicity and efficacy of replication incompetent genetic vaccines (DNA, MVA and adenovirus) in macaques that ignited interest in using the latter as an AIDS vaccine. Based, presumably, on the cellular immune response to the Gag protein incorporated into each of the vaccines, the recombinant adenovirus immunogen gave the best degree of virus load reduction after challenge with a SHIV. Adenoviral constructs were also shown to elicit immune responses in the mucosal compartments (Baig *et al.*, 2002; Lemiale *et al.*, 2003) and, most promising, an adenovirus prime/protein boost vaccination regime was shown to have a significant effect on

SIVmac viral loads following intrarectal challenge of immunized macaques (Patterson *et al.*, 2004). This combination of safety, immunogenicity and apparent efficacy against even mucosal challenge led quickly to the initiation of clinical trials in humans.

One of the greatest obstacles to using adenovirus-based vectors for immunization of humans is the seroprevalence of the general population. A large proportion of people in Europe and the USA (and indeed, Africa) have already been infected with the Ad5 strain of adenovirus (the strain most used for vector studies) and already have considerable immunity to the virus. This means that the vaccine vector is efficiently neutralized upon inoculation before it can deliver the genes of choice to the target cells. One obvious option is to re-engineer the vector system to one based on a less prevalent strain of adenovirus such as Ad7 (Nan *et al.*, 2003) or Ad35 (Vogels *et al.*, 2007). An alternative approach is to use the established Ad5-based vector in which the proteins recognized by existing antibodies are replaced with those from a less common form of adenovirus (Xin *et al.*, 2005, 2007).

As mentioned above, adenovirus-based vectors were used in a number of clinical trials, the most well-known of which is the Merck STEP trial involving 3000 volunteers at risk for HIV infection. The vaccine, a mixture of Ad5-based constructs expressing the *gag*, *pol* and *nef* genes of HIV-1, was designed to elicit protective T-cell responses and was administered three times. Late in 2007, however, it became clear that not only were the vaccinees showing no evidence of protection, those with pre-existing high levels of anti-adenovirus antibodies had an apparently increased risk of HIV infection compared to those in the placebo control group (Sekaly, 2008). The clinical trial was immediately halted and other similar trials were put on hold.

Due to the *post hoc* nature of the analyses, it is not possible to conclude conclusively that the vaccine had indeed increased the risk for infection in those recipients with a strong anti-vector immunity. It was suggested that the high levels of adenovirus-specific antibodies in the 'increased risk' group reflected a correspondingly high frequency of adenovirus-specific T-helper cells that, upon immunization with the Ad5/HIV constructs, became activated and expanded, offering an ideal pool of susceptible target cells for HIV. Until now, there are no actual data to support this scenario, and the unfortunate failure to include a group of volunteers receiving an 'empty' adenovirus control vector in the STEP trial plus the presence of confounding factors means that the decision to halt other vaccine trials, although understandable, may prove to be premature. However, the most critical outcome of the trial that is often forgotten in the discussion of possible enhancement of infection, is that this most promising of vaccine candidates failed to shown any degree of protective effect whatsoever. This, more than anything, has forced the scientific community to re-evaluate the approach to AIDS vaccine development (Robb, 2008; Sekaly, 2008; Watkins *et al.*, 2008).

Adeno-associated virus vectors

Adeno-associated virus (AAV) is a 20-nm icosahedral parvovirus that lacks the ability to replicate autonomously and is thought to be non-pathogenic in humans, despite up to 80% of the human population being seropositive for one of the 11 different serotypes of the virus. As the name suggests, AAV relies on genes provided by a helper adenovirus to replicate. Wild-type AAV, thanks to its *rep* gene, can integrate at a specific site in the host cell chromosome and this has no known detrimental effects. However, the recombinant form of the virus in which virtually the entire AAV genome, including the *rep* gene, is replaced by the gene of interest and its promoter, can integrate randomly, which could theoretically lead to the induction of cancer. However, experiments *in vitro*, in laboratory animals and clinical trials indicate that the likelihood of such side-effects is vanishingly small.

The virus has been used quite extensively as a vector for gene therapy thanks to its property of prolonged gene expression in even terminally differentiated and non-dividing cells. These factors make AAV a safe and attractive candidate for a vaccine vector, and indeed, a number of groups have developed the virus for this use in the AIDS field. Mice immunized systemically or orally with AAV recombinants, for example, were shown to develop strong cellular and humoral immune responses to the HIV proteins coded for by the vaccine vectors (Xin *et al.*, 2001, 2002) and macaques receiving a single shot of a recombinant AAV vector expressing SIV genes mounted antibody and T-cell responses and, upon challenge with pathogenic SIV, showed a significantly suppressed virus load (Johnson *et al.*, 2005).

Based on such results, clinical trials with AAV/HIV-based vaccines were carried out in Europe and India and, while safe, were found to be poorly immunogenic in humans (Mehendale *et al.*, 2008), although trials with multiple shots at higher doses are continuing in Africa. This disappointing result was compounded by data suggesting that the antigen-specific CD8$^+$ cytotoxic T-cells induced by the recombinant AAV vectors are, in mice at least, functionally impaired with regard to anti-HIV efficacy (Lin *et al.*, 2007; Lin *et al.*, 2007).

One of the 'holy grails' of AIDS vaccine research is the induction of broadly specific neutralizing antibodies able to block any incoming virus from infecting its target cells. Although the extreme variability of HIV's envelope glycoprotein renders most antibodies very type-specific with regards to neutralization, a small number of conserved epitopes have been identified, predominantly in the gp41 transmembrane glycoprotein, that can act as targets for broadly reactive neutralizing antibodies (Muster *et al.*, 1994; Stiegler *et al.*, 2001). Human monoclonal antibodies with such properties have been generated and have been shown to be effective in preventing infection when administered as a cocktail to macaques (Baba *et al.*, 2000; Hofmann-Lehmann *et al.*, 2001; Ferrantelli *et al.*, 2003; Ruprecht *et al.*, 2003) but all efforts to induce such antibodies by immunization have so far failed, due possibly in part to the unusual nature of the antibodies themselves. The human antibodies specific for defined, conserved regions of the HIV envelope transmembrane gp41 that do possess the broadly neutralizing activity desired appear to have the unusually long CDR H3 domains (Zwick *et al.*, 2004; Ou *et al.*, 2006) and may be actually a form of autoantibody (Alam *et al.*, 2007; Martinez *et al.*, 2009; Shen *et al.*, 2009). If so, the types of antibodies induced by immunization may not have the reach to bind to their epitopes in the confined space of the envelope/receptor complex that reveals them.

In a novel approach to overcoming this problem, AAV vectors have been produced that contain the gene coding for IgG1b12, one of these broadly reactive neutralizing antibodies (Lewis *et al.*, 2002). The somewhat counter-intuitive concept here is to circumvent the apparent difficulties involved in inducing broadly cross-reactive antibodies via immunization by providing the recipient with the genes necessary for producing the antibodies *in vivo* by gene transfer. Passively transferred antibodies of this type are known to be effective in macaques and in humans, and this new strategy of '*in situ* passive transfer' may allow the continuous production and maintenance of the high levels of antibodies needed to protect against infection.

Alphavirus vectors

These members of the family Togaviridae, such as Sindbis, Venezuelan equine encephalitis virus (VEEV) and Semliki Forest virus (SFV), offer a range of features desirable in a vaccine vector. Like AAV, recombinant alphavirus vectors can be produced by replacing essential viral genes with the 'vaccine' gene of choice. Expression of such constructs in cells that provide the missing genes in *trans* results in the production of single-cycle virus particles able to infect and replicate in target cells but not transmit to neighbouring cells. Vectors based on alphaviruses can infect a wide range of cells and have a tropism for monocytes and dendritic cells (MacDonald and Johnston, 2000; Ryman *et al.*, 2000). Target cells produce a high amount of the transgene product, albeit transiently, but as replication occurs exclusively in the cytoplasm there are none of the potential hazards posed by vectors that integrate into the host genome. Finally, antivector immunity, a significant problem for vaccines based on viruses such as adenovirus, is virtually non-existent in the population.

Experimental vaccines based on VEEV have been tested for immunogenicity and efficacy in macaques and have been shown to be able to reduce post-challenge viral loads (Davis *et al.*, 2000; Johnston *et al.*, 2005). Based on such studies, human trials of VEEV-based immunogens carrying the HIV-1 *gag* gene have been carried out. However, although the use of Semliki Forest virus based vectors as part of a multivector vaccination strategy in macaques showed some degree of efficacy against SIVmac challenge (Koopman *et al.*, 2004; Michelini *et al.*, 2004; Negri *et al.*, 2004; Stolte-Leeb *et al.*, 2006; Maggiorella *et al.*, 2007; Martinon *et al.*, 2008), no human trials involving this vector have so far been carried out.

Herpesvirus vectors

Herpes simplex virus (HSV)-1 is itself a relatively pathogenic virus that although usually associated with recurring lesions of the lips can cause fatal encephalitis. Non-replicating HSV vectors, however, offer a number of potential benefits, including a relatively large capacity for foreign genes. HSV is able to infect a very broad range of host cells, including non-dividing cells, with a high efficiency and is a strong activator of the innate immune system which may have a strong adjuvant effect (Dudek and Knipe, 2006). As with some of the other vectors, two basic approaches have been used to produce vaccine vectors based on HSV.

The first is to simply remove viral genes needed for productive infection (Samaniego *et al.*, 1998; Watanabe *et al.*, 2007) plus those that normally shut down host cell transcription and translation and cause downregulation of MHC. This allows a prolonged production and expression of the protein of choice. HSV-based vectors expressing the envelope and Nef proteins of SIVmac were shown to induce prolonged immune responses in macaques and to protect against high viral loads or even infection upon challenge with SIVmac (Murphy *et al.*, 2000). In an attempt to further improve on these results, a mixture of HSV vectors expressing the Gag, Env and Tat-Rev-Nef proteins of SIVmac was used to inoculate macaques (with and without prior DNA priming), and was shown to generate robust cellular immune responses and neutralizing antibody responses and, upon challenge with SIVmac, to reduce the virus loads in vaccinated animals (Kaur *et al.*, 2007).

The second approach to producing 'safe' herpesvirus vectors is to produce amplicons by which essential genes are provided in *trans* during particle production yielding a vector virus virtually devoid of HSV genes. One such construct expressing the HIV envelope gp120 was able to induce prolonged and robust cellular and humoral immune responses in mice after only a single inoculation (Hocknell *et al.*, 2002). Owing possibly to the technical challenges in producing HSV-based vectors at a large scale and the known high pre-existing immunity to HSV in the human population, there has been relatively little done to exploit this system as a basis for an AIDS vaccine. However, there is a very active field of developing replication defective HSV as a vaccine against HSV itself (Lu *et al.*, 2008), and it is likely that advances in this area will continue to be co-opted for AIDS vaccine development.

HSV is not the only member of the herpesvirus family to be used as a basis for an experimental AIDS vaccine. One of the more promising candidates has been varicella-zoster virus (VZV) because the attenuated form of the virus which is able to establish a persistent infection in humans is licensed and widely used as a safe vaccine against chicken-pox, even under conditions of immunosuppression. A recombinant form of the Oka VZV vaccine expressing the HIV-1 envelope glycoprotein was shown to induce both cellular and humoral immune responses against this protein in small laboratory animals (Shiraki *et al.*, 2001). A similar virus expressing SIV Env was therefore used to immunize rhesus macaques before challenge with SIVsmE660 (Staprans *et al.*, 2004). In the precise opposite of the result one would hope for, immunized animals were found to experience virus loads many orders of magnitude higher than those receiving 'empty' VSV vaccine or no vaccine at all. These dramatically increased virus loads were associated with severely depressed $CD4^+$ T-cell counts and an acceleration of disease progression. Careful analysis of the immune responses uncovered a failure of the vaccine to elicit a $CD8^+$ cytotoxic T-cell response. Furthermore, the antibodies specific for the SIV envelope glycoprotein, while fully able to bind the protein, were not able to neutralize the virus. This might suggest the induction of 'enhancing' antibodies, a phenomenon often described *in vitro* but rarely demonstrated *in vivo* for HIV and SIV. However, the fact that the enhancement of infection seemed to correlate directly with the early expansion of $CD4^+$ T-cells in the immunized macaques indicates that vaccination had primed SIV-specific T-helper cells to rapidly expand upon contact with the challenge virus, providing a large pool of readily infectible target cells for what is generally considered to be a relatively 'wimpy' challenge virus.

It should be noted, however, that this dramatic enhancement of infection and disease is probably an artefact of the animal model. Human VZV does not replicate well in macaques and this would explain the failure of the recombinant vaccine to induce a $CD8^+$ T-cell response. The situation would be considerably different in a susceptible host such as humans. To address this point, recombinant vaccines expressing the SIV envelope glycoprotein and Gag protein but based on simian varicella virus have been produced and shown to induce the desired anti-SIV cellular and humoral immune responses in primates (Ou *et al.*, 2007). It remains to be seen whether such responses are protective.

Measles virus vectors

As is the case with VZV, vaccines against measles have a long track record of safety and efficacy in the general population. An AIDS vaccine based on such a successful virus therefore has considerable appeal. An attenuated measles virus vector expressing the HIV-1 envelope glycoprotein was shown to induce in mice and in macaques antibodies able to neutralize both a SHIV carrying the corresponding *env* gene and a diverse range of primary HIV-1 isolates (Lorin *et al.*, 2004). Furthermore, high levels of both $CD4^+$ and $CD8^+$ T-cells specific for HIV-1 Env were induced. Recombinant viruses based on a licensed childhood measles vaccine and expressing a string of known human CTL epitopes were also shown to induce strong CTL responses in transgenic mice expressing the HLA-A0201 MHC-I molecule (Lorin *et al.*, 2005).

Since the advent of childhood vaccination against measles, most members of industrialized nations, at least, are seropositive for measles virus and it is therefore important that mice and macaques with a pre-existing immunity to measles virus could also react to vaccination with an HIV-specific immune response (Lorin *et al.*, 2004).

One of the fears in using a replication-competent recombinant virus, although highly attenuated, is that incorporation of HIV proteins into the vaccine virus might alter the cellular tropism and therefore the potential pattern of disease. It is therefore reassuring that recombinant measles virus vectors expressing SIV proteins show no evidence of having these foreign proteins incorporated into the virions themselves (Zuniga *et al.*, 2007).

Rhabdovirus vectors

The rhabdovirus vesicular stomatitis virus (VSV) has recently attracted much attention as a potential vehicle for delivery of HIV genes to the vaccine recipient (Clarke *et al.*, 2006). Apart from its very low seroprevalence in humans, a feature that avoids the problem of pre-existing immunity, VSV can infect a wide range of host cells, has a relatively large capacity for foreign gene expression and is suitable for large-scale growth in cells used for vaccine production.

VSV can be used in a number of ways as a vaccine vector. The 'classic' method is to use an attenuated form of the virus into which the HIV or SIV genes of choice have been inserted. This was initially done by Rose *et al.* (2001) who inoculated macaques with VSV recombinants expressing the HIV *env* and SIV *gag* genes. To avoid the anti-VSV immunity induced by the initial immunization, booster immunizations were performed using recombinant VSVs (rVSV) expressing glycoproteins from different serotypes but carrying the same HIV and SIV genes. Upon challenge with the appropriate pathogenic SHIV, vaccinated animals, although infected, showed only very low or undetectable virus loads and remained healthy for over a year. In contrast, most of the control animals progressed to AIDS within 6 months. A comparison of the route of delivery of these VSV-based vectors found that intranasal inoculation, despite having no effect on the levels of systemic and mucosal antibodies induced, was superior to intramuscular administration in terms of the virus-specific T-cell responses (Egan *et al.*, 2004). Both routes protected against intravaginal challenges with the pathogenic SHIV. The immune responses induced by these replicating VSV vectors was also shown to be enhanced by prior priming with plasmid DNA coding for the same HIV and SIV genes, even with intranasal delivery of the recombinant VSV vectors (Egan *et al.*, 2005). In this study, there was a clear correlation between the prechallenge levels of virus-specific T-cells and neutralizing antibodies. Similarly, boosting the responses induced with the rVSV vectors with MVA constructs expressing the same genes gave a better protective effect after challenge than did use of the rVSV vectors alone (Ramsburg *et al.*, 2004).

Despite these promising initial results, understandable concerns about the potential neurovirulence of replication-competent VSV vectors remain. In an attempt to address this potential problem, attenuated VSV has recently been further modified to dramatically reduce the inherent neurovirulence and, when used to deliver the HIV *gag* gene, to show enhanced immunogenicity (Cooper *et al.*, 2008).

To fully abrogate concerns of neurovirulence, the safest approach is to ensure that the vaccine virus cannot replicate. This was achieved by producing single-round vectors lacking the VSV G-protein (which was provided in *trans* during production). A side-by-side comparison of replication competent and single-cycle VSV recombinants showed, somewhat surprisingly perhaps, that the latter were as immunogenic as the former (Publicover *et al.*, 2005). To overcome the unlikely but possible generation of particles bearing the HIV Env protein in place of the VSV G-protein that could theoretically replicate, like HIV itself, in $CD4^+$ T-cells, recombinant single-cycle VSV particles were produced that coded for a secreted, rather than membrane-bound form of Env. Such constructs were found to be equally efficient at inducing Env-specific B-cell and T-cell responses.

However, another approach to utilize the beneficial properties of VSV, such as its tropism for antigen-presenting cells, is to make pseudovirions consisting of, for example, the HIV-1 Gag protein but expressing the VSV glycoprotein on their surface. Plasmid DNA constructs coding for such particles were used to immunize mice to allow *de novo* particle production *in vivo*, and were shown to be superior at inducing $CD8^+$ T-cell responses (Marsac *et al.*, 2002). Immunodeficiency virus-like particles pseudotyped with the VSV G-protein were shown to be highly immunogenic in mice (Racek *et al.*, 2006) and experiments in macaques using similar vectors showed clear (albeit limited) effects on SIVmac viral loads in immunized and challenged animals (Kuate *et al.*, 2006). Single-cycle SIV particles pseudotyped with VSV G protein and expressing macaque interferon-γ were also shown to be immunogenic and to be efficient at priming for anti-SIV T-cell responses when transduced into dendritic cells (Peng *et al.*, 2007).

VSV is not the only rhabdovirus being developed as a possible vector for vaccination against HIV infection. In what appears at first glance to be counter-intuitive, the virus causing one of the most feared of humans disease, rabies, has properties that make it attractive for such use. Schnell *et al.* (Schnell *et al.*, 2000) engineered a replication-competent rabies virus to express the HIV-1 envelope glycoprotein and showed that infection of mice with the recombinant virus and boosting with purified protein resulted in the induction of high titres of antibodies able to neutralize HIV. Vectors expressing HIV Gag could also induce very high levels of Gag-specific $CD8^+$ T-cells in mice (McGettigan *et al.*, 2001). To further allay understandable concerns about the safety of a replication competent rabies virus vector, the vaccine virus was further attenuated by selected mutations and deletions within the G protein (McGettigan *et al.*, 2003) and both the safety and immunogenicity of such recombinant rabies virus vectors could be enhanced by including the gene for IL-2 alongside the HIV genes (McGettigan *et al.*, 2006). Using recombinant rabies virus and VSV constructs expressing the same HIV gene in a heterologous prime-boost was found to elicit good humoral and cellular immune responses to HIV, including mucosal IgA, in mice (Tan *et al.*, 2005). Highly attenuated rabies virus vectors expressing HIV Env or SIVmac Gag were also used to immunize macaques but despite seroconversion to Env, no cellular immune responses could be detected (McKenna *et al.*, 2007). To overcome the problem that immunity to the original rabies virus made boosting with the same vaccines ineffective, new rabies virus vectors engineered to express the VSV G-protein were used to successfully re-immunize the macaques before challenge with the corresponding SHIV. Immunized animals were indeed shown to have reduced virus loads and protection from disease progression (McKenna *et al.*, 2007).

Poliovirus vectors

Coming only second to the smallpox vaccine in terms of global success, the Salk inactivated polio vaccine and the Sabin attenuated poliovirus vaccine have proven themselves to be both highly effective and safe. The live attenuated virus makes an especially attractive vaccine vector candidate because of its property of inducing mucosal immunity. Indeed, poliovirus vectors expressing the HIV p17 Gag and gp41 Env proteins, when administered by the intravenous and intranasal routes to macaques, induced rectal and vaginal HIV-specific IgA responses, long-lasting serum IgG responses and systemic proliferative and cytotoxic T-cell responses (Crotty *et al.*, 1999). One significant drawback of a vaccine vector based on poliovirus is, however, the relatively low capacity for expressing foreign DNA. This problem was overcome by Crotty *et al.* (2001), who made panels of many recombinant polioviruses expressing overlapping fragments of the SIV Gag, Pol, Env, Nef and Tat gene products and used these to immunize seven macaques before intravaginal challenge with pathogenic SIVmac. Two immunized animals appeared to have sterilizing immunity to challenge and another two showed severely suppressed post-challenge virus loads. Whereas all vaccinated monkeys remained healthy for the period of the study, half of the non-immunized control animals developed clinical AIDS.

Replication-incompetent poliovirus replicons expressing HIV Env and SIVmac Gag, when used in a prime-boost regime with purified Env protein, were also shown to rapidly clear the SHIV challenge virus and remain healthy whereas control animals had severe $CD4^{+}$ T-cell loss and died within 62 weeks (Fultz *et al.*, 2003).

One obvious drawback of an AIDS vaccine based on the virus used so successfully to protect humans from poliomyelitis is the fact that much of the current world population have already received the polio vaccine and would therefore have a strong pre-existing immunity to the putative AIDS vaccine. This could be overcome by using divergent serotypes of poliovirus, or indeed by combining vaccination against poliomyelitis and AIDS in one vaccine. It would be rather ironic if a successful AIDS vaccine were to be based on poliovirus given the unfounded (Cohen, 2001) but highly publicized accusations that the AIDS epidemic was initiated by the use of contaminated polio vaccines (Hooper, 2000, 2001).

Single-round infectious viruses

As mentioned earlier in the chapter, the most successful form of immunogen so far tested in the SIV animal model for AIDS is the live attenuated virus, but the fact that even these attenuated viruses can cause rapid AIDS in newborn monkeys and eventually even in adults rendered their use as a human vaccine impossible. However, assuming that the immune response induced by such live attenuated viruses is the reason for the protection induced (an assumption which still remains dubious because no such immune correlate of protection has yet been identified), an obvious approach would be to use an infectious virus that resembles HIV (or SIV) in every way except for its ability to replicate and cause disease.

Several groups have therefore generated and tested so-called 'single-cycle' viruses based on SIV for their safety and vaccine efficacy. One of the first was a nucleocapsid deletion mutant able to complete most steps in the infectious virus replication cycle but unable to produce progeny particles and that, while indistinguishable morphologically from wild-type virus, contain no viral RNA (Gorelick *et al.*, 2000). Macaques inoculated with DNA coding for such viruses, although not protected from challenge, were able to control the subsequent viraemia over a two-year period. A single-cycle immunodeficiency virus deficient in the *vif* gene was able to induce detectable levels of the defective vaccine virus in the plasma of inoculated macaques and stimulated cellular and humoral immune responses, although no protection against subsequent intravenous challenge with pathogenic SIVmac was achieved (Kuate *et al.*, 2003). Another approach was to introduce mutations into the SIV *Gag–Pol* frameshift site and others in the *pol* gene resulting (after co-transfection with a second plasmid providing *Gag–Pol in trans*) in the production of viruses able to infect and to undergo one round of replication but unable to initiate a productive infection (Evans *et al.*, 2004). Such viruses, further engineered with mutations in *nef* to avoid down-regulation of MHC-I, induced transient viraemia in inoculated macaques and protected some animals from high viral loads upon intravenous challenge with SIVmac239 (Evans *et al.*, 2005).

Single round viruses pseudotyped with the VSV G-protein to facilitate uptake and processing by dendritic cells were found to induce good cellular and humoral immune responses in macaques and to dramatically reduce plasma and cell-associated virus loads after SIVmac challenge (Kuate *et al.*, 2006). More recently, a single-round virus based on a molecular clone of SIVsm with engineered defects in *vif*, *nef* and *env* was also produced and pseudotyped with the VSV G-protein (Tang and Swanstrom, 2008).

An interesting variation of the single-round virus concept was used by Zheng *et al.* (Zheng *et al.*, 2008) who introduced a single mutation in the catalytic domain of the SIVsm integrase gene. Infection of macrophages with this virus resulted in the prolonged production of all viral proteins without genomic integration or the formation of infectious progeny virus. Although a single mutation would be insufficient to guarantee safety *in vivo*, the concept of a non-integrating single-round virus as a vaccine has significant potential.

Interestingly, single-round viruses have also been used to estimate the burst size of SIV *in vivo*. By infecting autologous cells at a known multiplicity of infection before reinfusion into macaques and measuring plasma viral loads, it was possible to determine that SIV produces approximately 5×10^4 progeny virus from every infected cell *in vivo* (Chen *et al.*, 2007).

Recombinant bacterial vectors

Until now, we have only discussed vaccine vectors based on naked DNA or on viruses. However, some strains of bacteria also offer potential as delivery vehicles for the genes of choice.

Salmonella vectors

Salmonella is of particular interest because attenuated salmonella vaccines are known to induce good mucosal immunity, important for a vaccine vector designed to prevent infection with a virus whose most common routes of entry are the rectal and vaginal mucosal surfaces. Recognizing this, recombinant constructs based on the vaccine strain of *Salmonella typhi* engineered to express the HIV Env, Gag and Nef proteins were produced as early as 1994 (Hone *et al.*, 1994) and a construct expressing gp120 was tested in phase I human trials by 1997 (Doepel, 1998; Gilbert *et al.*, 2003). Other constructs based on the *typhimurium* strain of *Salmonella* were shown to induce antibody and T-cell responses in mice (Berggren *et al.*, 1995; Fouts *et al.*, 1995) and intragastric delivery of a recombinant bacteria expressing HIV-1 Env induced specific $CD8^+$ T-cells in both systemic and mucosal lymphoid tissues (Shata *et al.*, 2001). Similarly, salmonella carrying DNA coding for an HIV-1 polyepitope CTL string could stimulate antibody, lymphoproliferative and cytotoxic T-cell responses in mice (Karpenko *et al.*, 2004).

Vaccination and challenge experiments have been carried out in macaques using both the *typhimurium* and *typhi* strains of *Salmonella* in which fragments of the SIV Gag protein are produced as part of proteins of the salmonella type III secretion system (Evans *et al.*, 2003). T-cell responses were transient and weak but were significantly higher after a booster immunization with recombinant MVA. However, despite the generation of these Gag-specific T-cells in both the peripheral blood and in the colon, there was no improvement of virus control after intrarectal challenge with SIVmac239. When tested in humans (Kotton *et al.*, 2006), similar constructs were found to be safe and disappointing with regards to immunogenicity. However, it is possible to modify the proteins to be expressed using this system to allow better secretion, presentation by the MHC-I and immunogenicity (Chen *et al.*, 2006).

As mentioned earlier, it is possible to optimize a gene to be used in a genetic vaccine for optimal codon usage in mammalian cells. For certain genes, particularly HIV or SIV *gag*, this can result in a huge increase in protein production and hence immunogenicity. In an interesting twist of this concept, it was found that modifying the sequence of *gag* for optimal codon usage in salmonella itself resulted in dramatic increase in the transgene expression (Tsunetsugu-Yokota *et al.*, 2007).

Finally, salmonella vectors in which the HIV *gag* gene is incorporated into the bacterial chromosome and the *env* gene is carried by a plasmid were shown, after intranasal inoculation, to induce high titres of serum IgG and mucosal IgA as well as systemic CTLs (Feng *et al.*, 2008). It seems clear that although the initial promise of a salmonella-based AIDS vaccine has yet to be fulfilled, there are many options for improving this form of recombinant vaccine.

Shigella vectors

Salmonella is not the only enteric bacteria that has been exploited as a delivery vehicle for HIV/SIV DNA. A single intranasal inoculation of mice with shigella carrying DNA coding for the HIV-1 gp120 was shown to induce strong CD8$^+$ T-cell responses and to afford significant protection against challenge with a recombinant vaccinia virus expressing gp120 (Shata and Hone, 2001). A side-by-side comparison of two salmonella and one highly attenuated shigella constructs indicated that the latter induced stronger T-cell responses and mucosal antibody responses (Vecino *et al.*, 2002). Intranasal inoculation with shigella carrying HIV *gag* DNA was as efficient as intramuscular injection of the naked DNA at stimulating Gag-specific cellular immune responses and was particularly efficient at boosting the response in mice primed with *gag* DNA (Xu *et al.*, 2003).

Listeria vectors

A third such bacterial vector is provided by *Listeria monocytogenes.* This Gram-positive intracellular pathogen is known to elicit strong cell-mediated immune responses against its own secreted proteins and, presumably, would do the same for any foreign protein expressed in the correct manner. Frankel *et al.* (Frankel *et al.*, 1995) showed this to be the case in mice using Listeria in which the genome had been modified to carry the HIV-1 *gag* gene. Hyperattenuation of the bacteria to diminish safety concerns by making it dependent on D-alanine for viability did not diminish immunogenicity (Friedman *et al.*, 2000) and immunized mice developed long-term CTL responses in a variety of tissues and were shown to be protected against challenge with vaccinia virus expressing Gag (Mata *et al.*, 2001; Rayevskaya and Frankel, 2001; Lieberman and Frankel, 2002), even when the challenge was performed via the vaginal route (Zhao *et al.*, 2006). Furthermore, the recombinant, attenuated bacteria were shown to be safe and immunogenic even in neonatal mice that have immature immune systems (Rayevskaya *et al.*, 2002). When used in a DNA-prime/*Listeria* oral boost vaccination regime in macaques (Neeson *et al.*, 2006), the recombinant bacteria carrying the SIV *gag* gene was shown to contribute to the induction of SIV Gag-specific CD8$^+$ T-cells in the peripheral blood, gut tissue, intraepithelial and lamina propria lymphocytes of the duodenum and ileum. Of particular note was the fact these cells expressed high levels of the gut-homing receptor α4β7 integrin and were predominantly of the effector memory phenotype – a positive characteristic for a vaccine designed to combat a virus such as HIV that initially replicates to high levels in the gut-associated lymphoid tissue (Centlivre *et al.*, 2007; Nilsson *et al.*, 2007). Using the recombinant bacteria alone to prime macaques via the oral route, it was shown that the route of a booster vaccination (oral or intramuscular) made a critical difference with regards to the type of immune response (Th1 vs, Th2) induced (Jiang *et al.*, 2007). Differences in the types of location of the cellular immune response to HIV-1 Gag was also shown in mice to be dependent of the route of booster immunization with replication-defective adenovirus HIV Gag constructs after *Listeria* priming (Li *et al.*, 2008).

Listeria monocytogenes produces a protein named listeriolysin O that is responsible for freeing the bacteria from the phagolysosome (via which it enters the cell by phagocytosis) allowing it to replicate in the cytoplasm. It is this property that makes Listeria so efficient at stimulating the CD8$^+$ T-cell response and incorporating the listeriolysin gene into bacteria that normally remain in the phagolysosome has been used to enhance their ability to induce CTLs (Dietrich *et al.*, 2001). Indeed, DNA constructs coding for a fusion protein consisting of HIV-1 Gag or Env plus a fragment of the listeriolysin O protein were shown to be superior to the unmodified constructs in inducing both humoral and cellular immunity (Bu *et al.*, 2003; Ye *et al.*, 2003).

BCG vectors

BCG (bacille Calmette–Guerin), an avirulent strain of *Mycobacterium bovis,* has been used as a vaccine against tuberculosis in billons of people over many decades and in terms of safety therefore has obvious potential as a vector for AIDS vaccine development. A method to insert foreign DNA into the mycobacterium was first published in 1987 (Jacobs *et al.*, 1987) and many groups since then have produced vectors expressing a variety of HIV and SIV genes and gene fragments. BCG carrying the HIV *gag*, *pol* and *env* genes were shown to induce cellular and humoral immune responses in mice (Fuerst *et al.*, 1991) and constructs expressing the SIV *gag* gene stimulated Gag-specific CTL in macaques (Yasutomi *et al.*, 1993) although animals receiving a BCG-based vaccine designed to stimulate a CTL response to a single epitope were not protected from challenge (Yasutomi *et al.*, 1995). BCG engineered to express the V3-loop of the HIV-1 gp120 was shown to induce neutralizing antibodies in guinea pigs and CTL in mice (Honda *et al.*, 1995), as did constructs expressing part of the SIV envelope glycoprotein (Lim *et al.*, 1997). Mice receiving, via a variety of mucosal routes, BCG engineered to express SIV genes responded with mucosal and systemic antibodies and systemic CTL (Lagranderie *et al.*, 1998) and similar results were achieved in macaques (Leung *et al.*, 2000). However, similar constructs failed to protect macaques against rectal SIVmac challenge (Mederle *et al.*, 2003). Similar to the strategy used for salmonella, optimization of codons used in the HIV *gag* gene for expression in BCG led to a dramatic increase in protein production and immunogenicity in mice (Kanekiyo *et al.*, 2005). Finally, in an effort to overcome the potential problem of the virtually global pre-existing immunity to BCG, recombinant vectors based on the related *Mycobacterium smegmatis* expressing the HIV *env* gene were produced and shown to stimulate HIV-specific T-cell responses in mice (Cayabyab *et al.*, 2006).

Combination vaccines

A recurring theme with many of the vaccine vectors being developed for potential use as an AIDS vaccine has been the need to overcome pre-existing immunity in the general population. Vaccines based on viruses that cause common infections, such as adenovirus, are particularly susceptible to such obstacles, as are attenuated vectors already used as vaccines against the pathogenic form of virus, such as poliovirus and measles virus. To some degree, these obstacles can be overcome by selecting virus serotypes not generally present in the target population or by engineering the vaccine virus to express surface proteins derived from rare serotypes.

However, most replication defective virus vaccines are relatively poor at stimulating a robust immune response after only one immunization and often require a series of booster immunizations to achieve the desired level immunogenicity. Once again, the problem of pre-existing anti-vector immunity can severely abrogate the efficacy of these booster shots. Although one solution is to use a series of vectors based on different strains of the vaccine virus, not all immune reactions are precisely type-specific or there may not be a series of naturally occurring strains available on which to base the different vectors.

The obvious solution to this dilemma is to use a series of immunizations based on entirely different forms of vectors but carrying and expressing the same retrovirus genes. Immunization with DNA, although often poorly immunogenic alone, is generally thought to be an excellent means of priming the immune response for subsequent boosting by other vectors. Part of DNA's attraction is the lack of any 'vector' proteins, either as part of the vaccine or produced *in situ*, that could stimulate an inappropriate or irrelevant immune response: the recipient is primed to respond to proteins coded for by the gene of choice alone.

Many different combinations of the genetic vaccines described so far in this chapter have been tested for immunogenicity and protective capacity with wildly differing degrees of success. However, systematic evaluation of different vaccination strategies using the same vaccine genes in different combinations of vectors are very rare. In one such large collaborative study comparing different combinations of the SIV *gag*, *pol*, *env*, *tat*, *rev* and *nef* genes delivered by DNA, SFV and MVA vectors, even the order of inoculation was found to have a profound effect on the outcome. For example, administration of the multigenic vaccine in a DNA, SFV, SFV, MVA regime resulted in full protection or a dramatic reduction of viral load (Michelini *et al.*, 2004) upon intrarectal challenge, whereas precisely the same batch and dose of vaccines given in a DNA, MVA, SFV, SFV regime had no discernible effect on challenge (Seibold *et al.*, manuscript in preparation). Other combinations of these vaccines had only sporadically beneficial effects, giving, for example, moderate reductions in viral loads and protection from profound $CD4^+$ T-cell loss (Stolte-Leeb *et al.*, 2006).

Molecular vaccines to other retroviruses

This chapter has so far concentrated on the intense efforts that have been made over the last 25 years to develop a vaccine able to prevent infection of humans with HIV. It will be clear that despite exploring virtually every conceivable avenue of vaccine development, the prospects for an AIDS vaccine in the foreseeable future are now no better than they were at the start of the epidemic. This may give the impression that vaccines against retroviruses in general are extremely difficult or even impossible to produce, but this is not the case. There is a wealth of literature describing, for example, experiments in mice in which the immunological mechanisms for protection against murine retroviruses are dissected in great detail (Hasenkrug and Dittmer, 2007), and such studies have provided a great deal of crucial information relevant for the development of vaccines against retroviruses infecting humans or domestic animals. Indeed, successful vaccines against a number of retroviruses of veterinary importance have been developed and are commercially available (Sparkes, 1997; Dunham, 2006; Hofmann-Lehmann *et al.*, 2008; Uhl *et al.*, 2008). Although superficially heartening, the apparent success of vaccines against some of the other retroviruses only serves to underscore the unexpected difficulties associated with the development of an AIDS vaccine.

Feline leukaemia virus (FeLV)

Attempts to develop vaccines against FeLV, a retrovirus causing immunosuppression in infected cats, have been ongoing since before the discovery of HIV. Indeed, the relative ease with which protection against disease can be achieved by vaccination against FeLV was a cause of early optimism for the prospects of an AIDS virus. Most commercial vaccines are based on inactivated virus or envelope subunits delivered with a strong adjuvant and were assumed to operate by the generation of neutralizing antibodies able to prevent the initial infection from occurring. However, it has recently been discovered using sensitive molecular assays, that vaccinated cats experiencing no antigenaemia and remaining free from disease nevertheless undergo limited viral replication and integration of the FeLV provirus into their cells (Hofmann-Lehmann *et al.*, 2007). Like all other vaccines, these vaccines do not therefore elicit sterilizing immunity and indeed, the initial plasma and cell-associated virus loads were found to be similar for 'protected' and 'infected' cats (Hofmann-Lehmann *et al.*, 2008).

As the commercial vaccines are not 100% effective at preventing disease, research on alternatives has continued using many of the same strategies described earlier for HIV. For example, inoculation of cats with plasmid DNA coding for the FeLV genes was shown to protect cats from challenge and this protection was correlated with the levels of FeLV-specific CTL induced (Flynn *et al.*, 2000). This protection could be enhanced by the inclusion of DNA coding for the cytokines IL-12 and IL-18 (Hanlon *et al.*, 2001) the effect of which was later shown to be due to the IL-18 (O'Donovan *et al.*, 2005).

Poxvirus vectors expressing FeLV genes have also been investigated for their vaccine potential, but the first such study using vaccinia virus itself expressing FeLV gp70, failed to induce FeLV-specific antibodies in vaccinated cats (Gilbert *et al.*, 1987). Later attempts using canarypox-based vectors were more successful, with constructs expressing both the *gag* and *env* genes inducing protection despite a failure to stimulate neutralizing antibodies (Tartaglia *et al.*, 1993; Poulet *et al.*, 2003). Vectors based on swinepox, a virus that, like the avian poxvirus vectors and MVA, is unable to replicate productively in mammalian cells, have also been developed (Winslow *et al.*, 2003; Winslow *et al.*, 2005) but their efficacy has not yet been demonstrated.

Feline immunodeficiency virus (FIV)

Feline immunodeficiency virus was first isolated in 1986 and was found to have so many similarities to HIV that many consider it a valid alternative model for AIDS vaccine development. Even without the HIV/AIDS connection, an FIV vaccine is of great commercial interest because a high percentage of sick cats are known to be seropositive for FIV (Malik *et al.*, 1997; Hartmann, 1998). Like HIV, one of the greatest obstacles to FIV vaccine development is the high degree of variability in Env (Olmsted *et al.*, 1989; Sodora *et al.*, 1994) which renders vaccines based on the induction of neutralizing antibodies alone somewhat type-specific. Nonetheless, and in contrast to the situation with SIV, whole-inactivated virus vaccines against FIV protect against infection and are even commercially available (Kusuhara *et al.*, 2005; Pu *et al.*, 2005), and despite concerns surrounding the validity of challenge studies carried out in the laboratory, these vaccines also appear to protect under conditions designed to simulate natural exposure to the virus (Matteucci *et al.*, 2000; Kusuhara *et al.*, 2005). Owing to uncertainty of the efficacy and breadth of protection induced by whole inactivated virus vaccines, efforts have been made to generate vaccines able to stimulate a cellular immune response to FIV, including those based on naked DNA. However, inoculation of DNA coding for the FIV gp120 was found to result in an enhancement, rather than suppression, of virus replication upon challenge (Cuisinier *et al.*, 1997; Richardson *et al.*, 1997). In contrast, an FIV multigene DNA vaccine was able to protect some cats against challenge in the absence of overt antibody responses (Hosie *et al.*, 1998) and both an almost full-length *vif*-deleted proviral DNA vaccine (Lockridge *et al.*, 2000) and a minimalistic DNA construct (Boretti *et al.*, 2000; Leutenegger *et al.*, 2000) were also shown to be effective. However, protection against a highly virulent, heterologous strain of FIV could not be achieved by DNA vaccination (Hosie *et al.*, 2000). As with FeLV, DNA vaccines expressing feline cytokines have also been tested with varying degrees of success for their protective capacity in cats (Dunham *et al.*, 2002; Gupta *et al.*, 2007).

Due possibly to a lack of suitable feline viruses relatively little has been done to develop FIV vaccines based on viral vectors, with only one report, for example, of an FIV/poxvirus vaccine in the literature (Tellier *et al.*, 1998). However, FIV genes have been inserted into *Listeria monocytogenes* and when used as a vaccine the bacteria was shown to reduce viral loads after vaginal FIV challenge (Stevens *et al.*, 2004) and subsequent studies showed that pre-existing immunity to the bacteria did not diminish the anti-FIV immune response in vaccinated cats (Stevens *et al.*, 2005). In contrast, cats immunized with attenuated *Salmonella* expressing the FIV *gag* gene experienced viral loads similar to those in controls upon challenge (Tijhaar *et al.*, 1997).

Finally, live attenuated FIV vaccines have been generated and have been shown to be effective in preventing infection or disease upon superinfection with the pathogenic wild-type virus (Broche-Pierre *et al.*, 2005; Pistello *et al.*, 2005).

Equine infectious anaemia virus (EIAV)

Like HIV, SIV and FIV, EIAV is a lentivirus causing chronic infection in its host and similar to HIV/SIV, early attempts to vaccinate horses against EIAV infection used inactivated virus or subunit immunogens. Indeed, although such vaccines were able to protect against challenge with homologous viruses, no such protection against a heterologous strain was seen (Issel *et al.*, 1992). In contrast, inoculation of ponies with purified EIAV envelope glycoprotein (gp90) produced using the recombinant baculovirus system resulted in an enhancement, rather than an amelioration of disease (Wang *et al.*, 1994) although in a second study, there was a spectrum of vaccine efficacy ranging from protection from disease to severe enhancement (Raabe *et al.*, 1998).

It is a little ironic that the first report of an EIAV vaccine that appeared in 1970 (Kono *et al.*, 1970) used a live attenuated form of the virus and it was not until 33 years later that the next report, using a virus attenuated by modifications of the S2 gene, appeared (Li *et al.*, 2003). Indeed, despite some work using vaccines based on lipopeptides (Ridgely *et al.*, 2003; Fraser *et al.*, 2005), plasmid DNA (Cook *et al.*, 2005; Mealey *et al.*, 2007) and even single round viruses (Liu and Wang, 2006), most activity in the EIAV vaccine field now concentrates on live attenuated viruses. Animals 'immunized' with the EIAV ΔS2 attenuated virus appeared to be totally resistant to intravenous challenge with the full-length wild-type virus (Li *et al.*, 2003). As is the case with deletion mutants of SIV in macaques, the efficacy of the live attenuated EIAV vaccines was shown to depend on the degree of attenuation and the timing of the challenge (Craigo *et al.*, 2005), the latter likely to be due to the need for immune response maturation. By immune suppression of protected animals using dexamethasone, it was possible to demonstrate that 50% of the horses had apparently been protected from actual infection with the challenge while the others had covert, undetectable infections that became full-blown in the absence of immune control (Craigo *et al.*, 2007). However, the efficacy of protection was shown to be highly dependent on the degree of homology between the vaccine and challenge viruses in the *env* gene (Craigo *et al.*, 2007) although this phenomenon was dependent on differences in CTL recognition rather, as one might have expected, than susceptibility to neutralizing antibodies (Tagmyer *et al.*, 2008). In contrast, protection afforded by a donkey-leukocyte attenuated live EIAV vaccine was associated with the levels of Th1 cytokines induced by the vaccine (Zhang *et al.*, 2007).

Conclusion

The aim of this chapter was to give a brief overview of the progress that has been made in the development of molecular vaccines against retroviral infections. Because the vast majority of the relevant research has been carried out in the area of AIDS vaccine development, most of the examples have been taken from this field. The huge variety of vectors and vaccination strategies that have been used can perhaps be seen to reflect the desperation permeating the AIDS vaccine community – virtually no stone has been left unturned in the hunt for a successful approach. And yet, despite these intense efforts on a global scale, we are today no closer to having an effective AIDS vaccine than we were 25 years ago. Indeed, one could even argue that having eliminated the 'classical' forms of vaccines such as whole-inactivated, live attenuated and subunit vaccines, the outlook is now bleaker than it was at the beginning.

Why, then, is HIV such a difficult target? As described earlier, vaccines against related veterinary retroviruses have been, to varying degrees, quite successful but HIV appears to possess a combination of unique factors that seem almost designed to hinder vaccines at every turn. The most commonly cited obstacle is the extraordinary high degree of variability of HIV. The degree of variation in one single infected individual is higher than the global variability of most other human pathogens (Fig. 12.3; Korber *et al.*, 2001; Walker and Korber, 2001). As much of this variability occurs in regions that act as targets for neutralizing antibodies the prospects for a broadly specific vaccine based, as many vaccines are, on the stimulation of such antibodies are poor. In this regard it is particularly frustrating that the induction by vaccination of antibodies resembling those known to recognize conserved epitopes of the gp41 envelope glycoprotein has so far proved impossible, despite intense effort.

However, it would unfortunately be wrong to think that variability is the major obstacle to vaccine development – there are more profound difficulties that have to be faced (Table 12.4). In the SIV/macaque animal model where the vaccine can be precisely matched to the sequence of the challenge virus, no useable vaccine has yet been shown to give robust protection. This is a basic hurdle that has to first be overcome before the problem of variability becomes relevant. At the intellectual level, one great difficulty is that the so-called 'correlates of immune protection' against HIV or SIV are not known. It is merely assumed that by stimulating high levels of cellular and humoral immunity by vaccination, infection or disease progression can be prevented. If we knew precisely what immune response against which protein is effective, vaccine development could focus on that one goal rather than trying everything that is available.

Another major problem is that HIV infects the very cells needed to combat the virus. Activated $CD4^+$ T-cells, primed by vaccination and stimulated by infection, probably offer an ideal target cell for the virus. If a vaccine does not succeed in preventing the initial establishment of infection, the cells needed for an effective immune response may by rapidly overwhelmed or eliminated.

Most viral infections are self-limiting, i.e. the immune system is able to eventually clear the virus from the body. The retroviral property of provirus integration makes clearance *per se* almost impossible, but the immune system appears even incapable of preventing the prolonged, active replication characteristic of HIV infection. We therefore have no definitive information about the types of immune responses able to control HIV and unfortunately there is no other

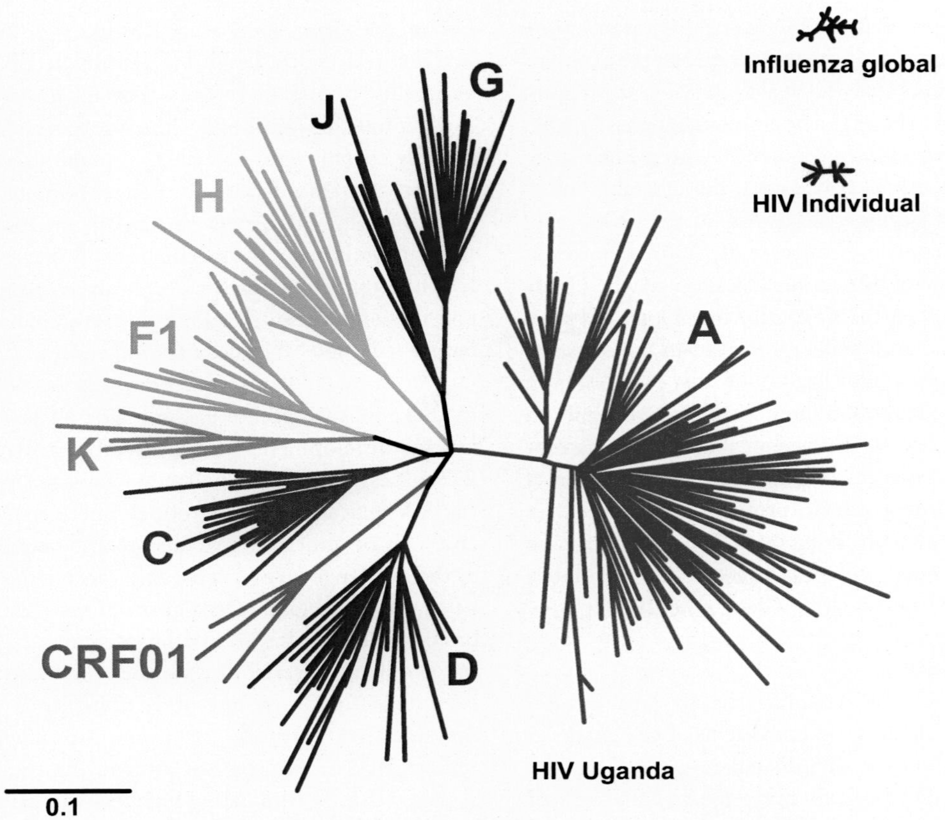

Figure 12.3 The problem of HIV variability. The degree of sequence diversity within a single HIV-infected individual is comparable to that for the entire global population of influenza viruses. The degree of diversity within Uganda alone is comparatively enormous. Diagrams represent evolutionary distances based on the HA1 domain of the influenza HA gene (all 96 isolates of H3N2 influenza from 1996) or the V2-C5 region of the HIV-1 *env* gene from one asymptomatic individual six years after infection or from 193 infected individuals from the Democratic Republic of Congo in 1997. (Adapted from Korber *et al.*, 2001.)

immunodeficiency virus that combines a high degree of variability with a tropism for cells of the immune system that could serve as guide in this endeavour.

We have to keep in mind that HIV is one of those rare pathogens that, like rabies virus, is almost 100% lethal in untreated humans and that, as it does for rabies virus, humankind represents a 'heterologous host' species for HIV. In the vast majority of viral and bacterial infections, morbidity and mortality are limited by the innate and adaptive immune systems and by genetic and other factors. Even with rabies, immunity is demonstrable in the natural host bat species and can be induced in heterologous hosts such as humans and dogs by vaccination. It therefore seems paradoxical that the natural hosts of SIV (over 30 different species of African primates known so far) have not, over the millennia, developed a similar state of protective immunity but have instead 'chosen to ignore' the life-long, high-level of virus replication occurring in their bodies. Is nature perhaps telling us that immunological eradication or control of SIV (and HIV) is not possible?

Like many viruses, HIV also has mechanisms to interfere with the immune response. For example, the envelope glycoprotein is 'shielded' to a great degree from antibody recognition by a massive amount of glycosylation and the viral Nef protein can down-regulate the expression of the host cell MHC-I, rendering it invisible to virus-specific CTLs.

These and other factors make HIV vaccine development very difficult and indeed, it is possible that in the long-run, an AIDS vaccine *per se*

Table 12.4 Difficulties in the development of AIDS vaccines

'Classic' vaccines are ineffective or too dangerous
No known correlate of immunity
Failure of the immune response to eliminate active infection
A reservoir of latently infected cells persists
Infection of $CD4^+$ cells
Down-regulation of MHC-I by viral proteins
Shielding of epitopes by conformation or glycosylation
Intra-host variability of HIV
Continuous development of escape mutants
Global diversity of HIV
Lack of a 'real' animal model
Failure to induce broadly reactive neutralizing antibodies
Huge organizational, financial and ethical problems with vaccine trials in humans
Failure of (most) pharmaceutical companies to become involved in vaccine development
High-profile failure of major clinical vaccine trials

Table 12.5 Reasons for optimism that an HIV vaccine is possible

Lessons being learned from long-term non-progressors and elite controllers
Correlation with MHC haplotypes
Recent successes in animal models
Drastic reductions of viral load and prevention of disease
Many vaccine candidates in the clinical trials pipeline
Refinement of promising approaches
Elimination of unsuccessful approaches
Increasingly deep knowledge of HIV and its interaction with the immune system
Innovative, targeted vaccine strategies being developed
Highly exposed persistently seronegative (HEPS) individuals
No infection despite frequent, unprotected sex with HIV-positive partners/customers
Protection against infection under field conditions is therefore possible

is not an achievable goal. However, a number of points give hope that this is not the case (Table 12.5). First, we now know more about HIV and the way in which it interacts with the host immune system than we do for any other human pathogen. This huge base of scientific knowledge means that strategies can be developed in an informed manner and the tools for measuring immune responses in exquisite detail are available. Second, although there is still no vaccine giving reproducible, robust protection in the SIV/macaque animal model, there are enough sporadic successes to indicate that we may be on the edge of achieving this goal. It is possible that we are 90% of the way there and only need the final tweaking of the vaccines to achieve protective immunity. Third, we know from studies of highly exposed, persistently seronegative (HEPS) individuals, most famously the cohorts of commercial sex workers working in areas of high HIV-seroprevalence, that protection against infection is possible under field conditions. If the underlying biological mechanisms protecting these individuals could be identified and reproduced by immunization, then an effective AIDS vaccine would be available. There are some indications that HEPS individuals have MHC-haplotypes relatively rare for their general population (Kaul *et al.*, 2001). This could mean that repeated exposure to many foreign MHC-molecules during unprotected sex might induce an 'anti-allo' response able to limit later exposure to HIV virions carrying such foreign MHC. It is known that immune responses to human MHC molecules can protect macaques from challenge with a virus grown in human cells (Chan *et al.*, 1992; Goldstein *et al.*, 1994; Norley *et al.*, 1998) and a similar mechanism might be contributing to the protection seen in the HEPS individuals.

Finally, it will not have escaped the reader's notice that the vast majority of molecular vaccines have been produced with safety as the first priority and efficacy as the second. Most vaccine candidates have therefore been replication-defective or highly attenuated vectors. The continuing failure of such vectors to give the level of immunogenicity and protection necessary, both in the animal model and in human efficacy trials, suggests that it might be time to re-evaluate our priorities and develop vaccines that, like the enormously successful smallpox and poliovirus vaccines, may carry a low but significant potential risk to some vaccine recipients. For people born and growing up in areas where the rates of HIV infection are in double-digit figures and where access to antiviral treatment is limited or non-existent, the risks posed by such vaccines may be entirely acceptable when balanced against the likelihood of HIV infection and AIDS.

HIV/AIDS is without doubt the biggest medical catastrophe of our age. Its impact both in terms of personal suffering and tragedy, and at the level of social and economic development globally, is simply staggering. Despite intense efforts in the areas of education and prevention, and despite the (unfortunately limited) availability of many antiretroviral drugs, the only realistic hope for controlling the pandemic lies in the production of an effective and affordable vaccine. Recent setbacks have cruelly driven home the fact that such a vaccine is presently nowhere in sight. However, the scale of the pandemic and the threat posed to future generations leaves us with no choice but to redouble our efforts to pursue that most elusive of goals, the development of an AIDS vaccine.

References

Aidoo, M., Otten, R.A., Rodriguez, V., Sariol, C.A., Martinez, M., Kraiselburd, E., Robinson, H., Folks, T., Butera, S., and Ellenberger, D. (2007). Absence of SHIV infection in gut and lymph node tissues in rhesus monkeys after repeated rectal challenges following HIV-1 DNA/MVA immunizations. Vaccine *25*, 6474–6481.

Alam, S.M., McAdams, M., Boren, D., Rak, M., Scearce, R.M., Gao, F., Camacho, Z.T., Gewirth, D., Kelsoe, G., Chen, P., and Haynes, B.F. (2007). The role of antibody polyspecificity and lipid reactivity in binding of broadly neutralizing anti-HIV-1 envelope human monoclonal antibodies 2F5 and 4E10 to glycoprotein 41 membrane proximal envelope epitopes. J. Immunol. *178*, 4424–4435.

Amara, R.R., Villinger, F., Altman, J.D., Lydy, S.L., O'Neil, S.P., Staprans, S.I., Montefiori, D.C., Xu, Y., Herndon, J.G., Wyatt, L.S., Candido, M.A., Kozyr, N.L., Earl, P.L., Smith, J.M., Ma, H.L., Grimm, B.D., Hulsey, M.L., Miller, J., McClure, H.M., McNicholl, J.M., Moss, B., and Robinson, H.L. (2001). Control of a mucosal challenge and prevention of AIDS by a multiprotein DNA/MVA vaccine. Science *292*, 69–74.

Baba, T.W., Liska, V., Hofmann-Lehmann, R., Vlasak, J., Xu, W., Ayehunie, S., Cavacini, L.A., Posner, M.R., Katinger, H., Stiegler, G., Bernacky, B.J., Rizvi, T.A., Schmidt, R., Hill, L.R., Keeling, M.E., Lu, Y., Wright, J.E., Chou, T.C., and Ruprecht, R.M. (2000). Human neutralizing monoclonal antibodies of the IgG1 subtype protect against mucosal simian-human immunodeficiency virus infection. Nat. Med. *6*, 200–206.

Baba, T.W., Liska, V., Khimani, A.H., Ray, N.B., Dailey, P.J., Penninck, D., Bronson, R., Greene, M.F., McClure, H.M., Martin, L.N., and Ruprecht, R.M. (1999). Live attenuated, multiply deleted simian immunodeficiency virus causes AIDS in infant and adult macaques. Nat. Med. *5*, 194–203.

Bagarazzi, M.L., Boyer, J.D., Ugen, K.E., Javadian, M.A., Chattergoon, M., Shah, A., Bennett, M., Ciccarelli, R., Carrano, R., Coney, L., and Weiner, D.B. (1998). Safety and immunogenicity of HIV-1 DNA constructs in chimpanzees. Vaccine *16*, 1836–1841.

Baig, J., Levy, D.B., McKay, P.F., Schmitz, J.E., Santra, S., Subbramanian, R.A., Kuroda, M.J., Lifton, M.A., Gorgone, D.A., Wyatt, L.S., Moss, B., Huang, Y., Chakrabarti, B.K., Xu, L., Kong, W.P., Yang, Z.Y., Mascola, J.R., Nabel, G.J., Carville, A., Lackner, A.A., Veazey, R.S., and Letvin, N.L. (2002). Elicitation of simian immunodeficiency virus-specific cytotoxic T-lymphocytes in mucosal compartments of rhesus monkeys by systemic vaccination. J. Virol. *76*, 11484–11490.

Barouch, D.H., Craiu, A., Kuroda, M.J., Schmitz, J.E., Zheng, X.X., Santra, S., Frost, J.D., Krivulka, G.R., Lifton, M.A., Crabbs, C.L., Heidecker, G., Perry, H.C., Davies, M.E., Xie, H., Nickerson, C.E., Steenbeke, T.D., Lord, C.I., Montefiori, D.C., Strom, T.B., Shiver, J.W., Lewis, M.G., and Letvin, N.L. (2000). Augmentation of immune responses to HIV-1 and simian immunodeficiency virus DNA vaccines by IL-2/Ig plasmid administration in rhesus monkeys. Proc. Natl. Acad. Sci. U.S.A. *97*, 4192–4197.

Barouch, D.H., Santra, S., Tenner-Racz, K., Racz, P., Kuroda, M.J., Schmitz, J.E., Jackson, S.S., Lifton, M.A., Freed, D.C., Perry, H.C., Davies, M.E., Shiver, J.W., and Letvin, N.L. (2002). Potent $CD4^+$ T-cell responses elicited by a bicistronic HIV-1 DNA vaccine expressing gp120 and GM-CSF. J. Immunol. *168*, 562–568.

Belshe, R.B., Gorse, G.J., Mulligan, M.J., Evans, T.G., Keefer, M.C., Excler, J.L., Duliege, A.M., Tartaglia, J., Cox, W.I., McNamara, J., Hwang, K.L., Bradney, A., Montefiori, D., and Weinhold, K.J. (1998). Induction of immune responses to HIV-1 by canarypox virus (ALVAC) HIV-1 and gp120 SF-2 recombinant vaccines in uninfected volunteers. NIAID AIDS Vaccine Evaluation Group. Aids *12*, 2407–2415.

Belyakov, I.M., Wyatt, L.S., Ahlers, J.D., Earl, P., Pendleton, C.D., Kelsall, B.L., Strober, W., Moss, B., and Berzofsky, J.A. (1998). Induction of a mucosal cytotoxic T-lymphocyte response by intrarectal immunization with a replication-deficient recombinant vaccinia virus expressing human immunodeficiency virus 89.6 envelope protein. J. Virol. *72*, 8264–8272.

Berggren, R.E., Wunderlich, A., Ziegler, E., Schleicher, M., Duke, R.C., Looney, D., and Fang, F.C. (1995). HIV gp120-specific cell-mediated immune responses in mice after oral immunization with recombinant Salmonella. J. Acquir. Immune Defic. Syndr. Hum. Retroviruses *10*, 489–495.

Berman, P.W., Gregory, T.J., Riddle, L., Nakamura, G.R., Champe, M.A., Porter, J.P., Wurm, F.M., Hershberg, R.D., Cobb, E.K., and Eichberg, J.W. (1990). Protection of chimpanzees from infection by HIV-1 after vaccination with recombinant glycoprotein gp120 but not gp160. Nature *345*, 622–625.

Bertley, F.M., Kozlowski, P.A., Wang, S.W., Chappelle, J., Patel, J., Sonuyi, O., Mazzara, G., Montefiori, D., Carville, A., Mansfield, K.G., and Aldovini, A. (2004). Control of simian/human immunodeficiency virus viremia and disease progression after IL-2-augmented DNA-modified vaccinia virus Ankara nasal vaccination in nonhuman primates. J. Immunol *172*, 3745–3757.

Berzofsky, J.A., Bensussan, A., Cease, K.B., Bourge, J.F., Cheynier, R., Lurhuma, Z., Salaun, J.J., Gallo, R.C., Shearer, G.M., and Zagury, D. (1988). Antigenic peptides recognized by T-lymphocytes from AIDS viral envelope-immune humans. Nature *334*, 706–708.

Boretti, F.S., Leutenegger, C.M., Mislin, C., Hofmann-Lehmann, R., Konig, S., Schroff, M., Junghans, C., Fehr, D., Huettner, S.W., Habel, A., Flynn, J.N., Aubert, A., Pedersen, N.C., Wittig, B., and Lutz, H. (2000). Protection against FIV challenge infection by genetic vaccination using minimalistic DNA constructs for FIV *env* gene and feline IL-12 expression. Aids *14*, 1749–1757.

Boyer, J.D., Robinson, T.M., Kutzler, M.A., Parkinson, R., Calarota, S.A., Sidhu, M.K., Muthumani, K., Lewis, M., Pavlakis, G., Felber, B., and Weiner, D. (2005). SIV DNA vaccine co-administered with IL-12 expression plasmid enhances CD8 SIV cellular immune responses in cynomolgus macaques. J. Med Primatol. *34*, 262–270.

Boyer, J.D., Ugen, K.E., Wang, B., Agadjanyan, M., Gilbert, L., Bagarazzi, M.L., Chattergoon, M., Frost, P., Javadian, A., Williams, W.V., Refaeli, Y., Ciccarelli, R.B., McCallus, D., Coney, L., and Weiner, D.B. (1997). Protection of chimpanzees from high-dose heterologous HIV-1 challenge by DNA vaccination. Nat. Med. *3*, 526–532.

Boyer, J.D., Wang, B., Ugen, K.E., Agadjanyan, M., Javadian, A., Frost, P., Dang, K., Carrano, R.A., Ciccarelli, R., Coney, L., Williams, W.V., and Weiner, D.B. (1996). In vivo protective anti-HIV immune responses in non-human primates through DNA immunization. J. Med Primatol. *25*, 242–250.

Broche-Pierre, S., Richardson, J., Moraillon, A., and Sonigo, P. (2005). Evaluation of live feline immunodeficiency virus vaccines with modified antigenic properties. J. Gen. Virol. *86*, 2495–2506.

Bruce, C.B., Akrigg, A., Sharpe, S.A., Hanke, T., Wilkinson, G.W., and Cranage, M.P. (1999). Replication-deficient recombinant adenoviruses expressing the human immunodeficiency virus Env antigen can induce both humoral and CTL immune responses in mice. J. Gen. Virol. *80 (Pt 10)*, 2621–2628.

Bu, Z., Ye, L., Skeen, M.J., Ziegler, H.K., Compans, R.W., and Yang, C. (2003). Enhancement of immune responses to an HIV *env* DNA vaccine by a C-terminal segment of listeriolysin O. AIDS Res. Hum. Retroviruses *19*, 409–420.

Buge, S.L., Richardson, E., Alipanah, S., Markham, P., Cheng, S., Kalyan, N., Miller, C.J., Lubeck, M., Udem, S., Eldridge, J., and Robert-Guroff, M. (1997). An adenovirus-simian immunodeficiency virus *env* vaccine elicits humoral, cellular, and mucosal immune responses in rhesus macaques and decreases viral burden following vaginal challenge. J. Virol. *71*, 8531–8541.

Carlson, J.R., McGraw, T.P., Keddie, E., Yee, J.L., Rosenthal, A., Langlois, A.J., Dickover, R., Donovan, R., Luciw, P.A., Jennings, M.B. , *et al.* (1990). Vaccine protection of rhesus macaques against simian immunodeficiency virus infection. AIDS Res. Hum Retroviruses *6*, 1239–1246.

Cayabyab, M.J., Hovav, A.H., Hsu, T., Krivulka, G.R., Lifton, M.A., Gorgone, D.A., Fennelly, G.J., Haynes, B.F., Jacobs, W.R., Jr. and Letvin, N.L. (2006). Generation of $CD8^+$ T-cell responses by a recombinant nonpathogenic Mycobacterium smegmatis vaccine vector expressing human immunodeficiency virus type 1 Env. J. Virol. *80*, 1645–1652.

Centlivre, M., Sala, M., Wain-Hobson, S., and Berkhout, B. (2007). In HIV-1 pathogenesis the die is cast during primary infection. Aids *21*, 1–11.

Chambers, V.C., and Ito, Y. (1964). Morphology of Shope papilloma virus associated with nucleic acid-induced tumors of cottontail rabbits. Virology 23, 434–436.

Chan, W.L., Rodgers, A., Hancock, R.D., Taffs, F., Kitchin, P., Farrar, G., and Liew, F.Y. (1992). Protection in simian immunodeficiency virus-vaccinated monkeys correlates with anti-HLA class I antibody response. J. Exp. Med. *176*, 1203–1207.

Chen, H.Y., Di Mascio, M., Perelson, A.S., Ho, D.D., and Zhang, L. (2007). Determination of virus burst size *in vivo* using a single-cycle SIV in rhesus macaques. Proc. Natl. Acad. Sci. U.S.A. *104*, 19079–19084.

Chen, L.M., Briones, G., Donis, R.O., and Galan, J.E. (2006). Optimization of the delivery of heterologous proteins by the Salmonella enterica serovar Typhimurium type III secretion system for vaccine development. Infect. Immun. *74*, 5826–5833.

Clarke, D.K., Cooper, D., Egan, M.A., Hendry, R.M., Parks, C.L., and Udem, S.A. (2006). Recombinant vesicular stomatitis virus as an HIV-1 vaccine vector. Springer Semin Immunopathol *28*, 239–253.

Cochi, S.L., and Kew, O. (2008). Polio today: are we on the verge of global eradication? JAMA *300*, 839–841.

Cohen, J. (2001). AIDS origins. Disputed AIDS theory dies its final death. Science *292*, 615.

Cook, R.F., Cook, S.J., Bolin, P.S., Howe, L.J., Zhou, W., Montelaro, R.C., and Issel, C.J. (2005). Genetic immunization with codon-optimized equine infectious anemia virus (EIAV) surface unit (SU) envelope protein gene sequences stimulates immune responses in ponies. Vet. Microbiol. *108*, 23–37.

Cooper, D., Wright, K.J., Calderon, P.C., Guo, M., Nasar, F., Johnson, J.E., Coleman, J.W., Lee, M., Kotash, C., Yurgelonis, I., Natuk, R.J., Hendry, R.M., Udem, S.A., and Clarke, D.K. (2008). Attenuation of recombinant vesicular stomatitis virus-human immunodeficiency virus type 1 vaccine vectors by gene translocations and g gene truncation reduces neurovirulence and enhances immunogenicity in mice. J. Virol. *82*, 207–219.

Cox, W.I., Tartaglia, J., and Paoletti, E. (1993). Induction of cytotoxic T-lymphocytes by recombinant canarypox (ALVAC) and attenuated vaccinia (NYVAC) viruses expressing the HIV-1 envelope glycoprotein. Virology *195*, 845–850.

Craigo, J.K., Durkin, S., Sturgeon, T.J., Tagmyer, T., Cook, S.J., Issel, C.J., and Montelaro, R.C. (2007). Immune suppression of challenged vaccinates as a rigorous assessment of sterile protection by lentiviral vaccines. Vaccine *25*, 834–845.

Craigo, J.K., Li, F., Steckbeck, J.D., Durkin, S., Howe, L., Cook, S.J., Issel, C., and Montelaro, R.C. (2005). Discerning an effective balance between equine infectious anemia virus attenuation and vaccine efficacy. J. Virol. *79*, 2666–2677.

Craigo, J.K., Zhang, B., Barnes, S., Tagmyer, T.L., Cook, S.J., Issel, C.J., and Montelaro, R.C. (2007). Envelope variation as a primary determinant of lentiviral vaccine efficacy. Proc. Natl. Acad. Sci. U.S.A. *104*, 15105–15110.

Crotty, S., Lohman, B.L., Lu, F.X., Tang, S., Miller, C.J., and Andino, R. (1999). Mucosal immunization of cynomolgus macaques with two serotypes of live poliovirus vectors expressing simian immunodeficiency virus antigens: stimulation of humoral, mucosal, and cellular immunity. J. Virol. *73*, 9485–9495.

Crotty, S., Miller, C.J., Lohman, B.L., Neagu, M.R., Compton, L., Lu, D., Lu, F.X., Fritts, L., Lifson, J.D., and Andino, R. (2001). Protection against simian immunodeficiency virus vaginal challenge by using Sabin poliovirus vectors. J. Virol. *75*, 7435–7452.

Cuisinier, A.M., Mallet, V., Meyer, A., Caldora, C., and Aubert, A. (1997). DNA vaccination using expression vectors carrying FIV structural genes induces immune response against feline immunodeficiency virus. Vaccine *15*, 1085–1094.

Daniel, M.D., Kirchhoff, F., Czajak, S.C., Sehgal, P.K., and Desrosiers, R.C. (1992). Protective effects of a live attenuated SIV vaccine with a deletion in the *nef* gene. Science *258*, 1938–1941.

Davis, N.L., Caley, I.J., Brown, K.W., Betts, M.R., Irlbeck, D.M., McGrath, K.M., Connell, M.J., Montefiori, D.C., Frelinger, J.A., Swanstrom, R., Johnson, P.R., and Johnston, R.E. (2000). Vaccination of macaques against pathogenic simian immunodeficiency virus with Venezuelan equine encephalitis virus replicon particles. J. Virol. *74*, 371–378.

Dietrich, G., Hess, J., Gentschev, I., Knapp, B., Kaufmann, S.H., and Goebel, W. (2001). From evil to good: a cytolysin in vaccine development. Trends Microbiol *9*, 23–28.

Doepel, L.K. (1998). Three new AIDS vaccine trials begin testing novel concepts. NIAID AIDS Agenda 8–9.

Dudek, T., and Knipe, D.M. (2006). Replication-defective viruses as vaccines and vaccine vectors. Virology *344*, 230–239.

Dunham, S.P. (2006). Lessons from the cat: development of vaccines against lentiviruses. Vet. Immunol. Immunopathol. *112*, 67–77.

Dunham, S.P., Flynn, J.N., Rigby, M.A., Macdonald, J., Bruce, J., Cannon, C., Golder, M.C., Hanlon, L., Harbour, D.A., Mackay, N.A., Spibey, N., Jarrett, O., and Neil, J.C. (2002). Protection against feline immunodeficiency virus using replication defective proviral DNA vaccines with feline interleukin-12 and -18. Vaccine *20*, 1483–1496.

Egan, M.A., Chong, S.Y., Megati, S., Montefiori, D.C., Rose, N.F., Boyer, J.D., Sidhu, M.K., Quiroz, J., Rosati, M., Schadeck, E.B., Pavlakis, G.N., Weiner, D.B., Rose, J.K., Israel, Z.R., Udem, S.A., and Eldridge, J.H. (2005). Priming with plasmid DNAs expressing interleukin-12 and simian immunodeficiency virus *gag* enhances the immunogenicity and efficacy of an experimental AIDS vaccine based on recombinant vesicular stomatitis virus. AIDS Res. Hum. Retroviruses *21*, 629–643.

Egan, M.A., Chong, S.Y., Rose, N.F., Megati, S., Lopez, K.J., Schadeck, E.B., Johnson, J.E., Masood, A., Piacente, P., Druilhet, R.E., Barras, P.W., Hasselschwert, D.L., Reilly, P., Mishkin, E.M., Montefiori, D.C., Lewis, M.G., Clarke, D.K., Hendry, R.M., Marx, P.A., Eldridge, J.H., Udem, S.A., Israel, Z.R., and Rose,

J.K. (2004). Immunogenicity of attenuated vesicular stomatitis virus vectors expressing HIV type 1 Env and SIV Gag proteins: comparison of intranasal and intramuscular vaccination routes. AIDS Res. Hum. Retroviruses *20*, 989–1004.

Egan, M.A., Pavlat, W.A., Tartaglia, J., Paoletti, E., Weinhold, K.J., Clements, M.L., and Siliciano, R.F. (1995). Induction of human immunodeficiency virus type 1 (HIV-1)-specific cytolytic T-lymphocyte responses in seronegative adults by a nonreplicating, host-range-restricted canarypox vector (ALVAC) carrying the HIV-1MN *env* gene. J. Infect. Dis. *171*, 1623–1627.

Ellenberger, D., Otten, R.A., Li, B., Aidoo, M., Rodriguez, I.V., Sariol, C.A., Martinez, M., Monsour, M., Wyatt, L., Hudgens, M.G., Kraiselburd, E., Moss, B., Robinson, H., Folks, T., and Butera, S. (2006). HIV-1 DNA/MVA vaccination reduces the per exposure probability of infection during repeated mucosal SHIV challenges. Virology *352*, 216–225.

Evans, D.T., Bricker, J.E., and Desrosiers, R.C. (2004). A novel approach for producing lentiviruses that are limited to a single cycle of infection. J. Virol. *78*, 11715–11725.

Evans, D.T., Bricker, J.E., Sanford, H.B., Lang, S., Carville, A., Richardson, B.A., Piatak, M., Jr., Lifson, J.D., Mansfield, K.G., and Desrosiers, R.C. (2005). Immunization of macaques with single-cycle simian immunodeficiency virus (SIV) stimulates diverse virus-specific immune responses and reduces viral loads after challenge with SIVmac239. J. Virol. *79*, 7707–7720.

Evans, D.T., Chen, L.M., Gillis, J., Lin, K.C., Harty, B., Mazzara, G.P., Donis, R.O., Mansfield, K.G., Lifson, J.D., Desrosiers, R.C., Galan, J.E., and Johnson, R.P. (2003). Mucosal priming of simian immunodeficiency virus-specific cytotoxic T-lymphocyte responses in rhesus macaques by the Salmonella type III secretion antigen delivery system. J. Virol. *77*, 2400–2409.

Feng, Y., Wang, S., Luo, F., Ruan, Y., Kang, L., Xiang, X., Chao, T., Peng, G., Zhu, C., Mu, Y., Zhu, Y., Zhang, X., and Wu, J. (2008). A novel recombinant bacterial vaccine strain expressing dual viral antigens induces multiple immune responses to the Gag and gp120 proteins of HIV-1 in immunized mice. Antiviral Res. *80*, 272–279.

Ferrantelli, F., Hofmann-Lehmann, R., Rasmussen, R.A., Wang, T., Xu, W., Li, P.L., Montefiori, D.C., Cavacini, L.A., Katinger, H., Stiegler, G., Anderson, D.C., McClure, H.M., and Ruprecht, R.M. (2003). Post-exposure prophylaxis with human monoclonal antibodies prevented SHIV89.6P infection or disease in neonatal macaques. Aids *17*, 301–309.

Ferrari, G., Humphrey, W., McElrath, M.J., Excler, J.L., Duliege, A.M., Clements, M.L., Corey, L.C., Bolognesi, D.P., and Weinhold, K.J. (1997). Clade B-based HIV-1 vaccines elicit cross-clade cytotoxic T-lymphocyte reactivities in uninfected volunteers. Proc. Natl. Acad. Sci. U.S.A. *94*, 1396–1401.

Flynn, J.N., Hanlon, L., and Jarrett, O. (2000). Feline leukaemia virus: protective immunity is mediated by virus-specific cytotoxic T-lymphocytes. Immunology *101*, 120–125.

Fouts, T.R., Tuskan, R.G., Chada, S., Hone, D.M., and Lewis, G.K. (1995). Construction and immunogenicity of Salmonella typhimurium vaccine vectors that express HIV-1 gp120. Vaccine *13*, 1697–1705.

Franchini, G., Robert-Guroff, M., Tartaglia, J., Aggarwal, A., Abimiku, A., Benson, J., Markham, P., Limbach, K., Hurteau, G., Fullen, J., *et al.* (1995). Highly attenuated HIV type 2 recombinant poxviruses, but not HIV-2 recombinant Salmonella vaccines, induce long-lasting protection in rhesus macaques. AIDS Res. Hum. Retroviruses *11*, 909–920.

Francis, D.P., Gregory, T., McElrath, M.J., Belshe, R.B., Gorse, G.J., Migasena, S., Kitayaporn, D., Pitisuttitham, P., Matthews, T., Schwartz, D.H., and Berman, P.W. (1998). Advancing AIDSVAX to phase 3. Safety, immunogenicity, and plans for phase 3. AIDS Res. Hum. Retroviruses *14 Suppl. 3*, S325–331.

Frankel, F.R., Hegde, S., Lieberman, J., and Paterson, Y. (1995). Induction of cell-mediated immune responses to human immunodeficiency virus type 1 Gag protein by using Listeria monocytogenes as a live vaccine vector. J. Immunol. *155*, 4775–4782.

Fraser, D.G., Leib, S.R., Zhang, B.S., Mealey, R.H., Brown, W.C., and McGuire, T.C. (2005). Lymphocyte proliferation responses induced to broadly reactive Th peptides did not protect against equine infectious anemia virus challenge. Clin. Diagn. Lab. Immunol. *12*, 983–993.

Friedman, R.S., Frankel, F.R., Xu, Z., and Lieberman, J. (2000). Induction of human immunodeficiency virus (HIV)-specific CD8 T-cell responses by Listeria monocytogenes and a hyperattenuated Listeria strain engineered to express HIV antigens. J. Virol. *74*, 9987–9993.

Fuerst, T.R., Stover, C.K., and de la Cruz, V.F. (1991). Development of BCG as a live recombinant vector system: potential use as an HIV vaccine. Biotechnol. Ther. *2*, 159–178.

Fuller, D.H., Murphey-Corb, M., Clements, J., Barnett, S., and Haynes, J.R. (1996). Induction of immunodeficiency virus-specific immune responses in rhesus monkeys following gene gun-mediated DNA vaccination. J. Med. Primatol. *25*, 236–241.

Fultz, P.N., Stallworth, J., Porter, D., Novak, M., Anderson, M.J., and Morrow, C.D. (2003). Immunogenicity in pig-tailed macaques of poliovirus replicons expressing HIV-1 and SIV antigens and protection against SHIV-89.6P disease. Virology *315*, 425–437.

Gao, F., Li, Y., Decker, J.M., Peyerl, F.W., Bibollet-Ruche, F., Rodenburg, C.M., Chen, Y., Shaw, D.R., Allen, S., Musonda, R., Shaw, G.M., Zajac, A.J., Letvin, N., and Hahn, B.H. (2003). Codon usage optimization of HIV type 1 subtype C gag, pol, *env*, and *nef* genes: *in vitro* expression and immune responses in DNA-vaccinated mice. AIDS Res. Hum. Retroviruses *19*, 817–823.

Gardner, M.B., Carlson, J.R., Jennings, M., Rosenthal, A., Langlois, A., Haynes, B., Bolognesi, D., and Palker, T.J. (1991). SIV vaccine protection of rhesus monkeys. Biotechnol. Ther. *2*, 9–19.

Gherardi, M.M., and Esteban, M. (2005). Recombinant poxviruses as mucosal vaccine vectors. J. Gen. Virol. *86*, 2925–2936.

Gilbert, J.H., Pedersen, N.C., and Nunberg, J.H. (1987). Feline leukemia virus envelope protein expression encoded by a recombinant vaccinia virus: apparent lack of immunogenicity in vaccinated animals. Virus Res. *7*, 49–67.

Gilbert, P.B., Chiu, Y.L., Allen, M., Lawrence, D.N., Chapdu, C., Israel, H., Holman, D., Keefer, M.C., Wolff, M., and Frey, S.E. (2003). Long-term safety analysis of preventive HIV-1 vaccines evaluated in AIDS vaccine evaluation group NIAID-sponsored Phase I and II clinical trials. Vaccine *21*, 2933–2947.

Girard, M., Kieny, M.P., Pinter, A., Barre-Sinoussi, F., Nara, P., Kolbe, H., Kusumi, K., Chaput, A., Reinhart, T., Muchmore, E. , *et al.* (1991). Immunization of chimpanzees confers protection against challenge with human immunodeficiency virus. Proc. Natl. Acad. Sci. U.S.A. *88*, 542–546.

Girard, M., van der Ryst, E., Barre-Sinoussi, F., Nara, P., Tartaglia, J., Paoletti, E., Blondeau, C., Jennings, M., Verrier, F., Meignier, B., and Fultz, P.N. (1997). Challenge of chimpanzees immunized with a recombinant canarypox-HIV-1 virus. Virology *232*, 98–104.

Goldstein, S., Elkins, W.R., London, W.T., Hahn, A., Goeken, R., Martin, J.E., and Hirsch, V.M. (1994). Immunization with whole inactivated vaccine protects from infection by SIV grown in human but not macaque cells. J. Med. Primatol *23*, 75–82.

Gomez, C.E., Najera, J.L., Jimenez, E.P., Jimenez, V., Wagner, R., Graf, M., Frachette, M.J., Liljestrom, P., Pantaleo, G., and Esteban, M. (2007). Head-to-head comparison on the immunogenicity of two HIV/AIDS vaccine candidates based on the attenuated poxvirus strains MVA and NYVAC co-expressing in a single locus the HIV-1BX08 gp120 and HIV-1(IIIB) Gag-Pol-Nef proteins of clade B. Vaccine *25*, 2863–2885.

Gorelick, R.J., Benveniste, R.E., Lifson, J.D., Yovandich, J.L., Morton, W.R., Kuller, L., Flynn, B.M., Fisher, B.A., Rossio, J.L., Piatak, M., Jr., Bess, J.W., Jr., Henderson, L.E., and Arthur, L.O. (2000). Protection of Macaca nemestrina from disease following pathogenic simian immunodeficiency virus (SIV) challenge: utilization of SIV nucleocapsid mutant DNA vaccines with and without an SIV protein boost. J. Virol. *74*, 11935–11949.

Guimaraes-Walker, A., Mackie, N., McCormack, S., Hanke, T., Schmidt, C., Gilmour, J., Barin, B., McMichael, A., Weber, J., Legg, K., Babiker, A., Hayes, P., Gotch, F., Smith, C., Dally, L., Dorrell, L., Cebere, I., Kay, R., Winstone, N., Moore, S., Goonetilleke, N., and Fast, P. (2008). Lessons from IAVI-006, a Phase I clinical trial to evaluate the safety and immunogenicity of the pTHr.HIVA DNA and MVA.HIVA vaccines in a prime-boost strategy to induce HIV-1 specific T-cell responses in healthy volunteers. Vaccine *26*, 6671–6677.

Gupta, S., Leutenegger, C.M., Dean, G.A., Steckbeck, J.D., Cole, K.S., and Sparger, E.E. (2007). Vaccination of cats with attenuated feline immunodeficiency virus proviral DNA vaccine expressing gamma interferon. J. Virol. *81*, 465–473.

Hanke, T., Blanchard, T.J., Schneider, J., Ogg, G.S., Tan, R., Becker, M., Gilbert, S.C., Hill, A.V., Smith, G.L., and McMichael, A. (1998). Immunogenicities of intravenous and intramuscular administrations of modified vaccinia virus Ankara-based multi-CTL epitope vaccine for human immunodeficiency virus type 1 in mice. J. Gen. Virol. *79*, 83–90.

Hanke, T., McMichael, A.J., Mwau, M., Wee, E.G., Ceberej, I., Patel, S., Sutton, J., Tomlinson, M., and Samuel, R.V. (2002). Development of a DNA-MVA/HIVA vaccine for Kenya. Vaccine *20*, 1995–1998.

Hanlon, L., Argyle, D., Bain, D., Nicolson, L., Dunham, S., Golder, M.C., McDonald, M., McGillivray, C., Jarrett, O., Neil, J.C., and Onions, D.E. (2001). Feline leukemia virus DNA vaccine efficacy is enhanced by coadministration with interleukin-12 (IL-12) and IL-18 expression vectors. J. Virol. *75*, 8424–8433.

Hanson, C.V. (1994). Measuring vaccine-induced HIV neutralization: report of a workshop. AIDS Res. Hum. Retroviruses *10*, 645–648.

Hartmann, K. (1998). Feline immunodeficiency virus infection: an overview. Vet. J. *155*, 123–137.

Hartung, S., Norley, S.G., Ennen, J., Cichutek, K., Plesker, R., and Kurth, R. (1992). Vaccine protection against SIVmac infection by high- but not low-dose whole inactivated virus immunogen. J. Acquir. Immune Defic. Syndr. *5*, 461–468.

Hasenkrug, K.J., and Dittmer, U. (2007). Immune control and prevention of chronic Friend retrovirus infection. Front Biosci. *12*, 1544–1551.

Haynes, J.R., Fuller, D.H., Eisenbraun, M.D., Ford, M.J., and Pertmer, T.M. (1994). Accell particle-mediated DNA immunization elicits humoral, cytotoxic, and protective immune responses. AIDS Res. Hum. Retroviruses *10 Suppl. 2*, S43–45.

Hirao, L.A., Wu, L., Khan, A.S., Satishchandran, A., Draghia-Akli, R., and Weiner, D.B. (2008). Intradermal/subcutaneous immunization by electroporation improves plasmid vaccine delivery and potency in pigs and rhesus macaques. Vaccine *26*, 440–448.

Hirsch, V.M., Fuerst, T.R., Sutter, G., Carroll, M.W., Yang, L.C., Goldstein, S., Piatak, M., Jr., Elkins, W.R., Alvord, W.G., Montefiori, D.C., Moss, B., and Lifson, J.D. (1996). Patterns of viral replication correlate with outcome in simian immunodeficiency virus (SIV)-infected macaques: effect of prior immunization with a trivalent SIV vaccine in modified vaccinia virus Ankara. J. Virol. *70*, 3741–3752.

Hirsch, V.M., Olmsted, R.A., Murphey-Corb, M., Purcell, R.H., and Johnson, P.R. (1989). An African primate lentivirus (SIVsm) closely related to HIV-2. Nature *339*, 389–392.

Hocknell, P.K., Wiley, R.D., Wang, X., Evans, T.G., Bowers, W.J., Hanke, T., Federoff, H.J., and Dewhurst, S. (2002). Expression of human immunodeficiency virus type 1 gp120 from herpes simplex virus type 1-derived amplicons results in potent, specific, and durable cellular and humoral immune responses. J. Virol. *76*, 5565–5580.

Hofmann-Lehmann, R., Cattori, V., Tandon, R., Boretti, F.S., Meli, M.L., Riond, B., and Lutz, H. (2008). How molecular methods change our views of FeLV infection and vaccination. Vet. Immunol. Immunopathol. *123*, 119–123.

Hofmann-Lehmann, R., Cattori, V., Tandon, R., Boretti, F.S., Meli, M.L., Riond, B., Pepin, A.C., Willi, B., Ossent, P., and Lutz, H. (2007). Vaccination against the feline leukaemia virus: outcome and response categories and long-term follow-up. Vaccine *25*, 5531–5539.

Hofmann-Lehmann, R., Vlasak, J., Rasmussen, R.A., Smith, B.A., Baba, T.W., Liska, V., Ferrantelli, F., Montefiori, D.C., McClure, H.M., Anderson, D.C., Bernacky, B.J., Rizvi, T.A., Schmidt, R., Hill, L.R., Keeling, M.E., Katinger, H., Stiegler, G., Cavacini, L.A., Posner, M.R., Chou, T.C., Andersen, J., and Ruprecht, R.M. (2001). Postnatal passive immunization of neonatal macaques with a triple combination of human monoclonal antibodies against oral simian-human immunodeficiency virus challenge. J. Virol. *75*, 7470–7480.

Honda, M., Matsuo, K., Nakasone, T., Okamoto, Y., Yoshizaki, H., Kitamura, K., Sugiura, W., Watanabe, K., Fukushima, Y., Haga, S., Katsura, Y., Tasaka, H., Komuro, K., Yamada, T., Asano, T., Yamazaki, A., and Yamazaki, S. (1995). Protective immune responses induced by secretion of a chimeric soluble protein from a recombinant *Mycobacterium bovis* bacillus Calmette-Guerin vector candidate vaccine for human immunodeficiency virus type 1 in small animals. Proc. Natl. Acad. Sci. U.S.A. *92*, 10693–10697.

Hone, D.M., Lewis, G.K., Beier, M., Harris, A., McDaniels, T., and Fouts, T.R. (1994). Expression of human immunodeficiency virus antigens in an attenuated Salmonella typhi vector vaccine. Dev. Biol. Stand. *82*, 159–162.

Hooper, E. (2000). Search for the origin of HIV and AIDS. Science *289*, 1140–1141.

Hooper, E. (2001). Experimental oral polio vaccines and acquired immune deficiency syndrome. Phil. Trans. R. Soc. Lond. B Biol. Sci. *356*, 803–814.

Hosie, M.J., Dunsford, T., Klein, D., Willett, B.J., Cannon, C., Osborne, R., Macdonald, J., Spibey, N., Mackay, N., Jarrett, O., and Neil, J.C. (2000). Vaccination with inactivated virus but not viral DNA reduces virus load following challenge with a heterologous and virulent isolate of feline immunodeficiency virus. J. Virol. *74*, 9403–9411.

Hosie, M.J., Flynn, J.N., Rigby, M.A., Cannon, C., Dunsford, T., Mackay, N.A., Argyle, D., Willett, B.J., Miyazawa, T., Onions, D.E., Jarrett, O., and Neil, J.C. (1998). DNA vaccination affords significant protection against feline immunodeficiency virus infection without inducing detectable antiviral antibodies. J. Virol. *72*, 7310–7319.

Hu, S.L., Abrams, K., Barber, G.N., Moran, P., Zarling, J.M., Langlois, A.J., Kuller, L., Morton, W.R., and Benveniste, R.E. (1992). Protection of macaques against SIV infection by subunit vaccines of SIV envelope glycoprotein gp160. Science *255*, 456–459.

Hu, S.L., Stallard, V., Abrams, K., Barber, G.N., Kuller, L., Langlois, A.J., Morton, W.R., and Benveniste, R.E. (1993). Protection of vaccinia-primed macaques against SIVmne infection by combination immunization with recombinant vaccinia virus and SIVmne gp160. J. Med. Primatol. *22*, 92–99.

Issel, C.J., Horohov, D.W., Lea, D.F., Adams, W.V., Jr., Hagius, S.D., McManus, J.M., Allison, A.C., and Montelaro, R.C. (1992). Efficacy of inactivated whole-virus and subunit vaccines in preventing infection and disease caused by equine infectious anemia virus. J. Virol. *66*, 3398–3408.

Jacobs, W.R., Jr., Tuckman, M., and Bloom, B.R. (1987). Introduction of foreign DNA into mycobacteria using a shuttle phasmid. Nature *327*, 532–535.

Jiang, S., Rasmussen, R.A., Nolan, K.M., Frankel, F.R., Lieberman, J., McClure, H.M., Williams, K.M., Babu, U.S., Raybourne, R.B., Strobert, E., and Ruprecht, R.M. (2007). Live attenuated Listeria monocytogenes expressing HIV Gag: immunogenicity in rhesus monkeys. Vaccine *25*, 7470–7479.

Johnson, P.R., Schnepp, B.C., Connell, M.J., Rohne, D., Robinson, S., Krivulka, G.R., Lord, C.I., Zinn, R., Montefiori, D.C., Letvin, N.L., and Clark, K.R. (2005). Novel adeno-associated virus vector vaccine restricts replication of simian immunodeficiency virus in macaques. J. Virol. *79*, 955–965.

Johnston, R.E., Johnson, P.R., Connell, M.J., Montefiori, D.C., West, A., Collier, M.L., Cecil, C., Swanstrom, R., Frelinger, J.A., and Davis, N.L. (2005). Vaccination of macaques with SIV immunogens delivered by Venezuelan equine encephalitis virus replicon particle vectors followed by a mucosal challenge with SIVsmE660. Vaccine *23*, 4969–4979.

Kanekiyo, M., Matsuo, K., Hamatake, M., Hamano, T., Ohsu, T., Matsumoto, S., Yamada, T., Yamazaki, S., Hasegawa, A., Yamamoto, N., and Honda, M. (2005). Mycobacterial codon optimization enhances antigen expression and virus-specific immune responses in recombinant Mycobacterium bovis bacille Calmette-Guerin expressing human immunodeficiency virus type 1 Gag. J. Virol. *79*, 8716–8723.

Karpenko, L.I., Nekrasova, N.A., Ilyichev, A.A., Lebedev, L.R., Ignatyev, G.M., Agafonov, A.P., Zaitsev, B.N., Belavin, P.A., Seregin, S.V., Danilyuk, N.K., Babkina, I.N., and Bazhan, S.I. (2004). Comparative analysis using a mouse model of the immunogenicity of artificial VLP and attenuated *Salmonella* strain carrying a DNA-vaccine encoding HIV-1 polyepitope CTL-immunogen. Vaccine *22*, 1692–1699.

Kaul, R., Dong, T., Plummer, F.A., Kimani, J., Rostron, T., Kiama, P., Njagi, E., Irungu, E., Farah, B., Oyugi, J., Chakraborty, R., MacDonald, K.S., Bwayo, J.J., McMichael, A., and Rowland-Jones, S.L. (2001). CD8(+) lymphocytes respond to different HIV epitopes in seronegative and infected subjects. J. Clin. Invest. *107*, 1303–1310.

Kaur, A., Sanford, H.B., Garry, D., Lang, S., Klumpp, S.A., Watanabe, D., Bronson, R.T., Lifson, J.D., Rosati, M., Pavlakis, G.N., Felber, B.K., Knipe, D.M., and Desrosiers, R.C. (2007). Ability of herpes simplex virus vectors to boost immune responses to DNA vectors and to protect against challenge by simian immunodeficiency virus. Virology *357*, 199–214.

Kestler, H.W., 3rd, Naidu, Y.N., Kodama, T., King, N.W., Daniel, M.D., Li, Y., and Desrosiers, R.C. (1989). Use of infectious molecular clones of simian immunodeficiency virus for pathogenesis studies. J. Med. Primatol. *18*, 305–309.

Kestler, H.W., 3rd, Ringler, D.J., Mori, K., Panicali, D.L., Sehgal, P.K., Daniel, M.D., and Desrosiers, R.C. (1991). Importance of the *nef* gene for maintenance of high virus loads and for development of AIDS. Cell *65*, 651–662.

Kim, J.J., Ayyavoo, V., Bagarazzi, M.L., Chattergoon, M.A., Dang, K., Wang, B., Boyer, J.D., and Weiner, D.B. (1997). In vivo engineering of a cellular immune response by coadministration of IL-12 expression vector with a DNA immunogen. J. Immunol. *158*, 816–826.

Kim, J.J., Yang, J.S., Montaner, L., Lee, D.J., Chalian, A.A., and Weiner, D.B. (2000). Coimmunization with IFN-gamma or IL-2, but not IL-13 or IL-4 cDNA can enhance Th1-type DNA vaccine-induced immune responses in vivo. J. Interferon Cytokine Res *20*, 311–319.

Kono, Y., Kobayashi, K., and Fukunaga, Y. (1970). Immunization of horses against equine infectious anemia (EIA) with an attenuated EIA virus. Natl Inst Anim Health Q (Tokyo) *10*, 113–122.

Koopman, G., Mortier, D., Hofman, S., Niphuis, H., Fagrouch, Z., Norley, S., Sutter, G., Liljestrom, P., and Heeney, J.L. (2004). Vaccine protection from $CD4^+$ T-cell loss caused by simian immunodeficiency virus (SIV) mac251 is afforded by sequential immunization with three unrelated vaccine vectors encoding multiple SIV antigens. J. Gen. Virol. *85*, 2915–2924.

Korber, B., Gaschen, B., Yusim, K., Thakallapally, R., Kesmir, C., and Detours, V. (2001). Evolutionary and immunological implications of contemporary HIV-1 variation. Br. Med. Bull. *58*, 19–42.

Kotton, C.N., Lankowski, A.J., Scott, N., Sisul, D., Chen, L.M., Raschke, K., Borders, G., Boaz, M., Spentzou, A., Galan, J.E., and Hohmann, E.L. (2006). Safety and immunogenicity of attenuated Salmonella enterica serovar Typhimurium delivering an HIV-1 Gag antigen via the Salmonella Type III secretion system. Vaccine *24*, 6216–6224.

Kuate, S., Stahl-Hennig, C., Stoiber, H., Nchinda, G., Floto, A., Franz, M., Sauermann, U., Bredl, S., Deml, L., Ignatius, R., Norley, S., Racz, P., Tenner-Racz, K., Steinman, R.M., Wagner, R., and Uberla, K. (2006). Immunogenicity and efficacy of immunodeficiency virus-like particles pseudotyped with the G protein of vesicular stomatitis virus. Virology *351*, 133–144.

Kuate, S., Stahl-Hennig, C., ten Haaft, P., Heeney, J., and Uberla, K. (2003). Single-cycle immunodeficiency viruses provide strategies for uncoupling *in vivo* expression levels from viral replicative capacity and for mimicking live-attenuated SIV vaccines. Virology *313*, 653–662.

Kusuhara, H., Hohdatsu, T., Okumura, M., Sato, K., Suzuki, Y., Motokawa, K., Gemma, T., Watanabe, R., Huang, C., Arai, S., and Koyama, H. (2005). Dual-subtype vaccine (Fel-O-Vax FIV) protects cats against contact challenge with heterologous subtype B FIV infected cats. Vet. Microbiol. *108*, 155–165.

Lagranderie, M., Winter, N., Balazuc, A.M., Gicquel, B., and Gheorghiu, M. (1998). A cocktail of Mycobacterium bovis BCG recombinants expressing the SIV Nef, Env, and Gag antigens induces antibody and cytotoxic responses in mice vaccinated by different mucosal routes. AIDS Res. Hum. Retroviruses *14*, 1625–1633.

Learmont, J.C., Geczy, A.F., Mills, J., Ashton, L.J., Raynes-Greenow, C.H., Garsia, R.J., Dyer, W.B., McIntyre, L., Oelrichs, R.B., Rhodes, D.I., Deacon, N.J., and Sullivan, J.S. (1999). Immunologic and virologic status after 14 to 18 years of infection with an attenuated strain of HIV-1. A report from the Sydney Blood Bank Cohort. N. Engl. J. Med. *340*, 1715–1722.

Lekutis, C., Shiver, J.W., Liu, M.A., and Letvin, N.L. (1997). HIV-1 *env* DNA vaccine administered to rhesus monkeys elicits MHC class II-restricted $CD4^+$ T helper cells that secrete IFN-gamma and TNF-alpha. J. Immunol. *158*, 4471–4477.

Lemiale, F., Kong, W.P., Akyurek, L.M., Ling, X., Huang, Y., Chakrabarti, B.K., Eckhaus, M., and Nabel, G.J. (2003). Enhanced mucosal immunoglobulin A response of intranasal adenoviral vector human immunodeficiency virus vaccine and localization in the central nervous system. J. Virol. *77*, 10078–10087.

Leung, N.J., Aldovini, A., Young, R., Jarvis, M.A., Smith, J.M., Meyer, D., Anderson, D.E., Carlos, M.P., Gardner, M.B., and Torres, J.V. (2000). The kinetics of specific immune responses in rhesus monkeys inoculated with live recombinant BCG expressing SIV Gag, Pol, Env, and Nef proteins. Virology *268*, 94–103.

Leutenegger, C.M., Boretti, F.S., Mislin, C.N., Flynn, J.N., Schroff, M., Habel, A., Junghans, C., Koenig-Merediz, S.A., Sigrist, B., Aubert, A., Pedersen, N.C., Wittig, B., and Lutz, H. (2000). Immunization of cats against feline immunodeficiency virus (FIV) infection by using minimalistic immunogenic defined gene expression vector vaccines expressing FIV gp140 alone or with feline interleukin-12 (IL-12), IL-16, or a CpG motif. J. Virol. *74*, 10447–10457.

Lewis, A.D., Chen, R., Montefiori, D.C., Johnson, P.R., and Clark, K.R. (2002). Generation of neutralizing activity against human immunodeficiency virus type 1 in serum by antibody gene transfer. J. Virol. *76*, 8769–8775.

Li, F., Craigo, J.K., Howe, L., Steckbeck, J.D., Cook, S., Issel, C., and Montelaro, R.C. (2003). A live attenuated equine infectious anemia virus proviral vaccine with a modified S2 gene provides protection from detectable infection by intravenous virulent virus challenge of experimentally inoculated horses. J. Virol. *77*, 7244–7253.

Li, Z., Zhang, M., Zhou, C., Zhao, X., Iijima, N., and Frankel, F.R. (2008). Novel vaccination protocol with two live mucosal vectors elicits strong cell-mediated immunity in the vagina and protects against vaginal virus challenge. J. Immunol. *180*, 2504–2513.

Lieberman, J., and Frankel, F.R. (2002). Engineered Listeria monocytogenes as an AIDS vaccine. Vaccine *20*, 2007–2010.

Lim, E.M., Lagranderie, M., Le Grand, R., Rauzier, J., Gheorghiu, M., Gicquel, B., and Winter, N. (1997). Recombinant Mycobacterium bovis BCG producing the N-terminal half of SIVmac251 Env antigen induces neutralizing antibodies and cytotoxic T-lymphocyte responses in mice and guinea pigs. AIDS Res Hum. Retroviruses *13*, 1573–1581.

Lin, J., Zhi, Y., Mays, L., and Wilson, J.M. (2007). Vaccines based on novel adeno-associated virus vectors elicit aberrant $CD8^+$ T-cell responses in mice. J. Virol. *81*, 11840–11849.

Lin, S.W., Hensley, S.E., Tatsis, N., Lasaro, M.O., and Ertl, H.C. (2007). Recombinant adeno-associated virus vectors induce functionally impaired transgene product-specific $CD8^+$ T-cells in mice. J. Clin Invest *117*, 3958–3970.

Liu, X.D., and Wang, B.Y. (2006). [Construction of recombinant lentivirus vaccine with single round replication]. Zhonghua Liu Xing Bing Xue Za Zhi *27*, 249–252.

Lockridge, K.M., Chien, M., Dean, G.A., Stefano Cole, K., Montelaro, R.C., Luciw, P.A., and Sparger, E.E. (2000). Protective immunity against feline immunodeficiency virus induced by inoculation with vif-deleted proviral DNA. Virology *273*, 67–79.

Lorin, C., Delebecque, F., Labrousse, V., Da Silva, L., Lemonnier, F., Brahic, M., and Tangy, F. (2005). A recombinant live attenuated measles vaccine vector primes effective HLA-A0201-restricted cytotoxic T-lymphocytes and broadly neutralizing antibodies against HIV-1 conserved epitopes. Vaccine *23*, 4463–4472.

Lorin, C., Mollet, L., Delebecque, F., Combredet, C., Hurtrel, B., Charneau, P., Brahic, M., and Tangy, F. (2004). A single injection of recombinant measles virus vaccines expressing human immunodeficiency virus (HIV) type 1 clade B envelope glycoproteins induces neutralizing antibodies and cellular immune responses to HIV. J. Virol. *78*, 146–157.

Lu, Z., Brans, R., Akhrameyeva, N.V., Murakami, N., Xu, X., and Yao, F. (2008). High-level expression of glycoprotein D by a dominant-negative HSV-1 virus augments its efficacy as a vaccine against HSV-1 infection. J. Invest. Dermatol. *129*, 1174–1184.

Lubeck, M.D., Natuk, R., Myagkikh, M., Kalyan, N., Aldrich, K., Sinangil, F., Alipanah, S., Murthy, S.C., Chanda, P.K., Nigida, S.M., Jr., Markham, P.D., Zolla-Pazner, S., Steimer, K., Wade, M., Reitz, M.S., Jr., Arthur, L.O., Mizutani, S., Davis, A., Hung, P.P., Gallo, R.C., Eichberg, J., and Robert-Guroff, M. (1997). Long-term protection of chimpanzees against high-dose HIV-1 challenge induced by immunization. Nat. Med. 3, 651–658.

MacDonald, G.H., and Johnston, R.E. (2000). Role of dendritic cell targeting in Venezuelan equine encephalitis virus pathogenesis. J. Virol. *74*, 914–922.

Maggiorella, M.T., Sernicola, L., Crostarosa, F., Belli, R., Pavone-Cossut, M.R., Macchia, I., Farcomeni, S., Tenner-Racz, K., Racz, P., Ensoli, B., and Titti, F. (2007). Multiprotein genetic vaccine in the SIV-Macaca animal model: a promising approach to generate sterilizing immunity to HIV infection. J. Med Primatol *36*, 180–194.

Malik, R., Kendall, K., Cridland, J., Coulston, S., Stuart, A.J., Snow, D., and Love, D.N. (1997). Prevalences of feline leukaemia virus and feline immunodeficiency virus infections in cats in Sydney. Aust. Vet J. *75*, 323–327.

Manrique, M., Micewicz, E., Kozlowski, P.A., Wang, S.W., Aurora, D., Wilson, R.L., Ghebremichael, M., Mazzara, G., Montefiori, D., Carville, A., Mansfield, K.G., and Aldovini, A. (2008). DNA-MVA vaccine protection after X4 SHIV challenge in macaques correlates with day-of-challenge antiviral $CD4^+$ cell-mediated immunity levels and postchallenge preservation of $CD4^+$ T-cell memory. AIDS Res. Hum. Retroviruses *24*, 505–519.

Marsac, D., Loirat, D., Petit, C., Schwartz, O., and Michel, M.L. (2002). Enhanced presentation of major histocompatibility complex class I-restricted human immunodeficiency virus type 1 (HIV-1) Gag-specific epitopes after DNA immunization with vectors coding for vesicular stomatitis virus glycoprotein-pseudotyped HIV-1 Gag particles. J. Virol. *76*, 7544–7553.

Marthas, M.L., Banapour, B., Sutjipto, S., Siegel, M.E., Marx, P.A., Gardner, M.B., Pedersen, N.C., and Luciw, P.A. (1989). Rhesus macaques inoculated with molecularly cloned simian immunodeficiency virus. J. Med. Primatol. *18*, 311–319.

Martinez, V., Diemert, M.C., Braibant, M., Potard, V., Charuel, J.L., Barin, F., Costagliola, D., Caumes, E., Clauvel, J.P., Autran, B., and Musset, L. (2009). Anticardiolipin antibodies in HIV infection are independently associated with antibodies to the membrane proximal external region of gp41 and with cell-associated HIV DNA and immune activation. Clin. Infect. Dis. *48*, 123–132.

Martinon, F., Brochard, P., Ripaux, M., Delache, B., Auregan, G., Vaslin, B., and Le Grand, R. (2008). Improved protection against simian immunodeficiency virus mucosal challenge in macaques primed with a DNA vaccine and boosted with the recombinant modified vaccinia virus Ankara and recombinant Semliki Forest virus. Vaccine *26*, 532–545.

Mascola, J.R., Snyder, S.W., Weislow, O.S., Belay, S.M., Belshe, R.B., Schwartz, D.H., Clements, M.L., Dolin, R., Graham, B.S., Gorse, G.J., Keefer, M.C., McElrath, M.J., Walker, M.C., Wagner, K.F., McNeil, J.G., McCutchan, F.E., and Burke, D.S. (1996). Immunization with envelope subunit vaccine products elicits neutralizing antibodies against laboratory-adapted but not primary isolates of human immunodeficiency virus type 1. The National Institute of Allergy and Infectious Diseases AIDS Vaccine Evaluation Group. J. Infect. Dis. *173*, 340–348.

Mata, M., Yao, Z.J., Zubair, A., Syres, K., and Paterson, Y. (2001). Evaluation of a recombinant Listeria monocytogenes expressing an HIV protein that protects mice against viral challenge. Vaccine *19*, 1435–1445.

Matteucci, D., Poli, A., Mazzetti, P., Sozzi, S., Bonci, F., Isola, P., Zaccaro, L., Giannecchini, S., Calandrella, M., Pistello, M., Specter, S., and Bendinelli, M. (2000). Immunogenicity of an anti-clade B feline immunodeficiency fixed-cell virus vaccine in field cats. J. Virol. *74*, 10911–10919.

Matthews, T.J. (1994). Dilemma of neutralization resistance of HIV-1 field isolates and vaccine development. AIDS Res. Hum. Retroviruses *10*, 631–632.

Matthews, T.J., Langlois, A.J., Robey, W.G., Chang, N.T., Gallo, R.C., Fischinger, P.J., and Bolognesi, D.P. (1986). Restricted neutralization of divergent human T-lymphotropic virus type III isolates by antibodies to the major envelope glycoprotein. Proc. Natl. Acad. Sci. U.S.A. *83*, 9709–9713.

McGettigan, J.P., Koser, M.L., McKenna, P.M., Smith, M.E., Marvin, J.M., Eisenlohr, L.C., Dietzschold, B., and Schnell, M.J. (2006). Enhanced humoral HIV-1-specific immune responses generated from recombinant rhabdoviral-based vaccine vectors co-expressing HIV-1 proteins and IL-2. Virology *344*, 363–377.

McGettigan, J.P., Pomerantz, R.J., Siler, C.A., McKenna, P.M., Foley, H.D., Dietzschold, B., and Schnell, M.J. (2003). Second-generation rabies virus-based vaccine vectors expressing human immunodeficiency virus type 1 *gag* have greatly reduced pathogenicity but are highly immunogenic. J. Virol. *77*, 237–244.

McGettigan, J.P., Sarma, S., Orenstein, J.M., Pomerantz, R.J., and Schnell, M.J. (2001). Expression and immunogenicity of human immunodeficiency virus type 1 Gag expressed by a replication-competent rhabdovirus-based vaccine vector. J. Virol. *75*, 8724–8732.

McKenna, P.M., Koser, M.L., Carlson, K.R., Montefiori, D.C., Letvin, N.L., Papaneri, A.B., Pomerantz, R.J., Dietzschold, B., Silvera, P., McGettigan, J.P., and Schnell, M.J. (2007). Highly attenuated rabies virus-based vaccine vectors expressing simian-human immunodeficiency virus89.6P Env and simian immunodeficiency virusmac239 Gag are safe in rhesus macaques and protect from an AIDS-like disease. J. Infect. Dis. *195*, 980–988.

Mealey, R.H., Stone, D.M., Hines, M.T., Alperin, D.C., Littke, M.H., Leib, S.R., Leach, S.E., and Hines, S.A. (2007). Experimental Rhodococcus equi and equine infectious anemia virus DNA vaccination in adult and neonatal horses: effect of IL-12, dose, and route. Vaccine *25*, 7582–7597.

Mederle, I., Le Grand, R., Vaslin, B., Badell, E., Vingert, B., Dormont, D., Gicquel, B., and Winter, N. (2003). Mucosal administration of three recombinant Mycobacterium bovis BCG-SIVmac251 strains to cynomolgus macaques induces rectal IgAs and boosts systemic cellular immune responses that are primed by intradermal vaccination. Vaccine *21*, 4153–4166.

Mehendale, S., van Lunzen, J., Clumeck, N., Rockstroh, J., Vets, E., Johnson, P.R., Anklesaria, P., Barin, B., Boaz, M., Kochhar, S., Lehrman, J., Schmidt, C., Peeters, M., Schwarze-Zander, C., Kabamba, K., Glaunsinger, T., Sahay, S., Thakar, M., Paranjape, R., Gilmour, J., Excler, J.L., Fast, P., and Heald, A.E. (2008). A phase 1 study to evaluate the safety and immunogenicity of a recombinant HIV type 1 subtype C adeno-associated virus vaccine. AIDS Res. Hum. Retroviruses *24*, 873–880.

Meyer, H., Sutter, G., and Mayr, A. (1991). Mapping of deletions in the genome of the highly attenuated vaccinia virus MVA and their influence on virulence. J. Gen. Virol. *72*, 1031–1038.

Michelini, Z., Negri, D.R., Baroncelli, S., Catone, S., Comini, A., Maggiorella, M.T., Sernicola, L., Crostarosa, F., Belli, R., Mancini, M.G., Farcomeni, S., Fagrouch, Z., Ciccozzi, M., Rovetto, C., Liljestrom, P., Norley, S., Heeney, J., and Titti, F. (2004). T-cell-mediated protective efficacy of a systemic vaccine approach in cynomolgus monkeys after SIV mucosal challenge. J. Med. Primatol. *33*, 251–261.

Montgomery, D.L., Shiver, J.W., Leander, K.R., Perry, H.C., Friedman, A., Martinez, D., Ulmer, J.B., Donnelly, J.J., and Liu, M.A. (1993). Heterologous and homologous protection against influenza A by DNA vaccination: optimization of DNA vectors. DNA Cell Biol. *12*, 777–783.

Mooij, P., Balla-Jhagjhoorsingh, S.S., Koopman, G., Beenhakker, N., van Haaften, P., Baak, I., Nieuwenhuis, I.G., Kondova, I., Wagner, R., Wolf, H., Gomez, C.E., Najera, J.L., Jimenez, V., Esteban, M., and Heeney, J.L. (2008). Differential $CD4^{+}$ versus $CD8^{+}$ T-cell responses elicited by different poxvirus-based human immunodeficiency virus type 1 vaccine candidates provide comparable efficacies in primates. J. Virol. *82*, 2975–2988.

Murphy, C.G., Lucas, W.T., Means, R.E., Czajak, S., Hale, C.L., Lifson, J.D., Kaur, A., Johnson, R.P., Knipe, D.M., and Desrosiers, R.C. (2000). Vaccine protection against simian immunodeficiency virus by recombinant strains of herpes simplex virus. J. Virol. *74*, 7745–7754.

Muster, T., Guinea, R., Trkola, A., Purtscher, M., Klima, A., Steindl, F., Palese, P., and Katinger, H. (1994). Cross-neutralizing activity against divergent human immunodeficiency virus type 1 isolates induced by the gp41 sequence ELDKWAS. J. Virol. *68*, 4031–4034.

Myagkikh, M., Alipanah, S., Markham, P.D., Tartaglia, J., Paoletti, E., Gallo, R.C., Franchini, G., and Robert-Guroff, M. (1996). Multiple immunizations with attenuated poxvirus HIV type 2 recombinants and subunit boosts required for protection of rhesus macaques. AIDS Res. Hum. Retroviruses *12*, 985–992.

Naidu, Y.M., Kestler, H.W., 3rd, Li, Y., Butler, C.V., Silva, D.P., Schmidt, D.K., Troup, C.D., Sehgal, P.K., Sonigo, P., Daniel, M.D., *et al.* (1988). Characterization of infectious molecular clones of simian immunodeficiency virus (SIVmac) and human immunodeficiency virus type 2: persistent infection of rhesus monkeys with molecularly cloned SIVmac. J. Virol. *62*, 4691–4696.

Nan, X., Peng, B., Hahn, T.W., Richardson, E., Lizonova, A., Kovesdi, I., and Robert-Guroff, M. (2003). Development of an Ad7 cosmid system and generation of an Ad7deltaE1deltaE3HIV(MN) *env/rev* recombinant virus. Gene Ther. *10*, 326–336.

Natuk, R.J., Chanda, P.K., Lubeck, M.D., Davis, A.R., Wilhelm, J., Hjorth, R., Wade, M.S., Bhat, B.M., Mizutani, S., Lee, S., *et al.* (1992). Adenovirus-human immunodeficiency virus (HIV) envelope recombinant vaccines elicit high-titered HIV-neutralizing antibodies in the dog model. Proc. Natl. Acad. Sci. U.S.A. *89*, 7777–7781.

Natuk, R.J., Lubeck, M.D., Chanda, P.K., Chengalvala, M., Wade, M.S., Murthy, S.C., Wilhelm, J., Vernon, S.K., Dheer, S.K., Mizutani, S. , *et al.* (1993). Immunogenicity of recombinant human adenovirus-human immunodeficiency virus vaccines in chimpanzees. AIDS Res. Hum. Retroviruses *9*, 395–404.

Neeson, P., Boyer, J., Kumar, S., Lewis, M.G., Mattias, L., Veazey, R., Weiner, D., and Paterson, Y. (2006). A DNA prime-oral Listeria boost vaccine in rhesus macaques induces a SIV-specific CD8 T-cell mucosal response characterized by high levels of alpha4beta7 integrin and an effector memory phenotype. Virology *354*, 299–315.

Negri, D.R., Baroncelli, S., Catone, S., Comini, A., Michelini, Z., Maggiorella, M.T., Sernicola, L., Crostarosa, F., Belli, R., Mancini, M.G., Farcomeni, S., Fagrouch, Z., Ciccozzi, M., Boros, S., Liljestrom, P., Norley, S., Heeney, J., and Titti, F. (2004). Protective efficacy of a multicomponent vector vaccine in cynomolgus monkeys after intrarectal simian immunodeficiency virus challenge. J. Gen. Virol. *85*, 1191–1201.

Nilsson, J., Kinloch-de-Loes, S., Granath, A., Sonnerborg, A., Goh, L.E., and Andersson, J. (2007). Early immune activation in gut-associated and peripheral lymphoid tissue during acute HIV infection. Aids *21*, 565–574.

Norley, S., Beer, B., Konig, H., Jensen, F., and Kurth, R. (1998). SIVmac vaccine studies using whole inactivated virus antigen sequentially depleted of viral proteins. J. Med. Primatol. *27*, 184–192.

O'Donovan, L.H., McMonagle, E.L., Taylor, S., Bain, D., Pacitti, A.M., Golder, M.C., McDonald, M., Hanlon, L., Onions, D.E., Argyle, D.J., Jarrett, O., and Nicolson, L. (2005). A vector expressing feline mature IL-18 fused to IL-1beta antagonist protein signal sequence is an effective adjuvant to a DNA vaccine for feline leukaemia virus. Vaccine 23, 3814–3823.

Olmsted, R.A., Hirsch, V.M., Purcell, R.H., and Johnson, P.R. (1989). Nucleotide sequence analysis of feline immunodeficiency virus: genome organization and relationship to other lentiviruses. Proc. Natl. Acad. Sci. U.S.A. *86*, 8088–8092.

Ou, W., Lu, N., Yu, S.S., and Silver, J. (2006). Effect of epitope position on neutralization by anti-human immunodeficiency virus monoclonal antibody 2F5. J. Virol. *80*, 2539–2547.

Ou, Y., Traina-Dorge, V., Davis, K.A., and Gray, W.L. (2007). Recombinant simian varicella viruses induce immune responses to simian immunodeficiency virus (SIV) antigens in immunized vervet monkeys. Virology *364*, 291–300.

Ourmanov, I., Brown, C.R., Moss, B., Carroll, M., Wyatt, L., Pletneva, L., Goldstein, S., Venzon, D., and Hirsch, V.M. (2000). Comparative efficacy of recombinant modified vaccinia virus Ankara expressing simian immunodeficiency virus (SIV) Gag–Pol and/or Env in macaques challenged with pathogenic SIV. J. Virol. *74*, 2740–2751.

Patterson, L.J., Malkevitch, N., Venzon, D., Pinczewski, J., Gomez-Roman, V.R., Wang, L., Kalyanaraman, V.S., Markham, P.D., Robey, F.A., and Robert-Guroff, M. (2004). Protection against mucosal simian immunodeficiency virus SIV(mac251) challenge by using replicating adenovirus-SIV multigene vaccine priming and subunit boosting. J. Virol. *78*, 2212–2221.

Peng, Y., Lin, F.C., Verardi, P.H., Jones, L.A., McChesney, M.B., and Yilma, T.D. (2007). Pseudotyped single-cycle simian immunodeficiency viruses expressing gamma interferon augment T-cell priming responses *in vitro*. J. Virol. *81*, 2187–2195.

Pialoux, G., Excler, J.L., Riviere, Y., Gonzalez-Canali, G., Feuillie, V., Coulaud, P., Gluckman, J.C., Matthews, T.J., Meignier, B., Kieny, M.P. , *et al.* (1995). A prime-boost approach to HIV preventive vaccine using a recombinant canarypox virus expressing glycoprotein 160 (MN) followed by a recombinant glycoprotein 160 (MN/LAI). The AGIS Group, and l'Agence Nationale de Recherche sur le SIDA. AIDS Res. Hum. Retroviruses *11*, 373–381.

Pistello, M., Bonci, F., Isola, P., Mazzetti, P., Merico, A., Zaccaro, L., Matteucci, D., and Bendinelli, M. (2005). Evaluation of feline immunodeficiency virus ORF-A mutants as candidate attenuated vaccine. Virology *332*, 676–690.

Pitisuttithum, P., Gilbert, P., Gurwith, M., Heyward, W., Martin, M., van Griensven, F., Hu, D., Tappero, J.W., and Choopanya, K. (2006). Randomized, double-blind, placebo-controlled efficacy trial of a bivalent recombinant glycoprotein 120 HIV-1 vaccine among injection drug users in Bangkok, Thailand. J. Infect. Dis. *194*, 1661–1671.

Pokorna, D., Rubio, I., and Muller, M. (2008). DNA-vaccination via tattooing induces stronger humoral and cellular immune responses than intramuscular delivery supported by molecular adjuvants. Genet. Vaccines Ther *6*, 4.

Poulet, H., Brunet, S., Boularand, C., Guiot, A.L., Leroy, V., Tartaglia, J., Minke, J., Audonnet, J.C., and Desmettre, P. (2003). Efficacy of a canarypox virus-vectored vaccine against feline leukaemia. Vet. Rec. *153*, 141–145.

Pu, R., Coleman, J., Coisman, J., Sato, E., Tanabe, T., Arai, M., and Yamamoto, J.K. (2005). Dual-subtype FIV vaccine (Fel-O-Vax FIV) protection against a heterologous subtype B FIV isolate. J. Feline Med. Surg. *7*, 65–70.

Publicover, J., Ramsburg, E., and Rose, J.K. (2005). A single-cycle vaccine vector based on vesicular stomatitis virus can induce immune responses comparable to those generated by a replication-competent vector. J. Virol. *79*, 13231–13238.

Putney, S.D., Matthews, T.J., Robey, W.G., Lynn, D.L., Robert-Guroff, M., Mueller, W.T., Langlois, A.J., Ghrayeb, J., Petteway, S.R., Jr., Weinhold, K.J. , *et al.* (1986). HTLV-III/LAV-neutralizing antibodies to an E. coli-produced fragment of the virus envelope. Science *234*, 1392–1395.

Raabe, M.L., Issel, C.J., Cook, S.J., Cook, R.F., Woodson, B., and Montelaro, R.C. (1998). Immunization with a recombinant envelope protein (rgp90) of EIAV produces a spectrum of vaccine efficacy ranging from lack of clinical disease to severe enhancement. Virology *245*, 151–162.

Racek, T., Jarmy, G., and Jassoy, C. (2006). Induction of humoral and cellular immune responses in mice by HIV-derived infectious pseudovirions. AIDS Res. Hum. Retroviruses *22*, 1162–1166.

Ramsburg, E., Rose, N.F., Marx, P.A., Mefford, M., Nixon, D.F., Moretto, W.J., Montefiori, D., Earl, P., Moss, B., and Rose, J.K. (2004). Highly effective control of an AIDS virus challenge in macaques by using vesicular stomatitis virus and modified vaccinia virus Ankara vaccine vectors in a single-boost protocol. J. Virol. *78*, 3930–3940.

Rayevskaya, M., Kushnir, N., and Frankel, F.R. (2002). Safety and immunogenicity in neonatal mice of a hyperattenuated Listeria vaccine directed against human immunodeficiency virus. J. Virol. *76*, 918–922.

Rayevskaya, M.V., and Frankel, F.R. (2001). Systemic immunity and mucosal immunity are induced against human immunodeficiency virus Gag protein in mice by a new hyperattenuated strain of Listeria monocytogenes. J. Virol. *75*, 2786–2791.

Redfield, R.R., Wright, D.C., James, W.D., Jones, T.S., Brown, C., and Burke, D.S. (1987). Disseminated vaccinia in a military recruit with human immunodeficiency virus (HIV) disease. N. Engl. J. Med. *316*, 673–676.

Richardson, J., Moraillon, A., Baud, S., Cuisinier, A.M., Sonigo, P., and Pancino, G. (1997). Enhancement of feline immunodeficiency virus (FIV) infection after DNA vaccination with the FIV envelope. J. Virol. *71*, 9640–9649.

Ridgely, S.L., Zhang, B., and McGuire, T.C. (2003). Response of ELA-A1 horses immunized with lipopeptide containing an equine infectious anemia virus ELA-A1-restricted CTL epitope to virus challenge. Vaccine *21*, 491–506.

Robb, M.L. (2008). Failure of the Merck HIV vaccine: an uncertain step forward. Lancet *372*, 1857–1858.

Rose, N.F., Marx, P.A., Luckay, A., Nixon, D.F., Moretto, W.J., Donahoe, S.M., Montefiori, D., Roberts, A., Buonocore, L., and Rose, J.K. (2001). An effective AIDS vaccine based on live attenuated vesicular stomatitis virus recombinants. Cell *106*, 539–549.

Ruprecht, R.M., Ferrantelli, F., Kitabwalla, M., Xu, W., and McClure, H.M. (2003). Antibody protection: passive immunization of neonates against oral AIDS virus challenge. Vaccine *21*, 3370–3373.

Ryman, K.D., Klimstra, W.B., Nguyen, K.B., Biron, C.A., and Johnston, R.E. (2000). Alpha/beta interferon protects adult mice from fatal Sindbis virus infection and is an important determinant of cell and tissue tropism. J. Virol. *74*, 3366–3378.

Samaniego, L.A., Neiderhiser, L., and DeLuca, N.A. (1998). Persistence and expression of the herpes simplex virus genome in the absence of immediate-early proteins. J. Virol. 72, 3307–3320.

Schnell, M.J., Foley, H.D., Siler, C.A., McGettigan, J.P., Dietzschold, B., and Pomerantz, R.J. (2000). Recombinant rabies virus as potential live-viral vaccines for HIV-1. Proc. Natl. Acad. Sci. U.S.A. *97*, 3544–3549.

Sekaly, R.P. (2008). The failed HIV Merck vaccine study: a step back or a launching point for future vaccine development? J. Exp. Med. *205*, 7–12.

Seth, A., Ourmanov, I., Kuroda, M.J., Schmitz, J.E., Carroll, M.W., Wyatt, L.S., Moss, B., Forman, M.A., Hirsch, V.M., and Letvin, N.L. (1998). Recombinant modified vaccinia virus Ankara-simian immunodeficiency virus *gag pol* elicits cytotoxic T-lymphocytes in rhesus monkeys detected by a major histocompatibility complex class I/peptide tetramer. Proc. Natl. Acad. Sci. U.S.A. *95*, 10112–10116.

Sharpe, S., Polyanskaya, N., Dennis, M., Sutter, G., Hanke, T., Erfle, V., Hirsch, V., and Cranage, M. (2001). Induction of simian immunodeficiency virus (SIV)-specific CTL in rhesus macaques by vaccination with modified vaccinia virus Ankara expressing SIV transgenes: influence of pre-existing anti-vector immunity. J. Gen. Virol. *82*, 2215–2223.

Shata, M.T., and Hone, D.M. (2001). Vaccination with a Shigella DNA vaccine vector induces antigen-specific CD8(+) T-cells and antiviral protective immunity. J. Virol. *75*, 9665–9670.

Shata, M.T., Reitz, M.S., Jr., DeVico, A.L., Lewis, G.K., and Hone, D.M. (2001). Mucosal and systemic HIV-1 Env-specific CD8(+) T-cells develop after intragastric vaccination with a Salmonella Env DNA vaccine vector. Vaccine *20*, 623–629.

Shen, X., Parks, R.J., Montefiori, D.C., Kirchherr, J.L., Keele, B.F., Decker, J.M., Blattner, W.A., Gao, F., Weinhold, K.J., Hicks, C.B., Greenberg, M.L., Hahn, B.H., Shaw, G.M., Haynes, B.F., and Tomaras, G.D. (2009). In Vivo gp41 Antibodies Targeting the 2F5 mAb Epitope Mediate HIV-1 Neutralization Breadth. J. Virol. *83*, 3617–3625.

Shiraki, K., Sato, H., Yoshida, Y., Yamamura, J.I., Tsurita, M., Kurokawa, M., and Kageyama, S. (2001). Construction of Oka varicella vaccine expressing human immunodeficiency virus *env* antigen. J. Med. Virol. *64*, 89–95.

Shiver, J.W., Davies, M.E., Yasutomi, Y., Perry, H.C., Freed, D.C., Letvin, N.L., and Liu, M.A. (1997). Anti-HIV *env* immunities elicited by nucleic acid vaccines. Vaccine *15*, 884–887.

Shiver, J.W., Fu, T.M., Chen, L., Casimiro, D.R., Davies, M.E., Evans, R.K., Zhang, Z.Q., Simon, A.J., Trigona, W.L., Dubey, S.A., *et al.* (2002). Replication-incompetent adenoviral vaccine vector elicits effective anti-immunodeficiency-virus immunity. Nature *415*, 331–335.

Shiver, J.W., Perry, H.C., Davies, M.E., Freed, D.C., and Liu, M.A. (1995). Cytotoxic T-lymphocyte and helper T-cell responses following HIV polynucleotide vaccination. Ann. N. Y. Acad. Sci. 772, 198–208.

Sodora, D.L., Shpaer, E.G., Kitchell, B.E., Dow, S.W., Hoover, E.A., and Mullins, J.I. (1994). Identification of three feline immunodeficiency virus (FIV) *env* gene subtypes and comparison of the FIV and human immunodeficiency virus type 1 evolutionary patterns. J. Virol. *68*, 2230–2238.

Sparkes, A.H. (1997). Feline leukaemia virus: a review of immunity and vaccination. J. Small Anim. Pract. *38*, 187–194.

Spear, G.T., Takefman, D.M., Sullivan, B.L., Landay, A.L., Jennings, M.B., and Carlson, J.R. (1993). Anti-cellular antibodies in sera from vaccinated macaques can induce complement-mediated virolysis of human immunodeficiency virus and simian immunodeficiency virus. Virology *195*, 475–480.

Staprans, S.I., Barry, A.P., Silvestri, G., Safrit, J.T., Kozyr, N., Sumpter, B., Nguyen, H., McClure, H., Montefiori, D., Cohen, J.I., and Feinberg, M.B. (2004). Enhanced SIV replication and accelerated progression to AIDS in macaques primed to mount a CD4 T-cell response to the SIV envelope protein. Proc. Natl. Acad. Sci. U.S.A. *101*, 13026–13031.

Stevens, R., Howard, K.E., Nordone, S., Burkhard, M., and Dean, G.A. (2004). Oral immunization with recombinant listeria monocytogenes controls virus load after vaginal challenge with feline immunodeficiency virus. J. Virol. *78*, 8210–8218.

Stevens, R., Lavoy, A., Nordone, S., Burkhard, M., and Dean, G.A. (2005). Pre-existing immunity to pathogenic Listeria monocytogenes does not prevent induction of immune responses to feline immunodeficiency virus by a novel recombinant *Listeria monocytogenes* vaccine. Vaccine *23*, 1479–1490.

Stiegler, G., Kunert, R., Purtscher, M., Wolbank, S., Voglauer, R., Steindl, F., and Katinger, H. (2001). A potent cross-clade neutralizing human monoclonal antibody against a novel epitope on gp41 of human immunodeficiency virus type 1. AIDS Res. Hum. Retroviruses *17*, 1757–1765.

Stolte-Leeb, N., Sauermann, U., Norley, S., Fagrouch, Z., Heeney, J., Franz, M., Hunsmann, G., and Stahl-Hennig, C. (2006). Sustained conservation of $CD4^+$ T-cells in multiprotein triple modality-immunized rhesus macaques after intrarectal challenge with simian immunodeficiency virus. Viral Immunol *19*, 448–457.

Stott, E.J., Chan, W.L., Mills, K.H., Page, M., Taffs, F., Cranage, M., Greenaway, P., and Kitchin, P. (1990). Preliminary report: protection of cynomolgus macaques against simian immunodeficiency virus by fixed infected-cell vaccine. Lancet *336*, 1538–1541.

Sutter, G., and Moss, B. (1992). Nonreplicating vaccinia vector efficiently expresses recombinant genes. Proc. Natl. Acad. Sci. U.S.A. *89*, 10847–10851.

Svanholm, C., Lowenadler, B., and Wigzell, H. (1997). Amplification of T-cell and antibody responses in DNA-based immunization with HIV-1 Nef by co-injection with a GM-CSF expression vector. Scand. J. Immunol. *46*, 298–303.

Tagmyer, T.L., Craigo, J.K., Cook, S.J., Even, D.L., Issel, C.J., and Montelaro, R.C. (2008). Envelope determinants of equine infectious anemia virus vaccine protection and the effects of sequence variation on immune recognition. J. Virol. *82*, 4052–4063.

Tan, G.S., McKenna, P.M., Koser, M.L., McLinden, R., Kim, J.H., McGettigan, J.P., and Schnell, M.J. (2005). Strong cellular and humoral anti-HIV Env immune responses induced by a heterologous rhabdoviral prime-boost approach. Virology *331*, 82–93.

Tang, Y., and Swanstrom, R. (2008). Development and characterization of a new single cycle vaccine vector in the simian immunodeficiency virus model system. Virology *372*, 72–84.

Tartaglia, J., Jarrett, O., Neil, J.C., Desmettre, P., and Paoletti, E. (1993). Protection of cats against feline leukemia virus by vaccination with a canarypox virus recombinant, ALVAC-FL. J. Virol. *67*, 2370–2375.

Tellier, M.C., Pu, R., Pollock, D., Vitsky, A., Tartaglia, J., Paoletti, E., and Yamamoto, J.K. (1998). Efficacy evaluation of prime-boost protocol: canarypoxvirus-based feline immunodeficiency virus (FIV) vaccine and inactivated FIV-infected cell vaccine against heterologous FIV challenge in cats. Aids *12*, 11–18.

Tijhaar, E.J., Siebelink, K.H., Karlas, J.A., Burger, M.C., Mooi, F.R., and Osterhaus, A.D. (1997). Induction of feline immunodeficiency virus specific antibodies in cats with an attenuated Salmonella strain expressing the Gag protein. Vaccine *15*, 587–596.

Titomirov, A.V., Sukharev, S., and Kistanova, E. (1991). In vivo electroporation and stable transformation of skin cells of newborn mice by plasmid DNA. Biochim. Biophys. Acta *1088*, 131–134.

Tsunetsugu-Yokota, Y., Ishige, M., and Murakami, M. (2007). Oral attenuated Salmonella enterica serovar Typhimurium vaccine expressing codon-optimized HIV type 1 Gag enhanced intestinal immunity in mice. AIDS Res. Hum. Retroviruses *23*, 278–286.

Ugen, K.E., Boyer, J.D., Wang, B., Bagarazzi, M., Javadian, A., Frost, P., Merva, M.M., Agadjanyan, M.G., Nyland, S., Williams, W.V., Coney, L., Ciccarelli, R., and Weiner, D.B. (1997). Nucleic acid immunization of chimpanzees as a prophylactic/immunotherapeutic vaccination model for HIV-1: prelude to a clinical trial. Vaccine *15*, 927–930.

Ugen, K.E., Nyland, S.B., Boyer, J.D., Vidal, C., Lera, L., Rasheid, S., Chattergoon, M., Bagarazzi, M.L., Ciccarelli, R., Higgins, T., Baine, Y., Ginsberg, R., Macgregor, R.R., and Weiner, D.B. (1998). DNA vaccination with HIV-1 expressing constructs elicits immune responses in humans. Vaccine *16*, 1818–1821.

Uhl, E.W., Martin, M., Coleman, J.K., and Yamamoto, J.K. (2008). Advances in FIV vaccine technology. Vet. Immunol. Immunopathol. *123*, 65–80.

Ulmer, J.B., Deck, R.R., DeWitt, C.M., Friedman, A., Donnelly, J.J., and Liu, M.A. (1994). Protective immunity by intramuscular injection of low doses of influenza virus DNA vaccines. Vaccine *12*, 1541–1544.

Vanbever, R., and Preat, V.V. (1999). In vivo efficacy and safety of skin electroporation. Adv. Drug Deliv. Rev. *35*, 77–88.

Vanichseni, S., Tappero, J.W., Pitisuttithum, P., Kitayaporn, D., Mastro, T.D., Vimutisunthorn, E., van Griensvan, F., Heyward, W.L., Francis, D.P., and Choopanya, K. (2004). Recruitment, screening and characteristics of injection drug users participating in the AIDSVAX B/E HIV vaccine trial, Bangkok, Thailand. Aids *18*, 311–316.

Vecino, W.H., Morin, P.M., Agha, R., Jacobs, W.R., Jr. and Fennelly, G.J. (2002). Mucosal DNA vaccination with highly attenuated Shigella is superior to attenuated Salmonella and comparable to intramuscular DNA vaccination for T-cells against HIV. Immunol. Lett. *82*, 197–204.

Verrier, B., Le Grand, R., Ataman-Onal, Y., Terrat, C., Guillon, C., Durand, P.Y., Hurtrel, B., Aubertin, A.M., Sutter, G., Erfle, V., and Girard, M. (2002). Evaluation in rhesus macaques of Tat and rev-targeted immunization as a preventive vaccine against mucosal challenge with SHIV-BX08. DNA Cell Biol *21*, 653–658.

Vogels, R., Zuijdgeest, D., van Meerendonk, M., Companjen, A., Gillissen, G., Sijtsma, J., Melis, I., Holterman, L., Radosevic, K., Goudsmit, J., and Havenga, M.J. (2007). High-level expression from two independent expression cassettes in replication-incompetent adenovirus type 35 vector. J. Gen. Virol. *88*, 2915–2924.

Walker, B.D., and Korber, B.T. (2001). Immune control of HIV: the obstacles of HLA and viral diversity. Nat. Immunol. *2*, 473–475.

Wang, B., Boyer, J., Srikantan, V., Coney, L., Carrano, R., Phan, C., Merva, M., Dang, K., Agadjanan, M., Gilbert, L., *et al.* (1993). DNA inoculation induces neutralizing immune responses against human immunodeficiency virus type 1 in mice and nonhuman primates. DNA Cell Biol. *12*, 799–805.

Wang, B., Boyer, J., Srikantan, V., Ugen, K., Gilbert, L., Phan, C., Dang, K., Merva, M., Agadjanyan, M.G., Newman, M., *et al.* (1995). Induction of humoral and cellular immune responses to the human immunodeficiency type 1 virus in nonhuman primates by *in vivo* DNA inoculation. Virology *211*, 102–112.

Wang, B., Merva, M., Dang, K., Ugen, K.E., Boyer, J., Williams, W.V., and Weiner, D.B. (1994). DNA inoculation induces protective *in vivo* immune responses against cellular challenge with HIV-1 antigen-expressing cells. AIDS Res. Hum. Retroviruses *10 Suppl. 2*, S35–41.

Wang, S.W., Bertley, F.M., Kozlowski, P.A., Herrmann, L., Manson, K., Mazzara, G., Piatak, M., Johnson, R.P., Carville, A., Mansfield, K., and Aldovini, A. (2004). An SHIV DNA/MVA rectal vaccination in macaques provides systemic and mucosal virus-specific responses and protection against AIDS. AIDS Res. Hum. Retroviruses *20*, 846–859.

Wang, S.Z., Rushlow, K.E., Issel, C.J., Cook, R.F., Cook, S.J., Raabe, M.L., Chong, Y.H., Costa, L., and Montelaro, R.C. (1994). Enhancement of EIAV replication and disease by immunization with a baculovirus-expressed recombinant envelope surface glycoprotein. Virology *199*, 247–251.

Watanabe, D., Brockman, M.A., Ndung'u, T., Mathews, L., Lucas, W.T., Murphy, C.G., Felber, B.K., Pavlakis, G.N., Deluca, N.A., and Knipe, D.M. (2007). Properties of a herpes simplex virus multiple immediate-early gene-deleted recombinant as a vaccine vector. Virology *357*, 186–198.

Watanaveeradej, V., Benenson, M.W., Souza, M.D., Sirisopana, N., Nitayaphan, S., Tontichaivanich, C., Amphaipit, R., Renzullo, P.O., Brown, A.E., McNeil, J.G., Robb, M.L., Birx, D.L., Tovanabutra, S., Carr, J.K., and McCutchan, F.E. (2006). Molecular epidemiology of HIV Type 1 in preparation for a Phase III prime-boost vaccine trial in Thailand and a new approach to HIV Type 1 genotyping. AIDS Res. Hum. Retroviruses *22*, 801–807.

Watkins, D.I., Burton, D.R., Kallas, E.G., Moore, J.P., and Koff, W.C. (2008). Nonhuman primate models and the failure of the Merck HIV-1 vaccine in humans. Nat. Med. *14*, 617–621.

Winslow, B.J., Cochran, M.D., Holzenburg, A., Sun, J., Junker, D.E., and Collisson, E.W. (2003). Replication and expression of a swinepox virus vector delivering feline leukemia virus Gag and Env to cell lines of swine and feline origin. Virus Res. *98*, 1–15.

Winslow, B.J., Kalabat, D.Y., Brown, S.M., Cochran, M.D., and Collisson, E.W. (2005). Feline B7.1 and B7.2 proteins produced from swinepox virus vectors are natively processed and biologically active: potential for use as nonchemical adjuvants. Vet Microbiol *111*, 1–13.

Xin, K.Q., Jounai, N., Someya, K., Honma, K., Mizuguchi, H., Naganawa, S., Kitamura, K., Hayakawa, T., Saha, S., Takeshita, F., Okuda, K., Honda, M., Klinman, D.M., and Okuda, K. (2005). Prime-boost vaccination with plasmid DNA and a chimeric adenovirus type 5 vector with type 35 fiber induces protective immunity against HIV. Gene Ther. *12*, 1769–1777.

Xin, K.Q., Ooki, T., Mizukami, H., Hamajima, K., Okudela, K., Hashimoto, K., Kojima, Y., Jounai, N., Kumamoto, Y., Sasaki, S., Klinman, D., Ozawa, K., and Okuda, K. (2002). Oral administration of recombinant adeno-associated virus elicits human immunodeficiency virus-specific immune responses. Hum Gene Ther. *13*, 1571–1581.

Xin, K.Q., Sekimoto, Y., Takahashi, T., Mizuguchi, H., Ichino, M., Yoshida, A., and Okuda, K. (2007). Chimeric adenovirus 5/35 vector containing the clade C HIV *gag* gene induces a cross-reactive immune response against HIV. Vaccine *25*, 3809–3815.

Xin, K.Q., Urabe, M., Yang, J., Nomiyama, K., Mizukami, H., Hamajima, K., Nomiyama, H., Saito, T., Imai, M., Monahan, J., Okuda, K., Ozawa, K., and Okuda, K. (2001). A novel recombinant adeno-associated virus vaccine induces a long-term humoral immune response to human immunodeficiency virus. Hum Gene Ther. *12*, 1047–1061.

Xu, F., Hong, M., and Ulmer, J.B. (2003). Immunogenicity of an HIV-1 *gag* DNA vaccine carried by attenuated *Shigella*. Vaccine *21*, 644–648.

Yasutomi, Y., Koenig, S., Haun, S.S., Stover, C.K., Jackson, R.K., Conard, P., Conley, A.J., Emini, E.A., Fuerst, T.R., and Letvin, N.L. (1993). Immunization with recombinant BCG-SIV elicits SIV-specific cytotoxic T-lymphocytes in rhesus monkeys. J. Immunol *150*, 3101–3107.

Yasutomi, Y., Koenig, S., Woods, R.M., Madsen, J., Wassef, N.M., Alving, C.R., Klein, H.J., Nolan, T.E., Boots, L.J., Kessler, J.A. , *et al.* (1995). A vaccine-elicited, single viral epitope-specific cytotoxic T-lymphocyte response does not protect against intravenous, cell-free simian immunodeficiency virus challenge. J. Virol. *69*, 2279–2284.

Yasutomi, Y., Robinson, H.L., Lu, S., Mustafa, F., Lekutis, C., Arthos, J., Mullins, J.I., Voss, G., Manson, K., Wyand, M., and Letvin, N.L. (1996). Simian immunodeficiency virus-specific cytotoxic T-lymphocyte induction through DNA vaccination of rhesus monkeys. J. Virol. *70*, 678–681.

Ye, L., Bu, Z., Skeen, M.J., Ziegler, H.K., Compans, R.W., and Yang, C. (2003). Enhanced immunogenicity of SIV Gag DNA vaccines encoding chimeric proteins containing a C-terminal segment of Listeriolysin O. Virus Res. *97*, 7–16.

Zagury, D. (1991). Anti-HIV cellular immunotherapy in AIDS. Lancet *338*, 694–695.

Zagury, D., Leonard, R., Fouchard, M., Reveil, B., Bernard, J., Ittele, D., Cattan, A., Zirimwabagabo, L., Kalumbu, M., Justin, W. , *et al.* (1987). Immunization against AIDS in humans. Nature *326*, 249–250.

Zagury, D., Bernard, J., Cheynier, R., Desportes, I., Leonard, R., Fouchard, M., Reveil, B., Ittele, D., Lurhuma, Z., Mbayo, K. , *et al.* (1988). A group specific anamnestic immune reaction against HIV-1 induced by a candidate vaccine against AIDS. Nature *332*, 728–731.

Zagury, D., Salaun, J.J., Bernard, J., Dechazal, L., Goussard, B., and Lurhuma, Z. (1988). [Immunization against the human immunodeficiency virus in Zaire]. Med Trop (Mars) *48*, 417–423.

Zhang, X., Wang, Y., Liang, H., Wei, L., Xiang, W., Shen, R., and Shao, Y. (2007). Correlation between the induction of Th1 cytokines by an attenuated equine infectious anemia virus vaccine and protection against disease progression. J. Gen. Virol. *88*, 998–1004.

Zhao, X., Zhang, M., Li, Z., and Frankel, F.R. (2006). Vaginal protection and immunity after oral immunization of mice with a novel vaccine strain of Listeria monocytogenes expressing human immunodeficiency virus type 1 gag. J. Virol. *80*, 8880–8890.

Zheng, Y., Ourmanov, I., and Hirsch, V.M. (2008). Persistent transcription of a nonintegrating mutant of simian immunodeficiency virus in rhesus macrophages. Virology *372*, 291–299.

Zuniga, A., Wang, Z., Liniger, M., Hangartner, L., Caballero, M., Pavlovic, J., Wild, P., Viret, J.F., Glueck, R., Billeter, M.A., and Naim, H.Y. (2007). Attenuated measles virus as a vaccine vector. Vaccine *25*, 2974–2983.

Zwick, M.B., Komori, H.K., Stanfield, R.L., Church, S., Wang, M., Parren, P.W., Kunert, R., Katinger, H., Wilson, I.A., and Burton, D.R. (2004). The long third complementarity-determining region of the heavy chain is important in the activity of the broadly neutralizing anti-human immunodeficiency virus type 1 antibody 2F5. J. Virol. *78*, 3155–3161.

Gammaretroviral and Lentiviral Vectors for Gene Delivery

13

Michael D. Mühlebach, Silke Schüle, Nina Gerlach, Matthias Schweizer, Christian J. Buchholz, Christine Hohenadl and Klaus Cichutek

Abstract

Gammaretroviral and lentiviral vectors for gene therapy have been developed that mediate stable genetic modification of treated cells by chromosomal integration of the transferred vector genomes. This is highly desired, not only for research use, but also for clinical gene therapy aiming at the long-term correction of genetic defects, e.g. in stem and progenitor cells. Retroviral vector particles with tropism for various target cells have been designed. Owing to split genome vector design the risk of replication-competent retrovirus formation has been minimized. Gammaretroviral and lentiviral vectors have so far been used in more than 300 clinical trials, addressing treatment options for various diseases. In some cases these trials resulted in benefit for treated patients suffering from life threatening disease. However, insertional mutagenesis due to vector integration in or next to cellular proto-oncogenes was concluded to be necessary for the lymphoproliferative disease observed in some patients treated with gammaretrovirally modified haematopoietic stem cells for X-linked severe combined immunodeficiency disease. These findings prompted the design of gammaretroviral vectors harbouring self-inactivating (SIN) long terminal repeats (LTRs), which current lentiviral vectors already have. SIN vectors may reduce the effect of insertional mutagenesis and proto-oncogene activation, thereby reducing the risk of oncogenesis. With a view to future clinical use, new developments such as cell entry targeting will further improve the safety and efficacy of retroviral vectors.

Introduction: retroviral vectors for gene transfer

Retroviral vectors are derived from retroviruses and are used to deliver heterologous genes into target cells. Since their introduction in the early 1980s (Mann *et al.*, 1983; Watanabe *et al.*, 1983; Wei *et al.*, 1981), replication-incompetent retroviral vectors have been widely used as gene delivery vehicles for various applications, including human gene therapy. Their widespread use is based on their capacity to actively integrate genetic material into the genome of infected target cells, thereby allowing long-term expression of the transferred gene in the genetically modified cells.

There are two main classes of retroviral vectors in current clinical use: gammaretroviral vectors and lentiviral vectors. Gammaretroviral vectors, which are derived from murine leukaemia virus (MLV) and its close relatives, allow gene transfer into proliferating cells. Replication-incompetent lentiviral vectors are derived from various lentiviruses, including human and simian immunodeficiency viruses (HIV and SIV, respectively). Lentiviral vectors derived from HIV-1 and equine infectious anaemia virus (EIAV) are currently used in clinical trials. An advantage of lentiviral vectors over gammaretroviral vectors is that the former enable gene transfer into non-proliferating cells. Replication-competent gammaretroviral vectors (RCRs) allow productive infection of progeny cells and thereby enable continuous transfer of contained heterologous genes.

Gammaretroviruses have a comparatively simple genome organization that facilitates the replacement of regions encoding retroviral structural (Gag, Env) and non-structural (Pol) proteins by heterologous genes. This is accomplished during construction of the retroviral transfer vector carrying the heterologous sequence, also termed transgene or therapeutic/preventive gene. The transgene may be a cDNA encoding a protein or a non-coding RNA with self-contained function (Coffin *et al.*, 1997; Miller *et al.*, 1993; Sinn *et al.*, 2005). Biotechnologically engineered mammalian packaging cells release replication-incompetent retroviral particles encompassing the transfer vector as their RNA genome, and additional factors and proteins necessary for cell entry, reverse transcription and chromosomal integration of the viral genome. During a single-round infection, the generated infectious virus particles transmit the retroviral transfer vector to a single cell. Replication-incompetent retroviral vectors have packaging capacities of about 10 kilobases (kb) and can transfer heterologous sequences of up to 8 kb. Lentiviral vector generation follows similar principles but requires additional technical considerations due to the presence of auxiliary genes in the lentiviral genomes. Details of vector and expression construct design and generation of packaging cells are described below.

Recent developments in vector technology include RCRs derived from gammaretroviruses encompassing all viral genes plus a heterologous gene (Finger *et al.*, 2005; Logg *et al.*, 2001; Solly *et al.*, 2003). RCRs may contain up to approximately 1.5 kb of extra genetic information. Because RCRs permit viral spread, they may improve the efficacy of gene transfer to infected tissues, which could be advantageous in immunotherapy or cancer gene therapy. However, their clinical use may have to await the development of additional features to abrogate virus spread to non-target tissue and to abrogate the risk of RCR-related tumour development.

The main safety issue in retroviral vector application is the possible incidence of insertional mutagenesis which may lead to transcriptional activation of cellular genes, including proto-oncogenes, adjacent to the integrated retroviral vector genome. In 2002, the first observation of insertional oncogenesis in mice caused by a replication-incompetent retroviral vector derived from MLV was reported (Li *et al.*, 2002). Between 2003 and 2008, this was followed by reports of lymphoproliferative diseases in two gene therapy clinical trials involving newborn patients suffering from lethal type X1 severe combined immunodeficiency disease. Patients had been administered with CD34-positive haematopoietic progenitor cells modified by similar replication-incompetent retroviral vectors derived from MLV. So-called self-inactivating (SIN) long terminal repeat (LTR)-containing vectors that may reduce the risk of insertional oncogenesis have been developed (described below). These vectors are designed to restrict retro- and lentiviral gene transfer to distinct target cells or tissues by restricting delivery at the level of cell entry or by limiting expression of the transferred genes to specific cells.

Vector design

Although retroviral vectors may be generated from members of all seven genera of the Retroviridae (International Committee on Taxonomy of Viruses, ICTVdB – The Universal Virus Database, version 4; http://www.ncbi.nlm.nih.gov/ICTVdb/ICTVdB), the most commonly used vectors are based on viruses characterized by a simple genome organization, such as gammaretroviruses (particularly MLV), or viruses that have the capacity to transduce non-dividing cells such as lentiviruses (e.g. HIV, SIV, or EIAV).

The retroviral vector particle basically consists of a structural core assembled from the Gag precursor protein which is processed into matrix (MA), capsid (CA), and nucleocapsid (NC) proteins (Coffin *et al.*, 1997). This core contains the viral non-structural proteins necessary for reverse transcription and integration, as well as two copies of the engineered RNA genome. After cell entry, the viral RNA is reverse transcribed into linear double-stranded DNA which is then translocated into the nucleus. After chromosomal integration, the modified vector genome – including the heterologous transgene – behaves like a cellular gene where expression may be driven by the original retroviral promoter contained in the 5′ LTR or, in case of SIN-LTRs, by a heterologous promoter located directly 5′ of the transferred gene.

Because transfer into the cell nucleus, which is followed by integration of gammaretroviral genomes into the host cell chromosomes, requires dissolution of the nuclear membrane during mitosis (Miller *et al.*, 1990; Roe *et al.*, 1993), gammaretroviral vectors mediate gene transfer only into proliferating cells. In contrast, lentiviral vectors can efficiently transduce fully differentiated quiescent hepatic, retinal, and muscle cells as well as non-stimulated progenitor cells such as $CD34^+$ stem cells (Lewis *et al.*, 1994; Miyoshi *et al.*, 1997; Naldini *et al.*, 1996b; Vigna *et al.*, 2000). The general basis for this capacity of lentiviral vectors is not fully understood but depends on complex interactions between viral and cellular proteins. Lentiviral Vpr, an accessory protein encoded by a spliced messenger RNA, appears to play a major role in this activity (De Noronha *et al.*, 2001). Lentiviral vector design and production (see below) is complicated by the presence of coding regions in addition to the main open reading frames (ORFs) encoding Gag, Pol and Env proteins.

Gammaretroviral vectors

The choice between gammaretroviral vs. lentiviral vectors largely depends on the target cells or tissues, in addition to various other considerations including the therapeutic setting. The construction of replication-incompetent gammaretroviral and lentiviral vectors involve similar underlying principles. For gammaretroviral vectors, packaging cells transfected with a distinct plasmid harbouring an expression construct encoding the capsid (Gag) and the non-structural enzymatic proteins (Pol), and a second plasmid encompassing the envelope (Env) protein (split genome approach), provide the basis for the generation of single-round infectious virus particles harbouring the respective retroviral/transfer vector RNA (Fig. 13.1B). Within the Gag/Pol expressing packaging construct and the Env expression construct, the *cis*-acting packaging signal psi (Ψ) required for packaging and transfer of the viral genome via virus particles is deleted (Fig. 13.1 A). The original viral promoter contained in the 5′LTR is replaced by a strong constitutively active heterologous promoter and endogenous viral transcription termination sequences are substituted with heterologous polyadenylation signals (Fig. 13.1 A). The retroviral transfer vector contains the packaging signal psi and all other viral *cis*-acting sequences required for efficient synthesis and packaging of vector RNA, its reverse transcription and integration. Expression of the included transgene in this basic version is controlled by the retroviral promoter located in the U3 region of the 5′LTR (Fig. 13.1 A).

The *cis*-acting sequences promoting transfer vector function include the 5′LTR containing the retroviral promoter and enhancer elements, the 5′ leader sequence including the tRNA primer binding site (PBS) essential for reverse transcription, and 360–400 bp of the *gag* gene representing the extended packaging signal (Ψ+; Fig. 13.2 A). Inclusion of Gag-encoding sequences clearly improves packaging of both gammaretroviral vectors and lentiviral vectors, resulting in increased virus titres (Linial *et al.*, 1990; Mangeot *et al.*, 2000; Zhao *et al.*, 2000). However, due to sequence homologies, this might promote unwanted recombination between the packaging construct and the transfer vector. The inserted transgene may be driven from a distinct promoter. The polypurine tract (PPT) and the 3′LTR, containing the only polyadenylation signal, are located downstream. The PPT is resistant to RNase H activity exerted by the reverse transcriptase and therefore facilitates second-strand DNA synthesis. Other commonly used heterologous *cis*-acting sequences are posttranscriptional regulatory elements (PRE) such as that of woodchuck hepatitis virus (WPRE), which enhance transgene expression and RNA stability and thereby improve virus titres (Zufferey *et al.*, 1999).

The split genome approach for generation of retroviral packaging cells significantly enhances biosafety since it practically excludes production of RCRs by homologous recombination. State of the art packaging cells are based on stably transfected cell lines, e.g. human embryonic kidney cells (HEK293), in which the packaging constructs are integrated in the genome, thus facilitating a controlled, long-term production of high titre vector viruses (Miller *et al.*, 1993). Expression of *gag/pol* and *env* might be either constitutive or inducible (Grignani *et al.*, 1998) in these cell lines.

Recent developments utilize DNA recombinases for targeted integration of vector constructs into defined loci of packaging cell chromosomes in order to provide well-defined high-titre producer

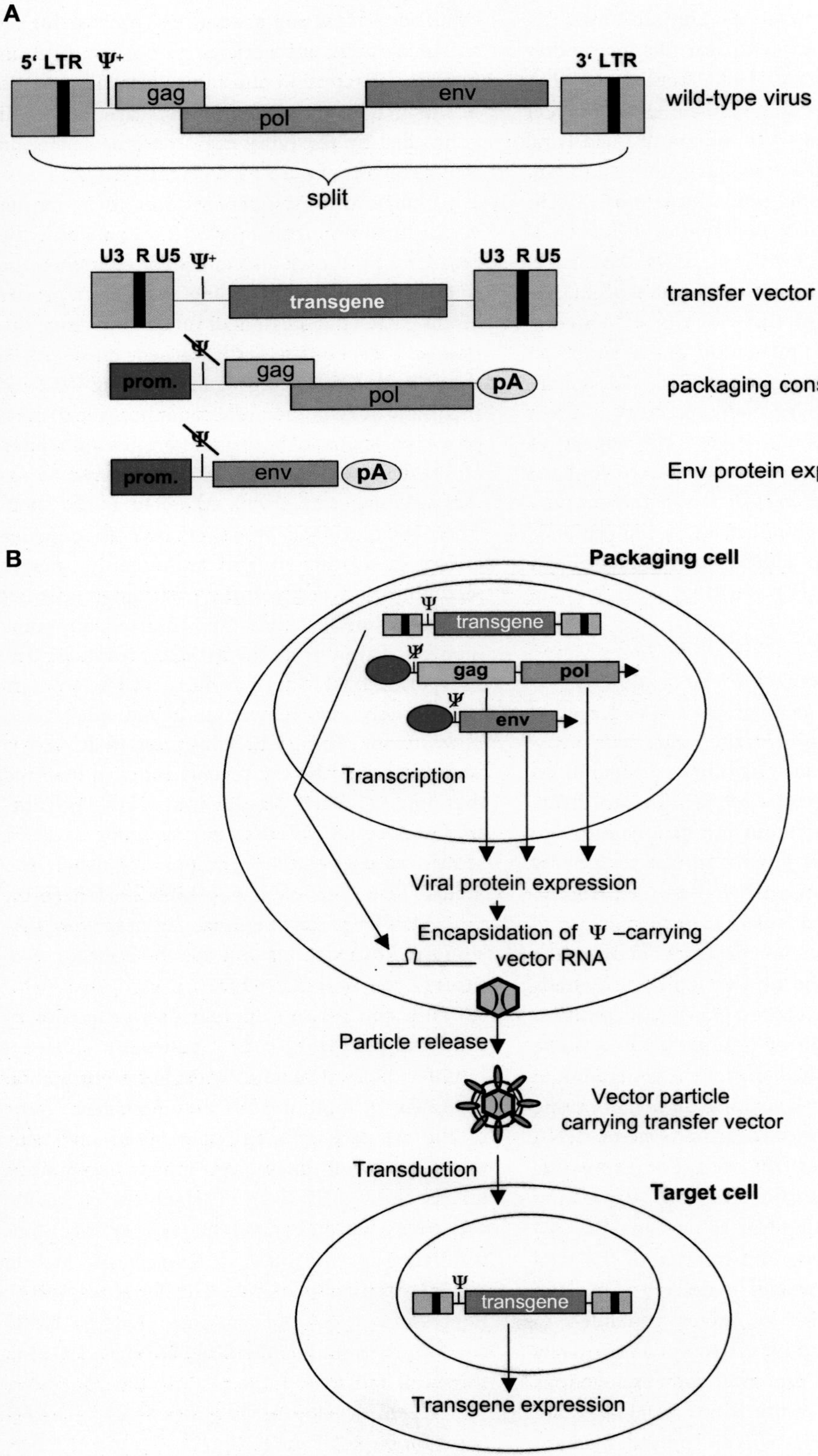
A
5' LTR
Ψ+
3' LTR
gag
env
pol
wild-type virus
split
U3 R U5
Ψ+
U3 R U5
transgene
transfer vector
Ψ
prom.
gag
pol
pA
packaging construct
Ψ
prom.
env
pA
Env protein expression construct
B
Packaging cell
Ψ
transgene
Ψ
gag
pol
Ψ
env
Transcription
Viral protein expression
Encapsidation of Ψ –carrying
vector RNA
Particle release
Vector particle
carrying transfer vector
Transduction
Target cell
Ψ
transgene
Transgene expression

cells. This system is based on the concerted action of Flp-recombinase derived from *Saccharomyces cerevisiae* and its respective recognition/binding sites (Flp-recombinase target, FRT) (Karreman *et al.*, 1996; Schucht *et al.*, 2006). For proof of concept, Flp293A, a HEK293-based retroviral producer cell line containing a single-copy retroviral vector integrated at a distinct chromosomal locus pre-selected for high transcriptional activity, was established. For targeted integration, the respective transfer vectors were flanked by non-interacting FRTs (Schucht *et al.*, 2006). Titres up to 2.5×10^7 were obtained for different vector constructs, indicating that a precise integration of viral vectors into a favourable chromosomal locus leads to substantial and predictable virus production. This method is probably compatible with other retroviral vectors, including the self-inactivating vectors described below.

Lentiviral vectors

Despite a common retroviral replication pathway, sequences are poorly conserved among the seven different retrovirus genera. In addition to Gag, Pol and Env proteins, lentiviruses encode 3–6 accessory proteins that contribute to virus replication and transcription (Seelamgari *et al.*, 2004). With the exception of Tat and Rev, the accessory proteins are non-essential for lentiviral vector production and may therefore be deleted from the Gag–Pol encompassing packaging construct (Vigna *et al.*, 2000). In comparison to gammaretroviruses, lentiviruses contain – besides the common *cis*-acting elements described above (Fig. 13.2 B) – an additional central polypurine tract (cPPT) and a central termination sequence (CTS) (Charneau *et al.*, 1992). This allows the synthesis of the lentiviral plus-strand DNA to be initiated at two origins, which increases the efficiency of reverse transcription and potentially also nuclear translocation of the pre-integration complex (Follenzi *et al.*, 2000). Moreover, splice donor (SD) and splice acceptor (SA) sites are retained, which further increases expression.

The lentivirus encoded Tat protein is a trans-activator of retroviral transcription driven by the endogenous promoter located within the 5′ LTR; Tat is necessary for efficient synthesis of lentiviral RNA. Rev is responsible for the efficient transport of unspliced and singly spliced mRNA from the nucleus into the cytoplasm. The respective RNA binding site, the Rev responsive element (RRE), is present within the *env* gene and must be included in the lentiviral transfer vector. In the absence of Rev, Gag/Pol encoding mRNA is targeted for degradation due to *cis*-repressive sequences – also called inhibitory sequences – that are present in the *gag–pol* coding region. Consequently, the Rev protein is required for efficient generation of vector particles and, depending on the stage of vector development, is expressed either directly from the packaging construct (Pavlakis *et al.*, 1990) or from a separate expression plasmid (Fig. 13.3).

Owing to biosafety issues, recent vector development approaches try to avoid inclusion of Tat and Rev encoding sequences within the packaging construct. This is achieved by providing one or more psi-negative *tat* and *rev* genes in the packaging cell in *trans* (Fig. 13.3 C). Moreover, the *trans*-acting function of Tat becomes dispensable if the viral promoter located in the 5′ U3 region of the transfer vector is replaced by a constitutively active promoter derived, e.g. from Rous sarcoma virus (RSV) (Dull *et al.*, 1998). In order to develop a Rev-independent lentiviral production system, the RRE in the transfer vector may be replaced by heterologous viral sequences that enhance export and stability of non-spliced

Figure 13.1 (opposite) Principal design of retroviral vector systems. (A) Representative genomic structure of a replication-competent retrovirus. All viral genes are included and flanked by the long terminal repeats (LTRs) containing the viral promoter (U3). To construct a replication-incompetent vector system, the viral components are split into three separate plasmids comprising the packaging constructs for expression of viral capsid (Gag) and non-structural (Pol), as well as envelope (Env) proteins. The transfer vector harbours the transgene and is marked by the presence of a packaging signal (Ψ+). To reduce sequence homologies, viral promoter and polyadenylation signals (pA) within the packaging constructs are replaced by heterologous elements displaying similar functions. (B) Retroviral packaging cell line, vector production and transduction of cells. Only the transfer vector encoding one or more heterologous transgenes is transferred and expressed in target cells.

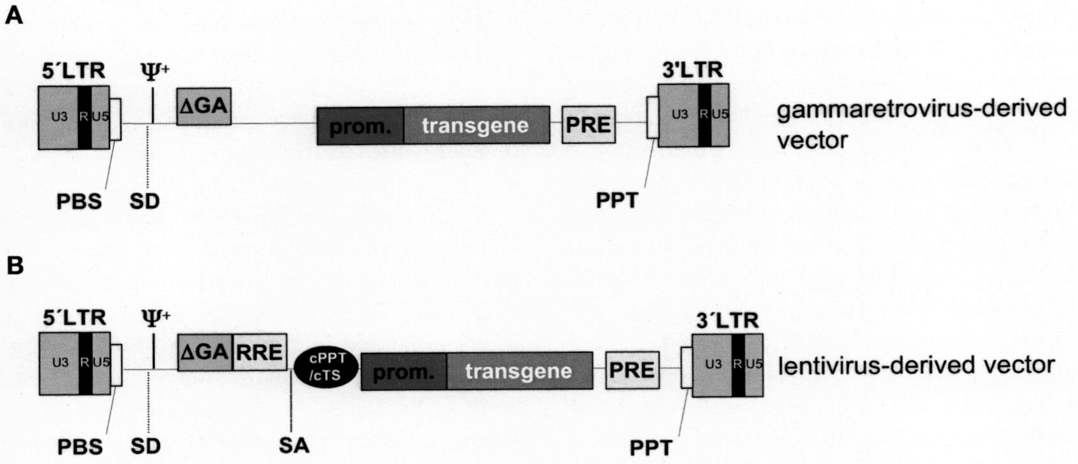

Figure 13.2 Schematic representation of gammaretrovirus- and lentivirus-derived transfer vector systems. (A) A gammaretroviral transfer vector includes the 5′ and 3′LTR, the 5′ leader sequence containing the tRNA primer binding site (PBS) and a splice donor (SD), followed by the extended packaging signal (Ψ+) overlapping with a part of the Gag-encoding region (GA). The encoded transgene may be driven by an internal promoter (prom). For enhanced transgene expression, posttranscriptional regulatory elements (PRE) may be included. The polypurine tract mediates second strand synthesis during reverse transcription. (B) Lentivirus-derived vectors also contain the Rev responsive element (RRE), which originates from the *env* gene, and a central polypurine tract (cPPT) facilitating transduction of non-dividing cells. Splice donor and splice acceptor (SA) sites are contained in the leader sequence.

transcripts, such as the constitutive transport elements (CTE) derived from Mason–Pfizer monkey virus (MPMV) (Gasmi *et al.*, 1999; Srinivasakumar *et al.*, 1999) or simian retrovirus type 1 (SRV-1) (Mautino *et al.*, 2000). Stability of the Gag/Pol-encoding RNA in the absence of Rev can be improved by codon optimization (Kotsopoulou *et al.*, 2000), i.e. adjusting codon usage to that of eukaryotic cells. This renders expression of *gag/pol* genes Rev-independent and also reduces sequence homologies between the transfer vector and the packaging construct (Fig. 13.3D).

Deletion of the *tat* gene and expression of *rev* from a separate expression plasmid resulted in the third generation of packaging constructs (Dull *et al.*, 1998). However, due to cytopathogenicity of some of the viral gene products, generation of stable packaging cell lines for production of lentiviral vector viruses is difficult. Up to now, only a limited number of stable lentiviral vector packaging cells have been described (Ikeda *et al.*, 2003; Klages *et al.*, 2000; Xu *et al.*, 2001). Thus, infectious lentivirus vector particles are usually produced by transient transfection of human 293T-cells with the respective transfer and packaging vector constructs, including an expression plasmid for synthesis of an envelope protein (Dull *et al.*, 1998).

Self-inactivating vectors

A further step towards development of safe and efficient retroviral gene delivery systems is the generation of so-called self-inactivating (SIN) vectors. In comparison to conventional retroviral vectors (Fig. 13.4A), SIN vectors are characterized by an almost complete deletion of the 3′ U3 region, which includes enhancer elements as well as the TATA box as a key element of the promoter (Fig. 13.4B). Since the 3′LTR region is the template used to generate both copies of the LTR in the integrated provirus, transduction of this vector results in the presence of two promoter-less and thus transcriptionally inactive LTRs (Yu *et al.*, 1986; Zufferey *et al.*, 1998). During construction of SIN vectors important sequences within the U3 region must be conserved: e.g. sequences regulating polyadenylation or integration (attachment site). For efficient synthesis of vector RNA within the packaging cell, a chimeric 5′LTR

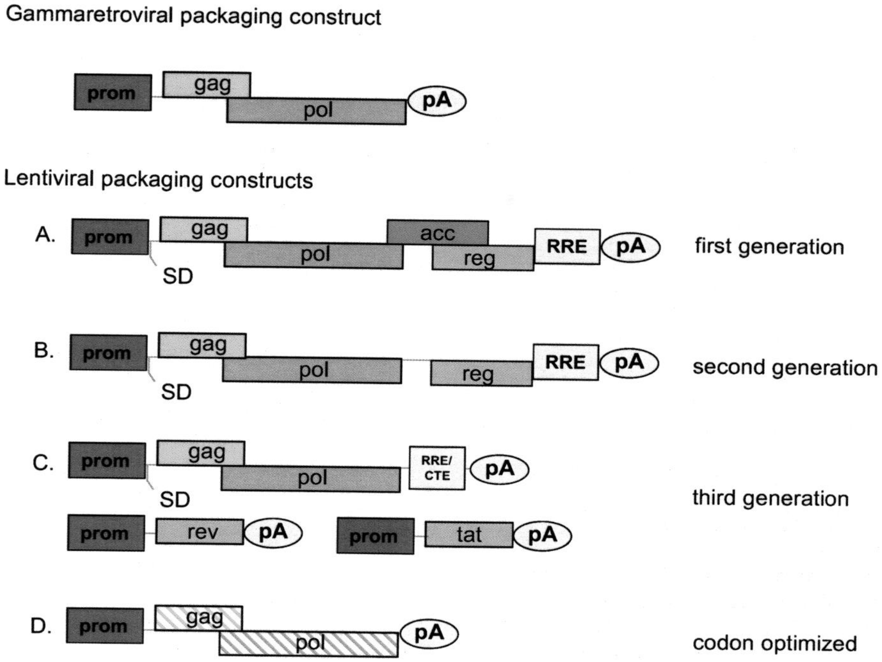

Figure 13.3 Schematic representation of typical first, second and third generation lentiviral packaging constructs. A strong constitutive promoter and polyadenylation signal (poly(A)) are used to replace the 5′ and 3′LTR, respectively. The packaging signal Ψ is deleted. In terms of development, the accessory genes (acc) and the regulatory genes (reg) are eliminated stepwise from the packaging construct. In third generation constructs the regulatory gene *rev* is expressed from a separate expression plasmid lacking homologous sequences. When using a RRE equivalent cytoplasmic transport element (CTE), *rev* might be deleted. Similarly, Tat might be expressed from a separate plasmid or, if a chimeric Tat-independent 3′ promoter is used, completely deleted.

containing a heterologous promoter (Naviaux *et al.*, 1996) may also be introduced (improved SIN vector, Fig. 13.4C). The SIN vector absolutely demands transcription of the inserted transgene from an internal, optionally transcription-targeted promoter.

Inactivation of both LTRs in transduced cells should reduce the risk of insertional mutagenesis resulting in transcriptional activation of cellular genes adjacent to the transfer vector integration site. Incorporation of so-called insulator sequences into the residual U3 region of the 3′LTR may also reduce the effects of insertional mutagenesis (Rivella *et al.*, 2000). Insulators are DNA sequences that can function as directional blocking elements, either by interfering with promoter–enhancer interactions when positioned in the intervening sequence, or by reducing position effects imparted on transgenes when flanking the integrated transcription units (Gaszner *et al.*, 2006; Recillas-Targa *et al.*, 2004). Although SIN vectors were established from both MLV-based and lentiviral vectors (Schambach *et al.*, 2006), they have only been widely adopted in the lentiviral vector field. In terms of transgene expression, lentiviral SIN vectors seem to perform similar to vectors with an intact 3′LTR. However, even with vectors encompassing large 3′LTR deletions, repair of the SIN deletion occurs at a significant rate (Lucke *et al.*, 2005; Modlich *et al.*, 2006). Although the concept of SIN vectors is powerful, further development and rigorous testing of this technology is required before it can be confidently used to address issues related to insertional mutagenesis (see below).

Multicistronic vector constructs

As described above, the gene transfer capacity of retroviral vectors is limited to about 8 kb, but the requirements for desired vectors may be

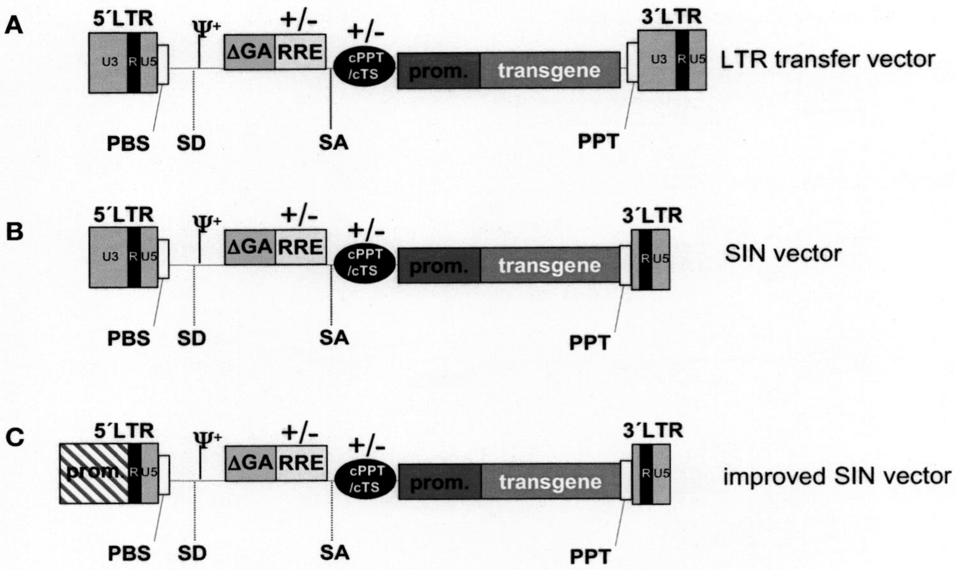

Figure 13.4 Development of self-inactivating retroviral- or lentiviral vectors. (A) The basic retroviral (– RRE and – cPPT) or lentiviral (+ RRE and + cPPT) transfer vector contains all sequences which are required for transfer and subsequent integration of the vector into the genome of the target cell. It is flanked by non-modified LTRs. (B) In contrast, a SIN vector carries a deletion in the U3 region of the 3′LTR. (C) Substitution of the 5′ U3 region with a constitutive heterologous promoter improves transcription of full-length viral RNA in virus producer cells. Upon infection and reverse transcription the deleted 3′ U3 region is copied to the 5′ end, rendering both LTRs transcriptionally inactive. PBS: primer binding site; Ψ packaging signal; SD: splice donor; GA: *gag* encoding region, RRE: Rev responsive element, cPPT/cTS: central polypurine tract/central termination sequence. SA: splice acceptor.

extensive. In addition to the therapeutic gene, genes encoding selection markers (e.g. enhanced green fluorescent protein or antibiotic resistance genes) might be included to enable identification or isolation of successfully transduced cells. Therefore, various methods were established to facilitate efficient and regulated (co-)expression of one (monocistronic vector), two (bicistronic vector) or multiple genes (polycistronic vectors) while simultaneously avoiding promoter interference that decreases expression levels (Ginn *et al.*, 2003).

In a basic retroviral transfer vector, transcription is driven by the viral promoter/enhancer elements present in the U3 region of the 5′LTR, allowing constitutive expression of the transferred gene in most cell types (Fig. 13.5A). Alternatively, especially for SIN vectors, transgene expression is mediated by a cell type-specific internal promoter (Fig. 13.5B and D). Expression of two individual genes may be achieved by using the promoter within the LTR in addition to the internal promoter (Fig. 13.5C); this concept is widely used for gammaretroviral gene transfer vectors. Introduction of an internal ribosomal entry site (IRES) located between two open reading frames (Fig. 13.5D) enables co-expression of two encoded proteins from a single RNA. The IRES is a *cis*-acting element derived from picornaviruses which allows translation of the downstream ORF without generation of a fusion protein (Martinez-Salas, 1999; Ngoi *et al.*, 2004). Similarly, the endopeptidase 2A-encoding sequence may be included for a co-translational separation of the two proteins (Fig. 13.5E) (Lengler *et al.*, 2005; Osborn *et al.*, 2005). In both cases the downstream protein should be a selectable marker; since expression levels of the second gene are significantly decreased compared to the first, this should guarantee expression of the first ORF.

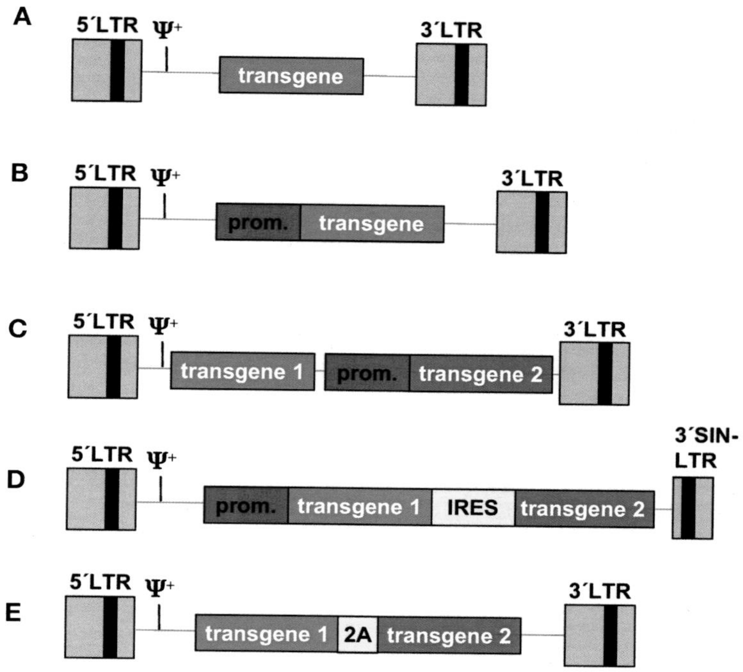

Figure 13.5 Schematic representation of different vector design strategies for transgene expression. (A) A single transgene is expressed via the viral promoter contained in the U3 region of the 5′LTR or (B) by a heterologous promoter located in front of the delivered transgene. (C) Combination of both facilitates expression of two genes in parallel. (D) Promoter interference is avoided by applying a SIN vector containing an internal promoter driving expression of two genes which are linked by an internal ribosomal binding site (IRES) or (E) the picornaviral 2A endopeptidase for individual translation.

Host range and cell targeting

The host range of retroviruses is determined by a number of factors that can influence any step in the viral replication cycle. With respect to targeted application of retroviral vectors, the most relevant factors are cell entry (i.e. receptor interaction), integration, and transgene expression.

The primary step in retroviral vector-mediated gene transfer is cell entry. For retrovirus-derived Env proteins, this involves the attachment of the viral particle to the target cell via a specific receptor molecule, followed by fusion of the viral envelope with the cellular membrane. This process is determined by the glycoprotein incorporated in(to) the viral envelope. The amphotropic MLV strains can infect a wide range of mammalian cells due to the interaction of the encoded Env protein with a ubiquitously expressed sodium-dependent phosphate symporter (Kavanaugh *et al.*, 1994a). In contrast, the ecotropic MLV uses a cationic amino acid transporter present exclusively on murine cells (Kavanaugh *et al.*, 1994b). The HIV-1 Env protein targets this virus specifically to T-lymphocytes and macrophages (Madani *et al.*, 1998; Siciliano *et al.*, 1999) via interaction with the CD4 molecule and associated chemokine receptors such as CCR5 and CXCR-4.

Depending on the target cell population intended to be transduced with the targeted retroviral vector, the respective wild-type virus Env protein could be replaced with glycoproteins from different viruses. This process (termed pseudotyping) is usually achieved by transfecting packaging cell lines with an expression construct encoding a heterologous envelope protein. Assuming efficient Env protein incorporation by the virus particles, pseudotyped viruses are obtained that reflect the host range of the heterologous Env protein (Cronin *et al.*, 2005; Kobinger *et al.*, 2001; Schnierle *et al.*, 1997). Although the rules of Env protein incorporation into viral envelopes are not yet fully understood (Sakalian *et al.*, 1998; Sandrin *et al.*, 2004), a high cell surface expression level is always critical for production of sufficient titres of infectious particles. For Env protein incorporation, modifications in the cytoplasmic tail of the heterologous Env protein may sometimes be required (Christodoulopoulos *et al.*, 2001; Merten *et al.*, 2005; Stitz *et al.*, 2000). The list of heterologous Env proteins used for pseudotyping MLV or HIV vectors is growing continuously. A very prominent example is the use of vesicular stomatitis virus glycoprotein (VSV-G), which facilitates generation of highly stable vector virus particles with broad host cell specificity (Yang *et al.*, 1995). Further examples include the gibbon ape leukaemia virus (GaLV) Env protein, the amphotropic MLV 10A1 Env protein, and the feline leukaemia virus RD114 Env protein, all of which have been applied for enhancing gene transfer into haematopoietic stem cells (HSCs) or T-lymphocytes (Muhlebach *et al.*, 2003; Uckert *et al.*, 2000; von Kalle *et al.*, 1994). Cell entry targeting of retroviral vectors to CD4-positive cells was achieved using a truncated variant of the HIV Env protein (Schnierle *et al.*, 1997), and airway epithelia cells were successfully transduced using the filovirus glycoprotein (Kobinger *et al.*, 2001).

For therapeutic applications *in vivo*, delivery of transferred genes exclusively to the target cells would be advantageous, especially if systemic administration is planned. Therefore, retroviral vector envelope proteins are currently optimized in order to target gene transfer at the level of cell entry. Three different strategies of Env protein engineering can be distinguished: (i) substitution of the complete envelope with a heterologous Env protein, (ii) addition or insertion of single chain antibodies (scAb) or polypeptide ligands into the Env protein, and (iii) engineering of protease cleavage sites into the Env protein (Fig. 13.6).

In contrast to envelope substitution, envelope engineering is not restricted by the availability of viral Env proteins evolved by nature. Therefore envelope engineering should theoretically enable gene transfer via many, if not all, cell surface molecules. To achieve this goal, Env proteins are modified with polypeptides derived from growth factors, chemokines or antibodies recognizing the cell surface receptor of interest. To enable direct evaluation of the modified molecules with respect to infectivity of respective vector particles for human cells, the modifying molecules are usually displayed on the ecotropic MLV Env protein. The list of polypeptides analysed has grown substantially during the past years and includes scAbs directed against various cell surface antigens, as well as chemokines and growth factors. However, in most cases low transduction efficiencies ($<10^2$

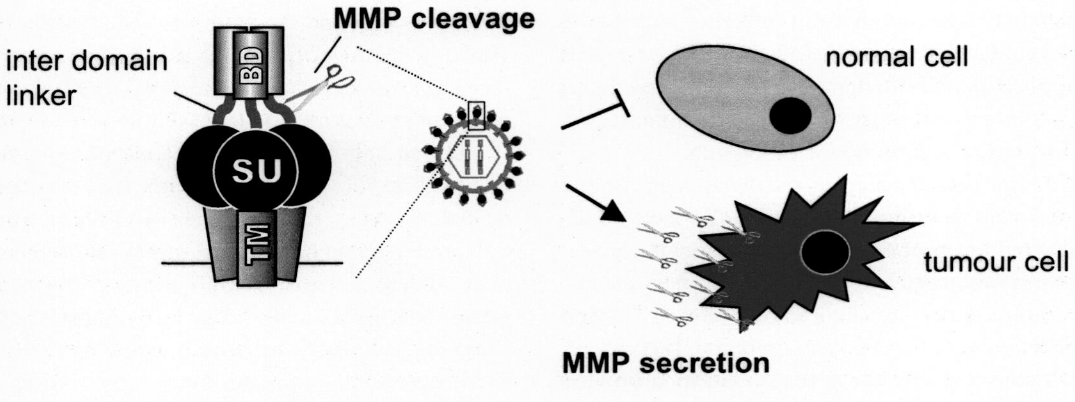

Figure 13.6 Schematic representation of tumour-selective MMP-mediated cell entry targeting. Receptor binding activity of retroviral envelope glycoproteins is blocked by N-terminal addition of a blocking domain (BD). The blocking domains are connected to the envelope proteins via a flexible interdomain linker containing a peptide stretch which represents a specific cleavage site for a tumour-associated matrix metalloprotease (MMP). Thus, infectivity is coupled to MMP expression, allowing selective infection of MMP-positive tumour cells. SU: surface domain of Env; TM: transmembrane domain of Env; BD: blocking domain; Scissors represent MMPs.

cfu/ml on target cells) were obtained, although efficient binding of the modified vector particles to the corresponding receptor molecules was demonstrated (reviewed in Buchholz *et al.*, 1999; Russell *et al.*, 1999).

A promising cell entry targeting strategy that uses modified Sindbis virus glycoproteins for pseudotyping was described for lentiviral vectors. This strategy involved mutation of residues in the Sindbis virus Env protein that are responsible for natural receptor interaction. Modification with the so-called Z domain, which binds the Fc part of immunoglobulins, allowed non-covalent binding of a monoclonal antibody directed against a melanoma-specific tumour antigen. These vector particles selectively transduced tumour tissue in a xenotransplantation tumour mouse model (Morizono *et al.*, 2005). Yang and co-workers (Yang *et al.*, 2006) incorporated a complete antibody molecule directed against the surface molecule CD20 in addition to the mutated Sindbis virus Env proteins into the envelope of vector particles, and demonstrated targeted gene transfer into CD20-positive B-lymphocytes.

Protease-activatable vectors are generated by inserting a protease-cleavable linker peptide between a so-called blocking domain and the N-terminus of the amphotropic MLV Env protein. The extracellular domain of CD40 ligand (CD40L), which prevents receptor recognition due to its trimeric structure, has frequently been used as a blocking domain. Exposure of such vector particles to the relevant protease releases the blocking domain and thereby restores membrane fusion capability and infectivity (Fig. 13.6). The system was initially established with a Factor Xa cleavage site, and was then extended to Env proteins activatable by matrix metalloproteinases (MMPs), which are extracellular proteases involved in tumour progression and metastasis (Nilson *et al.*, 1996; Peng *et al.*, 1997, 1999). MMP-activatable retroviral vectors became especially attractive after establishment of MLV-based protease substrate libraries, in which the amino acid residues of the linker peptide are combinatorially diversified to enable selection of suitable cleavage sites for a given tumour cell type and its specific MMP spectrum (Buchholz *et al.*, 1998; Hartl *et al.*, 2005; Schneider *et al.*, 2003). Vector particles equipped with linker peptides derived from this molecular evolution process were several orders of magnitude more efficient in gene transfer than those containing standard MMP cleavage sites. These selected cleavage sites were successfully transferred to other envelope proteins such as the influenza haemagglutinin (HA). When used for pseudotyping, the generated retroviral vector particles revealed the expected specificity for MMP-positive target cells (Szecsi *et al.*, 2006).

Following cell entry, integration of reverse transcribed vector genomes largely depends on the cell cycle status of the transduced target cell. As mentioned earlier, gammaretroviral vectors are unable to transfer genes into non-dividing cells (Miller *et al.*, 1990) since the breakdown of the nuclear membrane during mitosis is essential for entry of these viral genomes into the nucleus (Roe *et al.*, 1993). Since most tissues consisting of fully differentiated cells and non-replicating stem cells are not amenable to gene transfer using gammaretroviral vectors, *in vivo* applications in tumour gene therapy are limited. In *ex vivo* settings, isolated autologous cells may be stimulated with various cytokines to induce cell division and thereby enable retroviral transduction.

This restriction generally does not apply for lentiviral vectors. Although the precise mechanism whereby lentiviruses infect non-dividing cells is unknown, for HIV-1 this appears to be facilitated by the viral integrase, the matrix protein, and the accessory protein Vpr (Bukrinsky *et al.*, 1993; Gallay *et al.*, 1997; Heinzinger *et al.*, 1994; von Schwedler *et al.*, 1994). Vpr packaged into virus particles binds directly to the nuclear pore complex and affects the architecture and integrity of the nuclear membrane of infected cells (De Noronha *et al.*, 2001). The precise role of the other elements remains to be elucidated, especially since lentiviral vectors devoid of all accessory genes including *vpr* still infect non-dividing target cells (Zufferey *et al.*, 1997).

Thus, in contrast to gammaretroviral vectors, HIV-derived vectors can efficiently infect hepatic, retinal, glial, and muscle cells, as well as human HSCs (Kordower *et al.*, 2000; Miyoshi *et al.*, 1997; Naldini *et al.*, 1996a). However, certain resting/quiescent cells in the G0/G1a-phase of the cell cycle, such as non-stimulated T-lymphocytes, monocytes, or non-stimulated HSC, are still refractory to gene transfer by HIV-1-derived lentiviral vectors, and must be stimulated to facilitate efficient transduction. For monocytes and G0-arrested human diploid alpha-1 fibroblasts this hurdle can be overcome by utilizing vectors derived from a special SIV isolate, SIVsmmPBj1.9 (Muhlebach *et al.*, 2005), which exhibits a Vpx-dependent phenotype in monocytes (Wolfrum *et al.*, 2007). Alternatively, molecules such as IL-2 or IL-7 that exhibit chemokine activity may be fused to the N-terminus of the MLV Env protein and co-incorporated into vector particles along with the unmodified Env. Such vector particles stimulate resting T-cells during particle attachment, which enables transduction (Maurice *et al.*, 1999; Verhoeyen *et al.*, 2003).

In addition to targeting retroviral transduction to distinct cells at the level of cell entry, the use of cell-specific promoters may enhance the efficacy and safety of retroviral gene delivery. This so-called transcriptional targeting was successfully achieved using gammaretroviral and lentiviral vectors. Especially SIN vectors facilitated the use of internal cell- or tissue-specific promoters for the targeted expression of the encoded transgene in distinct cells, thereby avoiding promoter interference (Charrier *et al.*, 2007; De *et al.*, 2003; Follenzi *et al.*, 2004). For gammaretroviral vectors, promoter conversion was achieved by replacing the viral U3 region within the 3′LTR by cell-specific promoter elements (Diaz *et al.*, 1998; Mrochen *et al.*, 1997; Saller *et al.*, 1998). As mentioned earlier, upon infection and reverse transcription the 3′LTR including U3 is copied to the 5′ end, thereby replacing the original MLV promoter so that transgene expression is driven by the heterologous targeting promoter. Extensive vector improvements have yielded high titre vectors that facilitate efficient delivery and targeted expression of transgenes (Hlavaty *et al.*, 2004; Mavria *et al.*, 2005). Recent developments include the use of synthetic hybrid promoter/enhancer elements (Lipinski *et al.*, 2004; Metzl *et al.*, 2006) for targeting distinct types of tumours, e.g. colon or hepatocellular carcinomas, or for utilizing unique physiological conditions commonly encountered in solid tumours, e.g. hypoxia (Ingram *et al.*, 2005; Lee *et al.*, 2007; Lipnik *et al.*, 2006). Thus, it is possible to change vector tropism and increase transgene expression in specific tissues or cells.

Therapeutic applications

Due to their ability to stably integrate into the host cell chromosome, viral vectors derived from gammaretroviruses or lentiviruses are especially suited for long-term expression of the transferred gene in the target cell population. Possible applications therefore include substitution of mutated or missing genes in monogenic diseases.

Prominent examples are X-linked chronic granulomatous disease (X-CGD; Dinauer *et al.*, 1999), severe combined immunodeficiency diseases (SCID; Hirschhorn, 1990; Noguchi *et al.*, 1993), and Leber's congenital amaurosis (LCA; Bemelmans *et al.*, 2006). These hereditary diseases all have a defect in both alleles of a single respective gene that encodes a non-redundant protein and results in insufficient or non-detectable expression of the functional protein or synthesis of a dysfunctional version. For SCID-X1 and adenosine deaminase (ADA)-SCID (a metabolic form of the disease), retroviral vector-mediated gene therapy approaches have already proceeded to clinical trials. These trials demonstrated that successful retroviral gene transfer into haematopoietic cells may provide at least a transient benefit to treated patients. These studies also provide lessons for future applications of vectors for genetic modification of isolated non-differentiated stem cells that aim to repopulate diseased organs with functionally intact precursor cells (Fig. 13.7). For example, pioneering studies have investigated the use of genetically modified mesenchymal stem cells expressing vascular endothelial growth factor (VEGF) for treatment of ischaemic heart disease (Gao *et al.*, 2007; Usas *et al.*, 2007).

A second line of research concentrates on the use of retroviral vectors in cancer gene therapy using tumour suppressor genes such as p53, p21, p16, or BRCA1. These genes are expected to normalize cell cycle control in de-differentiated tumour cells, or may sensitize tumour cells to undergo programmed cell death (apoptosis) in response to chemotherapy or radiotherapy (Bai *et al.*, 2001; Hlavaty *et al.*, 2000; Kichina *et al.*, 2003; Roth *et al.*, 1996; Tait *et al.*, 1997). For cancer therapy, the use of gammaretroviral vectors offers

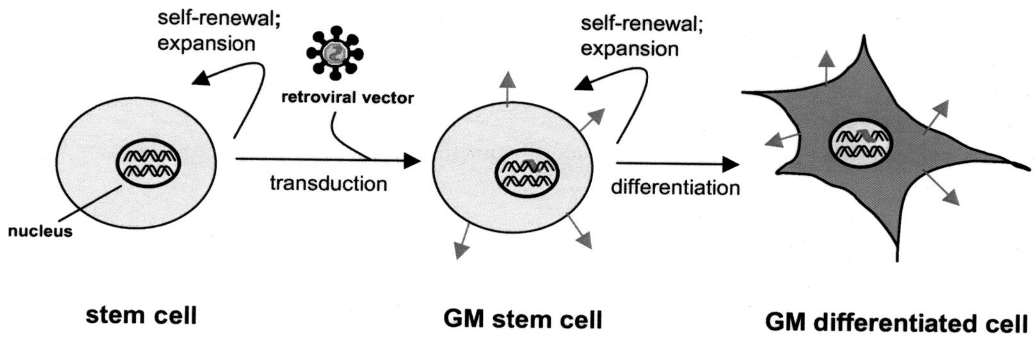

Figure 13.7 Schematic principle of retroviral gene transfer into stem cells. Stem cells of different origin can be transduced *ex* or *in vivo* using retroviral vectors and subsequently differentiated, with or without potential expansion or self-renewal, into different cell types, which still carry the transferred gene and express the respective protein of interest. GM: genetically modified; red helix: retroviral transfer vector; black helix: cellular genome; red arrows: gene of interest encoded by the retroviral vector.

the advantage that only dividing cells are transduced by these vectors, resulting of natural targeting to dividing cancer cells. Suicide genes on the corresponding transfer vectors encode proteins that facilitate conversion of non-toxic *pro*-drugs into cytotoxic compounds (Portsmouth *et al.*, 2007). Tumour cells transduced with the therapeutic vector and expressing the corresponding enzyme therefore become highly sensitive towards the applied drug while normal cells are relatively protected. The most common suicide gene currently used for this approach is the herpes simplex virus (HSV)-1 encoded thymidine kinase (HSV-Tk), used in combination with ganciclovir as a prodrug.

More recent approaches are designed to prevent or cure infectious diseases. For example, HIV infections may be treated by introducing so-called resistance genes, resulting in the expression of proteins that interfere with the retroviral life cycle. Resistance genes may encode dominant-negative mutants of HIV proteins such as RevM10 (Malim *et al.*, 1992), antisense-RNA against HIV proteins (Goodchild *et al.*, 1988), or siRNA designed to mediate downregulation of the HIV co-receptor CCR5 (Anderson *et al.*, 2007). Expression of a membrane-anchored form of the fusion inhibitor T20 (Egelhofer *et al.*, 2004) or proteins of the APOBEC3 and Trim5α families derived from non-human primates (Sakuma *et al.*, 2007; Stremlau *et al.*, 2004) were also investigated (for review, see Dropulic *et al.*, 2006). These genes may be transferred via retroviral vectors into either mature $CD4^+$ T-cells or the respective haematopoietic precursor cells in order to generate a target cell population resistant to HIV infections.

Another indication for retroviral vector-mediated gene transfer is genetic vaccination for treatment or prevention of infectious diseases or cancer. In this context, retroviral transfer vectors were used to deliver antigen encoding genes directly to antigen presenting cells (e.g. macrophages and dendritic cells), thereby triggering specific immune responses. For example, transfer of tumour-associated antigen (TAA) encoding genes such as the melanoma-specific tyrosinase-related protein (mTRP-2) (Metharom *et al.*, 2001) or ovalbumin (Dullaers *et al.*, 2006; He *et al.*, 2005) resulted in efficient TAA presentation and subsequent antitumour immunity. Transfer of pathogen-related antigen-encoding genes can result in efficient vaccination, and is a very promising approach for emerging diseases for which no conventional vaccine exists. The first successful results with this method involved the use of West Nile virus (WNV) envelope protein for vaccination (Iglesias *et al.*, 2006). A closely related approach involves the delivery of genes encoding *pro*-inflammatory cytokines (e.g. GM-CSF, IL-12) or immunologically relevant co-stimulatory molecules [e.g. B7.1 (CD80), B7.2] to tumour cells. For retroviral vectors the genes can be delivered *ex vivo*, i.e. via isolated autologous cells, or *in vivo*, i.e. via systemic virus application. For both approaches the goal is to enhance detection of tumour cells by the immune system, which facilitates the elimination of primary tumours and metastases (Chan *et al.*, 2005; Kitagawa *et al.*, 2003; Parney *et al.*, 2002; Wen *et al.*, 2001).

In addition to its therapeutic applications, retroviral gene transfer is a valuable tool in functional genomics. Vectors encoding selectable marker or reporter genes can be used to label distinct cell populations during *in vitro* experiments to enable screening of cDNA or retroviral display libraries (Hartl *et al.*, 2005; Merten *et al.*, 2005; Schneider *et al.*, 2003), or for generation of transgenic animals (Pfeifer *et al.*, 2002; van der Putten *et al.*, 1985). Lymphocytes transduced with a marker gene bearing retrovirus were initially used to address safety concerns regarding retroviral gene transfer technology. Using this method in a phase I clinical trial it was possible to follow the fate of genetically modified cells within a human body over prolonged periods of time (Rosenberg *et al.*, 1990).

Human clinical trials

According to Edelstein *et al.* (2007), retroviral vectors had been used until 2007 in 314 clinical trials, nine of which used lentiviral vectors. As summarized above, retroviral gene delivery may be aimed at restoring defective or missing genes, introducing new cellular functions, or inducing an immune response to prevent infectious diseases or cancer, or to eliminate tumour cells *in situ*.

More than 90% (286 of 314) of these trials were safety studies (phase I or phase I/II), 26 were phase II trials that focused on therapeutic efficacy, and three were phase III trials (Edelstein *et al.*, 2007). All phase III studies evaluated the therapeutic effects of a suicide gene for the treatment of glioblastoma. The transfer vectors employed in these studies were administered via intracranially implanted virus-producing cells. Therapeutic efficacy was disappointingly poor and appeared to be related to low transduction efficiencies of solid tumours *in vivo* (Rainov, 2000; Stockhammer *et al.*, 1997).

A total of 156 retroviral vectors have been approved for use in clinical trials for cancer gene therapy. Other indications for which retroviral vectors were approved for use in clinical trials include (in descending order by number of trials): gene marking studies, monogenic disease, infectious disease, neurological disease (including ocular diseases), cardiovascular disease and degenerative joint diseases. For some of these studies (Table 13.1) a beneficial effect of retroviral gene therapy was demonstrated, prompting further investigations. Strikingly, all of these trials targeted cells of haematopoietic origin.

Gene therapy for GvHD

More than 10 years ago Bonini *et al.* (1997) described the use of a suicide gene to successfully control graft versus host disease (GvHD) in leukaemia patients who received T-lymphocytes by adoptive transfer. The transfused allogeneic T-cells were genetically modified with MLV-derived vectors carrying the HSV-*tk* gene and an additional marker gene. Three of the eight patients treated with the genetically modified T-cells developed GvHD and were treated with ganciclovir exclusively. In two of the three ganciclovir-treated patients GvHD was eliminated, and the third showed partial reduction of GvHD. Recent follow-up studies confirmed that these patient statuses are still accurate (Bondanza *et al.*, 2006; Kornblau *et al.*, 2007).

Gene therapy for SCID

Promising results were also obtained from clinical trials to treat SCID. In SCID-X1, patients lack a functional gene for the gamma-chain of the interleukin(IL)-2 receptor. As a result, their T- and NK-cells remain undifferentiated and non-functional, leading to a total abrogation of the T- and NK-cell repertoire, which produces severe immune deficiency (Noguchi *et al.*, 1993). In ADA-SCID, patients have a mutation in the ADA gene, which causes accumulation of toxic metabolites and subsequent interference with lymphocyte function (Hirschhorn, 1990).

For both SCID-XI and ADA-SCID, transfer of the respective gene encoding a correct version of the IL2-R γc-chain or ADA into lymphocytes or their precursors may produce a functional lymphocyte repertoire, thereby curing the disease. Indeed, gammaretroviral transfer of the respective gene into isolated HSCs led to a cure of SCID-X1

Table 13.1 Examples of successful gene therapy clinical trials using retroviral vectors

Disease	Transferred gene	Target cells/vector	Outcome	References
Leukaemia, GvHD	Herpes simplex virus thymidine kinase (HSV-*tk*)	HSCs/retroviral vector	Successful GvHD treatment in all 3 GvHD patients	Bonini *et al.* (1997)
SCID-XI	Gamma-c-chain (IL-2R)	HSCs/retroviral vector	17 out of 20 infants cured[a]	Cavazzana-Calvo *et al.* (2000), Gaspar *et al.* (2004), Howe *et al.* (2008)
ADA-SCID	Adenosine deaminase gene (*ada*)	HSCs/retroviral vector	All 5 patients cured	Aiuti *et al.* (2002, 2007)
X-CGD	gp91*phox*	HSCs/retroviral vector	Both patients cured [b]	Ott *et al.* (2006)
Melanoma	TAA-specific T-cell receptors	T-lymphocytes/ retroviral vector	2 out of 17 patients with partial remission	Morgan *et al.* (2006)

a Five patients developed lymphoproliferative disease due to vector integration.

b One patient died, and death was considered unrelated to treatment.

GvHD, graft versus host disease; HSCs, haematopoietic stem cells; SCID, severe combined immunodeficiency disease; X-CDG, X-linked chronic granulomatous disease.

in 14 of treated 15 infants, enabling them to leave the confines of their germ-free environment (Cavazzana-Calvo *et al.*, 2000; Gaspar *et al.*, 2004). However, five treated patients[1] later developed a T-cell proliferative syndrome reminiscent of adult T-cell leukaemia, with large increases in the number of T-lymphocytes, accompanied by splenomegaly and anaemia. In the first trial, 4 of 10 treated patients developed leukaemia, but most of the leukaemic patients recovered after treatment with conventional chemotherapy; one patient died due to leukaemia (Cavazzana-Calvo *et al.*, 2007; Hacein-Bey-Abina *et al.*, 2003). T-cell proliferative syndrome has so far been observed in one patient 24 months post treatment in the second clinical trial (Howe *et al.*, 2008) and the reasons for this discrepancy, if it continues to hold true, need to be elucidated. Current hypotheses suggest that use of the Gibbon ape leukaemia virus (GaLV) Env in the second trial instead of the amphotropic MLV Env, which was used in the first trial to pseudotype the applied vectors used for transduction, and the inclusion of certain cytokines during cell propagation prior and during vector transduction, may have been responsible for this discrepancy (A. Thrasher and C. Baum, personal communication). Since the publication of these results, experts have concluded that the T-cell proliferative syndrome resulted from chromosomal integration of the gammaretroviral vector activating the cellular proto-oncogene *lmo2* or other proto-oncogenes (Cavazzana-Calvo *et al.*, 2007; Hacein-Bey-Abina *et al.*, 2003; Howe *et al.*, 2008).

In ADA-SCID, the patients' lack of a functional adenosine deaminase causes formation of toxic products during purine metabolism, and these products accumulate in the lymphocytes. *Ex vivo* transfer of the *ada* gene into isolated HSCs and subsequent administration to preconditioned patients produced a functional, multi-lineage haematopoietic reconstitution in all treated patients (Aiuti *et al.*, 2002; Aiuti *et al.*, 2007). No clonal selection or expansion of T-cells

1 Ten infants were treated in one SCID-X1 clinical trial and four of them developed T-cell proliferative syndrome. By 2008, one of the ten infants treated in a second SCID-X1 clinical trial had developed this syndrome.

was observed, although retroviral integration was again mapped to the *LMO2* locus (Aiuti *et al.*, 2007).

Gene therapy for X-linked CGD and Wiskott–Aldrich syndrome

Another type of immunodeficiency results from X-linked CGD. This disease was beneficially treated via transfer of a functional gp91phox subunit contained in the NADPH oxidase complex. Only two patients were treated in this trial and so far neither patient has shown clinical signs of a lymphoproliferative disease (Ott *et al.*, 2006). Selection of distinct T-cell clones revealed that close to 100% of retroviral integrations were within the *MDS1-EVI1*, *PRDM16* and *SETBP1* loci. Interestingly, the reported treatment successes in ADA-SCID and X-CGD were associated with a non-myeloablative pre-conditioning of the patients using busulfan, which promoted expansion of the re-infused genetically modified HSCs.

A third type of monogeneic immunodeficiency currently treated by the infusion of genetically modified HSCs is the Wiskott-Aldrich Syndrome. The correct version of the WASP gene was introduced into autologous CD34-positive cells by retroviral gene transfer, and both treated patients experienced polyclonal repopulation of the haematopoietic system for more than 12 months after treatment (Schwarzwaelder *et al.*, 2008).

Gene therapy with lentiviral vectors

Lentiviral vectors have only recently been used in human gene therapy trials. Their improved efficacy with respect to transduction of haematopoietic cells prompted the initiation of gene therapy for β-thalassaemia (Bank *et al.*, 2005). A SIN vector containing large elements of the β-globin locus control region, as well as insulators and other features that should prevent unintended events, was applied; no results are available from these trials.

Other lentiviral vector applications have primarily involved the use of HIV-1-based vectors for the treatment of HIV infections. The few trials so far approved concentrated on delivery of therapeutic RNA (antisense RNA, RNAi) designed to interfere with HIV expression in infected cells. For this purpose autologous T-cells were transduced *ex vivo* and subsequently re-infused into HIV-positive patients. Another approach targets expression of CCR5, a co-receptor of HIV infection, by introducing a CCR5 mRNA-specific ribozyme (Anderson *et al.*, 2007; Dropulic *et al.*, 2006).

Furthermore, clinical trials have been initiated that use lentiviral vector-based approaches to treat X-linked cerebral adrenoleucodystrophy, mucopolysaccharidosis type VII, Wiskott–Aldrich syndrome, or malignant melanoma. No results are available, yet.

In summary, treatment with retroviral vectors may provide advantages over conventional treatments, and these novel therapies could potentially be applied to a large number of diseases that are currently considered to be untreatable. However, certain inherent risks – namely insertional oncogenesis – must be evaluated using a case-by-case risk–benefit analysis. Hopefully, future generations of retroviral vectors will minimize these issues.

Safety concerns

The clinical studies described above show that undesired integration of retroviral vectors into the host cell genome is a major safety issue. Insertional mutagenesis caused by chromosomal integration of the vector may lead to destruction or alteration of an ORF or to alteration of cellular gene expression (Baum *et al.*, 2006). Moreover, adjacent sequences may be transcribed by the viral promoter within the 3′LTR of the vector. Hence, vector integration may lead to suppression of tumour suppressor genes or to transcriptional activation of proto-oncogenes, both of which could enable tumour induction. Based on a number of assumptions and theoretical calculations of probability, this risk was previously regarded to be negligible and indeed, no tumour induction was observed in clinical trials using retroviral vectors for more than a decade. However, as described above, this situation changed after several patients treated for SCID-X1 were diagnosed with leukaemia due to retroviral vector-mediated insertional mutagenesis that induced over-expression of the proto-oncogene *lmo2* and other proto-oncogenes. Although the special circumstances of SCID-X1 and non-SIN MLV vector use may have contributed to leukaemia induction (e.g. the young age of the patients, the physiological conditions due to SCID-X1 disease, the transgene product, the special vector pseudotype used, and/or the cell culture conditions and cytokines used), the outcome proved that insertional mutagenesis presents more than just a theoretical risk.

Various strategies for the enhancement of retroviral vector safety are under development. Some of them, such as the construction of SIN vectors to avoid transcription from the viral promoters and the insertion of insulators, were already mentioned. In addition, several groups are working on strategies to target retroviral vector integration into specific sites of the genome that are expected to be non-hazardous. Targeted integration has the potential to significantly enhance vector safety; however work in this area has been very limited up to now (Wu *et al.*, 2004).

To estimate the risk of insertional mutagenesis, improved molecular methods are being developed to analyse the safety of different vectors and their integration sites. These methods include linear amplification-mediated (LAM) PCR, combined with sophisticated computer analysis programs to assess expansion of clones (Modlich *et al.*, 2006; Schmidt *et al.*, 2003). Mouse models for analysis of the risk of leukaemia induction for different vectors, vector components, or transgenes have been described (Kustikova *et al.*, 2005; Li *et al.*, 2002; Will *et al.*, 2007).

In addition to the general problem of insertional mutagenesis, a major safety concern of retroviral vectors is the possible formation of replication-competent retroviruses (RCRs) in treated patients. Infection with lentiviruses or other retroviruses with pathogenic potential for humans must be avoided. Murine retroviruses like MLV were initially thought to be incapable of infecting humans due to inactivation by human serum (Welsh, Jr. *et al.*, 1975). However, vectors produced in human packaging cell lines are resistant to serum and may also resist complement-mediated inactivation (Cosset *et al.*, 1995). Moreover, transfusion of non-human primates with bone marrow cells infected with amphotropic MLV helper virus resulted in chronic infection and lymphoma induction (Donahue *et al.*, 1992; Purcell *et al.*, 1996; Vanin *et al.*, 1994). Therefore, preventing the formation of replication-competent viruses is crucial for all gene therapy trials using retroviral vectors (EMEA, 2001; US Federal Food and Drug Administration, 2000).

RCRs can be generated during the production process by recombination events between different components of the vectors or, at least theoretically, between vector components and endogenous or exogenous viral sequences. Indeed, for earlier generations of vector systems and packaging cells, RCRs were regularly detected. However, for present generations of vector systems that use split genome packaging cells bearing little or no sequence overlap between the vector components, formation of RCRs by recombination is extremely unlikely and the production of vector stocks without any RCR has become routine. Nevertheless, prior to use in human gene therapy, an absence of RCR must always be demonstrated for packaging cells, vector stocks, and cells transduced *ex vivo*.

Various assays have been used for detection of RCRs, such as detection of recombined sequences by PCR or virus detection in inoculated cell cultures. The latter can be achieved using marker rescue assays on cells carrying a selectable retroviral vector or using detection of RT activity in transduced cells. A very old but still relevant marker rescue assay is the S^+L^- assay, where a defective oncogenic virus is rescued by RCRs, and is detected after induction of foci on indicator cells (Bassin *et al.*, 1971). Altogether, great efforts are being made to protect the safety of patients participating in clinical trials that use retroviral vectors for gene therapy.

Conclusions

The field of retroviral vector development is evolving rapidly. Based on some early successes in which patients showed at least transient symptomatic relief from life-threatening inherited immunodeficiency diseases, and more recent successes showing retroviral vector-mediated transfer of an intact copy of the mutated gene, retroviral vectors are becoming an ever more important tool. Innovative technologies to improve the efficacy and safety of retroviral vectors are being developed.

Problems such as the insertional mutagenesis observed in one of the X-SCID clinical trials are being attacked via several different approaches. Non-integrating retroviral vectors in which the integrase was mutated were recently shown to allow long-term expression of therapeutic genes (Philippe *et al.*, 2006). Controlling the integration process to redirect integration of the vector into safe sites in the genome may become possible through the use of designer zinc finger endonucleases (Szczepek *et al.*, 2007). Variations in envelope proteins of retroviral vectors are being rapidly developed.

With all of these improvements, the array of available retroviral vectors – each equipped with a specific promoter and a distinct receptor specific for the target tissue or cell population of interest – will improve the specificity and safety of *in vivo* gene transfer strategies in the future. The insertion of tissue-specific cellular miRNA sequences into the vector genome can be used to ensure that the therapeutic gene is not expressed in non-targeted cells, thereby stabilizing expression and avoiding immune responses against the therapeutic protein (Brown *et al.*, 2006). This latter approach nicely illustrates the unique ability of retroviral vectors to enable the immediate transfer of basic mechanisms controlling gene expression into retrovectorology.

References

Aiuti, A., Cassani, B., Andolfi, G., Mirolo, M., Biasco, L., Recchia, A., Urbinati, F., Valacca, C., Scaramuzza, S., Aker, M., Slavin, S., Cazzola, M., Sartori, D., Ambrosi, A., Di, S.C., Roncarolo, M.G., Mavilio, F., and Bordignon, C. (2007). Multilineage hematopoietic reconstitution without clonal selection in ADA-SCID patients treated with stem cell gene therapy. J. Clin. Invest. *117*, 2233–2240.

Aiuti, A., Slavin, S., Aker, M., Ficara, F., Deola, S., Mortellaro, A., Morecki, S., Andolfi, G., Tabucchi, A., Carlucci, F., Marinello, E., Cattaneo, F., Vai, S., Servida, P., Miniero, R., Roncarolo, M.G., and Bordignon, C. (2002). Correction of ADA-SCID by stem cell gene therapy combined with nonmyeloablative conditioning. Science *296*, 2410–2413.

Anderson, J., and Akkina, R. (2007). Complete knockdown of CCR5 by lentiviral vector-expressed siRNAs and protection of transgenic macrophages against HIV-1 infection. Gene Ther. *14*, 1287–1297.

Bai, J., Zhu, X., Zheng, X., and Wu, Y. (2001). Retroviral vector containing human p16 gene and its inhibitory effect on Bcap-37 breast cancer cells. Chin. Med. J. (Engl.) *114*, 497–501.

Bank, A., Dorazio, R., and Leboulch, P. (2005). A phase I/II clinical trial of beta-globin gene therapy for beta-thalassemia. Ann. N. Y. Acad. Sci. *1054*, 308–316.

Bassin, R.H., Tuttle, N., and Fischinger, P.J. (1971). Rapid cell culture assay technic for murine leukaemia viruses. Nature *229*, 564–566.

Baum, C., Kustikova, O., Modlich, U., Li, Z., and Fehse, B. (2006). Mutagenesis and oncogenesis by chromosomal insertion of gene transfer vectors. Hum. Gene Ther. *17*, 253–263.

Bemelmans, A.P., Kostic, C., Crippa, S.V., Hauswirth, W.W., Lem, J., Munier, F.L., Seeliger, M.W., Wenzel, A., and Arsenijevic, Y. (2006). Lentiviral gene transfer of RPE65 rescues survival and function of cones in a mouse model of Leber congenital amaurosis. PLoS. Med. *3*, e347-

Bondanza, A., Valtolina, V., Magnani, Z., Ponzoni, M., Fleischhauer, K., Bonyhadi, M., Traversari, C., Sanvito, F., Toma, S., Radrizzani, M., La Seta-Catamancio, S., Ciceri, F., Bordignon, C., and Bonini, C. (2006). Suicide gene therapy of graft-versus-host disease induced by central memory human T-lymphocytes. Blood *107*, 1828–1836.

Bonini, C., Ferrari, G., Verzeletti, S., Servida, P., Zappone, E., Ruggieri, L., Ponzoni, M., Rossini, S., Mavilio, F., Traversari, C., and Bordignon, C. (1997). HSV-TK gene transfer into donor lymphocytes for control of allogeneic graft-versus-leukemia. Science *276*, 1719–1724.

Brown, B.D., Venneri, M.A., Zingale, A., Sergi, S.L., and Naldini, L. (2006). Endogenous microRNA regulation suppresses transgene expression in hematopoietic lineages and enables stable gene transfer. Nat. Med. *12*, 585–591.

Buchholz, C.J., Peng, K.W., Morling, F.J., Zhang, J., Cosset, F.L., and Russell, S.J. (1998). In vivo selection of protease cleavage sites from retrovirus display libraries. Nat. Biotechnol. *16*, 951–954.

Buchholz, C.J., Stitz, J., and Cichutek, K. (1999). Retroviral cell targeting vectors. Curr. Opin. Mol. Ther. *1*, 613–621.

Bukrinsky, M.I., Haggerty, S., Dempsey, M.P., Sharova, N., Adzhubel, A., Spitz, L., Lewis, P., Goldfarb, D., Emerman, M., and Stevenson, M. (1993). A nuclear localization signal within HIV-1 matrix protein that governs infection of non-dividing cells. Nature *365*, 666–669.

Cavazzana-Calvo, M., and Fischer, A. (2007). Gene therapy for severe combined immunodeficiency: are we there yet? J. Clin. Invest *117*, 1456–1465.

Cavazzana-Calvo, M., Hacein-Bey, S., de Saint Basile, G., Gross, F., Yvon, E., Nusbaum, P., Selz, F., Hue, C., Certain, S., Casanova, J.L., Bousso, P., Deist, F.L., and Fischer, A. (2000). Gene therapy of human severe combined immunodeficiency (SCID)-X1 disease. Science *288*, 669–672.

Chan, L., Hardwick, N., Darling, D., Galea-Lauri, J., Gaken, J., Devereux, S., Kemeny, M., Mufti, G., and Farzaneh, F. (2005). IL-2/B7.1 (CD80) fusagene transduction of AML blasts by a self-inactivating lentiviral vector stimulates T-cell responses *in vitro*: a strategy to generate whole cell vaccines for AML. Mol. Ther. *11*, 120–131.

Charneau, P., Alizon, M., and Clavel, F. (1992). A second origin of DNA plus-strand synthesis is required for optimal human immunodeficiency virus replication. J. Virol. *66*, 2814–2820.

Charrier, S., Dupre, L., Scaramuzza, S., Jeanson-Leh, L., Blundell, M.P., Danos, O., Cattaneo, F., Aiuti, A.,

Eckenberg, R., Thrasher, A.J., Roncarolo, M.G., and Galy, A. (2007). Lentiviral vectors targeting WASp expression to hematopoietic cells, efficiently transduce and correct cells from WAS patients. Gene Ther. *14*, 415–428.

Christodoulopoulos, I., and Cannon, P.M. (2001). Sequences in the cytoplasmic tail of the gibbon ape leukemia virus envelope protein that prevent its incorporation into lentivirus vectors. J. Virol. *75*, 4129–4138.

Coffin, J.M., Hughes, S.H., and Varmus, H.E. (1997). Retroviruses. *1997*, Cold Spring Harbor Laboratory Press, New York.

Cosset, F.L., Takeuchi, Y., Battini, J.L., Weiss, R.A., and Collins, M.K. (1995). High-titer packaging cells producing recombinant retroviruses resistant to human serum. J. Virol. *69*, 7430–7436.

Cronin, J., Zhang, X.Y., and Reiser, J. (2005). Altering the tropism of lentiviral vectors through pseudotyping. Curr. Gene Ther. *5*, 387–398.

De Noronha, C.M., Sherman, M.P., Lin, H.W., Cavrois, M.V., Moir, R.D., Goldman, R.D., and Greene, W.C. (2001). Dynamic disruptions in nuclear envelope architecture and integrity induced by HIV-1 Vpr. Science *294*, 1105–1108.

De, P.M., Venneri, M.A., and Naldini, L. (2003). In vivo targeting of tumor endothelial cells by systemic delivery of lentiviral vectors. Hum. Gene Ther. *14*, 1193–1206.

Diaz, R.M., Eisen, T., Hart, I.R., and Vile, R.G. (1998). Exchange of viral promoter/enhancer elements with heterologous regulatory sequences generates targeted hybrid long terminal repeat vectors for gene therapy of melanoma. J. Virol. *72*, 789–795.

Dinauer, M.C., Li, L.L., Bjorgvinsdottir, H., Ding, C., and Pech, N. (1999). Long-term correction of phagocyte NADPH oxidase activity by retroviral-mediated gene transfer in murine X-linked chronic granulomatous disease. Blood. *94*, 914–922.

Donahue, R.E., Kessler, S.W., Bodine, D., McDonagh, K., Dunbar, C., Goodman, S., Agricola, B., Byrne, E., Raffeld, M., Moen, R., *et al.* (1992). Helper virus induced T-cell lymphoma in nonhuman primates after retroviral mediated gene transfer. J. Exp. Med. *176*, 1125–1135.

Dropulic, B., and June, C.H. (2006). Gene-based immunotherapy for human immunodeficiency virus infection and acquired immunodeficiency syndrome. Hum. Gene Ther. *17*, 577–588.

Dull, T., Zufferey, R., Kelly, M., Mandel, R.J., Nguyen, M., Trono, D., and Naldini, L. (1998). A third-generation lentivirus vector with a conditional packaging system. J. Virol. *72*, 8463–8471.

Dullaers, M., Van, M.S., Heirman, C., Straetman, L., Bonehill, A., Aerts, J.L., Thielemans, K., and Breckpot, K. (2006). Induction of effective therapeutic antitumor immunity by direct *in vivo* administration of lentiviral vectors. Gene Ther. *13*, 630–640.

Edelstein, M.L., Abedi, M.R., and Wixon, J. (2007). Gene therapy clinical trials worldwide to 2007-an update. J. Gene Med. *9*, 833–842.

Egelhofer, M., Brandenburg, G., Martinius, H., Schult-Dietrich, P., Melikyan, G., Kunert, R., Baum, C., Choi, I., Alexandrov, A., and von Laer, D. (2004). Inhibition of human immunodeficiency virus type 1 entry in cells expressing gp41-derived peptides. J. Virol. *78*, 568–575.

E.M.E.A. (2001). CPMP/BWP/3088/99 Note for Guidance on the Quality, Preclinical and Clinical Aspects of Gene Transfer Medicinal Products. http://www.emea.eu.int/

Finger, C., Sun, Y., Sanz, L., Alvarez-Vallina, L., Buchholz, C.J., and Cichutek, K. (2005). Replicating retroviral vectors mediating continuous production and secretion of therapeutic gene products from cancer cells. Cancer Gene Ther. *12*, 464–474.

Follenzi, A., Ailles, L.E., Bakovic, S., Geuna, M., and Naldini, L. (2000). Gene transfer by lentiviral vectors is limited by nuclear translocation and rescued by HIV-1 *pol* sequences. Nat. Genet. *25*, 217–222.

Follenzi, A., Battaglia, M., Lombardo, A., Annoni, A., Roncarolo, M.G., and Naldini, L. (2004). Targeting lentiviral vector expression to hepatocytes limits transgene-specific immune response and establishes long-term expression of human antihemophilic factor IX in mice. Blood *103*, 3700–3709.

Gallay, P., Hope, T., Chin, D., and Trono, D. (1997). HIV-1 infection of nondividing cells through the recognition of integrase by the importin/karyopherin pathway. Proc. Natl. Acad. Sci. U.S.A. *94*, 9825–9830.

Gao, F., He, T., Wang, H., Yu, S., Yi, D., Liu, W., and Cai, Z. (2007). A promising strategy for the treatment of ischemic heart disease: Mesenchymal stem cell-mediated vascular endothelial growth factor gene transfer in rats. Can. J. Cardiol. *23*, 891–898.

Gasmi, M., Glynn, J., Jin, M.J., Jolly, D.J., Yee, J.K., and Chen, S.T. (1999). Requirements for efficient production and transduction of human immunodeficiency virus type 1-based vectors. J. Virol. *73*, 1828–1834.

Gaspar, H.B., Parsley, K.L., Howe, S., King, D., Gilmour, K.C., Sinclair, J., Brouns, G., Schmidt, M., von, K.C., Barington, T., Jakobsen, M.A., Christensen, H.O., Al, G.A., White, H.N., Smith, J.L., Levinsky, R.J., Ali, R.R., Kinnon, C., and Thrasher, A.J. (2004). Gene therapy of X-linked severe combined immunodeficiency by use of a pseudotyped gammaretroviral vector. Lancet *364*, 2181–2187.

Gaszner, M., and Felsenfeld, G. (2006). Insulators: exploiting transcriptional and epigenetic mechanisms. Nat. Rev. Genet. *7*, 703–713.

Ginn, S.L., Fleming, J., Rowe, P.B., and Alexander, I.E. (2003). Promoter interference mediated by the U3 region in early-generation HIV-1-derived lentivirus vectors can influence detection of transgene expression in a cell-type and species-specific manner. Hum. Gene Ther. *14*, 1127–1137.

Goodchild, J., Agrawal, S., Civeira, M.P., Sarin, P.S., Sun, D., and Zamecnik, P.C. (1988). Inhibition of human immunodeficiency virus replication by antisense oligodeoxynucleotides. Proc. Natl. Acad. Sci. U.S.A. *85*, 5507–5511.

Grignani, F., Kinsella, T., Mencarelli, A., Valtieri, M., Riganelli, D., Grignani, F., Lanfrancone, L., Peschle, C., Nolan, G.P., and Pelicci, P.G. (1998). High-efficiency gene transfer and selection of human hematopoietic progenitor cells with a hybrid EBV/retroviral vector expressing the green fluorescence protein. Cancer Res. *58*, 14–19.

Hacein-Bey-Abina, S., von Kalle, C., Schmidt, M., McCormack, M.P., Wulffraat, N., Leboulch, P., Lim, A., Osborne, C.S., Pawliuk, R., Morillon, E., Sorensen, R., Forster, A., Fraser, P., Cohen, J.I., de Saint Basile, G., Alexander, I., Wintergerst, U., Frebourg, T., Aurias, A., Stoppa-Lyonnet, D., Romana, S., Radford-Weiss, I., Gross, F., Valensi, F., Delabesse, E., Macintyre, E., Sigaux, F., Soulier, J., Leiva, L.E., Wissler, M., Prinz, C., Rabbitts, T.H., Le Deist, F., Fischer, A., and Cavazzana-Calvo, M. (2003). LMO2-associated clonal T-cell proliferation in two patients after gene therapy for SCID-X1. Science *302*, 415–419.

Hartl, I., Schneider, R.M., Sun, Y., Medvedovska, J., Chadwick, M.P., Russell, S.J., Cichutek, K., and Buchholz, C.J. (2005). Library-based selection of retroviruses selectively spreading through matrix metalloprotease-positive cells. Gene Ther. *12*, 918–926.

He, Y., Zhang, J., Mi, Z., Robbins, P., and Falo, L.D., Jr. (2005). Immunization with lentiviral vector-transduced dendritic cells induces strong and long-lasting T-cell responses and therapeutic immunity. J. Immunol. *174*, 3808–3817.

Heinzinger, N.K., Bukinsky, M.I., Haggerty, S.A., Ragland, A.M., Kewalramani, V., Lee, M.A., Gendelman, H.E., Ratner, L., Stevenson, M., and Emerman, M. (1994). The Vpr protein of human immunodeficiency virus type 1 influences nuclear localization of viral nucleic acids in nondividing host cells. Proc. Natl. Acad. Sci. U.S.A. *91*, 7311–7315.

Hirschhorn, R. (1990). Adenosine deaminase deficiency. Immunodefic. Rev. *2*, 175–198.

Hlavaty, J., Stracke, A., Klein, D., Salmons, B., Gunzburg, W.H., and Renner, M. (2004). Multiple modifications allow high-titer production of retroviral vectors carrying heterologous regulatory elements. J. Virol. *78*, 1384–1392.

Hlavaty, J., Tyukosova, S., Bies, J., Hlubinova, K., and Altaner, C. (2000). Retrovirus vector containing wild type p53 gene and its effect on human glioma cells. Neoplasma *47*, 204–211.

Howe, S.J., Mansour, M.R., Schwarzwaelder, K., Bartholomae, C., Hubank, M., Kempski, H., Brugman, M.H., Pike-Overzet, K., Chatters, S.J., de Ridder, D., Gilmour, K.C., Adams, S., Thornhill, S.I., Parsley, K.L., Staal, F.J., Gale, R.E., Linch, D.C., Bayford, J., Brown, L., Quaye, M., Kinnon, C., Ancliff, P., Webb, D.K., Schmidt, M., von Kalle, C., Gaspar, H.B., and Thrasher, A.J. (2008). Insertional mutagenesis combined with acquired somatic mutations causes leukemogenesis following gene therapy of SCID-X1 patients. J. Clin. Invest. *118*, 3143–3150.

Iglesias, M.C., Frenkiel, M.P., Mollier, K., Souque, P., Despres, P., and Charneau, P. (2006). A single immunization with a minute dose of a lentiviral vector-based vaccine is highly effective at eliciting protective humoral immunity against West Nile virus. J. Gene Med. *8*, 265–274.

Ikeda, Y., Takeuchi, Y., Martin, F., Cosset, F.L., Mitrophanous, K., and Collins, M. (2003). Continuous high-titer HIV-1 vector production. Nat. Biotechnol. *21*, 569–572.

Ingram, N., and Porter, C.D. (2005). Transcriptional targeting of acute hypoxia in the tumour stroma is a novel and viable strategy for cancer gene therapy. Gene Ther. *12*, 1058–1069.

Karreman, S., Hauser, H., and Karreman, C. (1996). On the use of double FLP recognition targets (FRTs) in the LTR of retroviruses for the construction of high producer cell lines. Nucleic Acids Res. *24*, 1616–1624.

Kavanaugh, M.P., Miller, D.G., Zhang, W., Law, W., Kozak, S.L., Kabat, D., and Miller, A.D. (1994a). Cell-surface receptors for gibbon ape leukemia virus and amphotropic murine retrovirus are inducible sodium-dependent phosphate symporters. Proc. Natl. Acad. Sci. U.S.A. *91*, 7071–7075.

Kavanaugh, M.P., Wang, H., Boyd, C.A., North, R.A., and Kabat, D. (1994b). Cell surface receptor for ecotropic host-range mouse retroviruses: a cationic amino acid transporter. Arch. Virol. Suppl. *9*, 485–494.

Kichina, J.V., Rauth, S., Das Gupta, T.K., and Gudkov, A.V. (2003). Melanoma cells can tolerate high levels of transcriptionally active endogenous p53 but are sensitive to retrovirus-transduced p53. Oncogene *22*, 4911–4917.

Kitagawa, T., Iwazawa, T., Robbins, P.D., Lotze, M.T., and Tahara, H. (2003). Advantages and limitations of particle-mediated transfection (gene gun) in cancer immuno-gene therapy using IL-10, IL-12 or B7–1 in murine tumor models. J. Gene Med. *5*, 958–965.

Klages, N., Zufferey, R., and Trono, D. (2000). A stable system for the high-titer production of multiply attenuated lentiviral vectors. Mol. Ther. *2*, 170–176.

Kobinger, G.P., Weiner, D.J., Yu, Q.C., and Wilson, J.M. (2001). Filovirus-pseudotyped lentiviral vector can efficiently and stably transduce airway epithelia *in vivo*. Nat. Biotechnol. *19*, 225–230.

Kordower, J.H., Emborg, M.E., Bloch, J., Ma, S.Y., Chu, Y., Leventhal, L., McBride, J., Chen, E.Y., Palfi, S., Roitberg, B.Z., Brown, W.D., Holden, J.E., Pyzalski, R., Taylor, M.D., Carvey, P., Ling, Z., Trono, D., Hantraye, P., Deglon, N., and Aebischer, P. (2000). Neurodegeneration prevented by lentiviral vector delivery of GDNF in primate models of Parkinson's disease. Science *290*, 767–773.

Kornblau, S.M., Aycox, P.G., Stephens, C., McCue, L.D., Champlin, R.E., and Marini, F.C. (2007). Control of graft-versus-host disease with maintenance of the graft-versus-leukemia effect in a murine allogeneic transplant model using retrovirally transduced murine suicidal lymphocytes. Exp. Hematol. *35*, 842–853.

Kotsopoulou, E., Kim, V.N., Kingsman, A.J., Kingsman, S.M., and Mitrophanous, K.A. (2000). A Rev-independent human immunodeficiency virus type 1 (HIV-1)-based vector that exploits a codon-optimized HIV-1 Gag–Pol gene. J. Virol. *74*, 4839–4852.

Kustikova, O., Fehse, B., Modlich, U., Yang, M., Dullmann, J., Kamino, K., von Neuhoff, N., Schlegelberger, B., Li, Z., and Baum, C. (2005). Clonal dominance of hematopoietic stem cells triggered by retroviral gene marking. Science *308*, 1171–1174.

Lee, C.H., Wu, C.L., and Shiau, A.L. (2007). Hypoxia-induced cytosine deaminase gene expression for cancer therapy. Hum. Gene Ther. *18*, 27–38.

Lengler, J., Holzmuller, H., Salmons, B., Gunzburg, W.H., and Renner, M. (2005). FMDV-2A sequence and protein arrangement contribute to functionality of CYP2B1-reporter fusion protein. Anal. Biochem. *343*, 116–124.

Lewis, P.F., and Emerman, M. (1994). Passage through mitosis is required for oncoretroviruses but not for the human immunodeficiency virus. J. Virol. *68*, 510–516.

Li, Z., Dullmann, J., Schiedlmeier, B., Schmidt, M., von Kalle, C., Meyer, J., Forster, M., Stocking, C., Wahlers, A., Frank, O., Ostertag, W., Kuhlcke, K., Eckert, H.G., Fehse, B., and Baum, C. (2002). Murine leukemia induced by retroviral gene marking. Science *296*, 497–497.

Linial, M.L., and Miller, A.D. (1990). Retroviral RNA packaging: sequence requirements and implications. Curr. Top. Microbiol. Immunol. *157*, 125–152.

Lipinski, K.S., Djeha, H.A., Gawn, J., Cliffe, S., Maitland, N.J., Palmer, D.H., Mountain, A., Irvine, A.S., and Wrighton, C.J. (2004). Optimization of a synthetic beta-catenin-dependent promoter for tumor-specific cancer gene therapy. Mol. Ther. *10*, 150–161.

Lipnik, K., Greco, O., Scott, S., Knapp, E., Mayrhofer, E., Rosenfellner, D., Gunzburg, W.H., Salmons, B., and Hohenadl, C. (2006). Hypoxia- and radiation-inducible, breast cell-specific targeting of retroviral vectors. Virology *349*, 121–133.

Logg, C.R., Tai, C.K., Logg, A., Anderson, W.F., and Kasahara, N. (2001). A uniquely stable replication-competent retrovirus vector achieves efficient gene delivery *in vitro* and in solid tumors. Hum. Gene Ther. *12*, 921–932.

Lucke, S., Grunwald, T., and Uberla, K. (2005). Reduced mobilization of Rev-responsive element-deficient lentiviral vectors. J. Virol. *79*, 9359–9362.

Madani, N., Kozak, S.L., Kavanaugh, M.P., and Kabat, D. (1998). gp120 envelope glycoproteins of human immunodeficiency viruses competitively antagonize signaling by coreceptors CXCR4 and CCR5. Proc. Natl. Acad. Sci. U.S.A. *95*, 8005–8010.

Malim, M.H., Freimuth, W.W., Liu, J., Boyle, T.J., Lyerly, H.K., Cullen, B.R., and Nabel, G.J. (1992). Stable expression of transdominant Rev protein in human T-cells inhibits human immunodeficiency virus replication. J. Exp. Med. *176*, 1197–1201.

Mangeot, P.E., Negre, D., Dubois, B., Winter, A.J., Leissner, P., Mehtali, M., Kaiserlian, D., Cosset, F.L., and Darlix, J.L. (2000). Development of minimal lentivirus vectors derived from simian immunodeficiency virus (SIVmac251) and their use for gene transfer into human dendritic cells. J. Virol. *74*, 8307–8315.

Mann, R., Mulligan, R.C., and Baltimore, D. (1983). Construction of a retrovirus packaging mutant and its use to produce helper-free defective retrovirus. Cell *33*, 153–159.

Martinez-Salas, E. (1999). Internal ribosome entry site biology and its use in expression vectors. Curr. Opin. Biotechnol. *10*, 458–464.

Maurice, M., Mazur, S., Bullough, F.J., Salvetti, A., Collins, M.K., Russell, S.J., and Cosset, F.L. (1999). Efficient gene delivery to quiescent interleukin-2 (IL-2)-dependent cells by murine leukemia virus-derived vectors harboring IL-2 chimeric envelope glycoproteins. Blood *94*, 401–410.

Mautino, M.R., Keiser, N., and Morgan, R.A. (2000). Improved titers of HIV-based lentiviral vectors using the SRV-1 constitutive transport element. Gene Ther. *7*, 1421–1424.

Mavria, G., Harrington, K.J., Marshall, C.J., and Porter, C.D. (2005). In vivo efficacy of HSV-TK transcriptionally targeted to the tumour vasculature is augmented by combination with cytotoxic chemotherapy. J. Gene Med. *7*, 263–275.

Merten, C.A., Stitz, J., Braun, G., Poeschla, E.M., Cichutek, K., and Buchholz, C.J. (2005). Directed evolution of retrovirus envelope protein cytoplasmic tails guided by functional incorporation into lentivirus particles. J. Virol. *79*, 834–840.

Metharom, P., Ellem, K.A., Schmidt, C., and Wei, M.Q. (2001). Lentiviral vector-mediated tyrosinase-related protein 2 gene transfer to dendritic cells for the therapy of melanoma. Hum. Gene Ther. *12*, 2203–2213.

Metzl, C., Mischek, D., Salmons, B., Gunzburg, W.H., Renner, M., and Portsmouth, D. (2006). Tissue- and tumor-specific targeting of murine leukemia virus-based replication-competent retroviral vectors. J. Virol. *80*, 7070–7078.

Miller, A.D., Miller, D.G., Garcia, J.V., and Lynch, C.M. (1993). Use of retroviral vectors for gene transfer and expression. Methods Enzymol. *217*, 581–599.

Miller, D.G., Adam, M.A., and Miller, A.D. (1990). Gene transfer by retrovirus vectors occurs only in cells that are actively replicating at the time of infection. Mol. Cell. Biol. *10*, 4239–4242.

Miyoshi, H., Takahashi, M., Gage, F.H., and Verma, I.M. (1997). Stable and efficient gene transfer into the retina using an HIV-based lentiviral vector. Proc. Natl. Acad. Sci. U.S.A. *94*, 10319–10323.

Modlich, U., Bohne, J., Schmidt, M., von Kalle, C., Knoss, S., Schambach, A., and Baum, C. (2006). Cell-culture assays reveal the importance of retroviral vector design for insertional genotoxicity. Blood *108*, 2545–2553.

Morgan, R.A., Dudley, M.E., Wunderlich, J.R., Hughes, M.S., Yang, J.C., Sherry, R.M., Royal, R.E., Topalian, S.L., Kammula, U.S., Restifo, N.P., Zheng, Z., Nahvi, A., de Vries, C.R., Rogers-Freezer, L.J., Mavroukakis, S.A., and Rosenberg S.A. (2006). Cancer regression in patients after transfer of genetically engineered lymphocytes. Science *314*, 126–129.

Morizono, K., Xie, Y., Ringpis, G.E., Johnson, M., Nassanian, H., Lee, B., Wu, L., and Chen, I.S. (2005). Lentiviral vector retargeting to P-glycoprotein on metastatic melanoma through intravenous injection. Nat. Med. *11*, 346–352.

Mrochen, S., Klein, D., Nikol, S., Smith, J.R., Salmons, B., and Gunzburg, W.H. (1997). Inducible expression of p21WAF-1/CIP-1/SDI-1 from a promoter conversion retroviral vector. J. Mol. Med. *75*, 820–828.

Muhlebach, M.D., Schmitt, I., Steidl, S., Stitz, J., Schweizer, M., Blankenstein, T., Cichutek, K., and Uckert, W.

(2003). Transduction efficiency of MLV but not of HIV-1 vectors is pseudotype dependent on human primary T-lymphocytes. J. Mol. Med. *81*, 801–810.

Muhlebach, M.D., Wolfrum, N., Schule, S., Tschulena, U., Sanzenbacher, R., Flory, E., Cichutek, K., and Schweizer, M. (2005). Stable transduction of primary human monocytes by simian lentiviral vector PBj. Mol. Ther. *12*, 1206–1216.

Naldini, L., Blomer, U., Gage, F.H., Trono, D., and Verma, I.M. (1996a). Efficient transfer, integration, and sustained long-term expression of the transgene in adult rat brains injected with a lentiviral vector. Proc. Natl. Acad. Sci. U.S.A. *93*, 11382–11388.

Naldini, L., Blomer, U., Gallay, P., Ory, D., Mulligan, R., Gage, F.H., Verma, I.M., and Trono, D. (1996b). In vivo gene delivery and stable transduction of nondividing cells by a lentiviral vector. Science *272*, 263–267.

Naviaux, R.K., Costanzi, E., Haas, M., and Verma, I.M. (1996). The pCL vector system: rapid production of helper-free, high-titer, recombinant retroviruses. J. Virol. *70*, 5701–5705.

Ngoi, S.M., Chien, A.C., and Lee, C.G. (2004). Exploiting internal ribosome entry sites in gene therapy vector design. Curr. Gene Ther. *4*, 15–31.

Nilson, B.H., Morling, F.J., Cosset, F.L., and Russell, S.J. (1996). Targeting of retroviral vectors through protease–substrate interactions. Gene Ther. *3*, 280–286.

Noguchi, M., Nakamura, Y., Russell, S.M., Ziegler, S.F., Tsang, M., Cao, X., and Leonard, W.J. (1993). Interleukin-2 receptor gamma chain: a functional component of the interleukin-7 receptor. Science. *262*, 1877–1880.

Osborn, M.J., Panoskaltsis-Mortari, A., McElmurry, R.T., Bell, S.K., Vignali, D.A., Ryan, M.D., Wilber, A.C., McIvor, R.S., Tolar, J., and Blazar, B.R. (2005). A picornaviral 2A-like sequence-based tricistronic vector allowing for high-level therapeutic gene expression coupled to a dual-reporter system. Mol. Ther. *12*, 569–574.

Ott, M.G., Schmidt, M., Schwarzwaelder, K., Stein, S., Siler, U., Koehl, U., Glimm, H., Kuhlcke, K., Schilz, A., Kunkel, H., Naundorf, S., Brinkmann, A., Deichmann, A., Fischer, M., Ball, C., Pilz, I., Dunbar, C., Du, Y., Jenkins, N.A., Copeland, N.G., Luthi, U., Hassan, M., Thrasher, A.J., Hoelzer, D., von, K.C., Seger, R., and Grez, M. (2006). Correction of X-linked chronic granulomatous disease by gene therapy, augmented by insertional activation of MDS1-EVI1, PRDM16 or SETBP1. Nat. Med. *12*, 401–409.

Parney, I.F., Farr-Jones, M.A., Kane, K., Chang, L.J., and Petruk, K.C. (2002). Human autologous *in vitro* models of glioma immunogene therapy using B7–2, GM-CSF, and IL12. Can. J. Neurol. Sci. *29*, 267–275.

Pavlakis, G.N., and Felber, B.K. (1990). Regulation of expression of human immunodeficiency virus. New Biol. *2*, 20–31.

Peng, K.W., Morling, F.J., Cosset, F.L., Murphy, G., and Russell, S.J. (1997). A gene delivery system activatable by disease-associated matrix metalloproteinases. Hum. Gene Ther. *8*, 729–738.

Peng, K.W., Vile, R., Cosset, F.L., and Russell, S. (1999). Selective transduction of protease-rich tumors by matrix-metalloproteinase-targeted retroviral vectors. Gene Ther. *6*, 1552–1557.

Pfeifer, A., Ikawa, M., Dayn, Y., and Verma, I.M. (2002). Transgenesis by lentiviral vectors: lack of gene silencing in mammalian embryonic stem cells and preimplantation embryos. Proc. Natl. Acad. Sci. U.S.A. *99*, 2140–2145.

Philippe, S., Sarkis, C., Barkats, M., Mammeri, H., Ladroue, C., Petit, C., Mallet, J., and Serguera, C. (2006). Lentiviral vectors with a defective integrase allow efficient and sustained transgene expression *in vitro* and *in vivo*. Proc. Natl. Acad. Sci. U.S.A. *103*, 17684–17689.

Portsmouth, D., Hlavaty, J., and Renner, M. (2007). Suicide genes for cancer therapy. Mol. Aspects Med. *28*, 4–41.

Purcell, D.F., Broscius, C.M., Vanin, E.F., Buckler, C.E., Nienhuis, A.W., and Martin, M.A. (1996). An array of murine leukemia virus-related elements is transmitted and expressed in a primate recipient of retroviral gene transfer. J. Virol. *70*, 887–897.

Rainov, N.G. (2000). A phase III clinical evaluation of herpes simplex virus type 1 thymidine kinase and ganciclovir gene therapy as an adjuvant to surgical resection and radiation in adults with previously untreated glioblastoma multiforme. Hum. Gene Ther. *11*, 2389–2401.

Recillas-Targa, F., Valadez-Graham, V., and Farrell, C.M. (2004). Prospects and implications of using chromatin insulators in gene therapy and transgenesis. Bioessays *26*, 796–807.

Rivella, S., Callegari, J.A., May, C., Tan, C.W., and Sadelain, M. (2000). The cHS4 insulator increases the probability of retroviral expression at random chromosomal integration sites. J. Virol. *74*, 4679–4687.

Roe, T., Reynolds, T.C., Yu, G., and Brown, P.O. (1993). Integration of murine leukemia virus DNA depends on mitosis. EMBO J. *12*, 2099–2108.

Rosenberg, S.A., Aebersold, P., Cornetta, K., Kasid, A., Morgan, R.A., Moen, R., Karson, E.M., Lotze, M.T., Yang, J.C., Topalian, S.L., *et al.* (1990). Gene transfer into humans– immunotherapy of patients with advanced melanoma, using tumor-infiltrating lymphocytes modified by retroviral gene transduction. N. Engl. J. Med. 323, 570–578.

Roth, J.A., Nguyen, D., Lawrence, D.D., Kemp, B.L., Carrasco, C.H., Ferson, D.Z., Hong, W.K., Komaki, R., Lee, J.J., Nesbitt, J.C., Pisters, K.M., Putnam, J.B., Schea, R., Shin, D.M., Walsh, G.L., Dolormente, M.M., Han, C.I., Martin, F.D., Yen, N., Xu, K., Stephens, L.C., McDonnell, T.J., Mukhopadhyay, T., and Cai, D. (1996). Retrovirus-mediated wild-type p53 gene transfer to tumors of patients with lung cancer. Nat. Med. *2*, 985–991.

Russell, S.J., and Cosset, F.L. (1999). Modifying the host range properties of retroviral vectors. J. Gene Med. *1*, 300–311.

Sakalian, M., and Hunter, E. (1998). Molecular events in the assembly of retrovirus particles. Adv. Exp. Med. Biol. *440*, 329–339.

Sakuma, R., Noser, J.A., Ohmine, S., and Ikeda, Y. (2007). Inhibition of HIV-1 replication by simian restriction factors, TRIM5alpha and APOBEC3G. Gene Ther. *14*, 185–189.

Saller, R.M., Ozturk, F., Salmons, B., and Gunzburg, W.H. (1998). Construction and characterization of a hybrid mouse mammary tumor virus/murine leukemia virus-based retroviral vector. J. Virol. *72*, 1699–1703.

Sandrin, V., Muriaux, D., Darlix, J.L., and Cosset, F.L. (2004). Intracellular trafficking of Gag and Env proteins and their interactions modulate pseudotyping of retroviruses. J. Virol. *78*, 7153–7164.

Schambach, A., Bohne, J., Chandra, S., Will, E., Margison, G.P., Williams, D.A., and Baum, C. (2006). Equal potency of gammaretroviral and lentiviral SIN vectors for expression of O6-methylguanine-DNA methyltransferase in hematopoietic cells. Mol. Ther. *13*, 391–400.

Schmidt, M., Glimm, H., Wissler, M., Hoffmann, G., Olsson, K., Sellers, S., Carbonaro, D., Tisdale, J.F., Leurs, C., Hanenberg, H., Dunbar, C.E., Kiem, H.P., Karlsson, S., Kohn, D.B., Williams, D., and von Kalle, C. (2003). Efficient characterization of retro-, lenti-, and foamyvector-transduced cell populations by high-accuracy insertion site sequencing. Ann. N. Y. Acad. Sci *996*, 112–121.

Schneider, R.M., Medvedovska, Y., Hartl, I., Voelker, B., Chadwick, M.P., Russell, S.J., Cichutek, K., and Buchholz, C.J. (2003). Directed evolution of retroviruses activatable by tumour-associated matrix metalloproteases. Gene Ther. *10*, 1370–1380.

Schnierle, B.S., Stitz, J., Bosch, V., Nocken, F., Merget-Millitzer, H., Engelstadter, M., Kurth, R., Groner, B., and Cichutek, K. (1997). Pseudotyping of murine leukemia virus with the envelope glycoproteins of HIV generates a retroviral vector with specificity of infection for CD4-expressing cells. Proc. Natl. Acad. Sci. U.S.A. *94*, 8640–8645.

Schucht, R., Coroadinha, A.S., Zanta-Boussif, M.A., Verhoeyen, E., Carrondo, M.J., Hauser, H., and Wirth, D. (2006). A new generation of retroviral producer cells: predictable and stable virus production by Flp-mediated site-specific integration of retroviral vectors. Mol. Ther. *14*, 285–292.

Schwarzwaelder, K., Schmidt, M., Boztug, K., Glimm, H., Lulay, C., Dewey, R.A., Diez, I.A., Diestelhorst, J., Naundorf, S., Kuhlcke, K., Kondratenko, I., Marodi, L., Welte, K., Klein, C., and von Kalle C. (2008). Polyclonal hematopoietic repopulation in clinical Wiskott-Aldrich Syndrome (WAS) gene therapy trial. Mol. Ther. *16*, S82.

Seelamgari, A., Maddukuri, A., Berro, R., de la, F.C., Kehn, K., Deng, L., Dadgar, S., Bottazzi, M.E., Ghedin, E., Pumfery, A., and Kashanchi, F. (2004). Role of viral regulatory and accessory proteins in HIV-1 replication. Front Biosci. *9, 2388–413.*, 2388–2413.

Siciliano, S.J., Kuhmann, S.E., Weng, Y., Madani, N., Springer, M.S., Lineberger, J.E., Danzeisen, R., Miller, M.D., Kavanaugh, M.P., DeMartino, J.A., and Kabat, D. (1999). A critical site in the core of the CCR5 chemokine receptor required for binding and infectivity of human immunodeficiency virus type 1. J. Biol. Chem. *274*, 1905–1913.

Sinn, P.L., Sauter, S.L., and McCray, P.B., Jr. (2005). Gene therapy progress and prospects: development of improved lentiviral and retroviral vectors– design, biosafety, and production. Gene Ther. *12*, 1089–1098.

Solly, S.K., Trajcevski, S., Frisen, C., Holzer, G.W., Nelson, E., Clerc, B., Abordo-Adesida, E., Castro, M., Lowenstein, P., and Klatzmann, D. (2003). Replicative retroviral vectors for cancer gene therapy. Cancer Gene Ther. *10*,

Srinivasakumar, N., and Schuening, F.G. (1999). A lentivirus packaging system based on alternative RNA transport mechanisms to express helper and gene transfer vector RNAs and its use to study the requirement of accessory proteins for particle formation and gene delivery. J. Virol. *73*, 9589–9598.

Stitz, J., Buchholz, C.J., Engelstadter, M., Uckert, W., Bloemer, U., Schmitt, I., and Cichutek, K. (2000). Lentiviral vectors pseudotyped with envelope glycoproteins derived from gibbon ape leukemia virus and murine leukemia virus 10A1. Virology *273*, 16–20.

Stockhammer, G., Brotchi, J., Leblanc, R., Bernstein, M., Schackert, G., Weber, F., Ostertag, C., Mulder, N.H., Mellstedt, H., Seiler, R., Yonekawa, Y., Twerdy, K., Kostron, H., De, W.O., Lambermont, M., Velu, T., Laneuville, P., Villemure, J.G., Rutka, J.T., Warnke, P., Laseur, M., Mooij, J.J., Boethius, J., Mariani, L., Gianella-Borradori, A., *et al.* (1997). Gene therapy for glioblastoma [correction of gliobestome] multiform: *in vivo* tumor transduction with the herpes simplex thymidine kinase gene followed by ganciclovir. J. Mol. Med. *75*, 300–304.

Stremlau, M., Owens, C.M., Perron, M.J., Kiessling, M., Autissier, P., and Sodroski, J. (2004). The cytoplasmic body component TRIM5alpha restricts HIV-1 infection in Old World monkeys. Nature *427*, 848–853.

Szczepek, M., Brondani, V., Buchel, J., Serrano, L., Segal, D.J., and Cathomen, T. (2007). Structure-based redesign of the dimerization interface reduces the toxicity of zinc-finger nucleases. Nat. Biotechnol. *25*, 786–793.

Szecsi, J., Drury, R., Josserand, V., Grange, M.P., Boson, B., Hartl, I., Schneider, R., Buchholz, C.J., Coll, J.L., Russell, S.J., Cosset, F.L., and Verhoeyen, E. (2006). Targeted retroviral vectors displaying a cleavage site-engineered hemagglutinin (HA) through HA–protease interactions. Mol. Ther. *14*, 735–744.

Tait, D.L., Obermiller, P.S., Redlin-Frazier, S., Jensen, R.A., Welcsh, P., Dann, J., King, M.C., Johnson, D.H., and Holt, J.T. (1997). A phase I trial of retroviral BRCA1sv gene therapy in ovarian cancer. Clin. Cancer Res. 3, 1959–1968.

US Federal Food and Drug Administration. (2000). Guidance for Industry: Supplemental guidance on testing for replication competent retrovirus in retroviral vector based gene therapy products and during follow-up of patients in clinical trials using retroviral vectors. http://www.fda.gov/cber/guidelines.htm

Uckert, W., Becker, C., Gladow, M., Klein, D., Kammertoens, T., Pedersen, L., and Blankenstein, T. (2000). Efficient gene transfer into primary human

CD8+ T-lymphocytes by MuLV-10A1 retrovirus pseudotype. Hum. Gene Ther. *11*, 1005–1014.

Usas, A., and Huard, J. (2007). Muscle-derived stem cells for tissue engineering and regenerative therapy. Biomaterials *28*, 5401–5406.

van der Putten, H., Botteri, F.M., Miller, A.D., Rosenfeld, M.G., Fan, H., Evans, R.M., and Verma, I.M. (1985). Efficient insertion of genes into the mouse germ line via retroviral vectors. Proc. Natl. Acad. Sci. U.S.A. *82*, 6148–6152.

Vanin, E.F., Kaloss, M., Broscius, C., and Nienhuis, A.W. (1994). Characterization of replication-competent retroviruses from nonhuman primates with virus-induced T-cell lymphomas and observations regarding the mechanism of oncogenesis. J. Virol. *68*, 4241–4250.

Verhoeyen, E., Dardalhon, V., Ducrey-Rundquist, O., Trono, D., Taylor, N., and Cosset, F.L. (2003). IL-7 surface-engineered lentiviral vectors promote survival and efficient gene transfer in resting primary T-lymphocytes. Blood *101*, 2167–2174.

Vigna, E., and Naldini, L. (2000). Lentiviral vectors: excellent tools for experimental gene transfer and promising candidates for gene therapy. J. Gene Med *2*, 308–316.

von Kalle C., Kiem, H.P., Goehle, S., Darovsky, B., Heimfeld, S., Torok-Storb, B., Storb, R., and Schuening, F.G. (1994). Increased gene transfer into human hematopoietic progenitor cells by extended *in vitro* exposure to a pseudotyped retroviral vector. Blood *84*, 2890–2897.

von Schwedler U., Kornbluth, R.S., and Trono, D. (1994). The nuclear localization signal of the matrix protein of human immunodeficiency virus type 1 allows the establishment of infection in macrophages and quiescent T-lymphocytes. Proc. Natl. Acad. Sci. U.S.A. *91*, 6992–6996.

Watanabe, S., and Temin, H.M. (1983). Construction of a helper cell line for avian reticuloendotheliosis virus cloning vectors. Mol. Cell. Biol. *3*, 2241–2249.

Wei, C.M., Gibson, M., Spear, P.G., and Scolnick, E.M. (1981). Construction and isolation of a transmissible retrovirus containing the src gene of Harvey murine sarcoma virus and the thymidine kinase gene of herpes simplex virus type 1. J. Virol. *39*, 935–944.

Welsh, R.M., Jr., Cooper, N.R., Jensen, F.C., and Oldstone, M.B. (1975). Human serum lyses RNA tumour viruses. Nature *257*, 612–614.

Wen, X.Y., Mandelbaum, S., Li, Z.H., Hitt, M., Graham, F.L., Hawley, T.S., Hawley, R.G., and Stewart, A.K. (2001). Tricistronic viral vectors co-expressing interleukin-12 (1L-12) and CD80 (B7-1) for the immunotherapy of cancer: preclinical studies in myeloma. Cancer Gene Ther. *8*, 361–370.

Will, E., Bailey, J., Schuesler, T., Modlich, U., Balcik, B., Burzynski, B., Witte, D., Layh-Schmitt, G., Rudolph, C., Schlegelberger, B., von Kalle, C., Baum, C., Sorrentino, B.P., Wagner, L.M., Kelly, P., Reeves, L., and Williams, D.A. (2007). Importance of murine study design for testing toxicity of retroviral vectors in support of phase I trials. Mol Ther *15*, 782–791.

Wolfrum, N., Muhlebach, M.D., Schule, S., Kaiser, J.K., Kloke, B.P., Cichutek, K., and Schweizer, M. (2007). Impact of viral accessory proteins of SIVsmmPBj on early steps of infection of quiescent cells. Virology *364*, 330–341.

Wu, X., and Burgess, S.M. (2004). Integration target site selection for retroviruses and transposable elements. Cell Mol Life Sci *61*, 2588–2596.

Xu, K., Ma, H., McCown, T.J., Verma, I.M., and Kafri, T. (2001). Generation of a stable cell line producing high-titer self-inactivating lentiviral vectors. Mol Ther *3*, 97–104.

Yang, L., Bailey, L., Baltimore, D., and Wang, P. (2006). Targeting lentiviral vectors to specific cell types *in vivo*. Proc. Natl. Acad. Sci. U.S.A. *103*, 11479–11484.

Yang, Y., Vanin, E.F., Whitt, M.A., Fornerod, M., Zwart, R., Schneiderman, R.D., Grosveld, G., and Nienhuis, A.W. (1995). Inducible, high-level production of infectious murine leukemia retroviral vector particles pseudotyped with vesicular stomatitis virus G envelope protein. Hum. Gene Ther. *6*, 1203–1213.

Yu, S.F., von Ruden, T., Kantoff, P.W., Garber, C., Seiberg, M., Ruther, U., Anderson, W.F., Wagner, E.F., and Gilboa, E. (1986). Self-inactivating retroviral vectors designed for transfer of whole genes into mammalian cells. Proc. Natl. Acad. Sci. U.S.A. *83*, 3194–3198.

Zhao, Y., Low, W., and Collins, M.K. (2000). Improved safety and titre of murine leukaemia virus (MLV)-based retroviral vectors. Gene Ther. *7*, 300–305.

Zufferey, R., Donello, J.E., Trono, D., and Hope, T.J. (1999). Woodchuck hepatitis virus posttranscriptional regulatory element enhances expression of transgenes delivered by retroviral vectors. J. Virol. *73*, 2886–2892.

Zufferey, R., Dull, T., Mandel, R.J., Bukovsky, A., Quiroz, D., Naldini, L., and Trono, D. (1998). Self-inactivating lentivirus vector for safe and efficient *in vivo* gene delivery. J. Virol. *72*, 9873–9880.

Zufferey, R., Nagy, D., Mandel, R.J., Naldini, L., and Trono, D. (1997). Multiply attenuated lentiviral vector achieves efficient gene delivery *in vivo*. Nat Biotechnol *15*, 871–875.

Non-primate Mammalian and Fish Retroviruses

14

Maribeth V. Eiden, Kathryn Radke, Joel Rovnak and Sandra L. Quackenbush

Abstract

The pioneering phase of the study of retroviruses resulted in the identification of viruses associated with diseases in chickens, mice and cats. Exogenous retroviruses have since been isolated from many vertebrate species, and classified into seven genera that can be grouped into two general categories. Alpharetroviruses, betaretroviruses and gammaretroviruses are genetically simple, encoding only nucleoprotein, matrix, capsid, reverse transcriptase, integrase, protease and envelope proteins. Deltaretroviruses, epsilonretroviruses, lentiviruses and spumaviruses are considered complex because they encode in addition to the proteins listed above, a number of ancillary proteins that often play an important role in gene regulation. In this chapter we review recent findings of representative simple mammalian gammaretroviruses and the complex piscine epsilonretroviruses and bovine leukaemia virus with the intent of illustrating how these viruses have shed light on the mechanisms of viral function, evolution and pathogenesis within the animal kingdom that hosts them.

Gammaretroviruses

Due to the widespread distribution of gammaretroviruses among their vertebrate hosts, especially non-human primates, extensive efforts were undertaken in the 1960s and 1970s to identify a gammaretrovirus associated with human disease. The discovery of human T-cell leukaemia viruses and human immunodeficiency viruses sidelined interest in gammaretroviruses as potential human pathogens. However, the identification of a gammaretrovirus in prostate cancer tissue (Urisman *et al.*, 2006; Dong *et al.*, 2007), the determination that a gammaretrovirus has recently endogenized into the genome of a mammalian species (Tarlinton *et al.*, 2006; Oliveira *et al.*, 2007) and evidence of recent interspecies transfer of a gammaretrovirus between gibbons and koalas (Hanger *et al.*, 2000) have renewed interest in gammaretroviruses as human pathogens and potential zoonotic agents. As mentioned above gammaretroviruses represent one of seven genera of retroviruses The relatedness of representative members of the different retroviral genera based on comparisons of amino acid alignment of seven conserved domains in reverse transcriptase shows clustering of members of each genera but no obvious clustering of simple viruses into a group apart for the complex viruses (Fig. 14.1).

Feline gammaretroviruses

Numerous feline leukaemia viruses (FeLVs) have been isolated from domestic cats. These isolates can be divided into acute transforming and chronic retroviruses. Acute transforming FeLVs induce disease after various periods of latency and are usually replication defective owing to the swapping of cellular sequences (transforming genes or viral oncogenes) for viral genes essential for viral replication. Chronic FeLVs are replication competent and induce disease after long periods of latency and do not contain viral oncogenes. There are four subgroups of chronic FeLV, subgroups A, B, C and T that have been characterized based on their interference, host range and receptor utilization properties

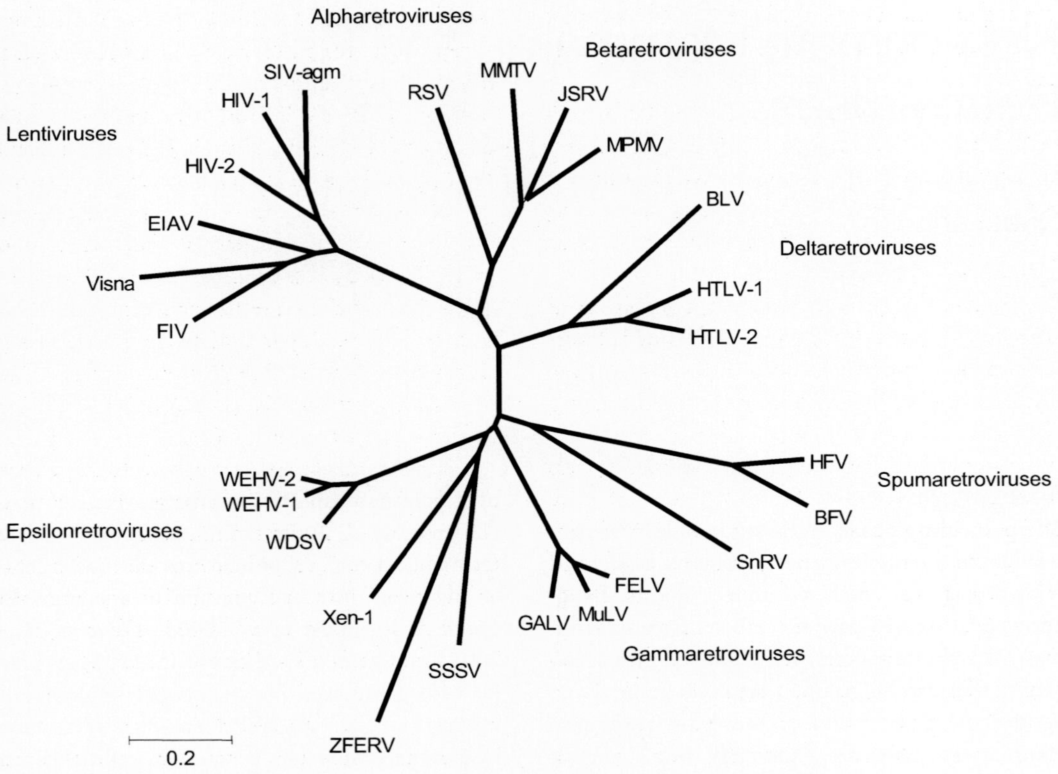

Figure 14.1 Unrooted phylogenetic tree of representative retroviruses based on an amino acid alignment of seven conserved domains in reverse transcriptase. Retroviruses are designated as follows: MPMV (Mason Pfizer monkey virus), JSRV (Jaagsiekte sheep retrovirus), MMTV (mouse mammary tumour virus), RSV (Rous sarcoma virus), EIAV (equine infectious anaemia virus), FIV (feline immunodeficiency virus), Visna (visna virus), HIV-2 (human immunodeficiency virus-2), HIV-1 (human immunodeficiency virus-1), SIV-agm (simian immunodeficiency virus-agm), HTLV-1 (human T-cell leukaemia virus type 1), HTLV-2 (human T-cell leukaemia virus 2), BLV (bovine leukaemia virus), SSSV (salmon swimbladder sarcoma virus), SnRV (snakehead retrovirus), BFV (bovine foamy virus), HFV (human foamy virus), FeLV (feline leukaemia virus), MuLV (murine leukaemia virus), GALV (gibbon ape leukaemia virus), WDSV (walleye dermal sarcoma virus), WEHV-1 (walleye epidermal hyperplasia virus type 1), WEHV-2 (walleye epidermal hyperplasia virus type 2), ZFERV (zebrafish endogenous retrovirus), Xen-1 (*Xenopus* endogenous retrovirus-1). Bootstrap values displayed at each branch point were determined from 100 replicates.

(Overbaugh *et al.*, 2001). Interference assays are based on the ability of cells productively infected by a virus to resist infection by viruses that use the same receptor to gain entry into the target cells, this phenomenon has been used to group viruses into different receptor groups. Host range refers to the types of cells derived from different species that viruses can productively infect *in vivo*.

FeLV-A is a weakly pathogenic virus that is commonly found in cats and is mainly transmitted horizontally through saliva, blood and other bodily fluids (Hardy *et al.*, 1973). FeLV can also be spread transplacentally from mother to offspring. Fleas may also be a potential vector for FeLV-A transmission (Vobis *et al.*, 2003). FeLV-A is an ecotropic virus present in all FeLV-infected cats and is associated with immunosuppression. All natural FeLV isolates from cats contain genetic segments derived from FeLV-A.

FeLV-B is isolated from FeLV-A infected cats and is associated with myeloproliferative disorders. It is now fairly well established that FeLV-B is a recombinant between FeLV-A and the FeLV-B envelope gene harboured as an endogenous element in the cat genome (Overbaugh *et al.*, 1988). It has also been recovered from the brain of cat inoculated with a molecular clone of FeLV-A that developed neurological disorders (Boomer *et al.*, 1994).

FeLV-C may have also originated via recombination and/or mutation but its origins are less clear then those proposed for FeLV-B (Overbaugh *et al.*, 2001). FeLV-C can induce aplastic anaemia, erythroid hypoplasia and immunosuppression (Riedel *et al.*, 1988; Abkowitz, 1991).

FeLV-T is a T-cell-tropic, cytopathic virus isolated from cats infected with a cloned, transmissible but nonvirulent form of FeLV-A (Rohn *et al.*, 1998). FeLV-T differs from the other FeLV subgroups in that it is not subject to superinfection interference and can reinfect T-lymphocytes productively infected with FeLV-T and replicate to a high copy number in these cells. This is in contrast to the resistance to infection by homologous virus that typically occurs in retroviral infections as a result of envelope-mediated receptor interference (Donahue *et al.*, 1991). The immunodeficient pathology associated with FeLV-T-mediated T-cell killing is probably a consequence of the failure to establish superinfection interference in infected cells (Donahue *et al.*, 1991).

RD114 is a replication-competent endogenous virus isolate not closely related to the FeLV subgroups A, B, C or T but, surprisingly, genetically related to the primate retrovirus baboon endogenous virus (BaEV) (Goldberg *et al.*, 1974). Originally obtained from human rhadomyosarcoma (RD) cells which had been passaged in neonatal cats (McAllister *et al.*, 1972), RD114 has been subsequently isolated from cell lines derived from domestic cats and shown to have a xenotropic host range (Fischinger *et al.*, 1973).

Feline gammaretrovirus receptors

The receptor for FeLV-A is presently unknown however a single cell surface protein of approximately 70kDa was proposed approximately 15 years ago to bind FeLV-A (Ghosh *et al.*, 1992). The receptor for FeLV-B is the ubiquitous inorganic phosphate transporter PiT1 (Takeuchi *et al.*, 1992) although FeLV-B can also use the feline orthologue of both PiT1 and the closely related Pit2 protein to infect feline cells (Anderson *et al.*, 2001). The receptor for FeLV-C has also been identified as a member of the major facilitator superfamily and, based on its relatedness to proteins found in bacteria and in *C. elegans*, may serve as a transporter for an organic anion or D-glucarate transporters (Tailor *et al.*, 1999b) (Quigley *et al.*, 2000). In 1990, Anderson and coworkers discovered that FeLV-T uses PiT1 as a receptor but requires an ancillary cellular co-factor termed FeLiX to gain entry into cells (Anderson *et al.*, 2000). FeLiX represents an endogenously expressed portion of the FeLV-B envelope protein.

RD114 is a member of the largest and most widely dispersed gammaretroviral interference group that includes BaEV as well as human endogenous virus type W (HERV-W) and Squirrel monkey retroviruses (SRVs). The retrovirus receptor for this large class of interfering viruses, ASCT2, has been found to be a Na^+-dependent transporter of neutral amino acids. A related transporter ASCT1 has been found to function as an ancillary receptor for BaEV and HERV-W (Torres-Zamorano *et al.*, 1998; Rasko *et al.*, 1999; Marin *et al.*, 2000; Lavillette *et al.*, 2002).

Murine gammaretroviruses

The earliest isolates of murine retroviruses were obtained from inbred mice. Later representative isolates from feral mice trapped in the wild were also identified and these viruses were placed into appropriate categories based on their interference, host range and receptor utilization properties. The main categories of murine retroviruses include ecotropic, amphotropic, polytropic and xenotropic.

Ecotropic murine viruses have an *in vitro* host range restricted to cells derived from rats and mice. One of the most extensively studied ecotropic murine retroviruses is Moloney murine leukaemia virus (MoMLV), originally isolated from cell free extract from the tumour of a neonatal mouse. It is pathologically associated with thymic leukaemia, disseminated lymphosarcoma and hepatosplenomegaly (Moloney, 1960a,b). MoMLV was the earliest of the gammaretroviruses to be cloned, sequenced and characterized (Sutcliffe *et al.*, 1980). The first retroviral vectors were based on MoMLV (Mann *et al.*, 1983) and most retroviral vectors used today as a gene transfer vehicles contain MoMLV-based genomes.

Amphotropic murine retroviruses (A-MLVs), originally isolated from feral mice trapped in geographically separate areas in southern California, have an extensive host range replicating in murine as well as a wide variety of cells derived from different species (Hartley and Rowe, 1976; Rasheed *et al.*, 1976). A-MLV sequences are not detected in the genome of feral or inbred mice, indicating that these viruses, in contrast to ecotropic viruses are not endogenous to mice (O'Neill *et al.*, 1987). A-MLVs cause mainly lymphoma in newborn inbred mice after a latent period of 6–12 months with a 1% incidence of hind limb paralysis and progressive central nervous system disease (Rasheed *et al.*, 1976). A-MLVs represent at least eight different field isolates. The newest isolate MuLV-1313 is the first A-MLV to be cloned and molecularly characterized in its entirety (Howard *et al.*, 2006). Howard and co-workers determined that the naturally occurring A-MLV MuLV-1313 is more closely related across its entire genome to the Cas-Br-E, an ecotropic retrovirus isolated from a Southern California wild mouse, than to any other MLV genome that has been sequenced, indicating that MuLV-1313 and Cas-Br-E share a common ancestor (Howard *et al.*, 2006).

10A1 MLV is a recombinant of A-MLV and an endogenous xenotropic virus of NIH Swiss mice obtained after passage of A-MLV in an NIH Swiss mouse (Rasheed *et al.*, 1982). 10A1 and A-MLV have 91% amino acid identity in their envelope proteins yet display different host ranges. For example, CHO-K1 cells are resistant to A-MLV but susceptible to infection by 10A1 (Ott *et al.*, 1990). More recently it has been resolved that as few as three residues unique to the 10A1 envelope when substituted for the corresponding residues of A-MLV confer 10A1 host range properties to the modified A-MLV (Han *et al.*, 1997).

Polytropic (P-MLVs) or duotropic MLVs are also designated mink cell focus forming viruses because of their ability to induce morphological alterations or focal formations in mink lung fibroblasts and are thought to be involved in wide range of leukaemias and lymphomas (Hartley *et al.*, 1977; Chattopadhyay *et al.*, 1981). Originally detected in the thymus glands of leukaemic and late preleukaemic AKR mice, they are neutralized by antisera to both ecotropic and xenotropic viruses, and can be interfered with by both viruses. Unlike either ecotropic or xenotropic MLVs, P-MLVs are not represented as replication competent endogenous retroviral elements but arise *de novo* via recombination between infectious ecotropic and P-MLV-related sequences. Their host range is quite broad excluding inbred murine cells but including some types of feral mice (Lyu *et al.*, 1999; Tailor *et al.*, 2003).

Xenotropic murine leukaemia viruses (X-MLVs) are endogenous viruses originally isolated from New Zealand Black mice wherein the increased expression of the virus was thought to result in autoimmune diseases (Levy, 1975). Xenotropic murine retroviruses, like polytropic murine retroviruses, are unable to infect their host species but are able to propagate efficiently in cells derived from other species (Levy, 1977). However, certain cell lines derived from European mice are resistant to X-MLVs yet are susceptible to P-MLVs (Kozak, 1985; Kozak and O'Neill, 1987).

Murine leukaemia virus receptors

The receptors for ecotropic, amphotropic, 10A1 MLV, polytropic and xenotropic MLVs have been cloned and identified as carrier facilitator proteins. Ecotropic MLVs utilize mCAT-1 also known as Atrc1, the transporter for cationic, basic amino acids, as a receptor (Albritton *et al.*, 1989). The inorganic phosphate transporter, PiT2 (now officially designated SLC20A2), functions as a receptor for A-MLV (Miller *et al.*, 1994; Zeijl *et al.*, 1994) and 10A1 MLV (Wilson *et al.*, 1995; Kavanaugh and Kabat, 1996). 10A1 can also employ the phosphate transporter PiT1 (now officially designated SLC20A1) as a receptor (Wilson *et al.*, 1995; Kavanaugh and Kabat, 1996). Xenotropic murine leukaemia viruses cannot infect cells from inbred mice due to the absence of a functional cell surface receptor required for virus entry, however, they have been shown to cross-interfere (Sommerfelt and Weiss, 1990) to various extents in non-inbred murine derived cell lines. Therefore it was not surprising that they can use a common receptor for infection of human cells (Battini *et al.*, 1999; Tailor *et al.*, 1999a). The murine orthologue of this protein present on the surface of cells derived from inbred mice functions solely as a P-MLV receptor while the human orthologue establishes infectivity by P-MLV as well as X-MLV, which do not infect laboratory mice (Yang *et al.*, 1999). The X-MLV, P-MLV receptor encodes a membrane protein related to the yeast Syg1 protein. In *Saccharomyces cerevisiae*, the mating pheromone-initiated signal is transduced by a G protein and normally results in transient cell cycle arrest and differentiation. A null allele of this G protein (GPA1/SCG1) subunit results in cell death. A truncated form of a Syg1 is a high copy suppressor of GPA1 lethality (Spain *et al.*, 1995).

Gammaretroviruses isolated from Asian feral mice

Gammaretroviruses obtained from feral Asia mice have only recently been molecularly characterized. It was not until 1996 that the first replication competent retrovirus *Mus dunni* endogenous retrovirus (MDEV) isolated from mice trapped on the Asian continent was molecularly characterized (Miller *et al.*, 1996; Miller and Wolgamot, 1997). MDEV is a replication competent but transcriptionally silent virus that can be activated by treatment of the cultured *Mus dunni* tail fibroblasts cells (Lander and Chattopadhyay, 1984) with hydrocortisone or 5-iodo-2′-deoxyuridine (Wolgamot and Miller, 1999). MDEV does not belong to any of the known inbred murine leukaemia virus groups by interference analysis but it appears to be genetically related in its coding regions to gibbon ape leukaemia virus (GALV). In addition, its long terminal repeats (LTRs) appear to derive from virus-like 30S elements, replication-defective retroelements that are similar in structure and replication cycle to retroviruses (Wolgamot and Miller, 1999).

In 2001, Prassolov *et al.* sequenced and characterized the M813 murine retrovirus isolated from the South-East Asian rodent *M. cervicolor*. It has been reported that two classes of replication competent gammaretroviruses are present in the genome of *M. cervicolor* (Callahan and Todaro, 1978). Type CI isolates have a xenotropic host range and appear closely related to GALV (Callahan and Todaro, 1978), however none of these isolates has been genetically characterized. The host range of the type CII isolates of *M. cervicolor* is limited to *M. musculus*. Type CII viruses appear to be closely related to MLVs isolated from *M. musculus* based on serological cross reactivity and genomic hybridization assays (Benveniste *et al.*, 1977) The cellular receptor for the M813 is the sodium-dependent myo-inositol transporter 1 (Hein *et al.*, 2003) and has been localized to mouse chromosome 2 (Prassolov *et al.*, 2001).

Koala retrovirus

Lymphoid neoplasias are a common cause of mortality among captive and free-ranging koalas in south-east Australia (Hanger *et al.*, 2003). Viral aetiology was suggested when virus particles were detected by electron microscopy of leukaemic animal tissues (Canfield *et al.*, 1988). Based on these observations, it was logical to suspect that a pathogenic virus could be the cause of haematopoietic neoplasias and related diseases observed in koalas. The sequence and molecular characterization of the koala retrovirus (KoRV) was reported in 2000 (Hanger *et al.*, 2000). Two surprising features of KoRV were the invariant presence of the virus among the tested koala populations and the similar viral DNA banding exhibited on Southern blots containing cellular DNA obtained from these animals. The DNA banding patterns suggested KoRV was heritable as an endogenous retrovirus (ERV). However, unlike other ERVs, KoRV remains biologically active. Electron microscopy of peripheral blood mononuclear cells derived from koalas with lymphoma found budding particles in 98% of examined koalas (Hanger *et al.*, 2000). These observations prompted further investigations that confirmed KoRV was present in the germ line of koalas (Tarlinton *et al.*, 2006). The presence of replicating virus particles and the considerable variability in the characterized proviruses in distinct populations of outbred animals implied that KoRV is in the initial stages of establishing itself as an ERV (Tarlinton *et al.*, 2006). This was further supported by the observation that all of the koalas on Kangaroo Island (representing the progeny of a handful of koalas removed from southern Australia in the 1920s to be used as foundation stock for the island) and several other koala populations in southern Australia are not infected by KoRV.

It was surprising when it was initially reported that KoRV is highly related to GALV and woolly monkey virus (WMV) (Hanger *et al.*, 2000). GALV and WMV were first identified in the early 1970s as exogenous retroviruses associated with leukaemias and lymphomas in captive primates (Kawakami *et al.*, 1972, 1973) – see

Chapter 15 for detailed descriptions of these primate retroviruses. The sequence similarity between KoRV, GALV and WMV is so close across all loci that these viruses can be considered to be conspecific (i.e. derive from a common ancestor virus) (Hanger *et al.*, 2000). Accurate estimation of the date of the origin of GALV is difficult because KoRV exists as both an endogenous and exogenous virus. Based on calibrations appropriate for a genomic endogenous virus, it is estimated that KoRV and GALV split millions of years ago, whereas calibration rates appropriate for exogenous non-heritable retroviruses indicate that GALV split from KoRV sometime within the last few decades (Bromham, 2002).

KoRV in vitro *and* in vivo *host range*

KoRV vectors have the broadest host range of any gammaretroviral vector identified to date. In addition to cells infectable by GALV (see Chapter 15), KoRV is capable of infecting a variety of murine and hamster derived cell lines that are resistant to GALV, FeLV-B and MLV vectors (Oliveira *et al.*, 2006). Interference assays were used to determine that human cells infected with GALV SEATO are resistant to KoRV vectors, suggesting KoRV and GALV both require the same receptor to infect human cells (Oliveira *et al.*, 2006). The *in vivo* host ranges of GALV SEATO and KoRV also differ. Unlike GALV, KoRV can be experimentally transmitted *in vivo* to rats (Fiebig *et al.*, 2006).

KoRV endogenization

KoRV is a newly endogenizing gammaretrovirus that shares 78% nucleotide identity across its entire genome with GALV SEATO (Hanger *et al.*, 2000; Tarlinton *et al.*, 2006). The finding that KoRV is *endogenous* in koalas while GALV and woolly monkey virus (WMV) are *exogenous* viruses suggests that KoRV predates GALV, and that primates and koalas acquired the virus at different times from a common source. The co-existence of conspecific exogenous and endogenous gammaretroviruses has allowed, for the first time, an in depth study of the retroviral genome for adaptive mutations acquired during early stages of retroviral endogenization. Despite their high degree of relatedness, the infectivity of KoRV vectors is substantially lower than that of GALV vectors (Oliveira *et al.*, 2006). Specific differences in the protein coding regions of KoRV that distinguish it from GALV, when introduced into the GALV genome, diminished GALV infectivity. A mutation was identified in an L domain of the KoRV matrix protein, as well as five residues present in the KoRV envelope which, when substituted for the corresponding residues of GALV, resulted in vectors exhibiting substantially reduced titres similar to those observed with KoRV vectors (Oliveira *et al.*, 2007). In addition, the KoRV envelope protein lacks an intact CETTG motif that has been newly identified as invariant among highly infectious gammaretroviruses (Oliveira *et al.*, 2007). Disruption of this motif in GALV results in vectors with reduced syncytia forming capabilities. Functional assessment of specific sequences that contribute to KoRV's attenuation from a highly infectious GALV-like progenitor virus has allowed the identification of specific modifications in the KoRV genome that correlate with its endogenization.

The origin of KoRV

As mentioned in the above section sequences related to GALV and KoRV have been identified in Asian rodents. This relatedness was based on the immunogenic cross reactivity between the viral proteins derived from inducible, replication-competent endogenous Asian mouse viruses and GALV or WMV (Lieber *et al.*, 1975), similarity of host range, liquid hybridization assays or viral interference assays (Benveniste *et al.*, 1977). Callahan and coworkers were able to induce a xenotropic virus from a cultured cell line derived from the Asian Palm mouse *V. oleracea*. The internal viral proteins of this induced virus were demonstrated to be immunologically related to respective GALV and WMV proteins (Callahan *et al.*, 1979). Liquid hybridization studies detected only a limited genetic relationship between the *V. oleracea* and GALV or WMV. The host range of *V. oleracea* is not similar to that of GALV, as canine and primate cells are resistant to infection with the *V. Oleracea* isolate. Thus, the endogenous virus present in the *V. oleracea* genome is an unlikely GALV progenitor candidate.

A more likely GALV/KoRV progenitor candidate was identified by M. Lieber *et al.*, who characterized an endogenous infectious retrovirus isolated from the Asian wild mouse

Mus caroli (Lieber *et al.*, 1975). They found that the induced virus has a xenotropic host range and interferes with GALV and WMV but not xenotropic MLVs and that antibodies to WMV reverse transcriptase, but not those to ecotropic MLVs, inhibited the *Mus caroli* virus polymerase activity (Lieber *et al.*, 1975). Finally, they determined that antibodies to the internal structural proteins and reverse transcriptase of the *M. caroli* isolate display immuno-cross-reactivity with GALV and WMV proteins but not inbred MLV proteins. Another group, led by Benveniste, performed liquid hybridization experiments with probes prepared from GALV or WMV segments and genomic DNA of *M. caroli;* their results showed genetic identity between these viruses (Benveniste *et al.*, 1977). More recently Southern blot and polymerase chain reaction analyses have verified that GALV related sequences are present in the genome of cells derived from *M. caroli* and *M. fragilicauda* (Eiden, unpublished data). *M. fragilicauda* is a recently discovered, although by no means new, mouse species that diverged from *M. musculus* at roughly the same time as *M. caroli.* However, *M. fragilicauda* was unknown when the studies undertaken in the 1970s were performed. Thus, inducible endogenous retroviruses similar to GALV have been described as present in the genome of two different species: *M. caroli and M. fragilicauda;* either one or both may have transmitted the virus to koalas and gibbons.

Cross-species transmission of the KoRV progenitor

The natural transmission of GALV among gibbons occurs via contact between an infected carrier gibbon and an uninfected animal, or prenatally, from a viraemic parent to its offspring (Kawakami *et al.*, 1978). Although it has not been definitively established, transmission of GALV could also occur by grooming, biting or exposure to faeces, urine (Kawakami *et al.*, 1977) or saliva (Reitz *et al.*, 1979). Koalas would be expected to transmit the virus in a similar manner. The ingestion of pap, a form of faeces that constitutes an important part of the young koala's diet, may also serve as a means of transmitting KoRV (Canfield, 1990).

How does trans-species transmission of the virus occur? Since KoRV has been isolated from both captive and wild koalas, while GALV has only been isolated from captive gibbons, iatrogenic infection of captive gibbons with a GALV-like virus is a possibility. Alternately, natural transmission could have occurred with an infected host species serving as the agent of transmission between the two continents. Rodents, fruit bats, and birds are all potential candidates, although no evidence of viruses has been observed in either of the latter two animals. Asian rodents are the most likely source animals, since some species of Asian mice contain GALV-like endogenous viruses. Evidence that rodents have migrated from South-East Asia to Australia on several occasions in the past, may indicate that Asian rodents served as vectors between the two continents (Martin *et al.*, 1999). Therefore, the most likely means of infection of gibbons and koalas with the murine progenitor virus involved horizontal transmission of the virus from Asian feral rodents to gibbon apes and koalas at different times.

Several questions arise when the hypothesis is posed that a virus present in rodents is capable of infecting gibbons. First, how does the Asian mouse virus circumvent key barriers in this species-jumping event? The MLVs implicated as progenitors to GALV or KoRV are xenotropic and don't propagate in most murine cells. However, cells derived from feral Asian rodents, unlike those derived from inbred mice, have been demonstrated to be susceptible to GALV. For example cells derived from the Japanese mouse, *M. musculus molossinus* are susceptible to GALV (Schneiderman *et al.*, 1996). It should additionally be noted that different cells derived from the same species vary in terms of susceptibility to GALV. For example, Chinese hamster ovary cells are resistant to GALV but cells derived from Chinese hamster lung are susceptible to GALV (Eglitis *et al.*, 1993). It is known that cell lines derived from tail fibroblasts of *M. caroli* and *M. fragilicauda* are resistant to GALV, but it remains unknown whether GALV or the GALV-progenitor virus can infect different tissues of its murine host *in vivo*.

Obviously, some of the most important questions regarding how an Asian MLV jumped species to infect primates and koalas may be answered through the isolation and molecular characterization of the progenitor MLV. Comparison of the progenitor virus host range, receptor utilization properties and differences in the key determinants between KoRV and GALV envelope proteins to those of the progenitor MLV may provide insights into this virus species-jumping capabilities.

Epsilonviruses

Several neoplastic and proliferative diseases in fish are associated with virus infection or exposure to chemical carcinogens. These diseases occur in wild populations of fish and higher incidence of disease is often observed in areas with greater densities of fish, such as aquaculture and hatchery facilities. Increased epizootics of neoplastic diseases in fish in farm-reared or hatchery facilities has increased the awareness of and prompted further investigation of the suspected viruses. Thirteen proliferative diseases of fish have been associated with the presence of retroviruses. The etiologic relationship of retroviruses with these diseases is primarily based on the observation of retrovirus-like particles, the presence of reverse transcriptase (RT) activity in neoplastic tissues and the ability to transmit disease with tumour extracts. Partial sequences have been obtained for some exogenous and endogenous fish retroviruses (Herniou, 1998; Gifford, 2003; Quackenbush and Casey, unpublished) and complete genomic sequence for five exogenous fish retroviruses from Atlantic salmon (*Salmo salar*), walleye (*Sander vitreus*), and snakehead (*Ophicephalus striatus*), and for the endogenous retrovirus of zebrafish (*Danio rerio*) have been determined (Holzschu *et al.*, 1995; Hart *et al.*, 1996; LaPierre *et al.*, 1999; Shen and Steiner, 2004; Paul *et al.*, 2006).

Isolation of the snakehead fish retrovirus in cell culture

Isolation of fish retroviruses in cell culture has been a particular challenge, and in many cases, has not been achieved. The presence of a retrovirus was noted during the characterization of cell lines established from three Southeast Asian fish species (Frerichs *et al.*, 1991). Electron microscopy of four different cell lines revealed the presence of type C virus particles, and supernatants from these cell lines contained high levels of RT activity. Of the four cell lines established from Asian freshwater fish, two of the lines (SSN-1 and SSN-3) were from striped snakehead (*Ophicephalus striatus*), one cell line, CP, was cultured from a climbing perch (*Anabas testudineus*) and the SGP cell line initiated from a snakeskin gourami (*Trichogaster pectoralis*). The snakehead retrovirus (SnRV) originally identified in one of the cell lines, SSN-1, was subsequently cloned by RT-PCR from purified virions (Hart *et al.*, 1996). The viral genome is 10.7 kb in length and includes an open reading frame (ORF) between *env* and the 3'LTR. The *env* start codon resides upstream of the major splice donor site and is predicted to fuse a 14 amino acid sequence in frame to *env*. The presence of an arginine tRNA primer binding site distinguishes SnRV from all other retroviruses. No endogenous copies of SnRV were identified in the snakehead fish genome.

The presence of SnRV in wild populations of fish has not been thoroughly examined. These fish were all healthy at the time the cell lines were derived and any association with disease is currently unknown.

Fish retroviruses associated with neoplastic and proliferative diseases

Retroviruses have been associated with different types of neoplastic and hyperplasias in fish, primarily due to observation of viral particles by EM and detection of RT activity in neoplastic tissues. Experimental transmissions with cell free extracts from nine retrovirus-associated diseases have been attempted, but disease development has only been achieved with six of them. Viruses from four of these have been cloned and sequenced.

Salmon swimbladder sarcoma-associated retrovirus

Neoplasia of the swimbladder in Atlantic salmon (*Salmo salar*) was first recognized in 1976 in salmon reared at a commercial fish farm in Scotland (Duncan, 1978; McKnight, 1978). Nodular masses on the surface of the swimbladder were observed in affected fish. Histological examination of the masses revealed that they arose at the junction of the inner smooth muscle and aerolar tissue zone. The tumours were well differentiated and comprising interlacing bundles of spindle cells with frequent mitotic cells and were classified as fibrosarcoma (Duncan, 1978) or leiomyosarcomas (McKnight, 1978). Examination of neoplastic tissue by electron microscopy detected budding retrovirus-like particles (Duncan, 1978).

A second outbreak of swimbladder sarcoma was reported in 1996 in Atlantic salmon collected from the Pleasant River in Maine and housed at the North Attleboro National Fish Hatchery in Massachusetts. These animals were housed at the hatchery as part of a native Atlantic salmon recovery program. Affected animals exhibited multifocal haemorrhages on the body and fins and multinodular masses on the swimbladder that were consistent with previous reports. By the spring of 1998, mortality was evident in 35% of the salmon population at the hatchery. The tumours comprised well-differentiated fibroblasts arranged in interlacing bundles and were classified as leiomyosarcomas.

Based on the previous observation of retrovirus-like particles in tumours, RT-PCR was used to amplify retroviral sequences from tumours with degenerate primers that targeted conserved amino acid sequences in the reverse transcriptase gene (VLPQG and YMDD) (Donehower *et al.*, 1990; Paul *et al.*, 2006). The complete proviral sequence of the salmon swimbladder sarcoma virus (SSSV) was obtained from a lambda library constructed from tumour DNA. SSSV is a simple retrovirus 10.9 kb in length with open reading frames that represent *gag, pol,* and *env* genes. It utilizes a methionine tRNA as a primer. Unlike many simple retroviruses, SSSV does not have related endogenous sequences in the host genome. Southern blot analysis of salmon swimbladder sarcomas showed that there was a very high proviral copy number (greater than 30 copies per cell) of SSSV with polyclonal integration (Paul *et al.*, 2006). The mechanisms that lead to the high copy number and their contribution to tumorigenesis are of interest for future research. Preliminary studies indicate that SSSV can be experimentally transmitted with cell free tumour extracts to naïve Atlantic salmon, however tumors were not observed (P. Bowser, personal communication).

Walleye retroviruses

Two different proliferative skin lesions in walleye discrete epidermal hyperplasia (WEH) and dermal sarcoma (WDS), are associated with distinct retroviruses. WEH and WDS were first recognized in walleye collected from Oneida Lake in New York in 1969 and have been reported to occur in other parts of North America (Walker, 1969; Yamamoto *et al.*, 1976). These diseases exhibit a seasonal cycle: the highest incidence of WEH and WDS occurs in the late fall through early spring (Bowser *et al.*, 1988). Lesions are rarely observed during the summer months.

Evidence for retroviral infection was first demonstrated by observation of type C particles with electron microscopy (Walker, 1969; Yamamoto *et al.*, 1985a,b). Sucrose density gradient centrifugation of homogenates from tumour and hyperplasia lesions yielded reverse transcriptase activity in fractions with a density of 1.18 g/ml and Mn^{2+} was the preferred cation (Martineau *et al.*, 1991; LaPierre *et al.*, 1998b). The retroviruses associated with WEH and WDS were subsequently cloned and sequenced (Martineau *et al.*, 1992; Holzschu *et al.*, 1995; LaPierre *et al.*, 1999). Currently, more is known about the pathogenesis of the walleye retroviruses than any of the other piscine retroviruses.

Walleye epidermal hyperplasia virus

Walleye epidermal hyperplasia lesions are broad, flat, translucent plaques of thickened epidermis with distinct boundaries. Microscopically, the plaques consist of a localized hyperplasia of Malpighian cells with frequent mitotic figures (Walker, 1969; Yamamoto *et al.*, 1985a).

Two independent retroviruses, walleye epidermal hyperplasia type 1 (WEHV1) and walleye epidermal hyperplasia type 2 (WEHV2), were identified in WEH lesions (LaPierre *et al.*, 1998b; LaPierre *et al.*, 1999). WEHV1 and WEHV2 were cloned by RT-PCR using degenerate *pol* primers followed by 5′ and 3′ RACE and isolation of genomic clones from a lambda library (Donehower *et al.*, 1990; LaPierre *et al.*, 1998b). Complete sequence analysis of WEHV1 and WEHV2 revealed that the proviruses were 12,999 bp and 13,125 bp in length, respectively (LaPierre *et al.*, 1999). The genome organization of WEHV consists of three open reading frames in addition to *gag, pol,* and *env* (Fig. 14.2) (Holzschu *et al.*, 1995; LaPierre *et al.*, 1999). Two ORFs, designated *orf a* and *orf b*, reside in the 3′ proximal region of the genome between *env* and the 3′LTR. The third open reading frame, *orf c*, is located between the 5′LTR and the start of *gag*. WEHV uses a histidyl-tRNA as

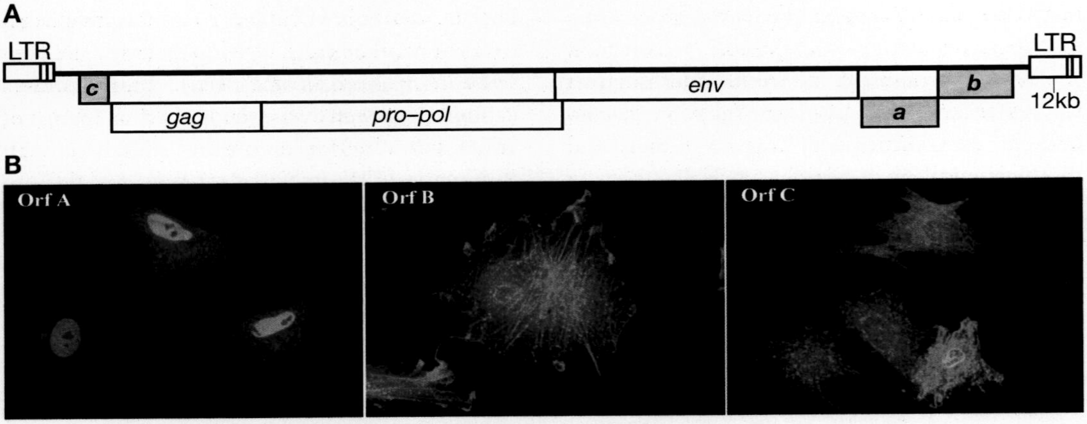

Figure 14.2 (A) Organization of the WDSV genome. (B) Immunofluorescent detection of endogenous WDSV accessory proteins in naturally infected cells derived from a spring regressing tumour.

the primer for minus-strand synthesis. Southern blot analysis indicates that hyperplasia lesions contain 1 to 3 copies of integrated proviral DNA (LaPierre *et al.*, 1998b). Northern blot analysis demonstrated that WEHV exhibits temporal gene expression in hyperplasia lesions collected in the fall and spring. Low levels of the viral transcripts that encode *orf a* are found in fall lesions whereas abundant levels of full-length genomic RNA, spliced *env*, and *orf a* and *orf b* transcripts are detected in spring lesions (LaPierre *et al.*, 1999).

Experimental transmission of WEH to walleye fingerlings has been achieved with cell-free filtrates from lesions collected in the spring (Bowser *et al.*, 1998). Ninety-seven per cent of inoculated fish developed WEH lesions that were grossly and microscopically consistent with those seen in adult wild caught fish. Analysis of lesions by PCR demonstrated the presence of WEHV2 in all lesions and WEHV1 in 69% of the lesions examined (Bowser *et al.*, 1998).

Walleye dermal sarcoma virus

Infection of walleye with walleye dermal sarcoma virus (WDSV) probably occurs through contact with virus in water or through direct contact with infected individuals when sexually mature walleye congregate during the spring spawning period (Bowser *et al.*, 1999). Tumours first appear on fish in the fall and throughout the fall and winter their number and size increase. The following spring during the spawning period WDS regresses (Bowser *et al.*, 1988; Bowser and Wooster, 1991). A study of different age classes of walleye conducted from 1995 through 2003 indicates that fish with tumours do not return with tumours during the spawning season in subsequent years. Data from these studies and experimental transmission studies suggest that most walleyes are likely to develop WDS in their lifetime (Bowser *et al.*, 1997; Bowser *et al.*, 1999; Getchell *et al.*, 2000, 2004).

Walleye dermal sarcoma is a cutaneous neoplasm that arises from the superficial surface of the scales (Walker, 1969; Yamamoto *et al.*, 1985b; Martineau *et al.*, 1990b). Microscopically the tumours are non-encapsulated, well-defined, nodular masses consisting of fibroblast cells that abut the epidermis. Mitotic figures are rare (Martineau *et al.*, 1990b). The overlaying epithelium of regressing tumours is often ulcerated. In some tumours perivascular accumulation of lymphocytes are observed as well as aggregates of inflammatory cells within the dermis surrounding the tumour (Martineau *et al.*, 1990b; Poulet *et al.*, 1994, 1995). Until recently, WDS was not found to be locally invasive in adult wild-caught fish. However, during the spring spawning run of 2000 three adult walleye with invasive tumours were identified (Bowser *et al.*, 2002).

To investigate a viral aetiology for walleye dermal sarcoma, 12-week-old walleye fingerlings were inoculated intramuscularly with cell-free tumour homogenates prepared from regressing spring tumours. By 14 weeks post infection 87%

of fish developed tumours that were grossly and histologically identical to those observed on adult walleyes (Martineau *et al.*, 1990a). Subsequent studies showed that tumour induction is very efficient using intramuscular, oral gavage, and topical application of tumour filtrates (Bowser *et al.*, 1997). Experimental infection by topical application and oral administration are more likely to simulate natural routes of infection. The most interesting finding regarding the pathogenesis of WDSV became evident when experimental transmission studies were conducted with filtrates derived from developing versus regressing tumours. In contrast to the efficient transmission of WDS with regressing tumours, filtrates from the developing, fall tumours were unable to transmit disease (Bowser *et al.*, 1996). The lack of transmission with these cell-free filtrates is due to the absence of infectious virus (Bowser *et al.*, 1996). Experimental transmission studies found that walleye fingerlings infected at 6 to 8 weeks of age developed invasive tumours (Earnest-Koons *et al.*, 1996). Histological examination found that the tumours were locally invasive and replaced normal tissue. Transmission of disease with material from naturally invasive tumours was poor (Bowser *et al.*, 2002). Quantification of the number of WDSV RNA copies in the invasive tumour homogenates was found to be significantly less than that in the standard regressing spring tumour preparations, suggesting that the reduced transmission rate is due to low levels of virus (Bowser *et al.*, 2002; Getchell *et al.*, 2002).

Experimental transmission of dermal sarcoma has been achieved in the closely related species, sauger (*Stizostedion canadense*) (Holzschu *et al.*, 1998). Dermal sarcoma developed in 97% of infected saugers and the tumours were grossly and histologically identical to those in walleye. Dermal sarcoma is also transmissible to yellow perch (*Perca flavescens*) and there is one report of a WDSV positive fibrosarcoma in yellow perch from an aquarium (Bowser *et al.*, 2001). These studies extend the host range of WDSV to include walleye, sauger, and yellow perch.

Further characterization of the etiologic agent of WDS relied on molecular techniques due to the lack of a cell culture system for propagation of virus at that time. The basis for ready transmission with regressing tumour homogenates became apparent when the extent of virus expression and viral DNA in developing and regressing tumours were evaluated. Developing tumours contain approximately 1 copy of viral DNA per cell, whereas cells of regressing tumours have 10 to 50 copies, most of which are unintegrated (unintegrated viral DNA or UVD) (Martineau *et al.*, 1991; Martineau *et al.*, 1992; Bowser *et al.*, 1996). Northern blot analysis of tumours showed that the developing fall tumours only express two subgenomic transcripts whereas the regressing spring tumours express subgenomic transcripts, spliced *env* transcript, and the full-length genomic RNA. The lack of genomic viral RNA in the developing tumours explains the inability to transmit disease (Bowser *et al.*, 1996; Quackenbush *et al.*, 1997).

WDSV was cloned from regressing tumour DNA and found to be 12.7 kb in length (Martineau *et al.*, 1991, 1992). Sequence analysis identified three open reading frames in addition to *gag, pol,* and *env*, designated *orf a*, *orf b*, and *orf c* (Holzschu *et al.*, 1995), an organization similar to WEHV. The *orf a* and *orf b* genes are located between *env* and the 3′LTR and *orf c* is found between the 5′LTR and gag. WDSV uses a histidyl-tRNA as the primer for minus-strand DNA synthesis. WDSV *gag, pro,* and *pol* are in the same reading frame and synthesize a polyprotein by termination suppression.

DNase I footprint analysis, electrophoretic mobility shift assays, and reporter assays were used to identify several *cis*-acting elements in the WDSV promoter important for transcription activation (Zhang *et al.*, 1999; Hronek *et al.*, 2004). Of particular interest was the differential binding of nuclear extracts from developing and regressing tumours to a region located between –82 and –32 (relative to the start of transcription) that contain three imperfect 15-bp repeats (Hronek *et al.*, 2004). Specific host or viral proteins that bind this region have not been identified, although it is likely that this element is critical for the induction of virus expression observed during tumour regression.

WDSV accessory protein function

Sequence comparisons of the Orf A proteins from walleye epidermal hyperplasia virus types 1 and 2 and WDSV revealed homology to D-type cyclins and the Orf A protein was thus named 'retroviral cyclin' or rv-cyclin (LaPierre *et al.*, 1998a). The rv-cyclins do not have the degree of homology to a cellular proto-oncogene that other retroviral oncogenes have, indicating they are not transduced cellular oncogenes. The homology of the rv-cyclins is limited to the cyclin box motif (19/29% and 17/30% identity/similarity to human and walleye D cyclins, respectively) (LaPierre *et al.*, 1998a). The cyclin box found in cellular cyclins and the herpesvirus v-cyclin forms an interface for interaction with cyclin-dependent kinases. Two critical residues, lysine and glutamate, necessary for such interactions are conserved in the walleye rv-cyclins (LaPierre *et al.*, 1998a).

The WDSV rv-cyclin was shown to rescue yeast (*Saccharomyces cerevisiae*) that are conditionally deficient in G1 cyclins (*Cln* genes) from growth arrest (LaPierre *et al.*, 1998a). In contrast, the WEHV1 and WEHV2 rv-cyclins were unable to complement G1 cyclin-deficient yeast. The structure of the WEHV1 and WEHV2 rv-cyclin proteins consists only of the cyclin box region whereas the WDSV rv-cyclin contains additional carboxy sequences in which an acidic transcriptional activation domain is located (Rovnak *et al.*, 2005).

Identification of WDSV rv-cyclin as a structural and functional homologue of cellular cyclins suggested a role in tumour development, and, when tested under the control of a keratin promoter in transgenic mice, rv-cyclin expression was associated with the induction of hyperplastic skin lesions in conjunction with wound healing (Lairmore *et al.*, 2000). The mice developed severe dermal hyperplasia at sites of injury such as tail clipping. These studies suggest that rv-cyclin is not oncogenic due to a single hit mechanism, but that additional genetic or epigenetic events are necessary for tumour development.

WDSV rv-cyclin is localized to the nucleus where it is concentrated in interchromatin granule clusters (IGCs) (or nuclear speckles), regions enriched in proteins necessary for transcription and mRNA processing (Rovnak *et al.*, 2001). rv-cyclin co-localizes and co-purifies with hyperphosphorylated forms of RNA polymerase II (Pol IIO) in IGCs, and is co-immunoprecipitated with antibody against Pol IIO. rv-cyclin was also found in a complex with cyclin-dependent kinase 8 (cdk8) and with its cognate partner, cyclin C (Rovnak and Quackenbush, 2002). cdk8 and cyclin C are components of the Mediator complex, which functions as a coactivator of transcription and is physically associated with the Pol II holoenzyme (Naar *et al.*, 2001). Cdk8 and cyclin C along with two additional proteins, Med12 and Med13, make up the cdk8 module of the Mediator complex, which functions to regulate transcription. The localization and physical association of WDSV rv-cyclin with components of transcription suggested that it has a functional role in transcription. Reporter gene assays evaluating WDSV and SV40 promoters demonstrated that the rv-cyclin has the capacity to both inhibit and enhance promoter activity in a cell-specific manner, similar to the actions of Mediator complexes (Rovnak and Quackenbush, 2002). Co-immunoprecipitation and GST pull down assays demonstrated an interaction of the rv-cyclin with p300/CBP, TBP, and the Mediator component, Sur2 (Rovnak *et al.*, 2005). Rv-cyclin physically interacts with cdk8 and mutations at K80 and E111, residues predicted to be important for cyclin/cdk interactions, reduced the ability of rv-cyclin to co-immunoprecipitate cdk8 (Rovnak and Quackenbush unpublished). In addition to the cyclin box, rv-cyclin contains an acidic transcription activation domain in its carboxy region (amino acids 240–270). The activation domain was identified by fusion of the rv-cyclin or the isolated activation domain to the GAL4 DNA binding domain and testing for activation of transcription from a GAL4 responsive luciferase reporter. Specific mutations at V260 within the activation domain eliminated activation of transcription (Rovnak *et al.*, 2005). Expression of rv-cyclin without a GAL4 DNA binding domain inhibits transcription from a WDSV promoter-driven luciferase reporter and mutation of residue V260 within the activation domain diminish this ability as well (Rovnak and Quackenbush, 2002; Rovnak *et al.*, 2005). Further studies demonstrated that the rv-cyclin activation domain directly contacts TATA binding protein-associated factor 9 (TAF9) in human and in walleye cells (Rovnak and Quackenbush, 2006). Mutation of residue V260 not only reduces the function of the activation domain it also prevents interaction with TAF9 and p300/CBP, but not with the Mediator component Sur2 (Rovnak and Quackenbush,

2006). This mutation had no effect on rv–cyclin/cdk8 interaction in immunoprecipitation assays (Rovnak and Quackenbush unpublished). Zhang and Martineau (Zhang and Martineau, 1999) also demonstrated Orf A (rv-cyclin) inhibition of promoter activity and further showed that rv-cyclin inhibits cell growth, however, they attributed this effect to the first 49 amino acids. This effect of rv-cyclin on cell growth has not been substantiated, and work by others indicates that the rv-cyclin exhibits growth promoting activity in a variety of cultured cells (LaPierre *et al.*, 1998a) (Rovnak unpublished).

The WDSV rv-cyclin protein negatively regulates the WDSV promoter in cultured walleye fibroblasts. Presumably, it is advantageous for the virus to have lower levels of gene expression during tumour development to avoid immune surveillance and the cytopathic effects associated with virus production. While rv-cyclin may function to reduce virus expression during tumour growth, the host, environmental, or viral signals that switch on full virus expression have yet to be determined.

The WDSV Orf B protein is also expressed during tumour formation. Orf B localizes to the plasma membrane in structures consistent with focal adhesions and in the cytoplasm with actin stress fibres in explanted tumour cells (Rovnak *et al.*, 2007). An interaction of Orf B with the Receptor for Activated C Kinase (RACK1) was first demonstrated by yeast two-hybrid analysis, then by co-immunoprecipitation (Daniels *et al.*, 2008). RACK1 is an adaptor protein that interacts with activated, conventional isoforms of PKC (α, βI, βII) and targets PKC to intracellular sites of action (McCahill *et al.*, 2002). The interaction of RACK1 with PKC is believed to maintain PKC in an active conformation. Walleye RACK1 exhibits 96% amino acid identity to human and mouse RACK1 and 97% identity to zebrafish RACK1, illustrating the highly conserved nature of this adaptor protein (Daniels *et al.*, 2008). RACK1 interacts with several cellular and viral proteins and is implicated in having an important role in the regulation of cell growth (McCahill *et al.*, 2002). In NIH3T3 cells activated PKCα is translocated to membranes only after PMA treatment. Orf B is found in a complex with PKCα in stable NIH3T3 transfects, and activated PKCα is found in membrane fractions without PMA stimulation and even under serum-deprived conditions (Daniels *et al.*, 2008). Constitutive expression of Orf B resulted in cell survival after treatment with staurosporine, a known inducer of apoptosis. In addition, Orf B-expressing cells were able to survive and proliferate when cultured under serum-deprived conditions (Daniels *et al.*, 2008). Treatment with the PKC inhibitor, bisindolymaleimide hydrochloride, reversed these effects, suggesting that activation of PKC signalling pathways contribute to tumour development.

WDSV Orf C was localized in mitochondria in transiently transfected cells, and this localization correlated with the induction of apoptosis (Nudson *et al.*, 2003). Orf C was also detected in regressing tumour explant cells and observed in regressing tumour sections by immunofluorescence assay (Rovnak *et al.*, 2007). Analysis of regressing tumours by terminal deoxynucleotidyl transferase biotin-dUTP nick end labelling (TUNEL) assay demonstrated that these cells are undergoing apoptosis (Rovnak, unpublished). Orf C colocalized with cytochrome c in mitochondria, which resulted in perinuclear clustering of mitochondria and the inability to retain MitoTracker Orange, a dye that accumulates in active mitochondria (Nudson *et al.*, 2003). Induction of apoptosis is probably due to disruption of the mitochondrial membrane potential in cells expressing Orf C (Nudson *et al.*, 2003). These data suggest that Orf C is targeted to the mitochondria in a process that is responsible for apoptosis in tumour cells and, ultimately, for tumour regression.

Establishment of a cell culture system for WDSV

Infectious virus is only produced during tumour regression in naturally infected fish. The establishment of productively infected walleye cells in culture has recently been achieved. A cell line was established from spring tumour explant cells (STEC) from a single dermal sarcoma collected during the spring spawning run of 2003. The three accessory proteins, rv-cyclin (Orf A), Orf B, and Orf C, and capsid protein were detected by IFA in these cells shortly after explant (Rovnak *et al.*, 2007).

WDSV production was confirmed by banding tissue culture supernatant pellets at 1.17

g/ml density on sucrose gradients and then examination of the banded material by TEM and qRT-PCR with WDSV *gag* primers (Rovnak *et al.*, 2007). The purified virus was also assayed for *in vitro* infection of walleye fibroblast cell line, W12. Proviral DNA was detected in infected cells by PCR with WDSV *gag* primers. Further analysis indicated that there is a significant amount of trypsin-resistant, intracellular, infectious virus. STEC provirus was cloned and sequenced in its entirety. This provirus harbours a total of 27 point mutations resulting in only five amino acid changes (one in gag, three in *env*, and one in orf b) when compared to the original WDSV clone that was isolated in 1990 (Martineau *et al.*, 1992; Rovnak *et al.*, 2007). Sequencing also identified a two bp insertion in *env*, which proved to be an error in the original clone sequence. This correction significantly alters the size and sequence of the large transmembrane domain (Holzschu *et al.*, 1995; Rovnak *et al.*, 2007). Perhaps the most significant conclusion is the overall stability of the WDSV genome over time, which suggests an extremely refined virus–host relationship. The walleye cell line, W12, has also been used to establish lines harbouring WDSV molecular clones (Rovnak and Quackenbush, unpublished). These cells produce infectious virus and also have significant quantities of infectious, intracellular particles.

Bovine leukaemia virus

Classification and genome

Bovine leukaemia virus (BLV) is the aetiological agent of enzootic bovine leukosis, a multicentric cancer of B-lymphocytes in cattle. BLV is a replication-competent, oncogenic member of the genus *Deltaretroviridae* along with primate T-lymphotropic virus (PTLV) 1 [human T-cell leukaemia virus (HTLV) 1 and simian T-cell leukaemia virus (STLV) 1], PTLV 2 (HTLV 2 and STLV 2) and PTLV 3 (HTLV 3 and STLV 3). Phylogenetic analysis of an integrase-encoding region of the *pol* gene indicates that BLV and PTLV diverged from each other millennia ago (Dube *et al.*, 1997). BLV is not derived from an endogenous retrovirus of cattle (Kettmann *et al.*, 1976), but may derive from a heretofore unidentified source.

BLV and HTLV have unusually long R regions that separate the polyadenylation signal located near the end of U3 from the polyadenylation site at the end of R. From its 5′ terminus, the BLV genome contains the *gag, pro, pol,* and *env* genes common to all replication-competent retroviruses. The arrangement of the 3′ end is unique to the deltaretroviruses: following *env* is a non-translated region (> 400 nt), followed in turn by downstream exons of orfs III and IV encoding accessory proteins, then the downstream exon of the regulatory *tax/rex* genes. *pro* and *pol* are translated after successive ribosome frameshifts from the *gag* reading frame. The non-translated region is conserved in length among BLV isolates, suggesting that it serves some important function, although deleting this sequence permits virus replication and tumorigenesis in sheep (Florins *et al.*, 2007). Complete proviral sequences have been obtained from tumours of Japanese, Belgian, and Australian cattle and from peripheral blood cells of an Argentine cow. The annotated NCBI reference genome sequence for BLV is NC_001414. A recent review presents the genome structure and mRNA splicing patterns of BLV, and comprehensively summarizes current knowledge about BLV gene expression and the functions of its encoded proteins (Gillet *et al.*, 2007).

Discovery and Isolation

Enzootic bovine leukosis was first described as a disease of mature dairy cattle in Germany in the 1870s (reviewed in Bendixen, 1965). By the early 1900s, the disease was known to be present continuously within herds but to affect only some of the animals. Hematologic observations revealed that cattle having abnormally high numbers of blood lymphocytes (lymphocytosis) outnumbered those developing leukosis. Patterns of natural and experimental transmission indicated that enzootic leukosis was caused by a blood-borne infectious agent, a conclusion reinforced in the early 1960s when lymphocytosis and leukosis occurred in Swedish dairy herds whose animals had been vaccinated with whole bovine blood to prevent babesiosis. In 1969, C-type retroviral particles were identified in cultures of mitogen-stimulated lymphocytes obtained from peripheral blood of cattle with lymphocytosis or leukosis, as well as from the blood of calves inoculated with tumour tissue or blood from such donors (reviewed in Miller and Van der Maaten, 1982). In 1972, BLV was shown to be infectious in cattle and sheep, and to be oncogenic in sheep. In 1974, reverse transcriptase activity was demonstrated in BLV virions, identifying it as a retrovirus.

The close relationship between BLV and HTLV 1, the cause of adult T-cell leukaemia and lymphoma, was discovered in the mid-1980s when their genomes were sequenced (reviewed in Radke, 1999; Gillet *et al.*, 2007). Although lymphocytes of different lineages are targeted (B-cells by BLV, $CD4^+$ T-cells by HTLV), the viruses share many features of pathogenesis. Both are agents of widespread, natural infection in outbred hosts and establish lifelong, largely latent infections accompanied by high-titre antiviral antibodies. Infected individuals can harbour many infected lymphocytes containing the proviral DNA genome integrated into host cell DNA, but most host cells rarely express viral genes (reviewed in Gillet *et al.*, 2007). Cancers occur after many years in only 1–5% of infected individuals.

Host range

Cattle are the natural hosts for BLV, but domesticated zebu and water buffalo, sheep, and wild capybara (a South American rodent) have also become infected. Sheep, which do not transmit BLV horizontally, are widely used as experimental subjects because of their sensitivity to infection and the high incidence of tumours in this species. Very few infected goats develop tumours. Rabbits can be persistently infected and become immunosuppressed but do not develop tumours. Chimpanzees, rhesus monkeys, deer, pigs, cats and rats can be infected, but do not develop disease. No direct evidence has yet demonstrated that BLV infects humans (reviewed in Gillet *et al.*, 2007), and no epidemiologic evidence has linked ingestion of raw milk from BLV-infected cows with human leukaemia, although unpasteurized milk contains cells capable of transmitting infection to sheep (Ferrer *et al.*, 1981) and BLV capsid protein can be immunostained in cells cultured from raw milk (Buehring *et al.*, 1994). Pasteurization destroys the infectivity of cells in milk.

Cells from many species including bovine, ovine, canine, feline, rodent, and human cells are susceptible to BLV infection in culture. However, most infected cells do not subsequently express viral genes. The cellular receptor for BLV, as yet unidentified, is distinct from receptors for a large number of other retroviruses, including HTLV.

Transmission

BLV infection is worldwide, affecting higher percentages of dairy than beef cattle. Infection of dairy cattle is widespread in the USA as well as many other countries, although a number of European countries have nearly eradicated BLV by culling infected animals. BLV is transmitted when provirus-carrying cells in colostrum, milk, or blood of an infected animal are transferred into the body of a naïve individual. Tracheal and bronchoalveolar washings, nasal secretions and saliva have been shown to contain infective cells, but not to transmit infection in experimental settings. BLV-infected cells are not found in the semen of infected bulls unless contaminating lymphocytes are present. Embryos and ova from BLV-infected cows do not harbour BLV (Hopkins and DiGiacomo, 1997). Nonetheless, many countries prohibit the importation of germplasm from animals in herds containing any BLV-infected animals.

Direct contact among animals facilitates BLV transmission. Sniffing, coughing, and sneezing could aerosolize lymphocytes and licking could transfer salivary lymphocytes to abraded skin, but experimental evidence supports only blood-borne transmission among adults. Preventing iatrogenic blood transfer during vaccination, tattooing, dehorning, and rectal palpation for pregnancy reduces transmission. Cattle are highly sensitive to intradermal infection, so large biting insects, especially in tropical areas, could transfer blood when interrupted during feeding. However, flies have not transmitted BLV in experiments testing normal feeding conditions.

Maternal to offspring transmission occurs, as might be expected from the long co-adaptation of BLV and Bovidae to ensure generation-to-generation propagation of the virus. In utero infection occurs at low frequencies (Hopkins and DiGiacomo, 1997). Perinatal transmission can occur upon exposure of the neonate to infected blood during or shortly after birth, or when it ingests colostrum or milk containing BLV-infected cells (Fig. 14.3) BLV antibodies that are present in colostrum at high concentrations help prevent detectable infection of the calf over several months (Nagy *et al.*, 2007). Long-term maintenance of maternal immune cells in the calf's body (maternal microchimerism) with subsequent activation of virus production is a potential mechanism of BLV transfer that remains to be explored. Sero-epidemiological evidence indicates that breastfeeding far outweighs in utero infection in importance as a mechanism of vertical transmission of HTLV (Furnia *et al.*, 1999). Interestingly, adult T-cell leukaemia occurs mainly in people who were infected with HTLV as infants.

Pathogenesis

Only a small proportion of BLV-infected cattle develop identifiable tumours but BLV infection perturbs the proliferation of many B-lymphocytes (reviewed in Gillet *et al.*, 2007). Most BLV-infected cows seem to be asymptomatic, but infection is detrimental to immune protection against infectious organisms (Trainin *et al.*, 1996; Trainin and Brenner, 2005). Subclinically infected cows produce less milk and are culled prematurely (reviewed in Rhodes *et al.*, 2003). About 30% of infected cattle develop persistent lymphocytosis in which their absolute peripheral blood lymphocyte count is increased stably and significantly above the average for age and breed. B-cells account for most of the excess and some are polyclonally infected with BLV. Not all cattle with persistent lymphocytosis go on to develop tumours and cattle can develop tumours without having persistent lymphocytosis. Host genes, some of which are part of the bovine major histocompatibility complex, influence susceptibility to infection, development of lymphocytosis and formation of tumours (reviewed in Mirsky *et al.*, 1998).

Tumour incidence is very high in sheep, the only experimental animal model for deltaretrovirus-induced tumorigenesis, so BLV infection has been intensively investigated in this species (Gillet *et al.*, 2007). Molecularly cloned DNA can infect sheep by direct injection (Willems *et al.*, 1992) and can induce tumours in sheep (Rovnak *et al.*, 1993), enabling *in vivo* testing of viral mutants (Willems *et al.*, 1993).

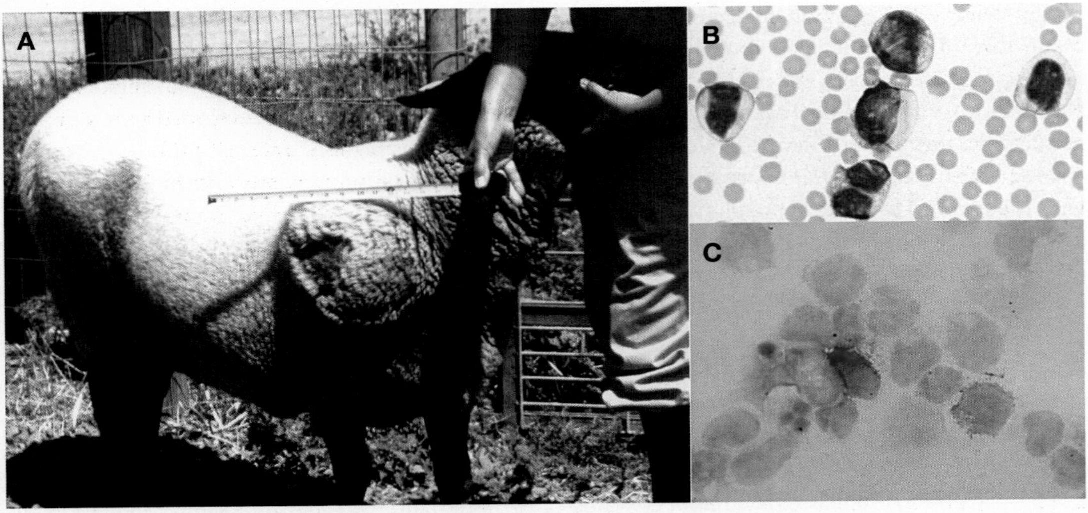

Figure 14.3 (A) Lymphosarcoma on the shoulder of a sheep infected 8 years earlier with BLV derived from the Bat2Cl6 cell line. The clone of lymphoblastic tumour cells also infiltrated multiple organs, including lymph nodes whose architecture was completely effaced by sheets of neoplastic cells. (B) The non-identical twin of sheep A, infected with the same inoculum, developed leukaemia 18 months after infection. (C) Colostral cells from a BLV-positive cow can express BLV in culture. Monocytes and large lymphocytes were enriched on a Percoll gradient, then were cultured overnight, fixed, and immunostained for the BLV capsid protein (red). Finding two positive cells in one field is unusual because only rare colostral cells contain BLV structural proteins after short-term culture.

Infection

When a naïve animal acquires BLV-infected lymphocytes, the allogeneic cells need to produce virus to infect cells of the new host. Where and how this happens during natural transmission remain unknown. When sheep are acutely infected by intradermal and subcutaneous injection, BLV-infected cells disseminate as early as day 4 in lymph emerging from the lymph node draining the region (Fulton *et al.*, 2006). Despite the appearance of virus-specific antibodies in lymph by day 6, BLV spreads to new host cells and is transmitted into a population of blood-borne B-cells that have high densities of surface IgM and do not recirculate through lymph. These $CD5^+$ $CD11b^+$ B-cells (Chevalier *et al.*, 1998) then begin to increase markedly in blood. Increased cell proliferation and survival contributes to transient B-cell lymphocytosis at about 3 weeks after infection, but only a minority of the cells harbour BLV. Serum antibodies become detectable at this time, as do BLV-specific cytotoxic T-lymphocytes (reviewed in Gillet *et al.*, 2007). Prior to and at the time of seroconversion, proviral load increases by virus replication and infection of new host cells. Then, immune selection against infected cells reduces the numbers of infectious centres (Radke *et al.*, 1990) and proviral insertion sites (Pomier *et al.*, 2008) among blood lymphocytes. During this whole time, only rare PBMCs show evidence of BLV expression *in vivo*. However, the persistence of antibodies specific for BLV structural proteins and the intermittent appearance of antibodies specific for the Rex regulatory protein (Powers *et al.*, 1991) indicate that some virus expression occurs throughout life-long infection (Gillet *et al.*, 2007).

Surface IgM-positive B-lymphocytes are the predominant host cells for BLV in the blood of infected sheep and cows. Most are $MHCII^+$ $CD5^+$ $CD11b^+$, although the provirus can also be found in $CD5^-$ $CD11b^-$ B-cells (reviewed in Gillet *et al.*, 2007). Blood monocytes, $CD8^+$ T-cells and granulocytes can contain provirus at very low levels (Schwartz *et al.*, 1994). Cell types other than B-lymphocytes may play important and yet unidentified roles in BLV propagation *in vivo*.

The factors triggering BLV gene expression *in vivo* are unknown. Molecules activating B-cells are effective in culture. BLV expression is governed by both repression and activation. The viral Tax protein, an early product of gene expression along with Rex, activates transcription from the LTR (Derse, 1987) by interacting with the cellular transcription factor CREB, which binds to three 21 bp imperfect repeats in U3 (Adam *et al.*, 1994). E box motifs overlapping the CREs repress basal transcription (Gillet *et al.*, 2007). CREMτ represses Tax interaction with CREB at the viral promoter (Nguyen *et al.*, 2007). Other regulatory sites in U3 are an NF-κB related binding site, a PU box, and a glucocorticoid-response-like element. An E box is located in the R region and an interferon regulatory factor binding site in U5. Epigenetic modifications also play a role in viral transcription (Gillet *et al.*, 2007). The early viral Rex protein regulates transport of unspliced and singly spliced BLV mRNA from the nucleus into cytoplasm (Derse, 1988), which is necessary for synthesis of the proteins that are incorporated into virions. Cellular roles of the accessory proteins are not well-defined. R3 is located in the nucleus and cellular membranes and G4, which interacts with farnesyl pyrophosphate synthetase, is located in the nucleus and in mitochondria (Lefebvre *et al.*, 2002). The genes encoding R3 and G4 are necessary for the development of high viral loads in sheep and for efficient tumour formation. Only 1 of 20 sheep injected with an R3/G4 deletion mutant developed a tumour after a long latency (Florins *et al.*, 2007).

Cytokine profiles are altered in cattle with persistent lymphocytosis. Reduced cell death assists the accumulation of BLV-infected B-cells in the blood (Debacq *et al.*, 2003). In contrast, increased cell proliferation causes lymphocytosis in sheep (Debacq *et al.*, 2002), when the BLV provirus is propagated mainly by division of infected cells (Moules *et al.*, 2005).

Tumours

Tumours develop in only 1–10% of infected cattle during their productive lives, with latencies of 4–8 years after infection. In contrast, most experimentally infected sheep develop tumours at 1 to >7 years (Fig. 14.3). Tumours, termed lymphomas or lymphosarcomas, are multicentric and infiltrating (Theilen and Madewell, 1987), with B-cell characteristics including rearranged Ig heavy chain genes and expression of class II

MHC antigens, plus IgG and CD5 in cows (Wu *et al.*, 1996: Vernau *et al.*, 1997), indicating a more mature phenotype, or IgM without CD5 in sheep (Murakami *et al.*, 1994). The cells resemble lymphoblasts, prolymphocytes, and mature lymphocytes. Histiocytes and macrophages are often found in tumour tissue. The presence of large numbers of neoplastic B-cells in the blood is termed leukaemia.

With tumour growth, cows present a variety of symptoms including decreased milk production, enlarged lymph nodes, loss of appetite and weight, depression, heart problems, fever, diarrhoea, bulging eyes, and posterior paralysis, depending upon which organs are infiltrated with tumour cells (Theilen and Madewell, 1987). Lymph nodes, abomasum, heart, uterus, spinal canal, and orbital cavities are commonly involved and tumour masses can become quite large. Tumours are inevitably fatal; animals usually live only a short time after diagnosis.

Tumours are clonal outgrowths of a single, BLV-infected cell containing one to several copies of the provirus integrated into host cell DNA (Gillet *et al.*, 2007). Integration sites vary in tumours from different individuals, indicating that the virus does not transform cells by insertional activation of a host cell gene. Tumours may contain full-length provirus, but some harbour incomplete proviruses retaining the 3′ tax/rex region of the genome. The long asymptomatic period between infection and development of clonal tumours, together with the low incidence of tumours among infected individuals, suggests that several genetic alterations are required for tumour formation.

Tax is necessary for tumorigenesis in sheep although it is no longer expressed in tumour cells of sheep or cows. Some data indicate that Tax expression must become completely suppressed at some stage during tumorigenesis in sheep (Merimi *et al.*, 2007). Tumorigenesis is initiated when the BLV Tax protein alters host cell gene expression; Tax may be expressed constantly or intermittently in provirus-positive cells (Gillet *et al.*, 2007). In blood-derived sheep B-lymphocytes and in a jejunal Peyer's patch-derived B-cell clone, constitutive Tax expression deregulates a number of genes involved in apoptosis, transcription, DNA repair, cell cycle regulation, and proto-oncogene function (Klener *et al.*, 2006). Tax can bind to cellular proteins (Gillet *et al.*, 2007). Alterations in cellular gene expression appear to render infected cells susceptible to mutations in host genes. Missense mutations in p53 have been identified in half of BLV-induced tumours in cows, but not in sheep, highlighting yet again the fact that BLV interacts differently with this experimental host than it does with its long-adapted bovine host. The events occurring throughout infection in sheep may illustrate an earlier stage of co-evolution of BLV and cattle that eventually led to a less pathogenic relationship.

Conclusions

Retroviruses are associated with a number of neoplastic diseases of vertebrate animals including chickens, fish, rodents, cats, ungulates, primates and, most recently, marsupials. Studies of these viruses will add to our understanding of retroviral biology, pathogenesis, and retroviral evolution as well as contribute to the development of strategies for the diagnosis and control of these pathogens.

Acknowledgements

M.V.E. is grateful for figures composed by Claudia Han. The Quackenbush Lab acknowledges Paul R. Bowser, James W. Casey, Greg Wooster and Rodman Getchell for providing materials and Volker Vogt for rabbit antisera against WDSV proteins and other members of the Quackenbush laboratory for their contributions. This research was supported in part by American Cancer Society grant RPG-00313-01-MBC to S.L.Q. and National Institutes of Health grant CA095056 to S.L.Q. K.R. thanks the members of the Radke laboratory for their many years of work, which was supported by grants from the American Cancer Society (MVV 199) National Institutes of Health (CA 40653 and 46374), and the University of California Cancer Research Coordinating Committee. We apologize to colleagues whose work was not cited in this brief review.

References

Abkowitz, J.L. (1991). Retrovirus-induced feline pure red blood cell aplasia: pathogenesis and response to suramin. Blood 77, 1442–1451.

Adam, E., Kerkhofs, P., Mammerickx, M., Kettmann, R., Burny, A., Droogmans, L., and Willems, L. (1994). Involvement of the cyclic AMP-responsive element binding protein in bovine leukemia virus expression *in vivo*. J. Virol. *68*, 5845–5853.

Albritton, L. M., Tseng, L., Scadden, D., and Cunningham, J. M. (1989). A putative murine ecotropic receptor gene encodes a multiple membrane-spanning protein and confers susceptibility to virus infection. Cell *57*, 659–666.

Anderson, M., Lauring, A. S., Robertson, S., Dirks, C., and Overbaugh, J. (2001). Feline Pit2 functions as a receptor for subgroup B feline leukemia viruses. J. Virol. *75*, 10563–10572.

Anderson, M. M., Lauring, A. S., Burns, C. C., and Overbaugh, J. (2000). Identification of a cellular cofactor required for infection by feline leukemia virus. Science *287*, 1828–1830.

Battini, J. L., Rasko, J. E., and Miller, A. D. (1999). A human cell-surface receptor for xenotropic and polytropic murine leukemia viruses: possible role in G protein-coupled signal transduction. Proc. Natl. Acad. Sci. U.S.A. *96*, 1385–1390.

Bendixen, H. J. 1965. Bovine enzootic leukosis. Adv. Vet. Sci. Comp. Med. *10*, 129–204.

Benveniste, R. E., Callahan, R., Scherr, C. J., Chapman, V., and Todaro, G. J. (1977). Two distinct endogenous type C viruses isolated from the asian rodent Mus cericolor: Conservation of virogene sequences in related reodent species. J. Virol. *21*, 849–862.

Boomer, S., Gasper, P., Whalen, L. R., and Overbaugh, J. (1994). Isolation of a novel subgroup B feline leukemia virus from a cat infected with FeLV-A. Virology *204*, 805–810.

Bowser, P., Wooster, G., Getchell, R., Chen, C.-Y., Sutton, C., and Casey, J. (2002). Naturally occurring invasive walleye dermal sarcoma and attempted experimental transmission of the tumor. J. Aquatic Anim. Hlth *14*, 288–293.

Bowser, P. R., Earnest-Koons, K. A., Wooster, G.A., LaPierre, L.A., Holzschu, D.L., and Casey, J. W. (1998). Experimental transmission of discrete epidermal hyperplasia in walleyes. J. Aquatic Anim. Hlth *10*, 282–286.

Bowser, P. R., Wolfe, M. J., Forney, J. L., and Wooster, G. A. (1988). Seasonal prevalence of skin tumors from walleye *(Stizostedion vitreum)* from Oneida Lake, New York. J. Wildlife Dis. *24*, 292–298.

Bowser, P. R., and Wooster, G. A. (1991). Regression of dermal sarcoma in adult walleyes *(Stizostedion vitreum)*. J. Aquatic Anim. Hlth 3, 147–150.

Bowser, P. R., Wooster, G. A., and Earnest-Koons, K. (1997). Effects of fish age and challenge route in experimental transmission of walleye dermal sarcoma in walleyes by cell-free tumor filtrates. J. Aquatic Anim. Hlth *9*, 274–278.

Bowser, P. R., Wooster, G. A., and Getchell, R. G. (1999). Transmission of walleye dermal sarcoma and lymphocystis via waterborne exposure. J. Aquatic Anim. Hlth *11*, 158–161.

Bowser, P. R., Wooster, G. A., Getchell, R. G., Paul, T. A., Casey, R. N., and Casey, J. W. (2001). Experimental transmission of walleye dermal sarcoma to yellow perch. J. Aquatic Anim. Hlth *13*, 214–219.

Bowser, P. R., Wooster, G. A., Quackenbush, S. L., Casey, R. N., and Casey, J. W. (1996). Comparison of fall and spring tumors as inocula for experimental transmission of walleye dermal sarcoma. J. Aquatic Anim. Hlth *8*, 78–81.

Bromham, L. D. (2002). The human zoo: endogenous retroviruses in the human genome. Trends Ecol. Evol. *17*, 91–97.

Buehring, G. C., P. M. Kramme, and R. D. Schultz. 1994. Evidence for bovine leukemia virus in mammary epithelial cells of infected cows. Lab. Inves.t.*71*, 359–365.

Callahan, R., Meade, C., and Todaro, G. J. (1979). Isolation of an endogenous type C virus related to the infectious primate type C viruses from the asian rodent *Vandeleuria oleracea*. J. Virol. *30*, 124–131.

Callahan, R., and Todaro, G. J. (1978). Four major endogenous retrovirus classes each genetically transmitted in various species of *mus*. In H.C. Morse, ed., Origins of Inbred Mice (Bethesda, MD: Academic Press), pp. 689–712.

Canfield, P. J. (1990). The Biology of the Koala. Sydney: Surrey Beatty and Sons.

Canfield, P. J., Sabine, J. M., and Love, D. N. (1988). Virus particles associated with leukaemia in a koala. Aust. Vet. J. *65*, 327–328.

Chattopadhyay, S. K., Lander, M. R., Gupta, S., Rands, E., and Lowy, D. R. (1981). Origin of mink cytopathic focus-forming (MCF) viruses: comparison with ecotropic and xenotropic murine leukemia virus genomes. Virology *113*, 465–483.

Chevallier, N., M. Berthelemy, D. Le Rhun, V. Laine, D. Levy, and I. Schwartz-Cornil. 1998. Bovine leukemia virus-induced lymphocytosis and increased cell survival mainly involve the CD11b+ B-lymphocyte subset in sheep. J. Virol. *72*, 4413–4420.

Daniels, C. C., Rovnak, J., and Quackenbush, S. L. (2008). Walleye dermal sarcoma virus Orf B functions through receptor for activated C kinase (RACK1) and protein kinase C. Virology *375*, 550–560.

Debacq, C., B. Asquith, P. Kerkhofs, D. Portetelle, A. Burny, R. Kettmann, and L. Willems. 2002. Increased cell proliferation, but not reduced cell death, induces lymphocytosis in bovine leukemia virus-infected sheep. Proc. Natl. Acad. Sci. U.S.A. *99*, 10048–10053.

Debacq, C., B. Asquith, M. Reichert, A. Burny, R. Kettmann, and L. Willems. 2003. Reduced cell turnover in bovine leukemia virus-infected, persistently lymphocytotic cattle. J. Virol. *77*, 13073–13083.

Derse, D. 1987. Bovine leukemia virus transcription is controlled by a virus-encoded *trans*-acting factor and by *cis*-acting response elements. J. Virol. *61*, 2462–2471.

Derse, D. 1988. *trans*-acting regulation of bovine leukemia virus mRNA processing. J. Virol. *62*, 1115–1119.

Donahue, P. R., Quackenbush, S. L., Gallo, M. V., de-Noronha, C. M., Overbaugh, J., Hoover, E. A., and Mullins, J. I. (1991). Viral genetic determinants of T-cell killing and immunodeficiency disease induction by the feline leukemia virus FeLV-FAIDS. J. Virol. *65*, 4461–4469.

Donehower, L. A., Bohannon, R. C., Ford, R. J., and Gibbs, R. A. (1990). The use of primers from highly conserved *pol* regions to identify uncharacterized retroviruses by the polymerase chain reaction. Journal of Virological Methods *28*, 33–46.

Dong, B., Kim, S., Hong, S., Das Gupta, J., Malathi, K., Klein, E. A., Ganem, D., Derisi, J. L., Chow, S. A., and Silverman, R. H. (2007). An infectious retrovirus susceptible to an IFN antiviral pathway from human prostate tumors. Proc. Natl. Acad. Sci. U.S.A. *104*, 1655–1660.

Dube, S., Bachman, S., Spicer, T., Love, J., Choi, D., Esteban, E., Ferrer, J.F. and Poiesz, B.J. (1997). Degenerate and specific PCR assays for the detection of bovine leukaemia virus and primate T-cell leukaemia/lymphoma virus *pol* DNA and RNA: phylogenetic comparisons of amplified sequences from cattle and primates from around the world. J. Gen. Virol. *78*, 1389–1398.

Duncan, I. B. (1978). Evidence for an oncovirus in the swim bladder fibrosarcoma of Atlantic salmon *Salmo salar* L. J. Fish Dis. *1*, 127–131.

Earnest-Koons, K., Wooster, G. A., and Bowser, P. R. (1996). Invasive walleye dermal sarcoma in laboratory-maintained walleyes, *Stizostedion vitreum*. Diseases of Aquatic Organisms *24*, 227–232.

Eglitis, M. A., Eiden, M. V., and Wilson, C. A. (1993). Gibbon ape leukemia virus and the amphotropic murine leukemia virus 4070A exhibit an unusual interference pattern on E36 chinese hamster cells. J. Virol. *67*, 5472–5477.

Ferrer, J. F., S. J. Kenyon, and P. Gupta. 1981. Milk of dairy cows frequently contains a leukemogenic virus. Science *213*, 1014–1016.

Fiebig, U., Hartmann, M. G., Bannert, N., Kurth, R., and Denner, J. (2006). Trans-species transmission of the endogenous koala retrovirus. J. Virol. *80*, 5651–5654.

Fischinger, P. J., Peebles, P. T., Nomura, S., and Haapala, D. K. (1973). Isolation of RD-114-like oncornavirus from a cat cell line. J. Virol. *11*, 978–985.

Florins, A., N. Gillet, M. Boxus, P. Kerkhofs, R. Kettmann, and L. Willems. 2007. Even attenuated bovine leukemia virus proviruses can be pathogenic in sheep. J. Virol. *81*, 10195–10200.

Frerichs, G. N., Morgan, D., Hart, D., Skerrow, C., Roberts, R. J., and Onions, D. E. (1991). Spontaneously productive C-type retrovirus infection of fish cell lines. J. Gen. Virol. 72, 2537–2539.

Fulton, B. E., Jr., M. Portella, and K. Radke. 2006. Dissemination of bovine leukemia virus-infected cells from a newly infected sheep lymph node. J. Virol. *80*, 7873–7884.

Furnia, A., R. Lal, E. Maloney, S. Wiktor, E. Pate, D. Rudolph, D. Waters, W. Blattner, and A. Manns. 1999. Estimating the time of HTLV-I infection following mother-to-child transmission in a breast-feeding population in Jamaica. J. Med. Virol. *59*, 541–546.

Getchell, R. G., Wooster, G. A., Rudstam, L. G., Van De Valk, A. J., Brooking, T. E., and Bowser, P. R. (2000). Prevalence of walleye dermal sarcoma by age-class in walleyes from Oneida Lake, New York. J. Aquatic Anim. Hlth *12*, 220–223.

Getchell, R. G., Wooster, G. A., Rudstam, L. G., Van De Valk, A. J., Brooking, T. E., and Bowser, P. R. (2004). Prevalence of walleye discrete epidermal hyperplasia by age-class in walleyes from Oneida Lake, New York. J. Aquatic Anim. Hlth *16*, 23–28.

Getchell, R. G., Wooster, G. A., Sutton, C. A., Casey, J. W., and Bowser, P. R. (2002). Dose Titration of Walleye Dermal Sarcoma (WDS) Tumor Filtrate. Journal of Aquatic Animal Health *14*, 247–253.

Ghosh, A. K., Bachmann, M. H., Hoover, E. A., and Mullins, J. I. (1992). Identification of a putative receptor for subgroup A feline leukemia virus on feline T-cells. J. Virol. *66*, 3707–3714.

Gillet, N., A. Florins, M. Boxus, C. Burteau, A. Nigro, F. Vandermeers, H. Balon, A. B. Bouzar, J. Defoiche, A. Burny, M. Reichert, R. Kettmann, and L. Willems. 2007. Mechanisms of leukemogenesis induced by bovine leukemia virus: prospects for novel anti-retroviral therapies in human. Retrovirology. *4*, 18.

Goldberg, R. J., Scolnick, E. M., Parks, W. P., Yakovleva, L. A., and Lapin, B. A. (1974). Isolation of a primate type-C virus from a lymphomatous baboon. Int J. Cancer *14*, 722–730.

Han, J.-H., Cannon, P. M., Lai, K.-M., Zhao, Y., Eiden, M. V., and Anderson, W. F. (1997). Identification of envelope protein residues required for the expanded host range of 10A1 murine leukemia virus. J. Virol. *71*, 8103–8108.

Hanger, J., McKee, J., Tarlinton, R., and Yates, A. (2003). Cancer and Haematological Disease in Koalas: A Clinical and Virological Update. Proceedings of the Annual Conference of the Australian Association of Veterinary Conservation Biologists: 19–37.

Hanger, J. J., Bromham, L. D., McKee, J. J., O'Brien, T. M., and Robinson, W. F. (2000). The nucleotide sequence of koala (*Phasocolarctos cinereus*) retrovirus: a novel type C endogenous virus related to gibbon ape leukemia virus. J. Virol. *74*, 4264–4272.

Hardy, W. D., Jr., Old, L. J., Hess, P. W., Essex, M., and Cotter, S. (1973). Horizontal transmission of feline leukaemia virus. Nature *244*, 266–269.

Hart, D., Frerichs, G. N., Rambaut, A., and Onions, D. E. (1996). Complete nucleotide sequence and transcriptional analysis of the snakehead fish retrovirus. J. Virol. *70*, 3606–3616.

Hartley, J. W., and Rowe, W. P. (1976). Naturally occurring murine leukemia viruses in wild mice: characterization of a new 'amphotropic' class. J. Virol. *19*, 19–25.

Hartley, J. W., Wolford, N. K., Old, L. J., and Rowe, W. P. (1977). A new class of murine leukemia virus associated with development of spontaneous lymphomas. Proc. Natl. Acad. Sci. U.S.A. *74*, 789–792.

Hein, S., Prassolov, V., Zhang, Y., Ivanov, D., Lohler, J., Ross, S. R., and Stocking, C. (2003). Sodium-dependent myo-inositol transporter 1 is a cellular receptor for *Mus cervicolor* M813 murine leukemia virus. J. Virol. *77*, 5926–5932.

Holzschu, D. L., Martineau, D., Fodor, S. K., Vogt, V. M., Bowser, P. R., and Casey, J. W. (1995). Nucleotide sequence and protein analysis of a complex piscine retrovirus, walleye dermal sarcoma virus. Journal of Virology *69*, 5320–5331.

Holzschu, D. L., Wooster, G. A., and Bowser, P. R. (1998). Experimental transmission of dermal sarcoma to the sauger *Stizostedion canadense*. Dis. Aquatic Organisms *32*, 9–14.

Hopkins, S. G., and R. F. DiGiacomo. 1997. Natural transmission of bovine leukemia virus in dairy and beef cattle. Vet. Clin. North Am. Food Anim Pract. *13*, 107–128.

Howard, T. M., Sheng, Z., Wang, M., Wu, Y., and Rasheed, S. (2006). Molecular and phylogenetic analyses of a new amphotropic murine leukemia virus (MuLV-1313). Virol J. 3, 101.

Hronek, B. W., Meagher, A., Rovnak, J., and Quackenbush, S. L. (2004). Identification and characterization of *cis*-acting elements residing in the walleye dermal sarcoma virus promoter. J. Virol. *78*, 7590–7601.

Kavanaugh, M. P., and Kabat, D. (1996). Identification and characterization of a widely expressed phosphate transporter/retrovirus receptor family. Kidney Int. *49*, 959–963.

Kawakami, T., Huff, S. D., Buckley, P., Dungworth, D. L., Snyder, S. P., and Gilden, R. V. (1972). C-type virus associated with gibbon lymphosarcoma. Nat. New Biol. *235*, 170–171.

Kawakami, T. G., Buckley, P., Huff, S., McKain, D., and Fielding, H. (1973). A comparative study *in vitro* of a simian virus isolated from spontaneous woolly monkey fibrosarcoma and of a known feline fibrosarcoma virus. Bibl. Haematol. *39*, 236–243.

Kawakami, T. G., Sun, L., and McDowell, T. S. (1977). Infectious primate type-C virus shed by healthy gibbons. Nature *268*, 448–450.

Kawakami, T. G., Sun, L., and McDowell, T. S. (1978). Natural transmission of gibbon leukemia virus. J. Natl. Cancer. I*nst*. *61*, 1113–1115.

Kettmann, R., D. Portetelle, M. Mammerickx, Y. Cleuter, D. Dekegel, M. Galoux, J. Ghysdael, A. Burny, and H. Chantrenne. 1976. Bovine leukemia virus: an exogenous RNA oncogenic virus. Proc. Natl. Acad. Sci. U.S.A. *73*, 1014–1018.

Klener, P., M. Szynal, Y. Cleuter, M. Merimi, H. Duvillier, F. Lallemand, C. Bagnis, P. Griebel, C. Sotiriou, A. Burny, P. Martiat, and A. Van den Broeke. 2006. Insights into gene expression changes impacting B-cell transformation: cross-species microarray analysis of bovine leukemia virus tax-responsive genes in bovine B cells. J. Virol. *80*, 1922–1938.

Kozak, C. A. (1985). Susceptibility of wild mouse cells to exogenous infection with xenotropic leukemia viruses: control by a single dominant locus on chromosome 1. J. Virol. *55*, 690–695.

Kozak, C. A., and O'Neill, R. R. (1987). Diverse wild mouse origins of xenotropic, mink cell focus-forming, and two types of ecotropic proviral genes. J. Virol. *61*, 3082–3088.

Lairmore, M. D., Stanley, J. R., Weber, S. A., and Holzschu, D. L. (2000). Squamous epithelial proliferation induced by walleye dermal sarcoma retrovirus cyclin in transgenic mice. Proc. Natl. Acad. Sci. U.S.A. *97*, 6114–6119.

Lander, M. R., and Chattopadhyay, S. K. (1984). A *Mus dunni* cell line that lacks sequences closely related to endogenous murine leukemia viruses and can be infected by ectropic, amphotropic, xenotropic, and mink cell focus-forming viruses. J. Virol. *52*, 695–698.

LaPierre, L. A., Casey, J. W., and Holzschu, D. L. (1998a). Walleye retroviruses associated with skin tumors and hyperplasias encode cyclin D homologs. Journal of Virology *72*, 8765–8771.

LaPierre, L. A., Holzschu, D. L., Bowser, P. R., and Casey, J. W. (1999). Sequence and transcriptional analyses of the fish retroviruses walleye epidermal hyperplasia virus types 1 and 2: Evidence for a gene duplication. J. Virol. *73*, 9393–9403.

LaPierre, L. A., Holzschu, D. L., Wooster, G. A., Bowser, P. R., and Casey, J. W. (1998b). Two closely related but distinct retroviruses are associated with walleye discrete epidermal hyperplasia. J. Virol. *72*, 3484–3490.

Lavillette, D., Marin, M., Ruggieri, A., Mallet, F., Cosset, F.-L., and Kabat, D. (2002). the envelope glycoprotein of human endogenous retrovirus type W Uses a divergent family of amino acid transporters/cell surface receptors. J. Virol. *76*, 6442–6452.

Lefebvre, L., Ciminale, V. Vanderplasschen, A., D'Agostino, D., Burny, A., Willems, L., and Kettmann, R. (2002). Subcellular localization of the bovine leukemia virus R3 and G4 accessory proteins. J. Virol. *76*, 7843–7854.

Levy, J. A. (1975). Xenotropic C-type viruses and autoimmune disease. J. Rheumatol. *2*, 135–148.

Levy, J. A. (1977). Murine xenotropic type C viruses III. Phenotypic mixing with avian leukosis and sarcoma viruses. Virology 77, 811–825.

Lieber, M., Scherr, C. J., Todaro, G. J., Benveniste, R. E., Callahan, R., and Coon, H. G. (1975). Isolation from the asian mouse Mus Caroli of an endogenous type C virus related to infectious primate type C viruses. Proc. Natl. Acad. Sci. U.S.A. *72*, 2315–2319.

Lyu, M. S., Nihrane, A., and Kozak, C. A. (1999). Receptor-mediated interference mechanism responsible for resistance to polytropic leukemia viruses in mus castaneus. J. Virol. *73*, 3733–3736.

McAllister, R. M., Nicolson, M., Gardner, M. B., Rongey, R. W., Rasheed, S., Sarma, P. S., Huebner, R. J., Hatanaka, M., Oroszlan, S., Gilden, R. V., *et al.* (1972). C-type virus released from cultured human rhabdomyosarcoma cells. Nat. New Biol. *235*, 3–6.

McCahill, A., Warwicker, J., Bolger, G. B., Houslay, M. D., and Yarwood, S. J. (2002). The RACK1 scaffold protein: a dynamic cog in cell response mechanisms. Mol Pharmacol *62*, 1261–1273.

McKnight, I. J. (1978). Sarcoma of the swim bladder of Atlantic salmon (*Salmo salar* L.). Aquaculture *13*, 55–60.

Mann, R., Mulligan, R. C., and Baltimore, D. (1983). Construction of a retrovirus packaging mutant and its use to produce helper-free defective retrovirus. Cell *33*, 153–159.

Marin, M., Tailor, C. S., Nouri, A., and Kabat, D. (2000). Sodium-dependent neutral amino acid transporter type 1 is an auxiliary receptor for baboon endogenous retrovirus. J. Virol. *74*, 8085–8093.

Martin, J., Herniou, E., Cook, J., O'Neill, R. W., and Tristem, M. (1999). Interclass transmission and phyletic host tracking in murine leukemia virus-related retroviruses. J. Virol. *73*, 2442–2449.

Martineau, D., Bowser, P. R., Renshaw, R. R., and Casey, J. W. (1992). Molecular characterization of a unique retrovirus associated with a fish tumor. Journal of Virology *66*, 596–599.

Martineau, D., Bowser, P. R., Wooster, G. A., and Armstrong, G. A. (1990a). Experimental transmission of a dermal sarcoma in fingerling walleyes (*Stizostedion vitreum vitreum*). Vet. Pathol. *27*, 230–234.

Martineau, D., Bowser, P. R., Wooster, G. A., and Forney, J. L. (1990b). Histologic and ultrastructural studies of dermal sarcoma of walleye (Pisces: *Stizostedion vitreum*). Vet. Pathol. *27*, 340–346.

Martineau, D., Renshaw, R., Williams, J. R., Casey, J. W., and Bowser, P. R. (1991). A large unintegrated retrovirus DNA species present in a dermal tumor of walleye *Stizostedion vitreum*. Diseases of Aquatic Organisms *10*, 153–158.

Merimi, M., P. Klener, M. Szynal, Y. Cleuter, C. Bagnis, P. Kerkhofs, A. Burny, P. Martiat, and A. Van den Broeke. 2007. Complete suppression of viral gene expression is associated with the onset and progression of lymphoid malignancy: observations in bovine leukemia virus-infected sheep. Retrovirology. *4*, 51.

Miller, A. D., Bonham, L., Alfano, J., Kiem, H. P., Reynolds, T., and Wolgamot, G. (1996). A novel murine retrovirus identified during testing for helper virus in human gene transfer trials. J. Virol. *70*, 1804–1809.

Miller, A. D., and Wolgamot, G. (1997). Murine retroviruses use at least six different receptors for entry into Mus dunni cells. J. Virol. *71*, 4531–4535.

Miller, D. G., Edwards, R. H., and Miller, A. D. (1994). Cloning of the cellular receptor for amphotropic murine retroviruses reveals homology to that for gibbon ape leukemia virus. Proc. Natl. Acad. Sci. U.S.A. *91*, 78–82.

Miller, J. M., and M. J. Van der Maaten. 1982. Bovine leukosis– its importance to the dairy industry in the United States. J. Dairy Sci. *65*, 2194–2203.

Mirsky, M. L., C. Olmstead, Y. Da, and H. A. Lewin. 1998. Reduced bovine leukaemia virus proviral load in genetically resistant cattle. Anim. Genet. *29*, 245–252.

Moloney, J. B. (1960a). Biological studies on a lymphoid-leukemia virus extracted from sarcoma 37. I. Origin and introductory investigations. J. Natl. Cancer. I*nst.* *24*, 933–951.

Moloney, J. B. (1960b). Properties of a leukemia virus. Natl .Cancer Inst. Monogr. *4*, 7–37.

Moules, V., C. Pomier, D. Sibon, A. S. Gabet, M. Reichert, P. Kerkhofs, L. Willems, F. Mortreux, and E. Wattel. 2005. Fate of premalignant clones during the asymptomatic phase preceding lymphoid malignancy. Cancer Res. *65*, 1234–1243.

Murakami, K., K. Okada, Y. Ikawa, and Y. Aida. 1994. Bovine leukemia virus induces CD5- B cell lymphoma in sheep despite temporarily increasing CD5$^+$ B cells in asymptomatic stage. Virology *202*, 458–465.

Naar, A. M., Lemon, B. D., and Tjian, R. (2001). Transcriptional coactivator complexes. Annual Reviews of Biochemistry *70*, 475–501.

Nagy, D. W., J. W. Tyler, and S. B. Kleiboeker. 2007. Decreased periparturient transmission of bovine leukosis virus in colostrum-fed calves. J. Vet. Intern. Med. *21*, 1104–1107.

Nguyen, T. L., S. de Walque, E. Veithen, A. Dekoninck, V. Martinelli, Y. de Launoit, A. Burny, R. Harrod, and C. Van Lint. 2007. Transcriptional regulation of the bovine leukemia virus promoter by the cyclic AMP-response element modulator tau isoform. J. Biol. Chem. *282*, 20854–20867.

Nudson, W. A., Rovnak, J., Buechner, M., and Quackenbush, S. (2003). Walleye dermal sarcoma virus Orf C is targeted to the mitochondria. J. Gen. Virol. *84*, 375–381.

O'Neill, R. R., Hartley, J. W., Repaske, R., and Kozak, C. A. (1987). Amphotropic proviral envelope sequences are absent from the Mus germ line. J. Virol. *61*, 2225–2231.

Oliveira, N. M., Farrell, K. B., and Eiden, M. V. (2006). In vitro characterization of a koala retrovirus. J. Virol. *80*, 3104–3107.

Oliveira, N. M., Satija, H., Kouwenhoven, I. A., and Eiden, M. V. (2007). Changes in viral protein function that accompany retroviral endogenization. Proc. Natl. Acad. Sci. U.S.A. *104*, 17506–17511.

Ott, D., Friedrich, R., and Rein, A. (1990). Sequence analysis of amphotropic and 10A1 murine leukemia viruses: close relationship to mink cell focus-inducing viruses. J. Virol. *64*, 757–766.

Overbaugh, J., Miller, A. D., and Eiden, M. V. (2001). Receptors and entry cofactors for retroviruses include single and multiple transmembrane-spanning proteins as well as newly described glycophosphatidylinositol-anchored and secreted proteins. Microbiol. Mol. Biol. Rev. *65*, 371–389.

Overbaugh, J., Reidel, N., Hoover, E. A., and Mullins, J. I. (1988). Transduction of endogenous envelope genes by feline leukemia virus *in vitro*. Nature *332*, 731–734.

Paul, T. A., Quackenbush, S. L., Sutton, C., Casey, R. N., Bowser, P. R., and Casey, J. W. (2006). Identification and characterization of an exogenous retrovirus from atlantic salmon swim bladder sarcomas. J. Virol. *80*, 2941–2948.

Pomier, C., M. T. Alcaraz, C. Debacq, A. Lancon, P. Kerkhofs, L. Willems, E. Wattel, and F. Mortreux. 2008. Early and transient reverse transcription during primary deltaretroviral infection of sheep. Retrovirology. *5*, 16.

Poulet, F. M., Bowser, P. R., and Casey, J. W. (1994). Retroviruses of fish, reptiles, and molluscs, In The Retroviridae, Levy, J. A. , ed. (New York: Plenum Press) pp. 1–38.

Poulet, F. M., Vogt, V. M., Bowser, P. R., and Casey, J. W. (1995). In situ hybridization and immunohistochemical study of walleye dermal sarcoma virus (WDSV) nucleic acids and proteins in spontaneous sarcomas of adult walleyes (*Stizostedion vitreum*). Vet. Pathol. 32, 162–172.

Powers, M. A., D. Grossman, L. C. Kidd, and K. Radke. 1991. Episodic occurrence of antibodies against the bovine leukemia virus Rex protein during the course of infection in sheep. J. Virol. *65*, 4959–4965.

Prassolov, V., Ivanov, D., Hein, S., Rutter, G., Munk, C., Lohler, J., and Stocking, C. (2001). The Mus cervicolor MuLV isolate M813 is highly fusogenic and induces a T-cell lymphoma associated with large multinucleated cells. Virology *290*, 39–49.

Quackenbush, S.L., Holzschu, D.L., Bowser, P.R., and Casey, J.W. (1997). Transcriptional analysis of walleye dermal sarcoma virus (WDSV). Virology *237*, 107–112.

Quigley, J., Burns, C., Anderson, M., Lynch, E., Sabo, K., Overbaugh, J., and Abkowitz, J. (2000). Cloning of the cellular receptor for feline leukemia virus subgroup C (FeLV-C), a retrovirus that induces red cell aplasia. Blood *95*, 1093–1099.

Radke, K., Grossman, D., and Kidd, L.C. (1990). Humoral immune response of experimentally infected sheep defines two early periods of bovine leukemia virus replication. Microb. Pathog. *9*, 159–171.

Radke, K. 1999. Bovine leukemia virus: general features; structure and molecular biology, pp. 191–198. *In* Webster, R.G. and Granoff, A. (eds.), Encyclopedia of Virology. Academic Press, London.

Rasheed, S., Gardner, M.B., and Chan, E. (1976). Amphotropic host range of naturally occuring wild mouse leukemia viruses. J. Virol. *19*, 13–18.

Rasheed, S., Pal, B.K., and Gardner, M. B. (1982). Characterization of a highly oncogenic murine leukemia virus from wild mice. Int J. Cancer *29*, 345–350.

Rasko, J.E.J., Battini, J.-L., Gottschalk, R.J., Mazo, I., and Miller, A.D. (1999). The RD114/simian type D retrovirus receptor is a neutral amino acid transporter. Proc. Natl. Acad. Sci. U.S.A. *96*, 2129–2134.

Reitz, M.S., Wong-Staal, F., Haseltine, W. A., Klein, D. G., Trainor, C.D., Gallagher, R. E., and Gallo, R.C. (1979). Gibbon ape leukemia virus-Hall's Island: Strain of gibbon ape leukemia virus. J. Virol. *29*, 395–400.

Rhodes, J.K., Pelzer, K.D., and Y.J. Johnson. 2003. Economic implications of bovine leukemia virus infection in mid-Atlantic dairy herds. J. Am. Vet. Med. Assoc. *223*, 346–352.

Riedel, N., Hoover, E., Dornsife, R.E., and Mullins, J.I. (1988). Pathogenic and host range determinants of the feline aplastic anemia virus. Proc. Natl. Acad. Sci. U.S.A. *85*, 2758–2762.

Rohn, J.L., Moser, M.S., Gwynn, S.R., Baldwin, D.N., and Overbaugh, J. (1998). In vivo evolution of a novel, syncytium-inducing and cytopathic feline leukemia virus variant. J. Virol. *72*, 2686–2696.

Rovnak, J., Boyd, A.L., Casey, J.W., Gonda, M.A., Jensen, W.A., and Cockerell, G.L. (1993). Pathogenicity of molecularly cloned bovine leukemia virus. J. Virol. *67*, 7096–7105.

Rovnak, J., Casey, J. W., and Quackenbush, S. L. (2001). Intracellular targeting of walleye dermal sarcoma virus Orf A (rv-cyclin). Virology *280*, 31–40.

Rovnak, J., Casey, R.N., Brewster, C.D., Casey, J.W., and Quackenbush, S.L. (2007). Establishment of productively infected walleye dermal sarcoma explant cells. J. Gen. Virol. *88*, 2583–2589.

Rovnak, J., Hronek, B.W., Ryan, S.O., Cai, S., and Quackenbush, S. L. (2005). An activation domain within the walleye dermal sarcoma virus retroviral cyclin protein is essential for inhibition of the viral promoter. Virology *342*, 240–251.

Rovnak, J., and Quackenbush, S.L. (2002). Walleye dermal sarcoma virus cyclin interacts with components of the Mediator complex and the RNA polymerase II holoenzyme. J. Virol. *76*, 8031–8039.

Rovnak, J., and Quackenbush, S L. (2006). Walleye dermal sarcoma virus retroviral cyclin directly contacts TAF9. J. Virol. *80*, 12041–12048.

Schneiderman, R. D., Farrell, K.B., Wilson, C.A., and Eiden, M.V. (1996). The Japanese feral mouse PiT1 and PiT2 homologs lack an acidic residue at position 550 but still function as gibbon ape leukemia virus receptors: Implications for virus binding motif. J. Virol. *70*, 6982–6986.

Schwartz, I., Bensaid, A., Polack, B., Perrin, B., Berthelemy, M., and Levy, D. (1994). In vivo leukocyte tropism of bovine leukemia virus in sheep and cattle. J. Virol. *68*, 4589–4596.

Shen, C. H., and Steiner, L. A. (2004). Genome structure and thymic expression of an endogenous retrovirus in zebrafish. J. Virol. *78*, 899–911.

Sommerfelt, M. A., and Weiss, R. A. (1990). Receptor interference groups of 20 retroviruses plating on human cells. Virology *176*, 58–69.

Spain, B H., Koo, D., Ramakrishnan, M., Dzudzor, B., and Colicelli, J. (1995). Truncated forms of a novel yeast protein suppress the lethality of a G protein alpha subunit deficiency by interacting with the beta subunit. J. Biol. Chem. *270*, 25435–25444.

Sutcliffe, J.G., Shinnick, T.M., Verma, I.M., and Lerner, R.A. (1980). Nucleotide sequence of Moloney leukemia virus: 3′ end reveals details of replication, analogy to bacterial transposons, and an unexpected gene. Proc. Natl. Acad. Sci. U.S.A. *77*, 3302–3306.

Tailor, C.S., Lavillette, D., Marin, M., and Kabat, D. (2003). Cell surface receptors for gammaretroviruses. Curr Top Microbiol Immunol *281*, 29–106.

Tailor, C.S., Nouri, A., Lee, C. G., Kozak, C., and Kabat, D. (1999a). Cloning and characterization of a cell surface receptor for xenotropic and polytropic murine leukemia viruses. Proc. Natl. Acad. Sci. U.S.A. *96*, 927–932.

Tailor, C. S., Willett, B. J., and Kabat, D. (1999b). A putative cell surface receptor for anemia-inducing feline leukemia virus subgroup C is a member of a transporter superfamily. J. Virol. *73*, 6500–6505.

Takeuchi, Y., Vile, R. G., Simpson, G., O'Hara, B., Collins, M. K., and Weiss, R. A. (1992). Feline leukemia virus subgroup B uses the same cell surface receptor as gibbon ape leukemia virus. J. Virol. *66*, 1219–1222.

Tarlinton, R. E., Meers, J., and Young, P. R. (2006). Retroviral invasion of the koala genome. Nature *442*, 79–81.

Theilen, G. H., and Madewell, B. R. 1987. Bovine hematopoietic neoplasms, sarcomas and related conditions, p. 408–430. *In* Theilen, G. H. and Madewell, B. R. (eds.), Veterinary Cancer Medicine. Lea & Febiger, Philadelphia.

Torres-Zamorano, V., Leibach, F. H., and Ganapathy, V. (1998). Sodium-dependent homo- and hetero-exchange of neutral amino acids mediated by the amino acid transporter ATB degree. Biochem. Biophys. Res. Commun. *245*, 824–829.

Trainin, Z., Brenner, J. , Meirom, R. , and Ungar-Waron, H. 1996. Detrimental effect of bovine leukemia virus (BLV) on the immunological state of cattle. Vet. Immunol. Immunopathol. *54*, 293–302.

Trainin, Z., and J. Brenner. 2005. The direct and indirect economic impacts of bovine leukemia virus infection on dairy cattle. Israel J. Vet. Med. *60*, 94–105.

Urisman, A., Molinaro, R. J., Fischer, N., Plummer, S. J., Casey, G., Klein, E. A., Malathi, K., Magi-Galluzzi, C., Tubbs, R. R., Ganem, D., *et al.* (2006). Identification of a novel Gammaretrovirus in prostate tumors of patients homozygous for R462Q RNASEL variant. PLoS Pathog. *2*, e25.

Vernau, W., R. M. Jacobs, V. E. Valli, and J. L. Heeney. 1997. The immunophenotypic characterization of bovine lymphomas. Vet. Pathol. *34*, 222–225.

Vobis, M., D'Haese, J., Mehlhorn, H., and Mencke, N. (2003). Evidence of horizontal transmission of feline leukemia virus by the cat flea (*Ctenocephalides felis*). Parasitol. Res. *91*, 467–470.

Walker, R. (1969). Virus associated with epidermal hyperplasia in fish. Natl. Cancer Instit. Monog. *31*, 195–207.

Willems, L., Kettmann, R., Dequiedt,.F., Portetelle, D., Vonèche, V., Cornil, I., Kerkhofs, P., Burny, A., and Mammerickx, M. (1993). In vivo infection of sheep by bovine leukemia virus mutants. J. Virol. *67*, 4078–4085.

Willems, L., Portetelle, D., Kerkhofs, P., Chen, G., Burny, A., Mammerickx, M., and Kettmann, R. (1992). In vivo transfection of bovine leukemia provirus into sheep. Virology *189*, 775–777.

Wilson, C.A., Eiden, M. V., Anderson, W.B., Lehel, C., and Olah, Z. (1995). The dual-function hamster receptor for amphotropic murine leukemia virus, 10A1 MuLV and gibbon ape leukemia virus is a phosphate symporter. J. Virol. *69*, 534–537.

Wolgamot, G., and Miller, A. D. (1999). Replication of *Mus dunni* endogenous retrovirus depends on promoter activation followed by enhancer multimerization. J. Virol. 73, 9803–9809.

Wu, D., K. Takahashi, K. Murakami, K. Tani, A. Koguchi, M. Asahina, M. Goryo, Y. Aida, and K. Okada. 1996. B-1a, B-1b and conventional B cell lymphoma from enzootic bovine leukosis. Vet. Immunol. Immunopathol. 55, 63–72.

Yamamoto, T., Kelly, R.K., and Nielsen, O. (1985a). Epidermal hyperplasia of walleye, *Stizostedion vitreum vitreum* (Mitchill), associated with retrovirus-like type-C particles: prevalence, histologic, and electron microscopic observations. J. Fish Dis. *19*, 425–436.

Yamamoto, T., Kelly, R.K., and Nielsen, O. (1985b). Morphological differentiation of virus-associated skin tumors of walleye (*Stizostedion vitreum vitreum*). Fish Pathol. *20*, 361–372.

Yamamoto, T., MacDonald, R.D., Gillespie, D.C., and Kelly, R.K. (1976). Viruses associated with lymphocystis and dermal sarcoma of walleye (*Stizostedion vitreum vitreum*). J. Fish Res. Board Can. 33, 2408–2419.

Yang, Y., Guo, L., Xu, S., Holland, C.A., Kitamura, T., Hunter, K., and Cunningham, J. (1999). Receptors for polytropic and xenotropic mouse leukemia viruses encoded by a single gene at RMC1. Nat. Genet. *21*, 216–219.

Zeijl, M.V., Johann, S.V., Cross, E., Cunningham, J., Eddy, R., Shows, T.B., and O'Hara, B. (1994). An amphotropic virus receptor is a second member of the gibbon ape leukemia virus receptor family. Proc. Natl. Acad. Sci. U.S.A. *91*, 1168–1172.

Zhang, Z., Kim, E., and Martineau, D. (1999). Functional characterization of a piscine retroviral promoter. J. Gen. Virol. *80*, 3065–3072.

Zhang, Z., and Martineau, D. (1999). Walley dermal sarcoma virus: OrfA N-terminal end inhibits the activity of a reporter gene directed by eukaryotic promoters and has a negative effect on the growth of fish and mammalian cells. J. Virol. 73, 8884–8889.

Simian Exogenous Retroviruses

Jonathan Luke Heeney and Ernst J. Verschoor

15

Abstract

Simians include diverse species of monkeys which are globally distributed predominantly in the southern hemisphere, and ape species that are restricted to the rainforests of central Africa and Southeast Asia. Types of simian retroviruses have been detected in almost all species of non-human primates which have been studied. As early identification and classification of primate retroviruses was largely possible due to the efforts to establish cell lines from various primate species, there were limitations. Firstly, the search for simian retroviruses was frequently restricted to only a few species cell lines. Secondly, not all retroviruses were permissive to the available cell lines due to species-specific restriction or cell-type specific factors. The earliest discoveries of simian retroviruses were often based on specific observations which arose from diagnostic workups of cases of then undefined illness or unusual cancers.

This chapter focuses on the four main groups of exogenous simian retroviruses; type D simian retroviruses (SRV), simian foamy viruses (SFV) commonly known as spumaviruses, the simian T-cell lymphotropic viruses (STLV), and the expanding group of simian immunodeficiency viruses (SIV). Initially viruses in these subgroups of simian retroviruses were identified based on the diseases from which they were associated with. Retroperitoneal fibrosarcomas and chronic wasting disease led to the early identification of the SRVs, found to be associated with a spectrum of diseases in different species of macaques. Other simian retroviruses are frequently asymptomatic in their natural hosts and in some cases simply cause cytopathic effect (CPE) in cell culture. In the last two decades molecular techniques have provided us with much more insight into the phylogeny of the wide variety of diverse retroviruses, knowledge which has enriched the seroprevalence evidence of widespread retroviral infections in many different primate species.

Overview of simian retroviruses

Classification

The family of *Retroviridae* consists of two subfamilies, the *Orthoretrovirinae* and the *Spumaretrovirinae*. The first contain six genera, *Alpha-, Beta, Gamma, Delta, Epsilonretroviruses*, and the *Lentiviruses*. The *Spumaretrovirinae* consist of only one genus, that of the *Spumaviruses*. The four exogenous simian retroviruses which form the focus of this chapter include the simian type D retroviruses (SRV), spumaviruses or foamy viruses (SFV), T-cell lymphotropic viruses (STLV), and the lentiviruses (SIV).

Within the *Retroviridae* family, there are yet no simian viruses which have been classified within the *Alpharetrovirus* and *Epsilonretrovirus* genera. The *Betaretrovirus* genus includes the simian type D retroviruses (SRV types 1 to 5), as well as the squirrel monkey retrovirus, and the langur retrovirus. Within the *Gammaretrovirus* genus, simian retroviruses include the gibbon ape leukaemia virus (GaLV), the baboon endogenous retrovirus (BaEV), and the woolly monkey sarcoma virus (WMSV) or simian sarcoma-associated virus (SSAV). These particular oncogenic retroviruses will be discussed only in a historical context.

Simian retroviruses within the *Deltaretrovirus* genus include the transacting T-cell lymphotropic viruses. These have previously been grouped with the oncogenic transacting C-type retroviruses, which were characterized by the oncogenic prototypes human T-cell lymphotropic viruses HTLV-1 and bovine leukaemia virus (BLV). HTLV and STLV are collectively known as primate T-cell lymphotropic viruses (PTLV). Formally there are three recognized groups of PTLV viruses (types 1–3), but new members have been recently proposed (Van Dooren *et al.*, 2007). In general, most STLVs have not been observed to be convincingly oncogenic in their natural hosts (Fultz, 1994), based on the definition of clonally integrated provirus in tumours (Heeney and Valli, 1990). The *Lentivirus* genus contains an expanding number of primate retroviruses which include the human immunodeficiency type 1 and type 2 viruses, as well as a diverse and growing list of simian immunodeficiency viruses (SIV). Primate lentiviruses are further categorized on the basis of their genetic organization. They represent the most well-studied of the primate retroviruses (Peeters *et al.*, 2002).

The *Spumavirus (foamy virus)* genus contains a growing number of non-human primate foamy viruses. This group includes the formerly termed 'human foamy virus (HFV)' which has now been reclassified as a chimpanzee foamy virus on the basis of sequence analysis (Brown *et al.*, 1978; Herchenroder *et al.*, 1994). This genus includes a large number of foamy viruses most of which have been identified in different primate species by polymerase chain reaction (PCR), but for which full genome sequence and culturable virus isolates have not yet been obtained.

Structure and morphology

Historically, as with other viruses, retroviruses were classified on the basis of their morphological characteristics as seen under the electron microscope. This included the morphological development of virions within infected cells and the extracellular virion structure once it had budded from and left an infected cell (Coffin, 1991).

The D-type simian retroviruses, or SRVs, differ characteristically in terms of intracellular morphogenesis. They possess 70- to 90-nm-diameter intracytoplasmic A-type particles. Interestingly, ring shaped structures bud at the cytoplasmic membrane where the envelope is acquired. This results in mature extracellular viral particles which are about 125 nm in diameter. The core differs from a lentivirus core by being less conical and more closely represents a cylinder within the envelope (Fine and Schochetman, 1978).

The simian foamy viruses (SFV) are distinguished by the prominent glycoprotein spikes which are even visible within intracytoplasmic vacuoles. The foamy viruses mature similarly to that of SRVs, but virus particles can also mature by budding through the cytoplasmic membrane as well as into intracytoplasmic vacuoles. They also possess a cylindrical core similar to the SRVs.

The STLVs and the simian lentiviruses have similarities in their morphology within an infected cell and have been morphologically classified as type C retroviruses because capsid assembly occurs on the inner side of the host cell membrane. As the viral capsid assembles, it interacts with the envelope proteins and buds from the cytoplasmic membrane in a thickened half moon or C shape to ultimately form extracellular virus. However, extracellularly these viruses differ in the shape of their core structure. For instance, mature HIV or SIV particles are 125–160 nm in diameter and have virion cores which are oblong and conical at one end. STLV particles on the other hand although similar in diameter, have a core which is morphologically distinct from lentiviruses (Fultz, 1994).

Genetic organization

In addition to morphological differences, the simian retroviruses can also be classified on the basis of the composition of their complex genomes. As with other retroviruses, they contain *gag, pol,* and *env* genes but also possess additional accessory genes (Fig. 15.1). The simplest of the four exogenous simian retroviruses in terms of its genetic composition are the SRVs of the *Betaretroviridae* genus. In addition to *gag, pol,* and *env* they possess a separate protease gene (*pro*) which is not an accessory gene. The only accessory gene they do possess is located at the 3′-end of the *env* gene and is called *sorf* (Fig. 15.1A). The second least complex, but with the longest genome are the simian foamy viruses which, in addition to *gag, pol,* and *env* genes, have two additional open

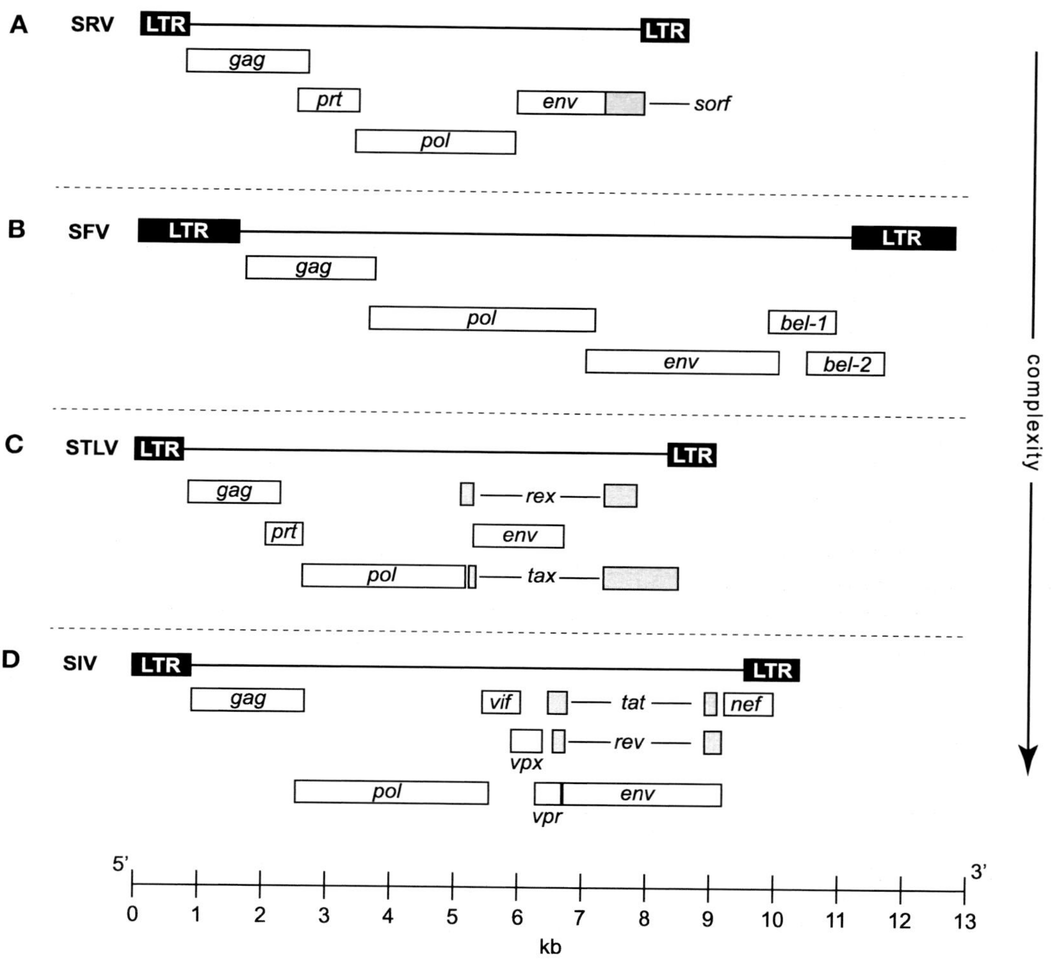

Figure 15.1 The genetic organization and relative genetic size of the proviruses of four main groups of exogenous primate retroviruses: (A) those of the *Betaretrovirus* genus include the simian retroviruses (SRV types 1 to 5), commonly referred to the D-type retroviruses; (B) *Spumavirus* genus (*foamy virus)* representing the largest genomes of the simian retroviruses; (C) the *Deltaretrovirus* genus representing the more complex genomes of the simian T-cell lymphotropic retroviruses; and (D) the complex genomes of *Lentivirus* genus representing the simian immunodeficiency like-viruses (SIVs) which are further sub-divided based on the composition of their accessory genes.

reading frames (ORFs) called *bel-1* and *bel-2* that are located 3′ to the *env* gene (Fig. 15.1B). The primate T-lymphotropic viruses also contain two accessory genes, known as *tax* and *rex*, both of which are required for virus replication. In addition, there are two other potential ORFs in the genome awaiting study and confirmation. Similar to the SRVs, STLVs encode a separate protease gene (*pro*) which is not encoded by the *pol* gene and is not an accessory gene. (Fig. 15.1C). The lentiviruses are the most genetically complex retroviruses with up to six accessory genes. It is the genetic composition of these different accessory genes which forms the basis for further classification of the primate lentiviruses.

Replication

The exogenous simian retroviruses all have accessory genes that function in the regulation of virus replication. Some of these accessory genes are absolutely essential for replication, for instance *tax* and *rev* of the STLVs. Others have accessory genes that enhance replication particularly in specific cell types. The complex genomes of the primate lentiviruses suggest that their replication is highly regulated for optimal replication kinetics in specific cell types or for modulating replication or for linking replication to specific differentiation states of certain cell types. For specific details readers are referred to earlier chapters in this series, and the following reviews (Bindhu *et al.*, 2004; Linial, 1999; Stevenson, 2003).

Early history of simian retroviruses

The first simian retrovirus that was characterized is the simian foamy virus. SFV was discovered in 1954 as a contaminant of primary monkey kidney cells, and was isolated as a virus-like agent in 1955 (Enders and Peebles, 1954; Rustigian *et al.*, 1955). The discovery that retroviruses could act as tumour-inducing agents in chickens (Ellerman and Bang, 1908; Rous, 1910, 1911), provoked considerable attention to retroviruses as potential causes of cancer in humans. The revised hunt for retroviruses and their link to cancer in the early 1970s led to the discovery of a number of type C viruses in non-human primates. These included the exogenous gibbon ape leukaemia virus (GaLV) (Kawakami *et al.*, 1972), the woolly monkey retrovirus (WMRV; SSAV) (Theilen, 1971), the endogenous type C retrovirus of baboons (BaEV) (Kalter *et al.*, 1973), and an endogenous D-type squirrel monkey retrovirus (SMRV) (Heberling *et al.*, 1977). The prototype of the morphologically defined D-type group was the Mason–Pfizer monkey virus (MPMV) isolated from a breast carcinoma of a rhesus monkey (Chopra *et al.*, 1970; Jensen *et al.*, 1970). Shortly after the discovery of the first human retrovirus, HTLV-1, in 1980 (Hinuma *et al.*, 1981; Poiesz *et al.*, 1980), a renewed search began for non-human primate retroviruses. It was in 1982 that an HTLV-1-related virus was identified by serology in Japanese macaques (*Macaca fuscata*) (Miyoshi *et al.*, 1982). Subsequently, STLV-1 has been identified in more than 25 different Old World monkey and ape species in Africa and Asia (Van Dooren *et al.*, 2007). A growing number of different lineages of the HTLV/STLV family have since been found to infect a large number of different primate species. Two new members of the human T-cell lymphotropic-related viruses have been identified, only one of which for now has no simian counterpart (Liegeois *et al.*, 2008; Mahieux and Gessain, 2009; Wolfe *et al.*, 2005).

The identification of AIDS in humans in the early 1980s triggered great interest in the simian acquired immunodeficiency syndrome (SAIDS) as a potential animal model for AIDS. Originally, this AIDS-like disease had been linked to the infection of rhesus macaques (*M. mulatta*) with the D-type retrovirus, SRV, at the California National Primate Research Center (CNPRC) (Marx *et al.*, 1985; Marx *et al.*, 1984; Maul *et al.*, 1986). Subsequently, SAIDS was also detected in rhesus macaques and other species of macaques in other primate centres (Lerche and Osborn, 2003). There were, however, significant virological, haematological, and pathological differences between this syndrome, and the pathogenesis of HIV-1-induced AIDS (Lerche *et al.*, 1987).

Overlapping with the occurrence of SRV-related SAIDS in US primate centres was a smouldering ongoing epidemic of another AIDS-like disease in rhesus macaques that was associated with lymphomas (Baskin *et al.*, 1986; Lowenstine *et al.*, 1992; Mansfield *et al.*, 1995). Shortly thereafter, an HIV-related lentivirus was isolated from these rhesus macaques (Daniel *et al.*, 1985). This virus, SIVmac, was subsequently linked to an asymptomatic natural SIV infection of sooty mangabeys (*Cercocebus atys*) (Murphey-Corb *et al.*, 1986). Leprosy studies performed at the Tulane National Primate Research Center in Louisiana were probably the cause of the unintentional iatrogenic transmission of asymptomatic SIVsm infection from sooty mangabeys to rhesus macaques (Gormus *et al.*, 2004, 1995a,b). The cause of the AIDS outbreaks due to SIVmac in rhesus macaques at the New England Primate Research Center, and to SIVstm infection in stump-tailed macaques (*M. arctoides*) in 1991 at the Yerkes National Primate Research Center (Khan *et al.*, 1991), could both be traced back to animals that had been imported from the CNPRC (Apetrei *et al.*, 2005; Daniel *et al.*, 1985; Gardner, 1996; Khan *et al.*, 1991; Mansfield *et al.*, 1995). More recently, new molecular and serological data have provided circumstantial evidence that these cases were caused by the accidental iatrogenically transmission of SIVsm during large scale Kuru experiments in these species (Apetrei *et al.*, 2006; Gibbs and Gajdusek, 1973).

D-type retroviruses

The prototypic virus of this group, the Mason–Pfizer monkey virus (MPMV) was first isolated from an adenocarcinoma from a female rhesus monkey (Chopra *et al.*, 1970). Seroepidemiological surveys have revealed a prevalence rate of up to 25% in macaque breeding colonies. The MPMV serologically correlates with only 1 of 5 distinct serotypes of SRV, with MPMV corresponding to the current designation of SRV-3. The SRV appear to be limited only to Asian macaque species, such as the rhesus macaques, the cynomolgous or long-tailed macaques (*M. fascicularis*), and pig-tailed macaques (*M. nemestrina*). The pathogenesis and pathological outcome depend considerably on the species and age of macaque when it became infected (Philipp-Staheli *et al.*, 2006). Interestingly, only one African non-human primate species has been reported to be seropositive for an SRV-like virus. This virus, SRV-Pc, was isolated from a yellow baboon (*Papio cynocephalus*), and limited sequence analysis showed a high homology with SRV types 1–3 (Grant *et al.*, 1995). Of the five recognized subtypes of SRV, only SRV-1, SRV-2, and SRV-3 have been molecularly characterized. The two most prevalent serotypes in captive breeding centres in the United States are SRV-1 and SRV-2. Both infections are correlated with simian AIDS, or SAIDS (Gardner *et al.*, 1994; Osborn *et al.*, 1984).

SRV-2 is commonly associated with retroperitoneal fibromas (RF). While 35% of SRV-2-infected macaques eventually develop this tumour, it is now suggested that a naturally occurring Kaposi's sarcoma-associated herpesvirus (KSHV) homologue in macaques, called RFHV, induces these RF-lesions due to chronic immunosuppression caused by SRV-2 (Philipp-Staheli *et al.*, 2006).

Other SRV-like viruses have recently described from Indian hanuman langurs (*Semnopithecus entellus*), and feral Indian rhesus macaques (Nandi *et al.*, 2000). These viruses, provisionally named SRV-6 and SRV-7, respectively, show relationship to SRVs, but also to simian endogenous retroviruses (SERV).

Foamy viruses

Foamy viruses (FV) are ubiquitous retroviruses found in a large number of mammalian species in addition to non-human primates. Specifically they are found in a wide range of non-human primates worldwide. Foamy viruses have been reported to infect humans but are not resident in the human population. To date they represent very rare acquired asymptomatic infections picked up by people who work with, or who have been in contact with non-human primates.

Foamy viruses were first described in the laboratory in 1954 as a contaminant of monkey kidney cells used for polio vaccine production (Enders and Peebles, 1954). Since then foamy viruses have been isolated from cattle, horses, cats, and from a wide range of non-human primates (Meiering and Linial, 2001). Rarely infection of humans with foamy viruses has been reported. However, based on sequence evidence FV are not native to humans and represent zoonotic infections (Achong *et al.*, 1971; Brooks *et al.*, 2002; Calattini *et al.*, 2007; Heneine *et al.*, 1998; Sandstrom *et al.*, 2000; Schweizer *et al.*, 1997; Wolfe *et al.*, 2004). While simian foamy viruses (SFV) commonly infect various simian species, specifically of Old World monkeys and apes, only one species of New World monkey species, the spider monkey, has been confirmed to be naturally infected (Thumer *et al.*, 2007).

Pathogenesis and disease symptoms

In contrast to the other retroviruses described in this chapter, foamy viruses are not associated with disease in their natural host, although infections with the so-called human foamy virus (HFV), have been associated with different human diseases, including Graves' disease, multiple sclerosis, thyroiditis de Quervain, and myasthenia gravis (Meiering and Linial, 2001). Most disease associations have been based on results obtained with a single diagnostic assay, without the necessary confirmatory tests. Claims by single surveys have yet to be confirmed (Debons-Guillemin *et al.*, 1992; Kuzmenok *et al.*, 2007; Schweizer *et al.*, 1994; Svenningsson *et al.*, 1992).

The most striking feature of foamy viruses is their ability to induce cytopathic effect (CPE) in cell culture. *In vitro,* foamy viruses have the capability to infect different cell types from a variety of species, causing syncytium formation, vacuolization and cell death in culture. Under magnification this CPE has a characteristic foamy appearance, hence the name foamy viruses or spumaviruses (foamy = spuma in Latin) (Linial, 1999; Meiering and Linial, 2001).

SFV can be detected in a wide number of tissues from infected hosts. However, active replication has only been shown in the oral mucosa of African green monkeys (Falcone *et al.*, 1999). Recently, it was demonstrated that SFV primarily replicates in superficial differentiated epithelial cells of the oral mucosa, and that abundant virus shedding occurs from that cell type (Murray *et al.*, 2008). This observation corresponds with the putative transmission by biting or grooming of SFV.

Genetic organization and viral proteins of foamy viruses.

Foamy viruses have the largest genomes of all retroviruses described to date. A typical FV proviral genome can be up to 13-kb in length (Fig. 15.1B). The viral genomes are flanked by LTRs, and contain, in addition to the basic retroviral *gag*, *pol*, and *env* genes, at least two open reading frames located 3′ to the end of *env*, designated *tas* and *bel2*. An additional ORF, *bel3*, has been described, but is not commonly found in all SFV (Flugel *et al.*, 1987).

Transcription of SFV initiates from two promoter regions, one located in the 5′ LTR, and in an internal promoter (IP) region located in the *env* gene (Lochelt *et al.*, 1993). Genome-length messenger RNAs, as well a subgenomic mRNA encoding the structural proteins are all transcribed from the LTR promoter, while both *tas* and *bel2* are both under control by IP.

In contrast to other retroviruses no Gag–Pol precursor polyprotein is translated in infected cells. Instead, the *pol*-precursor is translated from a spliced mRNA (Enssle *et al.*, 1996; Jordan *et al.*, 1996). Similarly, the envelope precursor is also formed by a single splicing event of the genome-length mRNA (Muranyi and Flugel, 1991).

Another feature that distinguishes foamy viruses from other retroviruses is that of the Gag precursor (p78). It is not efficiently cleaved by the viral protease, but instead is only cleaved once to generate p74 Gag. The Pol precursor, that in other retroviruses is cleaved into protease, reverse transcriptase-RnaseH, and integrase proteins, is cleaved only once in foamy viruses. The integrase protein is cleaved from the precursor protein, and a multifunctional protease- reverse transcriptase-RnaseH enzyme results (Flugel *et al.*, 1987). Envelope processing of foamy viruses is similar to that of other retroviruses. The precursor is cleaved by a cellular protease to mature surface and transmembrane proteins. The Tas protein, encoded by *tas* gene is a transactivator of transcription. Tas has binding sites on both the LTR as well as the IP (Rethwilm *et al.*, 1991). Another regulatory protein, Bet, is translated from a *tas-bel2*-spliced mRNA. Bet has recently been shown to act as an inhibitor of the APOBEC3 family (Lochelt *et al.*, 2005; Russell *et al.*, 2005).

Retroviral particles contain positive-stranded RNA molecules that after infection are reverse transcribed into DNA. Unlike other retroviruses, a large fraction of foamy virus particles (20%) contain DNA genomes (Yu *et al.*, 1999). These observations suggest that, uniquely in foamy viruses, reverse transcription takes place in the new virions before the virus infects new cells.

Molecular epidemiology

The nomenclature as recently proposed by Wolfe *et al.* (2004) designates simian foamy viruses as SFV followed by a three-letter identifier of the non-human primate species or subspecies from which it was isolated, similar to the nomenclature of the simian immunodeficiency viruses. However, other designations are also still in use.

The first simian foamy virus (SFV) that was fully genetically characterized, and fully sequenced was SFV-1. This foamy virus was a common asymptomatic infection of rhesus macaques (*M. mulatta*), and is thus renamed SFVmmu. The complete genomes are known from only a limited number of SFV, those infecting rhesus macaques (Kupiec *et al.*, 1991), African green monkeys (Renne *et al.*, 1992), chimpanzees (Herchenroder *et al.*, 1994), orang-utans (Verschoor *et al.*, 2003), and the spider monkey (Thumer *et al.*, 2007). For the majority of SFV only a small fragment of the integrase coding sequence has been molecularly characterized. Other simian viruses have only been detected in cell cultures or in serological assays, (including the putative SFV from marmosets (SFVmar), squirrel monkeys (SFV-4), capuchin monkeys (SFV-9), and the galago (SFV-5). Until full-length sequence data become available, and confirmed by PCR on uncultured tissues or fluids taken directly from animals in a sterile fashion, the characterization and exact phylogenetic origins and associations of these viruses will remain uncertain.

Integrase sequences from more than 50 primate species and subspecies have been deposited in GenBank. These include viruses from all ape species, from the five recognized baboon (sub) species, and eleven macaque species. In phylogenetic analyses, based on the aligned *int* sequences, SFVs infecting apes, and those of Old World monkeys, align to give distinct clusters. Within these clusters viruses from related hosts, or species tend to form subclusters (Calattini *et al.*, 2004; Switzer *et al.*, 2005; Verschoor *et al.*, 2004b). The pattern of the molecular phylogeny based on *int* sequence alignments with that of their host's mitochondrial DNA suggests that simian foamy viruses may to have co-evolved with their primate hosts for at least 30 million years (Switzer *et al.*, 2005).

A human foamy virus (HFV) was first isolated in 1971 from cell cultures from a Kenyan with a nasopharyngeal carcinoma (Achong *et al.*, 1971). This virus was thereafter designated human foamy virus, and is thus far the best-studied foamy virus. Later however, when sequence data became available from various isolates of chimpanzee SFV, it became apparent that this virus probably represented a zoonotic transmission of SFVcpz to humans. Since then, HFV has been referred to as SFVcpz(hu) (Brown *et al.*, 1978). Further characterization of SFVcpz(hu), and the genetic characterization of additional SFVcpz variants from the Eastern common chimpanzee (*Pan troglodytes schweinfurthii*) suggest that SFVcpz(hu) is in fact a variant of SFV from this chimpanzee subspecies. Other cases of zoonotic transmission of SFV have been reported by several groups (Brooks *et al.*, 2002; Calattini *et al.*, 2007; Heneine *et al.*, 1998; Schweizer *et al.*, 1997; Wolfe *et al.*, 2004), indicating that cross-species transmission of SFV occurs relatively frequently.

Simian T-cell lymphotropic viruses

In 1982 the existence of naturally occurring human T-cell lymphotropic virus (HTLV)-related viruses in non-human primates first became evident by the presence of cross-reactive anti-HTLV antibodies in Japanese macaques (Miyoshi *et al.*, 1982). Since then it has been determined that T-cell lymphotropic viruses form different lineages (STLV-1, STLV-2, STLV-3). STLV have been characterized from a large variety of Old World monkeys and great apes (Courgnaud *et al.*, 2004; Digilio, 1997; Giri *et al.*, 1994; Goubau *et al.*, 1994; Ibuki *et al.*, 1997; Ishikawa *et al.*, 1987; Koralnik *et al.*, 1994; Leendertz *et al.*, 2003, 2004; Liegeois *et al.*, 2008; Mahieux *et al.*, 1997, 1998a,b, 2000; Makuwa *et al.*, 2004a,b; Meertens and Gessain, 2003; Meertens *et al.*, 2001, 2002, 2003; Nerrienet *et al.*, 2001, 2004; Niphuis *et al.*, 2003; Saksena *et al.*, 1994; Takemura *et al.*, 2002; Van Dooren *et al.*, 2002, 2004, 2007; Vandamme *et al.*, 1996; Veazey *et al.*, 1998; Voevodin *et al.*, 1997).

Pathogenesis and disease symptoms

In humans HTLV-1 is associated with adult T-cell leukaemia (ATL) and HTLV-1-associated slowly progressive myelopathy (TSP/HAM) (Gessain and de The, 1996; Poiesz *et al.*, 1980). HTLV-2, though initially identified in a patient with hairy cell leukaemia, has been linked to neurological disorders (Roucoux and Murphy, 2004). In contrast, STLV is commonly associated with asymptomatic infections in their natural hosts. However, rare cases of STLV-associated leukaemia and/or lymphoma have been reported in African green monkeys, baboons, macaques and gorillas (Franchini and Reitz, 1994).

HTLVs are transmitted between sexually active partners (Schreiber *et al.*, 1997); from mother-to-child commonly through breast-feeding (Miyoshi *et al.*, 1982), and by blood-to-blood contact (Schreiber *et al.*, 1997). STLV infections are probably primarily sustained by blood contact as a consequence of fighting and hunting (Leendertz *et al.*, 2004; Nerrienet *et al.*, 1998; Niphuis *et al.*, 2003), whereas sexual transmission, or transmission of STLV-1 through breast-feeding have been documented in only very few cases (Fultz *et al.*, 1990; Georges-Courbot *et al.*, 1996; Niphuis *et al.*, 2003).

The prevalence of infection of PTLV is highly variable. Even in geographic regions were HTLV is endemic, infection rates are not uniform. In some regions HTLV-1 prevalence rates up to 12% are reported. The prevalence of STLV-1 infection in the wild varies greatly between 0 and 80%, depending on the species, habitat, age and territorial constraints examined (Niphuis *et al.*, 2003).

Genetic organization and viral proteins

The organizational structure of the STLV genome, and PTLV in general, is typical of retroviruses, but with a few characteristic differences (Fig. 15.1C). The genome contains the standard genes for retroviral structural proteins (*gag* and *env*), and replicatory enzymes (*pol*), but in addition contains a separate coding sequence for the viral protease (*pro*). It also possesses a unique region 3′ to the *env* gene. This 'X' region encodes the regulatory proteins (*tax* and *rex*), and has two additional open reading frames (ORFs I and II) (Poiesz *et al.*, 2003).

As with other retroviruses, the proviral genome is flanked by long terminal repeats (LTR). The LTRs contain short non-coding repeated sequences, designated R, U5, and U3, that play an essential role in viral replication and gene expression. The proviral DNA genome contains full-length R-U5-U3 LTR sequences at both the 5′ and 3′ends, resulting in a provirus of approximately 8.8 kb (Fig. 15.1C).

Transcription gives rise to three major messenger RNAs, which via splicing encode all viral proteins. The largest complete genome-length mRNA is utilized for the synthesis of the *gag*- and *pol*-encoded proteins, but is also packaged in newly formed viral particles. A second mRNA encodes the Tax and Rex proteins, while the third encodes the Env protein. Gag is translated as a precursor polyprotein, and is cleaved by the protease enzyme into the 19-kDa matrix protein (MA), the 24-kDa capsid protein(CA), and the 15-kDa nucleocapsid protein (NC) (Chen *et al.*, 1994).

Different from many other retroviruses, the viral protease (Pro) is not expressed as part of the Pol polyprotein, but instead is expressed by an open reading frame that overlaps the *Gag–Pol* genes. Pro is translated from the genome-length mRNA via ribosomal frame shifting (Nam *et al.*, 1993). Pro is responsible for the processing of the Gag polyprotein, and via self-maturation generates a mature form of itself (Hatanaka and Nam, 1989; Nam and Hatanaka, 1986). The other viral enzymes, reverse transcriptase (RT), the RNase H, and integrase are also expressed from the same mRNA, but expression involves two successive ribosomal frame-shifting events (Nam *et al.*, 1993).

The envelope glycoproteins, the 46-kDa surface glycoprotein (SU, gp46), and the 21-kDa transmembrane glycoprotein (TM, gp21) are synthesized from the envelope glycoprotein precursor by cleavage through cellular proteases (Hattori *et al.*, 1984). The envelope precursor is translated from a singly spliced subgenomic mRNA.

Multiple splicings of the same subgenomic mRNA result in minor mRNA species that are templates for the expression of the Tax and Rex regulatory proteins. Tax is a *trans*-acting activator of transcription and increases the transcription rate from the 5′LTR promoter (Cann *et al.*, 1985; Sodroski *et al.*, 1985). Tax activates viral and cellular transcription and promotes T-cell growth and malignant transformation (Copeland *et al.*, 1994). The Rex protein is involved in the regulation of gene expression (Ciminale *et al.*, 1995; Hidaka *et al.*, 1988). The exact roles of the putative ORFs I and II gene products in the PTLV life cycle has not yet been firmly established.

Molecular epidemiology

Simian T-cell lymphotropic viruses (STLV) naturally infect non-human primates and are genetically highly similar to the T-cell lymphotropic viruses that infect humans (HTLV-1 to -3). The terminology used to include simian and human deltaretroviruses are primate T-cell lymphotropic viruses, or PTLV. In contrast to foamy viruses, phylogenetic analyses indicate that HTLVs and STLVs do not form separate clusters of viruses (Saksena *et al.*, 1993). This suggests multiple cross-species transmissions in the distant as well as recent past. For that reason we do not limit ourselves in this section only to the simian viruses, but include all PTLVs.

On the basis of full genome sequence alignments, PTLV can be divided in three different lineages, namely PTLV-1, -2, and -3 (Goubau *et al.*, 1994). Within the large PTLV clusters subtypes can be distinguished. As cross-species transmissions not only occurred between simians and humans, but also frequently between different primate species that share the same regional distribution, clusters of viruses can be recognized according to the geographic range and ecosystem of the host (Fig. 15.2).

PTLV-1 form the largest group of T-cell lymphotropic retroviruses. HTLV-1 infections are found worldwide, but they most probably have their origins in Africa (Fig. 15.3). Within HTLV-1 six different lineages, or subtypes, can be recognized (A–F): the globally distributed, or so-called 'cosmopolitan' subtype A, the Central African subtypes B and D, and the Melanesian/Australian HTLV-1 subtype C. The subtypes E and F are relatively minor and contain HTLV from the Democratic Republic of Congo and Gabon.

The PTLV-1 group includes numerous non-human primate viruses also found in African and Asian Old World monkeys and apes. More than 25 different primate species belonging to the Old World monkeys and apes are infected with STLV-1 (Van Dooren *et al.*, 2007).

The PTLV-2 cluster of viruses comprises the HTLV-2 subtypes A, B, C, and D, and STLVs detected in the pygmy chimpanzee, or bonobo (*Pan paniscus*) (Giri *et al.*, 1994; Vandamme *et al.*, 1996). In contrast to HTLV-1, HTLV-2 is mainly restricted to African pygmies from Cameroon and Congo, and to Amerindian ethnic groups living in Latin America and the southern part of North America (Fig. 15.3). More recently HTLV-2 has spread to communities of intravenous drug users in America, Europe and Asia (Lee *et al.*, 1989). The STLV of *P. paniscus* belongs to the same lineage as a HTLV-2 subtype D virus strain from an Congolese Efe pygmy (Vandamme *et al.*, 1998). Because of the close geographic proximity of both

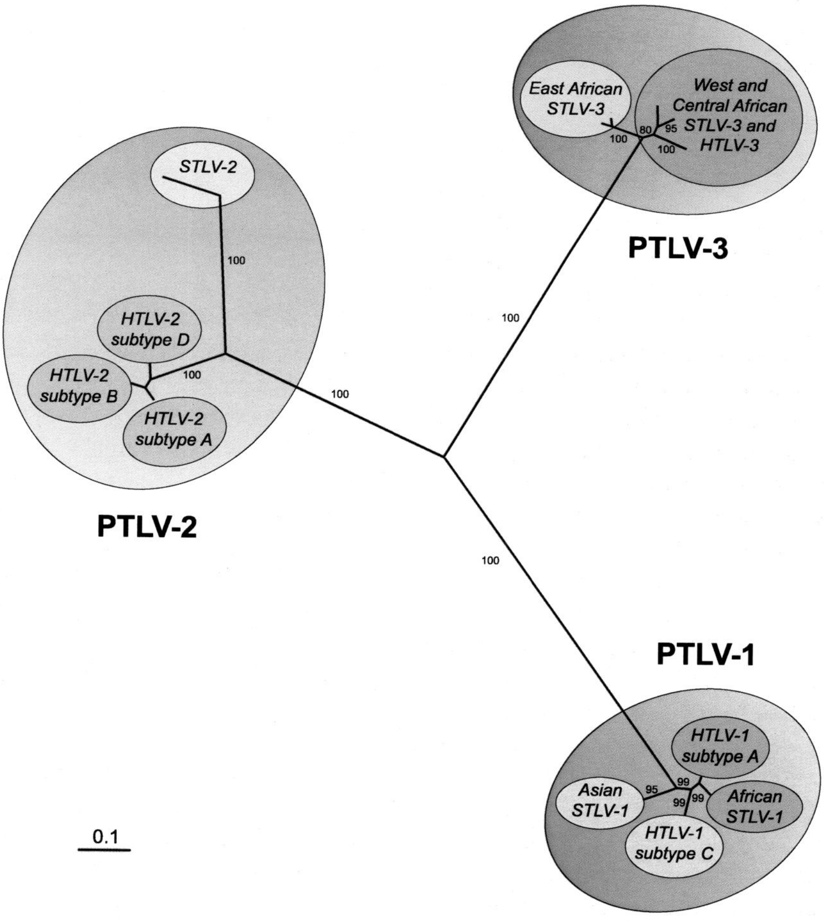

Figure 15.2 Phylogenetic tree of PTLV. Figure adapted from Calattini *et al.* (2006).

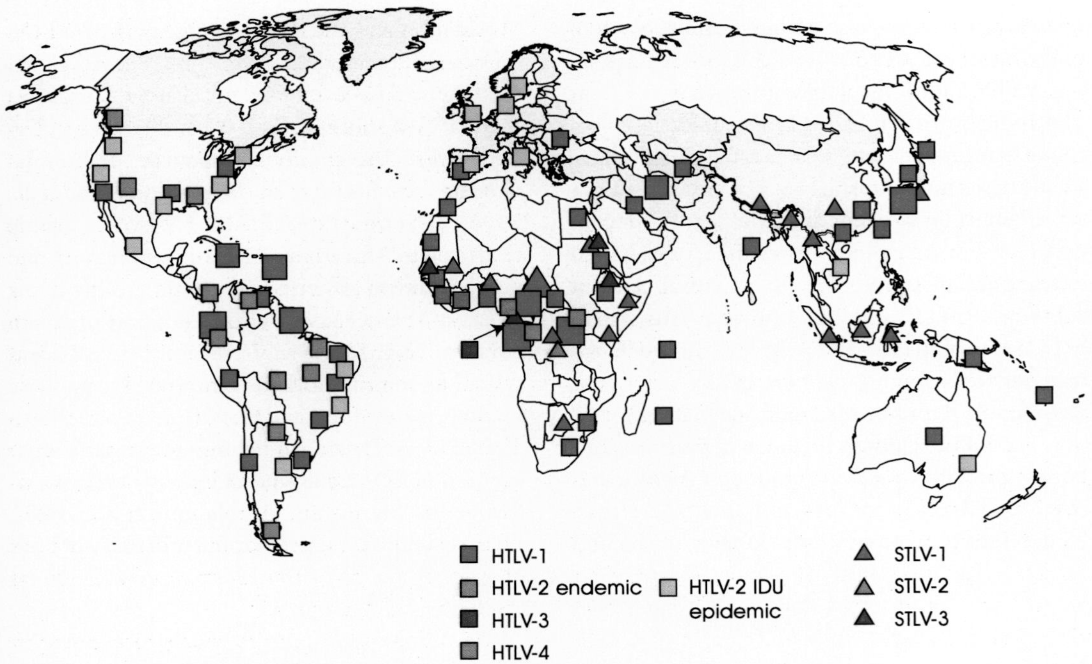

Figure 15.3 Global epidemiology of PTLV. Normal versus large-sized squares and triangles correspond to a seroprevalence of 0.3–3% and > 3% respectively. All different PTLV types can be found within Central Africa. Figure kindly provided by Sonia van Dooren, dissertation, University of Utrecht, 2005.

hosts, there is strong circumstantial evidence in favour of cross-species transmissions between bonobos and humans.

The first virus strain belonging to the PTLV-3 group was characterized from a captive hamadryas baboon from Eritrea (Goubau *et al.*, 1994). PTLV-3 includes other STLV-3s from a wide variety of non-human primates such as *Cercocebus agilis, Lophocebus albigena,* and *Cercopithecus nictitans* from Central Africa, geladas (*Theropithecus gelada*) and hamadryas baboons (*Papio hamadryas*) from East Africa, as well as genetically related viruses detected in West African monkeys (red-capped mangabeys, *Cercocebus torquatus torquatus*), and Guinea baboons (*Papio papio*) (Courgnaud *et al.*, 2004; Liegeois *et al.*, 2008; Meertens and Gessain, 2003; Meertens *et al.*, 2002; Takemura *et al.*, 2002; Van Dooren *et al.*, 2001, 2004). More recently, the PTLV-3 group of viruses has been expanded by the finding of HTLV-3. Human infections with a PTLV-3 have been found in people from Central Africa (Calattini *et al.*, 2005; Switzer *et al.*, 2006; Wolfe *et al.*, 2005) (Fig. 15.3). Genetically, the HTLV-3 strain Pyl43 from a Bakola pygmy from Cameroon is very closely related to a STLV-3 from a red-capped mangabey, and is probably another example of a zoonotic transmission. However, for strain 2026ND, which was isolated from a primate hunter from Cameroon, has to date no direct STLV-3 counterpart (Switzer *et al.*, 2006).

Of interest, Wolfe *et al.* (Wolfe *et al.*, 2005) recently described a new human virus designated HTLV-4. This virus is genetically different from other PTLVs and forms a separate lineage in the phylogenetic analyses. As such this would be the first PTLV lineage for which no simian counterparts have been described.

Simian lentiviruses

From more than 23 different non-human primate species there are more than 40 fully sequenced SIV genomes described, while partial sequences from a growing number of primate lentiviruses are becoming available. This situation is evolving regularly, and as there is serologic evidence of SIV infection for more primate species more sequence data will be published in the near future.

The nomenclature of simian lentiviruses is such that the species from which the isolate is derived is used to designate the type of SIV, given a three-letter abbreviation of the host primate species derived from. When subspecies are infected, the name of the subspecies is similarly included in the virus designation. Unfortunately there remain exceptions to this rule, for instance; subspecies of chimpanzees are infected with distinct variants of SIV, for example *Pan troglodytes troglodytes.* Here the geographic region or country from where the animal was located at the time the isolate was derived is used to further designate the virus isolate.

The primate lentiviruses form a genetically highly diverse group of viruses. In general, each primate species harbours its own species-specific virus. SIV can be detected in all major groups of African non-human primates, including apes, and the two tribes belonging to the subfamily of the *Cercopithecinae* (the *Papionini* and the *Cercopithecini*), and the *Colobinae* subfamily. However, though phylogenetic analysis is complicated by recombination and cross-species transmission events in the past, six major clusters of viruses can be distinguished. The largest cluster is formed by SIV from guenons (*Cercopithecus* monkeys), others clusters include viruses from African green monkeys (SIVagm), SIV from sooty mangabeys (SIVsm), including HIV-2 and SIVmac, the SIVlhoest group that contains viruses from L'Hoest's monkeys, sun-tailed guenons and mandrills (SIVlho; SIVsun; SIVmnd), and the lentiviruses from colobus monkeys (SIVcol). The sixth cluster is formed by SIV from chimpanzees (SIVcpz) and HIV-1 from humans. However, as sequence analysis has shown that SIVcpz is in fact a recombinant, it clusters with different SIV depending on the part of the genome used in the analysis. Using *pol* the cluster also includes viruses from mandrills (SIVmnd-2), drills (SIVdrl), and SIVrcm from red-capped mangabeys. In other parts of the genome SIVcpz clusters with the SIVgsn/SIVmon/SIVmus group of lentiviruses (for review see Gordon *et al.*, 2005).

Current data suggest that SIVcpz circulates in two out of four subspecies of chimpanzee, the Central African *Pan troglodytes troglodytes* and the Eastern African *Pan troglodytes schweinfurthii* (Bibollet-Ruche *et al.*, 2004b; Gao *et al.*, 1999; Keele *et al.*, 2006; Vanden Haesevelde *et al.*, 1996). More recently, SIVgor has been characterized from Western lowland gorillas *(Gorilla gorilla gorilla)* in Cameroon (Van Heuverswyn *et al.*, 2006). Phylogenetic sequence analysis from these great ape lentiviruses suggest that two of the three groups of HIV-1 were derived from cross-species transmission of SIVcpz from *P.t. troglodytes* to humans giving rise to HIV-1 groups M and N, and SIVgor giving rise to HIV-1 group 0. Exactly how these viruses were transmitted from great apes to humans is unknown, however local practices of hunting these species for bush meat in the countries where these animals are found, suggests that events involving hunting or preparing bush meat may have facilitated transmission of these great ape lentiviruses to humans.

The *Cercopithecinae* are widely distributed in sub-Saharan Africa. They are the largest reservoir species for SIV. All four African green monkey species (*Chlorocebus sp.*), vervets, grivets, tantalus, and sabaeus have high seroprevalence rates of subspecies-specific SIVagm. Other members of this group that carry their own variants of SIV include talapoin monkeys (*Miopithecus ogouensis*), monkeys belonging to the L'Hoest supergroup, and a large variety of *Cercopithecus* species (Beer *et al.*, 1999, 2000; Bibollet-Ruche *et al.*, 2004a; Courgnaud *et al.*, 2003a; Dazza *et al.*, 2005; Emau *et al.*, 1991; Hirsch *et al.*, 1999; Liegeois *et al.*, 2006; Osterhaus *et al.*, 1999; Verschoor *et al.*, 2004a).

Serological evidence of SIV infection other *Cercopithecinae* has been found for several mona monkey subspecies (*C. campbelli*, *C. pogonias*, and *C. lowei*), and Allen's swamp monkeys (*Allenopithecus nigroviridis*), but genetic characterization of these infections has not yet been reported (Beer *et al.*, 1999; Lowenstine *et al.*, 1986; Ohta *et al.*, 1988; Peeters *et al.*, 2002). There remain however members of this group from which seroprevalence data are not yet available (VandeWoude and Apetrei, 2006).

Sooty mangabeys *(Cercocebus atys)* are infected at high rate with SIVsm in the wild (Chen *et al.*, 1996; Marx *et al.*, 1991; VandeWoude and Apetrei, 2006). They belong to the greater *Papionini* radiation that also includes the red-capped mangabeys (*Cercocebus torquatus*) and the black mangabey (*Lophocebus aterrimus*), both of which are also infected with their own SIV (Beer *et al.*, 2001; Georges-Courbot *et al.*, 1998; Takemura *et al.*, 2005). For the grey-crested mangabey (*L. albigena*) serological evidence for SIV infection has been presented, but no SIV has yet been molecularly characterized from this species (Peeters *et al.*, 2002). Other members of the *Papionini* that naturally carry SIV infections include mandrills and drills (*Mandrillus sphinx and M. leucophaeus*) with mandrills being the natural host to two different SIVmnd (Hu *et al.*, 2003; Souquiere *et al.*, 2001; Takehisa *et al.*, 2001; Tsujimoto *et al.*, 1989). Very interestingly, there is currently no evidence that the wide-ranging baboon species naturally carry their own SIV infections. However, two baboon species have been shown to carry SIVagm from African green monkeys (Jin *et al.*, 1994; van Rensburg *et al.*, 1998).

Members of the *Colobinae* infected with their own species-associated variants of SIV include black-and-white, or mantled colobus monkeys (*Colobus guereza*), as well as three West African colobus species, the olive colobus (*Procolobus verus*), the Western red colobus (*Piliocolobus badius badius*), and Temminck's red colobus (*P. b. temminckii*) (Courgnaud *et al.*, 2001, 2003b; Locatelli *et al.*, 2008).

Phylogenetic analysis is complicated by recombination and cross-species transmission events, but reveals six major clusters of viruses. The largest cluster is formed solely by SIVs from guenons (*Cercopithecus* monkeys), others clusters include the viruses from African green monkeys (SIVagm), SIV from sooty mangabeys (SIVsm), including HIV-2 and SIVmac, the SIVlhoest group that contains viruses from L'Hoest's monkeys, sun-tailed guenons and mandrills (SIVlho; SIVsun; SIVmnd-1), and the lentiviruses from colobus monkeys (SIVcol). The sixth cluster was initially formed by SIV from chimpanzees (SIVcpz) and HIV-1 from humans. However, as sequence analysis has shown that SIVcpz is in fact a recombinant in clusters with different SIV depending on the part of the genome used in the analysis. Using *pol* the cluster also includes viruses from mandrills (SIVmnd-2), drills (SIVdrl), and SIVrcm from red-capped mangabeys. In other parts of the genome SIVcpz clusters with the SIVgsn/SIVmon/SIVmus group of lentiviruses (Gordon *et al.*, 2005).

All primate lentiviruses have five regulatory genes. The presence of two other regulatory genes (*vpx* and *vpu*) is variable (Tristem *et al.*, 1992; Tristem *et al.*, 1998). Only SIVs infecting the Papionini group of monkeys have *vpx* (Hu *et al.*, 2003; Takemura and Hayami, 2004). This appears to have occurred due to non-homologous recombination resulting in the duplication of the *vpr* gene. SIVs that infect the great apes contain *vpu* in contrast to SIVs infecting *Papionini* monkeys which contain *vpx*. Of the eight guenon species, only 3 have been shown to harbour *vpu* containing viruses (Barlow *et al.*, 2003; Courgnaud *et al.*, 2003a; Courgnaud *et al.*, 2002; Dazza *et al.*, 2005). Some authors argue that the *Cercopithicinae* are the major reservoir and the origin of SIVs since *vpu* first appeared in SIVs from these types of monkeys. They also appear to be a significant reservoir for viruses in the SIVcpz/HIV-1 lineage.

Examples of cross-species transmission of simian retroviruses

Based on analysis of the human, chimpanzee, and in part, other species genomes available, there is clearly 'genetic fossil' evidence that transmission of retroviruses has occurred across species barriers for millennia. As mentioned earlier, HIV-1 is most likely the result of several cross-species transmissions of SIVcpz from Central African common chimpanzees (Keele *et al.*, 2006) and SIVgor from lowland gorillas (Van Heuverswyn *et al.*, 2006). Sooty mangabeys naturally carry an asymptomatic SIVsm infection. Transmission of SIVsm variants to humans is the probable cause of HIV-2 infection in humans (Chen *et al.*, 1996), while transmission of SIVsm to Asian macaques gave rise to an AIDS-like disease (SIVmac). Importantly, not all cross-species transmissions of SIVsm resulted in virulent (pathogenic) variants of HIV-2, in fact some are relatively non-pathogenic. Similarly, only one particular SIVcpz*Ptt* lineage gave rise to the HIV-1 group M variants that have become the global epidemic strains, the others (HIV-1 'N' and 'O') have remained locally confined and are less pathogenic (Heeney *et al.*, 2006). Transmissions of SIV from African green monkeys have been documented in two baboon species, without any clear disease related to these events (Jin *et al.*, 1994; van Rensburg *et al.*, 1998). Another documented case of cross-species transmission is SIVdrl from drills to mandrills (van der Kuyl *et al.*, 2004), and SIVagm.Sab to patas monkeys (*Erythrocebus patas*) (Bibollet-Ruche *et al.*, 1996). In addition, several SIVs have been described that are probably recombinant viruses (SIVcpz, SIVmnd-2) (Courgnaud *et al.*, 2002; Leitner *et al.*, 2007; Souquiere *et al.*, 2001), implying previous transmission events.

Accidental transmissions of SFVs from simians to humans have also been reported (Boneva *et al.*, 2007; Brooks *et al.*, 2002; Calattini *et al.*, 2007; Heneine *et al.*, 1998; Sandstrom *et al.*, 2000; Schweizer *et al.*, 1997; Switzer *et al.*, 2004; Wolfe *et al.*, 2004). Foamy viruses can cause persistent and asymptomatic infections in humans. In an extensive study amongst 187 workers at primate keeping institutes, Switzer *et al.* (Switzer *et al.*, 2004) reported 5.3% (10/187) seropositive to SFV, and confirmed nine cases by PCR. No evidence was found for any disease signs connected to SFV infection, and human-to-human transmissions were reported, no documented cases of cross-species transmission have been reported.

As mentioned earlier, there is ample evidence of past and present cross-species transmissions between simian species, but also from non-human primates to humans of T-lymphotropic viruses (Calattini *et al.*, 2005; Salemi *et al.*, 2003; Slattery *et al.*, 1999; Voevodin *et al.*, 1997).

Changes in pathology following cross-species transmission

Infections which manifest with high morbidity or mortality are often investigated by pathologists, and if they cause similarly suspicious lesions in more than one individual these are brought to the attention of virologists. This is the diagnostic process by which many of the first non-human primate viruses were identified. However, not all cross-species infections of viruses result in successful contagions, and relatively fewer become successful pathogens. While chimpanzee foamy virus can be transmitted to humans, it does not cause any pathology that we can recognize, and is very inefficient at being transmitted from human to human (Sandstrom *et al.*, 2000; Schweizer *et al.*, 1997). Simian foamy viruses are asymptomatic and relatively common in their natural hosts. In fact scientists have not yet been able to identify any disease syndrome that is caused by a foamy virus. Other families of viruses are well known for the diseases they cause when transmitted to humans, such as Avian influenza, SARS coronavirus or the filoviruses such as Ebola (the last two are natural reservoirs in bats) (Heeney, 2006). Such infections are contagious and highly pathogenic. Although there are exceptions such as Jembrana disease in cattle, retrovirus infections, especially in primates have tended to be insidious. If they cause disease it is usually very chronic, manifesting years after infection as cancer or wasting disease such as AIDS.

We recognize natural lentiviruses as being common in African non-human primates and to date, naturally occurring infections in primates have been limited to those species found in Africa (Heeney *et al.*, 2006). Cross-species infection of SIVsm is recognized as HIV-2 in humans, while SIVcpz from chimpanzees has caused HIV-1 infection. Importantly, while both may cause AIDS in humans, they differ in virulence and their ability to be transmitted in human populations. A subtype (M) of HIV-1 has evolved as a predominant human lentivirus, being efficiently transmitted and the most common cause of AIDS. Other subtypes of HIV-1 and HIV-2 are less readily transmitted and do not cause as profound immunosuppressive disease in humans as the pandemic HIV-1 variants (clades C and CRF02_AG) (Heeney *et al.*, 2006).

It was commonly believed that in all African non-human primates lentiviruses caused asymptomatic infections, and only when they were transmitted to non-African species, such as the Asian macaques, that they would develop disease. Indeed AIDS develops in various macaque species when infected with SIVsm but with differing susceptibility which is related to the species or subspecies of macaque. The dogma that all African primates are susceptible to lentivirus infection but resistant to disease is not without important exception. While at least two subspecies of chimpanzees are infected with variants of SIVcpz, they are not completely immune to developing AIDS. We have termed their natural resistance, as relative resistance to reflect the observation that SIVcpz can have an insidious effect on their CD4 T-cell levels (Heeney *et al.*, 2006). Importantly, AIDS-like disease has been noted in chimpanzees infected with a particular recombinant of HIV-1 (Novembre *et al.*, 2001; O'Neil *et al.*, 2000). So while a larger population of African primates, in this case, chimpanzees, may be infected asymptomatically, and while they may be relatively resistant to disease, some individuals with either a susceptible genotype, or infected with a relatively virulent variant, may develop overt disease. It must be remembered that lentiviruses are continuously evolving, and although in many species they have appeared to have reached an apparent disease-free equilibrium, there are dynamics in infections and cross-species infections that occasionally distort an apparently finely balanced host-pathogen equilibrium. A case in point is the observation of AIDS-like disease in a Black mangabey infected with SIVsm (Apetrei *et al.*, 2004).

The pattern of human HTLV and non-human primate STLV infections perhaps represents a somewhat intermediate situation as compared to the global distribution of the lentiviruses and the foamy viruses and their disease causing potential. Evidence of cross-species transmissions are numerous, there are well-established human infections which are currently spreading, but slowly and not as global epidemics. Compared to the relatively young human HIV-1 epidemic, HTLV infections are much more limited and restricted to specific populations. The reservoir in non-human primates remains a continual source of potentially new human infections.

Conclusions

We are in an interesting period of scientific understanding of host susceptibility and resistance to retroviral infections. At least three distinct types of retroviral restriction factors have been identified affecting both pre-integration (TRIM5a), as well as post-transcription steps (APOBECs, Tetherins) (Neil *et al.*, 2008; Wolf and Goff, 2008). The mechanisms by which these factors seem to work is more apparent in many cases than their species-specific effects in protecting one species from certain retroviral infections better than others (Jern and Coffin, 2008). This said, this field of research is developing at unprecedented rates in science and important insights into species susceptibility and resistance to many of the different types of retroviruses and other viral families are likely to be gained in over the coming years.

References

Achong, B.G., Mansell, P.W.A., Epstein, M.A., and Clifford, P. (1971). An unusual virus in cultures from a human nasopharyngeal carcinoma. J. Natl. Cancer. Inst. *46*, 299–302.

Apetrei, C., Gormus, B., Pandrea, I., Metzger, M., ten Haaft, P., Martin, L. N., Bohm, R., Alvarez, X., Koopman, G., Murphey-Corb, M., *et al.* (2004). Direct inoculation of simian immunodeficiency virus from sooty mangabeys in black mangabeys (Lophocebus aterrimus): first evidence of AIDS in a heterologous African species and different pathologic outcomes of experimental infection. J. Virol. *78*, 11506–11518.

Apetrei, C., Kaur, A., Lerche, N.W., Metzger, M., Pandrea, I., Hardcastle, J., Falkenstein, S., Bohm, R., Koehler, J., Traina-Dorge, V., *et al.* (2005). Molecular epidemiology of simian immunodeficiency virus SIVsm in U.S. primate centers unravels the origin of SIVmac and SIVstm. J. Virol. *79*, 8991–9005.

Apetrei, C., Lerche, N. W., Pandrea, I., Gormus, B., Silvestri, G., Kaur, A., Robertson, D. L., Hardcastle, J., Lackner, A. A., and Marx, P. A. (2006). Kuru experiments triggered the emergence of pathogenic SIVmac. Aids *20*, 317–321.

Barlow, K.L., Ajao, A.O., and Clewley, J.P. (2003). Characterization of a novel simian immunodeficiency virus (SIVmonNG1) genome sequence from a mona monkey (Cercopithecus mona). J. Virol. *77*, 6879–6888.

Baskin, G.B., Martin, L.N., Rangan, S.R., Gormus, B.J., Murphey-Corb, M., Wolf, R.H., and Soike, K.F. (1986). Transmissible lymphoma and simian acquired immunodeficiency syndrome in rhesus monkeys. J. Natl. Cancer. Inst. *77*, 127–139.

Beer, B.E., Bailes, E., Dapolito, G., Campbell, B. J., Goeken, R. M., Axthelm, M. K., Markham, P.D., Bernard, J., Zagury, D., Franchini, G., *et al.* (2000). Patterns of genomic sequence diversity among their simian immunodeficiency viruses suggest that L'Hoest monkeys (Cercopithecus lhoesti) are a natural lentivirus reservoir. J. Virol. *74*, 3892–3898.

Beer, B.E., Bailes, E., Goeken, R., Dapolito, G., Coulibaly, C., Norley, S.G., Kurth, R., Gautier, J. P., Gautier-Hion, A., Vallet, D., *et al.* (1999). Simian immunodeficiency virus (SIV) from sun-tailed monkeys (*Cercopithecus solatus*): evidence for host-dependent evolution of SIV within the C. lhoesti superspecies. J. Virol. *73*, 7734–7744.

Beer, B.E., Foley, B.T., Kuiken, C. L., Tooze, Z., Goeken, R.M., Brown, C.R., Hu, J., St Claire, M., Korber, B.T., and Hirsch, V.M. (2001). Characterization of novel simian immunodeficiency viruses from red-capped mangabeys from Nigeria (SIVrcmNG409 and -NG411). J. Virol. *75*, 12014–12027.

Bibollet-Ruche, F., Bailes, E., Gao, F., Pourrut, X., Barlow, K.L., Clewley, J.P., Mwenda, J.M., Langat, D.K., Chege, G.K., McClure, H.M., *et al.* (2004a). New simian immunodeficiency virus infecting De Brazza's monkeys (Cercopithecus neglectus): evidence for a cercopithecus monkey virus clade. J. Virol. *78*, 7748–7762.

Bibollet-Ruche, F., Galat-Luong, A., Cuny, G., Sarni-Manchado, P., Galat, G., Durand, J.P., Pourrut, X., and Veas, F. (1996). Simian immunodeficiency virus infection in a patas monkey (*Erythrocebus patas*): evidence for cross-species transmission from African green monkeys (Cercopithecus aethiops sabaeus) in the wild. J. Gen. Virol. *77*, 773–781.

Bibollet-Ruche, F., Gao, F., Bailes, E., Saragosti, S., Delaporte, E., Peeters, M., Shaw, G.M., Hahn, B.H., and Sharp, P.M. (2004b). Complete genome analysis of one of the earliest SIVcpzPtt strains from Gabon (SIVcpzGAB2). AIDS Res. Hum. Retroviruses *20*, 1377–1381.

Bindhu, M., Nair, A., and Lairmore, M. D. (2004). Role of accessory proteins of HTLV-1 in viral replication, T-cell activation, and cellular gene expression. Front .Biosci. *9*, 2556–2576.

Boneva, R. S., Switzer, W.M., Spira, T. J., Bhullar, V.B., Shanmugam, V., Cong, M. E., Lam, L., Heneine, W., Folks, T.M., and Chapman, L.E. (2007). Clinical and virological characterization of persistent human infection with simian foamy viruses. AIDS Res. Hum. Retroviruses *23*, 1330–1337.

Brooks, J.I., Rud, E.W., Pilon, R. G., Smith, J.M., Switzer, W. M., and Sandstrom, P.A. (2002). Cross-species retroviral transmission from macaques to human beings. Lancet *360*, 387–388.

Brown, P., Nemo, G., and Gajdusek, D.C. (1978). Human foamy virus: further characterization, seroepidemiology, and relationship to chimpanzee foamy viruses. J. Infect. Dis. *137*, 421–427.

Calattini, S., Betsem, E. B., Froment, A., Mauclere, P., Tortevoye, P., Schmitt, C., Njouom, R., Saib, A., and Gessain, A. (2007). Simian foamy virus transmission from apes to humans, rural Cameroon. Emerg. Infect. Dis. *13*, 1314–1320.

Calattini, S., Chevalier, S.A., Duprez, R., Afonso, P., Froment, A., Gessain, A., and Mahieux, R. (2006). Human T-cell lymphotropic virus type 3: complete nucleotide sequence and characterization of the human tax3 protein. J. Virol. *80*, 9876–9888.

Calattini, S., Chevalier, S.A., Duprez, R., Bassot, S., Froment, A., Mahieux, R., and Gessain, A. (2005). Discovery of a new human T-cell lymphotropic virus (HTLV-3) in Central Africa. Retrovirology *2*, 30.

Calattini, S., Nerrienet, E., Mauclere, P., Georges-Courbot, M. C., Saib, A., and Gessain, A. (2004). Natural simian foamy virus infection in wild-caught gorillas, mandrills and drills from Cameroon and Gabon. J. Gen. Virol. *85*, 3313–3317.

Cann, A. J., Rosenblatt, J.D., Wachsman, W., Shah, N.P., and Chen, I. S. (1985). Identification of the gene responsible for human T-cell leukaemia virus transcriptional regulation. Nature *318*, 571–574.

Chen, Y. M., Jang, Y.J., Kanki, P.J., Yu, Q. C., Wang, J.J., Montali, R.J., Samuel, K.P., and Papas, T.S. (1994). Isolation and characterization of simian T-cell leukemia virus type II from New World monkeys. J. Virol. *68*, 1149–1157.

Chen, Z., Telfier, P., Gettie, A., Reed, P., Zhang, L., Ho, D.D., and Marx, P A. (1996). Genetic characterization of new West African simian immunodeficiency virus SIVsm: geographic clustering of household-derived SIV strains with human immunodeficiency virus type 2 subtypes and genetically diverse viruses from a single feral sooty mangabey troop. J. Virol. *70*, 3617–3627.

Chopra, H. C., Bogden, A. E., Zelljadt, I., and Jensen, E. M. (1970). Virus particles in a transplantable rat mammary tumor of spontaneous origin. Eur J. Cancer *6*, 287–290.

Ciminale, V., D'Agostino, D. M., Zotti, L., Franchini, G., Felber, B.K., and Chieco-Bianchi, L. (1995). Expression and characterization of proteins produced by mRNAs spliced into the X region of the human T-cell leukemia/lymphotropic virus type II. Virology *209*, 445–456.

Coffin, J. M. (1991). Retroviridae and their replication. In Fields, Bernard N Fields, DM Chanock *et al.* eds, Fundamentals of Virology, 2nd edn, pp. 645–708.

Copeland, K. F., Haaksma, A. G., Goudsmit, J., Krammer, P.H., and Heeney, J. L. (1994). Inhibition of apoptosis in T-cells expressing human T-cell leukemia virus type I Tax. AIDS Res. Hum. Retroviruses *10*, 1259–1268.

Courgnaud, V., Abela, B., Pourrut, X., Mpoudi-Ngole, E., Loul, S., Delaporte, E., and Peeters, M. (2003a). Identification of a new simian immunodeficiency virus lineage with a vpu gene present among different cercopithecus monkeys (*C. mona, C. cephus*, and *C. nictitans*) from Cameroon. J. Virol. *77*, 12523–12534.

Courgnaud, V., Formenty, P., Akoua-Koffi, C., Noe, R., Boesch, C., Delaporte, E., and Peeters, M. (2003b). Partial molecular characterization of two simian immunodeficiency viruses (SIV) from African colobids: SIVwrc from Western red colobus (Piliocolobus badius) and SIVolc from olive colobus (*Procolobus verus*). J. Virol. *77*, 744–748.

Courgnaud, V., Pourrut, X., Bibollet-Ruche, F., Mpoudi-Ngole, E., Bourgeois, A., Delaporte, E., and Peeters, M. (2001). Characterization of a novel simian immunodeficiency virus from guereza colobus monkeys (*Colobus guereza*) in Cameroon: a new lineage in the nonhuman primate lentivirus family. J. Virol. *75*, 857–866.

Courgnaud, V., Salemi, M., Pourrut, X., Mpoudi-Ngole, E., Abela, B., Auzel, P., Bibollet-Ruche, F., Hahn, B., Vandamme, A. M., Delaporte, E., and Peeters, M. (2002). Characterization of a novel simian immunodeficiency virus with a vpu gene from greater spot-nosed monkeys (*Cercopithecus nictitans*) provides new insights into simian/human immunodeficiency virus phylogeny. J. Virol. *76*, 8298–8309.

Courgnaud, V., Van Dooren, S., Liegeois, F., Pourrut, X., Abela, B., Loul, S., Mpoudi-Ngole, E., Vandamme, A., Delaporte, E., and Peeters, M. (2004). Simian T-cell leukemia virus (STLV) infection in wild primate populations in Cameroon: evidence for dual STLV type 1 and type 3 infection in agile mangabeys (*Cercocebus agilis*). J. Virol. *78*, 4700–4709.

Daniel, M.D., Letvin, N. L., King, N.W., Kannagi, M., Sehgal, P.K., Hunt, R.D., Kanki, P J., Essex, M., and Desrosiers, R.C. (1985). Isolation of T-cell tropic HTLV-III-like retrovirus from macaques. Science *228*, 1201–1204.

Dazza, M. C., Ekwalanga, M., Nende, M., Shamamba, K. B., Bitshi, P., Paraskevis, D., and Saragosti, S. (2005). Characterization of a novel vpu-harboring simian immunodeficiency virus from a Dent's Mona monkey (*Cercopithecus mona denti*). J. Virol. *79*, 8560–8571.

Debons-Guillemin, M.C., Valla, J., Gazeau, J., Wybier-Franqui, J., Giron, M. L., Toubert, M.E., Canivet, M., and Peries, J. (1992). No evidence of spumaretrovirus infection markers in 19 cases of De Quervain's thyroiditis. AIDS Res. Hum. Retroviruses *8*, 1547.

Digilio, L.G.A., Cho, N., Slattery, J., Markham, P., and Franchini, G. (1997). The simian T-lymphotropic/leukemia virus from *Pan paniscus* belongs to the type 2 family and infects Asian macaques. J. Virol. *71*, 3684–3692.

Ellerman, C., and Bang, O. (1908). Experimentelle Leukämie bei Hühnern. Centralbl Bakteriol *46*, 595–609.

Emau, P., McClure, H.M., Isahakia, M., Else, J.G., and Fultz, P. N. (1991). Isolation from African Sykes' monkeys (*Cercopithecus mitis*) of a lentivirus related to human and simian immunodeficiency viruses. J. Virol. *65*, 2135–2140.

Enders, J. F., and Peebles, T. C. (1954). Propagation in tissue cultures of cytopathogenic agents from patients with measles. Proc Soc Exp Biol Med *86*, 277–286.

Enssle, J., Jordan, I., Mauer, B., and Rethwilm, A. (1996). Foamy virus reverse transcriptase is expressed independently from the Gag protein. Proc. Natl. Acad. Sci. U.S.A. *93*, 4137–4141.

Falcone, V., Leupold, J., Clotten, J., Urbanyi, E., Herchenroder, O., Spatz, W., Volk, B., Bohm, N., Toniolo, A., Neumann-Haefelin, D., and Schweizer, M. (1999). Sites of simian foamy virus persistence in naturally infected African green monkeys: latent provirus is ubiquitous, whereas viral replication is restricted to the oral mucosa. Virology *257*, 7–14.

Fine, D., and Schochetman, G. (1978). Type D. primate retroviruses: a review. Cancer Res. *38*, 3123–3139.

Flugel, R.M., Rethwilm, A., Maurer, B., and Darai, G. (1987). Nucleotide sequence analysis of the *env* gene and its flanking regions of the human spumaretrovirus reveals two novel genes. EMBO J. *6*, 2077–2084.

Franchini, G., and Reitz, M.S., Jr. (1994). Phylogenesis and genetic complexity of the nonhuman primate retroviridae. AIDS Res. Hum. Retroviruses *10*, 1047–1060.

Fultz, P.N. (1994). STLV type-1. In J. Levy, ed. The Retroviridae. (New York: Plenum) *Vol. 3*, pp. 111–131.

Fultz, P.N., Gordon, T.P., Anderson, D.C., and McClure, H.M. (1990). Prevalence of natural infection with simian immunodeficiency virus and simian T-cell leukemia virus type I in a breeding colony of sooty mangabey monkeys. Aids *4*, 619–625.

Gao, F., Bailes, E., Robertson, D.L., Chen, Y., Rodenburg, C.M., Michael, S.F., Cummins, L.B., Arthur, L.O., Peeters, M., Shaw, G.M., *et al.* (1999). Origin of HIV-1 in the chimpanzee Pan troglodytes troglodytes. Nature *397*, 436–441.

Gardner, M.B. (1996). The history of simian AIDS. J. Med Primatol *25*, 148–157.

Gardner, M.B., Endres, M., and Barry, P. (1994). The simian retroviruses SIV and SRV. In: The Retroviridae Vol. 3, Levy JA (ed). New York: Plenum Press, pp. 133–1276.

Georges-Courbot, M.C., Lu, C.Y., Makuwa, M., Telfer, P., Onanga, R., Dubreuil, G., Chen, Z., Smith, S.M., Georges, A., Gao, F., *et al.* (1998). Natural infection of a household pet red-capped mangabey (Cercocebus torquatus torquatus) with a new simian immunodeficiency virus. J. Virol. *72*, 600–608.

Georges-Courbot, M.C., Moisson, P., Leroy, E., Pingard, A.M., Nerrienet, E., Dubreuil, G., Wickings, E. J., Debels, F., Bedjabaga, I., Poaty-Mavoungou, V., *et al.* (1996). Occurrence and frequency of transmission of naturally occurring simian retroviral infections (SIV, STLV, and SRV) at the CIRMF Primate Center, Gabon. J. Med Primatol *25*, 313–326.

Gessain, A., and de The, G. (1996). Geographic and molecular epidemiology of primate T lymphotropic retroviruses: HTLV-I, HTLV-II, STLV-I, STLV-PP, and PTLV-L. Adv Virus Res. *47*, 377–426.

Gibbs, C.J., Jr., and Gajdusek, D.C. (1973). Experimental subacute spongiform virus encephalopathies in primates and other laboratory animals. Science *182*, 67–68.

Giri, A., Markham, P., Digilio, L., Hurteau, G., Gallo, R.C., and Franchini, G. (1994). Isolation of a novel simian T-cell lymphotropic virus from *Pan paniscus* that is distantly related to the human T-cell leukemia/lymphotropic virus types I and II. J. Virol. *68*, 8392–8395.

Gordon, S., Pandrea, I., Dunham, R., Apetrei, C., and Silvestri, G. (2005). The call of the wild: what can we learn from studies of SIV infection of natural hosts? In Leitner, T., Foley, B., Hahn, B., Marx, P., McCutchan, F., Mellors, J., Wolinsky, S., and Korber, B. (eds) HIV Sequence Compendium 2005 Theoretical Biology and Biophysics Group, Los Alamos National Laboratory, NM, LA-UR 04-7420, 2–29.

Gormus, B.J., Martin, L.N., and Baskin, G.B. (2004). A brief history of the discovery of natural simian immunodeficiency virus (SIV) infections in captive sooty mangabey monkeys. Front. Biosci. *9*, 216–224.

Gormus, B.J., Xu, K., Baskin, G.B., Martin, L.N., Bohm, R.P., Blanchard, J.L., Mack, P.A., Ratterree, M.S., McClure, H.M., Meyers, W.M. , *et al.* (1995a). Experimental leprosy in monkeys. I. Sooty mangabey monkeys: transmission, susceptibility, clinical and pathological findings. Lepr. Rev. *66*, 96–104.

Gormus, B.J., Xu, K., Cho, S.N., Baskin, G.B., Bohm, R.P., Martin, L.N., Blanchard, J.L., Mack, P.A., Ratterree, M.S., Meyers, W.M., *et al.* (1995b). Experimental leprosy in monkeys. II. Longitudinal serological observations in sooty mangabey monkeys. Lep.r Rev. *66*, 105–125.

Goubau, P., Van Brussel, M., Vandamme, A.M., Liu, H.F., and Desmyter, J. (1994). A primate T-lymphotropic virus, PTLV-L, different from human T- lymphotropic viruses types I and II, in a wild-caught baboon

(*Papio hamadryas*). Proc. Natl. Acad. Sci. U.S.A. *91*, 2848–2852.

Grant, R.F., Windsor, S.K., Malinak, C.J., Bartz, C. R., Sabo, A., Benveniste, R.E., and Tsai, C.C. (1995). Characterization of infectious type D retrovirus from baboons. Virology *207*, 292–296.

Hatanaka, M., and Nam, S. H. (1989). Identification of HTLV-I *gag* protease and its sequential processing of the *gag* gene product. J. Cell Biochem. *40*, 15–30.

Hattori, S., Kiyokawa, T., Imagawa, K., Shimizu, F., Hashimura, E., Seiki, M., and Yoshida, M. (1984). Identification of *gag* and *env* gene products of human T-cell leukemia virus (HTLV). Virology *136*, 338–347.

Heberling, R.L., Barker, S.T., Kalter, S.S., Smith, G.C., and Helmke, R.J. (1977). Oncornavirus: isolation from a squirrel monkey (*Saimiri sciureus*) lung culture. Science *195*, 289–292.

Heeney, J.L. (2006). Zoonotic viral diseases and the frontier of early diagnosis, control and prevention. J. Intern. Med. *260*, 399–408.

Heeney, J.L., Dalgleish, A.G., and Weiss, R. A. (2006). Origins of HIV and the evolution of resistance to AIDS. Science *313*, 462–466.

Heeney, J.L., and Valli, V.E. (1990). Transformed phenotype of enzootic bovine lymphoma reflects differentiation-linked leukemogenesis. Lab. Invest. *62*, 339–346.

Heneine, W., Switzer, W.M., Sandstrom, P., Brown, J., Vedapuri, S., Schable, C.A., Khan, A.S., Lerche, N.W., Schweizer, M., Neumann-Haefelin, D., *et al.* (1998). Identification of a human population infected with simian foamy viruses. Nat. Med. *4*, 403–407.

Herchenroder, O., Renne, R., Loncar, D., Cobb, E. K., Murthy, K.K., Schneider, J., Mergia, A., and Luciw, P.A. (1994). Isolation, cloning, and sequencing of simian foamy viruses from chimpanzees (SFVcpz): high homology to human foamy virus (HFV). Virology *201*, 187–199.

Hidaka, M., Inoue, J., Yoshida, M., and Seiki, M. (1988). Post-transcriptional regulator (rex) of HTLV-1 initiates expression of viral structural proteins but suppresses expression of regulatory proteins. EMBO J. *7*, 519–523.

Hinuma, Y., Nagata, K., Hanaoka, M., Nakai, M., Matsumoto, T., Kinoshita, K.I., Shirakawa, S., and Miyoshi, I. (1981). Adult T-cell leukemia: antigen in an ATL cell line and detection of antibodies to the antigen in human sera. Proc. Natl. Acad. Sci. U.S.A. *78*, 6476–6480.

Hirsch, V.M., Campbell, B.J., Bailes, E., Goeken, R., Brown, C., Elkins, W. R., Axthelm, M., Murphey-Corb, M., and Sharp, P. M. (1999). Characterization of a novel simian immunodeficiency virus (SIV) from L'Hoest monkeys (*Cercopithecus l'hoesti*): implications for the origins of SIVmnd and other primate lentiviruses. J. Virol. *73*, 1036–1045.

Hu, J., Switzer, W. M., Foley, B. T., Robertson, D.L., Goeken, R.M., Korber, B.T., Hirsch, V.M., and Beer, B.E. (2003). Characterization and comparison of recombinant simian immunodeficiency virus from drill (*Mandrillus leucophaeus*) and mandrill (*Mandrillus sphinx*) isolates. J. Virol. *77*, 4867–4880.

Ibuki, K., Ido, E., Setiyaningsih, S., Yamashita, M., Agus, L. R., Takehisa, J., Miura, T., Dondin, S., and Hayami, M. (1997). Isolation of STLV-I from orangutan, a great ape species in Southeast Asia, and its relation to other HTLV-Is/STLV-Is. Jpn J. Cancer Res. *88*, 1–4.

Ishikawa, K., Fukasawa, M., Tsujimoto, H., Else, J.G., Isahakia, M., Ubhi, N.K., Ishida, T., Takenaka, O., Kawamoto, Y., Shotake, T., *et al.* (1987). Serological survey and virus isolation of simian T-cell leukemia/T-lymphotropic virus type I (STLV-I) in non-human primates in their native countries. Int J. Cancer *40*, 233–239.

Jensen, E.M., Zelljadt, I., Chopra, H.C., and Mason, M.M. (1970). Isolation and propagation of a virus from a spontaneous mammary carcinoma of a rhesus monkey. Cancer Res. *30*, 2388–2393.

Jern, P., and Coffin, J. M. (2008). Host-retrovirus arms race: trimming the budget. Cell Host Microbe *4*, 196–197.

Jin, M.J., Rogers, J., Phillips-Conroy, J.E., Allan, J. S., Desrosiers, R.C., Shaw, G.M., Sharp, P. M., and Hahn, B.H. (1994). Infection of a yellow baboon with simian immunodeficiency virus from African green monkeys: evidence for cross-species transmission in the wild. J. Virol. *68*, 8454–8460.

Jordan, I., Enssle, J., Guttler, E., Mauer, B., and Rethwilm, A. (1996). Expression of human foamy virus reverse transcriptase involves a spliced *pol* mRNA. Virology *224*, 314–319.

Kalter, S.S., Helmke, R.J., Panigel, M., Heberling, R.L., Felsburg, P.J., and Axelrod, L.R. (1973). Observations of apparent C-type particles in baboon (*Papio cynocephalus*) placentas. Science *179*, 1332–1333.

Kawakami, T.G., Huff, S.D., Buckley, P.M., Dungworth, D.L., Synder, S.P., and Gilden, R.V. (1972). C-type virus associated with gibbon lymphosarcoma. Nat New Biol *235*, 170–171.

Keele, B.F., Van Heuverswyn, F., Li, Y., Bailes, E., Takehisa, J., Santiago, M.L., Bibollet-Ruche, F., Chen, Y., Wain, L.V., Liegeois, F., *et al.* (2006). Chimpanzee reservoirs of pandemic and nonpandemic HIV-1. Science *313*, 523–526.

Khan, A.S., Galvin, T.A., Lowenstine, L.J., Jennings, M.B., Gardner, MB., and Buckler, C.E. (1991). A highly divergent simian immunodeficiency virus (SIVstm) recovered from stored stump-tailed macaque tissues. J. Virol. *65*, 7061–7065.

Koralnik, I.J., Boeri, E., Saxinger, W. C., Monico, A. L., Fullen, J., Gessain, A., Guo, H. G., Gallo, R.C., Markham, P., Kalyanaraman, V., *et al.* (1994). Phylogenetic associations of human and simian T-cell leukemia/lymphotropic virus type I strains: evidence for interspecies transmission. J. Virol. *68*, 2693–2707.

Kupiec, J J., Kay, A., Hayat, M., Ravier, R., Peries, J., and Galibert, F. (1991). Sequence analysis of the simian foamy virus type 1 genome. Gene *101*, 185–194.

Kuzmenok, O.I., Dvoryanchikov, G. A., Ponomareva, E.N., Goncharov, A.A., Fomin, I.K., Lee, S.T., Sanberg, P.R., and Potapnev, M.P. (2007). Myasthenia gravis accompanied by thymomas not related to foamy virus genome in Belarusian's patients. Int. J. Neurosci. *117*, 1603–1610.

Lee, H., Swanson, P., Shorty, V.S., Zack, J.A., Rosenblatt, J.D., and Chen, I.S. (1989). High rate of HTLV-II infection in seropositive i.v. drug abusers in New Orleans. Science *244*, 471–475.

Leendertz, F. H., Boesch, C., Junglen, S., Pauli, G., and Ellerbrok, H. (2003). Characterization of a new simian T-lymphocyte virus type 1 (STLV-1) in a wild living chimpanzee (Pan troglodytes verus) from Ivory Coast: evidence of a new STLV-1 group? AIDS Res. Hum. Retroviruses *19*, 255–258.

Leendertz, F. H., Junglen, S., Boesch, C., Formenty, P., Couacy-Hymann, E., Courgnaud, V., Pauli, G., and Ellerbrok, H. (2004). High variety of different simian T-cell leukemia virus type 1 strains in chimpanzees (Pan troglodytes verus) of the Tai National Park, Cote d'Ivoire. J. Virol. *78*, 4352–4356.

Leitner, T., Dazza, M. C., Ekwalanga, M., Apetrei, C., and Saragosti, S. (2007). Sequence diversity among chimpanzee simian immunodeficiency viruses (SIVcpz) suggests that SIVcpzPts was derived from SIVcpzPtt through additional recombination events. AIDS Res. Hum. Retroviruses *23*, 1114–1118.

Lerche, N.W., Marx, P.A., Osborn, K. G., Maul, D.H., Lowenstine, L.J., Bleviss, M. L., Moody, P., Henrickson, R.V., and Gardner, M.B. (1987). Natural history of endemic type D retrovirus infection and acquired immune deficiency syndrome in group-housed rhesus monkeys. J. Natl. Cancer. I*nst. 79*, 847–854.

Lerche, N.W., and Osborn, K. G. (2003). Simian retrovirus infections: potential confounding variables in primate toxicology studies. Toxicol. Pathol. *31 Suppl.*, 103–110.

Liegeois, F., Courgnaud, V., Switzer, W.M., Murphy, H.W., Loul, S., Aghokeng, A., Pourrut, X., Mpoudi-Ngole, E., Delaporte, E., and Peeters, M. (2006). Molecular characterization of a novel simian immunodeficiency virus lineage (SIVtal) from northern talapoins (*Miopithecus ogouensis*). Virology *349*, 55–65.

Liegeois, F., Lafay, B., Switzer, W. M., Locatelli, S., Mpoudi-Ngole, E., Loul, S., Heneine, W., Delaporte, E., and Peeters, M. (2008). Identification and molecular characterization of new STLV-1 and STLV-3 strains in wild-caught nonhuman primates in Cameroon. Virology *371*, 405–417.

Linial, M. L. (1999). Foamy viruses are unconventional retroviruses. J. Virol. *73*, 1747–1755.

Locatelli, S., Lafay, B., Liegeois, F., Ting, N., Delaporte, E., and Peeters, M. (2008). Full molecular characterization of a simian immunodeficiency virus, SIVwrcpbt from Temminck's red colobus (*Piliocolobus badius temminckii*) from Abuko Nature Reserve, The Gambia. Virology *376*, 90–100.

Lochelt, M., Muranyi, W., and Flugel, R.M. (1993). Human foamy virus genome possesses an internal, Bel-1-dependent and functional promoter. Proc. Natl. Acad. Sci. U.S.A. *90*, 7317–7321.

Lochelt, M., Romen, F., Bastone, P., Muckenfuss, H., Kirchner, N., Kim, Y.B., Truyen, U., Rosler, U., Battenberg, M., Saib, A., *et al.* (2005). The antiretroviral activity of APOBEC3 is inhibited by the foamy virus accessory Bet protein. Proc. Natl. Acad. Sci. U.S.A. *102*, 7982–7987.

Lowenstine, L.J., Lerche, N.W., Yee, J.L., Uyeda, A., Jennings, M.B., Munn, R., McClure, H.M., Anderson, D.C., Fultz, P.N., and Gardner, M.B. (1992). Evidence for a lentiviral etiology in an epizootic of immune deficiency and lymphoma in stump-tailed macaques (*Macaca arctoides*). J. Med Primatol *21*, 1–14.

Lowenstine, L.J., Pedersen, N.C., Higgins, J., Pallis, K.C., Uyeda, A., Marx, P., Lerche, N.W., Munn, R. J., and Gardner, M.B. (1986). Seroepidemiologic survey of captive Old-World primates for antibodies to human and simian retroviruses, and isolation of a lentivirus from sooty mangabeys (*Cercocebus atys*). Int J. Cancer *38*, 563–574.

Mahieux, R., Chappey, C., Georges-Courbot, M. C., Dubreuil, G., Mauclere, P., Georges, A., and Gessain, A. (1998a). Simian T-cell lymphotropic virus type 1 from Mandrillus sphinx as a simian counterpart of human T-cell lymphotropic virus type 1 subtype D. J. Virol. *72*, 10316–10322.

Mahieux, R., Chappey, C., Meertens, L., Mauclere, P., Lewis, J., and Gessain, A. (2000). Molecular characterization and phylogenetic analyses of a new simian T-cell lymphotropic virus type 1 in a wild-caught African baboon (*Papio anubis*) with an indeterminate STLV type 2-like serology. AIDS Res. Hum. Retroviruses *16*, 2043–2048.

Mahieux, R., and Gessain, A. (2009). The human HTLV-3 and HTLV-4 retroviruses: New members of the HTLV family. Pathol. Biol. *57*, 161–166.

Mahieux, R., Ibrahim, F., Mauclere, P., Herve, V., Michel, P., Tekaia, F., Chappey, C., Garin, B., Van Der Ryst, E., Guillemain, B., *et al.* (1997). Molecular epidemiology of 58 new African human T-cell leukemia virus type 1 (HTLV-1) strains: identification of a new and distinct HTLV-1 molecular subtype in Central Africa and in Pygmies. J. Virol. *71*, 1317–1333.

Mahieux, R., Pecon-Slattery, J., Chen, G.M., and Gessain, A. (1998b). Evolutionary inferences of novel simian T lymphotropic virus type 1 from wild-caught chacma (Papio ursinus) and olive baboons (*Papio anubis*). Virology *251*, 71–84.

Makuwa, M., Souquiere, S., Clifford, S.L., Telfer, P.T., Salle, B., Bourry, O., Onanga, R., Mouinga-Ondeme, A., Wickings, E.J., Abernethy, K.A., *et al.* (2004a). Two distinct STLV-1 subtypes infecting Mandrillus sphinx follow the geographic distribution of their hosts. AIDS Res. Hum. Retroviruses *20*, 1137–1143.

Makuwa, M., Souquiere, S., Telfer, P., Mouinga-Ondeme, A., Bourry, O., and Roques, P. (2004b). A New STLV-1 in a household pet Cercopithecus nictitans from Gabon. AIDS Res. Hum. Retroviruses *20*, 679–683.

Mansfield, K.G., Lerch, N.W., Gardner, M.B., and Lackner, A.A. (1995). Origins of simian immunodeficiency virus infection in macaques at the New England Regional Primate Research Center. J. Med. Primatol. *24*, 116–122.

Marx, P.A., Bryant, M.L., Osborn, K.G., Maul, D.H., Lerche, N.W., Lowenstine, L.J., Kluge, J.D., Zaiss, C.P., Henrickson, R.V., Shiigi, S. M., *et al.* (1985). Isolation of a new serotype of simian acquired immune deficiency syndrome type D retrovirus from Celebes black macaques (*Macaca nigra*) with immune deficiency and retroperitoneal fibromatosis. J. Virol. *56*, 571–578.

Marx, P.A., Li, Y., Lerche, N.W., Sutjipto, S., Gettie, A., Yee, J.A., Brotman, B.H., Prince, A.M., Hanson, A., Webster, R.G., *et al.* (1991). Isolation of a simian immunodeficiency virus related to human immunodeficiency virus type 2 from a west African pet sooty mangabey. J. Virol. *65*, 4480–4485.

Marx, P.A., Maul, D.H., Osborn, K.G., Lerche, N.W., Moody, P., Lowenstine, L.J., Henrickson, R.V., Arthur, L.O., Gilden, R. V., Gravell, M., *et al.* (1984). Simian AIDS: isolation of a type D retrovirus and transmission of the disease. Science *223*, 1083–1086.

Maul, D.H., Lerche, N.W., Osborn, K.G., Marx, P.A., Zaiss, C., Spinner, A., Kluge, J. D., MacKenzie, M. R., Lowenstine, L.J., Bryant, M. L., *et al.* (1986). Pathogenesis of simian AIDS in rhesus macaques inoculated with the SRV-1 strain of type D retrovirus. Am. J. Vet. *Res. 47*, 863–868.

Meertens, L., and Gessain, A. (2003). Divergent simian T-cell lymphotropic virus type 3 (STLV-3) in wild-caught Papio hamadryas papio from Senegal: widespread distribution of STLV-3 in Africa. J. Virol. *77*, 782–789.

Meertens, L., Mahieux, R., Mauclere, P., Lewis, J., and Gessain, A. (2002). Complete sequence of a novel highly divergent simian T-cell lymphotropic virus from wild-caught red-capped mangabeys (*Cercocebus torquatus*) from Cameroon: a new primate T-lymphotropic virus type 3 subtype. J. Virol. *76*, 259–268.

Meertens, L., Rigoulet, J., Mauclere, P., Van Beveren, M., Chen, G. M., Diop, O., Dubreuil, G., Georges-Goubot, M. C., Berthier, J. L., Lewis, J., and Gessain, A. (2001). Molecular and phylogenetic analyses of 16 novel simian T-cell leukemia virus type 1 from Africa: close relationship of STLV-1 from Allenopithecus nigroviridis to HTLV-1 subtype B strains. Virology *287*, 275–285.

Meertens, L., Shanmugam, V., Gessain, A., Beer, B. E., Tooze, Z., Heneine, W., and Switzer, W. M. (2003). A novel, divergent simian T-cell lymphotropic virus type 3 in a wild-caught red-capped mangabey (*Cercocebus torquatus torquatus*) from Nigeria. J. Gen. Virol. *84*, 2723–2727.

Meiering, C. D., and Linial, M. L. (2001). Historical perspective of foamy virus epidemiology and infection. Clin Microbiol Rev *14*, 165–176.

Miyoshi, I., Yoshimoto, S., Fujishita, M., Taguchi, H., Kubonishi, I., Niiya, K., and Minezawa, M. (1982). Natural adult T-cell leukemia virus infection in Japanese monkeys [letter]. Lancet *2*, 658.

Muranyi, W., and Flugel, R. M. (1991). Analysis of splicing patterns of human spumaretrovirus by polymerase chain reaction reveals complex RNA structures. J. Virol. *65*, 727–735.

Murphey-Corb, M., Martin, L. N., Rangan, S. R., Baskin, G. B., Gormus, B. J., Wolf, R. H., Andes, W. A., West, M., and Montelaro, R. C. (1986). Isolation of an HTLV-III-related retrovirus from macaques with simian AIDS and its possible origin in asymptomatic mangabeys. Nature *321*, 435–437.

Murray, S. M., Picker, L. J., Axthelm, M. K., Hudkins, K., Alpers, C. E., and Linial, M. L. (2008). Replication in a superficial epithelial cell niche explains the lack of pathogenicity of primate foamy virus infections. J. Virol. *82*, 6109–6119.

Nam, S. H., Copeland, T. D., Hatanaka, M., and Oroszlan, S. (1993). Characterization of ribosomal frameshifting for expression of *pol* gene products of human T-cell leukemia virus type I. J. Virol. *67*, 196–203.

Nam, S. H., and Hatanaka, M. (1986). Identification of a protease gene of human T-cell leukemia virus type I (HTLV-I) and its structural comparison. Biochem Biophys Res Commun *139*, 129–135.

Nandi, J. S., Bhavalkar-Potdar, V., Tikute, S., and Raut, C.G. (2000). A novel type D simian retrovirus naturally infecting the Indian Hanuman langur (Semnopithecus entellus). Virology *277*, 6–13.

Neil, S. J., Zang, T., and Bieniasz, P. D. (2008). Tetherin inhibits retrovirus release and is antagonized by HIV-1 Vpu. Nature *451*, 425–430.

Nerrienet, E., Amouretti, X., Muller-Trutwin, M. C., Poaty-Mavoungou, V., Bedjebaga, I., Nguyen, H. T., Dubreuil, G., Corbet, S., Wickings, E. J., Barre-Sinoussi, F., *et al.* (1998). Phylogenetic analysis of SIV and STLV type I in mandrills (*Mandrillus sphinx*): indications that intracolony transmissions are predominantly the result of male-to-male aggressive contacts. AIDS Res. Hum. Retroviruses *14*, 785–796.

Nerrienet, E., Meertens, L., Kfutwah, A., Foupouapouognigni, Y., Ayouba, A., and Gessain, A. (2004). Simian T-cell leukaemia virus type I subtype B in a wild-caught gorilla (*Gorilla gorilla gorilla*) and chimpanzee (*Pan troglodytes vellerosus*) from Cameroon. J. Gen. Virol. *85*, 25–29.

Nerrienet, E., Meertens, L., Kfutwah, A., Foupouapouognigni, Y., and Gessain, A. (2001). Molecular epidemiology of simian T-lymphotropic virus (STLV) in wild- caught monkeys and apes from Cameroon: a new STLV-1, related to human T- lymphotropic virus subtype F, in a *Cercocebus agilis*. J. Gen. Virol. *82*, 2973–2977.

Niphuis, H., Verschoor, E.J., Bontjer, I., Peeters, M., and Heeney, J. L. (2003). Reduced transmission and prevalence of simian T-cell lymphotropic virus in a closed breeding colony of chimpanzees (*Pan troglodytes verus*). J. Gen. Virol. *84*, 615–620.

Novembre, F. J., de Rosayro, J., Nidtha, S., O'Neil, S.P., Gibson, T.R., Evans-Strickfaden, T., Hart, C.E., and McClure, H.M. (2001). Rapid CD4(+) T-cell loss induced by human immunodeficiency virus type 1(NC) in uninfected and previously infected chimpanzees. J. Virol. *75*, 1533–1539.

O'Neil, S.P., Novembre, F. J., Hill, A.B., Suwyn, C., Hart, C.E., Evans-Strickfaden, T., Anderson, D. C., deRosayro, J., Herndon, J. G., Saucier, M., and McClure, H.M. (2000). Progressive infection in a subset of HIV-1-positive chimpanzees. J. Infect. Dis. *182*, 1051–1062.

Ohta, Y., Masuda, T., Tsujimoto, H., Ishikawa, K., Kodama, T., Morikawa, S., Nakai, M., Honjo, S., and Hayami, M. (1988). Isolation of simian immunodeficiency virus from African green monkeys and seroepidemiologic survey of the virus in various non-human primates. Int. J. Cancer *41*, 115–122.

Osborn, K.G., Prahalada, S., Lowenstine, L.J., Gardner, M.B., Maul, D.H., and Henrickson, R.V. (1984). The pathology of an epizootic of acquired immunodeficiency in rhesus macaques. Am J. Pathol. *114*, 94–103.

Osterhaus, A.D., Pedersen, N., van Amerongen, G., Frankenhuis, M. T., Marthas, M., Reay, E., Rose, T.M., Pamungkas, J., and Bosch, M.L. (1999). Isolation and partial characterization of a lentivirus from talapoin monkeys (*Myopithecus talapoin*). Virology *260*, 116–124.

Peeters, M., Courgnaud, V., Abela, B., Auzel, P., Pourrut, X., Bibollet-Ruche, F., Loul, S., Liegeois, F., Butel, C., Koulagna, D., *et al.* (2002). Risk to human health from a plethora of simian immunodeficiency viruses in primate bushmeat. Emerg. Infect. Dis. *8*, 451–457.

Philipp-Staheli, J., Marquardt, T., Thouless, M.E., Bruce, A.G., Grant, R.F., Tsai, C.C., and Rose, T.M. (2006). Genetic variability of the envelope gene of Type D simian retrovirus-2 (SRV-2) subtypes associated with SAIDS-related retroperitoneal fibromatosis in different macaque species. Virol. J. 3, 11.

Poiesz, B.J., Poiesz, M.J., and Choi, D. (2003). The human T-cell lymphoma/leukemia viruses. Cancer Invest. *21*, 253–277.

Poiesz, B.J., Ruscetti, F.W., Mier, J.W., Woods, A. M., and Gallo, R.C. (1980). T-cell lines established from human T-lymphocytic neoplasias by direct response to T-cell growth factor. Proc. Natl. Acad. Sci. U.S.A. *77*, 6815–6819.

Renne, R., Friedl, E., Schweizer, M., Fleps, U., Turek, R., and Neumann-Haefelin, D. (1992). Genomic organization and expression of simian foamy virus type 3 (SFV- 3). Virology *186*, 597–608.

Rethwilm, A., Erlwein, O., Baunach, G., Maurer, B., and ter Meulen, V. (1991). The transcriptional transactivator of human foamy virus maps to the bel 1 genomic region. Proc. Natl. Acad. Sci. U.S.A. *88*, 941–945.

Roucoux, D. F., and Murphy, E.L. (2004). The epidemiology and disease outcomes of human T-lymphotropic virus type II. AIDS Rev. *6*, 144–154.

Rous, P. (1910). A transmissible avian neoplasm (sarcoma of the common fowl). J. Exp. Med. *12*, 696–705.

Rous, P. (1911). A sarcoma of the fowl, transmissible by an agent seperable from the tumor cells. J. Exp. Med. *13*, 397–411.

Russell, R.A., Wiegand, H.L., Moore, M.D., Schafer, A., McClure, M.O., and Cullen, B.R. (2005). Foamy virus Bet proteins function as novel inhibitors of the APOBEC3 family of innate antiretroviral defense factors. J. Virol. *79*, 8724–8731.

Rustigian, R., Johnston, P., and Reihart, H. (1955). Infection of monkey kidney tissue cultures with virus-like agents. Proc. Soc. Exp. Biol. Med. *88*, 8–16.

Saksena, N.K., Herve, V., Durand, J.P., Leguenno, B., Diop, O. M., Digouette, J. P., Mathiot, C., Muller, M.C., Love, J.L., Dube, S., *et al.* (1994). Seroepidemiologic, molecular, and phylogenetic analyses of simian T- cell leukemia viruses (STLV-I) from various naturally infected monkey species from central and western Africa. Virology *198*, 297–310.

Saksena, N.K., Herve, V., Sherman, M.P., Durand, J.P., Mathiot, C., Muller, M., Love, J.L., LeGuenno, B., Sinoussi, F.B., Dube, D. K., *et al.* (1993). Sequence and phylogenetic analyses of a new STLV-I from a naturally infected tantalus monkey from Central Africa. Virology *192*, 312–320.

Salemi, M., De Oliveira, T., Courgnaud, V., Moulton, V., Holland, B., Cassol, S., Switzer, W.M., and Vandamme, A.M. (2003). Mosaic genomes of the six major primate lentivirus lineages revealed by phylogenetic analyses. J. Virol. *77*, 7202–7213.

Sandstrom, P.A., Phan, K.O., Switzer, W.M., Fredeking, T., Chapman, L., Heneine, W., and Folks, T.M. (2000). Simian foamy virus infection among zoo keepers. Lancet 355, 551–552.

Schreiber, G.B., Murphy, E.L., Horton, J.A., Wright, D.J., Garfein, R., Chien, H.C., and Nass, C.C. (1997). Risk factors for human T-cell lymphotropic virus types I and II (HTLV-I and -II) in blood donors: the Retrovirus Epidemiology Donor Study. NHLBI Retrovirus Epidemiology Donor Study. J. Acquir. Immune Defic. Syndr. Hum. Retrovirol. *14*, 263–271.

Schweizer, M., Falcone, V., Gange, J., Turek, R., and Neumann-Haefelin, D. (1997). Simian foamy virus isolated from an accidentally infected human individual. J. Virol. *71*, 4821–4824.

Schweizer, M., Turek, R., Reinhardt, M., and Neumann-Haefelin, D. (1994). Absence of foamy virus DNA in Graves' disease. AIDS Res. Hum. Retroviruses *10*, 601–605.

Slattery, J. P., Franchini, G., and Gessain, A. (1999). Genomic evolution, patterns of global dissemination, and interspecies transmission of human and simian T-cell leukemia/lymphotropic viruses. Genome Res. *9*, 525–540.

Sodroski, J., Rosen, C., Goh, W.C., and Haseltine, W. (1985). A transcriptional activator protein encoded by the x-lor region of the human T-cell leukemia virus. Science *228*, 1430–1434.

Souquiere, S., Bibollet-Ruche, F., Robertson, D.L., Makuwa, M., Apetrei, C., Onanga, R., Kornfeld, C., Plantier, J.C., Gao, F., Abernethy, K., *et al.* (2001). Wild *Mandrillus sphinx* are carriers of two types of lentivirus. J. Virol. 75, 7086–7096.

Stevenson, M. (2003). HIV-1 pathogenesis. Nat. Med. *9*, 853–860.

Svenningsson, A., Lycke, J., Svennerholm, B., Gronowitz, S., and Andersen, O. (1992). No evidence for spumavirus or oncovirus infection in relapsing-remitting multiple sclerosis. Ann. Neurol. 32, 711–714.

Switzer, W.M., Bhullar, V., Shanmugam, V., Cong, M. E., Parekh, B., Lerche, N. W., Yee, J. L., Ely, J. J., Boneva, R., Chapman, L. E., *et al.* (2004). Frequent simian foamy virus infection in persons occupationally exposed to nonhuman primates. J. Virol. *78*, 2780–2789.

Switzer, W.M., Qari, S. H., Wolfe, N.D., Burke, D.S., Folks, T.M., and Heneine, W. (2006). Ancient origin and molecular features of the novel human T-lymphotropic virus type 3 revealed by complete genome analysis. J. Virol. *80*, 7427–7438.

Switzer, W.M., Salemi, M., Shanmugam, V., Gao, F., Cong, M.E., Kuiken, C., Bhullar, V., Beer, B.E., Vallet, D., Gautier-Hion, A., *et al.* (2005). Ancient co-speciation of simian foamy viruses and primates. Nature *434*, 376–380.

Takehisa, J., Harada, Y., Ndembi, N., Mboudjeka, I., Taniguchi, Y., Ngansop, C., Kuate, S., Zekeng, L., Ibuki, K., Shimada, T., *et al.* (2001). Natural infection of wild-born mandrills (*Mandrillus sphinx*) with two different types of simian immunodeficiency virus. AIDS Res. Hum. Retroviruses *17*, 1143–1154.

Takemura, T., Ekwalanga, M., Bikandou, B., Ido, E., Yamaguchi-Kabata, Y., Ohkura, S., Harada, H., Takehisa, J., Ichimura, H., Parra, H.J., *et al.* (2005). A novel simian immunodeficiency virus from black mangabey (*Lophocebus aterrimus*) in the Democratic Republic of Congo. J. Gen. Virol. *86*, 1967–1971.

Takemura, T., and Hayami, M. (2004). Phylogenetic analysis of SIV derived from mandrill and drill. Front. Biosci. *9*, 513–520.

Takemura, T., Yamashita, M., Shimada, M.K., Ohkura, S., Shotake, T., Ikeda, M., Miura, T., and Hayami, M. (2002). High prevalence of simian T-lymphotropic virus type L in wild ethiopian baboons. J. Virol. *76*, 1642–1648.

Theilen, G.H. (1971). Continuing studies with transmissible feline fibrosarcoma virus in fetal and newborn sheep. J. Am. Vet. Med. Assoc. *158*, Suppl. *2*, 1040+.

Thumer, L., Rethwilm, A., Holmes, E.C., and Bodem, J. (2007). The complete nucleotide sequence of a New World simian foamy virus. Virology *369*, 191–197.

Tristem, M., Marshall, C., Karpas, A., and Hill, F. (1992). Evolution of the primate lentiviruses: evidence from vpx and vpr. EMBO J. *11*, 3405–3412.

Tristem, M., Purvis, A., and Quicke, D.L. (1998). Complex evolutionary history of primate lentiviral vpr genes. Virology *240*, 232–237.

Tsujimoto, H., Hasegawa, A., Maki, N., Fukasawa, M., Miura, T., Speidel, S., Cooper, R. W., Moriyama, E. N., Gojobori, T., and Hayami, M. (1989). Sequence of a novel simian immunodeficiency virus from a wild-caught African mandrill. Nature *341*, 539–541.

van der Kuyl, A.C., van den Burg, R., Hoyer, M.J., Gruters, R.A., Osterhaus, A.D., and Berkhout, B. (2004). SIVdrl detection in captive mandrills: are mandrills infected with a third strain of simian immunodeficiency virus? Retrovirology *1*, 36.

Van Dooren, S., Salemi, M., Pourrut, X., Peeters, M., Delaporte, E., Van Ranst, M., and Vandamme, A.M. (2001). Evidence for a second simian T-cell lymphotropic virus type 3 in *Cercopithecus nictitans* from Cameroon. J. Virol. *75*, 11939–11941.

Van Dooren, S., Shanmugam, V., Bhullar, V., Parekh, B., Vandamme, A.M., Heneine, W., and Switzer, W.M. (2004). Identification in gelada baboons (*Theropithecus gelada*) of a distinct simian T-cell lymphotropic virus type 3 with a broad range of Western blot reactivity. J. Gen. Virol. *85*, 507–519.

Van Dooren, S., Switzer, W. M., Heneine, W., Goubau, P., Verschoor, E., Parekh, B., De Meurichy, W., Furley, C., Van Ranst, M., and Vandamme, A. M. (2002). Lack of evidence for infection with simian immunodeficiency virus in bonobos. AIDS Res. Hum. Retroviruses *18*, 213–216.

Van Dooren, S., Verschoor, E. J., Fagrouch, Z., and Vandamme, A.M. (2007). Phylogeny of primate T lymphotropic virus type 1 (PTLV-1) including various new Asian and African non-human primate strains. Infect. Genet. Evol. *7*, 374–381.

Van Heuverswyn, F., Li, Y., Neel, C., Bailes, E., Keele, B.F., Liu, W., Loul, S., Butel, C., Liegeois, F., Bienvenue, Y., *et al.* (2006). Human immunodeficiency viruses: SIV infection in wild gorillas. Nature *444*, 164.

van Rensburg, E. J., Engelbrecht, S., Mwenda, J., Laten, J. D., Robson, B.A., Stander, T., and Chege, G.K. (1998). Simian immunodeficiency viruses (SIVs) from eastern and southern Africa: detection of a SIVagm variant from a chacma baboon. J. Gen. Virol. *79*, 1809–1814.

Vandamme, A. M., Liu, H. F., Van Brussel, M., De Meurichy, W., Desmyter, J., and Goubau, P. (1996). The presence of a divergent T-lymphotropic virus in a wild-caught pygmy chimpanzee (*Pan paniscus*) supports an African origin for the human T- lymphotropic/simian T-lymphotropic group of viruses. J. Gen. Virol. *77*, 1089–1099.

Vandamme, A.M., Salemi, M., Van Brussel, M., Liu, H. F., Van Laethem, K., Van Ranst, M., Michels, L., Desmyter, J., and Goubau, P. (1998). African origin of human T-lymphotropic virus type 2 (HTLV-2) supported by a potential new HTLV-2d subtype in Congolese Bambuti Efe Pygmies. J. Virol. *72*, 4327–4340.

Vanden Haesevelde, M.M., Peeters, M., Jannes, G., Janssens, W., van der Groen, G., Sharp, P.M., and Saman, E. (1996). Sequence analysis of a highly divergent HIV-1-related lentivirus isolated from a wild captured chimpanzee. Virology *221*, 346–350.

VandeWoude, S., and Apetrei, C. (2006). Going wild: lessons from naturally occurring T-lymphotropic lentiviruses. Clin. Microbiol. Rev. *19*, 728–762.

Veazey, R., DeMaria, M., Chalifoux, L.V., Shvetz, D. E., Pauley, D.R., Knight, H. L., Rosenzweig, M., Johnson, R. P., Desrosiers, R. C., and Lackner, A. A. (1998). Gastrointestinal tract as a major site of $CD4^+$ T-cell depletion and viral replication in SIV infection. Science *280*, 427–431.

Verschoor, E.J., Fagrouch, Z., Bontjer, I., Niphuis, H., and Heeney, J. L. (2004a). A novel simian immunodeficiency virus isolated from a Schmidt's guenon (*Cercopithecus ascanius schmidti*). J. Gen. Virol. *85*, 21–24.

Verschoor, E.J., Langenhuijzen, S., Bontjer, I., Fagrouch, Z., Niphuis, H., Warren, K. S., Eulenberger, K., and Heeney, J. L. (2004b). The phylogeography of orangutan foamy viruses supports the theory of ancient repopulation of Sumatra. J. Virol. *78*, 12712–12716.

Verschoor, E.J., Langenhuijzen, S., van den Engel, S., Niphuis, H., Warren, K. S., and Heeney, J. L. (2003). Structural and evolutionary analysis of an orangutan foamy virus. J. Virol. *77*, 8584–8587.

Voevodin, A., Samilchuk, E., Allan, J., Rogers, J., and Broussard, S. (1997). Simian T-lymphotropic virus type 1 (STLV-1) infection in wild yellow baboons

(*Papio hamadryas cynocephalus*) from Mikumi National Park, Tanzania. Virology *228*, 350–359.

Wolf, D., and Goff, S.P. (2008). Host restriction factors blocking retroviral replication. Annu. Rev. Genet. *42*, 143–163.

Wolfe, N.D., Heneine, W., Carr, J.K., Garcia, A.D., Shanmugam, V., Tamoufe, U., Torimiro, J. N., Prosser, A. T., Lebreton, M., Mpoudi-Ngole, E., *et al.* (2005). Emergence of unique primate T-lymphotropic viruses among central African bushmeat hunters. Proc. Natl. Acad. Sci. U.S.A. *102*, 7994–7999.

Wolfe, N.D., Switzer, W.M., Carr, J.K., Bhullar, V. B., Shanmugam, V., Tamoufe, U., Prosser, A. T., Torimiro, J. N., Wright, A., Mpoudi-Ngole, E., *et al.* (2004). Naturally acquired simian retrovirus infections in central African hunters. Lancet *363*, 932–937.

Yu, S., Sullivan, M.D., and Linial, M.L. (1999). Evidence that the human foamy virus genome is DNA. J. Virol. *73*, 1565–1572.

HTLV and HIV

16

Marvin S. Reitz, Jr and Robert C. Gallo

Abstract

For many years, retroviruses were known to be the cause of many kinds of animal leukaemias and hematopoietic tumours. In spite of the high expectation that this would also be true for humans, very little evidence for retroviral involvement in any human diseases was forthcoming. In the late 1970s, however, due to the development of sensitive and specific molecular methods to identify retroviruses and to produce large scale cultures of T-lymphocytes, HTLV-I was discovered and implicated as the cause of adult T-cell leukaemia, a particular and relatively infrequent leukaemia prevalent in southern Japan and parts of the Caribbean, and tropical spastic paraparesis, a demyelinating neuropathy similar to multiple sclerosis. The discovery that HTLV-I can be transmitted by breast milk has led to a significant decline in HTLV-I infections in Japan. Although no retroviruses have been identified to date in other human leukaemias or related diseases, the efforts that resulted in the discovery of HTLV-I were critical in isolating HIV-1 and identifying it as the cause of AIDS. The ability to grow the HIV-1 in quantity allowed the development of a blood test that has saved countless lives. The development of effective antiretroviral drugs has made HIV-1 infection a somewhat manageable chronic condition rather than a certain death sentence. Although vaccine trials thus far have been rather disappointing, an effective vaccine is one of our most important needs.

Background

Retroviruses were discovered early in the twentieth century. Ellerman and Bang showed that leukaemia could be transmitted in chickens by cell-free filtrates (Ellerman and Bang, 1908), and Rous used a similar technique to demonstrate transmission of sarcomas in chickens (Rous, 1910). These findings were replicated in mammals by Bittner for breast tumours in mice (Bittner *et al.*, 1945) and by Gross for mouse leukaemia (Gross, 1953). Jarrett showed that a retrovirus was responsible for cat leukaemia (Jarrett *et al.*, 1964a,b), which constituted the first demonstration of transmissible leukaemia in outbred species. Kawakami and Theilen and colleagues were the first to show that retroviruses could cause leukaemia in primates, specifically in gibbon apes and new world monkeys (Kawakami *et al.*, 1972; Theilen *et al.*, 1971).

The realization that retroviruses were the cause of leukaemias and lymphomas in chickens, mice, cats, and non-human primates led to the expectation that leukaemias were generally of retroviral aetiology. This in turn generated extensive efforts to identify analogous human viruses. An important fundamental discovery, the identification of reverse transcriptase in 1970, gave the field its first real molecular tool for the sensitive and relatively unambiguous detection of retroviruses. It also was the basis for naming these viruses retroviruses, as they reversed the normal flow of genetic information. Many laboratories devoted considerable efforts to refining this tool to identify human retroviruses, especially in leukaemias. Our own approach was two-fold. The first concentrated on the development of techniques with sufficient sensitivity and specificity to discriminate between reverse transcriptase and cellular DNA polymerases but with sufficient generality to detect reverse transcriptase from a variety of retroviruses. This would be particularly critical if a human retrovirus were to replicate at low levels. Although this was considered unlikely because animal retroviruses were expressed at very high levels, it was to prove the case with human T-cell leukaemia virus type I (HTLV-I). The second approach was to try to obtain and reproducibly grow sufficient quantities of the relevant hematopoietic cell types for detection of reverse transcriptase.

One of the first characteristics thought to be diagnostic for reverse transcriptase was the demonstration of RNAse-sensitive DNA synthesis in media fraction separated by density gradient, treated with non-ionic detergent and supplied with labelled DNA precursors (Schlom and Spiegelman, 1971). However, it became evident that often this activity was often due instead to RNA-primed but DNA-dependent DNA synthesized by mitochondrial DNA polymerase in membrane fragments banding with the approximate density of retroviral particles (Bobrow *et al.*, 1972, Reitz *et al.*, 1974). More rigorous criteria that were developed were the preference for synthetic polyribonucleotides over polydeoxyribonucleotides as templates by crude and purified reverse transcriptase (Goodman and Spiegelman, 1971, Robert *et al.*, 1972) and the ability of the purified enzyme to reverse transcribe natural RNA (Abrell *et al.*, 1975; Abrell and Gallo, 1973).

This second approach centred on the use of conditioned media as a source of cell growth factors. This work led to the discovery of interleukin 2 (IL-2), one of the first cytokine growth factors, and the recognition that it would support the growth of large quantities of T-cells over an extended time course (Morgan *et al.*, 1976; Ruscetti *et al.*, 1977). The discovery of IL-2 was of great significance in and of itself, but both techniques proved critical to the discovery of HTLV-I, because it is a poorly infectious virus that is not expressed at high levels.

Section I HTLV

Discovery of HTLV

These two avenues of work came to fruition in the late 1970s with the discovery of HTLV-I. The availability of IL-2 made it possible to culture large quantities of T-cells over an extended period of time, and a number of such cultures were established from T-cell malignancies, especially from patients diagnosed at the time with mycosis fungoides or Sezary syndrome, but now known to have had adult T-cell leukaemia/lymphoma (ATL) (Poiesz *et al.*, 1980b). Virus was detected from a cell line from a mycosis fungoides patient by reverse transcriptase assays and electron microscopy (Poiesz *et al.*, 1980a). Fig. 16.1A shows a typical HTLV-I viral particle. Subsequently, a similar virus was isolated from uncultured cells from a different patient diagnosed with Sezary syndrome (Poiesz *et al.*, 1981). The isolation and production of HTLV-I made nucleic acid and protein reagents available for its characterization and for clinical studies, including a blood test for the serum antibodies that signal infection. By nucleic acid and serologic techniques, HTLV-I was shown to be distinct from any known animal retroviruses (Kalyanaraman *et al.*, 1981b; Reitz *et al.*, 1981), to not be present in most types of leukaemia , and to be present specifically in the T-cells and not the B-cells of ATL patients (Gallo *et al.*, 1982). Miyoshi subsequently identified a retrovirus in ATL T-cells and in a cell line derived by co-cultivation of ATL T-cells with normal cord blood T-cells (Miyoshi *et al.*, 1981). The cell line was infected and was of the opposite sex to the ATL donor, demonstrating that the virus, later shown to be identical to HTLV-I, could both infect and transform target T-cells.

The identification of HTLV-I from a case of ATL was important in focusing on a possible causative association. ATL was recognized by Dr Takatsuki and his colleagues, who characterized it as being a distinct subset of leukaemias that was highly endemic in areas of southern Japan (Takatsuki *et al.*, 1976; Uchiyama *et al.*, 1977). Without this work, it would not have been a simple task to associate HTLV-I with leukaemia, as it is not found in other leukaemias and ATL is a relatively specific leukaemia. Fig. 16.1B shows the patient with ATL from whom HTLV-I was

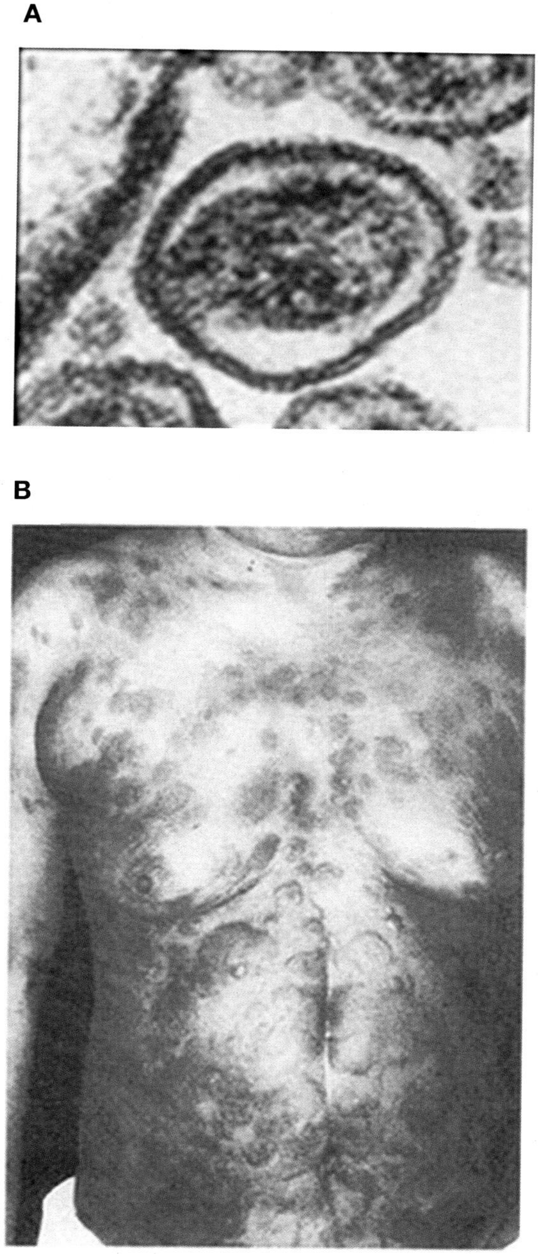

Figure 16.1 HTLV-I and adult T-cell leukaemia. (A) Electron micrograph of an extracellular HTLV-I virion. (B) Patient with cutaneous manifestations of adult T-cell leukaemia.

first isolated. The availability of an HTLV-I blood test allowed the clear identification of ATL in Japan, the Caribbean, and other areas (Blattner *et al.*, 1982; Kalyanaraman *et al.*, 1981a; Robert-Guroff *et al.*, 1981; Yoshida *et al.*, 1982) and led to the recognition that the virus also causes other diseases, most notably the multiple sclerosis-like tropical spastic paraparesis (TSP), also known as HTLV-associated myelopathy (Gessain *et al.*, 1985, Osame *et al.*, 1986). It was shown that HTLV-I was transmitted most efficiently by breast feeding (Ando *et al.*, 1987; Kinoshita *et al.*, 1987), making it possible to prevent a large fraction of transmissions. As a result, the incidence of ATL is declining precipitously in Japan.

The same techniques were used to isolate the second human retrovirus, HTLV-II, from a case of hairy cell leukaemia (Kalyanaraman *et al.*, 1982; Reitz *et al.*, 1983). HTLV-II, although relatively closely related to HTLV-I, has not so far been causally associated with any disease. Both viruses are widespread over various parts of the world and are clearly relatively ancient in human populations. Both also have counterparts, called simian T-cell leukaemia viruses types I and II (STLV-I and –II) in various Old World non-human primates (Chen *et al.*, 1994; Guo *et al.*, 1984; Watanabe *et al.*, 1985). More recently, a distinct but related pair of viruses called HTLV-III and STLV-III (Calattini *et al.*, 2005; Giri *et al.*, 1994; Liu *et al.*, 1994; Wolfe *et al.*, 2005) have been described in bush meat hunters and Old World non-human primates, respectively, although HTLV-III has not so far been associated with any diseases. Yet another member of the HTLV family, HTLV-IV, has also been identified from bush meat hunters (Wolfe *et al.*, 2005). It is likely that multiple cross-species transmissions from monkeys to humans represent the origins of the HTLV family, and it is possible that such transmissions are presently occurring.

HTLV virology

HTLV-I virions are poorly infectious, and transmission occurs most readily when it is cell to cell. It is likely that much viral replication depends upon duplication of integrated proviruses within infected cells as a consequence of cell division. Viral survival is thus partially dependent upon stimulation of infected cells to divide. Both $CD4^+$ and $CD8^+$ cells, especially from cord blood, can be infected *in vitro* (Markham *et al.*, 1984; Miyoshi *et al.*, 1981; Popovic *et al.*, 1983a,b), although $CD4^+$ cells are the main *in vivo* target. Infection results in a reduced dependence upon or an independence from exogenously added IL-2 and confers immortalization (Popovic *et al.*, 1983a). At first, the infected cells have a polyclonal integration site, but within a relatively short period of time an oligoclonal or monoclonal population emerges (Hahn *et al.*, 1984), suggesting that these cells have acquired a growth advantage. The specific integration sites change over time, probably through clonal evolution. This appears to also be the case in infected people over time, with the integrated viruses becoming increasingly defective, but generally retaining an LTR and the gene encoding the transactivating protein Tax. Generally, by the time of frank ATL, there is a monoclonal (although apparently random) integration site (Seiki *et al.*, 1984), suggesting that the virus is driving the clonal selection process. Interestingly, when ATL cells are put into culture, new clones appear and predominate that are not related to the leukemic cells (Jarrett *et al.*, 1986) and possibly represent newly infected normal cells.

HTLV genome

The genome of HTLV, shown schematically in Fig. 16.2, is about 8.7 kbp and is complex relative to most animal retroviruses (Seiki *et al.*, 1983). It contains an LTR at each end of the viral DNA. The LTRs contain the viral promoter region for RNA transcription, a Tax binding region, and a polyadenylation/transcription stop signal. The gag, protease, and polymerase genes are all encoded by a full-length transcript from different reading frames by translational frame shifting. The Gag proteins are synthesized as a polyprotein precursor and generated by cleavage by the viral protease. As shown schematically in Fig. 16.3, the polyprotein contains a p19 matrix protein, a p24 capsid protein and a p15 nucleocapsid protein. The matrix protein is myristoylated at position Gly2, targeting it to the plasma membrane. The protease, reverse transcriptase and integrase genes are likewise generated from polypeptide precursors. The envelope proteins are translated from a singly spliced transcript and the envelope is processed to a gp46 surface protein and a gp21 transmembrane protein (Fig. 16.4). These properties are broadly similar to the simple animal retroviruses. What was at the time unprecedented was the expression of a variety of transcripts encoding accessory or regulatory proteins that

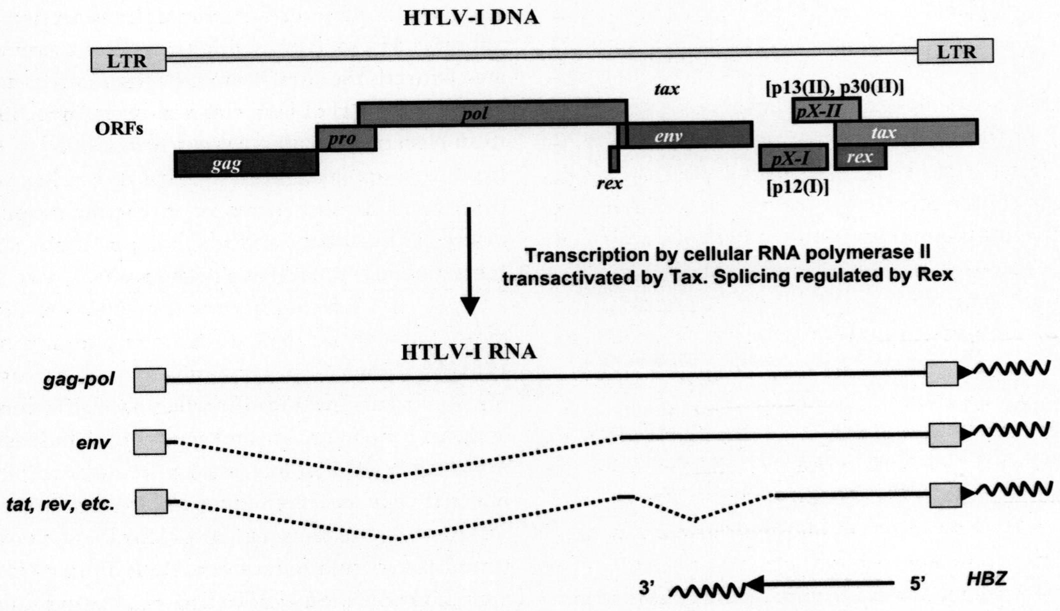

Figure 16.2 HTLV-I genome and transcripts. The genome of HTLV-I is represented with the open reading frames (ORFs) shown in their respective reading frames. LTR, long terminal repeat sequence. The squiggly lines on the RNA transcripts indicate poly(A) tracts, and the gray boxes show LTR-derived sequences. Arrowheads indicate the direction of transcription. Splices are represented by dashed lines.

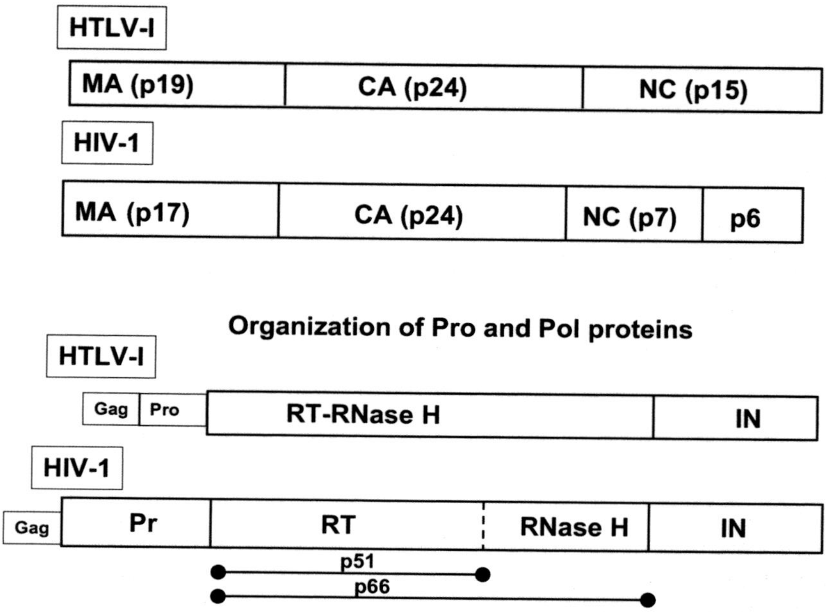

Figure 16.3 Organization of HTLV-I and HIV-1 Gag and Pol proteins. The organization of the HTLV-I and HIV-1 Gag and Pol polyproteins and their cleavage products is shown. The HTLV-I Pro protein is translated independently from its own reading frame. The HIV-1 RT is expressed in two forms, a 51-kDa form containing only RT and a 61-kDa form also containing RNase H.

contained three exons and were hence the products of multiple splicing (Arya *et al.*, 1985; Seiki *et al.*, 1985). Fig. 16.2 shows a representation of the viral transcripts.

HTLV regulatory proteins

The p40 Tax protein is encoded by exons 2 and 3 (Arya *et al.*, 1985; Seiki *et al.*, 1985), although exon 2 only contributes the initiation codon. Tax activates the transcription of viral RNA from the viral RNA polymerase promoter (Sodroski *et al.*, 1985b) and is necessary for viral expression. The p27 Rex protein is also encoded by exons 2 and 3 and regulates viral RNA splicing by enhancing the transport of unspliced and singly spliced viral transcripts into the cytoplasm (Inoue *et al.*, 1987). A smaller p21 form of Rex is translated from an alternative initiation codon within the third exon. Both Rev proteins are phosphorylated. Rev and Tax are both necessary for viral replication.

Transactivation of viral RNA transcription by Tax is mediated through its activation of multiple cellular transcription factors, including NFκB, CREB, AP-1 and SRF (Gitlin *et al.*, 1993; Lindholm *et al.*, 1990; Suzuki *et al.*, 1993a,b). Tax does not physically contact the viral promoter, but binds to transcription factors, increasing their ability to bind DNA and to recruit transcriptional co-activators such as CBP/p300. Tax disrupts p53 transcription function (Uittenbogaard *et al.*, 1995), probably leading to chromosomal instability and the aneuploidy that is a hallmark of ATL (Sanada *et al.*, 1985). Tax dysregulates cell cycle regulatory factors, including cyclin-dependent kinases and Rb (de La *et al.*, 2000; Haller *et al.*, 2002, Iwanaga *et al.*, 2001, Santiago *et al.*, 1999). It immortalizes T-cells, although they often remain at least somewhat dependent upon addition of exogenous IL-2. Cell transformation and at least some of the activation of transcriptional factors is dependent upon activation of the PI3 kinase (PI3K)-Akt (protein kinase B) pathway, although the mechanisms are complex and probably multifactorial. The activities of Tax go beyond activating transcription of viral RNA and include induction of expression of a wide variety of cellular genes including cytokines such as IL-2 and cell surface proteins such as the IL-2 receptor (Nimer *et al.*, 1989; Ruben *et al.*, 1988; Tschachler *et al.*, 1989; Wano *et al.*, 1988; Watanabe *et al.*, 1990).

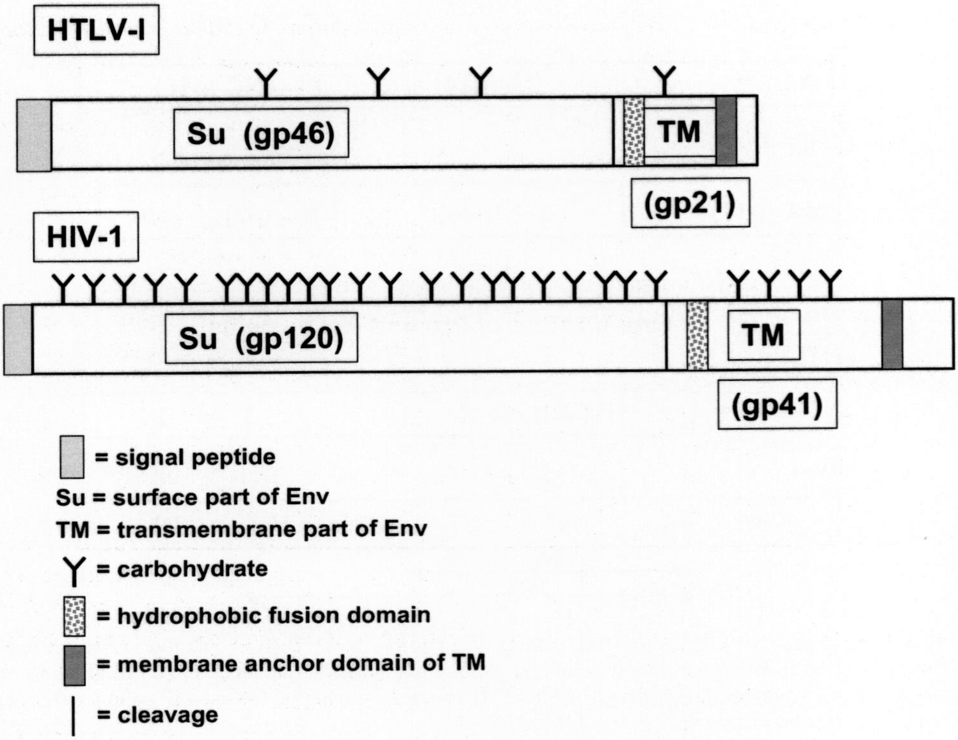

Figure 16.4 Organization of HTLV-I and HIV-1 Env proteins. The organization of the HTLV-I and HIV-1 Env polyproteins is shown, including glycosylation sites, the signal peptide/Su and Su/TM cleavage sites and the locations of the signal peptide and fusion and transmembrane anchor domains.

Rex is necessary for the export from the nucleus of unspliced and singly spliced mRNAs, which encode virion proteins (Inoue *et al.*, 1986; Inoue *et al.*, 1987). It functions by binding to a stem–loop structure of the viral RNA, the Rex responsive element (RxRE), encoded by the R/U5 element of the 5′LTR (Dillon *et al.*, 1990; Solomin *et al.*, 1990; Toyoshima *et al.*, 1990). Rex function involves binding to the cellular nuclear export protein CRM-1 (exportin-1) (Hakata *et al.*, 1998). In the absence of Rex, viral transcripts are processed to multiply spliced forms encoding only viral accessory proteins such as Rex and Tax or are degraded. Following nuclear export, Rex is cycled back into the nucleus following binding to importin-β (Palmeri and Malim, 1999).

Four other proteins, none of which is required for viral replication *in vitro*, are expressed from multiply spliced mRNAs. These include $p13^{II}$ and $p30^{II}$, encoded by pX ORF II, and $p12^{I}$ and $p27^{I}$, encoded by pX ORF I (Albrecht and Lairmore, 2002, Koralnik *et al.*, 1993). The two pX^{II} proteins, while not essential for viral replication in cell culture, are important for maintenance of viral loads in animal models (Bartoe *et al.*, 2000; Pique *et al.*, 2000). $p30^{II}$ appears to modulate transcription by modulating the activity of CREB-responsive promoters (Zhang *et al.*, 2001) and may repress viral expression (Albrecht and Lairmore, 2002). $p13^{II}$ alters mitochondrial membrane permeability and, interferes with cell proliferation and transformation, and promotes apoptosis, possibly through an influence on signal transduction by Ras (Hiraragi *et al.*, 2005; Silic-Benussi *et al.*, 2004). Little is known about the functions of $p27^{I}$. $p12^{I}$ appears to play a role in T-cell activation and to activate NF-AT through its binding to calcineurin (Albrecht *et al.*, 2002). $p12^{I}$ also activates STAT5 signalling (Nicot *et al.*, 2001).

Recent evidence indicates that a protein called HBZ is translated from the minus strand of viral RNA, transcribed from the 3′LTR, and is expressed in infected T-cells and in ATL cells as a 31-kDa protein. The HBZ protein heterodimerizes with cellular bZIP transcription factors including members of the CREB and Jun families (Mesnard *et al.*, 2006). It represses Tax-mediated viral transcription and supports proliferation of ATL cells.

HTLV pathogenesis

Infection with HTLV-I carries the risk of development of ATL, but with a relatively low efficiency. Far more people are infected that will develop ATL. As with animal leukaemias of retroviral origin, the latency period is very long, and infection in infancy is a strong risk factor (Murphy *et al.*, 1989). This suggests that other factors play a role in leukaemogenesis. These factors may be genetic (including immune response genes, leukaemia susceptibility genes, or sites of proviral integration) or environmental (including co-infections or chronic immune stimulation). Finding HTLV-I in all cases of ATL, however, strongly suggests that it is a necessary factor. Infection with HTLV-I is also predictive of disease, further supporting this idea. HTLV-I proviral DNA in ATL cells *in vivo* is mono- or oligoclonally integrated (Yoshida *et al.*, 1984), indicating that the cell from which the ATL originated was infected and suggesting causation in subsequent clonal outgrowth. Integration is semi-random, suggesting that leukaemia is not caused by insertional mutagenesis (Seiki *et al.*, 1984).

HTLV Tax protein transforms T-cells (Grassmann *et al.*, 1989), making it likely to be involved in ATL pathogenesis. Furthermore, although ATL cells generally contain partially deleted proviruses, Tax tends to be retained (Korber *et al.*, 1991), further suggesting its importance. On the other hand, viral genes are generally silent in ATL cells and the 5′LTR is methylated (Clarke *et al.*, 1984, Kitamura *et al.*, 1985), suggesting that by the time ATL is apparent, viral expression is no longer necessary. One caveat to this conclusion is that the HBZ protein described above appears to be expressed in ATL cells, and the 3′LTR, which contains the promoter for HBZ transcripts, is generally not methylated (Satou *et al.*, 2006). Given that Tax activities destabilize chromosomal structure (Liu *et al.*, 2003, Majone *et al.*, 2005), the likeliest scenario is that HTLV-I infection induces proliferation of infected T-cells. The host immune system eliminates most cells expressing virus, but clonal populations emerge that lack expression of one or more viral proteins. If Tax expression is maintained, another round of immune selection occurs, and this may continue for more rounds. During this time, genetic changes take place because of Tax or for other reasons. If, by the time that viral expression is no longer detected, the genetic changes are sufficient to make replication independent of Tax expression, ATL results. It is possible that HBZ plays a role at this point.

HIV

Early AIDS epidemic

The first reports of what became known as the acquired immunodeficiency syndrome (AIDS) were in 1981 (Friedman-Kien *et al.*, 1982; Gottlieb *et al.*, 1981; Hymes *et al.*, 1981; Masur *et al.*, 1981; Siegal *et al.*, 1981). They included reports of opportunistic infections, especially pneumocystis pneumonia, and Kaposi's sarcoma (KS), a previously rare and usually indolent form of skin cancer seen in elderly males of Mediterranean extraction. Identified risk groups, which first were restricted to homosexuals, included Haitians, heroin users, and haemophiliacs, and occurrence in the last group, as well as the identification of disease clusters, suggested an infectious origin. Many ideas of what was the causative agent were propounded. In 1982 our lab and that of Essex proposed that an HTLV-like virus was involved (Essex *et al.*, 1983, Gallo *et al.*, 1983, Gelmann *et al.*, 1983), since the disease affected $CD4^+$ T-cells, the target cell of HTLV, and since retroviruses, including feline leukaemia virus, could also cause immunodeficiency, and since HTLV-I causes modest immune impairment both *in vitro* and *in vivo*. Thus, the hypothesis of a new retrovirus related to but distinguishable from the known HTLVs was quite attractive.

Discovery of HIV-1 and its identification as the cause of AIDS

The first report of what is now called HIV-1 came in 1983 from the laboratory of Montagnier (Barre-Sinoussi *et al.*, 1983). Within the same year, our laboratory isolated many examples of what proved to be the same virus and put several into large scale production in late 1983-early 1984 (Gallo *et al.*, 1984; Popovic *et al.*, 1984; Sarngadharan *et al.*, 1984; Schupbach *et al.*, 1984). This permitted the development of serologic techniques that allowed the development of a blood test for the presence of the virus and identified it as the cause of AIDS. The blood test largely allowed the elimination of infected blood products and the consequent infection of haemophiliacs and transfusion recipients.

The availability of molecular probes for viral nucleic acids and proteins facilitated the discovery of a related human virus, called HIV-2, in people from western Africa (Barin *et al.*, 1985; Kanki *et al.*, 1986) and of a number of related viruses, called SIV, in Old World monkeys such as African green monkeys (Kanki *et al.*, 1985). Phylogenetic studies have suggested that HIV-2 was the result of cross-species transmission from monkeys to humans in Africa (Chen *et al.*, 1996; Sharp *et al.*, 1995), and that this has occurred recently and recurrently, presumably through the bush meat trade. More recently, chimpanzees (*Pan troglodytes troglodytes*) have been identified as the likely source of HIV-1 (Gao *et al.*, 1999; Keele *et al.*, 2006). The chimpanzee virus itself may have arisen by cross-species transmission as a recombinant between different SIV strains (Bailes *et al.*, 2003). Transmission appears to have occurred on at least three occasions, with one such transmission in south-eastern Cameroon being the origin of the most widespread HIV-1 group, HIV-1(M). Of course, it is likely that this has happened on innumerable other occasions, that the transmitted viruses failed to survive as infectious human agents, and that the three successful cross-species transmissions are the only ones that have been reported. Data has been recently presented (Gilbert *et al.*, 2007), suggesting that following its introduction into humans in Africa, HIV-1 was introduced into Haiti and subsequently into the US in the 1960s, from where it spread to many other countries, giving rise to clade B HIV-1(M), which is highly prevalent in the U.S and Europe as well as other locations. There are at least nine clades within HIV-1 group M, as well as many circulating recombinant forms, and it is likely that HIV will continue to evolve. Africa contains the widest variety of clades and inter-clade recombinants, with clade C being the most prevalent in many parts of Africa and in India.

HIV virology

HIV-1 belongs to the lentivirus group of retroviruses, which includes Maedi-visna virus, bovine and feline immunodeficiency viruses, equine infectious anaemia virus, and caprine arthritis-encephalomyelitis virus. A typical HIV-1 viral particle is shown in Fig. 16.5. Recently, an endogenous lentivirus in European rabbits, called RELIK, has been described that is estimated to have entered the rabbit genome seven million years ago and that could represent an ancestral form for all lentiviruses (Katzourakis *et al.*, 2007).

HIV-1 virions, unlike those of HTLV, are highly competent for infection of their target cells, which include primary $CD4^+$ T-cells, $CD4^+$ T-cell lines, and macrophage–monocytes. CD4 was identified as a primary host cell receptor for HIV-1 (Dalgleish *et al.*, 1984; Klatzmann *et al.*, 1984). HIV-1 isolates fell into two broad categories (Asjo *et al.*, 1986; Gartner *et al.*, 1986). One, called T-cell-tropic or syncytium inducing (SI), could infect primary T-cells and T-cell lines but was poorly infectious for macrophages and formed multicellular syncytia upon infection. The other, called macrophage-tropic or non-syncytium inducing (NSI), could infect macrophages and primary T-cells but not T-cell lines. A deeper understanding of the basis for this dichotomy was made possible by the identification of three β-chemokines, MIP-1α, MIP-1β, and RANTES, as natural host factors that strongly inhibited infection by NSI HIV-1 (Cocchi *et al.*, 1995). This was followed by the realization that SI isolates required a chemokine receptor now called CXCR4 (Feng *et al.*, 1996), which does not recognize MIP-1α, MIP-1β, or RANTES, as a second cell surface receptor, while NSI isolates instead required CCR5, for which all three chemokines are ligands (Alkhatib *et al.*, 1996; Berson *et al.*, 1996; Choe *et al.*, 1996; Deng *et al.*, 1996; Doranz *et al.*, 1996, Dragic *et al.*, 1996). Consequently, NSI and SI isolates are now called R5 and X4 isolates, respectively. Binding to CD4 is generally required to trigger a conformational change in the viral surface envelope protein that exposes the chemokine receptor binding site. Binding to the chemokine receptor triggers a second conformational change

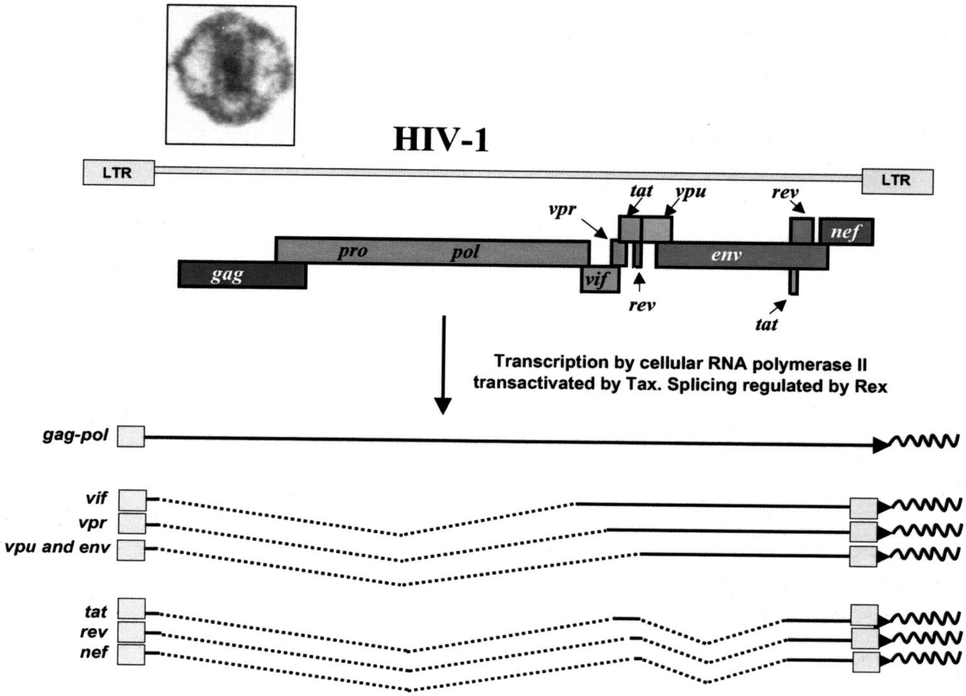

Figure 16.5 HIV-1 genome and transcripts. The genome of HIV-1 is represented with the coding regions for the indicated genes shown in their respective reading frames. LTR, long terminal repeat sequence. The squiggly lines on the RNA transcripts indicate poly (A) tracts, and the gray boxes show LTR-derived sequences. Arrowheads indicate the direction of transcription. Splices are represented by dashed lines. The inset shows an extracellular HIV-1 virion.

that exposes the membrane fusion domain of the transmembrane envelope protein and leads to cell entry. This is summarized in Fig. 16.6.

Transmission of HIV to a new host appears to occur primarily with R5 isolates. As a result, individuals who are homozygous for a deletion mutation, Δ32, that results in a CCR5-null phenotype are highly resistant to infection (Dean *et al.*, 1996; Huang *et al.*;1996, Liu *et al.*, 1996; Samson *et al.*, 1996). They do not suffer from obvious resultant pathologies, leading to the idea that targeting CCR5 may be a promising therapeutic (or even preventative) approach. Indeed, small molecules targeting CCR5 have been developed and are beginning to be used clinically.

Following entry into the cell, the preintegration complex (PIC) is released and reverse transcription takes place. The PIC is transported to the nucleus, where the viral DNA is integrated into host cell chromosomes. Unlike non-lentiviruses, the HIV-1 PIC is able to actively enter nuclei of non-dividing cells (Bukrinsky *et al.*, 1992; Di Marzio *et al.*, 1995; Gallay *et al.*, 1997; Mahalingam *et al.*, 1997). This is facilitated by nuclear targeting signals on the matrix, Vpr, and integrase proteins of the PIC (Bukrinsky *et al.*, 1993; Di Marzio *et al.*, 1995, Gallay *et al.*, 1997; Mahalingam *et al.*, 1997). This is shown schematically in Fig. 16.7.

HIV genome

The genome of HIV, represented in Fig. 16.2, is about 9750 bp (Muesing *et al.*, 1985; Ratner *et al.*, 1985; Sanchez-Pescador *et al.*, 1985; Wain-Hobson *et al.*, 1985). Although it bears only limited sequence homology to HTLV, structurally it bears a strong resemblance, in that it encodes numerous accessory proteins and undergoes a complex pattern of multiple RNA splicing, shown schematically in Figs. 16.2 and 16.8. As with other retroviruses, the LTRs contain the viral promoter/enhancer for cellular RNA polymerase and the polyadenylation/transcription stop signal. They also contain the transactivation response (TAR)

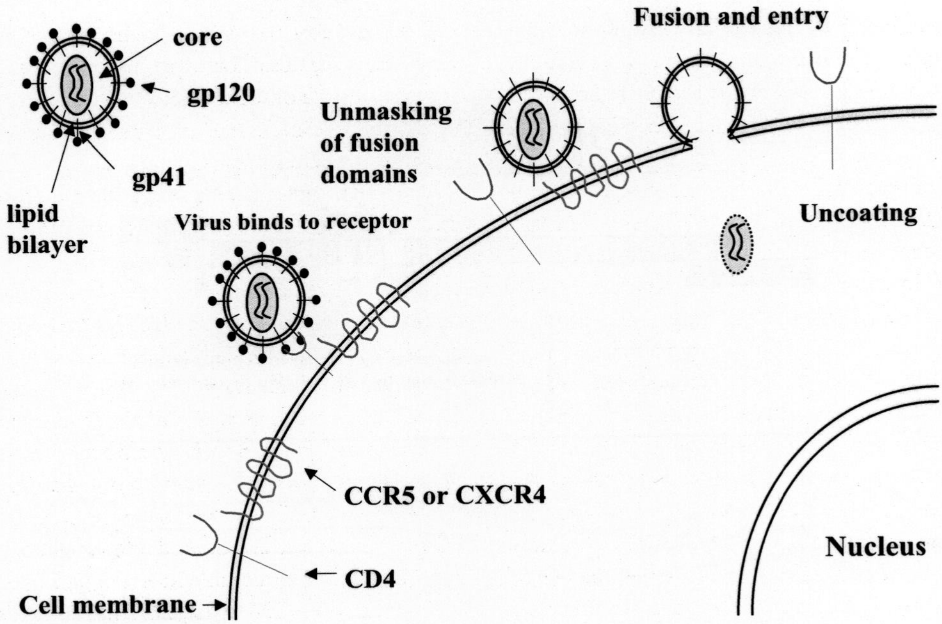

Figure 16.6 Early events in HIV infection: attachment and entry. Binding of gp120 to the CD4 and CCR5 receptors is the first step in infection, followed by fusion with the cell membrane, mediated by the transmembrane gp41 protein. Following fusion, the viral capsid is released into the cytoplasm.

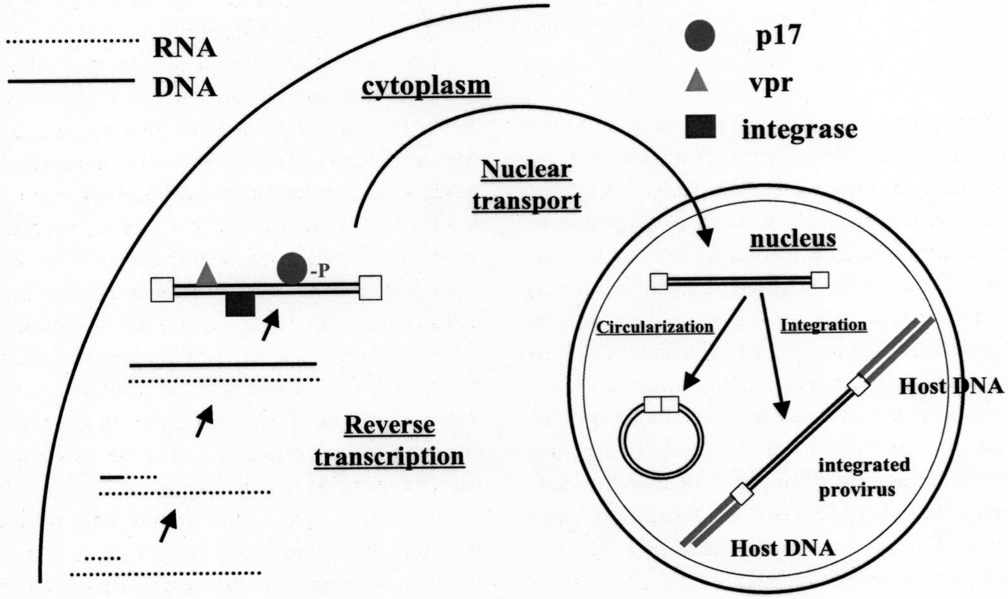

Figure 16.7 Early events in HIV infection: reverse transcription and integration. Following entry of the viral capsid into the cytoplasm, the capsid is partially disassembled and reverse transcription takes place. Viral DNA, p17 Gag, integrase and Vpr are contained in the preintegration complex. Nuclear import signals on these three proteins facilitate transport to and entry into the nucleus, where synthesis of the complete linear provirus is finished and it is integrated. Alternatively, the integrase reaction results in circularisation, which represents a dead end fate for viral DNA, or in other non-functional integrase-catalysed events.

region, a binding site for the HIV Tax functional homolog, Tat (Berkhout and Jeang, 1989; Hauber and Cullen, 1988; Muesing *et al.*, 1987; Sodroski *et al.*, 1985a). The Gag genes are translated from full-length viral RNA as a polypeptide precursor and processed by the viral protease into the myristoylated p17 matrix protein, the p24 capsid protein, the p7 nucleocapsid protein and a p6 protein (Gowda *et al.*, 1989; Mervis *et al.*, 1988). A translational frameshift permits translation of the Pol protein as a Gag–Pol precursor polypeptide (Jacks *et al.*, 1988; Wilson *et al.*, 1988), which is processed by the viral protease into a p32 integrase, a p66/51 reverse transcriptase/RNase H, and a p10 protease (Schulze *et al.*, 1991). The highly glycosylated envelope protein is translated from a singly spliced mRNA (Arrigo *et al.*, 1989; Hammarskjold *et al.*, 1989; Schwartz *et al.*, 1990b). After cleavage of the signal peptide, the gp120 surface protein and transmembrane gp41 protein are generated by cleavage with furin, a cellular protease (Hallenberger *et al.*, 1992).

HIV regulatory proteins

As with HTLV, the HIV Tat, Rev and Nef regulatory/accessory proteins are translated from mRNAs that contain more than two exons (Arrigo *et al.*, 1989; Robert-Guroff *et al.*, 1990; Schwartz *et al.*, 1990a). Vpu is translated from the same singly spliced transcript as Env (Schwartz *et al.*, 1990b). Tat and Rev are functional homologs of Tax and Rex of HTLV, and are necessary for viral replication. Tat transactivates viral RNA transcription (Sodroski *et al.*, 1985a,c), although its mechanism differs fundamentally from that of Tax. Tat is generally thought to facilitate RNA polymerase processivity of already initiated short transcripts (Kao *et al.*, 1987) by binding to p-TEFb, a complex of cyclin T1 and cdk9 (Wei *et al.*, 1998; Zhou *et al.*, 1998). The interaction allows binding to the TAR region of viral RNA and phosphorylation of the C terminal domain of RNA polymerase II, thereby increasing its processivity. More recently, it has been reported that Tat may also play a role in transcription initiation by facilitating the binding of TATAA binding protein (TBP) to the TATAA box, the earliest step in the assembly of transcriptional complexes (Raha *et al.*, 2005).

In addition to its role in transactivation of viral transcription, Tat appears to have other activities, some of which may be facilitated by two properties of Tat. First, it appears to be actively released by infected cells (Chang *et al.*, 1997; Ensoli *et al.*, 1993). Second, it is readily taken up by other cells by virtue of a basic protein transduction domain, allowing it to affect uninfected cells (Ensoli *et al.*, 1993, Frankel and Pabo, 1988). It appears to serve as a growth factor for KS-derived endothelial cells (Ensoli *et al.*, 1990)and increases tumour formation in a murine model of KS (Guo *et al.*, 2004). This may represent a direct cooperation of HIV-1 and HHV-8 in KS pathogenesis. It also dysregulates cytokine expression (Buonaguro *et al.*, 1992), and is immunosuppressive in mice (Agwale *et al.*, 2002; Cohen *et al.*, 1999), suggesting it may play a direct role in AIDS pathogenesis.

Rev, like HTLV-I Rex, is required to export full-length and singly spliced transcripts encoding the virion proteins from the nucleus to the cytoplasm (Feinberg *et al.*, 1986) (Fig. 16.8). It facilitates nuclear export by binding to Rev responsive elements (RREs) of the viral RNA (Daly *et al.*, 1989; Felber *et al.*, 1989; Itoh *et al.*, 1989; Zapp and Green, 1989). In its absence, viral transcripts are fully and multiply spliced to mRNA that encodes regulatory proteins. Like Rex, Rev is a shuttle protein that binds with cellular nuclear export protein CRM-1 (exportin-1) for transport to the cytoplasm (Askjaer *et al.*, 1998, Henderson and Percipalle, 1997, Neville *et al.*, 1997, Pasquinelli *et al.*, 1997) and with importin-β for transport back to the nucleus (Henderson and Percipalle, 1997). In the nucleus, Rev binding to importin-β allows binding to the Rev-responsive element (RRE) on viral RNA, and the resultant complex is transported to the nuclear pore complex, where nucleoporins bind to Rev and allow translocation across the nuclear membrane.

The remaining accessory proteins Nef, Vpr, Vif and Vpu have multiple, though less well defined, and sometimes apparently overlapping roles. The Nef protein, like the matrix protein, is myristoylated (Guy *et al.*, 1987) and is thereby targeted to the inner surface of the plasma membrane. Nef also appears to be incorporated into virions (Welker *et al.*, 1996), where it is complexed with the viral nucleoprotein complex.

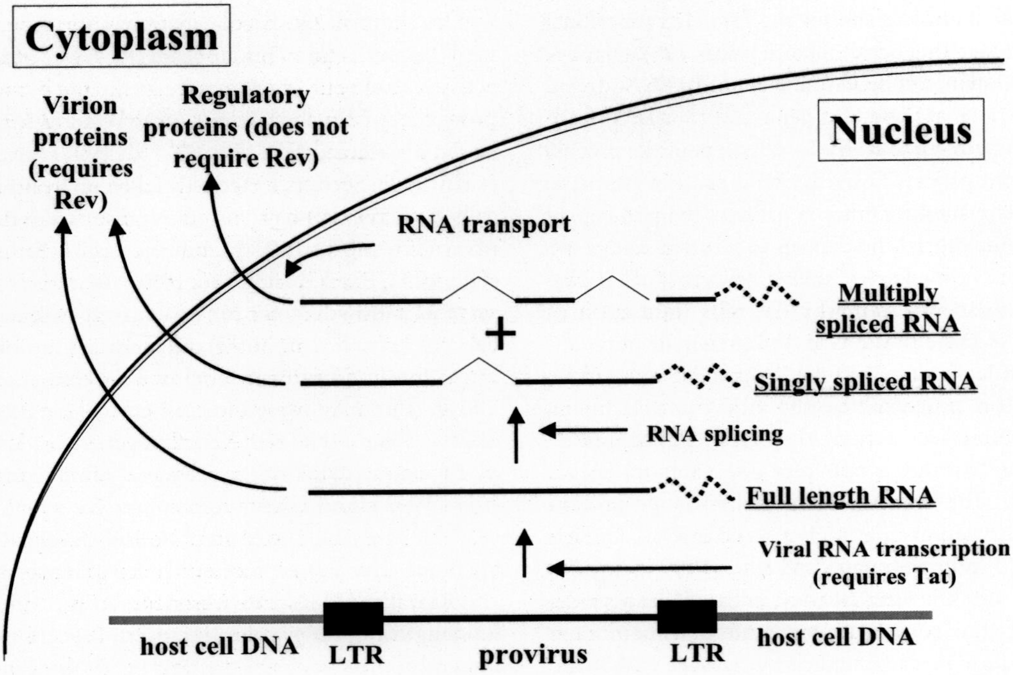

Figure 16.8 Later events in HIV infection: viral RNA synthesis, splicing and cytoplasmic export. Viral transcripts are synthesized from integrated viral DNA by cellular RNA polymerase. Transcriptional elongation is greatly facilitated by the viral Tat protein. RNA is spliced to multiply spliced forms by the cellular splicing machinery in the absence of Rev activity. Rev inhibits splicing and facilitates export from the nucleus of unspliced and singly spliced viral transcripts, allowing synthesis of virion proteins.

Although it is not required for viral replication *in vitro*, reverse transcription is attenuated in cells infected with Nef-defective HIV (Schwartz *et al.*, 1995). It appears to enhance viral replication *in vivo*, since lack of Nef function correlates with low viral loads in infected people (Deacon *et al.*, 1995, Kirchhoff *et al.*, 1995, Mariani *et al.*, 1996) and infection with Nef-negative SIV results in lower viral titers in infected monkeys (Schwartz *et al.*, 1995). The functions of Nef at the molecular level include down-regulation of cell surface CD4 (Garcia and Miller, 1991), CD28 (Swigut *et al.*, 2001) and MHC antigens(Schwartz *et al.*, 1996, Stumptner-Cuvelette *et al.*, 2001). Nef binds to the cytoplasmic tail of CD4 (Anderson *et al.*, 1994, Rhee and Marsh, 1994) and recruits adaptor protein complex (AP) proteins, which results in clathrin binding and endocytosis through clathrin-coated pits (Bresnahan *et al.*, 1998, Greenberg *et al.*, 1997, Mangasarian *et al.*, 1997, Piguet *et al.*, 1998). Further activities of Nef result in CD4 transport to lysosomes, where it is degraded. This may prevent viral envelope proteins from being trapped on the cell surface when virions are produced from infected cells. Down-regulation of MHC antigens occurs by broadly similar mechanisms and would help blunt the host immune response and facilitate viral immune escape.

Nef has been reported to also interact with a number of kinases important for cellular signal transduction, including ASK1 (Geleziunas *et al.*, 2001), PAK (Nunn and Marsh, 1996), and various Src family tyrosine kinases, including Lck (Collette *et al.*, 1996), Fyn (Arold *et al.*, 1997) and Hck (Moarefi *et al.*, 1997). Interactions with PAK2 and ASK1 may interfere with apoptosis of infected cells (Geleziunas *et al.*, 2001, Wolf *et al.*, 2001), keeping cells alive in order to maximize virion production. Interference with the Src family kinases likely interfere with signaling through the T-cell receptor and T-cell activation. These activities would reduce the host antiviral immune response.

Vpr is contained in virions. Its incorporation requires the viral p6 Gag protein (Paxton *et al.*, 1993). Among its functions are helping to target the preintegration complex (PIC) to the nucleus, which allows non-dividing cells to be infected (Heinzinger *et al.*, 1994) (Fig. 16.7), and arresting the cell cycle in G2 (He *et al.*, 1995, Jowett *et al.*, 1995, Re *et al.*, 1995), which may help maximize viral protein expression. Vpr increases binding of the nuclear localization signals on p17 matrix protein and the viral integrase to karyopherin-α, leading to transport to the nuclear envelope from which it enters the nucleus through the nuclear pore complex (Popov *et al.*, 1998). Vpr causes cell cycle arrest by inactivation of cdc25 and subsequent of cdk1/cyclin B complexes (Hrimech *et al.*, 2000). It is also thought to activate p21/Waf1, a cell cycle inhibitor (Chowdhury *et al.*, 2003).

Vpu, like Nef, down-regulates surface expression of CD4, reducing sequestration of viral gp160 into CD4/gp160 complexes during virion production (Willey *et al.*, 1992a,b). Vpu appears to target CD4 for ubiquitination and degradation via the E3 ubiquitin ligase complex (Schubert *et al.*, 1998).

The main function of Vif appears to be to inhibit the antiviral activity of APOBEC-3G (Mangeat *et al.*, 2003, Sheehy *et al.*, 2002), a cellular cytidine deaminase that carries out RNA editing and in the absence of Vif reportedly causes G to A mutations that result in hypermutation of the viral genome (Zhang *et al.*, 2003). Vif blocks the incorporation of APOBEC-3G into virions. For further discussion of APOBEC-3G function, see below.

HIV host cell interactions

There are two kinds of host cell interactions, based on their effects on the virus. Because it lacks much genetic capacity, the virus must rely on many cellular factors, a few of which are described above, to facilitate viral replication. These interactions represent potential targets for rational design of antiviral therapies. Recently, in a large-scale study using siRNA to knock down expression of all known human gene products, it was reported that more than 250 cellular genes are important for viral replication (Brass *et al.*, 2008). Although many of these may also be essential for the human host, it may eventually be possible to exploit some of them for antiviral therapy.

The second type of interaction is antiviral. Humoral and cellular immunity are two ways in which the host immune system attempts to eliminate HIV. Neutralizing antibodies are primarily directed against gp120, although a few neutralizing antibodies are directed against gp41. Antibodies against other viral proteins may participate in antibody-directed cellular cytotoxicity (ADCC), but their contribution is not clear. Vaccine approaches that rely on neutralizing antibodies face the difficulty that the gp120 is highly variable, and a few point mutations in the gp120 (or even the gp41) are sufficient to abrogate neutralization (diMarzo Veronese *et al.*, 1993; Kliks *et al.*, 1993; McKeating *et al.*, 1993; Reitz *et al.*, 1988). However, following CD4 binding, the CCR5 binding site of the envelope is exposed (Kwong *et al.*, 2000; Rizzuto *et al.*, 1998; Trkola *et al.*, 1996; Wu *et al.*, 1996; Wyatt and Sodroski, 1998). This may be a conformationally restrained site that cannot readily be changed without compromising envelope function. Approaches to make this site antigenic have included the use of gp120-CD4 proteins in which the two domains are connected by a peptide tether, and these appear to elicit antibodies with an unusually broad neutralization profile (DeVico *et al.*, 2007; Fouts *et al.*, 2002).

Cellular immunity is represented by CTL activity against viral proteins expressed in infected cells. Since some of these target proteins are less variable than the gp120, vaccine approaches have been designed to elicit CTL responses. A disadvantage is that cells must be infected for antiviral activity to occur (Gallo, 2005), and a large trial of such a vaccine has recently ended in failure (Cohen, 2007).

A third host antiviral interaction that has received increasing attention is innate immunity as opposed to adaptive immunity. Innate immunity is a somewhat loosely defined term that covers a wide spectrum of mechanisms, including natural killer cells, chemokines, cytokines, inflammatory responses, and intracellular mechanisms. New discoveries have greatly expanded our understanding of the wide breadth of innate immune mechanisms. For HIV-1, the first example of this kind of antiviral activity came with the recognition, cited above, that the cellular β chemokines MIP-1α, MIP-1β and RANTES were a potent antiviral factor inhibiting the interaction of gp120 with its co-receptor CCR5 (Cocchi *et al.*, 1995). Indeed, levels of these chemokines have been reported to correlate with resistance to infection in HIV-exposed people and with slow progression to AIDS in infected people (Cocchi *et al.*, 2000; Garzino-Demo *et al.*, 1999; Lehner *et al.*, 1996; Paxton *et al.*, 1998; Zagury *et al.*, 1998b), although it is possible that higher levels of chemokine expression are simply a marker for better preservation of the immune system. Among other components of the immune system, defensins may provide significant antiviral activity (Sun *et al.*, 2005).

The interaction of APOBEC-3G and related family members and the HIV-1 Vif protein is a recently discovered viral–host relationship. The APOBEC family may represent an ancient intracellular host defence mechanism against retroviruses; APOBEC inhibits replication of a wide variety of retroviruses. In the absence of Vif, APOBEC is incorporated into virions bound to viral RNA and becomes activated upon reverse transcription following infection of a new target cell (Kao *et al.*, 2003, Mariani *et al.*, 2003, Stopak *et al.*, 2003). Resting peripheral blood T-cells contain APOBEC in an active low molecular mass form (Chiu *et al.*, 2005). Activation of the T-cell generates the inactive high molecular mass form. Vif prevents incorporation of APOBEC into virions, apparently by depleting cytoplasmic APOBEC by a dual mechanism that includes inhibition of translation of APOBEC mRNA and facilitating its degradation by the 26S proteasome (Stopak *et al.*, 2003). It is not clear that the antiviral activity of APOBEC depends upon its cytidine deaminase activity, and a recent report suggest this may not be the case (Chiu *et al.*, 2005).

A second recently discovered intracellular antiviral mechanism is that mediated by the tripartite motif (TRIM) family (Stremlau *et al.*, 2004). The tripartite motif consists of a RING domain at the N-terminus, a B box-2 domain, and a coiled-coil domain (Reymond *et al.*, 2001). This also seems to represent an ancient antiretroviral innate host defence mechanism. TRIM5α appears to be analogous to one version of the Fv1 restriction element that regulates sensitivity of murine cells of different genetic backgrounds to infection with the so-called B-tropic and N-tropic murine leukaemia viruses (MuLV) and was identified forty years ago (Lilly, 1967). The target of both factors is the capsid protein, and the restriction by both human TRIM5α and murine Fv1-B of N-tropic MuLV infection depends upon residue 110 of the MuLV capsid protein (Perron *et al.*, 2004). HIV-1 is only partially sensitive to human TRIM5α (depending on allelic variability) (Javanbakht *et al.*, 2006a), perhaps helping to account for its successful cross-species transmission from chimpanzees, although macaque TRIM5α strongly restricts HIV-1 (Perron *et al.*, 2004; Stremlau *et al.*, 2004). TRIM5α binds to viral capsid hexamers (Stremlau *et al.*, 2006) and inhibits replication at a post-entry step. The precise mechanism is not clear and may be mediated at more than one level, but may include an accelerated and dysfunctional uncoating of the viral capsid (Stremlau *et al.*, 2006; Yap *et al.*, 2006). Trimerization of TRIM5α increases its antiviral activity (Javanbakht *et al.*, 2006b).

HIV pathogenesis

Various mechanisms for HIV pathogenesis leading to AIDS have been considered. These include direct mechanisms and indirect mechanisms. The direct mechanisms invoke killing of infected cells by the virus, either by syncytium formation, apoptosis, cell cycle arrest, or other mechanisms. At times this has been a prevalent view. However, although direct infected cell killing by the virus clearly occurs, infected cells are generally a small minority of total $CD4^+$ T-cells, and the scale of infection does not seem to be sufficient to explain the degree of immune destruction that occurs during infection. It certainly does not explain some of the pathologies common in HIV-infected people that do not directly involve cells infected with HIV, such as HIV-associated neuropathy or nephropathy or AIDS-KS, where HIV is absent from the tumour cells. More recently, it has been shown that initial HIV-1 and SIV infections result in a rapid infection and killing of $CD4^+$ T-cells in tissue of the gastrointestinal tract (Guadalupe *et al.*, 2003; Smit-McBride *et al.*, 1998), which could presumably be due to direct cell killing. Progression to AIDS is generally slow, however. To explain the slow course of the disease, it has been proposed that following rapid depletion of gut-associated lymphoid tissue, gastrointestinal integrity is compromised, leading to leakage of bacteria from the gut and a consequent chronic immune activation (Brenchley *et al.*, 2006), although this has not yet been widely accepted. This model of AIDS progression thus includes elements of direct and indirect mechanisms.

Killing of uninfected cells by indirect mechanisms could occur by chronic antigenic stimulation against HIV proteins, leading to clonal depletion and immune exhaustion, or could result from activities of circulating proteins, which could include circulating viral proteins, such as gp120 or Tat, or circulating cellular proteins such as cytokines whose expression is induced in infected cells. For example, IL-10 appears to be involved in facilitating functional exhaustion and allowing LCMV infection in mice to be chronic (Blackburn and Wherry, 2007), and IL-10 has been reported to be elevated in AIDS. Blockade of the IL-10 receptor in LCMV infected mice prevents both chronic infection and immune exhaustion (Blackburn and Wherry, 2007). Overproduction of α interferon, correlated with extracellular Tat and leading to reduced expression of antiviral β chemokines (Zagury *et al.*, 1998a), has been reported for T-cells from AIDS patients. Alternatively, cell surface molecules induced as a result of infection could result in aberrant cell-to-cell signaling. The accessory proteins of HIV, as discussed above, can interdict cellular signal transduction pathways, leading to inappropriate expression of cytokines and cell surface signal transduction proteins. It is likely, however, that HIV-mediated pathogenesis leading to AIDS is the result of multiple factors.

Prospects for future HIV therapies

The main hope for eradication of the virus lies in prevention of infection and realistically requires a protective vaccine. However, developing a vaccine has been quite difficult. As discussed above, the most recent vaccine trials have failed to show protection against infection and in a subset of people have actually resulted in increased infection. Retroviral infection, because of integration of viral DNA into host cell DNA, lasts the life of the cell. For all practical purposes, this means that once an individual is infected, it is virtually impossible to clear the body of all infected cells, especially when the virus can establish a latent infection of long-lived cells, such as has been demonstrated for HIV (Finzi *et al.*, 1997). This means infection is lifelong, and therefore treatment must be lifelong. As a consequence, toxicity, side effects and drug resistance (generated by the high genetic variability of HIV) become very serious issues.

Most currently used antiretroviral drugs target viral proteins and their functions, and have been developed by combinations of rational design based on protein structure and high throughput screening based on inhibition of function. The first class of antiretrovirals, stemming from work in Broder's laboratory at the National Cancer Institute, were the nucleoside analog reverse transcriptase (RT) inhibitors such as AZT (Yarchoan and Broder, 1987), followed by non-nucleoside RT inhibitors such as nevirapine (Merluzzi *et al.*, 1990). Shortly thereafter, molecules that inhibit the viral protease (Ashorn *et al.*, 1990), such as ritonavir, saquinavir and indinavir, were introduced. These and the RT inhibitors were shown to be effective in combination (Collier *et al.*, 1996). This therapy, now called highly active antiretroviral therapy (HAART), has proven quite effective in reducing viral burden and delaying progression to AIDS, although problems with toxicity, side effects and viral drug resistance commonly occur. More recently, inhibitors of fusion such as T20 (Fuzeon) (Wild *et al.*, 1993) and of CCR5 binding to envelope protein (Baba *et al.*, 1999) are beginning to be used clinically, although primarily within the context of people infected with HIV resistant to protease and RT inhibitors. These are collectively known as entry inhibitors. Side effects and toxicities remain a potential issue, and Fuzeon must be taken by injection. Although resistance to these drugs also occurs, a shift from R5 to X4 usage with the CCR5 inhibitors does not seem to readily occur, and the mechanism of resistance appears to instead involve altered usage of CCR5 (Trkola *et al.*, 2002). Raltegravir, an integrase inhibitor, represents another new class of antiretrovirals, and is being used in patients who harbour drug-resistant

HIV (Grinsztejn *et al.*, 2007), but only appears to be fully effective when used with other drugs. Beviramat, a derivative from a Chinese herb, is an inhibitor of viral maturation (Li *et al.*, 2003) that has recently entered clinical trials.

For the most part, the above drugs are targeted against viral proteins. Because HIV undergoes rather rapid genetic changes, resistant variants to these drugs occur fairly rapidly. As interactions of viral proteins with cellular proteins and the contributions of cellular proteins to viral replication become better understood, it may be possible to target cellular proteins, which would have only a limited possibility of change. If the cellular proteins were not critical for the health of the host, this might represent a specific and effective antiviral strategy. Indeed, this was the rationale behind the development of CCR5 antagonists. People homozygous for the Δ32 CCR5 mutation, although highly resistant to infection, do not have obvious health abnormalities (Dean *et al.*, 1996, Huang *et al.*, 1996, Liu *et al.*, 1996, Samson *et al.*, 1996), leading to the idea that targeting CCR5 would be a safe and effective therapy against HIV. In practice, however, changes in the HIV envelope can result in resistance to CCR5 inhibitors (Kuhmann *et al.*, 2004, Marozsan *et al.*, 2005, Trkola *et al.*, 2002), and these drugs are not without side effects, including cardiovascular events. Some drugs that affect cellular metabolism, such as hydroxyurea (which lowers intracellular DNA precursor pools) (Lori *et al.*, 1994) and rapamycin (which alters the cell cycle and causes a reduction in CCR5 expression) (Heredia *et al.*, 2003) have been reported to synergize with other antiretrovirals (RT inhibitors for hydroxyurea, CCR5 inhibitors for rapamycin), but also have associated toxicities.

Gene therapy approaches have included ribozymes, intracellular antibodies, antisense RNA, siRNA, and decoys for viral proteins. Gene therapy, however, has yet to deliver on its considerable promise because of difficulties that include achieving efficient and specific delivery to the proper cells and appropriate and durable expression. Since retroviral infection, as noted above, is lifelong, expression would need to be lifelong, which mandates the use of a vector that inserts the gene of interest into the host chromosomal DNA. This in turn raises a not inconsiderable risk of deleterious insertional mutagenesis, such as one resulting from insertion near or within an oncogene. This kind of event has already been reported in limited but partially successful gene therapy trials for severe combined immunotherapy X1 (Hacein-Bey-Abina *et al.*, 2003). One possibility to actually 'cure' infected cells would be therapy to specifically excise or destroy proviral DNA from them. Excision of HIV viral DNA from infected cells *in vitro* using an evolved recombinase has been reported, but any clinical use of such an approach is obviously far in the future. Given the ability of HIV to lie latent within long-lived T-cells, there will not likely be a single magic bullet for safe, effective and long-term treatment of infection. Progress will continue to be incremental, with the virus generating resistant mutants for each new approach, and the development of new approaches will continue to be necessary.

Conclusions

We review the discovery and characterization of HTLV and HIV, the first two known human retroviruses. During the 1970s, retroviruses were sought as the cause of human leukaemias. Following the development of sensitive and specific molecular methods to identify retroviruses and and methods for the large scale culture of T-lymphocytes, HTLV-I was discovered and identified as the cause of adult T-cell leukaemia, a type of leukaemia particularly prevalent in southern Japan and parts of the Caribbean, and tropical spastic paraparesis, a demyelinating neuropathy resembling multiple sclerosis. The discovery that HTLV-I is transmitted by breast milk has contributed to a significant decline in HTLV-I infections in Japan. Although no other retroviruses have been identified to date in human leukaemias or related diseases, the efforts that resulted in the discovery of HTLV-I were critical in isolating HIV-1 and identifying it as the etiologic agent in AIDS. The ability to grow the virus in quantity led to the development of a blood test that has saved countless lives. The development of effective antiretroviral drugs has resulted in HIV-1 infection being a somewhat manageable chronic condition rather than a certain death sentence. Although results with vaccines have been so far disappointing, development of an effective vaccine is currently one of the most important needs.

References

Abrell, J.W., and Gallo, R.C. (1973). Purification, characterization, and comparison of the DNA polymerases from two primate RNA tumor viruses. J. Virol. *12*, 431–439.

Abrell, J.W., Reitz, M.S., and Gallo, R.C. (1975). Transcription of 70S RNA by DNA polymerases from mammalian RNA viruses. J. Virol. *16*, 1566–1574.

Agwale, S.M., Shata, M.T., Reitz, M.S., Kalyanaraman, V.S., Gallo, R.C., Popovic, M., and Hone, D.M. (2002). A Tat subunit vaccine confers protective immunity against the immune-modulating activity of the human immunodeficiency virus type-1 Tat protein in mice. Proc. Natl. Acad. Sci. U.S.A. *99*, 10037–10041.

Albrecht, B., D'Souza, C.D., Ding, W., Tridandapani, S., Coggeshall, K.M., and Lairmore, M.D. (2002). Activation of nuclear factor of activated T-cells by human T-lymphotropic virus type 1 accessory protein p12(I). J. Virol. *76*, 3493–3501.

Albrecht, B., and Lairmore, M.D. (2002). Critical role of human T-lymphotropic virus type 1 accessory proteins in viral replication and pathogenesis. Microbiol. Mol. Biol. Rev. *66*, 396–406.

Alkhatib, G., Combadiere, C., Broder, C.C., Feng, Y., Kennedy, P.E., Murphy, P.M., and Berger, E.A. (1996). CC CKR5: a RANTES, MIP-1alpha, MIP-1beta receptor as a fusion cofactor for macrophage-tropic HIV-1. Science *272*, 1955–1958.

Anderson, S.J., Lenburg, M., Landau, N.R., and Garcia, J.V. (1994). The cytoplasmic domain of CD4 is sufficient for its down-regulation from the cell surface by human immunodeficiency virus type 1 Nef. J. Virol. *68*, 3092–3101.

Ando, Y., Nakano, S., Saito, K., Shimamoto, I., Ichijo, M., Toyama, T., and Hinuma, Y. (1987). Transmission of adult T-cell leukemia retrovirus (HTLV-I) from mother to child: comparison of bottle- with breast-fed babies. Jpn. J. Cancer Res. *78*, 322–324.

Arold, S., Franken, P., Strub, M.P., Hoh, F., Benichou, S., Benarous, R., and Dumas, C. (1997). The crystal structure of HIV-1 Nef protein bound to the Fyn kinase SH3 domain suggests a role for this complex in altered T-cell receptor signaling. Structure *5*, 1361–1372.

Arrigo, S.J., Weitsman, S., Rosenblatt, J.D., and Chen, I.S. (1989). Analysis of *rev* gene function on human immunodeficiency virus type 1 replication in lymphoid cells by using a quantitative polymerase chain reaction method. J. Virol. *63*, 4875–4881.

Arya, S.K., Guo, C., Josephs, S.F., and Wong-Staal, F. (1985). *Trans*-activator gene of human T-lymphotropic virus type III (HTLV-III). Science *229*, 69–73.

Ashorn, P., McQuade, T. J., Thaisrivongs, S., Tomasselli, A.G., Tarpley, W.G., and Moss, B. (1990). An inhibitor of the protease blocks maturation of human and simian immunodeficiency viruses and spread of infection. Proc. Natl. Acad. Sci. U.S.A. *87*, 7472–7476.

Asjo, B., Morfeldt-Manson, L., Albert, J., Biberfeld, G., Karlsson, A., Lidman, K., and Fenyo, E.M. (1986). Replicative capacity of human immunodeficiency virus from patients with varying severity of HIV infection. Lancet 2, 660–662.

Askjaer, P., Jensen, T.H., Nilsson, J., Englmeier, L., and Kjems, J. (1998). The specificity of the CRM1-Rev nuclear export signal interaction is mediated by RanGTP. J. Biol. Chem. *273*, 33414–33422.

Baba, M., Nishimura, O., Kanzaki, N., Okamoto, M., Sawada, H., Iizawa, Y., Shiraishi, M., Aramaki, Y., Okonogi, K., Ogawa, Y., Meguro, K., and Fujino, M. (1999). A small-molecule, nonpeptide CCR5 antagonist with highly potent and selective anti-HIV-1 activity. Proc. Natl. Acad. Sci. U.S.A. *96*, 5698–5703.

Bailes, E., Gao, F., Bibollet-Ruche, F., Courgnaud, V., Peeters, M., Marx, P.A., Hahn, B.H., and Sharp, P.M. (2003). Hybrid origin of SIV in chimpanzees. Science *300*, 1713.

Barin, F., M'Boup, S., Denis, F., Kanki, P., Allan, J.S., Lee, T.H., and Essex, M. (1985). Serological evidence for virus related to simian T-lymphotropic retrovirus III in residents of west Africa. Lancet 2, 1387–1389.

Barre-Sinoussi, F., Chermann, J.C., Rey, F., Nugeyre, M. T., Chamaret, S., Gruest, J., Dauguet, C., xler-Blin, C., Vezinet-Brun, F., Rouzioux, C., Rozenbaum, W., and Montagnier, L. (1983). Isolation of a T-lymphotropic retrovirus from a patient at risk for acquired immune deficiency syndrome (AIDS). Science *220*, 868–871.

Bartoe, J.T., Albrecht, B., Collins, N.D., Robek, M.D., Ratner, L., Green, P.L., and Lairmore, M.D. (2000). Functional role of pX open reading frame II of human T-lymphotropic virus type 1 in maintenance of viral loads *in vivo*. J. Virol. *74*, 1094–1100.

Berkhout, B., and Jeang, K.T. (1989). *trans* activation of human immunodeficiency virus type 1 is sequence specific for both the single-stranded bulge and loop of the *trans*-acting-responsive hairpin: a quantitative analysis. J. Virol. *63*, 5501–5504.

Berson, J.F., Long, D., Doranz, B.J., Rucker, J., Jirik, F.R., and Doms, R.W. (1996). A seven-transmembrane domain receptor involved in fusion and entry of T-cell-tropic human immunodeficiency virus type 1 strains. J. Virol. *70*, 6288–6295.

Bittner, J.J., Evans, C.A., and Green, R.G. (1945). Survival of the mammary tumor milk agents of mice. Science *101*, 95–97.

Blackburn, S.D., and Wherry, E.J. (2007). IL-10, T-cell exhaustion and viral persistence. Trends Microbiol. *15*, 143–146.

Blattner, W.A., Kalyanaraman, V. S., Robert-Guroff, M., Lister, T. A., Galton, D.A., Sarin, P.S., Crawford, M. H., Catovsky, D., Greaves, M., and Gallo, R.C. (1982). The human type-C retrovirus, HTLV, in Blacks from the Caribbean region, and relationship to adult T-cell leukemia/lymphoma. Int. J. Cancer *30*, 257–264.

Bobrow, S.N., Smith, R.G., Reitz, M.S., and Gallo, R.C. (1972). Stimulated normal human lymphocytes contain a ribonuclease-sensitive DNA polymerase distinct from viral RNA-directed DNA polymerase. Proc Natl Acad Sci U.S.A. *69*, 3228–3232.

Brass, A.L., Dykxhoorn, D. M., Benita, Y., Yan, N., Engelman, A., Xavier, R. J., Lieberman, J., and Elledge, S.J. (2008). Identification of Host Proteins Required for HIV Infection Through a Functional Genomic Screen. Science *319*, 921–926.

Brenchley, J.M., Price, D.A., Schacker, T.W., Asher, T.E., Silvestri, G., Rao, S., Kazzaz, Z., Bornstein, E., Lambotte, O., Altmann, D., Blazar, B.R., Rodriguez, B., Teixeira-Johnson, L., Landay, A., Martin, J.N., Hecht, F.M., Picker, L.J., Lederman, M.M., Deeks, S.G., and Douek, D.C. (2006). Microbial translocation is a cause of systemic immune activation in chronic HIV infection. Nat. Med. *12*, 1365–1371.

Bresnahan, P.A., Yonemoto, W., Ferrell, S., Williams-Herman, D., Geleziunas, R., and Greene, W. C. (1998). A dileucine motif in HIV-1 Nef acts as an internalization signal for CD4 downregulation and binds the AP-1 clathrin adaptor. Curr. Biol. *8*, 1235–1238.

Bukrinsky, M. I., Haggerty, S., Dempsey, M.P., Sharova, N., Adzhubel, A., Spitz, L., Lewis, P., Goldfarb, D., Emerman, M., and Stevenson, M. (1993). A nuclear localization signal within HIV-1 matrix protein that governs infection of non-dividing cells. Nature *365*, 666–669.

Bukrinsky, M.I., Sharova, N., Dempsey, M.P., Stanwick, T. L., Bukrinskaya, A.G., Haggerty, S., and Stevenson, M. (1992). Active nuclear import of human immunodeficiency virus type 1 preintegration complexes. Proc. Natl. Acad. Sci. U.S.A. *89*, 6580–6584.

Buonaguro, L., Barillari, G., Chang, H. K., Bohan, C.A., Kao, V., Morgan, R., Gallo, R. C., and Ensoli, B. (1992). Effects of the human immunodeficiency virus type 1 Tat protein on the expression of inflammatory cytokines. J. Virol. *66*, 7159–7167.

Calattini, S., Chevalier, S.A., Duprez, R., Bassot, S., Froment, A., Mahieux, R., and Gessain, A. (2005). Discovery of a new human T-cell lymphotropic virus (HTLV-3) in Central Africa. Retrovirology. *2*, 30.

Chang, H.C., Samaniego, F., Nair, B.C., Buonaguro, L., and Ensoli, B. (1997). HIV-1 Tat protein exits from cells via a leaderless secretory pathway and binds to extracellular matrix-associated heparan sulfate proteoglycans through its basic region. AIDS *11*, 1421–1431.

Chen, Y.M., Jang, Y. J., Kanki, P.J., Yu, Q.C., Wang, J.J., Montali, R.J., Samuel, K.P., and Papas, T.S. (1994). Isolation and characterization of simian T-cell leukemia virus type II from New World monkeys. J. Virol. *68*, 1149–1157.

Chen, Z., Telfier, P., Gettie, A., Reed, P., Zhang, L., Ho, D.D., and Marx, P. A. (1996). Genetic characterization of new West African simian immunodeficiency virus SIVsm: geographic clustering of household-derived SIV strains with human immunodeficiency virus type 2 subtypes and genetically diverse viruses from a single feral sooty mangabey troop. J. Virol. *70*, 3617–3627.

Chiu, Y.L., Soros, V.B., Kreisberg, J.F., Stopak, K., Yonemoto, W., and Greene, W.C. (2005). Cellular APOBEC3G restricts HIV-1 infection in resting $CD4^+$ T-cells. Nature *435*, 108–114.

Choe, H., Farzan, M., Sun, Y., Sullivan, N., Rollins, B., Ponath, P.D., Wu, L., Mackay, C.R., LaRosa, G., Newman, W., Gerard, N., Gerard, C., and Sodroski, J. (1996). The beta-chemokine receptors CCR3 and CCR5 facilitate infection by primary HIV-1 isolates. Cell *85*, 1135–1148.

Chowdhury, I. H., Wang, X. F., Landau, N.R., Robb, M.L., Polonis, V.R., Birx, D.L., and Kim, J.H. (2003). HIV-1 Vpr activates cell cycle inhibitor p21/Waf1/Cip1: a potential mechanism of G2/M cell cycle arrest. Virology *305*, 371–377.

Clarke, M.F., Trainor, C.D., Mann, D.L., Gallo, R. C., and Reitz, M.S. (1984). Methylation of human T-cell leukemia virus proviral DNA and viral RNA expression in short- and long-term cultures of infected cells. Virology *135*, 97–104.

Cocchi, F., DeVico, A.L., Garzino-Demo, A., Arya, S.K., Gallo, R.C., and Lusso, P. (1995). Identification of RANTES, MIP-1 alpha, and MIP-1 beta as the major HIV-suppressive factors produced by $CD8^+$ T-cells. Science *270*, 1811–1815.

Cocchi, F., DeVico, A.L., Yarchoan, R., Redfield, R., Cleghorn, F., Blattner, W. A., Garzino-Demo, A., Colombini-Hatch, S., Margolis, D., and Gallo, R. C. (2000). Higher macrophage inflammatory protein (MIP)-1alpha and MIP-1beta levels from $CD8^+$ T-cells are associated with asymptomatic HIV-1 infection. Proc. Natl. Acad. Sci. U.S.A. *97*, 13812–13817.

Cohen, J. (2007). AIDS research. Promising AIDS vaccine's failure leaves field reeling. Science *318*, 28–29.

Cohen, S.S., Li, C., Ding, L., Cao, Y., Pardee, A.B., Shevach, E.M., and Cohen, D.I. (1999). Pronounced acute immunosuppression *in vivo* mediated by HIV Tat challenge. Proc. Natl. Acad. Sci. U.S.A. *96*, 10842–10847.

Collette, Y., Dutartre, H., Benziane, A., Ramos, M., Benarous, R., Harris, M., and Olive, D. (1996). Physical and functional interaction of Nef with Lck. HIV-1 Nef-induced T-cell signaling defects. J. Biol. Chem. *271*, 6333–6341.

Collier, A.C., Coombs, R. W., Schoenfeld, D.A., Bassett, R.L., Timpone, J., Baruch, A., Jones, M., Facey, K., Whitacre, C., McAuliffe, V. J., Friedman, H.M., Merigan, T.C., Reichman, R. C., Hooper, C., and Corey, L. (1996). Treatment of human immunodeficiency virus infection with saquinavir, zidovudine, and zalcitabine. AIDS Clinical Trials Group. N. Engl. J. Med. *334*, 1011–1017.

Dalgleish, A. G., Beverley, P. C., Clapham, P. R., Crawford, D. H., Greaves, M.F., and Weiss, R.A. (1984). The CD4 (T4) antigen is an essential component of the receptor for the AIDS retrovirus. Nature *312*, 763–767.

Daly, T.J., Cook, K.S., Gray, G.S., Maione, T.E., and Rusche, J.R. (1989). Specific binding of HIV-1 recombinant Rev protein to the Rev-responsive element *in vitro*. Nature *342*, 816–819.

de La, F.C., Santiago, F., Chong, S.Y., Deng, L., Mayhood, T., Fu, P., Stein, D., Denny, T., Coffman, F., Azimi, N., Mahieux, R., and Kashanchi, F. (2000). Overexpression of p21(waf1) in human T-cell lymphotropic virus type 1-infected cells and its association with cyclin A/cdk2. J. Virol. *74*, 7270–7283.

Deacon, N.J., Tsykin, A., Solomon, A., Smith, K., Ludford-Menting, M., Hooker, D. J., McPhee, D. A., Greenway, A.L., Ellett, A., Chatfield, C., Lawson, V. A., Crowe, S., Maerz, A., Sonza, S., Learmont, J., Sullivan, J.S., Cunningham, A., Dwyer, D., Dowton, D., and Mills, J. (1995). Genomic structure of an attenuated quasi species of HIV-1 from a blood transfusion donor and recipients. Science *270*, 988–991.

Dean, M., Carrington, M., Winkler, C., Huttley, G. A., Smith, M.W., Allikmets, R., Goedert, J.J., Buchbinder, S.P., Vittinghoff, E., Gomperts, E., Donfield, S., Vlahov, D., Kaslow, R., Saah, A., Rinaldo, C., Detels, R., Hemophilia Growth and Development Study, Multicenter, A.I.D.S Cohort Study, Multicenter Hemophilia Cohort Study, San Francisco City Cohort, A.L.I.V.E Study, and O'Brien, S. J. (1996). Genetic restriction of HIV-1 infection and progression to AIDS by a deletion allele of the CKR5 structural gene. Science *273*, 1856–1862.

Deng, H., Liu, R., Ellmeier, W., Choe, S., Unutmaz, D., Burkhart, M., Di Marzio, P., Marmon, S., Sutton, R.E., Hill, C.M., Davis, C.B., Peiper, S.C., Schall, T. J., Littman, D R., and Landau, N.R. (1996). Identification of a major co-receptor for primary isolates of HIV-1. Nature *381*, 661–666.

DeVico, A., Fouts, T., Lewis, G.K., Gallo, R.C., Godfrey, K., Charurat, M., Harris, I., Galmin, L., and Pal, R. (2007). Antibodies to CD4-induced sites in HIV gp120 correlate with the control of SHIV challenge in macaques vaccinated with subunit immunogens. Proc. Natl. Acad. Sci. U.S.A. *104*, 17477–17482.

Di Marzio, P., Choe, S., Ebright, M., Knoblauch, R., and Landau, N. R. (1995). Mutational analysis of cell cycle arrest, nuclear localization and virion packaging of human immunodeficiency virus type 1 Vpr. J. Virol. *69*, 7909–7916.

diMarzo Veronese, F., V., Reitz, M.S., Gupta, G., Robert-Guroff, M., Boyer-Thompson, C., Louie, A., Gallo, R.C., and Lusso, P. (1993). Loss of a neutralizing epitope by a spontaneous point mutation in the V3 loop of HIV-1 isolated from an infected laboratory worker. J. Biol. Chem. *268*, 25894–25901.

Dillon, P. J., Nelbock, P., Perkins, A., and Rosen, C. A. (1990). Function of the human immunodeficiency virus types 1 and 2 Rev proteins is dependent on their ability to interact with a structured region present in *env* gene mRNA. J. Virol. *64*, 4428–4437.

Doranz, B.J., Rucker, J., Yi, Y., Smyth, R.J., Samson, M., Peiper, S. C., Parmentier, M., Collman, R.G., and Doms, R.W. (1996). A dual-tropic primary HIV-1 isolate that uses fusin and the beta- chemokine receptors CKR-5, CKR-3, and CKR-2b as fusion cofactors. Cell *85*, 1149–1158.

Dragic, T., Litwin, V., Allaway, G. P., Martin, S. R., Huang, Y., Nagashima, K.A., Cayanan, C., Maddon, P.J., Koup, R.A., Moore, J.P., and Paxton, W.A. (1996). HIV-1 entry into $CD4^+$ cells is mediated by the chemokine receptor CC-CKR-5. Nature *381*, 667–673.

Ellerman, V., and Bang, O. (1908). Experimentelle leukämie bei hühnern. Zentralbl Bakteriol Parasitenkd Infectionskr Hyg Abt Orig *46*, 595.

Ensoli, B., Barillari, G., Salahuddin, S.Z., Gallo, R.C., and Wong-Staal, F. (1990). Tat protein of HIV-1 stimulates growth of cells derived from Kaposi's sarcoma lesions of AIDS patients. Nature *345*, 84–86.

Ensoli, B., Buonaguro, L., Barillari, G., Fiorelli, V., Gendelman, R., Morgan, R.A., Wingfield, P., and Gallo, R.C. (1993). Release, uptake, and effects of extracellular human immunodeficiency virus type 1 Tat protein on cell growth and viral transactivation. J. Virol. *67*, 277–287.

Essex, M., McLane, M.F., Lee, T.H., Falk, L., Howe, C.W., Mullins, J I., Cabradilla, C., and Francis, D.P. (1983). Antibodies to cell membrane antigens associated with human T-cell leukemia virus in patients with AIDS. Science *220*, 859–862.

Feinberg, M.B., Jarrett, R.F., Aldovini, A., Gallo, R.C., and Wong-Staal, F. (1986). HTLV-III expression and production involve complex regulation at the levels of splicing and translation of viral RNA. Cell *46*, 807–817.

Felber, B.K., Derse, D., Athanassopoulos, A., Campbell, M., and Pavlakis, G.N. (1989). Cross-activation of the Rex proteins of HTLV-I and BLV and of the Rev protein of HIV-1 and nonreciprocal interactions with their RNA responsive elements. New Biol. *1*, 318–328.

Feng, Y., Broder, C.C., Kennedy, P.E., and Berger, E.A. (1996). HIV-1 entry cofactor: functional cDNA cloning of a seven-transmembrane, G protein-coupled receptor. Science *272*, 872–877.

Finzi, D., Hermankova, M., Pierson, T., Carruth, L.M., Buck, C., Chaisson, R.E., Quinn, T.C., Chadwick, K., Margolick, J., Brookmeyer, R., Gallant, J., Markowitz, M., Ho, D.D., Richman, D.D., and Siliciano, R. F. (1997). Identification of a reservoir for HIV-1 in patients on highly active antiretroviral therapy. Science *278*, 1295–1300.

Fouts, T., Godfrey, K., Bobb, K., Montefiori, D., Hanson, C. V., Kalyanaraman, V., DeVico, A., and Pal, R. (2002). Crosslinked HIV-1 envelope-CD4 receptor complexes elicit broadly cross-reactive neutralizing antibodies in rhesus macaques. Proc. Natl. Acad. Sci. U.S.A. *99*, 11842–11847.

Frankel, A.D., and Pabo, C.O. (1988). Cellular uptake of the *tat* protein from human immunodeficiency virus. Cell *55*, 1189–1193.

Friedman-Kien, A.E., Laubenstein, L.J., Rubinstein, P., Buimovici-Klein, E., Marmor, M., Stahl, R., Spigland, I., Kim, K.S., and Zolla-Pazner, S. (1982). Disseminated Kaposi's sarcoma in homosexual men. Ann. Intern. Med. *96*, 693–700.

Gallay, P., Hope, T., Chin, D., and Trono, D. (1997). HIV-1 infection of nondividing cells through the recognition of integrase by the importin/karyopherin pathway. Proc. Natl. Acad. Sci. U.S.A. *94*, 9825–9830.

Gallo, R.C. (2005). The end or the beginning of the drive to an HIV-preventive vaccine: a view from over 20 years. Lancet *366*, 1894–1898.

Gallo, R.C., Mann, D., Broder, S., Ruscetti, F. W., Maeda, M., Kalyanaraman, V. S., Robert-Guroff, M., and Reitz, M. S. (1982). Human T-cell leukemia-lymphoma virus (HTLV) is in T but not B-lymphocytes from a patient with cutaneous T-cell lymphoma. Proc Natl Acad Sci U.S.A. *79*, 5680–5683.

Gallo, R.C., Sarin, P. S., Gelmann, E. P., Robert-Guroff, M., Richardson, E., Kalyanaraman, V. S., Mann, D., Sidhu, G. D., Stahl, R. E., Zolla-Pazner, S., Leibowitch, J., and Popovic, M. (1983). Isolation of human T-cell leukemia virus in acquired immune deficiency syndrome (AIDS). Science *220*, 865–867.

Gallo, R.C., Salahuddin, S. Z., Popovic, M., Shearer, G. M., Kaplan, M., Haynes, B. F., Palker, T. J., Redfield, R., Oleske, J., and Safai, B. (1984). Frequent detection and isolation of cytopathic retroviruses (HTLV-III) from patients with AIDS and at risk for AIDS. Science *224*, 500–503.

Gao, F., Bailes, E., Robertson, D.L., Chen, Y., Rodenburg, C.M., Michael, S.F., Cummins, L.B., Arthur, L.O., Peeters, M., Shaw, G.M., Sharp, P.M., and Hahn, B.H. (1999). Origin of HIV-1 in the chimpanzee *Pan troglodytes troglodytes*. Nature *397*, 436–441.

Garcia, J.V., and Miller, A.D. (1991). Serine phosphorylation-independent downregulation of cell-surface CD4 by nef. Nature *350*, 508–511.

Gartner, S., Markovits, P., Markovitz, D.M., Betts, R.F., and Popovic, M. (1986). Virus isolation from and identification of HTLV-III/LAV-producing cells in brain tissue from a patient with AIDS. JAMA *256*, 2365–2371.

Garzino-Demo, A., Moss, R.B., Margolick, J.B., Cleghorn, F., Sill, A., Blattner, W.A., Cocchi, F., Carlo, D.J., DeVico, A.L., and Gallo, R.C. (1999). Spontaneous and antigen-induced production of HIV-inhibitory beta-chemokines are associated with AIDS-free status. Proc. Natl. Acad. Sci. U.S.A. *96*, 11986–11991.

Geleziunas, R., Xu, W., Takeda, K., Ichijo, H., and Greene, W.C. (2001). HIV-1 Nef inhibits ASK1-dependent death signalling providing a potential mechanism for protecting the infected host cell. Nature *410*, 834–838.

Gelmann, E.P., Popovic, M., Blayney, D., Masur, H., Sidhu, G., Stahl, R. E., and Gallo, R.C. (1983). Proviral DNA of a retrovirus, human T-cell leukemia virus, in two patients with AIDS. Science *220*, 862–865.

Gessain, A., Barin, F., Vernant, J.C., Gout, O., Maurs, L., Calender, A., and de, T.G. (1985). Antibodies to human T-lymphotropic virus type-I in patients with tropical spastic paraparesis. Lancet *2*, 407–410.

Gilbert, M.T., Rambaut, A., Wlasiuk, G., Spira, T.J., Pitchenik, A.E., and Worobey, M. (2007). The emergence of HIV/AIDS in the Americas and beyond. Proc. Natl. Acad. Sci. U.S.A. *104*, 18566–18570.

Giri, A., Markham, P., Digilio, L., Hurteau, G., Gallo, R.C., and Franchini, G. (1994). Isolation of a novel simian T-cell lymphotropic virus from Pan paniscus that is distantly related to the human T-cell leukemia/lymphotropic virus types I and II. J. Virol. *68*, 8392–8395.

Gitlin, S.D., Dittmer, J., Shin, R.C., and Brady, J.N. (1993). Transcriptional activation of the human T-lymphotropic virus type I long terminal repeat by functional interaction of Tax1 and Ets1. J. Virol. *67*, 7307–7316.

Goodman, N.C., and Spiegelman, S. (1971). Distinguishing reverse transcriptase of an RNA tumor virus from other known DNA polymerases. Proc. Natl. Acad. Sci. U.S.A. *68*, 2203–2206.

Gottlieb, M.S., Schroff, R., Schanker, H. M., Weisman, J.D., Fan, P.T., Wolf, R.A., and Saxon, A. (1981). *Pneumocystis carinii* pneumonia and mucosal candidiasis in previously healthy homosexual men: evidence of a new acquired cellular immunodeficiency. N. Engl. J. Med. *305*, 1425–1431.

Gowda, S.D., Stein, B. S., and Engleman, E.G. (1989). Identification of protein intermediates in the processing of the p55 HIV-1 *gag* precursor in cells infected with recombinant vaccinia virus. J. Biol. Chem. *264*, 8459–8462.

Grassmann, R., Dengler, C., Muller-Fleckenstein, I., Fleckenstein, B., McGuire, K., Dokhelar, M.C., Sodroski, J.G., and Haseltine, W.A. (1989). Transformation to continuous growth of primary human T-lymphocytes by human T-cell leukemia virus type I X-region genes transduced by a Herpesvirus saimiri vector. Proc. Natl. Acad. Sci. U.S.A. *86*, 3351–3355.

Greenberg, M.E., Bronson, S., Lock, M., Neumann, M., Pavlakis, G.N., and Skowronski, J. (1997). Co-localization of HIV-1 Nef with the AP-2 adaptor protein complex correlates with Nef-induced CD4 down-regulation. EMBO J. *16*, 6964–6976.

Grinsztejn, B., Nguyen, B.Y., Katlama, C., Gatell, J.M., Lazzarin, A., Vittecoq, D., Gonzalez, C.J., Chen, J., Harvey, C.M., and Isaacs, R.D. (2007). Safety and efficacy of the HIV-1 integrase inhibitor raltegravir (MK-0518) in treatment-experienced patients with multidrug-resistant virus: a phase II randomised controlled trial. Lancet *369*, 1261–1269.

Gross, L. (1953). Neck tumors, or leukemia, developing in adult C3H mice following inoculation, in early infancy, with filtered (Berkefeld N), or centrifugated (144,000 X g), Ak-leukemic extracts. Cancer *6*, 948–958.

Guadalupe, M., Reay, E., Sankaran, S., Prindiville, T., Flamm, J., McNeil, A., and Dandekar, S. (2003). Severe $CD4^+$ T-cell depletion in gut lymphoid tissue during primary human immunodeficiency virus type 1 infection and substantial delay in restoration following highly active antiretroviral therapy. J. Virol. *77*, 11708–11717.

Guo, H.G., Pati, S., Sadowska, M., Charurat, M., and Reitz, M. (2004). Tumorigenesis by human herpesvirus 8 vGPCR is accelerated by human immunodeficiency virus type 1 Tat. J. Virol. *78*, 9336–9342.

Guo, H.G., Wong-Staal, F., and Gallo, R.C. (1984). Novel viral sequences related to human T-cell leukemia virus in T-cells of a seropositive baboon. Science *223*, 1195–1197.

Guy, B., Kieny, M.P., Riviere, Y., Le, P.C., Dott, K., Girard, M., Montagnier, L., and Lecocq, J.P. (1987). HIV F/3′ orf encodes a phosphorylated GTP-binding protein resembling an oncogene product. Nature *330*, 266–269.

Hacein-Bey-Abina, S., Von, K.C., Schmidt, M., McCormack, M.P., Wulffraat, N., Leboulch, P., Lim, A., Osborne, C.S., Pawliuk, R., Morillon, E., Sorensen, R., Forster, A., *et al.* (2003). LMO2-associated clonal T-cell proliferation in two patients after gene therapy for SCID-X1. Science *302*, 415–419.

Hahn, B., Gallo, R.C., Franchini, G., Popovic, M., Aoki, T., Salahuddin, S.Z., Markham, P.D., and Staal, F.W. (1984). Clonal selection of human T-cell leukemia virus-infected cells *in vivo* and *in vitro*. Mol. Biol. Med. *2*, 29–36.

Hakata, Y., Umemoto, T., Matsushita, S., and Shida, H. (1998). Involvement of human CRM1 (exportin 1) in the export and multimerization of the Rex protein of human T-cell leukemia virus type 1. J. Virol. *72*, 6602–6607.

Hallenberger, S., Bosch, V., Angliker, H., Shaw, E., Klenk, H. D., and Garten, W. (1992). Inhibition of furin-mediated cleavage activation of HIV-1 glycoprotein gp160. Nature *360*, 358–361.

Haller, K., Wu, Y., Derow, E., Schmitt, I., Jeang, K. T., and Grassmann, R. (2002). Physical interaction of human T-cell leukemia virus type 1 Tax with cyclin-dependent kinase 4 stimulates the phosphorylation of retinoblastoma protein. Mol. Cell Biol. *22*, 3327–3338.

Hammarskjold, M. L., Heimer, J., Hammarskjold, B., Sangwan, I., Albert, L., and Rekosh, D. (1989). Regulation of human immunodeficiency virus *env* expression by the *rev* gene product. J. Virol. *63*, 1959–1966.

Hauber, J., and Cullen, B.R. (1988). Mutational analysis of the *trans*-activation-responsive region of the human immunodeficiency virus type I long terminal repeat. J. Virol. *62*, 673–679.

He, J., Choe, S., Walker, R., Di, M.P., Morgan, D.O., and Landau, N.R. (1995). Human immunodeficiency virus type 1 viral protein R (Vpr) arrests cells in the G2 phase of the cell cycle by inhibiting p34cdc2 activity. J. Virol. *69*, 6705–6711.

Heinzinger, N. K., Bukinsky, M. I., Haggerty, S. A., Ragland, A. M., Kewalramani, V., Lee, M. A., Gendelman, H. E., Ratner, L., Stevenson, M., and Emerman, M. (1994). The Vpr protein of human immunodeficiency virus type 1 influences nuclear localization of viral nucleic acids in nondividing host cells. Proc. Natl. Acad. Sci. U.S.A. *91*, 7311–7315.

Henderson, B.R., and Percipalle, P. (1997). Interactions between HIV Rev and nuclear import and export factors: the Rev nuclear localisation signal mediates specific binding to human importin-beta. J. Mol. Biol. *274*, 693–707.

Heredia, A., Amoroso, A., Davis, C., Le, N., Reardon, E., Dominique, J.K., Klingebiel, E., Gallo, R.C., and Redfield, R. R. (2003). Rapamycin causes down-regulation of CCR5 and accumulation of anti-HIV beta-chemokines: an approach to suppress R5 strains of HIV-1. Proc. Natl. Acad. Sci. U.S.A. *100*, 10411–10416.

Hiraragi, H., Michael, B., Nair, A., Silic-Benussi, M., Ciminale, V., and Lairmore, M. (2005). Human T-lymphotropic virus type 1 mitochondrion-localizing protein p13II sensitizes Jurkat T-cells to Ras-mediated apoptosis. J. Virol. *79*, 9449–9457.

Hrimech, M., Yao, X. J., Branton, P. E., and Cohen, E.A. (2000). Human immunodeficiency virus type 1 Vpr-mediated G(2) cell cycle arrest: Vpr interferes with cell cycle signaling cascades by interacting with the B subunit of serine/threonine protein phosphatase 2A. EMBO J. *19*, 3956–3967.

Huang, Y., Paxton, W.A., Wolinsky, S.M., Neumann, A.U., Zhang, L., He, T., Kang, S., Ceradini, D., Jin, Z., Yazdanbakhsh, K., *et al.*(1996). The role of a mutant CCR5 allele in HIV-1 transmission and disease progression. Nat. Med. *2*, 1240–1243.

Hymes, K.B., Cheung, T., Greene, J.B., Prose, N.S., Marcus, A., Ballard, H., William, D.C., and Laubenstein, L. J. (1981). Kaposi's sarcoma in homosexual men-a report of eight cases. Lancet *2*, 598–600.

Inoue, J., Seiki, M., and Yoshida, M. (1986). The second pX product p27 chi-III of HTLV-1 is required for *gag* gene expression. FEBS Lett. *209*, 187–190.

Inoue, J., Yoshida, M., and Seiki, M. (1987). Transcriptional (p40x) and post-transcriptional (p27x-III) regulators are required for the expression and replication of human T-cell leukemia virus type I genes. Proc. Natl. Acad. Sci. U.S.A. *84*, 3653–3657.

Itoh, M., Inoue, J., Toyoshima, H., Akizawa, T., Higashi, M., and Yoshida, M. (1989). HTLV-1 rex and HIV-1 *rev* act through similar mechanisms to relieve suppression of unspliced RNA expression. Oncogene *4*, 1275–1279.

Iwanaga, R., Ohtani, K., Hayashi, T., and Nakamura, M. (2001). Molecular mechanism of cell cycle progression induced by the oncogene product Tax of human T-cell leukemia virus type I. Oncogene *20*, 2055–2067.

Jacks, T., Power, M.D., Masiarz, F.R., Luciw, P.A., Barr, P.J., and Varmus, H.E. (1988). Characterization of ribosomal frameshifting in HIV-1 Gag–Pol expression. Nature *331*, 280–283.

Jarrett, R.F., Mitsuya, H., Mann, D.L., Cossman, J., Broder, S., and Reitz, M.S. (1986). Configuration and expression of the T-cell receptor beta chain gene in human T-lymphotrophic virus I-infected cells. J. Exp. Med. *163*, 383–399.

Jarrett, W.F., Crawford, E.M., Martin, W.B., and Davie, F. (1964a). A virus-like particle associated with leukemia (lymphosarcoma). Nature *202*, 567–569.

Jarrett, W.F., Martin, W.B., Crighton, G.W., Dalton, R.G., and Stewart, M.F. (1964b). Transmission experiments with leukemia (lymphosarcoma). Nature *202*, 566–567.

Javanbakht, H., An, P., Gold, B., Petersen, D. C., O'Huigin, C., Nelson, G. W., O'Brien, S. J., Kirk, G. D., Detels, R., Buchbinder, S., Donfield, S., Shulenin, S., Song, B., Perron, M. J., Stremlau, M., Sodroski, J., Dean, M., and Winkler, C. (2006a). Effects of human TRIM5alpha polymorphisms on antiretroviral function and susceptibility to human immunodeficiency virus infection. Virology *354*, 15–27.

Javanbakht, H., Yuan, W., Yeung, D.F., Song, B., az-Griffero, F., Li, Y., Li, X., Stremlau, M., and Sodroski, J. (2006b). Characterization of TRIM5alpha trimerization and its contribution to human immunodeficiency virus capsid binding. Virology *353*, 234–246.

Jowett, J. B., Planelles, V., Poon, B., Shah, N.P., Chen, M.L., and Chen, I.S. (1995). The human immunodeficiency virus type 1 vpr gene arrests infected T-cells in the G2 + M phase of the cell cycle. J. Virol. *69*, 6304–6313.

Kalyanaraman, V.S., Sarngadharan, M.G., Bunn, P. A., Minna, J. D., and Gallo, R.C. (1981a). Antibodies in human sera reactive against an internal structural protein of human T-cell lymphoma virus. Nature *294*, 271–273.

Kalyanaraman, V.S., Sarngadharan, M.G., Poiesz, B., Ruscetti, F.W., and Gallo, R.C. (1981b). Immunological properties of a type C retrovirus isolated from cultured human T-lymphoma cells and comparison to other mammalian retroviruses. J. Virol. *38*, 906–915.

Kalyanaraman, V.S., Sarngadharan, M. G., Robert-Guroff, M., Miyoshi, I., Golde, D., and Gallo, R. C. (1982). A new subtype of human T-cell leukemia virus (HTLV-II) associated with a T-cell variant of hairy cell leukemia. Science *218*, 571–573.

Kanki, P.J., Alroy, J., and Essex, M. (1985). Isolation of T-lymphotropic retrovirus related to HTLV-III/LAV from wild-caught African green monkeys. Science *230*, 951–954.

Kanki, P., Barin, F., M'Boup, S., Allan, J.S., Romet-Lemonne, J.L., Marlink, R., McLane, M.F., Lee, T.H., Arbeille, B., Denis, F., *et al.* (1986). New human T-lymphotropic retrovirus related to simian T-lymphotropic virus type III (STLV-IIIAGM). Science 232, 238–243.

Kao, S., Khan, M.A., Miyagi, E., Plishka, R., Buckler-White, A., and Strebel, K. (2003). The human immunodeficiency virus type 1 Vif protein reduces intracellular expression and inhibits packaging of APOBEC3G (CEM15), a cellular inhibitor of virus infectivity. J. Virol. *77*, 11398–11407.

Kao, S.Y., Calman, A.F., Luciw, P.A., and Peterlin, B.M. (1987). Anti-termination of transcription within the long terminal repeat of HIV-1 by *tat* gene product. Nature 330, 489–493.

Katzourakis, A., Tristem, M., Pybus, O.G., and Gifford, R.J. (2007). Discovery and analysis of the first endogenous lentivirus. Proc. Natl. Acad. Sci. U.S.A. *104*, 6261–6265.

Kawakami, T.G., Huff, S.D., Buckley, P. M., Dungworth, D.L., Synder, S.P., and Gilden, R.V. (1972). C-type virus associated with gibbon lymphosarcoma. Nat. New Biol. *235*, 170–171.

Keele, B.F., Van, H.F., Li, Y., Bailes, E., Takehisa, J., Santiago, M.L., Bibollet-Ruche, F., Chen, Y., Wain, L.V., Liegeois, F., Loul, S., Ngole, E.M., Bienvenue, Y., Delaporte, E., Brookfield, J. F., Sharp, P. M., Shaw, G.M., Peeters, M., and Hahn, B.H. (2006). Chimpanzee reservoirs of pandemic and nonpandemic HIV-1. Science *313*, 523–526.

Kinoshita, K., Amagasaki, T., Hino, S., Doi, H., Yamanouchi, K., Ban, N., Momita, S., Ikeda, S., Kamihira, S., Ichimaru, M., *et al.* (1987). Milk-borne transmission of HTLV-I from carrier mothers to their children. Jpn. J. Cancer Res. *78*, 674–680.

Kirchhoff, F., Greenough, T. C., Brettler, D. B., Sullivan, J. L., and Desrosiers, R. C. (1995). Brief report: absence of intact *nef* sequences in a long-term survivor with nonprogressive HIV-1 infection. N. Engl. J. Med. 332, 228–232.

Kitamura, T., Takano, M., Hoshino, H., Shimotohno, K., Shimoyama, M., Miwa, M., Takaku, F., and Sugimura, T. (1985). Methylation pattern of human T-cell leukemia virus *in vivo* and *in vitro*: pX and LTR regions are hypomethylated *in vivo*. Int. J. Cancer 35, 629–635.

Klatzmann, D., Champagne, E., Chamaret, S., Gruest, J., Guetard, D., Hercend, T., Gluckman, J. C., and Montagnier, L. (1984). T-lymphocyte T4 molecule behaves as the receptor for human retrovirus LAV. Nature *312*, 767–768.

Kliks, S. C., Shioda, T., Haigwood, N. L., and Levy, J. A. (1993). V3 variability can influence the ability of an antibody to neutralize or enhance infection by diverse strains of human immunodeficiency virus type 1. Proc. Natl. Acad. Sci. U.S.A. *90*, 11518–11522.

Koralnik, I. J., Fullen, J., and Franchini, G. (1993). The p12I, p13II, and p30II proteins encoded by human T-cell leukemia/lymphotropic virus type I open reading frames I and II are localized in three different cellular compartments. J. Virol. *67*, 2360–2366.

Korber, B., Okayama, A., Donnelly, R., Tachibana, N., and Essex, M. (1991). Polymerase chain reaction analysis of defective human T-cell leukemia virus type I proviral genomes in leukemic cells of patients with adult T-cell leukemia. J. Virol. *65*, 5471–5476.

Kuhmann, S. E., Pugach, P., Kunstman, K. J., Taylor, J., Stanfield, R. L., Snyder, A., Strizki, J. M., Riley, J., Baroudy, B. M., Wilson, I. A., Korber, B. T., Wolinsky, S. M., and Moore, J. P. (2004). Genetic and phenotypic analyses of human immunodeficiency virus type 1 escape from a small-molecule CCR5 inhibitor. J. Virol. *78*, 2790–2807.

Kwong, P. D., Wyatt, R., Sattentau, Q. J., Sodroski, J., and Hendrickson, W. A. (2000). Oligomeric modeling and electrostatic analysis of the gp120 envelope glycoprotein of human immunodeficiency virus. J. Virol. *74*, 1961–1972.

Lehner, T., Wang, Y., Cranage, M., Bergmeier, L.A., Mitchell, E., Tao, L., Hall, G., Dennis, M., Cook, N., Brookes, R., Klavinskis, L., Jones, I., Doyle, C., and Ward, R. (1996). Protective mucosal immunity elicited by targeted iliac lymph node immunization with a subunit SIV envelope and core vaccine in macaques. Nat. Med. 2, 767–775.

Li, F., Goila-Gaur, R., Salzwedel, K., Kilgore, N.R., Reddick, M., Matallana, C., Castillo, A., Zoumplis, D., Martin, D.E., Orenstein, J.M., Allaway, G.P., Freed, E.O., and Wild, C.T. (2003). PA-457: a potent HIV inhibitor that disrupts core condensation by targeting a late step in Gag processing. Proc. Natl. Acad. Sci. U.S.A. *100*, 13555–13560.

Lilly, F. (1967). Susceptibility to two strains of Friend leukemia virus in mice. Science *155*, 461–462.

Lindholm, P.F., Marriott, S.J., Gitlin, S.D., Bohan, C.A., and Brady, J.N. (1990). Induction of nuclear NF-kappa B DNA binding activity after exposure of lymphoid cells to soluble tax1 protein. New Biol. 2, 1034–1043.

Liu, B., Liang, M.H., Kuo, Y.L., Liao, W., Boros, I., Kleinberger, T., Blancato, J., and Giam, C.Z. (2003). Human T-lymphotropic virus type 1 oncoprotein tax promotes unscheduled degradation of Pds1p/securin and Clb2p/cyclin B1 and causes chromosomal instability. Mol. Cell Biol. 23, 5269–5281.

Liu, H.F., Vandamme, A.M., Van, B.M., Desmyter, J., and Goubau, P. (1994). New retroviruses in human and simian T-lymphotropic viruses. Lancet *344*, 265–266.

Liu, R., Paxton, W.A., Choe, S., Ceradini, D., Martin, S.R., Horuk, R., MacDonald, M.E., Stuhlmann, H., Koup, R.A., and Landau, N.R. (1996). Homozygous defect in HIV-1 coreceptor accounts for resistance of some multiply-exposed individuals to HIV-1 infection. Cell *86*, 367–377.

Lori, F., Malykh, A., Cara, A., Sun, D., Weinstein, J.N., Lisziewicz, J., and Gallo, R.C. (1994). Hydroxyurea as an inhibitor of human immunodeficiency virus-type 1 replication. Science *266*, 801–805.

Mahalingam, S., Ayyavoo, V., Patel, M., Kieber-Emmons, T., and Weiner, D.B. (1997). Nuclear import, virion incorporation, and cell cycle arrest/differentiation are mediated by distinct functional domains of human immunodeficiency virus type 1 Vpr. J. Virol. *71*, 6339–6347.

Majone, F., Luisetto, R., Zamboni, D., Iwanaga, Y., and Jeang, K.T. (2005). Ku protein as a potential human T-cell leukemia virus type 1 (HTLV-1) Tax target in clastogenic chromosomal instability of mammalian cells. Retrovirology. *2*, 45.

Mangasarian, A., Foti, M., Aiken, C., Chin, D., Carpentier, J.L., and Trono, D. (1997). The HIV-1 Nef protein acts as a connector with sorting pathways in the Golgi and at the plasma membrane. Immunity. *6*, 67–77.

Mangeat, B., Turelli, P., Caron, G., Friedli, M., Perrin, L., and Trono, D. (2003). Broad antiretroviral defence by human APOBEC3G through lethal editing of nascent reverse transcripts. Nature *424*, 99–103.

Mariani, R., Chen, D., Schrofelbauer, B., Navarro, F., Konig, R., Bollman, B., Munk, C., Nymark-McMahon, H., and Landau, N.R. (2003). Species-specific exclusion of APOBEC3G from HIV-1 virions by Vif. Cell *114*, 21–31.

Mariani, R., Kirchhoff, F., Greenough, T.C., Sullivan, J.L., Desrosiers, R.C., and Skowronski, J. (1996). High frequency of defective *nef* alleles in a long-term survivor with nonprogressive human immunodeficiency virus type 1 infection. J. Virol. *70*, 7752–7764.

Markham, P.D., Salahuddin, S.Z., Macchi, B., Robert-Guroff, M., and Gallo, R.C. (1984). Transformation of different phenotypic types of human bone marrow T-lymphocytes by HTLV-1. Int. J. Cancer 33, 13–17.

Marozsan, A.J., Kuhmann, S.E., Morgan, T., Herrera, C., Rivera-Troche, E., Xu, S., Baroudy, B.M., Strizki, J., and Moore, J.P. (2005). Generation and properties of a human immunodeficiency virus type 1 isolate resistant to the small molecule CCR5 inhibitor, SCH-417690 (SCH-D). Virology *338*, 182–199.

Masur, H., Michelis, M.A., Greene, J.B., Onorato, I., Stouwe, R.A., Holzman, R.S., Wormser, G., Brettman, L., Lange, M., Murray, H.W., and Cunningham-Rundles, S. (1981). An outbreak of community-acquired Pneumocystis carinii pneumonia: initial manifestation of cellular immune dysfunction. N. Engl. J. Med. *305*, 1431–1438.

McKeating, J.A., Bennett, J., Zolla-Pazner, S., Schutten, M., Ashelford, S., Brown, A.L., and Balfe, P. (1993). Resistance of a human serum-selected human immunodeficiency virus type 1 escape mutant to neutralization by CD4 binding site monoclonal antibodies is conferred by a single amino acid change in gp120. J. Virol. *67*, 5216–5225.

Merluzzi, V.J., Hargrave, K.D., Labadia, M., Grozinger, K., Skoog, M., Wu, J.C., Shih, C.K., Eckner, K., Hattox, S., Adams, J., Rosenthal, A.S., Faanes, R., Eckner, R.J., Koup, R.A., and Sullivan, J.L. (1990). Inhibition of HIV-1 replication by a nonnucleoside reverse transcriptase inhibitor. Science *250*, 1411–1413.

Mervis, R.J., Ahmad, N., Lillehoj, E.P., Raum, M.G., Salazar, F.H., Chan, H.W., and Venkatesan, S. (1988). The *gag* gene products of human immunodeficiency virus type 1: alignment within the *gag* open reading frame, identification of posttranslational modifications, and evidence for alternative *gag* precursors. J. Virol. *62*, 3993–4002.

Mesnard, J.M., Barbeau, B., and Devaux, C. (2006). HBZ, a new important player in the mystery of adult T-cell leukemia. Blood *108*, 3979–3982.

Miyoshi, I., Kubonishi, I., Yoshimoto, S., and Shiraishi, Y. (1981). A T-cell line derived from normal human cord leukocytes by co-culturing with human leukemic T-cells. Gann *72*, 978–981.

Moarefi, I., LaFevre-Bernt, M., Sicheri, F., Huse, M., Lee, C.H., Kuriyan, J., and Miller, W.T. (1997). Activation of the Src-family tyrosine kinase Hck by SH3 domain displacement. Nature *385*, 650–653.

Morgan, D.A., Ruscetti, F.W., and Gallo, R. (1976). Selective *in vitro* growth of T-lymphocytes from normal human bone marrows. Science *193*, 1007–1008.

Muesing, M.A., Smith, D.H., Cabradilla, C.D., Benton, C.V., Lasky, L.A., and Capon, D.J. (1985). Nucleic acid structure and expression of the human AIDS/lymphadenopathy retrovirus. Nature *313*, 450–458.

Muesing, M.A., Smith, D.H., and Capon, D.J. (1987). Regulation of mRNA accumulation by a human immunodeficiency virus *trans*-activator protein. Cell *48*, 691–701.

Murphy, E.L., Hanchard, B., Figueroa, J.P., Gibbs, W.N., Lofters, W.S., Campbell, M., Goedert, J.J., and Blattner, W.A. (1989). Modelling the risk of adult T-cell leukemia/lymphoma in persons infected with human T-lymphotropic virus type I. Int. J. Cancer *43*, 250–253.

Neville, M., Stutz, F., Lee, L., Davis, L.I., and Rosbash, M. (1997). The importin-beta family member Crm1p bridges the interaction between Rev and the nuclear pore complex during nuclear export. Curr. Biol. *7*, 767–775.

Nicot, C., Mulloy, J.C., Ferrari, M.G., Johnson, J.M., Fu, K., Fukumoto, R., Trovato, R., Fullen, J., Leonard, W.J., and Franchini, G. (2001). HTLV-1, p12(I) protein enhances STAT5 activation and decreases the interleukin-2 requirement for proliferation of primary human peripheral blood mononuclear cells. Blood *98*, 823–829.

Nimer, S.D., Gasson, J.C., Hu, K., Smalberg, I., Williams, J.L., Chen, I.S., and Rosenblatt, J.D. (1989). Activation of the GM-CSF promoter by HTLV-I and -II tax proteins. Oncogene *4*, 671–676.

Nunn, M.F., and Marsh, J.W. (1996). Human immunodeficiency virus type 1 Nef associates with a member of the p21-activated kinase family. J. Virol. *70*, 6157–6161.

Osame, M., Usuku, K., Izumo, S., Ijichi, N., Amitani, H., Igata, A., Matsumoto, M., and Tara, M. (1986). HTLV-I associated myelopathy, a new clinical entity. Lancet *1*, 1031–1032.

Palmeri, D., and Malim, M.H. (1999). Importin beta can mediate the nuclear import of an arginine-rich nuclear localization signal in the absence of importin alpha. Mol. Cell Biol. *19*, 1218–1225.

Pasquinelli, A.E., Powers, M.A., Lund, E., Forbes, D., and Dahlberg, J.E. (1997). Inhibition of mRNA export in vertebrate cells by nuclear export signal conjugates. Proc. Natl. Acad. Sci. U.S.A. *94*, 14394–14399.

Paxton, W., Connor, R.I., and Landau, N.R. (1993). Incorporation of Vpr into human immunodeficiency virus type 1 virions: requirement for the p6 region of *gag* and mutational analysis. J. Virol. *67*, 7229–7237.

Paxton, W.A., Liu, R., Kang, S., Wu, L., Gingeras, T.R., Landau, N.R., Mackay, C.R., and Koup, R.A. (1998). Reduced HIV-1 infectability of CD4[+] lymphocytes from exposed- uninfected individuals: association with low expression of CCR5 and high production of beta-chemokines. Virology *244*, 66–73.

Perron, M.J., Stremlau, M., Song, B., Ulm, W., Mulligan, R.C., and Sodroski, J. (2004). TRIM5alpha mediates the postentry block to N-tropic murine leukemia viruses in human cells. Proc. Natl. Acad. Sci. U.S.A. *101*, 11827–11832.

Piguet, V., Chen, Y.L., Mangasarian, A., Foti, M., Carpentier, J.L., and Trono, D. (1998). Mechanism of Nef-induced CD4 endocytosis: Nef connects CD4 with the mu chain of adaptor complexes. EMBO J. *17*, 2472–2481.

Pique, C., Ureta-Vidal, A., Gessain, A., Chancerel, B., Gout, O., Tamouza, R., Agis, F., and Dokhelar, M.C. (2000). Evidence for the chronic *in vivo* production of human T-cell leukemia virus type I Rof and Tof proteins from cytotoxic T-lymphocytes directed against viral peptides. J. Exp. Med. *191*, 567–572.

Poiesz, B.J., Ruscetti, F.W., Gazdar, A.F., Bunn, P.A., Minna, J.D., and Gallo, R.C. (1980a). Detection and isolation of type C retrovirus particles from fresh and cultured lymphocytes of a patient with cutaneous T-cell lymphoma. Proc Natl Acad Sci U.S.A. *77*, 7415–7419.

Poiesz, B.J., Ruscetti, F.W., Mier, J.W., Woods, A.M., and Gallo, R.C. (1980b). T-cell lines established from human T-lymphocytic neoplasias by direct response to T-cell growth factor. Proc. Natl. Acad. Sci. U.S.A. *77*, 6815–6819.

Poiesz, B.J., Ruscetti, F.W., Reitz, M.S., Kalyanaraman, V.S., and Gallo, R.C. (1981). Isolation of a new type C retrovirus (HTLV) in primary uncultured cells of a patient with Sezary T-cell leukaemia. Nature *294*, 268–271.

Popov, S., Rexach, M., Zybarth, G., Reiling, N., Lee, M.A., Ratner, L., Lane, C.M., Moore, M.S., Blobel, G., and Bukrinsky, M. (1998). Viral protein R regulates nuclear import of the HIV-1 pre-integration complex. EMBO J. *17*, 909–917.

Popovic, M., Lange-Wantzin, G., Sarin, P.S., Mann, D., and Gallo, R.C. (1983a). Transformation of human umbilical cord blood T-cells by human T-cell leukemia/lymphoma virus. Proc. Natl. Acad. Sci. U.S.A. *80*, 5402–5406.

Popovic, M., Sarin, P.S., Robert-Gurroff, M., Kalyanaraman, V.S., Mann, D., Minowada, J., and Gallo, R.C. (1983b). Isolation and transmission of human retrovirus (human t-cell leukemia virus). Science *219*, 856–859.

Popovic, M., Sarngadharan, M.G., Read, E., and Gallo, R.C. (1984). Detection, isolation, and continuous production of cytopathic retroviruses (HTLV-III) from patients with AIDS and pre-AIDS. Science *224*, 497–500.

Raha, T., Cheng, S.W., and Green, M.R. (2005). HIV-1 Tat stimulates transcription complex assembly through recruitment of TBP in the absence of TAFs. PLoS. Biol. 3, e44.

Ratner, L., Haseltine, W., Patarca, R., Livak, K.J., Starcich, B., Josephs, S.F., Doran, E.R., Rafalski, J.A., Whitehorn, E.A., Baumeister, K., *et al.* (1985). Complete nucleotide sequence of the AIDS virus, HTLV-III. Nature *313*, 277–284.

Re, F., Braaten, D., Franke, E.K., and Luban, J. (1995). Human immunodeficiency virus type 1 Vpr arrests the cell cycle in G2 by inhibiting the activation of p34cdc2-cyclin B. J. Virol. *69*, 6859–6864.

Reitz, M.S., Popovic, M., Haynes, B.F., Clark, S.C., and Gallo, R.C. (1983). Relatedness by nucleic acid hybridization of new isolates of human T-cell leukemia-lymphoma virus (HTLV) and demonstration of provirus in uncultured leukemic blood cells. Virology *126*, 688–672.

Reitz, M.S., Poiesz, B.J., Ruscetti, F.W., and Gallo, R.C. (1981). Characterization and distribution of nucleic acid sequences of a novel type C retrovirus isolated from neoplastic human T-lymphocytes. Proc Natl Acad Sci U.S.A. *78*, 1887–1891.

Reitz, M.S., Smith, R.G., Roseberry, E.A., and Gallo, R.C. (1974). DNA-directed and RNA-primed DNA synthesis in microsomal and mitochondrial fractions of normal human lymphocytes. Biochem. Biophys. Res. Commun. *57*, 934–948.

Reitz, M.S., Wilson, C., Naugle, C., Gallo, R.C., and Robert-Guroff, M. (1988). Generation of a neutralization-resistant variant of HIV-1 is due to selection for a point mutation in the envelope gene. Cell *54*, 57–63.

Reymond, A., Meroni, G., Fantozzi, A., Merla, G., Cairo, S., Luzi, L., Riganelli, D., Zanaria, E., Messali, S., Cainarca, S., Guffanti, A., Minucci, S., Pelicci, P.G., and Ballabio, A. (2001). The tripartite motif family identifies cell compartments. EMBO J. *20*, 2140–2151.

Rhee, S.S., and Marsh, J.W. (1994). Human immunodeficiency virus type 1 Nef-induced down-modulation of CD4 is due to rapid internalization and degradation of surface CD4. J. Virol. *68*, 5156–5163.

Rizzuto, C.D., Wyatt, R., Hernandez-Ramos, N., Sun, Y., Kwong, P.D., Hendrickson, W.A., and Sodroski, J. (1998). A conserved HIV gp120 glycoprotein structure involved in chemokine receptor binding. Science *280*, 1949–1953.

Robert, M.S., Smith, R.G., Gallo, R.C., Sarin, P.S., and Abrell, J.W. (1972). Viral and cellular DNA polymerase: comparison of activities with synthetic and natural RNA templates. Science *176*, 798–800.

Robert-Guroff, M., Popovic, M., Gartner, S., Markham, P., Gallo, R.C., and Reitz, M.S. (1990). Structure and expression of *tat*-, rev-, and nef-specific transcripts of human immunodeficiency virus type 1 in infected lymphocytes and macrophages. J. Virol. *64*, 3391–3398.

Robert-Guroff, M., Ruscetti, F.W., Posner, L.E., Poiesz, B.J., and Gallo, R.C. (1981). Detection of the human T-cell lymphoma virus p19 in cells of some patients with cutaneous T-cell lymphoma and leukemia using a monoclonal antibody. J. Exp. Med. *154*, 1957–1964.

Rous, P. (1910). A transmissible avian neoplasm (Sarcoma of the common fowl). J. Exp. Med. *12*, 696.

Ruben, S., Poteat, H., Tan, T.H., Kawakami, K., Roeder, R., Haseltine, W., and Rosen, C.A. (1988). Cellular transcription factors and regulation of IL-2 receptor gene expression by HTLV-I tax gene product. Science *241*, 89–92.

Ruscetti, F.W., Morgan, D.A., and Gallo, R.C. (1977). Functional and morphologic characterization of human T-cells continuously grown in vitro. J. Immunol *119*, 131–138.

Samson, M., Libert, F., Doranz, B.J., Rucker, J., Liesnard, C., Farber, C.M., Saragosti, S., Lapoumeroulie, C., Cognaux, J., Forceille, C., *et al.* (1996). Resistance to HIV-1 infection in caucasian individuals bearing mutant alleles of the CCR-5 chemokine receptor gene. Nature *382*, 722–725.

Sanada, I., Tanaka, R., Kumagai, E., Tsuda, H., Nishimura, H., Yamaguchi, K., Kawano, F., Fujiwara, H., and Takatsuki, K. (1985). Chromosomal aberrations in adult T-cell leukemia: relationship to the clinical severity. Blood *65*, 649–654.

Sanchez-Pescador, R., Power, M.D., Barr, P.J., Steimer, K.S., Stempien, M.M., Brown-Shimer, S.L., Gee, W.W., Renard, A., Randolph, A., Levy, J.A., *et al.* (1985). Nucleotide sequence and expression of an AIDS-associated retrovirus (ARV-2). Science *227*, 484–492.

Santiago, F., Clark, E., Chong, S., Molina, C., Mozafari, F., Mahieux, R., Fujii, M., Azimi, N., and Kashanchi, F. (1999). Transcriptional up-regulation of the cyclin D2 gene and acquisition of new cyclin-dependent kinase partners in human T-cell leukemia virus type 1-infected cells. J. Virol. *73*, 9917–9927.

Sarngadharan, M.G., Popovic, M., Bruch, L., Schupbach, J., and Gallo, R.C. (1984). Antibodies reactive with human T-lymphotropic retroviruses (HTLV-III) in the serum of patients with AIDS. Science *224*, 506–508.

Satou, Y., Yasunaga, J., Yoshida, M., and Matsuoka, M. (2006). HTLV-I basic leucine zipper factor gene mRNA supports proliferation of adult T-cell leukemia cells. Proc. Natl. Acad. Sci. U.S.A. *103*, 720–725.

Schlom, J., and Spiegelman, S. (1971). Simultaneous detection of reverse transcriptase and high molecular weight RNA unique to oncogenic RNA viruses. Science *174*, 840–843.

Schubert, U., Anton, L.C., Bacik, I., Cox, J.H., Bour, S., Bennink, J.R., Orlowski, M., Strebel, K., and Yewdell, J.W. (1998). CD4 glycoprotein degradation induced by human immunodeficiency virus type 1 Vpu protein requires the function of proteasomes and the ubiquitin-conjugating pathway. J. Virol. *72*, 2280–2288.

Schulze, T., Nawrath, M., and Moelling, K. (1991). Cleavage of the HIV-1, p66 reverse transcriptase/RNase H by the p9 protease *in vitro* generates active p15 RNase H. Arch. Virol. *118*, 179–188.

Schupbach, J., Sarngadharan, M.G., and Gallo, R.C. (1984). Antigens on HTLV-infected cells recognized by leukemia and AIDS sera are related to HTLV viral glycoprotein. Science *224*, 607–610.

Schwartz, O., Marechal, V., Danos, O., and Heard, J.M. (1995). Human immunodeficiency virus type 1 Nef increases the efficiency of reverse transcription in the infected cell. J. Virol. *69*, 4053–4059.

Schwartz, O., Marechal, V., Le, G.S., Lemonnier, F., and Heard, J.M. (1996). Endocytosis of major histocompatibility complex class I molecules is induced by the HIV-1 Nef protein. Nat. Med. *2*, 338–342.

Schwartz, S., Felber, B.K., Benko, D.M., Fenyo, E.M., and Pavlakis, G.N. (1990a). Cloning and functional analysis of multiply spliced mRNA species of human immunodeficiency virus type 1. J. Virol. *64*, 2519–2529.

Schwartz, S., Felber, B.K., Fenyo, E.M., and Pavlakis, G.N. (1990b). Env and Vpu proteins of human immunodeficiency virus type 1 are produced from multiple bicistronic mRNAs. J. Virol. *64*, 5448–5456.

Seiki, M., Eddy, R., Shows, T.B., and Yoshida, M. (1984). Nonspecific integration of the HTLV provirus genome into adult T-cell leukaemia cells. Nature *309*, 640–642.

Seiki, M., Hattori, S., Hirayama, Y., and Yoshida, M. (1983). Human adult T-cell leukemia virus: complete nucleotide sequence of the provirus genome integrated in leukemia cell DNA. Proc. Natl. Acad. Sci. U.S.A. *80*, 3618–3622.

Seiki, M., Hikikoshi, A., Taniguchi, T., and Yoshida, M. (1985). Expression of the pX gene of HTLV-I: general splicing mechanism in the HTLV family. Science *228*, 1532–1534.

Sharp, P.M., Robertson, D.L., and Hahn, B.H. (1995). Cross-species transmission and recombination of 'AIDS' viruses. Phil. Trans. R. Soc. Lond B Biol. Sci. *349*, 41–47.

Sheehy, A.M., Gaddis, N.C., Choi, J.D., and Malim, M.H. (2002). Isolation of a human gene that inhibits HIV-1 infection and is suppressed by the viral Vif protein. Nature *418*, 646–650.

Siegal, F.P., Lopez, C., Hammer, G.S., Brown, A.E., Kornfeld, S.J., Gold, J., Hassett, J., Hirschman, S.Z., Cunningham-Rundles, C., Adelsberg, B.R., *et al.* (1981). Severe acquired immunodeficiency in male homosexuals, manifested by chronic perianal ulcerative herpes simplex lesions. N. Engl. J. Med. *305*, 1439–1444.

Silic-Benussi, M., Cavallari, I., Zorzan, T., Rossi, E., Hiraragi, H., Rosato, A., Horie, K., Saggioro, D., Lairmore, M.D., Willems, L., Chieco-Bianchi, L., D'Agostino, D.M., and Ciminale, V. (2004). Suppression of tumor growth and cell proliferation by p13II, a mitochondrial protein of human T-cell leukemia virus type 1. Proc. Natl. Acad. Sci. U.S.A. *101*, 6629–6634.

Smit-McBride, Z., Mattapallil, J.J., McChesney, M., Ferrick, D., and Dandekar, S. (1998). Gastrointestinal T-lymphocytes retain high potential for cytokine responses but have severe CD4(+) T-cell depletion at all stages of simian immunodeficiency virus infection compared to peripheral lymphocytes. J. Virol. 72, 6646–6656.

Sodroski, J., Patarca, R., Rosen, C., Wong-Staal, F., and Haseltine, W. (1985a). Location of the *trans*-activating region on the genome of human T-cell lymphotropic virus type III. Science *229*, 74–77.

Sodroski, J., Rosen, C., Goh, W.C., and Haseltine, W. (1985b). A transcriptional activator protein encoded by the x-lor region of the human T-cell leukemia virus. Science *228*, 1430–1434.

Sodroski, J., Rosen, C., Wong-Staal, F., Salahuddin, S.Z., Popovic, M., Arya, S., Gallo, R.C., and Haseltine, W.A. (1985c). *Trans*-acting transcriptional regulation of human T-cell leukemia virus type III long terminal repeat. Science 227, 171–173.

Solomin, L., Felber, B.K., and Pavlakis, G.N. (1990). Different sites of interaction for Rev, Tev, and Rex proteins within the Rev-responsive element of human immunodeficiency virus type 1. J. Virol. *64*, 6010–6017.

Stopak, K., de, N.C., Yonemoto, W., and Greene, W.C. (2003). HIV-1 Vif blocks the antiviral activity of APOBEC3G by impairing both its translation and intracellular stability. Mol. Cell *12*, 591–601.

Stremlau, M., Owens, C.M., Perron, M.J., Kiessling, M., Autissier, P., and Sodroski, J. (2004). The cytoplasmic body component TRIM5alpha restricts HIV-1 infection in Old World monkeys. Nature *427*, 848–853.

Stremlau, M., Perron, M., Lee, M., Li, Y., Song, B., Javanbakht, H., az-Griffero, F., Anderson, D.J., Sundquist, W.I., and Sodroski, J. (2006). Specific recognition and accelerated uncoating of retroviral capsids by the TRIM5alpha restriction factor. Proc. Natl. Acad. Sci. U.S.A. *103*, 5514–5519.

Stumptner-Cuvelette, P., Morchoisne, S., Dugast, M., Le, G.S., Raposo, G., Schwartz, O., and Benaroch, P. (2001). HIV-1 Nef impairs MHC class II antigen presentation and surface expression. Proc. Natl. Acad. Sci. U.S.A. *98*, 12144–12149.

Sun, L., Finnegan, C.M., Kish-Catalone, T., Blumenthal, R., Garzino-Demo, P., La Terra Maggiore, G.M., Berrone, S., Kleinman, C., Wu, Z., Abdelwahab, S., Lu, W., and Garzino-Demo, A. (2005). Human beta-defensins suppress human immunodeficiency virus infection: potential role in mucosal protection. J. Virol. *79*, 14318–14329.

Suzuki, T., Fujisawa, J.I., Toita, M., and Yoshida, M. (1993a). The *trans*-activator tax of human T-cell leukemia virus type 1 (HTLV-1) interacts with cAMP-responsive element (CRE) binding and CRE modulator proteins that bind to the 21-base-pair enhancer of HTLV-1. Proc. Natl. Acad. Sci. U.S.A. *90*, 610–614.

Suzuki, T., Hirai, H., Fujisawa, J., Fujita, T., and Yoshida, M. (1993b). A *trans*-activator Tax of human T-cell leukemia virus type 1 binds to NF-kappa B p50 and serum response factor (SRF) and associates with enhancer DNAs of the NF-kappa B site and CArG box. Oncogene *8*, 2391–2397.

Swigut, T., Shohdy, N., and Skowronski, J. (2001). Mechanism for down-regulation of CD28 by Nef. EMBO J. *20*, 1593–1604.

Takatsuki, K., Uchiyama, T., Sagawa, K., and Yodoi, J. (1976). [Surface markers of malignant lymphoid cells in the classification of lymphoproliferative disorders, with special reference to adult T-cell leukemia (author's transl)]. Rinsho Ketsueki *17*, 416–421.

Theilen, G. H., Gould, D., Fowler, M., and Dungworth, D.L. (1971). C-type virus in tumor tissue of a woolly monkey (Lagothrix spp.) with fibrosarcoma. J. Natl Cancer Inst. *47*, 881–889.

Toyoshima, H., Itoh, M., Inoue, J., Seiki, M., Takaku, F., and Yoshida, M. (1990). Secondary structure of the human T-cell leukemia virus type 1 rex-responsive element is essential for rex regulation of RNA processing and transport of unspliced RNAs. J. Virol. *64*, 2825–2832.

Trkola, A., Dragic, T., Arthos, J., Binley, J.M., Olson, W.C., Allaway, G.P., Cheng-Mayer, C., Robinson, J., Maddon, P.J., and Moore, J.P. (1996). CD4-dependent, antibody-sensitive interactions between HIV-1 and its co-receptor CCR-5. Nature *384*, 184–187.

Trkola, A., Kuhmann, S.E., Strizki, J.M., Maxwell, E., Ketas, T., Morgan, T., Pugach, P., Xu, S., Wojcik, L., Tagat, J., Palani, A., Shapiro, S., Clader, J.W., McCombie, S., Reyes, G.R., Baroudy, B.M., and Moore, J.P. (2002). HIV-1 escape from a small molecule, CCR5-specific entry inhibitor does not involve CXCR4 use. Proc. Natl. Acad. Sci. U.S.A. *99*, 395–400.

Tschachler, E., Robert-Guroff, M., Gallo, R.C., and Reitz, M.S. (1989). Human T-lymphotropic virus I-infected T-cells constitutively express lymphotoxin *in vitro*. Blood 73, 194–201.

Uchiyama, T., Yodoi, J., Sagawa, K., Takatsuki, K., and Uchino, H. (1977). Adult T-cell leukemia: clinical and hematologic features of 16 cases. Blood *50*, 481–492.

Uittenbogaard, M.N., Giebler, H.A., Reisman, D., and Nyborg, J.K. (1995). Transcriptional repression of p53 by human T-cell leukemia virus type I Tax protein. J. Biol. Chem. *270*, 28503–28506.

Wain-Hobson, S., Sonigo, P., Danos, O., Cole, S., and Alizon, M. (1985). Nucleotide sequence of the AIDS virus, LAV. Cell *40*, 9–17.

Wano, Y., Feinberg, M., Hosking, J.B., Bogerd, H., and Greene, W.C. (1988). Stable expression of the tax gene of type I human T-cell leukemia virus in human T-cells activates specific cellular genes involved in growth. Proc. Natl. Acad. Sci. U.S.A. *85*, 9733–9737.

Watanabe, T., Seiki, M., Tsujimoto, H., Miyoshi, I., Hayami, M., and Yoshida, M. (1985). Sequence homology of the simian retrovirus genome with human T-cell leukemia virus type I. Virology *144*, 59–65.

Watanabe, T., Yamaguchi, K., Takatsuki, K., Osame, M., and Yoshida, M. (1990). Constitutive expression of parathyroid hormone-related protein gene in human T-cell leukemia virus type 1 (HTLV-1) carriers and adult T-cell leukemia patients that can be *trans*-activated by HTLV-1 tax gene. J. Exp. Med. *172*, 759–765.

Wei, P., Garber, M.E., Fang, S.M., Fischer, W.H., and Jones, K.A. (1998). A novel CDK9-associated C-type cyclin interacts directly with HIV-1 Tat and mediates its high-affinity, loop-specific binding to TAR RNA. Cell *92*, 451–462.

Welker, R., Kottler, H., Kalbitzer, H.R., and Krausslich, H.G. (1996). Human immunodeficiency virus type 1 Nef protein is incorporated into virus particles and specifically cleaved by the viral proteinase. Virology *219*, 228–236.

Wild, C., Greenwell, T., and Matthews, T. (1993). A synthetic peptide from HIV-1 gp41 is a potent inhibitor of virus-mediated cell–cell fusion. AIDS Res. Hum. Retroviruses *9*, 1051–1053.

Willey, R.L., Maldarelli, F., Martin, M.A., and Strebel, K. (1992a). Human immunodeficiency virus type 1 Vpu protein induces rapid degradation of CD4. J. Virol. *66*, 7193–7200.

Willey, R.L., Maldarelli, F., Martin, M.A., and Strebel, K. (1992b). Human immunodeficiency virus type 1 Vpu protein regulates the formation of intracellular gp160–CD4 complexes. J. Virol. *66*, 226–234.

Wilson, W., Braddock, M., Adams, S.E., Rathjen, P.D., Kingsman, S.M., and Kingsman, A.J. (1988). HIV expression strategies: ribosomal frameshifting is directed by a short sequence in both mammalian and yeast systems. Cell *55*, 1159–1169.

Wolf, D., Witte, V., Laffert, B., Blume, K., Stromer, E., Trapp, S., d'Aloja, P., Schurmann, A., and Baur, A.S. (2001). HIV-1 Nef associated PAK and PI3-kinases stimulate Akt-independent Bad-phosphorylation to induce anti-apoptotic signals. Nat. Med. *7*, 1217–1224.

Wolfe, N.D., Heneine, W., Carr, J. K., Garcia, A. D., Shanmugam, V., Tamoufe, U., Torimiro, J.N., Prosser, A. T., Lebreton, M., Mpoudi-Ngole, E., McCutchan, F.E., Birx, D.L., Folks, T. M., Burke, D.S., and Switzer, W.M. (2005). Emergence of unique primate T-lymphotropic viruses among central African bushmeat hunters. Proc. Natl. Acad. Sci. U.S.A. *102*, 7994–7999.

Wu, L., Gerard, N.P., Wyatt, R., Choe, H., Parolin, C., Ruffing, N., Borsetti, A., Cardoso, A.A., Desjardin, E., Newman, W., Gerard, C., and Sodroski, J. (1996). CD4-induced interaction of primary HIV-1 gp120 glycoproteins with the chemokine receptor CCR-5. Nature *384*, 179–183.

Wyatt, R., and Sodroski, J. (1998). The HIV-1 envelope glycoproteins: fusogens, antigens, and immunogens. Science *280*, 1884–1888.

Yap, M.W., Dodding, M.P., and Stoye, J.P. (2006). Trim-cyclophilin A fusion proteins can restrict human immunodeficiency virus type 1 infection at two distinct phases in the viral life cycle. J. Virol. *80*, 4061–4067.

Yarchoan, R., and Broder, S. (1987). Development of antiretroviral therapy for the acquired immunodeficiency syndrome and related disorders. A progress report. N. Engl. J. Med. *316*, 557–564.

Yoshida, M., Miyoshi, I., and Hinuma, Y. (1982). Isolation and characterization of retrovirus from cell lines of human adult T-cell leukemia and its implication in the disease. Proc. Natl. Acad. Sci. U.S.A. *79*, 2031–2035.

Yoshida, M., Seiki, M., Yamaguchi, K., and Takatsuki, K. (1984). Monoclonal integration of human T-cell leukemia provirus in all primary tumors of adult T-cell leukemia suggests causative role of human T-cell leukemia virus in the disease. Proc. Natl. Acad. Sci. U.S.A. *81*, 2534–2537.

Zagury, D., Lachgar, A., Chams, V., Fall, L.S., Bernard, J., Zagury, J. F., Bizzini, B., Gringeri, A., Santagostino, E., Rappaport, J., Feldman, M., Burny, A., and Gallo, R.C. (1998a). Interferon alpha and Tat involvement in the immunosuppression of uninfected T-cells and C-C chemokine decline in AIDS. Proc. Natl. Acad. Sci. U.S.A. *95*, 3851–3856.

Zagury, D., Lachgar, A., Chams, V., Fall, L. S., Bernard, J., Zagury, J.F., Bizzini, B., Gringeri, A., Santagostino, E., Rappaport, J., Feldman, M., O'Brien, S.J., Burny, A., and Gallo, R.C. (1998b). C-C chemokines, pivotal in protection against HIV type 1 infection. Proc. Natl. Acad. Sci. U.S.A. *95*, 3857–3861.

Zapp, M.L., and Green, M.R. (1989). Sequence-specific RNA binding by the HIV-1 Rev protein. Nature *342*, 714–716.

Zhang, H., Yang, B., Pomerantz, R. J., Zhang, C., Arunachalam, S. C., and Gao, L. (2003). The cytidine deaminase CEM15 induces hypermutation in newly synthesized HIV-1 DNA. Nature *424*, 94–98.

Zhang, W., Nisbet, J. W., Albrecht, B., Ding, W., Kashanchi, F., Bartoe, J. T., and Lairmore, M. D. (2001). Human T-lymphotropic virus type 1, p30(II) regulates gene transcription by binding CREB binding protein/p300. J. Virol. *75*, 9885–9895.

Zhou, Q., Chen, D., Pierstorff, E., and Luo, K. (1998). Transcription elongation factor P-TEFb mediates Tat activation of HIV-1 transcription at multiple stages. EMBO J. *17*, 3681–3691.

Index

A

B

R